AMIN R. ISMAIL
University of Dayton

VICTOR M. ROONEY
University of Dayton

DIGITAL CONCEPTS & APPLICATIONS

Saunders College Publishing
A Division of
Holt, Rinehart and Winston
Philadelphia Ft. Worth Chicago San Francisco
Montreal Toronto London Sydney Tokyo

Text Typeface: Times Roman
Compositor: General Graphic Services
Acquisitions Editor: Barbara Gingery
Developmental Editor: Alexa Barnes
Managing Editor: Carol Field
Project Editor: Maureen Iannuzzi
Manager of Art and Design: Carol Bleistine
Art and Design Coordinator: Doris Bruey
Text Designer: Tracy Baldwin
Cover Designer: Lawrence Didona & Associates
Text Artwork: Publication Services
Photo Research: Teri Stratford
Director of EDP: Tim Frelick
Production Manager: Charlene Squibb

Cover Credit: TSW–Click/Chicago

Printed in the United States of America

DIGITAL CONCEPTS AND APPLICATIONS

ISBN: 0-03-026628-9

Library of Congress Catalog Card Number: 89-042923

012 045 98765432

To my wife, Loretta, and granddaughters, Aarika and Melissa.

To my parents, Rashid and Salma, my grandfather, H. H. Ismail, and grandmother, Bernadette Carvalho.

Preface

This text covers the concepts and applications of digital circuits and is intended for college courses in two- and four-year programs in electronics technology, electrical engineering technology, and computer technology. The material in the book may also be well suited for digital circuit courses in computer science and some engineering programs. The prerequisites for this book are a basic knowledge of algebra and elementary electrical concepts; for later chapters a familiarity with basic electronics is beneficial but not required. The material covered in this book can be used in either a one- or two-semester course in digital circuit concepts and applications, depending on the depth of coverage.

Approach

The emphasis of this text is an understanding of concepts and applications of those concepts. A good understanding of the elementary concepts of digital circuits is vital to understanding the design of various application circuits. Therefore, much of this book is dedicated to building a strong conceptual foundation for use in the analysis and synthesis (design) of digital circuits. Application circuits covered in this book are first introduced by designing the circuit. Then equivalent commercially available circuits (in the form of ICs) are investigated through analysis and compared and contrasted with the circuits designed earlier. This technique makes it easier for the student to understand the function and operation of a digital application circuit.

Organization

The text deals with eight distinct areas of study—numbering systems, combinational logic, sequential and register-transfer logic, memory and I/O systems, digital computer and microprocessor architecture, programmable logic devices, logic technology, and interfacing.

Chapter 1 introduces numbering systems with an emphasis on the binary numbering system. Since we are accustomed to the decimal numbering system, it is used to explain some of the elementary concepts of other numbering systems. Also included is a detailed look at the octal and hexadecimal numbering systems since they play an important role in digital computers. Conversions between numbering systems and concepts of binary arithmetic and binary codes are also included. The chapter is written so that the sequence in which the different numbering systems are covered can be altered very easily to suit the instructor and course organization.

Chapters 2 through 4 cover the basic concepts of combinational logic and their application circuits. This material focuses on some of the application circuits that make up a digital computer as well as many other digital devices. The elementary concepts of logic are introduced first, followed by the basic logic gates, equations, truth tables, and simple circuits. Next the mathematics of logic—Boolean algebra—is covered in detail along with techniques for analyzing and synthesizing logic circuits. Finally these concepts are applied to the operation and design of specific application circuits such as adders, subtracters, comparators, and so on, and their hardware implementations in the form of integrated circuits. Techniques for implementing and troubleshooting combinational logic circuits in the laboratory are included throughout Chapters 2 through 4.

Sequential logic is covered in Chapters 5 and 6, and register-transfer logic is covered in Chapter 7. These chapters introduce the elementary concepts of sequential logic circuits and the building block of all sequential logic—the flip-flop. The operation of the different types of flip-flops is covered in detail, along with their applications in counters, registers, and register files. Also included in these chapters is an analysis of the various integrated circuits that implement the basic flip-flops, counters, registers, and so on in hardware as well as procedures to troubleshoot sequential logic circuits. Chapter 6 is oriented toward the design, analysis, and applications of various counters. Instructors who do not want to cover detailed design can omit the design procedures in Section 6–4 and use the circuits designed as classroom examples. Chapter 7 deals with the various types of registers and their applications.

The study of sequential logic and register files leads naturally to the design and operation of memory systems. The fundamental characteristics of memory are investigated in Chapter 8, along with the concepts of accessing memory and I/O and troubleshooting procedures for various circuits. These concepts and procedures are then used in Chapters 9, 10, and 13 in the design of memory and I/O systems for a digital computer.

The architecture of a digital computer integrates the first four areas of study—numbering systems, combinational logic, sequential logic, and memory and I/O systems. Due to their popularity, the microprocessor and microcomputer are used as models for the architecture of a typical digital computer. Chapter 10 covers this material and investigates the basic components and circuits that make up a digital computer. This treatment includes a detailed look at the CPU and microprocessor as well as the relationship between computer hardware and software. Techniques for troubleshooting such systems are also included.

Chapter 11 is a study of a related application of logic circuits—programmable logic devices (PLDs). PLDs are frequently used to replace many types of combinational and sequential logic circuits, and therefore some coverage is devoted to the operation and design of these devices.

The technology of logic circuits is covered in Chapter 12, including the electronics that makes up the basic logic gates. Chapter 12 begins with the application of the BJT and MOSFET as an electronic switch and its use in simple switching circuits. Next, the various electronic circuits that make up logic gates are analyzed. The switching and loading characteristics of the two major families of logic circuits (MOS and bipolar) are also investigated here. Laboratory procedures for troubleshooting the logic circuits of different families are included.

Chapter 13 covers the operation of various interface circuits. Circuits that interface various serial and parallel peripheral devices to a digital computer are examined here with their applications. Also covered in this chapter are the design and characteristics of such circuits as analog-to-digital converters and digital-to-analog converters, which provide the "link" between the analog and digital "worlds."

For convenience, the book also contains several appendices. Powers of 2 are listed in Appendix A to assist in various calculations. Appendix B contains a detailed description and explanation of the ANSI/IEEE 91-1984 logic symbols. However, due to their popularity, the distinctive shape symbols of the old standard have been retained for most of the logic circuits discussed in the book. Appendix C contains the manufacturer's data sheets of all the TTL integrated circuits used in the book, although selected portions of these data sheets have been reproduced throughout the text when necessary. A glossary of commonly used terms and definitions is included in Appendix D. Appendix E contains the answers to all odd-numbered end-of-chapter problems.

Flexibility

The material is organized so that it can be covered with or without a basic knowledge of electronics or electricity. The logic concepts introduced in Chapters 1 through 11 do not require an understanding of electronics or electricity. The investigations of various integrated circuits are separated in subsections so that they may be omitted if desired without any loss of continuity. The internal electronic construction of logic gates is not covered until Chapter 12. The electrical and switching characteristics of logic gates, however, are introduced early to complement the discussions on troubleshooting. Since this material is sectioned off, it also may be omitted if desired. Many academic programs introduce digital circuits early in their curriculum to allow for a larger selection of more advanced courses in microprocessors and related fields. The material in this book is organized to accommodate such programs since it can be covered without a prerequisite course in semiconductor devices. Some computer science programs may choose to omit the coverage of digital electronics entirely (Chapter 12) and possibly even the subsections that deal with integrated circuit logic, electrical and switching characteristics, and troubleshooting.

Learning aids

The organization of each chapter is designed to enhance the learning process of the student. Every chapter begins with an introduction that includes a list of chapter objectives and a chapter outline. This organization allows the reader to quickly identify the content of each chapter and to review the material thoroughly. The material is broken up into logical sections and contains an abundant supply of examples and illustrations to enhance the discussions and reinforce the concepts. Review questions are included at the end of every section and can be used as a self-test to determine if the concepts are understood. An end-of-chapter summary lists the key concepts and terms covered in the chapter in an easy-to-read form. At the end of every chapter are several problems and exercises marked by section, as well as troubleshooting case studies.

The following is a summary of the pedagogical and instructional features that are included in each chapter:

- **Chapter-opening photograph** of general interest relates to the subject matter. This helps to capture the student's interest in the material.
- **Chapter outline** identifies the sections and topics covered in the chapter.
- **List of objectives** assists in identifying material covered in the chapter.
- **Introduction** leads the student into the chapter's contents.
- **Two-color illustrations** make it easier to understand and analyze complex digital circuits and diagrams.
- **Review questions** at the end of each section are provided as a self-test.
- **Examples** and their solutions explain the various concepts and applications introduced in each chapter.
- **Troubleshooting case studies** introduce real-world examples of practical faults in digital circuits and procedures for correcting them.
- **Key terms** and **definitions** are identified throughout the text using italics for easy reference.
- **Chapter summaries** include a compiled list of key terms, definitions, and concepts covered in the chapter in an easy-to-read format.
- **End-of-chapter problems, questions,** and **case studies** are organized by section for easy reference.
- An eight-page four-color photographic insert on **careers in digital electronics** is included in the book. These photographs portray career opportunities open to students of digital electronics.

Ancillary package

A comprehensive ancillary package supports the textbook. The following is a list of its components, which can be used to assist greatly both the instructor and the student in preparing and studying the material covered in the book.

Instructor's manual with transparency masters and test bank

This manual contains the complete solutions and answers to all of the end-of-chapter problems. Also included is a competition comparison chart and a chart to coordinate the individual ancillary components with the material covered in the textbook. Transparency masters for approximately 100 tables and illustrations from the textbook are included to assist with class lectures. The Test Bank includes approximately 40 test items per chapter for use in preparing quizzes, tests, and exams. These test items are variations on the review questions and end-of-chapter problems from the textbook and include solutions.

Transparency acetates

Approximately 50 acetates are available to the instructor to complement and facilitate his or her lecture presentation. These selected illustrations can give the instructor a quick start into the course.

Laboratory manual

Approximately 30 class-tested lab experiments related to the material covered in the textbook are included in this manual. Lab experiments are organized with an introduction, lab objectives, equipment, and procedures section. Results and answers obtained from a lab experiment can be filled in by the student on separate pages and detached from the manual for submission to the instructor.

Instructor's laboratory manual

This manual contains the material in the Laboratory Manual along with answers and expected results for each experiment. Instructions on preparing the laboratory experiment are also included.

Computerized test bank

A compilation of all the Test Bank questions and problems is available on an IBM PC disk. The instructor can easily modify both the questions and the accompanying figures and diagrams to form new questions. The Computerized Test Bank provides a simple user interface and does not require detailed computer knowledge or experience to operate.

ACKNOWLEDGMENTS

We would like to individually acknowledge all the people who were involved with this project, but there are too many to mention in the little space that we have. However there are a few people that deserve special mention.

We appreciate the efforts of the following people in helping us prepare the manuscript by giving us their encouragement and honest criticisms during the review process:

Richard Anthony, Cuyahoga Community College
Jack Braun, Houston Community College
Herb Daugherty, Indiana Vocational Technical College, Muncie
Dick Foster, Grand Rapids Junior College
Harmit Kaur, Sinclair Community College
Clay Laster, San Antonio College
Mary McNamara, Lorrain County Community College
Frank Pugh, Santa Rosa Junior College
William Reed, DeVry Institute of Technology, Kansas City

We would also like to thank our Department Chairman, Joseph Farren, for allowing us to use the "raw" manuscript in our Digital Circuits course and for arranging our schedules so that we could teach the course on a regular basis. We would like to acknowledge the dedication and hard work of all the people at Saunders who were involved with the production of the book: Barbara Gingery, Electronics Technology Editor; Maureen Iannuzzi, Project Editor; Alexa Barnes, Developmental Editor; Laura Shur, Editorial Assistant; and Teri Stratford, photo researcher.

Last but certainly not the least, we would like to thank Loretta Rooney and Rhonda Long for helping us prepare the manuscript, and the many students at the University of Dayton who helped "iron-out" the "wrinkles" in the manuscript.

A. R. I.
V. M. R.

Contents

The abacus is often considered to be the first computer invented. It was used to assist in calculations and was constructed with pebbles or, in Latin, "calculus," which is where the word "calculate" comes from (The Bettman Archive).

OUTLINE

OBJECTIVES

The objectives of this chapter are to

- introduce the basic concepts that relate to the decimal numbering system.
- explore the procedures for performing arithmetic using only the process of addition.
- apply the decimal numbering system concepts to the octal, hexadecimal, and binary numbering systems.
- study techniques of converting between different numbering systems.
- study the various procedures for performing binary arithmetic.
- investigate the different types of binary codes and their applications.

1 NUMBERING SYSTEMS

1–1 INTRODUCTION

Numbering systems have been in existence for thousands of years and have provided us with the means to assign value, determine quantity, calculate distances, measure sizes, etc. Besides serving as a very important tool in people's everyday lives, numbering systems play an important part in the design and operation of digital circuits.

The decimal numbering system has been in existence for over 2000 years and is simply based on a series of ten unique digits that were originally used to represent the ten fingers of the hand (hence the use of the term "digit," derived from the latin word "digitus," meaning "finger"). This implies that a numbering system does not have to be based on ten digits but can be based on a larger or smaller number. There are a few other numbering systems that are more suitable in the study of digital computers and their associated circuits, and therefore it is important to be able to understand some of the basic concepts of numbering systems so that these concepts can be applied to other numbering systems that are important in the study of digital circuits. Because of our familiarity with the decimal numbering system, this chapter will begin by introducing us to some of the basic numbering system concepts using the decimal numbering system as a model and then will apply those concepts to three other numbering systems—octal, hexadecimal, and binary. The octal and hexadecimal numbering systems will be covered first since they are closest to the decimal numbering system. Once we have developed a good understanding of our own (decimal) numbering system and two other closely related numbering systems we shall then study the binary numbering system. This study will include the binary counting sequence; conversions between binary numbers and decimal, octal, and hexadecimal numbers; and representation of binary numbers, binary arithmetic, and binary codes. We will then apply many of the concepts and procedures developed in this chapter to the design and operation of various digital circuits to be covered in subsequent chapters.

1–2 Decimal numbering system concepts

The decimal numbering system is based on a set of ten digits, each digit representing a value. These ten digits in the order of value can be listed as follows:

0 (zero)
1 (one)
2 (two)
3 (three)
4 (four)
5 (five)
6 (six)
7 (seven)
8 (eight)
9 (nine)

The first digit 0 (zero) has the lowest value (no value) and the tenth digit 9 (nine) has the highest value. If there is a need to have larger values represented by the decimal numbering system we start using *combinations* of the basic ten digits. Therefore after 9, the next highest values would be

10 (ten)
11 (eleven)
.
.
.
etc.

Note that "10" (ten), "11" (eleven), etc. are not digits but combinations of digits that are assigned increasing values. If we continue assigning values to these combinations we obtain a *count*. If we obtain a count of all possible decimal digits taken singly we have a count of 10:

0, 1, 2, 3, 4, 5, 6, 7, 8, 9

If we obtain a count of all possible numbers taking two digits at a time, we have a count of 100:

00, 01, 02, 03, . . . , 10, 11, 12, . . . , 20, 21, 22, . . . , 97, 98, 99

That is, there are 100 combinations of digits between 0 and 99 when the combinations are taken two at a time.

Similarly, there are 1000 combinations of digits between 0 and 999 when the combinations are taken three at a time:

000, 001, 002, . . . , 010, 011, . . . , 098, 099, 100, 101, . . . , 999

In general, for any numbering system we can determine the number of combinations of n digits by using the following relationship:

$$B^n$$

where B is the *base* of the numbering system and n is the number of digits to be combined. Since the decimal numbering system is a base-ten numbering system (i.e., there are ten digits in the numbering system), the number of combinations of n decimal digits is

$$10^n$$

Thus, there are 10^2 or 100 combinations of two digits, 10^3 or 1000 combinations of three digits, 10^4 or 10,000 combinations of four digits, and so on. When these combinations are organized into an orderly sequence of increasing values, we obtain a count.

EXAMPLE 1–1

How many combinations would there be of five decimal digits and what would the largest value in the count be?

SOLUTION The number of combinations would be 10^5 or 100,000. The count would begin at 00000 and end at 99999. Therefore, the largest value in the count would be 99999.

A decimal number such as 6375 is simply a combination of the four decimal digits 6, 3, 7, and 5. However, the place, or *position,* of each digit in the number is of significance. For example, even though the number 7653 has the same digits as 6375, the two numbers have different values. Each digit in a decimal number is assigned a *positional weight.* For example

$$\begin{array}{rcccc} \text{number} \longrightarrow & 6 & 3 & 7 & 5 \\ & \uparrow & \uparrow & \uparrow & \uparrow \\ \text{positional weight} \longrightarrow & 10^3 & 10^2 & 10^1 & 10^0 \end{array}$$

In the number 6375, the digit 5 has a weight of 10^0, or 1, the digit 7 has a weight of 10^1, or 10, the digit 3 has a weight of 10^2, or 100, and the digit 6 has a weight of 10^3, or 1000. We can express the value of the number 6375 as follows:

$$\begin{array}{rcl} 6 \times 10^3 = 6 \times 1000 & = & 6000 \\ 3 \times 10^2 = 3 \times 100 & = & 300 \\ 7 \times 10^1 = 7 \times 10 & = & 70 \\ 5 \times 10^0 = 5 \times 1 & = & \underline{5} \; + \\ & & 6375 \end{array}$$

Thus each digit in a decimal number is assigned a position (beginning with 0) and a weight that is a power of 10. The rightmost digit of any number is called the *least significant digit* (LSD) and the leftmost digit is called the *most significant digit* (MSD). The LSD always has a position of 0 associated with it while the position of the MSD will depend on the size of the number. Therefore for the number 6375

$$\begin{array}{rcccc} & 6 & 3 & 7 & 5 \\ & \uparrow & \uparrow & \uparrow & \uparrow \\ \text{position:} & 3 & 2 & 1 & 0 \\ & \uparrow & & & \uparrow \\ & \text{MSD} & & & \text{LSD} \end{array}$$

the LSD is 5 (position 0) and the MSD is 6 (position 3).

EXAMPLE 1–2

Determine the value of the number 90,812 in terms of its positional weights.

SOLUTION

$$\begin{array}{rcl} 9 \times 10^4 = 9 \times 10{,}000 & = & 90{,}000 \\ 0 \times 10^3 = 0 \times 1000 & = & 0000 \\ 8 \times 10^2 = 8 \times 100 & = & 800 \\ 1 \times 10^1 = 1 \times 10 & = & 10 \\ 2 \times 10^0 = 2 \times 1 & = & \underline{2} \; + \\ & & 90{,}812 \end{array}$$

In general, for a number in any numbering system, the value of any digit in the number is given by the expression

$$D \times B^p$$

where D is the digit, B is the base of the numbering system, and p is the position of the digit. We will apply this relationship to other numbering systems in later sections.

We can also develop a technique for working the positional weight system in reverse. That is, given a decimal number that represents a value, we can obtain the digits that make up the number by a series of repetitive divisions by the base of the numbering system. For example, to determine the digits that make up the decimal number 125, we can first divide 125 by 10

$$125 \div 10 = 12 \text{ with a remainder of } 5$$

Notice that the remainder 5 is the LSD.
Next we take the result 12 and again divide by 10

$$12 \div 10 = 1 \text{ with a remainder of } 2$$

Notice that the remainder 2 is the next digit.
Since 1 cannot be divided any further, it is the final remainder that is the MSD.

To illustrate this procedure further, consider the following example.

EXAMPLE 1–3

Determine the digits that make up the decimal number 8475 by using the repetitive division procedure.

SOLUTION

10	8475	5	⟵ remainder
10	847	7	⟵ remainder
10	84	4	⟵ remainder
		8	⟵ remainder

The digits that make up the decimal number 8475 are given by the remainders (reading bottom to top) 8, 4, 7, 5.

This procedure will also be applied to the other numbering systems discussed later in this chapter to facilitate conversions from decimal to other numbering systems.

Almost all mathematical calculations can be performed through the use of the four basic arithmetic operations—addition, subtraction, multiplication, and division. Furthermore, addition is particularly important since it can be used to perform the other three arithmetic operations. For example, one can subtract two numbers by adding a negative number to a positive number or vice versa. Multiplication can be performed by repetitive addition, and division by repetitive subtractions. Many digital computers perform all arithmetic operations through the process of addition, and therefore the techniques by which these operations are conducted are worth investigating.

Decimal addition

The simple process of adding decimal numbers is often taken for granted in everyday life. However, since we will be applying the concepts of the decimal numbering system and decimal arithmetic operations to various other numbering systems, it is important to identify certain procedures and terms used when performing these operations.

Consider the addition of two decimal numbers, 7632 and 6489. The addition is performed by adding each pair of digits in both numbers starting with the LSDs, obtaining a sum of the two LSDs, keeping the LSD of this sum as the LSD of the final sum, carrying over the MSD of the sum for addition to the next pair of digits. The process is then repeated for the pairs that follow until all pairs of digits have been added. We can illustrate this procedure in a less complicated manner as follows:

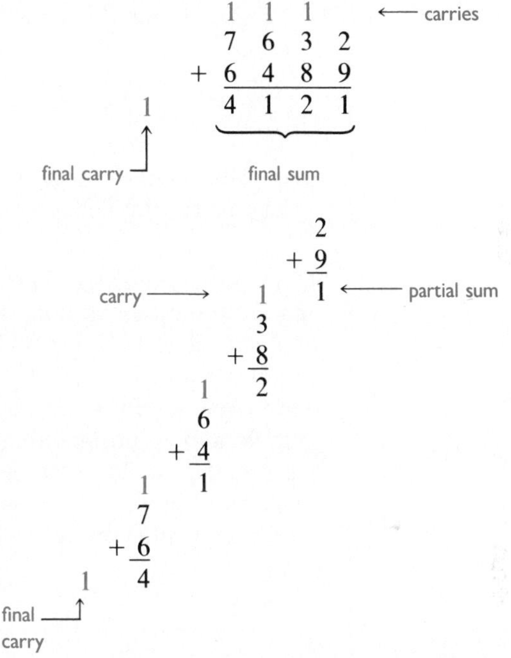

Notice that since we have performed the addition of 2 four-digit numbers, the final sum is considered to be the four-digit number 4121 with a final carry of 1.

Decimal subtraction—9's complement

The *9's complement* technique of subtracting decimal numbers allows us to perform subtraction by adding a positive number to a negative number and vice versa. In order to apply this technique of subtraction, we must first define the 9's complement of a decimal number.

The *9's complement* of a decimal number can be used to represent a negative number. It is obtained by subtracting each digit in the number from 9. For example, the 9's complement of the number 2 is 7, and the 9s complement of the number 6 is 3. A *signed* negative decimal number can be represented as an *unsigned* 9's complement representation. For example, the 9's complement representation of the number −5 is 4. Here 4 is not a positive number but is an unsigned representation of the number −5. The unsigned number 4 and the signed number −5 are *equivalent* and *complementary*. It can be seen that it is very easy to confuse negative unsigned numbers represented in 9's complement form with positive numbers. The 9's complement of multidigit decimal numbers can be obtained by subtracting each digit in the number from 9.

EXAMPLE 1–4

Represent the following signed negative numbers in 9's complement form:
(a) −1 (b) −8 (c) −4 (d) −9 (e) −72 (f) −4218

SOLUTIONS
(a) 9 − 1 = 8
Therefore, the 9's complement representation of −1 is 8.
(b) 9 − 8 = 1
Therefore, the 9's complement representation of −8 is 1.
(c) 9 − 4 = 5
Therefore, the 9's complement representation of −4 is 5.
(d) 9 − 9 = 0
Therefore, the 9's complement representation of −9 is 0.
(e) 99 − 72 = 27
Therefore, the 9's complement representation of −72 is 27.
(f) 9999 − 4218 = 5781
Therefore, the 9's complement representation of −4218 is 5781.

The word *complement* in the term "9's complement" implies that we can apply the same operation in reverse. For example, if the 9's complement of 6 is 3, then the 9's complement of 3 is 6. Similarly, if the 9's complement of 72 is 27, then the 9's complement of 27 is 72. This allows one to very easily switch back and fourth between a signed negative number and an unsigned negative number in 9's complement representation. For example, if the number 38,743 is the 9's complement representation of some negative number, then in order to obtain its actual (or signed) value you can obtain its 9's complement:

$$\begin{array}{r} 99{,}999 \\ -\ \underline{38{,}743} \\ 61{,}256 \end{array}$$

Therefore the actual value of the number is −61,256.

EXAMPLE 1–5

The following is a list of negative numbers expressed in 9's complement form. Determine the signed representation of each number.
(a) 937,569 (b) 8 (c) 54 (d) 734,958,457

SOLUTIONS
(a) 9's complement of 937,569 is 999,999 − 937,569 = 62,430.
Therefore the signed number is −62,430.
(b) 9's complement of 8 is 9 − 8 = 1.
Therefore the signed number is −1.
(c) 9's complement of 54 is 99 − 54 = 45
Therefore the signed number is −45.
(d) 9's complement of 734,958,457 is
999,999,999 − 734,958,457 = 265,041,542
Therefore the signed number is −265,041,542.

To obtain the difference between two numbers using the 9's complement subtraction technique, the following steps are performed:

1. Obtain the 9's complement representation of the negative number.
2. Add the 9's complement representation of the negative number (from step 1) to the positive number.
3. Add the final carry obtained in step 2 to the final sum obtained by adding the two numbers; this is known as the *end-around* carry. The sum obtained will be the final result (the difference between the two numbers).

To illustrate the procedure used in 9's complement arithmetic let us consider the following examples:

EXAMPLE 1–6

Perform the following subtractions using the 9's complement arithmetic technique:
(a) 6 − 4 = 2 (b) 9 − 8 = 1 (c) 42 − 39 = 3
(d) 125 − 67 = 58

SOLUTIONS

(a) The 9's complement representation of −4 is 5.

```
                         6
                       + 5
end-around carry ──→ 1   1
                     └─→ 1
                         2 ←── final result
```

(b) The 9's complement representation of −8 is 1.

```
                         9
                       + 1
end-around carry ──→ 1   0
                   + └─→ 1
                         1 ←── final result
```

(c) The 9's complement representation of −39 is 60.

```
                        42
                      + 60
end-around carry ──→ 1  02
                   + └─→ 1
                         3 ←── final result
```

(d) In this problem since the largest number being operated on is a three-digit number, we must assume that both numbers are three digits long. Thus 125 − 067 = 058.

The 9's complement representation of −067 is 932.

```
                       125
                     + 932
end-around carry ──→ 1 057
                   + └──→ 1
                       058 ←── final result
```

Notice in all the problems in Example 1–6 that the results of the subtraction were positive numbers and that every addition produced an end-around carry of 1 that was added to the sum to obtain the final result. This is true in *all* cases of 9's complement arithmetic that yield positive results. However, when the 9's complement technique is performed on operations that yield negative results, the carry will always be 0, and the result will always be in *unsigned* 9's complement representation. Because the 9's complement representation of a negative number is unsigned, the carry is the only means by which one can identify the final result as being positive or negative. Furthermore, if the result is negative, the result must be recomplemented in order to obtain its signed value. To illustrate this fact let us consider the following examples.

EXAMPLE 1–7

Perform the following subtractions using the 9's complement arithmetic technique:
(a) $4 - 6 = -2$ (b) $8 - 9 = -1$ (c) $39 - 42 = -3$
(d) $67 - 125 = -58$ (e) $7 - 7 = 0$

SOLUTIONS
(a) The 9's complement representation of -6 is 3.

```
                          4
                        + 3
end-around carry ——→   0  7
(negative result)    + └→ 0
                          7 ←— final result in 9's complement representation
```

The 9's complement of 7 is 2, therefore the signed result is -2.
(b) The 9's complement representation of -9 is 0.

```
                          8
                        + 0
end-around carry ——→   0  8
(negative result)    + └→ 0
                          8 ←— final result in 9's complement representation
```

The 9's complement of 8 is 1, therefore the signed result is -1.
(c) The 9's complement representation of -42 is 57.

```
                         39
                       + 57
end-around carry ——→   0 96
(negative result)    + └→ 0
                         96 ←— final result in 9's complement representation
```

The 9's complement of 96 is 3, therefore the signed result is -3.
(d) In this problem since the largest number being operated on is a three-digit number, we must assume that both numbers are three digits long. Thus $067 - 125 = -058$.
The 9's complement representation of -125 is 874.

```
                         067
                       + 874
end-around carry ——→   0 941
(negative result)    + └→  0
                         941 ←— final result in 9's complement representation
```

The 9's complement of 941 is 58, therefore the signed result is −58.
(e) The 9's complement representation of −7 is 2.

```
                                7
                              + 2
end-around carry  ──→      0    9
(negative result)        + └──→ 0
                                9  ←── final result in 9's complement representation
```

The 9's complement of 9 is 0, therefore the signed result is −0. Since a sign is not generally associated with the digit 0, the negative sign can be dropped in this case.

Notice in all the above examples that even though the end-around carry is 0 in all the cases, you still go ahead and add it in so that the same procedure is followed for both types of results. In Chapter 4 we will see that the circuit that exactly implements this procedure will always add in the end-around carry regardless of its value.

In all the examples worked so far, notice that the 9's complement technique can be used to obtain the difference between two numbers without having to perform a conventional subtraction, but simply by adding two numbers (if we ignore the fact that we performed a conventional subtraction to obtain the 9's complement representation of the negative number). The advantage of performing subtraction in this manner will be seen in Section 1–6 and in Chapter 4.

Decimal subtraction—10's complement

The 9's complement technique is just one of two techniques that may be used to obtain the difference between two numbers through the process of addition. Another technique may be used to obtain the same results. This technique is known as 10's complement subtraction.

The *10's complement* of a decimal number can also be used to represent a negative number. It is obtained by subtracting the number from 10. For example, the 10's complement of the number 2 is 8, and the 10's complement of the number 6 is 4. A signed negative decimal number can be represented as an unsigned 10's complement representation. For example, the 10's complement representation of the number −4 is 6. Here 6 is not a positive number but is an unsigned representation of the number −4. Another way of determining the 10's complement of a number is to obtain its 9's complement and then add 1 to it; this is the preferred way of obtaining the 10's complement since it is more applicable to the technique used in binary arithmetic (examined in Section 1–6) and in digital computer circuits.

EXAMPLE 1–8

Represent the following signed negative numbers in 10's complement form:
(a) −1 (b) −8 (c) −4 (d) −9 (e) −72 (f) −4218

SOLUTIONS
(a) The 9's complement representation of −1 is 8.
Therefore its 10's complement representation is 8 + 1, or 9.

(b) The 9's complement representation of −8 is 1.
Therefore its 10's complement representation is 1 + 1, or 2.
(c) The 9's complement representation of −4 is 5.
Therefore its 10's complement representation is 5 + 1, or 6.
(d) The 9's complement representation of −9 is 0.
Therefore its 10's complement representation is 0 + 1, or 1.
(e) The 9's complement representation of −72 is 27.
Therefore its 10's complement representation is 27 + 1, or 28.
(f) The 9's complement representation of −4218 is 5781.
Therefore its 10's complement representation is 5781 + 1, or 5782.

As mentioned before, the word "complement" in the term "10's complement" implies that the same operation can be applied in reverse. For example, if the 10's complement of 6 is 4, then the 10's complement of 4 is 6. Similarly, if the 10's complement of 72 is 28, then the 10's complement of 28 is 72. We can also switch back and forth between a signed negative number and an unsigned negative number in 10's complement representation. For example, if the number 38,743 is the 10's complement representation of some negative number, then in order to obtain its actual (or signed) value we can obtain its 10's complement.

$$\begin{array}{r} 99{,}999 \\ -\ \underline{38{,}743} \\ 61{,}256 \\ +\ \underline{\qquad 1} \\ 61{,}257 \end{array}$$

Therefore the actual value of the number is −61,257.

EXAMPLE 1–9

The following is a list of negative numbers expressed in 10's complement form. Determine the signed representation of each number.
(a) 937,569 (b) 8 (c) 54 (d) 734,958,457

SOLUTIONS
(a) 999,999 − 937,569 = 62,430 (9's complement)
62,430 + 1 (10's complement)
Therefore the signed number is −62,431.
(b) 9 − 8 = 1 (9's complement)
1 + 1 (10's complement)
Therefore the signed number is −2.
(c) 99 − 54 = 45 (9's complement)
45 + 1 (10's complement)
Therefore the signed number is −46.
(d) 999,999,999 − 734,958,457 = 265,041,542 (9's complement)
265,041,542 + 1 (10's complement)
Therefore the signed number is −265,041,543.

In order to obtain the difference between two numbers using the 10's complement subtraction technique, we must perform the following steps.

1. Obtain the 10's complement representation of the negative number.

2. Add the 10's complement representation of the negative number (from step 1) to the positive number.
3. Ignore the final carry obtained from step 2, but interpret the meaning of the carry. That is, a carry of 1 will indicate a positive result, and a carry of 0 will indicate a negative result. The sum obtained from step 2 will be the final result (the difference between the two numbers). If the result is negative it will be represented in unsigned 10's complement form.

To illustrate the procedure used in 10's complement arithmetic let us consider the following examples:

EXAMPLE 1–10

Perform the following subtractions using the 10's complement arithmetic technique:
(a) $6 - 4 = 2$ (b) $9 - 8 = 1$ (c) $42 - 39 = 3$ (d) $125 - 67 = 58$
(e) $7 - 7 = 0$

SOLUTIONS
(a) The 10's complement representation of -4 is 6.

$$\begin{array}{r} 6 \\ +\ 6 \\ \hline 1\quad 2 \end{array}$$

ignore carry ⟶ 1, 2 ⟵ final result

(b) The 10's complement representation of -8 is 2.

$$\begin{array}{r} 9 \\ +\ 2 \\ \hline 1\quad 1 \end{array}$$

ignore carry ⟶ 1, 1 ⟵ final result

(c) The 10's complement representation of -39 is 61.

$$\begin{array}{r} 42 \\ +\ 61 \\ \hline 1\quad 03 \end{array}$$

ignore carry ⟶ 1, 03 ⟵ final result

(d) In this problem since the largest number being operated on is a three-digit number, we must assume that both numbers are three digits long. Thus $125 - 067 = 058$.
The 10's complement representation of -067 is 933.

$$\begin{array}{r} 125 \\ +\ 933 \\ \hline 1\quad 058 \end{array}$$

ignore carry ⟶ 1, 058 ⟵ final result

(e) The 10's complement representation of -7 is 3.

$$\begin{array}{r} 7 \\ +\ 3 \\ \hline 1\quad 0 \end{array}$$

ignore carry ⟶ 1, 0 ⟵ final result

EXAMPLE 1–11

Perform the following subtractions using the 10's complement arithmetic technique:
(a) $4 - 6 = -2$ (b) $8 - 9 = -1$ (c) $39 - 42 = -3$
(d) $67 - 125 = -58$

SOLUTIONS

(a) The 10's complement representation of -6 is 4.

$$\begin{array}{r} 4 \\ +\ 4 \\ \hline 0\ \ 8 \end{array}$$

ignore carry ⟶ 0 (negative result); 8 ⟵ final result in 10's complement representation

The 10's complement of 8 is 2, therefore the signed result is -2.

(b) The 10's complement representation of -9 is 1.

$$\begin{array}{r} 8 \\ +\ 1 \\ \hline 0\ \ 9 \end{array}$$

ignore carry ⟶ 0 (negative result); 9 ⟵ final result in 10's complement representation

The 10's complement of 9 is 1, therefore the signed result is -1.

(c) The 10's complement representation of -42 is 58.

$$\begin{array}{r} 39 \\ +\ 58 \\ \hline 0\ \ 97 \end{array}$$

ignore carry ⟶ 0 (negative result); 97 ⟵ final result in 10's complement representation

The 10's complement of 97 is 3, therefore the signed result is -3.

(d) In this problem since the largest number being operated on is a three-digit number, we must assume that both numbers are three digits long. Thus $067 - 125 = -058$.

The 10's complement representation of -125 is 875.

$$\begin{array}{r} 067 \\ +\ 875 \\ \hline 0\ \ 942 \end{array}$$

ignore carry ⟶ 0 (negative result); 942 ⟵ final result in 10's complement representation

The 10's complement of 942 is 58, therefore the signed result is -58.

Decimal multiplication and division

Just as we investigated two different techniques for calculating the difference between two numbers using only the process of addition, we can now examine techniques to perform multiplication and division through the processes of repetitive addition and subtraction, respectively.

The process of multiplying two numbers is simply a process by which one number, the *multiplicand*, is repetitively added to itself a certain number of times (specified by the *multiplier*) to obtain a *product*. For example, to perform the following multiplication:

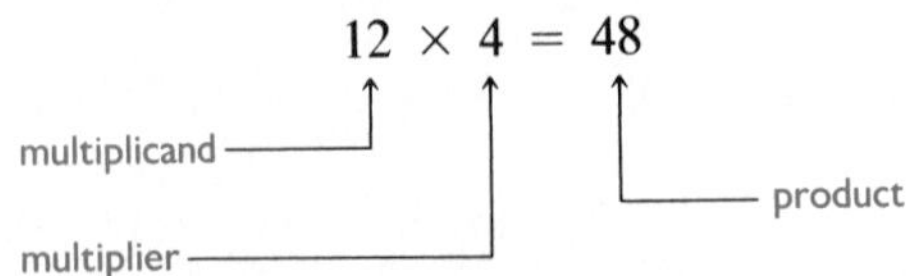

we can add the multiplicand, 12, four (multiplier = 4) times to itself to obtain a total sum of 48 (product):

$$\begin{array}{r} 12 \\ 12 \\ 12 \\ +\ \underline{12} \\ 48 \end{array}$$

The longhand technique of multiplication can also be used, but the repetitive addition technique is more applicable to how a digital computer would perform multiplication using the binary numbering system (to be discussed in Section 1–6).

Dividing two decimal numbers can be accomplished by repetitive subtraction, and since subtraction can be accomplished by addition (9's or 10's complement subtraction), the entire division process can be done through addition. For example, consider dividing the number 24 (*dividend*) by 3 (*divisor*) to obtain a result of 8 (*quotient*).

```
                    8    ← quotient
divisor ⟶   3 ⌋ 24    ← dividend
                   24
                   --
                    0    ← remainder
```

The operation can also be performed by repetitively subtracting the divisor (3) from the dividend (24) until a remainder of 0 or a remainder that is less than the divisor (3) is obtained. The quotient will be a count of the number of subtractions that have taken place:

		quotient (tally)	
dividend:	24		
divisor:	− 3	1	
remainder:	21		
divisor:	− 3	1	
remainder:	18		
divisor:	− 3	1	
remainder:	15		
divisor:	− 3	1	
remainder:	12		
divisor:	− 3	1	
remainder:	9		
divisor:	− 3	1	
remainder:	6		
divisor:	− 3	1	
remainder:	3		
divisor:	− 3	+ 1	
final remainder ⟶	0	8	← quotient

To illustrate this procedure with numbers that yield a remainder when divided, consider the following example.

EXAMPLE 1–12

Divide the number 96 by 9 using the repetitive subtraction procedure:

		10	⟵ quotient
divisor ⟶	9	96	⟵ dividend
		90	
		6	⟵ remainder

SOLUTION

		quotient (tally)
dividend:	96	
divisor:	− 9	1
remainder:	87	
divisor:	− 9	1
remainder:	78	
divisor:	− 9	1
remainder:	69	
divisor:	− 9	1
remainder:	60	
divisor:	− 9	1
remainder:	51	
divisor:	− 9	1
remainder:	42	
divisor:	− 9	1
remainder:	33	
divisor:	− 9	1
remainder:	24	
divisor:	− 9	1
remainder:	15	
divisor:	− 9	+ 1
final remainder (less than 9) ⟶	6	10 ⟵ quotient

This section has dealt with some of the concepts of the decimal numbering system and decimal arithmetic, all of which will be applied to other numbering systems discussed in this chapter. The techniques of subtracting, multiplying, and dividing decimal numbers through the process of intrinsic addition was also investigated, and will be applied to the binary numbering system and binary arithmetic covered in Section 1–6. Since the decimal numbering system is one with which we are most familiar, a good understanding of these concepts will be of assistance in understanding other numbering systems as well as some of the circuits to be designed and analyzed later in this book.

Review questions

1. What is meant by the *base* of a numbering system?
2. Why are the decimal numbers 23 and 32 different in value when they both have the same digits?
3. How can we represent a negative number without a sign?
4. In 9's and 10's complement subtraction, how is the end-around carry interpreted?

> **5.** Why is addition considered to be the most important arithmetic operation as far as digital computers are concerned?

1–3 The octal numbering system

The octal numbering system is a *base-eight* numbering system. That is, there are only eight digits in the numbering system. The word *octal* is derived from the latin word *octa* or "eight."

0 (zero)
1 (one)
2 (two)
3 (three)
4 (four)
5 (five)
6 (six)
7 (seven)

The digits, 8 (eight) and 9 (nine), do not exist in this system. Therefore if we were to count in octal, we would follow the same procedure as we did in decimal—start combining digits after we reach a count of 7:

. . . 7, 10, 11, 12, . . . , 15, 16, 17, 20, 21, . . . , 75, 76, 77, 100

Notice that if we were to examine a count of all possible combinations of two octal digits, the largest number would be 77. That is, the count would end at 77. Using the relationship introduced in Section 1–1, B^n, the total number of combinations for two octal digits would be

$$8^2$$

or

$$64$$

Therefore there are 64 (decimal) counts between the octal numbers 00 and 77. The *value* of the octal number 77 would actually be 63 (in decimal). Thus octal numbers are always larger than decimal numbers representing the same value; this is because the octal numbering system has fewer digits. Table 1–1 lists the relationship between the ten decimal digits and their octal equivalents.

Table 1–1
Decimal digits and their octal equivalents

Decimal	Octal
0	0
1	1
2	2
3	3
4	4
5	5
6	6
7	7
8	10
9	11

A note on terminology— if the decimal combination 10 is referred to as "ten," and the combination 11 as "eleven," etc., should the octal combination 10 be referred to as "ten" and the octal combination 11 as "eleven," or should these combinations be referred to as "eight" and "nine"? The answer is neither! The octal combination 10 should be referred to as "one-zero," 11 as "one-one," 12 as "one-two," and so on, since the terms "eight," "nine," "ten," "eleven," etc. are used to describe decimal digits and combinations that do not exist or are not the same in the octal numbering system. Furthermore, we have now been introduced to a numbering system that uses the same digits as the decimal numbering system although not all similar digits and combinations in both systems have the same value. Therefore it is important to distinguish between octal numbers and decimal numbers. The notation used to identify numbers in different numbering systems is to append to the number a *subscript* that identifies its base. For example, the following are octal numbers:

$$645_8,\ 7364510_8,\ 7_8,\ 13007_8$$

and the following are decimal numbers:

$$2746_{10},\ 92648_{10},\ 25347_{10},\ 4_{10}$$

At this stage it is important to be consistent in identifying numbers in different bases to eliminate confusion when dealing with different numbering systems. For example, if a number is specified as 72645, we would not know if it were an octal number or a decimal number. However, the number 223648 is definitely a decimal number since it contains the digit 8 which is not part of the octal set. Still, it is recommended that the base of the numbering system be identified when dealing with numbers from now on.

Since the decimal numbering system is the most frequently used, we often drop the subscript base specifier (10) when identifying decimal numbers and assume decimal as the *default*. Thus if a number such as 17263 is specified (without any base specifier) it will be assumed to be decimal.

Conversions

Converting a number from one numbering system to another is simply a process by which the value of a number in one system (or base) is converted to its equivalent value in another system (or base). For example, with reference to Table 1–1, the value of the octal number 11 is 9 in decimal. Therefore the decimal equivalent of 11_8 is 9_{10}. Extending the count in Table 1–1 to 77_8 and 63_{10}, we can obtain the octal and decimal equivalents. However, this is a very inconvenient way of performing conversions, especially with very large numbers.

Recall from Section 1–2 that the value of a decimal number could be obtained by taking the sum of the products of each digit and its positional weight. The positional weight of each digit was given by the relationship B^p, where B is the base of the numbering system and p is the position of the digit. This technique can also be applied to the octal numbering system to determine the (decimal) values of octal numbers. For example, consider the number 76251_8. The value of this number in decimal would be

$$(7 \times 8^4) + (6 \times 8^3) + (2 \times 8^2) + (5 \times 8^1) + (1 \times 8^0)$$
$$= (7 \times 4096) + (6 \times 512) + (2 \times 64) + (5 \times 8) + (1 \times 1)$$
$$= 28672 + 3072 + 128 + 40 + 1$$
$$= 31913_{10}$$

Again, notice that it takes a smaller decimal number (31913) to represent the same value as a larger octal number (76251).

EXAMPLE 1–13

A person represents herself as being 102_8 years old. What is her actual age? That is, what is her age in decimal?

SOLUTION

$$102_8 = (1 \times 8^2) + (0 \times 8^1) + (2 \times 8^0)$$
$$= 64 + 0 + 2$$
$$= 66_{10}$$

The person is thus 66 years old.

EXAMPLE 1–14

Convert the following numbers from octal to decimal.
(a) 6351_8 (b) 24073_8 (c) 791_8

SOLUTION

(a) $6351_8 = 6 \times 8^3 + 3 \times 8^2 + 5 \times 8^1 + 1 \times 8^0$
$= 6 \times 512 + 3 \times 64 + 5 \times 8 + 1 \times 1$
$= 3072 + 192 + 40 + 1$
$= 3305_{10}$

(b) $24{,}073_8 = 2 \times 8^4 + 4 \times 8^3 + 0 \times 8^2 + 7 \times 8^1 + 3 \times 8^0$
$= 2 \times 4096 + 4 \times 512 + 0 \times 64 + 7 \times 8 + 3 \times 1$
$= 8192 + 2048 + 0 + 56 + 3$
$= 10{,}299_{10}$

(c) No solution possible. 791_8 is not a valid octal number since "9" is not an octal digit.

To convert from decimal to octal, the process is similar to the one discussed in Section 1–2. That is, you must obtain the individual digits that make up the number by repetitively dividing the decimal number by 8 (the base of the octal numbering system). The remainders for each successive division will make up the octal number. The following example illustrates this procedure.

EXAMPLE 1–15

Convert the decimal number (a) 3305 to octal, (b) 10299 to octal.

SOLUTION

(a)

8	3305	1	← remainder
8	413	5	← remainder
8	51	3	← remainder
	6		← remainder

Reading the remainders from bottom to top, the octal equivalent is 6351_8.

(b)

8	10299	3	⟵ remainder
8	1287	7	⟵ remainder
8	160	0	⟵ remainder
8	20	4	⟵ remainder
		2	⟵ remainder

Reading the remainders from bottom to top, the octal equivalent is 24073_8.

Arithmetic

Arithmetic operations with octal numbers are performed in much the same way as arithmetic operations with decimal numbers, except that it is important for us to remember that the counting sequence for octal numbers is different and that the digits 8 and 9 do not exist. For example, in octal, 7 + 1 is not equal to 8 but is equal to 10, and 20 − 5 is not equal to 15 but is equal to 13.

Overall, in comparing the octal numbering system with the decimal numbering system many similarities can be seen. All the concepts of the decimal numbering system developed in Section 1–2 apply to octal; the differences only lie in the numbers being manipulated. Even though a numbering system such as octal, with its reduced repertoire of digits, seems disadvantageous at first, we shall see in Section 1–5 that the octal numbering system provides us with a very convenient way to represent numbers in the most important numbering system—binary.

Review questions

1. How many digits does the octal numbering system have?
2. What is the largest possible three-digit octal number?
3. Compare the octal and decimal numbering systems in terms of their representation of a value (such as 75).

1–4 The hexadecimal numbering system

The hexadecimal numbering system is a *base-sixteen* numbering system. That is, there are 16 digits in the numbering system. The word *hexadecimal* is derived from the latin word *hexa,* or "six," since it is used to describe a numbering system that has six additional digits over the decimal numbering system. Since the decimal numbering system uses the graphic symbols 0, 1, 2, 3, 4, 5, 6, 7, 8, 9 to represent the 10 basic digits, we must therefore "invent" six "new" graphic symbols to represent the additional six hexadecimal digits. However, instead of designing new graphic symbols for these additional six digits, we choose to use something that we are more familiar with—the letters A, B, C, D, E, F. Note, however, that when dealing with hexadecimal numbers these are no longer considered to be letters of the alphabet but are now treated as digits in exactly the same way as are the basic 10 digits 0 through 9.

0	(zero)
1	(one)
2	(two)
3	(three)
4	(four)
5	(five)
6	(six)
7	(seven)
8	(eight)
9	(nine)
A	(ay)
B	(bee)
C	(see)
D	(dee)
E	(ee)
F	(ef)

Now if one were to count in hexadecimal, one would follow the same procedure as one did in decimal and octal—start combining digits after a count of *F* is reached.

. . . F, 10, 11, 12, . . . , 19, 1A, 1B, 1C, 1D, 1E, 1F, 20, 21, . . . ,
99, 9A, . . . , 9F, A0, A1, . . . , FD, EF, FF, 100

Notice that if we were to examine a count of all possible combinations of two hexadecimal digits, the largest number would be FF. That is, the count would end at FF. Using the relationship introduced in Section 1–1, B^n, the total number of combinations for two hexadecimal digits would be

$$16^2$$

or

$$256$$

Therefore there are 256 (decimal) counts between the hexadecimal numbers 00 and FF. The *value* of the hexadecimal number FF would actually be 255 (in decimal). Thus hexadecimal numbers are always smaller than decimal numbers representing the same value; this is because the hexadecimal numbering system has more digits. Table 1–2 lists the relationship between the 16 hexadecimal digits and their decimal equivalents.

As before, the subscript 16 must be appended to a number if we consider that number to be hexadecimal. For example, the following are hexadecimal numbers:

$$6A45_{16},\ 7364510_{16},\ F7_{16},\ 13007_{16}$$

Table 1–2
Hexadecimal digits and their decimal equivalents

Hexadecimal	Decimal
0	0
1	1
2	2
3	3
4	4
5	5
6	6
7	7
8	8
9	9
A	10
B	11
C	12
D	13
E	14
F	15

Conversions

One can convert from hexadecimal to decimal by applying the concepts studied in Section 1–2. The procedure used is similar to the octal-to-decimal conversion procedure developed in Section 1–3. To convert a number from hexadecimal to decimal we take the sum of the products of each digit and its positional weight. The positional weight of each digit in the hexadecimal numbering system is 16^p, where p is the position of each digit in the number to be converted. For example, consider the number 6251_{16}. The value of this number in decimal would be

$$\begin{aligned} &(6 \times 16^3) + (2 \times 16^2) + (5 \times 16^1) + (1 \times 16^0) \\ &= (6 \times 4096) + (2 \times 256) + (5 \times 16) + (1 \times 1) \\ &= 24{,}576 + 512 + 80 + 1 \\ &= 25{,}169_{10} \end{aligned}$$

Again, notice that it takes a larger decimal number (25,169) to represent the same value as a smaller hexadecimal number (6251).

EXAMPLE 1–16

A person represents herself as being 23_{16} years old. What is her actual age? That is, what is her age in decimal?

SOLUTION

$$\begin{aligned} 23_{16} &= (2 \times 16^1) + (3 \times 16^0) \\ &= 32 + 3 \\ &= 35_{10} \end{aligned}$$

The person is 35 years old.

EXAMPLE 1–17

Convert the following numbers from hexadecimal to decimal.
(a) $6A51_{16}$ (b) $240F3_{16}$ (c) BAD_{16}

SOLUTIONS

(a) $6A51_{16} = 6 \times 16^3 + A \times 16^2 + 5 \times 16^1 + 1 \times 16^0$
$= 6 \times 4096 + A \times 256 + 5 \times 16 + 1 \times 1$

(*Note:* since A has a decimal value of 10, A × 256 is evaluated as 10 × 256.)

$= 24{,}576 + 2560 + 80 + 1$
$= 27{,}217_{10}$

(b) $240F3_{16} = 2 \times 16^4 + 4 \times 16^3 + 0 \times 16^2 + F \times 16^1 + 3 \times 16^0$
$= 2 \times 65{,}536 + 4 \times 4096 + 0 \times 256 + F \times 16 + 3 \times 1$

(*Note:* since F has a decimal value of 15, F × 16 is evaluated as 15 × 16.)

$= 131072 + 16384 + 0 + 240 + 3$
$= 147699_{10}$

(c) $BAD_{16} = B \times 16^2 + A \times 16^1 + D \times 16^0$
$= 11 \times 256 + 10 \times 16 + 13 \times 1$
$= 2816 + 160 + 13$
$= 2989_{10}$

To convert from decimal to hexadecimal, the process is similar to the one discussed in Sections 1–2 and 1–3. That is, we must obtain the individual digits that make up the number by repetitively dividing the decimal number by 16 (the base of the hexadecimal numbering system). The remainders for each successive division will make up the hexadecimal number. The following example illustrates this procedure.

EXAMPLE 1–18

Convert the decimal number (a) 3305 to hexadecimal, (b) 10,299 to hexadecimal.

SOLUTIONS

(a)

16	3305	9	← remainder
16	206	14	← remainder
		12	← remainder

Note: The hexadecimal digit C represents the decimal number 12, and the hexadecimal digit E represents the decimal number 14. Therefore the last remainder 12 will be expressed as the hexadecimal digit C and the second remainder 14 will be expressed as the hexadecimal digit E.

Reading the remainders from bottom to top, the hexadecimal equivalent is $CE9_{16}$.

(b)

16	10,299	11	← remainder
16	643	3	← remainder
16	40	8	← remainder
		2	← remainder

Reading the remainders from bottom to top, the hexadecimal equivalent is $283B_{16}$.

Arithmetic

Arithmetic operations with hexadecimal numbers are performed in much the same way as arithmetic operations with decimal and octal numbers, except that it is important to remember that the counting sequence for hexadecimal numbers is different, and that there are six additional digits that follow the digit 9. For example, in hexadecimal, 9 + 1 is not equal to 10 but is equal to A, and 20 − 5 is not equal to 15 but is equal to 1B.

Again, many similarities are apparent among the three numbering systems discussed so far, and all the concepts of the decimal numbering system developed in Section 1–2 apply to hexadecimal; the differences only lie in the numbers being manipulated. Besides allowing larger values to be expressed with smaller numbers, the most important advantage of the hexadecimal numbering system will become apparent in the following section and in later chapters.

Review questions

1. How many digits does the hexadecimal numbering system have?
2. What is the largest possible three-digit hexadecimal number?
3. Compare the hexadecimal and decimal numbering systems in terms of their representation of a value (such as 75).
4. Compare the octal and hexadecimal numbering systems in terms of their representation of a value (such as 75).

1–5 The binary numbering system

The binary numbering system is the most important numbering system in the study of digital circuits and their applications. It is this numbering system that provides the basis for all digital computer circuits, and is a natural numbering system for the description of the logic circuits to be studied later in this book.

The binary numbering system is a *base-two* numbering system. That is, there are only two digits in the numbering system. The word *binary* is used to describe something constituted of two things or parts, and when used in this context describes a numbering system that is composed of only two digits, 0 and 1. No other digits exist in the binary numbering system. Therefore if one were to count in binary, the same procedure would be followed as in decimal—one would start combining digits after a count of 1 is reached.

0, 1, 10, 11, 100, 101, 110, 111, 1000,
1001, 1010, 1011, 1100, 1101, 1110, 1111

If we start counting from 0, the next count is 1. However, notice that after 1, the next count is 10 since the digits 2 through 9 do not exist in the binary numbering system. The next count is 11. However, after 11 the next count is 100 since the numbers 12 through 99 do not exist in the binary numbering system. The count then proceeds to 101, 110, 111, and so on.

Notice that if we are to examine a count of all possible combinations of four binary digits, the largest number would be 1111. That is, the count would end at 1111. Using the relationship introduced in Section 1–1, B^n, the total number of combinations for four binary digits would be,

$$2^4$$

or

$$16$$

Therefore, there are 16 (decimal) counts between the binary numbers 0000 and 1111. The *value* of the binary number 1111 would actually be 15 (in decimal). Thus binary numbers are always larger than decimal numbers representing the same value; this is because the binary numbering system has only two digits. Table 1–3 lists the relationship between the ten decimal digits and their binary equivalents. Notice in Table 1–3 that binary numbers are often operated on in groups. A single binary digit (0 or 1) is called a *bit*, a group of 4 bits is called a *nibble*, a group of 8 bits is called a *byte*, and a group of 16 bits is called a *word*. Since the largest number in the table is 9, and the binary number used to represent 9 requires 4 bits, we add leading zeros to the other numbers in the table to make them 4 bits long. Of course, the addition of leading zeros has absolutely no effect on the values of the numbers.

The binary counting sequence is of particular importance for it follows a definite pattern that is important in the design and analysis of digital circuits. Table 1–4 lists all possible combinations of 4 bits in the counting sequence, along with their equivalent decimal, hexadecimal, and octal digits or numbers.

Notice in Table 1–4 that the count of all possible combinations of 4 bits follows a pattern similar to the counting sequence in any other numbering system. That is, the least significant digit (LSD) cycles through every possible digit in sequence while the other digits in positions to the left increment once every time a cycle from the digit at the right is completed. Since there are only two possible digits (bits) in the binary numbering system, the least significant bit (LSB) toggles back and forth between a 0 and a 1. The next bit toggles once for every cycle of the LSB. The next bit position toggles once for every cycle of the bit to its right; or once for every two cycles of the LSB. Finally, the *most significant bit* (MSB) toggles once for every cycle of the bit to its right; or once for every four cycles of the LSB.

Table 1–4 also illustrates two very important characteristics of the octal and hexadecimal numbering systems. Notice that every 4-bit binary combination can be uniquely represented by a hexadecimal digit. That is, there is one hexadecimal digit available for every 4-bit binary number. This also applies to octal digits and 3-bit binary numbers. Every digit in

Table 1–3
Decimal digits and their binary equivalents

Decimal	Binary
0	0000
1	0001
2	0010
3	0011
4	0100
5	0101
6	0110
7	0111
8	1000
9	1001

Table 1–4
Binary, decimal, hexadecimal, and octal equivalence table

Binary	Hexadecimal	Decimal	Octal
0000	0	0	0
0001	1	1	1
0010	2	2	2
0011	3	3	3
0100	4	4	4
0101	5	5	5
0110	6	6	6
0111	7	7	7
1000	8	8	10
1001	9	9	11
1010	A	10	12
1011	B	11	13
1100	C	12	14
1101	D	13	15
1110	E	14	16
1111	F	15	17

the octal set can uniquely represent every 3-bit binary combination; there is one octal digit for every 3-bit binary number. As will be seen later in this section, this perfect representation makes conversion among hexadecimal, octal, and binary numbers very simple. Unfortunately, the relationship between binary and decimal is not that perfect. Notice in Table 1–4 that even though every decimal digit can be represented by a unique 4-bit binary number, the set of decimal digits cannot represent all 4-bit combinations since there are 16 combinations and only 10 decimal digits. Furthermore, even though a decimal digit can be used to represent each unique 3-bit combination, all the digits in the decimal set cannot be represented by 3-bit combinations since there are only eight combinations. Therefore 3-bit or 4-bit numbers are not a perfect ''fit'' for the decimal digits.

As before, we must append the subscript 2 to a number if we consider that number to be binary. For example, the following are binary numbers:

$$1110_2,\ 1010110_2,\ 10_2,\ 10001_2$$

Notice that binary numbers tend to be very long. For example, the binary representation of the decimal number 60,000 would be a 16-bit binary number, 1110101001100000. It is not uncommon to encounter binary numbers this size (and even larger) during the study of digital circuits and computers. We therefore need a more convenient means of representing these large binary numbers to somewhat ease the task of working with them. At first thought, decimal numbers would be the ideal choice to represent binary numbers—decimal numbers are shorter in size and we are most familiar with the decimal numbering system. Binary numbers could be converted to decimal, manipulated, and then reconverted back to binary. However, the decimal numbering system is not very well suited to representing binary numbers, and conversions are relatively more time consuming and complicated. As shall be seen later in this section, the octal and hexadecimal numbering systems are the ideal choices for representing binary numbers.

Decimal-to-binary and binary-to-decimal conversions

Once again, we can convert from binary to decimal by applying the concepts studied in Section 1–2. The procedure used is similar to the octal-to-decimal and hexadecimal-to-decimal conversion procedures developed in Sections 1–3 and 1–4. To convert a number from binary to decimal we take the sum of the products of each bit and its positional weight. The positional weight of each bit in the binary numbering system is 2^p, where p is the position of each bit in the number to be converted. For example, consider the number 11011_2. The value of this number in decimal would be

$$
\begin{aligned}
&(1 \times 2^4) + (1 \times 2^3) + (0 \times 2^2) + (1 \times 2^1) + (1 \times 2^0) \\
&= (1 \times 16) + (1 \times 8) + (0 \times 4) + (1 \times 2) + (1 \times 1) \\
&= 16 + 8 + 0 + 2 + 1 \\
&= 27_{10}
\end{aligned}
$$

Again, notice that it takes a smaller decimal number (27) to represent the same value as a larger binary number (11011).

EXAMPLE 1–19

A person represents himself as being 11000_2 years old. What is his actual age? That is, what is his age in decimal?

SOLUTION

$$
\begin{aligned}
11000_2 &= (1 \times 2^4) + (1 \times 2^3) + (0 \times 2^2) + (0 \times 2^1) + (0 \times 2^0) \\
&= 16 + 8 + 0 + 0 + 0 \\
&= 24_{10}
\end{aligned}
$$

The person is 24 years old.

The conversions from binary to decimal can be greatly simplified by simply adding up the powers of 2 for bit positions that are 1's. For example, consider the binary number 10100110_2.

number:	1	0	1	0	0	1	1	0
	↑	↑	↑	↑	↑	↑	↑	↑
position:	7	6	5	4	3	2	1	0
weight:	2^7	2^6	2^5	2^4	12^3	2^2	2^1	2^0
	128	64	32	16	8	4	2	1

To convert the number 10100110_2 to decimal we simply add up all the weights corresponding to the 1 bits (since if the bit is 0 then the product of the bit and the weight is 0)

$$128 + 32 + 4 + 2 = 166_{10}$$

Table 1–5 lists the powers of 2 for bit positions 0 through 15 as a convenience in the conversion of binary numbers up to 16 bits long. Appendix A lists all powers of 2 from 0 to 63.

Table 1–5
Powers of 2 from 0 to 15

n	2^n
0	1
1	2
2	4
3	8
4	16
5	32
6	64
7	128
8	256
9	512
10	1024
11	2048
12	4096
13	8192
14	16,384
15	32,768

EXAMPLE 1–20

Convert the following numbers from binary to decimal using Table 1–5:
(a) 1001010010_2 (b) 1111_2 (c) 01001011_2

SOLUTIONS

(a)

```
1  0  0  1  0  1  0  0  1  0
↑        ↑     ↑        ↑
512      64    16       2
```

$512 + 64 + 16 + 2 = 594_{10}$

(b)

```
1  1  1  1
↑  ↑  ↑  ↑
8  4  2  1
```

$8 + 4 + 2 + 1 = 15_{10}$

(c)

```
0  1  0  0  1  0  1  1
   ↑        ↑     ↑  ↑
   64       8     2  1
```

$64 + 8 + 2 + 1 = 75_{10}$

To convert from decimal to binary, the process is similar to the one discussed in Section 1–2, and also similar to the decimal-to-octal and decimal-to-hexadecimal conversions covered in Sections 1–3 and 1–4, respectively. That is, one must obtain the individual digits that make up the number by repetitively dividing the decimal number by 2 (the base of the binary numbering system). The remainders for each successive division will make up the binary number. The following example illustrates this procedure.

EXAMPLE 1–21

Convert the decimal number (a) 33 to binary, (b) 624 to binary.

SOLUTIONS

(a)

2	33	1	← remainder
2	16	0	← remainder
2	8	0	← remainder
2	4	0	← remainder
2	2	0	← remainder
	1		← remainder

Reading the remainders from bottom to top, the binary equivalent is 100001_2.

(b)

2	624	0	← remainder
2	312	0	← remainder
2	156	0	← remainder
2	78	0	← remainder
2	39	1	← remainder
2	19	1	← remainder
2	9	1	← remainder
2	4	0	← remainder
2	2	0	← remainder
	1		← remainder

Reading the remainders from bottom to top, the binary equivalent is 1001110000_2.

Binary-to-octal and octal-to-binary conversions

It was mentioned earlier in this section that the hexadecimal and octal numbering systems are ideal for representing groups of binary numbers since the digits in each system can perfectly represent all possible combinations of 3 bits (octal) or all possible combinations of 4 bits (hexadecimal); this fact was illustrated in Table 1–4. The capability of the hexadecimal and octal numbering systems to perfectly represent groups of binary bits also makes conversions very simple. For example, consider the binary number, 100111010011. If we break up this number into groups of 3 bits, starting at the LSB, and then substitute the octal digit that represents each group, we will obtain the octal equivalent of the binary

Table 1–6
Binary representation of the octal digits

Binary	Octal
000	0
001	1
010	2
011	3
100	4
101	5
110	6
111	7

number. That is, we have converted from binary to octal. Table 1–6 lists the eight octal digits and their 3-bit binary equivalents.

$$\underbrace{1\ 0\ 0}_{4}\quad\underbrace{1\ 1\ 1}_{7}\quad\underbrace{0\ 1\ 0}_{2}\quad\underbrace{0\ 1\ 1}_{3}$$

Therefore 100111010011_2 is equivalent to 4723_8.

We can verify that the conversion is valid by converting the binary number 100111010011_2 to decimal

$$\begin{aligned}100111010011_2 &= 1 + 2 + 16 + 64 + 128 + 256 + 2048\\ &= 2515_{10}\end{aligned}$$

and then converting the octal number 4723_8 to decimal

$$\begin{aligned}4723_8 &= 4 \times 8^3 + 7 \times 8^2 + 2 \times 8^1 + 3 \times 8^0\\ &= 2515_{10}\end{aligned}$$

Since both numbers yield the same decimal equivalent, the two numbers are equivalent.

EXAMPLE 1–22

Convert the following binary numbers to octal:
(a) 110001010111 (b) 10100110 (c) 1 (d) 0001011000

SOLUTIONS

(a)

$$\underbrace{1\ 1\ 0}_{6}\ \underbrace{0\ 0\ 1}_{1}\ \underbrace{0\ 1\ 0}_{2}\ \underbrace{1\ 1\ 1}_{7}$$

Therefore $110001010111_2 = 6127_8$.

(b)

leading zero added

$$\rightarrow \underbrace{0\ 1\ 0}_{2}\ \underbrace{1\ 0\ 0}_{4}\ \underbrace{1\ 1\ 0}_{6}$$

Therefore $10100110_2 = 246_8$.

(c) $1_2 = 1_8$

(d) 0 0 0 1 0 1 1 0 0 0 → 1 3 0

Therefore $0001011000_2 = 130_8$.

To convert from octal to binary, the process is simply reversed. That is, we substitute the 3-bit binary code for each octal digit. For instance, see Example 1–23.

EXAMPLE 1–23

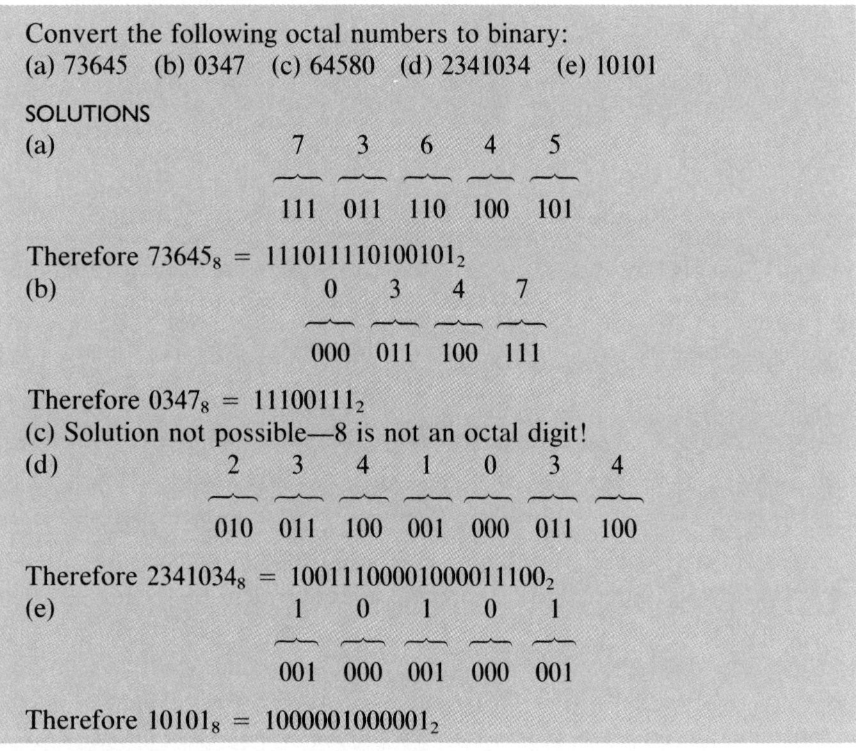

Convert the following octal numbers to binary:
(a) 73645 (b) 0347 (c) 64580 (d) 2341034 (e) 10101

SOLUTIONS

(a) 7 3 6 4 5 → 111 011 110 100 101

Therefore $73645_8 = 111011110100101_2$

(b) 0 3 4 7 → 000 011 100 111

Therefore $0347_8 = 11100111_2$

(c) Solution not possible—8 is not an octal digit!

(d) 2 3 4 1 0 3 4 → 010 011 100 001 000 011 100

Therefore $2341034_8 = 10011100001000011100_2$

(e) 1 0 1 0 1 → 001 000 001 000 001

Therefore $10101_8 = 1000001000001_2$

Binary-to-hexadecimal and hexadecimal-to-binary conversions

For the same reasons stated earlier, we can apply the same techniques to convert from binary to hexadecimal and from hexadecimal to binary. However, since there are 16 hexadecimal digits and since this set perfectly represents all possible combinations of 4 bits, as shown in Table 1–7, we must perform the conversion with 4-bit groups. For example, consider the binary number, 100111010011. If we break up this number into groups of 4 bits, starting at the LSB, and then substitute the hexadecimal digit that represents each group, we obtain the hexadecimal equivalent of the binary number. That is, we have converted from binary to hexadecimal.

1 0 0 1 1 1 0 1 0 0 1 1 → 9 D 3

Therefore 100111010011_2 is equivalent to $9D3_{16}$.

Table 1–7
Binary representation of the hexadecimal digits

Binary	Hexadecimal
0000	0
0001	1
0010	2
0011	3
0100	4
0101	5
0110	6
0111	7
1000	8
1001	9
1010	A
1011	B
1100	C
1101	D
1110	E
1111	F

We can verify that the conversion is valid by converting the binary number 100111010011_2 to decimal

$$
\begin{aligned}
100111010011_2 &= 1 + 2 + 16 + 64 + 128 + 256 + 2048 \\
&= 2515_{10}
\end{aligned}
$$

and then converting the hexadecimal number $9D3_{16}$ to decimal

$$
\begin{aligned}
9D3_{16} &= 9 \times 16^2 + D \times 16^1 + 3 \times 16^0 \\
&= 2515_{10}
\end{aligned}
$$

Since both numbers yield the same decimal equivalent, the two numbers are equivalent.

EXAMPLE 1–24

Convert the following binary numbers to hexadecimal:
(a) 110001010111 (b) 10100110 (c) 1 (d) 0001011000

SOLUTIONS

(a)

$$\underbrace{1\;1\;0\;0}_{C}\;\;\underbrace{0\;1\;0\;1}_{5}\;\;\underbrace{0\;1\;1\;1}_{7}$$

Therefore $110001010111_2 = C57_{16}$

(b)

$$\underbrace{1\;0\;1\;0}_{A}\;\;\underbrace{0\;1\;1\;0}_{6}$$

Therefore $10100110_2 = A6_{16}$

(c)

$$1_2 = 1_{16}$$

(d)

$$0\;0\;\underbrace{0\;1\;0\;1}_{5}\;\;\underbrace{1\;0\;0\;0}_{8}$$

Therefore $0001011000_2 = 58_{16}$

To convert from hexadecimal to binary, the process is simply reversed. That is, we substitute the 4-bit binary code for each hexadecimal digit (see Example 1–25).

EXAMPLE 1–25

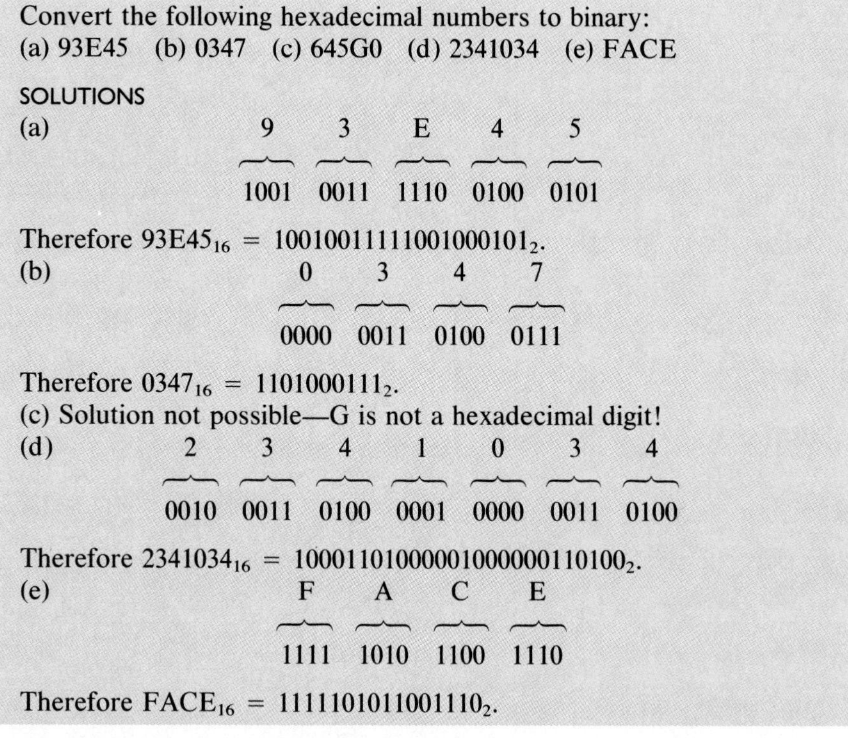

Convert the following hexadecimal numbers to binary:
(a) 93E45 (b) 0347 (c) 645G0 (d) 2341034 (e) FACE

SOLUTIONS

(a)

9	3	E	4	5
1001	0011	1110	0100	0101

Therefore $93E45_{16} = 10010011111001000101_2$.

(b)

0	3	4	7
0000	0011	0100	0111

Therefore $0347_{16} = 1101000111_2$.

(c) Solution not possible—G is not a hexadecimal digit!

(d)

2	3	4	1	0	3	4
0010	0011	0100	0001	0000	0011	0100

Therefore $2341034_{16} = 1000110100000100000110100_2$.

(e)

F	A	C	E
1111	1010	1100	1110

Therefore $FACE_{16} = 1111101011001110_2$.

Octal-to-hexadecimal and hexadecimal-to-octal conversions

The simplest way to convert between the octal and hexadecimal numbering systems is to perform a binary conversion as an intermediate step. That is, first convert from one base to binary, and then regroup the bits and convert to the other base. To illustrate this procedure, consider the following examples.

EXAMPLE 1–26

Convert the following octal numbers to hexadecimal:
(a) 51 (b) 64320 (c) 100011

SOLUTIONS

(a)

5	1
101	001

Regrouping the number 101001_2 into groups of 4 bits and adding two leading zeros we have

0010	1001
2	9

Therefore $51_8 = 29_{16}$.

(b)

6	4	3	2	0
110	100	011	010	000

Regrouping the number 110100011010000_2 into groups of 4 bits and adding one leading zero we have

$$\underbrace{0110}_{6}\ \underbrace{1000}_{8}\ \underbrace{1101}_{D}\ \underbrace{0000}_{0}$$

Therefore $64320_8 = 68D0_{16}$

(c)

$$\overbrace{001}^{1}\ \overbrace{000}^{0}\ \overbrace{000}^{0}\ \overbrace{000}^{0}\ \overbrace{001}^{1}\ \overbrace{001}^{1}$$

Regrouping the number 001000000000001001_2 into groups of 4 bits and discarding the two leading zero we have

$$\underbrace{1000}_{8}\ \underbrace{0000}_{0}\ \underbrace{0000}_{0}\ \underbrace{1001}_{9}$$

Therefore $100011_8 = 8009_{16}$.

EXAMPLE 1–27

Convert the following hexadecimal numbers to octal:
(a) 29 (b) 68D0 (c) 8009

SOLUTIONS

(a)

$$\overbrace{0010}^{2}\ \overbrace{1001}^{9}$$

Regrouping the number 00101001_2 into groups of 3 bits and discarding the two leading zeros we have

$$\underbrace{101}_{5}\ \underbrace{001}_{1}$$

Therefore $29_{16} = 51_8$.

(b)

$$\overbrace{0110}^{6}\ \overbrace{1000}^{8}\ \overbrace{1101}^{D}\ \overbrace{0000}^{0}$$

Regrouping the number 0110100011010000_2 into groups of 3 bits and discarding one leading zero we have

$$\underbrace{110}_{6}\ \underbrace{100}_{4}\ \underbrace{011}_{3}\ \underbrace{010}_{2}\ \underbrace{000}_{0}$$

Therefore $68D0_{16} = 64320_8$.

(c)

$$\overbrace{1000}^{8}\ \overbrace{0000}^{0}\ \overbrace{0000}^{0}\ \overbrace{1001}^{9}$$

Regrouping the number 1000000000001001_2 into groups of 3 bits and adding 2 leading zeros we have

$$\underbrace{001}_{1}\ \underbrace{000}_{0}\ \underbrace{000}_{0}\ \underbrace{000}_{0}\ \underbrace{001}_{1}\ \underbrace{001}_{1}$$

Therefore $8009_{16} = 100011_8$.

The main purpose of the octal and hexadecimal numbering system is in the *representation* of binary numbers. Because of the ease with which we can switch back and forth between binary and octal/hexadecimal numbers, we have a very convenient means of representing these rather large binary numbers. For example, instead of having to work with numbers such as

$$100010100100111100101001_2$$

we can now work with equivalents such as $8A4F29_{16}$ or 42447451_8 which are much more convenient to "carry around" and are less prone to error. However, it is important to remember that these numbers are only convenient representations of binary numbers. Using decimal numbers to represent binary numbers is possible but is a time-consuming process which involves a lot of arithmetic. The selection of either the octal or hexadecimal representation of binary numbers is a matter of choice. However, some digital circuit applications that process binary numbers are more suited to octal representation while others are better suited to hexadecimal representation.

Binary coded decimal

At this point it is important to note that the substitution technique used to convert from octal and hexadecimal to binary and vice versa cannot be used for decimal-to-binary and binary-to-decimal conversions. As previously stated, this is because the set of decimal digits cannot perfectly represent all possible combinations of 3 or 4 bits. There are ten decimal digits, whereas there are 2^3 or 8 combinations of 3 bits, and 2^4, or 16 combinations of 4 bits. It may however *appear* to work in some cases, but the results are erroneous. For example, consider the binary number, 10010111. Using the substitution technique we have

$$\underbrace{1\ 0\ 0\ 1}_{9}\quad \underbrace{0\ 1\ 1\ 1}_{7}$$

or the number 97, which is actually a hexadecimal number. The decimal equivalent of 10010111,

$$10010111_2 = 1 + 2 + 4 + 16 + 128 = 151$$

is 151 and not 97.

Furthermore, notice that there absolutely is no way to convert a number such as 11001110_2 to decimal using the substitution technique, since there isn't any *single* decimal digit that can represent the binary combinations 1100 and 1110. In fact, there are no decimal digits to represent any of the six combinations, 1010, 1011, 1100, 1101, 1110, 1111. Using two decimal digits to represent each combination is not permitted

Table 1–8
Binary representation of the decimal digits (BCD)

Binary	Decimal
0000	0
0001	1
0010	2
0011	3
0100	4
0101	5
0110	6
0111	7
1000	8
1001	9

since it will not be possible to reconvert the number back to its original value:

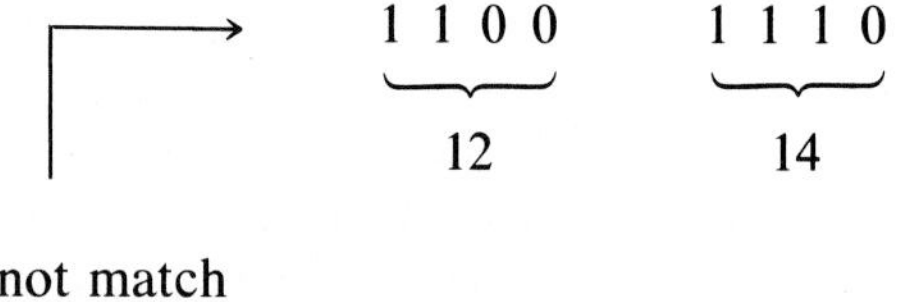

does not match

1 2 1 4

0001 0010 0001 0100

This substitution technique can, however, be used to *represent* decimal digits (but not to *convert* from decimal to binary) in binary form. When a 4-bit binary number is substituted for a decimal digit, this 4-bit binary number is called a *binary coded decimal* (BCD) number. Table 1–8 lists the 10 decimal digits and their BCD values. The following examples illustrate the procedures used for converting from decimal to BCD and from BCD to decimal.

EXAMPLE 1–28

Convert the following decimal numbers to their BCD representations:
(a) 63547 (b) 12 (c) 10101 (d) 0398

SOLUTIONS

(a)

6 3 5 4 7

0110 0011 0101 0100 0111

Therefore 63547_{10} = 1100011010101000111 (BCD).

(b)

1 2

0001 0010

Therefore 12_{10} = 10010 (BCD).

(c)

1 0 1 0 1

0001 0000 0001 0000 0001

Therefore 10101_{10} = 10000000100000001 (BCD).

(d) 0 3 9 8

0000 0011 1001 1000

Therefore 0398_{10} = 1110011000 (BCD)

EXAMPLE 1–29

Convert the following BCD numbers to decimal:
(a) 100100010000 (b) 110001101010100111 (c) 11100101100

SOLUTIONS

(a) 1001 0001 0000

9 1 0

Therefore 100100010000 (BCD) = 910_{10}

(b) 0110 0011 0101 0100 0111

6 3 5 4 7

Therefore 110001101010100111 (BCD) = 63547_{10}

(c) 0111 0010 1100

7 2 ?

No solution is possible! The number 11100101100 is not a valid BCD number and cannot be converted to decimal since 1100 cannot be represented by a decimal digit.

Digital computers use two techniques to store decimal numbers in memory. The first involves converting a decimal number to binary and then storing the binary representation of the decimal number. An alternative technique involves converting the decimal number to BCD (using the substitution technique) and then storing the BCD representation of the decimal number. The former technique is more complicated since it involves a lot of arithmetic manipulation, the latter technique is preferable since it involves only a simple substitution process. For example, a digital computer can store the decimal number 7395 as

1110011100011

after converting from decimal to binary using the repetitive division procedure, or it can store the number as

111001110010101

after converting from decimal to BCD using the substitution procedure.

There are advantages and disadvantages in both procedures. BCD numbers generally require more storage, that is, they are usually larger than their binary equivalents. Performing arithmetic on BCD numbers is more complicated as will be seen in Section 1–6. However, conversions to and from BCD and decimal are much simpler than conversions to and from binary and decimal.

This section has introduced us to the concepts of binary numbers and their relationship to decimal, octal, and hexadecimal numbers. The binary numbering system may seem very cumbersome to work with and manipulate due to its very limited set of digits. However, this *binary* set lends itself very well to the operation of the digital circuits (many of which are designed to manipulate binary numbers) to be examined in the rest of this book. The octal and hexadecimal numbering systems are very well suited to representing binary numbers since the conversions back and forth are very simple. As far as the decimal numbering system is concerned—we live in a decimal world and must therefore be able to switch back and forth between the decimal numbering system and the others when dealing with digital circuits that interface to the "real world."

Review questions

1. How many digits does the binary numbering system have?
2. What is the largest possible eight-digit binary number?
3. Compare the binary and decimal numbering systems in terms of their representation of a value (such as 75).
4. Compare the octal and binary numbering systems in terms of their representation of a value (such as 75).
5. Compare the binary and hexadecimal numbering systems in terms of their representation of a value (such as 75).
6. Why can't we use the substitution procedure to convert from decimal to binary and binary to decimal?
7. What is the difference between the binary representation of decimal numbers and the BCD representation of decimal numbers?

1–6 Binary arithmetic

One of the functions of a digital circuit is to perform the four basic arithmetic operations on binary numbers. It was stated earlier that the process of addition is the most important since most digital circuits perform the other three operations (subtraction, multiplication, and division) through various techniques of addition; we saw examples of these in Section 1–2 using the decimal numbering system as an example. In this section we will examine methods of performing addition, subtraction, multiplication, and division with binary numbers using techniques that are very similar to those developed in Section 1–2.

Addition

The binary addition procedure is similar to addition in any other numbering system. The only thing we should remember is that there are only two digits (bits) in this numbering system, and therefore the results obtained when adding these bits are quite different. Table 1–9 lists the sums obtained by adding all possible combinations of 2 bits. Notice in Table 1–9 that we have extended the sum (result) of the two numbers to 2 bits since the largest sum is 10 (2). Also notice that 1 + 1 is equal to 2_{10} which is 10_2. Table 1–10 extends the arithmetic performed in Table 1–9 to 3 bits. Again notice in Table 1–10, that since the largest sum is 11 (3), the

Table 1–9
Sums of all possible combinations of 2 bits

0	+	0	=	0	0
0	+	1	=	0	1
1	+	0	=	0	1
1	+	1	=	1	0

Table 1–10
Sums of all possible combinations of 3 bits

0	+	0	+	0	=	0	0
0	+	0	+	1	=	0	1
0	+	1	+	0	=	0	1
0	+	1	+	1	=	1	0
1	+	0	+	0	=	0	1
1	+	0	+	1	=	1	0
1	+	1	+	0	=	1	0
1	+	1	+	1	=	1	1

result of adding 1 + 1 + 1, we only need two bits to represent the results. Notice in Tables 1–9 and 1–10 that in order to list all possible combinations of 2 or 3 bits we simply count in binary. Also, the most significant bit of the sum is referred to as the *carry bit*.

EXAMPLE 1–30

Construct a table similar to Tables 1–9 and 1–10 that will list the sums of all possible combinations of 4 bits.

SOLUTION Since the largest sum will be 1 + 1 + 1 + 1, or 100 (4) we need to extend the result to 3 bits.

0 + 0 + 0 + 0 = 0 0 0
0 + 0 + 0 + 1 = 0 0 1
0 + 0 + 1 + 0 = 0 0 1
0 + 0 + 1 + 1 = 0 1 0
0 + 1 + 0 + 0 = 0 0 1
0 + 1 + 0 + 1 = 0 1 0
0 + 1 + 1 + 0 = 0 1 0
0 + 1 + 1 + 1 = 0 1 1
1 + 0 + 0 + 0 = 0 0 1
1 + 0 + 0 + 1 = 0 1 0
1 + 0 + 1 + 0 = 0 1 0
1 + 0 + 1 + 1 = 0 1 1
1 + 1 + 0 + 0 = 0 1 0
1 + 1 + 0 + 1 = 0 1 1
1 + 1 + 1 + 0 = 0 1 1
1 + 1 + 1 + 1 = 1 0 0

So far we have looked at single-bit addition. We now need to extend this process to the addition of multibit binary numbers. Recall from Section 1–2 that when we add decimal numbers, we simply add the LSDs of each number, obtain a single-digit sum and carry, and then add the carry to the next set of digits in sequence, continuing this process to the MSD. The same procedure is used to add multibit binary numbers. For example, consider the addition of the binary number 1001 (9) and 0101 (5) to produce a result of 1110 (14 or E_{16}):

```
      0 0 1     ← carries
    1 0 0 1     ← number 9
+   0 1 0 1     ← number 5
0   1 1 1 0     ← result 14
↑   └──┬──┘
final carry  sum
```

Starting with the LSB, 1 + 1 produces 10, or 0 with a carry of 1. This carry is then added to the next position, 1 + 0 + 0, to obtain 01 (or 1 with a carry of 0). The carry along with the next two bits in position are then added 0 + 0 + 1, to obtain a result of 01 (or 1 with a carry of 0). Finally, this carry is then added to the two MSBs, 0 + 1 + 0, to obtain the MSB of the result, 1, and a final carry of 0. Notice that the largest number being added is 9 (1001) which is a 4-bit number and therefore we have extended the addition to 4 bits. However, the result is a 5-bit number when the final carry is included. For the addition of two 4-bit numbers the result will never be greater than 5 bits. To illustrate this, consider the addition of the two largest possible 4-bit numbers 1111 (15 or F_{16}),

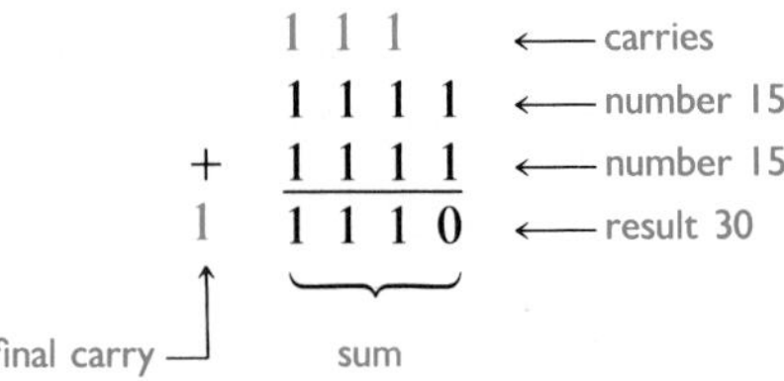

The result of the addition of 1111 (15) and 1111 (15) produces the binary number 11110 (30). The following examples further illustrate the technique of binary addition.

EXAMPLE 1–31

Perform the following additions in binary arithmetic:
(a) 7 + 2 (b) 20 + 32 (c) $3E_{16} + 75_{16}$ (d) $324_8 + 106_8$

SOLUTIONS

(a)

```
             1 0       ← carries
   7         1 1 1
 + 2       + 0 1 0
   9       1 0 0 1
```

(b)

```
             0 0 0 0 0       ← carries
   20        0 1 0 1 0 0
 + 32      + 1 0 0 0 0 0
   52      0 1 1 0 1 0 0
```

(c)

```
             1 1 1 1 0 0       ← carries
   3E        0 1 1 1 1 1 0
 + 75      + 1 1 1 0 1 0 1
   B3      1 0 1 1 0 0 1 1
```

(d)

```
             1 0 0 0 1 0 0       ← carries
   324       1 1 0 1 0 1 0 0
 + 106     + 0 1 0 0 0 1 1 0
   432     1 0 0 0 1 1 0 1 0
```

Subtraction–1's complement

The 1's complement technique for binary subtraction is similar to the 9's complement technique for decimal subtraction discussed in Section 1–2. The *1's complement* of a binary number is obtained by subtracting each bit in the number from 1. However, observe that,

$$1 - 1 = 0 \quad \text{and} \quad 1 - 0 = 1$$

which indicates that the 1's complement of 1 is 0 and the 1's complement of 0 is 1. Therefore a more direct (or easier) approach to obtaining the 1's complement of a binary number is to *invert* each bit in the number. Here the term *invert* refers to the process of changing a 0 to a 1, and a 1 to a 0. For example, the 1's complement of the binary number, 10010110 is 01101001. Notice that this completely eliminates the need for subtraction! Also, the process of obtaining the inverse of a bit can be performed very easily by a digital circuit called an inverter (to be seen in Chapter 2). To subtract two binary numbers using the 1's complement technique we must do the following:

1. Obtain the 1's complement representation of the negative number.
2. Add the positive number to the 1's complement representation of the negative number.
3. Add the end-around (final) carry obtained from step 2 to the sum obtained from step 2.
4. The final sum is the difference between the two numbers.
5. If the end-around carry is a 1 then the result is positive. If the end-around carry is 0 then the result is negative and is represented in unsigned 1's complement form; it must be recomplemented in order to obtain its signed representation.

The following examples illustrate this technique:

EXAMPLE 1–32

Subtract the following numbers using 1's complement binary arithmetic:
(a) 9 − 5 (b) 85 − 32 (c) $F3_{16} - 8B_{16}$ (d) $63_8 - 12_8$

SOLUTIONS
(a) 9 − 5 = 4
9 = 1001
5 = 0101 Therefore −5 = 1010 (in 1's complement representation).
Note that 5 (101) must be extended to 4 bits (0101) since the largest number being operated on, 9 (1001), is 4 bits long.

```
     1 0 0 1
   + 1 0 1 0
  1  0 0 1 1
+ └──────→ 1
     0 1 0 0
```

The result is 0100 or 4.
(b) 85 − 32 = 53
85 = 1010101
32 = 0100000
Therefore −32 = 1011111 (in 1's complement representation).

```
     1 0 1 0 1 0 1
   + 1 0 1 1 1 1 1
  1  0 1 1 0 1 0 0
+ └────────────→ 1
     0 1 1 0 1 0 1
```

The result is 0110101 or 53.

(c) $F3_{16} - 8B_{16} = 68_{16}$
F3 = 11110011
8B = 10001011
Therefore −8B = 01110100 (in 1's complement representation).

```
         1 1 1 1 0 0 1 1
       + 0 1 1 1 0 1 0 0
     1   0 1 1 0 0 1 1 1
   + └──────────────────→1
         0 1 1 0 1 0 0 0
```

The result is 01101000 or 68_{16}.
(d) $63_8 - 12_8 = 51_8$
63 = 110011
12 = 001010
Therefore − 12 = 110101 (in 1's complement representation).

```
         1 1 0 0 1 1
       + 1 1 0 1 0 1
     1   1 0 1 0 0 0
   + └──────────────→1
         1 0 1 0 0 1
```

The result is 101001 or 51_8.

EXAMPLE 1–33

Subtract the following numbers using 1's complement binary arithmetic:
(a) 5 − 9 (b) 32 − 85 (c) $8B_{16} - F3_{16}$ (d) $12_8 - 63_8$

SOLUTIONS
(a) 5 − 9 = −4
5 = 0101
9 = 1001
Therefore −9 = 0110 (in 1's complement representation).
Note that 5 (101) must be extended to 4 bits (0101) since the largest number being operated on, 9 (1001), is 4 bits long.

```
         0 1 0 1
       + 0 1 1 0
     0   1 0 1 1
   + └──────────→0
         1 0 1 1
```

Since the carry is 0, the result is 1011, or −4, in 1's complement representation. The signed result is the 1's complement of 1011, or −0100.
(b) 32 − 85 = −53
32 = 0100000
85 = 1010101
Therefore −85 = 0101010 (in 1's complement representation).

```
         0 1 0 0 0 0 0
       + 0 1 0 1 0 1 0
     0   1 0 0 1 0 1 0
   + └────────────────→0
         1 0 0 1 0 1 0
```

Since the carry is 0, the result is 1001010, or −53, in 1's

complement representation. The signed result is the 1's complement of 1001010, or −0110101.

(c) $8B_{16} - F3_{16} = -68_{16}$

8B = 10001011

F3 = 11110011

Therefore −F3 = 00001100 (in 1's complement representation).

```
        1 0 0 0 1 0 1 1
      + 0 0 0 0 1 1 0 0
    0   1 0 0 1 0 1 1 1
  + └──────────────────→0
        1 0 0 1 0 1 1 1
```

Since the carry is 0, the result is 10010111, or −68, in 1's complement representation. The signed result is the 1's complement of 10010111, or −01101000.

(d) $12_8 - 63_8 = -51_8$

12 = 001010

63 = 110011

Therefore −63 = 001100 (in 1's complement representation).

```
        0 0 1 0 1 0
      + 0 0 1 1 0 0
    0   0 1 0 1 1 0
  + └──────────────→0
        0 1 0 1 1 0
```

Since the carry is 0, the result is 010110, or −51, in 1's complement representation. The signed result is the 1's complement of 010110, or −101001.

Subtraction—2's complement

The 2's complement technique for binary subtraction is similar to the 10's complement technique for decimal subtraction discussed in Section 1–2. The 2's complement of a binary number is obtained by adding 1 to the 1's complement of a number. For example, if the 1's complement of the number 10010110 is 01101001, then the 2's complement of the number 10010110 will be

```
   01101001   1's complement
 +        1   plus 1
   01101010   2's complement
```

To subtract two binary numbers using the 2's complement technique, we must do the following:

1. Obtain the 2's complement representation of the negative number.
2. Add the positive number to the 2's complement representation of the negative number.
3. Discard the final carry.
4. The final sum is the difference between the two numbers.
5. If the final carry is a 1 then the result is positive. If the final carry is 0 then the result is negative and is represented in unsigned 2's complement form; it must be recomplemented in order to obtain its signed representation.

The following examples illustrate this technique.

EXAMPLE 1–34

Subtract the following numbers using 2's complement binary arithmetic:
(a) 9 − 5 (b) 85 − 32 (c) $F3_{16} - 8B_{16}$ (d) $63_8 - 12_8$

SOLUTIONS

(a) 9 − 5 = 4
9 = 1001
5 = 0101
Therefore −5 = 1010 + 1 = 1011 (in 2's complement representation).

```
                       1 0 0 1
                     + 1 0 1 1
                     ---------
discard ⟶        1     0 1 0 0
carry
```

The result is 0100 or 4.

(b) 85 − 32 = 53
85 = 1010101
32 = 0100000
Therefore −32 = 1011111 + 1 = 1100000 (in 2's complement representation).

```
                       1 0 1 0 1 0 1
                     + 1 1 0 0 0 0 0
                     ---------------
discard ⟶        1     0 1 1 0 1 0 1
carry
```

The result is 0110101 or 53.

(c) $F3_{16} - 8B_{16} = 68_{16}$
F3 = 11110011
8B = 10001011
Therefore −8B = 01110100 + 1 = 01110101 (in 2's complement representation).

```
                       1 1 1 1 0 0 1 1
                     + 0 1 1 1 0 1 0 1
                     -----------------
discard ⟶        1     0 1 1 0 1 0 0 0
carry
```

The result is 01101000 or 68_{16}.

(d) $63_8 - 12_8 = 51_8$
63 = 110011
12 = 001010
Therefore −12 = 110101 + 1 = 110110 (in 2's complement representation).

```
                       1 1 0 0 1 1
                     + 1 1 0 1 1 0
                     -------------
discard ⟶        1     1 0 1 0 0 1
carry
```

The result is 101001 or 51_8.

EXAMPLE 1–35

Subtract the following numbers using 2's complement binary arithmetic:
(a) 5 − 9 (b) 32 − 85 (c) $8B_{16} - F3_{16}$ (d) $12_8 - 63_8$

SOLUTIONS

(a) $5 - 9 = -4$
$5 = 0101$
$9 = 1001$
Therefore $-9 = 0110 + 1 = 0111$ (in 2's complement representation).

$$\begin{array}{lr} & 0\,1\,0\,1 \\ & +\,\underline{0\,1\,1\,1} \\ \text{discard carry} \rightarrow 0 & 1\,1\,0\,0 \end{array}$$

Since the carry is 0 the result is 1100, or −4, in 2's complement representation. The signed result is the 2's complement of 1100, or −0100.

(b) $32 - 85 = -53$
$32 = 0100000$
$85 = 1010101$
Therefore $-85 = 0101010 + 1 = 0101011$ (in 2's complement representation).

$$\begin{array}{lr} & 0\,1\,0\,0\,0\,0\,0 \\ & +\,\underline{0\,1\,0\,1\,0\,1\,1} \\ \text{discard carry} \rightarrow 0 & 1\,0\,0\,1\,0\,1\,1 \end{array}$$

Since the carry is 0 the result is 1001011, or −53, in 2's complement representation. The signed result is the 2's complement of 1001011, or −0110101.

(c) $8B_{16} - F3_{16} = -68_{16}$
$8B = 10001011$
$F3 = 11110011$
Therefore $-F3 = 00001100 + 1 = 00001101$ (in 2's complement representation).

$$\begin{array}{lr} & 1\,0\,0\,0\,1\,0\,1\,1 \\ & +\,\underline{0\,0\,0\,0\,1\,1\,0\,1} \\ \text{discard carry} \rightarrow 0 & 1\,0\,0\,1\,1\,0\,0\,0 \end{array}$$

Since the carry is 0 the result is 10011000, or −68, in 2's complement representation. The signed result is the 2's complement of 10011000, or −01101000.

(d) $12_8 - 63_8 = -51_8$
$12 = 001010$
$63 = 110011$
Therefore $-63 = 001100 + 1 = 001101$ (in 2's complement representation).

$$\begin{array}{lr} & 0\,0\,1\,0\,1\,0 \\ & +\,\underline{0\,0\,1\,1\,0\,1} \\ \text{discard carry} \rightarrow 0 & 0\,1\,0\,1\,1\,1 \end{array}$$

Since the carry is 0 the result is 010111, or −51, in 2's complement representation. The signed result is the 2's complement of 010111, or −101001.

Multiplication

There are two techniques used to perform binary multiplication. The first is the process of repetitive additions similar to the procedure discussed in Section 1–2 in which the multiplicand is added to itself a certain number of times specified by the multiplier. The second is a technique that uses a "shift and add" scheme to perform the multiplication. Both techniques will be investigated here.

The simplest technique of binary multiplication is the repetitive addition procedure. For example, if we were required to multiply the binary number 1001 (9) by the binary number 101 (5) we could add the number 1001 to itself 5 (101) times.

```
     9        1001
  +  9        1001  +
    18       10010
  +  9        1001  +
    27       11011
  +  9        1001  +
    36      100100
  +  9        1001  +
    45      101101
```

The final result is 101101 or 45_{10}.

A more practical way to perform binary multiplication is to apply some of the concepts of longhand multiplication. If we were to multiply the two binary numbers 1001 and 101 using conventional long multiplication techniques, we would have

```
     1001   <- multiplicand
  ×   101   <- multiplier
  -------
     1001
    0000
   1001
  -------
   101101   <- product
```

The final result is 101101 or 45_{10}.

Notice that in order to perform the multiplication all we have to do is shift and add the multiplicand. Both these operations are easily performed by digital circuits. The following are examples of binary multiplication.

EXAMPLE 1–36

Multiply the following numbers using the shift-and-add binary multiplication technique (longhand multiplication).
(a) 15 × 12 (b) $3A_{16}$ × 2 (c) 10_8 × 10_8

SOLUTIONS
(a) 15 × 12 = 180
In binary, 1111 × 1100 = 10110100.

```
       1111
     × 1100
     ------
       0000
      0000
     1111
    1111
   --------
   10110100
```

The final result is 10110100 or 180_{10}.

(b) $3A_{16} \times 2 = 74_{16}$
In binary, 111010 × 10 = 1110100.

```
      1 1 1 0 1 0
  ×           1 0
  ---------------
      0 0 0 0 0 0
    1 1 1 0 1 0
  ---------------
    1 1 1 0 1 0 0
```

The final result is 1110100 or 74_{16}.
(c) $10_8 \times 10_8 = 100_8$
In binary, 1000 × 1000 = 1000000.

```
            1 0 0 0
          × 1 0 0 0
  -----------------
            0 0 0 0
          0 0 0 0
        0 0 0 0
      1 0 0 0
  -----------------
      1 0 0 0 0 0 0
```

The final result is 1000000 or 100_8.

Division

Binary division is performed by repetitive subtractions. The technique of division by repetitive subtractions was described in Section 1–2 with reference to the decimal numbering system. Recall that this process simply involves repetitively subtracting the divisor from the dividend while keeping a count of how many subtractions have taken place. When the remainder of the subtractions reach a value that is less than the divisor or equal to zero, the division is complete and the quotient is equal to the count of the number of subtractions that have taken place. For example, if we are to divide the binary number 11110 (30) by the binary number 101 (5) we can perform the division by adding the 2's complement of the divisor 101 to the quotient repetitively until we obtain a remainder of zero. It is preferable to use 2's complement subtraction since this does not involve adding in the end-around carry each time. The following examples illustrate the procedure.

EXAMPLE 1–37

Divide the number 30 by 5 using the repetitive binary subtraction procedure.

SOLUTION
The dividend is $30_{10} = 11110_2$.
The divisor is $5 = 101_2$.
Since we will be subtracting the divisor from the dividend we need to obtain the 2's complement of 5 (101) extended to 5 bits (00101) since the dividend 30 (11110) is a 5-bit number.

2's complement of 00101 = 11010 + 1 = 11011.

dividend:	30		1 1 1 1 0	quotient (tally)
divisor:	− 5		+ 1 1 0 1 1	1
remainder:	25	1	1 1 0 0 1	
divisor:	− 5		+ 1 1 0 1 1	1
remainder:	20	1	1 0 1 0 0	
divisor:	− 5		+ 1 1 0 1 1	1
remainder:	15	1	0 1 1 1 1	
divisor:	− 5		+ 1 1 0 1 1	1
remainder:	10	1	0 1 0 1 0	
divisor:	− 5		+ 1 1 0 1 1	1
remainder:	5	1	0 0 1 0 1	
divisor:	− 5		+ 1 1 0 1 1	1 +
final remainder (less than 5) ⟶	0	1	0 0 0 0 0	6 ← quotient

EXAMPLE 1–38

Divide the number 10 by 3 using the repetitive binary subtraction procedure.

SOLUTION

The dividend is $10_{10} = 1010_2$.
The divisor is $3 = 11_2$.
Since we will be subtracting the divisor from the dividend we need to obtain the 2's complement of 3 (11) extended to 4 bits (0011) since the dividend 10 (1010) is a 4-bit number.
2's complement of 0011 = 1100 + 1 = 1101.

dividend:	10		1 0 1 0	quotient (tally)
divisor:	− 3		+ 1 1 0 1	1
remainder:	7	1	0 1 1 1	
divisor:	− 3		+ 1 1 0 1	1
remainder:	4	1	0 1 0 0	
divisor:	− 3		+ 1 1 0 1	1 +
final remainder (less than 3) ⟶	1	1	0 0 0 1	3 ← quotient

BCD addition

Section 1–5 introduced us to the representation of decimal numbers in a form known as binary coded decimal (BCD). These BCD numbers were made up of 4-bit binary codes that represented each digit in a decimal number. Recall that BCD numbers are *not* the binary equivalent of decimal numbers, they are merely a representation. In Section 1–5 it was stated that a digital computer uses one of two techniques to store a decimal number in memory—convert the number into its binary

ROBOTICS

Robots are no longer science fiction! Computer controlled robots are used today in manufacturing, hazardous environments, research, and very soon in our homes to assist in domestic chores.

◀ An experimental "personal robot" being controlled by a remote computer.

© Charles/Gamma Liaison

A mobile robot that is being used to defuse a phony suitcase bomb in front of the Irish Embassy in Brussels, Belgium. ▶

Photo News/Gamma Liaison

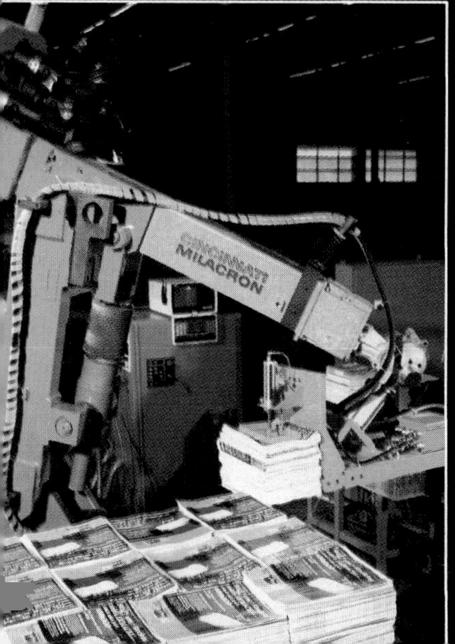

◀ An industrial robot used in automated packaging that is programmed to stack magazines.

A series of industrial robots used in an automated manufacturing process—an automobile assembly line. ▶

COMPUTER AIDED DESIGN & COMPUTER ASSISTED ENGINEERING

Digital computers simplify the process of design and development of mechanical and electronic equipment. CAD/CAE systems today are used to design automobiles, highways, spacecraft, and electronic microchips.

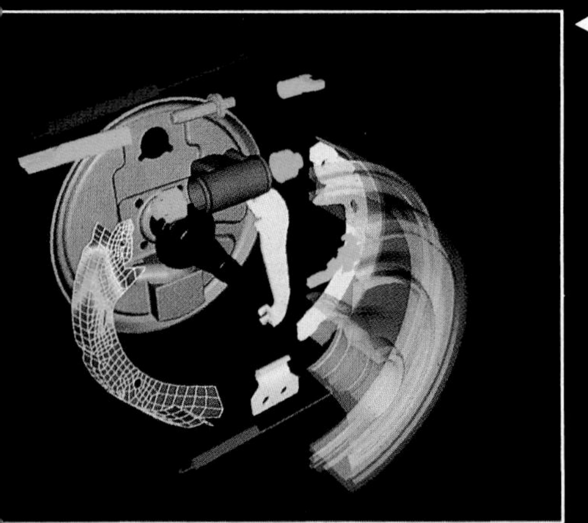
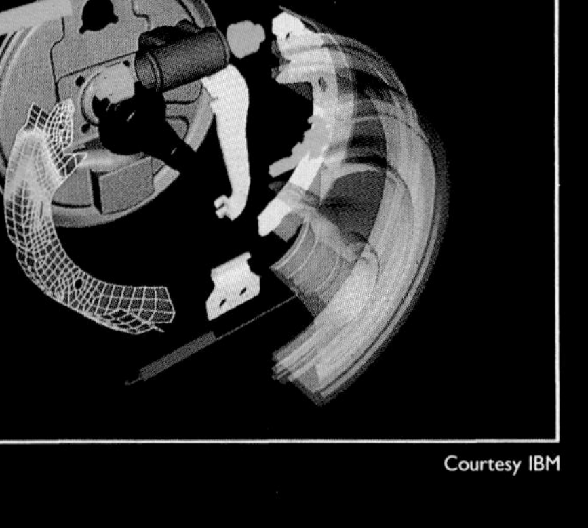

◀ An exploded view of a brake assembly designed and displayed on a CAD system.

Courtesy IBM

Photo courtesy Hewlett-Packard Company

▲ The Model 9845 from Hewlett-Packard has a high resolution color graphics display for CAD/CAE applications.

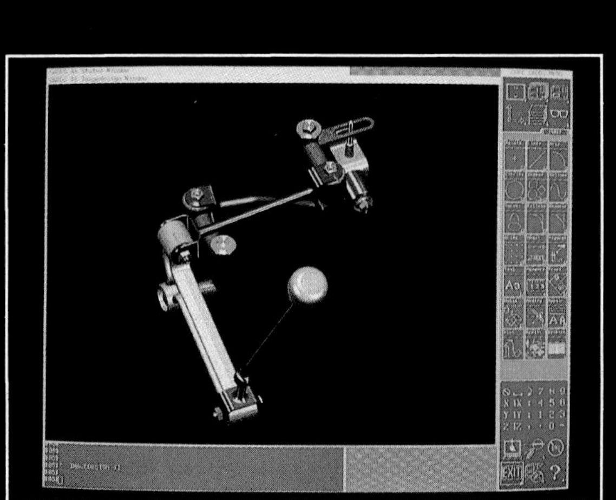

Courtesy Computer Vision

▲ A 3-dimensional CAD software package works with a CAD workstation to assist in complex mechanical designs.

Courtesy TRW, Inc.

A light pen is used to ▶ make changes on the design layout of a microelectronic integrated circuit.

TELECOMMUNICATIONS

Telecommunications Technology has influenced almost every aspect of our everyday lives—from telephone to television. Digital Electronics have played a major role in its evolution.

An artist's concept of a Navstar ▶ Satellite in Earth Orbit to assist in Navigation.

Courtesy NASA

Courtesy NASA

◀ Mission Control Center at the Johnson Space Center contains all the communications equipment necessary to remotely monitor and control all activities on spacecraft and other space equipment.

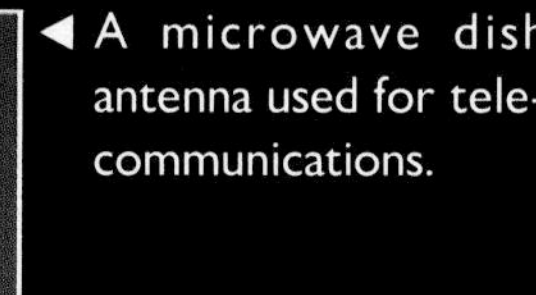

◀ A microwave dish antenna used for telecommunications.

Courtesy Hughes Aircraft

MEDICINE

Accurate diagnosis of abnormalities in the human body is made possible through the use of advanced digital computer circuits. Computer controlled equipment is also being used in the treatment of previously untreatable diseases.

Liane Enkells/Stock, Boston

◀ A computerized heart monitoring system used to monitor and record data from a patient.

Charles Gupton/Stock, Boston

An Electroencephalograph (EEG) Monitor displays the brain activity of a human subject. ▶

John Coletti/Stock, Boston

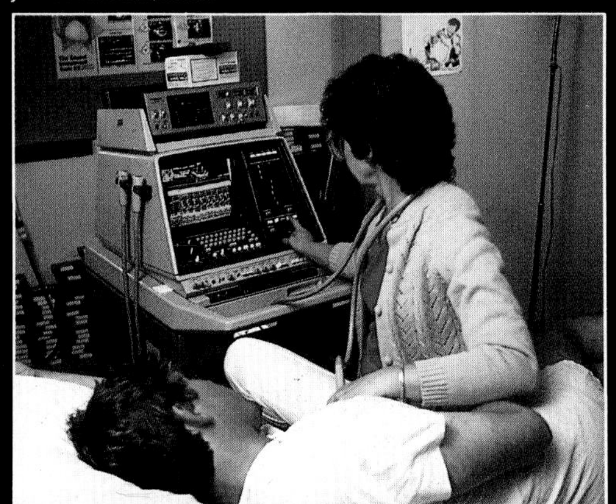

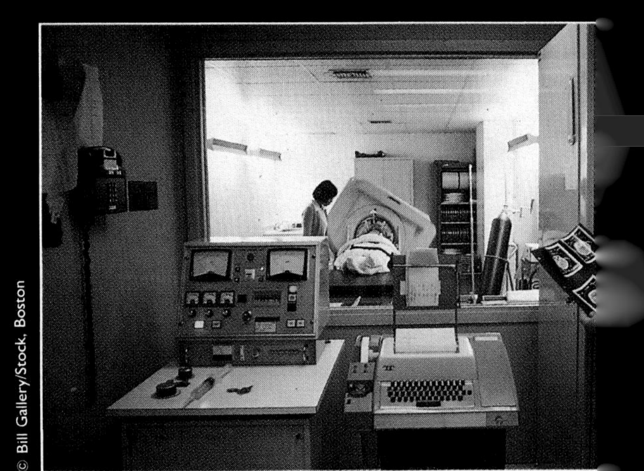

© Bill Gallery/Stock, Boston

▲ A Computer Assisted Tomography (CAT) scan-

AUTOMOBILES

The applications of digital computers in automobiles are approaching those used in space and aircraft. Engine timing, exhaust emissions, fuel flow, and other vital functions are now monitored and controlled by an on-board computer.

Courtesy Chrysler

▲ A futuristic automobile "cockpit"—The Chrysler Millennium.

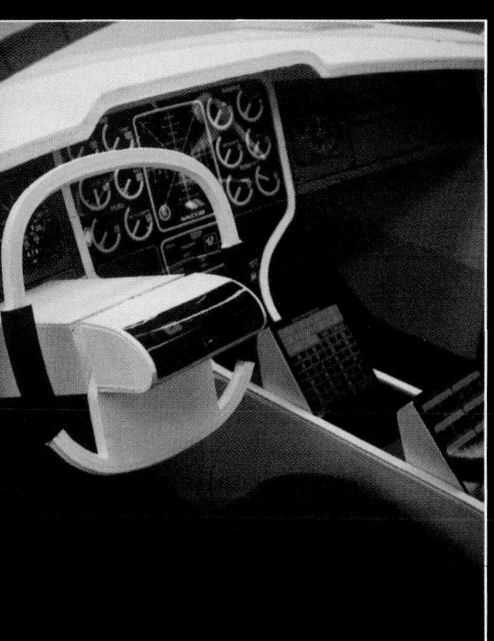

Courtesy Ford Motor Company

◀ An instrument panel of an experimental automobile showing the computer controlled navigation system.

Courtesy Ford Motor Company

Digital instrumentation ▶ in a current model Ford "Aerostar."

DEFENSE & SPACE

Much of the advancement in Digital Technology can be attributed to research done in Space and Defense Systems. The microprocessor evolved to a certain extent from the need in the space and defense industry for smaller and faster digital computers.

Courtesy Department of Defense

▲ The computer controlled cockpit of a fighter aircraft.

Courtesy NASA

▲ A view of the Apollo II Lunar Module as it separates from the Command Module and heads towards the Moon's surface.

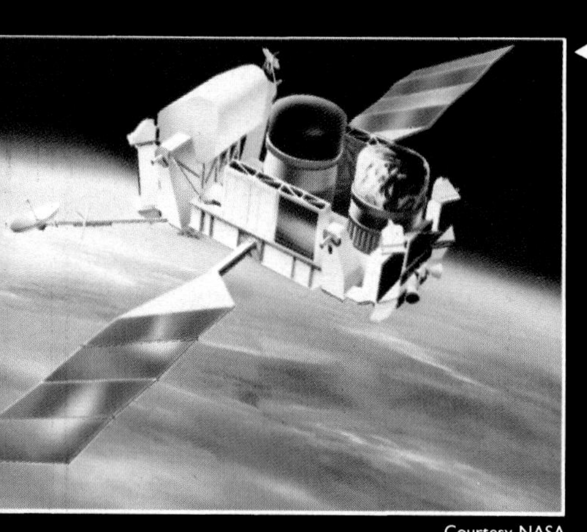

Courtesy NASA

◀ The Gamma Ray Observatory (GRO) to be launched in late 1990 weighs 15 tons, and measures 70 feet between the tips of the solar arrays.

Courtesy NASA

The space shuttle Atlantis ▶ is launched from the Kennedy Space Center on a Department of Defense dedicated mission.

COMPUTERS

Digital Electronics has made its greatest impact in Computer Technology. Computers today are used in almost every field of Engineering. Digital computers that once occupied large rooms are now being used in pocket-size appliances for high-speed computations.

▲ One of the fastest computers in the world! The CRAY-2 System is a supercomputer whose circuits are immersed in a fluorocarbon liquid that dissipates the heat generated by the densely packed electronic components.

▲ The personal computer (PC) is used today for entertainment, business, bookkeeping, and other routine tasks that were once done manually.

Paul Shambroom/Cray Research, Inc.

◀ The CRAY X-MP uses multiple processors to increase the execution speed of its instructions and consequently its computational performance.

Office Automation— ▶ The Personal Productivity Center makes using the most advanced office automation system features as easy as using a PC.

Photo courtesy Hewlett-Packard Company

ENTERTAINMENT

Digital Computers are always behind the scenes when it's time to entertain us. They are found in television sets, and video games, and are vital to the production of special effects in movies and television programs.

▲ The SONY XBR-PRO television set. A high technology microprocessor controlled TV of the future that provides features that are not currently found in TV sets.

Miro Vintopiv/Stock, Boston

▲ Kids playing computer video games on a home television set in the living room.

◀ Video games at a video arcade. These games make use of sophisticated microprocessors to handle the complex graphics that are displayed and manipulated on their screens.

© Teri Stratford, 1988

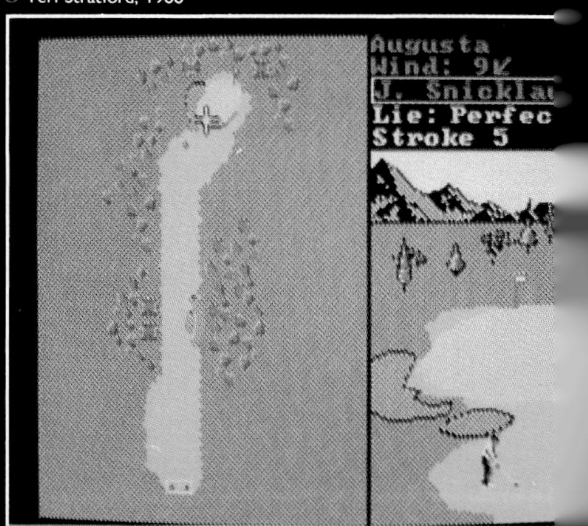

▲ Through the use of computer software a P can be used to provide hours of enterta ment in the form of challenging games.

equivalent before storage, or convert the number to its BCD representation before storage.

If the digital computer stores decimal numbers as their binary equivalents, arithmetic performed on these binary numbers is simple and can be done using the standard binary addition procedure discussed earlier in this section. The result can then be reconverted back from binary to decimal before retrieval. For example, if the two decimal numbers 27 and 34 were to be added, they would first be converted to binary, 011011 and 100010, respectively, and then added as follows:

```
  27       0 1 1 0 1 1
+ 34     + 1 0 0 0 1 0
----     -------------
  61       1 1 1 1 0 1
```

The binary sum obtained is 111101 which when converted back to decimal is 61, the desired result.

However, if the digital computer stored the two numbers, 27 and 34, in BCD representation, 00100111 and 00110100, respectively, the result would be incorrect if these two numbers were added using the standard binary addition procedure used in the previous case

```
  27       0 0 1 0 0 1 1 1
+ 34     + 0 0 1 1 0 1 0 0
----     -----------------
  5?       0 1 0 1 1 0 1 1
           \_____/ \_____/
              5       ?
```

Notice that the result of the addition yields not only the wrong MSD, but also an invalid BCD number 1011. This is because binary addition takes into account the *six* "missing" combinations 1010, 1011, 1100, 1101, 1110, and 1111 that are invalid in BCD. To compensate for these six digits during addition one must perform an operation known as *decimal adjust* on the result of a binary addition of two BCD numbers. The rules for BCD addition and for performing this decimal adjust operation are as follows:

1. Add the two BCD numbers using regular binary addition.
2. Check each group of 4 bits (nibble) of the result. If the value of the nibble is greater than 9 (1001), add 6 (0110) to the nibble. Also add 6 if there was a carry-out generated from the nibble when the two BCD numbers were added.

We can apply these rules to the addition of the two numbers, 27 and 34, attempted earlier as follows:

```
  27       0 0 1 0 0 1 1 1
+ 34     + 0 0 1 1 0 1 0 0
----     -----------------
  5?       0 1 0 1 1 0 1 1
   6     +         0 1 1 0
----     -----------------
  61       0 1 1 0 0 0 0 1
```

Notice that no carry-out was generated out of the least significant nibble or the most significant nibble. However, since the least significant nibble has a value that is greater than 9, we have added 6 to the least significant nibble to obtain the desired result 61_{10}.

The following examples of BCD addition illustrate other variations in the use of the rules for BCD addition.

EXAMPLE 1–39

Perform the addition of the following decimal numbers using BCD addition.
(a) 59 + 39 (b) 98 + 89 (c) 75 + 56 (d) 3928 + 4165

SOLUTIONS
(a) 59 + 39 = 98

```
                 1 <-------- carry-out
  59        0 1 0 1  1 0 0 1
+ 39      + 0 0 1 1  1 0 0 1
----      ------------------
  ??        1 0 0 1  0 0 1 0
```

A carry-out was generated from the least significant nibble, therefore we must add 6 to the least significant nibble:

```
  ??        1 0 0 1  0 0 1 0
+  6      +          0 1 1 0
----      ------------------
  98        1 0 0 1  1 0 0 0
```

The final sum is 10011000 (BCD) or 98_{10}.
(b) 98 + 89 = 187

```
                 1 <-------- carry-out
  98        1 0 0 1  1 0 0 0
+ 89      + 1 0 0 0  1 0 0 1
----      ------------------
  ??      1 0 0 1 0  0 0 0 1
          ^
          carry-out
```

A carry-out was generated from the least significant nibble and the most significant nibble, therefore we must add 6 to the least significant nibble and the most significant nibble:

```
  ??      1 0 0 1 0  0 0 0 1
+ 66      +  0 1 1 0  0 1 1 0
----      ------------------
 187      1 1 0 0 0  0 1 1 1
```

The final sum is 110000111 (BCD) or 187_{10}.
(c) 75 + 56 = 131

```
  75        0 1 1 1  0 1 0 1
+ 56      + 0 1 0 1  0 1 1 0
----      ------------------
  ??        1 1 0 0  1 0 1 1
```

The least significant nibble and the most significant nibble are both greater than 9, therefore we must add 6 to both the least significant and most significant nibbles.

```
  ??        1 1 0 0  1 0 1 1
+ 66      + 0 1 1 0  0 1 1 0
----      ------------------
 131      1 0 0 1 1  0 0 0 1
```

The final sum is 100110001 (BCD) or 131_{10}.

(d) 3928 + 4165 = 8093

3928	0 0 1 1 1 0 0 1 0 0 1 0 1 0 0 0
+ 4165	+ 0 1 0 0 0 0 0 1 0 1 1 0 0 1 0 1
????	0 1 1 1 1 0 1 0 1 0 0 0 1 1 0 1

The least significant nibble and the third nibble (from the right) are greater than 9, therefore 6 must be added to both these nibbles:

????	0 1 1 1 1 0 1 0 1 0 0 0 1 1 0 1
+ 0606	+ 0 1 1 0 0 1 1 0
8093	1 0 0 0 0 0 0 0 1 0 0 1 0 0 1 1

The final sum is 1000000010010011 (BCD) or 8093_{10}.

In this section we have investigated the techniques used to perform the basic arithmetic functions of addition, subtraction, multiplication, and division on binary numbers. We observed that these techniques were very similar to the techniques developed for decimal arithmetic in Section 1–2. The 1's and 2's complement subtraction procedure is analogous to the 9's and 10's complement subtraction procedure, respectively, for decimal numbers. Both techniques can be used to perform binary subtraction and there are advantages and disadvantages of one over the other; the choice of which technique to use generally depends on the application. Binary multiplication can be performed using multiple additions or by using the longhand procedure of shift-and-adds. Binary division is performed by the repetitive subtraction procedure. The addition of BCD numbers is similar to the addition of binary numbers except that it requires a decimal adjust to ensure that the result of the addition is a valid BCD number.

All binary arithmetic procedures studied in this section have made use of addition or a series of additions to obtain the required results. The basic digital circuit that performs all arithmetic operations in most digital computers is the adder, whose design and operation will be studied in Chapter 4.

Review questions

1. How do we identify the sign of the result of a binary subtraction?
2. Why is a decimal adjust required during BCD addition?

1–7 Binary codes

At this point it is apparent to us that many of the digital circuits to be covered later in this book will operate on binary numbers. Besides performing arithmetic on binary numbers, many digital circuits interpret binary numbers as instructions, while others interpret binary numbers as information. Information transfer between digital devices plays a very important part in digital communications.

In a digital communication scheme, if two devices want to transfer information between themselves (i.e., communicate) they do so by transmitting and receiving binary numbers. In much the same way as words make up the vocabulary of a spoken language, these binary numbers make up the "vocabulary" of a basic digital communications scheme that allows information to be transferred. The information transferred between two digital communicating devices are binary numbers that represent all the symbols necessary to accomplish the intelligible transfer of information. Thus each letter in the English alphabet (as well as some international alphabets), each digit in our numbering system, commonly used symbols for punctuation, and other necessary graphic characters are assigned unique binary numbers (codes) which are then transferred back and forth in a digital communications system.

A good example of a simple digital communicating system is the keyboard connected to a computer. When a key is pressed on the keyboard, a device known as an *encoder* inside the keyboard detects which key was pressed and sends a unique binary number (code) to the computer. The computer receives the code, and *decodes* it to determine which key was pressed. Standardization of these codes is extremely important. That is, if the keyboard transmits the code 1000001 when the "A" key was pressed, the computer should be able to identify that this code corresponds to the letter *A* and not any other character. Just as in a spoken language, all persons involved must speak the same language to prevent a communications breakdown, in a digital communications system, all devices must use the same code assignments for the symbols used in the communications vocabulary.

Over the years there have been several different sets of codes used in digital communications. However, the most popular and universally accepted code is the American Standard Code for Information Interchange (ASCII). Table 1–11 lists the ASCII codes and the symbols they represent.

Notice in Table 1–11 that even though each ASCII code is represented by an 8-bit number, since the most significant bit of each code is 0, the ASCII code is really a 7-bit code. Since there are 2^7 or 128 possible combinations of 7-bit binary numbers, the ASCII code can represent a maximum of 128 symbols. The symbols that the ASCII code can represent are of two types—graphic and nongraphic. Graphic ASCII symbols (codes 20_{16} through $7F_{16}$) are those that can be displayed or printed. Examples of such are the letters, numbers, punctuation, etc. Nongraphic ASCII symbols (codes 00_{16} through 19_{16}) are those that cannot be printed or displayed and are used to control the transmission of information or activate certain special features on the communicating devices. For example, when the RETURN (or ENTER) key is pressed on an ASCII keyboard (a keyboard that generates ASCII codes) no graphic character is actually displayed on the screen, instead the digital computer simply advances the cursor to the next line. Most of the nongraphic codes are generated through the use of the CTRL (CONTROL) key on an ASCII keyboard. Table 1–11 lists only the function abbreviations of these nongraphic codes; however, the definitions and key sequences for these nongraphic ASCII codes are shown in Table 1–12.

An extension of the ASCII code developed by IBM and called the extended ASCII code used 8 bits, thus allowing for up to 2^8 or 256 symbols to be represented. This code is shown in Table 1–13. The symbols assigned to the codes 00000000 through 01111111 are exactly the same as the basic ASCII code. The symbols assigned to the codes

Table 1–11
The ASCII Code

Hex	Binary	Symbol	Hex	Binary	Symbol	Hex	Binary	Symbol	Hex	Binary	Symbol
00	00000000	NUL	20	00100000	SP	40	01000000	@	60	01100000	
01	00000001	SOH	21	00100001	!	41	01000001	A	61	01100001	a
02	00000010	STX	22	00100010	"	42	01000010	B	62	01100010	b
03	00000011	ETX	23	00100011	#	43	01000011	C	63	01100011	c
04	00000100	EOT	24	00100100	$	44	01000100	D	64	01100100	d
05	00000101	ENQ	25	00100101	%	45	01000101	E	65	01100101	e
06	00000110	ACK	26	00100110	&	46	01000110	F	66	01100110	f
07	00000111	BEL	27	00100111	′	47	01000111	G	67	01100111	g
08	00001000	BS	28	00101000	(	48	01001000	H	68	01101000	h
09	00001001	HT	29	00101001	)	49	01001001	I	69	01101001	i
0A	00001010	LF	2A	00101010	*	4A	01001010	J	6A	01101010	j
0B	00001011	VT	2B	00101011	+	4B	01001011	K	6B	01101011	k
0C	00001100	FF	2C	00101100	,	4C	01001100	L	6C	01101100	l
0D	00001101	CR	2D	00101101	–	4D	01001101	M	6D	01101101	m
0E	00001110	SO	2E	00101110	.	4E	01001110	N	6E	01101110	n
0F	00001111	SI	2F	00101111	/	4F	01001111	O	6F	01101111	o
10	00010000	DLE	30	00110000	0	50	01010000	P	70	01110000	p
11	00010001	DC1	31	00110001	1	51	01010001	Q	71	01110001	q
12	00010010	DC2	32	00110010	2	52	01010010	R	72	01110010	r
13	00010011	DC3	33	00110011	3	53	01010011	S	73	01110011	s
14	00010100	DC4	34	00110100	4	54	01010100	T	74	01110100	t
15	00010101	NAK	35	00110101	5	55	01010101	U	75	01110101	u
16	00010110	SYN	36	00110110	6	56	01010110	V	76	01110110	v
17	00010111	ETB	37	00110111	7	57	01010111	W	77	01110111	w
18	00011000	CAN	38	00111000	8	58	01011000	X	78	01111000	x
19	00011001	EM	39	00111001	9	59	01011001	Y	79	01111001	y
1A	00011010	SUB	3A	00111010	:	5A	01011010	Z	7A	01111010	z
1B	00011011	ESC	3B	00111011	;	5B	01011011	[	7B	01111011	{
1C	00011100	FS	3C	00111100	<	5C	01011100	\	7C	01111100	|
1D	00011101	GS	3D	00111101	=	5D	01011101	]	7D	01111101	}
1E	00011110	RS	3E	00111110	>	5E	01011110	^	7E	01111110	~
1F	00011111	US	3F	00111111	?	5F	01011111	_	7F	01111111	DEL

10000000 through 11111111 make up the extended portion of the set and are used to represent various graphic characters and symbols.

Many digital communicating devices transfer binary codes with *parity*. Parity generating and checking is a scheme by which any errors that occur during transfer of these binary codes can be detected. There are two types of parity used during the transfer of binary codes—even and odd. To illustrate the concept of even and odd parity, consider the basic ASCII codes shown in Table 1–11. All 128 codes listed in Table 1–11 can be classified as those that have *even parity* and those that have *odd parity*. The codes that have an even number of 1 bits have even parity, while the codes that have an odd number of 1 bits have odd parity. For example, the following codes have even parity:

"!" 010 0001— two 1-bits
"M" 100 1101— four 1-bits
"w" 111 0111— six 1-bits

Table 1–12
Nongraphic ASCII code definitions and key sequences

Code	Key	Definition	Code	Key	Definition
NUL	CTRL @	Null	DLE	CTRL P	Data Link Escape
SOH	CTRL A	Start of Heading	DC1	CTRL Q	Direct Control 1
STX	CTRL B	Start Text	DC2	CTRL R	Direct Control 2
ETX	CTRL C	End Text	DC3	CTRL S	Direct Control 3
EOT	CTRL D	End of Transmission	DC4	CTRL T	Direct Control 4
ENQ	CTRL E	Enquiry	NAK	CTRL U	Negative Acknowledge
ACK	CTRL F	Acknowledge	SYN	CTRL V	Synchronous idle
BEL	CTRL G	Bell	ETB	CTRL W	End Transmission Block
BS	CTRL H	Backspace	CAN	CTRL X	Cancel
HT	CTRL I	Horizontal Tab	EM	CTRL Y	End of Medium
LF	CTRL J	Line Feed	SUB	CTRL Z	Substitute
VT	CTRL K	Vertical Tab	ESC	CTRL [	Escape
FF	CTRL L	Form Feed	FS	CTRL \	Form Separator
CR	CTRL M	Carriage Return	GS	CTRL]	Group Separator
SO	CTRL N	Shift Out	RS	CTRL ^	Record Separator
SI	CTRL O	Shift In	US	CTRL _	Unit Separator

while the following have odd parity:

"@" 100 0000— one 1-bit
"Q" 101 0001— three 1-bits
"O" 100 1111— five 1-bits

In an even parity transfer of information, the transmitter will add an additional bit (most significant 8 bit) to each 7-bit ASCII code called the *parity bit*. This bit would be set to a 1 if the code had odd parity so as to make it even or set to a 0 if the code had even parity. Thus all codes transmitted would have even parity by forcing the parity bit to a 1 or a 0. The receiver would be conditioned to check for even parity. That is, when the code is received the receiver checks to see if the total number of 1 bits is even. If the total number of 1 bits is even, the code was transmitted properly, otherwise the code was corrupted during transfer. An odd parity data communications system basically works in the same manner except that the parity bit of each code transmitted is set to a 1 or a 0 so as to make the total number of 1 bits an odd number. The receiver then counts the number of 1 bits received and signals an error if this number is not odd.

EXAMPLE 1–40

Determine the ASCII codes for the following characters if they were transmitted with, no parity, even parity, and odd parity:
(a) "5" (b) "$" (c) "E" (d) ">"

SOLUTIONS

(a) "5" = 35_{16} = 011 0101
With no parity, 011 0101
With even parity, 0011 0101
With odd parity, 1011 0101

(b) "$" = 24_{16} = 010 0100
With no parity, 010 0100
With even parity, 0010 0100
With odd parity, 1010 0100

Table 1–13
The IBM extended ASCII code

Hex	Binary	Symbol	Hex	Binary	Symbol	Hex	Binary	Symbol	Hex	Binary	Symbol
80	10000000	Ç	A0	10100000	á	C0	11000000	└	E0	11100000	α
81	10000001	ü	A1	10100001	í	C1	11000001	┴	E1	11100001	β
82	10000010	é	A2	10100010	ó	C2	11000010	┬	E2	11100010	Γ
83	10000011	â	A3	10100011	ú	C3	11000011	├	E3	11100011	π
84	10000100	ä	A4	10100100	ñ	C4	11000100	─	E4	11100100	Σ
85	10000101	à	A5	10100101	Ñ	C5	11000101	┼	E5	11100101	σ
86	10000110	å	A6	10100110	ª	C6	11000110	╞	E6	11100110	μ
87	10000111	ç	A7	10100111	º	C7	11000111	╟	E7	11100111	γ
88	10001000	ê	A8	10101000	¿	C8	11001000	╚	E8	11101000	Φ
89	10001001	ë	A9	10101001	⌐	C9	11001001	╔	E9	11101001	θ
8A	10001010	è	AA	10101010	¬	CA	11001010	╩	EA	11101010	Ω
8B	10001011	ï	AB	10101011	½	CB	11001011	╦	EB	11101011	δ
8C	10001100	î	AC	10101100	¼	CC	11001100	╠	EC	11101100	∞
8D	10001101	ì	AD	10101101	¡	CD	11001101	═	ED	11101101	ϕ
8E	10001110	Ä	AE	10101110	$\ll$	CE	11001110	╬	EE	11101110	ϵ
8F	10001111	Å	AF	10101111	$\gg$	CF	11001111	╧	EF	11101111	$\cap$
90	10010000	É	B0	10110000	░	D0	11010000	╨	F0	11110000	$\equiv$
91	10010001	æ	B1	10110001	▒	D1	11010001	╤	F1	11110001	$\pm$
92	10010010	Æ	B2	10110010	▓	D2	11010010	╥	F2	11110010	$\geq$
93	10010011	ô	B3	10110011	│	D3	11010011	╙	F3	11110011	$\leq$
94	10010100	ö	B4	10110100	┤	D4	11010100	╘	F4	11110100	⌠
95	10010101	ò	B5	10110101	╡	D5	11010101	╒	F5	11110101	⌡
96	10010110	û	B6	10110110	╢	D6	11010110	╓	F6	11110110	$\div$
97	10010111	ù	B7	10110111	╖	D7	11010111	╫	F7	11110111	$\approx$
98	10011000	ÿ	B8	10111000	╕	D8	11011000	╪	F8	11111000	°
99	10011001	Ö	B9	10111001	╣	D9	11011001	┘	F9	11111001	•
9A	10011010	Ü	BA	10111010	║	DA	11011010	┌	FA	11111010	·
9B	10011011	¢	BB	10111011	╗	DB	11011011	█	FB	11111011	$\sqrt{}$
9C	10011100	£	BC	10111100	╝	DC	11011100	▄	FC	11111100	^{n}
9D	10011101	¥	BD	10111101	╜	DD	11011101	▌	FD	11111101	^{2}
9E	10011110	₧	BE	10111110	╛	DE	11011110	▐	FE	11111110	■
9F	10011111	ƒ	BF	10111111	┐	DF	11011111	▀	FF	11111111	

(c) "E" = 45_{16} = 100 0101
With no parity, 100 0101
With even parity, 1100 0101
With odd parity, 0100 0101

(d) ">" = $3E_{16}$ = 011 1110
With no parity, 011 1110
With even parity, 1011 1110
With odd parity, 0011 1110

The binary codes discussed in this section are used mainly for information transfers as well as for representing the alphabets that make up our spoken language. Textual matter is usually stored in a computer's memory in ASCII representation. This also includes the other graphic symbols that we use in everyday writing. Symbolic representation is therefore another application of the binary numbering system without which the processing of information by digital computers cannot be accomplished.

Review questions

1. What are binary codes used for?
2. Why is there a need for a standard code?
3. What is the purpose of parity?

1–8 SUMMARY

- The decimal numbering system is a *base-ten* numbering system and has ten digits.
- A *count* in any numbering system is simply an ordered sequence of combinations of digits.
- Almost all mathematical operations can be reduced to the four basic arithmetic operations—addition, subtraction, multiplication, and division.
- Subtraction, multiplication, and division can be accomplished through various techniques of addition.
- The 9's and 10's complement of a decimal number can be used to represent a negative number without a sign.
- The *9's complement* of a number is obtained by subtracting each digit in the number from 9.
- The *10's complement* of a number can be obtained by adding 1 to its 9's complement.
- The 9's and 10's complement subtraction procedures can be used to subtract two decimal numbers through the process of addition.
- The *final carry* in 9's and 10's complement subtraction can be used to identify a negative or positive result.
- The end-around carry is the final carry that is added to the sum in 9's complement arithmetic and discarded in 10's complement arithmetic.
- The *octal* numbering system is a *base-eight* numbering system and has a set of only eight digits.
- The *hexadecimal* numbering system is a *base-16* numbering system and has a set of 16 digits.
- The binary numbering system is the numbering system that is used by all digital circuits.
- The *binary* numbering system is a *base-two* numbering system and has only two digits.
- A single binary digit is called a *bit*, a group of 4 bits make a *nibble*, 2 nibbles make up a *byte*, and 2 bytes make up a *word*.
- The octal and hexadecimal numbering systems are ideal for *representing* binary numbers.
- *Binary coded decimal* (BCD) numbers are binary numbers that represent decimal digits.
- The BCD value of a decimal number is only a representation of the decimal number and not its binary equivalent (value).

- The BCD system is a convenient means of representing decimal numbers since it does not require complicated conversions and can be accomplished through the substitution procedure.
- Binary subtraction is performed through the 1's and 2's complement techniques.
- The *1's complement* of a binary number is obtained by reversing (inverting) each bit in the number.
- The *2's complement* of a binary number is obtained by adding 1 to its 1's complement.
- The 1's and 2's complement form can be used to represent a negative binary number without a sign.
- The final carry in 1's and 2's complement subtraction can be used to identify a negative or positive result.
- The end-around carry is the final carry that is added to the sum in 1's complement arithmetic and discarded in 2's complement arithmetic.
- The addition of BCD numbers involves an extra step called *decimal adjust* to account for the six combinations that cannot be represented by any decimal digit.
- *Binary codes* are used in digital communications for transferring information.
- The *ASCII* code is a standard set of 7-bit binary numbers that represent the various symbols used in information transfer.
- The *extended ASCII* code simply adds another 128 combinations and symbols to the basic ASCII code by extending it to 8 bits.
- *Parity* generation and checking is done to ensure that data transmission errors can be detected.
- Two types of parity schemes are used—*even* and *odd* parity.
- Binary numbers transmitted with even parity have an even number of 1 bits while binary numbers transmitted with odd parity have an odd number of 1 bits.

PROBLEMS

Section 1–2 Decimal numbering system concepts

1. How many combinations would there be of 8 decimal digits, and what would the largest value in the count be?
2. Determine the value of the number 17,563 in terms of its positional weights.
3. Determine the digits that make up the decimal number 9140 by using the repetitive division procedure.
4. Represent the following signed negative numbers in 9's complement form:
   a. −4
   b. −12
   c. −6
   d. −640
   e. −98
   f. −4657

5. The following is a list of negative numbers expressed in 9's complement form. Determine the signed representation of each number.
a. 734,658
b. 238
c. 12
d. 948,505,734
e. 7

6. Perform the following subtractions using the 9's complement arithmetic technique:
a. $9 - 4 = 5$
b. $7 - 3 = 4$
c. $65 - 11 = 54$
d. $143 - 76 = 67$

7. Perform the following subtractions using the 9's complement arithmetic technique:
a. $2 - 4 = -2$
b. $3 - 8 = -5$
c. $24 - 87 = -63$
d. $43 - 205 = -162$
e. $1 - 643 = -642$

8. Represent the following signed negative numbers in 10's complement form:
a. -3
b. -25
c. -7
d. -10
e. -99
f. -7310

9. The following is a list of negative numbers expressed in 10's complement form. Determine the signed representation of each number.
a. 671,239
b. 9
c. 22
d. 819,735,021
e. 912

10. Perform the following subtractions using the 10's complement arithmetic technique:
a. $3 - 1 = 2$
b. $7 - 3 = 4$
c. $75 - 5 = 70$
d. $105 - 8 = 97$
e. $9 - 9 = 0$

11. Perform the following subtractions using the 10's complement arithmetic technique:
a. $4 - 8 = -4$
b. $1 - 7 = -6$
c. $9 - 19 = -10$
d. $7 - 512 = -505$

12. Divide the number 128 by 14 using the repetitive subtraction procedure.

```
                  9    ←quotient
divisor →  14 ) 128    ←dividend
                126
                ---
                  2    ←remainder
```

Section 1–3 The octal numbering system

13. Obtain an octal count sequence for the following decimal sequences:
a. 0 to 10
b. 100 to 108
c. 25 to 35
d. 55 to 66

14. Convert the following numbers from octal to decimal:
a. 1234_8
b. 03776_8
c. 812_8
d. 10_8
e. 7777_8
f. 3057672_8

15. Convert the following decimal numbers to octal:
a. 9731_{10}
b. $18{,}325_{10}$
c. 64_{10}
d. 8_{10}

Section 1–4 The hexadecimal numbering system

16. Obtain a hexadecimal count sequence for the following decimal sequences:
a. 0 to 12
b. 100 to 108
c. 25 to 35
d. 245 to 257

17. Convert the following numbers from hexadecimal to decimal:
a. 5691_{16}
b. $1B6A9_{16}$
c. $FACE_{16}$
d. 10_{16}
e. $FFFF_{16}$
f. 16789_{16}

18. Convert the following decimal numbers to hexadecimal:
a. 2916_{10}
b. 12087_{10}
c. 256_{10}
d. 16_{10}

Section 1–5 The binary numbering system

19. Obtain a binary count sequence for the following sequences:
a. 0 to 12
b. 100 to 108
c. 25_{16} to 30_{16}
d. 245_8 to 260_8

20. Convert the following numbers from binary to decimal using Table 1–5:
a. 1101101101_2
b. 1010_2
c. 10110101_2
d. 100021101_2
e. 100001_2
f. 11110_2

21. Convert the following decimal numbers to binary:
a. 10
b. 987
c. 801
d. 76

22. Convert the following binary numbers to octal:
a. 101010101010
b. 11100101
c. 10110
d. 0110110
e. 0000001
f. 11
g. 101000000

23. Convert the following octal numbers to binary:
a. 63543
b. 7777
c. 06354
d. 1110110
e. 53692
f. 1546
g. 2645

24. Convert the following binary numbers to hexadecimal:
a. 111101011001
b. 01011011
c. 101
d. 10010111101
e. 11111101111111
f. 010000001
g. 0000101010010

25. Convert the following hexadecimal numbers to binary:
a. ABCDE
b. 6F62B
c. 27549
d. 73AC645
e. BA32
f. 100101
g. 88

26. Convert the following octal numbers to hexadecimal:
a. 77
b. 10111
c. 352440
d. 6235213
e. 10321
f. 12

27. Convert the following hexadecimal numbers to octal:
a. AC
b. 6345
c. 10010
d. 7DA64
e. 02365
f. F

28. Convert the following decimal numbers to their BCD representations:
a. 90,017
b. 65
c. 63,515
d. 1010
e. 5
f. 236,021

29. Convert the following BCD numbers to decimal:
a. 010110011000
b. 0100010010101001000
c. 10100101110
d. 100100010101
e. 0100100010000010
f. 00010

30. The number 2103 is a base-four number. Convert this number to:
a. decimal
b. octal
c. hexadecimal
d. binary

31. Convert the following numbers to their base-four equivalents:
a. 75
b. 634_8
c. $F9_{16}$
d. 10101001_2

Section 1–6 Binary arithmetic

32. Perform the following additions in binary arithmetic:
a. $9 + 7$
b. $56 + 95$
c. $F5_{16} + 10_{16}$
d. $107_8 + 632_8$

33. Subtract the following numbers using 1's complement binary arithmetic:
a. $7 - 3$
b. $92 - 21$
c. $E7_{16} - 6D_{16}$
d. $172_8 - 65_8$

34. Subtract the following numbers using 1's complement binary arithmetic:
a. $6 - 8$
b. $62 - 75$
c. $A7_{16} - E9_{16}$
d. $25_8 - 76_8$

35. Subtract the following numbers using 2's complement binary arithmetic:
a. $8 - 3$
b. $43 - 12$
c. $EF_{16} - AB_{16}$
d. $42_8 - 33_8$

36. Subtract the following numbers using 2's complement binary arithmetic:
    a. 3 − 7
    b. 76 − 98
    c. $29_{16} - CB_{16}$
    d. $56_8 - 72_8$

37. Multiply the following numbers using the shift-and-add binary multiplication technique (longhand multiplication):
    a. 10 × 9
    b. $2C_{16} \times 3_{16}$
    c. $17_8 \times 13_8$

38. Divide the number 70 by 14 using the repetitive binary subtraction procedure.

39. Divide the number 32 by 10 using the repetitive binary subtraction procedure.

40. Perform the addition of the following decimal numbers using BCD addition.
    a. 23 + 56
    b. 48 + 12
    c. 90 + 77
    d. 7364 + 1238

Section 1–7 Binary codes

41. Determine the ASCII codes for the following characters if they were transmitted with no parity, even parity, and odd parity:
    a. "W"
    b. "9"
    c. "@"
    d. ETX

42. Determine the ASCII codes for the following characters if they were transmitted with no parity, even parity, and odd parity:
    a. "4"
    b. "="
    c. BEL
    d. "H"

Digital circuit boards are usually made up of many logic circuits in the form of ICs. The photograph shows a person troubleshooting such a circuit board (Photo courtesy Hewlett-Packard Company).

OUTLINE

OBJECTIVES

The objectives of this chapter are to:

- introduce the basic concepts of logic circuits, their operation, and their analysis.
- introduce the general operation of a logic gate and its hardware implementation in ICs.
- examine the operation of the four basic logic gates—the buffer, the inverter, the *AND* gate, and the *OR* gate.
- introduce the distinctive shape symbols and the ANSI/IEEE uniform shape symbols for each gate.
- introduce the various ICs that implement the four basic gates in hardware.
- examine the operation of the *NAND* and *NOR* gates, and introduce their IC implementations.
- introduce simple logic circuits that are constructed with the basic gates and develop a technique for analyzing and constructing these circuits.
- examine the procedure to set up logic circuit hardware.
- study the electrical characteristics of a logic gate.
- study the various troubleshooting procedures to test and debug logic gates and circuits.

2 BASIC LOGIC GATES AND CIRCUITS

2–1 INTRODUCTION

Logic circuits are the building blocks of all digital computers. The operation of these circuits is described by the principles of logic rather than the principles of conventional mathematics. Logic circuits are also known as *switching circuits* since the basic circuit element of early logic circuits was the electromechanical *switch* known as the *relay*. Also, the characteristics of logic circuits to *switch* from one state to another has led to continued use of the term to describe modern logic circuits. Since digital computers are made up of various logic circuits, and logic circuits are often used to process *digits* and numbers, we often refer to these circuits as *digital circuits*.

The basic element of modern logic circuits is the *gate* whose function is to electronically implement the various logic *primitives* or *functions* that are used in the design of logic circuits. Gates and logic circuits, accept binary numbers as *input*, process or manipulate these numbers, and produce binary numbers as *output*. We can broadly classify these logic circuits into two categories—*combinational logic circuits* and *sequential logic circuits*. Combinational logic circuits (discussed in this chapter and in Chapters 3 and 4) are circuits whose outputs are dependent only on the inputs of the circuit. However, the outputs of a sequential logic circuit (discussed in Chapters 5 through 7) are dependent not only on its inputs but also on the past history of the outputs; that is, sequential logic circuits include a *memory element*, or feedback.

This chapter introduces the reader to the basic logic gates that will be used in all logic circuits. The standard *distinctive shape symbols* as well as the new ANSI/IEEE *uniform shape symbols* are introduced with each gate. Due to their popularity, however, we will use the distinctive shape symbols in most of the logic circuits discussed in the book. A complete explanation of the ANSI/IEEE logic symbols is included in Appendix B. We will then study the principle of operation of each gate along with the actual hardware implementation of the gate in the form of an *integrated circuit* (IC). Also included in this chapter is an introduction to the electrical characteristics of logic gates and the interpretation of data sheets for digital integrated circuits. The application of the basic gates in simple logic circuits is then examined along with procedures to set up the logic circuit hardware. Chapter 3 continues the study to include procedures to design and analyze logic circuits. At the end of this chapter we will examine various procedures to troubleshoot the basic logic gates and circuits in a laboratory environment. These procedures can then be applied to many of the other circuits included in the rest of this book.

2–2 Basic logic concepts

This section introduces us to the basic concepts used in the analysis and synthesis of logic gates and circuits. These concepts are developed using simple switching circuit analogies and later in this chapter they are applied to specific gates and circuits. It should therefore be noted that the switching circuits developed in this section are not representative of the circuits to be covered in the rest of the book but are simply conceptual models that are used to illustrate the principles of logic and the operation of logic circuits.

The basic element of a switching circuit is a device known as a *switch*. A switch is used to control the flow of current through a circuit and is a *two-state device*; that is, the switch can exist in only one of two possible states—ON or OFF. Figure 2–1 illustrates these two possible states. In Figure 2–1a the switch is shown with its contacts open thus preventing the passage of current though the switch; this will be referred to as the OFF state. In Figure 2–1b, the switch is now closed and allows the passage of current through it; this state will be referred to as the ON state.

Now consider a simple switching cricuit such as the one shown in Figure 2–2. The circuit consists of a battery (or some power supply) connected in series with a switch and a lamp (light bulb). We will assume that the battery maintains a fixed voltage and current for the circuit. We have identified the switch and the lamp with the variables S and L, respectively. The lamp, L, is also a two-state device; that is, the lamp can either be ON or OFF. For both the lamp and the switch there are only two possible states—ON or OFF, with nothing in between these two states. Notice in Figure 2–2 that the state of the lamp, L, is *dependent* on the state of the switch, S. However, the state of the switch, S is *independent* of the state of the lamp, L. In other words, turning the switch ON or OFF will cause the lamp to turn ON or OFF, but turning the lamp ON or OFF will not cause the switch to turn ON or OFF. The switch is therefore the controlling factor, whereas the lamp is the controlled factor. In the study of logic circuits, L is called the *dependent variable* and S is called the *independent variable*.

Figure 2–1
The two states of a switch.

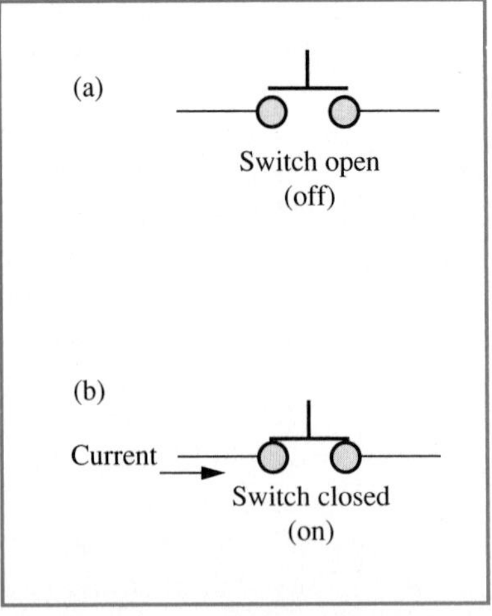

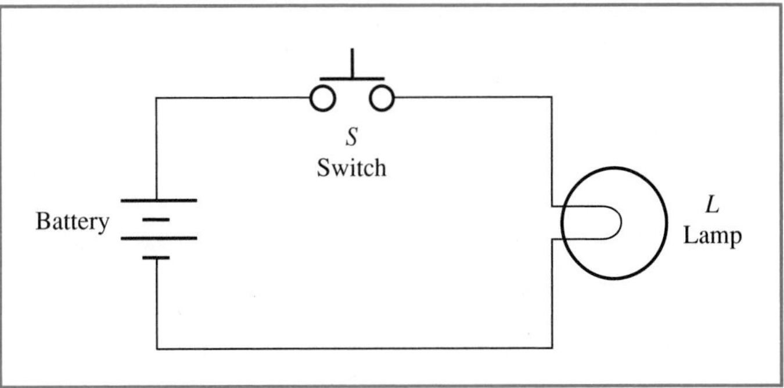

Figure 2–2
A simple switching circuit.

In order to describe the operation of the circuit shown in Figure 2–2, we can now construct a table that lists all possible states of the dependent variable, L, as a function of the independent variable, S. Such a table is called a *truth table* and is shown in Table 2–1.

Notice in Table 2–1 that the state of the lamp, L, corresponds exactly to the state of the switch, S. That is, if the switch is ON then the lamp is ON, and if the switch is OFF then the lamp is OFF. The circuit in Figure 2–2 and the truth table in Table 2–1 can also be described by the *logic equation*

$$L = S$$

which when expressed in words simply means that the state of the lamp is the same as (equal to) the state of the switch. Observe that the dependent variable is always placed on the left-hand side of the logic equation.

The operation of the circuit shown in Figure 2–2 can be determined by simple logical deduction, hence we often refer to such a circuit as a *logic circuit* since the operation of such circuits are often governed by the simple rules of logic. For reasons mentioned in Section 2–1, we also refer to these circuits as *digital circuits*. All digital circuits are two-state circuits. That is, the variables in digital circuits can only have one of two possible states. The counterpart of a digital circuit is often referred to as an *analog circuit*. For example, if the circuit in Figure 2–2 were modified such that a variable resistor were used to control the state of the lamp instead of the switch, then the lamp would have several states of intensity between being completely ON and completely OFF; the circuit could no longer be classified as a digital circuit but would now be considered an analog circuit.

To illustrate the concepts of logic circuits, truth tables, and logic equations further, consider the circuit shown in Figure 2–3.

In Figure 2–2, the switch was placed in series with the lamp so that the switch directly controlled the current flowing through the lamp. In Figure 2–3, we have now placed the switch in parallel with the lamp (i.e., across the lamp) so that the switch now controls the flow of current *away* from the lamp. Of course, we should not experimentally implement this circuit without a suitable current-limiting power supply since the switch

Table 2–1
Truth table for Figure 2–2

S	**L**
OFF	OFF
ON	ON

Figure 2–3
An inverting circuit.

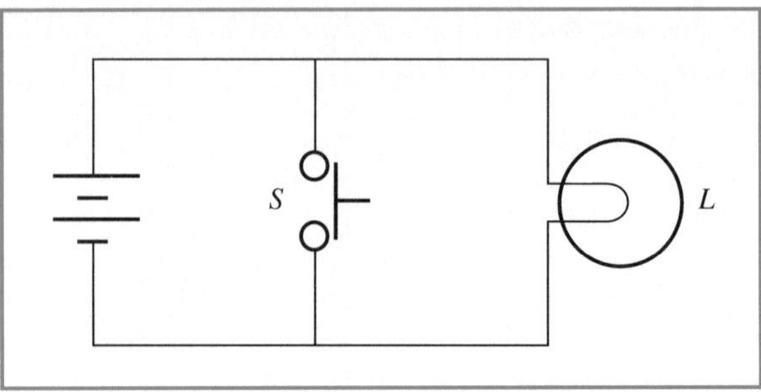

effectively short-circuits the power supply. The logic of the circuit is quite apparent. If the switch is open (OFF) then all the current flows from the battery to the lamp and the lamp is ON. However, if the switch is closed, all the current flows through the switch (remember, current always follows the path of least resistance) and the lamp is OFF. Therefore, the state of the lamp is the *inverse* of the state of the switch. The circuit is known as an *inverting circuit* or more simply as an *inverter* and its truth table is shown in Table 2–2. Notice in Table 2–2 that the state of the lamp, L, at any given instant is the opposite of the state of the switch, S. We refer to L as the *complement* of S, and to S as the complement of L. In the study of logic circuits, the word "complement" is identified with "inversion" rather than with "equivalence." The logic equation for the circuit of Figure 2–3 and its truth table is

$$L = NOT\ S$$

which when expressed in words means that the state of L is not the state of S. Since there are only two possible states, the term "NOT" identifies the inverse state. For example if S = ON,

$$\text{then } L = NOT\ S = NOT\ \text{ON} = \text{OFF}$$

if S = OFF,

$$\text{then } L = NOT\ S = NOT\ \text{OFF} = \text{ON}$$

Instead of using the word *NOT* to identify the inverse state we use a horizontal bar over the letter of the variable. Thus the same logic equation can be represented as

$$L = \overline{S}$$

The equation is read as "L equals S not" or "L equals not S" and identifies the variable L and its complement, S.

Table 2–2
Truth table for Figure 2–3

S	L
OFF	ON
ON	OFF

EXAMPLE 2–1

The variable A is the complement of the variable B, and the variable B is the complement of the independent variable C. Construct a truth table for the three variables, A, B, and C. What conclusion can be drawn from the relationship between the variables A and C?

SOLUTION Since B is the complement of C

$$B = \overline{C}$$

Since A is the complement of B

$$A = \overline{B}$$

Thus looking at all possible states of the independent variable C and determining the states of A and B from the preceding equations we obtain the following truth table:

C	B	A
ON	OFF	ON
OFF	ON	OFF

From the truth table we can conclude that the state of A is the same as the state of C. Therefore

$$A = C$$

We can also reach this conclusion mathematically as follows:
If $B = \overline{C}$
Then $\overline{B} = C$
But $A = \overline{B}$
Therefore $A = C$.

So far we have examined circuits that had one dependent variable and one independent variable. Let us now extend the study of logic concepts to circuits that contain more than one independent variable. Figure 2–4 expands the circuit of Figure 2–2 to include two switches, A and B, connected in series with the lamp, L. As before, notice that the state of L is dependent on the states of both A and B, but not vice versa. Therefore L is the dependent variable and A and B are the independent variables.

We can describe the operation of the circuit in Figure 2–4 in terms of the conditions necessary for the lamp to turn ON. This is known as *positive logic* and will be used throughout this book. Note that if the operation of the circuit was described in terms of the conditions necessary to turn the lamp OFF, we would be dealing with *negative logic*. The lamp, L, will be ON if and only if *both* switches A *and* B are ON. The circuit is known as an "AND" circuit since the dependent variable L

Figure 2–4
An *AND* circuit.

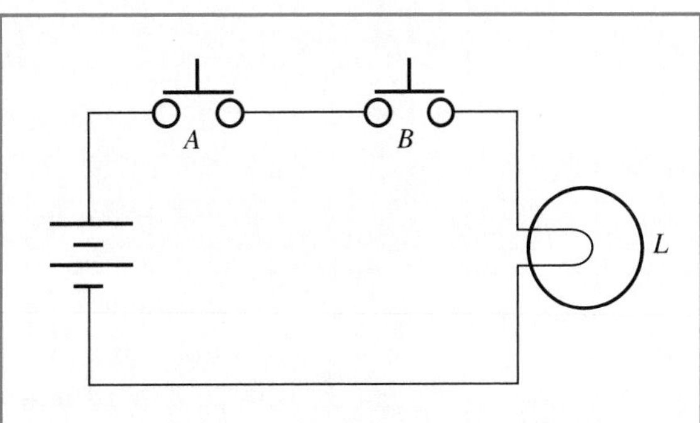

Table 2–3
Truth table for Figure 2–4

A	B	L
OFF	OFF	OFF
OFF	ON	OFF
ON	OFF	OFF
ON	ON	ON

will be in the ON state if the independent variables A and B are both in the ON state. If any of the independent variables are OFF, the dependent variable will be OFF. The truth table shown in Table 2–3 illustrates this relationship.

Notice in Table 2–3 that we have looked at all possible combinations of the independent variables A and B, and for each combination, determined the state of the dependent variable L. Also note that if the binary digit 1 were substituted for ON and a 0 for OFF, we would have a binary count of all possible combinations of 2 bits in the columns representing the two independent variables A and B. This format for constructing truth tables will be used for all the logic circuits to follow.

The logic equation for the circuit in Figure 2–4 and its truth table is expressed as follows:

$$L = A \text{ } AND \text{ } B$$

However, in the study of logic circuits, we choose not to use the word "AND" but rather to replace it with the symbol "·" which will be used to represent the *AND* function. Thus

$$L = A \cdot B$$

The equation is read as "L equals A and B." The concept of the *AND function* can be extended to more than one independent variable as shown in the following example.

EXAMPLE 2–2

Construct a truth table for the dependent variable X for the circuit shown in Figure 2–5, and determine the logic equation for the circuit.

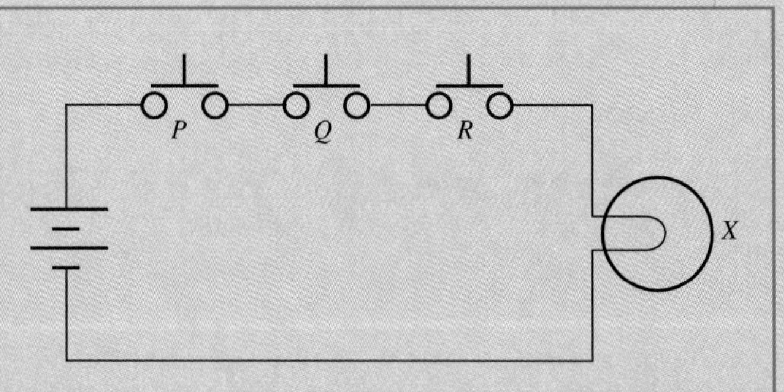

Figure 2–5
AND circuit with three switches.

SOLUTION Regardless of the number of switches in the circuit, the lamp X will be ON, if and only if all three switches are ON.

Therefore the truth table for the circuit is as follows:

P	Q	R	X
OFF	OFF	OFF	OFF
OFF	OFF	ON	OFF
OFF	ON	OFF	OFF
OFF	ON	ON	OFF
ON	OFF	OFF	OFF
ON	OFF	ON	OFF
ON	ON	OFF	OFF
ON	ON	ON	ON

The logic equation for the circuit is

$$X = P \cdot Q \cdot R$$

Again, observe in Example 2–2 that we have used the binary counting sequence for all possible combinations of three bits in order to obtain all possible states of the three independent variables P, Q, and R. Also notice that the logic of the AND function is the same irrespective of the number of independent variables; therefore the logic can be extended to as many variables as needed.

The circuit in Figure 2–6 represents another type of logic function that is commonly used in digital circuits. The circuit is known as an *OR* circuit since either switch A *or* switch B must be ON in order for the lamp L to turn ON. The truth table for the *OR* circuit is shown in Table 2–4.

Notice in Table 2–4 that if any of the independent variables is ON, the dependent variable will be ON.

The logic equation for the circuit in Figure 2–6 and its truth table is expressed as follows:

$$L = A \text{ } OR \text{ } B$$

However, in the study of logic circuits, we choose not to use the word "OR" but rather to replace it with the symbol "+" which will be used to represent the *OR* function. Thus

$$L = A + B$$

The equation is read as "L equals A or B."

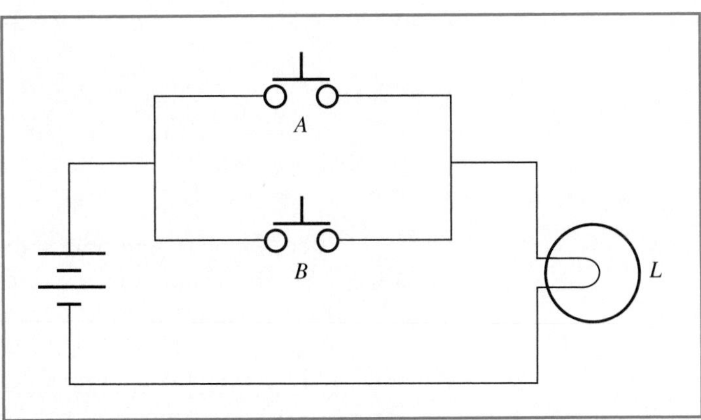

Figure 2–6
An *OR* circuit.

Table 2–4
Truth table for Figure 2–6

A	B	L
OFF	OFF	OFF
OFF	ON	ON
ON	OFF	ON
ON	ON	ON

The concept of the *OR* function can be extended to more than one independent variable as shown in the following example.

EXAMPLE 2–3

Construct a truth table for the dependent variable X for the circuit shown in Figure 2–7, and determine the logic equation for the circuit.

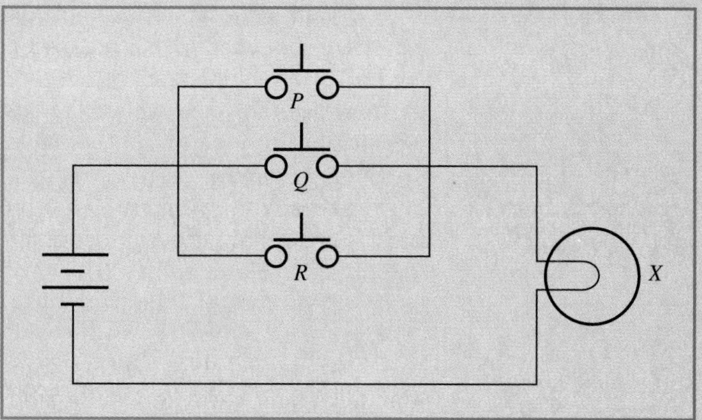

Figure 2–7
OR circuit with three switches.

SOLUTION Regardless of the number of switches in the circuit, the lamp X will be ON if any of the three switches are ON. Therefore the truth table for the circuit is as follows:

P	Q	R	X
OFF	OFF	OFF	OFF
OFF	OFF	ON	ON
OFF	ON	OFF	ON
OFF	ON	ON	ON
ON	OFF	OFF	ON
ON	OFF	ON	ON
ON	ON	OFF	ON
ON	ON	ON	ON

The logic equation for the circuit is

$$X = P + Q + R$$

As before, Example 2–3 shows us that the logic of the *OR* function is the same irrespective of the number of independent variables; therefore the logic can be extended to as many variables as needed.

We have so far used the symbols "·" and "+" to represent the *AND* and *OR* functions, respectively, in logic equations. It is apparent that these are the same symbols used to represent the arithmetic operations of multiplication and addition, respectively. In Chapter 3, logic equations containing these functions will be manipulated in much the same way as we manipulate algebraic expressions, and since there are a lot of algebraic similarities between the arithmetic and logical functions, we choose to use the same symbols.

A logic circuit can contain combinations of the various logic functions discussed earlier. Example 2–4 illustrates such a circuit and its analysis.

EXAMPLE 2–4

Construct a truth table for the dependent variable Z for the circuit shown in Figure 2–8, and determine the logic equation for the circuit.

Figure 2–8

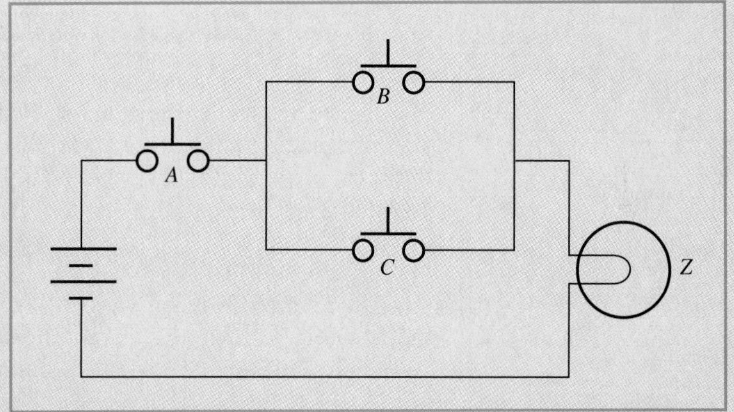

SOLUTION We can describe the logic of the circuit as follows. The lamp Z will be ON if and only if switch A is ON, and either switch B or C is ON. This logic can be expressed as a logic equation

Equation (2–1)

$$Z = A \cdot (B + C)$$

We can also describe the logic of the circuit as follows. The lamp Z will be ON if and only if switch A AND switch B are ON, OR switch A AND switch C are ON. This logic can also be expressed as an equivalent logic equation

Equation (2–2)

$$Z = (A \cdot B) + (A \cdot C)$$

We can now construct a truth table that looks at all possible states of A, B, and C and for each combination the state of Z can be determined as follows:

A	B	C	Z
OFF	OFF	OFF	OFF
OFF	OFF	ON	OFF
OFF	ON	OFF	OFF
OFF	ON	ON	OFF
ON	OFF	OFF	OFF
ON	OFF	ON	ON
ON	ON	OFF	ON
ON	ON	ON	ON

In Example 2–4 we obtained two equivalent logic equations for the circuit in Figure 2–8. Equations 2–1 and 2–2 both described the logic of the circuit in different forms. We can also obtain a logic equation from the truth table of the circuit by analyzing the last three conditions in the table. Notice that the last three combinations cause the lamp Z to turn ON. Expressed in words this simply means that in order for the lamp Z to turn ON, one of the following three conditions have to be satisfied:

1. Switch A must be ON, switch B must be OFF, and switch C must be ON.

or

2. Switch A must be ON, switch B must be ON, and switch C must be OFF.

or

3. Switch A must be ON, switch B must be ON, and switch C must be ON.

Since we can express an OFF state as the complement of the ON state by placing a bar over a variable, we can express these conditions in the form of the following logic equation that is equivalent to Equations 2–1 and 2–2.

Equation (2–3)

$$Z = (A \cdot \overline{B} \cdot C) + (A \cdot B \cdot \overline{C}) + (A \cdot B \cdot C)$$

In Chapter 3, using various laws defined by a field of logical mathematics known as *Boolean algebra*, it will be proved that Equations 2–1, 2–2, and 2–3 are logically equivalent.

Review questions

1. Why is a switching circuit considered to be a two-state device?
2. What are the two types of variables that exist in a switching circuit and how do they differ from each other?
3. What is the difference between a digital and an analog circuit?
4. What does an inverter do?
5. What are the conditions necessary for the dependent variable of an *AND* circuit to be ON?
6. What are the conditions necessary for the dependent variable of an *OR* circuit to be ON?

2–3 The logic gate

At this point we have examined the three basic *logic functions* known as the logic inversion (*NOT*), the logic *AND*, and the logic *OR*. We have used simple switching circuits to illustrate the concepts of logic functions and the equations and tables that describe them. However, these switching circuits are only conceptual models and are not used in modern digital devices. Instead we use circuits that operate on the same principles discussed previously but have different configurations. These circuits are known as *logic gates* or more simply, *gates*.

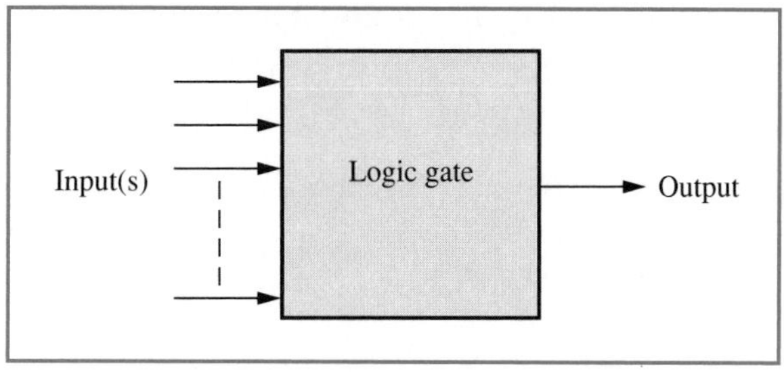

Figure 2–9
The model of a logic gate.

Figure 2–9 illustrates the general model that represents all logic gates. A logic gate can be viewed as a "black box" that has one or more inputs and a single output. The inputs and output to the box are 1's and 0's (binary numbers). The gate accepts binary numbers at its input and produces a 1 or a 0 at its output depending on its *function*. The output of the gate is therefore considered to be the dependent variable because it depends on the binary combinations at the input(s) of the gate (independent variables). Like the switching circuits discussed earlier, the operation of a gate can be described by its truth table and logic equation that relate the dependent and independent variables.

Within the box is an electronic circuit (to be studied in Chapter 12) that makes the gate function in the appropriate manner. At this stage we will view the gate as a "black box" with signals going in and coming out, with its inner workings essentially invisible. The binary digits 1 and 0 are applied to the gate's input(s) and measured at the output in the form of voltages. Typically +5 volts represents a logic 1 (also referred to as a logic *high level*) and 0 volts (ground potential) represents a logic 0 (also referred to as a logic *low level*). We will examine the details of the electrical characteristics of these gates in Section 2–9 and Chapter 12.

In the previous section we obtained a truth table for a switching circuit by examining the state of the dependent variable (lamp) for all possible combinations of the states of the independent variables (switches). We saw that these combinations followed a binary counting sequence. The truth tables that describe the operation of logic gates follow the same format. That is, we will look at all possible combinations of the independent variables (inputs) by listing a binary count. Recall from Chapter 1 that the number of binary combinations of n bits is given by the relationship 2^n. Therefore the truth table for a single input gate would have 2^1 or two entries

Input
0
1

The truth table for a two-input gate would have 2^2 or four entries

Inputs	
0	0
0	1
1	0
1	1

The truth table for a three-input gate would have 2^3 or eight entries

Inputs		
0	0	0
0	0	1
0	1	0
0	1	1
1	0	0
1	0	1
1	1	0
1	1	1

And the truth table for a four-input gate would have 2^4 or 16 entries

Inputs			
0	0	0	0
0	0	0	1
0	0	1	0
0	0	1	1
0	1	0	0
0	1	0	1
0	1	1	0
0	1	1	1
1	0	0	0
1	0	0	1
1	0	1	0
1	0	1	1
1	1	0	0
1	1	0	1
1	1	1	0
1	1	1	1

Integrated circuit logic

A logic gate is usually packaged in the form of an IC. One of the most popular and most accepted *series* of digital ICs is the *74xxx TTL series*. These ICs provide the digital circuit designer with a wide variety of gates and application circuits that can be used in most applications. Each IC in the series is given a two- or three-digit number ranging from 00 to (currently) 670 that identifies the function of the IC. The numbers are prefixed with the digits 74 (or 54) to identify the series, and hence we refer to the series as the 74xxx series, where xxx identifies a specific IC in the series. More will be said about this series in Chapter 12.

Integrated circuits are manufactured in a variety of sizes that are classified according to the number of *pins* or connections. Figure 2–10a shows typical IC packages that are used for logic gates in 14- and 16-pin sizes. Figure 2–10b includes the pin layout when each IC (commonly referred to as a *chip*) is viewed from above. A notch on one end of the IC usually identifies the top. Notice that the pins are numbered sequentially starting from the upper left-hand side and proceeding down to the end of one side. The count is then continued from the bottom right-hand side to the top right-hand side of the IC. Many ICs include a small dot on the upper left-hand side to identify pin 1. Generally two pins on an IC are reserved for applying power to the internal circuitry: +5 volts and ground

(a)

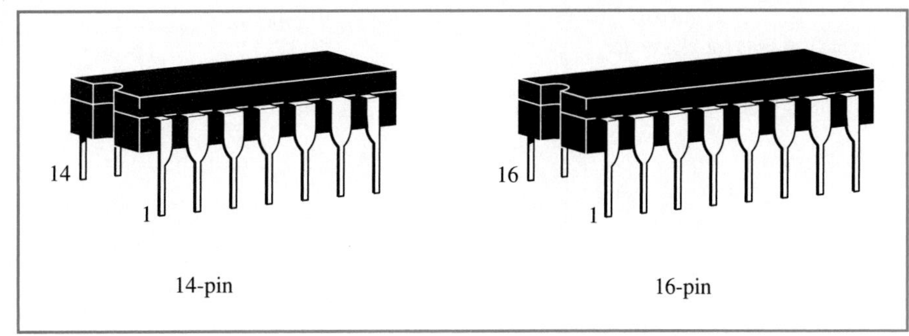

(b)

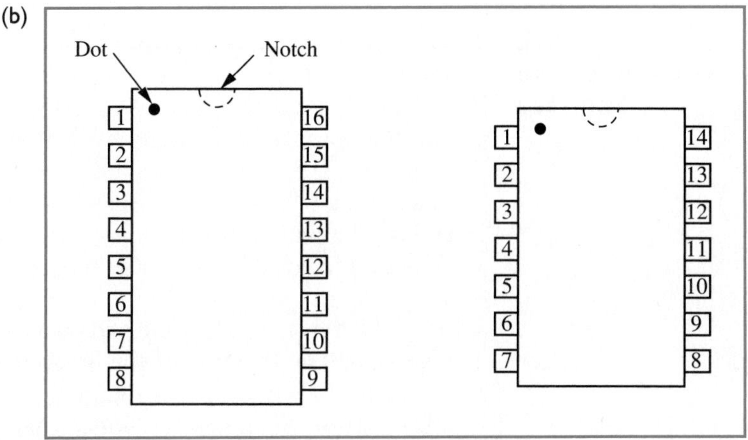

Figure 2–10
14- and 16-pin ICs. (a) Pictorial view; (b) pin layout (top view).

potential. These two pins are labeled V_{CC} and GND (ground), respectively and are usually (but not always) assigned to pins 14 and 7, respectively, in a 14-pin IC, and pins 8 and 16, respectively, in a 16-pin IC. The other pins on the IC allow access to the input(s) and output(s) of the gate(s) in the IC.

The next few sections in this chapter will examine the various gates that implement the basic functions discussed in this section. The operation and analysis of each gate examined will also include the specific 74xxx IC(s) that implement the function of the gate in hardware (i.e., electronically). Appendix C includes the complete data sheets for the ICs discussed in the book, which include the pin configurations, symbols, and electrical characteristics.

Review questions

1. How is a logic gate similar to a simple switching circuit?
2. What form do the dependent and independent variables of a logic gate take—conceptually and electronically?
3. How do we identify the pins on an IC?
4. What is the purpose of the V_{CC} and GND pins on an IC?

(a)

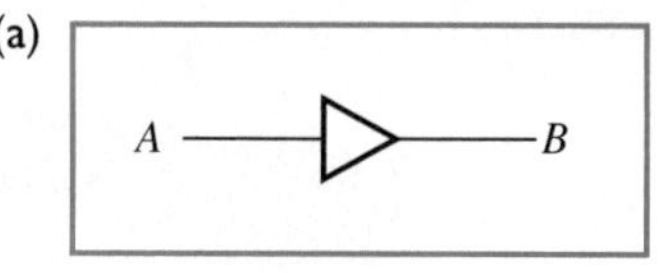

(b)

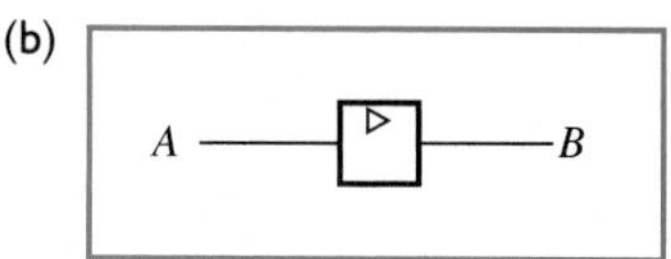

Figure 2–11
Logic symbols for the buffer. (a) Distinctive shape symbol; (b) ANSI/IEEE Std. 91-1984 symbol.

2–4 The buffer and inverter

The *buffer* is a gate that has only one input and one output. The logic symbols for the buffer are shown in Figure 2–11. The buffer, like the switching circuit in Figure 2–2, provides no logic function. This means that the output of the buffer (B) will always be at the same state as the input (A). The buffer, however, does provide a very important electronic function as will be seen in Section 2–9 and in Chapter 12. The truth table and logic equation for the buffer are shown in Table 2–5. Notice the similarity in logic between the buffer and the switching circuit of Figure 2–2.

The *inverter*, also referred to as the *NOT* gate, implements logical inversion. The function of the inverter is similar to the function of the switching circuit shown in Figure 2–3. The logic symbols for the inverter are shown in Figure 2–12. Notice that the symbols for the inverter are almost identical to that of a buffer except that a "bubble" (○) has been added to the symbol. It is this bubble that indicates that a logic inversion has been performed and is referred to as the *negation indicator*. The placement of the bubble at the input of the inverter or the output of the inverter has no logic significance; it is only significant as far as the electronic implementation of the gate is concerned since it indicates whether the inversion is done at the input or at the output of the gate. The ANSI/IEEE (American National Standards Institute/Institute of Electrical and Electronics Engineers) version of the symbol may also represent inversion by means of the *polarity indicator* which is shown as a triangle (◺) at either the input or the output of the gate. We will use the negation indicator throughout this textbook to represent inversion.

The output of the inverter (B) is always equal to the complement of the input (A). The truth table and logic equation for the inverter shown in Table 2–6 illustrate its operation.

Since the logic state at the input of an inverter is simply complemented, two inverters connected in series will simply function logically as a buffer. The following example illustrates this fact.

Table 2–5
Truth table and logic equation for the buffer

A	B
0	0
1	1

Logic equation: $B = A$

Figure 2–12
Logic symbols for the inverter. (a) Distinctive shape symbol; (b) ANSI/IEEE Std. 91-1984 symbol.

(a)

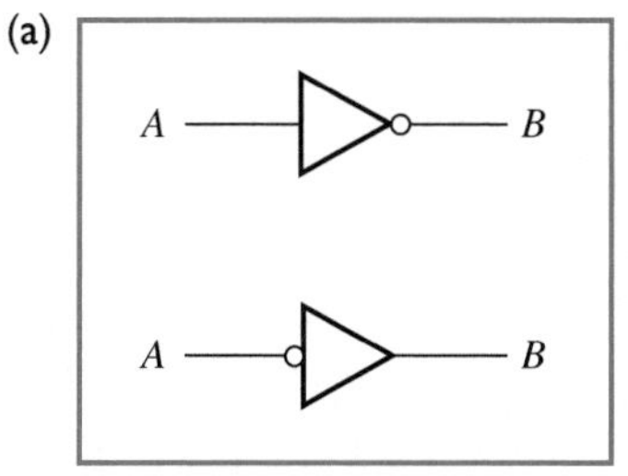

(b)

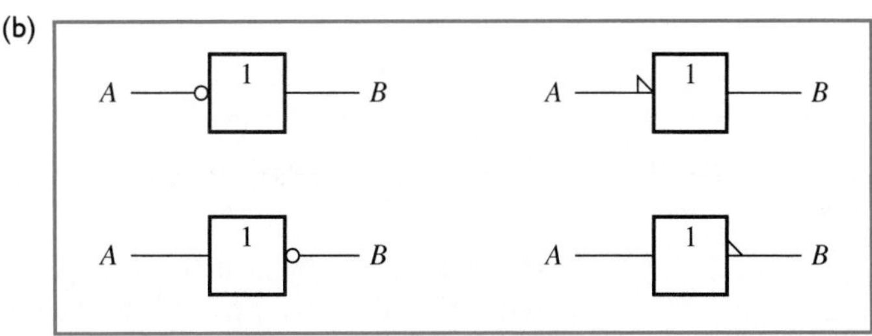

Table 2–6
Truth table and logic equation for the inverter

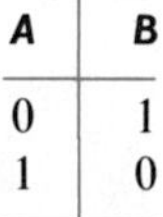

A	B
0	1
1	0

Logic equation: $B = \overline{A}$

EXAMPLE 2–5

Construct a truth table for the dependent variable X for the circuit shown in Figure 2–13, and determine the logic equation for the circuit.

Figure 2–13
Cascaded inverters.

SOLUTION

Y	Output of gate 1	Output of gate 2	X
0	1	1	0
1	0	0	1

Therefore from the truth table, the logic equation for the circuit is

$$X = Y$$

Integrated circuit logic

The buffer and inverter are packaged in several 74xxx ICs, two of which are shown in Figure 2–14. The 7404 is known as a ''hex inverter'' IC since it contains six (hexa) inverters in one package. The inputs to the

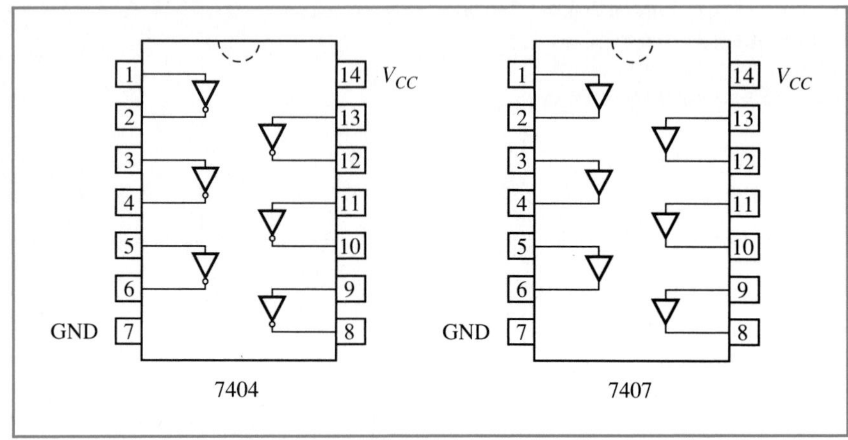

Figure 2–14 Pin configuration of the 7404 inverter and the 7407 buffer.

inverters are accessed through pins 1, 3, 5, 9, 11, and 13, while the outputs are derived from pins 2, 4, 6, 8, 10, and 12, respectively. Pins 7 and 14 are used to supply power to the IC. Similarly, the 7407 is known as a ''hex buffer'' and is used in much the same way as the 7404.

Review questions

1. Logically, what is the difference between a buffer and an inverter?
2. In the symbols that represent an inverter, what actually identifies the process of logical inversion?

2–5 The *AND* gate

The *AND* gate implements the logic *AND* function discussed in Section 2–2. The function of the AND gate is similar to the operation of the *AND* circuit shown in Figure 2–4. This means that the output of the *AND* gate

(a)

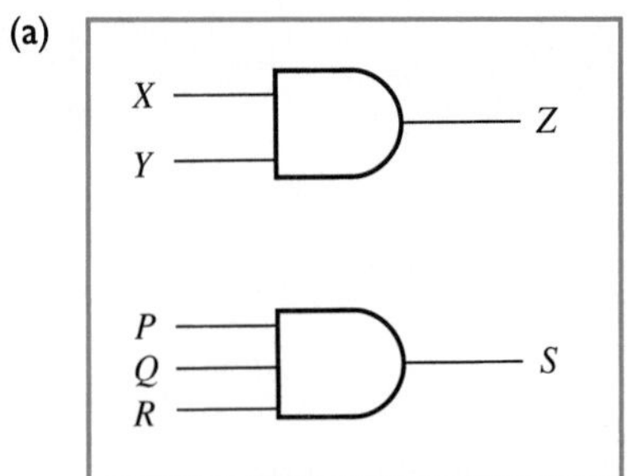

(b)

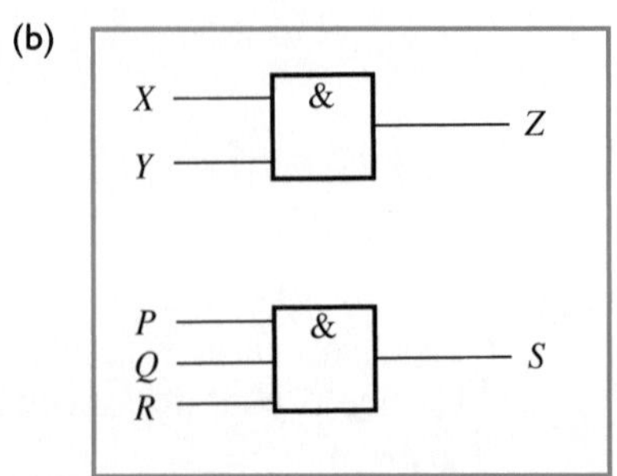

Figure 2–15 Logic symbols for the *AND* gate. (a) Distinctive shape symbol; (b) ANSI/IEEE Std. 91-1984 symbol.

Table 2–7
Truth and table and logic equation for a two-input *AND* gate

X	Y	Z
0	0	0
0	1	0
1	0	0
1	1	1

Logic equation: $Z = X \cdot Y$

Table 2–8
Truth table and logic equation for a three-input *AND* gate

P	Q	R	S
0	0	0	0
0	0	1	0
0	1	0	0
0	1	1	0
1	0	0	0
1	0	1	0
1	1	0	0
1	1	1	1

Logic equations: $S = P \cdot Q \cdot R$

(dependent variable) will be a logic 1 if and only if *all* inputs (independent variables) are at the logic 1 state. The symbols for the AND gate are shown in Figure 2–15. *AND* gates can have several inputs but they must have at least two inputs to properly implement the AND function. The truth tables and logic equations for the two-input and three-input AND gates shown in Figure 2–15 are given in Tables 2–7 and 2–8, respectively.

EXAMPLE 2–6

The five-input *AND* gate shown in Figure 2–16 has one of its inputs permanently tied high (logic 1 state). Obtain the truth table and the logic equation for the gate. What would happen to the gate if one of its inputs were permanently tied low (logic 0 state)?

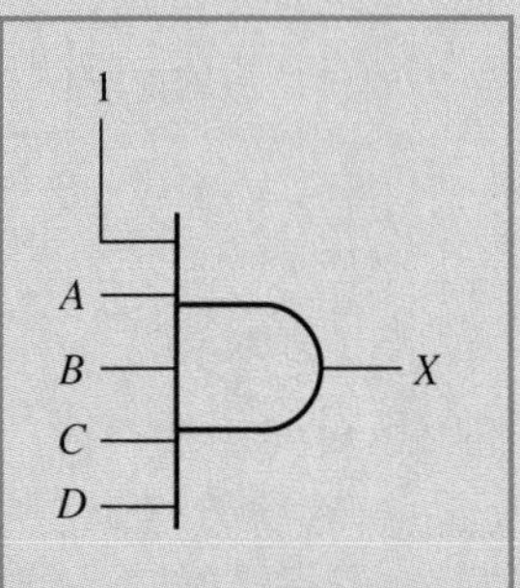

Figure 2–16
A 5-input *AND* gate used as a 4-input *AND* gate.

SOLUTION Since one of the five inputs is permanently tied to the logic 1 state, the five-input *AND* gate now functions like a four-input AND gate. This is because the output of the gate (X) now depends on whether the states of the four remaining inputs (A, B, C, D) are all 1s, or if any one of the inputs is a 0.

Therefore its truth table and logic equation is

A	B	C	D	X
0	0	0	0	0
0	0	0	1	0
0	0	1	0	0
0	0	1	1	0
0	1	0	0	0
0	1	0	1	0
0	1	1	0	0
0	1	1	1	0
1	0	0	0	0
1	0	0	1	0
1	0	1	0	0
1	0	1	1	0
1	1	0	0	0
1	1	0	1	0
1	1	1	0	0
1	1	1	1	1

Logic equation: $X = A \cdot B \cdot C \cdot D$

In any *AND* gate, if any one of the inputs is a logic 0, the output will be a logic 0 and will not be affected by the states of the other inputs.

Integrated circuit logic

The *AND* gate is available in several IC packages as shown in Figure 2–17. The 7408 is known as the "quad 2-input *AND* gate" and contains four (quad) two-input *AND* gates in one IC. The 7411 is known as the

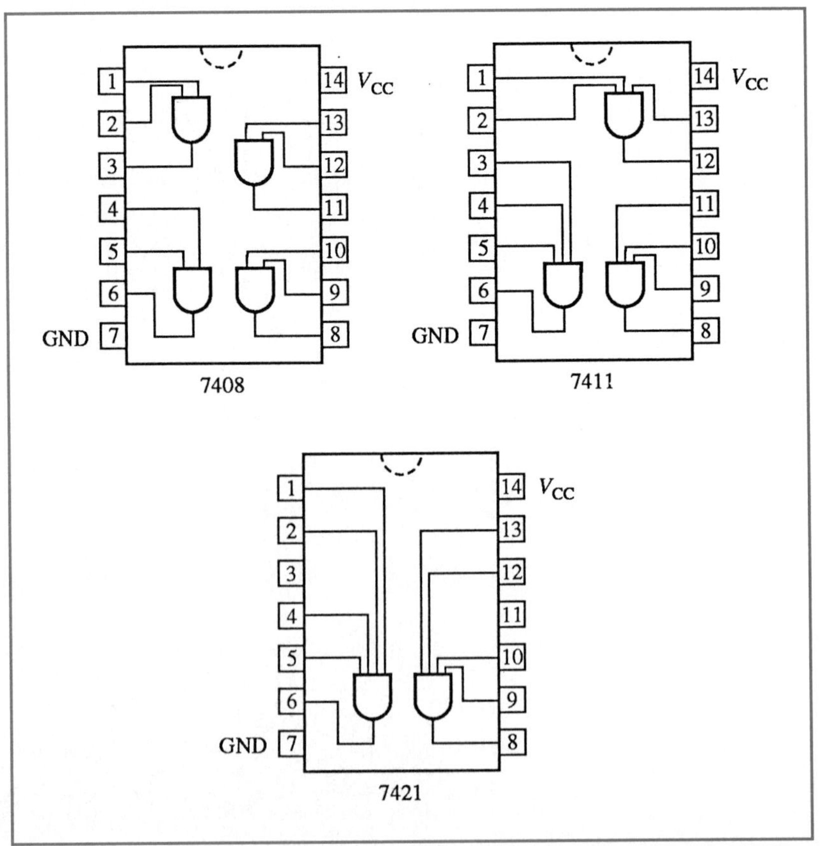

Figure 2–17 Pin configurations of various *AND* gate ICs.

"triple 3-input *AND* gate" and contains three (triple) 3-input *AND* gates in one IC. The 7421 is known as the "dual 4-input *AND* gate" and contains two (dual) 4-input *AND* gates in one IC. As before, notice that power is supplied to pins 7 (GND) and 14 (V_{CC}) on each IC.

Review questions

1. What is the difference between an *AND* gate and an inverter?
2. Is it possible to make a four-input *AND* gate function logically like a buffer? If yes, how can it be done?
3. How can we connect two 2-input *AND* gates to function as a three-input *AND* gate?

2–6 The *OR* gate

The *OR* gate implements the logic OR function discussed in Section 2–2. The function of the OR gate is similar to the operation of the *OR* circuit shown in Figure 2–6. This means that the output of the OR gate (dependent variable) will be a logic 1 if *any* of its inputs (independent variables) are at the logic 1 state. The symbols for the *OR* gate are shown in Figure 2–18. *OR* gates can have several inputs but they must have at least two inputs to properly implement the *OR* function. The truth tables and logic equations for the two- and three-input *OR* gates shown in Figure 2–18 are given in Tables 2–9 and 2–10, respectively.

(a)

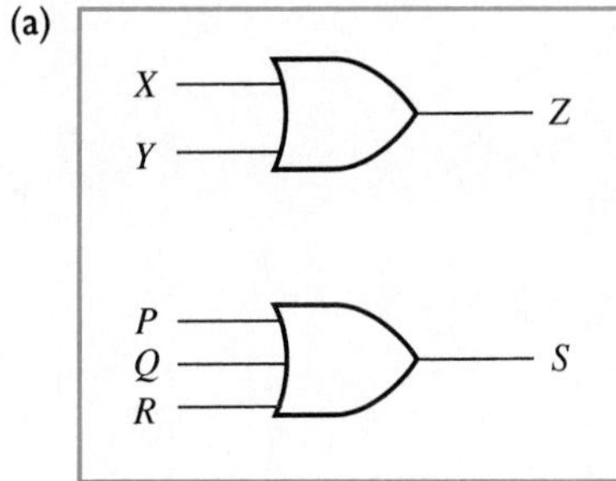

(b)

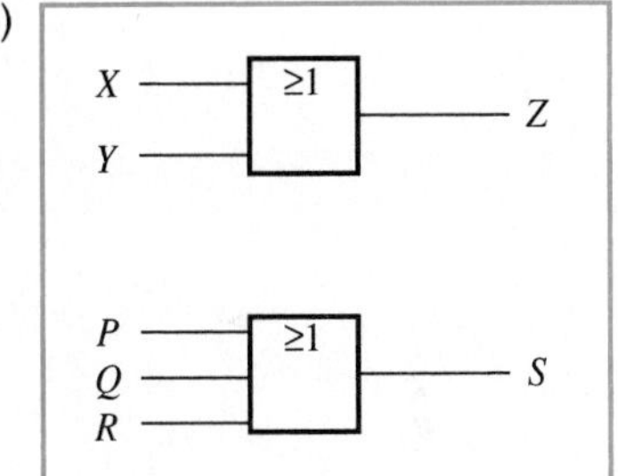

Figure 2–18
Logic symbols for the *OR* gate. (a) Distinctive shape symbol; (b) ANSI/IEEE Std. 91-1984 symbol.

Table 2–9
Truth table and logic equation for a two-input *OR* gate

X	Y	Z
0	0	0
0	1	1
1	0	1
1	1	1

Logic equation: $Z = X + Y$

Table 2–10
Truth table and logic equation for a three-input *OR* gate

P	Q	R	S
0	0	0	0
0	0	1	1
0	1	0	1
0	1	1	1
1	0	0	1
1	0	1	1
1	1	0	1
1	1	1	1

Logic equation: $S = P + Q + R$

EXAMPLE 2–7

The five-input *OR* gate shown in Figure 2–19 has one of its inputs permanently tied low (logic 0 state). Obtain the truth table and the logic equation for the gate. What would happen to the gate if one of its inputs were permanently tied high (logic 1 state)?

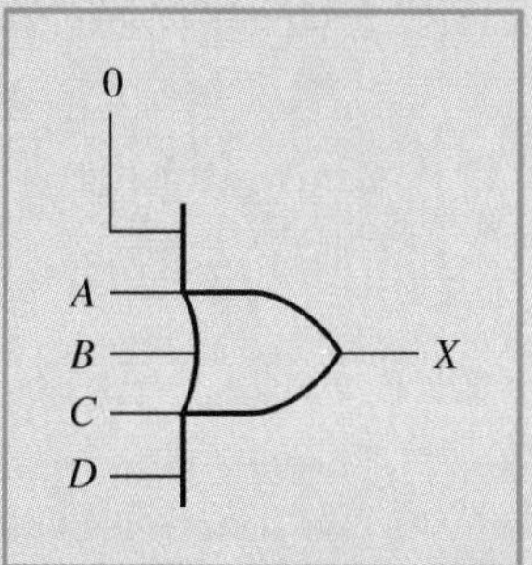

Figure 2–19
A five-input *OR* gate used as a four-input *OR* gate.

SOLUTION Since one of the five inputs is permanently tied to the logic 0 state, the five-input *OR* gate now functions like a four-input *OR* gate. This is because the output of the gate (X) now depends on whether the states of the four remaining inputs (A, B, C, D) are all 0's, or if any one of the inputs is a 1.
Therefore its truth table and logic equation is

A	B	C	D	X
0	0	0	0	0
0	0	0	1	1
0	0	1	0	1
0	0	1	1	1
0	1	0	0	1
0	1	0	1	1
0	1	1	0	1
0	1	1	1	1
1	0	0	0	1
1	0	0	1	1
1	0	1	0	1
1	0	1	1	1
1	1	0	0	1
1	1	0	1	1
1	1	1	0	1
1	1	1	1	1

Logic equation: $X = A + B + C + D$

In any *OR* gate, if any one of the inputs is a logic 1, the output will be a logic 1 and will not be affected by the states of the other inputs.

Integrated circuit logic

The *OR* gate is only available in one IC package as shown in Figure 2–20. The 7432 is known as the "quad 2-input *OR* gate" and contains four (quad) 2-input *OR* gates in one IC. Having only one IC that provides only two-input *OR* gates in not a limitation in any way. As will be seen in the next section and in Chapter 3, there are several procedures that we can use to emulate the function of *OR* gates having more than two inputs, or to construct *OR* gates with several inputs.

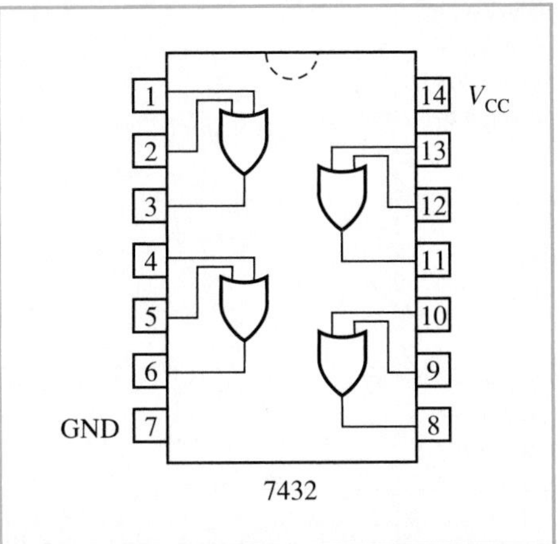

Figure 2–20
Pin configuration of a 7432 OR-gate IC.

Review questions

1. Compare and contrast the *AND* and *OR* gates.
2. Is it possible to make a four-input *OR* gate function logically like a buffer? If yes, how can it be done?
3. How could we connect two 2-input *OR* gates to function like a three-input *OR* gate?

2–7 The *NAND* and *NOR* gates

There are two other basic logic gates that are frequently used in digital circuits—the *NAND* and *NOR* gates. Both these gates are actually combinations of *AND* gates, *OR* gates, and inverters but are treated as discrete logic gates.

The *NAND* gate is a combination of an *AND* gate and a *NOT* gate (inverter). Hence the letter *N* in the word "NAND" indicates that the *AND* function is inverted. This inversion is done at the output as shown in Figure 2–21.

Figure 2–21a and b show the symbols that represent two- and three-input NAND gates. As before, these symbols can be modified to

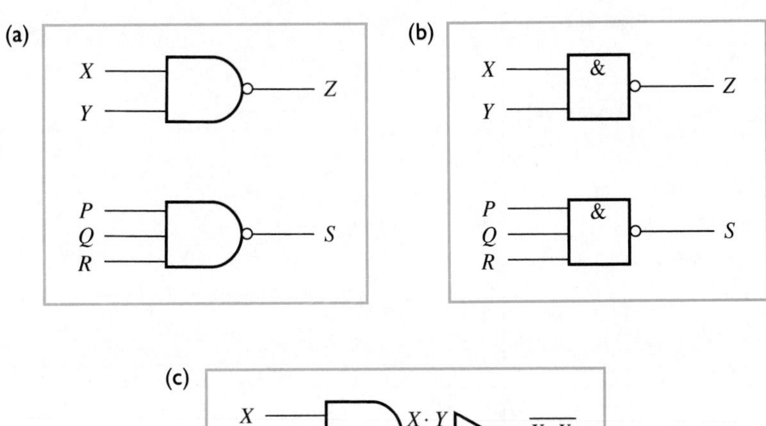

Figure 2–21
The *NAND* gate. (a) Distinctive shape symbol; (b) ANSI/IEEE Std. 91-1984 symbol; (c) equivalent circuit.

Table 2–11
Truth table and logic equation for a two-input *NAND* gate

X	Y	Z
0	0	1
0	1	1
1	0	1
1	1	0

Logic equation: $Z = \overline{X \cdot Y}$

Table 2–12
Truth table and logic equation for a three-input *NAND* gate

P	Q	R	S
0	0	0	1
0	0	1	1
0	1	0	1
0	1	1	1
1	0	0	1
1	0	1	1
1	1	0	1
1	1	1	0

Logic equation: $S = \overline{P \cdot Q \cdot R}$

represent more inputs. Notice that the symbols for the *NAND* gate is very similar to the symbol of an *AND* gate except that the output line has a "bubble." Recall from Section 2–3 that the "bubble" symbol indicates inversion, and therefore this means that the output of the gate is internally inverted. The circuit in Figure 2–21c represents a two-input *NAND* gate as an *AND* gate with an inverter connected to its output. The output of a *NAND* gate therefore is exactly the opposite of the output of an *AND* gate. This means that the output of a *NAND* gate (dependent variable) will be a logic 0 if and only if *all* inputs (independent variables) are at the logic 1 state. The truth tables and logic equations for the two- and three-input NAND gates shown in Figure 2–21 are given in Tables 2–11 and 2–12, respectively.

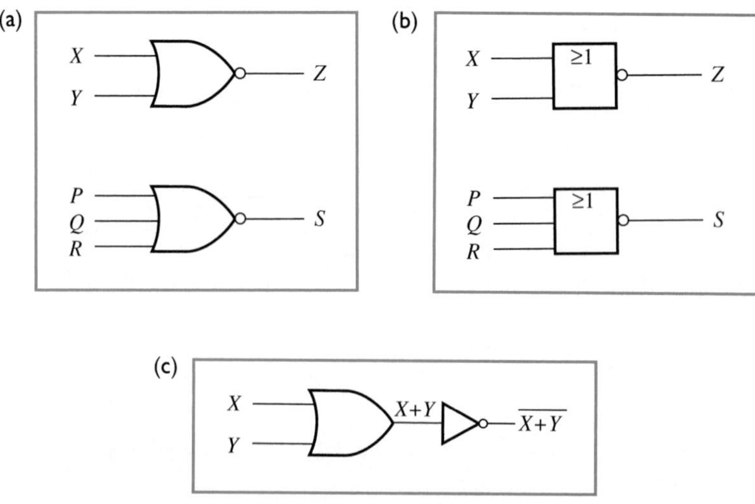

Figure 2–22
The *NOR* gate. (a) Distinctive shape symbol; (b) ANSI/IEEE Std. 91-1984 symbol; (c) equivalent circuit.

Table 2–13
Truth table and logic equation for a two-input *NOR* gate

X	Y	Z
0	0	1
0	1	0
1	0	0
1	1	0

Logic equation: $Z = \overline{X + Y}$

Table 2–14
Truth table and logic equation for a three-input *NOR* gate

P	Q	R	S
0	0	0	1
0	0	1	0
0	1	0	0
0	1	1	0
1	0	0	0
1	0	1	0
1	1	0	0
1	1	1	0

Logic equation: $S = \overline{P + Q + R}$

Notice that placement of the bar over the entire term on the right-hand side of each logic equation indicates that the *AND* function is inverted to produce the *NAND* function.

The *NOR* gate is a combination of an *OR* gate and a *NOT* gate (inverter). Hence the letter *N* in the word "*NOR*" indicates that the *OR* function is inverted. Like the *NAND* gate, this inversion is done at the output as shown in Figure 2–22.

Figure 2–22a and b show the symbols that represent two- and three-input *NOR* gates. As before, these symbols can be modified to represent more inputs. Notice that the symbols for the *NOR* gate is very similar to the symbol for an *OR* gate except that the output line has a "bubble." Again, since the bubble symbol indicates inversion, this means that the output of the gate is internally inverted. The circuit in Figure 2–21c represents a two-input *NOR* gate as an *OR* gate with an inverter connected to its output. The output of a *NOR* gate therefore is exactly the opposite of the output of an OR gate. This means that the output of the *NOR* gate (dependent variable) will be a logic 0 if *any* of its inputs (independent variables) are at the logic 1 state. The truth tables and logic equations for the two- and three-input *NOR* gates shown in Figure 2–22 are given in Tables 2–13 and 2–14, respectively.

As in the *NAND* gate, the placement of the bar over the entire term on the right-hand side of each logic equation indicates that the *OR* function is inverted to produce the *NOR* function.

EXAMPLE 2–8

The five-input *NAND* gate shown in Figure 2–23 has one of its inputs permanently tied high (logic 1 state). Obtain the truth table and the logic equation for the gate. What would happen to the gate if one of its inputs were permanently tied low (logic 0 state)?

SOLUTION Since one of the five inputs is permanently tied to the logic 1 state, the five-input *NAND* gate now functions like a four-

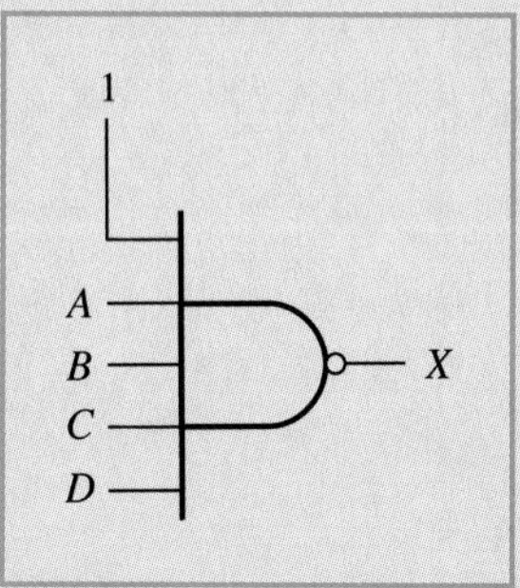

Figure 2–23
A five-input *NAND* gate used as a four-input *NAND* gate.

input *NAND* gate. This is because the output of the gate (X) now depends on whether the states of the four remaining inputs (A, B, C, D) are all 1's, or if any one of the inputs is a 0.
Therefore its truth table and logic equation is

A	B	C	D	X
0	0	0	0	1
0	0	0	1	1
0	0	1	0	1
0	0	1	1	1
0	1	0	0	1
0	1	0	1	1
0	1	1	0	1
0	1	1	1	1
1	0	0	0	1
1	0	0	1	1
1	0	1	0	1
1	0	1	1	1
1	1	0	0	1
1	1	0	1	1
1	1	1	0	1
1	1	1	1	0

Logic equation: $X = \overline{A \cdot B \cdot C \cdot D}$

In any *NAND* gate, if any one of the inputs is a logic 0, the output will be a logic 1 and will not be affected by the states of the other inputs.

EXAMPLE 2–9

The five-input *NOR* gate shown in Figure 2–24 has one of its inputs permanently tied low (logic 0 state). Obtain the truth table and the

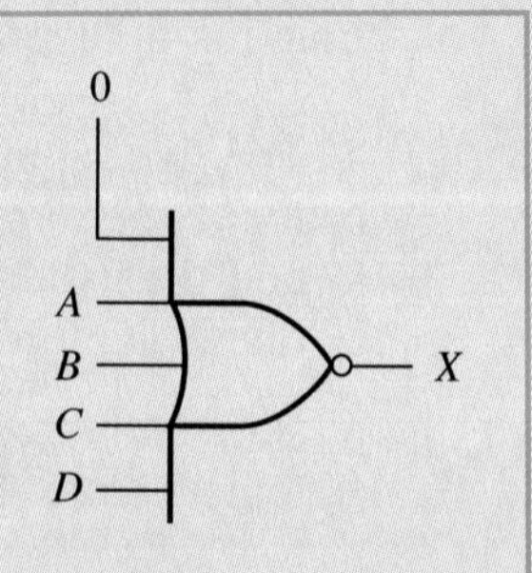

Figure 2–24
A five-input *NOR* gate used as a four-input *NOR* gate.

logic equation for the gate. What would happen to the gate if one of its inputs were permanently tied high (logic 1 state)?

SOLUTION Since one of the five inputs is permanently tied to the logic 0 state, the five-input *NOR* gate now functions like a four-input *NOR* gate. This is because the output of the gate (X) now depends on whether the states of the four remaining inputs (A, B, C, D) are all 0's, or if any one of the inputs is a 1.
Therefore its truth table and logic equation is

A	B	C	D	X
0	0	0	0	1
0	0	0	1	0
0	0	1	0	0
0	0	1	1	0
0	1	0	0	0
0	1	0	1	0
0	1	1	0	0
0	1	1	1	0
1	0	0	0	0
1	0	0	1	0
1	0	1	0	0
1	0	1	1	0
1	1	0	0	0
1	1	0	1	0
1	1	1	0	0
1	1	1	1	0

Logic equation: $X = \overline{A + B + C + D}$

In any *NOR* gate, if any one of the inputs is a logic 1, the output will be a logic 0 and will not be affected by the states of the other inputs.

The *NAND* and *NOR* gates are very useful in logic circuits since they can greatly reduce the complexity of a logic circuit. The techniques used to do this will be examined in the next chapter. *NAND* and *NOR* gates can also be used as *AND* and *OR* gates, respectively, by inverting the outputs. Figure 2–25 shows the circuits required to do this.

Integrated circuit logic

The 74xxx series includes a variety of ICs that package the *NAND* and *NOR* gates. Figure 2–26 shows the pinouts of several *NAND* ICs—the

Figure 2–25
(a) *AND* gate; (b) *OR* gate.

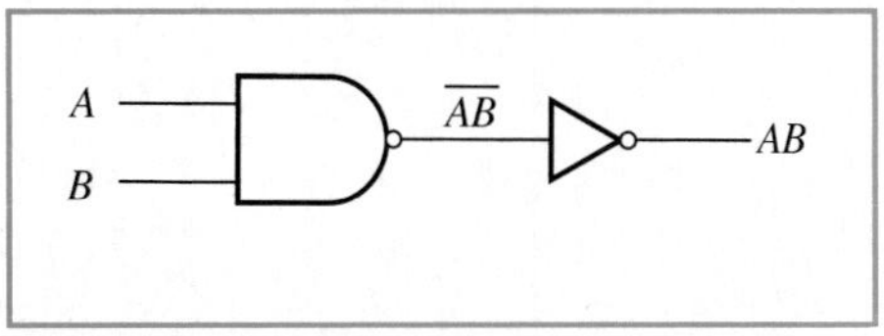

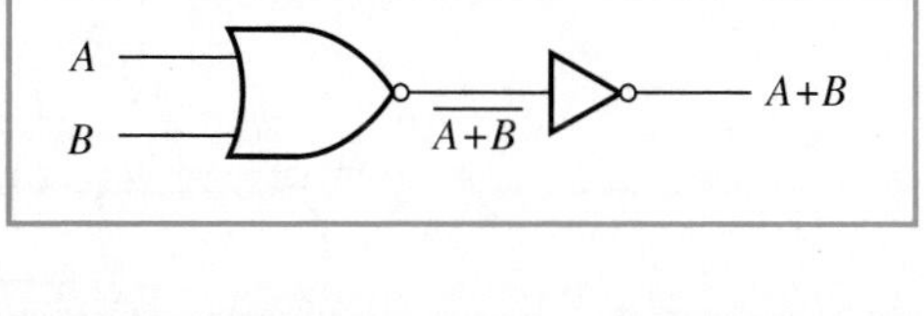

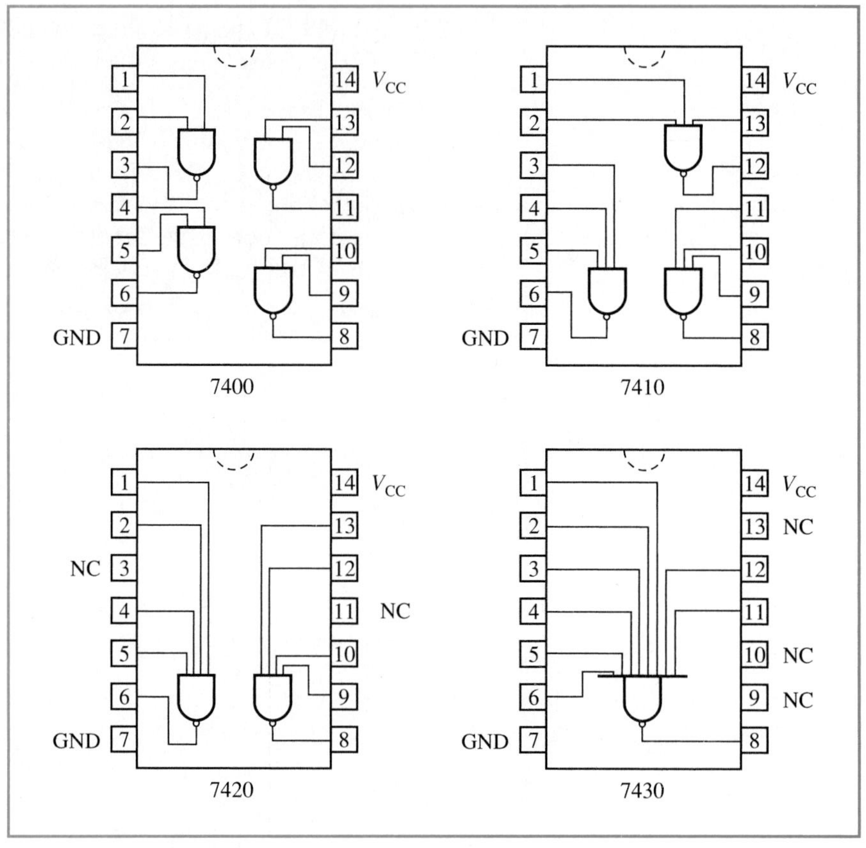

Figure 2–26 Pin configurations of various *NAND* gate ICs.

7400 quad 2-input *NAND* gate, the 7410 triple 3-input *NAND* gate, the 7420 dual 4-input *NAND* gate, and the 7430 eight-input *NAND* gate.

There are two ICs that package the *NOR* gates. These are shown in Figure 2–27—The 7402 quad 2-input NOR gate, and the 7427 triple 3-input NOR gate.

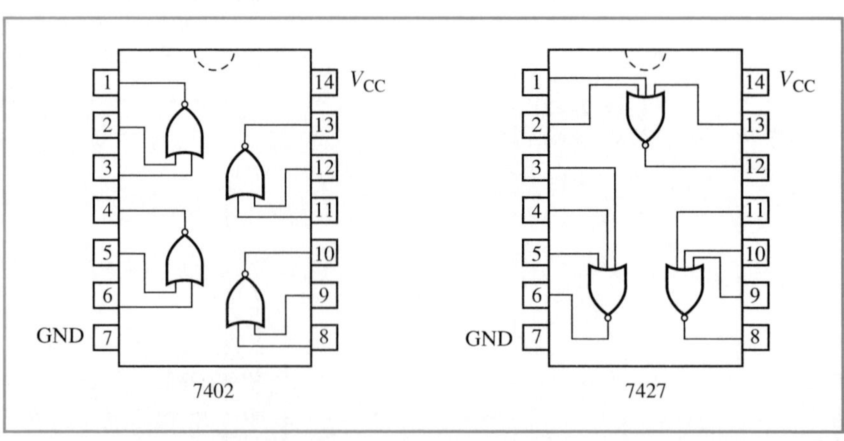

Figure 2–27 Pin configurations of various *NOR* gate ICs.

Review questions

1. Is it possible to make a four-input *NAND* gate or *NOR* gate function as an inverter? If yes, how can it be done?
2. What is the significance of the bubble in the logic symbols that are used to represent the NAND and NOR gates?
3. Is it possible to connect two 2-input *NOR* gates to function like a three-input *NOR* gate?
4. Is it possible to connect two 2-input *NAND* gates to function like a three-input *NAND* gate?

2–8 Logic circuits

Like logic gates, logic circuits can be viewed as "black boxes" with binary inputs (independent variables) and binary outputs (dependent variables). Logic circuits, however, can have several outputs and are made up of the various logic gates discussed in this chapter and other gates that will be introduced in following chapters. The basic model of a logic circuit is shown in Figure 2–28.

Just as a truth table was used to describe the operation of a logic gate, so can it be used to describe the operation of a logic circuit. Similarly a logic equation can also be used to relate the output of a logic circuit to its inputs. To illustrate this fact, consider the following examples.

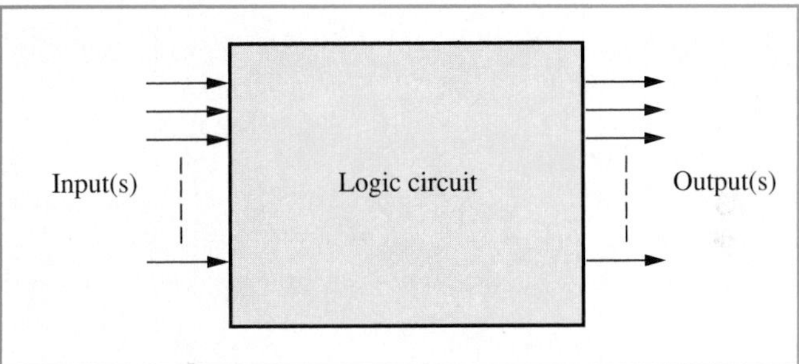

Figure 2–28
The model of a logic circuit.

EXAMPLE 2–10

Obtain the logic equations and truth tables for the circuits shown in Figure 2–29a and b.

SOLUTION (a) In Figure 2–29a, we have obtained the logic equation for the circuit by tracing the independent variables A, B, and C through each gate in the circuit. Since the input to the *OR* gate is B and C, the output of the *OR* gate is $B + C$. One of the inputs of the *AND* gate is A, while the other input is connected to the output of the *OR* gate, $B + C$. Therefore the output of the *AND* gate and the logic equation for the entire circuit is

$$Z = \mathrm{A} \cdot (\mathrm{B} + \mathrm{C})$$

Notice that we have grouped the term $B + C$ in parentheses to indicate that it is evaluated before the *AND* function. Enclosing the term $B + C$ in parentheses allows us to treat the term as if it were a

Figure 2–29

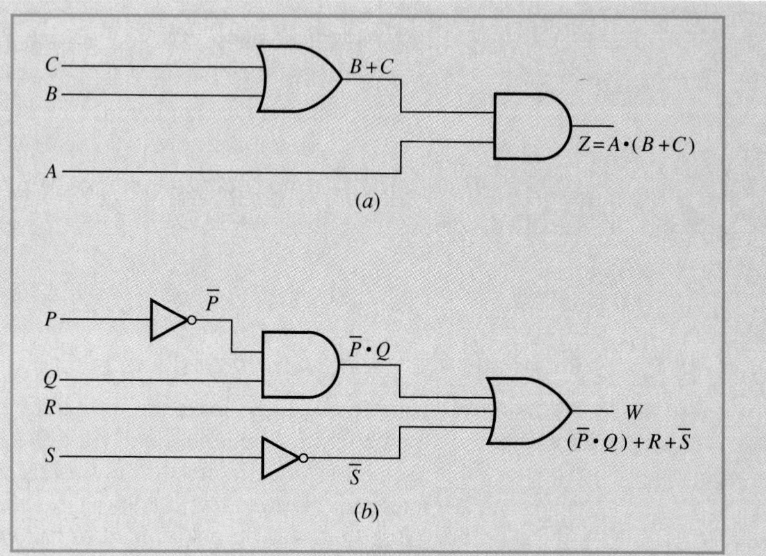

single variable. The truth table for the logic equation can be obtained by analyzing each gate in the circuit separately as follows:

A	B	C	B + C	A · (B + C)
0	0	0	0	0
0	0	1	1	0
0	1	0	1	0
0	1	1	1	0
1	0	0	0	0
1	0	1	1	1
1	1	0	1	1
1	1	1	1	1

In order to obtain the output of the *OR* gate in terms of the variables *B* and *C*, we simply "OR" (apply the *OR* function) the columns for *B* and *C* using the basic truth table for an *OR* gate. To obtain the output of the *AND* gate (the final output of the circuit) we now "AND" (apply the *AND* function) the columns for *A* and $B + C$ using the basic truth table for an *AND* gate.

(b) In Figure 2–29b we have applied the same procedure as we did for Figure 2–29a to determine the logic equation for the circuit

$$W = (\overline{P} \cdot Q) + R + \overline{S}$$

Again, $(\overline{P} \cdot Q)$ are grouped together since this term is evaluated before the "OR" function and is treated like a single variable at the input of the *OR* gate. As before, the truth table can be determined by tracing all possible combinations of the independent variables *P*, *Q*, *R*, and *S* through each gate in the circuit.

P	Q	R	S	$\overline{P}$	$\overline{S}$	$\overline{P} \cdot Q$	W $(\overline{P} \cdot Q) + R + \overline{S}$
0	0	0	0	1	1	0	1
0	0	0	1	1	0	0	0
0	0	1	0	1	1	0	1
0	0	1	1	1	0	0	1
0	1	0	0	1	1	1	1
0	1	0	1	1	0	1	1
0	1	1	0	1	1	1	1
0	1	1	1	1	0	1	1
1	0	0	0	0	1	0	1
1	0	0	1	0	0	0	0
1	0	1	0	0	1	0	1
1	0	1	1	0	0	0	1
1	1	0	0	0	1	0	1
1	1	0	1	0	0	0	0
1	1	1	0	0	1	0	1
1	1	1	1	0	0	0	1

To obtain the $\overline{P}$ and $\overline{S}$ columns we simply invert the P and S columns. The $(\overline{P} \cdot Q)$ column is obtained by applying the *AND* function to the $\overline{P}$ and Q columns. The last column represents the final output W and is obtained by applying the *OR* function to the R, $\overline{S}$, and $(\overline{P} \cdot Q)$ columns.

If a logic circuit has more than one output, then the circuit is described by a series of logic equations, each one representing a single output. Each logic equation can be treated independently in order to obtain the truth table for the circuit. The following example illustrates this procedure.

EXAMPLE 2–11

Obtain the logic equation and truth table for the circuit shown in Figure 2–30.

Figure 2–30

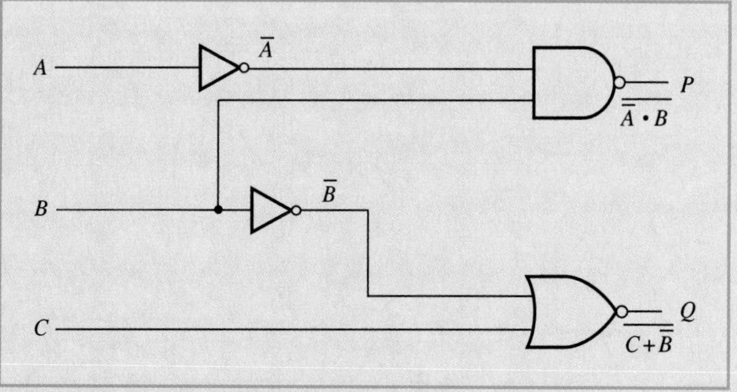

SOLUTION The outputs of the logic circuit are described by the following logic equations:

Equation (2–4)
$$P = \overline{\overline{A} \cdot B}$$

Equation (2–5)
$$Q = \overline{C + \overline{B}}$$

The truth table of the circuit is obtained by independently determining the values of P and Q for all possible combinatins of A, B, and C. That is, we can think of the circuit in Figure 2–30 as being made up of two circuits, one that is described by Equation 2–4, and the other by Equation 2–5.

					P	Q
A	B	C	$\overline{A}$	$\overline{B}$	$\overline{\overline{A} \cdot B}$	$\overline{C + \overline{B}}$
0	0	0	1	1	1	0
0	0	1	1	1	1	0
0	1	0	1	0	0	1
0	1	1	1	0	0	0
1	0	0	0	1	1	0
1	0	1	0	1	1	0
1	1	0	0	0	1	1
1	1	1	0	0	1	0

It is also possible to work backward and draw the circuit for a given logic equation. The order in which each term in the equation is drawn is the same as the order in which each term is evaluated, with terms in parentheses treated as single variables. Example 2–12 illustrates these procedures for a few simple logic equations. More examples will follow in Chapter 3 after the various mathematical techniques of manipulating these equations have been introduced.

EXAMPLE 2–12

Draw the logic circuits for the following logic equations:

(a) $T = (A \cdot B) + C$

(b) $X = \overline{P} \cdot Q \cdot (R + \overline{Q})$

(c) $F = (\overline{X + Y + Z}) + (W \cdot \overline{X})$

SOLUTIONS

(a)

Figure 2–31

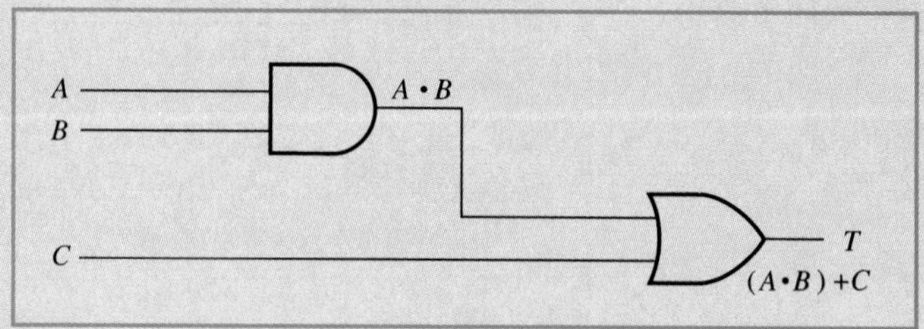

(b)

Figure 2–32

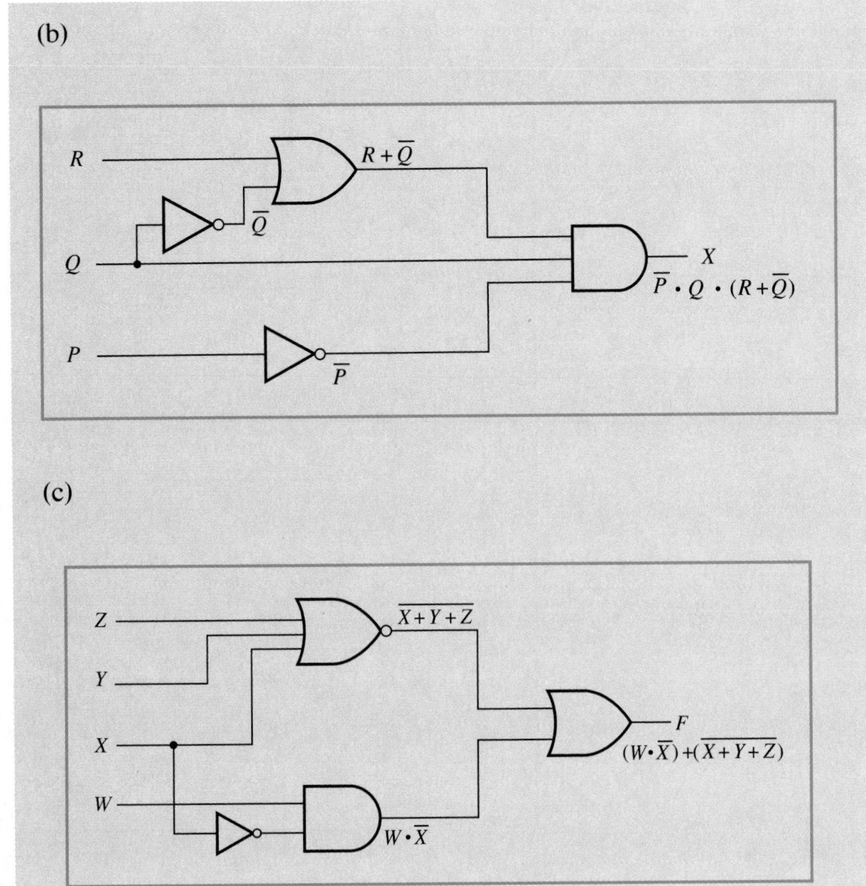

Figure 2–33

Constructing logic circuits with ICs

In order to construct a logic circuit in the laboratory we must make use of the various 74xxx ICs that implement the basic logic gates. For example, consider the construction of the logic circuit shown in Figure 2–34.

The first step would be to draw a *schematic diagram* that identifies the chip numbers (74xxx IC numbers) and pin numbers for each gate in the circuit. The resulting diagram is shown in Figure 2–35.

Notice in Figure 2–35 that three ICs are needed—a 7404, a 7408, and a 7432—to implement the logic circuit. We have identified the chip number for each gate and also the pin numbers for the inputs and output of each gate. Since we have not used all the gates available in each IC

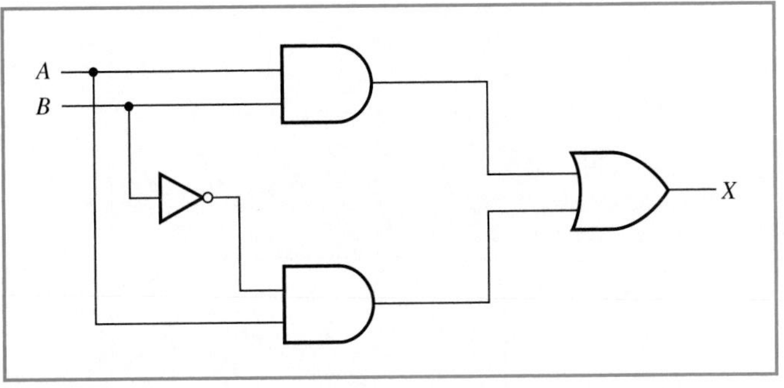

Figure 2–34
Logic circuit to be constructed.

Figure 2–35
Schematic diagram.

7408
A
B
7432
X
7404
7408

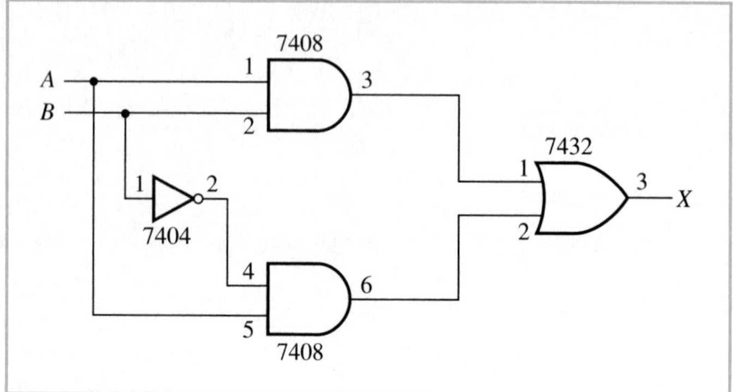

Figure 2–36
Layout and wiring diagram.

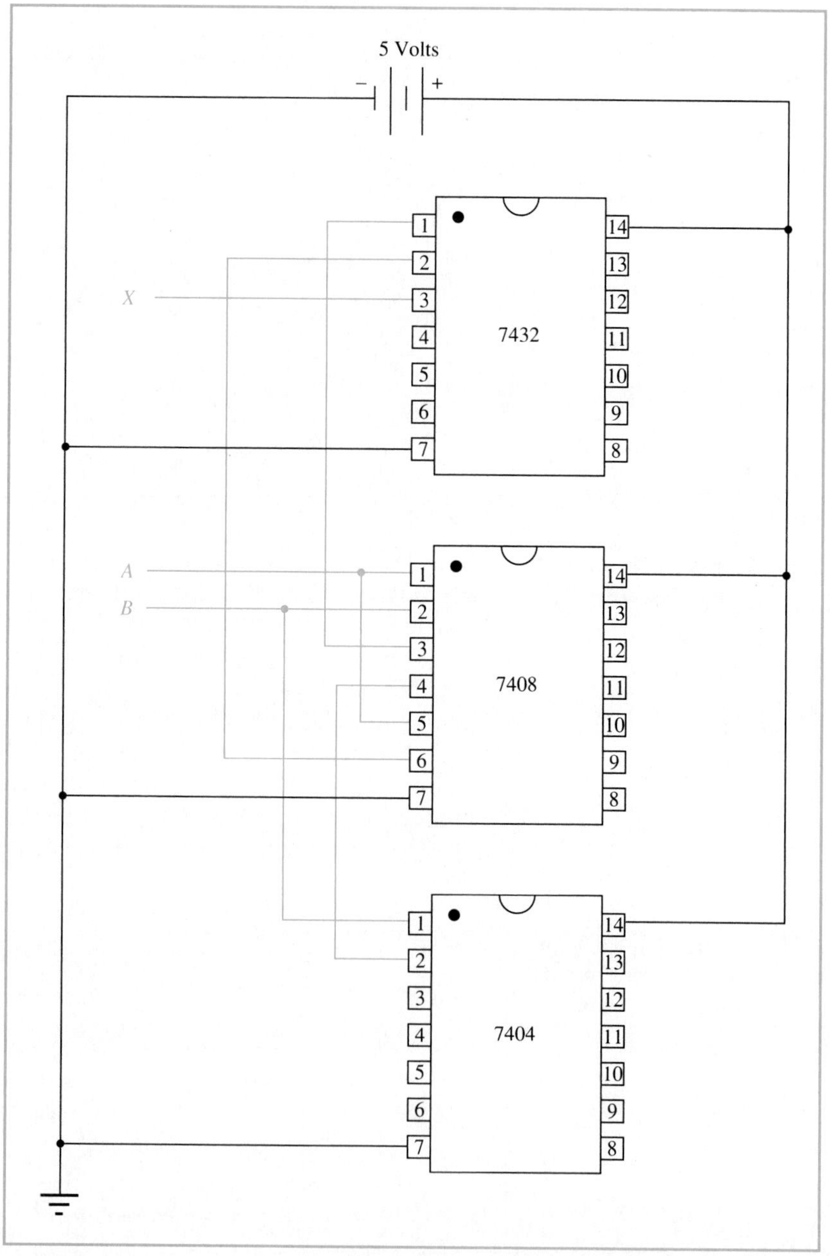

5 Volts
X
7432
A
B
7408
7404

there will be five "spare" inverters in the 7404, two "spare" *AND* gates in the 7408, and three "spare" *OR* gates in the 7432.

The next step would be to draw a layout and wiring diagram that show how the ICs are laid out on a breadboard and wired together. This diagram is shown in Figure 2–36 and represents the actual logic circuit that would be set up in the laboratory. Generally before the circuit is set up it is good practice to wire-up the V_{CC} and GND pins of all the ICs to the 5-volt power supply as shown in Figure 2–36.

It is important to remember that any unused inputs to a gate that is being used in a logic circuit must be *conditioned* to a logic high or logic low as appropriate. For example, if a three-input *AND* gate is being used to function as a two-input *AND* gate, then the unused third input must be tied to a logic high (V_{CC}) to disable it (see Figure 2–16). Similarly if a three-input *NOR* gate is being used to function as a two-input *NOR* gate, then the unused third input must be tied to a logic low (GND) to disable it (see Figure 2–24). We should never assume that since an unconnected input has *no voltage* on it, it is equivalent to having a logic low; a logic low is a voltage—ground potential! In reality, the inputs to a 74xxx TTL logic gate are internally *pulled-up* to a logic high and therefore if left unconnected will assume the logic high state; however this is not recommended since unconditioned (or floating) inputs have a tendency to pick up noise that can affect the operation of the logic circuit.

The outputs of two or more logic gates should never be connected together. Especially if one output is at a logic high state and another is at a logic low state. Logically, the level at this connection (node) is undefined since there are two logic levels applied to it, and electrically, this sort of connection could cause damage to the outputs of the gates since we are effectively shorting a 5-volt potential to ground through the output of the gate. There are two exceptions to this rule which will be examined in Chapters 4 and 12.

EXAMPLE 2–13

Draw a schematic diagram for the logic circuit shown in Figure 2–30 showing the chip and pin numbers for each gate. Also draw a layout and wiring diagram that represents the laboratory set-up of the logic circuit.

Schematic diagram

Figure 2–37

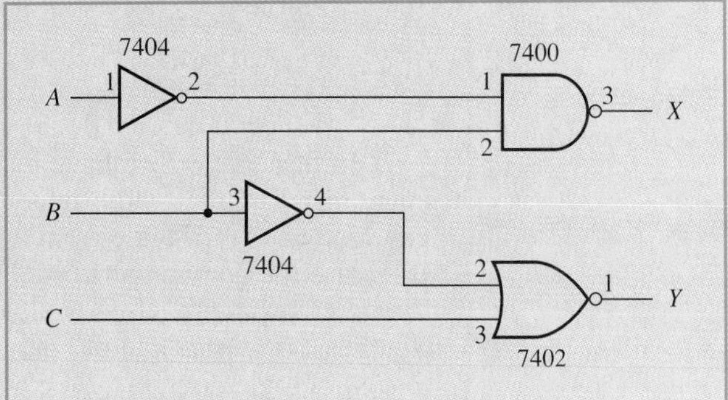

Figure 2–38

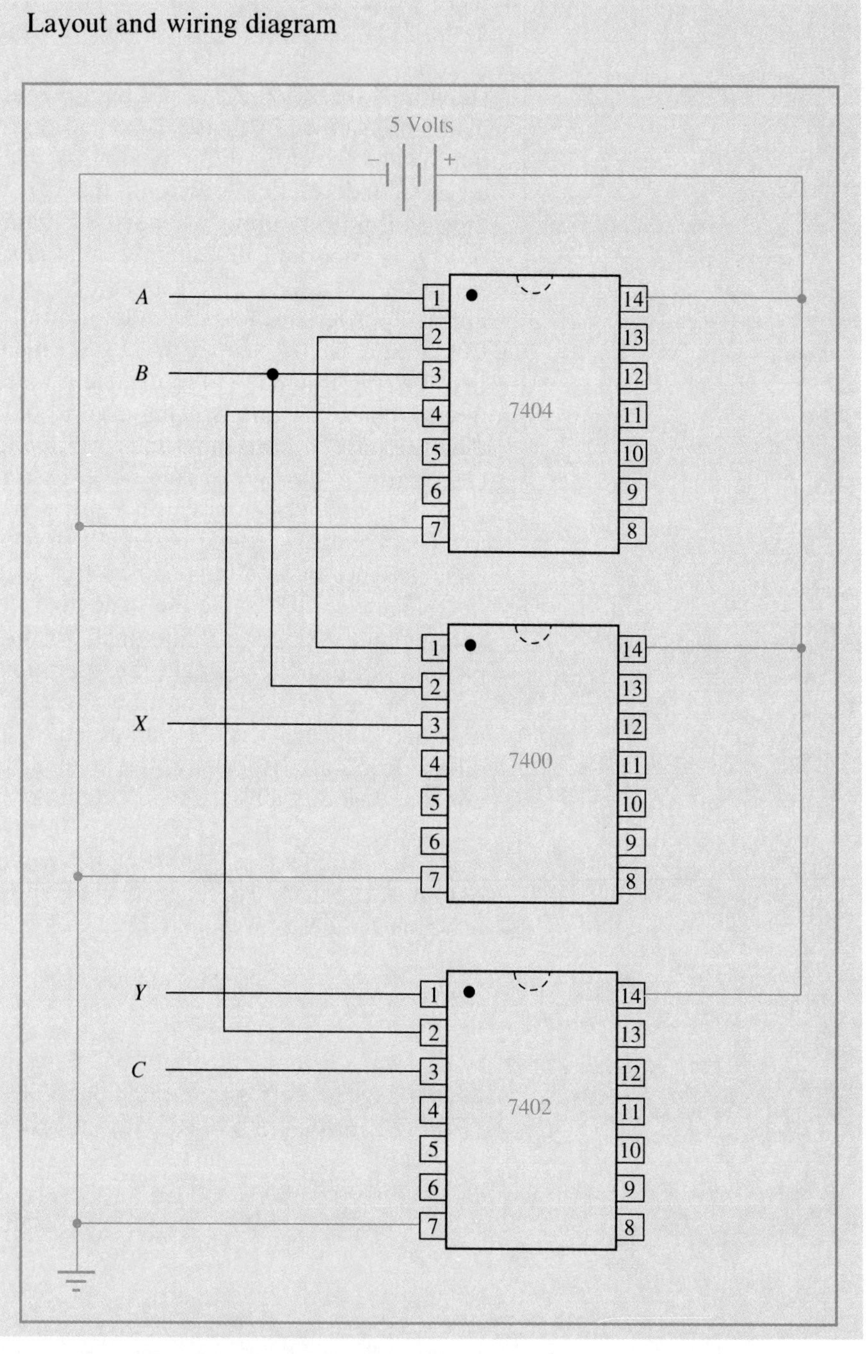

In this section we have examined simple logic circuits, their equations, and their truth tables. The process of obtaining a logic equation and truth table from a given circuit is known as *analysis* since these are the components of a logic circuit that describe its operation. The study of digital circuits also involves the *synthesis* of logic circuits. Synthesis (the opposite of analysis) involves the putting together or *design* of logic circuits, given a specification, truth table, or logic equation. This section has included simple examples of both analysis and synthesis of logic circuits. However, in order to study the different procedures of analysis and synthesis of logic circuits, and apply them to the design of more complex circuits, we must first study the mathematical concepts of logic that will allow us to understand the basic laws of logic functions and their interrelationships. This, and other techniques of designing and analyzing digital circuits will be discussed in Chapter 3.

Review questions

1. What are the similarities and differences between a basic logic gate and a logic circuit?
2. What does the analysis of logic circuits involve?
3. What does the synthesis of logic circuits involve?
4. What is the purpose of drawing a schematic diagram for a logic circuit?
5. How does a layout and wiring diagram differ from the schematic diagram of a logic circuit?
6. Why is it important to condition all unconnected inputs of a logic gate to the high or low state?

2–9 Electrical characteristics of logic gates

In the previous sections we saw that the inputs and outputs of logic gates are the binary numbers 0 and 1 that are represented by the voltages 0 and +5 volts, respectively. These voltages correspond to the power supply potentials for the IC, which for the 74xxx series are referred to as GND and V_{CC}, respectively. In electrical terminology, a logic 1 is referred to as a high voltage level (or simply a "high") and a logic 0 is referred to as a low voltage level (or simply a "low"). Because of the effects of internal resistances within the gates and loading effects, an output may not necessarily produce exactly +5 volts for a logic high and 0 volts for a logic low. Since the output of a gate may have to drive several inputs, the input of a logic gate must be able to accept a *range* of voltages that are acceptable for interpretation of a logic high and a logic low. These are some of gate specifications that are included in the *electrical characteristics* of the IC manufacturer's *data sheets*. Some of the important characteristics included in the manufacturer's data sheets are shown in Figure 2–39.

The electrical characteristics shown in Figure 2–39 are for the *standard 74xxx TTL series*. There are other *families* of ICs that have different electrical characteristics that will be examined in Chapter 12. This section will examine some of the parameters given in manufacturers' data sheets so that we can interpret the various data sheets for various digital ICs and also be aware of the characteristics, requirements, and limitations of these circuits. As stated in the data sheet shown in Figure 2–39, all parameters are specified for an operating temperature range of 0 to 70°C. This is the recommended temperature range for reliable operation of the IC.

Power requirements

Notice that the data sheet shown in Figure 2–39 includes the power supply voltage range for V_{CC}. This indicates that the voltage applied to power the IC can be in the range 4.75 to 5.25 volts but is typically 5 volts.

Recommended operating conditions

Parameter	Limits			Units
	Minimum	*Typical*	*Maximum*	
Supply voltage (V_{CC})	4.75	5.0	5.25	V
Operating free-air temperature range	0	25	70	°C
HIGH-level output current (I_{OH})			−400	μA
LOW-level output current (I_{OL})			16	mA

Electrical characteristics over operating temperature range (unless otherwise noted)

Parameter	Limits			Units	Test conditions[a]
	Minimum	*Typical*[b]	*Maximum*		
HIGH-level input voltage (V_{IH})	2.0			V	
LOW-level input voltage (V_{IL})			0.8	V	
HIGH-level output voltage (V_{OH})	2.4	3.4		V	V_{CC} = min, I_{OH} = 0.4 mA V_{IN} = 0.8 V
LOW-level output voltage (V_{OL})		0.2	0.4	V	V_{CC} = min, I_{OL} = 16 mA V_{IN} = 2.0 V
HIGH-level input current (I_{IH})			40	μA	V_{CC} = max, V_{IN} = 2.4 V
LOW-level input current (I_{IL})			−1.6	mA	V_{CC} = max, V_{IN} = 0.4 V
Short-circuit output current[c] (I_{OS})	−18		−55	mA	V_{CC} = max
Total supply current with outputs high (I_{CCH})		4.0	8.0	mA	V_{CC} = max
Total supply current with outputs low (I_{CCL})		12	22	mA	V_{CC} = max

Switching characteristics (T_A = 25°C)

Parameter	Limits			Units	Test conditions
	Minimum	*Typical*	*Maximum*		
Propagation delay time, LOW-to-HIGH output (t_{PLH})		11	22	ns	V_{CC} = 5.0 V C_{LOAD} = 15 pF R_{LOAD} = 400 ohms
Propagation delay time, HIGH-to-LOW output (t_{PHL})		7.0	15	ns	

NOTES:

[a]For conditions shown as min or max, use the appropriate value specified under recommended operating conditions for the applicable device type.
[b]Typical limits are at V_{CC} = 5.0 V, 25°C.
[c]Not more that one output should be shorted at a time. Duration of short not to exceed 1 s.

Figure 2–39
Manufacturer's data sheet showing the electrical characteristics of the 74xxx ICs.

The total amount of current drawn by the IC is given by two parameters, I_{CCH} and I_{CCL}. I_{CCL} is the total amount of current drawn from the power supply V_{CC} when *all* the outputs are in the *low* state. I_{CCH} is the total amount of current drawn from the power supply V_{CC} when *all* the outputs are in the *high* state. Notice that I_{CCL} is usually higher than I_{CCH} (the reason for this will be seen later).

Therefore the average current drawn from V_{CC} is

$$I_{CC} = \frac{I_{CCH} + I_{CCL}}{2}$$

and the average power dissipated by the IC is

$$P_{AVG} = I_{CC} \times V_{CC}$$

Input characteristics

When a logic high is applied to the input of a gate, the voltage can be in the range 2 to 5 volts. This is specified by the parameter V_{IH} (the IH stands for ''high input voltage'') which is a *minimum* of 2 volts. This means that any voltage less than 2 volts will not be interpreted as a logic high by the gate. Since a positive voltage is applied to the input of the gate, the gate input represents a *load* and draws current. This current is typically 40 μA when the positive input voltage is 2.4 volts. This input current at the logic high state is referred to as I_{IH}. The relationship between the input voltage and I_{IH} is shown in Figure 2–40a. The reason why 2.4 volts is chosen for the input voltage will be seen later.

Figure 2–40
Input characteristics of a 74xxx logic gate.

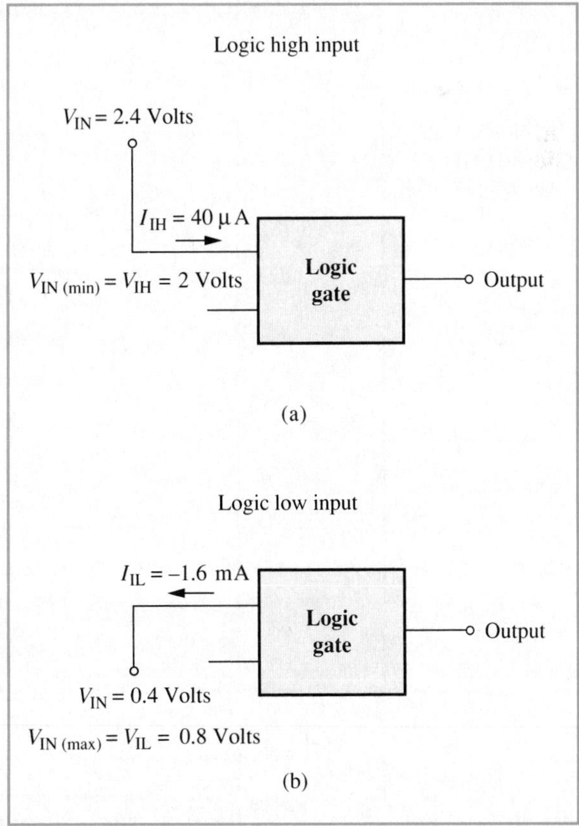

When a logic low is applied to the input of a gate, current flows out of the input. By convention we identify this as a negative current. A low is generally 0 volts or ground potential but can be a *maximum* of 0.8 volts as specified by the parameter V_{IL} in the data sheet. This means that a logic gate will interpret any voltage in the range 0 to 0.8 volts as a logic low but any voltage greater than 0.8 volts will not be considered a logic low. In Figure 2–40b, when 0.4 volts is applied to the input of the gate (the reason for picking this value will be seen later) the current flowing out of the gate, I_{IL}, is -1.6 mA as specified in the data sheet.

Thus the input of a logic gate will not respond correctly to any voltage between 0.8 and 2 volts since this area does not define a logic high or a logic low state. A logic gate should therefore never be operated in this *undefined* area.

Values for the input voltages and currents V_{IH}, V_{IL}, I_{IH}, and I_{IL} are given in the data sheet for the 74xxx TTL ICs shown in Figure 2–39.

Output characteristics

We have so far examined the *input characteristics* of a logic gate. Now let us examine the *output characteristics*.

When the output of a logic gate is at a logic high its voltage can be anywhere from 2.4 to 5 volts. This is specified by the parameter V_{OH}. The manufacturer guarantees that V_{OH} will be a *minimum* of 2.4 volts. The value of the output voltage will change depending on how much current is drawn from the output. Typically, the output of a logic gate will be connected to (drive) the inputs of other logic gates and therefore the amount of current that is drawn from an output in the logic high state will depend on the number of loads (or inputs) attached to the output. Figure 2–41 illustrates the relationship between the output voltage, V_{OUT}, and

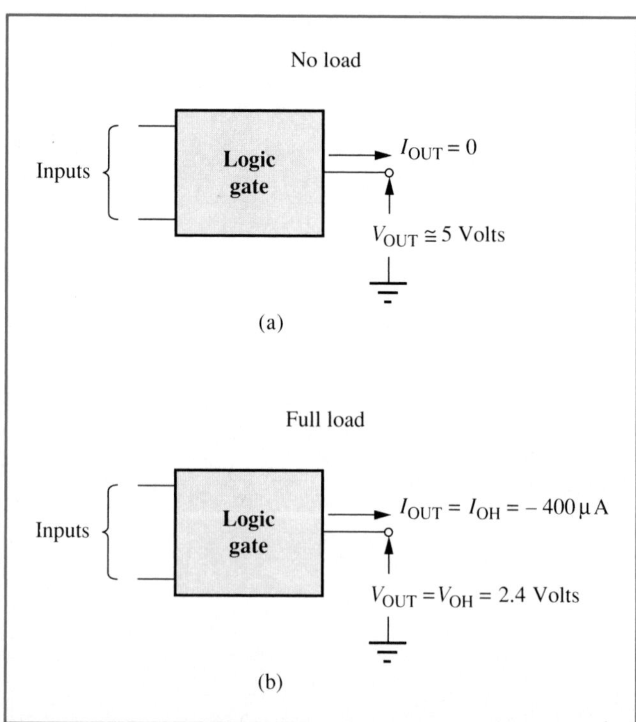

Figure 2–41 Output characteristics of a 74xxx logic gate at the high state.

the output current, I_{OUT}. If there is no load attached to the output of the gate, then the current flowing out of the gate is 0 and V_{OUT} has a value very close to 5 volts as shown in Figure 2–41a. When a load is attached to the output it draws current which causes V_{OUT} to decrease (due to internal gate resistances). The greater the loading, the more the current drawn and the smaller the value of V_{OUT}. However, we do not want to load the output to a point where V_{OUT} reaches a value less than 2.4 volts (V_{OH}). While maintaining $V_{OUT} = V_{OH}$ at 2.4 volts, the *maximum* amount of current that can be drawn from the output of the gate is $-400\ \mu A$ (by convention, the negative sign indicates that the current flow is out of the gate) as shown in Figure 2–41b. This current is known as I_{OH}. Any further increase in I_{OUT} will cause the output voltage to drop below V_{OH} (2.4 volts). The current I_{OH} is known as *source current*.

When the output of a logic gate is at a logic low its voltage can be anywhere from 0 to 0.4 volts. This is specified by the parameter V_{OL}. The manufacturer guarantees that V_{OL} will be a *maximum* of 0.4 volts. The value of the output voltage will change depending on how much current is drawn from the load attached to the output. Typically, the output of a logic gate will be connected to (drive) the inputs of other logic gates and therefore the amount of current that is drawn from an input in the logic low state will depend on the number of loads (or inputs) attached to the output.

When the output of a logic gate is at the logic low state it draws current from the load. This current flows into the output of the logic gate. If no load is attached to the output as shown in Figure 2–42a, then the current is 0 and the output voltage V_{OUT} is very close to 0 since the

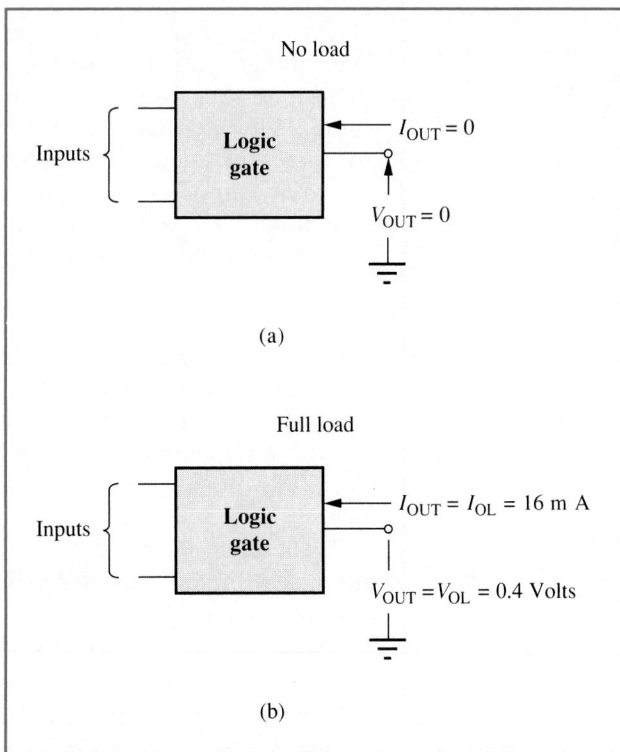

Figure 2–42
Output characteristics of a 74xxx logic gate at the low state.

internal resistance of the gate does not affect the output voltage. If a load is attached to the output of the gate when at the logic low state, the flow of current into the gate's output causes a voltage drop inside the gate and increases the output voltage. The greater the loading, or the larger the number of loads, the larger the output voltage. However, we do not want to load the output to a point where V_{OUT} reaches a value greater than 0.4 volts. While maintaining V_{OUT} at 0.4 volts (V_{OL}), the *maximum* amount of current that can be drawn from the load(s) is 16 mA and is known as I_{OL}. Any further increase in I_{OUT} will cause the output voltage to increase above V_{OL} (0.4 volts). The current I_{OL} is known as *sink current.*

Therefore a logic output sources current when in the high state and sinks current when in the low state.

Values for the input voltages and currents V_{OH}, V_{OL}, I_{OH}, and I_{OL} are given in the data sheet for the 74xxx TTL ICs shown in Figure 2–39. Notice that the data sheet also lists the short-circuit output current I_{OS}. This is the total amount of current that the output can source when short-circuited; of course the output voltage will be 0 volts.

Fan-in and fan-out

We have seen so far that 40 μA of current (I_{IH}) will flow into the input of a logic gate when a logic high is applied to it, and -1.6 mA of current (I_{IL}) will flow out of the input of a logic gate when a logic low is applied to it. These currents represent 1 *unit load* (UL) to the output of a logic circuit.

Since an output can source -400 μA of current in the high state (I_{OH}), the total number of inputs that an output can drive in the high state will be

$$\frac{I_{OH}}{I_{IH}} = \frac{400\ \mu\text{A}}{40\ \mu\text{A}} = 10$$

Thus an output can drive ten inputs or 10 ULs when in the high state. We refer to this number as the *fan-out* (*high*) of an output and it is illustrated in Figure 2–43a. Notice that the addition of any more loads will cause the output voltage to decrease below V_{OH} and therefore could place the voltage in the undefined area between a logic high and a logic low.

Since an output can *sink* 16 mA of current in the low state (I_{OL}), the total number of inputs that an output can drive in the low state will be

$$\frac{I_{OL}}{I_{IL}} = \frac{16\ \text{mA}}{1.6\ \text{mA}} = 10$$

Thus an output can also drive ten inputs or 10 ULs when in the low state. We refer to this number as the *fan-out* (*low*) of an output and it is illustrated in Figure 2–43b. Notice that the addition of any more loads will cause the output voltage to increase above V_{OL} and therefore could place the voltage in the undefined area between a logic low and a logic high.

Therefore the *fan-out* of a standard 74xxx TTL logic gate is 10. This means that the output can safely drive ten inputs (unit loads). If an attempt is made to make the output drive more than this number, unpredictable results could occur. Note that the term *fan-in* is sometimes used to refer to a unit load (UL) since these inputs are usually connected to the outputs (that have fan-outs). Most standard 74xxx TTL logic gates have a fan-out of 10, but some gates in this family, such as *buffers* and

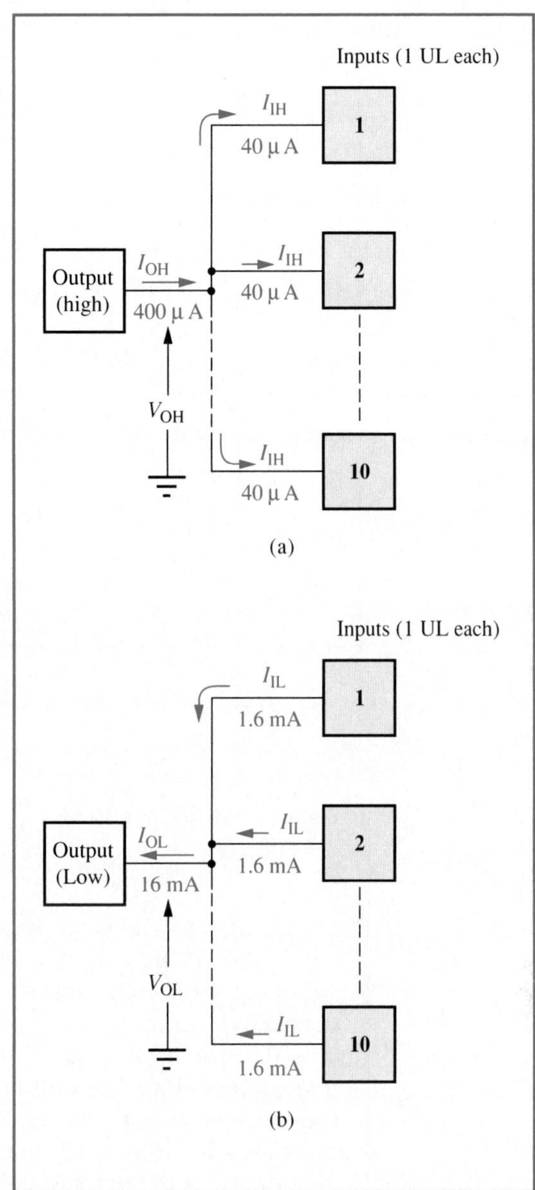

Figure 2–43
(a) Fan-out in the high state; (b) fan-out in the low state.

open-collector ICs, have larger fan-outs; these will be examined in Chapter 12. Similarly most standard 74xxx TTL logic gates have inputs that represent 1 UL; there are however variations of these ICs that have inputs that represent more and even less than 1 UL.

Noise margin

When we examine the input and output characteristics of logic gates, particularly the voltages, we can make the following observations during *worst case* conditions—that is, when the gate is fully loaded and operating at its thresholds. Let us assume that the output of a 74xxx TTL gate is connected to ten inputs so that it is fully loaded (fan-out = 10).

The input of a logic gate requires a *minimum* voltage of 2 volts for a *logic high*—V_{IH}. The output of a logic gate when at the high state and fully loaded will produce a voltage of 2.4 volts—V_{OH}. There is therefore a

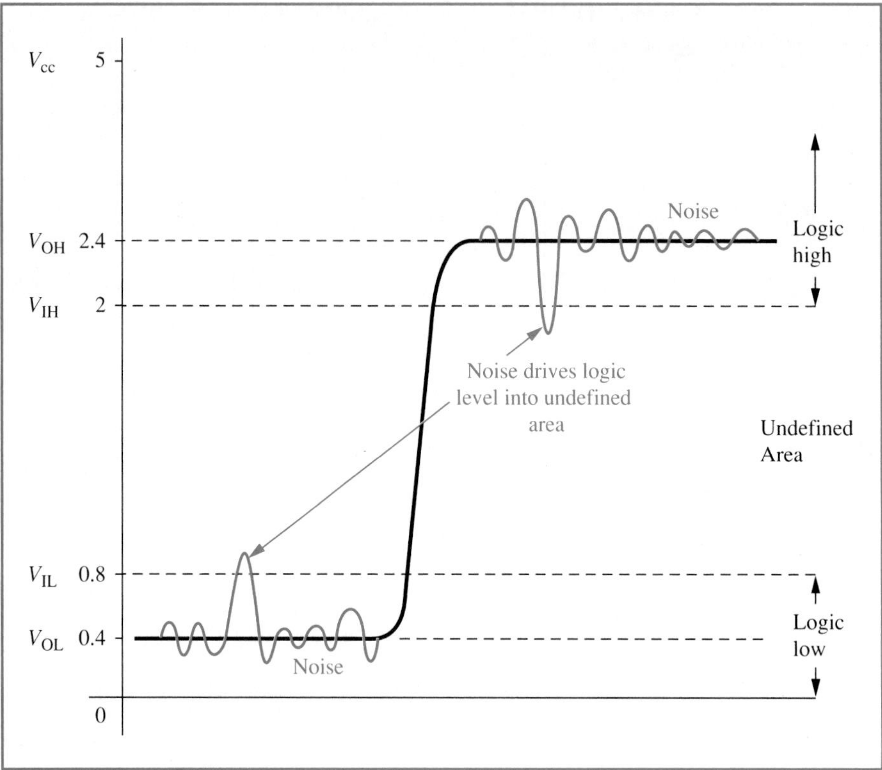

Figure 2–44 Effects of noise on a logic level.

margin of 0.4 volts when the gate is operating under worst case conditions. If the output voltage drops below this margin of 0.4 volts, it will no longer be considered a logic high by the inputs of the gates connected to it.

The input of a logic gate requires a *maximum* voltage of 0.8 volts for a *logic low*—V_{IL}. The output of a logic gate when at the low state and fully loaded will produce a voltage of 0.4 volts—V_{OL}. Again, there is a margin of 0.4 volts when the gate is operating under worst case conditions. If the output voltage rises above this margin of 0.4 volts, it will no longer be considered a logic low by the inputs of the gates connected to it.

Figure 2–44 illustrates the effects of noise on logic gates operating under worst case conditions. Assuming that the logic high level is right at V_{OH} (2.4 volts) and the logic low level is right at V_{OL} (0.4 volts), any electrical noise that is picked up over the interconnecting wires will ''ride'' on these logic levels and will not affect the operation of the gates as long as the amplitude of the noise is less than 0.4 volts. If the amplitude of the noise exceeds 0.4 volts (peak voltage), then the logic level could be driven into the undefined area.

This value of 0.4 volts is known as the *noise margin* since it establishes the amount of noise that can exist on a logic level under worst case conditions. The noise margin is therefore defined as the difference between worst case output voltage and worst case input voltage.

Thus the noise margin in the high state is

$$V_{NH} = V_{OH} - V_{IH} = 2.4 - 0.4 = 0.4 \text{ volts}$$

and the noise margin in the low state is

$$V_{NL} = V_{IL} - V_{OL} = 0.8 - 0.4 = 0.4 \text{ volts}$$

Propagation delay

When the input of a gate changes states and this change causes the output to change state, the change of state at the output is not instantaneous. There is a certain amount of delay time associated with the output. This delay time is known as *propagation delay*.

Figure 2–45 shows the effects of propagation delay on an inverter both under ideal conditions and actual conditions. Ideally, when the input of the inverter switches from a low (L) to a high (H), the output of the inverter will instantaneously switch from a high to a low. However, in practice there is a delay associated with a logic level that switches from a low to a high (*rise time*) and a delay associated with a logic level that switches from a high to a low (*fall time*). The propagation delay time is the time between a change of state that occurs at the input and a change

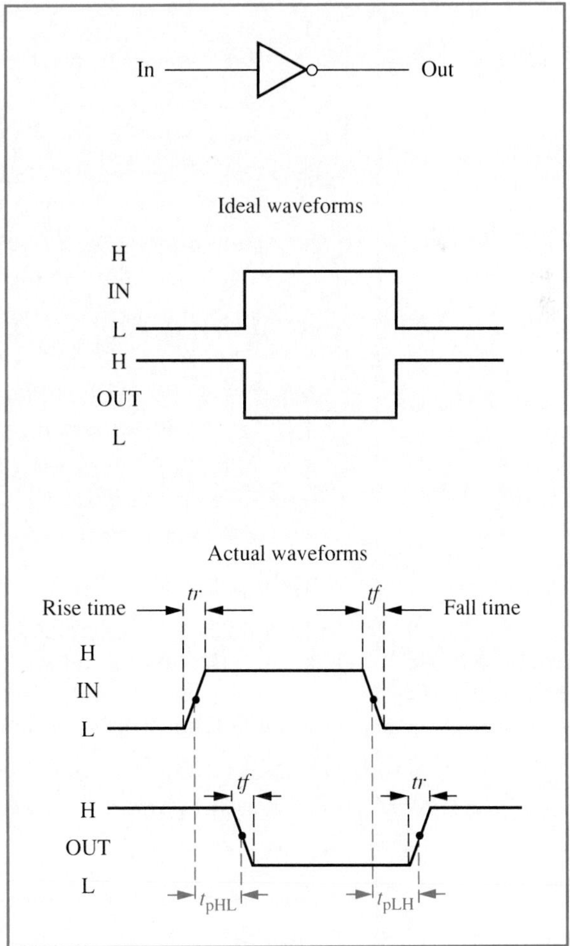

Figure 2–45
Propagation delay in a logic gate.

of state that occurs at the output. Manufacturer data sheets often specify two propagation delay times as shown in Figure 2–39:

t_{pHL} = propagation delay time when the *output* changes from high to a low state

t_{pLH} = propagation delay time when the *output* changes from low to a high state

These two values are usually equal, but if they are not then the *average propagation delay time* (t_{pd}) is

$$t_{pd} = \frac{t_{pHL} + t_{pLH}}{2}$$

and the worst case propagation delay time is the larger of the two values.

Propagation delays are an indication of the switching speed of the gate. A slow gate has a longer propagation delay time and therefore has a slow switching speed while a fast gate has a smaller propagation delay time and a fast switching speed. The effects of gate propagation delays are usually felt on sequential circuits and will be examined in Chapter 5.

Review questions

1. Explain why the current drawn from the power supply V_{CC} is more when all the outputs are at the low state.
2. Define the range of voltages that are recognized as a logic high and a logic low at the input of a gate.
3. Describe the currents that flow at the input of a gate when a logic high and a logic low are applied to it.
4. Define the range of voltages that are produced by the output of a gate when at the high state and at the low state.
5. Describe the terms *source* and *sink currents* with reference to the output of a gate.
6. Define the terms *fan-out* and *fan-in*.
7. Define *noise margin*.
8. What is propagation delay. What effect does it have on the speed of a gate?

2–10 Troubleshooting logic gates and circuits

In Section 2–8 we examined the procedures to set-up a logic circuit with ICs using the layout and wiring diagram. Once the circuit is constructed on a breadboard there is often a good possibility that the circuit will not work as intended due to one of the following reasons:

1. Faulty power supply
2. Defective IC
3. Wiring mistake

This section will deal with the procedures used to identify and correct such faults in a logic circuit. These procedures are known as *troubleshooting procedures*.

Troubleshooting equipment

One of the advantages in troubleshooting digital circuits over analog circuits is that since digital circuits are two-stated we only have to determine if the inputs and outputs of a circuit are at a logic 1 (high) or a logic 0 (low) state. In troubleshooting analog circuits we generally need to measure continuously varying voltages and currents that have to be within fairly close tolerances. This characteristic of digital circuits allows us to test and troubleshoot logic circuits without the need for expensive equipment and lengthy procedures. The most commonly used equipment for troubleshooting digital circuits can be listed as follows in the order of importance:

1. Logic probe
2. Logic pulser
3. Current tracer
4. Voltmeter and ammeter
5. Oscilloscope
6. Logic analyzer
7. Signature analyzer

This section will examine the use of the logic probe, logic pulser, current tracer, voltmeter, and ammeter in troubleshooting combinational logic circuits. Later chapters will discuss the application of the oscilloscope, logic analyzer, and signature analyzer for the *dynamic troubleshooting* of various application circuits and sequential logic circuits.

The logic probe can be used to detect 90% of all problems with simple combinational logic circuits. The logic probe simply indicates whether a logic level is at the logic high state, logic low state, or changing states. The logic probe consists of a thin needle-like probe tip connected to a housing that has three light-emitting diodes (LEDs). A set of leads are used to supply power (usually 5 volts) to the logic probe. To determine the logic level at a particular pin of an IC, we touch the probe tip to the pin and observe the LEDs. One LED indicates a logic high state, the other indicates a logic low state, and the third indicates if the logic state at the pin is changing rapidly from a high to a low or vice versa (pulsing). If neither the high LED or the low LED illuminate, then the logic level is neither high nor low but is floating, or in a *high-impedance state* (more will be said about this state in Chapter 4). Many logic probes may also have a switch that is used to select between the various families of logic ICs.

The logic pulser is similar in construction to the logic probe. The logic pulser is used to *inject* a pulse into a logic circuit. The logic pulser has a pushbutton switch that is used to apply a short-duration pulse to a circuit node; this pulse will cause a node that is at a logic high state to momentarily go low and a node that is at a logic low state to momentarily go high. The logic pulser can be used to inject a pulse into a circuit node without the need to isolate the node from the rest of the circuit. It cannot produce a voltage pulse on a node that is directly shorted to ground or V_{CC}, it can however produce a current pulse at such a node.

The current tracer is often used in conjunction with the logic pulser to detect a *changing* current in a wire or connection between nodes. The current tracer is also similar in construction to the logic probe but contains an insulated tip that can sense the presence of a magnetic field caused by a changing current. An LED or lamp indicates the presence of current flow in the circuit.

Voltmeters and ammeters are useful instruments for measuring logic levels to ensure that the voltages and currents are within the ranges specified for the IC. The oscilloscope and logic analyzer are generally used to examine dynamic logic states in sequential logic circuits where logic levels are constantly changing. The signature analyzer is a useful tool in quickly pinpointing a faulty component in a circuit without having to go through elaborate diagnostic procedures. More will be said about the logic analyzer and signature analyzer in Chapters 8 and 9.

Causes of faults in ICs

Digital ICs often contain several gates in one package. Often a single gate may malfunction while the other gates operate properly, or in some cases all the gates in the IC may malfunction due to an internal problem with the power supply. The reason why a single gate may not work as it should could have to do with a number of problems related to the internal circuitry of that gate. Typical problems include

1. Damage of internal components such as transistors and resistors due to overloading or driving outputs and inputs
2. Inputs or outputs that are internally shorted to V_{CC} or ground
3. Inputs or outputs that are internally open-circuited

Since it is impossible to repair any of the above three situations, the IC that exhibits one or more of these symptoms is generally replaced. It is therefore fruitless for us to devote much troubleshooting time in determining the reason for the defective operation of the gate or IC, instead let us concentrate our efforts on troubleshooting logic circuits and narrowing the problem down to a defective IC that can simply be replaced.

Troubleshooting procedures

The first step in any troubleshooting procedure is to make sure that each IC in the circuit is powered-up properly. One of the most common faults in a logic circuit is a defective power supply, or, in cases where the circuit is being breadboarded, unconnected or misconnected power supply pins on the IC(s). In cases where no logic levels are detectable in the circuit it is obvious that an inoperable power supply is at fault. In all other cases, a marginally operating power supply could cause erratic operation of the circuit. Power supply potential should be checked with a voltmeter to make sure that it is between the recommended 4.75- to 5.25-volt range. The closer the voltage to 5 volts the better since (depending on the quality of the power supply) the voltage could change a lot when the outputs of gates in the logic circuit switch to the logic low state. If the power supply voltages are acceptable, then the next step is to try to isolate the fault to a defective IC assuming that there are no wiring errors. If the circuit is being breadboarded, then the wiring should be checked for proper connections.

Let us consider a logic circuit such as the one shown in Figure 2–46 that contains a fault. The expected values for the output of the circuit and the recorded values are shown in the truth table in Table 2–15.

Figure 2–46
Logic circuit containing a fault.

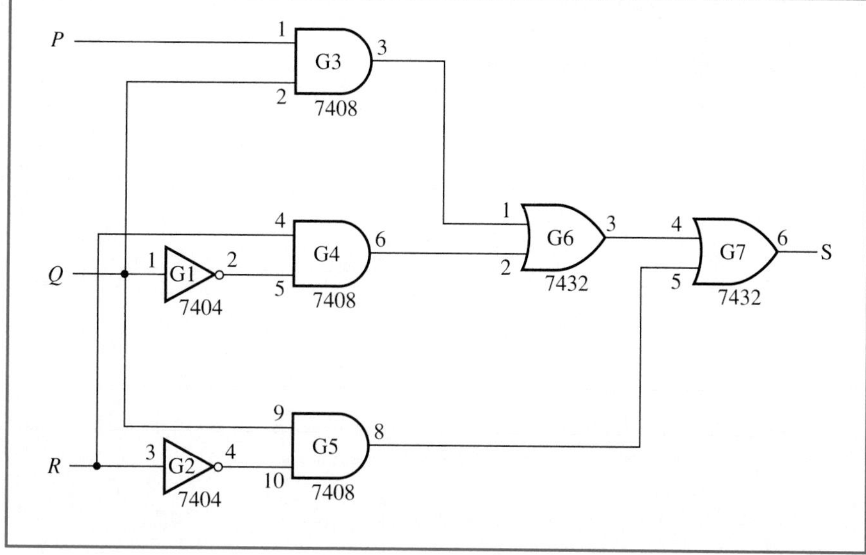

Notice in Table 2–15 that the last combination does not produce the expected output of a logic high; instead, it produces a logic low. All other combinations work as expected. In order to troubleshoot this circuit we apply this combination (111) to the inputs of the circuit. The logic levels that we can expect from the output of each gate is shown in Figure 2–47. Using the logic probe we then verify that each gate produces the correct output as shown in Figure 2–47. However, the logic probe indicates that the output of gate G3 is at the logic low (L) state when it should really be at a logic high (H) state. Since the output of G3 is connected to the input of G6 we have isolated the fault to one of two possible causes

1. The output of gate G3 is shorted to GND thereby pulling the node down to the logic low state.
2. The input of gate G6 is shorted to GND thereby pulling the node down to the logic low state.

To isolate the defective gate we can (if possible) disconnect the connection between the output of G3 and the input of G6 and measure the logic level at the output of G3 as shown in Figure 2–48. If the output of G3 now indicates a logic high level (Figure 2–48a), the defective gate is

Table 2–15

P	Q	R	Expected S	Recorded S	
L	L	L	L	L	
L	L	H	H	H	
L	H	L	H	H	
L	H	H	L	L	
H	L	L	L	L	
H	L	H	H	H	
H	H	L	H	H	
H	H	H	H	L	⟵ fault

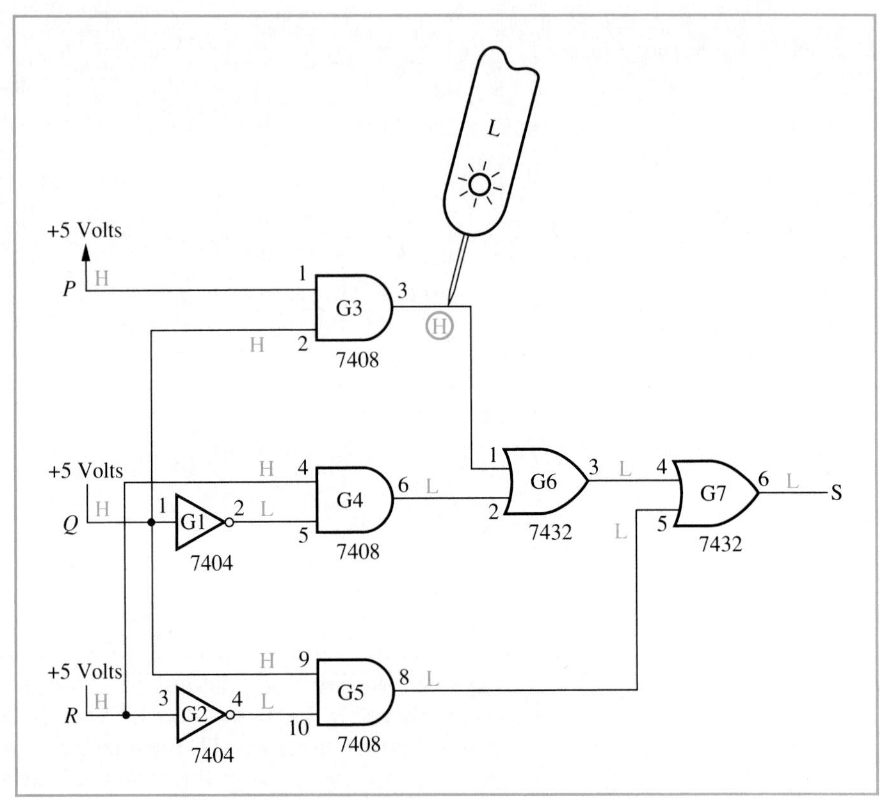

Figure 2–47
Identifying the fault in a circuit by using the logic probe.

G6 due to an internally shorted input to ground. If the output of G3 is still a logic low, then gate G3 is defective (Figure 2–48b) due to an internally shorted output to ground.

Figure 2–48
Isolation of the fault in the circuit using a logic probe.

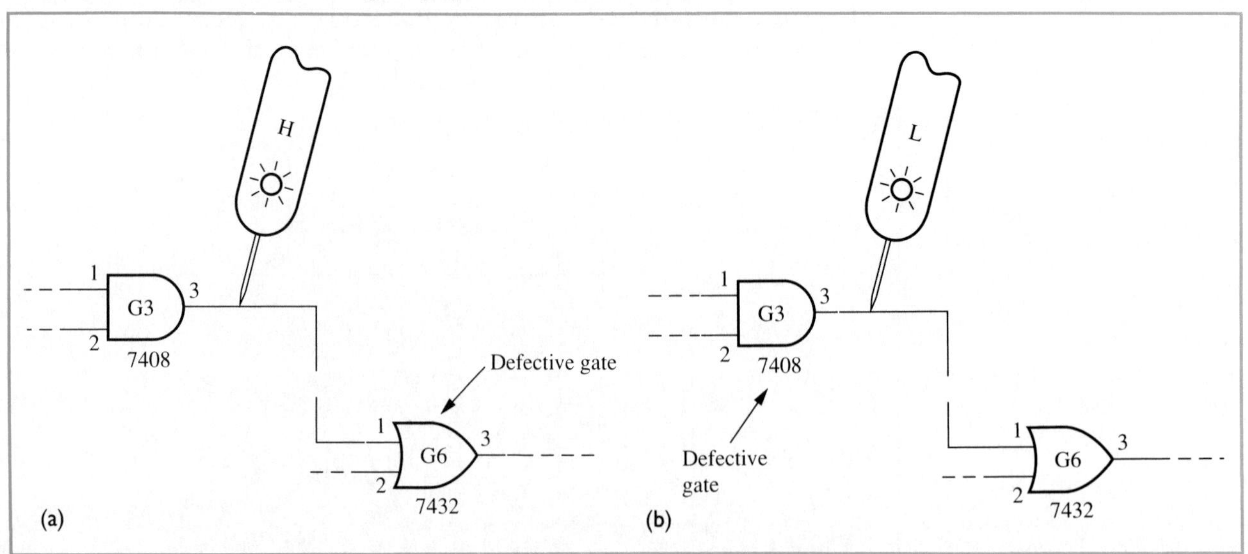

If it is not possible to disconnect the connection between the output of G3 and the input of G6 [such as in a printed circuit (PC) board], then the current pulser and current tracer can be used to isolate the defective gate. The pulser is used to inject a pulse of current into the node while the current tracer is used to determine if the current pulse flows into the input of G6 or into the output of G3. The current will always flow in the direction of the short. In Figure 2–49a the pulser is used to inject a pulse of current into the node and the tracer is placed near the input of G6. If the tracer detects the current pulse, then the input of G6 is shorted to ground and therefore G6 is the defective gate. In Figure 2–49b the tracer is placed near the output of G3. If the tracer detects the current pulse, the output of G3 is shorted to ground and therefore G3 is the defective gate.

The problem has now been isolated to a single gate and can be corrected by replacing the IC.

Figure 2–49
Isolation of the fault in the circuit using a pulser and current tracer.

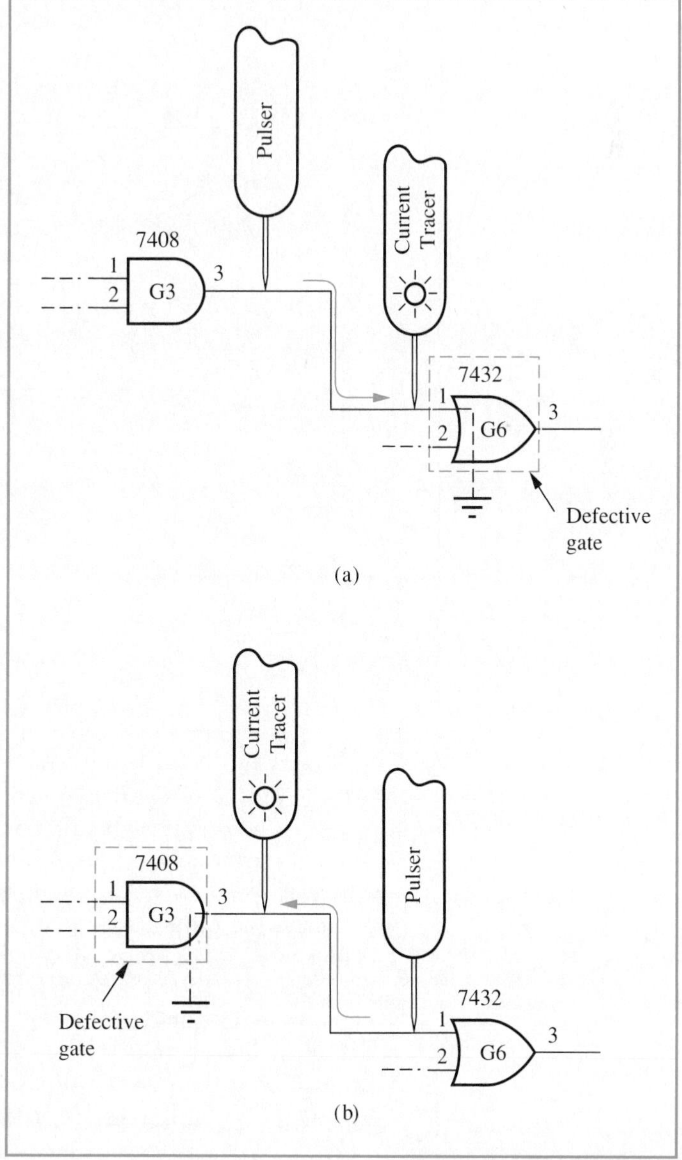

The following example illustrates another case study of a similar procedure used to troubleshoot a logic circuit.

EXAMPLE 2–14

The circuit shown in Figure 2–50 has a fault as shown in Table 2–16. The recorded output of the circuit indicates that the output remains at a logic high state and does not respond to any combination applied to the input. After making sure that the ICs have proper power supply, we must now determine which one of the gates in the circuit has a fault. The best approach is to begin at the output of the circuit since if the output of G7 is shorted to V_{CC} it would remain high regardless of its input levels.

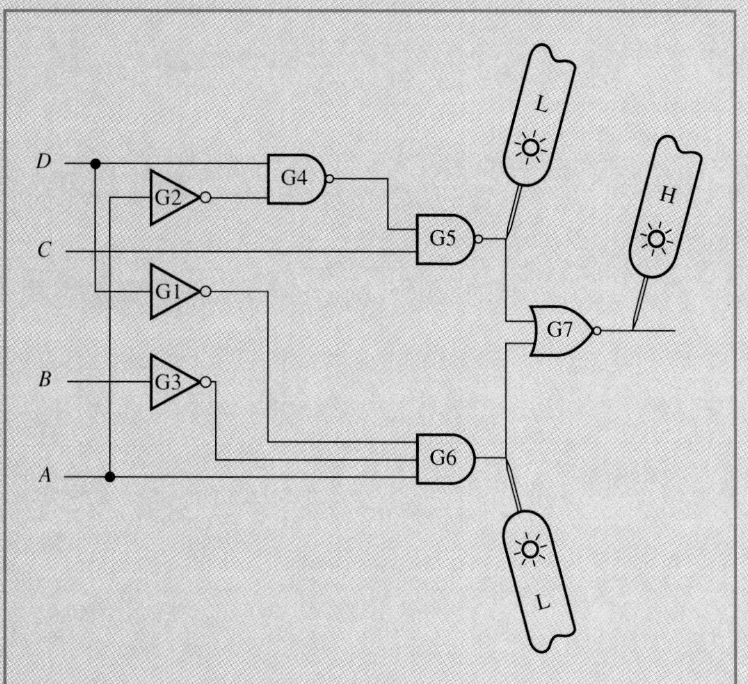

Figure 2–50
Circuit containing a fault.

An examination of the inputs of G7 with the logic probe indicates that both inputs are at the low state. This could account for the output of G7 being high (since if both inputs of a *NOR* gate are low the output is high). To make sure that the inputs of G7 are not shorted to ground, we inject a high going pulse (a pulse that changes from a low to a high state) to one input using the pulser and examine the output with a logic probe as shown in Figure 2–51a. The same procedure is repeated with the other input of G7 as shown in Figure 2–51b. In both cases the output goes low when the pulse is applied indicating that gate G7 is not at fault.

It is quite unlikely that the outputs of G5 and G6 are shorted to ground since the pulser would not have been able to produce a voltage pulse on the inputs of G7 if this were the case. To see if it is

Table 2–16

A	B	C	D	Expected X	Recorded X
L	L	L	L	L	H
L	L	L	H	L	H
L	L	H	L	H	H
L	L	H	H	L	H
L	H	L	L	L	H
L	H	L	H	L	H
L	H	H	L	H	H
L	H	H	H	L	H
H	L	L	L	L	H
H	L	L	H	L	H
H	L	H	L	L	H
H	L	H	H	H	H
H	H	L	L	L	H
H	H	L	H	L	H
H	H	H	L	H	H
H	H	H	H	H	H

possible to change the output of G5 to a logic high we ground (apply a logic low) the *C* input as shown in Figure 2–52a; the logic probe attached to the output of G5 still indicates a logic low. This indicates that gate G5 is defective due to an internal malfunction since it should have produced a logic high at its output.

To see if the output of gate G6 changes to a logic high, the *A* input is connected high, the *B* input is connected low, and the *D* input is connected low so that all the inputs of G6 are high as shown in Figure 2–52b. The output of gate G6 remains low. However, the output of G3 indicates a low state when it should be high. Use of the current tracer as shown in Figure 2–52c indicates that the output of G3 is shorted to ground.

Replacement of G5 (7400 IC) and G3 (7404 IC) produce recorded results that match the expected results of Table 2–16.

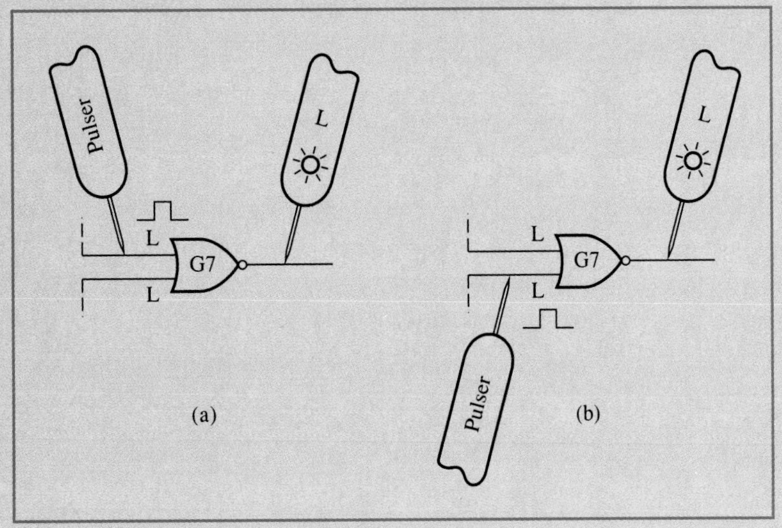

Figure 2–51 Testing a gate for proper operation.

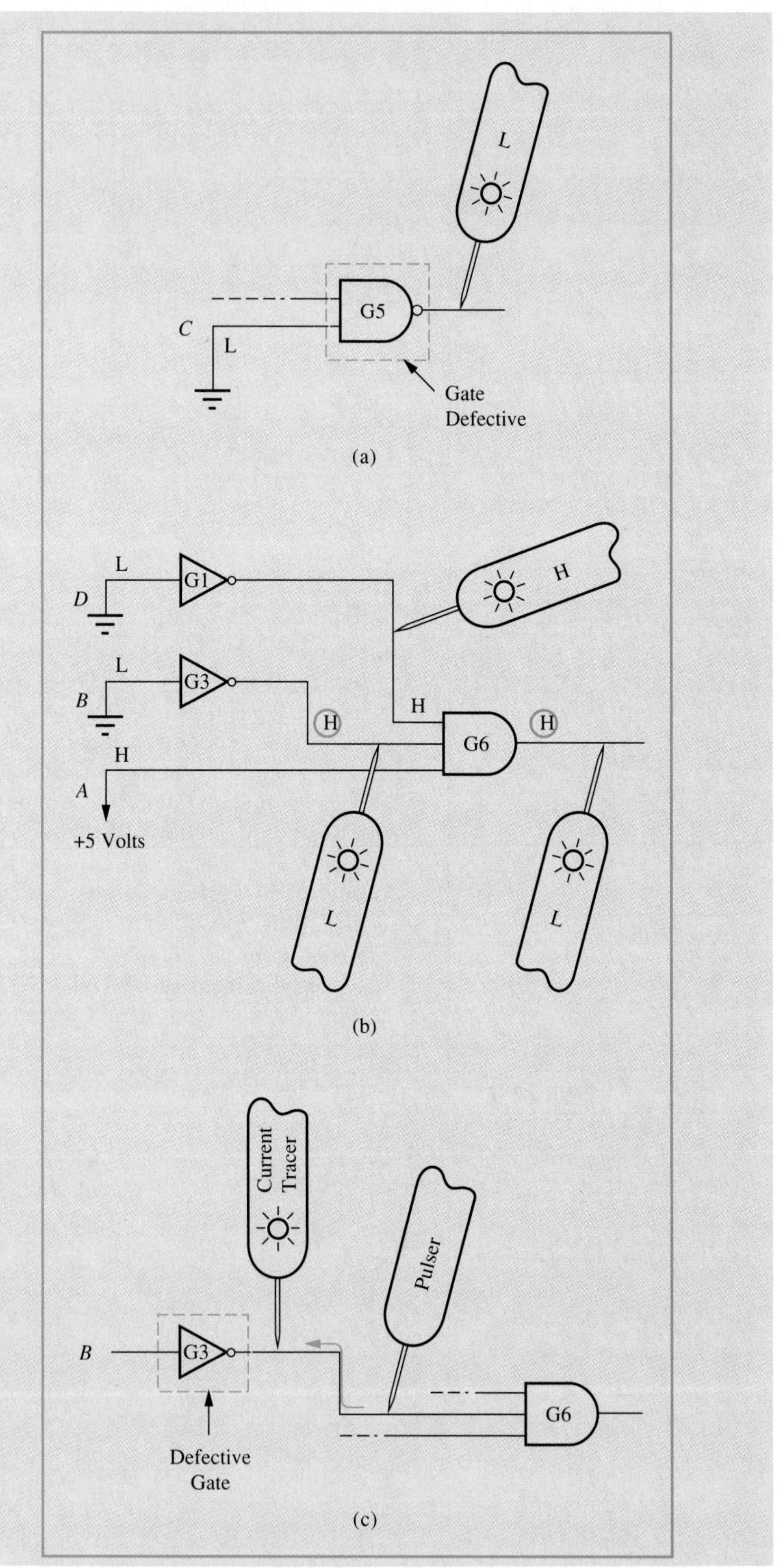

Figure 2–52
Isolation of the fault.

The troubleshooting procedures examined in this section provide a systematic approach to isolating and correcting faults in a logic circuit. A major portion of troubleshooting time is spent on isolation of the fault. Correction of the fault is usually as simple as replacement of an IC or in some cases removal of an external short such as a solder bridge. The logic probe is a versatile tool to isolate most common problems. However, the gates being tested with a logic probe generally have to be isolated from the rest of the circuit to make measurements. The pulser/current tracer combination is excellent for in-circuit tests of logic gates. Since most combinational logic circuits are similar to the ones studied in this chapter, the procedures and equipment introduced in this section can be applied to troubleshoot almost any logic circuit.

Review questions

1. What is troubleshooting? What does it involve?
2. Describe the use of the logic probe, the pulser, and the current tracer to troubleshoot logic circuits
3. What are the typical faults that could exist in a logic gate?

2–11 SUMMARY

- Logic circuits are also called *digital circuits* and *switching circuits*.
- The basic element of a modern logic circuit is a device known as the *logic gate*.
- A gate is a device that electronically implements the various *logic functions* that are used in the design of logic circuits.
- Gates are implemented in hardware in the form of an electronic *integrated circuit* (IC) also known as a *chip*.
- Logic circuits are broadly classified into two categories—*combinational logic circuits* and *sequential logic circuits*.
- The variables in a switching circuit can have only one of two states.
- *Digital circuits* are *two-state* circuits, whereas *analog circuits* can have many more states.
- The variables in a digital circuit are generally classified as *dependent* or *independent*.
- The states of dependent variables depend on other factors such as the state of the independent variables while the states of independent variables do not depend on anything.
- *Logical inversion* is a process or function that reverses the state of a variable.
- The *complement* of a variable is its opposite state.
- A logic gate can have several inputs (independent variable) and one output (dependent variable) whose value depends on the binary combinations at the inputs and the function of the gate.
- The inputs and outputs of a gate are represented conceptually by binary numbers and electronically by voltages.

- The 74xxx TTL series of ICs is a family of ICs that implement various logic gates and digital circuits.
- A *buffer* does not provide any logic function—the logic output is equal to the logic input.
- The output of an *inverter* (*NOT gate*) is the complement of its input.
- The output of the *AND* gate will be a logic 1 if and only if all its inputs are at the logic 1 state.
- The output of the *OR* gate will be a logic 1 if any of its inputs is at the logic 1 state.
- *AND* and *OR* gates can have any number of inputs but must have at least two inputs.
- The output of the *NAND* gate will be a logic 0 if and only if all its inputs are at the logic 1 state.
- The output of the *NOR* gate will be a logic 0 if any of its inputs is at the logic 1 state.
- NAND and *NOR* gates can have any number of inputs but must have at least two inputs.
- Logic circuits are constructed from logic gates and can have several inputs and outputs.
- *Truth tables* and *logic equations* are used to describe the operation of logic gates and circuits.
- The *analysis* of logic circuits involves determining the operation of the circuit by obtaining its truth table and logic equation.
- The *synthesis* of logic circuits involves the design of the circuit given its logic equation, truth table, or a specification.
- In electrical terminology, a logic 1 is referred to as a *high* and a logic 0 is referred to as a *low*.
- A logic output *sinks* current in the low state and *sources* current in the high state.
- A *unit load* (UL) describes the amount of loading an input puts on an output and is also known as *fan-in*.
- *Fan-out* is a measure of how many inputs (ULs) an output can drive under worst case conditions.
- *Noise margin* is the difference between worst case output voltage and worst case input voltage.
- *Propagation delay* is the amount of delay that occurs from the time the input of a gate changes states to the time the output changes states.
- The propagation delay of a gate is a direct indication of the switching speed of the gate; a slow gate has a long propagation delay time while a fast gate has a short propagation delay time.

PROBLEMS

Section 2–2 Basic logic concepts

1. Obtain a truth table for each of the switching circuits shown in Figure 2–53.
2. Obtain a truth table for each of the switching circuits shown in Figure 2–54.
3. Obtain the logic equations for each of the switching circuits shown in Figure 2–53.
4. Obtain the logic equations for each of the switching circuits shown in Figure 2–54.

Figure 2–53

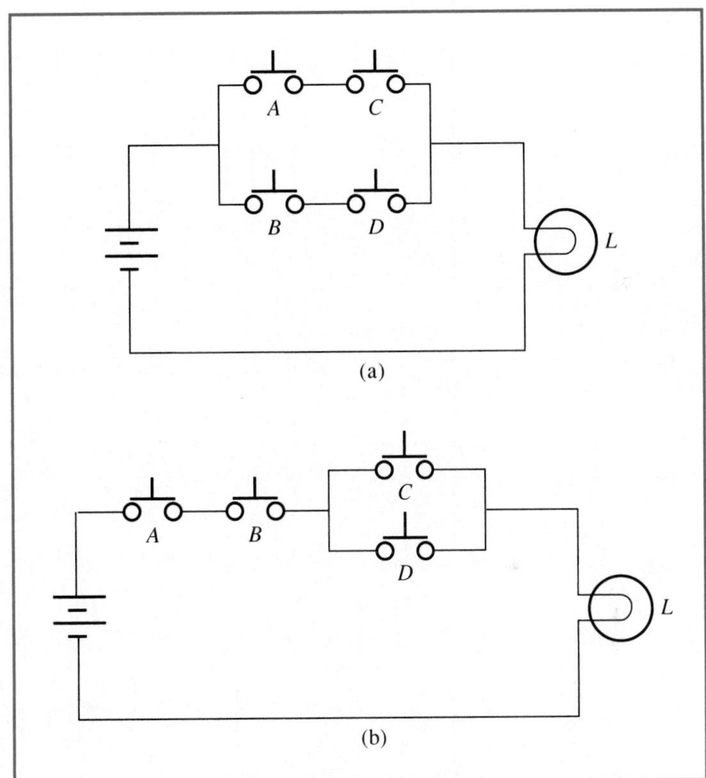

Section 2–4 The buffer and inverter

5. Trace through the circuit shown in Figure 2–55 and determine the logic states of B through H for all possible values of the independent variable A.
6. From the circuit shown in Figure 2–55, determine the logic equations for the dependent variables B through H in terms of the independent variable A.

Section 2–5 The AND gate

7. Obtain a logic equation and truth table for the circuit shown in Figure 2–56. What can you conclude about this circuit?
8. Construct a 16-input *AND* gate using only three-input *AND* gates.

Figure 2–54

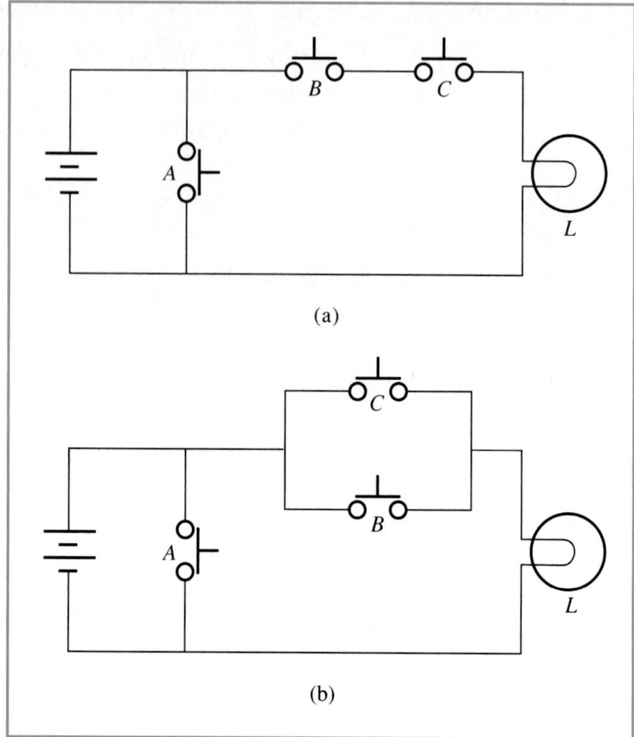

Figure 2–55

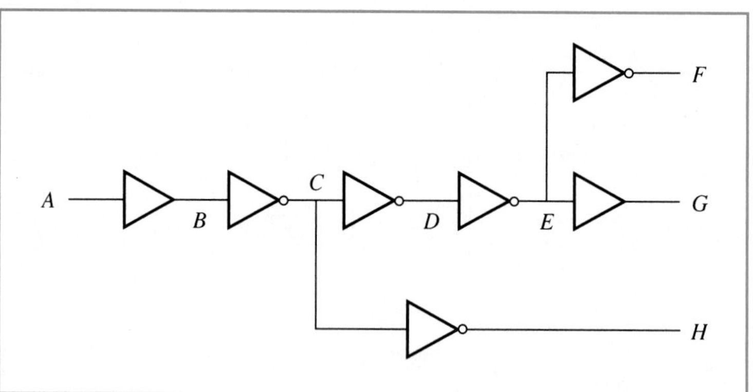

Figure 2–56

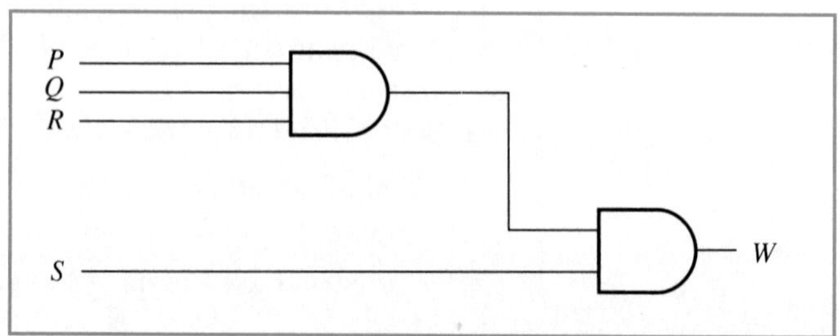

Section 2–6 The OR gate

9. Obtain a logic equation and truth table for the circuit shown in Figure 2–57. What can you conclude about this circuit?

10. Construct a 16-input *OR* gate using only three-input *OR* gates.

Figure 2–57

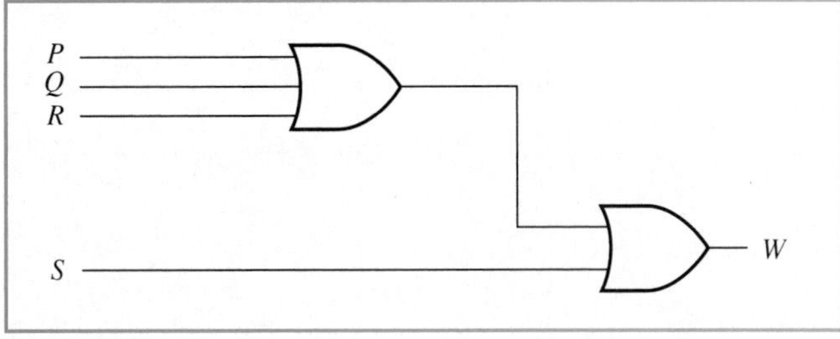

Section 2–7 The NAND and NOR gates

11. Obtain a truth table for each of the circuits shown in Figure 2–58. What conclusions can you reach about these circuits?

12. Obtain a truth table and logic equation for each of the circuits shown in Figure 2–59. What conclusions can you reach about these circuits?

13. Obtain a truth table and logic equation for each of the circuits shown in Figure 2–60.

14. Construct a 16-input *NAND* gate using only commercially available (74xxx) *AND* gates and inverters.

15. Construct a 16-input *NOR* gate using only commercially available (74xxx) *OR* gates and inverters.

Section 2–8 Logic circuits

16. Analyze each of the circuits shown in Figure 2–61, and obtain their truth tables.

17. Analyze each of the circuits shown in Figure 2–62 and obtain their truth tables.

18. Obtain the logic equations for each of the circuits shown in Figure 2–61.

Figure 2–58

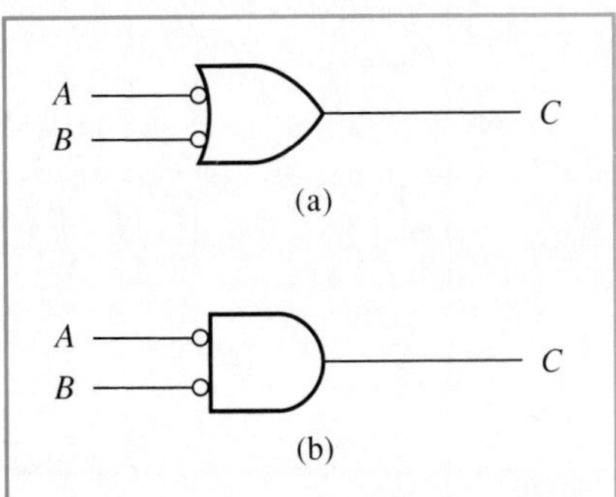

Figure 2–59

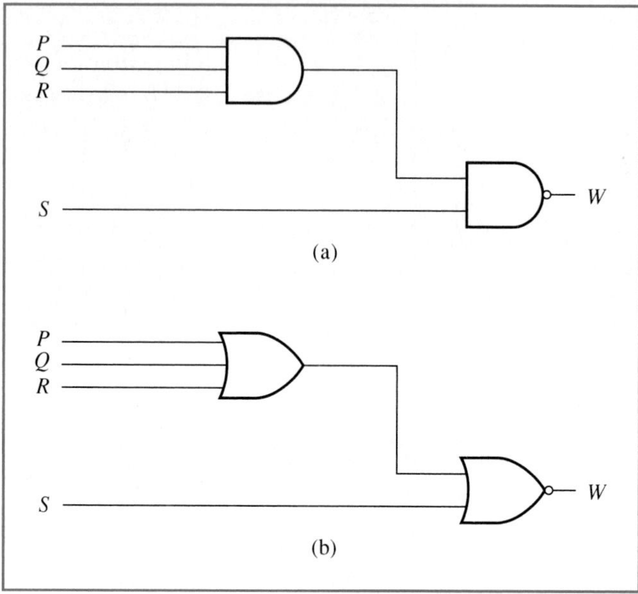

19. Obtain the logic equations for each of the circuits shown in Figure 2–62.

20. Draw the logic circuits that represent the following logic equations:
(a) $A = \overline{P} \cdot (Q + \overline{R}) \cdot S$
(b) $Z = (A \cdot \overline{\overline{B}} \cdot C) + (\overline{A} \cdot \overline{B} \cdot \overline{C})$
(c) $T = \overline{X + Y} + \overline{Y + Z}$
(d) $W = \overline{A} \cdot \overline{B \cdot C} \cdot \overline{D}$

21. Draw the logic circuits that represent the following logic equations:
(a) $X = \overline{W} + \overline{Z} + (\overline{P} \cdot Z)$
(b) $Z = \overline{A \cdot B} \cdot \overline{C \cdot D}$

Figure 2–60

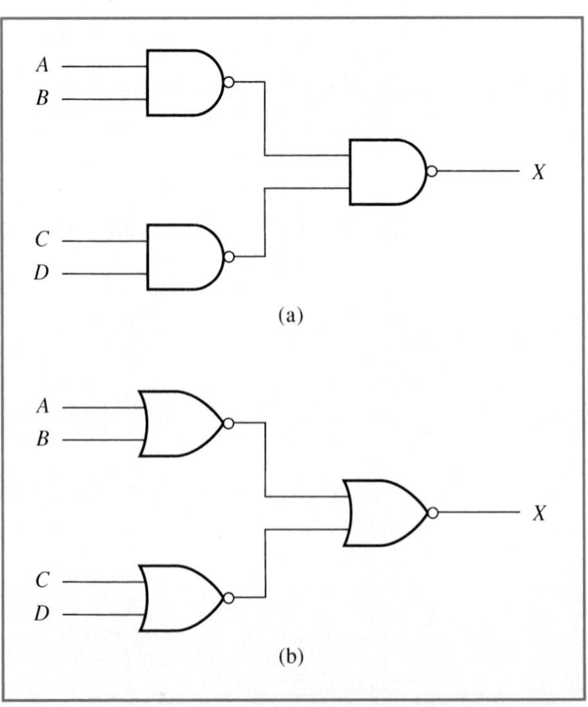

Figure 2–61

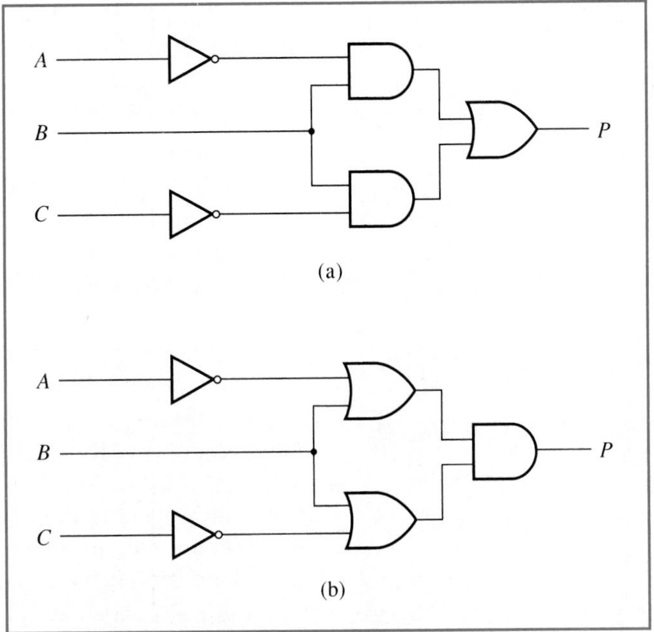

(c) $W = \overline{P + (Q \cdot R)}$
(d) $A = (X \cdot Y) + \overline{X} + \overline{Y}$

22. Draw a schematic diagram for each of the circuits shown in Figure 2–61 using the appropriate 74xxx ICs.
23. Draw a schematic diagram for each of the circuits shown in Figure 2–62 using the appropriate 74xxx ICs.
24. Draw a layout and wiring diagram for each of the circuits shown in Figure 2–61.
25. Draw a layout and wiring diagram for each of the circuits shown in Figure 2–62.

Section 2–9 Electrical characteristics of logic gates

26. An IC belonging to the 74LS family of logic integrated circuits has the following electrical parameters:

Figure 2–62

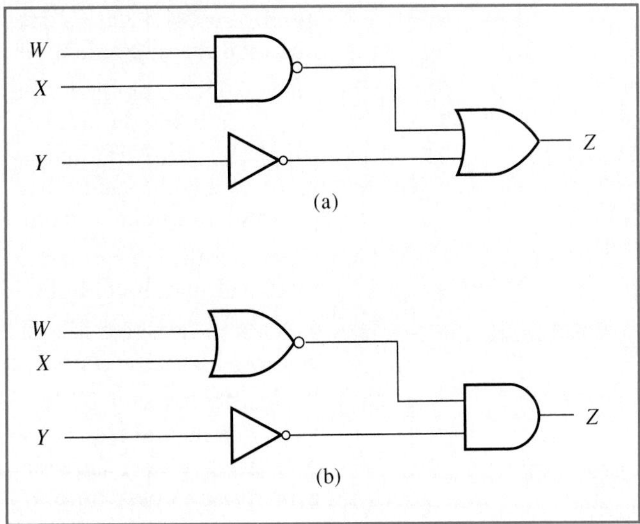

$I_{CCH} = 18$ mA, $I_{CCL} = 19$ mA, and $V_{CC} = 5$ volts
Calculate the average power dissipated by the IC.

27. An IC belonging to the 74S family of logic integrated circuits has the following electrical parameters:
$I_{CCH} = 9$ mA, $I_{CCL} = 30$ mA, and $V_{CC} = 5$ volts
Calculate the average power dissipated by the IC.

28. The 74LS family of logic integrated circuits can detect at their inputs all voltages in the range 2 to 5 volts as a logic high and all voltages in the range 0 to 0.8 volts as a logic low. The outputs are guaranteed to produce a voltage in the range 2.7 to 5 volts in the high state and a voltage of 0 to 0.5 volts in the low state. Determine the voltages V_{OH}, V_{OL}, V_{IH}, V_{IL} for this family.

29. The 74AS family of logic integrated circuits is guaranteed to produce a minimum of 3 volts at its output when at the logic high state and a maximum of 0.5 volts when at the logic low state. The minimum amount of voltage required at the input of a logic high is 2 volts and the maximum amount of voltage at the input for a logic low is 0.8 volts. Determine the voltages V_{OH}, V_{OL}, V_{IH}, and V_{IL} for this family.

30. The output of a 74LS gate can source 0.4 mA of current in the high state and sink 8 mA in the low state. A current of 20 μA flows into the input when a logic high is applied to it and 400 μA flows out of the input when a logic low is applied to it. Determine the currents I_{IL}, I_{IH}, I_{OL}, and I_{OH}.

31. The output of a 74AS gate can source 2 mA of current in the high state and sink 20 mA in the low state. A current of 20 μA flows into the input when a logic high is applied to it, and 500 μA flows out of the input when a logic low is applied to it. Determine the current I_{IL}, I_{IH}, I_{OL}, and I_{OH}.

32. Calculate the fan-out of the 74LS family of ICs from the electrical parameters obtained from Problem 30.

33. Calculate the fan-out of the 74AS family of ICs from the electrical parameters obtained from Problem 31.

34. Since 1 UL is defined with reference to the standard 74xxx TTL family as being an input that draws 40 μA in the high state and supplies 1.6 mA in the low state, determine how many ULs the input of a 74LS gate corresponds to with reference to the parameters obtained from Problem 30.

35. Since 1 UL is defined with reference to the standard 74xxx TTL family as being an input that draws 40 μA in the high state and supplies 1.6 mA in the low state, determine how many ULs the input of a 74AS gate corresponds to with reference to the parameters obtained from Problem 31.

36. Calculate the noise margin of the 74LS family of ICs from the electrical parameters obtained in Problem 28.

37. Calculate the noise margin of the 74AS family of ICs from the electrical parameters obtained in Problem 29.

38. A 74LS05 inverter IC has the following propagation delay times: $t_{pLH} = 6$ ns and $t_{pHL} = 3$ ns. Draw a diagram of the input and output waveform showing these delays, and calculate the average propagation delay time, t_{pd}.

Figure 2–63

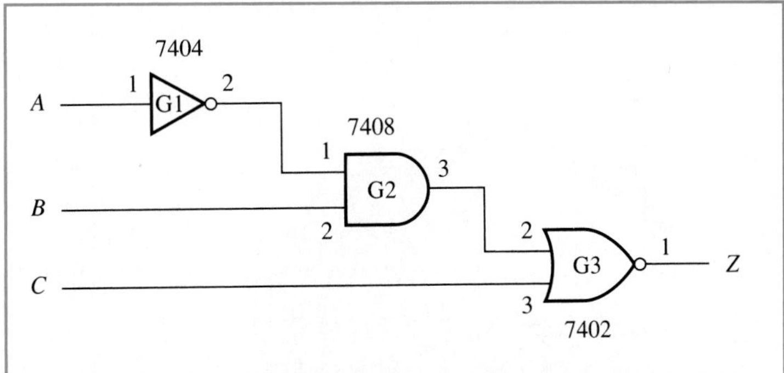

39. A 74L00 NAND gate IC has the following propagation delay times: $t_{pLH} = 35$ ns and $t_{pHL} = 31$ ns. Draw a diagram of the input and output waveform showing these delays, and calculate the average propagation delay time, t_{pd}. Assume that one of the inputs to the NAND gate is permanently tied high.

Troubleshooting

Section 2–10 Troubleshooting logic gates and circuits

40. The output of gate G2 in Figure 2–63 has an internal open circuit. What would a logic probe indicate at the output of G2?

41. What would the output of the circuit (Z) shown in Figure 2–63 be if the output of G2 was an open circuit and the three inputs to the circuit (A, B, and C) are logic lows.

42. If the output of gate G2 (refer to Figure 2–63) is an open circuit, discuss procedures that could be used to detect this fault.

43. The output of gate G3 in Figure 2–64 is always at the logic low state and is not affected by any change in logic levels at the inputs A and B. Where could the possible fault lie?

44. Discuss a troubleshooting procedure that could be used to identify the defective gate in Problem 43 without having to disconnect the connection between G3 and G2.

45. Discuss how a logic probe alone could be used to isolate the fault in the malfunctioning circuit described in Problem 43.

Figure 2–64

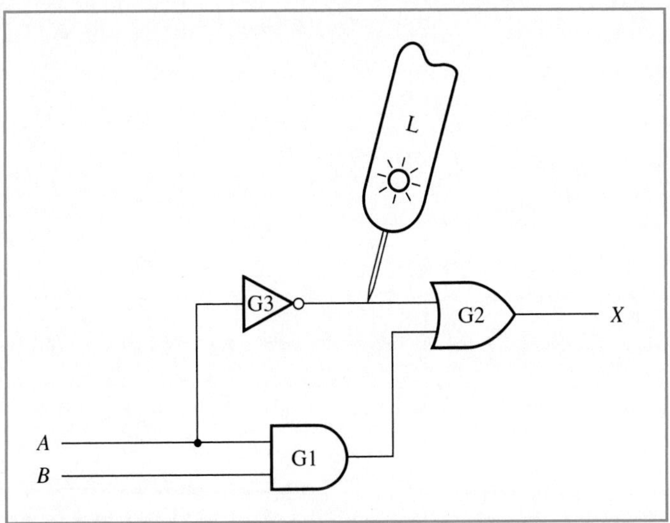

The English mathematician, George Boole (1815–1864) was the developer of Boolean Algebra, the mathematics of logic, which is the basis of all digital computers in existence today (The Bettman Archive).

OUTLINE

OBJECTIVES

The objectives of this chapter are to

- study the various techniques of analyzing logic circuits to obtain their logic equations and truth tables.
- introduce the concepts of minterms and maxterms.
- design logic circuits using the sum of minterms (sum of products) and product of maxterms (product of sums) forms.
- introduce and verify the basic laws of Boolean algebra.
- apply the laws of Boolean algebra to the simplification of logic equations and circuits.
- apply DeMorgan's law to convert from *AND-OR* logic circuits to *NAND* and *NOR* circuits.
- study the Karnaugh map techniques for simplifying logic equations and circuits.

3 THE ANALYSIS AND SYNTHESIS OF LOGIC CIRCUITS

3–1 INTRODUCTION

The study of digital circuits includes two areas that provide its foundations. These two areas of study are the analysis and synthesis of logic circuits. Techniques of analyzing logic circuits can be used to determine their function and to assist in electronic troubleshooting procedures. The synthesis of logic circuits encompasses techniques to design logic circuits for specific applications. The procedures for analysis and synthesis of logic circuits have opposite paths since the objective of synthesis is to obtain a circuit, given its specification, while the objective of analysis is to obtain its function (and indirectly its specifications), given the circuit.

The analysis and synthesis of logic circuits are based on the elementary principles and functions of logic circuits discussed in Chapter 2. We have seen that the behavior of gates and circuits can be described by mathematical logic equations. The manipulation of these equations also plays a vital part in the analysis and synthesis of logic circuits, and includes the study of a branch of discrete mathematics known as *Boolean algebra*. Using Boolean algebraic techniques we can greatly simplify the procedures for designing and analyzing logic circuits.

This chapter continues the study of logic circuits introduced in Chapter 2 to include other techniques of analysis and synthesis of logic circuits. The mathematics of Boolean algebra along with its application in design and analysis procedures is also included in this chapter. The material covered in this chapter then serves as a foundation for the study of the various application circuits that are designed and analyzed in Chapter 4.

3–2 Analysis of logic circuits

Chapter 2 introduced us to simple logic circuits and procedures for analyzing them in order to obtain their logic equations and truth tables. This section continues the study of the analysis of similar logic circuits using more formal techniques. A good understanding of the basic logic gates and their functions is an important prerequisite to the study of these techniques.

In order to obtain the logic equation for any logic circuit we must trace the basic logic functions through each gate in the circuit and then build the logic equation from these components. The manner in which this is done can be seen in Figure 3–1.

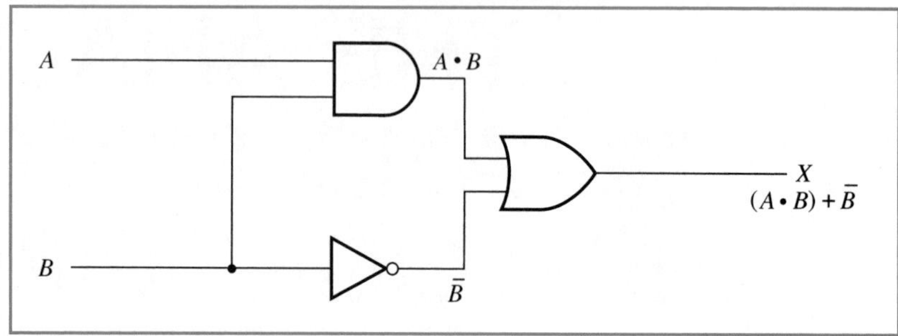

Figure 3–1
Trace to determine the logic equation of a circuit.

In order to obtain the logic equation for the circuit in Figure 3–1, we begin at the inputs of the circuit, and trace the independent variables through each gate. For example, since A and B are the inputs of the *AND* gate, the output of the *AND* gate is

$$A \cdot B$$

Similarly the output of the inverter is

$$\overline{B}$$

Notice that since the inputs to the *OR* gate are connected to the outputs of the *AND* gate and the inverter, the output of the *OR* gate and the output of the circuit is

$$(A \cdot B) + \overline{B}$$

Also, notice that since the term $(A \cdot B)$ is treated like a single variable that makes up one of the inputs of the *OR* gate, we choose to enclose the term in parentheses. In Section 3–4 we shall see that this is not always necessary. The parentheses enclosing the *AND* function also indicates that the *AND* function is evaluated before the *OR* function, that is, the *AND* gate will appear *before* the *OR* gate in the logic circuit. The logic equation for the circuit in Figure 3–1 is

$$X = (A \cdot B) + \overline{B}$$

Once the logic equation for the circuit has been determined we can now use one of two techniques to obtain its truth table. The first technique involves the application of the basic logic functions (*AND, OR, NOT*) to the logic equations. This is done by breaking down the logic equation into its basic logic components, obtaining the logic states of each one of these components as a function of the independent variables, and then determining the logic states of the dependent variable by combining these logic components appropriately. This procedure is illustrated in Table 3–1. The first two columns in Table 3–1 list all possible combinations of the independent variables, A and B. Since the logic equation for the circuit is made up of the two basic components or terms, $\overline{B}$ and $(A \cdot B)$, we have a column for each one of these terms. Finally since the logic equation for the circuit is simply the logical *OR* of these two terms, the last column represents the dependent variable X that represents the logic equation for the circuit.

To complete the truth table, we simply determine the logic states in each column starting with the left-most column. The column for $\overline{B}$ is obtained by complementing the column for B. The column for $A \cdot B$ is obtained by "*AND*ing" (applying the *AND* function to) the A and B columns. Finally, the column for $(A \cdot B) + \overline{B}$ (the dependent variable X) is obtained by "*OR*ing" (applying the *OR* function to) the columns for $\overline{B}$ and $(A \cdot B)$. The completed truth table is shown in Table 3–2.

An alternative technique for obtaining the truth table of a logic circuit is to apply all possible binary combinations to the inputs of the circuit and trace these logic states through the circuit. Figure 3–2 shows the logic states at the outputs of each gate in the circuit of Figure 3–1 for all possible combinations of inputs. The construction of the truth table for each trace is also included in Figure 3–2. Notice that we do not have to obtain the logic equation for the circuit first, as was the case with the previous technique of analysis.

The technique of obtaining a truth table for a logic circuit by tracing through all possible combinations of the inputs can often be quite time-consuming, especially when the circuit is fairly complex and has more than two input variables. However, in many logic circuits this process of analysis can be accelerated by recognizing certain configurations and applying the general rules of the basic *AND, OR,* and *NOT* functions. For example, the circuit in Figure 3–3 has four inputs and therefore we must trace through 16 combinations to obtain the truth table. However, notice that the output of the circuit, W, is actually the output of the *AND* gate, G4, and if any of the inputs of an *AND* gate is a logic 0, its output will be a logic 0 regardless of the states of the other inputs. Since one of the inputs of the *AND* gate G4, is connected to the input variable A, through the inverter, G3, the output of the *AND* gate G4 (which is also the output of the circuit) will be a 0 whenever the variable A is a 1. We can therefore complete 8 of the 16 truth table entries as follows:

	A	B	C	D	W
	0	0	0	0	
	0	0	0	1	
	0	0	1	0	
	0	0	1	1	
	0	1	0	0	
	0	1	0	1	
	0	1	1	0	
	0	1	1	1	
⟶	1	0	0	0	0
⟶	1	0	0	1	0
⟶	1	0	1	0	0
⟶	1	0	1	1	0
⟶	1	1	0	0	0
⟶	1	1	0	1	0
⟶	1	1	1	0	0
⟶	1	1	1	1	0

Similarly we can apply the same rule to the other inputs of the *AND* gate, G4. Since the second input of the *AND* gate, G4, is connected to the output of the *OR* gate, G2, this input will be a logic 0 if and only if both B and C are at the logic 0 state. Thus the output W will be a 0 if B and C are both 0s. We can therefore fill in all *remaining* truth table combinations where B and C are both at the logic 0 states:

Table 3–1
Truth table analysis for Figure 3–1

A	B	$\overline{B}$	A · B	X (A · B) + $\overline{B}$
0	0			
0	1			
1	0			
1	1			

Table 3–2
Completed truth table for Figure 3–1

A	B	$\overline{B}$	A · B	X (A · B) + $\overline{B}$
0	0	1	0	1
0	1	0	0	0
1	0	1	0	1
1	1	0	1	1

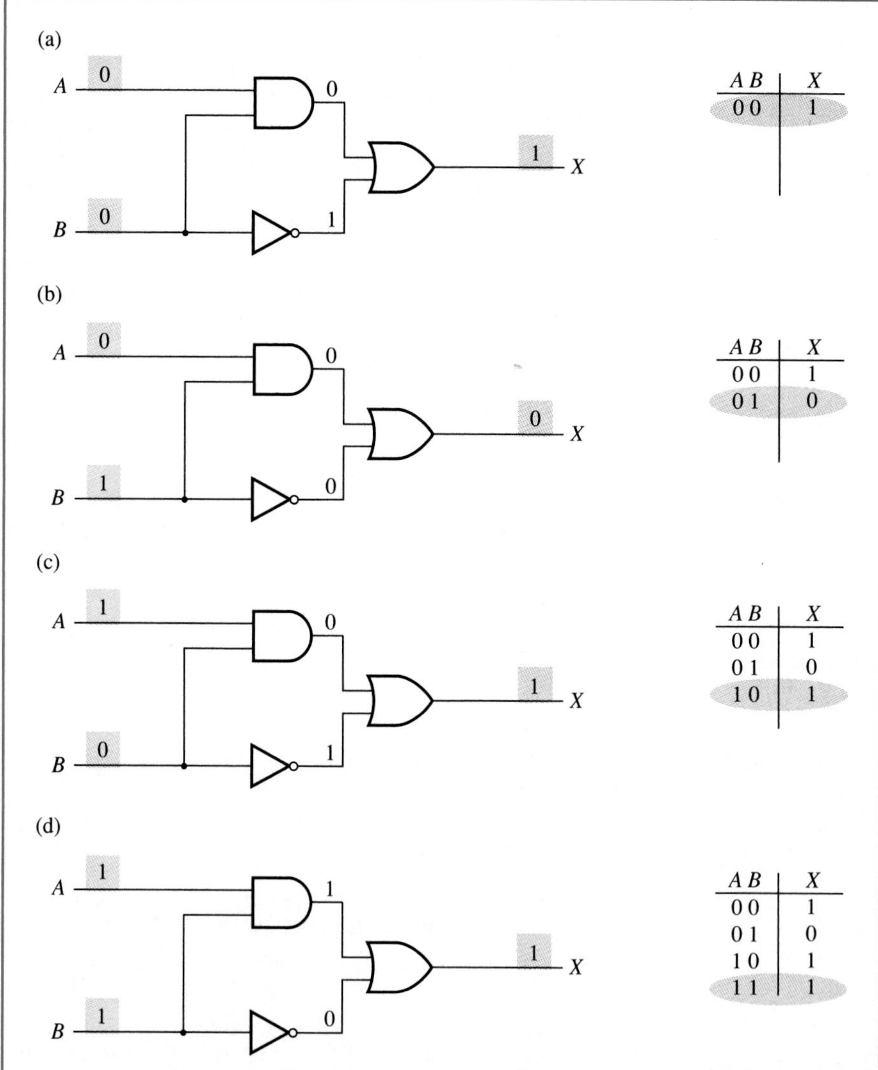

Figure 3–2
Tracing logic states through a circuit.

Figure 3–3

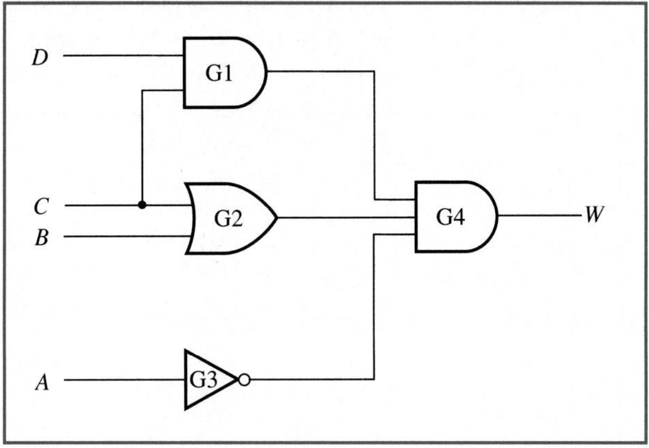

A	B	C	D	W
→ 0	0	0	0	0
→ 0	0	0	1	0
0	0	1	0	
0	0	1	1	
0	1	0	0	
0	1	0	1	
0	1	1	0	
0	1	1	1	
1	0	0	0	0
1	0	0	1	0
1	0	1	0	0
1	0	1	1	0
1	1	0	0	0
1	1	0	1	0
1	1	1	0	0
1	1	1	1	0

Finally, the third input of the *AND* gate, G4, will be a logic 0 if either *D* or *C* are at the logic 0 state since *D* and *C* are *AND*ed together by G1. We can therefore complete all remaining combinations where *D* or *C* are at the logic 0 state:

A	B	C	D	W
0	0	0	0	0
0	0	0	1	0
→ 0	0	1	0	0
0	0	1	1	
→ 0	1	0	0	0
→ 0	1	0	1	0
→ 0	1	1	0	0
0	1	1	1	
1	0	0	0	0
1	0	0	1	0
1	0	1	0	0
1	0	1	1	0
1	1	0	0	0
1	1	0	1	0
1	1	1	0	0
1	1	1	1	0

Since we have now looked at all possible conditions necessary to produce a 0 at one or more of the inputs of the *AND* gate, G4, the remaining conditions will produce logic 1s on all the inputs of G4; this can be verified by tracing the remaining two combinations through the circuit. The completed truth table is therefore

A	B	C	D	W
0	0	0	0	0
0	0	0	1	0
0	0	1	0	0
0	0	1	1	1
0	1	0	0	0
0	1	0	1	0
0	1	1	0	0
0	1	1	1	1
1	0	0	0	0
1	0	0	1	0
1	0	1	0	0
1	0	1	1	0
1	1	0	0	0
1	1	0	1	0
1	1	1	0	0
1	1	1	1	0

Thus we can use either of these techniques to obtain the truth table for a logic circuit. The first procedure obtains the truth table from the logic equation derived from the circuit, while the second procedure obtains the truth table directly from the circuit by tracing through various logic combinations. The decision as to which procedure to use depends on the configuration of the circuit and its complexity. However, both procedures can be used to check the results of the analysis. The following examples illustrate both techniques of analysis for various circuits.

EXAMPLE 3–1

The logic equation for the circuit shown in Figure 3–4 can be obtained by tracing the logic functions through each gate in the circuit. The logic function outputs of each gate are also shown in the circuit.

The output of the *AND* gate is

$$A \cdot \overline{B}$$

Figure 3–4

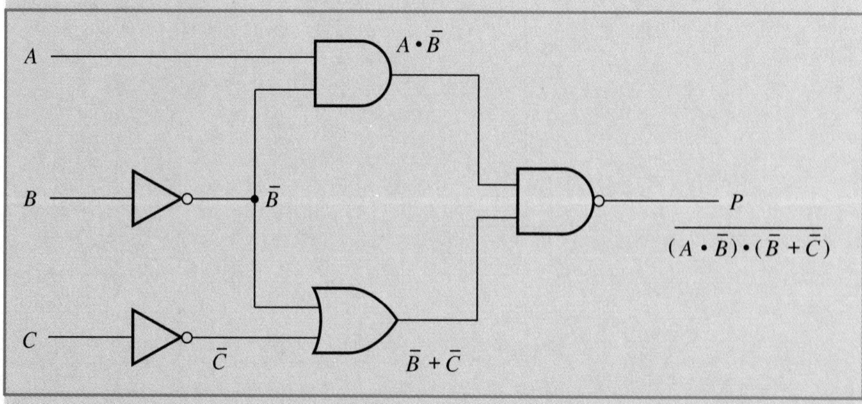

and the output of the *OR* gate is

$$\overline{B} + \overline{C}$$

Since these two terms (functions) are fed into the input of the *NAND* gate, they are treated like single variables (grouped in parentheses) and the logic equation for the circuit is

$$P = \overline{(A \cdot \overline{B}) \cdot (\overline{B} + \overline{C})}$$

We can now obtain the truth table for the circuit by breaking down its logic equation into its basic logic components as shown in Table 3–3.

Table 3–3
Truth table for Figure 3–4

A	B	C	$\overline{B}$	$\overline{C}$	$A \cdot \overline{B}$	$\overline{B} + \overline{C}$	P $\overline{(A \cdot \overline{B}) \cdot (\overline{B} + \overline{C})}$
0	0	0	1	1	0	1	1
0	0	1	1	0	0	1	1
0	1	0	0	1	0	1	1
0	1	1	0	0	0	0	1
1	0	0	1	1	1	1	0
1	0	1	1	0	1	1	0
1	1	0	0	1	0	1	1
1	1	1	0	0	0	0	1

We can also obtain the truth table for the circuit in Figure 3–4 by tracing through all possible binary combinations of the input variables *A*, *B*, and *C*. However, to simplify the process of analysis, we can make certain judgments based on the rules of the basic logic functions.

Notice that since the output of the circuit is a *NAND* gate, if any of its inputs is at the logic 0 state, the output will be a logic 1. Since one of the inputs of the *NAND* gate is connected to the output of the *OR* gate (whose inputs are connected to *B* and *C* through inverters), we can conclude that the output of the *OR* gate (also the input of the *NAND* gate) will be a logic 0 if and only if both *B* and *C* are at the logic 1 states. Thus the output, *P*, will be a logic 1 if both *B* and *C* are at the logic 1 states. We can therefore complete all entries in the following truth table that satisfy this condition

	A	B	C	P
	0	0	0	
	0	0	1	
	0	1	0	
⟶	0	1	1	1
	1	0	0	
	1	0	1	
	1	1	0	
⟶	1	1	1	1

We can also conclude that the remaining input of the *NAND* gate will be a logic 0 if either *A* is a logic 0 or if *B* is a logic 1, since this input is connected to an *AND* gate whose inputs are A and $\overline{B}$. Therefore we can complete all remaining entries in the truth table where P will be a logic 1

	A	B	C	P
⟶	0	0	0	1
⟶	0	0	1	1
⟶	0	1	0	1
	0	1	1	1
	1	0	0	
	1	0	1	
⟶	1	1	0	1
	1	1	1	1

Since we have satisfied all conditions necessary for the variable P to be a logic 1, the remaining combinations for A, B, and C will cause the output of the circuit to be a logic 0. This can be verified by tracing the remaining two combinations (100, 101) through the circuit. The final truth table is shown in Table 3–4.

Notice that the truth tables shown in Table 3–3 and Table 3–4 are identical with respect to the dependent variable P.

Table 3–4
Truth table for Figure 3–4

A	B	C	P
0	0	0	1
0	0	1	1
0	1	0	1
0	1	1	1
1	0	0	0
1	0	1	0
1	1	0	1
1	1	1	1

EXAMPLE 3–2

The logic equation for the circuit shown in Figure 3–5 can be obtained by tracing the logic functions through each gate in the circuit. The logic function outputs of each gate are also shown in the circuit.

The inputs to the four-input *OR* gate, G7, are

$$(\overline{P} \cdot Q), (\overline{Q + R}), \overline{R}, (\overline{R} + \overline{S})$$

As before, we have grouped the terms in parentheses since each term is treated as a single variable to obtain the final logic equation for the circuit

$$Z = (\overline{P} \cdot Q) + (\overline{Q + R}) + \overline{R} + (\overline{R} + \overline{S})$$

We can now obtain the truth table for the circuit by breaking down its logic equation into its basic logic components as shown in Table 3–5.

We can also obtain the truth table for the circuit in Figure 3–5 by tracing through all possible binary combinations of the input variables P, Q, R, and S. However, to simplify the process of analysis, we can make certain judgments based on the rules of the basic logic functions.

Figure 3–5

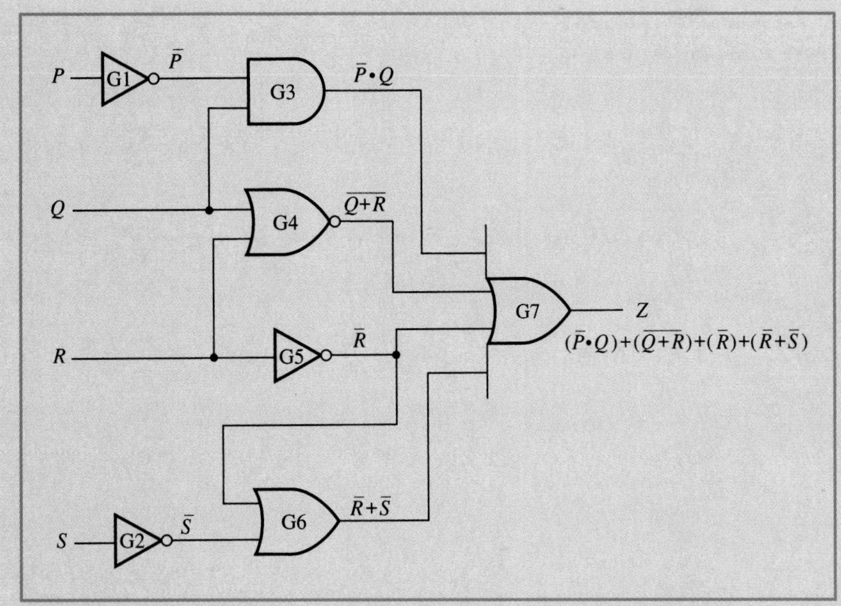

Notice that since the output of the circuit is the *OR* gate, G7, if any of its inputs are at the logic 1 state, the output will be a logic 1. Since one of the inputs of G7 is connected to the output of the *AND* gate, G3 (whose inputs are connected to $\bar{P}$ and Q), we can conclude that the output of the *AND* gate, G3 (also the input of the *OR* gate), will be a logic 1 if and only if P is at the logic 0 state and Q is a logic 1. Thus the output, Z, will be a logic 1 if P is a logic 0 and Q is a logic 1. We can therefore complete all entries in the following truth table that satisfy this condition:

	P	Q	R	S	Z
	0	0	0	0	
	0	0	0	1	
	0	0	1	0	
	0	0	1	1	
⟶	0	1	0	0	1
⟶	0	1	0	1	1
⟶	0	1	1	0	1
⟶	0	1	1	1	1
	1	0	0	0	
	1	0	0	1	
	1	0	1	0	
	1	0	1	1	
	1	1	0	0	
	1	1	0	1	
	1	1	1	0	
	1	1	1	1	

The output of the *NOR* gate, G4, also feeds into the input of the *OR* gate, G7. This *NOR* gate will produce a logic 1 at its output if and only if both Q and R are at the logic 0 state. Therefore the output of the circuit, Z, will be a logic 1 when both Q and R are logic 0s.

Table 3–5
Truth table for Figure 3–5

P	Q	R	S	$\overline{P}$	$\overline{S}$	$\overline{P} \cdot Q$	$\overline{Q + R}$	$\overline{R}$	$\overline{R} + \overline{S}$	Z
0	0	0	0	1	1	0	1	1	1	1
0	0	0	1	1	0	0	1	1	1	1
0	0	1	0	1	1	0	0	0	1	1
0	0	1	1	1	0	0	0	0	0	0
0	1	0	0	1	1	1	0	1	1	1
0	1	0	1	1	0	1	0	1	1	1
0	1	1	0	1	1	1	0	0	1	1
0	1	1	1	1	0	1	0	0	0	1
1	0	0	0	0	1	0	1	1	1	1
1	0	0	1	0	0	0	1	1	1	1
1	0	1	0	0	1	0	0	0	1	1
1	0	1	1	0	0	0	0	0	0	0
1	1	0	0	0	1	0	0	1	1	1
1	1	0	1	0	0	0	0	1	1	1
1	1	1	0	0	1	0	0	0	1	1
1	1	1	1	0	0	0	0	0	0	0

	P	Q	R	S	Z
→	0	0	0	0	1
→	0	0	0	1	1
	0	0	1	0	
	0	0	1	1	
	0	1	0	0	1
	0	1	0	1	1
	0	1	1	0	1
	0	1	1	1	1
→	1	0	0	0	1
→	1	0	0	1	1
	1	0	1	0	
	1	0	1	1	
	1	1	0	0	
	1	1	0	1	
	1	1	1	0	
	1	1	1	1	

The third input of the *OR* gate, G7, is connected to R through the inverter, G5. Therefore the output of the circuit, Z, will be a logic 1 if R is a logic 0:

	P	Q	R	S	Z
	0	0	0	0	1
	0	0	0	1	1
	0	0	1	0	
	0	0	1	1	
	0	1	0	0	1
	0	1	0	1	1
	0	1	1	0	1
	0	1	1	1	1
	1	0	0	0	1
	1	0	0	1	1
	1	0	1	0	
	1	0	1	1	
→	1	1	0	0	1
→	1	1	0	1	1
	1	1	1	0	
	1	1	1	1	

Finally, notice that the fourth input of the *OR* gate, G7, is connected to another *OR* gate, G6, whose output will be a logic 1 if either *R* or *S* is at the logic 0 state (since *R* and *S* are inverted before being connected to the input of this *OR* gate)

	P	Q	R	S	Z
	0	0	0	0	1
	0	0	0	1	1
→	0	0	1	0	1
	0	0	1	1	
	0	1	0	0	1
	0	1	0	1	1
	0	1	1	0	1
	0	1	1	1	1
	1	0	0	0	1
	1	0	0	1	1
→	1	0	1	0	1
	1	0	1	1	
	1	1	0	0	1
	1	1	0	1	1
→	1	1	1	0	1
	1	1	1	1	

Since we have satisfied all conditions necessary for the output of the circuit, *Z*, to be a logic 1, the remaining combinations for *P*, *Q*, *R*, and *S* (0011, 1011, 1111) will cause the output of the circuit to be a logic 0. This can be verified by tracing the remaining three combinations through the circuit. The final truth table is shown in Table 3–6.

Notice that the truth table shown in Tables 3–5 and 3–6 are identical with respect to the dependent variable *Z*.

Table 3–6
Truth table for Figure 3–5

P	Q	R	S	Z
0	0	0	0	1
0	0	0	1	1
0	0	1	0	1
0	0	1	1	0
0	1	0	0	1
0	1	0	1	1
0	1	1	0	1
0	1	1	1	1
1	0	0	0	1
1	0	0	1	1
1	0	1	0	1
1	0	1	1	0
1	1	0	0	1
1	1	0	1	1
1	1	1	0	1
1	1	1	1	0

This section has examined procedures for analyzing various logic circuits. The objective of analysis is to determine the function or operation of a logic circuit by obtaining its truth table, or logic equation, or both. The analysis of logic circuits plays an important role in the study of digital circuits since these procedures also allow us to simplify the design of existing logic circuits and to troubleshoot logic circuits. The analysis procedures can also be applied to check the proper design of logic circuits, as will be seen in the next section.

Review questions

1. What is the purpose of analyzing logic circuits?
2. What does the analysis of a logic circuit involve?
3. Describe the two techniques used in the analysis of logic circuits.
4. Why do we use parentheses in some of the logic equations obtained from a circuit?

3–3 Synthesis of logic circuits

The synthesis of logic circuits involves designing or "putting together" logic circuits from a given specification, truth table, or logic equation. Synthesis, therefore, is the opposite of analysis. In Chapter 2 we examined several simple examples of synthesizing logic circuits from logic equations. In this section we will investigate the procedures for obtaining logic equations from truth tables; these logic equations will then be used to produce the logic circuits that represent the equations and truth tables.

Before we attempt to learn the techniques for designing logic circuits, let us first identify certain characteristics of truth tables. Table 3–7 includes all possible combinations of two independent variables, X and Y. The first column (labeled "number") in the table lists the decimal equivalent of each combination.

The fourth column in Table 3–7 lists the *minterms* for each combination. Each minterm in the table will produce a logic 1 when its corresponding combination is applied to it, and a logic 0 for all other combinations. This can be verified by drawing the logic circuits for each of the minterms shown in Table 3–7 and then applying all possible combinations to each circuit. Figure 3–6 illustrates these circuits.

Figure 3–6a represents the circuit for the minterm, $\overline{X} \cdot \overline{Y}$. Notice that the output of the circuit will be a logic 1 only when the combination 00 is applied to it. All other combinations will produce a logic 0 at the output. Similarly, the circuit in Figure 3–6b represents the circuit for the minterm $\overline{X} \cdot Y$, whose output will be a logic 1 if and only if the combination 01 is applied to it. Figure 3–6c and d show the circuits for the remaining two minterms $X \cdot \overline{Y}$ and $X \cdot Y$, whose outputs will be a logic 1 for the combinations 10 and 11, respectively.

A *minterm* is therefore a logic expression (or term) that will produce a logic 1 for one and only one combination. Notice that this combination very closely resembles the minterm. For example, in Table 3–7, notice that the variables in each minterm are *barred* (complemented) if its value

Table 3–7
Two-variable minterms and maxterms

Number	X	Y	Minterms	Maxterms
0	0	0	$\overline{X} \cdot \overline{Y}$	$X + Y$
1	0	1	$\overline{X} \cdot Y$	$X + \overline{Y}$
2	1	0	$X \cdot \overline{Y}$	$\overline{X} + Y$
3	1	1	$X \cdot Y$	$\overline{X} + \overline{Y}$

Figure 3–6
Minterm circuits.

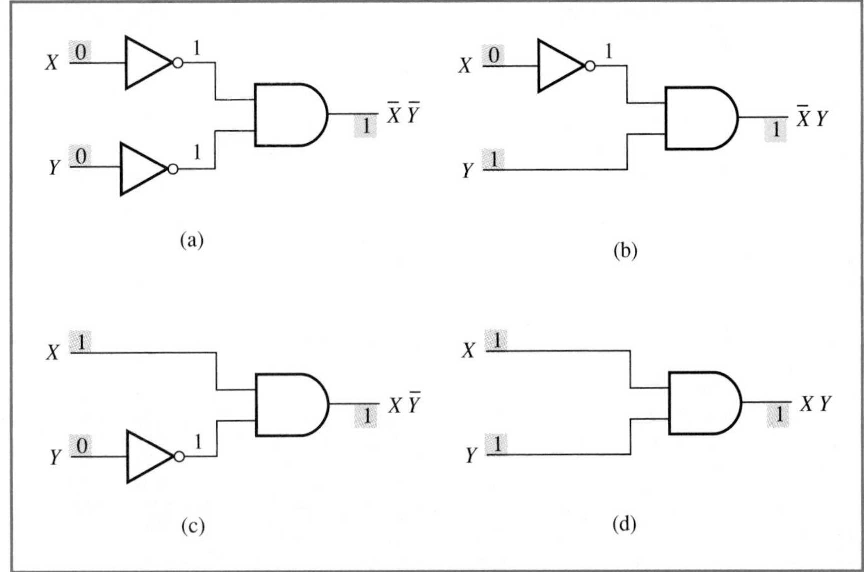

in its corresponding combination is a logic 0

minterm:	$\overline{X} \cdot \overline{Y}$	$\overline{X} \cdot Y$	$X \cdot \overline{Y}$	$X \cdot Y$
combination:	0 0	0 1	1 0	1 1

A minterm can therefore be used to *detect* a certain combination by producing a logic 1 for that combination only, and a logic 0 for all other combinations. We often refer to minterms by number. This number is the decimal (or hexadecimal) equivalent of the combination that the minterm represents. For example, with reference to Table 3–7, minterm 0 is $\overline{X} \cdot \overline{Y}$, minterm 1 is $\overline{X} \cdot Y$, and so on. Because we use the "·" to represent the multiplication operator in conventional algebra, minterms are often referred to as *logical products*.

A *maxterm* is the complement of a minterm. This fact will be proved in the next section. A maxterm will produce a *logic 0* for one and only one combination. The fifth column in Table 3–7 lists the maxterms for all possible combinations of X and Y. Figure 3–7 shows the four circuits that represent the maxterms along with the unique combinations that produce a logic 0 for each circuit.

Notice in Figure 3–7a that if any combination other than a 00 is applied to the inputs of the circuit, the output will be a logic 1. Therefore the circuit will produce a logic 0 if and only if the combination applied to its inputs is 00. Thus the maxterm $X + Y$ represents the combination 00. Similarly, the circuits in Figure 3–7b, c, and d represent the maxterms $X + \overline{Y}$, $\overline{X} + Y$, and $\overline{X} + \overline{Y}$, respectively, which correspond to the combinations 01, 10, and 11, respectively.

A maxterm is therefore a logic expression (or term) that will produce a logic 0 for one and only one combination. Notice that this combination very closely resembles the maxterm. For example, in Table 3–7 notice that the variables in each maxterm are *barred* (complemented) if its value in its corresponding combination is a logic 1

maxterm:	$X + Y$	$X + \overline{Y}$	$\overline{X} + Y$	$\overline{X} + \overline{Y}$
combination:	0 0	0 1	1 0	1 1

Figure 3–7
Maxterm circuits.

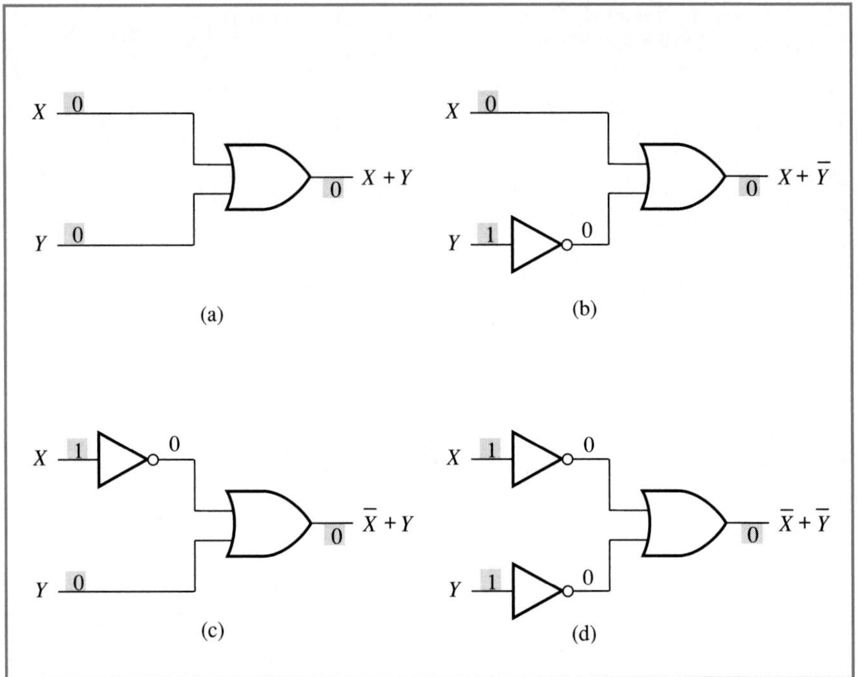

Like a minterm, a maxterm can also be used to *detect* a certain combination by producing a logic 0 for that combination only, and a logic 1 for all other combinations. Maxterms can also be referred to by a number that is the decimal (or hexadecimal) equivalent of the combination that the maxterm represents. For example, with reference to Table 3–7, maxterm 0 is $X + Y$, maxterm 1 is $X + \overline{Y}$, and so on. Since we use the "+" to represent the addition operator in conventional algebra, maxterms are often referred to as *logical sums*.

We can obtain minterms and maxterms for combinations of any size. Tables 3–8 and 3–9 list the minterms and maxterms for all possible combinations of 3 and 4 bits, respectively.

The logic equation for a truth table can be obtained in two forms—*sum-of-products* (SOP) and *product-of-sums* (POS). For example, consider the truth table shown in Table 3–10. Since the combinations 01 and 10 produce logic 1's for the dependent variable X, we can obtain the logic equation for the circuit by "*OR*ing" the minterms for these two combinations

Equation (3–1)

$$X = (\overline{A} \cdot B) + (A \cdot \overline{B})$$

Table 3–8
Three-variable minterms and maxterms

Number	X	Y	Z	Minterms	Maxterms
0	0	0	0	$\overline{X} \cdot \overline{Y} \cdot \overline{Z}$	$X + Y + Z$
1	0	0	1	$\overline{X} \cdot \overline{Y} \cdot Z$	$X + Y + \overline{Z}$
2	0	1	0	$\overline{X} \cdot Y \cdot \overline{Z}$	$X + \overline{Y} + Z$
3	0	1	1	$\overline{X} \cdot Y \cdot Z$	$X + \overline{Y} + \overline{Z}$
4	1	0	0	$X \cdot \overline{Y} \cdot \overline{Z}$	$\overline{X} + Y + Z$
5	1	0	1	$X \cdot \overline{Y} \cdot Z$	$\overline{X} + Y + \overline{Z}$
6	1	1	0	$X \cdot Y \cdot \overline{Z}$	$\overline{X} + \overline{Y} + Z$
7	1	1	1	$X \cdot Y \cdot Z$	$\overline{X} + \overline{Y} + \overline{Z}$

Table 3–9
Four-variable minterms and maxterms

Number	W	X	Y	Z	Minterms	Maxterms
0	0	0	0	0	$\overline{W}\cdot\overline{X}\cdot\overline{Y}\cdot\overline{Z}$	$W + X + Y + Z$
1	0	0	0	1	$\overline{W}\cdot\overline{X}\cdot\overline{Y}\cdot Z$	$W + X + Y + \overline{Z}$
2	0	0	1	0	$\overline{W}\cdot\overline{X}\cdot Y\cdot\overline{Z}$	$W + X + \overline{Y} + Z$
3	0	0	1	1	$\overline{W}\cdot\overline{X}\cdot Y\cdot Z$	$W + X + \overline{Y} + \overline{Z}$
4	0	1	0	0	$\overline{W}\cdot X\cdot\overline{Y}\cdot\overline{Z}$	$W + \overline{X} + Y + Z$
5	0	1	0	1	$\overline{W}\cdot X\cdot\overline{Y}\cdot Z$	$W + \overline{X} + Y + \overline{Z}$
6	0	1	1	0	$\overline{W}\cdot X\cdot Y\cdot\overline{Z}$	$W + \overline{X} + \overline{Y} + Z$
7	0	1	1	1	$\overline{W}\cdot X\cdot Y\cdot Z$	$W + \overline{X} + \overline{Y} + \overline{Z}$
8	1	0	0	0	$W\cdot\overline{X}\cdot\overline{Y}\cdot\overline{Z}$	$\overline{W} + X + Y + Z$
9	1	0	0	1	$W\cdot\overline{X}\cdot\overline{Y}\cdot Z$	$\overline{W} + X + Y + \overline{Z}$
A	1	0	1	0	$W\cdot\overline{X}\cdot Y\cdot\overline{Z}$	$\overline{W} + X + \overline{Y} + Z$
B	1	0	1	1	$W\cdot\overline{X}\cdot Y\cdot Z$	$\overline{W} + X + \overline{Y} + \overline{Z}$
C	1	1	0	0	$W\cdot X\cdot\overline{Y}\cdot\overline{Z}$	$\overline{W} + \overline{X} + Y + Z$
D	1	1	0	1	$W\cdot X\cdot\overline{Y}\cdot Z$	$\overline{W} + \overline{X} + Y + \overline{Z}$
E	1	1	1	0	$W\cdot X\cdot Y\cdot\overline{Z}$	$\overline{W} + \overline{X} + \overline{Y} + Z$
F	1	1	1	1	$W\cdot X\cdot Y\cdot Z$	$\overline{W} + \overline{X} + \overline{Y} + \overline{Z}$

Notice that since we have *OR*ed the two minterms, the variable X will only be a 1 when either the combination 01 or the combination 10 is applied to the circuit.

Since minterms are known as logical products and since we have taken the logical sum of these products, this equation is said to be in *sum-of-products form*. The circuit for Equation 3–1 is shown in Figure 3–8.

We can prove that Equation 3–1 represents Table 3–10 by analyzing the circuit in Figure 3–8 as follows:

A	B	$\overline{A}$	$\overline{B}$	$\overline{A}\cdot B$	$A\cdot\overline{B}$	X
0	0	1	1	0	0	0
0	1	1	0	1	0	1
1	0	0	1	0	1	1
1	1	0	0	0	0	0

It is also possible to obtain the logic equation and circuit for the truth table in Table 3–10 by "*AND*ing" the maxterms for combinations that produce logic 0s for the dependent variable X. Since the two combinations 00 and 11 produce logic 0's, the logic equation for the circuit could also be expressed as

Equation (3–2)
$$X = (A + B)\cdot(\overline{A} + \overline{B})$$

Notice that since we have *AND*ed the two maxterms, the variable X will only be a 0 when either the combination 00 or the combination 11 is applied to the circuit.

Since maxterms are known as logical sums and since we have taken the logical product of these sums, this equation is said to be in *product-of-sums form*. The circuit for Equation 3–2 is shown in Figure 3–9.

Table 3–10

A	B	X
0	0	0
0	1	1
1	0	1
1	1	0

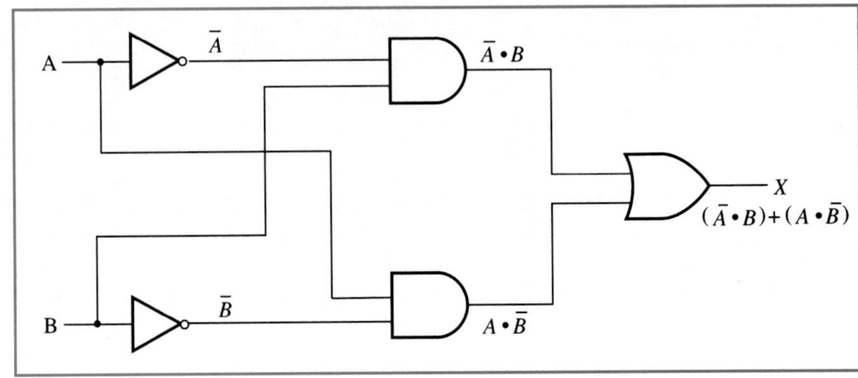

Figure 3–8 Circuit obtained from sum-of-products equation.

We can also prove that Equation 3–2 represents Table 3–10 by analyzing the circuit in Figure 3–9 as follows:

A	B	$\bar{A}$	$\bar{B}$	A + B	$\bar{A} + \bar{B}$	X
0	0	1	1	0	1	0
0	1	1	0	1	1	1
1	0	0	1	1	1	1
1	1	0	0	1	0	0

Both the sum-of-products form (Equation 3–1) and the product-of-sums form (Equation 3–2) yield the same truth table, thus proving that the circuits of Figures 3–8 and 3–9 are logically equivalent. In this example, both the circuits are the same in complexity (but not in configuration) and have the same number of gates. However, this may not be true for all truth tables as can be seen in the following examples.

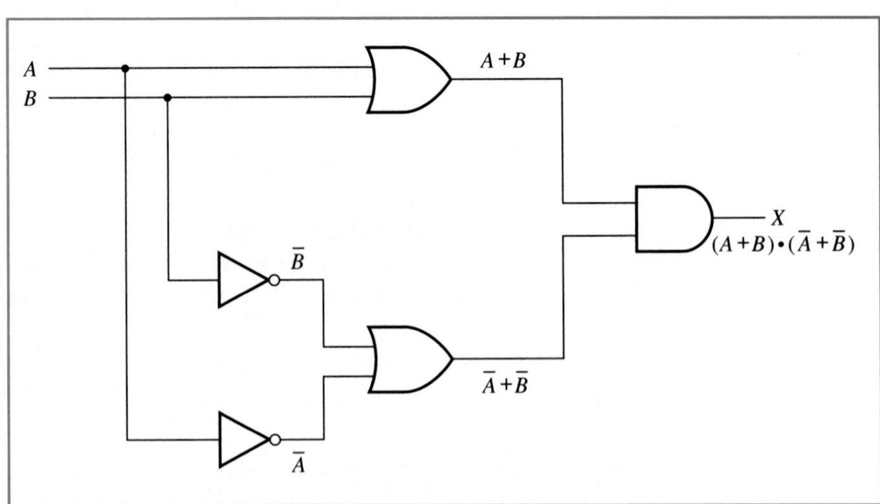

Figure 3–9 Circuit obtained from product-of-sums equation.

EXAMPLE 3–3

Obtain the logic equation and logic circuit for the following truth table in (a) sum-of-products form and (b) product-of-sums form.

P	Q	R	S
0	0	0	1
0	0	1	0
0	1	0	0
0	1	1	0
1	0	0	1
1	0	1	1
1	1	0	1
1	1	1	1

SOLUTION

(a) The logic equation for the truth table in sum-of-products form is

Equation (3–3)

$$S = (\overline{P} \cdot \overline{Q} \cdot \overline{R}) + (P \cdot \overline{Q} \cdot \overline{R}) + (P \cdot \overline{Q} \cdot R) + (P \cdot Q \cdot \overline{R}) + (P \cdot Q \cdot R)$$

The logic circuit for Equation 3–3 is shown in Figure 3–10.

(b) The logic equation for the truth table in product-of-sums form is

Equation (3–4)

$$S = (P + Q + \overline{R}) \cdot (P + \overline{Q} + R) \cdot (P + \overline{Q} + \overline{R})$$

The logic circuit for Equation 3–4 is shown in Figure 3–11.

The equations obtained in Example 3–3 (Equations 3–3 and 3–4) are logically equivalent and therefore the circuits in Figures 3–10 and 3–11 are also logically equivalent. However, notice that the circuit in Figure 3–10 is more complex in construction and requires many more gates than the circuit in Figure 3–11.

Figure 3–10

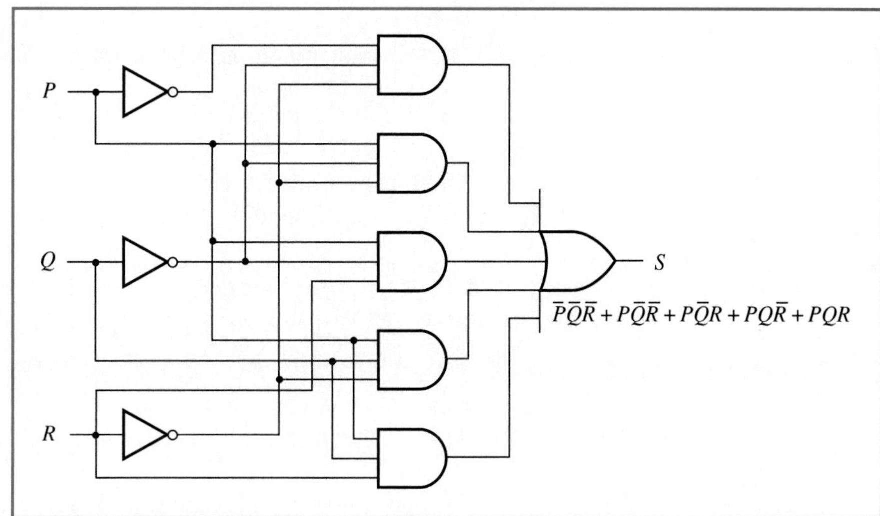

Figure 3–11

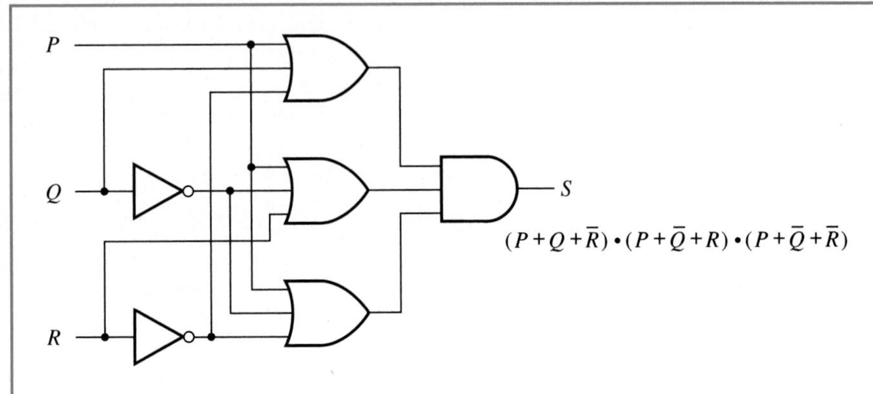

EXAMPLE 3–4

Obtain the logic equation and logic circuit for the following truth table in (a) sum-of-products form and (b) product-of-sums form.

A	B	C	D	X
0	0	0	0	0
0	0	0	1	1
0	0	1	0	0
0	0	1	1	1
0	1	0	0	0
0	1	0	1	0
0	1	1	0	1
0	1	1	1	0
1	0	0	0	0
1	0	0	1	1
1	0	1	0	1
1	0	1	1	0
1	1	0	0	0
1	1	0	1	0
1	1	1	0	0
1	1	1	1	1

SOLUTION

(a) The logic equation for the truth table in sum-of-products form is

Equation (3–5)

$$X = (\overline{A}\cdot\overline{B}\cdot\overline{C}\cdot D) + (\overline{A}\cdot\overline{B}\cdot C\cdot D) + (\overline{A}\cdot B\cdot C\cdot\overline{D}) + (A\cdot\overline{B}\cdot\overline{C}\cdot D) + (A\cdot\overline{B}\cdot C\cdot\overline{D}) + (A\cdot B\cdot C\cdot D)$$

The logic circuit for Equation 3–5 is shown in Figure 3–12.

(b) The logic equation for the truth table in product-of-sums form is

Equation (3–6)

$$X = (A + B + C + D)\cdot(A + B + \overline{C} + D)\cdot(A + \overline{B} + C + D)\cdot (A + \overline{B} + C + \overline{D})\cdot(A + \overline{B} + \overline{C} + \overline{D})\cdot (\overline{A} + B + C + D)\cdot(\overline{A} + B + \overline{C} + \overline{D})\cdot (\overline{A} + \overline{B} + C + D)\cdot(\overline{A} + \overline{B} + C + \overline{D})\cdot (\overline{A} + \overline{B} + \overline{C} + D)$$

The logic circuit for Equation 3–6 is shown in Figure 3–13.

Again, notice in Example 3–4 that even though Equations 3–5 and 3–6 and the circuits in Figures 3–12 and 3–13 are logically equivalent, the

Figure 3–12

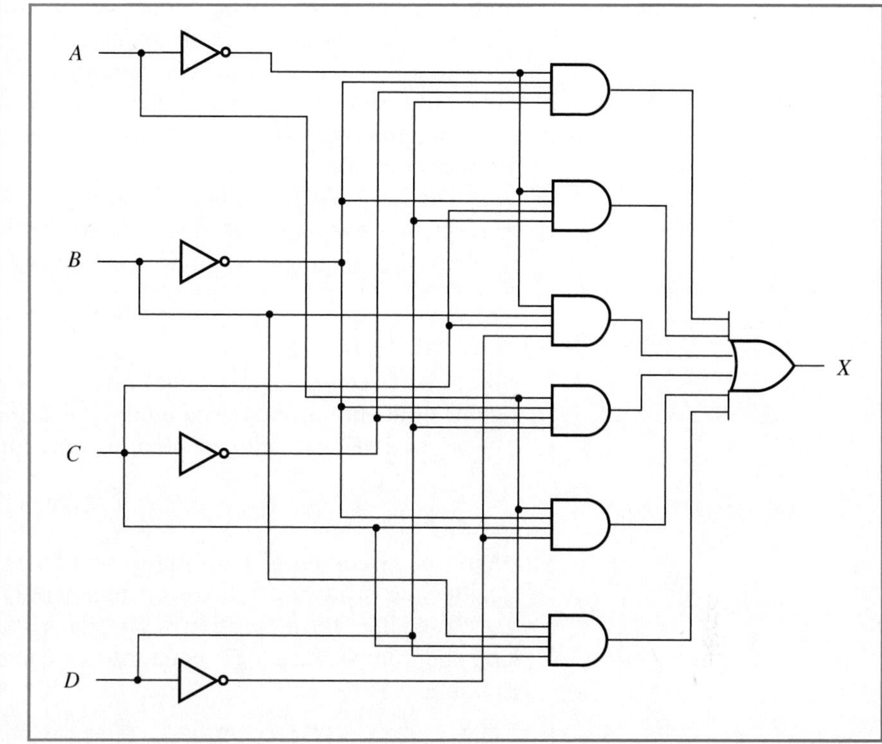

Figure 3–13

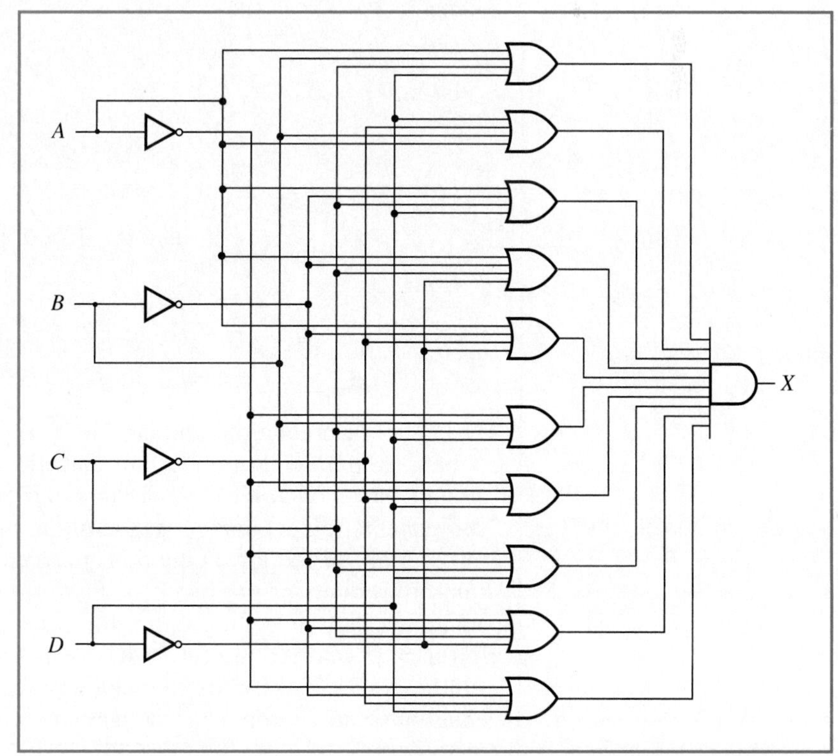

circuit in Figure 3–13 is more complex in construction than the circuit in Figure 3–12. It is therefore apparent that when the dependent variable in a truth table has more logic 0's than logic 1's, the sum-of-products form will yield a simpler circuit. However, when the dependent variable in a truth table has more logic 1's than logic 0's, the product-of-sums form will produce a simpler circuit.

The logic equations derived so far can also be expressed in a more convenient *short-form* notation in terms of the minterm and maxterm numbers. For example, Equation 3–1 can be expressed as

$$X = \Sigma_m\ 1, 2$$

where the Greek letter Σ (sigma) represents a sum and the letter m signifies the sum of minterms numbered 1 and 2.

Equation 3–2 is represented in short-form notation as follows:

$$X = \Pi_M\ 0, 3$$

where the Greek letter Π (pi) represents a product and the letter M signifies the product of maxterms numbered 0 and 3.

The short-form notation is particularly useful for representing very long and complicated logic equations as can be seen in the following example.

EXAMPLE 3–5

The short-form notation for Equations 3–3 through 3–6 can be obtained by determining the appropriate minterm and maxterm numbers from their truth tables that produce 1's and 0's, respectively. Note that the minterm and maxterm numbers are shown in hexadecimal, but can also be expressed in decimal.

Equation 3–3:

$$S = \Sigma_m\ 0, 4, 5, 6, 7$$

Equation 3–4:

$$S = \Pi_M\ 1, 2, 3$$

Equation 3–5:

$$X = \Sigma_m\ 1, 3, 6, 9, A, F$$

Equation 3–6:

$$X = \Pi_M\ 0, 2, 4, 5, 7, 8, B, C, D, E$$

The logic equation and circuits derived in this section are considered to be *unsimplified*. An important consideration in the design of digital circuits is to construct logic circuits with a minimum number of components (gates) and consequently a minimum number of ICs. Most (but not all) of the logic equations derived directly from a truth table in sum-of-products or product-of-sums form can be *simplified* into equivalent equations that yield the same truth table and produce circuits that have fewer gates and require fewer ICs. It is therefore important to extend the synthesis of logic circuits to include procedures to simplify logic equations and circuits. The remainder of this chapter deals with such procedures.

Review questions

1. What is the difference between the procedures of analysis and synthesis of logic circuits?
2. What does the synthesis of a logic circuit involve?
3. What are minterms and maxterms?
4. What two forms can the equations derived from a truth table take?
5. What is the purpose of having a short-form notation for expressing logic equations?

3–4 Boolean algebra

In the previous sections and in Chapter 2 we can see that the logic equations that represent logic circuits very closely resemble conventional algebraic equations. In fact, these logic equations contain some of the same operators (+ and ·) that we use in conventional algebra. The reason for this similarity in operators and format is because we can manipulate these logic equations and expressions in much the same way as we manipulate conventional algebraic equations and expressions. The mathematics of manipulating logic equations and expressions is called *Boolean algebra* and was developed by the English mathematician George Boole around 1847. Boolean algebra, or the "algebra of logic" is a very natural system for representing logic circuits even though it was developed much before the invention of electronic switching circuits. In this section we will study the various laws, theorems, and postulates of Boolean algebra that will allow us to manipulate and simplify logic equations and expressions so that the resulting circuits can be constructed with a minimum number of gates and ICs.

In Boolean algebra there are certain assumptions that we must make based on the basic logic functions discussed previously—*AND*, *OR*, and *NOT*. These assumptions are known as *postulates* and will be described for each one of the three basic logic functions.

There are two postulates for the *NOT* function

$$\overline{1} = 0 \quad \text{and} \quad \overline{0} = 1$$

The two postulates simply describe the operation of the inverter and read "*NOT* 1 is equal to 0" and "*NOT* 0 is equal to 1." In other words, just as we can complement a variable, we can also complement a logic state, since a variable actually represents a logic state.

A double inverted variable will always produce the variable. Thus

$$\overline{\overline{X}} = X$$

There are four postulates for the *AND* function. These four postulates are often referred to as *logical multiplication* since the *AND operator* " · " is also the multiplication operator in conventional mathematics.

$$0 \cdot 0 = 0 \qquad 0 \cdot 1 = 0$$
$$1 \cdot 0 = 0 \qquad 1 \cdot 1 = 1$$

Notice that the four *AND* postulates are simply derived from the truth table of the *AND* gate which simply states that "anything" *AND*ed with a 0 will produce a 0.

There are also four postulates for the *OR* function. These four postulates are often referred to as *logical addition* since the *OR operator* " + " is also the addition operator in conventional mathematics.

$$0 + 0 = 0 \qquad 0 + 1 = 1$$
$$1 + 0 = 1 \qquad 1 + 1 = 1$$

Notice that the four *OR* postulates are simply derived from the truth table of the *OR* gate, which simply states that "anything" *OR*ed with a 1 will produce a 1.

In conventional arithmetic and algebra we use certain rules of precedence in evaluating expressions. For example

$$3 \cdot 2 + 5 = 11$$

since we multiply 3 and 2 *first* and then add the result (6) to 5 for a final result of 11. Therefore the multiplication function has a higher precedence than the addition function. If we wanted to add 2 and 5 and multiply the result (7) by 3 for a final result of 21 we would use parentheses to force precedence of evaluation in the expression as follows:

$$3 \cdot (2 + 5) = 21$$

We use the same rules of precedence for the logical operators *AND* (·) and *OR* (+), as for the arithmetic operators *MULTIPLY* (·) and *ADD* (+). That is, in a logic expression containing both types of operators, the *AND* function is always evaluated first. The same rule applies to the construction of logic circuits. Thus the following logic equation:

Equation (3–7)

$$X = A \cdot B + C$$

is the same as

Equation (3–8)

$$X = (A \cdot B) + C$$

since the *AND* function has a higher precedence than the *OR* function. In the logic circuit for Equation 3–7, *A* and *B* would be "*AND*ed" first and then "*OR*ed" with *C* as shown in Figure 3–14. Parentheses can be used to force precedence of evaluation. For example, in Equation 3–7, if we wanted to "*OR*" *B* and *C* first and then "*AND*" the result with *A*, the logic equation would be

Equation (3–9)

$$X = A \cdot (B + C)$$

Notice in Equation 3–9 that we must enclose $B + C$ in parentheses to force precedence since the *OR* function has a lower precedence than the

Figure 3–14
Circuit for Equation 3–7.

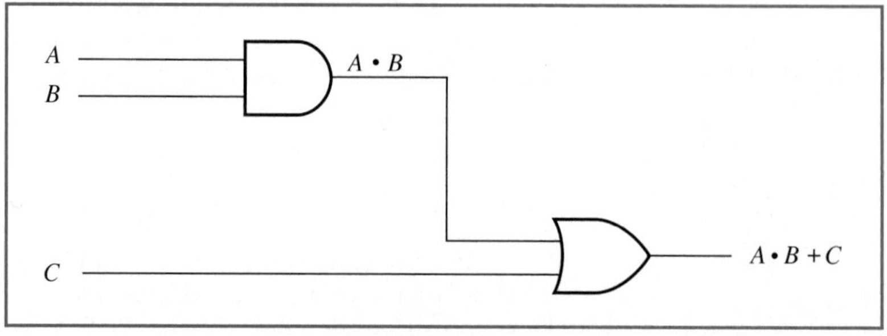

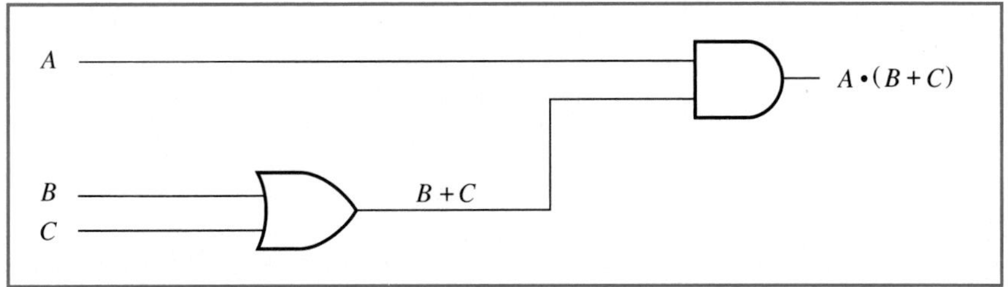

Figure 3–15
Circuit for Equation 3–9.

AND function. The circuit for Equation 3–9 is shown in Figure 3–15. Equation 3–9 is *not* logically equivalent to Equation 3–7.

Terms that are enclosed in parentheses can be treated as single variables. For example, if we define the term $(B + C)$ in Equation 3–9 to be represented by the variable, P, then the equation becomes

$$X = A \cdot P$$

which identifies the output gate (outermost gate) of the circuit as an *AND* gate.

The *NAND* and *NOR* functions have the same precedence as terms enclosed in parentheses. Thus the expression

$$\overline{A \cdot B} + \overline{A + B}$$

is equivalent to

$$(\overline{A \cdot B}) + (\overline{A + B})$$

The *NOT* operator always has the highest precedence. That is, inversions are always done first, before the *AND* and *OR* functions.

As in conventional algebra for convenience, we often choose to drop the " · " operator when dealing with logic expressions. Thus the following logic equation

$$Z = P \cdot \overline{Q} \cdot R + \overline{P} \cdot Q \cdot \overline{R}$$

can also be written as

$$Z = P\overline{Q}R + \overline{P}Q\overline{R}$$

We will be using the latter form for most of the logic equations that follow.

EXAMPLE 3–6

In each of the following logic equations, determine the value of the dependent variable for the following values of the independent variables:

$$A = 0 \quad B = 1 \quad C = 1 \quad D = 0$$

(a) $X = A\overline{B} + AD$ (b) $Y = A(B + \overline{C})$ (c) $Z = B(\overline{A} + \overline{C})C$
(d) $P = 1 + \overline{D}B + \overline{A}$

SOLUTIONS

(a)
$$\begin{aligned} X &= A \cdot \overline{B} + A \cdot D \\ &= 0 \cdot \overline{1} + 0 \cdot 0 \\ &= 0 \cdot 0 + 0 \cdot 0 \\ &= 0 + 0 \\ &= 0 \end{aligned}$$

(b)
$$\begin{aligned} Y &= A \cdot (B + \overline{C}) \\ &= 0 \cdot (1 + \overline{1}) \\ &= 0 \cdot (1 + 0) \\ &= 0 \cdot 1 \\ &= 0 \end{aligned}$$

(c) $Z = B \cdot (\overline{A} + \overline{C}) \cdot C$
$= 1 \cdot (\overline{0} + \overline{1}) \cdot 1$
$= 1 \cdot (1 + 0) \cdot 1$
$= 1 \cdot 1 \cdot 1$
$= 1$

(d) $P = 1 + \overline{D} \cdot B + \overline{A}$
$= 1 + \overline{0} \cdot 1 + \overline{0}$
$= 1 + 1 \cdot 1 + 1$
$= 1 + 1 + 1$
$= 1$

There are ten basic Boolean algebra theorems that will allow us to simplify logic equations and minimize the number of gates in logic circuits. These ten theorems or rules are known as the *laws of equivalence* and are listed in Table 3–11.

Each rule listed in Table 3–11 has two interpretations, the *OR* interpretation (on the left-hand side) and the *AND* interpretation (on the right-hand side). The remainder of this section will deal with the explanation and verification of each rule.

The idempotent law

The idempotent law can be verified by constructing a circuit for each interpretation as shown in Figure 3–16. In Figure 3–16a a two-input *OR* gate has both its inputs connected to the independent variable, A. The output of the circuit therefore is $A + A$. If $A = 0$ then the output of the *OR* gate is 0, and if $A = 1$ then the output of the *OR* gate is 1. Therefore

Table 3–11
Boolean algebra laws of equivalence

Rule	Law	OR interpretation	AND interpretation
Rule 1	Idempotent law	$A + A = A$	$A \cdot A = A$
Rule 2	Associative law	$(A + B) + C = A + (B + C)$	$(A \cdot B) \cdot C = A \cdot (B \cdot C)$
Rule 3	Commutative law	$A + B = B + A$	$A \cdot B = B \cdot A$
Rule 4	Distributive law	$A + (B \cdot C) = (A + B) \cdot (A + C)$	$A \cdot (B + C) = (A \cdot B) + (A \cdot C)$
Rule 5	Identity law	$A + 0 = A$	$A \cdot 1 = A$
Rule 6	Dominance law	$A + 1 = 1$	$A \cdot 0 = 0$
Rule 7	Complementarity law	$A + \overline{A} = 1$	$A \cdot \overline{A} = 0$
Rule 8	Absorbtion law	$A + (A \cdot B) = A$	$A \cdot (A + B) = A$
Rule 9	DeMorgan's law	$\overline{A + B} = \overline{A} \cdot \overline{B}$	$\overline{A \cdot B} = \overline{A} + \overline{B}$
Rule 10	Miscellaneous	$\overline{A} + (A \cdot B) = \overline{A} + B$	$A + (\overline{A} \cdot B) = A + B$

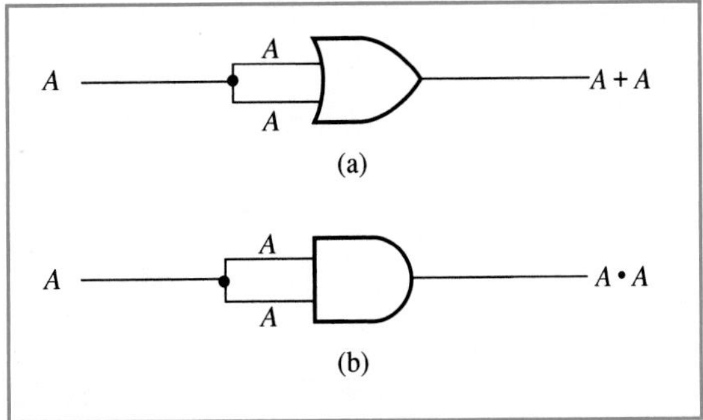

Figure 3–16
Circuits to represent idempotence.

the output of the *OR* gate will always be at the same state as *A*. The following truth table confirms this fact:

A	A + A = A
0	0 + 0 = 0
1	1 + 1 = 1

Similarly, in Figure 3–16b, if $A = 0$ then the output of the *AND* gate is 0, and if $A = 1$ then the output of the *AND* gate is 1. The idempotent law can also be verified by the following truth table using the Boolean postulates:

A	A · A = A
0	0 · 0 = 0
1	1 · 1 = 1

The idempotent law can be extended to more than two occurrences. Thus

$$A \cdot A \cdot A = A \quad \text{and} \quad A + A + A = A$$
$$A \cdot A \cdot A \cdot A = A \quad \text{and} \quad A + A + A + A = A$$

etc.

The associative law

The associative law simply specifies that there are no rules of precedence when evaluating logical expressions that contain the same operators. Therefore the circuits in Figure 3–17a through c are all equivalent. This can be also be proved by applying the Boolean postulates to the logic expressions as shown in Table 3–12.

Similarly, the associative law can also be applied to the *AND* function, and therefore the circuits shown in Figure 3–17d through f are equivalent. Table 3–13 verifies the associative law for the *AND* function.

Observe that the associative laws can be applied to extend the number of inputs in *AND* and *OR* gates. For example, the circuits in Figure 3–17a and b are equivalent to the three-input *OR* gate in Figure 3–17c. Similarly, the circuits in Figure 3–17d and e are equivalent to the three-input *AND* gate in Figure 3–17f.

The commutative law

The commutative law simply states that the position of the operands of a logical operator have no bearing on the logic of a logic function. This means that the function of a logic gate will be the same regardless of the order in which the inputs are applied, as can be seen in Figure 3–18a and b.

Figure 3–17 Circuits to represent association.

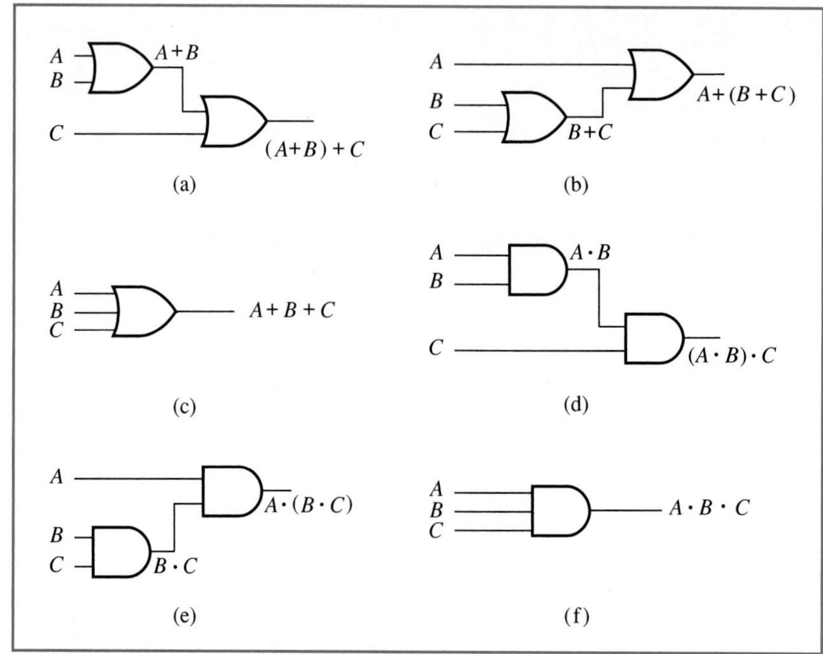

The distributive law

The distributive law is one of the most important laws because it is used frequently in the reduction of logic equations. Figure 3–19a and b show the two circuits that represent the left-hand and right-hand side of the *OR* interpretation of the distributive law. The two circuits can be proved to be equivalent by analyzing the two circuits and obtaining a truth table for the outputs. The results of the analysis are shown in Table 3–14.

Similarly, Figure 3–19c and d are equivalent as shown by the analysis in Table 3–15.

The identity law

The identity law can be verified by constructing the circuit for each interpretation as shown in Figure 3–20. Notice in Figure 3–20a that since one of the inputs of the *OR* gate is tied permanently to a logic 0, if $A = 1$ then the output of the gate will be a 1, but if $A = 0$, the output will be a 0. Thus the state of the output is equal to the state of A. The following truth table illustrates this analysis:

A	A + 0 = A
0	0 + 0 = 0
1	1 + 0 = 1

Table 3–12

A	B	C	A + B	B + C	A + (B + C)	(A + B) + C
0	0	0	0	0	0	0
0	0	1	0	1	1	1
0	1	0	1	1	1	1
0	1	1	1	1	1	1
1	0	0	1	0	1	1
1	0	1	1	1	1	1
1	1	0	1	1	1	1
1	1	1	1	1	1	1

Table 3–13

A	B	C	A · B	B · C	A · (B · C)	(A · B) · C
0	0	0	0	0	0	0
0	0	1	0	0	0	0
0	1	0	0	0	0	0
0	1	1	0	1	0	0
1	0	0	0	0	0	0
1	0	1	0	0	0	0
1	1	0	1	0	0	0
1	1	1	1	1	1	1

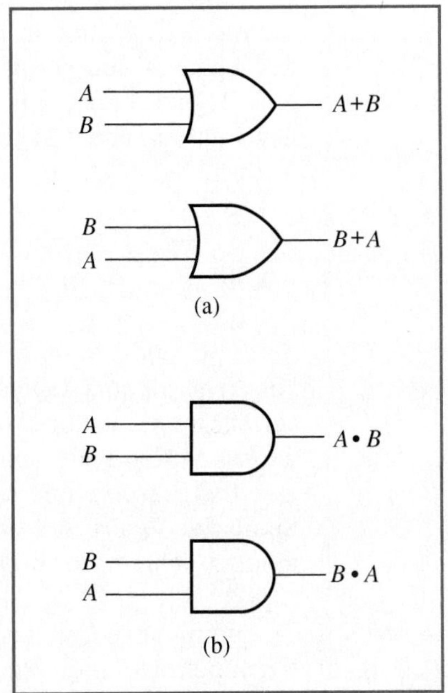

Figure 3–18 Circuits to represent commutation.

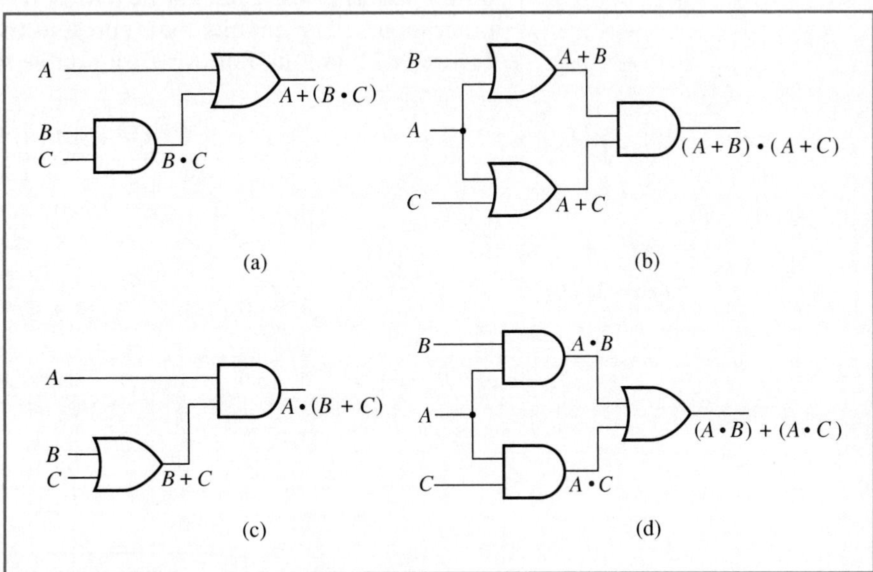

Figure 3–19 Circuits to represent distribution.

Table 3–14

A	B	C	A + (B · C)	(A + B) · (A + C)
0	0	0	0	0
0	0	1	0	0
0	1	0	0	0
0	1	1	1	1
1	0	0	1	1
1	0	1	1	1
1	1	0	1	1
1	1	1	1	1

The identity law when applied to the *AND* function produces the same result. Notice in Figure 3–20b that since one of the inputs to the *AND* gate is tied permanently to a logic 1, the output will be a logic 1 if $A = 1$, and a logic 0 if $A = 0$. Therefore the state of the output of the *AND* gate is equal to the state of A. This is illustrated in the following truth table:

A	A · I = A
0	0 · 1 = 0
1	1 · 1 = 1

The identity law is very useful when it is necessary to disable certain inputs of the *AND* and *OR* gates. For example, in Figure 3–21a we have effectively converted a four-input *OR* gate to a three-input *OR* gate by *disabling* one of the inputs by tying it at the logic 0 state. Similarly, Figure 3–21b reduces the number of effective inputs in a four-input *AND* gate to three.

The dominance law

The dominance law describes the basic characteristics of the *AND* and *OR* functions (and gates). That is, if any of the inputs of an *OR* gate is a logic 1, the output of the gate will be a logic 1 regardless of the states of the other inputs. Similarly, if any of the inputs of an *AND* gate is a logic 0, the output of the gate will be a logic 0 regardless of the states of the other inputs. The circuits that represent the dominance laws are shown in Figure 3–22 and the following truth table verifies the laws:

A	A + I = I	A · 0 = 0
0	0 + 1 = 1	0 · 0 = 0
1	1 + 1 = 1	1 · 0 = 0

Table 3–15

A	B	C	A · (B + C)	(A · B) + (A · C)
0	0	0	0	0
0	0	1	0	0
0	1	0	0	0
0	1	1	0	0
1	0	0	0	0
1	0	1	1	1
1	1	0	1	1
1	1	1	1	1

Figure 3–20
Circuits to represent identity.

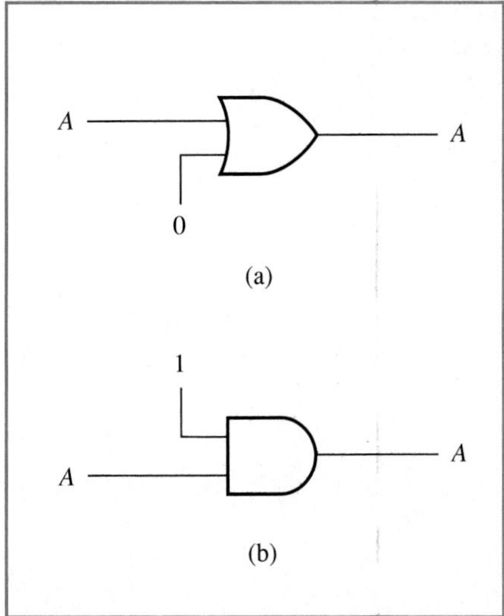

Notice that we can apply the identity law and the dominance law to a two-input gate to effectively "pass-through" or "block-out" the logic level of a variable.

The complementarity law

The complementarity law is another important law that is used very frequently in the simplification of logic circuits. Figure 3–23 includes the two circuits that represent this law. Notice in Figure 3–23a that the output of the *OR* gate will always be a logic 1 since one input is the complement of the other and therefore for any value of A, one of the inputs will always be a logic 1. Similarly, in Figure 3–23b, the output of the *AND* gate will always be a logic 0 since the logic states at the two inputs will always be complementary, and therefore for any value of A one of the inputs will always be a logic 0. The following truth table verifies the complementarity law:

A	$\bar{A}$	$A + \bar{A} = 1$	$A \cdot \bar{A} = 0$
0	1	$0 + 1 = 1$	$0 \cdot 1 = 0$
1	0	$1 + 0 = 1$	$1 \cdot 0 = 0$

Figure 3–21
Reducing the number of effective inputs in a gate. (a) *OR* gate; (b) *AND* gate.

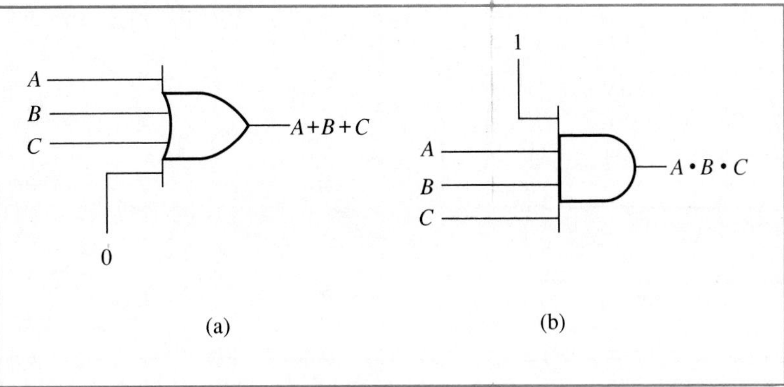

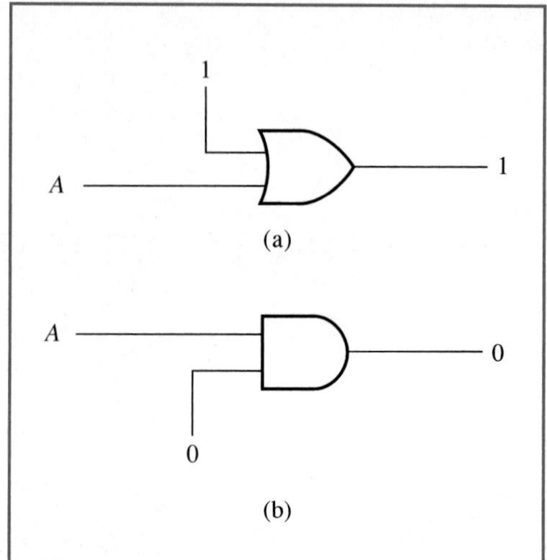

Figure 3–22
Circuits to represent dominance.

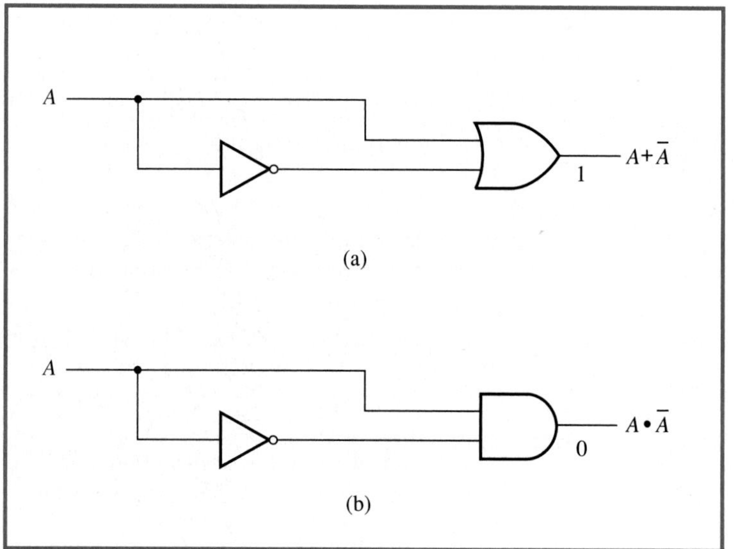

Figure 3–23
Circuits to represent complementarity.

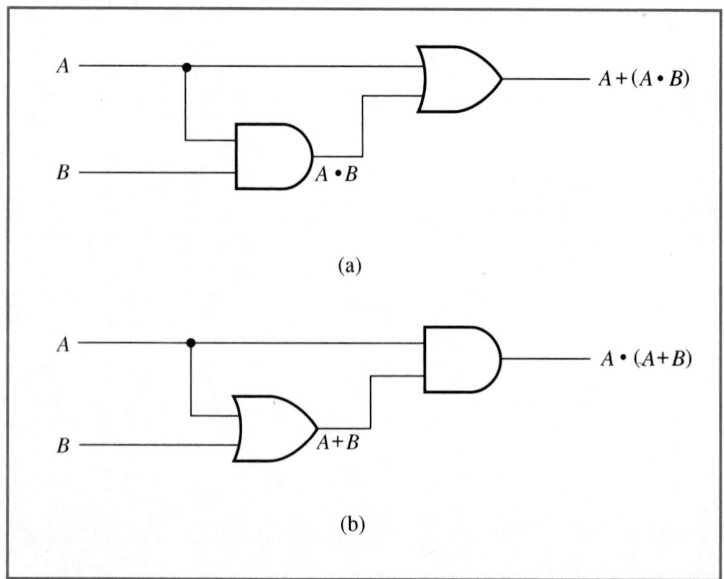

Figure 3–24
Circuits to represent absorbtion.

Table 3–16

A	B	A · B	A + B	A + (A · B)	A · (A + B)
0	0	0	0	0	0
0	1	0	1	0	0
1	0	0	1	1	1
1	1	1	1	1	1

The absorbtion law

The circuits that represent the absorbtion laws are snown in Figure 3–24. Notice in Figure 3–24a that if *A* is a logic 1 then the output of the *OR* gate will be a logic 1, and if *A* is a logic 0 the output of the *OR* gate will be a logic 0 since both its inputs are at logic 0 state; therefore the output of the circuit is equal to the state of *A*. In Figure 3–24b if *A* is a logic 0 then the output of the *AND* gate is a logic 0, and if *A* is a logic 1 the output of the *AND* gate will be a logic 1 since both its inputs are at logic 1 state; therefore the output of the circuit is equal to the state of *A*. The truth table shown in Table 3–16 verifies the absorbtion law.

DeMorgan's law

DeMorgan's law (also known as DeMorgan's theorem) relates the three basic functions—*AND*, *OR*, and *NOT*—to the *NAND* and *NOR* functions. Its importance in changing the configuration of a circuit will be seen in Section 3–6.

DeMorgan's law states that the complement of the logical sum of two or more variables equals the logical product of the complements of the variables. Conversely, the complement of the logical products of two or more variables equals the logical sum of the complements of the variables.

Figure 3–25 includes the four circuits that represent DeMorgan's law. Figure 3–25a is a simple two-input *NOR* gate, while Figure 3–25b is its equivalent circuit constructed with an *AND* gate and inverters. Recall that the output of a *NOR* gate will be a logic 0 if any of its inputs is a logic 1. Notice that this logic also applies to the circuit in Figure 3–25b since if either *A* or *B* is a logic 1, the output of the *AND* gate will be a logic 0. The truth table shown in Table 3–17 verifies DeMorgan's law for the circuits shown in Figure 3–25a and b.

Similarly, Figure 3–25c is a simple two-input *NAND* gate, while

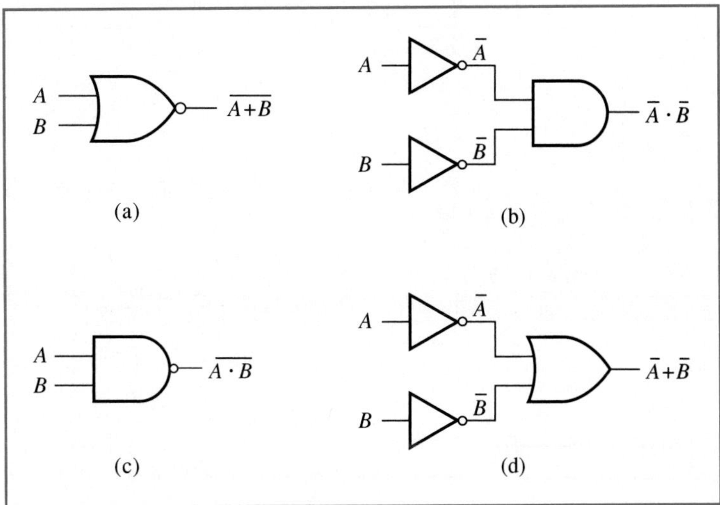

Figure 3–25
Circuits to represent DeMorgan's law.

Table 3–17

A	B	$\overline{A}$	$\overline{B}$	$\overline{A + B}$	$\overline{A} \cdot \overline{B}$
0	0	1	1	1	1
0	1	1	0	0	0
1	0	0	1	0	0
1	1	0	0	0	0

Figure 3–25d is its equivalent circuit constructed with an *OR* gate and inverters. Recall that the output of a *NAND* gate will be a logic 1 if any of its inputs is a logic 0. Notice that this logic also applies to the circuit in Figure 3–25d since if either *A* or *B* is a logic 0, the output of the *OR* gate will be a logic 1. The truth table shown in Table 3–18 verifies DeMorgan's law for the circuits shown in Figure 3–25c and d.

DeMorgan's law can also be used to convert minterms to their corresponding maxterms and vice versa. Recall that minterms and maxterms are complements of each other. For example, we can convert minterm no. 3 in Table 3–8 to its corresponding maxterm (no. 3) as follows:

minterm no. 3: $\overline{X} \cdot Y \cdot Z$

applying DeMorgan's law: $\overline{\overline{X} \cdot Y \cdot Z}$

$= \overline{\overline{X}} + \overline{Y} + \overline{Z}$

maxterm no. 3: $= X + \overline{Y} + \overline{Z}$

Similarly, we can convert maxterm no. 3 to its complementary minterm (no. 3) as follows:

maxterm no. 3: $X + \overline{Y} + \overline{Z}$

applying DeMorgan's law: $\overline{X + \overline{Y} + \overline{Z}}$

$= \overline{X} \cdot \overline{\overline{Y}} \cdot \overline{\overline{Z}}$

minterm no. 3: $= \overline{X} \cdot Y \cdot Z$

DeMorgan's law allows the symbols for the *NAND* and *NOR* gates to be shown in terms of *AND* and *OR* gates. For example, since a *NAND* gate is equivalent to an *OR* gate with inverted inputs, it can also be represented as

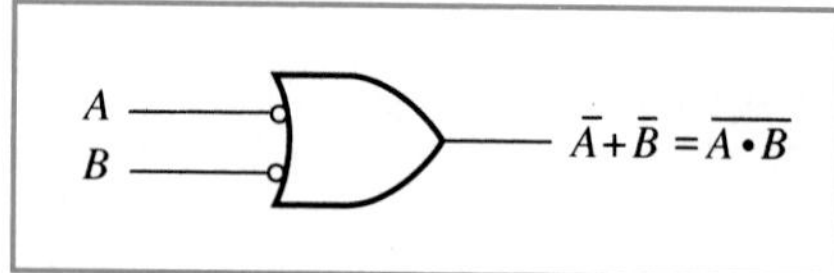

Recall that the "bubble" in a logic symbol indicates inversion.

Table 3–18

A	B	$\overline{A}$	$\overline{B}$	$\overline{A \cdot B}$	$\overline{A} + \overline{B}$
0	0	1	1	1	1
0	1	1	0	1	1
1	0	0	1	1	1
1	1	0	0	0	0

Similarly, a *NOR* gate is equivalent to an *AND* gate with inverted inputs and can be represented as

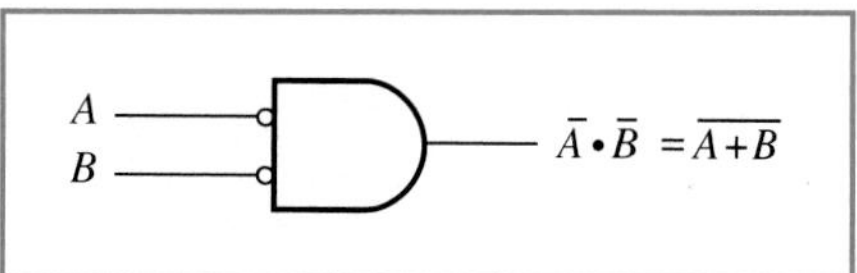

Miscellaneous rules

The two miscellaneous rules listed in Table 3–11 are not part of the Boolean algebra laws but are useful in the reduction of logic equations. The circuits that represent these rules are shown in Figure 3–26—according to the first rule, the circuits in Figure 3–26a and b are equivalent, and according to the second rule, the circuits in Figure 3–26c and d are equivalent. The verification of the two rules are shown in

Figure 3–26
Circuits to represent the miscellaneous rules.

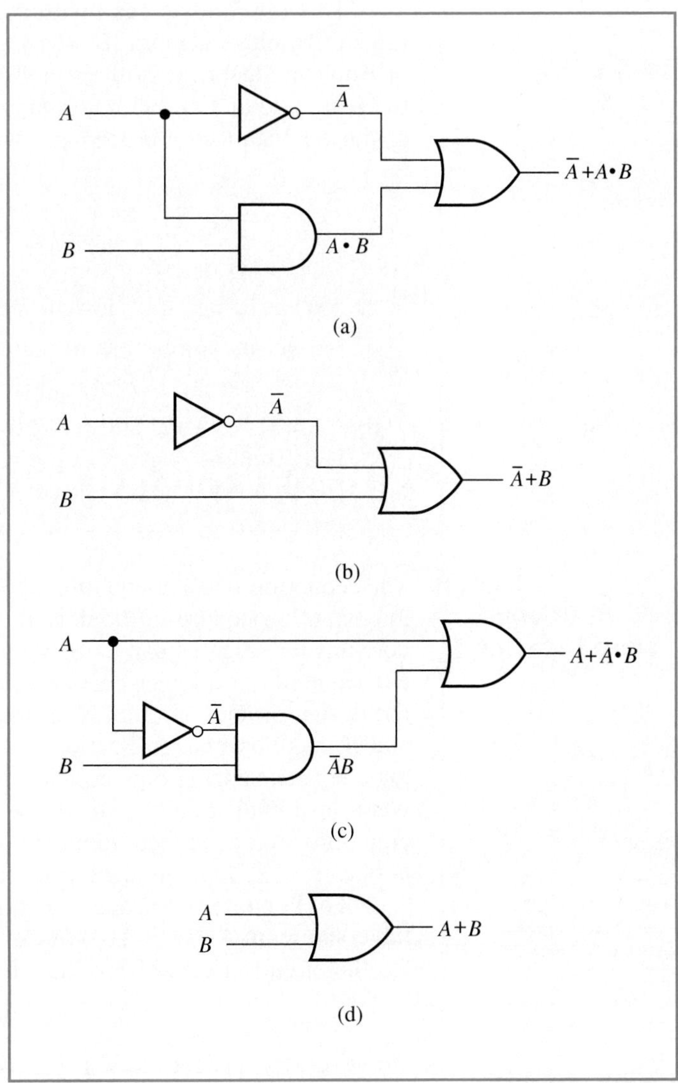

Table 3–19

A	B	$\bar{A}$	$A \cdot B$	$\bar{A} + (A \cdot B)$	$\bar{A} + B$
0	0	1	0	1	1
0	1	1	0	1	1
1	0	0	0	0	0
1	1	0	1	1	1

Table 3–20

A	B	$\bar{A}$	$\bar{A} \cdot B$	$A + (\bar{A} \cdot B)$	$A + B$
0	0	1	0	0	0
0	1	1	1	1	1
1	0	0	0	1	1
1	1	0	0	1	1

Tables 3–19 and 3–20 by obtaining truth tables for each set of circuits through analysis.

This section has dealt with the basic postulates, laws, theorems, and rules of Boolean algebra. It was stated earlier that one of the applications of Boolean algebra is in the simplification (or reduction) of logic equations to yield logic circuits with a minimum number of gates. The applications of these laws in such procedures will be seen in the next section.

Review questions

1. What is Boolean algebra? What is its purpose in the study of digital circuits?
2. What is meant by logical multiplication and addition?
3. What is the purpose of using parentheses in logic expressions?
4. What are the Boolean algebra laws of equivalence used for?
5. Are the *NAND* and *NOR* functions associative? That is, can the *AND* and *OR* gates in the circuits of Figure 3–17 be replaced with *NAND* and *NOR* gates, respectively?

3–5 Boolean simplification

The reduction of logic equations and minimization of logic circuits is an important procedure in the design of digital circuits since a circuit that contains fewer gates can usually be implemented with fewer ICs, and consequently at a lower cost and smaller size. The objective, therefore, in the design of logic circuits is to obtain a required circuit with as few gates and ICs as possible. In Section 3–3 we developed a procedure to obtain logic equations from a given truth table. It was stated that these equations were unsimplified and in most cases could be reduced to equivalent equations that produced much simpler circuits. This section deals with the application of Boolean algebra in the reduction of logic circuits.

To illustrate the process of Boolean simplification, consider the truth table shown in Table 3–21. We can obtain the *unsimplified* logic equation for the circuit in sum-of-products form as

Equation (3–10)

$$Z = \bar{X}\,\bar{Y} + X\,\bar{Y}$$

However, by inspection of Table 3–21 notice that the dependent variable, Z, is the complement of the independent variable Y. Therefore the *simplified* logic equation for the circuit is

Equation (3–11)

$$Z = \overline{Y}$$

In most cases the simplified logic equation cannot be obtained so easily by inspection, and we must apply the rules of Boolean algebra (see Table 3–11) to reduce the equation to its simplest form.

To begin the reduction of Equation 3–10, we must first apply the distributive law, to factor out the $\overline{Y}$

$$Z = \overline{Y}(\overline{X} + X)$$

The complementarity law can now be applied to the term $(\overline{X} + X)$ to obtain the following equation

$$Z = \overline{Y} \cdot 1$$

Finally, we apply the identity law to obtain the final simplified equation

$$Z = \overline{Y}$$

When we can no longer apply any of the Boolean algebra laws, the equation is in its simplest form.

It is also possible to apply the laws of Boolean algebra to reduce the unsimplified product-of-sums equation for Table 3–21

$$Z = (X + \overline{Y})(\overline{X} + \overline{Y})$$
$$Z = \overline{X}(X + \overline{Y}) + \overline{Y}(X + \overline{Y}) \quad \text{(distributive)}$$
$$Z = \overline{X}X + \overline{X}\,\overline{Y} + \overline{Y}X + \overline{Y}\,\overline{Y} \quad \text{(distributive)}$$
$$Z = 0 + \overline{X}\,\overline{Y} + \overline{Y}X + \overline{Y}\,\overline{Y} \quad \text{(complementarity)}$$
$$Z = \overline{X}\,\overline{Y} + \overline{Y}X + \overline{Y}\,\overline{Y} \quad \text{(identity)}$$
$$Z = \overline{X}\,\overline{Y} + \overline{Y}X + \overline{Y} \quad \text{(idempotent)}$$
$$Z = \overline{Y}(\overline{X} + X) + \overline{Y} \quad \text{(distributive)}$$
$$Z = \overline{Y} \cdot 1 + \overline{Y} \quad \text{(complementarity)}$$
$$Z = \overline{Y} + \overline{Y} \quad \text{(identity)}$$
$$Z = \overline{Y} \quad \text{(idempotent)}$$

It can be proved that the simplified and unsimplified logic equations and their circuits are equivalent by constructing truth tables for each equation or circuit and comparing their truth tables. For example, we can reduce the unsimplified Equation 3–3 (from Section 3–3) as follows:

$$S = \overline{P}\,\overline{Q}\,\overline{R} + P\overline{Q}\,\overline{R} + P\overline{Q}R + PQ\overline{R} + PQR$$
$$S = \overline{Q}\,\overline{R}(\overline{P} + P) + P\overline{Q}R + PQ(\overline{R} + R) \quad \text{(distributive)}$$
$$S = \overline{Q}\,\overline{R}1 + P\overline{Q}R + PQ1 \quad \text{(complementarity)}$$
$$S = \overline{Q}\,\overline{R} + P\overline{Q}R + PQ \quad \text{(identity)}$$
$$S = \overline{Q}(\overline{R} + PR) + PQ \quad \text{(distributive)}$$
$$S = \overline{Q}(\overline{R} + P) + PQ \quad \text{(miscellaneous)}$$
$$S = \overline{Q}\,\overline{R} + \overline{Q}P + PQ \quad \text{(distributive)}$$
$$S = \overline{Q}\,\overline{R} + P(\overline{Q} + Q) \quad \text{(distributive)}$$
$$S = \overline{Q}\,\overline{R} + P1 \quad \text{(complementarity)}$$
$$S = \overline{Q}\,\overline{R} + P \quad \text{(identity)}$$

Equation (3–12)

$$S = \overline{Q}\,\overline{R} + P$$

Table 3–21

X	Y	Z
0	0	1
0	1	0
1	0	1
1	1	0

Table 3–22

P	Q	R	$\bar{Q}$	$\bar{R}$	$\bar{Q}\bar{R}$	S
0	0	0	1	1	1	1
0	0	1	1	0	0	0
0	1	0	0	1	0	0
0	1	1	0	0	0	0
1	0	0	1	1	1	1
1	0	1	1	0	0	1
1	1	0	0	1	0	1
1	1	1	0	0	0	1

We can prove that the reduced Equation (3–12) is equivalent to the unsimplified Equation (3–3) by constructing its truth table (Table 3–22). Notice that the truth table shown in Table 3–22 is identical to the truth table of Equation 3–3 (see Example 3–3).

The circuit for Equation 3–12 is shown in Figure 3–27. Notice that this logic circuit is considerably simpler than the circuit in Figure 3–10, requiring a total of only four gates as compared to nine gates for Figure 3–10. Futhermore, the IC implementation of Figure 3–10 would require four ICs—two 7411s for the three-input *AND* gates, one 7432 for the *OR* gate, and one 7404 for the inverters. The circuit in Figure 3–27 would only require three ICs—one 7408 for the two-input *AND* gate, one 7432 for the two-input *OR* gate, and one 7404 for the inverters.

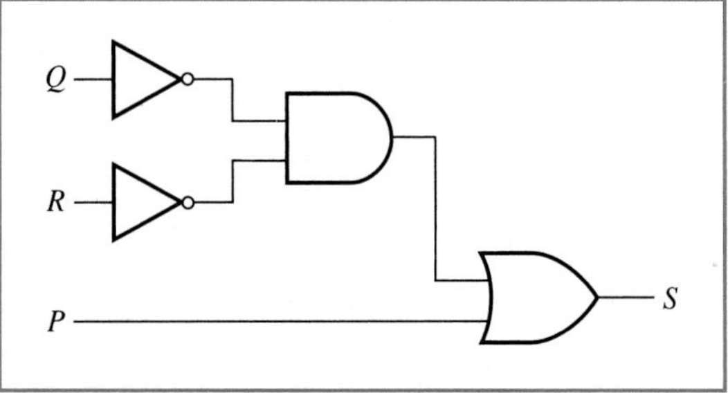

Figure 3–27 Figure 3–10 reduced by applying Boolean algebra.

EXAMPLE 3–7

Simplify Equation 3–5 using Boolean algebra and draw the logic circuit for the simplified equation. Compare the simplified circuit with the unsimplified circuit (Figure 3–12) in terms of the number of gates and ICs and verify that the two circuits are equivalent.

SOLUTION

$$X = \bar{A}\bar{B}\bar{C}D + \bar{A}\bar{B}CD + \bar{A}BC\bar{D} + A\bar{B}\bar{C}D + A\bar{B}C\bar{D} + ABCD$$

$$X = \bar{A}\bar{B}\bar{C}D + \bar{A}\bar{B}CD + A\bar{B}\bar{C}D + \bar{A}BC\bar{D} + A\bar{B}C\bar{D} + ABCD \quad \text{(commutative)}$$

$$X = \bar{A}\bar{B}D(\bar{C} + C) + A\bar{B}\bar{C}D + \bar{A}BC\bar{D} + A\bar{B}C\bar{D} + ABCD \quad \text{(distributive)}$$

$$X = \bar{A}\bar{B}D1 + A\bar{B}\bar{C}D + \bar{A}BC\bar{D} + A\bar{B}C\bar{D} + ABCD \quad \text{(complementarity)}$$

$$X = \bar{A}\bar{B}D + A\bar{B}\bar{C}D + \bar{A}BC\bar{D} + A\bar{B}C\bar{D} + ABCD \quad \text{(identity)}$$

$$X = \bar{B}D(\bar{A} + A\bar{C}) + \bar{A}BC\bar{D} + A\bar{B}C\bar{D} + ABCD \quad \text{(distributive)}$$

$$X = \bar{B}D(\bar{A} + \bar{C}) + \bar{A}BC\bar{D} + A\bar{B}C\bar{D} + ABCD \quad \text{(miscellaneous)}$$

$$X = \bar{A}\bar{B}D + \bar{B}\bar{C}D + \bar{A}BC\bar{D} + A\bar{B}C\bar{D} + ABCD \quad \text{(distributive)}$$

The circuit in Figure 3–12 has a total of 11 gates as compared to the simplified circuit shown in Figure 3–28 consisting of 10 gates. To implement the circuit in Figure 3–12 we need six ICs—two 7432s, three 7421s, and one 7404. To implement the circuit in Figure 3–28 we need five ICs—one 7432, three 7421s, and one 7404.

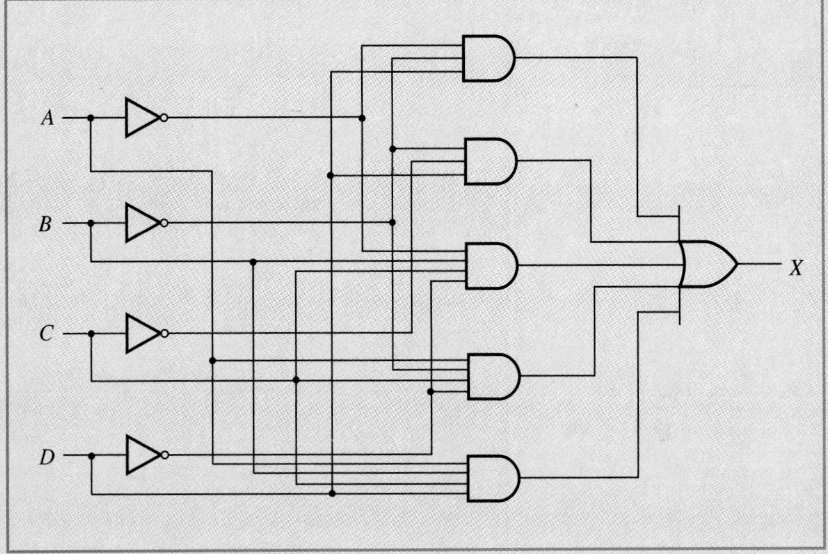

Figure 3–28
Figure 3–12 reduced by applying Boolean algebra.

It can be proved that the simplified circuit in Figure 3–28 is equivalent to the unsimplified circuit in Figure 3–12 by constructing a truth table for the circuit in Figure 3–28 and comparing it to the truth table for the circuit in Figure 3–12 (shown in Example 3–4):

A	B	C	D	X
0	0	0	0	0
0	0	0	1	1
0	0	1	0	0
0	0	1	1	1
0	1	0	0	0
0	1	0	1	0
0	1	1	0	1
0	1	1	1	0
1	0	0	0	0
1	0	0	1	1
1	0	1	0	1
1	0	1	1	0
1	1	0	0	0
1	1	0	1	0
1	1	1	0	0
1	1	1	1	1

It can be seen that since the two tables are identical, their circuits and equations are equivalent.

The rules of Boolean algebra do not always work in the simplification of logic equations and circuits. There are some logic equations that cannot be simplified by applying the laws of Boolean algebra, while there are others that even when simplified will not provide much of a benefit in terms of gate and IC reduction. However, in most cases Boolean algebra can be used to provide the designer with a circuit that contains the smallest number of gates and ICs. The following examples illustrate the use of Boolean algebra in the reduction of different types of logic equations.

EXAMPLE 3–8

Simplify the following logic equations using Boolean algebra:

(a) $H = \overline{X}\,\overline{Y} + Y\overline{Z} + X\overline{Y}\,\overline{Z} + X\overline{Y}Z + X(Y + \overline{X}) + X$

(b) $P = \overline{A}\,\overline{B}\,\overline{C} + \overline{A}B\overline{C} + \overline{A}BC + A\overline{B}\,\overline{C} + A\overline{B}C + AB\overline{C}$

(c) $X = R(P + \overline{P}Q + \overline{Q})(\overline{Q} + \overline{R}P)$

(d) $Z = (A\overline{B}\,\overline{C} + A\overline{B}C + ABC + AB\overline{C})(A + B)$

(e) $W = \overline{AC + A\overline{D}} + \overline{\overline{B}C}$

(f) $A = P + \overline{P}Q\overline{R} + \overline{Q + R}$

SOLUTIONS

(a)

$H = \overline{X}\,\overline{Y} + Y\overline{Z} + X\overline{Y}\,\overline{Z} + X\overline{Y}Z + X(Y + \overline{X}) + X$	
$H = \overline{X}\,\overline{Y} + Y\overline{Z} + X\overline{Y}(\overline{Z} + Z) + X(Y + \overline{X}) + X$	(distributive)
$H = \overline{X}\,\overline{Y} + Y\overline{Z} + X\overline{Y}1 + X(Y + \overline{X}) + X$	(complementarity)
$H = \overline{X}\,\overline{Y} + Y\overline{Z} + X\overline{Y} + X(Y + \overline{X}) + X$	(identity)
$H = Y\overline{Z} + \overline{X}\,\overline{Y} + X\overline{Y} + X(Y + \overline{X}) + X$	(commutative)
$H = Y\overline{Z} + \overline{Y}(\overline{X} + X) + X(Y + \overline{X}) + X$	(distributive)
$H = Y\overline{Z} + \overline{Y}1 + X(Y + \overline{X}) + X$	(complementarity)
$H = Y\overline{Z} + \overline{Y} + X(Y + \overline{X}) + X$	(identity)
$H = \overline{Z} + \overline{Y} + X(Y + \overline{X}) + X$	(miscellaneous)
$H = \overline{Z} + \overline{Y} + XY + X\overline{X} + X$	(distributive)
$H = \overline{Z} + \overline{Y} + XY + 0 + X$	(complementarity)
$H = \overline{Z} + \overline{Y} + XY + X$	(identity)
$H = \overline{Z} + \overline{Y} + X$	(absorbtion)

(b)

$P = \overline{A}\,\overline{B}\,\overline{C} + \overline{A}B\overline{C} + \overline{A}BC + A\overline{B}\,\overline{C} + A\overline{B}C + AB\overline{C}$	
$P = \overline{A}\,\overline{C}(\overline{B} + B) + \overline{A}BC + A\overline{B}(\overline{C} + C) + AB\overline{C}$	(distributive)
$P = \overline{A}\,\overline{C} + \overline{A}BC + A\overline{B} + AB\overline{C}$	(complementarity and identity)
$P = \overline{A}(\overline{C} + BC) + A(\overline{B} + B\overline{C})$	(distributive)
$P = \overline{A}(\overline{C} + B) + A(\overline{B} + \overline{C})$	(miscellaneous)
$P = \overline{A}\,\overline{C} + \overline{A}B + A\overline{B} + A\overline{C}$	(distributive)
$P = \overline{A}\,\overline{C} + A\overline{C} + A\overline{B} + \overline{A}B$	(associative)
$P = \overline{C}(\overline{A} + A) + A\overline{B} + \overline{A}B$	(distributive)
$P = \overline{C} + A\overline{B} + \overline{A}B$	(complementarity and identity)

(c)

$X = R(P + \overline{P}Q + \overline{Q})(\overline{Q} + \overline{R}P)$	
$X = R(P + Q + \overline{Q})(\overline{Q} + \overline{R}P)$	(miscellaneous)
$X = R(P + 1)(\overline{Q} + \overline{R}P)$	(complementarity)
$X = R\,1(\overline{Q} + \overline{R}P)$	(dominance)
$X = R(\overline{Q} + \overline{R}P)$	(identity)

$X = R\bar{Q} + R\bar{R}P$	(distributive)
$X = R\bar{Q} + 0P$	(complementarity)
$X = R\bar{Q} + 0$	(dominance)
$X = R\bar{Q}$	(identity)

(d)

$Z = (A\bar{B}\bar{C} + A\bar{B}C + ABC + AB\bar{C})\,(A + B)$	
$Z = (A\bar{B}\,(\bar{C} + C) + AB\,(C + \bar{C})\,)\,(A + B)$	(distributive)
$Z = (A\bar{B} + AB)\,(A + B)$	(complementarity and identity)
$Z = (A\,(\bar{B} + B))\,(A + B)$	(distributive)
$Z = A\,(A + B)$	(complementarity and identity)
$Z = A$	(absorbtion)

(e)

$W = \overline{AC + A\bar{D}} + \overline{\bar{B}C}$	
$W = \overline{AC} \cdot \overline{A\bar{D}} + \overline{\bar{B}C}$	(DeMorgan's law)
$W = (\bar{A} + \bar{C}) \cdot (\bar{A} + \bar{\bar{D}}) + (\bar{\bar{B}} + \bar{C})$	(DeMorgan's law)
$W = (\bar{A} + \bar{C}) \cdot (\bar{A} + D) + (B + \bar{C})$	($\bar{\bar{X}} = X$)
$W = \bar{A}\,(\bar{A} + D) + \bar{C}\,(\bar{A} + D) + (B + \bar{C})$	(distributive)
$W = \bar{A}\bar{A} + \bar{A}D + \bar{C}\bar{A} + \bar{C}D + (B + \bar{C})$	(distributive)
$W = \bar{A} + \bar{A}D + \bar{C}\bar{A} + \bar{C}D + (B + \bar{C})$	(idempotent)
$W = \bar{A} + \bar{C}\bar{A} + \bar{C}D + (B + \bar{C})$	(absorbtion)
$W = \bar{A} + \bar{C}D + (B + \bar{C})$	(absorbtion)
$W = \bar{A} + \bar{C}D + B + \bar{C}$	(associative)
$W = \bar{C} + \bar{C}D + B + \bar{A}$	(commutative)
$W = \bar{C} + B + \bar{A}$	(absorbtion)

(f)

$A = P + \bar{P}Q\bar{R} + \overline{Q + R}$	
$A = P + Q\bar{R} + \overline{Q + R}$	(miscellaneous)
$A = P + Q\bar{R} + \bar{Q}\bar{R}$	(DeMorgan's law)
$A = P + \bar{R}\,(Q + \bar{Q})$	(distributive)
$A = P + \bar{R}$	(complementarity and identity)

This section has examined the application of the Boolean algebra laws, theorems, and postulates in the simplification of logic equations and circuits. In many of the examples of Boolean simplification it can be seen that the number of gates and ICs can be greatly reduced by obtaining an equivalent simplified logic equation. However, it is difficult to determine when a logic equation has been simplified to a point where no further simplification is possible. As a general rule, one can consider a logic equation completely simplified when no other rule can be applied to further simplify it. However, there is always a possibility that we may not obtain the simplest form because the application of a particular law or rule is not apparent. In Section 3–7 a technique will be examined that (if used properly) will allow one to obtain the simplest possible form using a more direct approach. Section 3–6 will also deal with a technique of reducing the size of some logic circuits by applying DeMorgan's law.

Review questions

1. What is the objective of Boolean simplification?
2. When is a logic equation considered to be completely simplified?
3. Can all logic equations be simplified using Boolean algebra?

3–6 *NAND* and *NOR* logic

The unsimplified sum-of-products and product-of sums equations that were synthesized in Section 3–3 and their simplified equivalents were all implemented using *AND* gates, *OR* gates, and inverters. These circuits are therefore known as *AND-OR-INVERT* circuits or more simply *AND-OR* logic. The hardware implementation of these circuits, no matter how simple, required a minimum of three ICs if the circuit contained *AND* gates, *OR* gates, and inverters, since each IC usually contains gates having the same function.

NAND and *NOR* gates have a special characteristic that allow them to replace *AND-OR* logic circuits with circuits that contain only *NAND* gates or only *NOR* gates. In some circumstances the number of ICs required to construct a circuit can be reduced if only *NAND* gates or *NOR* gates are used. *NAND* gate circuits have another advantage in that the manufacturing cost of *NAND* gates is lower than other gates since fewer electronic components are used to make up a *NAND* gate than any other 74xxx gate; this also means that *NAND* gate circuits are more compact than those of any other gate. *NAND* gate circuits are therefore preferable in the integration of complex logic circuits on an IC. This section will therefore deal with the techniques of converting logic equations from *AND-OR* logic to *NAND* and *NOR* logic to further minimize the number of ICs required to implement the circuit rather than the number of gates in the circuit.

To illustrate the universal properties of the *NAND* and *NOR* gates, first consider the circuits shown in Figure 3–29. The circuit shown in Figure 3–29a is an inverter that is constructed from a two-input *NAND* gate. We can prove that the circuit functions as an inverter by applying the idempotent law:

Equation (3–13)

$$A \cdot A = A$$

If we invert the left-hand side of Equation 3–13 we must also invert the right-hand side to balance the equation

Equation (3–14)

$$\overline{A \cdot A} = \overline{A}$$

Notice that the left-hand side of Equation 3–14 represents a two-input

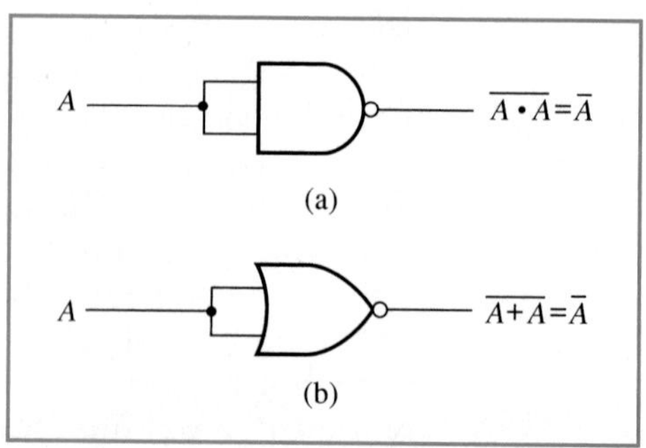

Figure 3–29
(a) *NAND* gate used as an inverter; (b) *NOR* gate used as an inverter.

NAND gate with both its inputs connected to the same independent variable *A*, and the right-hand side of Equation 3–14 represents the complement of *A*.

Similarly, Equation 3–15 proves by idempotence that the circuit in Figure 3–29b is an inverter that is constructed from a *NOR* gate.

Equation (3–15)
$$\overline{A + A} = \overline{A}$$

We can also use *NAND* and *NOR* gates with more than two inputs to function as inverters by tying *all* inputs together.

Figure 3–30 shows circuits that implement the *AND* function. Figure 3–30a is a circuit that functions as an *AND* gate but is constructed exclusively with *NAND* gates. Notice that we have simply inverted the output of a two-input *NAND* gate to obtain the *AND* function. To implement the *AND* function using only *NOR* gates is more complex, as can be seen in Figure 3–30b. Notice that the output of the circuit is

Equation (3–16)
$$\overline{\overline{A} + \overline{B}}$$

Applying DeMorgan's theorem to Equation 3–16 we have

Equation (3–17)
$$\overline{\overline{A}} \cdot \overline{\overline{B}}$$

Equation 3–17 further simplifies into the equation for an *AND* gate

Equation (3–18)
$$A \cdot B$$

Note that we can also implement a *NAND* gate using only *NOR* gates by inverting the output of Figure 3–30b. As stated before, these circuits can be extended to more than two inputs.

Figure 3–31 shows circuits that implement the *OR* function. Figure 3–31b is a circuit that functions as an *OR* gate but is constructed exclusively with *NOR* gates. Notice that we have simply inverted the output of a two-input *NOR* gate to obtain the *OR* function. To implement the *OR* function using only *NAND* gates is more complex as can be seen in Figure 3–31a. Notice that the output of the circuit is

Equation (3–19)
$$\overline{\overline{A} \cdot \overline{B}}$$

Applying DeMorgan's theorem to Equation 3–19 we have

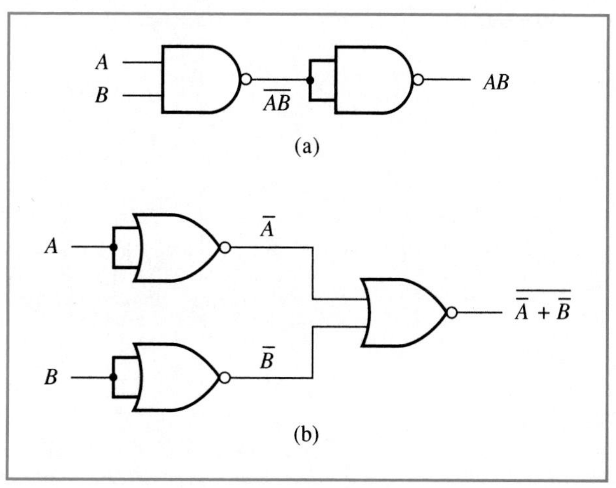

Figure 3–30
(a) *AND* gate using only *NAND* gates; (b) *AND* gate using only *NOR* gates.

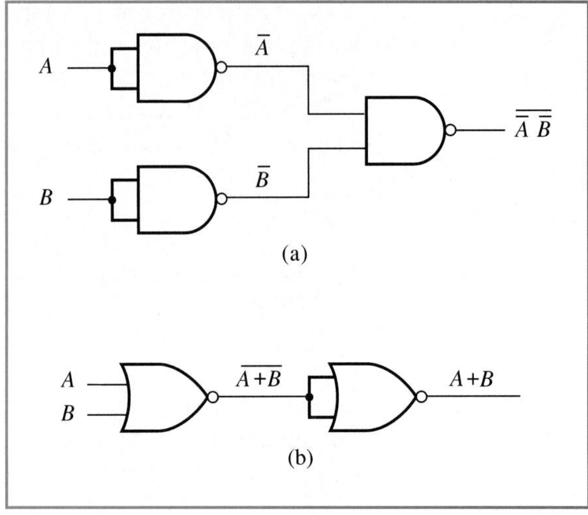

Figure 3–31 (a) *OR* gate using only *NAND* gates; (b) *OR* gate using only *NOR* gates.

Equation (3–20)

$$\overline{\overline{A}} + \overline{\overline{B}}$$

Equation 3–20 further simplifies into the equation for an *OR* gate

Equation (3–21)

$$A + B$$

Note that we can also implement a *NOR* gate using only *NAND* gates by inverting the output of Figure 3–31a. As stated before, these circuits can also be extended to more than two inputs.

By inspecting the circuits shown in Figures 3–29 through 3–31 it may appear that the *NAND* and *NOR* logic implementation of the basic *AND*, *OR*, and Invert functions would increase the complexity of a logic circuit rather than reduce it. However, this is not the case, as can be seen in the following examples.

EXAMPLE 3–9

Figure 3–32a is a circuit that represents the *AND-OR* logic implementation of the expression

Equation (3–22)

$$AB + CD$$

Figure 3–32 (a) *AND-OR* implementation; (b) *NAND* implementation.

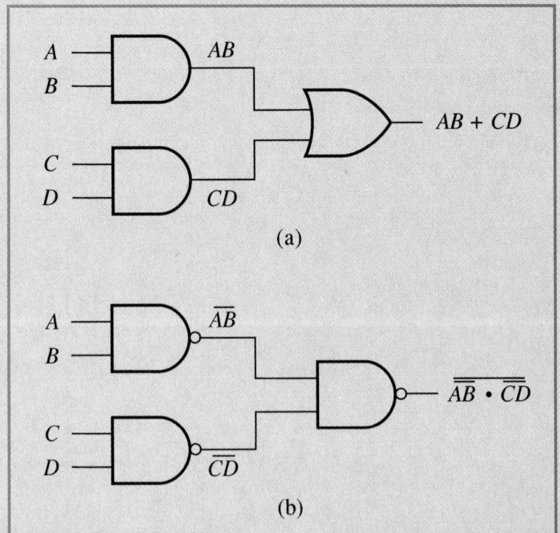

The circuit contains three gates and its hardware implementation will contain two ICs—one 7432 and one 7408.

Figure 3–32b is the *NAND* logic implementation of the circuit shown in Figure 3–32a. The logic expression for the circuit is

Equation (3–23)

$$\overline{\overline{AB} \cdot \overline{CD}}$$

We can prove that the two circuits are equivalent by applying DeMorgan's theorem to Equation 3–23

Equation (3–24)

$$\overline{\overline{AB}} + \overline{\overline{CD}}$$

Simplifying Equation 3–24 further we have

Equation (3–25)

$$AB + CD$$

Notice that since Equation 3–25 and Equation 3–22 are identical, the two circuits are equivalent.

Even though the circuit in Figure 3–32b has the same number of gates as the circuit in Figure 3–32a, it will require only one 7400 IC as compared to the two-IC implementation of Figure 3–32a.

EXAMPLE 3–10

Figure 3–33a is a circuit that represents the *AND-OR* logic implementation of the expression

Equation (3–26)

$$(A + B) \cdot (C + D)$$

The circuit contains three gates and its hardware implementation will contain two ICs—one 7408 and one 7432.

Figure 3–33b is the *NOR* logic implementation of the circuit shown in Figure 3–33a. The logic expression for the circuit is

Equation (3–27)

$$\overline{\overline{A + B} + \overline{C + D}}$$

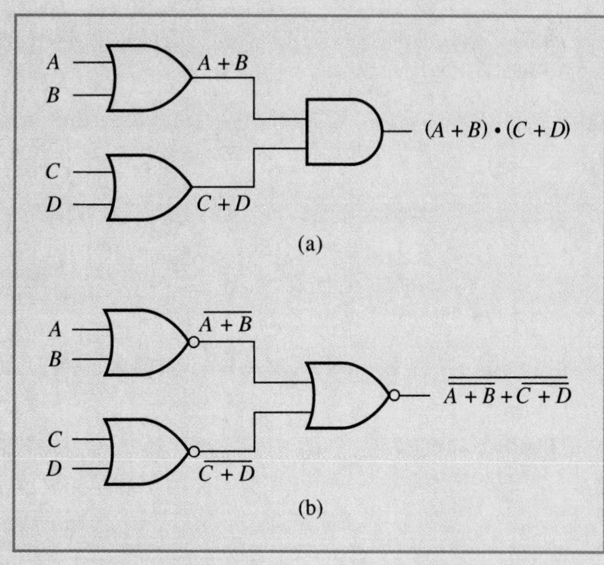

Figure 3–33
(a) *AND-OR* implementation; (b) *NOR* implementation.

We can prove that the two circuits are equivalent by applying DeMorgan's theorem to Equation 3–27

Equation (3–28)

$$\overline{\overline{A + B}} \cdot \overline{\overline{C + D}}$$

Simplifying Equation 3–28 further we have

Equation (3–29)
$$(A + B) \cdot (C + D)$$

Notice that since Equation 3–29 and Equation 3–26 are identical, the two circuits are equivalent.

Even though the circuit in Figure 3–33b has the same number of gates as the circuit in Figure 3–33a, it will require only one 7402 IC as compared to the two-IC implementation of Figure 3–33a.

Examples 3–9 and 3–10 have shown us that *AND-OR* circuits when converted to *NAND* and *NOR* circuits do not reduce the number of gates in the circuit but can reduce the number of ICs required to implement the circuit in hardware. Furthermore, it can be seen that when an *AND-OR* circuit is in sum-of-products form it is easier to convert the circuit to *NAND* logic, and when the *AND-OR* circuit is in product-of-sums form it is easier to convert it to *NOR* logic. We shall now investigate the techniques used to convert *AND-OR* logic circuits to *NAND* and *NOR* logic.

Consider the following *AND-OR* logic equation in sum-of-products form:

Equation (3–30)
$$X = \overline{A}B + AC$$

The circuit for Equation 3–30 is made up of four gates and will require three ICs (7404, 7408, 7432) as shown in Figure 3–34. We can convert this circuit to *NAND* logic by applying the following procedure to Equation 3–30.

Both sides of the equation are first inverted

Equation (3–31)
$$\overline{X} = \overline{\overline{A}B + AC}$$

We then apply DeMorgan's law to the right-hand side of the equation

Equation (3–32)
$$\overline{X} = \overline{\overline{A}B} \cdot \overline{AC}$$

Finally, to restore the equation to its original value in terms of the variable X we invert both sides again

Equation (3–33)
$$\overline{\overline{X}} = \overline{\overline{\overline{A}B} \cdot \overline{AC}}$$

The final *NAND* equation is

Equation (3–34)
$$X = \overline{\overline{\overline{A}B} \cdot \overline{AC}}$$

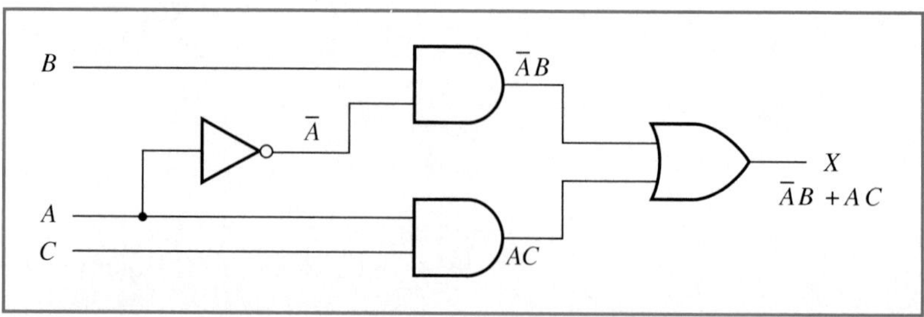

Figure 3–34
AND-OR circuit for Equation 3–30.

Figure 3–35
NAND circuit for Equation 3–30.

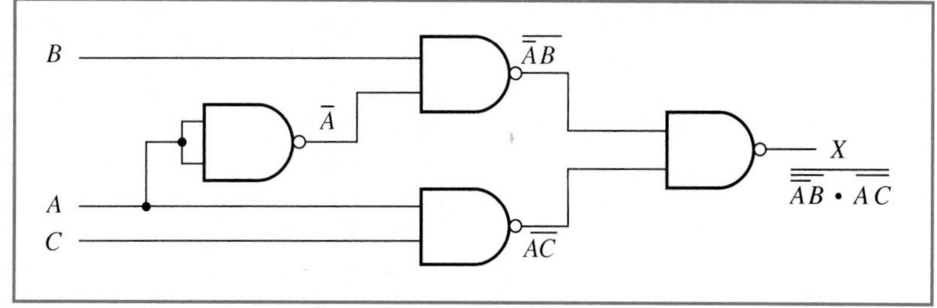

The circuit for Equation 3–34 is shown in Figure 3–35. Notice that like its equivalent *AND-OR* circuit shown in Figure 3–34, the *NAND* circuit also has four gates but only requires one IC (7400) for its hardware implementation.

To convert the *AND-OR* circuit in Figure 3–34 to *NOR* logic we must first convert Equation 3–30 into product-of-sums form. Converting from sum-of-products to product-of-sums can be done by applying various Boolean algebra rules to Equation 3–30, but the procedure involves trial and error and is time-consuming. A preferred technique will be studied in Section 3–7. For the present we will just use the product-of-sums equation without being concerned about its derivation. Equation 3–35 is the product-of-sums equivalent for Equation 3–30.

Equation (3–35)

$$X = (A + B) \cdot (\overline{A} + C)$$

We must now apply the same procedure as before to convert Equation 3–35 to *NOR* logic. Both sides of the equation are inverted

Equation (3–36)

$$\overline{X} = \overline{(A + B) \cdot (\overline{A} + C)}$$

Next, DeMorgan's law is applied to the right-hand side of Equation 3–36

Equation (3–37)

$$\overline{X} = \overline{(A + B)} + \overline{(\overline{A} + C)}$$

Finally, both sides of the equation are inverted again to restore the equation to its original value in terms of X

Equation (3–38)

$$\overline{\overline{X}} = \overline{\overline{(A + B)} + \overline{(\overline{A} + C)}}$$

The final *NOR* equation is

Equation (3–39)

$$X = \overline{\overline{(A + B)} + \overline{(\overline{A} + C)}}$$

The circuit for Equation 3–39 is shown in Figure 3–36. Notice that like its equivalent *AND-OR* circuit shown in Figure 3–34 and its equivalent *NAND* circuit in Figure 3–35, the *NOR* circuit also has four

Figure 3–36
NOR circuit for Equation 3–30.

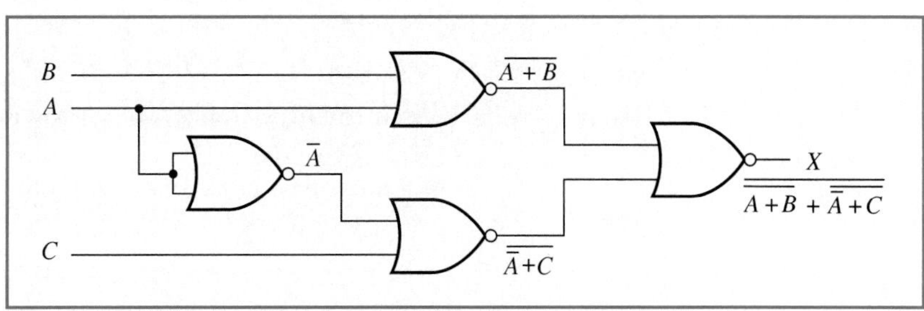

Table 3–23

A	B	C	X (Figure 3–34)	X (Figure 3–35)	X (Figure 3–36)
0	0	0	0	0	0
0	0	1	0	0	0
0	1	0	1	1	1
0	1	1	1	1	1
1	0	0	0	0	0
1	0	1	1	1	1
1	1	0	0	0	0
1	1	1	1	1	1

gates. However, like the *NAND* circuit, it only requires one IC (7402) for its hardware implementation. We can prove by truth tables that the three circuits are equivalent as shown in Table 3–23.

It should be noted that before the previously discussed procedures are applied to *AND-OR* logic equations (in sum-of-products or product-of-sums form) they should be simplified completely using either the Boolean algebra laws or through the Karnaugh map (K-map) procedure to be discussed in the next section. To convert an *AND-OR* equation to *NAND* logic the equation must be in sum-of-products form, and to convert an *AND-OR* equation to *NOR* logic the equation must be in product-of-sums form. As was stated earlier in this section, conversions between logic equations in sum-of-products and product-of-sums form will be examined in the next section.

EXAMPLE 3–11

Convert the following simplified *AND-OR* logic equations to *NAND* logic and draw the *NAND* circuit for each equation:

(a) $T = XY\overline{Z} + \overline{X}\,\overline{Y} + \overline{X}Z$

(b) $X = P\,(Q + \overline{R}S)$

(c) $Z = (\overline{A + B})\,C + A\,(\overline{B + C})$

SOLUTIONS

(a) $T = XY\overline{Z} + \overline{X}\,\overline{Y} + \overline{X}Z$

Inverting both sides

$$\overline{T} = \overline{XY\overline{Z} + \overline{X}\,\overline{Y} + \overline{X}Z}$$

Applying DeMorgan's law

$$\overline{T} = \overline{XY\overline{Z}} \cdot \overline{\overline{X}\,\overline{Y}} \cdot \overline{\overline{X}Z}$$

Inverting both sides again

$$\overline{\overline{T}} = \overline{\overline{XY\overline{Z}} \cdot \overline{\overline{X}\,\overline{Y}} \cdot \overline{\overline{X}Z}}$$

The final *NAND* equation is

Equation (3–40)

$$T = \overline{\overline{XY\overline{Z}} \cdot \overline{\overline{X}\,\overline{Y}} \cdot \overline{\overline{X}Z}}$$

The circuit for Equation 3–40 is shown in Figure 3–37.

(b) $X = P\,(Q + \overline{R}S)$

We must first convert the equation to sum-of-products form using the distributive law

$$X = PQ + P\overline{R}S$$

Figure 3–37

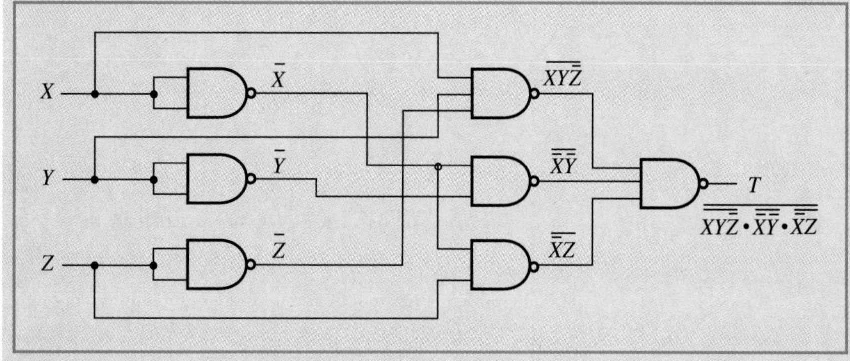

Both sides of the equation are now inverted

$$\overline{X} = \overline{PQ + P\overline{R}S}$$

Applying DeMorgan's law

$$\overline{X} = \overline{PQ} \cdot \overline{P\overline{R}S}$$

Inverting both sides again

$$\overline{\overline{X}} = \overline{\overline{PQ} \cdot \overline{P\overline{R}S}}$$

The final *NAND* equation is

Equation (3–41)

$$X = \overline{\overline{PQ} \cdot \overline{P\overline{R}S}}$$

The circuit for Equation 3–41 is shown in Figure 3–38.

Figure 3–38

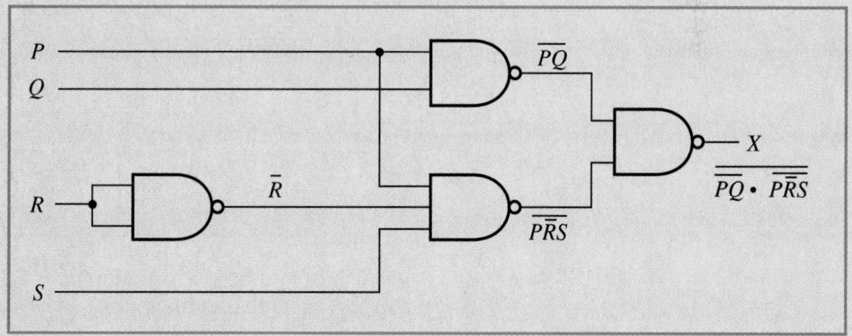

(c) $Z = \overline{(A + B)}\, C + A\, \overline{(B + C)}$

We must first convert the equation to sum-of-products form using DeMorgan's law

$$Z = (\overline{A} \cdot \overline{B})\, C + A\, (\overline{B} \cdot \overline{C})$$

$$Z = \overline{A}\overline{B}C + A\overline{B}\overline{C}$$

Inverting both sides of the equation

$$\overline{Z} = \overline{\overline{A}\overline{B}C + A\overline{B}\overline{C}}$$

Applying DeMorgan's law

$$\bar{Z} = \overline{\bar{A}\bar{B}C} \cdot \overline{A\bar{B}\bar{C}}$$

Inverting both sides again

$$\bar{\bar{Z}} = \overline{\overline{\bar{A}\bar{B}C} \cdot \overline{A\bar{B}\bar{C}}}$$

The final *NAND* equation is

Equation (3–42)

$$Z = \overline{\overline{\bar{A}\bar{B}C} \cdot \overline{A\bar{B}\bar{C}}}$$

The circuit for Equation 3–42 is shown in Figure 3–39.

Figure 3–39

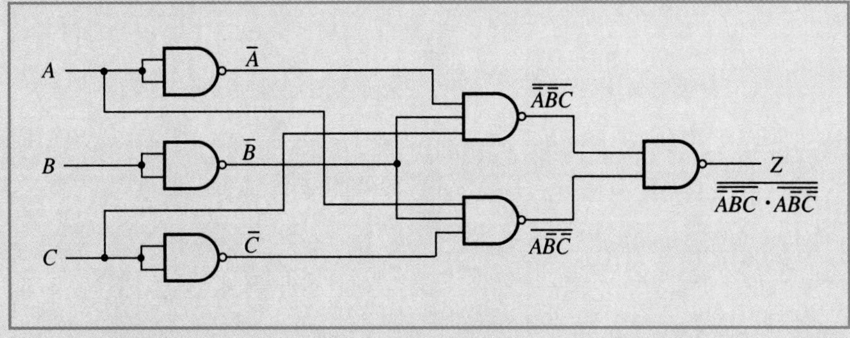

EXAMPLE 3–12

Convert the following simplified *AND-OR* logic equations to *NOR* logic and draw the *NOR* circuit for each equation.

(a) $A = (\bar{X} + Y + \bar{Z})(X + Y)(X + Z)$

(b) $Z = A + B\bar{C}$

(c) $S = (\overline{PQ} + R)(P + \overline{QR})$

SOLUTIONS

(a) $A = (\bar{X} + Y + \bar{Z})(X + Y)(X + Z)$

Inverting both sides

$$\bar{A} = \overline{(\bar{X} + Y + \bar{Z})(X + Y)(X + Z)}$$

Applying DeMorgan's law

$$\bar{A} = (\overline{\bar{X} + Y + \bar{Z}}) + (\overline{X + Y}) + (\overline{X + Z})$$

Inverting both sides again

$$\bar{\bar{A}} = \overline{(\overline{\bar{X} + Y + \bar{Z}}) + (\overline{X + Y}) + (\overline{X + Z})}$$

The final *NOR* equation is

Equation (3–43)

$$A = \overline{(\overline{\bar{X} + Y + \bar{Z}}) + (\overline{X + Y}) + (\overline{X + Z})}$$

The circuit for Equation 3–43 is shown in Figure 3–40.

Figure 3–40

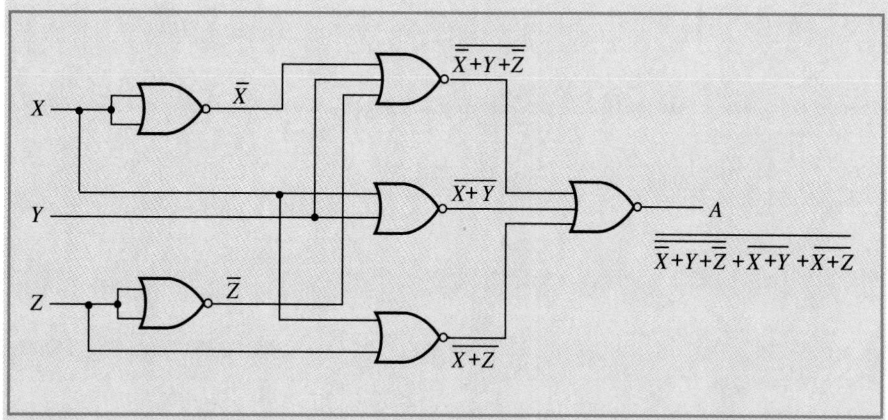

(b) $Z = A + B\overline{C}$

We must first convert the equation to product-of-sums form using the distributive law

$$Z = (A + B) \cdot (A + \overline{C})$$

Both sides of the equation are now inverted

$$\overline{Z} = \overline{(A + B) \cdot (A + \overline{C})}$$

Applying DeMorgan's law

$$\overline{Z} = \overline{(A + B)} + \overline{(A + \overline{C})}$$

Inverting both sides again

$$\overline{\overline{Z}} = \overline{\overline{(A + B)} + \overline{(A + \overline{C})}}$$

The final *NOR* equation is

Equation (3–44)

$$Z = \overline{\overline{(A + B)} + \overline{(A + \overline{C})}}$$

The circuit for Equation 3–44 is shown in Figure 3–41.

Figure 3–41

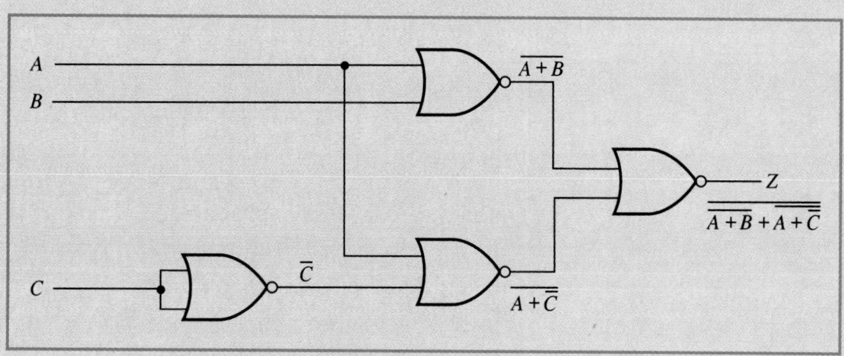

(c) $S = (\overline{PQ} + R)(P + \overline{QR})$

We must first convert the equation to product-of-sums form using DeMorgan's law

$$S = (\bar{P} + \bar{Q} + R)(P + \bar{Q} + \bar{R})$$

Inverting both sides of the equation

$$\bar{S} = \overline{(\bar{P} + \bar{Q} + R)(P + \bar{Q} + \bar{R})}$$

Applying DeMorgan's law

$$\bar{S} = \overline{(\bar{P} + \bar{Q} + R)} + \overline{(P + \bar{Q} + \bar{R})}$$

Inverting both sides again

$$\bar{\bar{S}} = \overline{\overline{(\bar{P} + \bar{Q} + R)} + \overline{(P + \bar{Q} + \bar{R})}}$$

The final *NOR* equation is

Equation (3–45)

$$S = \overline{\overline{(\bar{P} + \bar{Q} + R)} + \overline{(P + \bar{Q} + \bar{R})}}$$

The circuit for Equation 3–45 is shown in Figure 3–42.

Figure 3–42

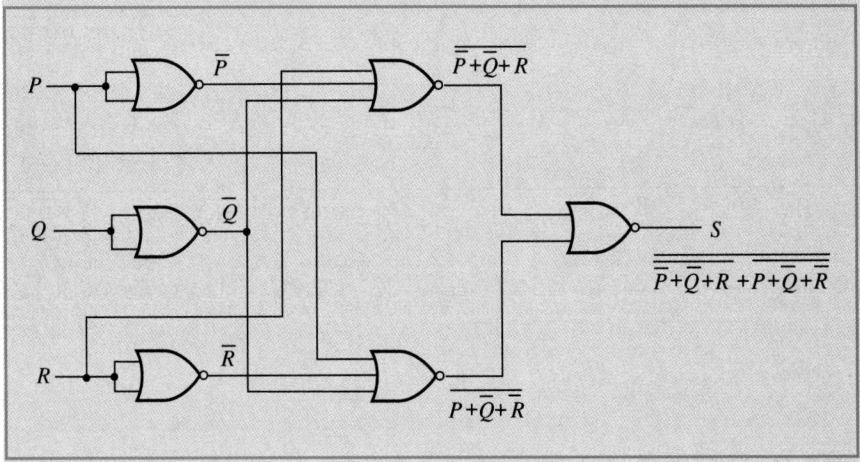

Review questions

1. What is an *AND-OR* logic circuit? How does it differ from a *NAND* and *NOR* logic circuit?
2. What are the advantages of *NAND* and *NOR* circuits over *AND-OR* circuits?
3. Why are *NAND* and *NOR* gates considered to have a "universal" property?
4. What form should an *AND-OR* logic equation be in before it is converted to *NAND* logic?
5. What form should an *AND-OR* logic equation be in before it is converted to *NOR* logic?

3–7 Karnaugh map simplification

The simplification of logic equations using the laws of Boolean algebra requires that one be proficient in applying these laws appropriately to obtain an equivalent logic equation in its simplest form. In many cases, Boolean simplification must be performed on a trial-and-error basis and consequently can be quite time-consuming. The *Karnaugh map* (K-map) simplification technique is a more direct and mechanical technique that (if used properly) can obtain the simplest logic equation from the truth table of an unsimplified logic equation. Furthermore, the K-map can also be used to obtained simplified logic equations in both sum-of-products and product-of-sums form using the same procedures.

The Karnaugh map, also referred to as a *K-map*, is a modified version of a truth table. Like a truth table, it maps the dependent variable as a function of the independent variables, but its arrangement is such that we can tell by inspection if simplification is possible or not, and if simplification is possible its configuration allows us to obtain the simplest possible logic equation. K-maps are generally classified according to the number of independent variables in the logic equation, circuit, or truth table. We generally deal with two-variable, three-variable, and four-variable K-maps. K-maps with more than four variables are possible but can get very complicated and therefore other techniques must be used. This section will examine the configuration of these K-maps and their use in the simplification of logic equations. This technique will be used in the simplification of logic equations for most of the examples covered in the rest of this book and therefore it is important that its usage be thoroughly understood.

Two-variable Karnaugh maps

The general configuration of a two-variable K-map is shown in Figure 3–43 to represent the independent variables A and B and the dependent variable X. The K-map is made up of four *cells* (corresponding to the four combinations of two independent variables, $2^2 = 4$). For reference purposes only, each cell has been numbered from 0 to 3 to identify the binary combination of A and B that identifies the cell (assuming that A is the MSB and B is the LSB). Therefore cell no. 0 corresponds to $A = 0$ and $B = 0$, and cell no. 1 corresponds to $A = 0$ and $B = 1$, cell no. 2 corresponds to $A = 1$ and $B = 0$, and cell no. 3 corresponds to $A = 1$ and $B = 1$. The bits that will be placed in each cell will be the values of the dependent variable X. For example, Figure 3–44 shows a two-variable truth table and its corresponding K-map configuration. Notice in Figure

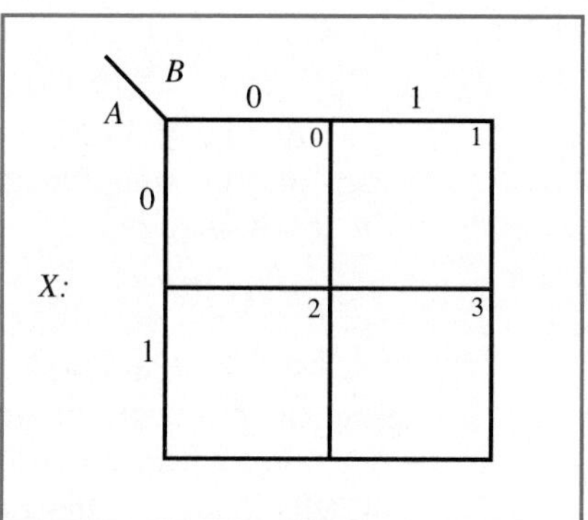

Figure 3–43
General configuration of a two-variable K-map.

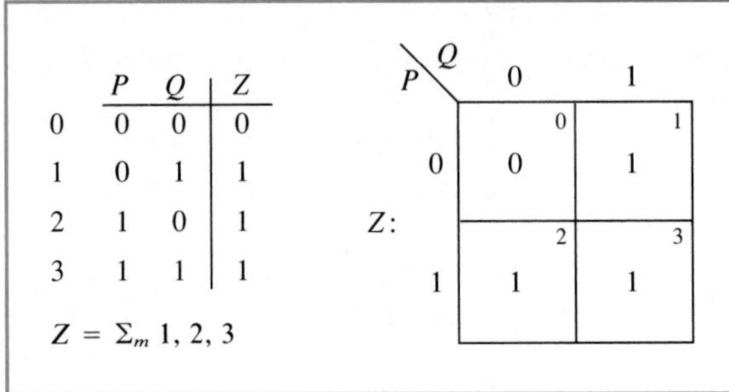

Figure 3–44
A two-variable truth table and its K-map configuration.

3–44 that the value inside each cell is the value of the dependent variable Z, and the coordinates of each cell are the values of the independent variables P and Q.

Before we can obtain the logic equation from the K-map we must first define the *adjacency* of two cells in a K-map. Two cells are said to be *adjacent* if *only one independent variable changes in going from one cell to the other*. For example, in Figure 3–43, cell no. 0 and cell no. 1 are adjacent because the variable B changes from a 0 to a 1 (or a 1 to a 0) but the variable A remains the same (0) in both cells. Similarly, cell no. 1 and cell no. 3 are adjacent because A changes from a 0 to a 1 (or a 1 to a 0) but the variable B remains the same (1) in both cells. Notice that cell no. 1 and cell no. 2 are *not* adjacent because both variables A and B change in going from one cell to the other—in going from cell no. 1 to cell no. 2, B changes from a 1 to a 0 and A changes from a 0 to a 1. Similarly, cell no. 0 and cell no. 3 are not adjacent.

There are two methods of using a K-map. The first involves combining all the cells containing 1's (1-cells) to obtain a simplified sum-of-products equation, and the other involves combining all the cells containing 0's (0-cells) to obtain the product-of-sums equation. Notice that this procedure is similar to the procedure used to obtain the *unsimplified* sum-of-products and product-of-sums equation from a truth table. We shall first examine the sum-of-products procedure and then apply the concepts developed to the product-of-sums procedure discussed at the end of this section.

Figure 3–45a shows a K-map representation of a truth table containing two 1's for the dependent variable P. Notice that these two 1's are located in cells that are *not* adjacent. This means that simplification is *not possible*, and we must take the sum of the minterms corresponding to these two 1's. The *unsimplified* equation is

$$P = \overline{X}\overline{Y} + XY$$

Similarly, in Figure 3–45b, the two 1-cells are not adjacent and therefore no simplification is possible, and the unsimplified logic equation for the K-map is

$$P = X\overline{Y} + \overline{X}Y$$

If a K-map contains a single 1-cell then the equation will contain only the minterm for that cell.

Simplification is only possible in K-maps that have logic 1's located in cells that are adjacent. For example, in Figure 3–46a, since there are

Figure 3–45
Nonadjacent 1-cells.

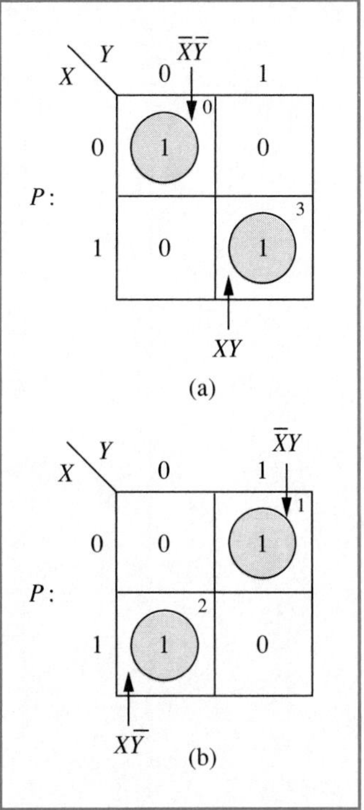

two adjacent 1-cells, and since the variable X remains the same (1) in both cells, the logic equation for the K-map is

$$P = X$$

Notice in the K-map shown in Figure 3–46b that the variable X also remains the same in both cells but its value is a 0. Therefore the logic equation for the K-map is

$$P = \overline{X}$$

In Figure 3–46c, the variable Y remains constant in both adjacent 1-cells, therefore the equation is

$$P = Y$$

Similarly in Figure 3–46d, the variable $\overline{Y}$ remains constant in both adjacent 1-cells (since its value is 0) and therefore the equation for the K-map is

$$P = \overline{Y}$$

In K-maps where there are more than one set of adjacent 1's we must *OR* the variables obtained from each set. For example, in Figure 3–47a since adjacent cells no. 1 and no. 3 produce the variable Y, and adjacent cells no. 2 and no. 3 produce the variable X, the logic equation for the K-map is

$$P = X + Y$$

Similarly, for the K-map in Figure 3–47b

$$P = \overline{X} + Y$$

Figure 3–46
Groups of two adjacent 1-cells.

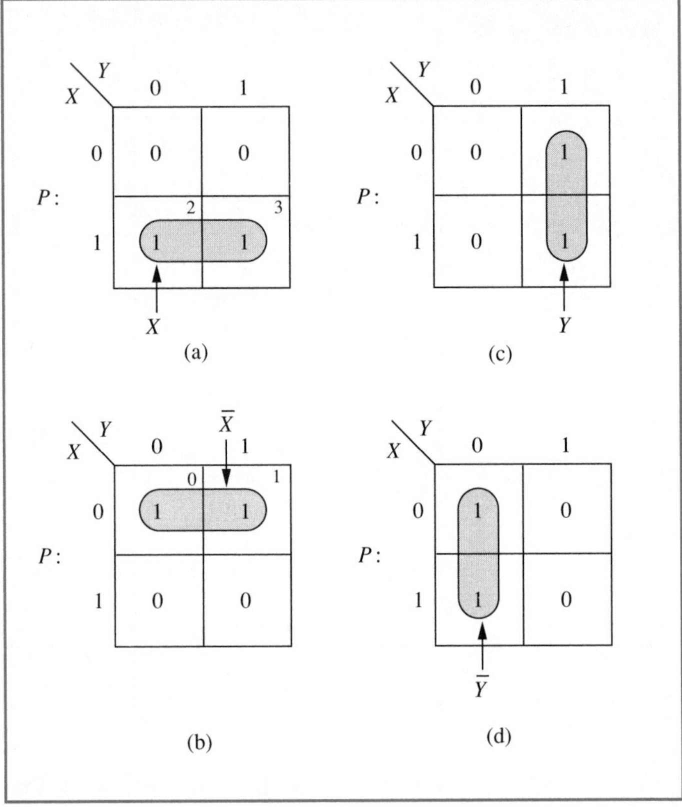

Figure 3–47
Multiple groups of two adjacent 1-cells.

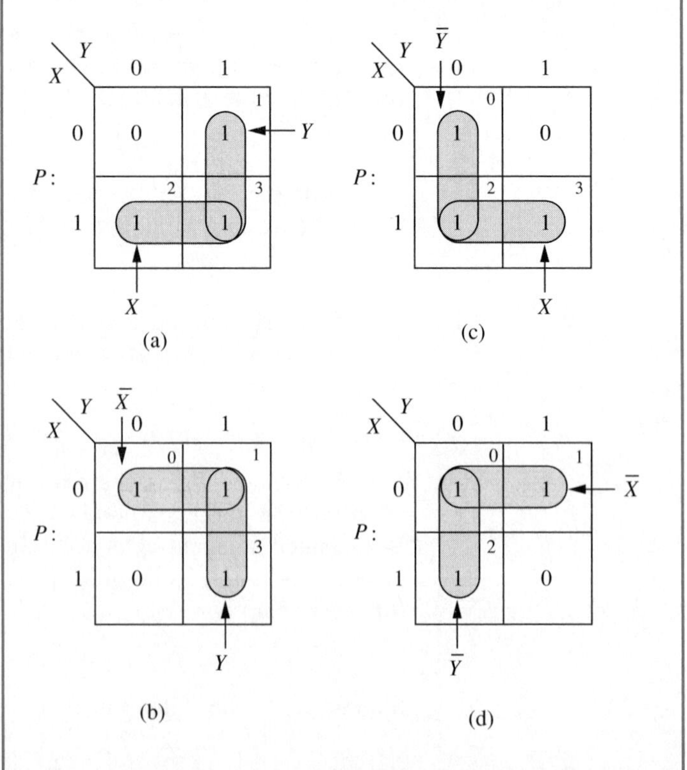

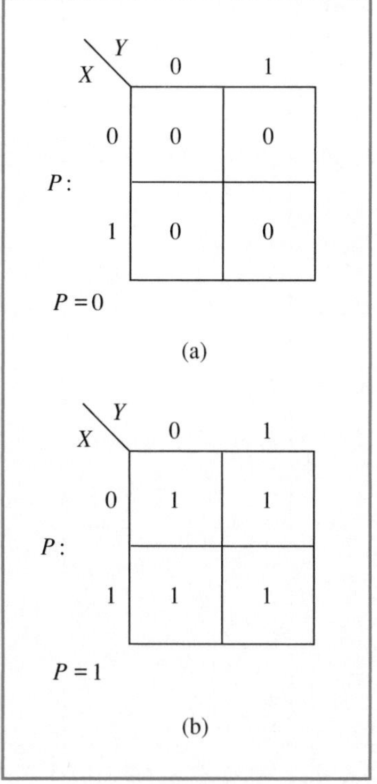

Figure 3–48
(a) K-map with all 0-cells; (b) K-map with all 1-cells.

and for the K-maps in Figure 3–47c and d, $P = X + \overline{Y}$ and $P = \overline{X} + \overline{Y}$, respectively.

If a K-map contains all 0's as shown in Figure 3–48a, then it is obvious that the dependent variable has a value of 0 permanently and does not depend on the states of the independent variables. Similarly in Figure 3–48b the dependent variable P has a value of 1 regardless of the independent variables X and Y.

Two-variable K-maps are not very beneficial in the simplification of logic equations since in most cases the simplified equation for a two-variable truth table can be obtained by inspection or by applying some elementary Boolean algebra rules. However, the advantage of using K-maps becomes evident when they are used to obtain simplified equations for three- and four-variable truth tables.

Three-variable Karnaugh maps

Three-variable K-maps map a dependent variable against three independent variables. Therefore the number of cells in a three-variable K-map will be 2^3 or 8. Since the number of independent variables is an odd number, there are two ways in which a three-variable K-map can be drawn as shown in Figure 3–49a and b. Again, notice that each cell is assigned a number corresponding to the binary combination of the independent variables (A, B, and C), assuming that A is the MSB and C is the LSB. Also, notice that the cells are arranged such that in going from one cell to another both vertically and horizontally, only one variable changes. Therefore, cell no. 0 is adjacent to cells nos. 1, 4, and 2; cell no. 1 is adjacent to cells nos. 3, 5, and 0; cell no. 3 is adjacent to cells nos. 2, 7, and 1; cell no. 2 is adjacent to cells nos. 0, 3, and 6; cell no. 4 is adjacent to cells nos. 0, 5, and 6, and so on. As in two-variable K-maps,

Figure 3–49
Two different configurations of a three-variable K-map.

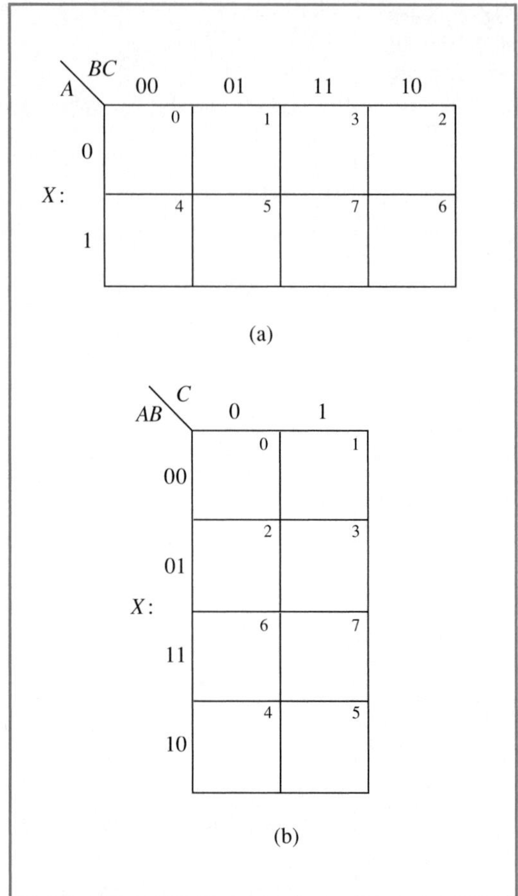

cells that are diagonally opposite each other are not adjacent since more than one variable changes in going from one cell to another. Thus, cells nos. 4 and 1 are *not* adjacent, cells nos. 5 and 3 are *not* adjacent, etc.

Because of the manner in which the cells in a three-variable K-map are numbered, care must be taken when filling in the values of the dependent variable into each cell in going from a truth table to a K-map. Figure 3–50 shows a three-variable truth table and its three-variable K-map representations. Notice that the values in each cell of the K-map are the values of the dependent variable Z, and the "coordinates" of each cell are the values of the independent variables P, Q, and R.

As stated earlier, there are two methods used to obtain a simplified equation from a K-map—combining cells containing 1's into adjacent groups to obtain a simplified sum-of-products equation, and combining cells containing 0's into adjacent groups to obtain a simplified product-of-sums equation. We will examine the first method now and then apply the concepts to the second method at the end of this section.

As stated earlier, if a K-map has nonadjacent cells containing 1's, then no simplification is possible. Figure 3–51 illustrates such a K-map. In cases such as this the logic equation is simply the sum of the minterms for each cell

$$P = \overline{A}\overline{B}\overline{C} + \overline{A}BC + A\overline{B}C$$

Figure 3–50
A three-variable truth table and its K-map.

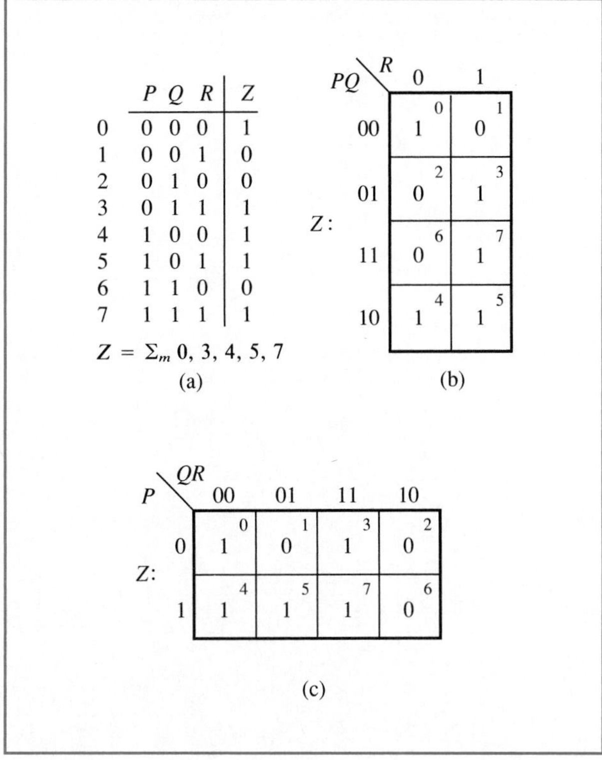

	P	Q	R	Z
0	0	0	0	1
1	0	0	1	0
2	0	1	0	0
3	0	1	1	1
4	1	0	0	1
5	1	0	1	1
6	1	1	0	0
7	1	1	1	1

$Z = \Sigma_m\, 0, 3, 4, 5, 7$

The following examples illustrate other K-map configurations in which no simplification is possible.

K-maps that have cells containing 1's that are adjacent in groups of two will produce simplified logic equations. For example, in Figure 3–53 we have K-maps with five cells containing 1's. Notice that Figure 3–53a and b illustrate the *same* K-map, but we have simply grouped the adjacent 1-cells differently. In Figure 3–53a and b we have combined cells nos. 0 and 4, and cells nos. 3 and 7, into two groups. Cell no. 1 is adjacent to both cell no. 0 and cell no. 3, so we can combine it with cell no. 0 (as shown in Figure 3–53a) or with cell no. 3 (as shown in Figure 3–53b).

With reference to the group made up of cell no. 0 and cell no. 4, notice that the variables *B* and *C* remain constant (the variable *A*

Figure 3–51
Nonadjacent 1-cells.

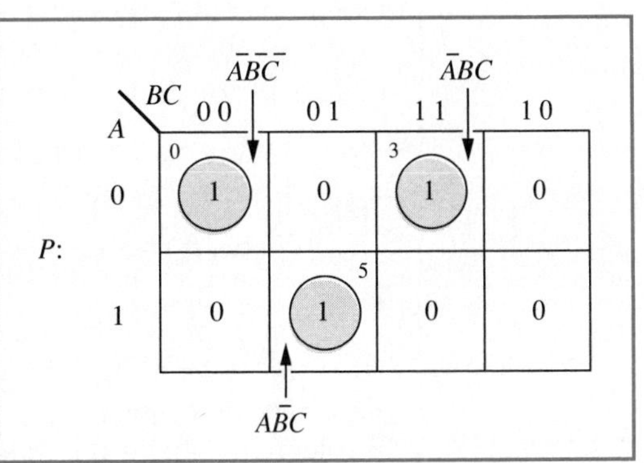

EXAMPLE 3–13

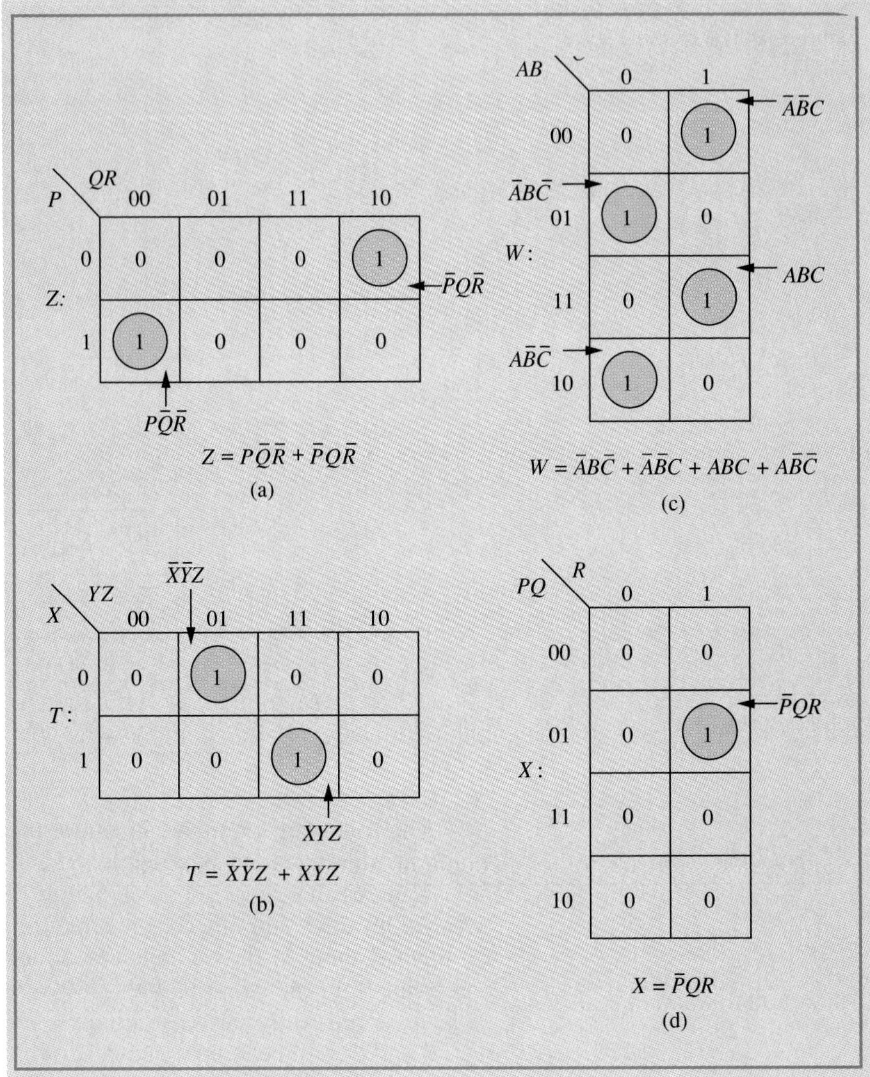

Figure 3–52
Examples of K-maps that cannot be simplified.

changes). Similarly, the variables B and C remain constant in the group made up of cells nos. 3 and 7. Since the values of B and C correspond to 00 and 11 for each group, respectively, the common variables are

$\overline{B}\overline{C}$ for adjacent cells nos. 0 and 4

and

BC for adjacent cells nos. 3 and 7

Since we will be obtaining the simplified *sum-of-products* equation from the K-map, each group will produce the *logical product* of the variables that remain constant, and we will take the *logical sum* of these products to obtain the final logic equation.

In Figure 3–53a, the group made up of adjacent cells nos. 0 and 1 yields the term

$$\overline{A}\overline{B}$$

since the variable C changes, but A and B remain constant and their values are 0. The simplified logic equation for the K-map in Figure 3–53a can now be obtained as

Equation (3–46)

$$P = \overline{B}\overline{C} + \overline{A}\overline{B} + BC$$

In Figure 3–53b, the group made up of adjacent cells nos. 1 and 3 yields the term

$$\overline{A}C$$

since the variable B changes but A and C remain constant at values of 0 and 1, respectively. The simplified logic equation for the K-map in Figure 3–53b can now be obtained as

Equation (3–47)

$$P = \overline{B}\overline{C} + \overline{A}C + BC$$

Figure 3–53
Groups of two adjacent 1-cells.

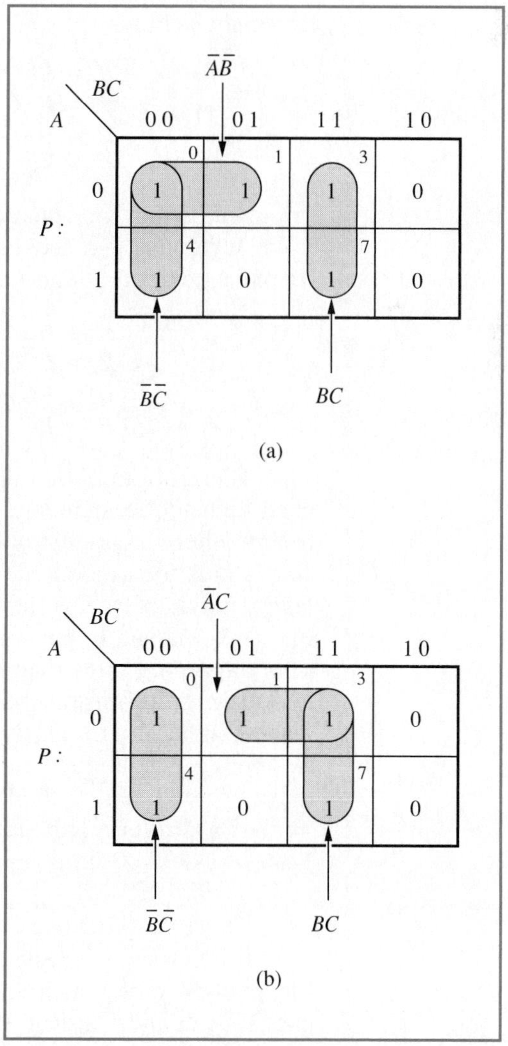

(a)

(b)

Equations 3–46 and 3–47 are logically equivalent and can be considered the same as far as simplification is concerned since both equations are made up of three terms, each term containing two variables. Notice that combining 1-cells into adjacent groups of two produces terms that have only two variables. If, for example, we did not combine the 1-cells into groups, we would have to take each cell individually and would have the following logic equation

Equation (3–48)

$$P = \bar{A}\bar{B}\bar{C} + \bar{A}\bar{B}C + \bar{A}BC + A\bar{B}\bar{C} + ABC$$

It is apparent that Equation 3–48 is unsimplified and is more complex than Equations 3–46 and 3–47 since it contains five terms, each term having three variables.

To further emphasize the importance of combining 1-cells into groups, let us obtain the equation for either of the K-maps in Figure 3–53 by *not* including cell no. 1 in any group

Equation (3–49)

$$P = \bar{B}\bar{C} + \bar{A}\bar{B}C + BC$$

Notice that Equation 3–49 is logically equivalent to Equations 3–46 and 3–47, but is more complex since it contains one term with three variables. We can prove that Equation 3–49 is not *completely* simplified by using Boolean algebra

$$\begin{aligned} P &= \bar{B}\bar{C} + \bar{A}\bar{B}C + BC & & \\ P &= \bar{B}(\bar{C} + \bar{A}C) + BC & & \text{(distributive)} \\ P &= \bar{B}(\bar{C} + \bar{A}) + BC & & \text{(miscellaneous)} \\ P &= \bar{B}\bar{C} + \bar{A}\bar{B} + BC & & \text{(distributive)} \end{aligned}$$

Notice that the equation we have obtained is the same as Equation 3–46. We could also use a different approach to the Boolean simplification and obtain Equation 3–47

$$\begin{aligned} P &= \bar{B}\bar{C} + \bar{A}\bar{B}C + BC & & \\ P &= \bar{B}\bar{C} + C(\bar{A}\bar{B} + B) & & \text{(distributive)} \\ P &= \bar{B}\bar{C} + C(\bar{A} + B) & & \text{(miscellaneous)} \\ P &= \bar{B}\bar{C} + \bar{A}C + BC & & \text{(distributive)} \end{aligned}$$

Therefore to obtain the simplest possible logic equation an attempt must be made to include a 1-cell in a group. 1-cells that are not adjacent to any other 1-cell must be taken alone (i.e., as minterms). However, all 1-cells must be accounted for in the final equation either as groups or taken individually. Overlapping groups are possible as can be seen in Figure 3–53a and b, but if all 1-cells have been accounted for either alone or in adjacent groups then it would not be appropriate to combine 1-cells into other groups again. For example, in Figure 3–53a combining cell no. 1 and cell no. 3 into *another* group would not be appropriate since these two cells have already been included in other groups; if we did combine these two cells again, the logic equation obtained (Equation 3–50) would have an extra term although it would be logically equivalent to Equation 3–46

Equation (3–50)

$$P = \bar{B}\bar{C} + \bar{A}\bar{B} + BC + \bar{A}C$$

It is possible for K-maps to have *four* adjacent cells containing 1's. Figure 3–54 shows such a K-map. Notice in Figure 3–54 that since cells nos. 1, 3, 7, and 5 are adjacent (in that order), only one variable, R,

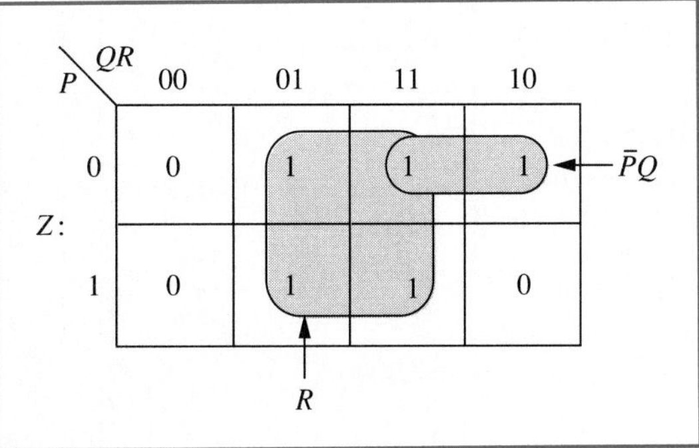

Figure 3–54
K-map with a group of four adjacent 1-cells.

remains constant in that group of four cells. Therefore the term produced by the group of four cells is R. The logic equation for the K-map in Figure 3–54 is

Equation (3–51)

$$Z = R + \bar{P}Q$$

It should now be apparent that the larger the group of adjacent 1-cells, the smaller the resulting term, and consequently the simpler the resulting logic equation. In a three-variable K-map the first attempt should be to try to find groups of four adjacent 1-cells, then groups of two adjacent 1-cells, and finally single 1-cells. The following example illustrates other K-map configurations and their logic equations.

Four-variable Karnaugh maps

Four-variable K-maps have 2^4, or 16, cells, each cell representing a unique combination of the four independent variables. As stated earlier, the value in each cell corresponds to the value of the dependent variable. Figure 3–56 shows the general configuration of a four-variable K-map that maps the dependent variable X against the independent variables A, B, C, and D. For convenience we have numbered each cell (in hexadecimal) with numbers that correspond to the binary values of the variables A, B, C, and D (assuming that A is the MSB and D is the LSB). Notice that the cells are arranged in the same manner as a three-variable K-map so that only one variable changes in going from one adjacent cell to another. As before, cells located diagonally opposite each other are *not* adjacent. Notice that cells in all rows and columns are adjacent and that the cells in each corner of the K-map (0, 2, A, 8) are adjacent.

As stated earlier care must be taken when filling in the cells of the K-map with the values of the dependent variable from a truth table, since the cells are not arranged in the same order as the entries in the truth table. Figure 3–57 shows a four-variable truth table with its K-map representation.

To obtain the simplified sum-of-products equation from a four-variable K-map we must again combine the adjacent 1-cells in groups that are as large as possible. In a four-variable K-map, the largest possible group is a group of eight adjacent 1-cells as shown in Figure 3–58. A group of eight adjacent 1-cells (cell numbers 0, 1, 4, 5, C, D, 8, 9) will only have one variable ($\bar{C}$ in this case) that remains constant. A group of eight adjacent 1-cells will therefore yield a term that has a single variable. The next largest group of cells in a four-variable K-map is 4. In Figure 3–58, cell nos. 0, 1, 3, and 2, make up a group of four adjacent 1-cells

EXAMPLE 3–14

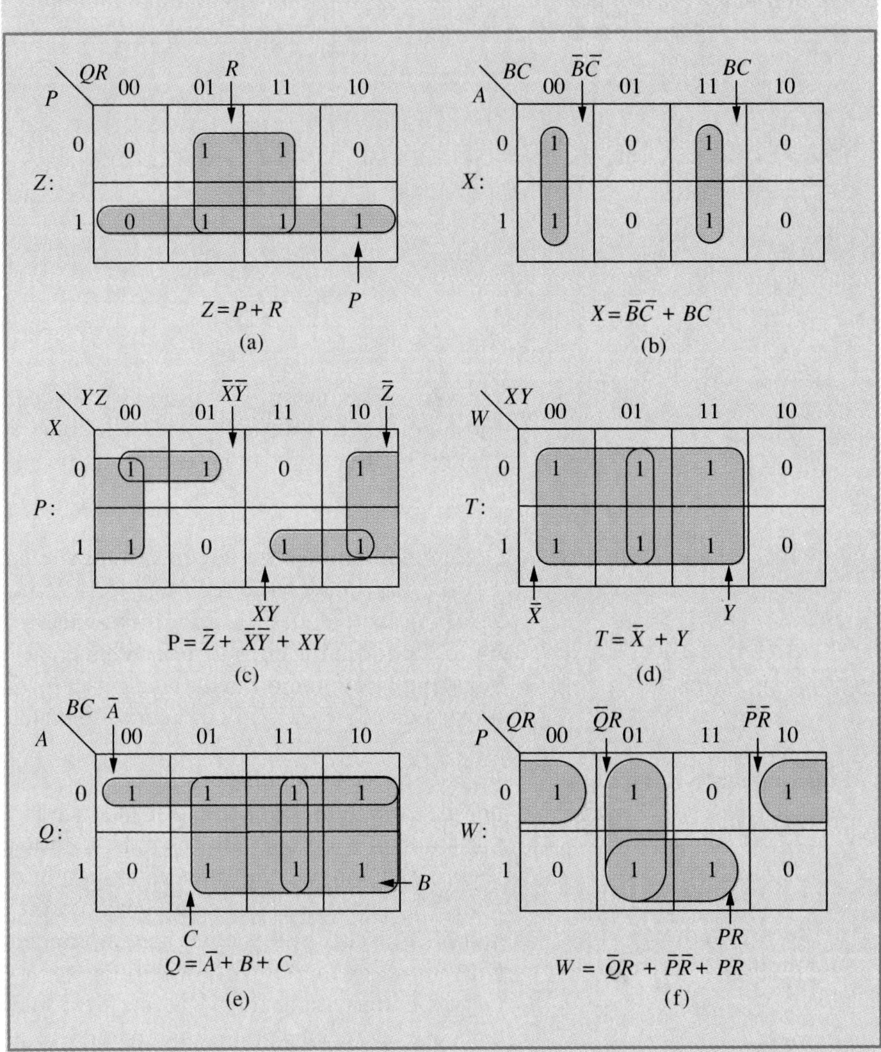

Figure 3–55 Examples of three-variable K-map simplification.

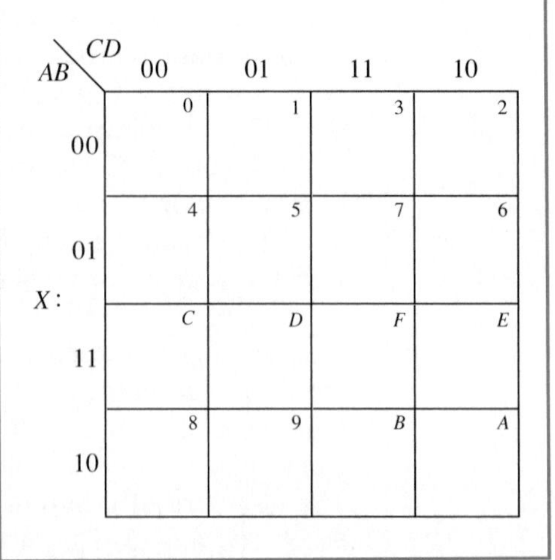

Figure 3–56 General configuration of a four-variable K-map.

Figure 3–57
A four-variable truth table and its K-map.

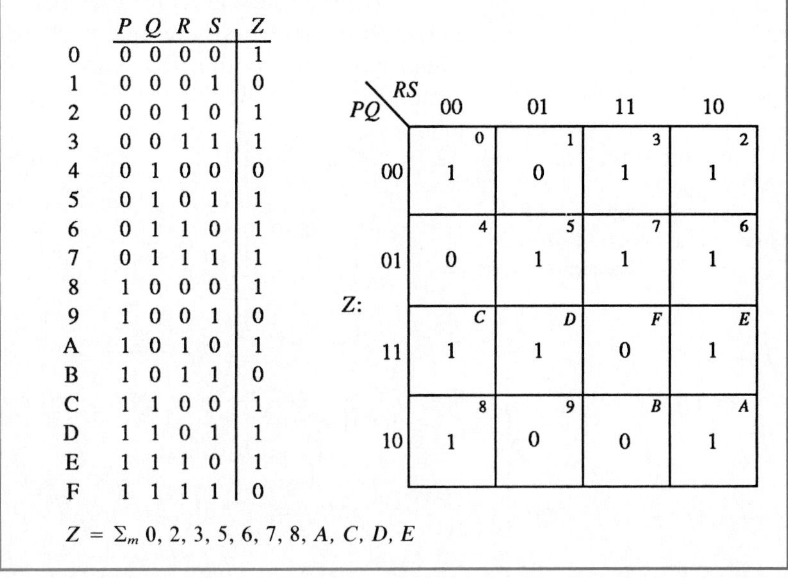

	P	Q	R	S	Z
0	0	0	0	0	1
1	0	0	0	1	0
2	0	0	1	0	1
3	0	0	1	1	1
4	0	1	0	0	0
5	0	1	0	1	1
6	0	1	1	0	1
7	0	1	1	1	1
8	1	0	0	0	1
9	1	0	0	1	0
A	1	0	1	0	1
B	1	0	1	1	0
C	1	1	0	0	1
D	1	1	0	1	1
E	1	1	1	0	1
F	1	1	1	1	0

$Z = \Sigma_m\ 0, 2, 3, 5, 6, 7, 8, A, C, D, E$

Z:

PQ \ RS	00	01	11	10
00	1 (0)	0 (1)	1 (3)	1 (2)
01	0 (4)	1 (5)	1 (7)	1 (6)
11	1 (C)	1 (D)	0 (F)	1 (E)
10	1 (8)	0 (9)	0 (B)	1 (A)

that yield the term $\overline{A}\overline{B}$. Finally, the smallest group is a group of two adjacent 1-cells. In Figure 3–58, cell no. D and cell no. F make up a group of two adjacent 1-cells that yield the term ABD. As stated earlier, 1-cells that are not adjacent to any other 1-cells must be taken singly, as minterms. There are no single 1-cells in Figure 3–58. The simplified logic equation for the K-map in Figure 3–58 can be obtained by taking the logical sum of the simplified terms (products) produced by each group of adjacent 1-cells:

Equation (3–52)

$$X = \overline{C} + \overline{A}\overline{B} + ABD$$

Figure 3–58
Combining various groups of 1-cells in a four-variable K-map.

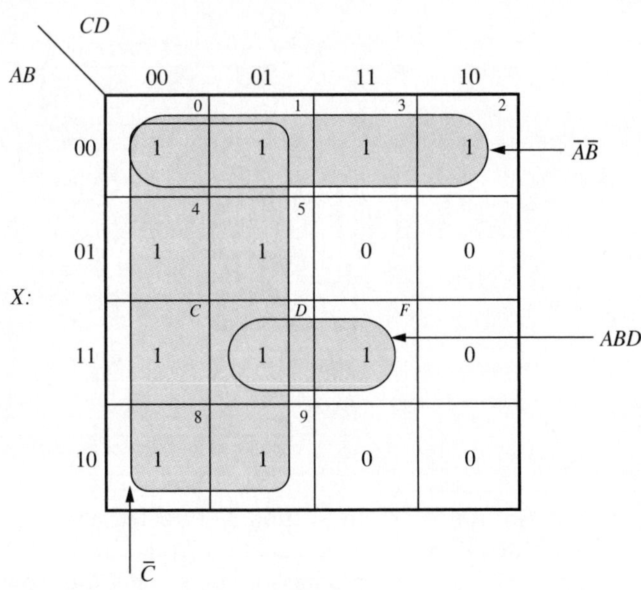

Observe that the larger the group of adjacent 1-cells, the smaller its corresponding term in the final logic equation. The following example illustrates various other four-variable K-map configurations and the simplified logic equations obtained from them.

EXAMPLE 3–15

Figure 3–59 Examples of four-variable K-map simplification.

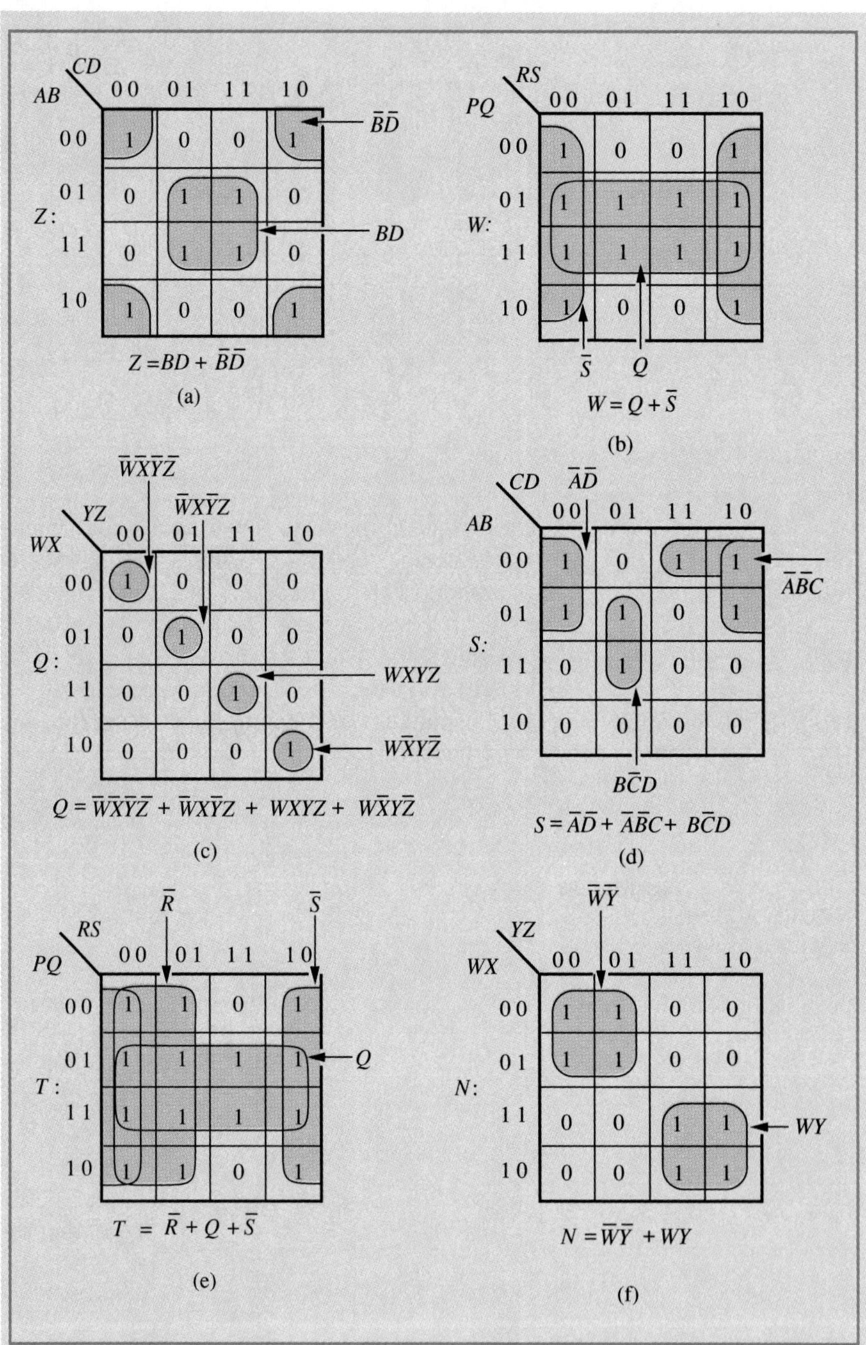

Product-of-sums simplification

In Section 3–3 we learned that it was possible to obtain an unsimplified logic equation from a truth table in one of two forms—sum-of-products or product-of-sums. This was done by obtaining the logical sum of the minterms (sum-of-products) for which the dependent variable was a 1 or the logical product of the maxterms (product-of-sums) for which the dependent variable was a 0. Earlier in this section we combined groups of adjacent 1-cells in K-maps to obtain simplified *products* (terms) that were logically *added* together to produce the final simplified logic equation. Similarly we can also combine groups of adjacent 0-cells in K-maps to obtain simplified *sums* (terms) that are logically *multiplied* together to produce the final simplified logic equation. Recall that for minterms, a variable that represented a 0 was barred and a variable that represented a 1 was unbarred, but for maxterms a barred variable represented a 1 and an unbarred variable represented a 0.

To illustrate the procedure for obtaining a simplified product-of-sums equation from a K-map, consider the K-map shown in Figure 3–54. For convenience, the K-map has been redrawn and shown in Figure 3–60 with the two groups of adjacent 0-cells identified. The variables P and R remain constant in the group made up of cell nos. 4 and 6, and since their values are 1 and 0, respectively, the term that corresponds to this group is

$$(\bar{P} + R)$$

Notice that since we are combining 0's, the term is a *sum* rather than a product with variables that have values of 0 left unbarred and variables that have values of 1 barred. Similarly, for the group made up of cell nos. 0 and 4, the variables Q and R remain constant with values of 0, and therefore this group yields the term

$$(Q + R)$$

The simplified logic equation for the K-map in Figure 3–60 is obtained by taking the product of the simplified terms (sums)

Equation (3–53)

$$Z = (\bar{P} + R) \cdot (Q + R)$$

Equation 3–53 and the simplified sum-of-products equation (Equation 3–51) obtained from the K-map in Figure 3–54 are logically equivalent. This can be verified by applying the distributive law to Equation 3–51

$$Z = R + \bar{P}Q$$
$$Z = (R + \bar{P})(R + Q)$$

The following example illustrates various other K-map configurations and their simplified product-of-sums equations.

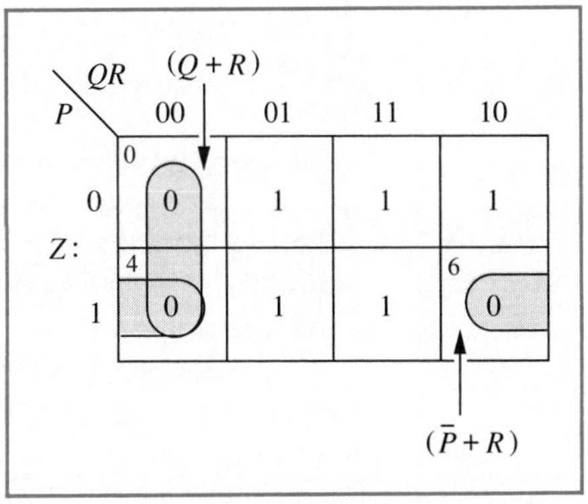

Figure 3–60
Combining 0-cells in a K-map simplification.

P \ QR	00	01	11	10
0	0 (cell 0)	1	1	1
1	0 (cell 4)	1	1	0 (cell 6)

Z: group of cells 0 and 4: $(Q+R)$; group of cells 4 and 6: $(\bar{P}+R)$

EXAMPLE 3–16

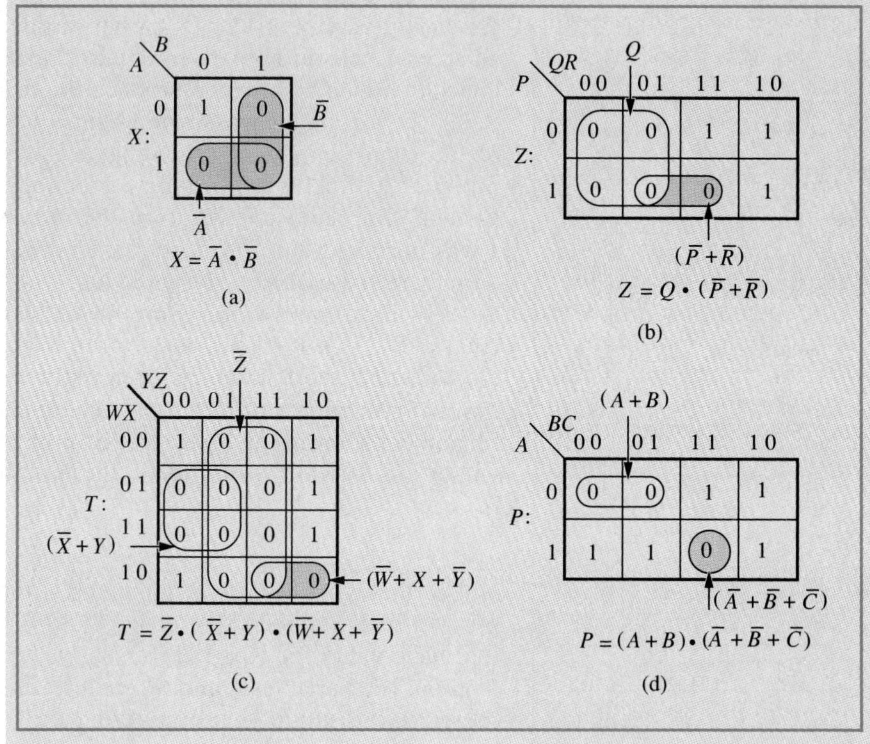

Figure 3–61 Examples of K-map simplification to yield equations in product-of-sums form.

Note that the procedure for obtaining the simplified product-of sums equation from a K-map is very similar to the procedure for obtaining the simplified sum-of-products equation, except that everything is now reversed. K-maps also provide a very convenient means of converting a logic equation from simplified sum-of-products form to simplified product-of-sums form and vice versa. In order to do this we must first obtain a truth table for the given equation, construct a K-map, and then solve the K-map to obtain the equation in its complementary form. The following examples illustrate this procedure.

EXAMPLE 3–17

Convert the following equations into their complementary forms: (a) $Z = A\overline{B} + \overline{A}C$ and (b) $X = (P + Q)(R + \overline{S})$

SOLUTIONS

(a) To convert the equation

$$Z = A\overline{B} + \overline{A}C$$

into product-of-sums form, the equation's truth table shown in Table 3–24 is first obtained. Next, the truth table values are entered into the K-map shown in Figure 3–62. Note that we could also go directly from the equation to a K-map since a K-map is effectively a truth table.

From the K-map in Figure 3–62 we can now obtain the simplified product-of-sums equation

$$Z = (A + C)(\overline{A} + \overline{B})$$

Table 3–24

A	B	C	Z
0	0	0	0
0	0	1	1
0	1	0	0
0	1	1	1
1	0	0	1
1	0	1	1
1	1	0	0
1	1	1	0

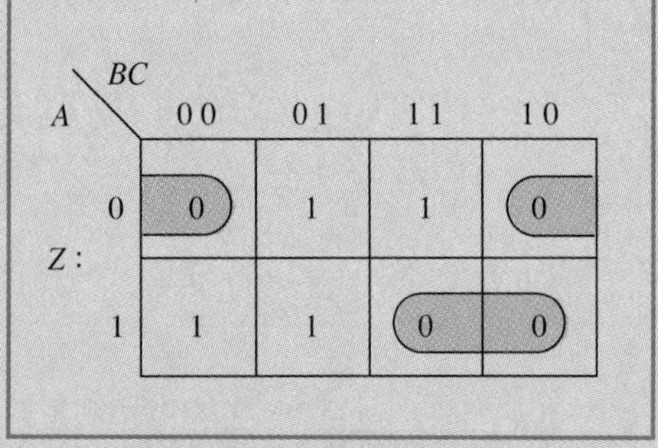

Figure 3–62
Using a K-map to convert from sum-of-products to product-of-sums form.

(b) To convert the equation

$$X = (P + Q)(R + \overline{S})$$

into sum-of-products form, the equation's truth table shown in Table 3–25 is first obtained. Next the truth table values are entered into the K-map shown in Figure 3–63.

From the K-map in Figure 3–63 we can now obtain the simplified sum-of-products equation

$$Z = PR + QR + P\overline{S} + Q\overline{S}$$

Table 3–25

P	Q	R	S	X
0	0	0	0	0
0	0	0	1	0
0	0	1	0	0
0	0	1	1	0
0	1	0	0	1
0	1	0	1	0
0	1	1	0	1
0	1	1	1	1
1	0	0	0	1
1	0	0	1	0
1	0	1	0	1
1	0	1	1	1
1	1	0	0	1
1	1	0	1	0
1	1	1	0	1
1	1	1	1	1

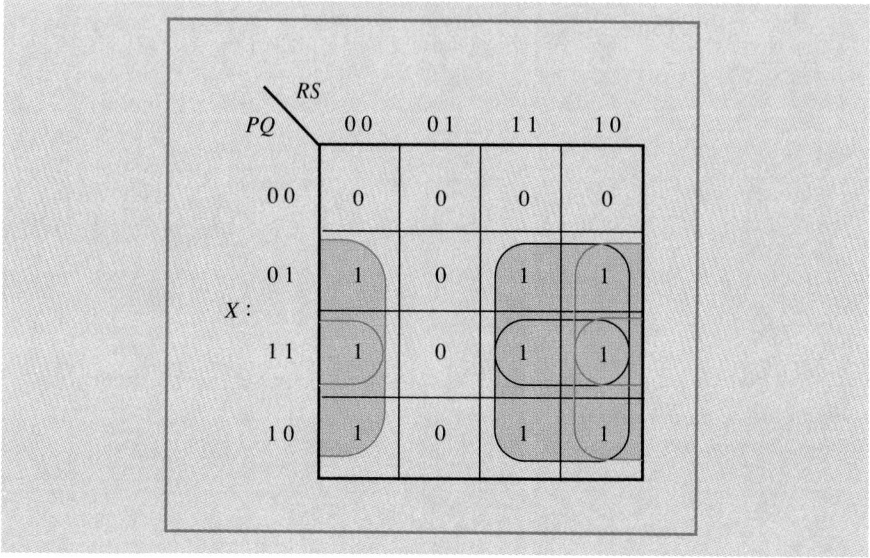

Figure 3–63
Using a K-map to convert from product-of-sums to sum-of-products form.

This section has examined an important technique used in the simplification of logic equations and circuits—the Karnaugh map technique. The Karnaugh map simplification is not designed to be a complete replacement of the Boolean algebra techniques of simplification but is a quick and effective substitute in many cases. Unlike Boolean simplification, the K-map approach can directly indicate whether simplification is possible and if so provide the simplest logic equation. Boolean algebra can still be an effective means of reducing a logic circuit even beyond the capabilities of a K-map. Besides, the function of a K-map is based on the laws of Boolean algebra, and therefore a good understanding of the laws of Boolean algebra is an important prerequisite to the efficient use of a Karnaugh map. The techniques of simplification examined in this chapter will be applied to many of the circuits that will be designed and analyzed in subsequent chapters; the specific simplification procedure used however, will always be the one that best fits the application.

Review questions

1. In what way is a K-map similar to a truth table?
2. What determines adjacency between cells in a K-map?
3. What is the number of cells in a K-map based on?
4. How does one recognize that there is no simplification possible in a K-map?
5. What are the largest group of adjacent 1- and 0-cells that can exist in two-, three-, and four-variable K-maps?
6. What are the differences between the sum-of-products and the product-of-sums procedures of using a K-map?

3–8 Integrated circuit logic and troubleshooting applications

Many of the rules of Boolean algebra introduced in this chapter can be (and have been) applied to the construction of IC logic circuits and the troubleshooting of logic gates and circuits.

For example, in Figure 3–64a we can test the operation of an *AND* gate by applying the identity law ($A \cdot 1 = A$). One input of the *AND* gate is tied high while the other input is pulsed low. If the output pulses low, then the gate is functioning properly. But if the output indicates a permanent high or a low (shorted output) or if there is no logic level detected (open output), then the gate is defective. Similarly in Figure 3–64b we can test the operation of the *OR* gate by applying the identity

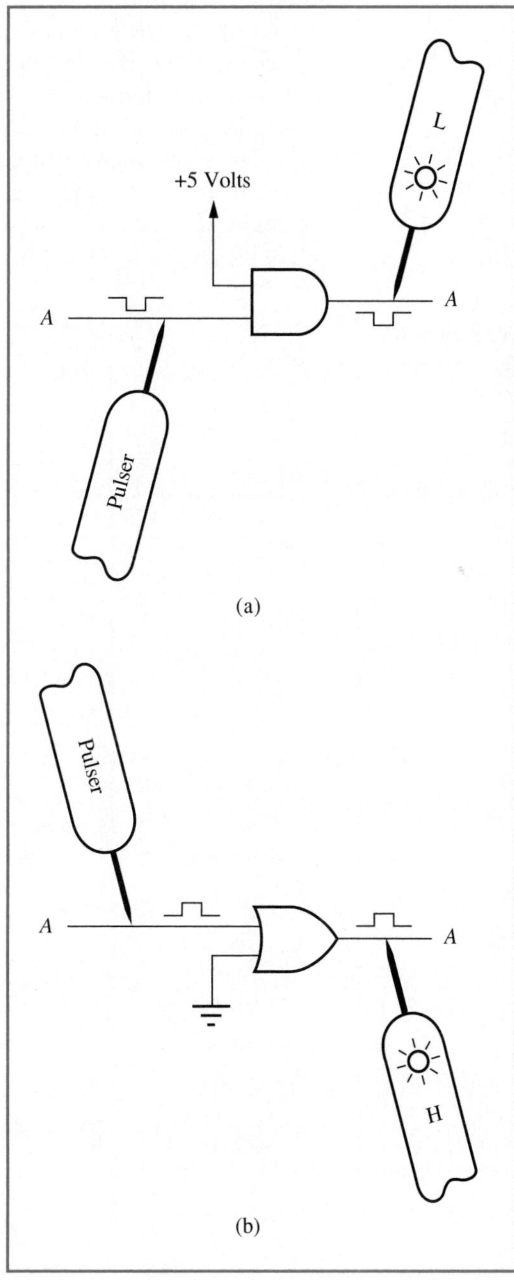

Figure 3–64
Application of the identity law in troubleshooting a logic circuit.

law ($A + 0 = A$). One input of the *OR* gate is tied low while the other input is pulsed high. If the output pulses high, then the gate is functioning properly. But if the output indicates a permanent high or a low (shorted output) or if there is no logic level detected (open output), then the gate is defective. The same procedures can also be applied to test *NAND* and *NOR* gates.

In isolating various faults in logic gates and circuits we generally apply the dominance law. For example, if the output of an *AND* gate is stuck at the logic low state as shown in Figure 3–65a and if we have determined that the output is not shorted to ground, then by logical deduction the dominance law tells us that one of the inputs could be internally shorted to ground ($A \cdot 0 = 0$). Similarly, in Figure 3–65b if the output of the *OR* gate is stuck at the logic high state and if we have determined that the output is not shorted to V_{CC}, then by logical deduction the dominance law tells us that one of the inputs could be internally shorted to V_{CC} ($A + 1 = 1$).

There are many instances when the design of a logic circuit requires a type of logic gate that may not be available in the form of an IC. For example, if a logic circuit requires a three-input *OR* gate, which is not available in the form of an IC, we can apply the associative law and

Figure 3–65
Application of the dominance law in troubleshooting a logic circuit.

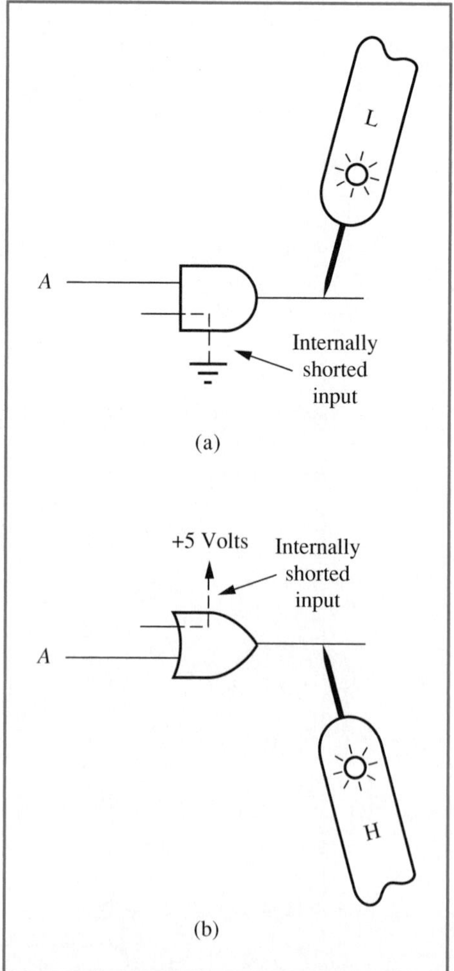

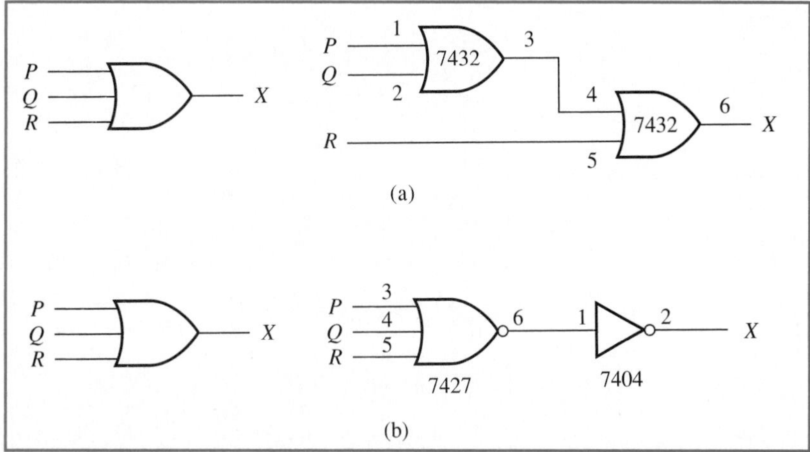

Figure 3–66 Use of Boolean algebra in the implementation of logic circuits.

implement the circuit using two 2-input *OR* gates from a 7432 IC as shown in Figure 3–66a. Or we could apply one of the Boolean postulates (double negation) and use a 7427 three-input *NOR* gate along with a 7404 inverter as shown in Figure 3–66b. DeMorgan's law is used extensively to convert from *NAND* and *NOR* logic to *AND-OR* logic and vice versa can greatly reduce the gate and IC count (and consequently the cost) of a logic circuit. Examples of this were seen in Section 3–6. In the design and implementation of logic circuits we often have many "spare" gates left over in the ICs that make up the circuit. These gates can often be efficiently utilized through the proper application of Boolean algebra to produce a more efficient and compact circuit.

The technique of analysis introduced in Section 3–2 is used extensively to troubleshoot logic circuits. This technique allows us to determine the outputs of a circuit by tracing each combination through the circuit from inputs to outputs. For example, if we were troubleshooting the circuit shown in Figure 3–1, we would construct a table similar to the one shown in Table 3–2 and record the measured logic levels at the outputs of each gate. The recorded data table can then be compared to the expected data table (Table 3–2) and any discrepancy in logic level can be attributed to a specific gate or a pair of gates. The following example illustrates this procedure.

EXAMPLE 3–18

The following table lists the recorded data obtained from Figure 3–67 after a logic probe was used to record the logic outputs of each

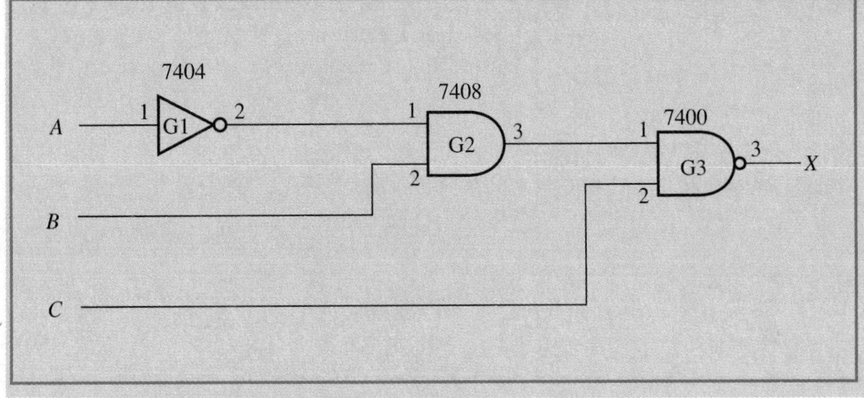

Figure 3–67 Circuit containing a fault.

gate for all possible combinations of the input. The circuit has a fault in it. Identify the fault in the circuit using this data.

A	B	C	Outputs G1	G2	G3	
L	L	L	H	H	H	← fault
L	L	H	H	H	L	← fault
L	H	L	H	H	H	
L	H	H	H	H	L	
H	L	L	L	L	H	
H	L	H	L	L	H	
H	H	L	L	L	H	
H	H	H	L	L	H	

To determine the fault in the circuit, the circuit must be analyzed by tracing through all possible combinations and determining the output of each gate for each combination. The resulting table (expected results) is shown below for the logic equation $X = \overline{\overline{A}B \cdot C}$

A	B	C	Outputs $\overline{A}$	$\overline{A}B$	$\overline{\overline{A}B \cdot C}$	
L	L	L	H	L	H	← fault
L	L	H	H	L	H	← fault
L	H	L	H	H	H	
L	H	H	H	H	L	
H	L	L	L	L	H	
H	L	H	L	L	H	
H	H	L	L	L	H	
H	H	H	L	L	H	

Notice that the first two entries in the above table of expected results (highlighted) do not match the first two (corresponding) entries in the table of recorded results. The fault cannot be in the output of G3 since the low output is produced because both of its inputs are in the high state. Besides, the output of G3 is correct for all other combinations. Since the output of G1 appears to be normal, it would appear that gate G2 has a fault possibly because pin 2 is internally shorted to V_{CC}.

It is therefore important for us to have a good knowledge of the rules of Boolean algebra when it comes to troubleshooting logic gates and circuits. A basic understanding of the analysis procedures is also important since it enables us to quickly and efficiently narrow down the fault (or faults) in a logic circuit to a specific gate or IC.

Review questions

1. Explain the ways in which some of the laws of Boolean algebra can be applied to the construction of logic circuits with ICs.
2. Discuss some of the applications of Boolean algebra in troubleshooting logic circuits.
3. How can we apply the procedure for analyzing logic circuits to troubleshooting?

3–9 SUMMARY

- The *analysis* of a logic circuit involves procedures to obtain its truth table, logic equation, or both in order to determine the function of the circuit.
- The *synthesis* of a logic circuit involves procedures to obtain (or design) a logic circuit from a truth table, logic equation, or specification.
- *Minterms* are logical products that produce a 1 for one and only one combination of the independent variables.
- *Maxterms* are logical sums that produce a 0 for one and only one combination of the independent variables.
- Minterms and maxterms are complements of each other.
- Logic equations derived from a truth table can be expressed in two equivalent forms—*sum-of-products* (SOP) and *product-of-sums* (POS).
- The sum-of-products and product-of-sums equations obtained directly from a truth table are generally unsimplified.
- *Boolean algebra* is also known as the algebra of logic.
- Boolean algebra is used in the simplification of logic equations and circuits.
- The objective of Boolean simplification is to reduce the number of gates and ICs required to implement the circuit in hardware.
- A logic equation is considered to be completely simplified when no rule of Boolean algebra can be used to reduce it any further.
- Some logic equations cannot be simplifed using Boolean algebra.
- *AND-OR logic* circuits are logic circuits that are made up of *AND* gates, *OR* gates, and inverters.
- *NAND logic* circuits are logic circuits that are made up of *NAND* gates only.
- *NOR logic* circuits are logic circuits that are made up of *NOR* gates only.
- *NAND* and *NOR* logic circuits can greatly reduce the number of ICs required to implement a circuit in hardware.
- Any of three basic functions—*AND*, *OR*, and *NOT*—can be implemented with *NAND* and *NOR* gates.
- The *Karnaugh map* (*K-map*) is a type of truth table that maps the dependent variable against all possible combinations of the independent variables.
- *Adjacent cells* in a K-map are cells in which only one independent variable changes in going from one cell to another.
- The cells in a K-map are organized such that all cells located in rows and columns are adjacent while cells that are located diagonally opposite each other are not.
- The number of cells in a K-map is equal to the number of combinations of the independent variables.

- The value entered in a K-map cell is the value of the dependent variable while the coordinates of the cell are the combination of the independent variables.
- We can obtain a simplified sum-of-products equation from a K-map by combining all *1-cells* (cells containing logic 1's).
- We can obtain a simplified product-of-sums equation from a K-map by combining all *0-cells* (cells containing logic 0's).
- No simplification is possible if 1-cells or 0-cells are not adjacent in the K-map.
- We combine adjacent cells in a K-map in groups of 8, 4, and 2, with each group producing a simplified term.
- The larger the group of adjacent cells, the smaller the resulting term.
- Cells that are not adjacent to any other cells must be taken as minterms (1-cells) or maxterms (0-cells).
- The K-map can also be used to obtain an equivalent simplified product-of-sums equation, given a sum-of-products equation, and vice versa.

PROBLEMS

Section 3–2 Analysis of logic circuits

1. Obtain the logic equations for the circuits shown in Figure 3–68.
2. Obtain the logic equations for the circuits shown in Figure 3–69.
3. Obtain the truth tables for the circuits shown in Figure 3–68 by tracing through all possible combinations of the input (independent) variables.
4. Obtain the truth tables for the circuits shown in Figure 3–69 by tracing through all possible combinations of the input (independent) variables.

Figure 3–68

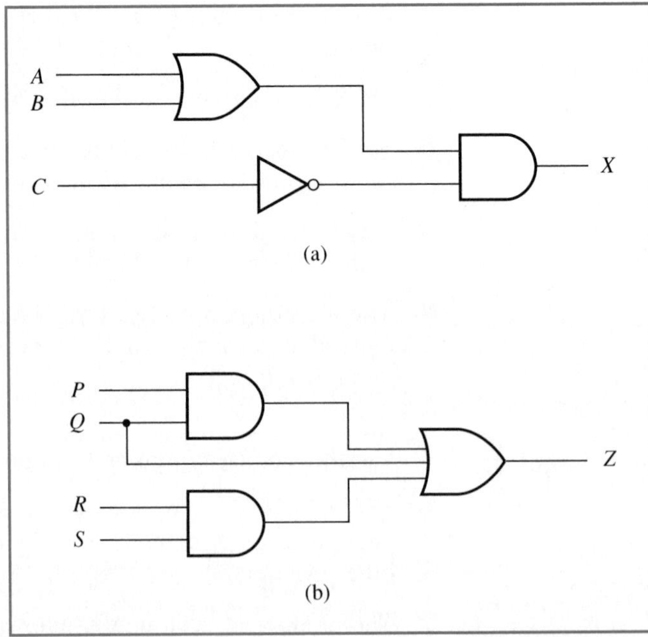

Figure 3–69

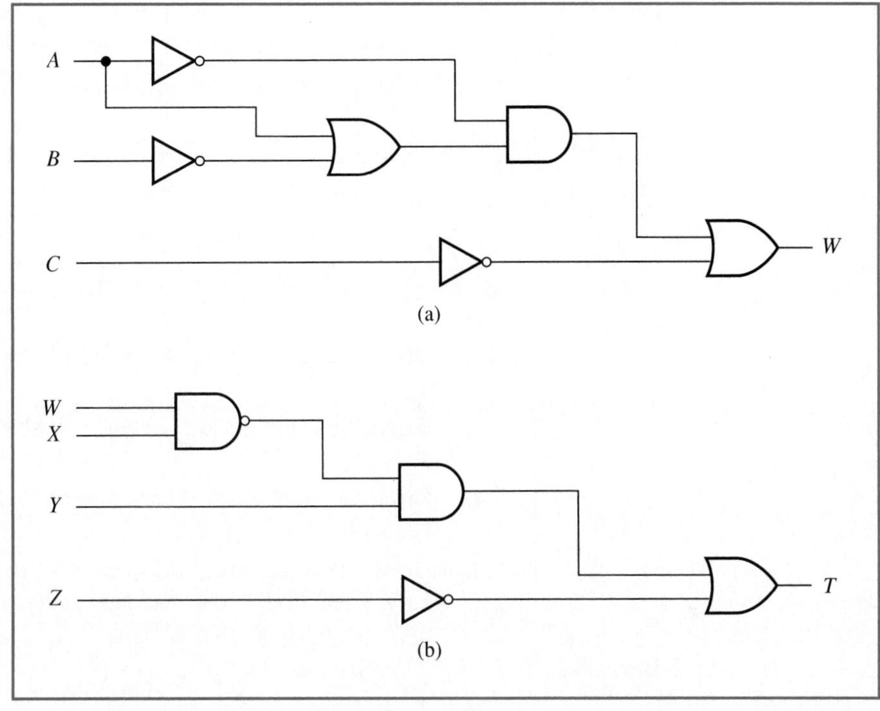

5. Obtain the truth tables for the circuits shown in Figure 3–68 by breaking down their logic equations into their basic functions and constructing truth tables like the ones shown in Tables 3–2, 3–3, etc.
6. Obtain the truth tables for the circuits shown in Figure 3–69 by breaking down their logic equations into their basic functions and constructing truth tables like the ones shown in Tables 3–2, 3–3, etc.

Section 3–3 Synthesis of logic circuits

7. Derive the unsimplified SOP equations from the truth tables shown in Tables 3–26 and 3–27.
8. Derive the unsimplified SOP equations from the truth tables shown in Tables 3–28 and 3–29.
9. Derive the unsimplified POS equations from the truth tables shown in Tables 3–26 and 3–27.
10. Derive the unsimplified POS equations from the truth tables shown in Tables 3–28 and 3–29.
11. Draw the logic circuit for each of the equations obtained in Problem 7.

Table 3–26

X	Y	Z
0	0	1
0	1	1
1	0	0
1	1	0

Table 3–27

P	Q	R	S
0	0	0	0
0	0	1	1
0	1	0	0
0	1	1	0
1	0	0	1
1	0	1	1
1	1	0	0
1	1	1	1

12. Draw the logic circuit for each of the equations obtained in Problem 8.

13. Draw the logic circuit for each of the equations obtained in Problem 9.

14. Draw the logic circuit for each of the equations obtained in Problem 10.

***15.** Draw the logic circuits for each of the following logic equations:
a. $X = A(\overline{B} + C) + A\overline{C}(B + D)$
b. $Y = \overline{PQR} + P(\overline{R} + \overline{Q}S)$
c. $Z = A\,\overline{B + C}\,B + AC$
d. $W = Q(R + \overline{Q})ST$

16. Rewrite each of the equations obtained in Problem 8 in short-form notation.

17. Rewrite each of the equations obtained in Problem 7 in short-form notation.

18. Rewrite each of the equations obtained in Problem 10 in short-form notation.

19. Rewrite each of the equations obtained in Problem 9 in short-form notation.

Table 3–28

A	B	C	D	X
0	0	0	0	1
0	0	0	1	1
0	0	1	0	1
0	0	1	1	1
0	1	0	0	0
0	1	0	1	1
0	1	1	0	0
0	1	1	1	1
1	0	0	0	0
1	0	0	1	0
1	0	1	0	0
1	0	1	1	0
1	1	0	0	0
1	1	0	1	1
1	1	1	0	0
1	1	1	1	1

*The answer to this problem is not given.

Table 3–29

P	Q	R	S
0	0	0	1
0	0	1	1
0	1	0	1
0	1	1	0
1	0	0	1
1	0	1	1
1	1	0	1
1	1	1	0

Section 3–4 Boolean algebra

20. In each of the following logic equations, determine the value of the dependent variable for the following values of the independent variable:

$$X = 1 \quad Y = 0 \quad Z = 0$$

a. $T = \overline{Y} + Z\overline{X} + ZY$
b. $A = Y(\overline{Y} + X)\,\overline{Z}$
c. $C = Z\overline{X} + (Y + 1)$
d. $S = X\overline{Y}Z + X(Y \cdot 0)$

21. In each of the following logic equations, determine the value of the dependent variable for the following values of the independent variable:

$$A = 1 \quad B = 0 \quad C = 1 \quad D = 1$$

a. $X = A\overline{B}(C + \overline{B}) + \overline{A}C(D + A)$
b. $Y = B + DC + B\overline{D}C$
c. $Z = ABCD(0 + A\overline{B}CD)$
d. $Q = \overline{A + B} \cdot \overline{CD}$

22. Determine the Boolean algebra law that is applicable in the following logic equations:
a. $PQ(RS + \overline{R}S) = PQRS + PQ\overline{R}S$
b. $\overline{AB + CD} = \overline{AB} \cdot \overline{CD}$
c. $XY + XYZ = XY$
d. $(P + Q) + (\overline{P + Q}) = 1$

23. Prove by truth tables that the right-hand side of each of the equations given in Problem 22 is equal to the left-hand side.

Section 3–5 Boolean simplification

24. Simplify the equations obtained from Problem 8 using Boolean algebra and verify that the simplified equation is equivalent to the unsimplified equation. Compare the unsimplified and simplified equations in terms of the number of logic gates and ICs required to implement them.

25. Simplify the equations obtained from Problem 7 using Boolean algebra and verify that the simplified equation is equivalent to the unsimplified equation. Compare the unsimplified and simplified equations in terms of the number of logic gates and ICs required to implement them.

26. Simplify the equations obtained from Problem 10 using Boolean algebra and verify that the simplified equation is equivalent to the unsimplified equation. Compare the unsimplified and simplified equations in terms of the number of logic gates and ICs required to implement them.

27. Simplify the equations obtained from Problem 9 using Boolean algebra and verify that the simplified equation is equivalent to the unsimplified equation. Compare the unsimplified and simplified equations in terms of the number of logic gates and ICs required to implement them.

28. Simplify the following logic expressions using Boolean algebra:
a. $\overline{A}\,\overline{B} + \overline{A}B + A\overline{B}$
b. $\overline{P}\,\overline{Q}\,\overline{R} + \overline{P}Q + P\overline{Q} + \overline{P}R$
c. $(P + Q + R)\,(P + \overline{Q} + R)$
d. $\overline{A}\,\overline{B}C + \overline{A}B\overline{C} + ABC$
e. $X\overline{Y} + \overline{W}\,\overline{X}Y + XY + W\overline{X}Y$

29. Simplify the following logic expressions using Boolean algebra:
a. $A\,(A + B) + B + \overline{A}\,(A + AB)$
b. $\overline{A} + AB + A + A\overline{B}$
c. $\overline{P\overline{Q}\,\overline{R} + PR}$
d. $\overline{(P + R) \cdot (\overline{P} + R)}$
e. $\overline{\overline{A} + B + \overline{C}} + \overline{(A + \overline{B}) \cdot (\overline{A} + \overline{D})} + A\overline{B}\,\overline{C}\,\overline{D}$

Section 3–6 NAND and NOR logic

30. Convert each of the simplified SOP equations obtained from Problem 24 to *NAND* logic by applying DeMorgan's theorem. Then verify that the *NAND* logic equations obtained are equivalent to their original *AND-OR* logic equations.

31. Convert each of the simplified SOP equations obtained from Problem 25 to *NAND* logic by applying DeMorgan's theorem. Then verify that the *NAND* logic equations obtained are equivalent to their original *AND-OR* logic equations.

32. Construct the logic circuits for each of the *NAND* equations obtained in Problem 30 and compare these circuits with their *AND-OR* equivalent circuits in terms of the number of gates and ICs required to implement them.

33. Construct the logic circuits for each of the *NAND* equations obtained in Problem 31 and compare these circuits with their *AND-OR* equivalent circuits in terms of the number of gates and ICs required to implement them.

34. Convert the following simplified *AND-OR* logic equations to *NAND* logic and draw the *NAND* logic circuit for each equation:

a. $Z = \overline{P}Q + P(\overline{Q} + RS)$
b. $X = \overline{A}\overline{B} + AB + BC$
c. $G = \overline{Y} + \overline{W}Z + WX\overline{Z}$
d. $F = \overline{A\overline{B}} + \overline{D}\overline{C}$
e. $A = \overline{\overline{P} + \overline{R}} + \overline{R}(\overline{P} + QS)$

35. Convert the following simplified *AND-OR* logic equations to *NOR* logic and draw the *NOR* logic circuit for each equation:

a. $Z = (\overline{P} + Q) \cdot (P + \overline{R})$
b. $T = \overline{X} + YZ$
c. $S = A \cdot (B + \overline{D}) \cdot (B + C + D)$
d. $W = \overline{\overline{P}Q\overline{R}} \cdot \overline{P\overline{Q}S}$
e. $Z = \overline{AD + BC} \cdot (\overline{C} + D)$

Section 3–7 Karnaugh map simplification

36. Using a K-map, obtain the simplified sum-of-products equations for Tables 3–26 through 3–29.

37. Using a K-map, obtain the simplified product-of-sums equations for Tables 3–26 through 3–29.

38. Simplify each of the logic expressions given in Problem 28 into sum-of-products form by using the K-map technique.

39. Simplify each of the logic expressions given in Problem 28 into product-of-sums form by using the K-map technique.

40. Simplify each of the logic expressions given in Problem 29 into POS form by using the K-map technique.

41. Simplify each of the logic expressions given in Problem 29 into SOP form by using the K-map technique.

42. Convert the following logic equations to their equivalent POS form by using a K-map:

a. $Q = \overline{A}\overline{B} + AB$
b. $X = \overline{Q} + PR$
c. $T = \overline{X}\overline{Z} + YZ + XZ$
d. $W = \overline{P}\overline{Q}\overline{R} + \overline{P}QR + PQ\overline{R} + P\overline{Q}R$
e. $Z = \overline{A} + \overline{C} + BD + A\overline{B}\overline{D}$

43. Convert the following logic equations to their equivalent SOP form by using a K-map:

a. $Z = (P + Q)(\overline{P} + \overline{Q})$
b. $X = (\overline{A} + B + C)(A + \overline{C})(\overline{B} + \overline{C})$
c. $S = (P + Q)(R + \overline{Q})(\overline{P} + \overline{Q})(\overline{R} + \overline{P})$
d. $A = (W + X + Y + Z)(\overline{X} + \overline{Z})(\overline{W} + X + \overline{Y} + Z)$
e. $T = R(P + \overline{S})(\overline{P} + Q + S)$

Troubleshooting

Section 3–8 Integrated circuit logic and troubleshooting applications

44. The logic circuit shown in Figure 3–70 has a fault. The recorded outputs of the logic gates G1, G2, G3, and G4 are shown below for all possible combinations of the inputs B and C. Use the analysis procedure to troubleshoot the circuit and isolate the fault.

		Outputs			
B	**C**	G1	G2	G3	G4
L	L	H	H	H	L
L	H	H	L	H	L
H	L	L	H	L	L
H	H	L	L	H	L

Figure 3–70

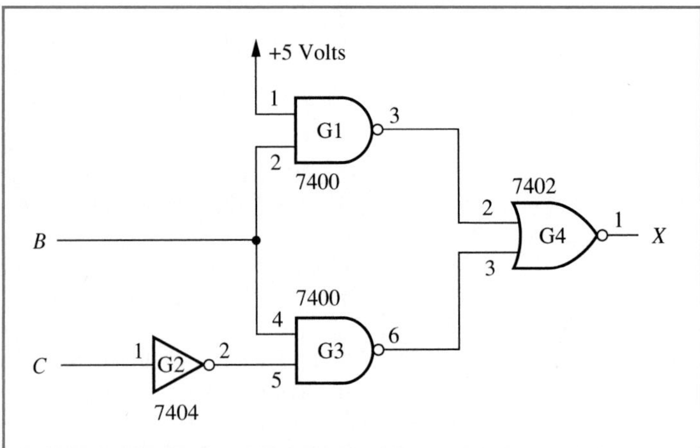

45. The logic circuit shown in Figure 3–71 has a fault. The recorded outputs of the logic gates G1, G2, and G3 are shown below for all possible combinations of the inputs P, Q, and R. Use the analysis procedure to troubleshoot the circuit and isolate the fault.

Figure 3–71

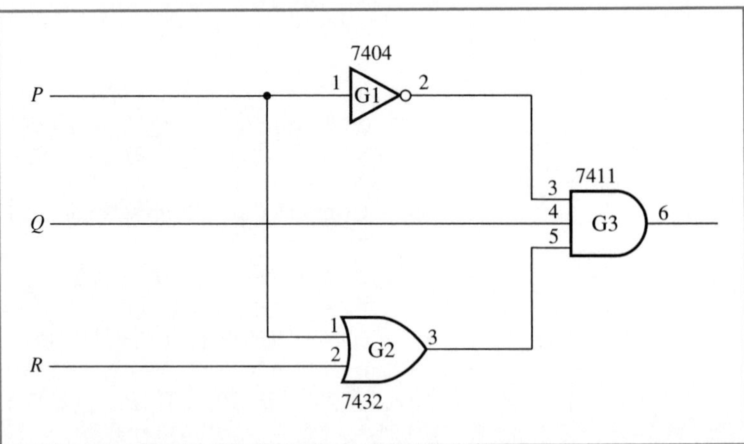

P	***Q***	***R***	**Outputs**		
			G1	G2	G3
L	L	L	H	L	L
L	L	H	H	L	L
L	H	L	H	L	L
L	H	H	H	L	L
H	L	L	L	H	L
H	L	H	L	H	L
H	H	L	L	H	L
H	H	H	L	H	L

One of the most common applications of digital circuits is the calculator. The calculator contains various logic circuits to perform everything from simple arithmetic to complex mathematical operations (Photo courtesy Hewlett-Packard Company).

OUTLINE

OBJECTIVES

The objectives of this chapter are to

- study the basic procedures for designing application circuits from a given specification.
- investigate the design and operation of various types of adder circuits such as half-adders, full-adders, parallel-adders, etc.
- introduce the operation of the Exclusive-*OR* gate and its application in adder circuits.
- apply the concepts of 1's and 2's complement binary subtraction to the design and operation of subtracter circuits.
- design and analyze various circuits that are used to compare binary numbers.
- study the applications of encoding and decoding binary numbers and the design of encoders and decoders.
- apply the concepts of decoding to code converters such as seven-segment decoders, and introduce the operation of seven-segment LEDs.
- investigate the design and operation of multiplexers and demultiplexers.
- introduce the concepts of tristate logic, tristate logic gates, and their applications in multiplexers and demultiplexers.
- design and analyze various types of circuits to generate and check the parity of binary codes.
- incorporate various application circuits into the design of an arithmetic logic unit (ALU).

4 APPLICATIONS OF COMBINATIONAL LOGIC

4–1 INTRODUCTION

Logic circuits are used for many applications in the field of digital computers and electronics. Combinational logic circuits are often designed to respond to certain input combinations by producing 1's and 0's at their outputs as required by the application. There are different applications of combinational logic circuits. Many of these circuits are used in digital computers to process binary numbers. For example, circuits such as adders are used to add binary numbers while comparators are used to compare binary numbers. This chapter investigates the design and analysis of these application circuits.

The procedures for designing and analyzing application circuits have been studied in Chapter 3. The synthesis procedures will be used to design the various application circuits from a given specification while the analysis procedures will be used to determine the function of various application circuits. Therefore the procedures that will be used to design and analyze the application circuits covered in this chapter will be identical except for a few variations introduced for certain applications to improve efficiency.

This chapter also introduces another gate that is used frequently in adders, comparators, and parity circuits—the Exclusive-*OR* gate. Also introduced in conjunction with the study of multiplexers and demultiplexers is the concept of tristate logic and logic gates with tristate outputs. Even though this chapter concludes the study of combinational logic circuits, many of the concepts and circuits developed in this and previous chapters will be applied to other circuits and concepts covered in later chapters since this material is often considered to be the foundation of the study of digital circuits.

4–2 Application circuit design

To design a logic circuit for a specific application we generally begin with a specification that describes the function of the circuit. For example, consider the following specification:

"Design a logic circuit that will be used to detect three binary numbers. The circuit should produce a logic 1 at its output whenever the binary equivalent of the number 2, 5, or 6 is applied at its input."

The first step is to draw a *logic symbol* for the circuit that will represent the inputs and outputs of the circuit in the form of a *block diagram*. Figure 4–1 shows the logic symbol for the circuit to be designed.

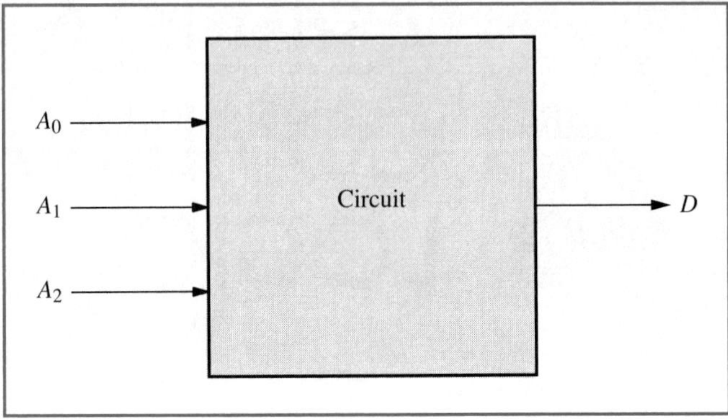

Figure 4–1
Logic symbol for circuit to be designed.

Notice that the logic symbol for the circuit contains three inputs labeled A_2, A_1, and A_0 since the inputs must be able to accommodate the numbers 2 (10_2), 5 (101_2), and 6 (110_2), and the largest number, 6, is represented by a 3-bit binary number, 110. Also notice that we have labeled the inputs with the same variable name A but we have used the subscripts 2, 1, and 0 to identify each bit that makes up the 3-bit binary number A. The subscript 0 will be used to identify the LSB and the highest subscript (2 in this example) will be used to identify the MSB. This scheme of symbolically identifying binary numbers will be used throughout the book. Thus we can say that the input to the circuit of Figure 4–1 is the 3-bit binary number A which is made of bits A_2, A_1, and A_0. The output of the circuit is D.

Since the specification usually includes all the information necessary to construct a truth table for the circuit, the next step is to construct a truth table (Table 4–1) that represents the operation of the circuit.

Notice how the input variable A is ordered in Table 4–1—the LSB (A_0) is placed in the rightmost column and the MSB (A_2) is placed in the leftmost column. Since the circuit is to produce a 1 at its output (D) when the number 010, 101, or 110 is applied to its input, the output variable D in Table 4–1 is a logic 1 for these combinations only.

We can now obtain the logic equation for the circuit by using one of two techniques discussed in Chapter 3—obtain the unsimplified (sum-of-products or product-of-sums) logic equation and then reduce it by using Boolean algebra, or obtain the simplified equation directly from a K-map. We shall use the latter technique for this and most of the other examples to follow. The K-map for the circuit is shown in Figure 4–2.

Table 4–1
Truth table for specification

A_2	A_1	A_0	D
0	0	0	0
0	0	1	0
0	1	0	1
0	1	1	0
1	0	0	0
1	0	1	1
1	1	0	1
1	1	1	0

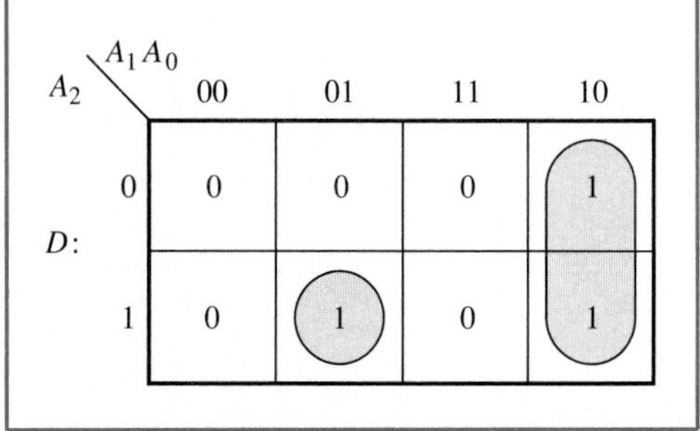

Figure 4–2
K-map for the circuit to be designed.

From the K-map shown in Figure 4–2 we can combine the two adjacent 1-cells and the single isolated 1-cell to produce the following simplified logic equation in sum-of-products form:

Equation (4–1)

$$D = A_2\overline{A}_1A_0 + A_1\overline{A}_0$$

The design can now be completed by constructing the logic circuit for Equation 4–1 as shown in Figure 4–3. Notice that the circuit in Figure 4–3 fills in the "box" that represents the logic symbol for the circuit (Figure 4–1).

Many application circuits have more than one output. For such circuits, the design procedures remains the same except that each output is designed independently. The following example illustrates the design of an application circuit that has three outputs.

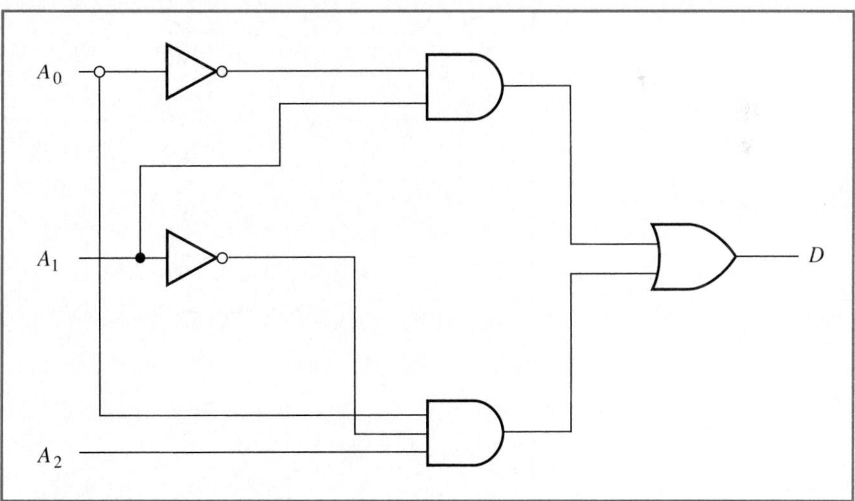

Figure 4–3
Logic circuit for the specification.

EXAMPLE 4–1

Design a logic circuit that will accept all possible combinations of 3-bit binary numbers as input, pass through only the odd combinations, and filter out all even combinations. For example, if any of the odd numbers, 001, 011,101, or 111 are applied to its input, the output will be the same as the input, but if any of the even numbers 000, 010, 100, or 110 are applied to its input, the output will be a 000.

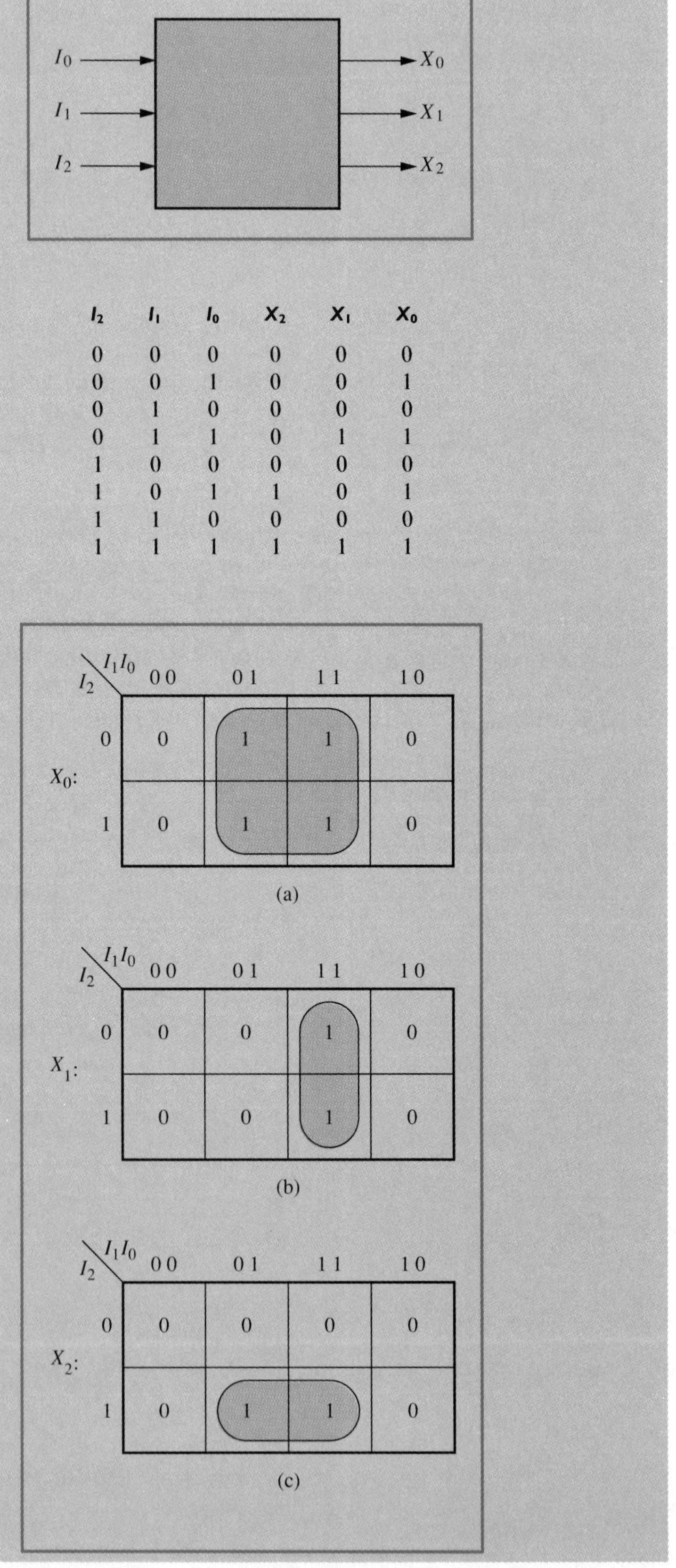

Figure 4–4
Logic symbol for "filter" circuit.

Table 4–2
Truth table for "filter" circuit

I_2	I_1	I_0	X_2	X_1	X_0
0	0	0	0	0	0
0	0	1	0	0	1
0	1	0	0	0	0
0	1	1	0	1	1
1	0	0	0	0	0
1	0	1	1	0	1
1	1	0	0	0	0
1	1	1	1	1	1

Figure 4–5
K-maps for the "filter" circuit.

The logic symbol for the circuit is shown in Figure 4–4. The variables I represent the input number, and the variables X represent the output number.

The truth table for the specification is shown in Table 4–2. Since Table 4–2 represents the values for the three output variables X_0, X_1, and X_2, it can be treated as three independent truth tables for the purpose of synthesis.

We can now obtain a K-map for each output as shown in Figure 4–5. From Figure 4–5a, the simplified logic equation for X_0 is

Equation (4–2)

$$X_0 = I_0$$

Similarly, from Figure 4–5b

Equation (4–3)

$$X_1 = I_1 I_0$$

and from Figure 4–5c

Equation (4–4)

$$X_2 = I_2 I_0$$

Notice that the simplicity of this truth table also enables us to obtain Equations 4–2 through 4–4 by inspection.

The final step is to construct the circuits for Equations 4–2 through 4–4. Even though these equations have been derived independently, to produce three independent circuits, the three circuits function together as one application circuit as shown in Figure 4–6.

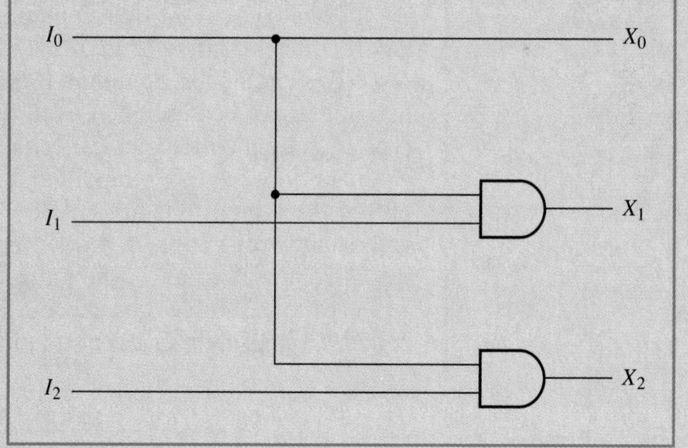

Figure 4–6 Logic circuit that "filters" out all even numbers.

The design procedures outlined in this section will be used to design most of the logic circuit applications discussed in the remainder of this chapter and the rest of the book. This procedure can be used to design any logic circuit from a given specification. We shall refer to this approach as the *conventional approach* to designing application circuits. We shall also see that even though this may be the most direct and efficient design procedure for most application circuits, there are a few applications where this approach can get complicated and does not necessarily produce the most efficient logic circuit.

Review questions

1. What does a logic symbol represent?
2. Briefly outline the steps involved in designing a circuit using the conventional approach.

4–3 Adders

Adders are logic circuits that are used to perform binary addition. It was stated in Chapter 1 that most digital computers reduce all arithmetic operations to the process of addition, and therefore adders play an important role in the architecture of a digital computer. This section deals with the design and construction of the building blocks of adders and their applications in adder circuits.

One of the basic elements of binary addition is a circuit known as a *half-adder*. The logic symbol of a half-adder is shown in Figure 4–7. A half-adder is simply a logic circuit that is used to add 2 bits (A, B) and produce a single-bit sum (S) and a single-bit carry (C). The process of adding all possible combinations of 2 bits was discussed in Chapter 1 and is shown in Table 1–9. The truth table for the half-adder can therefore be constructed as shown in Table 4–3. Notice in Table 4–3 that the equation for the output variable C is the equation for the *AND* function and that no simplification is possible for the output variable S. The logic equations for the half-adder circuit are therefore

Equation (4–5)
$$C = AB$$

Equation (4–6)
$$S = \overline{A}B + A\overline{B}$$

The logic circuit that implements Equations 4–5 and 4–6 to produce a half-adder is shown in Figure 4–8.

The circuit for Equation 4–6 and consequently the circuit for the half-adder can be simplified even further by replacing it with a special type of gate known as an *Exclusive-OR gate*. An Exclusive-*OR* (XOR) gate can only have two inputs and will produce a logic 0 output if and only if the 2 input bits are *equal*. The truth table for the Exclusive-*OR* gate is shown in Table 4–4 and its two logic symbols and equivalent *AND-OR* logic circuit is shown in Figure 4–9a, b, and c respectively.

Notice in Table 4–4 that the Exclusive-*OR* gate produces the sum of X and Y (ignoring carries) at its output Z.

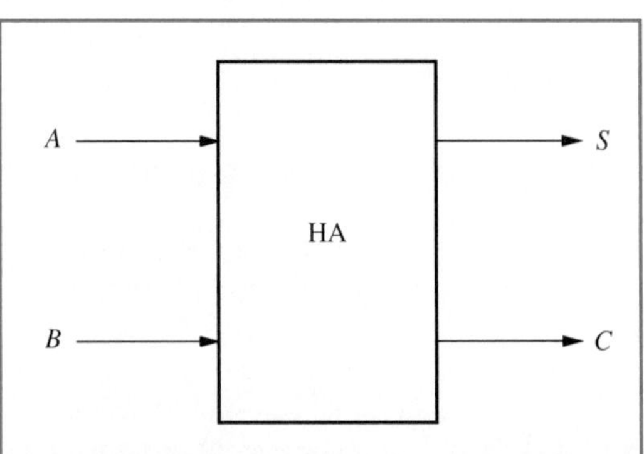

Figure 4–7
Logic symbol for a half-adder.

Table 4–3
Truth table for a half-adder

A	B	C	S
0	0	0	0
0	1	0	1
1	0	0	1
1	1	1	0

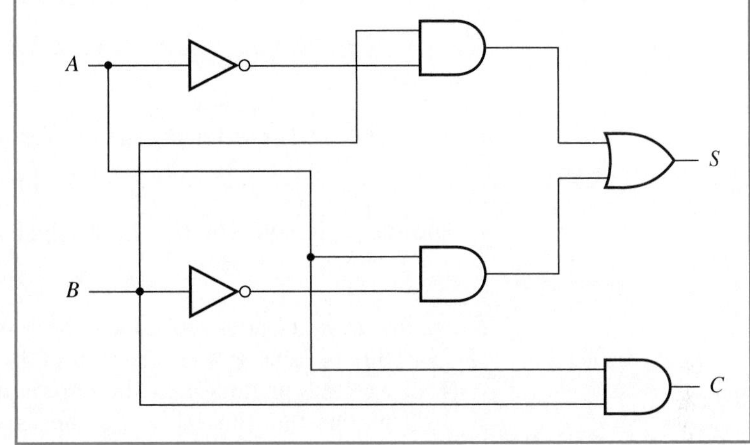

Figure 4–8
Half-adder circuit.

Table 4–4
Truth table for an Exclusive-*OR* gate

X	Y	Z
0	0	0
0	1	1
1	0	1
1	1	0

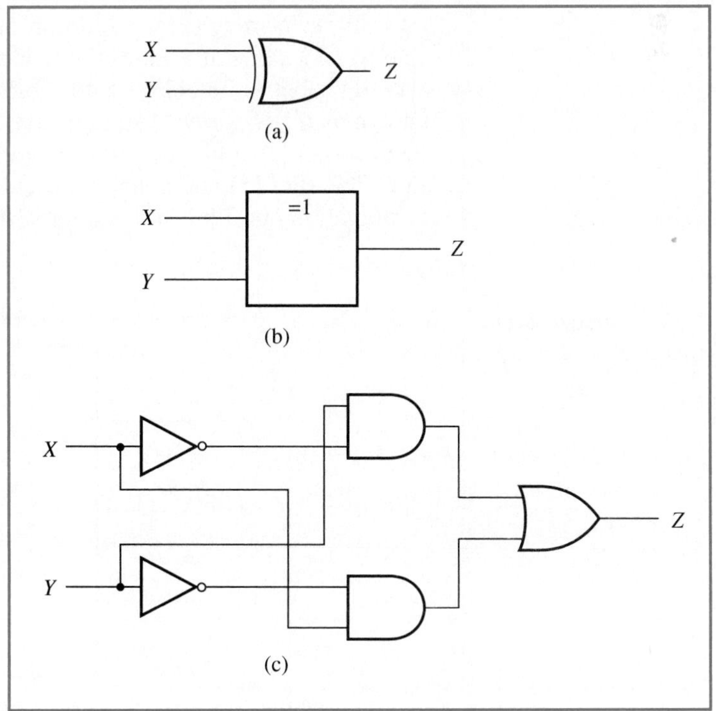

Figure 4–9
Logic symbols for an Exclusive-*OR* gate. (a) Distinctive shape symbol; (b) ANSI/IEEE Std. 91-1984 symbol; (c) equivalent *AND-OR* circuit.

Figure 4–10
Simplified half-adder circuit.

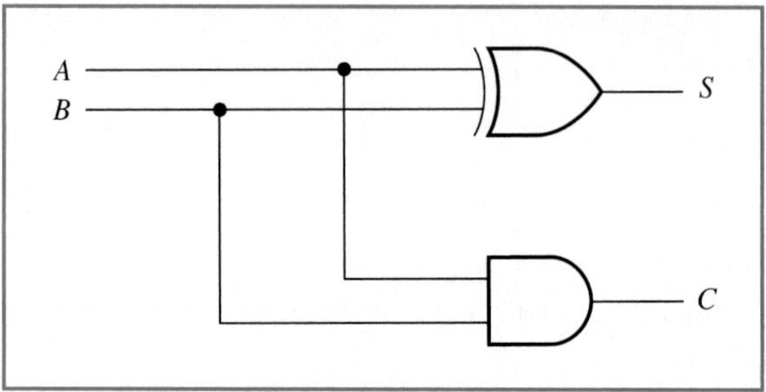

The *AND-OR* logic equation for Table 4–4 is

Equation (4–7)

$$Z = \overline{X}Y + X\overline{Y}$$

and the logic equation for the Exclusive-*OR* gate is

Equation (4–8)

$$Z = X \oplus Y$$

where the symbol $\oplus$ represents the Exclusive-*OR* function.

Since Table 4–4 is represented by both Equation 4–7 and Equation 4–8, we can conclude that they are equivalent.

Notice that the values for the variable S in Table 4–3 represent the Exclusive-*OR* function:

Equation (4–9)

$$S = A \oplus B$$

We can now redesign the half-adder circuit shown in Figure 4–8 by replacing the circuit for Equation 4–6 with the circuit for Equation 4–9 and reconstructing the circuit as shown in Figure 4–10.

Notice in Figure 4–10 that we have reduced the original half-adder circuit from six gates to two by incorporating the Exclusive-*OR* gate into the circuit.

In performing binary addition of numbers it is often desirable to be able to add 2 bits at a time; the half-adder accomplishes this task. It is also desirable to be able to add 3 bits at a time; a circuit that does this is known as a *full-adder*. The logic symbol for the full-adder is shown in Figure 4–11. A full-adder adds 3 input bits (A, B, C_i) and also produces a single-bit sum (S) and a single-bit carry (C_o). The addition of all possible combinations of 3 bits was covered in Chapter 1 and illustrated in Table

Figure 4–11
Logic symbol for a full-adder.

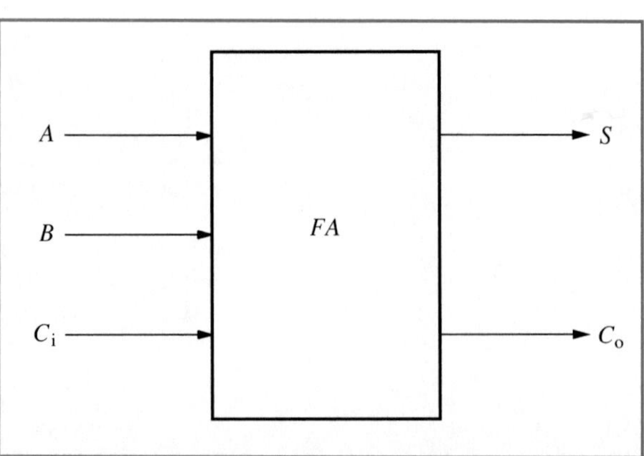

Table 4–5
Truth table for a full-adder

A	B	C_i	C_o	S
0	0	0	0	0
0	0	1	0	1
0	1	0	0	1
0	1	1	1	0
1	0	0	0	1
1	0	1	1	0
1	1	0	1	0
1	1	1	1	1

1–10. The variable C_i is called the *carry-in* and the variable C_o is called the *carry-out* and their functions will be examined later in this section. Table 4–5 shows the truth table for a full-adder.

The K-maps for the output variables S and C_o shown in Figure 4–12 produce the following simplified equations:

Equation (4–10)
$$C_o = AB + BC_i + AC_i$$

Equation (4–11)
$$S = \overline{A}\overline{B}C_i + \overline{A}B\overline{C}_i + ABC_i + A\overline{B}\overline{C}_i$$

The complete circuit for the full-adder is finally constructed from Equations 4–10 and 4–11 as shown in Figure 4–13.

The full-adder circuit shown in Figure 4–13 can be greatly simplified if the full-adder circuit is constructed with half-adders as shown in Figure 4–14. Notice that the circuit in Figure 4–14 requires only five gates (since each half-adder contains two gates), whereas the circuit in Figure 4–13 requires 12 gates.

The circuit in Figure 4–14 adds the 3 input bits A, B, and C_i by first adding the 2 bits, A and B using HA1. The sum of these 2 bits (S_1) is then added to the third input C_i using HA2 to produce the final sum output (S). The final carry output (C_o) is obtained by "*OR*ing" the carry outputs (C_1 and C_2) of both half-adders, HA1 and HA2. We can prove that

Equation (4–12)
$$C_o = C_1 + C_2$$

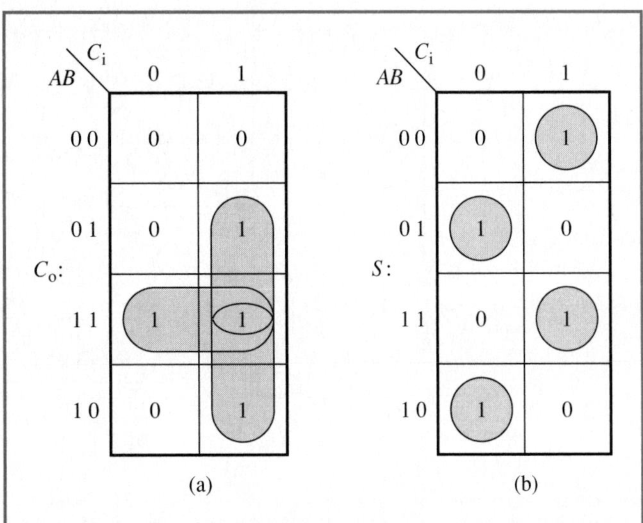

Figure 4–12
K-maps for the design of the full-adder.

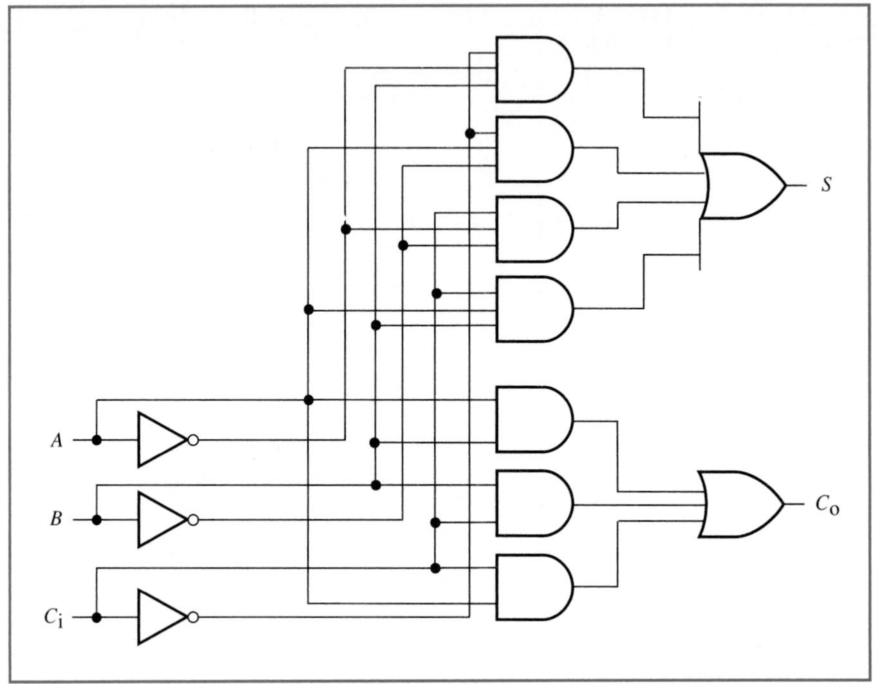

Figure 4–13
Full-adder circuit.

as follows
Since

$$C_1 = AB$$

and

$$C_2 = S_1 C_i$$

Equation 4–12 becomes

Equation (4–13)

$$C_o = AB + S_1 C_i$$

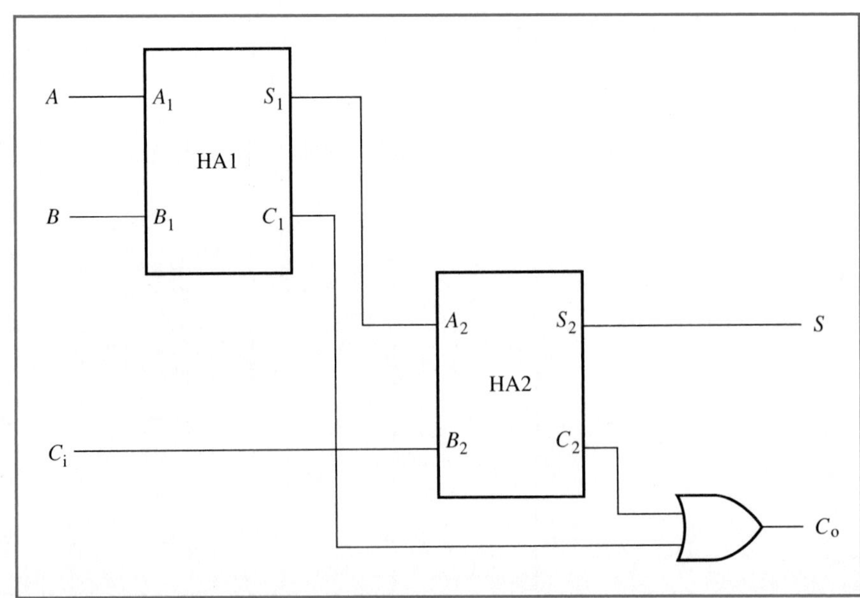

Figure 4–14
Full-adder circuit constructed from half-adders.

However

$$S_1 = A \oplus B$$

Therefore Equation 4–13 becomes

$$\begin{aligned}
C_o &= AB + (A \oplus B)\,C_i \\
&= AB + (A\overline{B} + \overline{A}B)\,C_i \\
&= AB + A\overline{B}C_i + \overline{A}BC_i \\
&= A\,(B + \overline{B}C_i) + \overline{A}BC_i \\
&= A\,(B + C_i) + \overline{A}BC_i \\
&= AB + AC_i + \overline{A}BC_i \\
&= AB + C_i\,(A + \overline{A}B) \\
&= AB + C_i\,(A + B) \\
&= AB + AC_i + BC_i
\end{aligned}$$

Equation (4–14)

Notice that Equation 4–14 is the same as Equation 4–10.

It is possible to construct other adders that are capable of adding 4 bits or more using the same procedures discussed in this section. However, there are not many applications for such circuits. The main applications of half- and full-adders are in the addition of multibit binary numbers using circuits known as *parallel binary adders*.

Parallel binary adders are classified according to the size of the numbers that are processed (added) by the circuit. A parallel binary adder that adds two 2-bit numbers is referred to as a *2-bit parallel binary adder* while one that adds two 4-bit numbers is called a *4-bit parallel binary adder*. Parallel binary adders can be of any size, and function in the same manner regardless of the size of the numbers processed. For example, consider the 2-bit parallel binary adder whose logic symbol is shown in Figure 4–15. The circuit adds two 2-bit numbers A (A_1A_0) and B (B_1B_0) and produces a 2-bit sum S (S_1S_0) and a carry C. We can use the conventional approach to designing this circuit—that is, we can construct a truth table of all possible combinations of the inputs A_1, A_0, B_1, B_0, and determine the logic equations and circuits for each output, S_1, S_0, and C. However, it is much easier to design the circuit by interconnecting various half- and full-adders to accomplish the addition on a bit-by-bit basis.

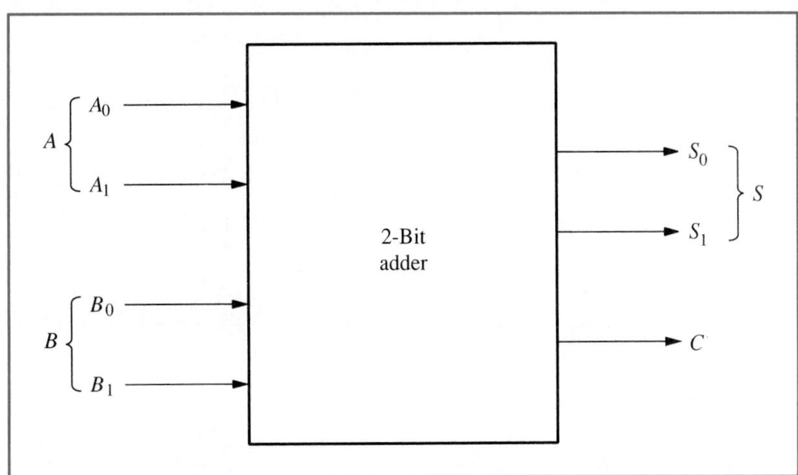

Figure 4–15
Logic symbol for a 2-bit parallel binary adder.

Recall from Chapter 1 that the binary addition of the numbers *A* and *B* is accomplished by adding each bit in each number starting with the LSBs and carrying over any overflow from one position to the next up to the MSB. For example

$$\begin{array}{rrr} & C_0 & \\ & A_1 & A_0 \\ & +\ B_1 & B_0 \\ \hline C & S_1 & S_0 \end{array}$$

This addition can be accomplished by using a half-adder

$$\begin{array}{rr} & A_0 \\ & +\ B_0 \\ \hline C_0 & S_0 \end{array}$$

and a full-adder

$$\begin{array}{rr} & C_0 \\ & A_1 \\ & +\ B_1 \\ \hline C & S_1 \end{array}$$

The circuit that implements the 2-bit parallel binary adder is shown in Figure 4–16. Notice that the circuit in Figure 4–16 can be very easily extended to add numbers of any size simply but adding more full-adders to the circuit. The following examples illustrate the concept of *expanding* the size of a parallel binary adder.

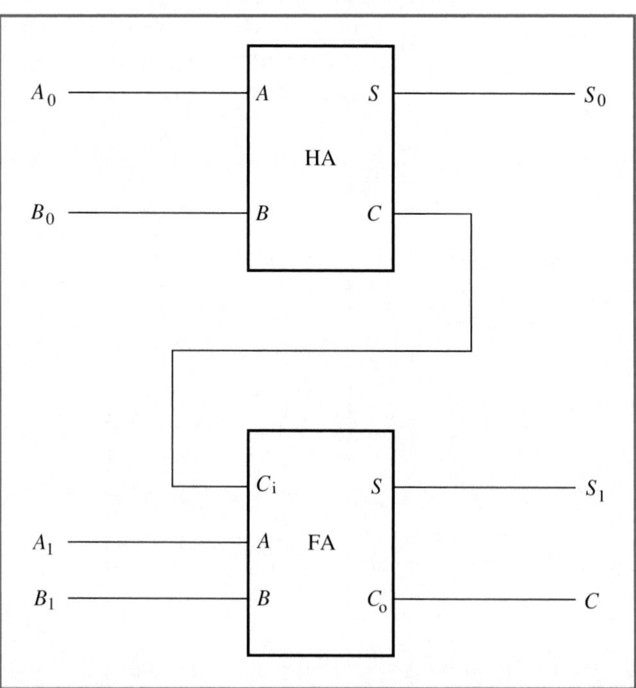

Figure 4–16
Logic circuit for a 2-bit parallel binary adder.

EXAMPLE 4–2

Design a 4-bit parallel binary adder whose logic symbol is shown in Figure 4–17. Notice that the circuit should incorporate a single-bit carry input (C_i) that will be added to the numbers A and B to facilitate expansion.

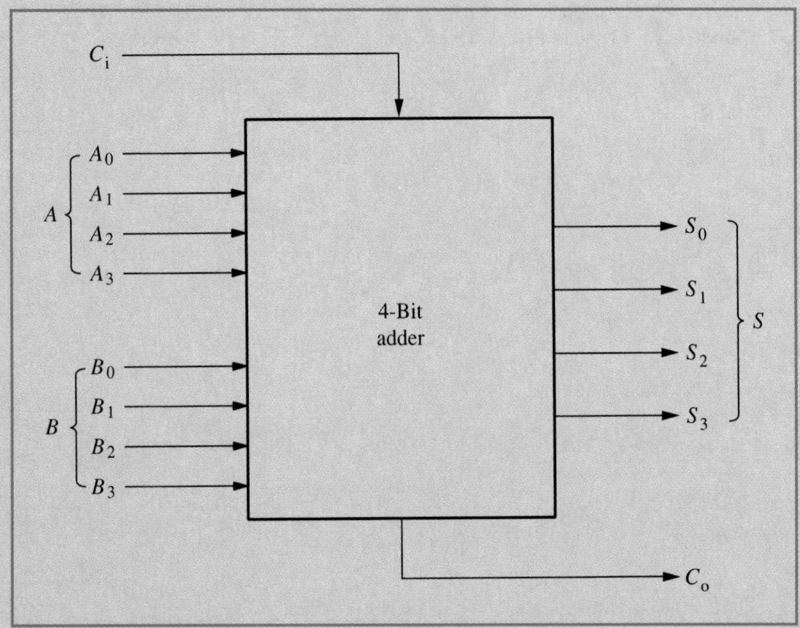

Figure 4–17 Logic symbol for a 4-bit parallel binary adder.

The design of the 4-bit parallel binary adder is shown in Figure 4–18. Notice that the only difference between this circuit and the circuit in Figure 4–16 is that two more full-adders have been added to expand the size of each number to 4 bits instead of 2 bits and that the least significant half-adder (in Figure 4–16) has been replaced by a full-adder to accommodate a carry input.

EXAMPLE 4–3

Expand the 4-bit parallel binary adder shown in Figure 4–17 to add two 8-bit numbers as illustrated by the logic symbol shown in Figure 4–19.

The logic symbol shown in Figure 4–19 represents each set of eight input lines as a double arrow, sometimes referred to as a *bus* (more will be said about this in Section 4–11). To add the two 8-bit numbers A and B and obtain an 8-bit sum S, we must first add the least significant nibbles (4 bits) of each number, and then add the most significant nibbles of both numbers along with the carry generated from the addition of the least significant nibbles. The resulting circuit is shown in Figure 4–20.

$$\begin{array}{cccc} & C_i \leftarrow & & \\ & A_7\;A_6\;A_5\;A_4 & & A_3\;A_2\;A_1\;A_0 \\ & \underline{B_7\;B_6\;B_5\;B_4} & & \underline{B_3\;B_2\;B_1\;B_0} \\ C_o & S_7\;S_6\;S_5\;S_4 & C_o & S_3\;S_2\;S_1\;S_0 \end{array}$$

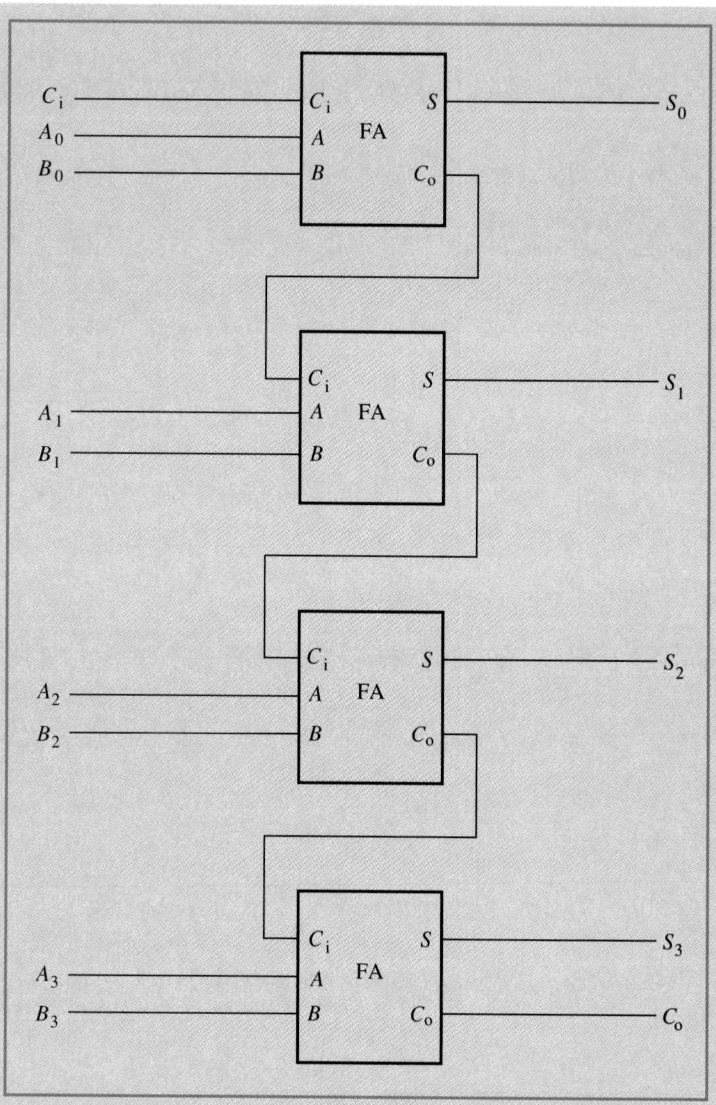

Figure 4–18
Logic circuit for a 4-bit parallel binary adder.

Figure 4–19
Logic symbol for an 8-bit parallel binary adder.

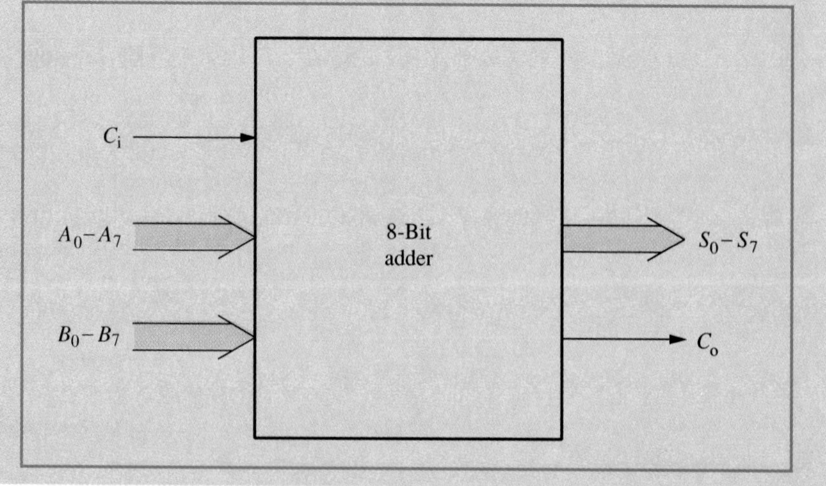

Figure 4–20
Logic circuit for an 8-bit parallel binary adder.

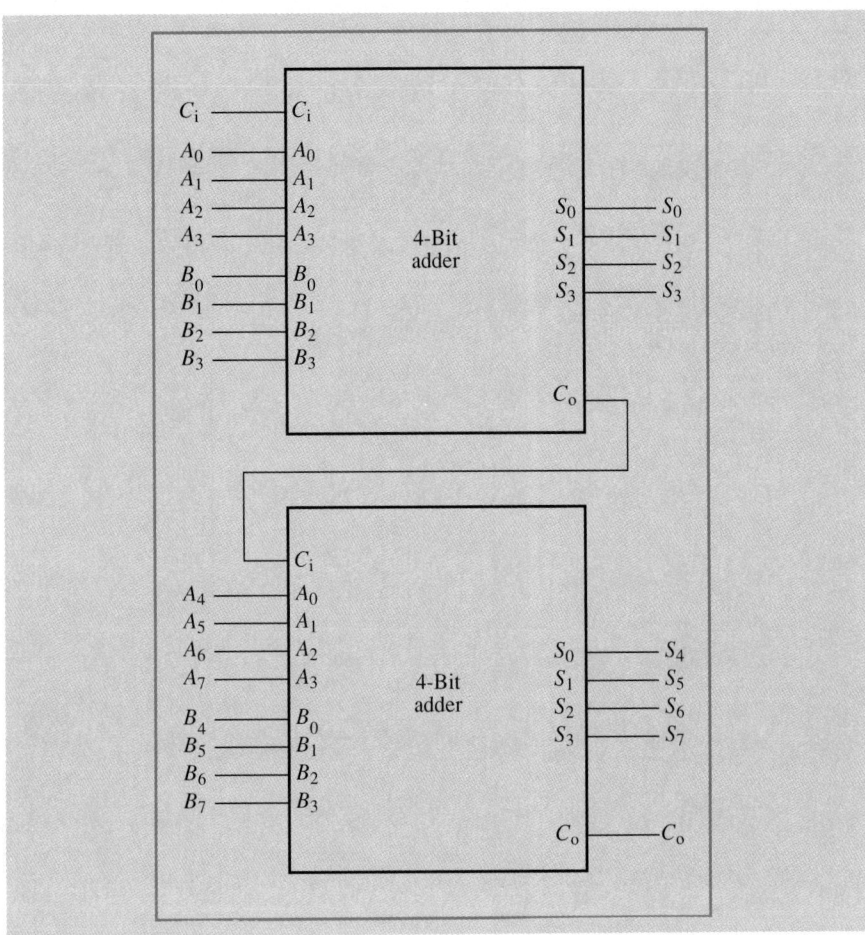

Integrated circuit logic

The 74xxx series includes two ICs that provide the functions of parallel binary addition with expansion capabilities. The logic symbol and pin configurations of the two ICs are shown in Figure 4–21. The 7482 is a 2-bit parallel binary adder and is similar in function to the circuit designed in Figure 4–16. Σ_1 and Σ_2 are the outputs that represent the sum of the numbers A (A_2A_1) and B (B_2B_1) while C_{in} is the carry input and C_2 is the carry output.

The 7483 is a 4-bit parallel binary adder and is similar in function to the circuit designed in Figure 4–18. The 4-bit inputs A_3–A_0 and B_3–B_0 are added together with a carry input C_0 to produce a 4-bit sum, S_3–S_0, and a carry output, C_4.

The 7486 is an IC that packages four 2-input Exclusive-*OR* gates as shown in Figure 4–22. These gates can be used in the construction of the basic half- and full-adders discussed earlier in this section as well as in some of the other application circuits to be discussed in this chapter and in others.

Review questions

1. What is a half-adder? How many bits can it add?
2. What is a full-adder? How many bits can it add?

3. Describe the function of an Exclusive-*OR* gate.
4. Describe the operation of a parallel binary adder.
5. What is the purpose of the C_i and C_o input and output in a parallel binary adder?

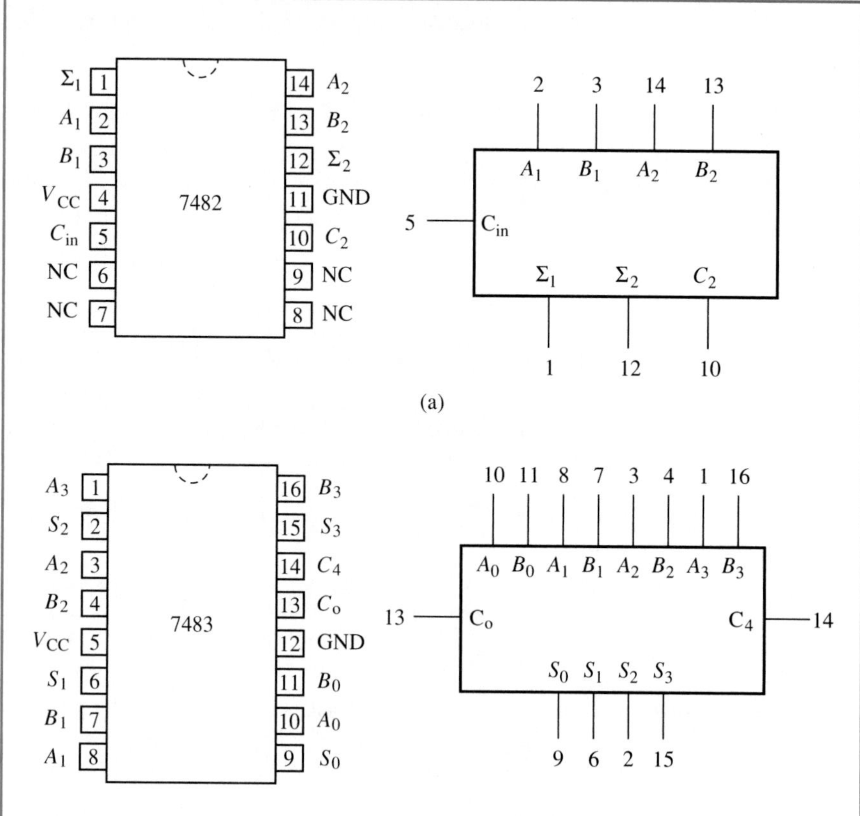

Figure 4–21 Logic symbols and pin configurations of (a) the 7482 2-bit parallel adder and (b) the 7483 4-bit parallel adder.

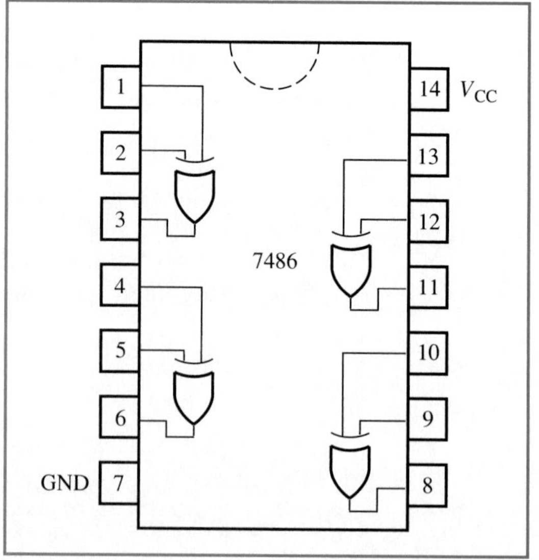

Figure 4–22 Pin configuration of the 7486 quad 2-input Exclusive-*OR* gate.

4–4 Subtracters

Subtracter circuits are used to perform binary subtraction on two numbers. In Chapter 1 we saw that binary subtraction is performed by *adding* a negative number (in 1's or 2's complement representation) to a positive number. We can therefore make a few simple modifications to the adder circuits discussed in Section 4–3 to allow these circuits to perform binary subtraction. Figure 4–23 shows the logic symbol for a 4-bit subtracter which will accept as input the two 4-bit numbers A $(A_3A_2A_1A_0)$ and B $(B_3B_2B_1B_0)$ and obtain the difference between the two numbers D $(D_3D_2D_1D_0)$ using either the 1's or 2's complement technique. The *sign bit*, S, identifies whether the result is negative or positive.

1's complement subtracter

Consider the subtraction of two 4-bit numbers A and B to obtain the difference between the two numbers, D

$$\begin{array}{r} A_3\ A_2\ A_1\ A_0 \\ -\ B_3\ B_2\ B_1\ B_0 \\ \hline D_3\ D_2\ D_1\ D_0 \end{array}$$

Notice above that the number B is subtracted from the number A to obtain a negative or positive difference, D.

If we were to perform this subtraction using the 1's complement technique, we would first obtain the 1's complement representation of the negative number B by *inverting* each bit in the number

$$\overline{B}_3\ \overline{B}_2\ \overline{B}_1\ \overline{B}_0$$

Next, we add the 1's complement representation of the negative number, $\overline{B}$, to the positive number, A, to obtain a partial sum, P, and an end-around carry, C.

$$\begin{array}{rr} & A_3\ A_2\ A_1\ A_0 \\ & +\ \overline{B}_3\ \overline{B}_2\ \overline{B}_1\ \overline{B}_0 \\ \hline C & P_3\ P_2\ P_1\ P_0 \end{array}$$

The end-around carry, C, is then added to the partial sum, P, to obtain the final result D, which is the difference between A and B. The value of

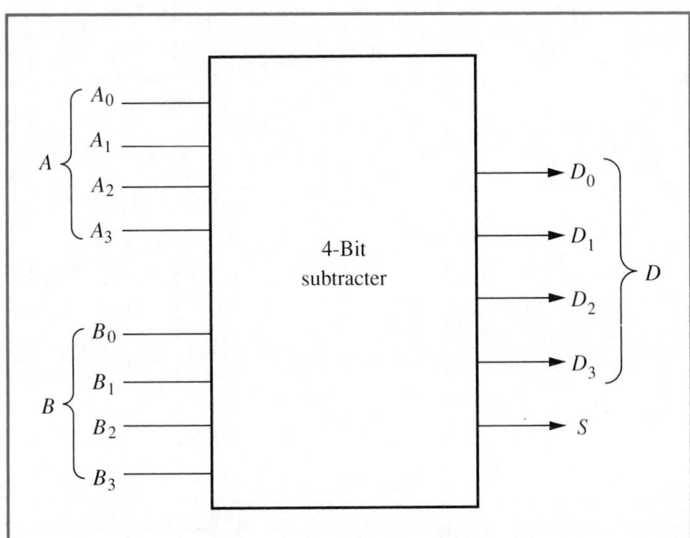

Figure 4–23 Logic symbol for a 4-bit subtracter.

C determines whether the result is positive ($C = 1$) or negative ($C = 0$)

$$\begin{array}{rcccc} & P_3 & P_2 & P_1 & P_0 \\ + & & & & C \\ \hline & D_3 & D_2 & D_1 & D_0 \end{array}$$

The circuit to implement 1's complement subtraction as described above is shown in Figure 4–24. Notice that the 1's complement of the number B is first obtained by passing each bit in the number through an inverter. A simple parallel binary adder (identified as Stage 1) is used to add the numbers A and $\overline{B}$ and obtain the partial sum, P. The carry output C of the most significant full-adder in Stage 1 is the sign bit that is added to the LSB of the partial sum, P_0 (Stage 2). The three remaining half-adders in Stage 2 are included to take care of any carry that may be generated during the addition of each set of bits since C must be added to

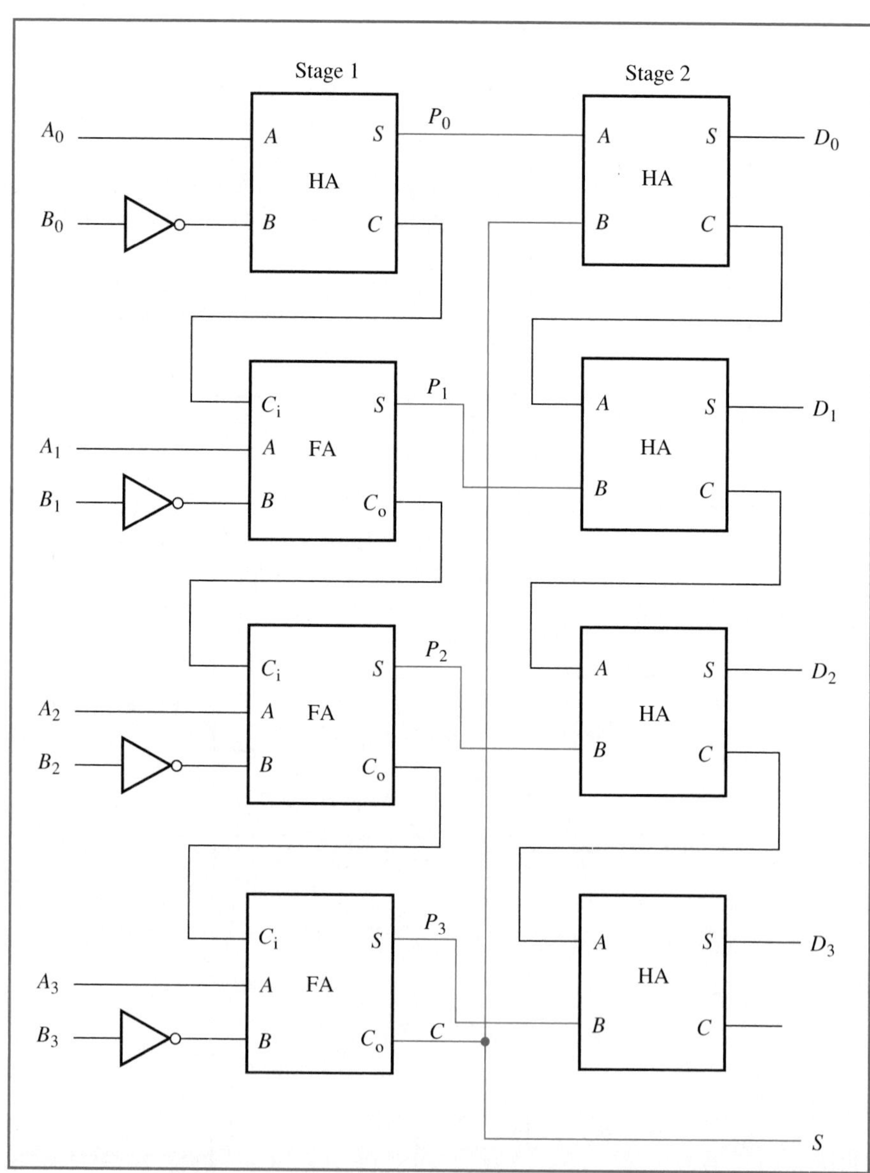

Figure 4–24
A 4-bit 1's complement subtracter circuit.

the entire number P rather than just P_0. The outputs D of the half-adders in Stage 2 represent the final result. Notice that the carry output of the most significant half-adder in Stage 2 can be ignored since it will always be a logic 0 because the result of any 4-bit subtraction will never require more than 4 bits of representation.

2's complement subtracter

Again, consider the subtraction of two 4-bit numbers A and B to obtain the difference between the two numbers, D

$$\begin{array}{r} A_3\,A_2\,A_1\,A_0 \\ -\ B_3\,B_2\,B_1\,B_0 \\ \hline D_3\,D_2\,D_1\,D_0 \end{array}$$

As before, notice above that the number B is subtracted from the number A to obtain a negative or positive difference, D. If we were to perform this subtraction using the 2's complement technique, we would first obtain the 2's complement representation of the negative number B by *inverting* each bit in the number and then adding 1 to the final result:

$$\begin{array}{r} \overline{B}_3\,\overline{B}_2\,\overline{B}_1\,\overline{B}_0 \\ +\qquad\qquad 1 \end{array}$$

Next, we add the 2's complement representation of the negative number to the positive number, A, to obtain the difference D and a final carry, C.

$$\begin{array}{r} A_3\,A_2\,A_1\,A_0 \\ \left.\begin{array}{r}\overline{B}_3\,\overline{B}_2\,\overline{B}_1\,\overline{B}_0 \\ +\qquad\qquad 1\end{array}\right\}\ \text{2's complement of } B \\ \hline C\ \ D_3\,D_2\,D_1\,D_0 \end{array}$$

As before, the value of C determines whether the result is positive ($C = 1$) or negative ($C = 0$) and is therefore considered to be the sign bit.

The circuit to implement 2's complement subtraction as described above is shown in Figure 4–25. Notice that the 1's complement of the number B is first obtained by passing each bit in the number through an inverter. However, instead of immediately obtaining the 2's complement of the number B by adding 1 to its 1's complement and then adding the 2's complement of B to the number A, the circuit adds the number A, the 1's complement of B, and a logic 1, using a parallel binary adder to effectively produce the same result with a minimum amount of components. The least significant full-adder has its carry-input tied to a logic 1 to effectively convert the 1's complement of B into its 2's complement. The difference between the two numbers is produced directly at the output of the adder and the carry-output of the most significant full-adder represents the sign bit of the subtracter.

Both the 1's complement and the 2's complement subtracters can be extended to subtract binary numbers of any size simply by adding more components to accommodate the additional bits; the procedure is therefore similar to that discussed in Section 4–3 to expand the size of parallel binary adders. It is apparent that the design of the 2's complement subtracter is less complex than the design of the 1's complement subtracter and is therefore the preferred method of binary subtraction in most digital computers. The conventional approach can also be used to design subtracters, but the resulting circuits would be

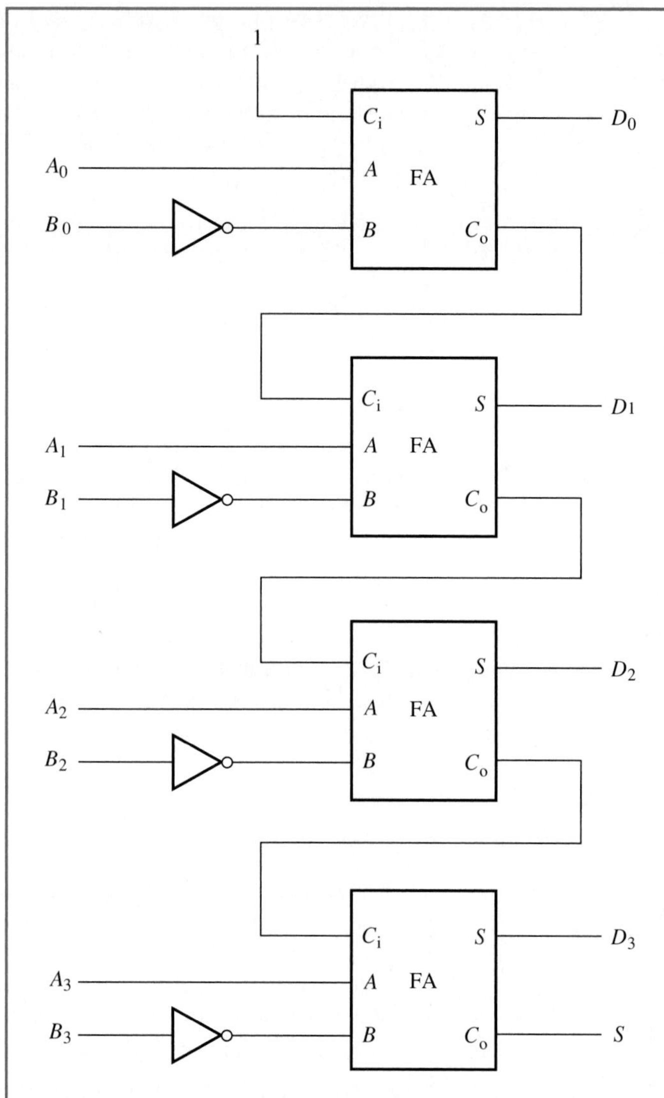

Figure 4–25 A 4-bit 2's complement subtracter circuit.

much more complex than the circuits obtained by using parallel binary adders. Since binary subtracters can be designed using parallel binary adders, there are no special ICs available that perform binary subtraction.

Review questions

1. Describe the operation of a 1's complement subtracter.
2. Describe the operation of a 2's complement subtracter.
3. Why is a 2's complement subtracter preferred over a 1's complement subtracter?

4–5 Comparators

Comparators are logic circuits that are used to compare two binary numbers. The result of the comparison are single-bit outputs that indicate whether the numbers being compared are equal to each other, or if one number is greater or less than another. Figure 4–26 shows the logic symbol for a comparator (also referred to as a magnitude comparator) that

Figure 4–26
Logic symbol for a 2-bit magnitude comparator.

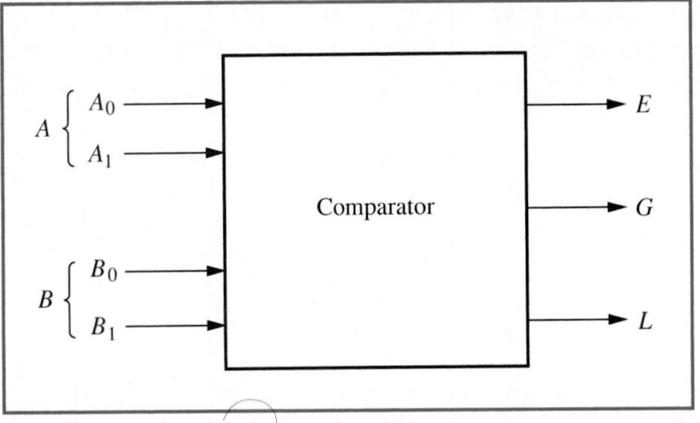

compares two 2-bit binary numbers A (A_1A_0) and B (B_1B_0) and activates one of three outputs, E, G, or L, if A is equal to B, A is greater than B, or A is less than B, respectively. We can use the conventional approach to design the comparator circuit as follows.

The truth table for the 2-bit magnitude comparator is shown in Table 4–6. The truth table includes the decimal equivalents of the numbers A and B for convenience in comparing their values. The outputs of the circuits are *active high,* which means that the state of an output will be a logic 1 (*high*) when *activated*; for example, whenever the two numbers are equal, the output E will be (activated to) a logic 1 to indicate equality, otherwise its state will be a logic 0.

We can now obtain the simplified logic equations for each of the outputs, E, G, and L from the truth table shown in Table 4–6 by constructing K-maps for each output as shown in Figure 4–27.

Equation (4–15)
$$E = \overline{A}_1\overline{A}_0\overline{B}_1\overline{B}_0 + \overline{A}_1A_0\overline{B}_1B_0 + A_1A_0B_1B_0 + A_1\overline{A}_0B_1\overline{B}_0$$

Equation (4–16)
$$G = A_1\overline{B}_1 + A_0\overline{B}_1\overline{B}_0 + A_1A_0\overline{B}_0$$

Equation (4–17)
$$L = \overline{A}_1B_1 + \overline{A}_1\overline{A}_0B_0 + \overline{A}_0B_1B_0$$

Even though Equation 4–15 cannot be simplified, we can obtain an equivalent circuit that will be less complex by using the Exclusive-*OR*

Table 4–6
Truth table for a 2-bit magnitude comparator

A		B	A_1	A_0	B_1	B_0	E	G	L
0	=	0	0	0	0	0	1	0	0
0	<	1	0	0	0	1	0	0	1
0	<	2	0	0	1	0	0	0	1
0	<	3	0	0	1	1	0	0	1
1	>	0	0	1	0	0	0	1	0
1	=	1	0	1	0	1	1	0	0
1	<	2	0	1	1	0	0	0	1
1	<	3	0	1	1	1	0	0	1
2	>	0	1	0	0	0	0	1	0
2	>	1	1	0	0	1	0	1	0
2	=	2	1	0	1	0	1	0	0
2	<	3	1	0	1	1	0	0	1
3	>	0	1	1	0	0	0	1	0
3	>	1	1	1	0	1	0	1	0
3	>	2	1	1	1	0	0	1	0
3	=	3	1	1	1	1	1	0	0

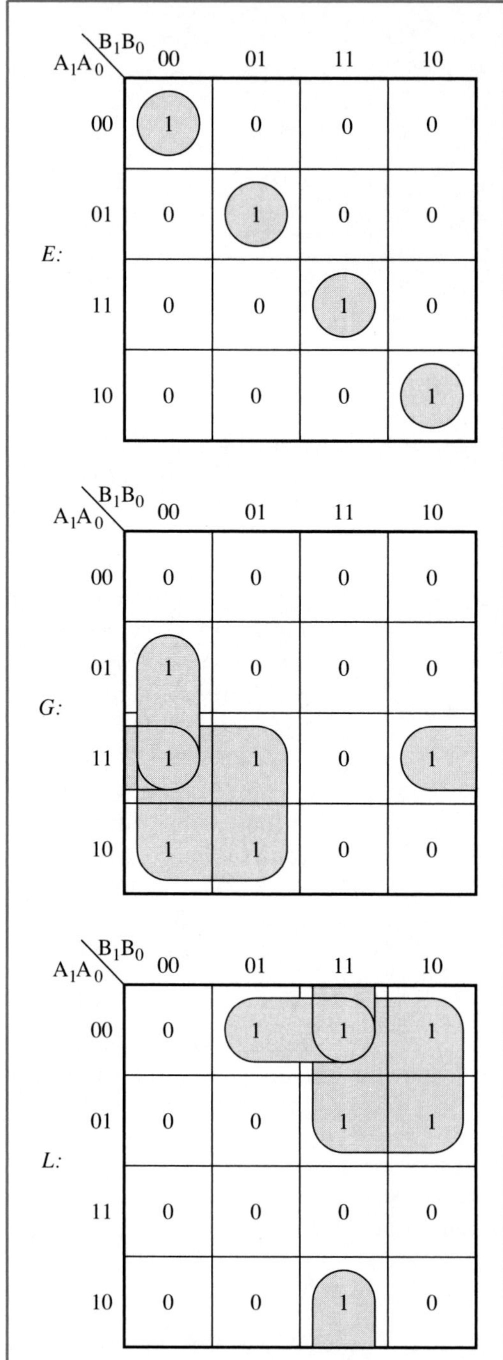

Figure 4–27
K-maps for the design of the 4-bit magnitude comparator.

gate to test a pair of bits for equality. The manner in which this is done can be explained as follows.

For the two numbers A and B to be equal, the following conditions must be satisfied:

$$A_0 = B_0$$

and

$$A_1 = B_1$$

Since the output of an Exclusive-*OR* gate will be a logic 0 if and only if both inputs are equal, the following conditions are true:

Equation (4–18)
$$A_0 \oplus B_0 = 0$$

and

Equation (4–19)
$$A_1 \oplus B_1 = 0$$

if the numbers *A* and *B* are equal.

Therefore to generate the comparator output *E* we must "*NOR*" Equations 4–18 and 4–19 so that *E* is a logic 1 if and only if $A_0 \oplus B_0$ and $A_1 \oplus B_1$ produce logic 0's. The simplified equation for E then becomes

Equation (4–20)
$$E = \overline{(A_0 \oplus B_0) + (A_1 \oplus B_1)}$$

Notice that the circuit for Equation 4–20 will require fewer gates than the circuit for Equation 4–15.

The complete circuit for the 2-bit magnitude comparator can now be constructed with Equations 4–16, 4–17, and 4–20, as shown in Figure 4–28.

Integrated circuit logic

The 7485 is similar in operation to the 2-bit magnitude comparator designed in this section, except that the 7485 compares two 4-bit numbers and has the capability to extend the comparison to numbers of any size. The 7485 is called a 4-bit magnitude comparator because of the size of the

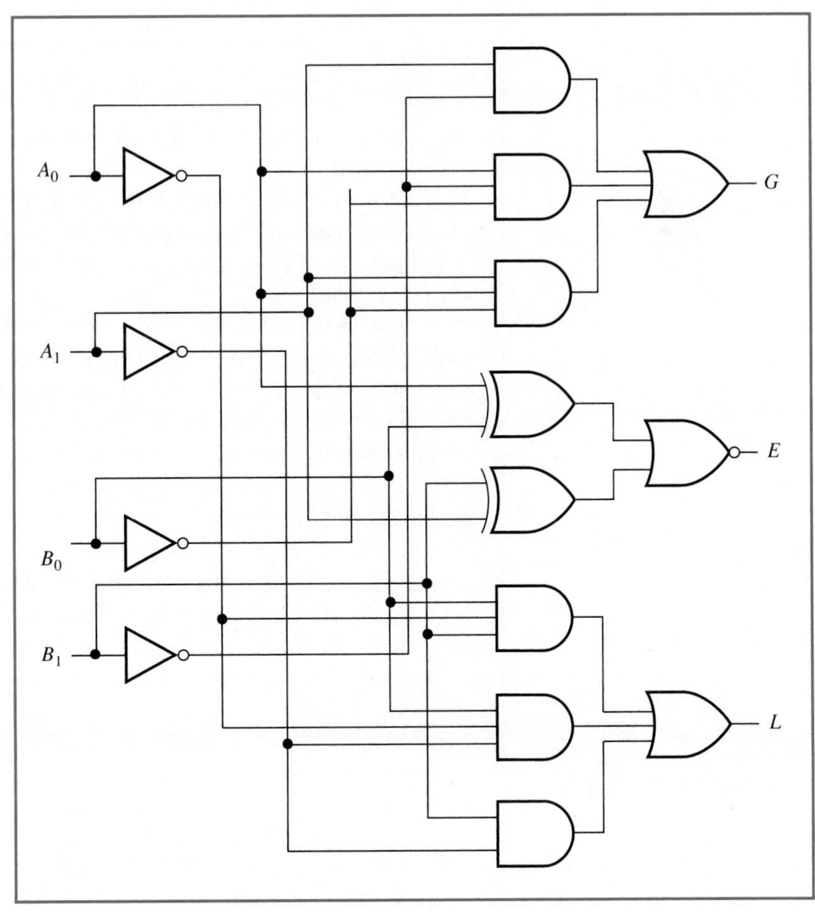

Figure 4–28
The 4-bit magnitude comparator circuit.

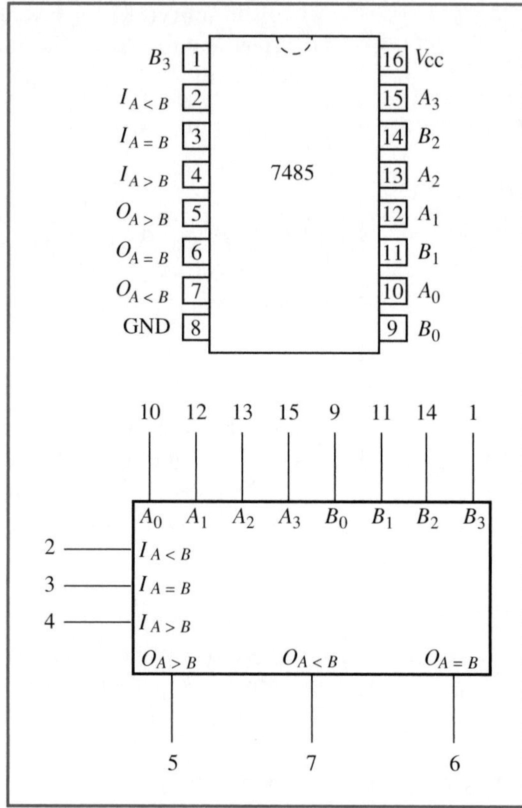

Figure 4–29
Logic symbol and pin configuration of the 7485 4-bit magnitude comparator.

numbers it compares. The logic symbol and pin configuration of the 7485 are shown in Figure 4–29.

The two 4-bit binary numbers being compared are applied to the A (A_0–A_3) and B (B_0–B_3) inputs of the 7485 and the result of the comparison is indicated by activating (to a logic 1) one of the three outputs, O, as shown in Table 4–7.

The three *expansion inputs*, $I_{A>B}$, $I_{A<B}$, and $I_{A=B}$, are used to expand the size of the numbers being compared by connecting together several 7485 ICs. If the size of the numbers being compared is less than or equal to 4 bits, that is, if only a single 7485 IC is being used with no expansion, then the $I_{A=B}$ input must be tied to a logic 1; the states of the other two expansion inputs are not significant to the operation of the 7485 under these conditions.

To compare numbers greater than 4 bits, several 7485 ICs can be connected serially. Figure 4–30a shows how two 7485s can be connected together to function as an 8-bit comparator. Notice that IC-1 compares

Table 4–7
Truth table for the 7485 magnitude comparator

Inputs	$O_{A>B}$	$O_{A<B}$	$O_{A=B}$
$A > B$	1	0	0
$A < B$	0	1	0
$A = B$	0	0	1

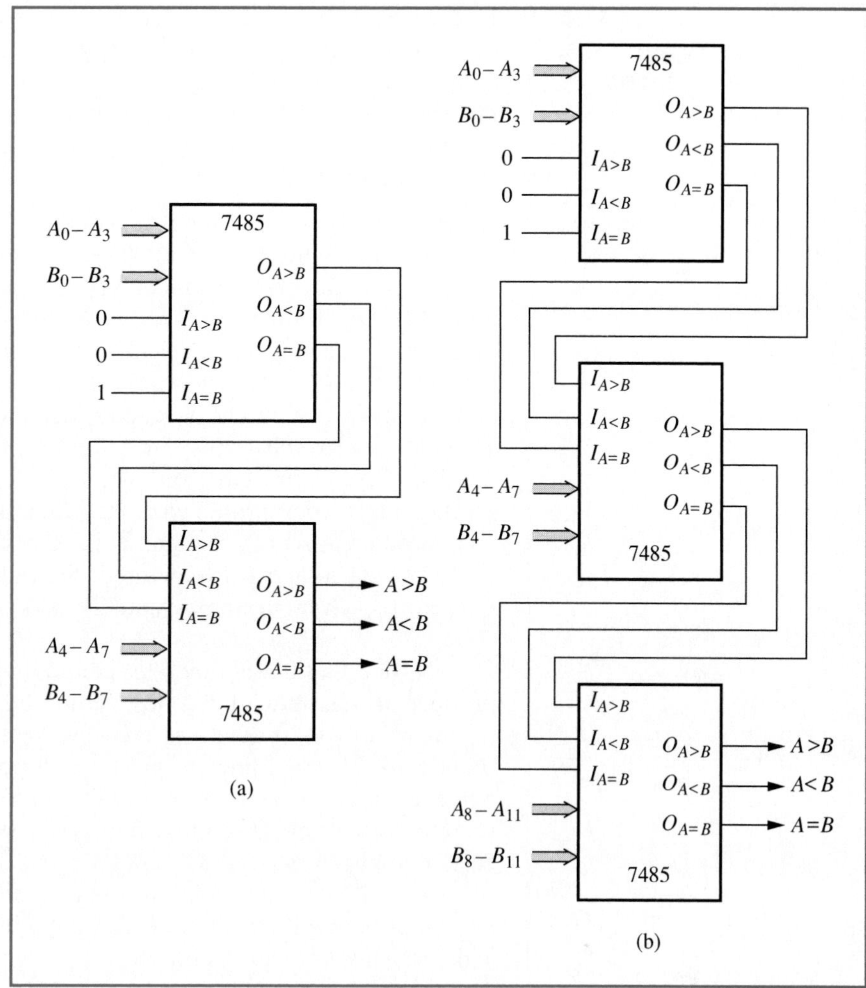

Figure 4–30 Cascading comparators. (a) Comparing two 8-bit numbers; (b) comparing two 12-bit numbers.

the least significant nibble of both numbers while IC-2 compares the most significant nibble. The expansion inputs $I_{A>B}$ and $I_{A<B}$ of IC-1 are tied to a logic 0 state while $I_{A=B}$ is tied to a logic 1. The outputs of IC-1 are connected to the expansion inputs of IC-2 so that the final result of the comparison will be based on both parts of each number. The three outputs of IC-2 provide the results of the comparison of the entire 8-bit number. Figure 4–30b shows three 7485s connected to compare two 12-bit numbers. Notice that the circuit is very similar to the circuit in Figure 4–30a with another 7485 added in series to accommodate the 4 extra bits. Thus a pair of binary numbers of any size can be compared by using the expansion inputs of the 7485 as shown in the circuits of Figure 4–30.

Review questions

1. Describe the operation of a comparator.
2. What is the purpose of the expansion inputs of the 7485 IC?

4–6 Encoders and decoders

Encoders and decoders are digital circuits that are used to generate and detect digital codes. Besides playing an important part in digital communications, these circuits are also used in digital computers to interface many of the various components that make up the digital computer. Applications of encoders and decoders will be seen in this and other chapters in this book.

Encoders

An encoder has several input lines that each produce a unique binary code when activated. Encoders are generally categorized according to the size of the codes they produce. Typical encoders produce 3-bit (octal) or 4-bit (BCD or hexadecimal) codes. To illustrate the operation and design of a simple encoder, consider the 2-bit encoder whose logic symbol is shown in Figure 4–31. The circuit has four inputs labeled I_0 through I_3, and a 2-bit binary output X_1X_0 provides the required code. The input lines are labeled with subscripts that identify the binary code produced when the line is activated. For example, when line I_0 is activated (set to a logic 1), the output code produced is 00, when line I_1 is activated, the output code produced is 01, when line I_2 is activated, the output code produced is 10, and when line I_3 is activated, the output code produced is 11. Therefore each input line essentially tells the encoder to putout its unique code.

Notice that a 2-bit encoder generally requires 4 (2^2) input lines to produce all possible 2-bit codes. Similarly, a 3-bit encoder will require 8 (2^3) input lines to produce all possible 3-bit codes, and a 4-bit encoder will require 16 (2^4) input lines to produce all possible 4-bit codes. Note, however, that encoders do not have to produce all possible codes and therefore the number of input lines may be less than the maximum for certain applications.

To complete the design of the encoder shown in Figure 4–31 we must first obtain its truth table as shown in Table 4–8 and then obtain the logic equations for each output.

The function of an encoder is to produce a unique binary code when *one* of its input lines is activated since (as stated earlier) activating an input line signals the encoder to output the binary code corresponding to that line. If more than one line is activated we are essentially "telling" the encoder to output more than one code on the same set of output lines

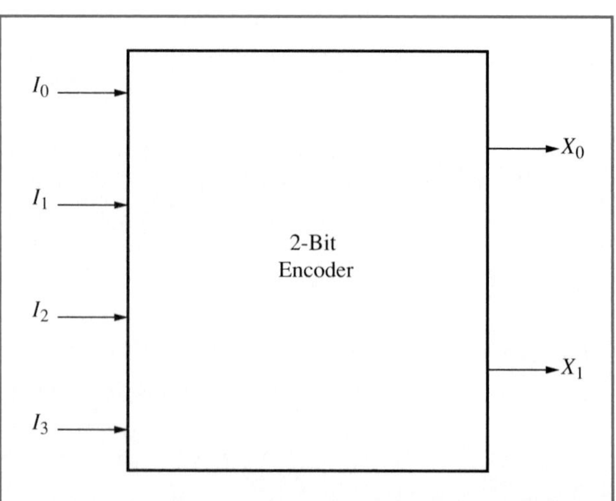

Figure 4–31 Logic symbol for a 2-bit encoder.

Table 4–8
Truth table for a 2-bit encoder

I_0	I_1	I_2	I_3	X_1	X_0	
0	0	0	0	x	x	
0	0	0	1	1	1	←
0	0	1	0	1	0	←
0	0	1	1	x	x	
0	1	0	0	0	1	←
0	1	0	1	x	x	
0	1	1	0	x	x	
0	1	1	1	x	x	
1	0	0	0	0	0	←
1	0	0	1	x	x	
1	0	1	0	x	x	
1	0	1	1	x	x	
1	1	0	0	x	x	
1	1	0	1	x	x	
1	1	1	0	x	x	
1	1	1	1	x	x	

which is logically impossible. Therefore in the design of the encoder we must make the assumption that only one of the input lines will be activated at a given instant and if more than one line is activated we *don't care* what the output will be. These *don't care conditions* are identified by the letter x in the truth table shown in Table 4–8. Since an x identifies an insignificant logic state, we can use it as a 1 or a 0 when convenient. We also don't care what the output of the encoder is if none of the inputs are activated; therefore the first entry in the truth table also has x states for the outputs. Note that if none of the input lines are activated we would ideally like to have no code produced at the output (i.e., no logic levels). However, at this time we will assume that this is not possible since the output of any logic gate or circuit must be at one of two logic states, 1 or 0, and therefore when none of the input lines are activated there will be some insignificant code produced at the outputs.

The K-maps for the output variables X_1 and X_0 are shown in Figure 4–32. Since we must include all the logic 1's in the K-map in the largest possible group and with as few groups as possible, in each K-map we have considered two of the don't care states to be logic 1's and the rest of the don't care states to be logic 0's. The presence of don't care states in a K-map can in most cases assist greatly in producing simplified equations as long as their use is limited to increasing the size of groups of adjacent 1-cells (or 0-cells for product-of-sums simplification). The logic equations obtained from the K-maps in Figure 4–32 are as follows:

Equation (4–21)
$$X_1 = \overline{I}_0\overline{I}_1$$

Equation (4–22)
$$X_0 = \overline{I}_0\overline{I}_2$$

We can now construct the logic circuit for the 2-bit encoder from Equations 4–21 and 4–22 as shown in Figure 4–33. Notice that the outputs of the circuit appear to be independent of the input I_3 which is not connected to anything. This is because we specified don't care outputs for the first entry in the truth table and because of the manner in which the two x cells were combined in each K-map, the output of the circuit when none of the inputs are activated is a 11. Therefore if I_3 is

Figure 4–32
K-maps for the design of the 2-bit encoder.

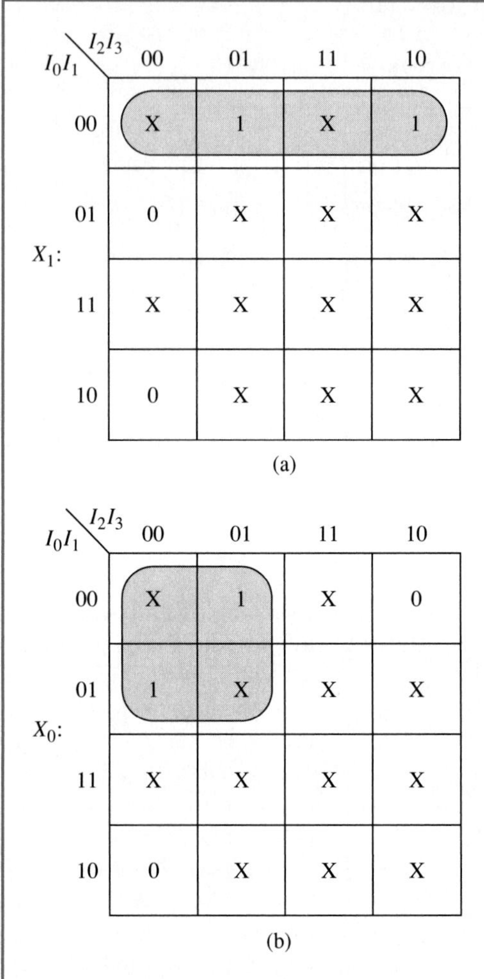

Figure 4–33
Logic circuit for a 2-bit encoder.

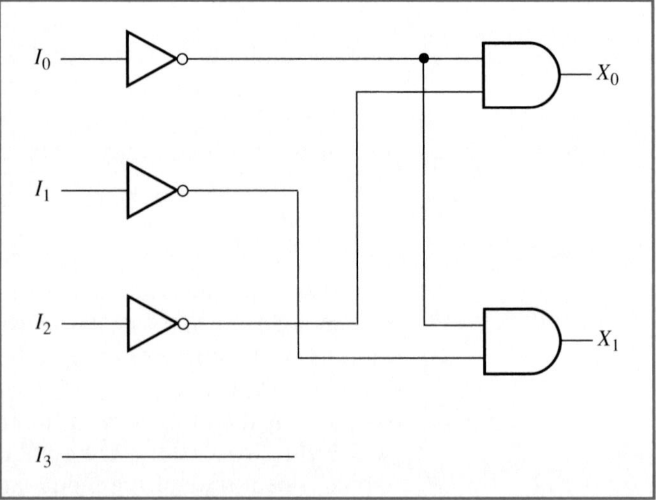

active or inactive the output code will be a 11. This condition does not pose a problem since the correct code is produced when I_3 is activated even though it is not connected to anything.

An encoder that takes into consideration the possibility of more than one input being activated at the same time is known as a *priority encoder*. A priority encoder therefore resolves all the don't care conditions in Table 4–8 by assigning priorities to each input line. Most priority encoders are *high-order data line encoded*. This means that if two or more inputs are activated at the same time, the line with the higher code gets priority. For example, if I_1 and I_3 are activated at the same time, the code for I_3 is put out. Table 4–9 is a modification of Table 4–8 to include high-order priority encoding of the inputs.

Notice in Table 4–9 that the first entry in the truth table is still shown with x outputs since this entry corresponds to no active input. The K-maps for each of the outputs is shown in Figure 4–34, and the logic equations obtained from the K-maps are

Equation (4–23)
$$X_1 = I_2 + I_3$$

Equation (4–24)
$$X_0 = I_3 + I_1\overline{I_2}$$

The logic circuit for the 2-bit priority encoder shown in Figure 4–35 is obtained from Equations 4–23 and 4–24. Again notice that one of the input lines, I_0, is not connected to anything because in this case if none of the input lines are activated, the output will be a 00.

Encoders are used mainly in keyboards and keypads to produce unique binary codes for each key pressed. For example, an ASCII keyboard on a computer uses an ASCII encoder circuit to generate a 7-bit ASCII code for each of the keys or key-combinations pressed. Figure 4–36 shows a typical encoder application with a telephone-style keypad.

The keypad simply consists of an arrangement of pushbutton switches, each of which outputs a logic 1 on a separate line (labeled 0 through *) whenever pressed. The 12 switch outputs are connected to a 4-bit encoder that will produce a unique 4-bit code for each of the keys in the keypad as shown in Table 4–10.

Table 4–9
Truth table for a 2-bit priority encoder

I_0	I_1	I_2	I_3	X_1	X_0
0	0	0	0	x	x
0	0	0	1	1	1
0	0	1	0	1	0
0	0	1	1	1	1
0	1	0	0	0	1
0	1	0	1	1	1
0	1	1	0	1	0
0	1	1	1	1	1
1	0	0	0	0	0
1	0	0	1	1	1
1	0	1	0	1	0
1	0	1	1	1	1
1	1	0	0	0	1
1	1	0	1	1	1
1	1	1	0	1	0
1	1	1	1	1	1

Figure 4–34
K-maps for the design of a 2-bit priority encoder.

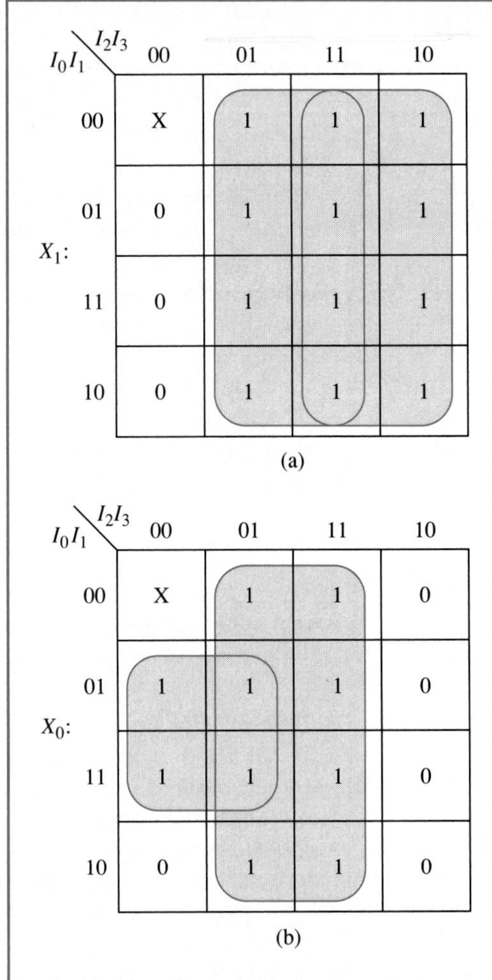

Since a 4-bit encoder has the capability to produce 16 codes (2^4) from a set of 16 input lines, 4 of these 16 lines are unused in the keypad circuit since there are only 12 keys. The four unused codes (1100 through 1111) are shown in Table 4–10.

BCD encoders are also encoders that generate 4-bit codes but do not generate all 16 combinations. BCD encoders are circuits that output the

Figure 4–35
Logic circuit for a 2-bit priority encoder.

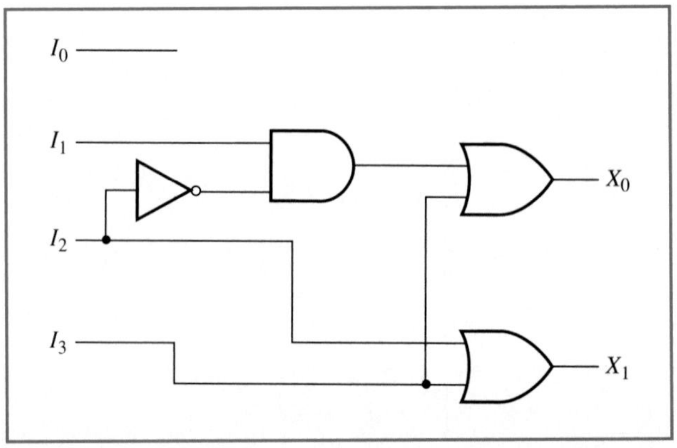

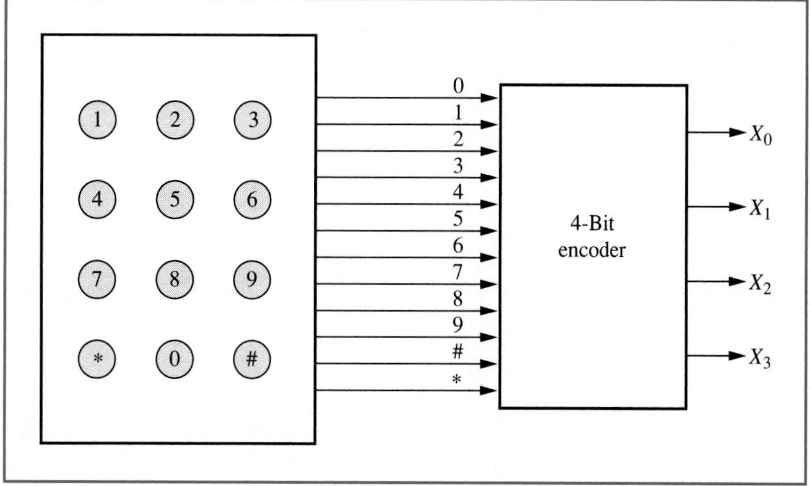

Figure 4–36
Encoded keypad circuit.

ten BCD codes (0000 through 1001) and have only ten input lines corresponding to each of these codes. BCD encoders are used in applications where the ten keys of a decimal keypad (such as in a calculator) have to be encoded to BCD.

Decoders

The function of a decoder is the opposite of the function of an encoder. An encoder *produces* a unique binary code for each of its input lines while a decoder *detects* a unique binary code by activating one of several output lines. Figure 4–37 shows the logic symbol for a 1-of-4 decoder. The inputs to the circuit, X_1X_0, represent a 2-bit code while the outputs of the circuit, O_0 through O_3, are used to identify the 2-bit code that is applied to the inputs; the subscript used in each output variable identifies the input code that activates it. For example, if the code 00 is applied to the input, output line O_0 is activated; if the code 01 is applied to the input, O_1 is activated; if the code 10 is applied to the input, O_2 is activated; and if the code 11 is applied to the input, O_3 is activated. Thus

Table 4–10
Key codes for encoded keypad circuit

X_3	X_2	X_1	X_0	Key
0	0	0	0	0
0	0	0	1	1
0	0	1	0	2
0	0	1	1	3
0	1	0	0	4
0	1	0	1	5
0	1	1	0	6
0	1	1	1	7
1	0	0	0	8
1	0	0	1	9
1	0	1	0	#
1	0	1	1	*
1	1	0	0	Not used
1	1	0	1	Not used
1	1	1	0	Not used
1	1	1	1	Not used

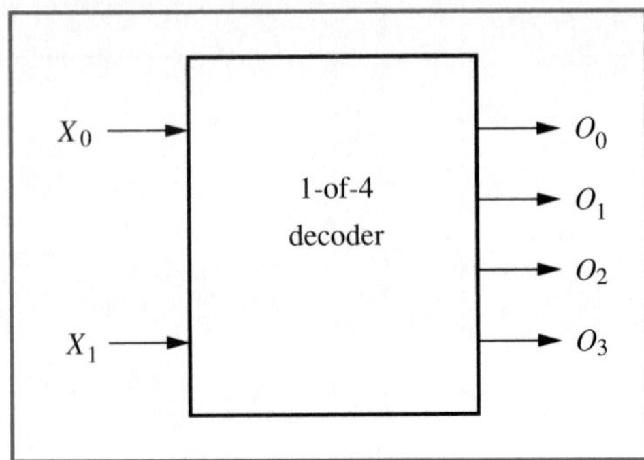

Figure 4–37
Logic symbol for a 1-of-4 decoder.

at any given instant, only *one of four* lines will be active when a code is present at the input and therefore the decoder is called a *1-of-4 decoder*. Table 4–11 shows the truth table for the 1-of-4 decoder. The logic equations for each one of the outputs of the decoder can be obtained directly from Table 4–11 (since no simplification is necessary) as follows:

Equation (4–25)
$$O_0 = \overline{X}_1\overline{X}_0$$

Equation (4–26)
$$O_1 = \overline{X}_1X_0$$

Equation (4–27)
$$O_2 = X_1\overline{X}_0$$

Equation (4–28)
$$O_3 = X_1X_0$$

Finally, the logic circuit for the decoder can be constructed as shown in Figure 4–38 from Equations 4–25 through 4–28.

Table 4–11
Truth table for a 1-of-4 decoder

X_1	X_0	O_0	O_1	O_2	O_3
0	0	1	0	0	0
0	1	0	1	0	0
1	0	0	0	1	0
1	1	0	0	0	1

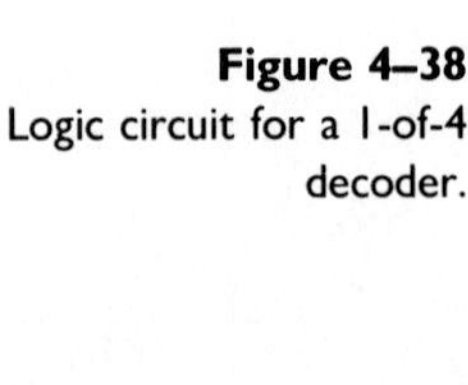

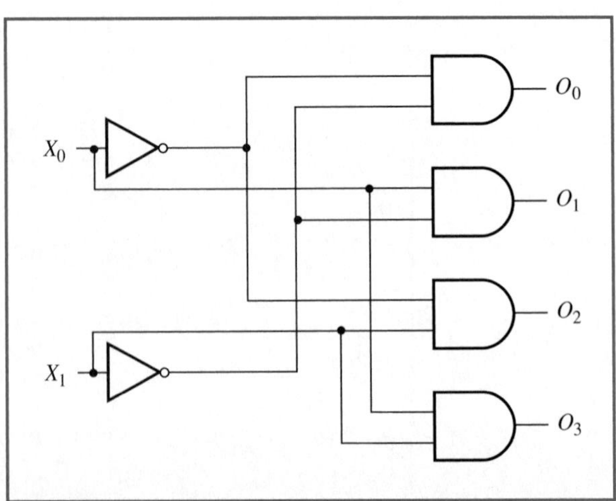

Figure 4–38
Logic circuit for a 1-of-4 decoder.

EXAMPLE 4–4

Design a 1-of-8 decoder whose logic symbol is shown in Figure 4–39.

Figure 4–39
Logic symbol for a 1-of-8 decoder.

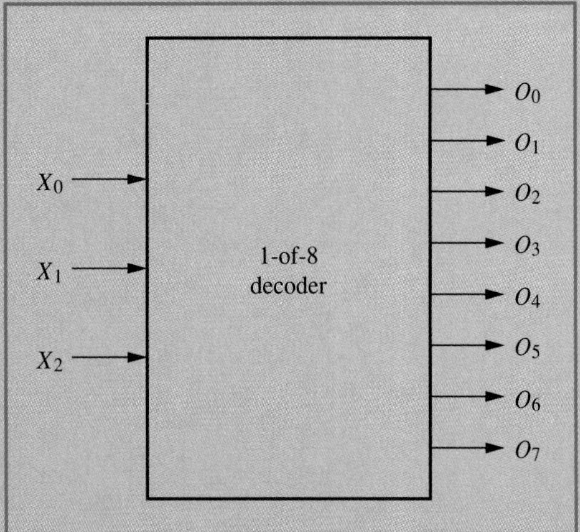

Table 4–12
Truth table for a 1-of-8 decoder

X_2	X_1	X_0	O_0	O_1	O_2	O_3	O_4	O_5	O_6	O_7
0	0	0	1	0	0	0	0	0	0	0
0	0	1	0	1	0	0	0	0	0	0
0	1	0	0	0	1	0	0	0	0	0
0	1	1	0	0	0	1	0	0	0	0
1	0	0	0	0	0	0	1	0	0	0
1	0	1	0	0	0	0	0	1	0	0
1	1	0	0	0	0	0	0	0	1	0
1	1	1	0	0	0	0	0	0	0	1

SOLUTION The truth table for the 1-of-8 decoder is shown in Table 4–12. The logic equations for O_0 through O_7 are obtained directly from Table 4–12 as follows:

Equation (4–29)
$$O_0 = \overline{X}_2\overline{X}_1\overline{X}_0$$

Equation (4–30)
$$O_1 = \overline{X}_2\overline{X}_1X_0$$

Equation (4–31)
$$O_2 = \overline{X}_2X_1\overline{X}_0$$

Equation (4–32)
$$O_3 = \overline{X}_2X_1X_0$$

Equation (4–33)
$$O_4 = X_2\overline{X}_1\overline{X}_0$$

Equation (4–34)
$$O_5 = X_2\overline{X}_1X_0$$

Equation (4–35)
$$O_6 = X_2X_1\overline{X}_0$$

Equation (4–36)
$$O_7 = X_2X_1X_0$$

The logic circuit for the 1-of-8 decoder shown in Figure 4–40 is obtained from Equations 4–29 through 4–36.

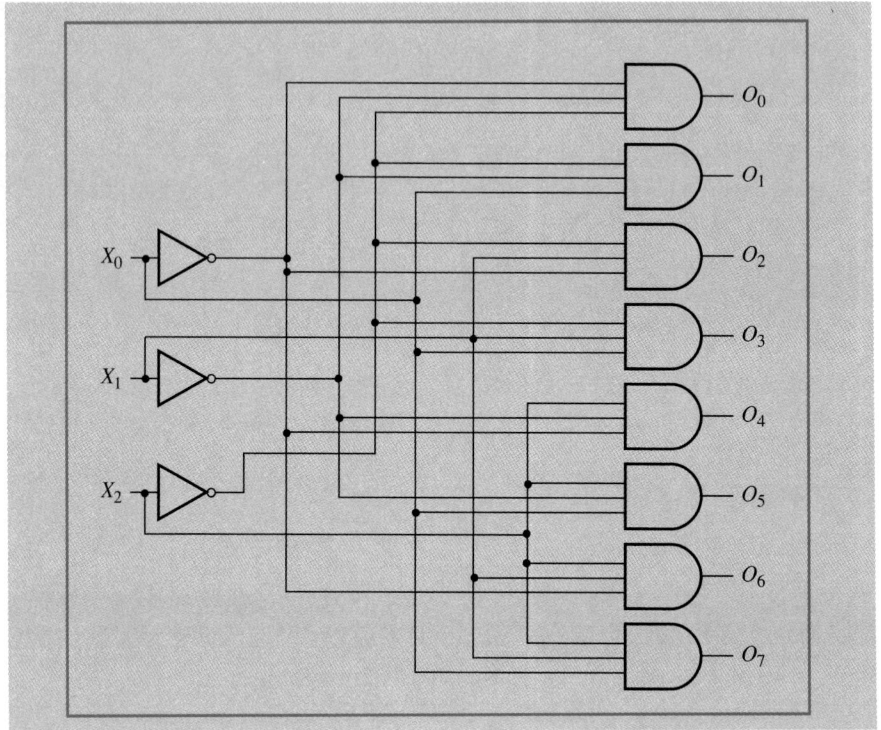

Figure 4–40
Logic circuit for a 1-of-8 decoder.

Encoders and decoders are very frequently used in digital communication circuits to transfer information between digital devices. For example, the keys on a keyboard are usually used to transfer information to a digital computer or circuit by encoding the keys and transmitting the key codes rather than transmitting the signals generated by each individual key. The codes are then decoded at the receiving end into their original form so that the process of encoding and decoding appears transparent to the transfer of information. Figure 4–41 illustrates such a communications scheme where each input line I will indirectly activate the corresponding output line O. For example, activating I_5 will indirectly activate O_5 through the encoding and decoding process. The encoding and decoding circuits make it appear that there is a direct

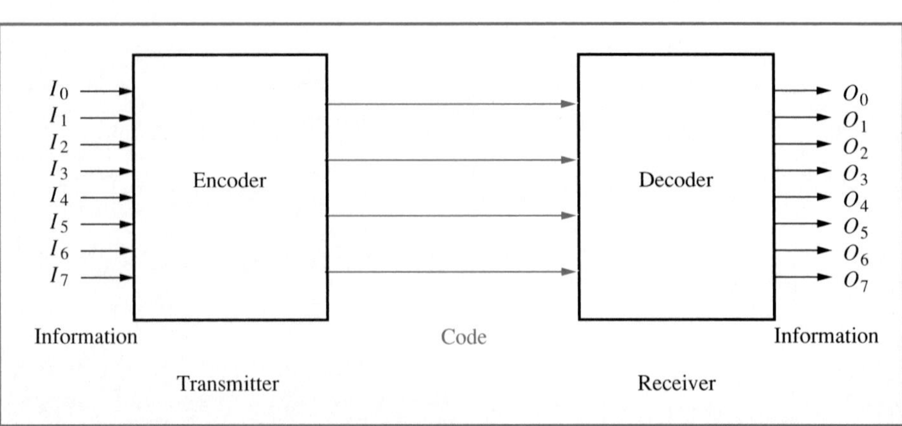

Figure 4–41
Digital information transfer using an encoder and decoder.

connection between the input (I) lines and the output (O) lines with the advantage of having fewer lines connected between the transmitter and receiver.

Integrated circuit logic

Priority encoders are available in the form of two ICs—the 74147 and the 74148. The logic symbol and pin configuration of the 74147 and the 74148 are shown in Figure 4–42a and b, respectively.

The 74147 is a 4-bit BCD priority encoder that encodes ten input lines into a 4-bit BCD code. The truth table for the 74147 is shown in Table 4–13. Notice in Table 4–13 that the inputs ($\overline{1}$ through $\overline{9}$) are *active low*. This means that the encoder will encode a particular input line when that line is (activated) at the *low state*. Active low inputs and outputs are represented by placing a bar over the variable and a bubble in the logic symbol (note: some manufacturers use either the bar *or* the bubble but not both to identify active low signals). The bubble simply indicates that a logic low is internally inverted to produce a high at the input and a logic high is internally inverted to produce a logic low at the output. Also notice that the output code produced by the encoder is complemented. For example, when line $\overline{9}$ is activated the output is 0110 (1001 complemented), and when line $\overline{1}$ is activated the output is 1110 (0001 complemented). The 74147 incorporates highest-order data line encoding and therefore the output of the encoder will always correspond to the highest-order input line if more than one input line is active. For example, in Table 4–13 the second entry indicates that if input line $\overline{9}$ is activated,

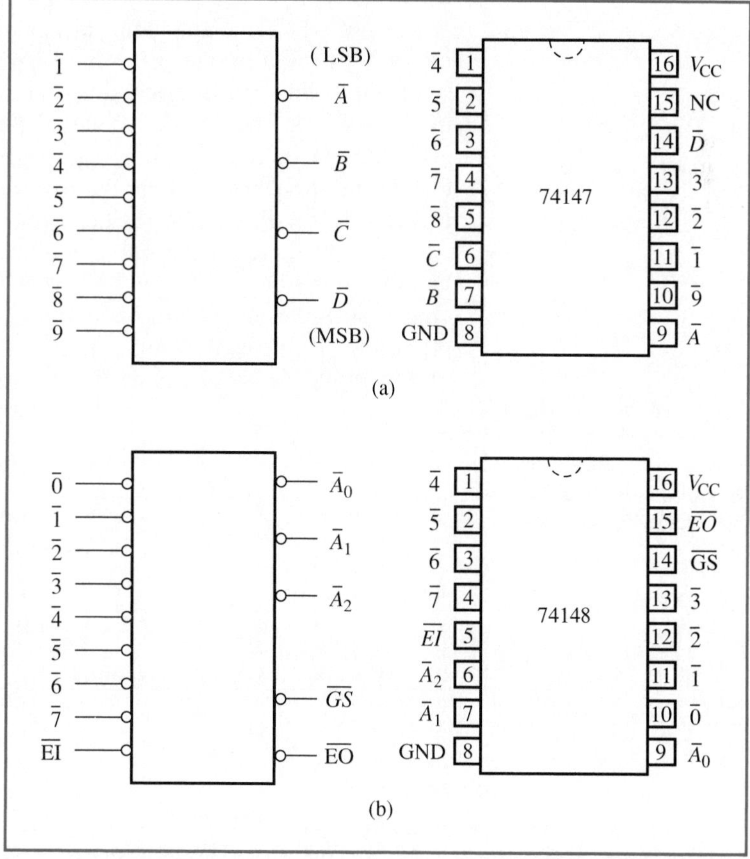

Figure 4–42
Logic symbols and pin configurations of (a) the 74147 priority encoder and (b) the 74148 priority encoder.

Table 4–13
Truth table for the 74147 priority encoder

Inputs									Outputs			
$\overline{1}$	$\overline{2}$	$\overline{3}$	$\overline{4}$	$\overline{5}$	$\overline{6}$	$\overline{7}$	$\overline{8}$	$\overline{9}$	$\overline{D}$	$\overline{G}$	$\overline{B}$	$\overline{A}$
1	1	1	1	1	1	1	1	1	1	1	1	1
X	X	X	X	X	X	X	X	0	0	1	1	0
X	X	X	X	X	X	X	0	1	0	1	1	1
X	X	X	X	X	X	0	1	1	1	0	0	0
X	X	X	X	X	0	1	1	1	1	0	0	1
X	X	X	X	0	1	1	1	1	1	0	1	0
X	X	X	0	1	1	1	1	1	1	0	1	1
X	X	0	1	1	1	1	1	1	1	1	0	0
X	0	1	1	1	1	1	1	1	1	1	0	1
0	1	1	1	1	1	1	1	1	1	1	1	0

the output code will be a 0110 (1001 complemented), the code for line $\overline{9}$, regardless of the states of the other input lines (as indicated by the don't cares—X). Similarly, the fifth entry in Table 4–13 indicates that if line $\overline{6}$ is activated, the output of the encoder will be 1001 (0110 complemented) regardless of the states of lines $\overline{1}$ through $\overline{5}$ since line $\overline{6}$ has a higher priority; however, since lines $\overline{7}$ through $\overline{9}$ have a higher priority than line $\overline{6}$ these line must be inactive (high state). Also notice in the logic symbol of the 74147 in Figure 4–42a and its truth table in Table 4–13 that input line $\overline{0}$ is missing. This is because the code for this line, 1111 (0000 complemented) is produced when none of the other input lines are active (first entry in Table 4–13). This implied condition therefore does not require any input line.

The 74148 is a 3-bit priority encoder that encodes eight input lines into a 3-bit (octal) code. Like the 74147, the 74148 incorporates highest-order input line priority encoding. The truth table for the 74148 is shown in Table 4–14. The 74148 encodes eight input lines ($\overline{0}$ through $\overline{7}$) into a 3-bit code at the output lines ($\overline{A}_2\overline{A}_1\overline{A}_0$). The 74148 is similar to the 74147 in that the input lines are active-low and the output code is represented in inverted form. The 74148 includes two *enable* lines that allow the IC to be expanded to increase the number of input lines encoded. The enable input line ($\overline{EI}$) must be active (low) in order for the 74148 to function. The enable output line $\overline{EO}$ will be at the low state if none of the encoders input lines ($\overline{0}$ through $\overline{7}$) are active and the $\overline{EI}$ line is active. An example of expanding the 74148 using these lines follows. The $\overline{GS}$ output line is

Table 4–14
Truth table for the 74148 priority encoder

Inputs									Outputs				
$\overline{EI}$	$\overline{0}$	$\overline{1}$	$\overline{2}$	$\overline{3}$	$\overline{4}$	$\overline{5}$	$\overline{6}$	$\overline{7}$	$\overline{A}_2$	$\overline{A}_1$	$\overline{A}_0$	$\overline{GS}$	$\overline{EO}$
1	X	X	X	X	X	X	X	X	1	1	1	1	1
0	1	1	1	1	1	1	1	1	1	1	1	1	0
0	X	X	X	X	X	X	X	0	0	0	0	0	1
0	X	X	X	X	X	X	0	1	0	0	1	0	1
0	X	X	X	X	X	0	1	1	0	1	0	0	1
0	X	X	X	X	0	1	1	1	0	1	1	0	1
0	X	X	X	0	1	1	1	1	1	0	0	0	1
0	X	X	0	1	1	1	1	1	1	0	1	0	1
0	X	0	1	1	1	1	1	1	1	1	0	0	1
0	0	1	1	1	1	1	1	1	1	1	1	0	1

active (low) when the $\overline{EI}$ line is active and when any of the input lines ($\overline{0}$ through $\overline{7}$) are active. This line can therefore be used to indicate if the output of the encoder contains a valid code. For example, if a keypad were connected to the 74148, the closure of any key would activate this line. The following example illustrates how the $\overline{EI}$, $\overline{EO}$, and $\overline{GS}$ lines can be used to expand the capabilities of the 74148.

EXAMPLE 4–5

Figure 4–43b illustrates how two 74148 ICs can be connected together to make up a 4-bit encoder that encodes 16 input lines into a 4 bit code. The logic symbol for such an encoder is shown in Figure 4–43a.

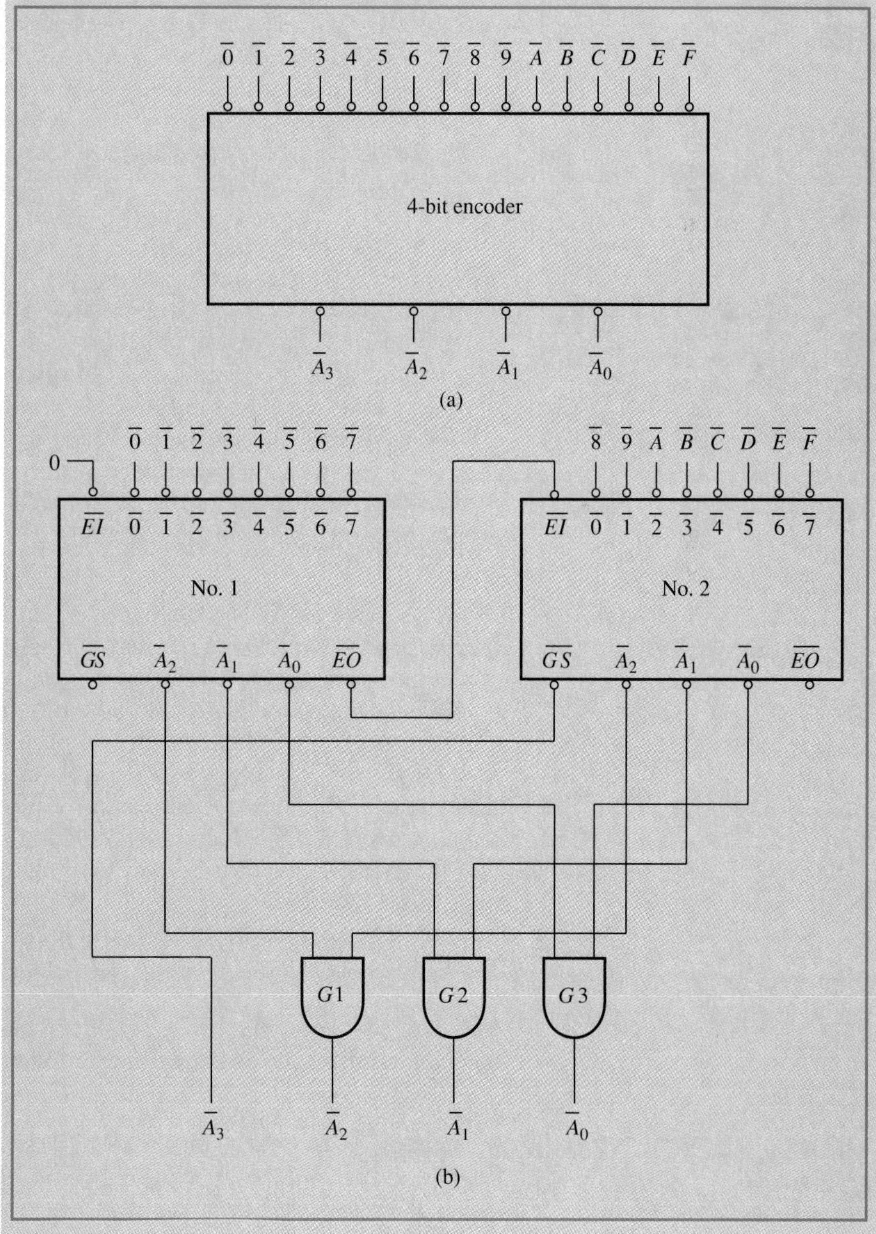

Figure 4–43
(a) Logic symbol for a 4-bit encoder; (b) connecting two 74148s to function as a 4-bit encoder.

Encoder no. 1 has its $\overline{EI}$ line permanently tied low to enable the encoder at all times. If any of the input lines ($\overline{0}$ through $\overline{7}$) of encoder no. 1 is activated, the enable output line $\overline{EO}$ will be a logic 1 and therefore disable encoder no. 2 since this line is connected to $\overline{EI}$ of encoder no. 2. The outputs of encoder no. 2 ($\overline{A}_2\overline{A}_1\overline{A}_0$) will therefore be high (111), and therefore the outputs of encoder no. 1 will be passed through the *AND* gates (*G*1, *G*2, and *G*3) as the output code ($\overline{A}_3\overline{A}_2\overline{A}_1\overline{A}_0$). The $\overline{GS}$ output of encoder no. 2 represents the MSB of the output code ($\overline{A}_3$) and will be a logic 1 if any of the input lines ($\overline{0}$ through $\overline{7}$) of encoder no. 1 are activated. Thus encoder no. 1 will produce the codes 1111 through 1000 (0000 through 0111 complemented) at the output lines $\overline{A}_3\overline{A}_2\overline{A}_1\overline{A}_0$ when any of the input lines $\overline{0}$ through $\overline{7}$ are activated. If none of the input lines ($\overline{0}$ through $\overline{7}$) of encoder no. 1 are activated, its $\overline{EO}$ output is low, and therefore encoder no. 2 is enabled. Since the output lines $\overline{A}_2\overline{A}_1\overline{A}_0$ of encoder no. 1 are now at the high state (because its input lines $\overline{0}$ through $\overline{7}$ are inactive), the output lines $\overline{A}_2\overline{A}_1\overline{A}_0$ of encoder no. 2 will be passed through the *AND* gates (*G*1, *G*2, and *G*3) as the output code ($\overline{A}_3\overline{A}_2\overline{A}_1\overline{A}_0$) if any of the input lines ($\overline{8}$ through $\overline{F}$) of encoder no. 2 are activated. The $\overline{GS}$ line of encoder no. 2 is a logic 0 when the encoder's input lines ($\overline{8}$ through $\overline{F}$) are active, and therefore activating any of the input lines of encoder no. 2 will produce the codes 0111 through 0000 (1000 through 1111 complemented) at the output lines $\overline{A}_3\overline{A}_2\overline{A}_1\overline{A}_0$.

Note that this circuit does not incorporate highest-order data input priority encoding between the input lines connected to encoder no. 1 ($\overline{0}$ through $\overline{7}$) and the input lines connected to encoder no. 2 ($\overline{8}$ through $\overline{F}$). For example, if line $\overline{2}$ and line $\overline{A}$ are activated at the same time, the output of the circuit $\overline{A}_3\overline{A}_2\overline{A}_1\overline{A}_0$ will be 1101 (0010 complemented).

The 74138 is a 1-of-8 decoder IC that decodes a 3-bit binary code at its input and activates one of eight output lines. The logic symbol and pin configuration of the 74138 is shown in Figure 4–44.

The 74138 decodes a 3-bit code at its input lines $A_2A_1A_0$ and activates one of eight (active-low) output lines $\overline{O}_0$ through $\overline{O}_7$. The truth table for the 74138 is shown in Table 4–15. Notice that the 74138 has three input enable lines, $\overline{E}_1$, $\overline{E}_2$, and E_3. These three enable lines must be active in order for the decoder to function. Since $\overline{E}_1$ and $\overline{E}_2$ are active-low, these two lines must be tied to a logic 0, while the active-high enable input line E_3 must be tied to a logic 1. If any one of these three enable lines are inactive, the decoder's output will be inactive regardless of the states of the other inputs as illustrated in the first three entries of Table 4–15.

The enable inputs of the 74138 can be used to expand the decoding capabilities of the IC as shown in the following example.

EXAMPLE 4–6

The 1-of-16 decoder whose logic symbol is shown in Figure 4–45a can be implemented by using two 74138 decoders as shown in Figure 4–45b.

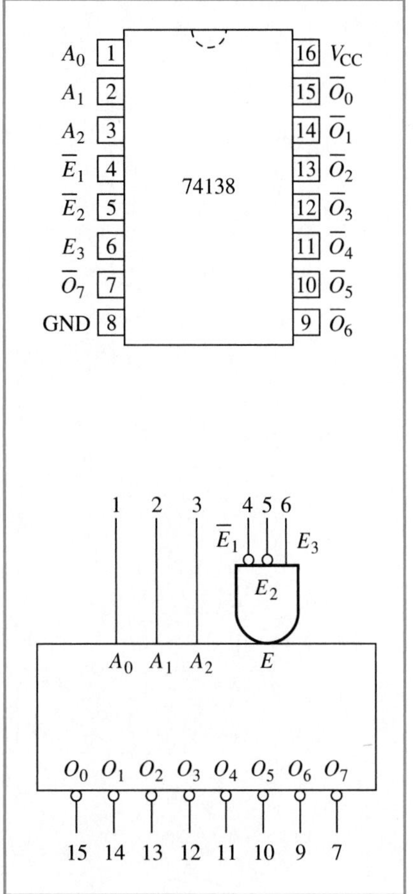

Figure 4–44
Logic symbol and pin configuration of the 74138 1-of-8 decoder.

Table 4–15
Truth table for the 74138 decoder

Inputs						Outputs							
$\overline{E}_1$	$\overline{E}_2$	E_3	A_2	A_1	A_0	$\overline{O}_0$	$\overline{O}_1$	$\overline{O}_2$	$\overline{O}_3$	$\overline{O}_4$	$\overline{O}_5$	$\overline{O}_6$	$\overline{O}_7$
1	X	X	X	X	X	1	1	1	1	1	1	1	1
X	1	X	X	X	X	1	1	1	1	1	1	1	1
X	X	0	X	X	X	1	1	1	1	1	1	1	1
0	0	1	0	0	0	0	1	1	1	1	1	1	1
0	0	1	0	0	1	1	0	1	1	1	1	1	1
0	0	1	0	1	0	1	1	0	1	1	1	1	1
0	0	1	0	1	1	1	1	1	0	1	1	1	1
0	0	1	1	0	0	1	1	1	1	0	1	1	1
0	0	1	1	0	1	1	1	1	1	1	0	1	1
0	0	1	1	1	0	1	1	1	1	1	1	0	1
0	0	1	1	1	1	1	1	1	1	1	1	1	0

Notice in Figure 4–45b that the input lines A_0, A_1, and A_2 of the circuit are connected to the input lines ($A_2A_1A_0$) of both the decoders. However input line A_3 will either enable decoder no. 1 (when at the logic 0 state) or enable decoder no. 2 (when at the logic 1 state), and therefore decoder no. 1 will activate its output lines $\overline{O}_0$ through $\overline{O}_7$ whenever the codes 0000 through 0111 appear at the

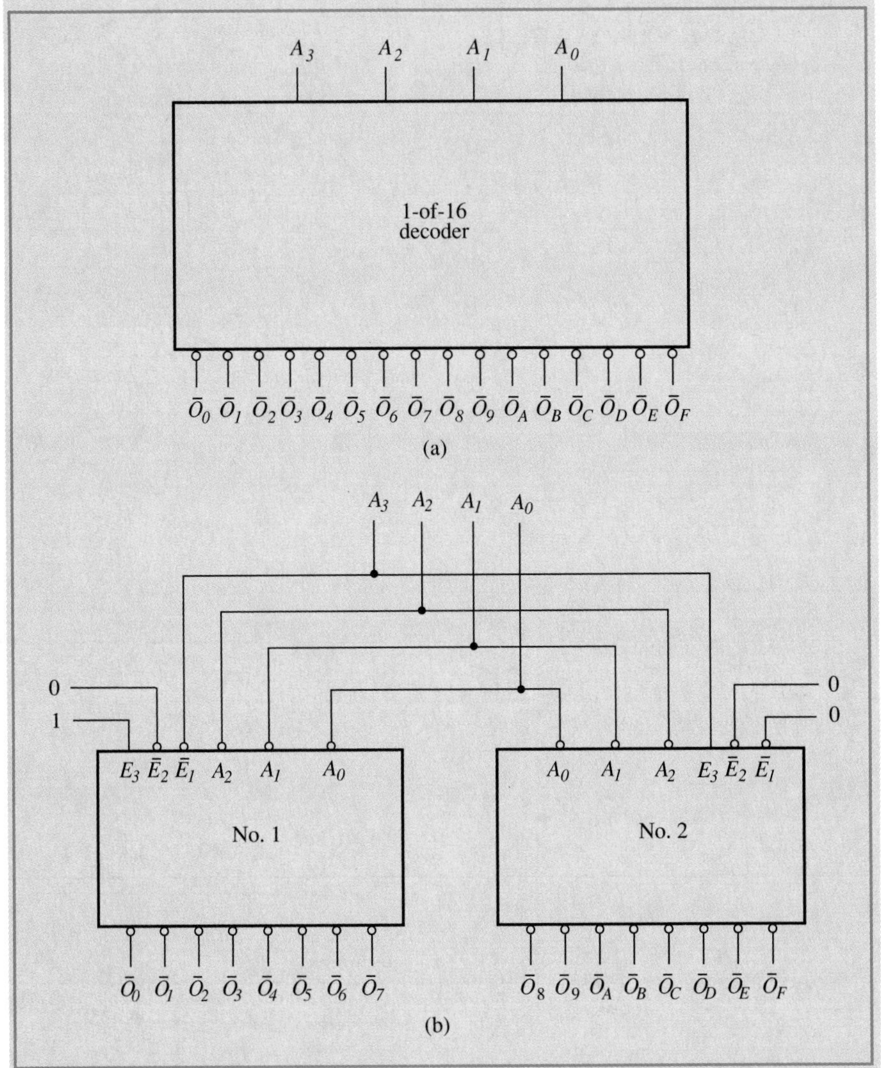

Figure 4–45 (a) logic symbol of a 1-of-16 decoder; (b) two 74138s connected to function as a 1-of-16 decoder.

circuit's input lines $A_3A_2A_1A_0$ while decoder no. 2 will activate its output lines $\overline{O}_8$ through $\overline{O}_F$ whenever the codes 1000 through 1111 appear at the circuit's input lines $A_3A_2A_1A_0$. Notice that the remaining enable lines of decoders nos. 1 and 2 are unused and therefore tied active.

The 74154 is a 1-of-16 decoder that functions in a manner very similar to the circuit designed in Example 4–6. The logic symbol and pin configuration of the 74154 is shown in Figure 4–46. The 74154 decodes a 4-bit code appearing on the input lines $A_3A_2A_1A_0$ and activates one of 16 active-low outputs, $\overline{O}_0$ through $\overline{O}_{15}$. The 74154 has two active-low enable inputs, $\overline{E}_0$ and $\overline{E}_1$, that must both be at the low state in order for the IC to function.

The 74139 is a dual 1-of-4 decoder IC that incorporates two independent 1-of-4 decoders in one package. The logic symbol and pin configuration for the 74139 is shown in Figure 4–47.

Figure 4–46
Logic symbol and pin configuration of the 74154 1-of-16 decoder.

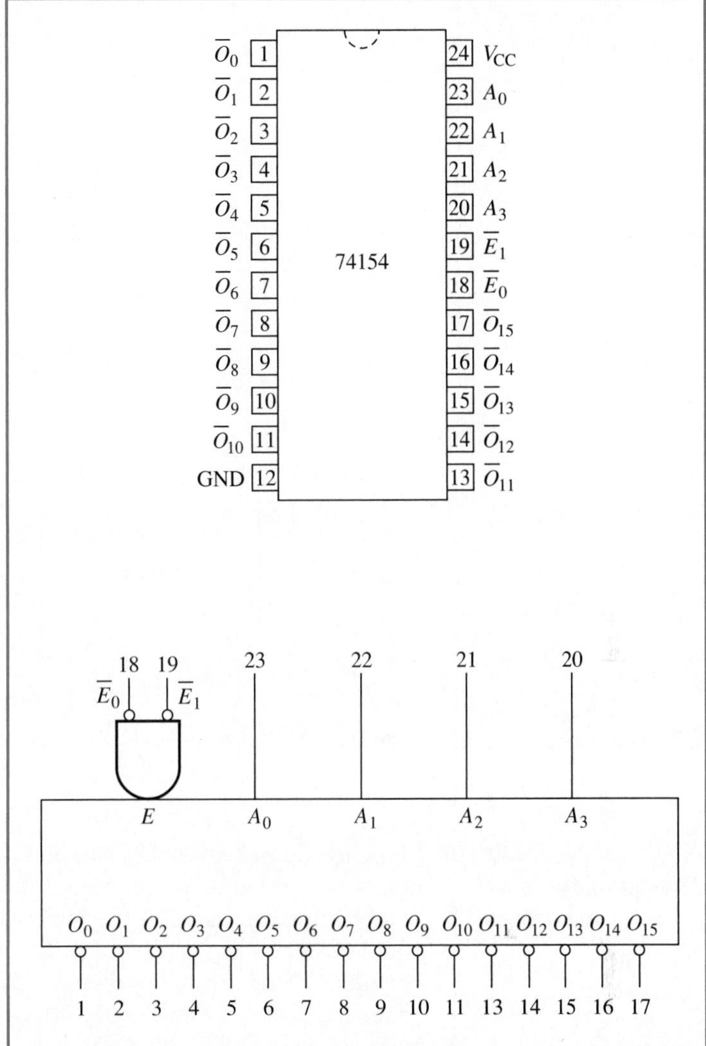

Each decoder in the 74139 IC decodes a 2-bit code appearing on the input lines A_1A_0 and activates one of 4 active-low output lines $\overline{O}_0$ through $\overline{O}_3$. The decoders are very similar in operation to the circuit designed in Figure 4–38. Each decoder also has an active-low enable line, $\overline{E}$, which must be tied to the logic 0 state in order to enable the circuit.

Review questions

1. Describe the operation of an encoder.
2. What is a priority encoder?
3. What is meant by high-order data line encoding?
4. Describe an application of an encoder.
5. Describe the function of a decoder.
6. Why are most decoders known as 1-of-*n* decoders?

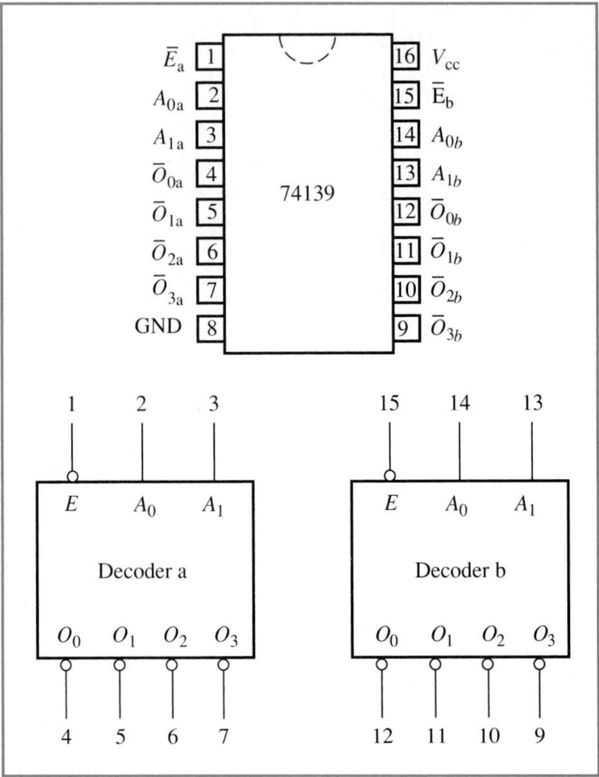

Figure 4–47 Logic symbol and pin configuration of the 74139 dual 1-of-4 decoder.

4–7 Seven-segment decoders

The decoders discussed in Section 4–6 detected binary codes at their inputs by activating a unique output line whenever a particular binary combination appeared at the input. The output line that was activated could then be used to identify the binary code present at the input of the decoder. There are many digital applications where it is necessary to decode a binary code directly into a form that represents that code. For example, a typical application may require that a binary number be decoded into a form that can be displayed on a seven-segment light-emitting diode (LED). This section deals with decoders that are used to drive seven-segment LEDs. These decoders are actually known as *code converters* since they effectively convert one code into another.

A seven-segment LED consists of a set of seven bar-shaped LEDs arranged in the shape of the number 8 as shown in Figure 4–48a. The seven segments are labeled *a* through *g* for identification purposes. There are two common configurations available, the *common cathode* (CC) configuration (Figure 4–48c) and the *common anode* (CA) configuration (Figure 4–48b). Notice in Figure 4–48c that all the cathodes of the seven-segment LED are tied together at one point (cc) that is grounded (logic 0). Since an LED must be forward-biased (i.e., current must pass from anode to cathode) in order for it to emit light, a *logic 1* on any of the segment connections *a* through *g* will illuminate the respective segment. In Figure 4–48b since all the anodes of the seven-segment LED are tied together at one point (ca) that is tied to a positive voltage V^+, a *logic 0* on any of the segment connections *a* through *g* will forward bias the respective diode and illuminate the segment. Thus, common cathode seven-segment LEDs require a logic 1 to illuminate a segment while common anode seven-

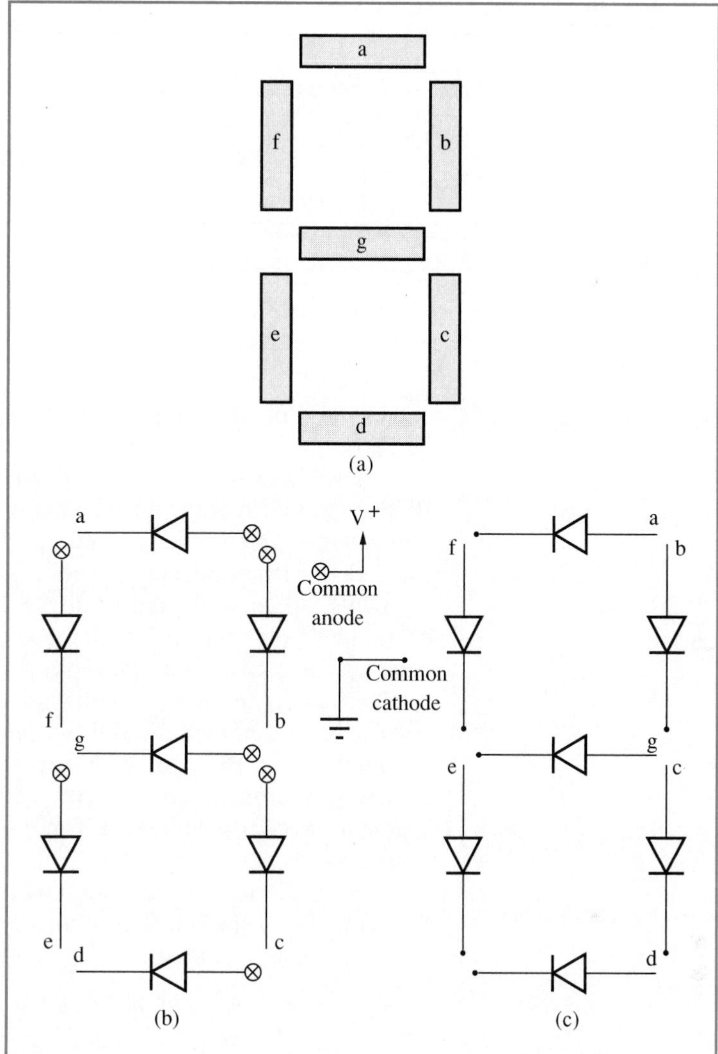

Figure 4–48
Seven-segment LEDs. (a) Configuration; (b) common anode; (c) common cathode.

segment LEDs require a logic 0 to illuminate a segment. There are certain design considerations necessary to supply or limit the flow of current through the LED segments, but these considerations will be examined in a later chapter.

We can now investigate the binary codes necessary to display all the hexadecimal digits (0 through F) on a common cathode seven-segment LED. Since the hexadecimal digit set includes all the binary, octal, and decimal digits, these codes can be used in any application requiring a display in any one of the four numbering systems. The display format for the hexadecimal digits is shown in Figure 4–49. Notice that the digits B and D must be displayed as lowercase letters b and d, respectively. The pictorial representation of each digit is then translated into the binary codes required to activate the appropriate segments for each digit as shown in Table 4–16. For example, to display the digit 5 on the seven-segment LED, segments *a*, *c*, *d*, *f*, and *g* must be illuminated and therefore in Table 4–16 these segments have been set to logic 1's for the digit 5. Note that the codes shown in Table 4–16 are common cathode

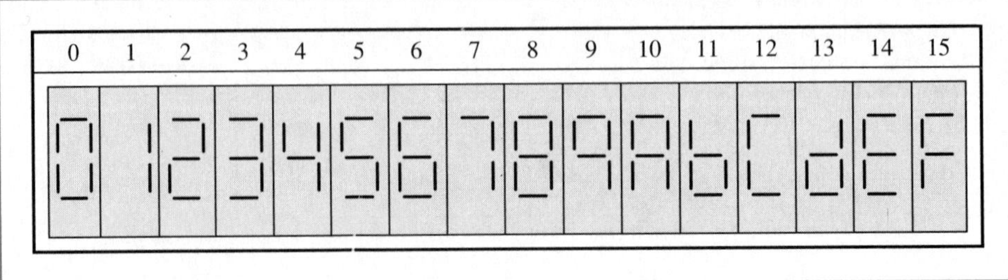

Figure 4–49
Seven-segment display of hexadecimal digits.

codes and must be complemented for common anode seven-segment LEDs.

A BCD to seven-segment decoder decodes (actually converts) a 4-bit BCD code that represents the decimal digits 0 through 9 into the (first ten) seven-segment codes listed in Table 4–16 so that when the output of the decoder is connected to a seven-segment LED, the appropriate segments will be activated to display the decimal digit corresponding to each one of the ten BCD codes. A hexadecimal to seven-segment decoder decodes (or converts) all possible combinations of 4 bits into the seven-segment codes to display the hexadecimal digits that represent the 4-bit codes. Since a hexadecimal to seven-segment decoder has the capability to function as a BCD to seven-segment decoder, we shall investigate the design of the former circuit. The truth table for a hexadecimal to seven-segment decoder is shown in Table 4–17 and its logic symbol is shown in Figure 4–50.

From Table 4–17 we can now obtain the simplified logic equations for each one of the outputs, *a* through *g*, through the use of the K-maps shown in Figure 4–51.

Equation (4–37)
$$a = A_2A_1 + \overline{A}_3A_1 + A_3\overline{A}_0 + \overline{A}_2\overline{A}_0 + \overline{A}_3A_2A_0 + A_3\overline{A}_2\overline{A}_1$$

Equation (4–38)
$$b = \overline{A}_3\overline{A}_2 + A_3\overline{A}_1A_0 + \overline{A}_2\overline{A}_0 + \overline{A}_3\overline{A}_1\overline{A}_0 + \overline{A}_3A_1A_0$$

Table 4–16
Common cathode seven-segment codes

Digit	a	b	c	d	e	f	g
0	1	1	1	1	1	1	0
1	0	1	1	0	0	0	0
2	1	1	0	1	1	0	1
3	1	1	1	1	0	0	1
4	0	1	1	0	0	1	1
5	1	0	1	1	0	1	1
6	1	0	1	1	1	1	1
7	1	1	1	0	0	0	0
8	1	1	1	1	1	1	1
9	1	1	1	0	0	1	1
A	1	1	1	0	1	1	1
B	0	0	1	1	1	1	1
C	1	0	0	1	1	1	0
D	0	1	1	1	1	0	1
E	1	0	0	1	1	1	1
F	1	0	0	0	1	1	1

Table 4–17
Truth table for a hexadecimal to seven-segment decoder

Digit	Hex code $A_3A_2A_1A_0$	Seven-segment code						
		a	*b*	*c*	*d*	*e*	*f*	*g*
0	0 0 0 0	1	1	1	1	1	1	0
1	0 0 0 1	0	1	1	0	0	0	0
2	0 0 1 0	1	1	0	1	1	0	1
3	0 0 1 1	1	1	1	1	0	0	1
4	0 1 0 0	0	1	1	0	0	1	1
5	0 1 0 1	1	0	1	1	0	1	1
6	0 1 1 0	1	0	1	1	1	1	1
7	0 1 1 1	1	1	1	0	0	0	0
8	1 0 0 0	1	1	1	1	1	1	1
9	1 0 0 1	1	1	1	0	0	1	1
A	1 0 1 0	1	1	1	0	1	1	1
B	1 0 1 1	0	0	1	1	1	1	1
C	1 1 0 0	1	0	0	1	1	1	0
D	1 1 0 1	0	1	1	1	1	0	1
E	1 1 1 0	1	0	0	1	1	1	1
F	1 1 1 1	1	0	0	0	1	1	1

Equation (4–39)
$$c = A_3\overline{A}_2 + \overline{A}_3A_2 + \overline{A}_1A_0 + \overline{A}_3\overline{A}_1 + \overline{A}_3A_0$$

Equation (4–40)
$$d = A_3\overline{A}_1\overline{A}_0 + A_2\overline{A}_1A_0 + A_2A_1\overline{A}_0 + \overline{A}_2A_1A_0 + \overline{A}_3\overline{A}_2\overline{A}_0$$

Equation (4–41)
$$e = \overline{A}_2\overline{A}_0 + A_3A_2 + A_1\overline{A}_0 + A_3A_1$$

Equation (4–42)
$$f = \overline{A}_1\overline{A}_0 + A_3\overline{A}_2 + A_2\overline{A}_0 + A_3A_1 + \overline{A}_3A_2\overline{A}_1$$

Equation (4–43)
$$g = A_3\overline{A}_2 + A_3A_0 + \overline{A}_2A_1 + A_1\overline{A}_0 + \overline{A}_3A_2\overline{A}_1$$

The logic circuit for the hexadecimal to seven-segment decoder is obtained by implementing Equations 4–37 through 4–43 as shown in Figure 4–52.

It is therefore apparent that seven-segment decoders are actually code converters for they convert binary codes from one form to another. The circuits discussed in this section are not limited only to seven-segment LED displays but can be modified or redesigned for more

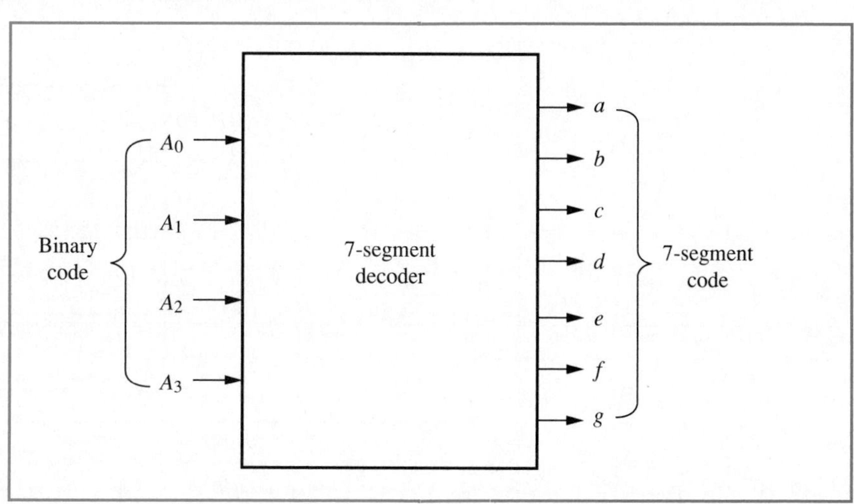

Figure 4–50
Logic symbol for a hexadecimal to seven-segment decoder.

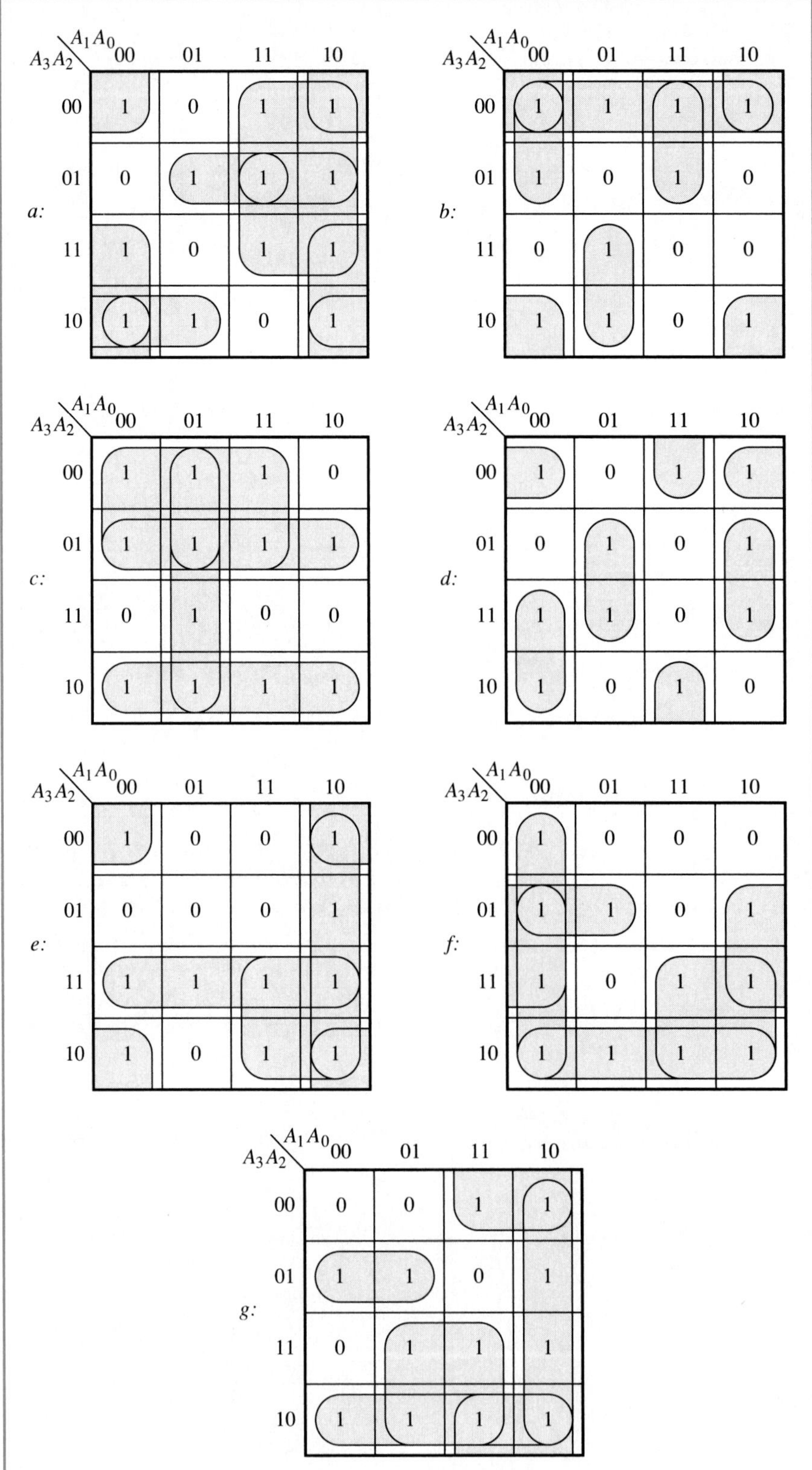

Figure 4–51
K-maps for the design of the seven-segment decoder.

Figure 4–52
Seven-segment decoder circuit.

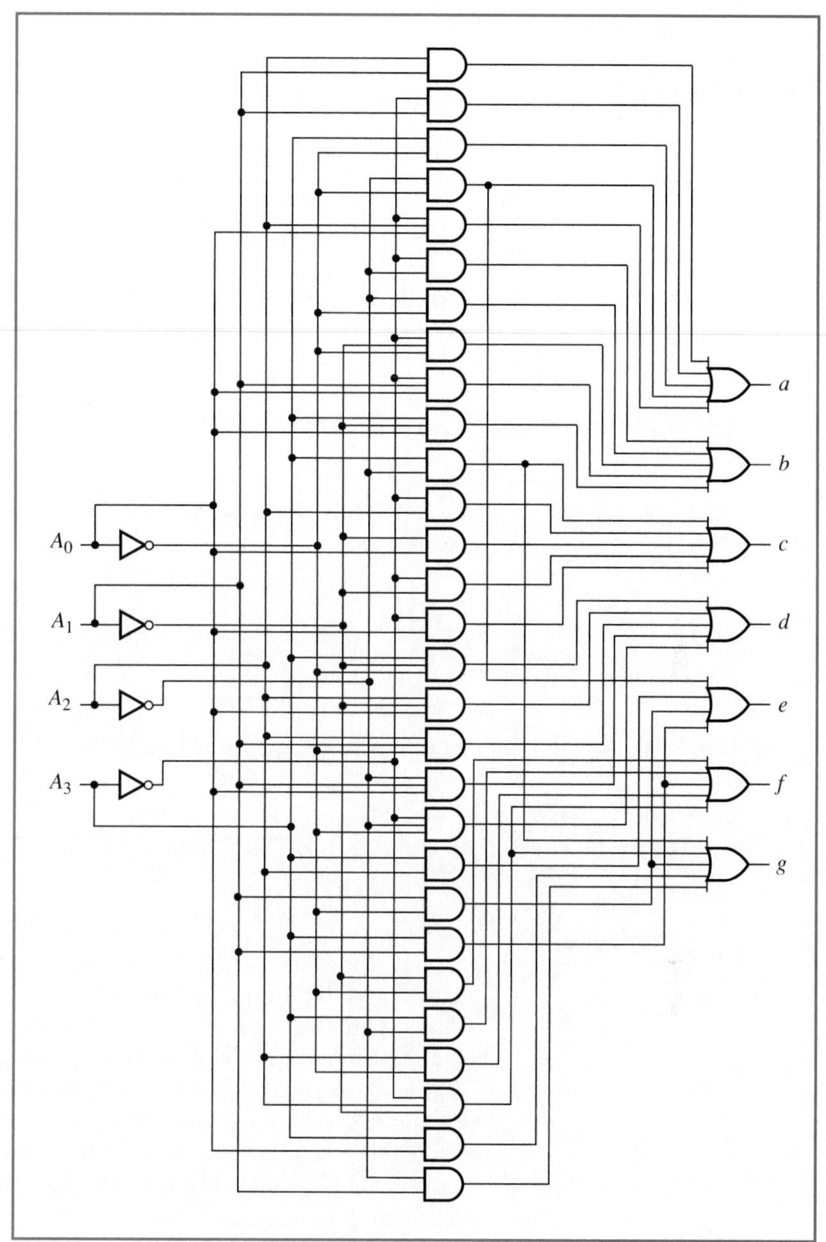

complex LED designs, including dot-matrix LEDs and print-heads for printers, cathode-ray tube (CRT) character generation, and liquid crystal displays (LCDs).

Integrated circuit logic

The 9368 seven-segment decoder/driver/latch is an IC that incorporates a hexadecimal to seven-segment decoder and is designed to drive common cathode LEDs. The logic symbol and pin configuration of the 9368 is shown in Figure 4–53.

Notice that the 9368 is very similar in operation to the circuit designed in Figure 4–52 except that it incorporates a few additional features. The 9368 accepts as input a 4-bit binary code ($A_3A_2A_1A_0$) and

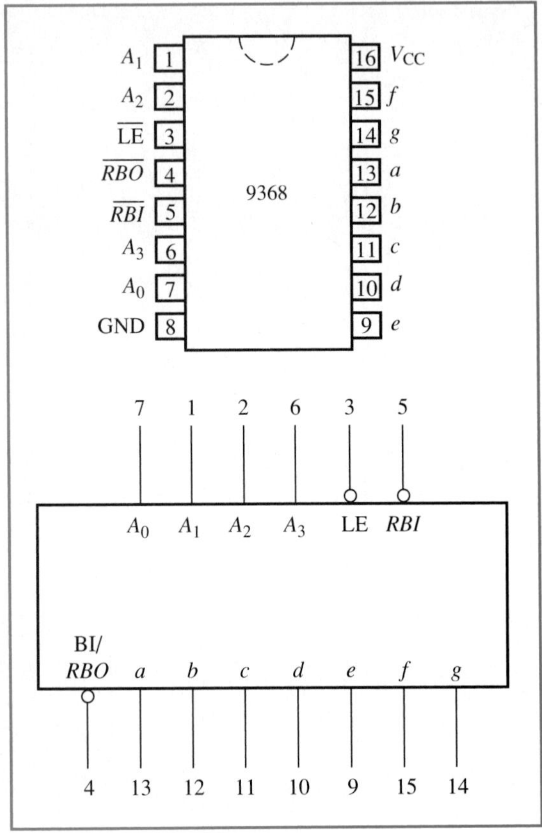

Figure 4–53
Logic symbol and pin configuration of the 9368 seven-segment decoder/driver/latch.

decodes it into a seven-segment code (a through g) to display the hexadecimal digit corresponding to the input binary number. The 9368 has a ripple blanking input ($\overline{RBI}$) and a ripple blanking output ($\overline{RBO}$) that are used for leading zero suppression in circuits containing several 9368 ICs. *Leading zero suppression* is a technique that is used to blank out leading zeros in LED displays; for example, the number 0645 is displayed as 645 with leading zero suppression. If the binary number 0000 is applied to the input of the 9368 and $\overline{RBI}$ is at the low state, the LED is *blanked* by deactivating all segments and the $\overline{RBO}$ output is set to a logic 0. If $\overline{RBI}$ is at the high state when the binary number 0000 is applied to the input of the 9368, the number 0 is displayed on the LEDs and the $\overline{RBO}$ output is set to a logic 1. Thus $\overline{RBI}$ and $\overline{RBO}$ can be connected such that leading zeros are blanked out on a multidigit display containing several 9368 ICs. In order to incorporate leading zero suppression in multidigit displays, the $\overline{RBI}$ input of the MSD is tied low so that the MSD is blanked when 0000 is applied to the 9368. The $\overline{RBO}$ output of the MSD is connected to the $\overline{RBI}$ input of the next significant digit which will only be blanked if the MSD has been blanked (since $\overline{RBO}$ of the MSD = 0) and the number at the input is 0000. The next significant digit will not be blanked if the MS digit is displaying a nonzero since $\overline{RBO}$ will be at the logic 1 state. The $\overline{RBI}$ and $\overline{RBO}$ lines are connected in this "daisy chain" manner for all the other digits in sequence up to the LSD. The 9368 also has an active-low

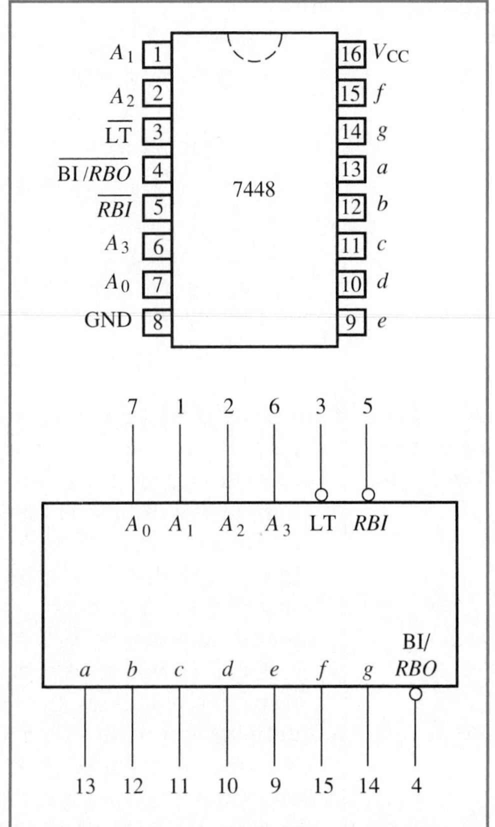

Figure 4–54
Logic symbol and pin configuration of the 7448 BCD to seven-segment decoder.

latch enable input, $\overline{E}_L$, that must be at the logic 0 state in order for the IC to function. This input controls the operation of an internal storage system known as a *latch* (to be discussed in Chapter 5) that is used to store the binary number being displayed.

The 7448 BCD to seven-segment decoder IC (Figure 4–54) is similar in operation to the 9368 except that the 7448 decodes a 4-bit BCD code in the range 0000 to 1001 into the seven-segment codes required to display the decimal digits 0 through 9, respectively. If an attempt is made to decode the codes 1010 to 1111 the resulting seven-segment codes produce unintelligible displays on the seven-segment LED connected to the 7448. Instead of a latch enable line, $\overline{E}_L$, the 7448 has an active-low *lamp test*, $\overline{L}_T$, line that is used to test the segments of the seven-segment LED connected to the 7448. When this line is low, all seven segments are activated and thus any defective segments can be identified. The $\overline{RBI}$ and $\overline{RBO}$ lines function in the same manner as the 9368 and are used for ripple blanking and leading zero suppression.

Notice that the 9368 and 7448 are interchangeable. That is, their pin configurations are identical. Since the 9368 can display all the decimal digits, it can substitute for the 7448 but not vice versa since the 7448 cannot display the hexadecimal digits A through F. There are also identical ICs available that are designed with inverted outputs (*a* through *g*) to allow them to interface to common anode–type LEDs.

Review questions

1. What is the difference between a standard decoder (such as the ones discussed in Section 4–6) and a seven-segment decoder?
2. What is the difference between a common cathode and a common anode seven-segment LED?
3. Why is a seven-segment decoder also known as a code converter?
4. What is the purpose of the ripple blanking inputs and outputs in a seven-segment decoder IC?

4–8 Multiplexers and tristate logic

In many digital applications, particularly in digital computers, it is often necessary to route or channel data in the form of binary numbers from different circuits for processing. The circuit that provides this routing or *multiplexing* of data is known as a *data selector* or *multiplexer*. A multiplexer can be thought of as a multiposition switch as shown in Figure 4–55 that can be used to connect one of several input *channels* (*A* through *F*) to a common *OUTPUT*. Each channel can represent a binary number that can be a single bit or several bits in size.

The circuit shown in Figure 4–55 provides data selection by mechanically moving the switch to the desired channel. A digital multiplexer provides data selection by means of a binary code that identifies a particular channel. Figure 4–56 shows the logic symbol for a two-channel multiplexer that multiplexes two input channels, *A* and *B*, to the output, *OUT*. The input channels (and consequently, the output) are 1 bit wide. The multiplexer has a *SELECT* line which is used to select the channel being multiplexed. We will assume that if *SELECT* is a logic 1, then channel *A* is multiplexed (connected) to the output, and if *SELECT* is a logic 0 then channel *B* is multiplexed (connected) to the output. Using this information we can construct a truth table for the multiplexer as shown in Table 4–18.

Notice in Table 4–18 that when $SELECT = 0$, the output, *OUT*, of the multiplexer is equal to the logic states of channel *B*; it appears as if the output *OUT* is directly connected to *B*. Similarly when $SELECT = 1$,

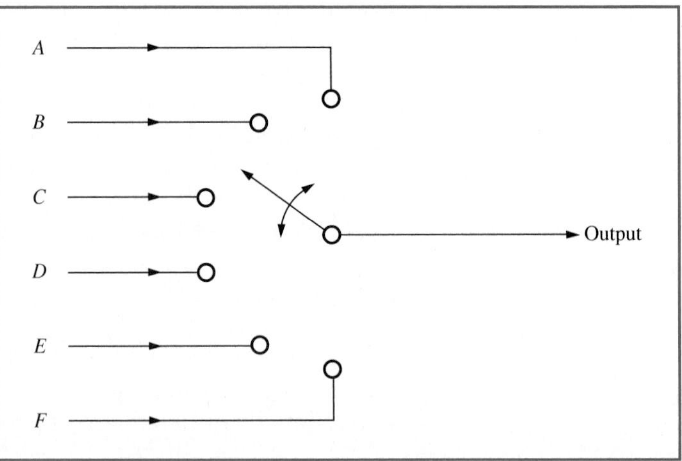

Figure 4–55
Conceptual multiplexer circuit.

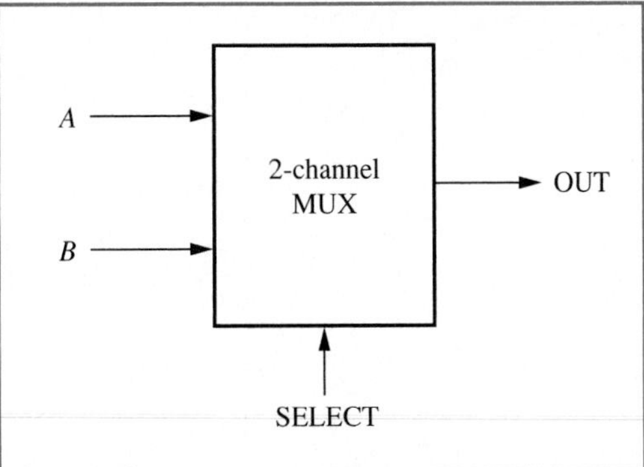

Figure 4–56
Logic symbol for a two-channel 1-bit multiplexer.

the output *OUT* is equal to the logic states of channel *A* making it seem as if *OUT* is directly connected to *A*.

We can now use the conventional approach to obtain the logic equations and logic circuit to implement the truth table for the multiplexer shown in Figure 4–56. The K-map for the output of the multiplexer is shown in Figure 4–57 and its logic equation is as follows:

Equation (4–44)

$$OUT = \overline{SELECT} \cdot B + SELECT \cdot A$$

The logic circuit for Equation 4–44 shown in Figure 4–58 functions as a two-channel, 1-bit multiplexer. Notice in Figure 4–58 that the logic levels of *A* and *B* are passed through the *AND* gates by applying a logic 1 on the second input. For example, when $SELECT = 1$, since one of the inputs of the *AND* gate *G*1 is a logic 1, the output will be equal to the logic state of *A* (identity law). However when $SELECT = 1$, one of the inputs of the *AND* gate *G*2 is a logic 0 and therefore the output will be a logic 0 (dominance law). Since one of the inputs of the *OR* gate is a logic

Table 4–18
Truth table for a two-channel 1-bit multiplexer

SELECT	*A*	*B*	*OUT*
0	0	0	0
0	0	1	1
0	1	0	0
0	1	1	1
1	0	0	0
1	0	1	0
1	1	0	1
1	1	1	1

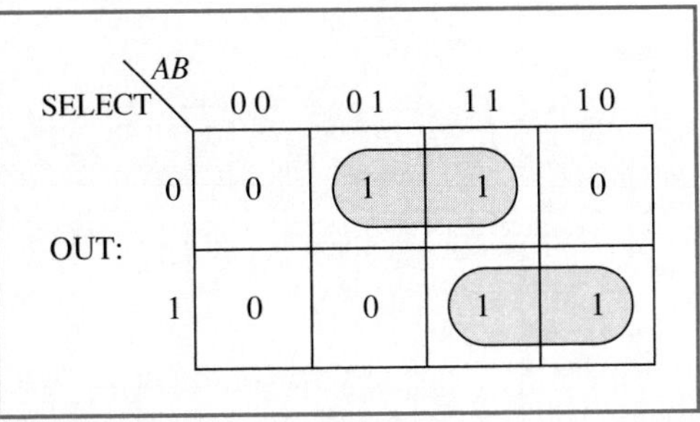

Figure 4–57
K-map for the design of the two-channel 1-bit multiplexer.

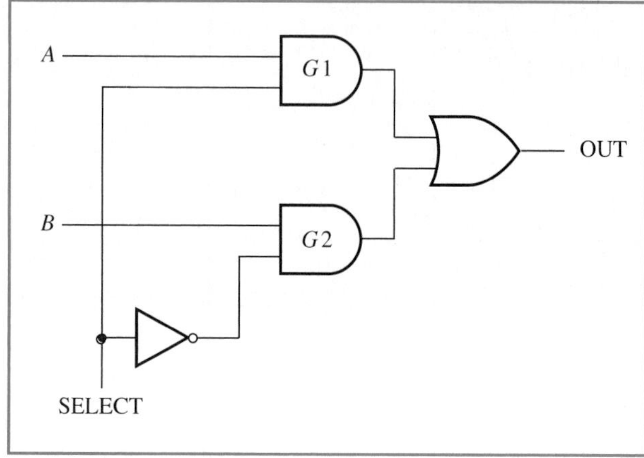

Figure 4–58
Logic circuit for the two-channel 1-bit multiplexer.

0, the output will be equal to the state of the other input (identity law)—in this case, the state of *A*. Similarly when *SELECT* = 0 gate *G*1 effectively *blocks A* and *passes B* by applying a logic 0 and a logic 1 at the second inputs of gates *G*1 and *G*2, respectively. Thus we can multiplex as many single-bit channels as desired simply by using a two-input *AND* gate for each channel and applying a logic 1 to one of its inputs to pass the logic level of the channel through the gate and by applying a logic 0 to one of its inputs to prevent the logic level of the channel from passing through the gate. The outputs of each of these *AND* gates can then be *OR*ed together to obtain the multiplexed output. The following example illustrates the design of a four-channel, 1-bit multiplexer.

EXAMPLE 4–7

The logic symbol for a four-channel multiplexer is shown in Figure 4–59. The circuit multiplexes one of four input channels (*A* through *D*) to the output, *X*. Since one of four channels is being multiplexed, the circuit requires two select lines (S_1S_0) so that each 2-bit combination applied to the select lines will select a particular channel to be multiplexed as shown in Table 4–19.

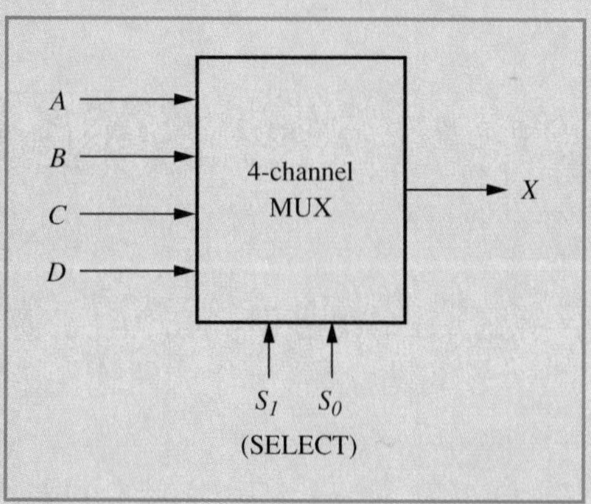

Figure 4–59
Logic symbol for a four-channel 1-bit multiplexer.

Table 4–19
Truth table for a four-channel, 1-bit multiplexer

S_1	S_0	X
0	0	*A*
0	1	*B*
1	0	*C*
1	1	*D*

Notice in Table 4–19 that if $S_1S_0 = 00$ then the logic state of the output *X* is equal to channel *A*, if $S_1S_0 = 01$ then the logic state of the output *X* is equal to channel *B*, and so on. Instead of using the conventional approach to design the circuit to implement the multiplexer shown in Figure 4–59, we can apply the operation of the circuit in Figure 4–58 along with our knowledge of decoders (from Section 4–6) to obtain the circuit shown in Figure 4–60.

Notice that the circuit in Figure 4–60 is similar to the circuit in Figure 4–58 as far as the arrangement of the *AND* gates and the *OR* gate is concerned. That is, as stated earlier we have used *AND* gates for each input channel and the outputs of the *AND* gates are ORed together to provide the final multiplexed output, *X*. However, notice that in order to "enable" each *AND* gate (to pass the logic level of the channel connected to it) by applying a logic 1 at its other input, we have used a 1-of-4 decoder. The decoder accepts one of four possible binary codes at its inputs S_1S_0 and activates (logic 1) a single output which is used to enable the respective *AND* gate and the appropriate channel.

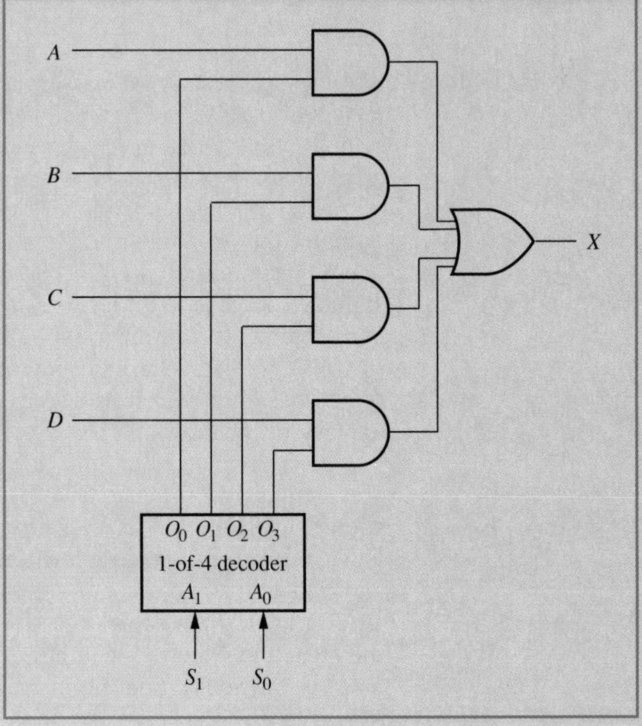

Figure 4–60
Logic circuit for a four-channel 1-bit multiplexer.

Figure 4–61
Logic symbol for a two-channel 2-bit multiplexer.

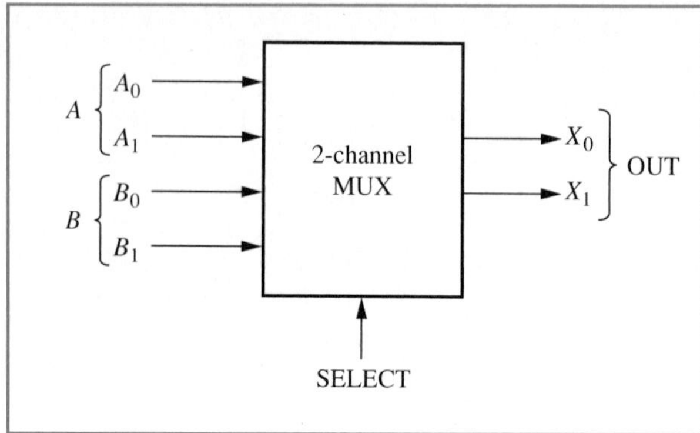

Many digital applications require larger binary numbers to be routed to different circuits using multiplexers. We can design multiplexers to handle channels larger than 1 bit by combining the single-bit multiplexers investigated so far. For example, Figure 4–61 shows the logic symbol for a two-channel, 2-bit multiplexer that multiplexes two channels A and B to the output, OUT. However, notice that channels A and B are actually 2-bit numbers, A_1A_0 and B_1B_0, respectively, and the output is a 2-bit output, X_1X_0. The truth table for the multiplexer is shown in Table 4–20. Instead of using the conventional approach to implement the two-channel, 2-bit multiplexer, we can more efficiently implement the circuit by using two 2-channel, 1-bit multiplexers (from Figure 4–56) as shown in Figure 4–62. Notice in Figure 4–62 that each multiplexer is used to multiplex a single bit that makes up the binary number and therefore the circuit can be extended to multiplex two numbers of any size. Similarly, we can increase the number of channels by using single-bit multiplexers with the desired number of channels.

Tristate logic

At this stage it may be appropriate to introduce the concept of tristate logic and tristate logic gates. These gates can greatly simplify the design of multiplexers and are also used extensively in other digital computer circuits that will be examined in later chapters.

The logic gates we have examined so far have all had *two-stated* outputs, that is, the output of a logic gate is either a logic 1 (5 volts) or a logic 0 (ground potential, 0 volts) and must exist in one of these two states. A logic gate with *three-stated* outputs, also referred to as a *tri-state* logic gate, can have its output exist in one of three states—a logic 1 (5 volts), a logic 0 (ground potential), or no logic level (no voltage). This third state is often referred to as a *high-impedance* state and can be explained as follows. If we view the output of a two-stated logic gate as a two-position switch that provides either a logic 1 or a logic 0 by connecting the output to 5 volts or ground potential, respectively, as shown in Figure 4–63a, then the output of a tristated logic gate can be viewed as a three-position switch (Figure 4–63b) that provides a logic 1 and a logic 0 as before but includes a third position that is neither a logic 1 nor a logic 0 but represents *no voltage*. Notice that the third position of the switch in Figure 4–63b is not connected to any logic state but is *floating* (or open), and therefore the impedance or resistance of the output

Table 4–20
Truth table for a two-channel, 2-bit multiplexer

Select	X_1X_0
0	B_1B_0
1	A_1A_0

in this state is infinite or very very high. This state is therefore referred to as the high-impedance, or Z, state.

Tristate logic is often incorporated into the outputs of many of the basic logic gates. However, one of the more common gates that is available with a tristate output is the buffer. Figure 4–64a shows the logic symbol for a tristate buffer. Notice that the buffer has an additional input labeled E that is used to *enable* or *disable* the output of the buffer. When the output of the buffer is enabled ($E = 1$) as shown in Figure 4–64b, the buffer functions normally and the states of the output B are equal to the input A. However, when the buffer is disabled ($E = 0$) as shown in Figure 4–64c, the output of the buffer is in a high-impedance state (Z) and independent of the input A. The tristate buffer can therefore be viewed as a switch that is opened or closed as shown in Figure 4–64b and c, respectively—when closed, the output is connected to the input ($B = A$), and when open, the output is floating at a high-impedance state (Z). Table 4–21 is the truth table for the tristate buffer shown in Figure 4–64a.

Figure 4–65 illustrates some of the other basic gates with tristate outputs. All the gates shown in Figure 4–65 operate normally (i.e., according to their functions) when the output is enabled ($E = 1$). However, when the output is disabled ($E = 0$) it is effectively cut off from the gate and is in a high-impedance state.

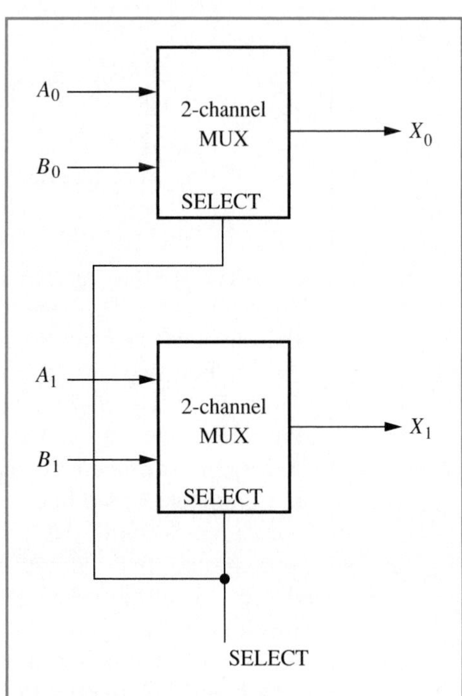

Figure 4–62
Logic circuit for a two-channel 2-bit multiplexer.

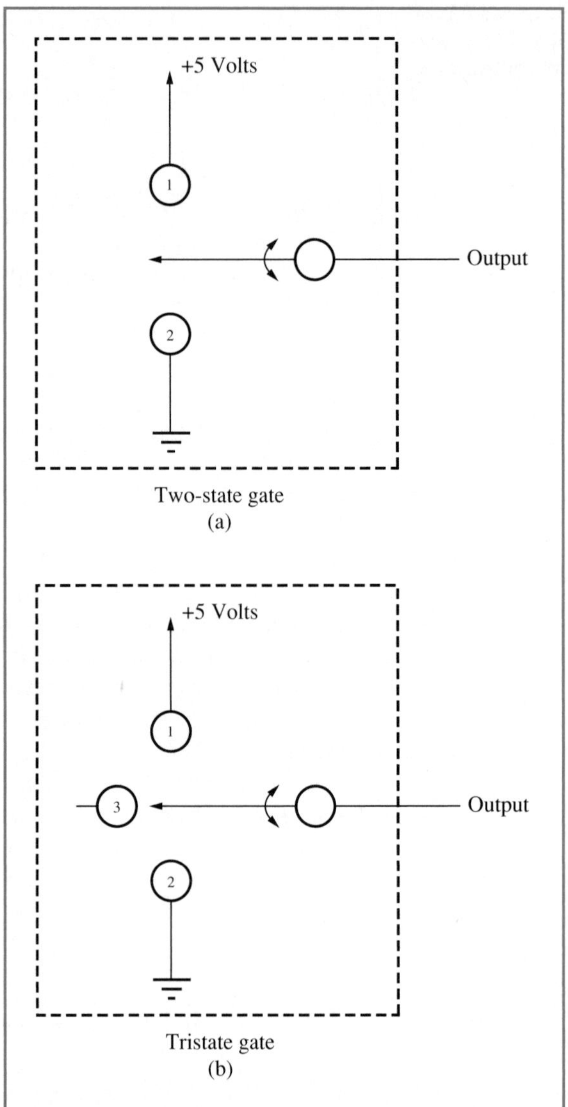

Figure 4–63
Conceptual circuits for two-state and tristate outputs.

The tristate gates investigated so far have active-high enable lines (E). However, there are many gates that have active-low enable lines ($\overline{E}$) that must be tied to a logic 0 in order to enable the output.

When two or more outputs of logic gates are connected together there exists a *conflict of logic levels*, especially when the outputs are at different logic states. This condition cannot be defined logically since there are no precedence rules for logic states. For example, if two outputs are connected together and the two outputs are at different logic states, we cannot assume that a logic 0 state will override a logic 1 state or vice versa. Also, from an electronics perspective this means that the 5 volts (logic 1) is effectively shorted to ground (logic 0), and even though the resulting state would be a logic 0 the connection could damage the internal electronic circuitry of the gates (there are exceptions to this rule which will be covered in Chapter 12) and should be avoided. Keeping this

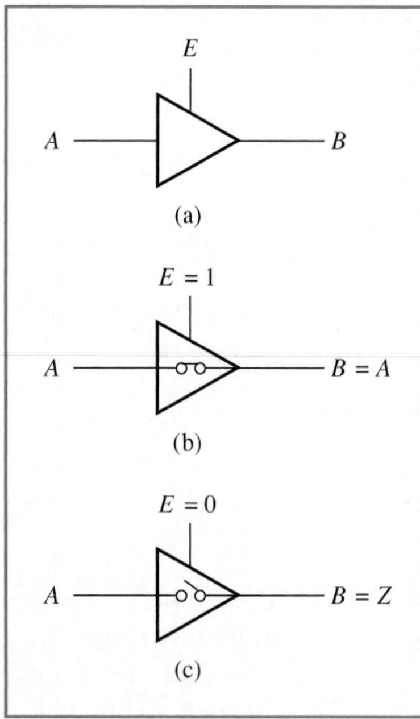

Figure 4–64
The tristate buffer. (a) Logic symbol; (b) output enabled; (c) output disabled.

Table 4–21
Truth table for a tristate buffer

E	A	B
0	0	Z
0	1	Z
1	0	0
1	1	1

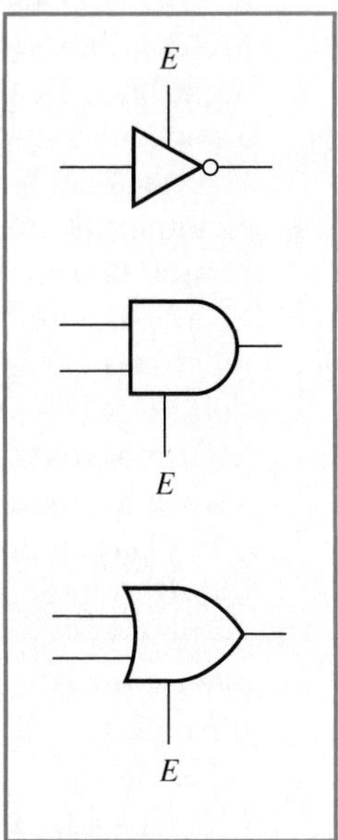

Figure 4–65
Other types of gates with tristate outputs.

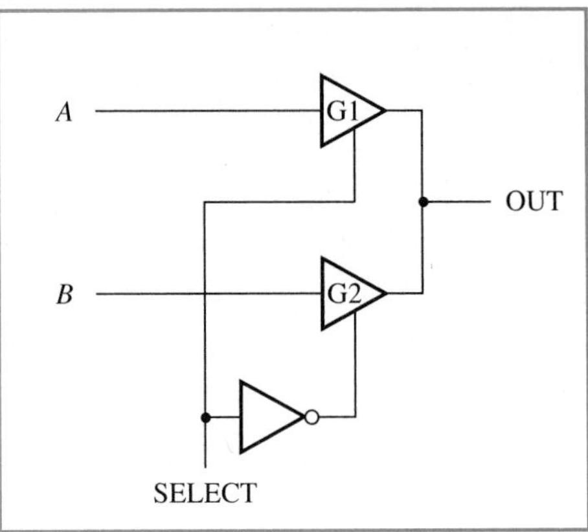

Figure 4–66
A two-channel 1-bit multiplexer implemented with tristate logic.

in mind we can now redesign the multiplexer in Figure 4–56 using tristate buffers as shown in Figure 4–66.

Observe in Figure 4–66 that we have connected the A and B channels together through a set of tristate buffers. When $SELECT = 1$, the buffer $G1$ is enabled and $G2$ is disabled. The output of $G2$ is therefore at a high impedance state effectively cutting off B from the output OUT, while the output of $G1$ is equal to A. Therefore input A is multiplexed to OUT. When $SELECT = 0$, $G2$ is now enabled and $G1$ is disabled, causing A to be isolated from the output and B to be connected to the output; B is therefore multiplexed to OUT. It is therefore apparent that tristate buffers can simplify the design of a multiplexer when the equivalent circuits shown in Figures 4–66 and 4–58 are compared. Example 4–8 illustrates another design of a multiplexer using tristate logic.

Integrated circuit logic

The 74151 is an eight-channel, 1-bit multiplexer that multiplexes 8 single-bit channels to a single output. The logic symbol and pin configuration of the 74151 is shown in Figure 4–68. The 74151 has three select lines, $S_2S_1S_0$, that select 1 of 8 channels to be multiplexed. Each one of the input lines I can be multiplexed to the output Z when the binary code for a particular line is applied to the select lines $S_2S_1S_0$. The subscript that identifies each input line also identifies the binary code that selects it. For example, to multiplex I_3 to Z, the code at the select lines must be 011. Notice that the 74151 also provides the output Z in complementary form, $\overline{Z}$, as a convenience. The 74151 also has an active-low enable line, $\overline{E}$, that must be tied low in order for the multiplexer to function; if $\overline{E} = 1$ the output of the multiplexer Z is held permanently low. The enable line $\overline{E}$ is used to expand the 74151 (i.e., increase the number of input channels) as shown in Figure 4–69.

The circuit in Figure 4–69 illustrates the use of the 74151's enable line $\overline{E}$ to interconnect two ICs to double the number of input channels multiplexed to 16. The 16 inputs to be multiplexed (I_0 through I_{15}) are evenly broken up between 74151 no. 1 (I_0 through I_7) and 74151 no. 2 (I_8 through I_{15}). Since the circuit has a total of 16 channels it requires four select lines, $S_3S_2S_1S_0$, to select one of 16 channels. The three select lines $S_2S_1S_0$ select the basic eight channels of each IC, while the most significant select line S_3 selects either IC no. 1 or IC no. 2 when at the 0

EXAMPLE 4–8

Implement the four-channel, 1-bit multiplexer whose logic symbol is shown in Figure 4–59 using tristate buffers.

SOLUTION The logic circuit for the multiplexer is shown in Figure 4–67. The design is similar to the circuit in Figure 4–66 except that the 1-of-4 decoder is used to enable one of the four tristate buffers to multiplex the input to the output.

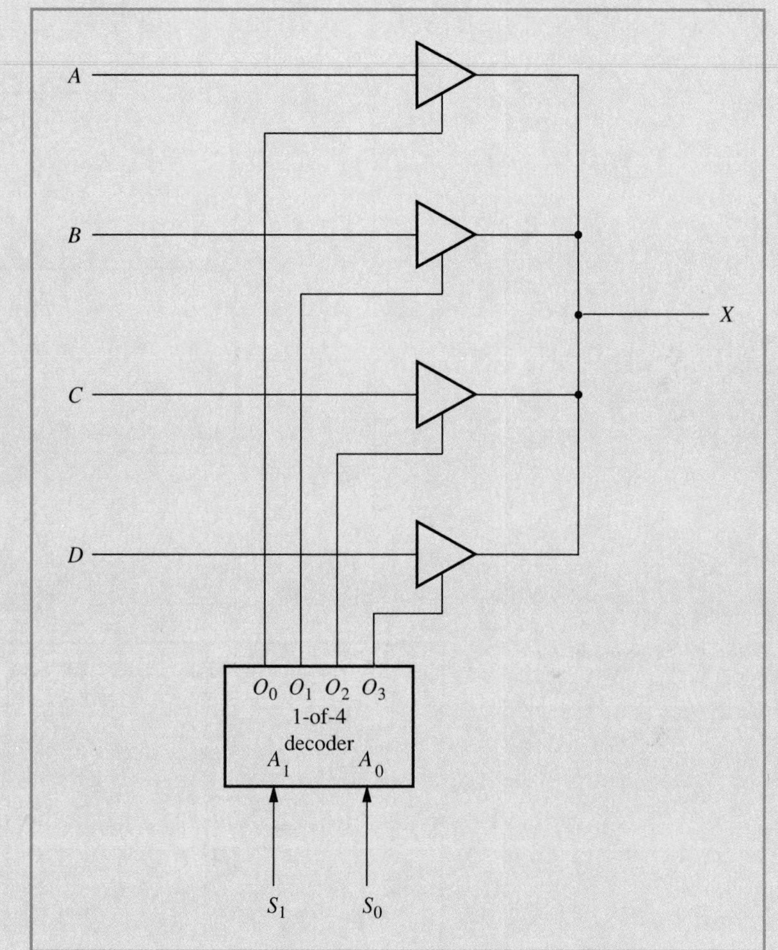

Figure 4–67
A four-channel 1-bit multiplexer implemented with tristate logic.

and 1 states, respectively, since this line is connected to the enable lines $\overline{E}$ of IC no. 1 and IC no. 2. Thus when S_3 is a logic 0, IC no. 1 will be enabled and the eight combinations of $S_2S_1S_0$ will select the first eight inputs (I_0 through I_7). Also when S_3 is a logic 0, the tristate buffer $G1$ is enabled (while $G2$ is disabled) and the output Z of IC no. 1 is passed to OUT. When S_3 is a logic 1, IC no. 2 is enabled and the eight combinations of $S_2S_1S_0$ will select the next eight inputs (I_8 through I_{15}). Also when S_3 is a logic 1, the tristate buffer $G2$ is enabled (while $G1$ is disabled) and the output Z of IC no. 2 is passed to OUT.

The 74150 IC is a 16-input multiplexer and effectively provides the same function as the circuit in Figure 4–69. The logic symbol and pin configuration of the 74150 are shown in Figure 4–70. Like the 74151, the

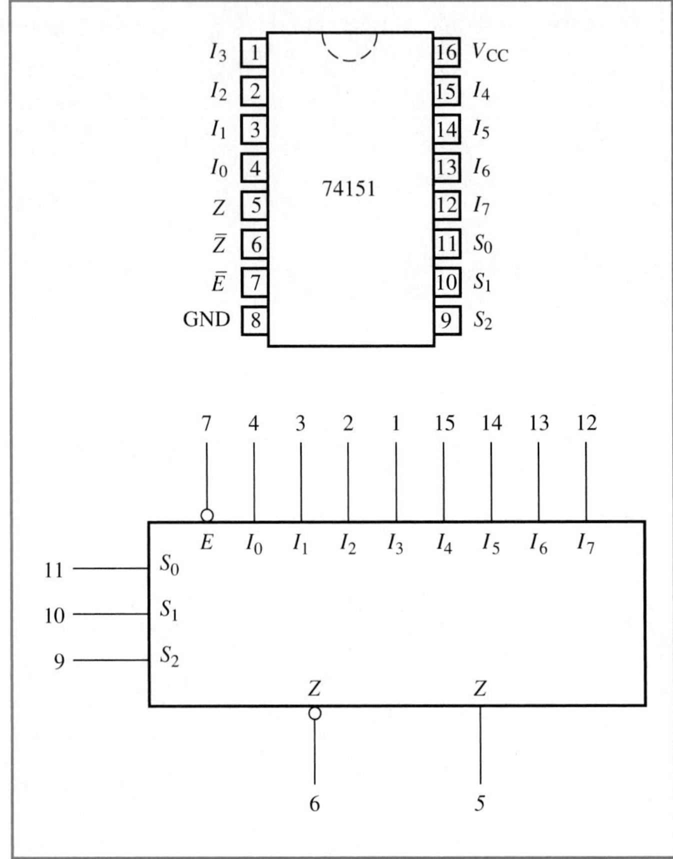

Figure 4–68 Logic symbol and pin configuration of the 74151 eight-input multiplexer.

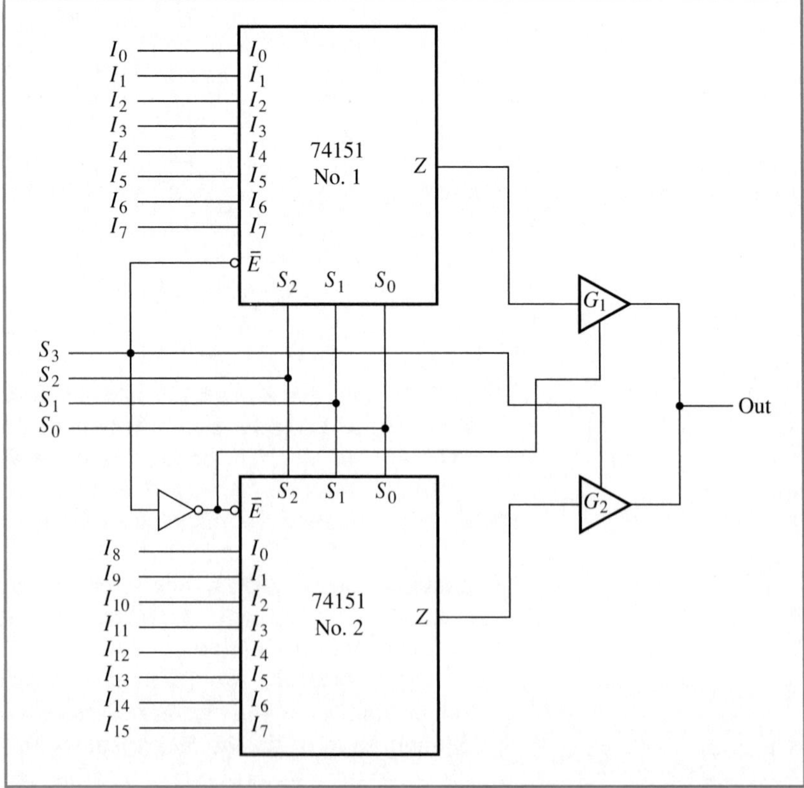

Figure 4–69 16-channel multiplexer constructed from two 74151s.

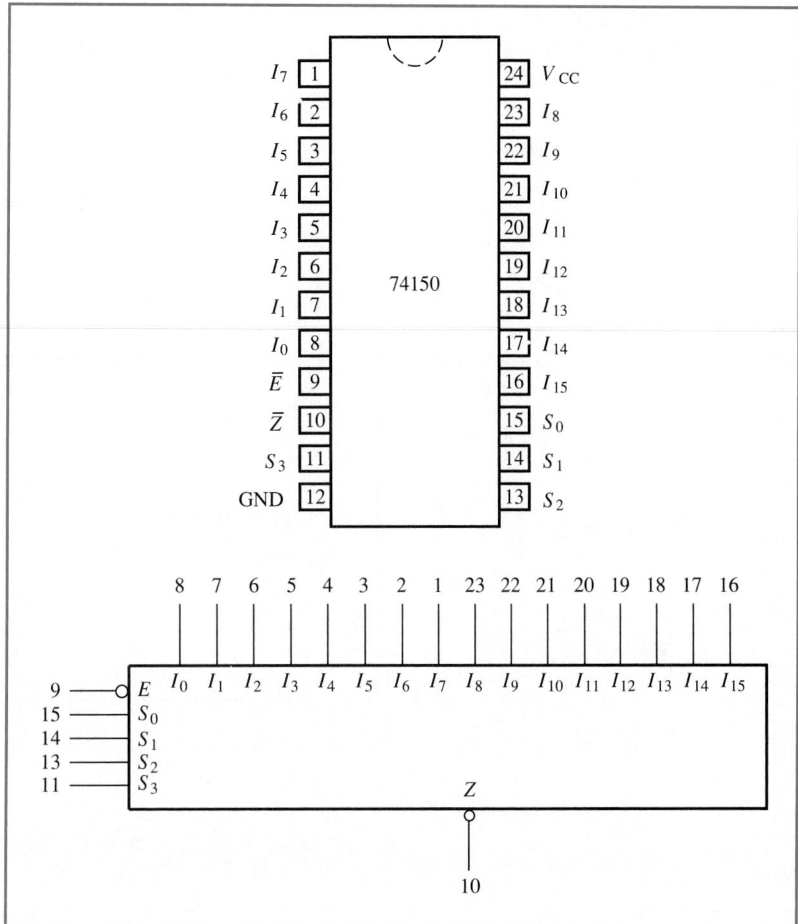

Figure 4–70
Logic symbol and pin configuration of the 74150 16-input multiplexer.

74150 can be expanded to multiplex any number of input channels by using its enable line $\overline{E}$.

The 74153 is a dual 4-input multiplexer IC that contains two 4-channel multiplexers in one package. The logic symbol and pin configuration of the 74153 are shown in Figure 4–71. The two 4-channel multiplexers of the 74153 multiplex four input lines I_0 through I_3 to the output Z. The letters a and b identify the two multiplexers that make up the 74153. The two multiplexers function independently and are enabled separately by the active-low enable lines, $\overline{E}$. However, only one set of select lines S_1S_0 selects the input channels to be multiplexed. For example, when $S_1S_0 = 01$, I_{1a} and I_{1b} will be multiplexed to outputs Z_a and Z_b, respectively. The 74153 can therefore be used as a four-channel, 2-bit multiplexer as well.

The 74157 quad 2-input multiplexer is similar to the 74153 except that the 74157 contains four separate two-channel multiplexers. The logic symbol and pin configuration for the 74157 are shown in Figure 4–72. The four channels of the 74157 are identified by the letters a, b, c, and d. Like the 74153, a single select line is used to select one of two input channels of all four multiplexers together. For example, if $S = 1$ then $Z_a = I_{1a}$ and $Z_b = I_{1b}$ and $Z_c = I_{1c}$ and $Z_d = I_{1d}$. Thus the 74157 can also be used as a two-channel, 4-bit multiplexer. Unlike the 74153, the 74157 only has a single active-low enable line that enables the entire IC.

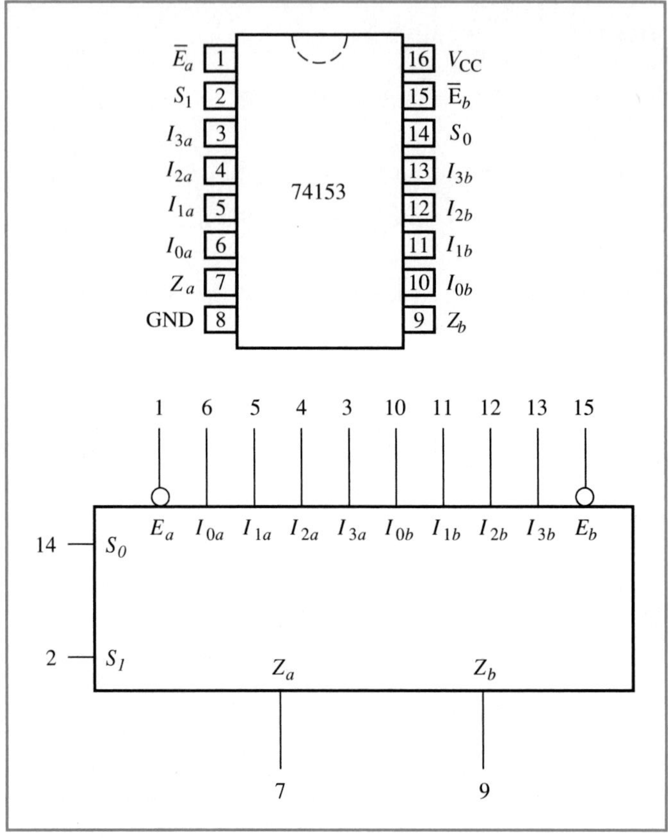

Figure 4–71 Logic symbol and pin configuration of the 75153 dual 4-input multiplexer.

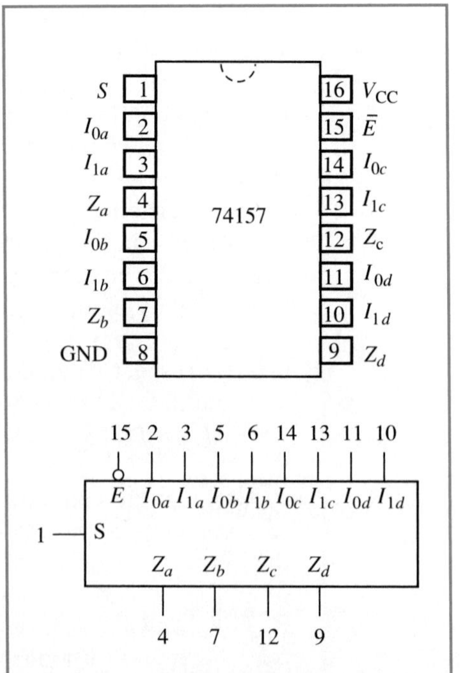

Figure 4–72 Logic symbol and pin configuration of the 74157 quad 2-input multiplexer.

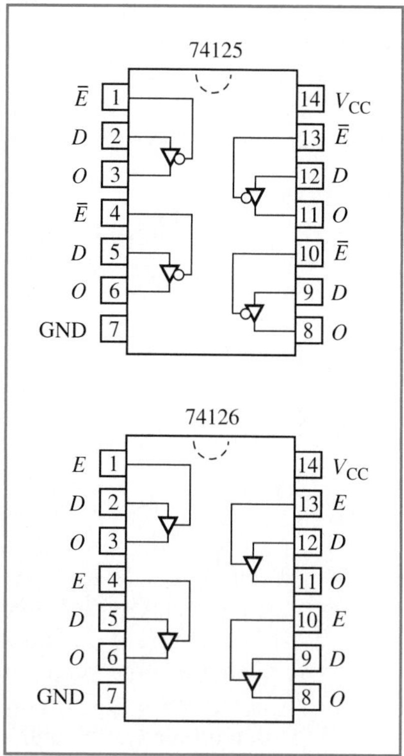

Figure 4–73
Pin configurations of the 74125 and 74126 quad buffers.

The tristate buffers introduced in this section are available in the form of two ICs—the 74125 and 74126. Figure 4–73 shows the pin configurations of the two ICs. The 74126 is a package containing 4 tristate buffers with active-high enable lines while the 74125 is a package containing 4 tristate buffers with active-low enable lines.

Review questions

1. Describe the operation of a multiplexer.
2. State a possible application of a multiplexer.
3. Explain tristate logic and the "high-impedance state."
4. Why should not two (two-state) logic outputs be connected together?
5. Why is it all right for two tristate logic outputs to be connected together? What precaution should be observed when doing this?

4–9 Demultiplexers

Demultiplexers reverse the process of multiplexing. A multiplexer channels several input lines onto one output line while a demultiplexer distributes the data from a single input line over several output lines. The conceptual model of a demultiplexer is shown in Figure 4–74. In comparing the model shown in Figure 4–74 with the conceptual model of a multiplexer shown in Figure 4–55, it is apparent that the function of a multiplexer and a demultiplexer are the opposite. The demultiplexer (represented by a multiposition switch) connects the single *INPUT* line to one of six output lines, *A* through *F*, by changing the position of the

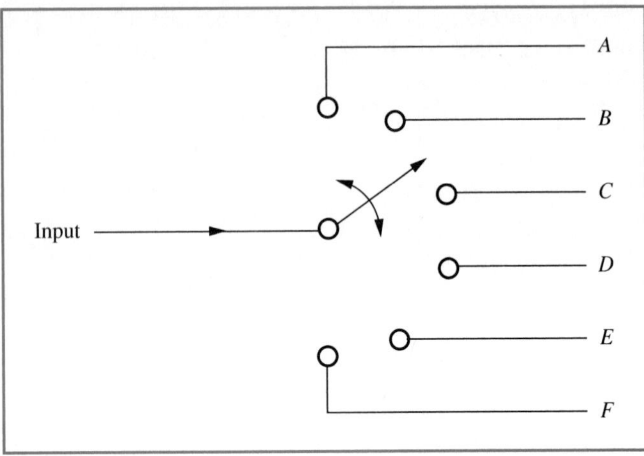

Figure 4–74
Conceptual demultiplexer circuit.

switch. Of course, a logic circuit would not make use of a mechanical switch to select the output channel but would use logic gates to provide the demultiplexing function.

Figure 4–75 shows the logic symbol for a two-channel, single-bit demultiplexer. Notice that the demultiplexer shown in Figure 4–75 reverses the function of the multiplexer shown in Figure 4–56. The demultiplexer takes the logic state of the input *IN* and channels it to the output *A* if *SELECT* = 1, or channels it to the output *B* if *SELECT* = 0. The truth table for the demultiplexer is shown in Table 4–22. Notice in Table 4–22 that when *SELECT* = 0 the state of *B* is equal to the state of *IN* and we don't care about the state of *A*. Similarly, when *SELECT* = 1, the state of *A* is equal to the state of *IN* and we don't care about the state of *B*. We can now obtain the simplified logic equations for the outputs *A*

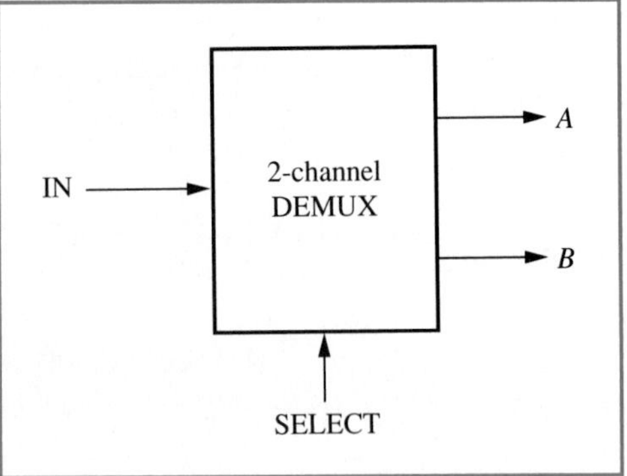

Figure 4–75
Logic symbol for a two-channel 1-bit demultiplexer.

Table 4–22
Truth table for a two-channel demultiplexer

SELECT	*IN*	A	B
0	0	X	0
0	1	X	1
1	0	0	X
1	1	1	X

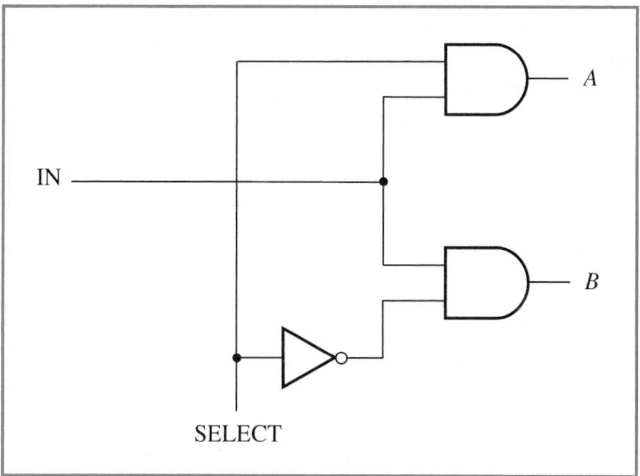

Figure 4–76
Logic circuit for a two-channel, 1-bit demultiplexer.

and B directly from Table 4–22 by assuming that the don't care conditions are zeros

Equation (4–45)
$$A = SELECT \cdot IN$$

Equation (4–46)
$$B = \overline{SELECT} \cdot IN$$

The logic circuit for the demultiplexer can be constructed by implementing Equations 4–45 and 4–46 as shown in Figure 4–76. Again notice that a logic 1 appearing at one of the inputs of each *AND* gate (connected to *SELECT* or $\overline{SELECT}$) in Figure 4–76 is used to pass the logic level at the second input (*IN*) to the output of the *AND* gate (identity law). Using this concept we can design a four-channel demultiplexer without going through the conventional approach. The logic symbol and circuit for a four-channel, single-bit demultiplexer are shown in Figure 4–77.

The demultiplexer in Figure 4–77 requires two select lines to select one of four output channels. The logic circuit for the demultiplexer works by applying *IN* to one of the inputs of all four *AND* gates. However, *IN* will be passed to the output of a specific *AND* gate if and only if the second input of the *AND* gate is a logic 1. The 1-of-4 decoder uses the 2-bit code at its input, S_1S_0, to select one of four *AND* gates (or output channels) to pass the input *IN*. Thus when $S_1S_0 = 00$, $A = IN$ and when $S_1S_0 = 01$, $B = IN$, and so on.

The demultiplexers shown in Figure 4–76 and 4–77 can also be constructed using tristate buffers as illustrated by the following example.

EXAMPLE 4–9

The two-channel demultiplexer whose logic symbol is shown in Figure 4–75 is implemented using tristate buffers as shown in Figure 4–78.

The circuit in Figure 4–78 uses two tristate buffers ($G1$ and $G2$) to connect the input line *IN* to either one of the outputs *A* or *B*. When $SELECT = 1$, $G1$ is enabled ($G2$ is disabled since it has an active-low enable line) and $A = IN$. When $SELECT = 0$, $G2$ is enabled while $G1$ is disabled and $B = IN$.

Figure 4–79 illustrates the use of tristate buffers to implement the four-channel demultiplexer whose logic symbol is shown in Figure 4–77a. Again, the tristate buffers are used to connect one of

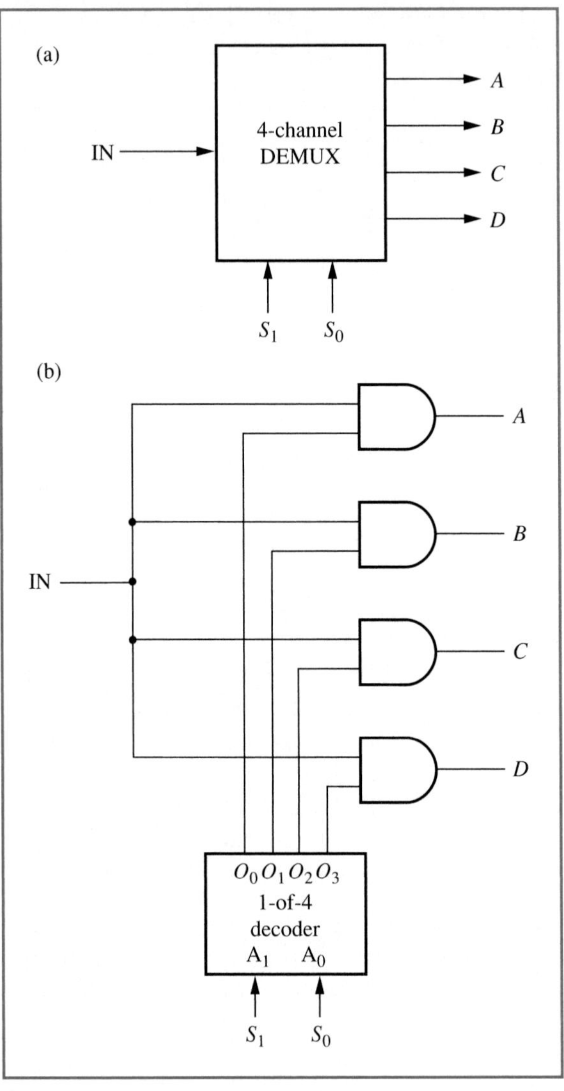

Figure 4–77
A four-channel 1-bit demultiplexer. (a) Logic symbol; (b) logic circuit.

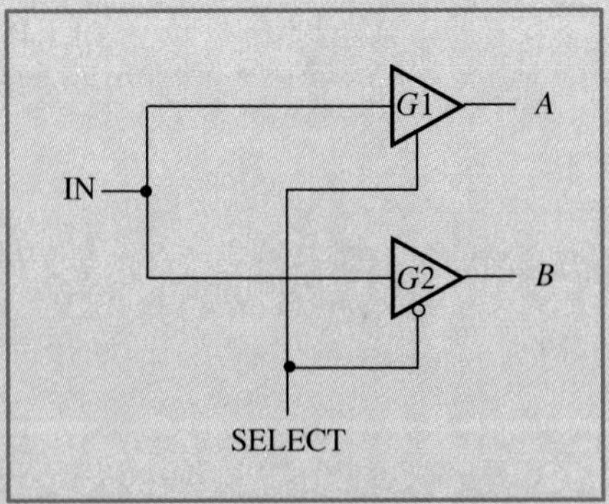

Figure 4–78
Logic circuit for a two-channel 1-bit demultiplexer using tristate logic.

four outputs, A through D, to the input, IN. The select lines S_1S_0 when decoded by the 1-of-4 decoder determine which one of the four tristate buffers are enabled.

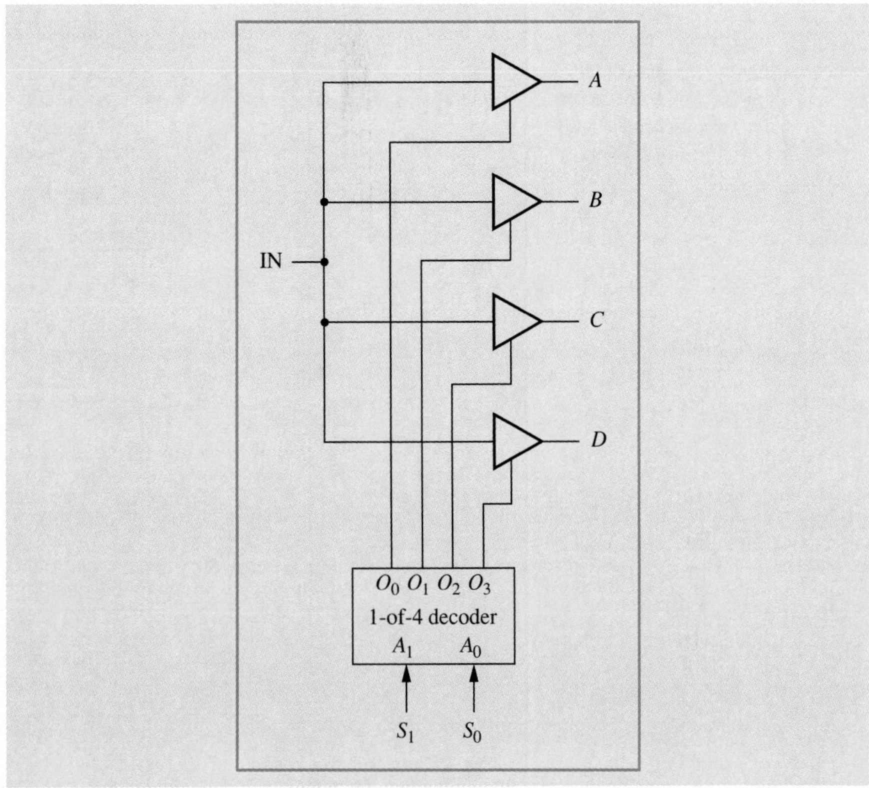

Figure 4–79
Logic circuit for a four-channel 1-bit demultiplexer using tristate logic.

Just as we could connect several multiplexers together to increase the size of the channels being multiplexed, we can also connect several demultiplexers to increase the size of the channels being demultiplexed. Figure 4–80a shows the logic symbol for a two-channel demultiplexer that demultiplexes a 2-bit input channel I_1I_0 to one of two output channels A_1A_0 and B_1B_0. The logic circuit for the demultiplexer is shown in Figure 4–80b and is constructed by using two 2-channel, 1-bit demultiplexers (Figure 4–75), one for I_0 and another for I_1. Since the *SELECT* lines of both the demultiplexers are connected together, when $SELECT = 0$, I_1I_0 will be connected to B_1B_0 and when $SELECT = 1$, I_1I_0 will be connected to A_1A_0.

Integrated circuit logic

The 74xxx series does not include any ICs that are dedicated to the demultiplexing function since many *decoders* can be configured for demultiplexing. Hence many of the decoder ICs available in the 74xxx series are often referred to as *decoders/demultiplexers*.

Figure 4–81 illustrates the use of the 74138 1-of-8 decoder (logic symbol and pin configuration shown in Figure 4–44) as an eight-channel, 1-bit demultiplexer. The circuit in Figure 4–81 connects the input line *IN* to one of eight output channels, *A* through *H*, when selected by one of eight possible binary combinations (codes) appearing on $S_2S_1S_0$. Notice that the two enable lines of the decoder, $\overline{E_2}$ and E_3, are permanently enabled by connection to ground and +5 volts, respectively. However, the demultiplexer input line *IN* is connected to the third active-low enable line $\overline{E_1}$, so that when $\overline{E_1}$ is a logic 0, the 74138 is enabled, and depending

Figure 4–80 A two-channel 2-bit demultiplexer. (a) Logic symbol; (b) logic circuit constructed from two 2-channel 1-bit demultiplexers.

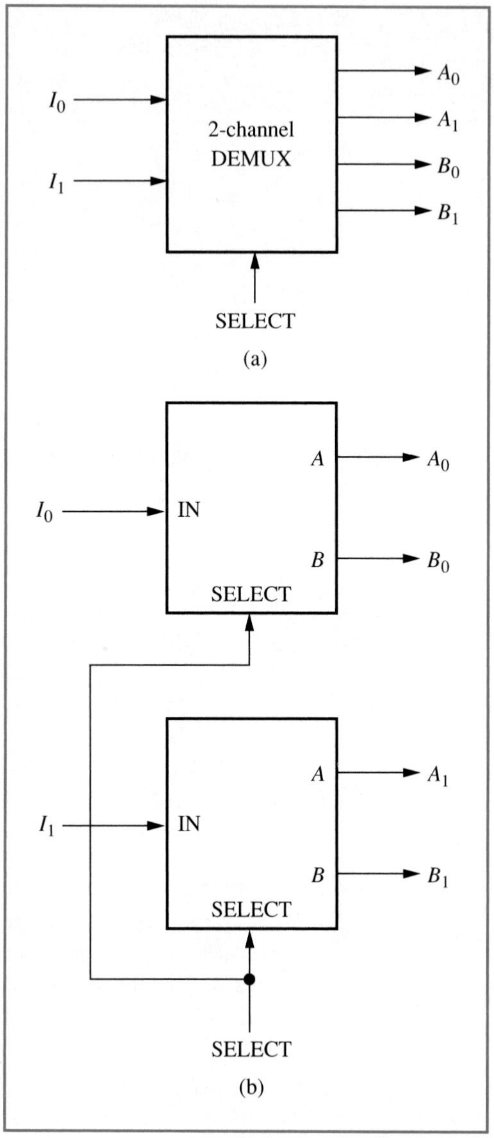

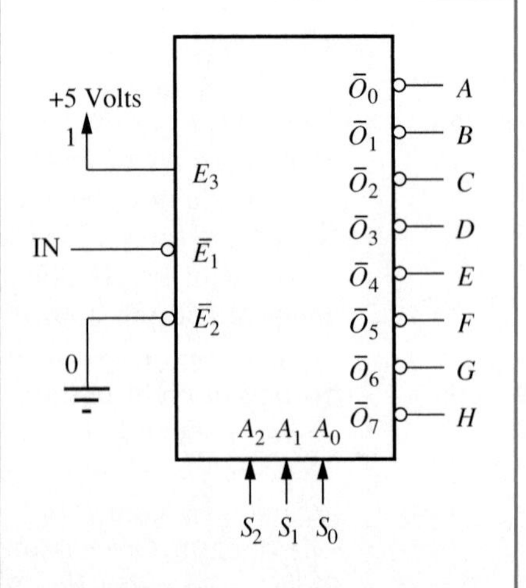

Figure 4–81 74138 configured as an eight-channel 1-bit demultiplexer.

on the code at $A_2A_1A_0$ (demultiplexer lines $S_2S_1S_0$) one of the eight output lines $\overline{O}_0$ through $\overline{O}_7$ (demultiplexer channels A through H) is activated (logic 0). When the input of the demultiplexer IN ($\overline{E_1}$) is a logic 1, the 74138 is disabled and all its outputs $\overline{O}_0$ through $\overline{O}_7$ (demultiplexer lines A through H) are inactive at the logic 1 state. However, since a logic 1 appears at all the output channels, it appears as if the input IN has been directed to the appropriate output channel, since IN is at the logic 1 state.

Review questions

1. Describe the operation of a demultiplexer.
2. State a possible application of a demultiplexer.
3. Why are decoders also known as demultiplexers?

4–10 Parity circuits

In Chapter 1, the use of binary codes and their applications in digital communication systems were introduced. Recall that many digital communication systems transmit binary codes with a parity bit for error checking. The type of parity used can be even or odd. In odd parity systems all binary codes are transferred with an odd number of 1-bits and therefore if the number of 1-bits in the code is an odd number, the parity bit is set to a logic 0 to keep the total number of 1-bits transferred an odd number; if the number of 1-bits in the code is an even number, the parity bit is set to a logic 1 to make the total number of 1-bits transferred an odd number. In even parity systems all binary codes are transferred with an even number of 1-bits and therefore if the number of 1-bits in the code is an even number, the parity bit is set to a logic 0 to keep the total number of 1-bits transferred an even number; if the number of 1-bits in the code is an odd number, the parity bit is set to a logic 1 to make the total number of 1-bits transferred an even number. This section deals with an application circuit that is used to generate the parity bit for a binary code and also to determine the parity of a binary code.

Figure 4–82 shows the logic symbol for a 4-bit *parity generator circuit*. The circuit accepts as input a 4-bit number $D_3D_2D_1D_0$ and activates one of two active-low outputs $\overline{P}_o$ or $\overline{P}_e$, depending on the parity of the code applied to the inputs. The truth table for the parity generator circuit of Figure 4–82 is shown in Table 4–23.

Notice in Table 4–23 that the reason for active-low outputs, $\overline{P}_e$ and $\overline{P}_o$, is because they represent the parity bit for even and odd parity

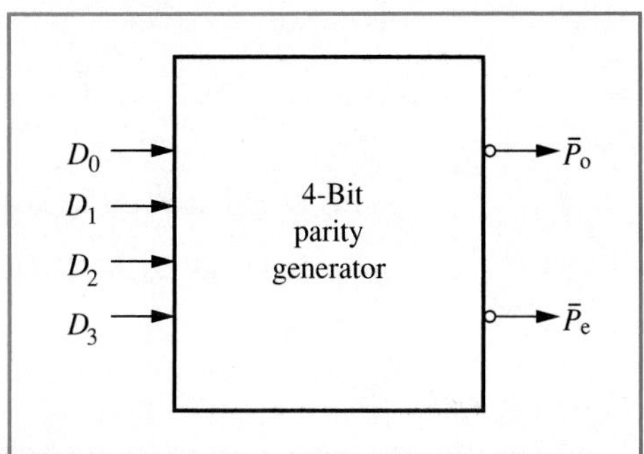

Figure 4–82 Logic symbol for a 4-bit parity generator.

Table 4–23
Truth table for a 4-bit parity generator

Number of 1-bits	D_3	D_2	D_1	D_0	$\overline{P}_e$	$\overline{P}_o$
0	0	0	0	0	0	1
1	0	0	0	1	1	0
1	0	0	1	0	1	0
2	0	0	1	1	0	1
1	0	1	0	0	1	0
2	0	1	0	1	0	1
2	0	1	1	0	0	1
3	0	1	1	1	1	0
1	1	0	0	0	1	0
2	1	0	0	1	0	1
2	1	0	1	0	0	1
3	1	0	1	1	1	0
2	1	1	0	0	0	1
3	1	1	0	1	1	0
3	1	1	1	0	1	0
4	1	1	1	1	0	1

generation, respectively, and must be set to a logic 0 if the total number of bits (data bits and parity bit) is even or odd, respectively. For example, the number 0111 with even parity is

$$\underset{\overline{P}_e}{1}\ 0111$$

The number 1100 with even parity is

$$\underset{\overline{P}_e}{0}\ 1100$$

The number 0111 with odd parity is

$$\underset{\overline{P}_o}{0}\ 0111$$

The number 1100 with odd parity is

$$\underset{\overline{P}_o}{1}\ 1100$$

Figure 4–83 shows the K-map obtained from Table 4–23 for the output $\overline{P}_e$. Since no cells are adjacent no simplification is possible, and therefore the logic equation for $\overline{P}_e$ from the K-map is as follows:

Equation (4–47)

$$\overline{P}_e = \overline{D}_3\overline{D}_2\overline{D}_1D_0 + \overline{D}_3\overline{D}_2D_1\overline{D}_0 + \overline{D}_3D_2\overline{D}_1\overline{D}_0 + \overline{D}_3D_2D_1D_0 + D_3D_2\overline{D}_1D_0 + D_3D_2D_1\overline{D}_0 + D_3\overline{D}_2\overline{D}_1\overline{D}_0 + D_3\overline{D}_2D_1D_0$$

Using the distributive law on Equation 4–47 we have

Equation (4–48)

$$\overline{P}_e = \overline{D}_3\overline{D}_2(\overline{D}_1D_0 + D_1\overline{D}_0) + \overline{D}_3D_2(\overline{D}_1\overline{D}_0 + D_1D_0) + D_3D_2(\overline{D}_1D_0 + D_1\overline{D}_0) + D_3\overline{D}_2(\overline{D}_1\overline{D}_0 + D_1D_0)$$

Notice in Equation 4–48 that

$$\overline{D}_1D_0 + D_1\overline{D}_0 = D_1 \oplus D_0 \text{ (Exclusive-}OR\text{ function)}$$

Figure 4–83
K-map for the design of the 4-bit parity generator.

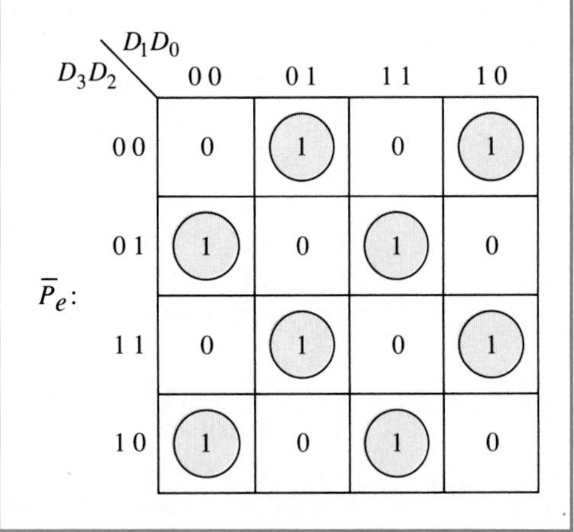

and

$$\overline{D}_1\overline{D}_0 + D_1D_0 = \overline{D_1 \oplus D_0} \text{ (this proof is left as an exercise)}$$

Therefore Equation 4–48 now becomes

$$\overline{P}_e = \overline{D}_3\overline{D}_2(D_1 \oplus D_0) + \overline{D}_3D_2(\overline{D_1 \oplus D_0}) + D_3D_2(D_1 \oplus D_0) + D_3\overline{D}_2(\overline{D_1 \oplus D_0})$$

Using the distributive law again

$$\overline{P}_e = (D_1 \oplus D_0)(\overline{D}_3\overline{D}_2 + D_3D_2) + (\overline{D_1 \oplus D_0})(\overline{D}_3D_2 + D_3\overline{D}_2)$$

Now applying the Exclusive-*OR* function

Equation (4–49)
$$\overline{P}_e = (D_1 \oplus D_0)(\overline{D_3 \oplus D_2}) + (\overline{D_1 \oplus D_0})(D_3 \oplus D_2)$$

Notice that Equation 4–49 is in the *AND-OR* form that is suitable for conversion to Exclusive-*OR* form

Equation (4–50)
$$\overline{P}_e = (D_1 \oplus D_0) \oplus (D_3 \oplus D_2)$$

Since $\overline{P}_o$ is the complement of $\overline{P}_e$, the logic equation for $\overline{P}_o$ is

Equation (4–51)
$$\overline{P}_o = \overline{\overline{P}}_e = P_e$$

The logic circuit for the 4-bit parity generator shown in Figure 4–84 is implemented from Equations 4–50 and 4–51.

Notice in Figure 4–84 that the circuit effectively *adds* the four input bits D_0 through D_3 since each Exclusive-*OR* gate is capable of adding 2 bits (without a carry). It can also be observed (from Table 4–23) that the

Figure 4–84
Logic circuit for the 4-bit parity generator.

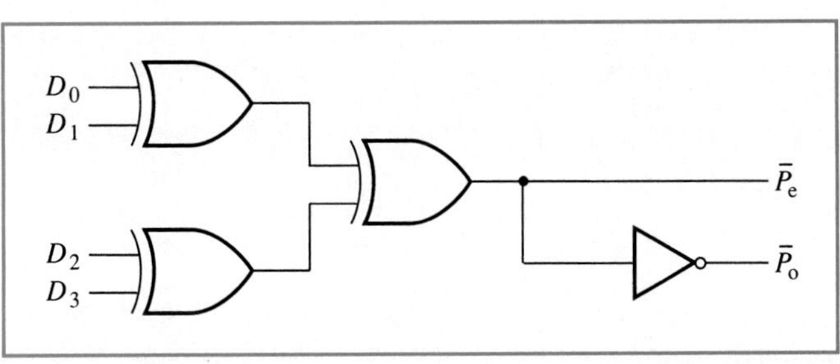

sum of all the input bits will be a logic 0 if there are an even number of 1-bits and a logic 1 if there are an odd number of 1-bits. Using this information we can now design a circuit that can be used to check the parity of a 4-bit code.

A *parity checking circuit* must be able to accept as input the data bits (code) as well as the parity bit, and determine if the parity of all the input bits is even or odd. Figure 4–85 shows the logic symbol for a 4-bit *parity generator/checker*. The circuit has a 4-bit input $D_3D_2D_1D_0$ for the code, and a set of inputs OI and EI for the parity bit of the 4-bit code. If even parity checking is being performed by the circuit, the parity bit of the code is applied to EI (EVEN INPUT) but if odd parity checking is being performed by the circuit, the parity bit of the code is applied to OI (ODD INPUT). The active-low outputs $\overline{P}_e$ and $\overline{P}_o$ as before indicate whether the input code has even or odd parity. Notice that the circuit can operate as a parity generator as well as a checker; the manner in which this is done will be seen shortly.

Figure 4–86 shows the logic circuit for the 4-bit parity generator/checker. Notice that the portion of the circuit that is made up of gates $G1$, $G2$, and $G3$ is similar to the circuit shown in Figure 4–84 and simply sums (without carries) the input code (bits D_0 through D_3). Thus the output of this subcircuit S will be a logic 0 if the parity of $D_3D_2D_1D_0$ is even and a logic 1 if the parity of $D_3D_2D_1D_0$ is odd.

For even parity checking the output of the gate $G4$ will be a logic 0 (indicating even parity) if S is a logic 0 and the parity bit (EI) is a logic 0. If, however, S is a logic 1, indicating that $D_3D_2D_1D_0$ has odd parity but the parity bit (EI) is a logic 1 to force it even, then the output of gate $G4$ will be a logic 0. All other conditions at the input of $G4$ will produce a logic 1 out, indicating a parity error.

For odd parity checking the output of gate $G5$ will be a logic 0 (indicating odd parity) if S is a logic 1 and the parity bit (OI) is a logic 0. If, however, S is a logic 0 indicating that $D_3D_2D_1D_0$ has even parity but the parity bit (OI) is a logic 1 to force it odd then the output of gate $G5$ will be a logic 0. All other conditions at the input of $G5$ will produce a logic 1 out, indicating a parity error.

Thus for even parity checking, a logic 0 at $\overline{P}_e$ indicates that there is no parity error while a logic 1 at $\overline{P}_e$ identifies a parity error. Similarly for

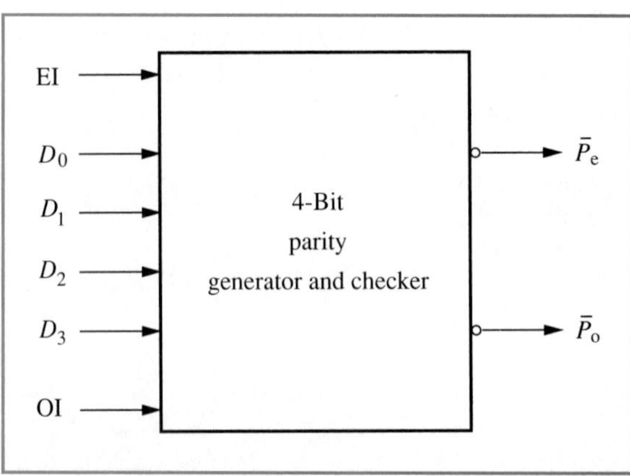

Figure 4–85
Logic symbol for a 4-bit parity generator/checker.

Figure 4–86
Logic circuit for a 4-bit parity generator/checker.

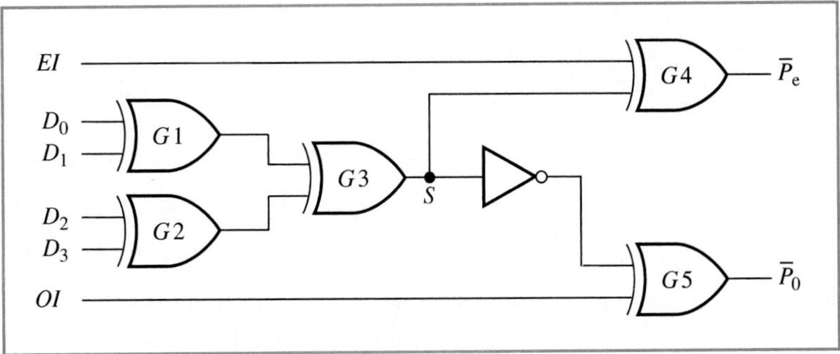

odd parity checking, a logic 0 at $\overline{P}_o$ indicates that there is no parity error while a logic 1 at $\overline{P}_o$ identifies a parity error.

The 4-bit parity generator/checker circuit shown in Figure 4–86 can be used to generate the parity bit for the 4-bit input code $D_3D_2D_1D_0$ by tying EI to a logic 0 (even parity generation) or by tying OI to a logic 0 (odd parity generation). In Figure 4–86 if EI is a logic 0 then $\overline{P}_e$ will be a logic 0 when S is a logic 0, and $\overline{P}_e$ will be a logic 1 when S is a logic 1. However, S will be a logic 0 (and therefore $\overline{P}_e$ will be a logic 0) if the code $D_3D_2D_1D_0$ has even parity. Similarly, if OI is a logic 0 then $\overline{P}_o$ will be a logic 0 when S is a logic 1, and $\overline{P}_o$ will be a logic 1 when S is a logic 0. However, S will be a logic 1 (and therefore $\overline{P}_o$ will be a logic 0) if the code $D_3D_2D_1D_0$ has odd parity. Thus $\overline{P}_e$ will be a logic 0 when $D_3D_2D_1D_0$ has even parity, and $\overline{P}_o$ will be a logic 0 when $D_3D_2D_1D_0$ has odd parity. For parity generation, the circuit therefore operates like the circuit in Figure 4–84.

Integrated circuit logic

The 74180 8-bit parity generator/checker IC is similar in concept to the circuit shown in Figure 4–86 except that it processes an 8-bit number. The logic symbol and pin configuration of the 74180 is shown in Figure 4–87. The operation of the 74180 is slightly different from the circuit in Figure 4–86 as illustrated by its truth table shown in Table 4–24.

For even parity checking, the 8-bit code $I_7I_6I_5I_4I_3I_2I_1I_0$ is applied to the IC and the parity bit is applied to EI. If the parity of all 9 bits is even then the Σ_E output will be a logic 0. For odd parity checking, the parity bit is applied to OI, and if the parity of all 9 bits is odd then Σ_E will be a logic 0. Thus the output Σ_E is used to identify the parity for both even and odd parity checking configurations, unlike the circuit in Figure 4–86 that uses two outputs.

Table 4–24
Truth table for the 74180 parity generator/checker

Inputs			Outputs	
Sum of 1's at I_0 through I_7	*EI*	*OI*	Σ_E	Σ_O
EVEN	1	0	1	0
ODD	1	0	0	1
EVEN	0	1	0	1
ODD	0	1	1	0
X	1	1	0	0
X	0	0	1	1

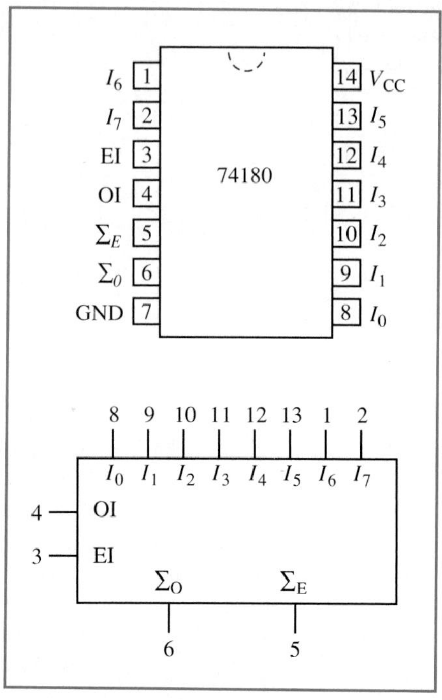

Figure 4–87 Logic symbol and pin configuration of the 74180 8-bit parity generator/checker.

Parity generation in the 74180 is also slightly different from the parity generation produced by the circuit in Figure 4–86. For even parity generation, the *EI* line is tied low and the *OI* line is tied high. An 8-bit code at $I_7I_6I_5I_4I_3I_2I_1I_0$ with even parity will then produce a 0 parity bit at Σ_E and a 1 parity bit at Σ_E if the code has odd parity. For odd parity generation, the *EI* line is tied high and the *OI* line is tied low. An 8-bit code at $I_7I_6I_5I_4I_3I_2I_1I_0$ with odd parity will then produce a 0 parity bit at Σ_E and a 1 parity bit at Σ_E if the code has even parity. Thus for even or odd parity generation the output Σ_E supplies the required parity bit. Since Σ_O is the complement of Σ_E it can be used if the complement of the parity bit is desired.

Review questions

1. How can one determine the parity of a binary number?
2. Describe the operation of a parity generator circuit.
3. Describe the operation of a parity checking circuit.

4–11 Arithmetic logic units

This section incorporates several of the application circuits designed in this chapter into a logic circuit known as an *arithmetic logic unit* (ALU). ALUs provide an important function in the operation of a digital computer and are designed to provide the computer with the basic arithmetic and logical operations that are applied to binary numbers. Figure 4–88 shows the logic symbol for a typical ALU that operates on two 4-bit numbers. The truth table for the ALU is shown in Table 4–25.

The ALU shown in Figure 4–88 accepts two 4-bit numbers *A* ($A_3A_2A_1A_0$) and *B* ($B_3B_2B_1B_0$) and produces a 4-bit result *F* ($F_3F_2F_1F_0$) as output. The *MODE* input, *M*, to the ALU selects it for either an

Figure 4–88
Logic symbol for a 4-bit ALU.

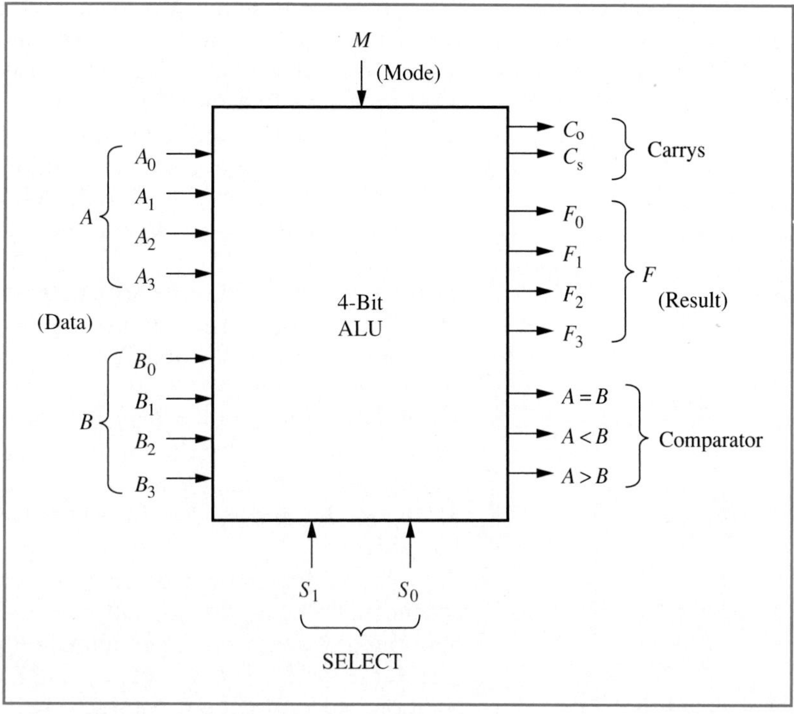

arithmetic (0) or a logic (1) operation. The *SELECT* inputs S_1S_0 select one of four arithmetic or logic operations depending on the state of M. Notice in the truth table shown in Table 4–25 that the four arithmetic operations produce as a result the sum of A and B, the difference between A and B, the negated value (in 2's complement representation) of B, and the incremented value of A. The outputs C_o and C_s shown in Figure 4–88 provide the carry-outs from addition and subtraction, respectively. The four logic operations are performed on the numbers A and B on a *bit-by-bit* basis and include the logical sum and product of A and B, the complement of A, and A exclusively *OR*ed with B. For example, if A = 0101 and B = 0110, then the logic product (A · B) of A and B would be 0100 if each bit that makes up the numbers A and B is *AND*ed on a bit-by-bit basis—$A_0 \cdot B_0$, $A_1 \cdot B_1$, $A_2 \cdot B_2$, $A_3 \cdot B_3$. The ALU also incorporates a comparator that compares the data inputs A and B and produces the results of the comparison at the outputs. $A = B$, $A < B$, and $A > B$ as discussed in Section 4–5.

Table 4–25
Truth table for a 4-bit ALU

Function select		Function	
S_1	S_0	M = 1 (logic)	M = 0 (arithmetic)
0	0	$A \cdot B$	A plus B
0	1	$A + B$	A minus B
1	0	$A \oplus B$	Minus B (2's complement)
1	1	$\overline{A}$	A plus 1

Since the design of the logic circuit for the ALU shown in Figure 4–88 can get fairly complex, we will take another approach to obtain the circuit. We will first examine the internal *architecture*, or layout, of the ALU in terms of various *function blocks*. The operation and purpose of each function block will then be investigated. Finally, these function blocks will be implemented using various logic circuits, some of which have already been developed in this chapter.

Figure 4–89 shows the internal architecture of the ALU whose logic symbol is shown in Figure 4–88. The ALU can be viewed as having two main processing units—the arithmetic unit that is responsible for providing the four basic arithmetic functions listed in Table 4–25, and the logic unit that is responsible for providing the four basic logic functions also listed in Table 4–25. The data inputs for the two 4-bit numbers, A and B, are represented in Figure 4–89 as *busses* for simplicity. Recall that a *bus* is simply a collection of lines grouped together to represent a binary number. Thus the double arrows are busses that represent the data lines, A_0–A_3, B_0–B_3, and F_0–F_3; these busses will be referred to as A, B, and F, respectively.

The arithmetic unit accepts as input the numbers A and B and produces the number F_A as output (result). The actual operation that is performed depends on the combination present at the select inputs S_1S_0 (see Table 4–25). The outputs C_o and C_s are the carry outputs from addition and subtraction, respectively.

Figure 4–89
Internal architecture of a 4-bit ALU.

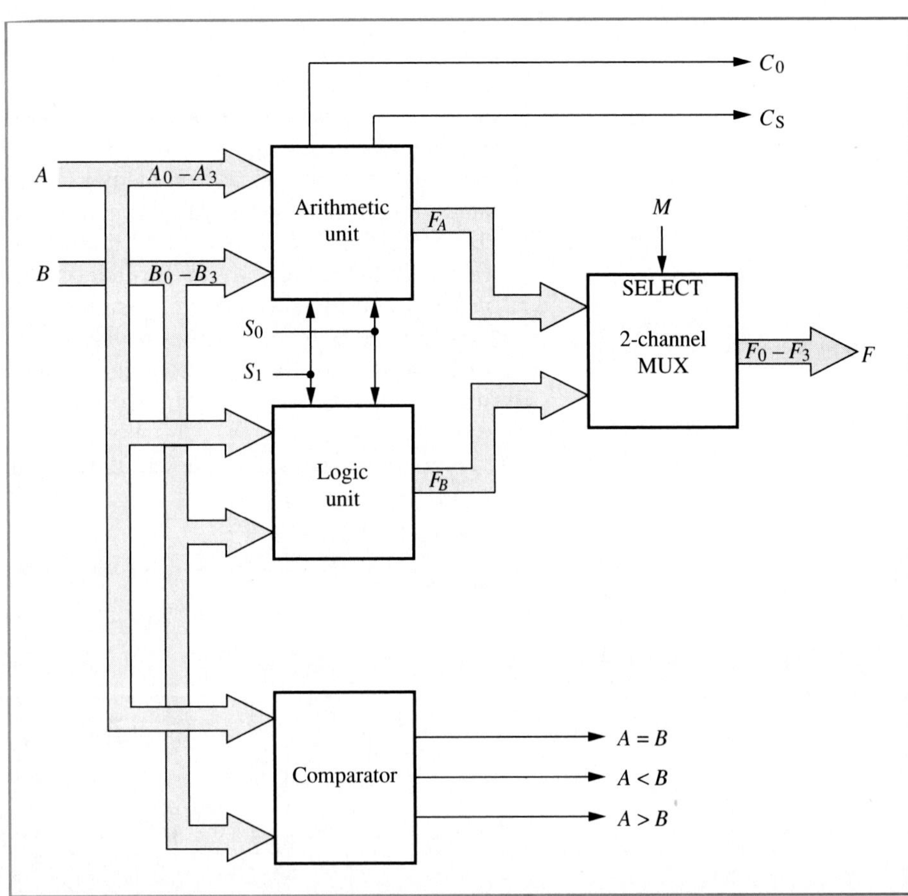

The logic unit accepts as input the numbers A and B and produces the number F_L as output (result). Again, the actual operation that is performed depends on the combination present at the select inputs S_1S_0 (see Table 4–25).

The internal operation of the arithmetic and logic function blocks shown in Figure 4–89 will be discussed later.

The 4-bit outputs of the arithmetic unit and logic unit are connected to a two-channel, 4-bit multiplexer (discussed in Section 4–8) that will multiplex F_A to the output (F) when M ($MODE$) is at the logic 0 state and F_L to the output (F) when M ($MODE$) is at the logic 1 state. Notice in Table 4–25 that the $MODE$ (M) input selects between logic and arithmetic functions. The data inputs A and B are also fed into a 4-bit magnitude comparator (discussed in Section 4–5) that produces three outputs $A = B$, $A > B$, and $A < B$ to indicate the results of the comparison.

The internal architecture of the arithmetic unit function block is shown in Figure 4–90. Each one of the four arithmetic functions listed in Table 4–25 is implemented by a separate circuit. The adder and subtracter function like the adders and subtracters discussed in Sections 4–3 and 4–4, respectively. The adder and subtracter circuits accept as input the 4-bit numbers A and B and produce their results at $F0$ and $F1$, respectively, with carries C_o and C_s, respectively.

Figure 4–90
Architecture of the arithmetic unit.

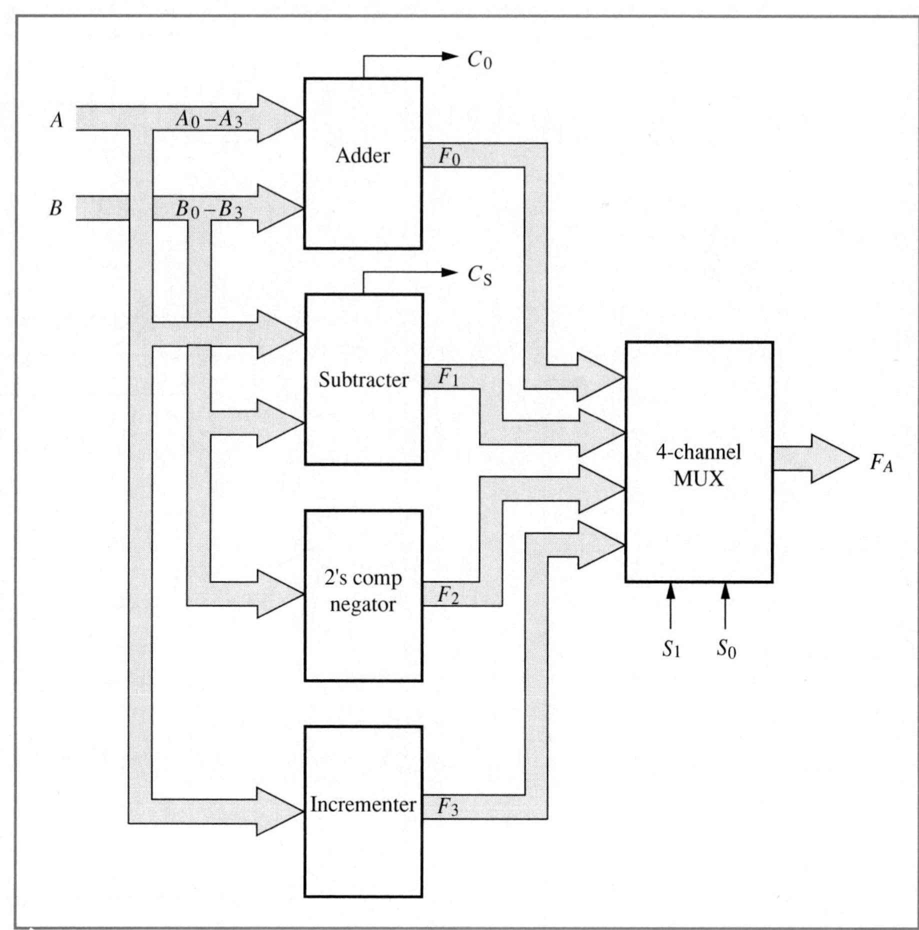

The 2's complement negator circuit obtains the 2's complement of the 4-bit number *B* by first complementing each bit that makes up the number *B* and then adding 1 to the resulting number using a series of half-adders as shown in Figure 4–91. The output of the circuit, *F*2, is the 2's complement of the number *B*.

The incrementer circuit adds 1 to the 4-bit number *A* (increments *A*) by using a parallel adder circuit similar to the one shown in Figure 4–91. The incrementer circuit is shown in Figure 4–92. Notice in Figure 4–92 that the series of half-adders simply adds 1 to the LSB of *A* and provides for the carries to ripple through the other bits for addition if necessary. The incremented value of *A* is made available at the output of the circuit, *F*3.

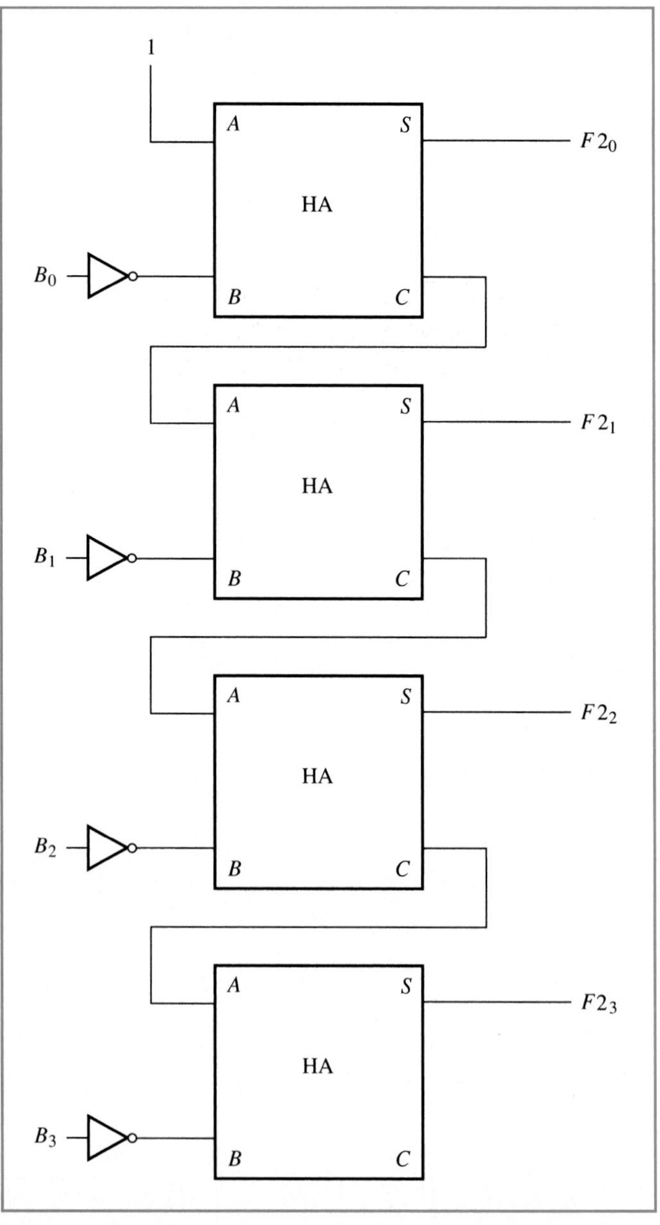

Figure 4–91
2's complement negator circuit.

Figure 4–92
Incrementer circuit.

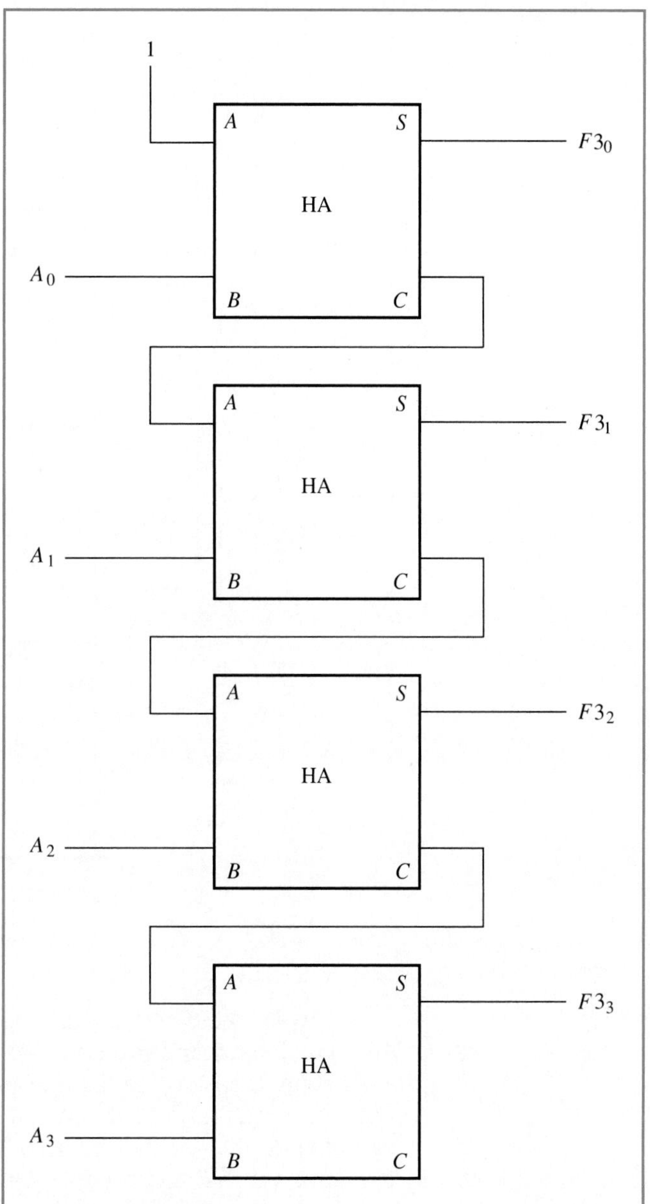

The outputs of the four circuits that make up the arithmetic unit represent the 4-bit results of each function listed in Table 4–25. Since the output of the ALU, f_A, should only supply the arithmetic operation specified by the select lines S_1S_0, a four-channel, 4-bit multiplexer is used to multiplex F_0 through F_3 onto the output lines F_A of the ALU.

The internal architecture of the logic unit function block is shown in Figure 4–93. Like the arithmetic unit, each one of the four logic functions listed in Table 4–25 is implemented by a separate circuit. The 4-bit data inputs to the ALU, A and B, are applied to the inputs of each circuit that makes up the logic unit. Each circuit is responsible for performing a particular logical operation on either A or B or both and providing the result at one of the outputs F_4 through F_7.

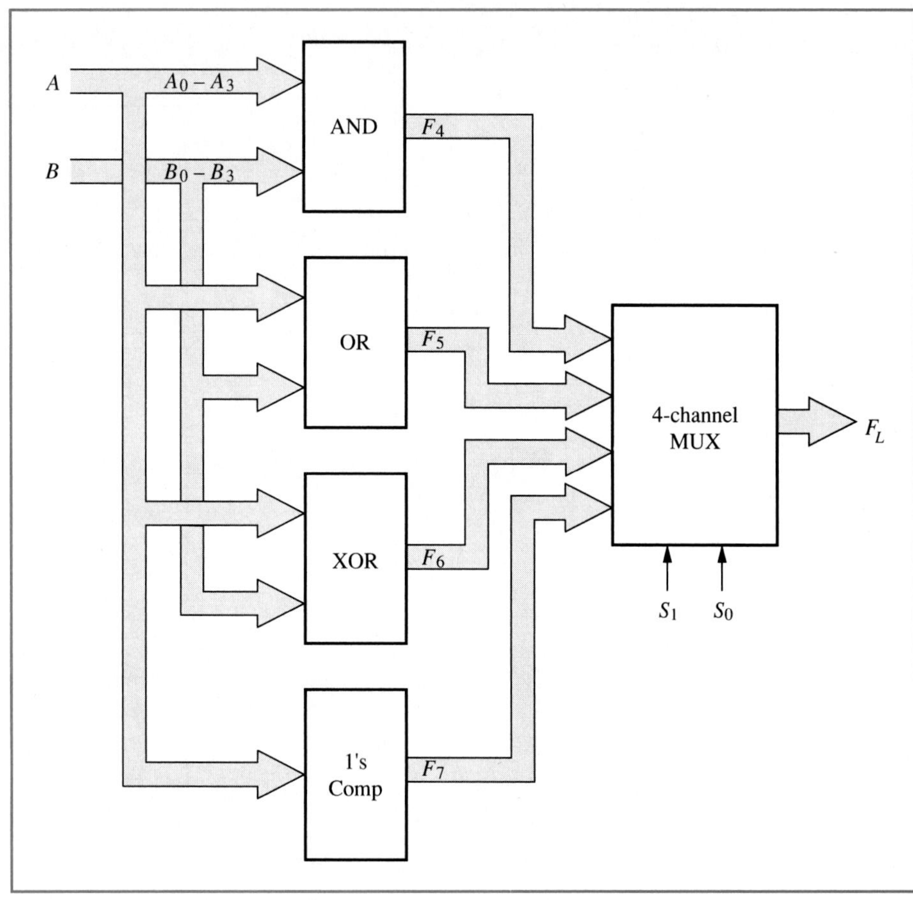

Figure 4–93
Architecture of the logic unit.

It was stated earlier that logical operations are performed on the 4-bit numbers A and B on a bit-by-bit basis. The circuits shown in Figure 4–94 show the four circuits that provide the logic unit with its four basic functions and illustrate the manner in which the two numbers are operated on in a bit-by-bit basis. The circuit in Figure 4–94a "*AND*s" the numbers A and B and produces the result at the outputs F_4. Similarly, the circuit shown in Figure 4–94b "*OR*s" the numbers A and B and produces the result at the outputs F_5, while the circuit shown in Figure 4–94c "*XOR*s" the numbers A and B and produces the result at the outputs F_6. The circuit in Figure 4–94d obtains the 1's complement of the number A at the output F_7 by simply inverting each bit in the number.

Notice in Figure 4–93 that the 4-bit outputs of each one of the four circuits shown in Figure 4–94 are connected to a four-channel, 4-bit multiplexer that channels the appropriate result to the output of the logic unit F_L, depending on the combination at the select inputs S_1S_0.

The design of the ALU described in this section is a conceptual (but functional) design. An actual ALU could be more efficiently designed by eliminating many of the redundancies inherent in this conceptual model. However, studying the operation of a complex circuit such as an ALU is more easily done by viewing it as a collection of function blocks rather than as a single entity.

Figure 4–94
Logic unit (a) "*AND*" circuit; (b) "*OR*" circuit; (c) "*XOR*" circuit; (d) 1's complement circuit.

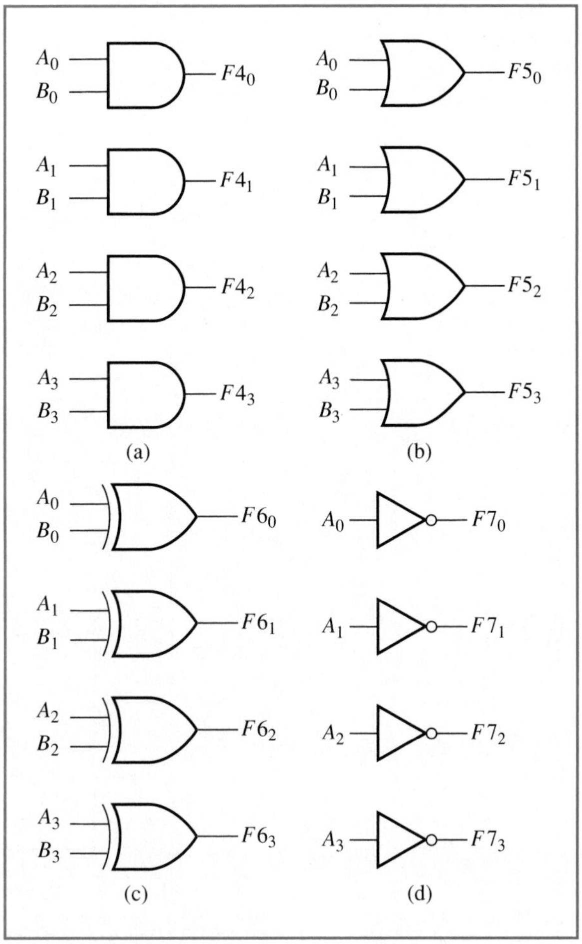

Integrated circuit logic

The 74181 4-bit ALU is an IC that is capable of performing 16 logic operations on two 4-bit input numbers plus a variety of arithmetic operations. The logic symbol and pin configuration of the 74181 are shown in Figure 4–95 and its truth table is shown in Table 4–26.

There are a lot of similarities and differences between the 74181 ALU and the ALU designed in this section (Figure 4–88). Both ALUs process two 4-bit data inputs A ($A_3A_2A_1A_0$) and B ($B_3B_2B_1B_0$) and provide the results at the output F ($F_3F_2F_1F_0$). The *MODE* select input M selects either the arithmetic unit or the logic unit in both ALUs. However, notice that the 74181 ALU has four function select lines $S_3S_2S_1S_0$ that allow it to (theoretically) select 16 arithmetic and 16 logic functions as listed in Table 4–26. The 74181 also incorporates a comparator that compares the numbers A and B but provides only one output, $A = B$, to indicate whether the two numbers are equal. The 74181's carry outputs P and G are for addition and subtraction operations performed by its arithmetic unit. The carry input, C_n, and carry output, C_{n+4}, can be used to expand the capabilities of the 74181 besides serving other functions. The 74181 ALU also has the capability to process active-low data at the inputs A and B and produce active-low results at the output F; in this mode, however, the functions assigned to each one of the codes appearing at S are in a different order and are not as shown in Table 4–26.

Figure 4–95
Logic symbol and pin configuration of the 74181 4-bit arithmetic logic unit.

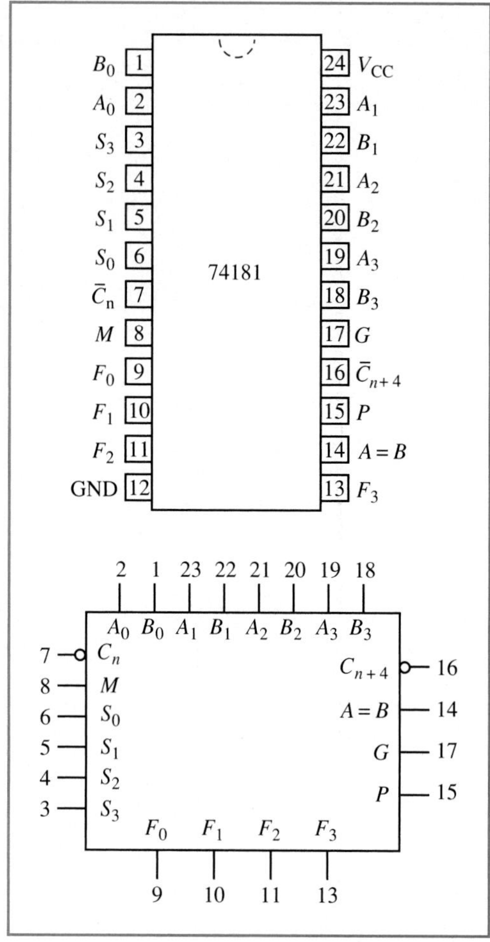

Table 4–26
Truth table for the 74181 4-bit ALU

Function select				**Function**	
S_3	S_2	S_1	S_0	Logic $M = 1$	Arithmetic $M = 0$
0	0	0	0	$\overline{A}$	A
0	0	0	1	$\overline{A + B}$	$A + B$
0	0	1	0	$\overline{A}B$	$A + \overline{B}$
0	0	1	1	0	Minus 1
0	1	0	0	$\overline{AB}$	$A + A\overline{B}$
0	1	0	1	$\overline{B}$	$(A + B)$ plus $A\overline{B}$
0	1	1	0	$A \oplus B$	A minus B minus 1
0	1	1	1	$A\overline{B}$	AB minus 1
1	0	0	0	$\overline{A} + B$	A plus AB
1	0	0	1	$\overline{A \oplus B}$	A plus B
1	0	1	0	B	$(A + \overline{B})$ plus AB
1	0	1	1	AB	AB minus 1
1	1	0	0	1	A plus A
1	1	0	1	$A + \overline{B}$	$(A + B)$ plus A
1	1	1	0	$A + B$	$(A + \overline{B})$ plus A
1	1	1	1	A	A minus 1

Review questions

1. Describe the operation of an ALU.
2. Which ALU line selects between an arithmetic or logic operation?
3. How are the actual arithmetic or logic functions selected?
4. What purpose does the multiplexer shown in Figure 4–89 serve?
5. What purpose do the multiplexers shown in Figures 4–90 and 4–93 serve?

4–12 Troubleshooting techniques

This chapter has introduced us to many application-specific ICs that implement the more complex logic functions of adding, subtracting, encoding, decoding, etc. The troubleshooting procedures for these ICs are similar to the procedures used to troubleshoot logic circuits containing the basic logic gates. This is because these ICs are actually circuits constructed from the basic logic gates. For troubleshooting purposes, the operation of an application-specific IC can be verified by using its truth table in much the same way as we would verify the operation of a simple logic gate or a logic circuit by making sure that it implements its truth table correctly. Application-specific ICs that are defective also exhibit the same symptoms as logic gates—inputs that are shorted to V_{CC} or GND, outputs that are shorted to V_{CC} or GND, open inputs, and open outputs.

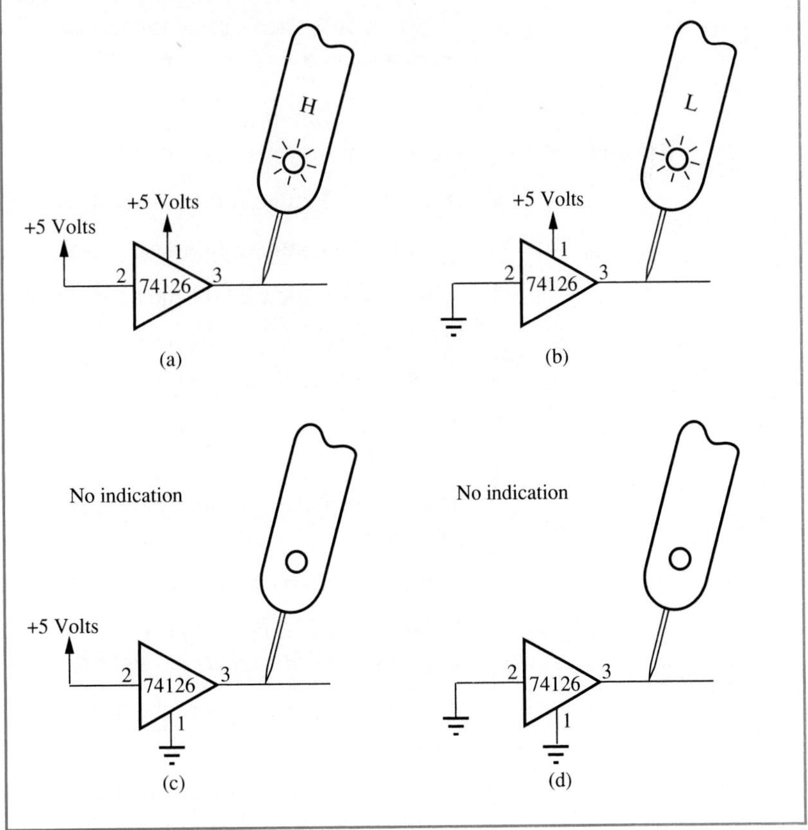

Figure 4–96
Identifying the outputs of a tristate logic gate.

Therefore the same troubleshooting procedures that we used for simple logic circuits can be applied to troubleshoot the more complex logic circuits introduced in this chapter.

Section 4–8 introduced us to the concept of tristate logic and the logic gate with tristate outputs. When we troubleshoot a logic circuit that contains gates with tristate outputs we should remember that a gate whose output is in a high-impedance state will exhibit the symptoms of an open output. For example, in Figure 4–96a, the 74126 tristate buffer gate is enabled by connecting the enable input to a logic high (H). The logic probe connected to the output indicates a logic high state since the input of the gate is connected to a logic high. If the input is connected low (L) as shown in Figure 4–96b then the logic probe indicates a low state at the output. However, if the enable input of the 74126 gate is tied low (to disable the gate) as shown in Figure 4–96c and d, then the output of the gate is in a high-impedance state (open circuited) and therefore the logic probe indicates no logic level. Notice that if a gate with a two-stated output produced this condition the gate would be considered defective due to an open output.

Review questions

1. Why do the inputs and outputs of application-specific ICs exhibit the same faults as the simple logic gates?
2. Why is it possible to apply the troubleshooting procedures for simple logic circuits to complex circuits that contain application-specific ICs?
3. How does the logic probe indicate that the output of a logic gate is in a high-impedance state?

4–13 SUMMARY

- The *logic symbol* for a circuit represents the circuit as a block and identifies its inputs and outputs.
- *Adders* are logic circuits that are used to add binary numbers.
- A *half-adder* adds 2 bits and produces a single-bit sum and carry.
- A *full-adder* adds 2 bits plus a carry input and produces a single-bit sum and carry-out.
- An *Exclusive-OR gate* can have only two inputs.
- The Exclusive-*OR* gate produces a logic 1 at its output if both inputs are not equal.
- The output of an Exclusive-*OR* gate is the sum (ignoring carries) of its two inputs.
- A *parallel binary adder* adds multibit binary numbers.
- *Subtracter* circuits use parallel binary adders and can be of two types—1's complement and 2's complement.
- A 1's complement subtracter subtracts two numbers by using the 1's complement binary subtraction technique and requires two addition steps.
- A 2's complement subtracter subtracts two numbers by using the 2's complement binary subtraction technique and requires one addition step.

- A *comparator* is a logic circuit that compares the magnitude of two binary numbers.
- *Encoders* are logic circuits that produce a unique binary code when one of their input lines is activated.
- A *priority encoder* allows more than one of its inputs to be activated by assigning priorities to each input.
- Most priority encoders use *high-order data line encoding* which causes the encoder to output the code of the line that has the highest code if more than one input line is activated at a time.
- A *decoder* detects a unique binary code by activating one of several output lines to identify the code.
- Most decoders are known as *1-of-n* decoders since they will only activate one of several (*n*) output lines in response to an input code.
- *Seven-segment decoders* convert a binary code into a code suitable for activating the segments of a *seven-segment LED*.
- A *ripple blanking* circuit is used to blank out leading zeros in a seven-segment LED display system.
- *Multiplexers* channel binary data from one of several input channels onto a single output channel.
- *Tristate* logic gates have outputs that can exist in one of three states—1, 0, or *Z* (high impedance).
- A *demultiplexer* reverses the process of multiplexing.
- Decoders can be configured to provide demultiplexing and therefore decoders are also known as *decoders/demultiplexers*.
- *Parity generators* are designed to produce a parity bit for an input binary number.
- The parity of a binary number can be obtained by adding all the bits in the number (ignoring carries) and examining the single-bit sum for a 0 (even number of 1-bits) or a 1 (odd number of 1-bits).
- *Parity checkers* determine if a number has even or odd parity.
- The parity generating and checking functions can usually be combined into one circuit called a *parity generator/checker*.
- *Arithmetic logic units* (*ALUs*) are logic circuits that provide the basic arithmetic and logic functions for operation on two binary numbers.

PROBLEMS

Section 4–2 Application circuit design

1. Design a logic circuit that will accept all possible combinations of 3-bit binary numbers as input, pass through only the even combinations, and filter out all the odd combinations. For example, if any of the even numbers 000, 010, 100, or 110 are applied to its input, the output will be the same as the input, but if any of the odd numbers 001, 011, 101, or 111 are applied to its input, the output will be 111.

2. The pH scale is often used to determine the acidity or alkalinity of water and is given by a number in the range 0 through 14. A pH of 7 is neutral while a pH less than 7 is acidic and a pH greater than 7 is alkaline. A logic circuit is to be designed to monitor and control the pH level of a swimming pool. The pH level of the water is applied to the input of the circuit as a binary number and the circuit is to maintain the pH level between 6 and 7 by activating a valve to release acid if the pH is greater than 7 or by activating another valve to release base if the pH is less than 6. The circuit should also have four LED indicators that are activated under the following conditions:
   (a) If the pH is between 6 and 7, a green LED should be illuminated to indicate the optimum setting.
   (b) If the pH is greater than 7, a blue LED should be illuminated to indicate alkalinity.
   (c) If the pH is less than 7, a yellow LED should be illuminated to indicate acidity.
   (d) If the pH is greater than 9 or if the pH is less than 5, a red LED should be illuminated as a warning.

Section 4–3 Adders

3. Prove that the following equation is true:

$$\overline{X \oplus Y} = X \cdot Y + \overline{X} \cdot \overline{Y}$$

4. Design a 2-bit parallel binary adder that will add two 2-bit numbers A_1A_0 and B_1B_0 and produce a sum S_1S_0 and a carry-out C_o. Use the conventional approach. Compare this circuit with the circuit shown in Figure 4–16 in terms of the number of gates required.
5. Design a parallel binary adder that will add two 16-bit numbers. Use only 4-bit adders in your design.
6. Construct a schematic diagram of a 4-bit adder using 7482 ICs.
7. Construct a schematic diagram of an 8-bit adder using 7483 ICs.

Section 4–4 Subtracters

8. Design a 2-bit subtracter that will subtract the 2-bit number A_1A_0 from B_1B_0 and produce the difference D_1D_0 and a sign S. Use the conventional approach.
9. Design an 8-bit 2's complement subtracter using a series of parallel binary adders.
10. Design an 8-bit 1's complement subtracter using a series of parallel binary adders.

Section 4–5 Comparators

11. Design a comparator that will compare all numbers from 0 to 7 with the number 5. If the number is less than 5 then the output of the circuit will be a logic 1, otherwise the output of the circuit will be a logic 0. Use the conventional approach.
12. Design a 4-bit magnitude comparator by using two 2-bit magnitude comparators whose logic symbol is shown in Figure 4–26.

Section 4–6 Encoders and decoders

13. The encoder circuit shown in Figure 4–33 was designed with no priorities. However, when more than one line is activated, the output will be at some logic state that will represent some code. Therefore there will be some priorities built into the circuit by default. Analyze the circuit and determine the priorities of the input lines; that is, obtain the complete truth table for the circuit through analysis.
14. Design a BCD encoder (no priorities) that will encode ten input lines into the BCD code. The inputs and outputs of the encoder should be active high.
15. Redesign the circuit from Problem 14 to incorporate *low-order* data line encoding.
16. Design a 1-of-8 decoder with active-low outputs.
17. Incorporate an active-low enable line for the decoder designed in Problem 16.
18. Expand the decoder designed in Problem 17 to function as a 1-of-16 decoder.
19. Determine the priorities of all 16 input lines of the encoder circuit shown in Figure 4–43.

Section 4–7 Seven-segment decoders

20. Design a BCD to seven-segment decoder to drive a common anode seven-segment LED.
21. Incorporate a ripple blanking input $\overline{RBI}$ and a ripple blanking output $\overline{RBO}$ into the circuit designed in Problem 20.
22. The *Gray code* is a 4-bit code that has its combinations organized such that only 1 bit changes from one code to the next in sequence

    0 0 0 0
    0 0 0 1
    0 0 1 1
    0 0 1 0
    0 1 1 0
    0 1 1 1
    0 1 0 1
    0 1 0 0
    1 1 0 0
    1 1 0 1
    1 1 1 1
    1 1 1 0
    1 0 1 0
    1 0 1 1
    1 0 0 1
    1 0 0 0

    Design a code conversion circuit that will convert the Gray code to the 4-bit binary code.
23. Design a code conversion circuit that will convert the 4-bit binary code to the Gray code described in Problem 22.

24. The excess-3 code is obtained by adding 3 (0011) to each 4-bit BCD code as shown in the following table. When we obtained the 1's complement of the excess-3 code of a decimal digit, we get the excess-3 code of the 9's complement of the decimal digit. For example, the excess-3 code for 5 is 1000; the 1's complement of 1000 is 0111, which is the excess-3 code for 4. 4 is the 9's complement of 5.

Decimal	BCD	Excess-3
0	0000	0011
1	0001	0100
2	0010	0101
3	0011	0110
4	0100	0111
5	0101	1000
6	0110	1001
7	0111	1010
8	1000	1011
9	1001	1100

Design a code conversion circuit that will convert the BCD code to the excess-3 code.

25. Design a code conversion circuit that will convert the excess-3 code described in Problem 24 to the BCD code.

Section 4–8 Multiplexers and tristate logic

26. Design a three-channel multiplexer that will channel three 2-bit input numbers onto a single 2-bit output channel. Use regular (two-state) logic in your design.

27. Implement the circuit designed in Problem 26 using tristate logic gates.

28. Analyze the circuit shown in Figure 4–97 by constructing a truth table for the circuit. Use the letter Z to identify all high-impedance states.

29. Design a two-channel, 4-bit multiplexer using regular (two-state) logic.

30. Redesign the circuit in Problem 29 using tristate logic gates. Compare the two circuits.

Figure 4–97

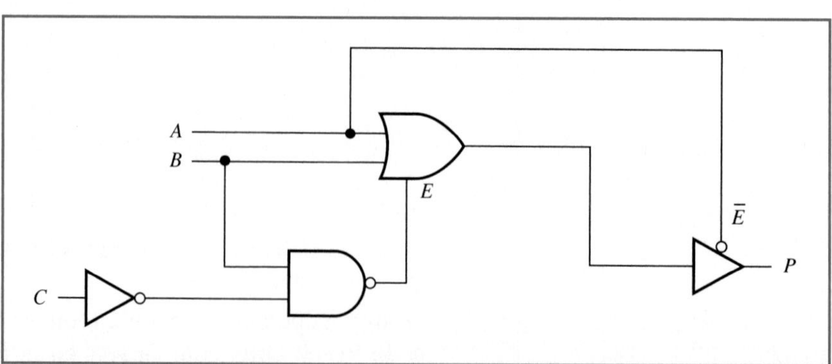

Figure 4–98

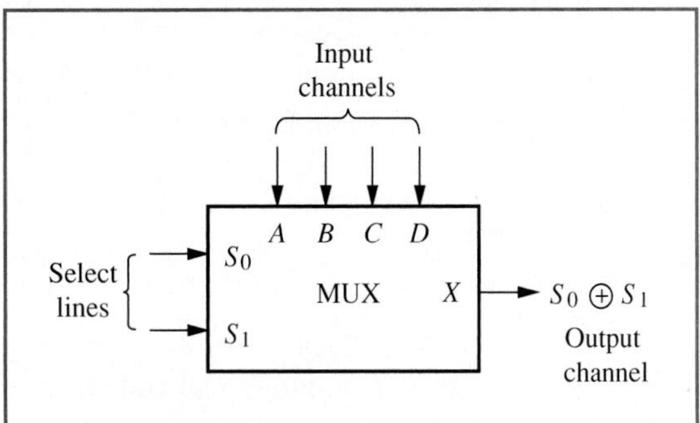

31. A single-bit, multichannel multiplexer can very easily be used to implement a truth table by using the select lines as the independent variables and the output line as the dependent variable. The input channel lines are set to the values of the dependent variable so that these values are multiplexed to the output as each combination is applied to the select inputs.

Make the multiplexer shown in Figure 4–98 emulate the function of an Exclusive-*OR* gate by appropriately connecting the input channels to logic 1's or 0's.

32. Using the information provided in Problem 31, make the 74150 16-input multiplexer implement the following truth table:

A	B	C	D	X
0	0	0	0	1
0	0	0	1	0
0	0	1	0	0
0	0	1	1	1
0	1	0	0	0
0	1	0	1	1
0	1	1	0	1
0	1	1	1	0
1	0	0	0	0
1	0	0	1	1
1	0	1	0	1
1	0	1	1	0
1	1	0	0	1
1	1	0	1	0
1	1	1	0	0
1	1	1	1	1

33. Compare the circuit obtained in Problem 32 with the circuit that would be obtained if the logic equation for the truth table were implemented. Compare in terms of the number of ICs required.

Section 4–9 Demultiplexers

34. Design a two-channel, 4-bit demultiplexer using two-channel, 1-bit demultiplexers whose logic symbol is shown in Figure 4–75.

35. Design a four-channel, 2-bit demultiplexer using a four-channel, 1-bit demultiplexer whose logic symbol is shown in Figure 4–77.

36. Configure the 74154 decoder IC (see Figure 4–46) to function as a 16-channel 1-bit demultiplexer.

Section 4–10 Parity circuits

37. Design an 8-bit parity generator circuit that will accept an 8-bit input number and indicate whether the number has even or odd parity by activating a *single* output that will be a logic 1 if the parity is even or a logic 0 if the parity is odd.

38. Expand the parity generator/checker circuit in Figure 4–86 to check an 8-bit input number.

Section 4–11 Arithmetic logic units

39. If the following two numbers were applied to the 74181 ALU:

$$A = 1010 \quad B = 0110$$

what would the output of the circuit F be for the following conditions:
(a) $M = 1, S = 0001$
(b) $M = 0, S = 1001$
(c) $M = 1, S = 0111$
(d) $M = 0, S = 1111$

40. Design a 4-bit ALU to implement the following truth table:

S_2	S_1	S_0	Function
0	0	0	$\overline{A \cdot B}$
0	0	1	$\overline{A + B}$
0	1	0	$\overline{A \oplus B}$
0	1	1	$\overline{B}$
1	0	0	A plus A
1	0	1	A minus 1
1	1	0	B minus A
1	1	1	Minus A (2's complement)

Troubleshooting

Section 4–12 Troubleshooting techniques

41. The circuit shown in Figure 4–43b uses two 74148 encoders to encode 16 inputs into a 4-bit active-low output code. When the circuit is set up and tested, the input lines 0 through 7 produce the correct codes 1111 through 1000. However, when the input lines 8 through F are activated the output of the circuit appears to be stuck at 1111. Troubleshoot the circuit and suggest possible faults that could produce this situation.

42. The logic circuit shown in Figure 4–99 has a fault. The recorded outputs of the logic gates $G1$, $G2$, and $G3$ are shown below for all possible combinations of the inputs P, Q, and R. Use the analysis procedure to troubleshoot the circuit and isolate the fault.

Figure 4–99

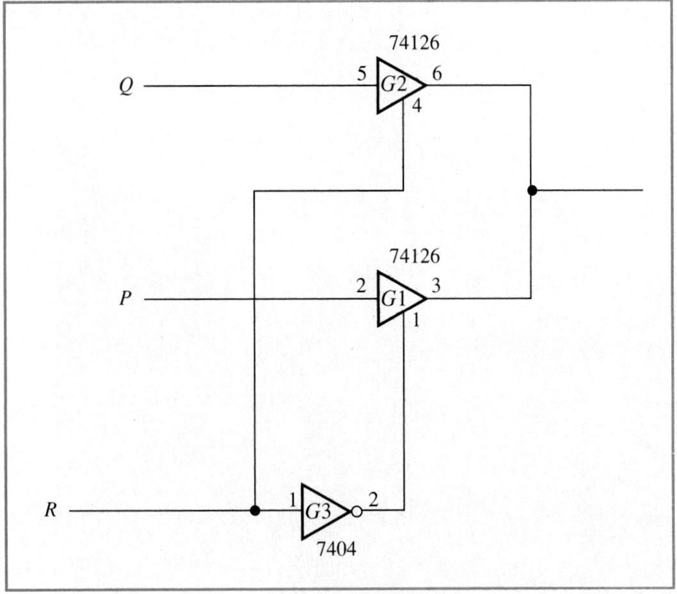

P	Q	R	Outputs G1	G2	G3
L	L	L	L	L	H
L	L	H	Z	Z	L
L	H	L	L	L	H
L	H	H	Z	Z	L
H	L	L	H	H	H
H	L	H	Z	Z	L
H	H	L	H	H	H
H	H	H	Z	Z	L

Note that the outputs of $G1$ and $G2$ were connected together as shown in Figure 4–99 when the above data was recorded.

The basic unit of storage in early computers were the magnetic toroids shown in the photograph. Today semiconductor logic circuits have completely eliminated the need for these devices in digital computers (Courtesy IBM).

OUTLINE

OBJECTIVES

The objectives of this chapter are to

- study the design and operation of a simple ungated and gated latch.
- introduce the timing diagram technique of analyzing sequential logic circuits.
- introduce the basic operation of a flip-flop as compared to a gated latch.
- study the operation of the *S-R* flip-flop and develop its state transition table and equation.
- study the operation of the *T* flip-flop and develop its state transition table and equation.
- study the operation of the *D* flip-flop and develop its state transition table and equation.
- investigate the various ICs that implement the *D* flip-flop in hardware.
- study the operation of the *J-K* flip-flop and develop its state transition table and equation.
- investigate the various ICs that implement the *J-K* flip-flop in hardware.
- examine the characteristics and operation of master-slave flip-flops.
- examine the versatility of the *D* and *J-K* flip-flops by designing circuits to convert from one type to another.
- study the design and operation of the monostable and astable multivibrator.
- examine the switching characteristics of flip-flops and interpret data sheet parameters.
- investigate the procedures to test and troubleshoot flip-flop circuits and identify the typical faults that could exist in a sequential logic circuit.

5 INTRODUCTION TO SEQUENTIAL LOGIC

5–1 INTRODUCTION

The logic circuits studied so far are known as *combinational logic circuits* since the output(s) for the circuits depend only on various binary combinations applied to the input(s). This chapter introduces *sequential logic circuits*. Like combinational logic circuits, sequential logic circuits have output(s) that could depend on various combinations applied to the input(s), but the outputs also depend on the *past history* of the state of the outputs. Sequential logic circuits therefore incorporate a *memory element* since these circuits must "remember" the past states of their outputs in order to produce an appropriate output. One of the characteristics of sequential logic circuits is *feedback*—a technique by which an output logic level is treated just like another input to the circuit. Circuits that incorporate feedback are classified as sequential logic circuits while circuits that don't are classified as combinational logic circuits.

Sequential logic circuits play an important role in the operation of a digital computer. Besides serving as the primary means of storage, these circuits are also used for timing and counting. This chapter introduces the basic unit of a sequential logic circuit—the flip-flop. Chapters 6 and 7 will then investigate the various applications of flip-flops and sequential logic circuits. Since the design and operation of the flip-flop is based on the basic logic gates and concepts covered in earlier chapters, a thorough understanding of this elementary material is vital to the study of sequential logic circuits.

5–2 The latch

A *latch* is a sequential logic circuit that can exist in one of two possible states—*SET* or *RESET*. When a latch is *SET,* its output is at the logic 1 state and when the latch is *RESET* its output is at the logic 0 state. The term *CLEAR* is also used to represent a *RESET* condition. Figure 5–1a shows the logic symbol of a *SET-RESET* (*S-R*) latch. Notice that the latch has two inputs labeled *SET* (S) and *RESET* (R) and two outputs labeled Q and $\overline{Q}$ that are always complements of each other. The state of the latch is always described in terms of the Q output, that is, if $Q=1$ then the latch is said to be *SET* and if $Q=0$ the latch is said to be *RESET*. The $\overline{Q}$ output is made available as a convenience only.

In order to set the *S-R* latch we must activate the S input by applying a logic 1 to it. Only one of the two inputs must be active at a time and therefore the R input must be (inactive) at a logic 0. Similarly, to reset the *S-R* latch we must activate the R input by applying a logic 1 to it

Figure 5–1
The *S-R NOR* latch. (a) logic symbol; (b) logic circuit.

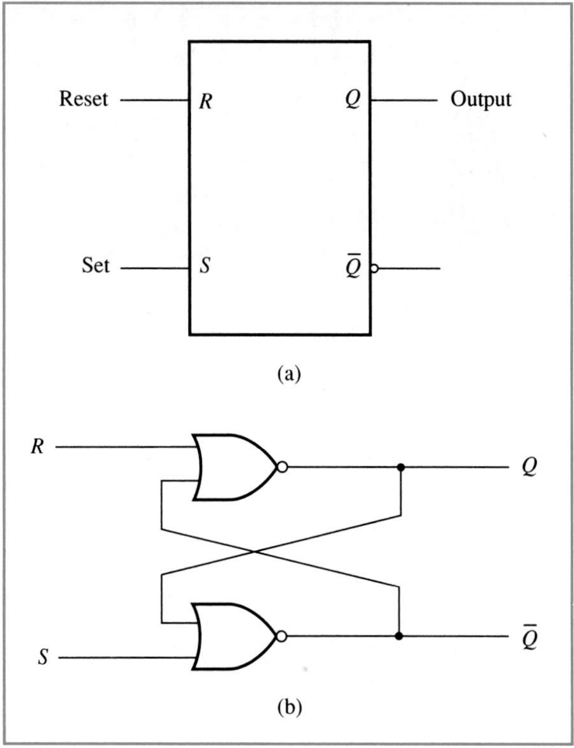

while keeping the S input inactive at the logic 0 state. Figure 5–2 illustrates the *timing diagram* for the *S-R* latch. A timing diagram describes the operation of a sequential logic circuit by graphically showing the output states of the circuits as the inputs change. In the timing diagram of Figure 5–2 we have traced the state of the latch for several possible (arbitrarily picked) states of the inputs S and R. Each

Figure 5–2
Timing diagram for the *S-R NOR* latch.

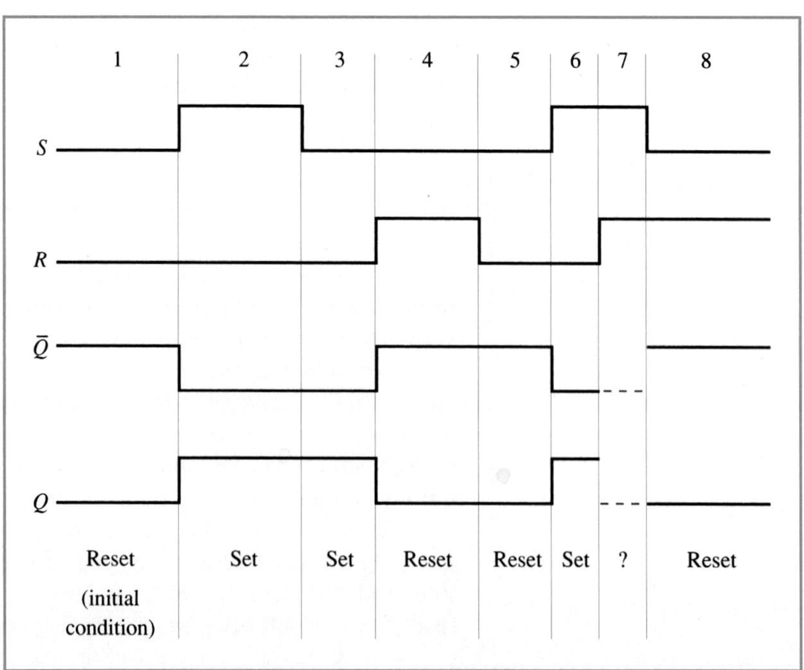

time *S* or *R* changes state we must determine if the output state of the latch has changed. The intervals during which *S* and *R* remain constant are labeled in Figure 5–2 with the numbers 1 through 8 for reference purposes in the following analysis.

1. Both *S* and *R* are inactive, and the initial state of the output of the latch (*Q*) is *RESET*. Therefore the latch remains *RESET*. This is a *no-change* condition.
2. *S* is active and commands the latch to *SET*. The output of the latch is therefore *SET*.
3. *S* is inactive now (while *R* remains inactive) but the latch *remains SET* since this is a no-change condition.
4. *R* is now active and commands the latch to *RESET*. The output of the latch is therefore *RESET*.
5. *R* is inactive now (while *S* remains inactive) but the latch remains *RESET* since this is a no-change condition.
6. *S* is activated again and the latch changes to the *SET* state.
7. Both *S* and *R* are now active and this causes *both* outputs to go to the logic 0 state. This effectively commands the latch to *SET* and *RESET* at the same time. Since this is impossible the state of the latch is unknown. This is an *undefined* or *illegal* state.
8. *S* is now inactive while *R* is still active, thus commanding the latch to *RESET*.

Notice in the timing diagram shown in Figure 5–2 that the $\overline{Q}$ output is the complement of the *Q* output. Also notice from the analysis that the *S* and *R* inputs function as command inputs that instruct the latch to either *SET* or *RESET* its output. The *memory element* characteristic of a sequential logic circuit is apparent in intervals 3 and 5 of the timing diagram. In interval 2 the latch was *SET* by activating the *S* input. However, when the *S* input was returned back to its inactive state in interval 3, the latch remained *SET*. That is, the output was *latched* or held at the *SET* state even though the stimulus that produced it was removed. Therefore by momentarily activating the *S* input we have *stored* a logic 1 at the output of the latch. Similarly, in interval 4 the *R* input was activated to *RESET* the latch. In interval 5, the latch remained *RESET* even though *R* returned to the inactive state. Therefore by momentarily activating the *R* input we have stored a logic 0 at the output of the latch. To understand how the latch accomplishes its function we must analyze its logic circuit.

The logic circuit for the *S-R* latch is shown in Figure 5–1b. Since this circuit is made up of two *NOR* gates it is sometimes referred to as a *NOR latch*. Notice that the circuit incorporates the feedback from output to input that is typical of most sequential logic circuits. In order to analyze the *NOR* latch circuit we must first assume some initial state for the latch. Let us assume the same initial conditions specified in interval 1 of the timing diagram in Figure 5–2 and trace through each of the following intervals to verify its validity.

The circuits in Figure 5–3a through 5–3h show the trace of the logic levels for each of eight intervals of the timing diagram in Figure 5–2. Recall that if any input of a *NOR* gate is a logic 1 the output is a logic 0, and if both inputs of the *NOR* gate are logic 0s the output is a logic 1. Figure 5–3a shows the logic circuit of the *S-R* latch during interval 1 (initial conditions). Since the latch is *RESET* ($Q=0$) and both R and S are inactive ($R=0$, $S=0$) the output of gate $G2$ ($\bar{Q}$) is a logic 1 which is fed back to the input of gate $G1$ producing a logic 0 at its output (Q) and therefore keeping the latch *RESET*. During interval 2, the S line changes to a logic 1 immediately causing the output of $G2$ to change to a logic 0 and consequently the output of $G1$ to change to a logic 1. The latch is now *SET* as shown in Figure 5–3b. During interval 3 the S input changes back to a logic 0, but the output of $G2$ does not change since its other input is still held at the logic 1 state by Q. The latch remains *SET* as shown in Figure 5–3c. During interval 4 the R input changes to a logic 1

Figure 5–3
Trace of the various states of the *NOR* latch.

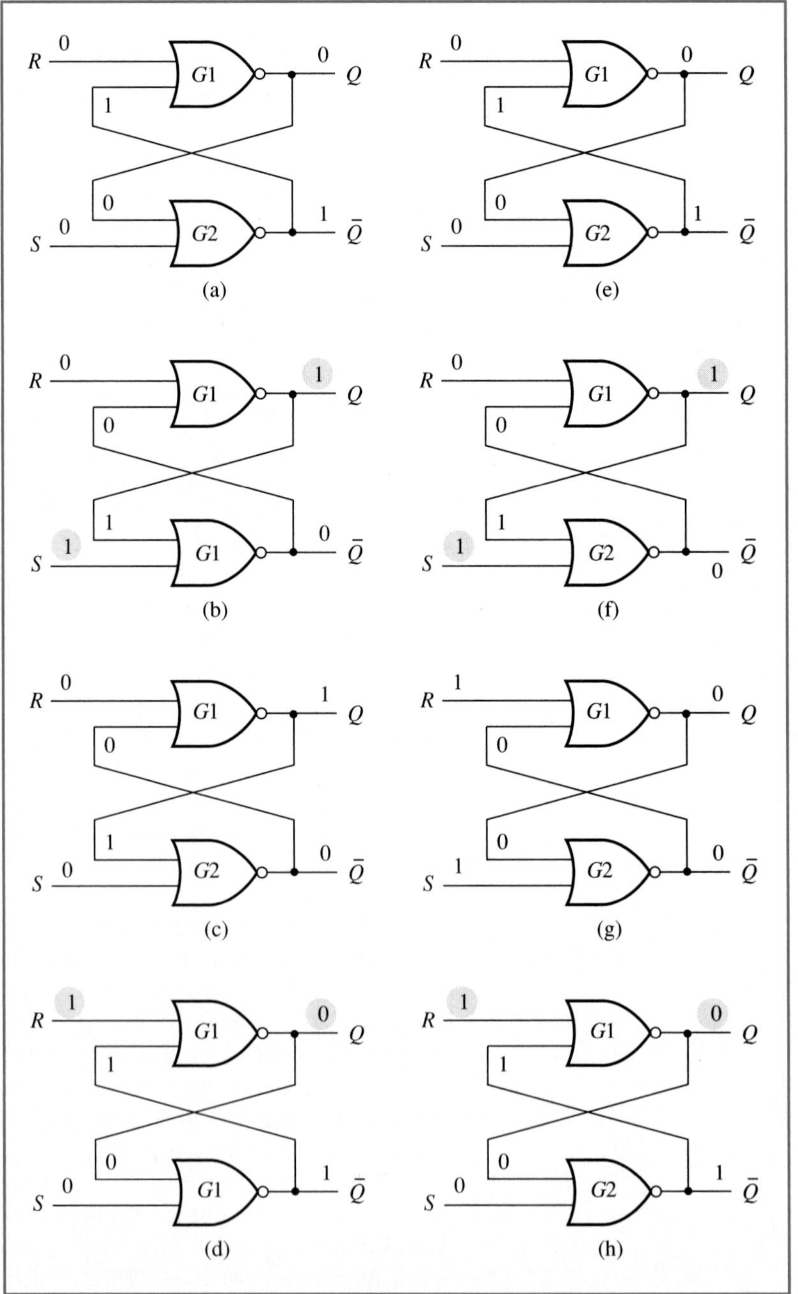

state. This causes the output of $G1$ to change to a logic 0 and the output of $G2$ to a logic 1. The latch is now *RESET* as shown in Figure 5–3d. During interval 5 the R input returns to the logic 0 state, but the output of $G1$ does not change since its other input is held high by $\overline{Q}$. The latch therefore remains *RESET* as shown in Figure 5–3e. During interval 6 the S input again changes to a logic 1 state, which causes the output of $G2$ to change to a logic 0 and the output of $G1$ to change to a logic 1, thus setting the latch as shown in Figure 5–3f. During interval 7 while S is active at the logic 1 state, the input R now changes to a logic 1 which causes the output of $G1$ to change to a logic 0. However, the output of $G2$ remains at a logic 0 state since S is a logic 1. Both the outputs of the latch are at the logic 0 state as shown in Figure 5–3g. This is an invalid condition since the outputs of the latch are defined to be complements of each other. During interval 8 the input S returns to the logic 0 state, while R remains active at the logic 1 state. This causes the output of $G2$ to change to a logic 1 and the output of $G1$ to change to a logic 0, thus resetting the latch.

An *S-R* latch can also be designed with *active-low* inputs. The operation of such a latch is identical to the *active-high* input *S-R* latch shown in Figure 5–1 except that a logic 0 on the S and R inputs is required to set and reset the latch, respectively. The logic symbol and circuit for the active-low input *S-R* latch is shown in Figure 5–4a and b, respectively. Notice that due to its *NAND* gates, the latch is sometimes referred to as a *NAND latch*.

The timing diagram for the *S-R NAND* latch is shown in Figure 5–5. Again we have analyzed the state of the latch for various (arbitrarily picked) values of $\overline{S}$ and $\overline{R}$. The timing diagram is divided into five time intervals during which $\overline{S}$ and $\overline{R}$ do not change.

Figure 5–4
The *S-R NAND* latch. (a) Logic symbol; (b) logic circuit.

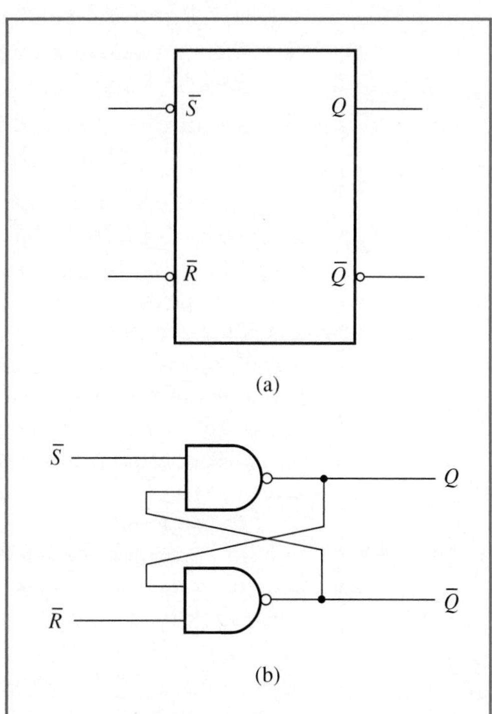

Figure 5–5
Timing diagram for the *NAND* latch.

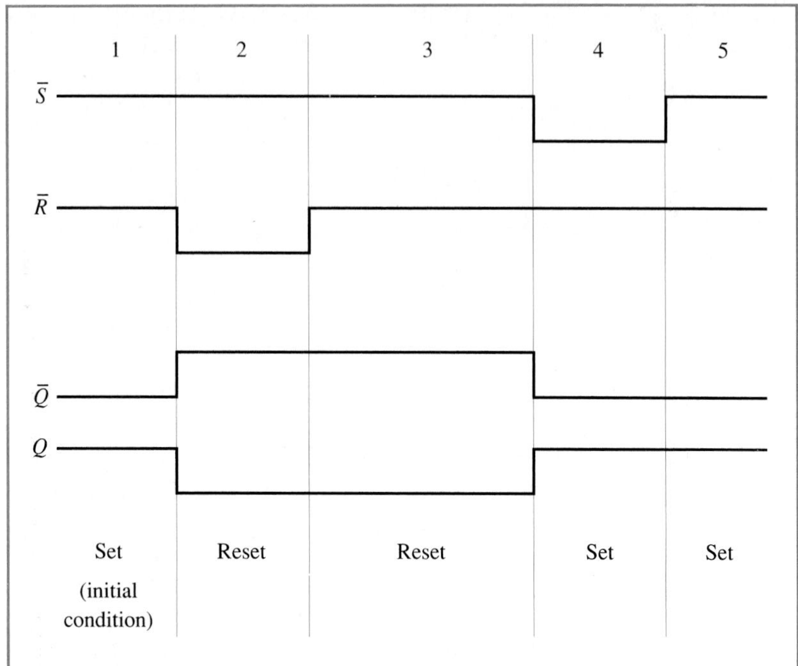

1. During this interval we have assumed that the latch is currently *SET* ($Q=1$) and that the $\overline{S}$ and $\overline{R}$ inputs are inactive at the logic 1 state. This is the initial condition of the latch.
2. $\overline{R}$ is now active (0) while $\overline{S}$ is inactive (1), and therefore the latch is *RESET* and the output Q changes to the logic 0 state.
3. $\overline{R}$ returns to the inactive state (1) but the latch remains *RESET* ($Q=0$) since $\overline{S}$ is also inactive (1). This is a no-change condition.
4. $\overline{S}$ is now active (0) while $\overline{R}$ is inactive (1), and therefore the latch is now *SET* and the output Q changes to the logic 1 state.
5. $\overline{S}$ returns to the inactive state (1) but the latch remains *SET* ($Q=1$) since $\overline{R}$ is also inactive (1). This is also a no-change condition.

In order to understand the operation of the *NAND* latch we can analyze its logic circuit shown in Figure 5–4b for each of the five time intervals of the timing diagram shown in Figure 5–5.

The circuits in Figure 5–6a through e show the trace of the logic levels for each of the five intervals of the timing diagram in Figure 5–5. Recall that if any input of a *NAND* gate is a logic 0 the output is a logic 1, and if both inputs of the *NAND* gate are logic 1's the output is a logic 0. Figure 5–6a shows the logic circuit of the *S-R* latch during interval 1 (initial conditions). Since the latch is *SET* ($Q=1$) and both $\overline{R}$ and $\overline{S}$ are inactive ($\overline{R}=1$, $\overline{S}=1$), the output of gate *G*2 ($\overline{Q}$) is a logic 0, which is fed back to the input of gate *G*1 producing a logic 1 at its output (Q) and therefore keeping the latch *SET*. During interval 2, the $\overline{R}$ line changes to a logic 0, immediately causing the output of *G*2 to change to a logic 1 and consequently the output of *G*1 to change to a logic 0. The latch is now *RESET* as shown in Figure 5-6b. During interval 3 the $\overline{R}$ input changes back to a logic 1, but the output of *G*2 does not change since its other input is still held at the logic 0 state by Q. The latch remains *RESET* as

Figure 5–6
Trace of the various states of the *NAND* latch.

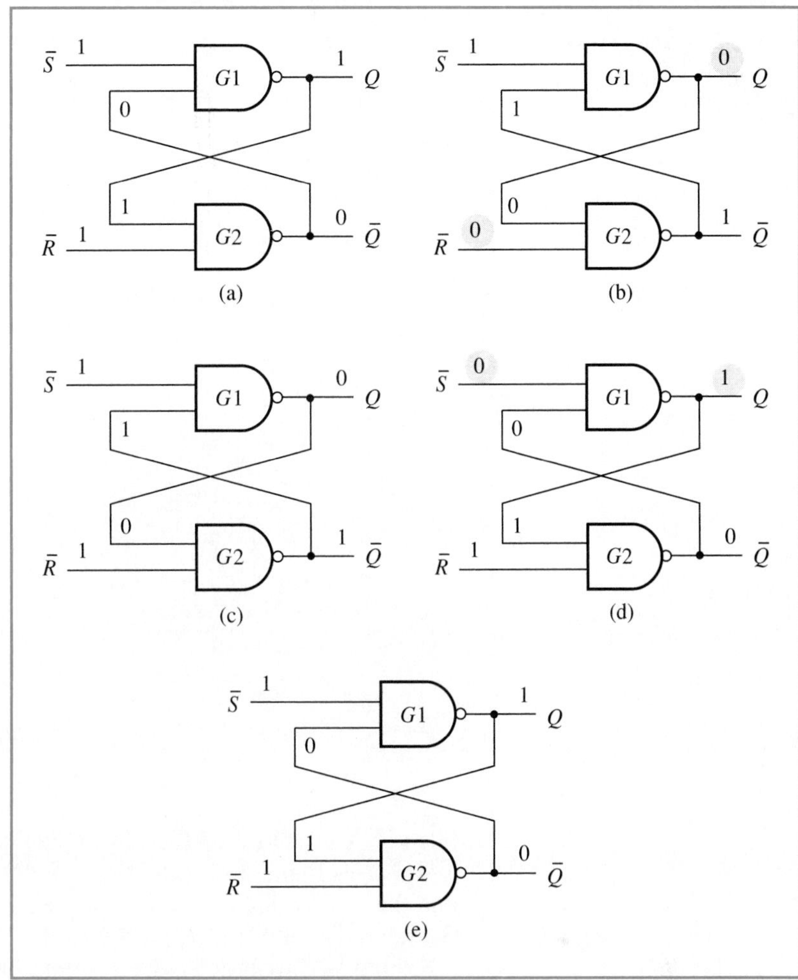

shown in Figure 5–6c. During interval 4 the $\overline{S}$ input changes to a logic 0 state. This causes the output of *G*1 to change to a logic 1 and the output of *G*2 to a logic 0. The latch is now *SET* as shown in Figure 5–6d. During interval 5 the $\overline{S}$ input returns to the logic 1 state, but the output of *G*1 does not change since its other input is held low by $\overline{Q}$. The latch therefore remains *SET* as shown in Figure 5–6e.

The *S-R* latches discussed so far could be set or reset by activating the *S* and *R* inputs, respectively. Activating either input caused the latch to change states immediately. A *gated S-R* latch has another input called the *gate* that determines *when* the latch will change states. The logic symbol and circuit for the gated *S-R* latch is shown in Figure 5–7. The *S* and *R* inputs function as described earlier and are used to *SET* and *RESET* the latch. However, the latch will not change states until a logic 1 is applied to the gate input *G*. The gate *G* effectively blocks (0) or passes (1) the *S* and *R* inputs by means of the two *NAND* gates *G*1 and *G*2 as shown in Figure 5–7b. If $G=0$ then the outputs of *G*1 and *G*2 are 1's, regardless of the states of *S* and *R* (dominance law). Since *G*3 and *G*4 make up an active-low *NAND* latch, the logic 1's on the inputs of *G*3 and *G*4 make the latch inactive, and therefore there is no change in state. If $G=1$, the outputs of *G*1 and *G*2 are equal to $\overline{S}$ and $\overline{R}$, respectively

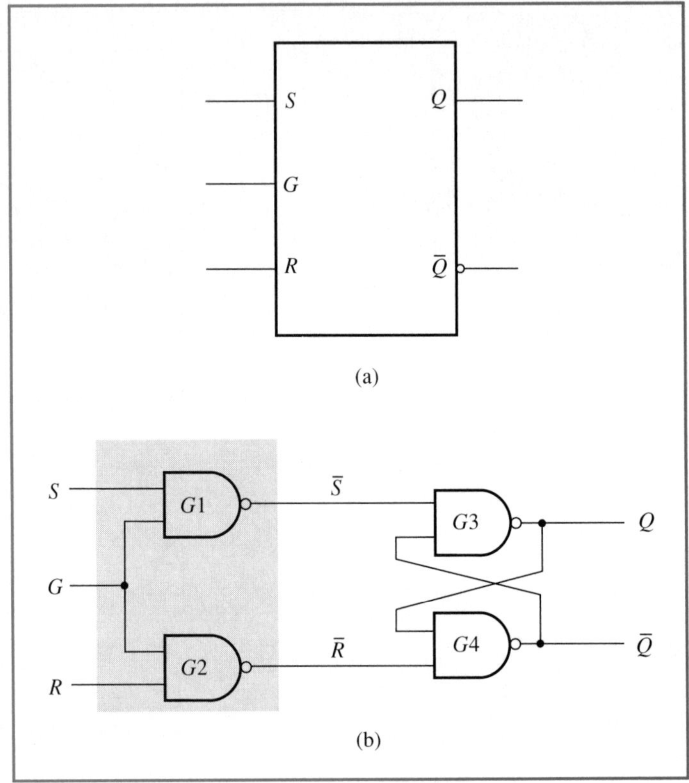

Figure 5–7
The gated *S-R* (*NAND*) latch. (a) Logic symbol; (b) logic circuit.

(identity law) and thus the *NAND* latch changes states as dictated by the logic levels applied to *S* and *R*. The *G* input thus functions like an *enable* line that either allows *S* and *R* to control the state of the latch or disables the influence of *S* and *R* on the latch. The timing diagram for the gated *S-R* latch is shown in Figure 5–8 and has nine time intervals identified for the following analysis.

1. The latch is initially *RESET* but will not change states even though *S* changes to a logic 1 during this interval since *G* is a logic 0.
2. *G* is now a logic 1 and the latch inputs *S* and *R* are enabled. Since *S* is a logic 1 during this interval, the latch is now *SET*.
3. *G* is now a logic 0 and the latch inputs are disabled. However, since the previous state is latched, there is no change and the latch remains *SET*. *S* and *R* change during this interval but it has no effect on the state of the latch since $G = 0$.
4. *G* is now a logic 1 while *R* is active thus causing the latch to *RESET*.
5. *G* is now a logic 0 and therefore any change in *R* and *S* do not affect the state of the latch. The latch remains *RESET*.
6. *G* is now a logic 1 but *S* and *R* are inactive (no-change command) and therefore the latch remains *RESET*.
7. *S* is now active while $G = 1$ which causes the latch to *SET*.
8. *S* and *R* are inactive (no change) while $G = 1$, and therefore the latch remains *SET*.
9. *R* is now active while $G = 1$ which causes the latch to *RESET* and remain at that state.

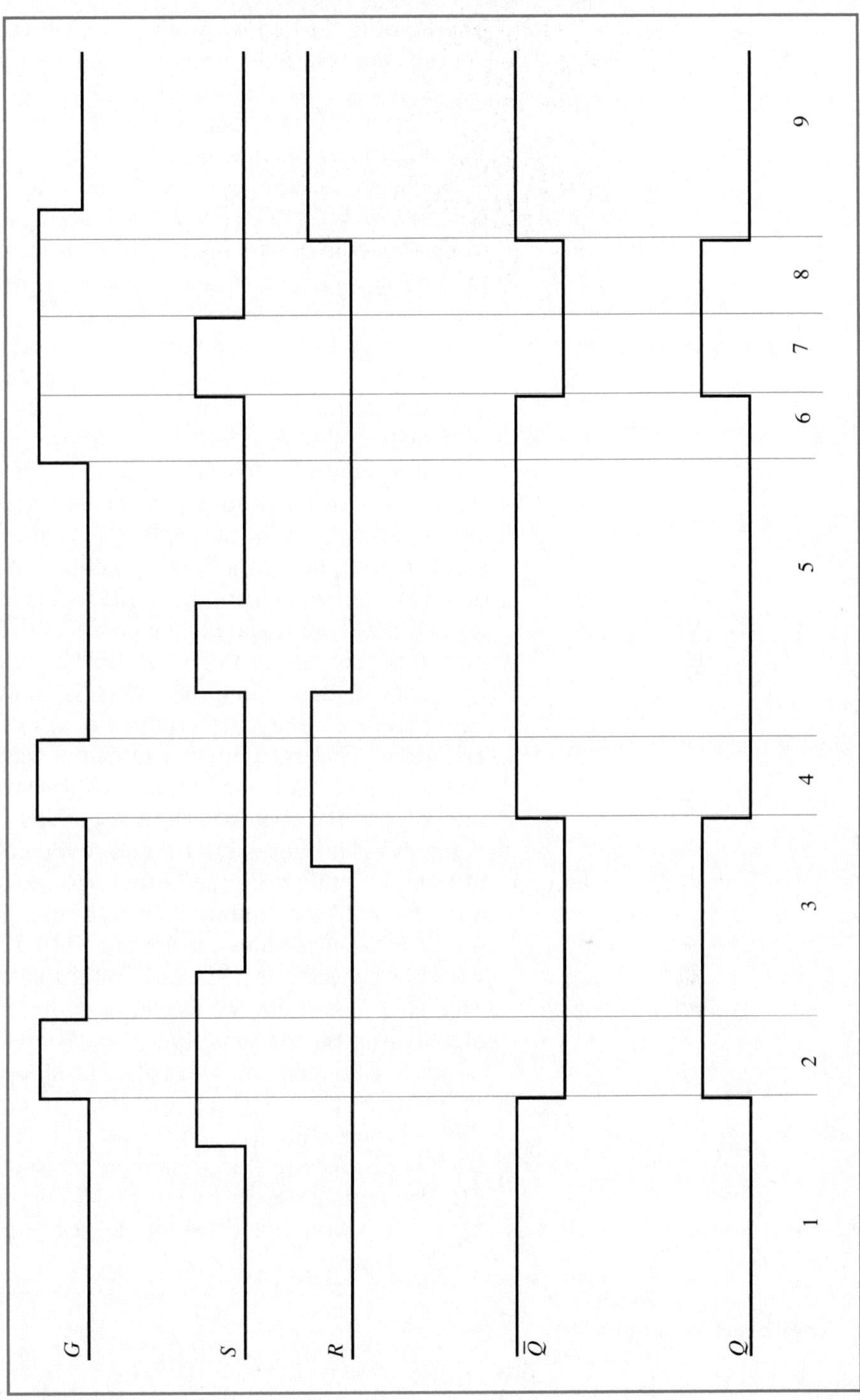

Figure 5–8
Timing diagram for the gated *S-R* latch.

Therefore in a gated latch, the S and R inputs function as command or control inputs and determine if the latch will be *SET* or *RESET*, respectively, *only when the gate is enabled* (*1*).

The *S-R NOR* latch of Figure 5–1b can also be converted to a gated *S-R NOR* latch as shown in Figure 5–9. Again, the two *AND* gates will either pass the logic states of R and S to the *NOR* latch if $G=1$ (identity law) or block the logic states of R and S to the *NOR* latch if $G=0$ (dominance law). The operation of the circuit in Figure 5–9 is identical to the operation of the circuit in Figure 5–7b.

An application of a latch

One of the most common applications of an ungated latch is in a circuit known as a *debounce circuit*. A debounce circuit is a circuit that is used to eliminate the effects of *contact bounce* which occurs when the contacts of a switch open or close. For example, consider the simple switch circuit shown in Figure 5–10a. The switch in the circuit is a momentary pushbutton switch whose contacts are normally open. The output of the circuit is therefore at the high state (logic 1) when the switch is open since the output is at a 5-volt potential. When the pushbutton switch is depressed momentarily, the contacts close, the output of the circuit is at ground potential (logic 0), the contacts then open and the output of the switch is back to the logic 1 state. The voltage pulse generated by momentarily depressing the switch is shown in Figure 5–10b. The circuit can therefore be used to produce a *pulse*. Figure 5–10b indicates an ideal condition. That is, the pulse shown is what we would like the circuit to produce. However, the contacts of the switch are mechanical and when the poles of the switch strike the contact, the contact actually vibrates or bounces. This causes intermittent contact to be made for a short duration of time, and therefore the output voltage oscillates between a logic 0 and a logic 1 until the contact comes to rest. The effect of this contact bounce produces a pulse shown in Figure 5–10c. It is apparent that the circuit does not generate the "clean" pulse that we expect. This actual pulse produced by such a circuit can often be unacceptable to many types of digital circuits, and therefore a solution is needed to eliminate the effect of contact bounce and attempt to produce a pulse that is ideal. The circuit in Figure 5–11a eliminates the effect of contact bounce and produces a "clean" pulse like the one shown in Figure 5–10b. An active-low *NAND* latch is used along with a momentary single-pole, double-throw switch. The switch is normally at position 1. When the switch is depressed, it goes to position 2, and when released goes back to position 1. Since the

Figure 5–9
Logic circuit for the gated *S-R* (*NOR*) latch.

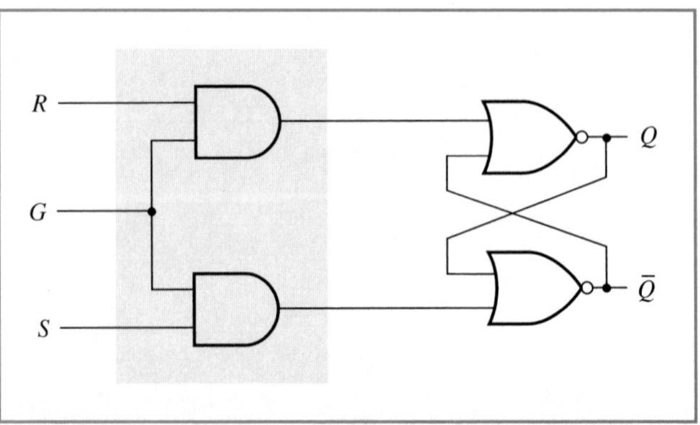

Figure 5–10
The pulse generated by a switch. (a) Switch circuit; (b) ideal output assuming no contact bounce; (c) actual output with contact bounce.

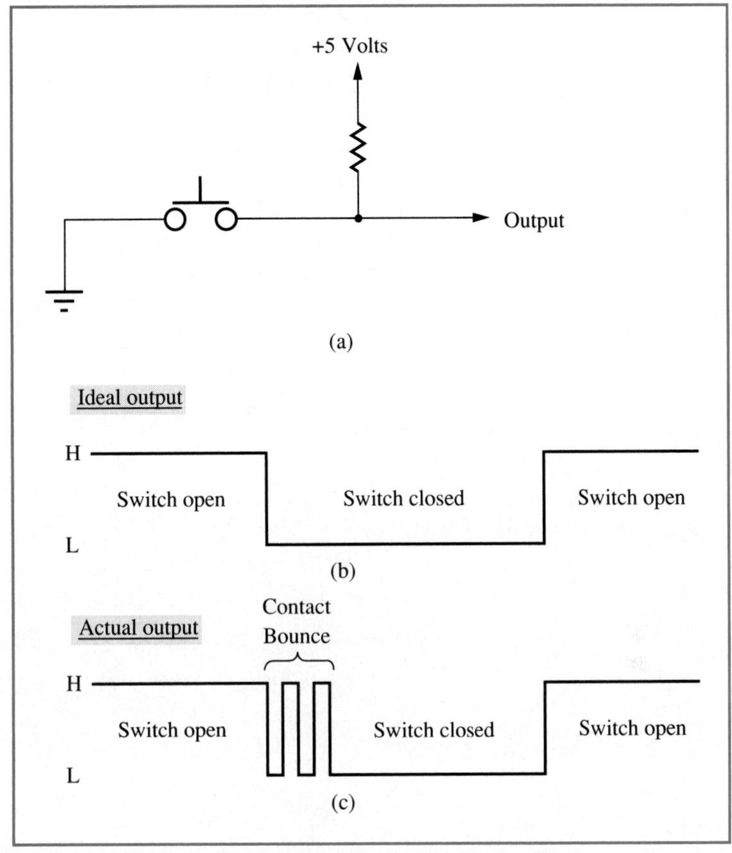

Figure 5–11
(a) The debounce circuit; (b) output waveform.

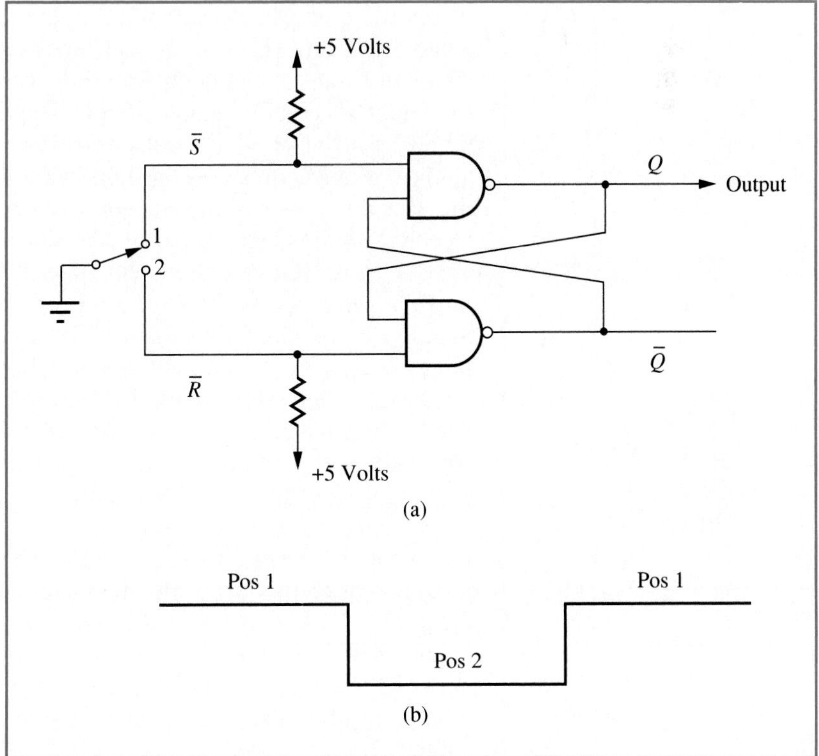

switch is at position 1, the $\overline{S}$ input to the latch has a ground potential (logic 0) applied to it while the $\overline{R}$ input is at the 5-volt level (logic 1). This causes the output of the latch to be *SET* as shown in Figure 5–11b. When the switch is depressed, the contact moves to position 2 and applies a logic 0 on the $\overline{R}$ input; the $\overline{S}$ input is automatically pulled up to 5 volts. The latch now *RESETS*. Now if there is any contact bounce, it will produce oscillations on the $\overline{R}$ input of the latch and continuously try to reset the latch. This will have no effect on the latch since the latch is already reset. When the contact returns to position 1, the latch will be set but again there will be contact bounce that will produce oscillations on the $\overline{S}$ input of the latch, attempting to repetitively set it. But since the latch is already set, it will remain at that state. The resulting pulse produced and shown in Figure 5–11b is an exact reproduction of the ideal pulse.

Review questions

1. Define the terms *SET, RESET,* and *CLEAR*.
2. Describe the operation of a latch.
3. What is the purpose of a timing diagram?
4. Describe the operation of a gated latch.

5–3 The flip-flop

A *flip-flop* is a sequential logic circuit that functions like a gated latch. The only difference between a flip-flop and a latch is in the manner in which the control or command inputs (such as *S* and *R*) are enabled. In a gated *S-R* latch, the gate input *G* enables the *S* and *R* control inputs to affect the state of the latch. In other words, it is the gate (*G*) that causes the latch to actually change states. This is done by applying a logic *level* (logic 1 for the previously discussed circuits) at the *G* input to cause the latch to change states. A flip-flop has a *clock* input *C* which functions like the gate input to a latch but requires a *transition* in order for the flip-flop to change states as dictated by the settings of the control inputs.

A *transition* is a change in state from one logic level to another. A *positive transition* is a transition from a logic 0 to a logic 1 while a *negative transition* is a transition from a logic 1 to a logic 0. Examples of negative and positive transitions are shown in Figure 5–12. The waveforms shown in Figure 5–12 are known as *clock pulses* and are often used to *trigger* or change the state of a flip-flop. Figure 5–12a illustrates the positive transitions (identified by the arrows) of two types of clock pulses. Notice that since the positive edge of the pulses appears to rise, these edges are also referred to as *rising edges*. Figure 5–12b illustrates the negative transitions (identified by the arrows) of two types of clock pulses. Notice that since the negative edge of the pulses appears to fall, these edges are also referred to as *falling edges*. Flip-flops are designed to change states on either transition. A flip-flop that changes states on a negative transition of the clock is called a *negative edge-triggered flip-flop* while a flip-flop that changes states on a positive transition of the clock is called a *positive edge-triggered flip-flop*. The logic circuits for flip-flops

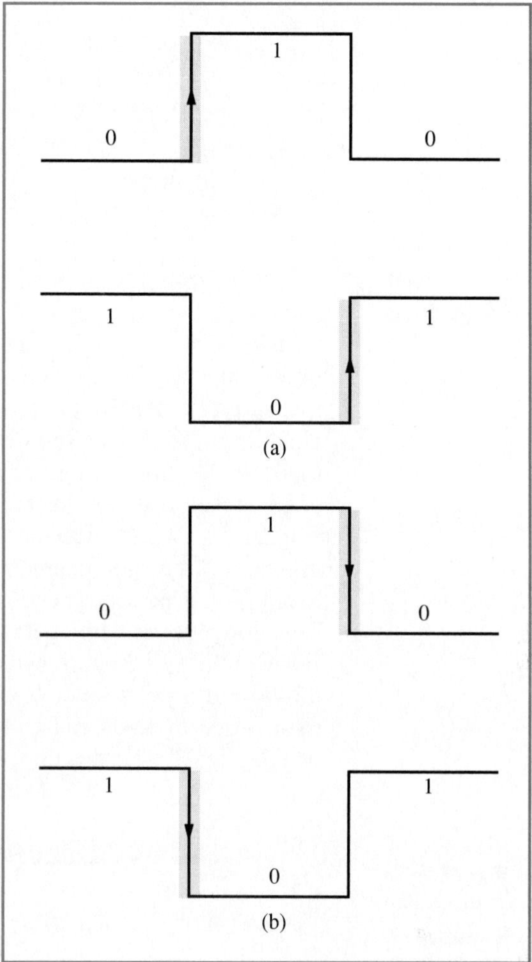

Figure 5–12
Clock transitions used to trigger a flip-flop. (a) Positive edge; (b) negative edge.

are similar to the logic circuits for gated latches except that a circuit to detect a transition on the clock input is added to the circuit. Because of the difference in the manner in which flip-flops and latches change states, the timing diagrams *may* be quite different. For example, in Figure 5–8 if the *G* input of the *S-R* latch was positive edge-triggered, the state of the latch during intervals 6, 7, 8, and 9 would be *RESET*, since the positive transition of *G* occurs during the beginning of interval 6 when *S* and *R* are logic 0's (no change) and the current state of the latch is *RESET*.

Flip-flops are the basic logic elements of sequential logic circuits, and therefore the remainder of this chapter is dedicated to the study of the various types of flip-flops available and their operational characteristics. Four different types of flip-flops will be introduced—the *S-R* flip-flop, the *T* flip-flop, the *D* flip-flop, and the *J-K* flip-flop. The *S-R* and *T* flip-flops are conceptual flip-flops. That is, there are no commercial implementations (in the form of 74xxx ICs) for these flip-flops, but they merely serve as models to illustrate the basic concepts of flip-flop operation. The *D* and *J-K* flip-flops are available commercially and can emulate the functions of the *S-R* and *T* flip-flops as will be seen in Sections 5–6 and 5–7. The specific applications of flip-flops in various types of sequential circuits will be examined in Chapters 6 and 7.

Review questions

1. How does a flip-flop differ from a latch?
2. What does the term *trigger* mean with reference to the operation of flip-flops?
3. What are the different ways in which a flip-flop can be triggered?

5–4 The *S-R* flip-flop

The *S-R* flip-flop functions like the *S-R* latch. The *S* and *R* control inputs *command* the latch to *SET* or *RESET* when the clock *C* makes a transition. Figure 5–13a and b shows the logic symbols for the two types of *S-R* flip flops—positive edge-triggered and negative edge-triggered, respectively. The *dynamic input indicator* shown as a "triangle" at the clock input identifies the device as being edge-triggered. The bubble at the input of the clock (Figure 5–13b) indicates that the flip-flop is negative edge-triggered while no bubble indicates a positive edge-triggered flip-flop (Figure 5–13a). Notice that the bubble symbol at the clock input identifies inversion, and this means that a negative edge-triggered flip-flop is essentially a positive edge-triggered flip-flop with its clock pulse inverted. This concept can thus be used to convert one type of flip-flop to another. For example, in Figure 5–14 we have converted a negative edge-triggered flip-flop to a positive edge-triggered flip-flop by inverting the clock. The rising edge of clock pulse *A* is converted to a falling edge in clock pulse *B* which triggers the flip-flop. Similarly, in Figure 5–15 a positive edge-

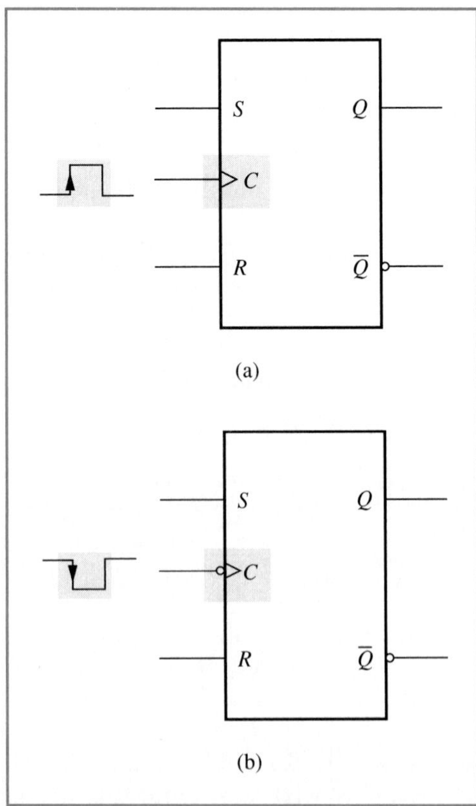

Figure 5–13 Logic symbols for an *S-R* flip-flop. (a) Positive edge-triggered; (b) negative edge-triggered.

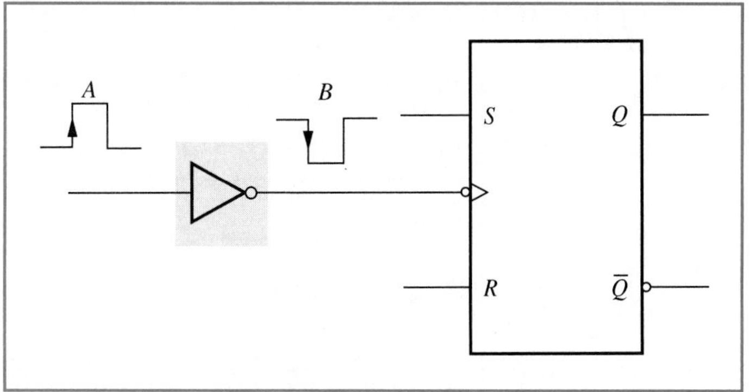

Figure 5–14
Converting a negative edge-triggered flip-flop to a positive edge-triggered flip-flop.

triggered flip-flop is converted to a negative edge-triggered flip-flop by converting the falling edge of input pulse *A* to a rising edge in pulse *B* that triggers the flip-flop.

Figure 5–16 shows the timing diagram for a positive edge-triggered *S-R* flip-flop. The clock input is fed by a continuous train of periodic pulses, and each positive transition of a clock pulse causes the flip-flop to change states. The state of *R* and *S* at the instant the positive transition of the clock takes place dictates the state of the flip-flop. Notice in interval 3 that the flip-flop remains *RESET* because *S* changed from a 0 to a 1 *after* the positive transition of the second clock pulse took place. During interval 4 the flip-flop is *SET*, since *S* is a logic 1 when the third clock pulse makes a positive transition.

To illustrate the effects of changing the edge at which a flip-flop triggers, we have analyzed the timing diagram of Figure 5–16 for a negative edge-triggered *S-R* flip-flop. The revised timing diagram is shown in Figure 5–17.

Notice in Figure 5–17 that each negative transition of the clock pulse now causes the flip-flop to change states. The state of *R* and *S* at the instant the negative transition of the clock takes place dictates the state of the flip-flop. Notice in interval 3 that this time the flip-flop is *SET* because *S* changed from a 0 to a 1 *before* the negative transition of the second clock pulse took place. During interval 4 the flip-flop is still *SET* since *S* is a logic 0 (and *R* is also a logic 0) when the third clock pulse makes a negative transition. The resulting output *Q* in Figure 5–17 is

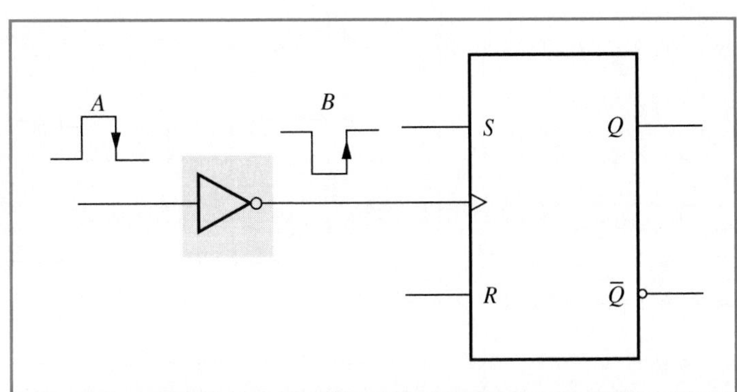

Figure 5–15
Converting a positive edge-triggered flip-flop to a negative edge-triggered flip-flop.

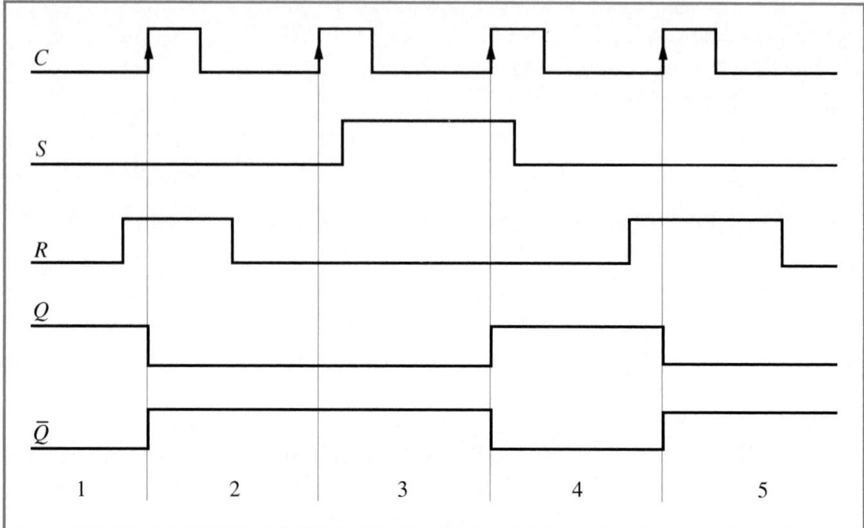

Figure 5–16
Timing diagram for a positive edge-triggered flip-flop.

considerably different from the output Q in Figure 5–16, even though the same clock and inputs to S and R were used.

Just as truth tables were used to describe the operation of the basic gates and combinational logic circuits, *state transition tables* are used to describe the operation of flip-flops and sequential logic circuits. A state transition table simply identifies the *next state* of a flip-flop given the *present* conditions. The next state of a flip-flop is the state of the flip-flop *after* the clock pulse makes a transition and triggers the flip-flop. Table 5–1 shows the *summarized* state transition table for an S–R flip-flop. The next state of the S-R flip-flop in Table 5–1 is identified by Q^{n+1}.

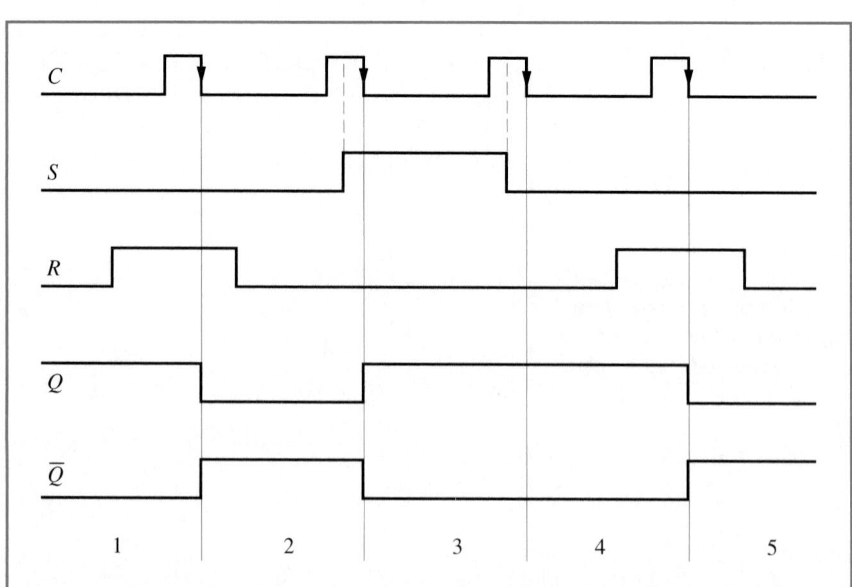

Figure 5–17
Timing diagram for a negative edge-triggered flip-flop.

Table 5–1
Summarized state transition table for an *S-R* flip-flop

S	R	Q^{n+1}	
0	0	Q^n	No change
0	1	0	*RESET*
1	0	1	*SET*
1	1	x	Invalid

Notice that the letter Q represents the output of the flip-flop, and the superscript $n+1$ indicates that it is the output of the flip-flop *after* the clock pulse triggers it. A *complete* state transition table for a flip-flop is often more appropriate for describing its operation. The complete state transition table, or simply the state transition table, shows the next state of the flip-flop Q^{n+1} as a function of S, R, and the *present state*. The present state of the flip-flop, Q^n, is the state of the flip-flop *before* the clock pulse triggers. The state transition table for the *S-R* flip-flop is shown in Table 5–2. Like a truth table, the state transition table of Table 5–2 lists all possible combinations of S, R, and Q^n. The value of Q^{n+1} can then be determined for each combination from the summarized state transition table. For example, when $S=0$ and $R=0$ there is no change in states when the clock pulse triggers the flip-flop. In other words, the next state of the flip-flop, Q^{n+1}, is equal to the present state, Q^n. When $S=0$ and $R=1$, the next state of the flip-flop is a logic 0 (*RESET*) regardless of the present state. Similarly when $S=1$ and $R=0$, the next state of the flip-flop is a logic 1 (*SET*) regardless of the present state. The *S-R* flip-flop does not support the condition when $S=1$ and $R=1$, and therefore we don't care what the next state of the flip-flop is under these conditions since they should never be used for an *S-R* flip-flop.

From Table 5–2 we can obtain a K-map for the output variable Q^{n+1} as shown in Figure 5–18 and the following logic equation:

Equation (5–1)

$$Q^{n+1} = S + \overline{R}Q^n$$

Equation 5–1 is called the *state transition equation* for the *S-R* flip-flop and its application in the analysis of sequential logic circuits will be examined in the next chapter. This equation can be used to determine the next state of the *S-R* flip-flop given the logic levels of S, R, and its present state, Q^n.

Table 5–2
State transition table for an *S-R* flip-flop

S	R	Q^n	Q^{n+1}	
0	0	0	0	No change
0	0	1	1	
0	1	0	0	Reset
0	1	1	0	
1	0	0	1	Set
1	0	1	1	
1	1	0	x	Invalid
1	1	1	x	

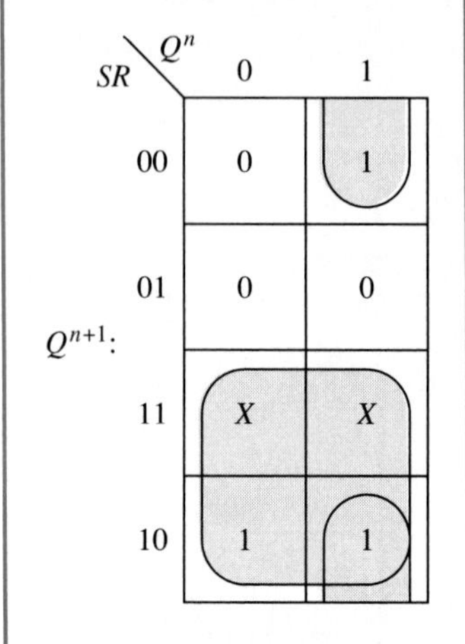

Figure 5–18
K-map for the *S-R* flip-flop.

Review questions

1. What does the dynamic input indicator identify?
2. What symbol identifies the flip-flop as being negative edge-triggered?
3. Describe the operation of the *S-R* flip-flop.
4. Why is $R=1$, $S=1$ an invalid condition for the *S-R* flip-flop?
5. What is the purpose of a state transition table?

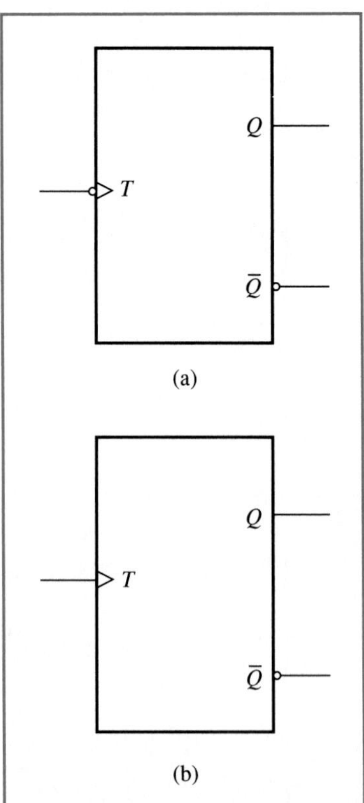

Figure 5–19
Logic symbols for a *T* flip-flop. (a) Negative edge-triggered; (b) positive edge-triggered.

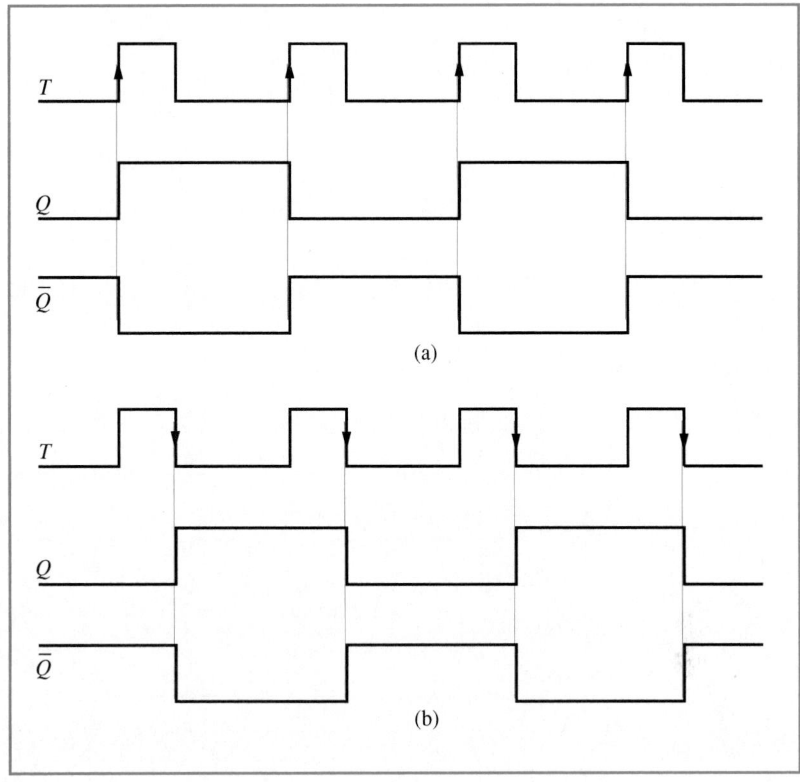

Figure 5–20
Timing diagrams for a *T* flip-flop. (a) Positive edge-triggered; (b) negative edge-triggered.

5–5 The *T* flip-flop

The *T* or *toggle* flip-flop has no control inputs like the *S-R* flip-flop but only contains a clock input labeled *T* as shown in Figure 5–19. Every time the clock pulse triggers the flip-flop it *toggles*. That is, it changes to the complement of the present state. For example, if the present state is a logic 1, when the clock pulse makes a transition the next state will be a logic 0, and if the present state is a logic 0 the next state will be a logic 1 when the clock pulse makes a transition.

Figure 5–20a and b show the timing diagrams for positive edge-triggered and negative edge-triggered *T* flip-flops, respectively. Notice that the outputs of both types of flip-flops are identical. Also notice that the *T* flip-flop effectively divides the input clock frequency at the *T* input by 2. The following example illustrates how several *T* flip-flops can be connected in series to reduce the frequency of a clock pulse. More examples of frequency division will be seen in Chapter 6.

EXAMPLE 5–1

Construct a timing diagram for the circuit shown in Figure 5–21 showing the outputs Q_A, Q_B, and Q_C as a function of nine *CLOCK* pulses. Also determine the frequency of the pulses at each of the outputs if the clock frequency is 800 Hz.

SOLUTION The timing diagram for Figure 5–21 is shown in Figure 5–22, assuming that Q_A, Q_B, and Q_C start off at the logic 0 states. Notice in the timing diagram of Figure 5–22 that Q_A divides the *CLOCK* frequency (800 Hz) by 2 to produce a frequency of 400 Hz.

Figure 5–21
Frequency division using T flip-flops.

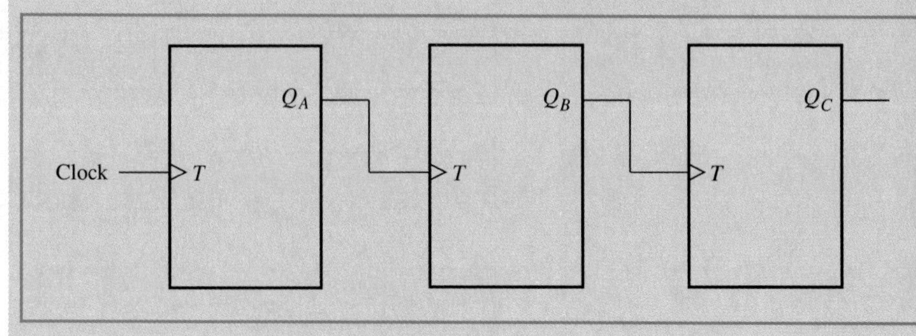

Figure 5–22
Timing diagram for Figure 5–21.

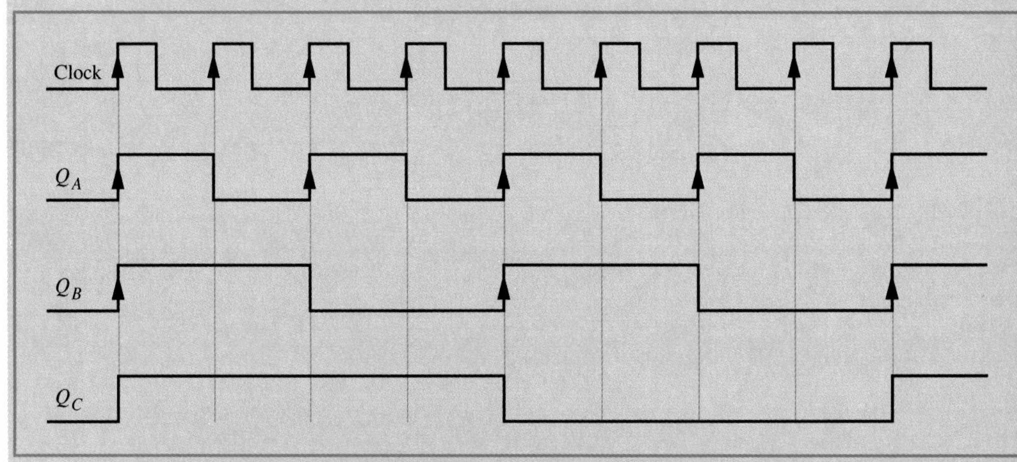

Q_B divides Q_A by 2 to produce a frequency of 200 Hz, and Q_C divides Q_B by 2 to produce a frequency to 100 Hz. Therefore

$$Q_A = CLOCK \div 2$$
$$Q_B = CLOCK \div 4$$
$$Q_C = CLOCK \div 8$$

The state transition table for the T flip-flop is shown in Table 5–3. Since there are no control inputs, the only input variable in the table is the present state Q^n. We can now obtain the state transition equation for the T flip-flop directly from Table 5–3

Equation (5–2)
$$Q^{n+1} = \overline{Q}^n$$

Equation 5–2 simply states that the next state of a T flip-flop is equal to the complement of the present state—a toggle condition.

Review questions

1. Describe the operation of the T flip-flop.
2. How does the operation of the T flip-flop compare with the operation of the S-R flip-flop?
3. What does the word *toggle* mean?
4. How does the T flip-flop accomplish frequency division?

Table 5–3
State transition table for a *T* flip-flop

Q^n	Q^{n+1}
0	1
1	0

5–6 The *D* flip-flop

The *D* or *data*-type flip-flop has a single control input *D* and a positive edge-triggered or negative-edge triggered clock input as shown in Figure 5–23. The logic level at the *D* input determines the next state of the flip-flop. For example, if *D* is a logic 1 before the clock pulse makes a transition the next state of the flip-flop will be a logic 1, and if *D* is a logic 0 before the clock pulse makes a transition, the next state of the flip-flop will be a logic 0. The *D* flip-flop can therefore be used to *store* 1 bit of data, and hence it is known as a *D* or data flip-flop.

Figure 5–24a and b show the timing diagram for positive and negative edge-triggered *D* flip-flops, respectively. Notice that the outputs of the two flip-flops are different even though *D* and *C* are the same in both timing diagrams. This is because in Figure 5–24a *D* is a logic 1 when the third clock pulse makes a positive transition and in Figure 5–24b *D* is a logic 0 when the third clock pulse makes a negative transition. Thus, the outputs of negative and positive edge-triggered *D* flip-flops will be

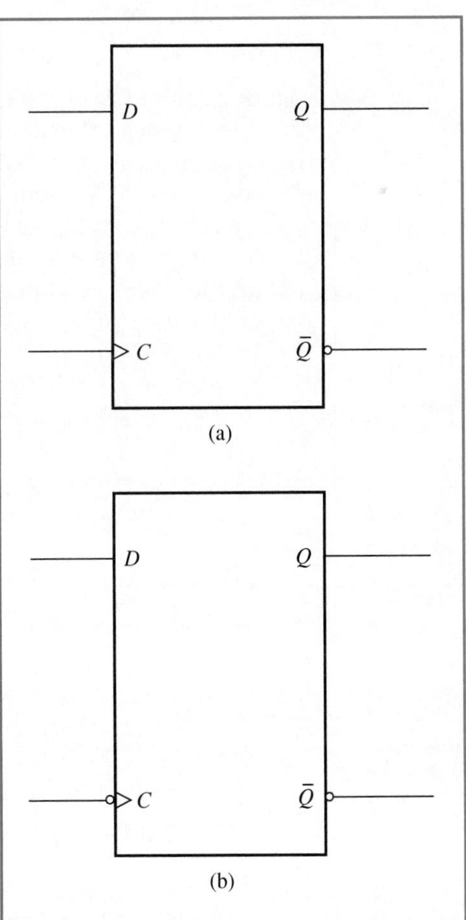

Figure 5–23
Logic symbols for a *D* flip-flop. (a) Positive edge-triggered; (b) negative edge-triggered.

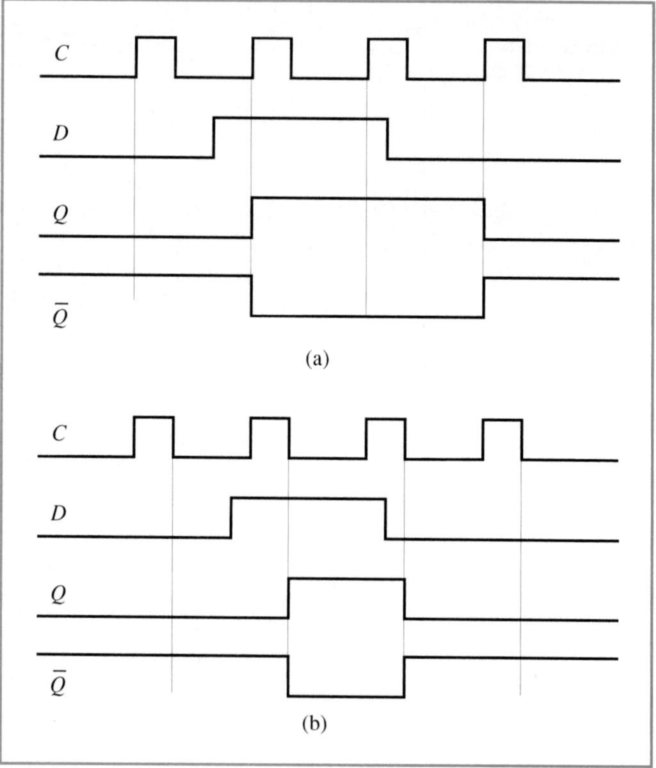

Figure 5–24
Timing diagrams for a *D* flip-flop. (a) Positive edge-triggered; (b) negative edge-triggered.

different only if D changes between the positive and negative transitions of a clock pulse. The summarized state transition table for the D flip-flop is shown in Table 5–4. Notice that the next state of the flip-flop is not dependent on the present state but is only dependent on the value of D.

Even though the summarized state transition table shown in Table 5–4 is adequate to describe the operation of the D flip-flop, we can also obtain the complete state transition table as shown in Table 5–5. We can now obtain the state transition equation for the D flip-flop directly from Table 5–4 or Table 5–5

Equation (5–3)

$$Q^{n+1} = D$$

Table 5–4
Summarized state transition table for a *D* flip-flop

D	Q^{n+1}
0	0
1	1

Table 5–5
State transition table for a *D* flip-flop

D	Q^n	Q^{n+1}
0	0	0
0	1	0
1	0	1
1	1	1

Equation 5–3 simply states that the next state of the D flip-flop will be equal to the present state of D.

When the D flip-flop is used in the synthesis or design of sequential logic circuits (see Chapter 6), it is often more convenient to view the state transition table from a different perspective. The state transition table in Table 5–5 tells us what the next state of the D flip-flop would be, given its present state and the value of D. From Table 5–5 we can also determine what D has to be in order for the flip-flop to change from one state to another. This information is shown in Table 5–6. Table 5–6 lists all possible transitions that a flip-flop can make in going from one state to another. These transitions are listed in the first two columns Q^n and Q^{n+1} The third column lists the values of D required for the flip-flop to make the transition. For example, in order for the output of a D flip-flop to change from a 0 to a 1, D must be a 1. Similarly, in order for a D flip-flop to change from a 1 to a 0, D must be a 0. For the D flip-flop to "change" from a 0 to a 0 (no change), D must be a 0. Similarly, in order for the D flip-flop to "change" from a 1 to a 1 (no change), D must be a 1. Notice that for two of the entries in the table, even though no change (or transition) actually takes place, we still consider those two conditions to be transitions.

Integrated circuit logic

The 7474 dual D flip-flop contains two independent positive edge-triggered D-type flip-flops in one IC package. The logic symbol and pin configuration for the IC is shown in Figure 5–25. Notice that each flip-flop has the standard D and C (labeled CP) inputs for the data and clock, respectively, and the standard Q and $\overline{Q}$ outputs. However, notice that each flip-flop also has two additional inputs labeled $\overline{S}_D$ (*SET*) and $\overline{C}_D$ (*CLEAR*). The D input to a D flip-flop is known as a *synchronous* input since it is synchronized with the transition of the clock pulse. This means that if the logic level of D changes, the flip-flop will not change states unless the clock pulse triggers the flip-flop. The $\overline{S}_D$ and $\overline{C}_D$ inputs to the 7474 D flip-flops are known as *asynchronous* inputs since they are used to *directly SET* or *CLEAR* (*RESET*) the output of the flip-flop and do not depend on a clock pulse transition. Therefore their effect is not synchronized with the clock—it is, asynchronous. Notice that both $\overline{S}_D$ and $\overline{C}_D$ are active-low inputs and therefore a logic 0 on $\overline{S}_D$ will *SET* the flip-flop and a logic 0 on $\overline{C}_D$ will *CLEAR* the flip-flop regardless of the states of D or CP. Obviously both asynchronous inputs cannot be connected to a logic 0 at the same time. If these inputs are not used in the D flip-flops they must be connected to the inactive logic 1 state. Since the $\overline{C}_D$ and $\overline{S}_D$ inputs to the flip-flop can directly *CLEAR* or *SET* the flip-flop respectively, they are often referred to as the *direct set* and *direct clear* inputs.

Table 5–6
Modified state transition table for a *D* flip-flop

Q^n	Q^{n+1}	D
0	0	0
0	1	1
1	0	0
1	1	1

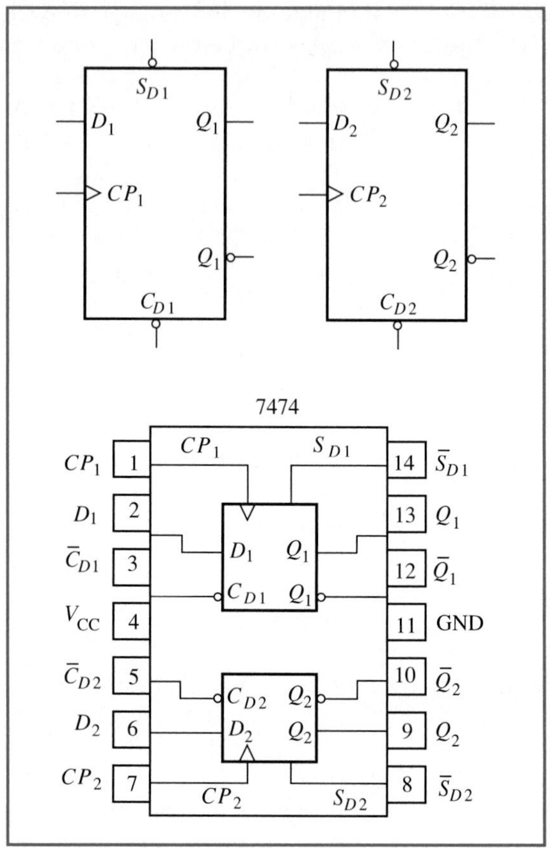

Figure 5–25 Logic symbol and pin configuration of the 7474 dual D-type positive edge-triggered flip-flop.

Review questions

1. Describe the operation of the *D* flip-flop.
2. Compare the *SET* and *RESET* operations of the *D* flip-flop with the *SET* and *RESET* operations of the *S-R* flip-flop.
3. What are asynchronous inputs? How do they differ from synchronous inputs?

5–7 The *J-K* flip-flop

The *J-K* flip-flop emulates the operation of the *S-R* flip-flop. The *J* input corresponds to *SET* and the *K* input corresponds to *RESET*. There is no significance to the letters *J* and *K* used to identify the control inputs of the flip-flop. There is one difference between the *J-K* and *S-R* flip-flops. In an *S-R* flip-flop, if $S=1$ and $R=1$, the condition is defined as invalid and the next state of the flip-flop is undefined. However, in a *J-K* flip-flop, if $J=1$ and $K=1$, the next state of the flip-flop is equal to the complement of its present state; in other words, the flip-flop toggles. The logic symbols for a negative edge-triggered and a positive edge-triggered *J-K* flip-flop are shown in Figure 5–26a and b, respectively. Figure 5–27 shows the timing diagram for a positive edge-triggered *J-K* flip-flop. During the first transition of the clock pulse, $J=1$ and $K=0$ and therefore the output *Q* of the flip-flop is *SET* during interval 1. At the second transition of the clock pulse $J=1$ and $K=1$ and the present output of the flip-flop is a logic 1, therefore the flip-flop toggles and the next state of the

Figure 5–26
Logic symbols for a *J-K* flip-flop. (a) Negative edge-triggered; (b) positive edge-triggered.

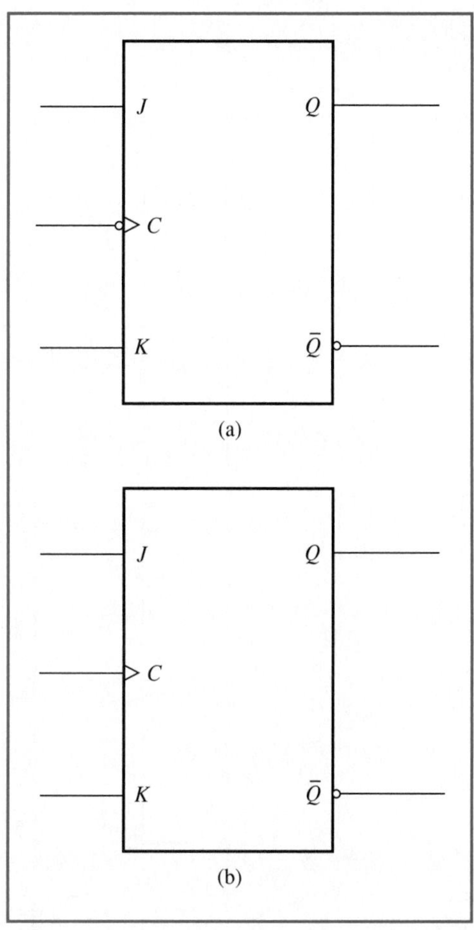

flip-flop is *RESET* (logic 0) during interval 2. At the third transition of the clock pulse, $J=0$ and $K=1$, which ‘‘tells’’ the flip-flop to *RESET;* however, the flip-flop is already *RESET,* so there is no change during interval 3. At the fourth transition of the clock, $J=0$ and $K=0$, indicating a no-change condition, and therefore the flip-flop remains *RESET* during interval 4. At the fifth transition of the clock $J=1$ and $K=0$ and the flip-flop is *SET* during interval 5. At the sixth transition of the clock $J=1$ and $K=1$ (toggle) and therefore the next state of the flip-flop is the complement of the present state; the flip-flop is *RESET* during interval 6.

The summarized state transition table for the *J-K* flip-flop is shown in Table 5–7. Notice that the next state of the *J-K* flip-flop is dependent on *J, K,* and the present state Q^n. To thoroughly describe the operation of the *J-K* flip-flop and obtain its state transition equation we must construct the complete state transition table as shown in Table 5–8. Using Table 5–8, the state transition equation for the *J-K* flip-flop can be obtained from the K-map shown in Figure 5–28

Equation (5–4)

$$Q^{n+1} = J\overline{Q}^n + \overline{K}Q^n$$

Notice that Equation 5–4 indicates that the next state of the *J-K* flip-flop depends on *J, K,* and the present state Q^n of the flip-flop.

It was stated in Section 5–6 that it is sometimes convenient to reorganize the state transition table to allow us to determine what the

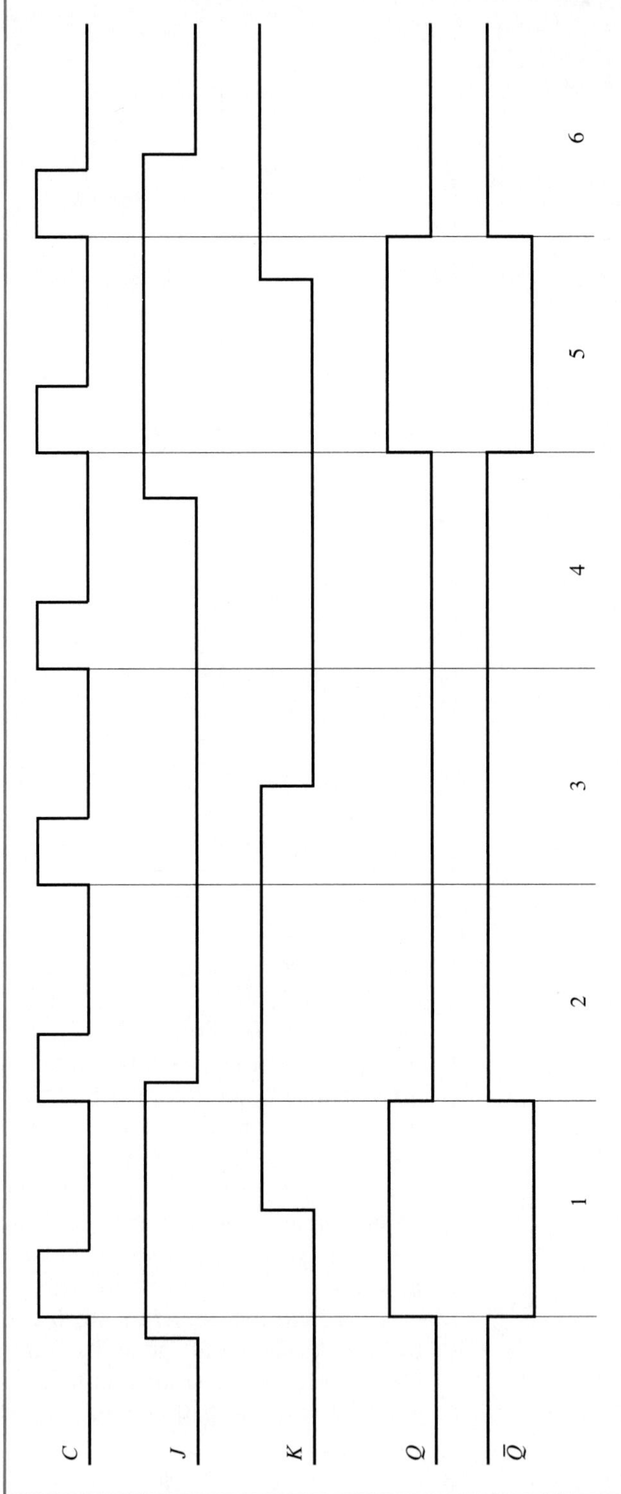

Figure 5–27
Timing diagram for a positive edge-triggered *J-K* flip-flop.

Table 5–7
Summarized state transition table for a J-K flip-flop

J	K	Q^{n+1}	
0	0	Q^n	No change
0	1	0	*RESET*
1	0	1	*SET*
1	1	$\bar{Q}^n$	Toggle

Table 5–8
State transition table for a J-K flip-flop

	J	K	Q^n	Q^{n+1}	
0	0	0	0	0	No change
1	0	0	1	1	
2	0	1	0	0	*RESET*
3	0	1	1	0	
4	1	0	0	1	*SET*
5	1	0	1	1	
6	1	1	0	1	Toggle
7	1	1	1	0	

control input(s) should be for a particular transition to take place at the output of a flip-flop. Notice that we have numbered the eight entries for reference purposes in Table 5–8 (0 through 7) and that entries 3 and 7 show the output of the *J-K* flip-flop making a transition from a 1 (Q^n) to a 0 (Q^{n+1}). For these two entries K must be a logic 1 but J can be either a 0 or a 1. In other words, for the output of a *J-K* flip-flop to change from a 1 to a 0, K must be a 1 and we don't care what J is

Q^n		Q^{n+1}	J	K
1	$\longrightarrow$	0	x	1

Similarly, with reference to entries 4 and 6, the *J-K* flip-flop changes from a 0 (Q^n) to a 1 (Q^{n+1}). For these two entries J must be a logic 1 but K can be either a 0 or a 1. In other words, for the output of a *J-K* flip-flop to

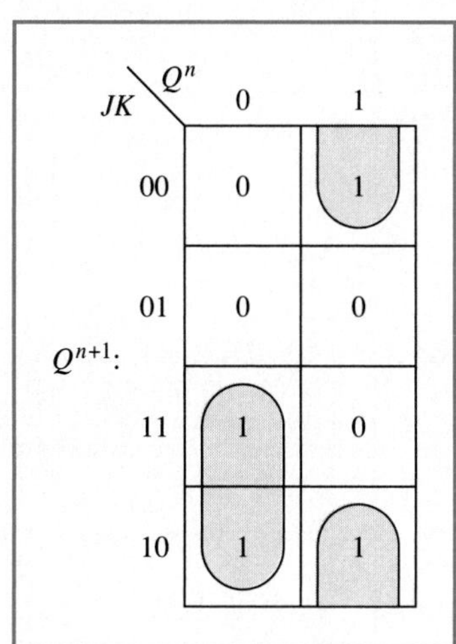

Figure 5–28
K-map for the J-K flip-flop.

change from a 0 to a 1, *J* must be a 1 and we don't care what *K* is

Q^n		Q^{n+1}	*J*	*K*
0	⟶	1	1	X

There are two possibilities for no change to occur in the output of the *J-K* flip-flop. Entries 0 and 2 indicate that the flip-flop changes from a 0 to a 0. For these two entries *J* must be a logic 0 but *K* can be either a 0 or a 1. In other words, for the output of a *J-K* flip-flop to remain constant at a 0, *J* must be a 0 and we don't care what *K* is

Q^n		Q^{n+1}	*J*	*K*
0	⟶	0	0	X

Similarly, entries 1 and 5 indicate that the flip-flop "changes" from a 1 to a 1. For these two entries *K* must be a logic 0 but *J* can be either a 0 or a 1. In other words, for the output of a *J-K* flip-flop to remain constant at a 1, *K* must be a 0 and we don't care what *J* is

Q^n		Q^{n+1}	*J*	*K*
1	⟶	1	X	0

From this analysis we can now construct a modified state transition table for the *J-K* flip-flop as shown in Table 5–9.

Integrated circuit logic

The 74112 dual *J-K* flip-flop contains two independent negative edge-triggered *J-K* type flip-flops in one IC package. The logic symbol and pin configuration for the IC are shown in Figure 5–29. Like the 7474 *D* flip-flop discussed in the previous section, the 74112 flip-flop has the regular synchronous inputs *J* and *K* as well as the asynchronous active-low inputs $\overline{S}_D$ and $\overline{C}_D$ for each flip-flop. A logic 0 on $\overline{S}_D$ will *SET* the output of the flip-flop and a logic 0 at $\overline{C}_D$ will *CLEAR* (reset) the flip-flop without requiring a clock transition. If the asynchronous inputs are not being used, they should be connected to the inactive logic 1 state. The *CP* input accepts the clock pulses that work in conjunction with the *J* and *K* inputs.

Table 5–9
Modified state transition table for a *J-K* flip-flop

Q^n	Q^{n+1}	*J*	*K*
0	0	0	X
0	1	1	X
1	0	X	1
1	1	X	0

Review questions

1. Describe the operation of the *J-K* flip-flop.
2. How does the *J-K* flip-flop emulate the operation of the *S-R* flip-flop?
3. How is the *J-K* flip-flop set up to toggle?
4. How is the *J-K* flip-flop set up to *SET* and *RESET*?

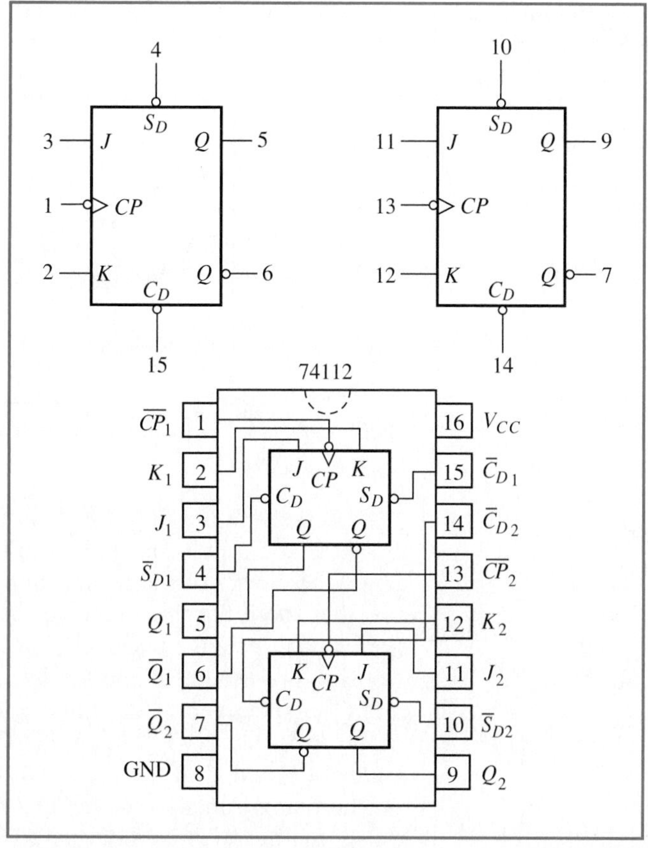

Figure 5–29
Logic symbol and pin configuration of the 74112 dual *J-K* negative edge-triggered flip-flop.

5–8 Master-slave flip-flops

The flip-flops discussed so far changed states on either a positive transition or negative transition of a clock pulse and were therefore known as *edge-triggered* flip-flops. All that was required to trigger the flip-flop was the edge of a clock pulse, and not a complete clock pulse. There are many applications that require flip-flops to change states on receiving a complete clock pulse—that is, a clock pulse that makes both a negative and a positive transition. A flip-flop that changes states on receiving a complete pulse is known as a *master-slave* or *pulse-triggered* flip-flop.

The most common type of master-slave flip-flop available is the *J-K* flip-flop although the master-slave design can be applied to the *D, S-R,* and *T* flip-flops also. Since the design of the master-slave *S-R* flip-flop is the simplest to understand, we shall use it as a model. Figure 5–30 shows the logic symbol for a master-slave *S-R* flip-flop. Notice that the symbol does not have a dynamic input indicator since the flip-flop is not edge-triggered. Instead it has a *postponed output* indicator (⌝) at each output since the output of the flip-flop does not change until the entire pulse (both transitions) has appeared at the clock input.

The master-slave *S-R* flip-flop actually consists of two edge-triggered *S-R* flip-flops connected in series as shown in Figure 5–31. The *S, R,* and *C* inputs to the master-slave flip-flops are applied to the *S, R,* and *C* inputs of a positive edge-triggered *S-R* flip-flop known as the *master,* and the Q and $\overline{Q}$ outputs of the master-slave *S-R* flip-flop are obtained from a negative edge-triggered flip-flop known as the *slave*.

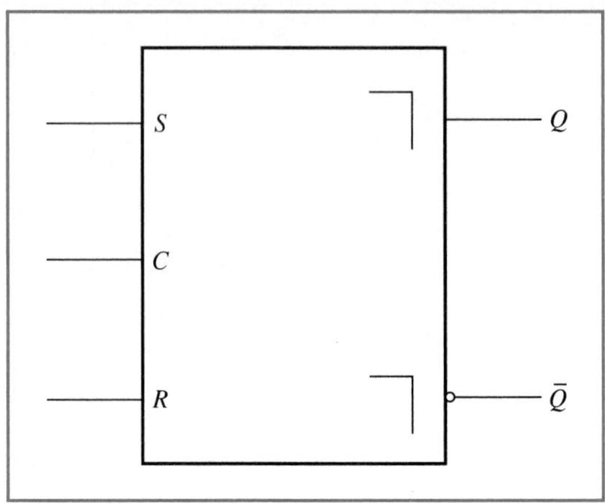

Figure 5–30
Logic symbol for a master-slave *S-R* flip-flop.

To set the flip-flop, the S input is set to a logic 1 and the R input to a logic 0 and a pulse is applied to the clock C as shown in Figure 5–32. Since the clocks of the master and the slave flip-flops are connected together, both flip-flops receive the pulse. However, since the master flip-flop is positive edge-triggered, it will change to the *SET* state on the rising transition of the pulse and the logic levels of S and R will be latched to Q and $\overline{Q}$, respectively. But Q and $\overline{Q}$ of the master flip-flop are connected to S and R of the slave flip-flop and thus we have effectively transferred the states of S and R of the master flip-flop to S and R of the slave flip-flop on the rising transition of the clock pulse. On the falling edge of the clock pulse, the negative edge-triggered slave flip-flop is now *SET* and its outputs (which are the outputs of the master-slave flip-flop) are at the same state as the outputs of the master. Therefore the master flip-flop only responds to the rising edge and the slave flip-flop only responds to the falling edge of the clock pulse.

Similarly, to *RESET* the flip-flop, the S input is *SET* to a logic 0 and the R input to a logic 1. On the positive edge of the clock pulse, the master flip-flop changes states and its Q and $\overline{Q}$ outputs are 0 and 1, respectively. Since the slave flip-flop's S and R inputs are at the same

Figure 5–31
Equivalent circuit for a master-slave *S-R* flip-flop.

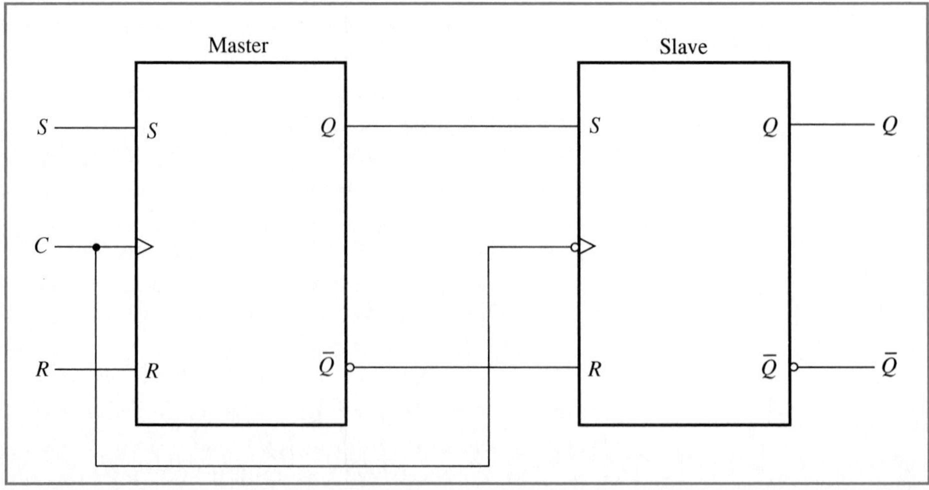

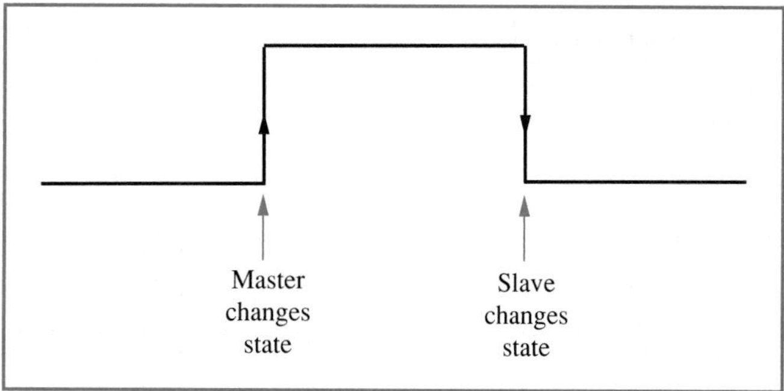

Figure 5–32 Pulse used to trigger a master-slave flip-flop.

logic states as the master flip-flop's Q and $\overline{Q}$ outputs, respectively, on the negative transition of the clock pulse, the slave flip-flop is *RESET*.

For a no-change condition, $R=0$ and $S=0$, and thus the output of the master flip-flop does not change on the positive transition of the clock pulse. The output of the master is the same as the output of the slave. Therefore on the negative transition if the slave is in the reset state it is *RESET* again, and if the slave is in the set state it is *SET* again, thereby producing no change at the output of the slave.

Since the output of a master-slave flip-flop changes only after the entire pulse appears at its clock, its timing diagram is exactly the same as the timing diagram of a negative edge-triggered flip-flop and therefore in many applications is interchangeable with such flip-flops.

Integrated circuit logic

The 7476 dual *J-K* flip-flop contains two independent master-slave *J-K* flip-flops in one IC package. The logic symbol and pin configuration for the IC are shown in Figure 5–33.

The 7476 *J-K* flip-flops function like the 74112 *J-K* flip-flops (discussed in the Section 5–7) except that a complete clock pulse is required to make the outputs of the flip-flops change states. Notice that each flip-flop has the clock pulse *CP*, a set of active-low synchronous inputs *J* and *K*, and a set of asynchronous inputs $\overline{S}_D$ and $\overline{C}_D$. The bubble shown at *CP* is sometimes used to indicate that the outputs of the flip-flop will change states on the negative edge of the clock pulse.

Review questions

1. How does one identify a master-slave flip-flop?
2. How does the operation of a master-slave flip-flop differ from the operation of an edge-triggered flip-flop?
3. State another name for a master-slave flip-flop.

5–9 Flip-flop conversions

It was stated earlier that the only commercially available flip-flops are the *D* and *J-K* types. This is because these two flip-flops can very easily emulate the functions of the *S-R* and *T* flip-flops and can also be configured to emulate the functions of each other. Since adapting one flip-flop to function as another can be very helpful in the design of digital circuits, this section deals with the configuration of the *D* and *J-K* types of flip-flops to emulate the functions of other types of flip-flops.

Figure 5–33
Logic symbol and pin configuration of the 7476 dual *J-K* master-slave flip-flop.

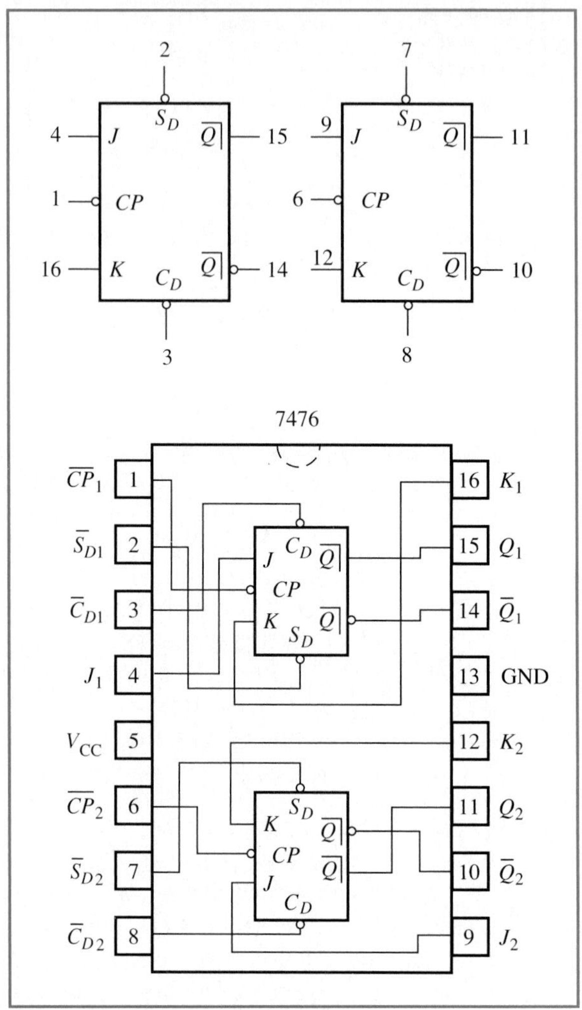

J-K flip-flop configurations

Since the *J-K* flip-flop functions like the *S-R* flip-flop for the first three binary combinations of *J* and *K*, we can simply define the *J* input to be the *S* input and the *K* input to the *R* input as shown in Figure 5–34. The condition where $S = 1$ and $R = 1$ is invalid for an *S-R* flip-flop and therefore will not be used in the circuit shown in Figure 5–34.

Figure 5–34
J-K flip-flop configured as an *S-R* flip-flop.

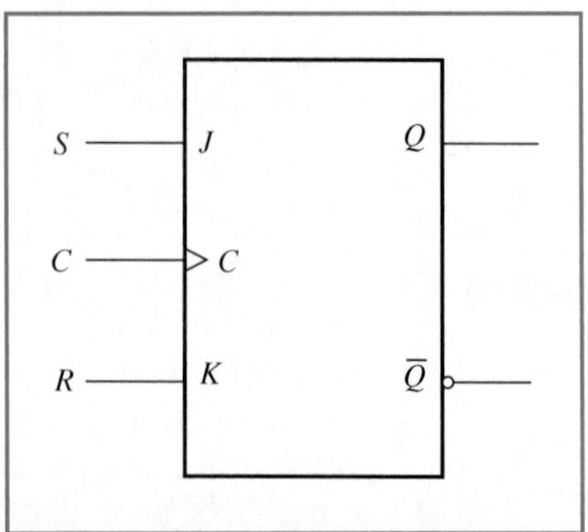

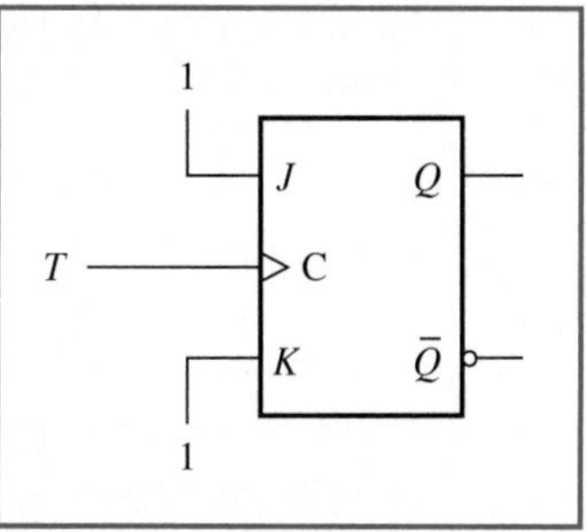

Figure 5–35
J-K flip-flop configured as a *T* flip-flop.

To make the J-K flip-flop function like a T-type flip-flop we have to connect the *J* and *K* inputs to a logic 1 as shown in Figure 5–35. Recall that when $J=1$ and $K=1$, the next state of the *J-K* flip-flop is equal to the complement of the present state and the flip-flop toggles.

The circuit to make the *J-K* flip-flop function like a *D* flip-flop is shown in Figure 5–36. Notice that the input of the *D* flip-flop is directly connected to *J* and is indirectly connected to *K* via an inverter. Thus if *D* is a logic 1, then $J=1$ and $K=0$ and the *J-K* flip-flop is *SET*, and therefore its output is a logic 1 (equal to the state of D). When *D* is a logic 0, then $J=0$ and $K=1$ and the *J-K* flip-flop is *RESET*, and therefore its output is a logic 0 (equal to the state of *D*).

The versatility in configuring the *J-K* flip-flop to function like the three other flip-flops is quite apparent. This is due to the capability of the *J-K* flip-flop to incorporate the functions of the *S-R, T,* and *D* flip-flops simply by changing the logic levels at the *J* and *K* control inputs. The *D* flip-flop can also be configured to function like the three other flip-flops, but it requires a considerable amount of additional circuitry.

D flip-flop configurations

To configure the *D* flip-flop to function as an *S-R* flip-flop we must first examine the transition equations for the *D* and *S-R* flip-flops given in Equations 5–3 and 5–1, respectively, and rewritten here as follows:
For the *D* flip-flop

$$Q^{n+1} = D$$

and for the *S-R* flip-flop

$$Q^{n+1} = S + \bar{R}Q^n$$

Since the two transition equations represent the next states of the *D* and *S-R* flip-flops and since we would like to make the next state of the *D* flip-

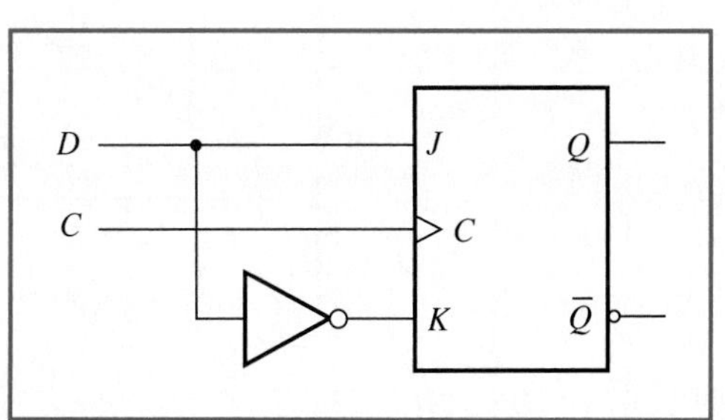

Figure 5–36
J-K flip-flop configured as a *D* flip-flop.

flop the same as the next state of the *S-R* flip-flop, we must equate the two equations as follows:

$$D = Q^{n+1} = S + \bar{R}Q^n$$

Therefore

Equation (5–5)

$$D = S + \bar{R}Q^n$$

The circuit to implement Equation 5–5 is shown in Figure 5–37. The circuit in Figure 5–37 will therefore function like an *S-R* flip-flop.

The circuit to configure the *D* flip-flop to function like a *T* flip-flop is shown in Figure 5–38. Notice that the *D* input is connected to the $\bar{Q}$ output so that each time the clock pulse makes a transition, the logic level at *D* toggles, and therefore the logic level at *Q* toggles. For example, if *Q* is initially a logic 1, $\bar{Q}$ is a logic 0 and therefore *D* is a logic 0. When the clock pulse triggers the flip-flop, the *Q* output changes to a logic 0 (the state of *D*) and $\bar{Q}$ (and *D*) change to a logic 1. On the next transition of the clock pulse *Q* is back to a logic 1 and the process repeats itself. Mathematically, using the state transition equations of the *D* and *T* flip-flops (Eqs. 5–3 and 5–2, respectively)

$$Q^{n+1} = D$$

and

$$Q^{n+1} = \bar{Q}^n$$

we must equate the two in order to make the next state of the *D* flip-flop the same as the next state of the *T* flip-flop

$$D = Q^{n+1} = \bar{Q}^n$$

Therefore

Equation (5–6)

$$D = \bar{Q}^n$$

The circuit shown in Figure 5–38 is obtained from Equation 5–6. In order to configure the *D* flip-flop to function as a *J-K* flip-flop, the procedure is similar to the one used to configure the *D* flip-flop to function like an *S-R* flip-flop.

We must first examine the transition equations for the *D* and *J-K* flip-flops given in Equations 5–3 and 5–4, respectively, and rewritten here as follows.

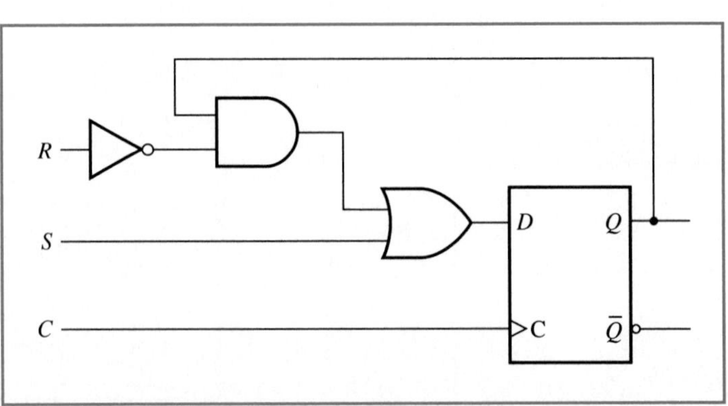

Figure 5–37
D flip-flop configured as an S-R flip-flop.

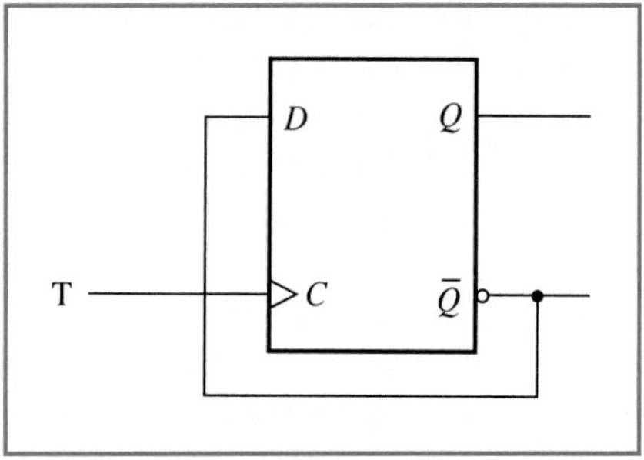

Figure 5–38
D flip-flop configured as a *T* flip-flop.

For the *D* flip-flop

$$Q^{n+1} = D$$

and for the *J-K* flip-flop

$$Q^{n+1} = J\bar{Q}^n + \bar{K}Q^n$$

Since the two transition equations represent the next states of the *D* and *J-K* flip-flops and since we would like to make the next state of the *D* flip-flop the same as the next state of the *J-K* flip-flop, we must equate the two equations as follows:

$$D = Q^{n+1} = J\bar{Q}^n + \bar{K}Q^n$$

Therefore

Equation (5–7)

$$D = J\bar{Q}^n + \bar{K}Q^n$$

The circuit to implement Equation 5–7 is shown in Figure 5–39. The circuit in Figure 5–39 will therefore function like a *J-K* flip-flop.

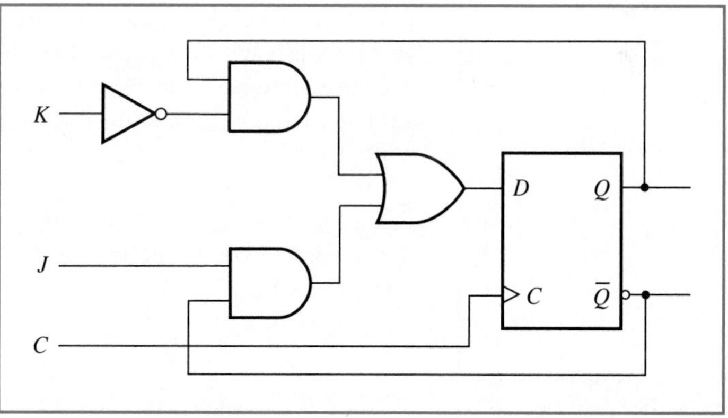

Figure 5–39
D flip-flop configured as a *J-K* flip-flop.

Review questions

1. Discuss the versatility of the *J-K* flip-flop in functioning like an *S-R* and *T* flip-flop.
2. Discuss the versatility of the *D* flip-flop in functioning like an *S-R* and *T* flip-flop.
3. Why is it important for the *J-K* and *D* flip-flops to be able to function like the *T* and *S-R* flip-flops?

5–10 Multivibrators

The flip-flop is often called a *bistable multivibrator* since it can exist in one of two (bi) possible states—*SET* or *RESET*—and it is capable of switching (vibrating) between these two states. There are two other types of multivibrators that are often used in digital circuits—the *monostable multivibrator* and the *astable multivibrator*.

The monostable multivibrator

The monostable multivibrator, also known as the *one-shot,* has only one (mono) stable state. When the one-shot is activated or triggered, it switches from its stable state to an *unstable state* and remains there for a set period of time, after which it returns back to its stable state. The logic symbol for the one-shot is shown in Figure 5–40.

The one-shot is triggered by applying a very short duration pulse at its T input. When this occurs the one-shot goes into its unstable (or *quasistable*) state and remains there for a period of time that is established by the values chosen for the external resistor R_x and capacitor C_x. Figure 5–41 shows the timing diagram of a typical one-shot. The duration (or width) of the output pulse t_W will depend on the RC time constant of the external resistor and capacitor.

There are two types of one-shots available—*nonretriggerable* and *retriggerable*. Once a nonretriggerable one-shot has been triggered and is in its quasistable state, it will not respond to any more pulses at its T input but will return to its stable state after a time period equal to t_W. A retriggerable one-shot can be triggered while it is in its quasistable state. When this occurs the output pulse width is extended for another duration of time period t_W.

The 74121 monostable multivibrator is a nonretriggerable one-shot. The pin configuration and logic symbol for the 74121 are shown in Figure 5–42. The one-shot is triggered by applying a trigger pulse at input B with either the $\overline{A}_1$ or the $\overline{A}_2$ input tied low; the one-shot is then triggered on the positive edge of the pulse. Alternatively, a trigger pulse can be applied to the $\overline{A}_1$ or $\overline{A}_2$ inputs with the B input tied high, in which case the one-shot will be triggered on the negative edge of the pulse. An external capacitor and resistor are connected to the R_xC_x and C_x inputs and establish the duration of the output pulse according to the following relationship:

$$t_W = 0.69\ R_xC_x$$

The output pulse is obtained from the Q and $\overline{Q}$ (complemented) outputs of the one-shot.

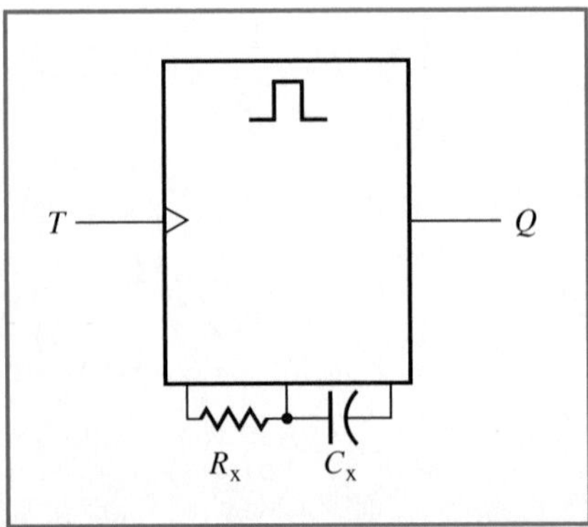

Figure 5–40
Logic symbol for a monostable multivibrator (one-shot).

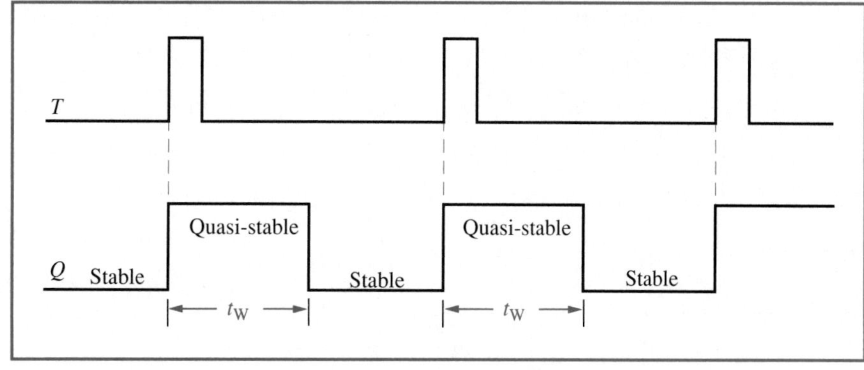

Figure 5–41
Timing diagram for a one-shot.

The 74122 is a retriggerable, resettable monostable multivibrator. The pin configuration and logic symbol for the 74122 are shown in Figure 5–43. The trigger inputs to the 74122 work in the same manner as the trigger inputs to the 74121 except that the 74122 has two B inputs, B_1 and B_2, instead of a single B input like the 74121. Since B_1 and B_2 are "*AND*ed" together, they must both be high if one of the A lines is being used to trigger the one-shot, or one of the B lines must be high if the other is being used to trigger the one-shot. The 74122 also provides a *Direct Clear* input ($\overline{C}_D$) that can be used to reset the one-shot to its stable state after it has been triggered into the quasistable state. The duration of the output pulse is established by the following relationship for the 74122:

$$t_W = 0.32\ R_xC_x\left[1 + \frac{0.7}{R_x}\right]$$

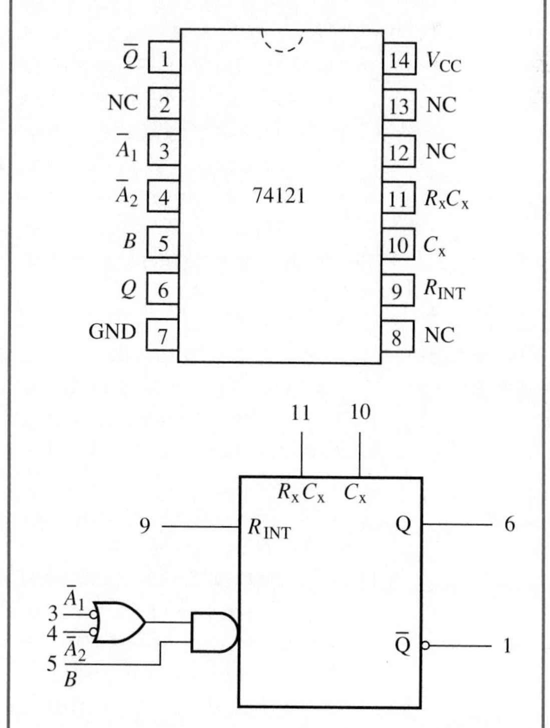

Figure 5–42
Logic symbol and pin configuration of the 74121 monostable multivibrator.

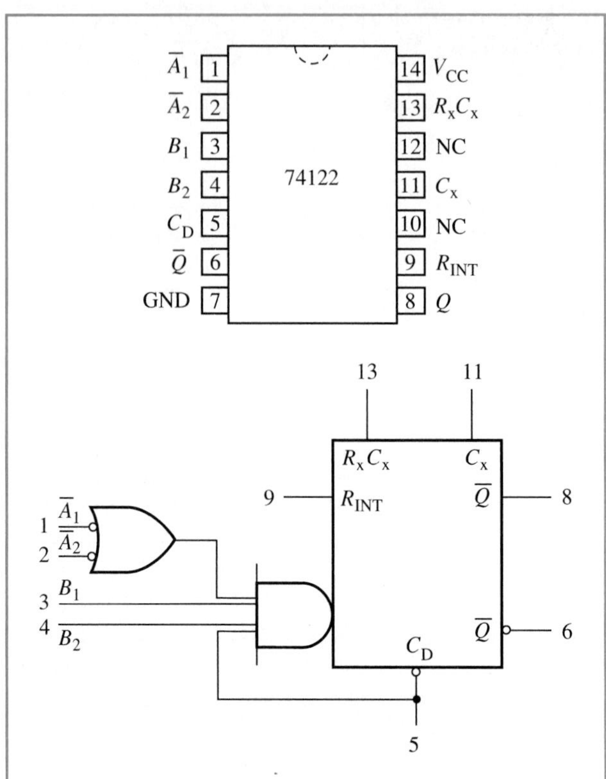

Figure 5–43 Logic symbol and pin configuration of the 74122 retriggerable, resettable multivibrator.

The 74123 is also a retriggerable one-shot but contains two independent one-shots in one IC package as shown in Figure 5–44. Like the 74122, the 74123 can also be directly cleared while it is in its quasistable state by applying a low to the $\overline{C}_D$ input. However, each one-shot has only two trigger inputs—*A* and *B*. The negative edge of a pulse applied to the *A* input (while *B* is high) will trigger the one-shot, or the positive edge of a pulse applied to the *B* input (while *A* is low) can also be used to trigger the one-shot. The output pulse duration can be individually set for each one-shot by attaching the appropriate *RC* circuit.

Most of the applications of one-shots are in timing applications and applications that require extremely short-duration pulses to be detected or "stretched-out."

The astable multivibrator

The astable multivibrator is also known as the *free-running* multivibrator. The astable multivibrator is *astable,* that is, it does not have any stable state. This means that the astable multivibrator will oscillate continuously between two states. Astable multivibrators are therefore used as oscillators in various digital circuits and in application circuits that require a periodic string of clock pulses for timing purposes. Examples of this will be seen in Chapter 6.

Figure 5–45 illustrates how two one-shots can be connected together to produce a free-running multivibrator. Since the 74123 contains two one-shots the circuit shown in Figure 5–45 could easily be set up using one IC. The operation of the circuit can be explained with reference to the timing diagram shown in Figure 5–46 as follows.

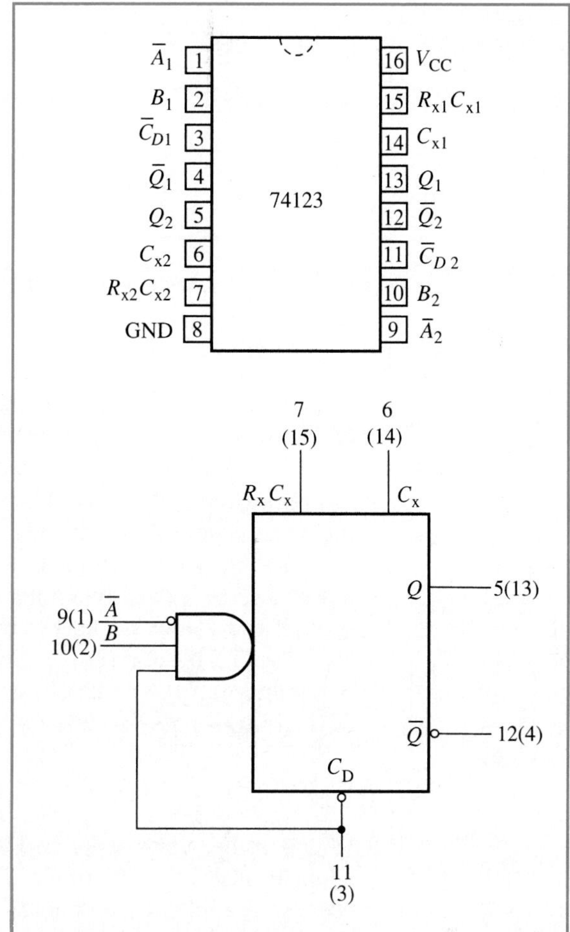

Figure 5–44 Logic symbol and pin configuration of the 74123 dual retriggerable, resettable multivibrator.

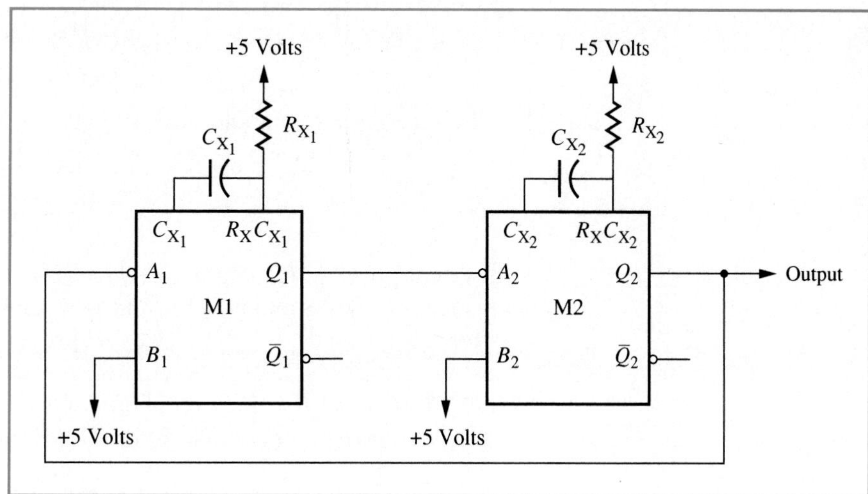

Figure 5–45 Two one-shots connected together to function as a free-running (astable) multivibrator.

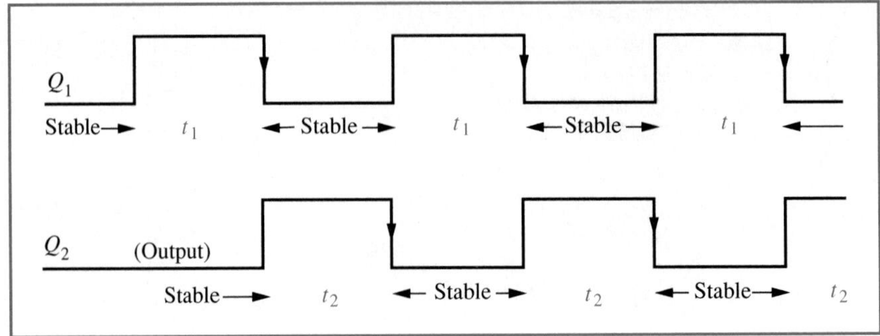

Figure 5–46
Timing diagram for an astable multivibrator.

Assuming that $M1$ is triggered on power-up, after a period t_1 when $M1$ returns to its stable state, it will trigger $M2$ on the negative edge of the pulse. $M2$ will go to the quasistable state for a period t_2. When $M2$ returns to the stable state it will retrigger $M1$ on the negative edge, and the process repeats itself over and over again.

The *duty cycle* (high and low duration) of the output stream of pulses can be varied by changing the values of R_x and C_x for each one-shot. If the values of these components are equal, then the output pulse will have a 50% duty cycle (the high and low durations will be equal). Assuming that the circuit is set up for a 50% duty cycle, the frequency (f) of the output pulse stream will be

$$f = \frac{1}{t_1 + t_2}$$

Another very versatile IC that is used for many digital and analog applications is the 555 timer. The 555 timer can be configured for a variety of applications, one of which is the free-running mode. Because of its complexity, and because it is beyond the scope of this text, we shall only examine the configuration of the 555 for astable operation and will not attempt to investigate the internal operation of the IC.

Figure 5–47 shows the 555 timer configured to operate as an astable multivibrator. The output pulse stream is obtained from pin 3. The pulse widths (both high and low) are controlled by the values of the external components R_A, R_B, and C as follows:

$$t_1 = 0.693\ R_B\mathrm{C}$$
$$t_2 = 0.693\ (R_A + R_B)C$$

where resistor R_A should have a value greater than or equal to 500 ohms. The output frequency of the pulse stream would then be

$$f = \frac{1}{t_1 + t_2}$$

It is not possible to obtain a perfect 50% duty cycle with the 555 since by inspecting the formulas for t_1 and t_2 we can see that R_A must be 0 in order to make t_1 equal to t_2. However, to prevent the circuit from drawing excess current, R_A must be greater than or equal to 500 ohms. We can, however, make R_B much much greater in value than R_A so that the value of t_1 is approximately equal to t_2. The formula then approximates to

$$f \cong \frac{1.44}{2\ R_B C}$$

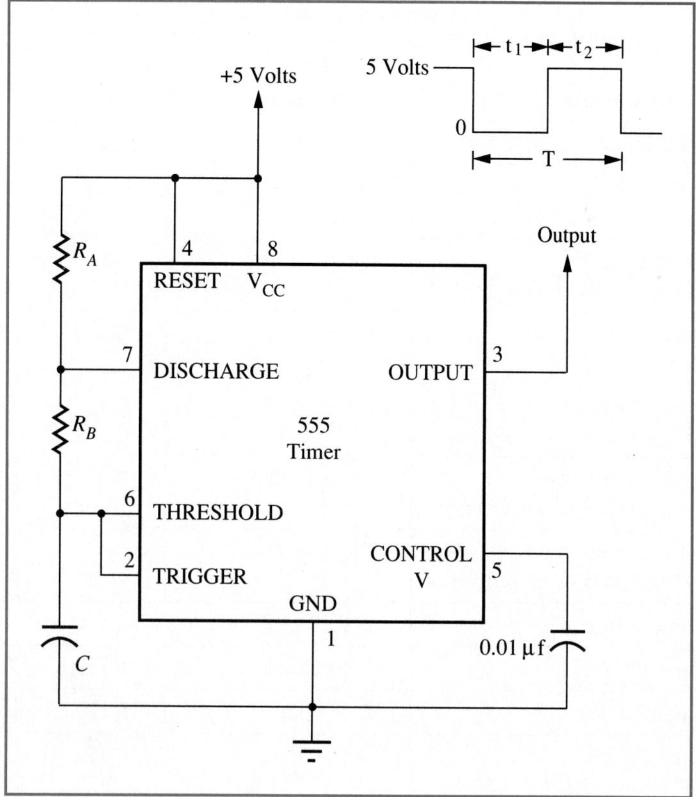

Figure 5–47
555 timer configuration for astable operation.

Review questions

1. Why is the flip-flop known as a bistable multivibrator?
2. Describe the operation of a monostable multivibrator.
3. What is the difference between a nonretriggerable and retriggerable one-shot?
4. What is the difference between the monostable and an astable multivibrator?

5–11 Switching characteristics of flip-flops

Flip-flops and other multivibrators exhibit the same input and output electrical characteristics as the basic logic gates. The same is true for loading rules and noise margins. These parameters were discussed in Chapter 2 and apply to all the multivibrators in the 74xxx TTL series of ICs. However, it is important for us to be able to understand some of the other characteristics that are typical of flip-flops, specifically, the switching characteristics.

Integrated circuit data books often include the switching or AC characteristics of flip-flops along with the other electrical and logic information about the IC. Figure 5–48 shows the switching characteristics of a 7474 *D*-type flip-flop obtained from the manufacturer's data book. Notice that most of these parameters relate to the timing of the flip-flop.

Maximum clock frequency

The data sheet includes the maximum frequency of the clock pulses that are applied at the clock input, f_{MAX}. If this frequency is exceeded, the

Switching Characteristics at V_{CC} = 5 volts and T_A = 25°C

Parameter	From (input) to (output)	Symbol	C_L = 15 pF, R_L = 2K		Units
			Minimum	*Maximum*	
Maximum clock frequency		f_{MAX}	25		MHz
Propagation delay time LOW to HIGH level output	Clock to Q or $\overline{Q}$	t_{PLH}		25	ns
Propagation delay time HIGH to LOW level output	Clock to Q or $\overline{Q}$	t_{PHL}		30	ns
Propagation delay time LOW to HIGH level output	Preset to Q	t_{PLH}		25	ns
Propagation delay time HIGH to LOW level output	Preset to $\overline{Q}$	t_{PHL}		30	ns
Propagation delay time LOW to HIGH level output	Clear to $\overline{Q}$	t_{PLH}		25	ns
Propagation delay time HIGH to LOW level output	Clear to Q	t_{PHL}		30	ns
Setup time HIGH	D to Clock	t_s (H)	20		ns
Hold time HIGH	D to Clock	t_h (H)	5		ns
Setup time LOW	D to Clock	t_s (L)	20		ns
Hold time LOW	D to Clock	t_h (L)	5		ns
Clock pulse width		t_w (H) t_w (L)	30 37		ns
Clear or preset pulse width (LOW)		t_w (L)	30		ns

Figure 5–48
Manufacturer's data sheet showing the switching characteristics of the 7474 D-type flip-flop.

flip-flop will not respond properly to the clock pulses and therefore its operation will be impaired.

Propagation delays

Notice in the data sheet that the flip-flop has several propagation delays associated with its operation. These parameters have the same symbol t_P, but describe delays between different inputs and outputs of the flip-flop.

The propagation delay time between the clock pulse and the output of the flip-flop is measured from the time the clock pulse makes a transition to the time the Q or $\overline{Q}$ output changes states. Two values are specified for this propagation delay—t_{PLH}, the propagation delay when the output changes from a *LOW* to a *HIGH*, as shown in Figure 5–49a, and t_{PHL}, the propagation delay when the output changes from a *HIGH* to a *LOW* as shown in Figure 5–49b. Note that propagation delays are measured from the midpoint of each transition. The waveforms in Figure 5–49 assume a positive edge-triggered flip-flop.

The other propagation delays given in the data sheet are the delay times between the direct set (preset) and direct clear (reset) asynchronous inputs of the flip-flop and the Q or $\overline{Q}$ outputs. The delays are shown in Figure 5–50. t_{PLH} is the propagation delay from the time the direct set ($\overline{S}$) or direct clear ($\overline{C}$) input is activated (to a *LOW*) to the time the (Q or $\overline{Q}$) output changes state from a *LOW* to a *HIGH*. t_{PHL} is the propagation

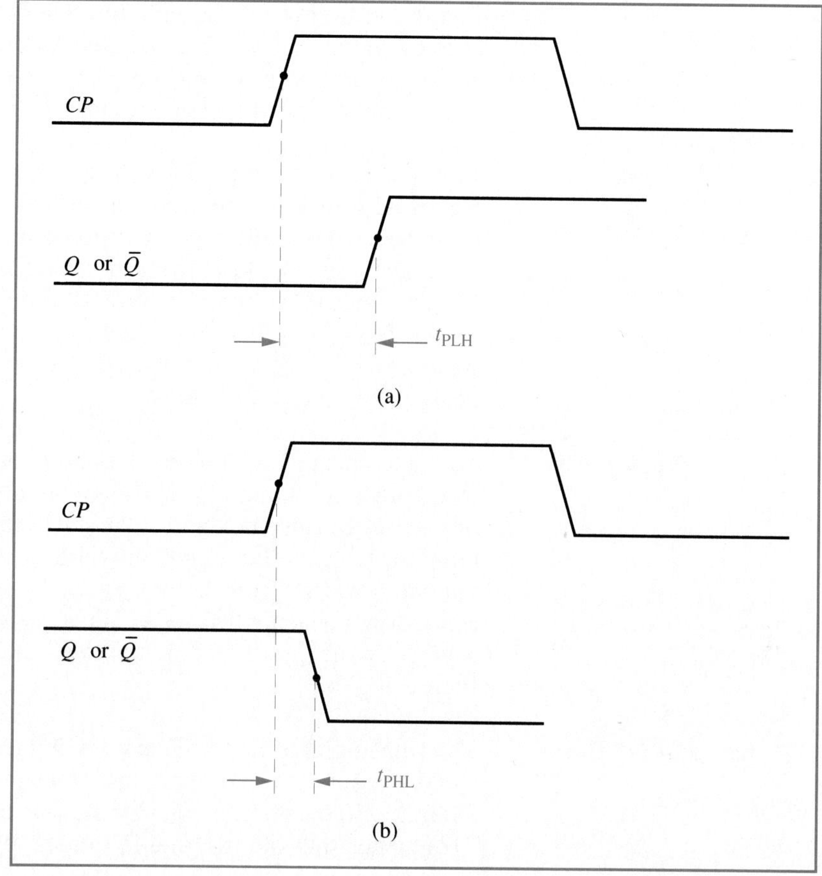

Figure 5–49 Propagation delays between the clock pulse and the outputs of a flip-flop. (a) LOW to HIGH; (b) HIGH to LOW.

delay from the time the $\overline{S}$ or $\overline{C}$ input is activated to the time the (Q or $\overline{Q}$) output changes state from a *HIGH* to a *LOW*.

Setup time

The setup time of a flip-flop is the minimum amount of time before the clock makes its transition that the control input(s) (D in this example) have to be stable. t_s(H) is the time interval between the application of a *HIGH* at the D input and the transition of the clock pulse as shown in

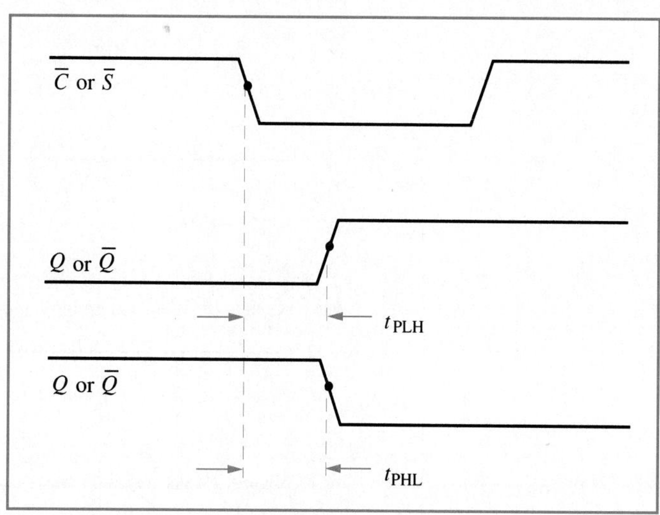

Figure 5–50 Propagation delays between the asynchronous control inputs and the outputs of a flip-flop.

Figure 5–51a. $t_s(L)$ is the time interval between the application of a *LOW* at the *D* input and the transition of the clock pulse as shown in Figure 5–51b. Again, both waveforms for the clock pulse (*CP*) shown in Figure 5–51 assume a positive edge-triggered clock.

Hold time

The *hold time* of a flip-flop is the minimum amount of time after the clock makes its transition that the control input(s) (*D* in this example) have to be stable. $t_h(H)$ is the time interval between the transition of the clock pulse and a change in *D* from the *HIGH* state as shown in Figure 5–52a. $t_h(L)$ is the time interval between the transition of the clock pulse and a change in *D* from the *LOW* state as shown in Figure 5–52b. Again, both waveforms for the clock pulse (*CP*) shown in Figure 5–52 assume a positive edge-triggered clock.

Pulse widths

The data sheet also includes the minimum pulse widths required at the clock pulse (*CP*) and asynchronous inputs ($\overline{S}$ and $\overline{C}$) for reliable operation. For asymmetric clock pulses applied to the *CP* input of the flip-flop, $t_w(H)$ is the minimum width of the *HIGH* cycle and $t_w(L)$ is the minimum width of the *LOW* cycle as shown in Figure 5–53a. For an active-low pulse applied to the direct set ($\overline{S}$) and direct clear ($\overline{C}$) asynchronous inputs, the minimum width of the pulse is $t_w(L)$ as shown in Figure 5–53b.

Rise and fall times

The transitions of the clock pulses applied to the clock inputs of flip-flops usually require an extremely fast transition from the *HIGH* to *LOW* or *LOW* to *HIGH* state. Typically the transition has to be around 6 to 13 ns. This means that the rise and fall times of clock pulses must be extremely

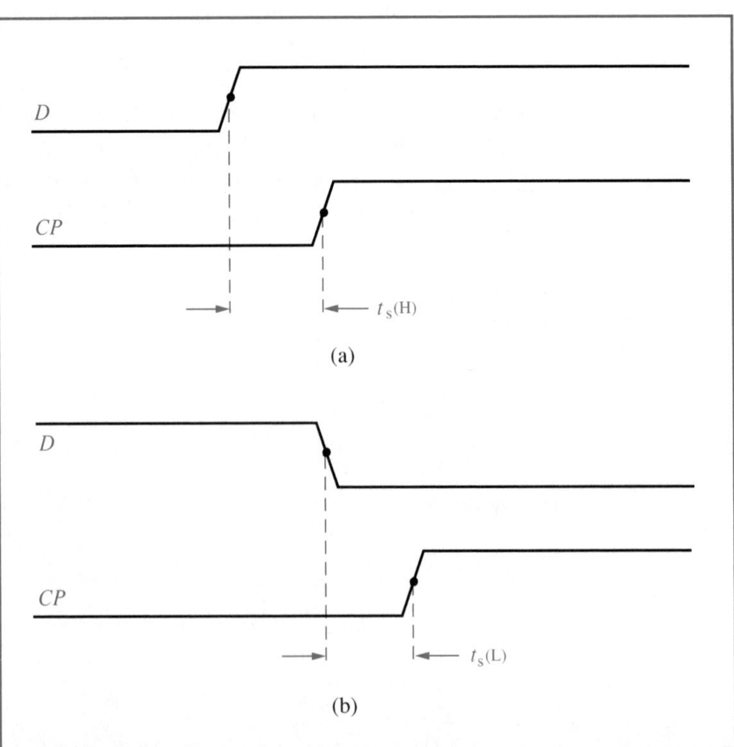

Figure 5–51
Setup times for a flip-flop. (a) *D* goes HIGH; (b) *D* goes LOW.

Figure 5–52
Hold times for a flip-flop. (a) *D* goes LOW; (b) *D* goes HIGH.

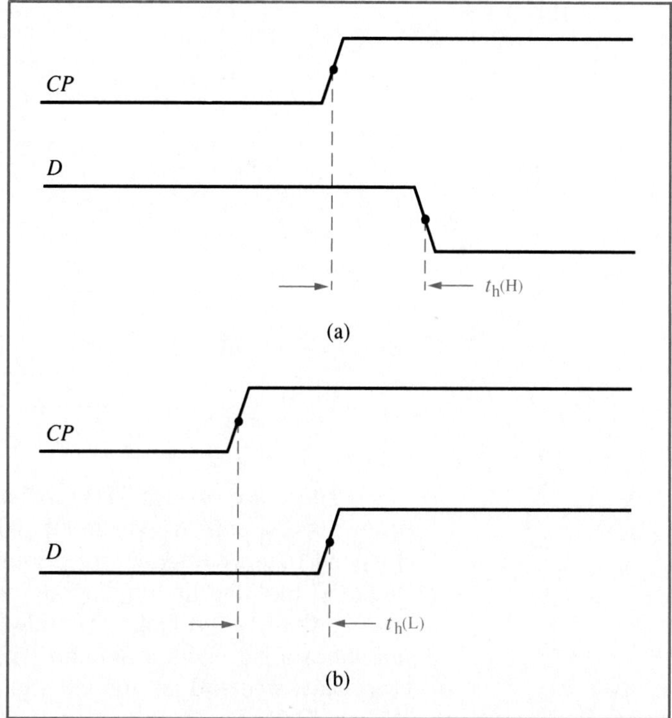

small to properly trigger the flip-flop. If the rise and fall times of the pulse(s) applied to the clock input of the flip-flop is too long, the flip-flop often produces undesirable oscillations as it passes through the undefined area between a *HIGH* and a *LOW*. Most pulse generator circuits such as astable multivibrators do produce pulses with quick rise and fall times, but due to the effect of capacitance in a circuit the pulses could be distorted by longer rise and fall times.

Figure 5–53
Pulse width measurements. (a) Clock input; (b) asynchronous control inputs.

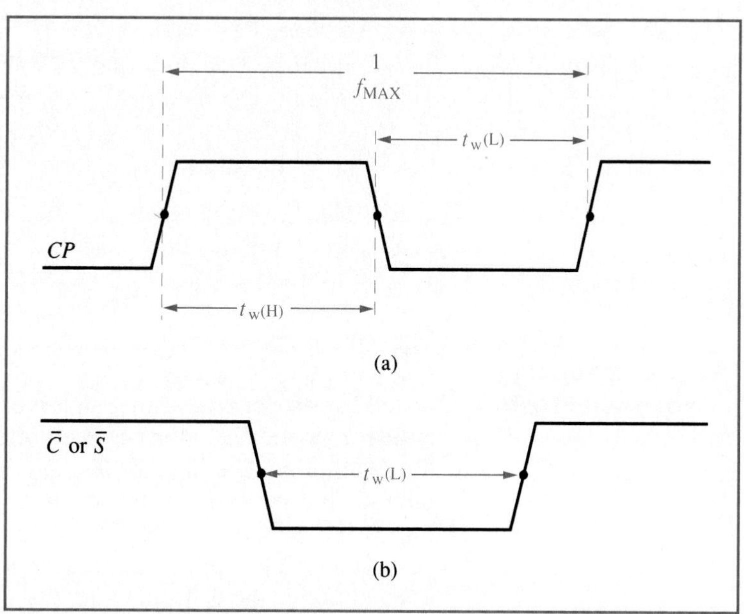

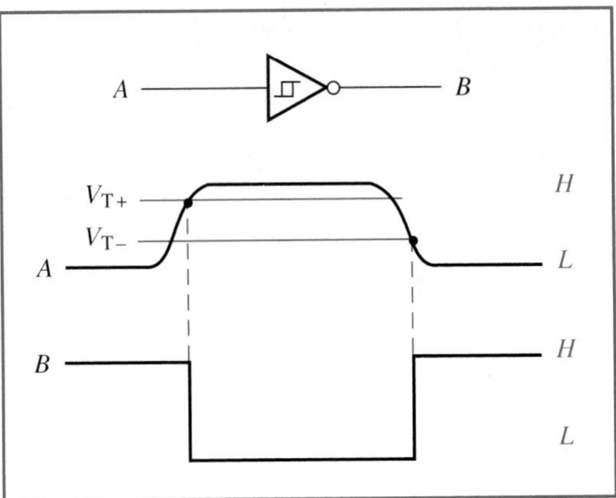

Figure 5–54
A Schmitt-trigger inverter. (a) Logic symbol; (b) timing diagram.

There are several 74xxx IC gates that have special types of inputs known as *Schmitt trigger* inputs. Gates with these inputs are designed to respond reliably to slow transitions. For example, Figure 5–54 shows the logic symbol and timing diagram for a 7414 hex Schmitt trigger inverter. Notice that the symbol (⎍) inside the logic symbol for the inverter identifies a gate with a Schmitt trigger input. Assume that a pulse with a very slow rise and fall time is applied to input *A* of the inverter shown in Figure 5–54. The output of the inverter changes from a *HIGH* to a *LOW* when the input reaches the threshold voltage V_{T+} (typically around 1.7 volts) and changes from a *LOW* to a *HIGH* when the input voltage reaches the threshold voltage V_{T-}. Once the threshold voltage has been detected at the input and the output changes states, the output will remain at that state even if the input voltage fluctuates slightly. Therefore we can convert a pulse with a slow transition to a pulse with a fast transition using a gate with a Schmitt trigger input.

Review questions

1. How are the different types of propagation delays measured for a flip-flop.
2. What is the setup time for a flip-flop?
3. What is the hold time for a flip-flop?
4. What effect does a slow rise or fall time generally have on the clock input of a flip-flop?
5. How can a gate with a Schmitt trigger input quicken the rise and fall times of a pulse?

5–12 Troubleshooting sequential logic

The logic probe and pulser can be used to test flip-flops and other multivibrators for faults. Figure 5–55 illustrates how a *D* flip-flop can be tested for proper operation using these two test instruments. The *D* input is tied high and the pulser is used to inject a pulse into the clock input of the *D* flip-flop. The logic probe can then be used to examine the Q and $\overline{Q}$ outputs and verify that they are in the high and low state, respectively. Similarly, we can test the *D* flip-flop with a low applied to the *D* input.

One of the most common problems with flip-flop circuits deal with unconditioned asynchronous inputs. If the direct set ($\overline{S}$) and clear ($\overline{C}$)

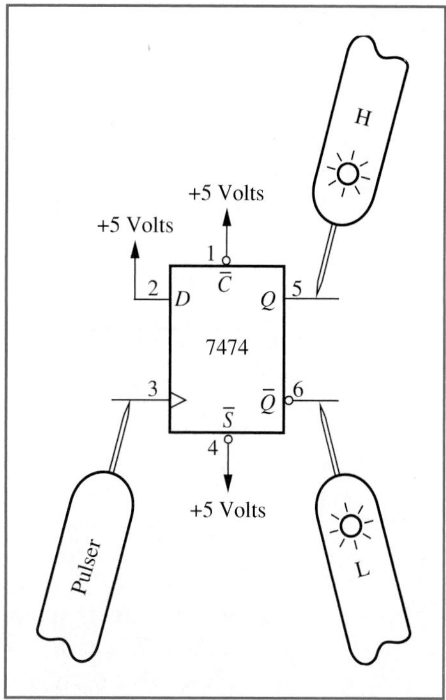

Figure 5–55
Testing a *D* flip-flop for proper operation.

inputs are not being used, they should be tied inactive at the *HIGH* state. If these inputs are left floating, they are susceptible to picking up noise and could cause the flip-flop to set and reset erratically.

Besides the faults that are common in combinational logic circuits (shorted inputs, shorted outputs, floating inputs, etc.), sequential logic circuits (i.e., circuits containing flip-flops) can be affected by faults due to timing problems. These problems are generally due to the effects of propagation delays at high operating frequencies although many of the faults due to timing problems could also be attributed to improper setup and hold times for flip-flop circuits.

Propagation delays in sequential logic circuits become exaggerated at high frequencies. For example, consider a standard 7474 *D*-type flip-flop being clocked at its maximum frequency of 25 MHz (see data sheet in Figure 5–48). The period of the clock at 25 MHz would be 40 ns. The propagation delay t_{PHL} from the clock to the output would be *more than half the period of the clock* as shown in Figure 5–56. This could lead to problems in many circuits.

To illustrate the effects of propagation delays at high frequencies consider the timing diagram of the *D* flip-flop that has been configured as a *T* flip-flop as shown in Figure 5–38. Again we will assume that the *D* flip-flop is being clocked at its maximum clock frequency of 25 MHz. Now the output *Q* of the flip-flop will toggle from a *LOW* to a *HIGH* 25 ns (t_{PLH}) after the clock makes its transition and will toggle from a *HIGH* to a *LOW* 30 ns (t_{PHL}) after the clock makes the next transition. The timing diagram is shown in Figure 5–57. The $\overline{Q}$ output of the flip-flop will have the same delays. Since the $\overline{Q}$ output is connected to the *D* input of the flip-flop, the setup time t_s(L) will be only 15 ns and the setup time t_s(H) will be only 10 ns. Notice that the data sheet in Figure 5–48

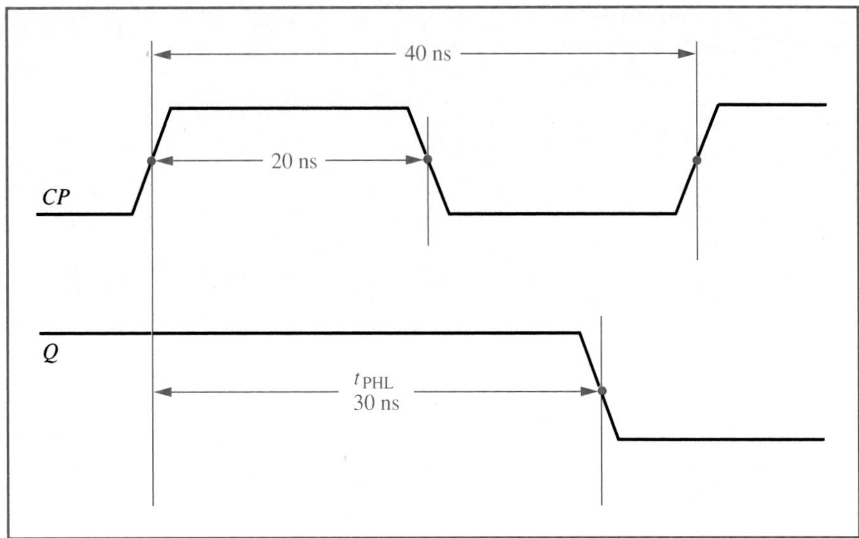

Figure 5–56
Propagation delay at high frequency.

specifies minimum setup times of 20 ns for both t_s(H) and t_s(L). This would therefore cause unreliable operation of the toggle flip-flop.

The only solution to this type of a problem is to use a *D*-type flip-flop with a smaller propagation delay or use a *J-K* flip-flop configured as a toggle flip-flop. More examples of such faults and their troubleshooting procedures will be examined in Chapter 6.

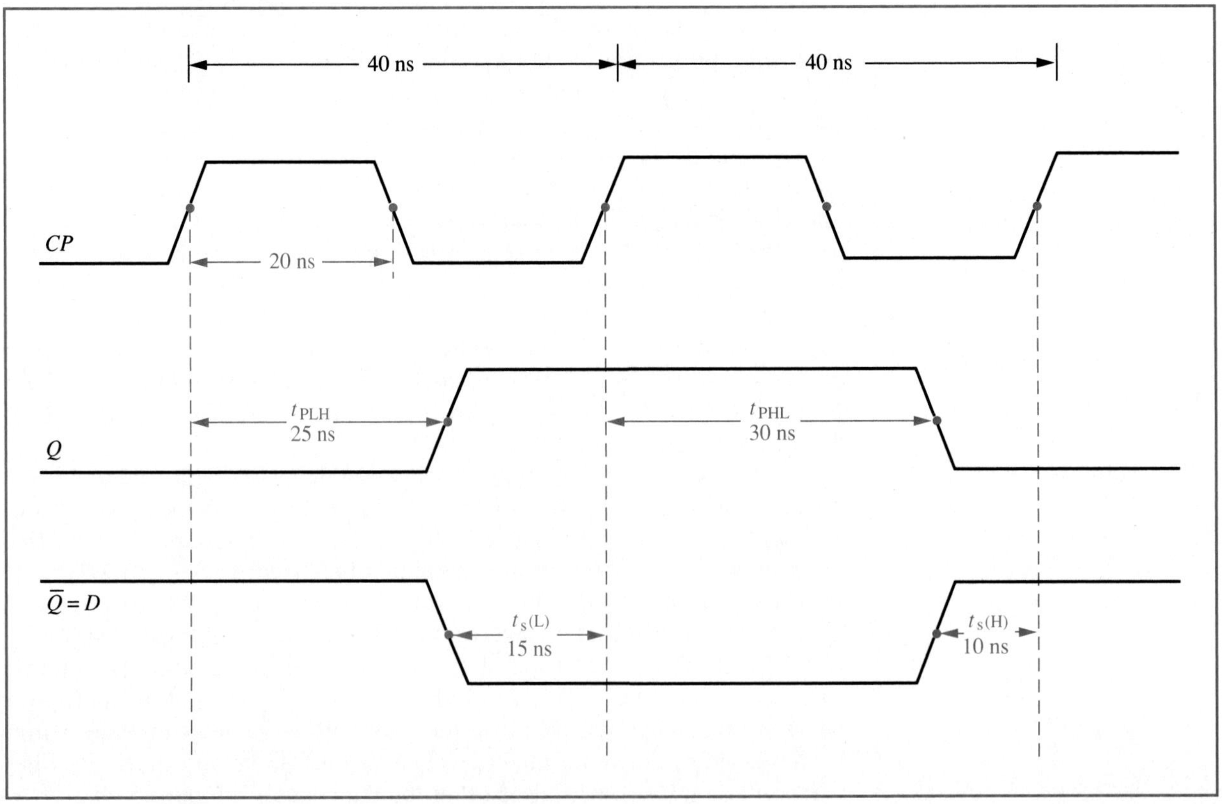

Figure 5–57
Effects of propagation delays.

Review questions

1. Why is it important to condition the direct set and clear inputs of a flip-flop?
2. When do propagation delays generally affect the operation of a sequential logic circuit?

5–13 SUMMARY

- Combinational logic circuits are circuits whose outputs depend only on binary combinations applied to the inputs.
- Sequential logic circuits are circuits whose outputs depend on binary combinations applied to the inputs as well as on the past history of the outputs.
- A *latch* is a sequential logic circuit that can exist in either the set or reset states.
- A *set* state indicates that the output of a sequential logic circuit is a logic 1.
- A *reset* or *clear* state indicates that the output of a sequential logic circuit is a logic 0.
- A *timing diagram* describes the operation of a sequential logic circuit by graphically showing the relationship between the inputs and outputs.
- A *gated latch* has a *gate input* that determines when the latch will change states.
- A *flip-flop* is a gated latch that changes states on the transition of a clock pulse applied to the gate.
- The gate input of a flip-flop is known as the *clock input.*
- Flip-flops can be made to change states (trigger) by a negative or positive transition of a clock pulse.
- The *negative transition* of the clock pulse is the *falling edge* or the transition from a logic 1 to a logic 0.
- The *positive transition* of the clock pulse is the *rising edge* or the transition from a logic 0 to a logic 1.
- A *negative edge-triggered* flip-flop is a flip-flop that changes states on the negative edge of a clock pulse.
- A *positive edge-triggered* flip-flop is a flip-flop that changes states on the positive edge of a clock pulse.
- A negative edge-triggered flip-flop is identified by a bubble at its clock input while a positive edge-triggered flip-flop does not have a bubble at its clock input.
- A *state transition table* describes the operation of a flip-flop by identifying its *next state* given its *present state* and the logic levels at its *control inputs*.
- The T or *toggle* flip-flop has no control inputs but only has a single clock input.

- The next state of a T flip-flop is always equal to the complement of its present state.
- The D or *data* flip-flop has a single control input that determines the next state of the flip-flop.
- The next state of a D flip-flop is equal to the value of D before the clock pulse triggers the flip-flop.
- Flip-flops can have *synchronous* control inputs that cause the flip-flop to change states only when the clock pulse makes its transition, and *asynchronous* control inputs that cause the flip-flop to change states without requiring a clock pulse transition.
- The asynchronous inputs to a flip-flop are generally used to directly set or clear the flip-flop and are therefore known as the *direct* set and clear inputs.
- The J-K flip-flop is similar in operation to the S-R flip-flop except that it does not have an invalid combination for the two control inputs.
- The J input of the J-K flip-flop sets the flip-flop while the K input resets the flip-flop.
- A *master-slave* flip-flop is also known as a *pulse-triggered* flip-flop.
- In order for a master-slave flip-flop to change states, a *complete* pulse must appear at its clock input.
- The flip-flop is also known as a *bistable multivibrator* because it can exist in one of two possible stable states.
- A multivibrator that can exist in only one stable state is known as a *monostable multivibrator* or *one-shot*.
- A multivibrator that does not have any stable states is known as a *free-running* or *astable* multivibrator.
- The propagation delay time of a flip-flop is usually measured from the time the clock triggers the flip-flop to the time the output changes state.
- The propagation delay time can also be measured from the time the direct set or clear inputs are activated to the time the output changes state.
- The *setup time* of the flip-flop is the minimum amount of time that the control input of a flip-flop is to remain stable before the clock makes its transition.
- The *hold time* of the flip-flop is the minimum amount of time that the control input of the flip-flop is to remain stable after the clock makes its transition.
- A gate with a *Schmitt trigger* input can respond to a slow transition without producing undesirable oscillations at the output.
- Propagation delays generally affect the operation of sequential logic circuits at high operating frequencies.

PROBLEMS

Section 5–2 The latch

1. Complete the timing diagram for the circuit shown in Figure 5–58.
2. Complete the timing diagram for the circuit shown in Figure 5–59.

Figure 5–58

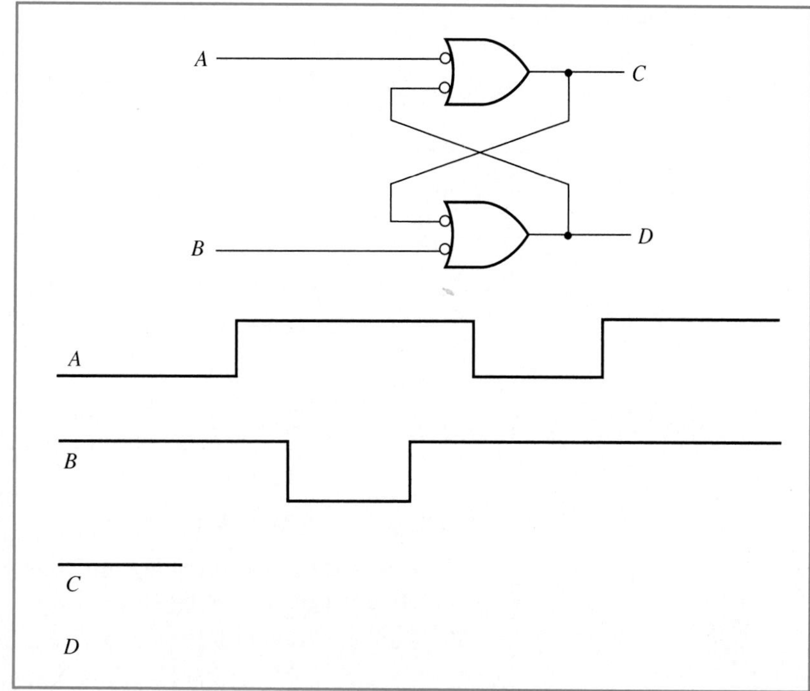

3. Identify the Q and $\overline{Q}$ outputs of the circuit in Figure 5–58 and the inputs to *SET* and *RESET* the latch.
4. Identify the Q and $\overline{Q}$ outputs of the circuit in Figure 5–59 and the inputs to *SET* and *RESET* the latch.

Figure 5–59

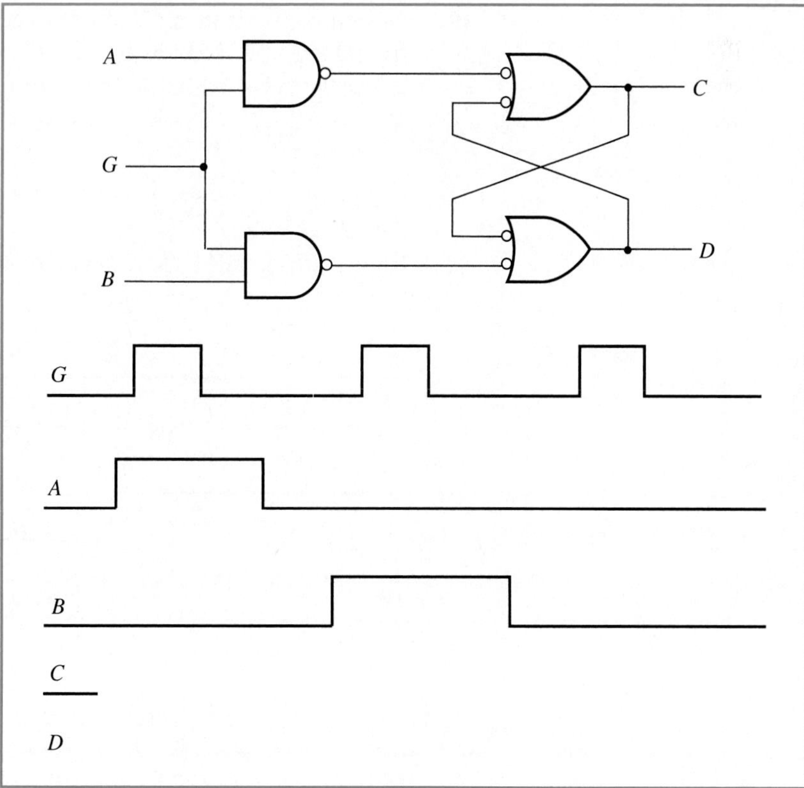

Figure 5–60

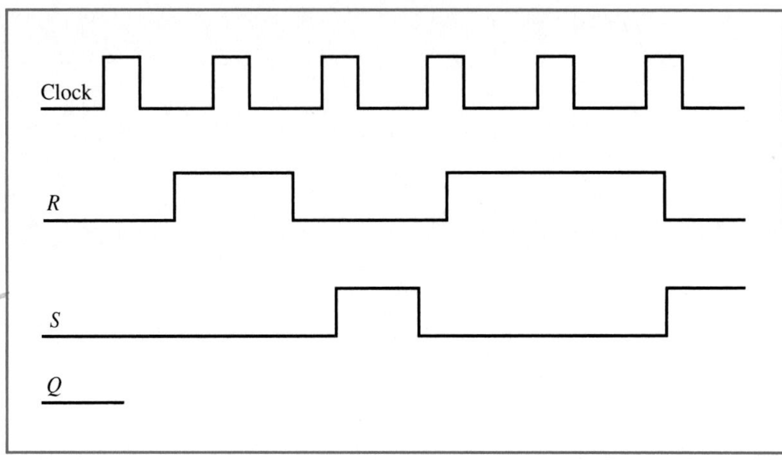

Section 5–4 The S-R flip-flop

5. Complete the timing diagram shown in Figure 5–60 for a positive edge-triggered *S-R* flip-flop.
6. Repeat Problem 5 for a negative edge-triggered flip-flop.
7. Complete the timing diagram shown in Figure 5–61 for a negative edge-triggered *S-R* flip-flop.
8. Repeat Problem 7 for a positive edge-triggered flip-flop.

Section 5–5 The T flip-flop

9. Draw a timing diagram for the circuit shown in Figure 5–62 illustrating the outputs *X* and *Y* as a function of the clock.
10. Design a circuit using *T* flip-flops to accept a clock pulse having a frequency of 224 Hz as input and producing a clock pulse having a frequency of 7 Hz as output.

Section 5–6 The D flip-flop

11. Complete the timing diagram shown in Figure 5–63 for a positive edge-triggered *D* flip-flop.
12. Repeat Problem 11 for a negative edge-triggered flip-flop.

Figure 5–61

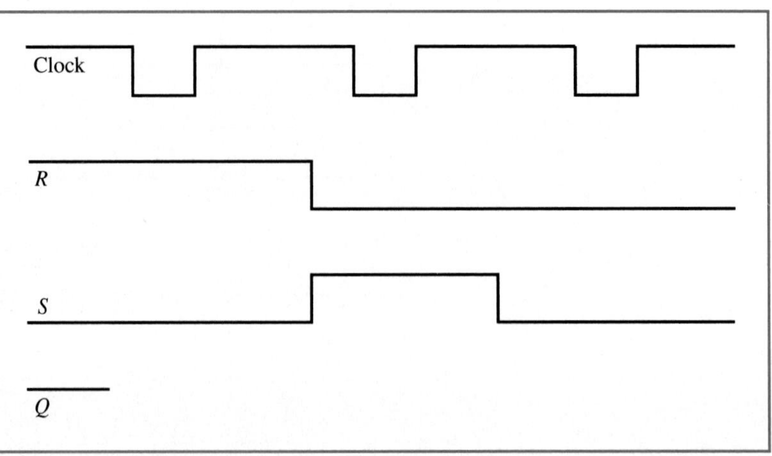

Figure 5–62

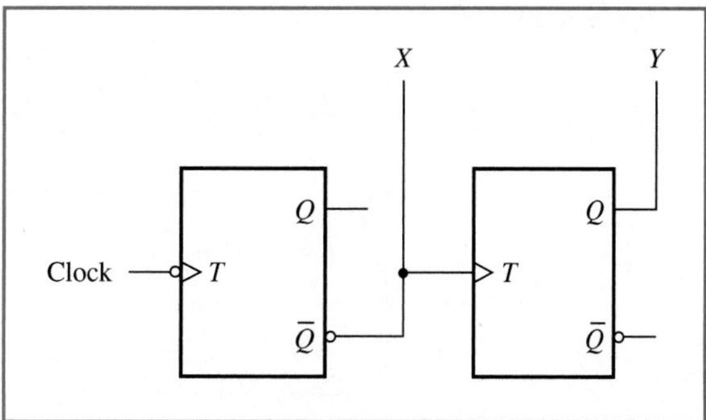

Figure 5–63

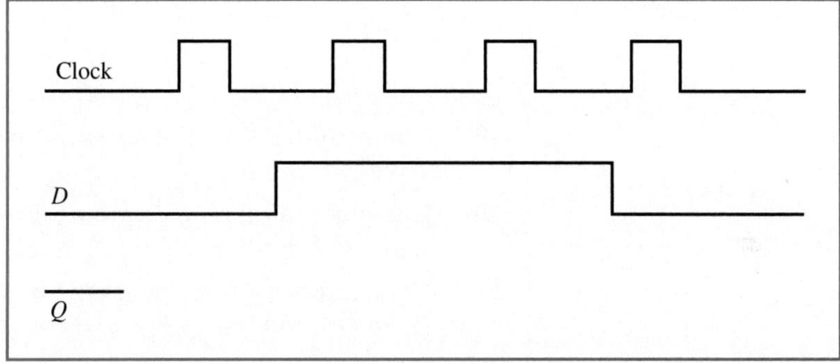

13. Complete the timing diagram shown in Figure 5–64 for a negative edge-triggered *D* flip-flop.
14. Repeat Problem 13 for a positive edge-triggered flip-flop.

Section 5–7 The J-K flip-flop

15. Complete the timing diagram shown in Figure 5–65 for a positive edge-triggered *J-K* flip-flop.
16. Repeat Problem 15 for a negative edge-triggered flip-flop.
17. Complete the timing diagram shown in Figure 5–66 for a negative edge-triggered *J-K* flip-flop.
18. Repeat Problem 17 for a positive edge-triggered flip-flop.

Figure 5–64

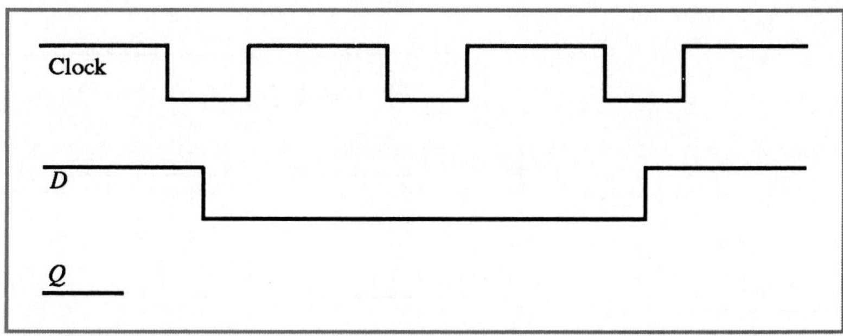

Figure 5–65

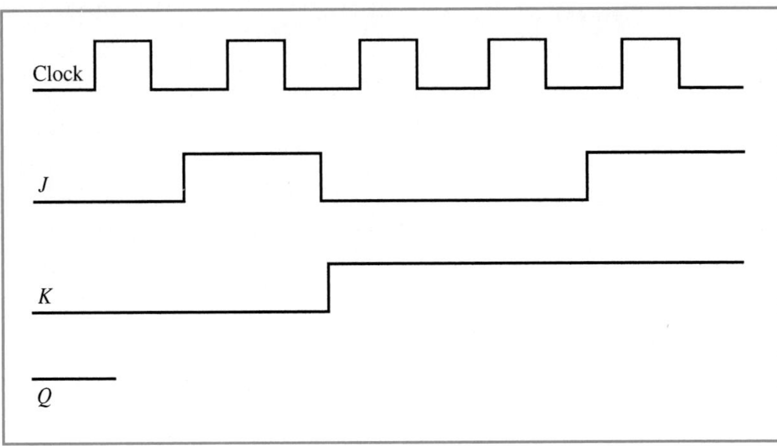

Section 5–8 Master-slave flip-flops

19. Complete the timing diagram shown in Figure 5–60 for a master-slave *S-R* flip-flop.
20. Complete the timing diagram shown in Figure 5–61 for a master-slave *S-R* flip-flop.
21. Complete the timing diagram shown in Figure 5–63 for a master-slave *D* flip-flop.
22. Complete the timing diagram shown in Figure 5–64 for a master-slave *D* flip-flop.
23. Complete the timing diagram shown in Figure 5–65 for a master-slave *J-K* flip-flop.
24. Complete the timing diagram shown in Figure 5–66 for a master-slave *J-K* flip-flop.
25. Draw a timing diagram for a master-slave *T* flip-flop.

Section 5–9 Flip-flop conversions

26. Configure the *S-R* flip-flop to function like a *J-K* flip-flop.
27. Configure the *S-R* flip-flop to function like a *D* flip-flop.

Figure 5–66

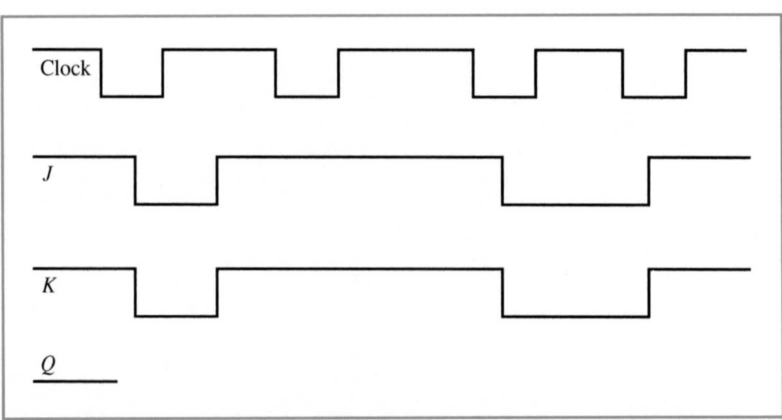

28. Configure the *S-R* flip-flop to function like a *T* flip-flop.

29. Analyze the circuit shown in Figure 5–37 by constructing a state transition table showing Q^{n+1} as a function of *S*, *R*, Q^n, and *D*.

30. Analyze the circuit shown in Figure 5–39 by constructing a state transition table showing Q^{n+1} as a function of *J*, *K*, Q^n, $\overline{Q}^n$, and *D*.

31. The operation of the *X-Y* flip-flop whose logic symbol is shown in Figure 5–67 is described by the following summarized state transition table.

X	Y	Q^{n+1}
0	0	1
0	1	Q^n
1	0	$\overline{Q}^n$
1	1	0

Construct a complete state transition table and obtain a state transition equation for the *X-Y* flip-flop.

32. Convert the *X–Y* flip-flop from Problem 31 to a *D*-type flip-flop.

33. Convert the *X-Y* flip-flop from Problem 31 to a *T*-type flip-flop.

34. Convert the *X-Y* flip-flop from Problem 31 to a *J-K*–type flip-flop.

35. Convert the *X-Y* flip-flop from Problem 31 to a *S-R*–type flip-flop.

36. The transition equation for the *L-M* flip-flop is

$$Q^{n+1} = \overline{L}Q^n + ML$$

Construct a transition table for the flip-flop.

Section 5–10 Multivibrators

37. Design a 74121 one-shot circuit to produce an active-high pulse having a duration of 1 ms. You may pick any appropriate values of R_x and C_x.

38. Design a 74122 one-shot circuit to produce an active-low pulse having a duration of 0.5 ms. You may pick any appropriate values of R_x and C_x.

Figure 5–67

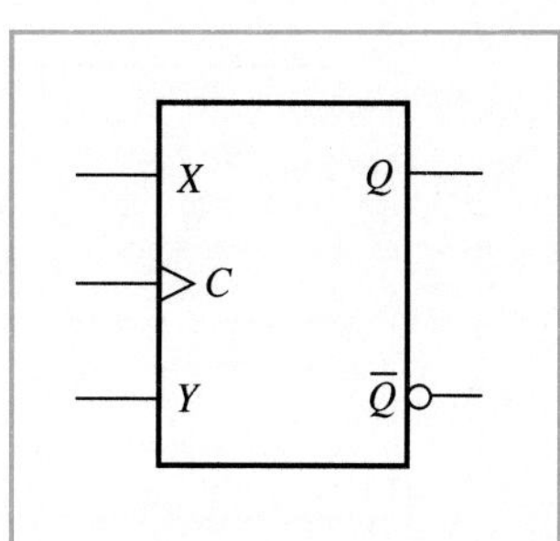

Figure 5–68

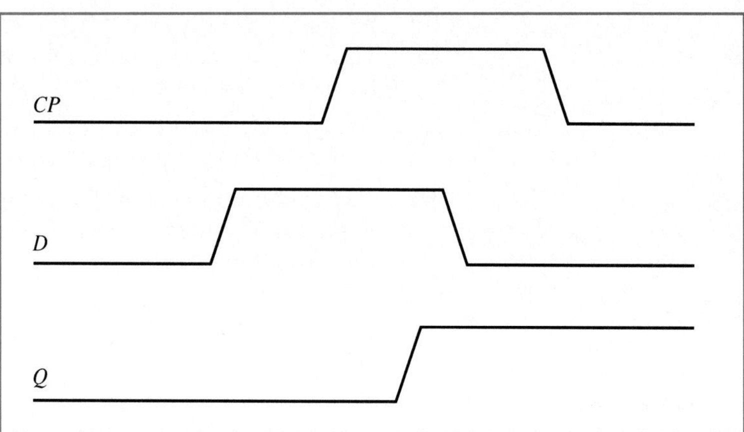

39. Determine the values of R_x and C_x in Figure 5–45 to obtain a free-running multivibrator having a symmetric square-wave output with a frequency of 1000 Hz.
40. Determine the values of R_A, R_B, and C for the 555 timer circuit in Figure 5–47 to produce a free-running multivibrator having a symmetric square-wave output with a frequency of 1000 Hz.

Section 5–11 Switching characteristics of flip-flops

41. Figure 5–68 shows a timing diagram of a positive edge-triggered *D* flip-flop. Identify the following on the diagram:
    a. The propagation delay t_{PHL}
    b. The setup time t_s(H)
    c. The hold time t_h(H)
42. Figure 5–69 shows a timing diagram of a negative edge-triggered *D* flip-flop. Identify the following on the diagram:
    a. The propagation delay t_{PHL}
    b. The setup time t_s(L)
    c. The hold time t_h(L)

Figure 5–69

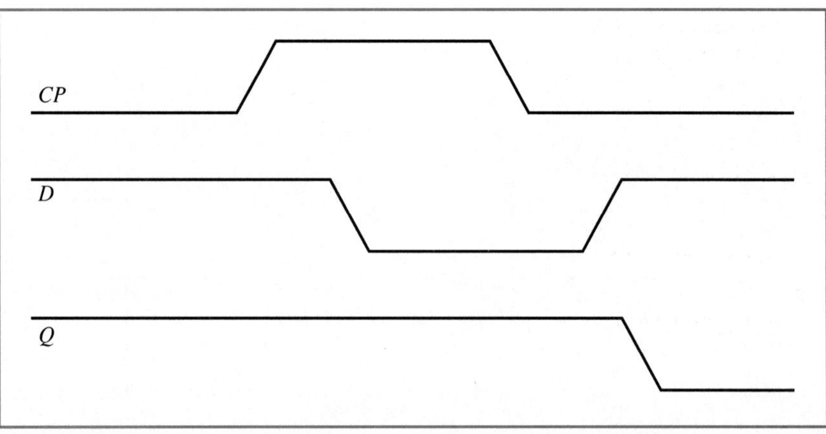

Troubleshooting

Section 5–12 Troubleshooting sequential logic

43. Construct a circuit to test a *J-K* flip-flop. Discuss the procedure that could be used to verify its operation.

44. Referring to the timing diagram of Figure 5–57, calculate a safe frequency to operate the flip-flop while maintaining the setup times at the minimum of 20 ns.

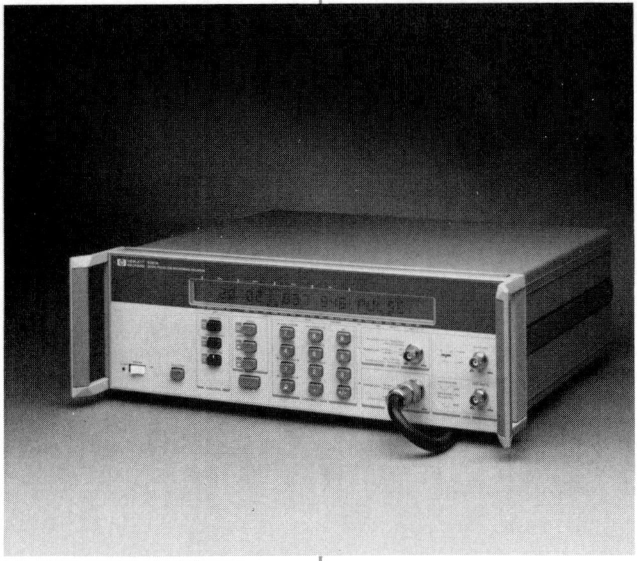

The dynamic operation of digital circuits often requires the use of a digital frequency counter (shown in the photograph) for testing and troubleshooting procedures (Photo courtesy Hewlett-Packard Company).

OUTLINE

OBJECTIVES

The objectives of this chapter are to

- study the design and operation of asynchronous binary counters and their analysis using timing diagrams.
- introduce synchronous counters and their analysis using timing diagrams.
- learn the design procedures (synthesis) used to design various types of synchronous counters.
- develop a formal technique for analyzing synchronous binary counters.
- introduce an application of counters by designing a 24-hour time-of-day digital clock.
- study the application of counters for frequency division.
- investigate the operation and applications of the various IC implementations of counters.

6 COUNTERS

6–1 INTRODUCTION

The basic flip-flop introduced in Chapter 5 has many applications in sequential logic circuits, one of which is in the design of *counters*. A *counter* is a circuit that produces a numeric count each time an input clock pulse makes a transition. The general logic symbol for a typical counter is shown in Figure 6–1. The clock input to the counter is used to change the state of the counter in much the same way as the clock input to a flip-flop is used to change the state of the flip-flop; the difference is that a counter has several outputs lines (labeled X in Figure 6–1) that can exist in several states (called *counts*) while a flip-flop has only one output (not including $\overline{Q}$) that can exist in one of two states. The maximum number of states that a counter can have depends on the relationship 2^n where n is the number of outputs. Thus a 3-bit counter will have 8 possible states and a 4-bit counter will have 16. During each state, the output of the counter has a different value that identifies the state of the counter. The maximum number of states that a counter counts through is known as its *modulus*. The number of outputs dictate the size of the counter (2-bit, 3-bit, 4-bit, etc.), and the actual sequence in which the counter counts (BCD, binary, etc.) at these outputs depends on its design. A counter may have one or more control inputs (labeled C in Figure 6–1) that could be used to modify the counting sequence of the counter.

This chapter studies the operation, design, and analysis of the various types of counters used in digital circuit applications. The counters introduced in this chapter are classified into two broad categories—synchronous and asynchronous. Asynchronous counters are characterized by very simple logic circuits as long as the count they produce is simple. For more complex counting sequences synchronous counters can be very easily designed using the techniques of synthesis covered in Section 6–4. The analysis of these counters is also included in this chapter. The timing diagram is a versatile tool for analyzing all types of counters, but for synchronous counters the analysis techniques discussed in Section 6–5 is more direct. Sections 6–6 and 6–7 include the complete design of a counter application—a time-of-day digital clock—and the application of counters in frequency division. Finally, Section 6–8 investigates the operation and applications of various counters implemented in the form of ICs.

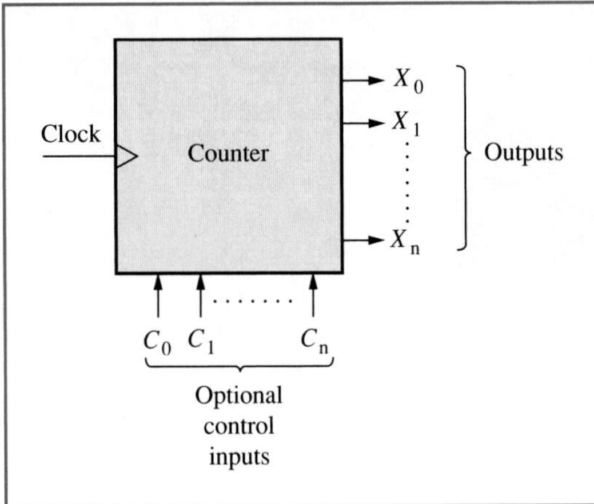

Figure 6–1 General logic symbol for a counter.

6–2 Asynchronous counters

An *asynchronous counter* is made up of a set of *T* flip-flops connected in series as shown in Figure 6–2. Note that the use of the *T* flip-flop is symbolic and could be replaced with the *J-K* or *D* implementations designed in Figures 5–35 and 5–38, respectively. The counter has three outputs labeled $X_2X_1X_0$ and a *CLOCK* input that is used to cycle the counter through its various states. The counter is therefore a 3-bit counter and will have 2^3 or 8 possible states. The inputs and outputs of each flip-flop are labeled with the subscripts 0, 1, and 2 to identify the significance of each bit. Notice that the *CLOCK* input is connected to flip-flop F0 which will toggle each time a *positive transition* of the clock pulse takes place. The output of *F*0 is connected to the *T* input of *F*1 and serves as the clock for *F*1; *F*1 will therefore toggle each time the output of *F*0 makes a positive transition. Similarly the output of *F*1 is the clock for *F*2 and therefore *F*2 will toggle each time the output of *F*1 makes a positive transition. Thus it is apparent that all flip-flops will not change states at the same time, and hence this type of counter is called an *asynchronous* counter since the term *asynchronous* is used to define events that do not occur at the same time. This type of counter is also called a *ripple counter* since the clock pulse appears to *ripple* through each flip-flop connected in series.

Figure 6–3 shows the timing diagram for the 3-bit ripple counter of Figure 6–2. The timing diagram shows the outputs of the counter (which

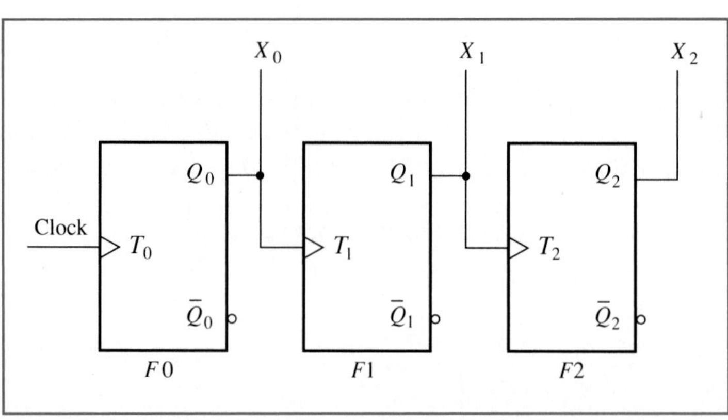

Figure 6–2 A 3-bit asynchronous down counter.

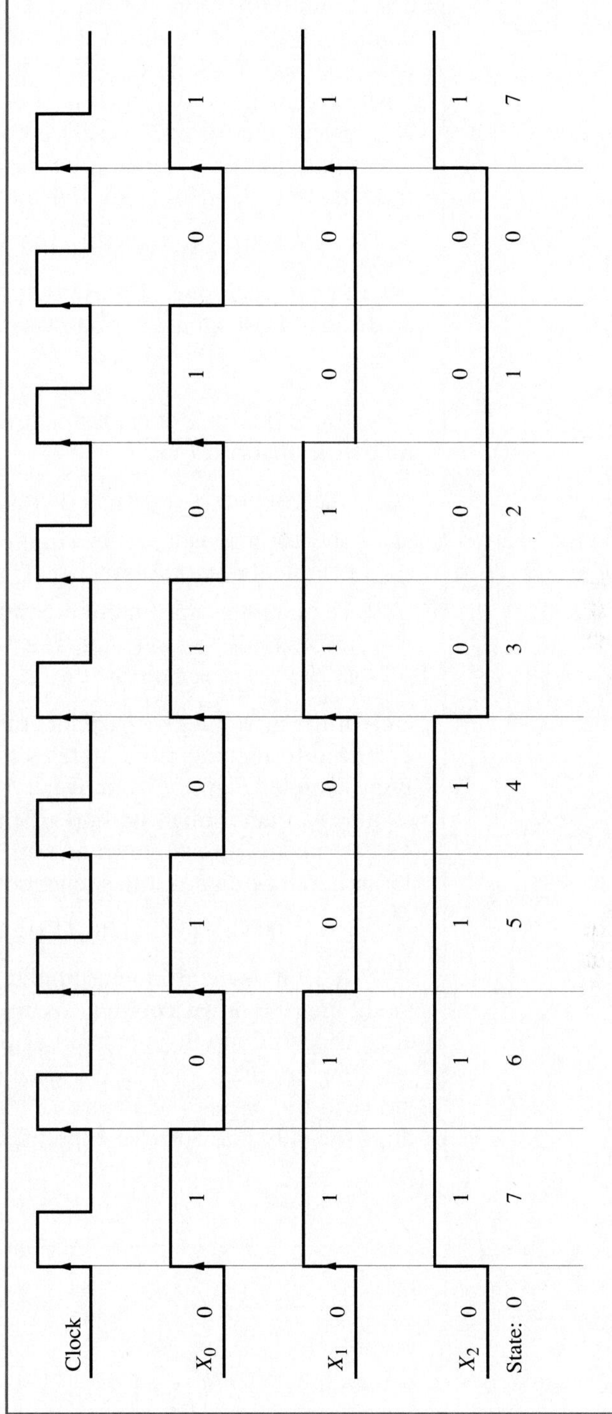

Figure 6–3
Timing diagram for a 3-bit asynchronous down counter.

are the outputs of the flip-flops) as a function of the *CLOCK* input of the counter. We have assumed that the counter is initially at a state at which its output is 000 (State 0). As each clock pulse of *CLOCK* makes a positive transition, X_0 toggles. Similarly, for each positive transition of X_0, X_1 toggles, and for each positive transition of X_1, X_2 toggles. If we determine the binary code present at the output of the counter for each transition of the main clock pulse *CLOCK*, the sequence is

000, 111, 110, 101, 100, 011, 010, 001, 000, 111, . . .

We refer to each one of these outputs as *states* and list them respectively as decimal equivalents as follows:

0, 7, 6, 5, 4, 3, 2, 1, 0, 7, . . .

From the timing diagram shown in Figure 6–3 we can make the following observations:

1. The counter is a *down counter* since it counts down from 7 to 0.
2. The counter is a *binary counter* since it counts through all possible combinations of 3 bits.
3. The counter will function continuously. In other words once it *cycles* through all possible states, it will *recycle* through those same states *in the same sequence*.

One of the advantages of ripple counters is that they can easily be extended to include more states simply by adding more flip-flops in series. For example, Figure 6–4 shows a 4-bit binary ripple down counter that includes an additional flip-flop *F*3 that will toggle each time the output of *F*2 (X_2) makes a positive transition. The counter will have 16 states 0 to 15 and count down in the sequence

0000, 1111, 1110, 1101, 1100, 1011, 1010, . . . , 0000

A 3-bit asynchronous ripple up counter can be constructed by modifying the down counter circuit in Figure 6–2 to contain negative edge-triggered flip-flops instead of positive edge-triggered flip-flops. The resulting circuit is shown in Figure 6–5. Notice that there is no difference between the circuits of Figures 6–2 and 6–5 other than the type of flip-flops (negative or positive edge-triggered) used. The operation of the up

Figure 6–4
A 4-bit binary ripple down counter.

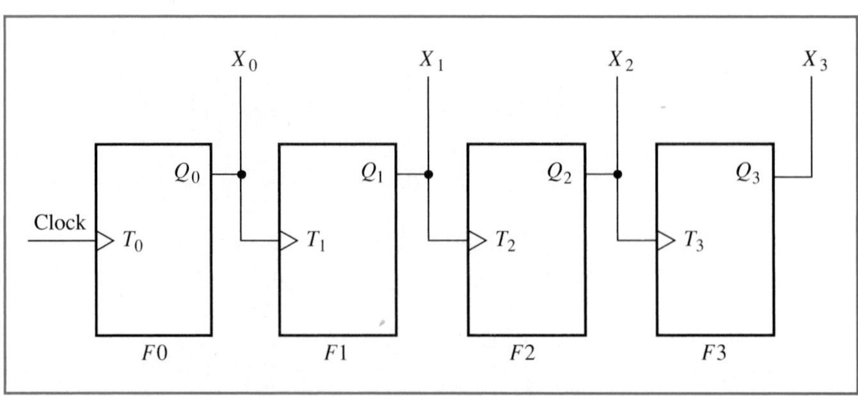

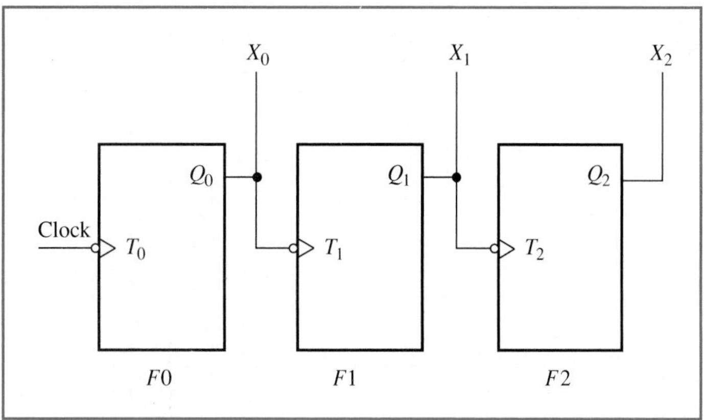

Figure 6–5
A 3-bit binary ripple up counter.

counter can be verified by analyzing the circuit and obtaining a timing diagram as shown in Figure 6–6. The timing diagram analysis in Figure 6–6 is conducted in much the same way as the analysis for the down counter, except that the output of flip-flop $F0$ (X_0) toggles on the negative edge of the clock pulse. Similarly, the output of $F1$ (X_1) toggles on the negative edge of X_0, and the output of $F2$ (X_2) toggles on the negative edge of X_1. Therefore for each transition of the *CLOCK*, the sequence of states that the counter goes through is as follows:

$$0, 1, 2, 3, 4, 5, 6, 7, 0, 1, \ldots$$

Notice that the count sequence is an up count since it increases from 0 to 7 and then recycles back to 0.

Recall from Chapter 5 that a negative edge-triggered flip-flop is actually a positive edge-triggered flip-flop with an inverted clock input. Therefore to convert the down counter circuit in Figure 6–2 to an up counter, we do not have to replace the positive edge-triggered flip-flops with negative edge-triggered flip-flops (as shown in Figure 6–5) but can simply invert the clock pulses entering flip-flops $F1$ and $F2$. The type of flip-flop used for $F0$ is not significant since it is triggered by the main *CLOCK*. Instead of adding inverters to the circuit we can use the $\overline{Q}$ outputs, effectively producing the same results as shown in Figure 6–7.

The timing diagram for the up counter in Figure 6–7 is shown in Figure 6–8 (p. 364). The analysis for this counter is a little more complex because each flip-flop in series is not triggered by the output of the previous flip-flop but is triggered by the complement of its output. In Figure 6–8 since $F0$ is directly triggered by the *CLOCK* it changes states on the positive transition of each clock pulse. $F1$ is triggered on the positive edge of $\overline{Q}_0$ which is the complement of X_0 as shown in the timing diagram; thus X_1 toggles on the positive edge of $\overline{Q}_0$ which is actually the negative edge of X_0. Similarly X_2 toggles on the positive edge of $\overline{Q}_1$ which is the negative edge of X_1. The resulting timing diagram is similar to the timing diagram of the up counter shown in Figure 6–6.

It is possible for us to incorporate the design of Figures 6–2 and 6–7 into a counter known as an *up/down counter*. The logic symbol for such a counter is shown in Figure 6–9a (p. 365). Notice that the counter has a *control input, S,* that is used to select the desired counting sequence—if

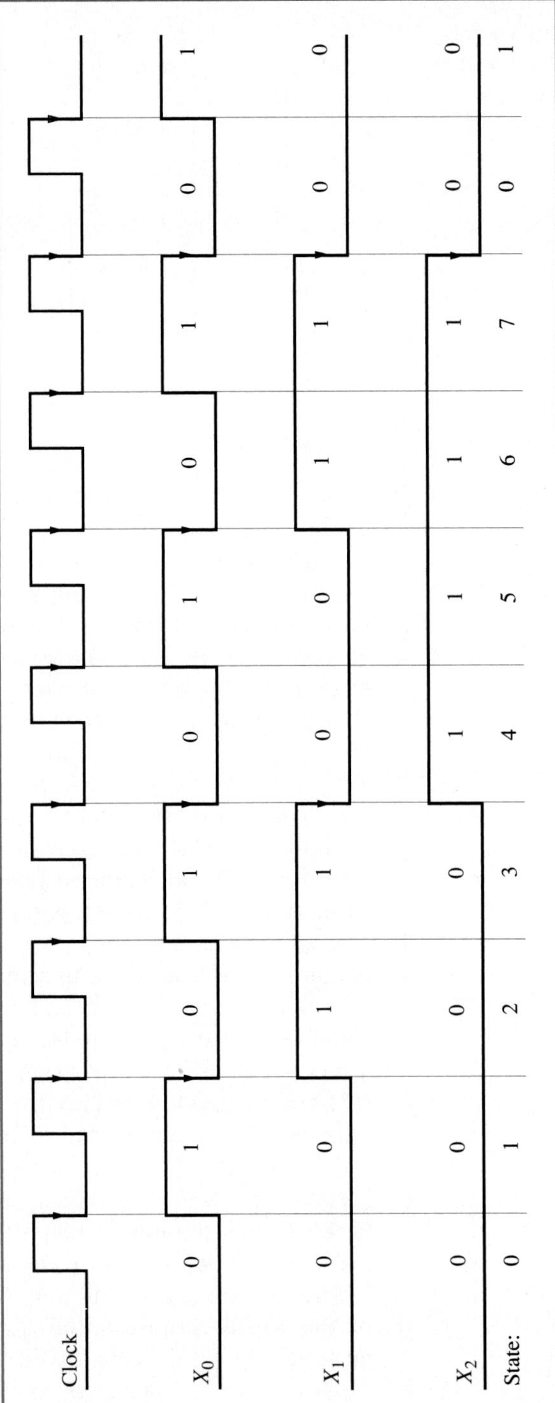

Figure 6–6
Timing diagram for a 3-bit binary ripple up counter.

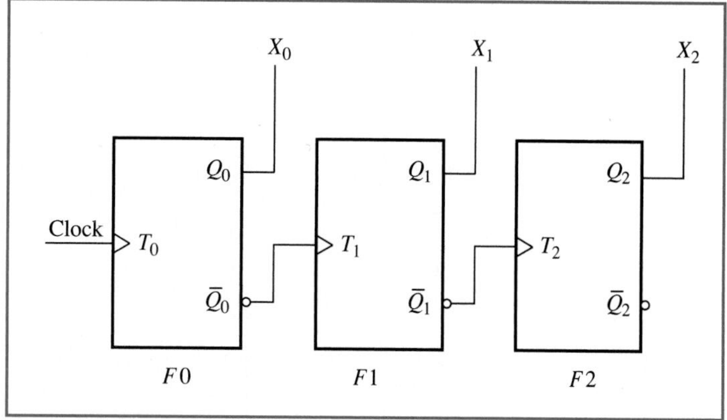

Figure 6–7
A 3-bit binary ripple up counter using positive edge-triggered flip-flops.

$S = 0$ then the counter counts *UP*, and if $S = 1$ then the counter counts *DOWN*. The logic circuit for the up/down counter shown in Figure 6–9b uses two-channel, single-bit multiplexers to configure the circuit as a down counter or an up counter. When $S = 0$, Channel B of each multiplexer is multiplexed to the output; this effectively "connects" the $\bar{Q}$ output of each flip-flop to the clock of the next flip-flop in series, configuring the circuit like the up counter shown in Figure 6–7. When $S = 1$, Channel A of each multiplexer is multiplexed to the output; this effectively "connects" the Q output of each flip-flop to the clock of the next flip-flop in series, configuring the circuit like the down counter shown in Figure 6–2.

The manner in which the flip-flops are connected together and the types of flip-flops used (negative or positive edge-triggered) determine the counting sequence of an asynchronous counter. The following example illustrates the analysis of a 4-bit asynchronous counter that counts in a disordered sequence.

EXAMPLE 6–1

Analyze the 4-bit ripple counter shown in Figure 6–10 by obtaining a timing diagram for the outputs $X_3X_2X_1X_0$. Since the counter will have 16 possible states, the timing diagrams should be analyzed with at least 17 clock pulses to verify that the counter recycles. Assume that the counter starts off at state 13 (1101).

The timing diagram for Figure 6–10 is shown in Figure 6–11 (p. 366). Notice that X_3 toggles on each positive transition of the main *CLOCK*. X_2 toggles on each positive transition of $\bar{Q}_3$ (the complement of X_3). X_1 toggles on each positive transition of X_2. X_0 toggles on each negative transition of $\bar{Q}_1$ (the complement of X_1). The sequence of states through which the counter cycles is also shown in the timing diagram.

A counter does not necessarily have to cycle through all possible states. For example, in order to implement a BCD counter that cycles through ten states, 0 through 9, we must use a 4-bit counter (to accommodate the numbers 1000 and 1001) that has a maximum of 16 possible states. Therefore a BCD counter is a 4-bit counter implemented with a modulus of ten (uses only 10 of the 16 possible states). Since a BCD counter has ten counts, it is often referred to as a *decade* counter.

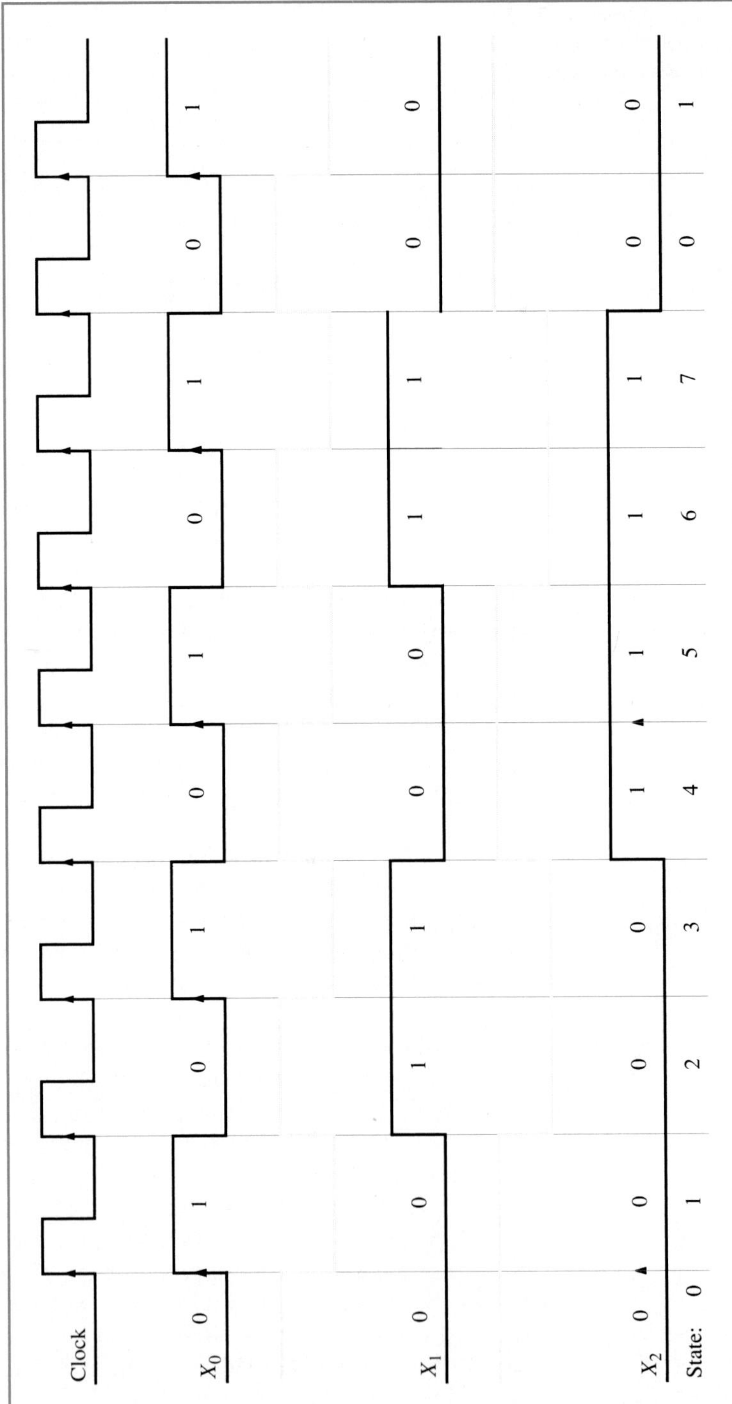

Figure 6–8
Timing diagram for Figure 6–7.

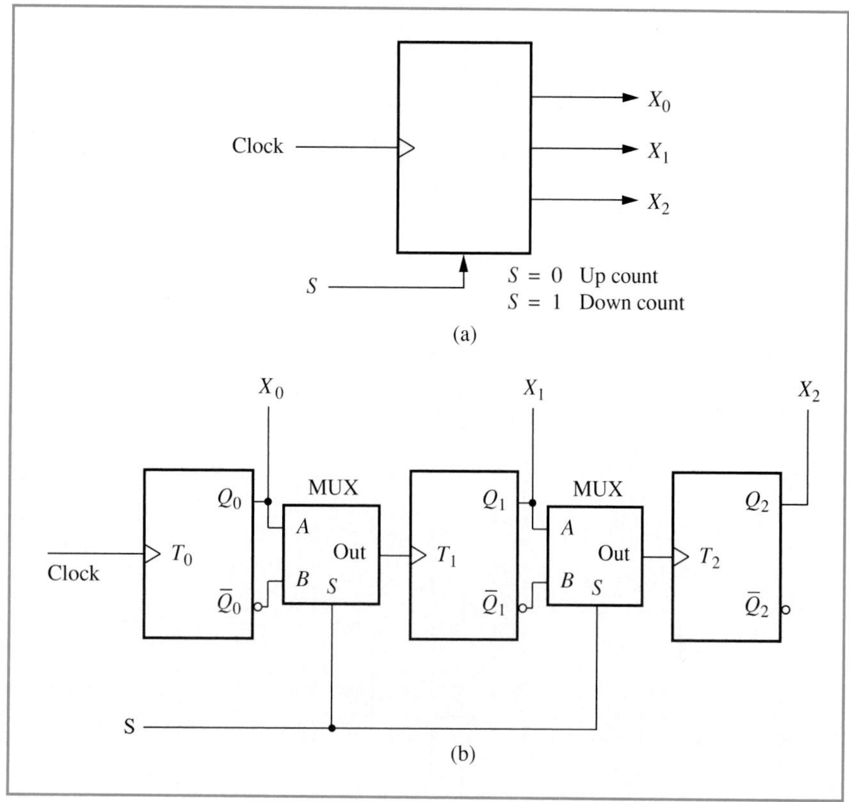

Figure 6–9
An asynchronous up/down counter. (a) Logic symbol; (b) circuit.

To design a BCD ripple counter we must first construct a 4-bit binary counter (up or down) and then modify the circuit to automatically reset to 0000 after the count reaches 1001 (9). Figure 6–12 shows the logic circuit for a BCD up counter. Notice that we have used *J-K* flip-flops in this circuit with the *K* inputs to each flip-flop permanently connected to a logic 1 state. The *J* inputs of all the flip-flops are connected together to the output of a *NAND* gate. The output of this *NAND* gate will be a logic 1 for all counts from 0000 to 1000 (states 0 to 8) since either of its inputs X_0 or X_3 will be a logic 0 during these counts. Since the *J* inputs (and also the *K* inputs) to all the flip-flops are logic 1's, the flip-flops function as toggle flip-flops during states 0 through 8. When the counter reaches state

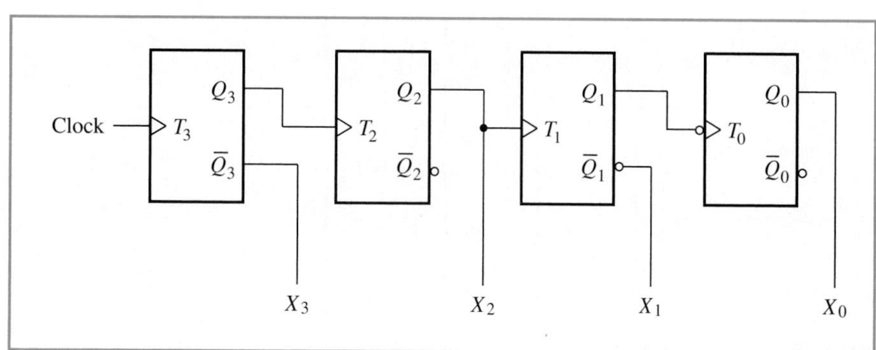

Figure 6–10

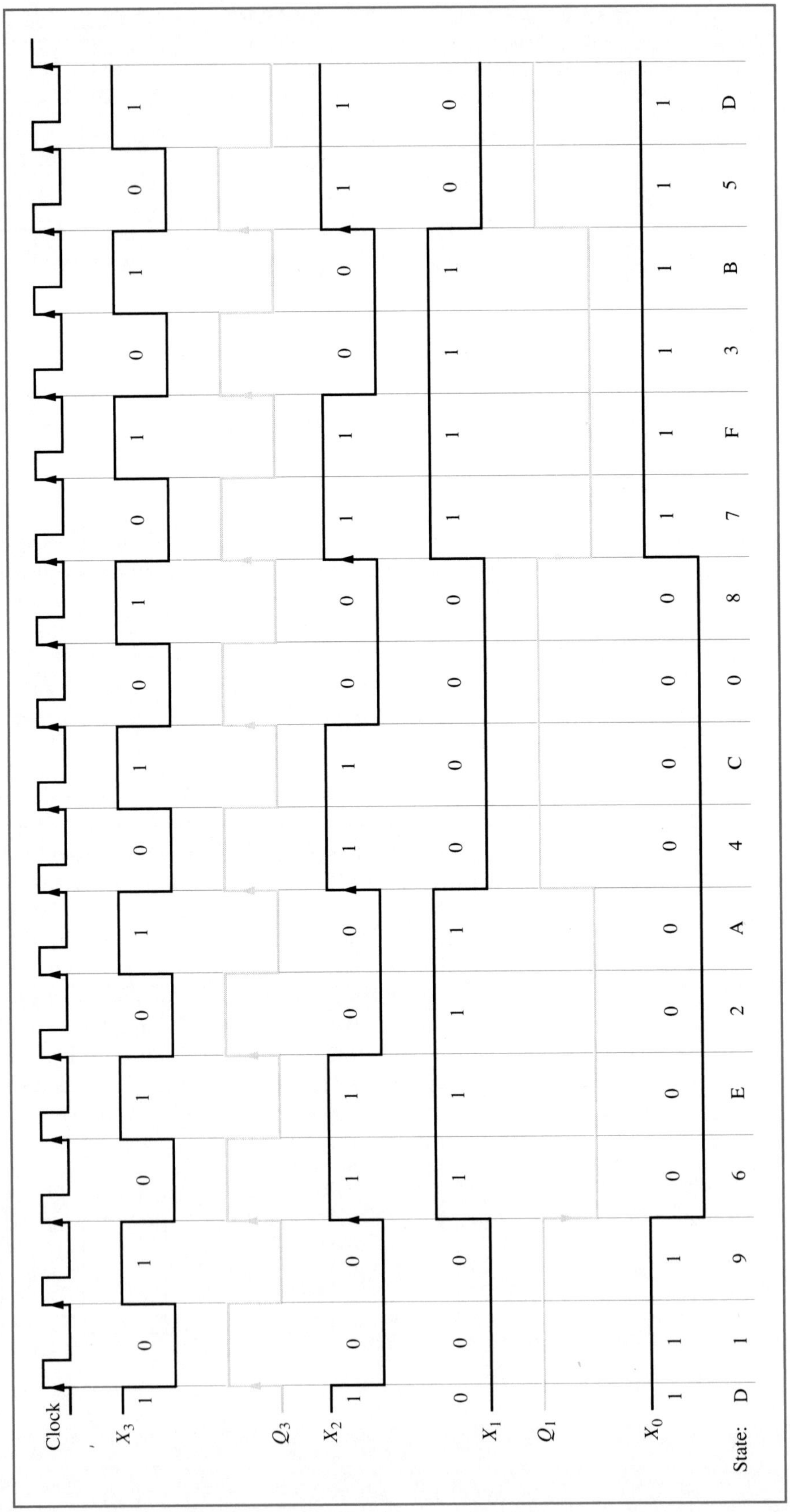

Figure 6–11

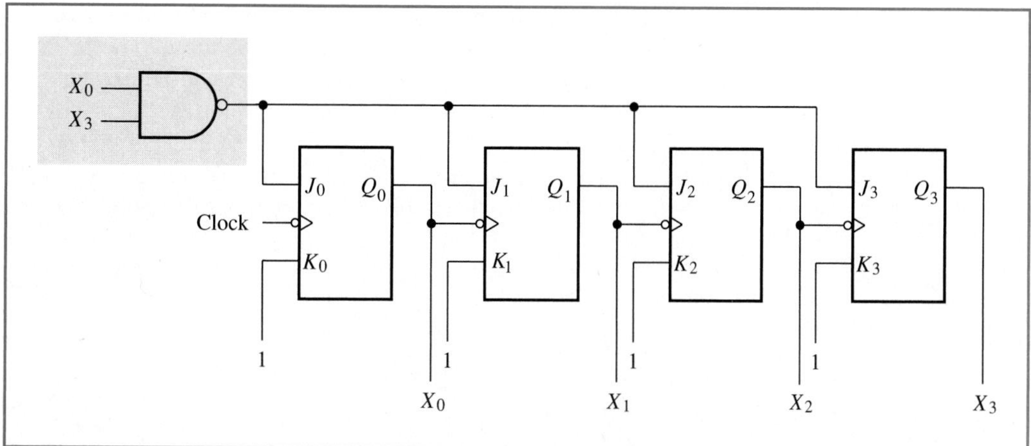

Figure 6–12
An asynchronous BCD up counter.

9 and its output is 1001, X_0 and X_3 are both logic 1's, and the output of the *NAND* gate is a logic 0. Since all the *J* inputs are now logic 0's (and all *K* inputs are logic 1's), on the next transition of the clock pulse all the flip-flops reset, the output of the counter is 0000 (state 0), and the counter recycles. Figure 6–13 shows the timing diagram for the BCD counter.

The BCD counter just described uses a technique known as *counter decoding*. This process uses a combinational logic circuit (or gate) to *monitor* the output of the counter. When the output of the counter reaches a certain value, the combinational logic circuit resets the counter.

Figure 6–13
Timing diagram for the BCD counter.

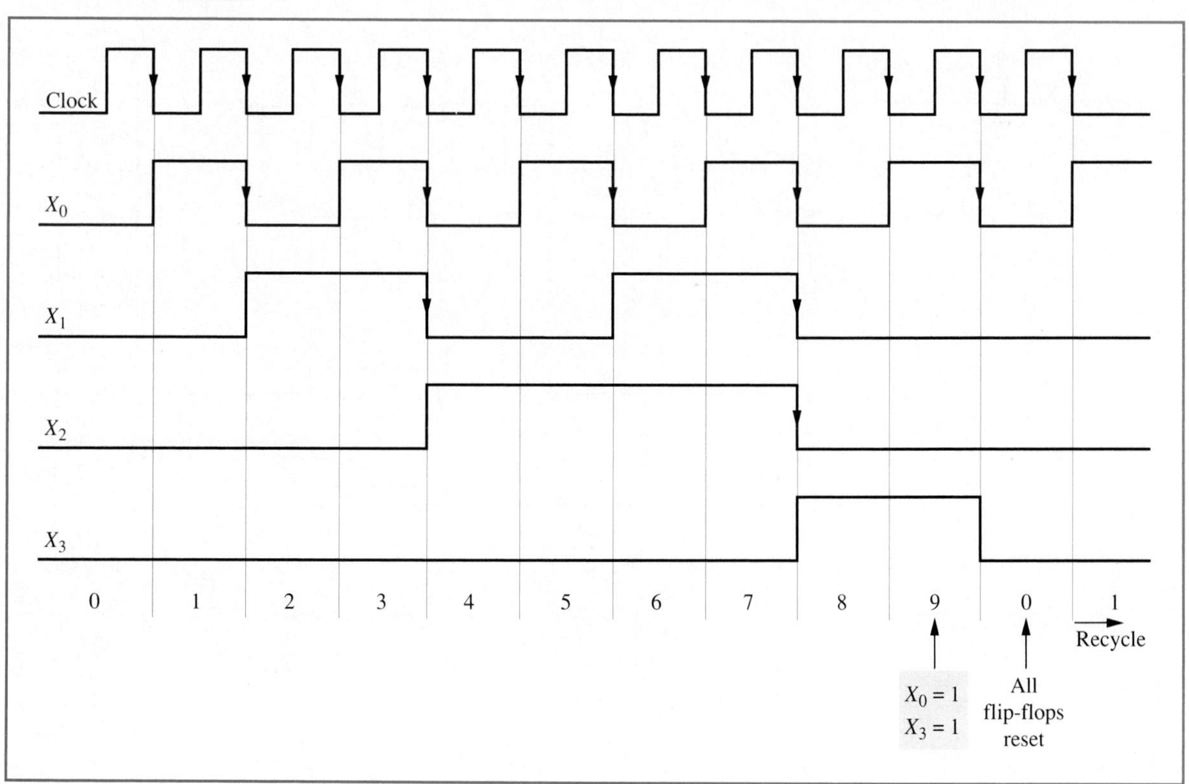

The direct set and clear (asynchronous) inputs of the flip-flops can also be used in conjunction with counter decoding. The following examples illustrate the design of a counter that uses the asynchronous inputs of the flip-flops to change the counting sequence.

EXAMPLE 6–2

Figure 6–14 shows a 3-bit asynchronous counter that counts from 0 to 5 (000 to 101) and then recycles to 0 (000). We have used *J-K* flip-

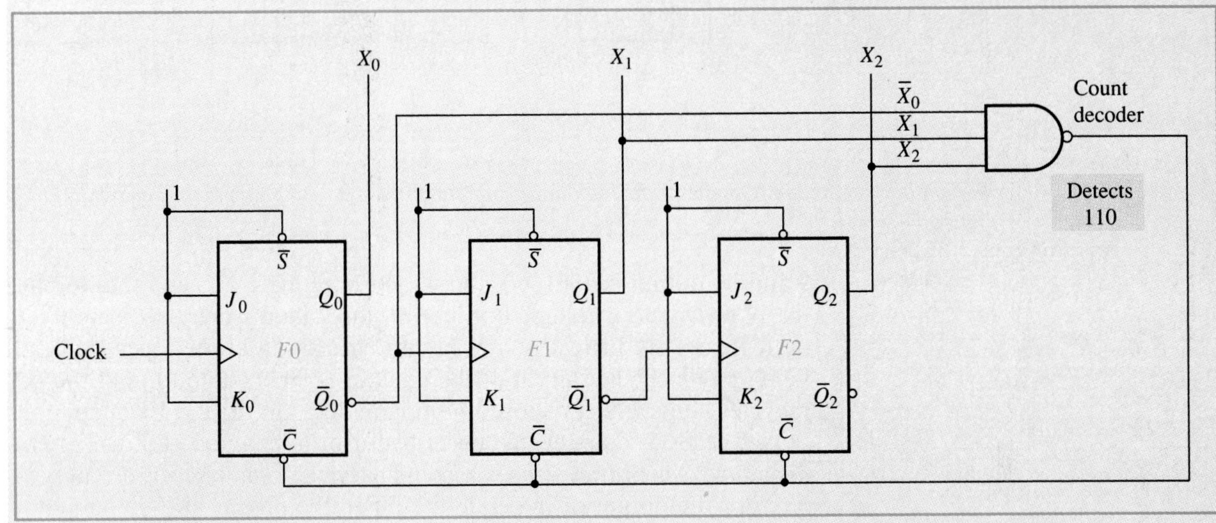

Figure 6–14
Counter implemented with a count decoder.

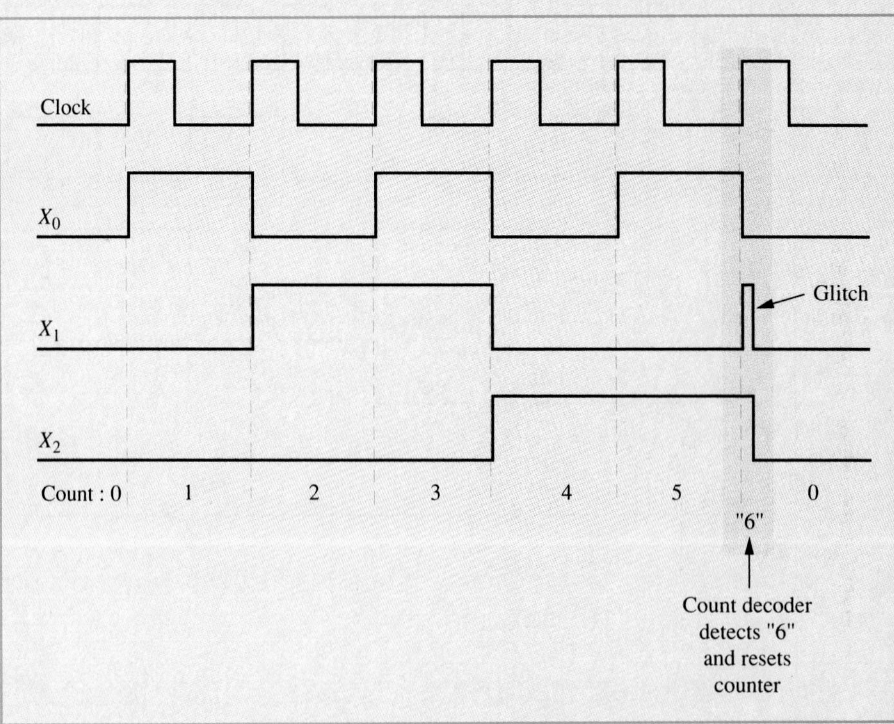

Figure 6–15
Timing diagram for Figure 6–14.

flops configured as *T* flip-flops. Since we are not using the direct *SET* inputs, $\overline{S}$, we have tied them inactive at the logic 1 state. The direct *CLEAR* inputs, $\overline{C}$, are connected together to the output of a *NAND* gate that is called the *count decoder*. The output of the counter $X_2X_1\overline{X}_0$ is applied to the input of the *NAND* gate, and when the count reaches 6 (110) the output of the *NAND* gate produces a logic 0 which (almost instantaneously) *CLEARS* all the flip-flops. The counter therefore recycles to 000 after a count of 5 (101).

Figure 6–15 shows the timing diagram of the 0-5 counter. Notice that after a count of 5 (101), for a very short duration, the output of the counter does change to 6 (110) but the count decoder detects the 110 combination during this state and resets the counter. However, we can see that a spurious output, sometimes called a glitch, is produced on the X_1 output. This could be a problem in some applications as will be seen in Section 6–9.

The use of counter decoding does not have to be limited to simply resetting the counter to the 0 state. We can also use counter decoding to recycle the counter to any other state as shown in the following example.

EXAMPLE 6–3

Figure 6–16 shows a 3-bit asynchronous counter that counts from 3 to 7 (011 to 111) and then recycles to 3 (011). This time we have used *D* flip-flops configured as *T* flip-flops although *J-K* flip-flops configured as *T* flip-flops could have been used as well.

The count decoder will produce a logic 0 when the counter reaches a count of 000. The output of the count decoder is used to *SET* flip-flop no. 0 and flip-flop no. 1, and to *RESET* flip-flop no. 2; this will preset the counter to 011. Therefore when the counter reaches a count of 7 (111) on the next count, 0 (000), the count decoder will cause the outputs of the flip-flops to recycle to 3 (011). The timing diagram of the counter is shown in Figure 6–17. Again, notice that since the counter does momentarily recycle to 000 after a count of 111, glitches are produced at the X_0 and X_1 outputs.

Figure 6–16
Counter implemented with a count decoder.

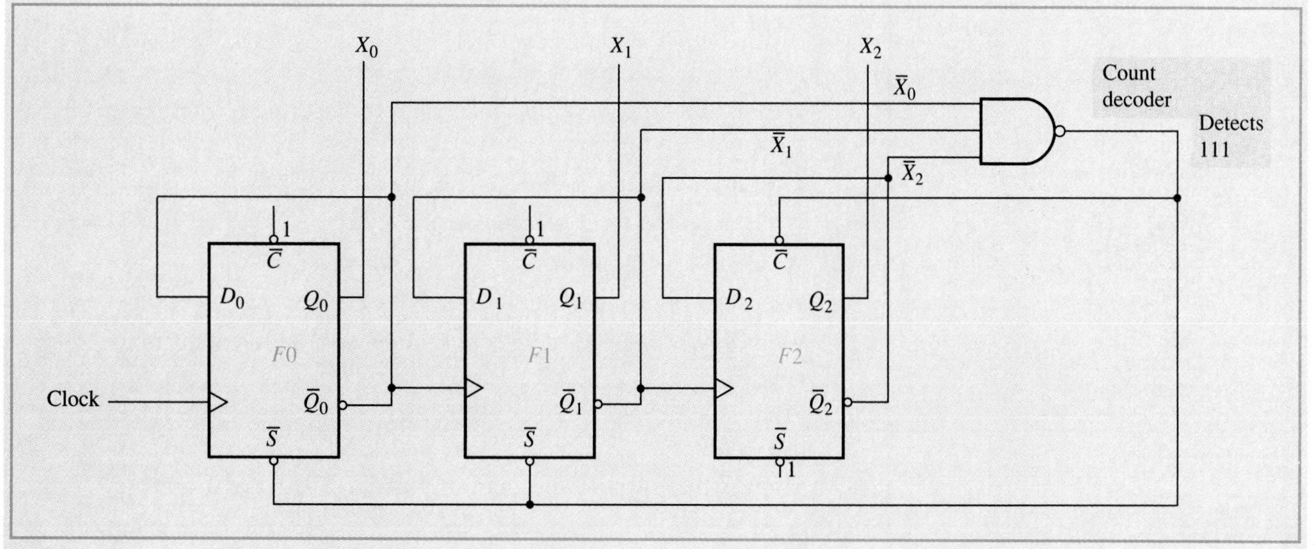

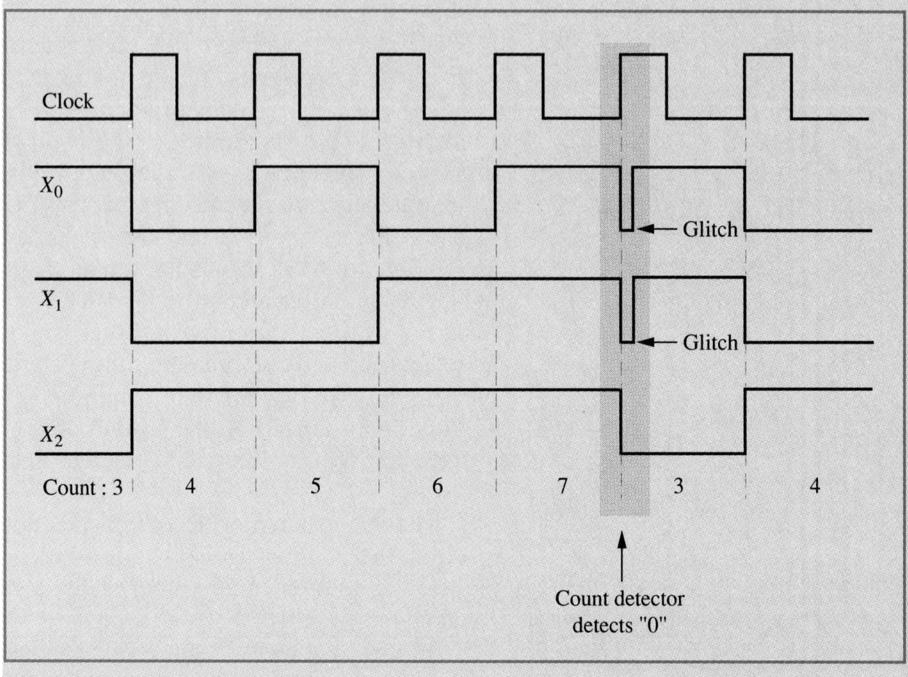

Figure 6–17
Timing diagram for Figure 6–16.

Asynchronous counters are fairly simple logic circuits and can be analyzed very easily though the use of timing diagrams, provided that the circuits are fairly simple. However, the design of asynchronous counters with unusual count sequences is quite difficult since there is no *formal* procedure and therefore the design must be conducted on a trial-and-error basis. Asynchronous counters are limited to *T* flip-flops or *D* and *J-K* flip-flops configured as *T* flip-flops. The next few sections deal with the design and analysis of synchronous counters. Unlike asynchronous counters, there is a formal procedure that can be used to design practically any type of count sequence for a synchronous counter from a given specification. The same procedure can be used in reverse to analyze synchronous counters.

Review questions

1. Why is the term *asynchronous* used to describe asynchronous counters?
2. Why are asynchronous counters known as ripple counters?
3. What does the term *modulus* mean?
4. What is an up counter and a down counter?
5. What is a binary counter?
6. What determines the counting sequence of an asynchronous counter?
7. What is a BCD counter? What other terms are used to identify this counter?
8. What is the purpose of counter decoding? How is it done?

6–3 Synchronous counters

A synchronous counter is made up of a set of flip-flops that are triggered simultaneously (in parallel) by the main clock of the counter. Since all the flip-flops change states when the clock makes its transition, the term *synchronous* is used to describe this type of counter. Unlike asynchronous counters, all the flip-flops used in the counters are not always *T* flip-flops (or configured as *T* flip-flops). Furthermore, changing the type of flip-flops used in a synchronous counter (negative or positive edge-triggered) does not affect the counting sequence as long as all flip-flops are of the same type.

Figure 6–18 shows the logic circuit and timing diagram for a 2-bit synchronous binary up counter. Notice in the logic circuit shown in Figure 6–18a that the clocks of both the flip-flops are connected together. Also note that flip-flop *F*0 is configured as a toggle flip-flop, but flip-flop *F*1 is not. The timing diagram for the counter is shown in Figure 6–18b for five clock pulses; the intervals between clocks pulses have been numbered 0 through 6 for identification purposes. To analyze the circuit and construct the timing diagram shown in Figure 6–18b we first complete the timing for the output X_0 since *F*0 will toggle each time a positive transition of the clock pulse takes place. To trace the output X_1 we must determine the logic levels at J_1 and K_1 *before* each clock pulse makes a transition and causes flip-flop F_1 to change states. The following analysis traces the logic level of X_1 during each interval.

Figure 6–18
A 2-bit synchronous binary up counter. (a) Logic circuit; (b) timing diagram.

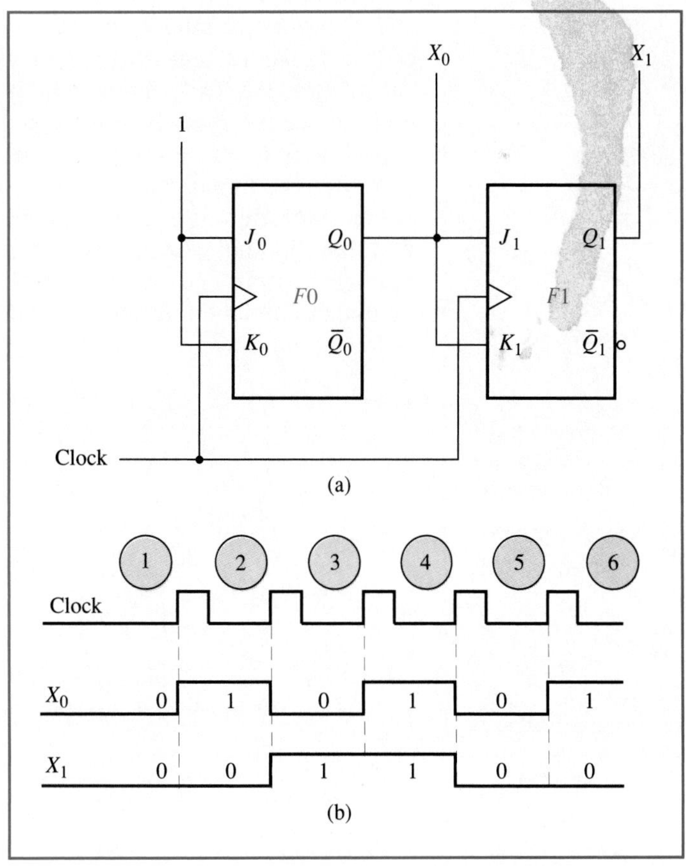

① This is the initial condition. $X_1 = 0$ and $X_0 = 0$. Since X_0 is connected to J_1 and K_1, and since X_0 is a 0 ($X_0 = J_1 = K_1 = 0$), when the next transition of the clock takes place, $F1$ will not change states.

② $X_0 = 1$ and X_1 remains at a 0. Again, $J_1 = K_1 = 1$ (since $X_0 = 1$). Therefore, when the next transition takes place, $F1$ will toggle to a 1.

③ $X_0 = 0$ and X_1 toggles to a 1. Since $X_0 = J_1 = K_1 = 0$, when the next transition takes place, $F1$ will not change states.

④ $X_0 = 1$ and X_1 remains at a 1. Since $X_0 = J_1 = K_1 = 1$, when the next transition takes place, $F1$ will toggle to a 0.

⑤ $X_0 = 0$ and X_1 toggles to a 0. Since $X_0 = J_1 = K_1 = 0$, when the next transition takes place, $F1$ will not change states.

⑥ $X_0 = 1$ and X_1 remains at a 0. Since $X_0 = J_1 = K_1 = 1$, when the next transition takes place, $F1$ will toggle to a 1.

The logic circuits and procedure used to analyze synchronous counters get more complex as the size of the counter increases. Synchronous counters typically contain many of the basic logic gates and have few toggle flip-flops. Since the control inputs of the flip-flops are dynamically configured as the counter cycles through its states, this adds to the complexity of analysis.

Figure 6–19 shows the logic circuit for a 3-bit synchronous binary down counter. The function of the circuit is exactly the same as the asynchronous counter shown in Figure 6–2 but its operation is quite different. Notice that the main clock is connected to all three flip-flops, $F0$, $F1$, and $F2$. Only flip-flop $F0$ is a toggle flip-flop. The outputs of the counter are the outputs of the flip-flops and are labeled X_2, X_1, and X_0.

Figure 6–20 shows the timing diagram for the 3-bit synchronous counter. The timing diagram analysis of synchronous counters is somewhat different from the analysis of asynchronous counters because all three flip-flops change states together and their states must be analyzed together, before and after the clock pulse makes a transition. The timing diagram is divided into ten intervals, labeled 1 through 10, for the following analysis.

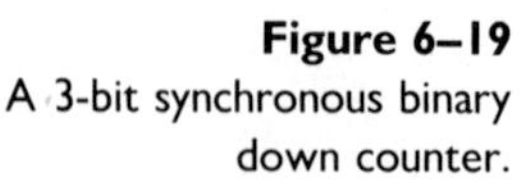
Figure 6–19
A 3-bit synchronous binary down counter.

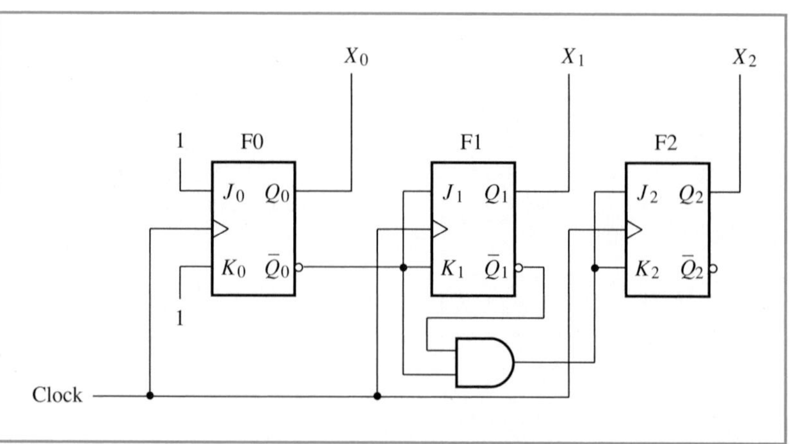

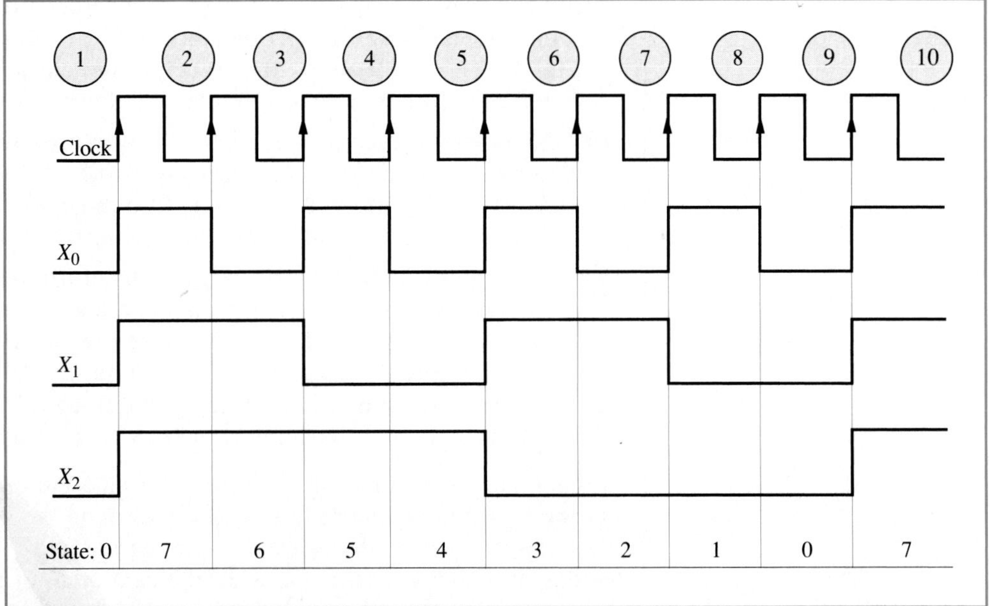

Figure 6–20
Timing diagram for the 3-bit synchronous binary down counter.

(1) This is the initial condition. We have assumed that the counter is at the 0 state.

(2) Before the clock makes a transition $X_0 = 0$ and therefore during this interval X_0 toggles and is a logic 1. Before the clock makes a transition $J_1 = K_1 = \overline{Q}_0 = 1$ since Q_0 (X_0) is a logic 0. Therefore, flip-flop $F1$ is configured to toggle and does so when the clock makes its transition. Since the previous state of X_1 was a 0, its state during this interval is a 1. Before the clock makes a transition $J_2 = K_2 = \overline{Q}_1\overline{Q}_0 = 1 \cdot 1 = 1$ since Q_0 and Q_1 are logic 0s. Flip-flop $F2$ is also configured to toggle and does so when the clock makes its transition. Since the previous state of X_2 was a logic 0, its state during this interval is a logic 1.

(3) X_0 toggles to a 0. $J_1 = K_1 = \overline{Q}_0 = 0$ before the clock triggers (no change) and therefore X_1 remains at a logic 1 during this interval. $J_2 = K_2 = \overline{Q}_1\overline{Q}_0 = 0 \cdot 0 = 0$ before the clock triggers (no change) and therefore X_2 remains at a logic 1 during this interval.

(4) X_0 toggles to a 1. $J_1 = K_1 = \overline{Q}_0 = 1$ before the clock triggers (toggle) and therefore X_1 toggles to a logic 0 during this interval. $J_2 = K_2 = \overline{Q}_1\overline{Q}_0 = 0 \cdot 1 = 0$ before the clock triggers (no change) and therefore X_2 remains at a logic 1 during this interval.

(5) X_0 toggles to a 0. $J_1 = K_1 = \overline{Q}_0 = 0$ before the clock triggers (no change) and therefore X_1 remains at a logic 0 during this interval. $J_2 = K_2 = \overline{Q}_1\overline{Q}_0 = 1 \cdot 0 = 0$ before the clock triggers (no change) and therefore X_2 remains at a logic 1 during this interval.

(6) X_0 toggles to a 1. $J_1 = K_1 = \overline{Q}_0 = 1$ before the clock triggers (toggle) and therefore X_1 toggles to a logic 1 during this interval. $J_2 = K_2 = \overline{Q}_1\overline{Q}_0 = 1 \cdot 1 = 1$ before the clock triggers (toggle) and therefore X_2 toggles to a logic 0 during this interval.

⑦ X_0 toggles to a 0. $J_1 = K_1 = \overline{Q}_0 = 0$ before the clock triggers (no change) and therefore X_1 remains at a logic 1 during this interval. $J_2 = K_2 = \overline{Q}_1\overline{Q}_0 = 0 \cdot 0 = 0$ before the clock triggers (no change) and therefore X_2 remains at a logic 0 during this interval.

⑧ X_0 toggles to a 1. $J_1 = K_1 = \overline{Q}_0 = 1$ before the clock triggers (toggle) and therefore X_1 toggles to a logic 0 during this interval. $J_2 = K_2 = \overline{Q}_1\overline{Q}_0 = 0 \cdot 1 = 0$ before the clock triggers (no change) and therefore X_2 remains at a logic 0 during this interval.

⑨ X_0 toggles to a 0. $J_1 = K_1 = \overline{Q}_0 = 0$ before the clock triggers (no change) and therefore X_1 remains at a logic 0 during this interval. $J_2 = K_2 = \overline{Q}_1\overline{Q}_0 = 1 \cdot 0 = 0$ before the clock triggers (no change) and therefore X_2 remains at a logic 0 during this interval. During this interval the counter is back to state 0 (interval 1) and the counter will now recycle the count as shown in interval 10.

Thus it can be seen that in order to determine the outputs of the counter during each timing interval, we must determine the states of J and K for each flip-flop *before* the clock pulse triggers it and then, based on the setting of J and K, determine what the next state of the flip-flop is going to be; that is, the state of the flip-flop *after* the clock pulse triggers, or the state of the flip-flop during the interval.

Recall from our study of asynchronous counters that we could simply reverse the count of the counter by using the opposite type of flip-flops. For example, an asynchronous down counter could be converted to an asynchronous up counter simply by using negative edge-triggered T flip-flops instead of positive edge-triggered T flip-flops. This is not true for synchronous counters since synchronous counters are not sensitive to the edge at which the flip-flops trigger. Thus the design of a 3-bit synchronous *up* counter as shown in Figure 6–21 is considerably different from the 3-bit synchronous *down* counter in Figure 6–19.

Again notice that the 3-bit synchronous up counter in Figure 6–21 has a common clock for all its flip-flops and that the output of the counter is taken from the output of the flip-flops. This is true for all synchronous counters. The timing diagram for the up counter is shown in Figure 6–22. Again, for the purpose of the analysis that follows, we have identified the ten intervals, labeled 1 through 10, in the timing diagram.

Figure 6–21
A 3-bit synchronous binary up counter.

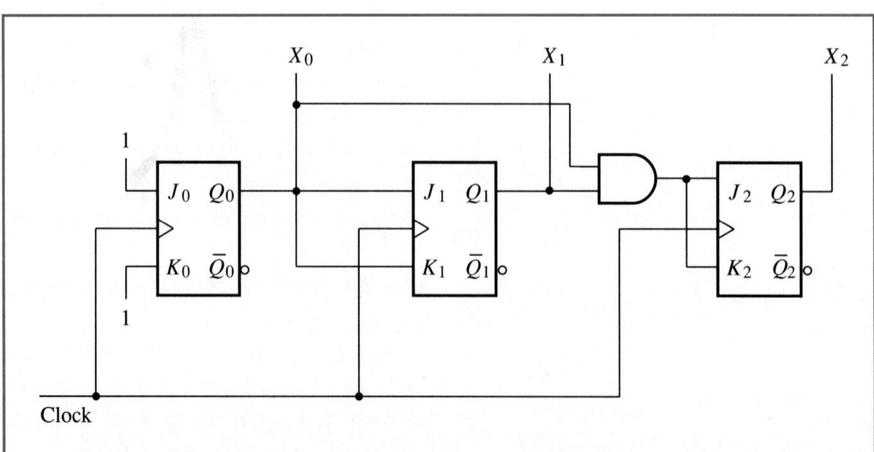

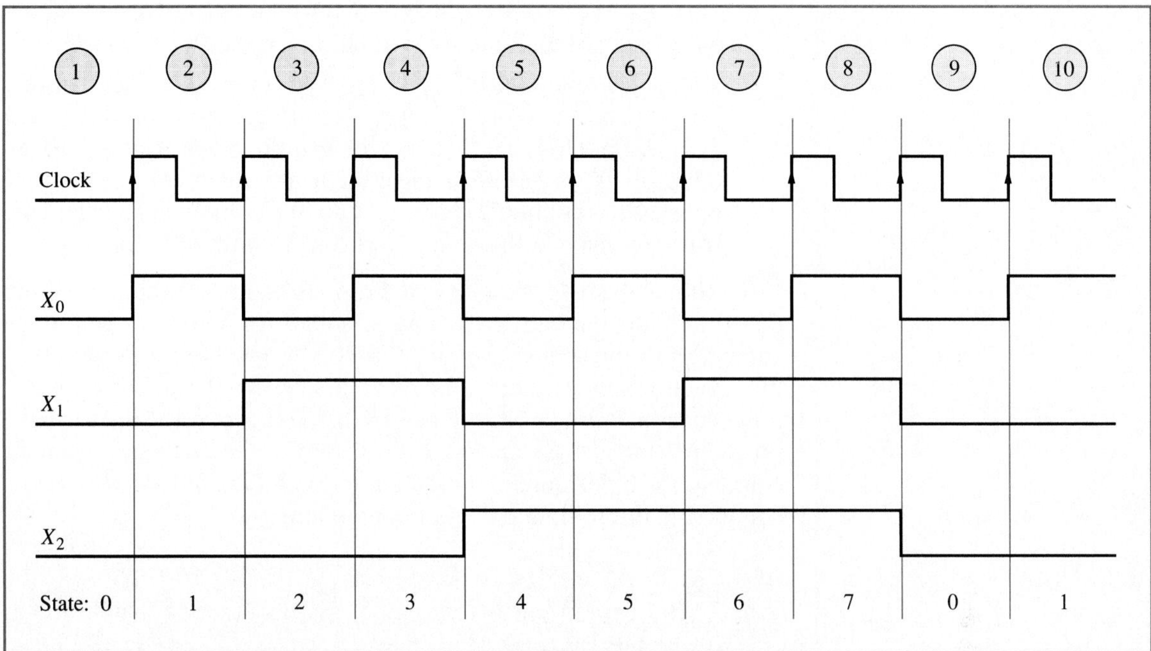

Figure 6–22
Timing diagram for the 3-bit synchronous binary up counter.

① This is the initial condition. We have assumed that the counter is at the 0 state.

② X_0 toggles to a 1. $J_1 = K_1 = Q_0 = 0$ before the clock triggers (no change) and therefore X_1 remains at a logic 0 during this interval. $J_2 = K_2 = Q_1Q_0 = 0 \cdot 0 = 0$ before the clock triggers (no change) and therefore X_2 remains at a logic 0 during this interval.

③ X_0 toggles to a 0. $J_1 = K_1 = Q_0 = 1$ before the clock triggers (toggle) and therefore X_1 toggles to a logic 1 during this interval. $J_2 = K_2 = Q_1Q_0 = 0 \cdot 1 = 0$ before the clock triggers (no change) and therefore X_2 remains at a logic 0 during this interval.

④ X_0 toggles to a 1. $J_1 = K_1 = Q_0 = 0$ before the clock triggers (no change) and therefore X_1 remains at a logic 1 during this interval. $J_2 = K_2 = Q_1Q_0 = 1 \cdot 0 = 0$ before the clock triggers (no change) and therefore X_2 remains at a logic 0 during this interval.

⑤ X_0 toggles to a 0. $J_1 = K_1 = Q_0 = 1$ before the clock triggers (toggle) and therefore X_1 toggles to a logic 0 during this interval. $J_2 = K_2 = Q_1Q_0 = 1 \cdot 1 = 1$ before the clock triggers (toggle) and therefore X_2 toggles to a logic 1 during this interval.

⑥ X_0 toggles to a 1. $J_1 = K_1 = Q_0 = 0$ before the clock triggers (no change) and therefore X_1 remains at a logic 0 during this interval. $J_2 = K_2 = Q_1Q_0 = 0 \cdot 0 = 0$ before the clock triggers (no change) and therefore X_2 remains at a logic 1 during this interval.

⑦ X_0 toggles to a 0. $J_1 = K_1 = Q_0 = 1$ before the clock triggers (toggle) and therefore X_1 toggles to a logic 1 during this interval. $J_2 = K_2 = Q_1Q_0 = 0 \cdot 1 = 0$ before the clock triggers (no change) and therefore X_2 remains at a logic 1 during this interval.

⑧ X_0 toggles to a 1. $J_1 = K_1 = Q_0 = 0$ before the clock triggers (no change) and therefore X_1 remains at a logic 1 during this interval.

$J_2 = K_2 = Q_1Q_0 = 1 \cdot 0 = 0$ before the clock triggers (no change) and therefore X_2 remains at a logic 1 during this interval.

(9) X_0 toggles to a 0. $J_1 = K_1 = Q_0 = 1$ before the clock triggers (toggle) and therefore X_1 toggles to a logic 0 during this interval. $J_2 = K_2 = Q_1Q_0 = 1 \cdot 1 = 1$ before the clock triggers (toggle) and therefore X_2 toggles to a logic 0 during this interval. During this interval the counter is back to state 0 (interval 1) and the counter will now recycle the count as shown in interval 10.

From the analysis of the two synchronous counter circuits it is apparent that the timing diagram procedure is quite involved and complex when compared to the timing diagram analysis of asynchronous counters. The two synchronous counters analyzed so far were designed using a formal procedure that will be introduced in the next section. This formal procedure can also be applied (in reverse) to the analysis of synchronous counters (to be examined in Section 6–5) and is considerably less complex than the timing diagram approach.

Review questions

1. Why is the term *synchronous* used to describe synchronous counters?
2. Compare the construction of a synchronous counter with that of an asynchronous counter.
3. Compare the counting sequence of a synchronous counter with that of an asynchronous counter.

6–4 Synthesis of synchronous counters

The *synthesis*, or design, of synchronous counters involves a formal six-step procedure. Using this procedure one can design any type of counter from a given specification. For example, consider the following procedure to design a 2-bit synchronous binary up counter using D flip-flops.

Step 1: state diagram

The *state diagram* is a graphic representation of the states of the counter and the connections among its states. The state diagram for the 2-bit binary up counter is shown in Figure 6–23. Notice that each state of the

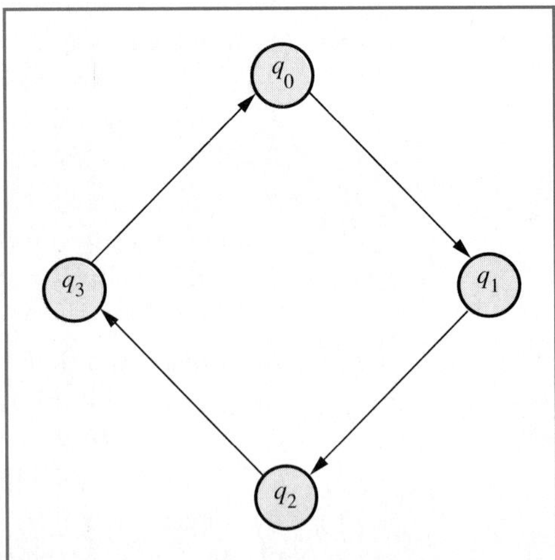

Figure 6–23
State diagram for a synchronous 2-bit binary up counter.

Table 6–1
State table for a 2-bit binary up counter

q^n	q^{n+1}
q_0	q_1
q_1	q_2
q_2	q_3
q_3	q_0

counter is identified by a circle and the variable q_n where n is the number of the state and corresponds to the output of the counter when in that state. There are four states shown in Figure 6–23 since the size of the counter is 2 bits. Notice also that we use lowercase q to identify the state of the counter (uppercase Q is used to identify the state (or output) of a flip-flop). The connection between the states is shown by arrows and depends on the counting sequence of the counter. For example, for an up counter, if the present state is q_3, the next state is q_0, if the present state is q_0 the next state is q_1, and so on. A transition from one state to another occurs when a clock pulse triggers the counter. The state diagram therefore provides almost the same information as a timing diagram since it identifies the sequence in which the counter counts.

Step 2: state table

The *state table* is simply a translation of the state diagram into tabular form. The state table for the 2-bit binary up counter is shown in Table 6–1. The first column in the state table lists the *present state* of the counter q^n. The second column of the state table lists the *next state* of the counter, q^{n+1}, that is, the state of the counter after the clock pulse makes a transition. Note the use of the n and $n + 1$ notation to identify the present and next states, respectively; this notation was used in the state transition tables of flip-flops to identify their present and next states.

To construct a state table we first list all possible present states in column 1 in order starting with state q_0. We then refer to the state diagram and determine what the next state of the counter will be when the clock pulse triggers the counter. For example, if the present state is q_0, the next state is q_1 as shown in the first entry on the state table. Similarly, if the present state is q_1, the next state is q_2, if the present state is q_2, the next state is q_3, and so on.

Step 3: transition table

The *transition table* is a translation of the state table and relates each state of the counter to the individual outputs of each flip-flop. The circuit will have two flip-flops which we shall refer to as flip-flop no. 0 (LSB) and flip-flop no. 1 (MSB). The output of flip-flop no. 0 is Q_0 and the output of flip-flop no. 1 is Q_1. The transition table for the 2-bit binary up counter is shown in Table 6–2. Observe that in constructing the transition table we have simply converted each state of the counter (from the state table) to a 2-bit binary number that represents the outputs of the two flip-flops that make up the counter. For example, when the counter is in state q_0 the

Table 6–2
Transition table for a 2-bit binary up counter

$(Q_1\ Q_0)^n$	$(Q_1\ Q_0)^{n+1}$
0 0	0 1
0 1	1 0
1 0	1 1
1 1	0 0

outputs of the flip-flops (which are the outputs of the counter) are 00, and therefore the outputs of the individual flip-flops are

$$Q_1 = 0, \qquad Q_0 = 0$$

Similarly, when the counter is in state q_1 the outputs of the flip-flops are 01, and therefore the outputs of the individual flip-flops are

$$Q_1 = 0, \qquad Q_0 = 1$$

and so on.

The present state of the outputs of the flip-flops is identified in the transition table by the heading

$$(Q_1 Q_0)^n$$

for convenience and could also be written as

$$Q_1{}^n Q_0{}^n$$

The next state of the outputs of the flip-flops is identified in the transition table by the heading

$$(Q_1 Q_0)^{n+1}$$

for convenience and could also be written as

$$Q_1{}^{n+1} Q_0{}^{n+1}$$

The transition table tells us the state transition of each flip-flop as the counter changes from one state to the next. For example, if we extract the present and next states of flip-flop no. 0 (whose output is Q_0) from Table 6–2, we have the following:

$Q_0{}^n$	$\longrightarrow$	$Q_0{}^{n+1}$
0	$\longrightarrow$	1
1	$\longrightarrow$	0
0	$\longrightarrow$	1
1	$\longrightarrow$	0

Notice that flip-flop no. 0 toggles. Similarly, we can determine how flip-flop no. 1 changes states as the counter counts through its sequence by extracting the $Q_1{}^n$ and $Q_1{}^{n+1}$ columns from Table 6–2

$Q_1{}^n$	$\longrightarrow$	$Q_1{}^{n+1}$
0	$\longrightarrow$	0
0	$\longrightarrow$	1
1	$\longrightarrow$	1
1	$\longrightarrow$	0

Step 4: excitation maps

The previous three steps in the synthesis were not dependent on the type of flip-flop that is to be used in the counter circuit. Before this step is started we must determine the type of flip-flop, *J-K* or *D*, that the counter is to have. Since the specification for this example requires *D* flip-flops, we shall proceed with that assumption.

Excitation maps are used to determine what the control inputs of each flip-flop (*J* and *K*, or *D*) has to be in order for the flip-flop to make the transitions specified in the transition table. The excitation maps for the D input of each flip-flop are shown in Figure 6–24. These maps are essentially K-maps and are used in the same manner.

The excitation map for flip-flop no. 0 is shown in Figure 6–24a. The map represents the control input D_0 of the flip-flop, and therefore the value in each cell is the value of D_0 needed in order for the flip-flop to change from its present state to its next state. To determine the values of D_0 required for the flip-flop to change from one state to another we must refer to the modified transition table for the D flip-flop shown in Table 5–6 and reproduced here as follows:

Q^n	Q^{n+1}	D
0	0	0
0	1	1
1	0	0
1	1	1

Recall that the table shown above tells us what D has to be in order for a particular transition to take place.

Since each cell in Figure 6–24a is mapped by Q_1Q_0, the cell number corresponds to the present state of the counter. For example, in cell no. 0, since the present state of the counter is 0 ($Q_1 = 0$, $Q_0 = 0$) and, according to the transition table (refer to extracted $Q_0{}^n$ and $Q_0{}^{n+1}$ columns), the flip-flop must change from a 0 to a 1 (first entry in the transition table), D_0 must be a 1 (see reproduction of Table 5–6). Similarly, in cell no. 1, since the present state of the counter is 1 ($Q_1 = 0$, $Q_0 = 1$) and the flip-flop must change from a 1 to a 0 (second entry in transition table), D_0 must be a 0. In cell no. 2, since the present state of the counter is 2 ($Q_1 = 1$, $Q_0 = 0$) and the flip-flop must change from a 0 to a 1 (third entry in transition table), D_0 must be a 1. Finally, in cell no. 3, since the present state of the counter is 3 ($Q_1 = 1$, $Q_0 = 1$) and the

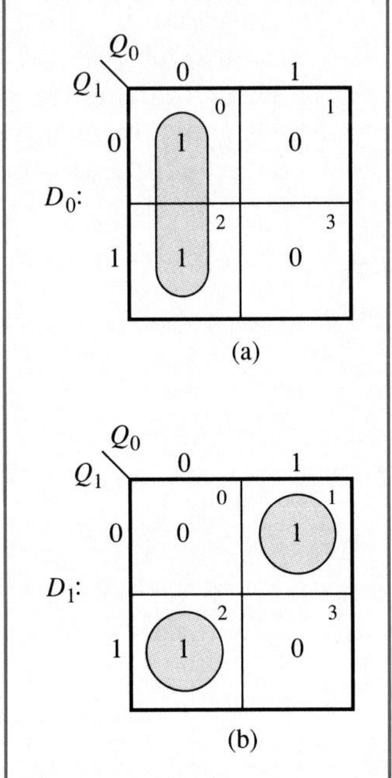

Figure 6–24
Excitation maps for the synchronous 2-bit binary up counter using D flip-flops.

flip-flop must change from a 1 to a 0 (last entry in transition table), D_0 must be a 0.

The excitation map for flip-flop no. 1 can be filled in in the same manner as shown in Figure 6–24b. Refer to the transition table shown in Table 6–2 (or the extracted $Q_1{}^n$ and $Q_1{}^{n+1}$ columns shown earlier) and the reproduction of Table 5–6.

Cell no. 0: The present state of the counter is 0 and the flip-flop changes from a 0 to a 0. For this to occur D_1 must be a 0.

Cell no. 1: The present state of the counter is 1 and the flip-flop changes from a 0 to a 1. For this to occur D_1 must be a 1.

Cell no. 2: The present state of the counter is 2 and the flip-flop changes from a 1 to a 1. For this to occur D_1 must be a 1.

Cell no. 3: The present state of the counter is 3 and the flip-flop changes from a 1 to a 0. For this to occur D_1 must be a 0.

Step 5: excitation equations

The simplified logic equations obtained from the excitation maps are known as the *excitation equations* for the counter. Thus from the excitation maps of the 2-bit binary up counter shown in Figure 6–24, the excitation equations are as follows:

Equation (6–1)

$$D_0 = \overline{Q}_0$$

$$D_1 = Q_1\overline{Q}_0 + \overline{Q}_1 Q_0$$

Equation (6–2)

$$\therefore \quad D_1 = Q_1 \oplus Q_0$$

Notice that the excitation equations identify the connections that are to be made between the outputs of the flip-flops (Q_0 and Q_1) and the inputs of the flip-flops (D_0 and D_1).

Step 6: logic circuit

The logic circuit for the counter is implemented from the excitation equations. Thus for the 2-bit binary up counter, Equations 6–1 and 6–2 produce the circuit for the counter shown in Figure 6–25. Notice that the clocks of both the flip-flops are connected to a common *CLOCK* that is used to cycle the counter through each of its states. The type of flip-flop, negative or positive edge-triggered, is not significant to the operation of the circuit. Also notice that the outputs of the counter (X_1X_0) are the

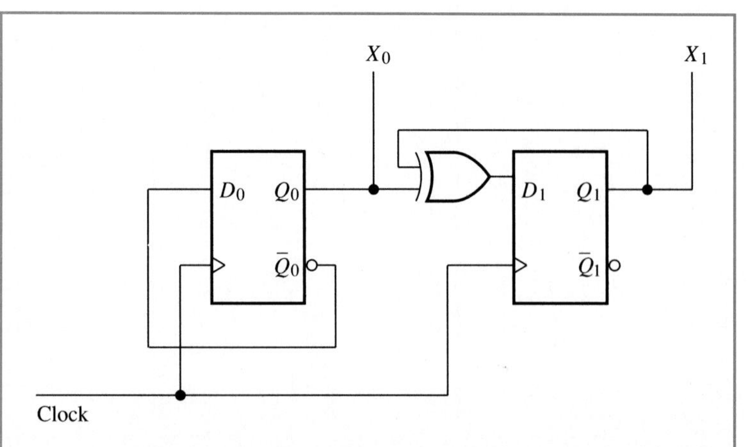

Figure 6–25 Logic circuit for the synchronous 2-bit binary up counter using *D* flip-flops.

outputs of the flip-flops (Q_1Q_0) and that flip-flop no. 0 will function as a T flip-flop.

The counter designed in Figure 6–25 can very easily be redesigned to use J-K flip-flops by modifying only steps 4 and 5 of the synthesis as shown in the following example.

EXAMPLE 6–4

To redesign the 2-bit binary up counter shown in Figure 6–25 to use J-K flip-flops instead of D flip-flops, we must obtain the excitation maps for the two J-K flip-flops in step 4 of the synthesis:

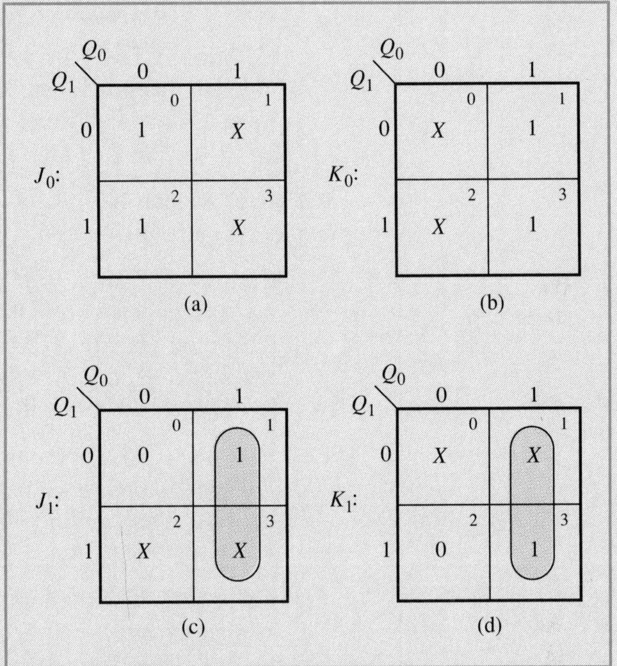

Figure 6–26 Excitation maps for the synchronous 2-bit binary up counter using J-K flip-flops.

Step 4: excitation maps
Since each J-K flip-flop has two control inputs, J and K, we must have excitation maps for each of the inputs of each flip-flop as shown in Figure 6–26. Referring to the modified transition table for a J-K flip-flop shown in Table 5–9 and reproduced here as follows:

Q^n	Q^{n+1}	J	K
0	0	0	x
0	1	1	x
1	0	x	1
1	1	x	0

we can fill in each set of cells for J_0 and K_0 (shown in the maps of Figure 6–26a and b, respectively) as follows:

Cell no. 0: The present state of the counter is 0 and flip-flop number 0 changes from a 0 to a 1 (refer to the transition table shown in Table 6–2 or the extracted $Q_0{}^n$ and $Q_0{}^{n+1}$ columns). For this to occur J must be a 1 and K is an X. Thus cell no. 0

for the J_0 map has a 1 filled in and cell no. 0 for the K_0 map has an X filled in.

Cell no. 1: The present state of the counter is 1 and the flip-flop changes from a 1 to a 0. For this to occur J is an X and K must be a 1. Thus cell no. 1 for the J_0 map has an X filled in and cell no. 1 for the K_0 map has a 1 filled in.

Cell no. 2: The present state of the counter is 2 and the flip-flop changes from a 0 to a 1. For this to occur J must be a 1 and K is an X. Thus cell no. 2 for the J_0 map has a 1 filled in and cell no. 2 for the K_0 map has an X filled in.

Cell no. 3: The present state of the counter is 3 and the flip-flop changes from a 1 to a 0. For this to occur J is an X and K must be a 1. Thus cell no. 3 for the J_0 map has an X filled in and cell no. 3 for the K_0 map has a 1 filled in.

Similarly, we can fill in each set of cells for J_1 and K_1 (shown in the maps of Figure 6–26c and d, respectively) as follows:

Cell no. 0: The present state of the counter is 0 and flip-flop no. 1 changes from a 0 to a 0 (refer to Table 6–2 or the extracted ${Q_1}^n$ and ${Q_1}^{n+1}$ columns). For this to occur J must be a 0 and K is an X. Thus cell no. 0 for the J_1 map has a 0 filled in and cell no. 0 for the K_1 map has an X filled in.

Cell no. 1: The present state of the counter is 1 and the flip-flop changes from a 0 to a 1. For this to occur J must be a 1 and K is an X. Thus cell no. 1 for the J_1 map has a 1 filled in and cell no. 1 for the K_1 map has an X filled in.

Cell no. 2: The present state of the counter is 2 and the flip-flop changes from a 1 to a 1. For this to occur J is an X and K must be a 0. Thus cell no. 2 for the J_1 map has an X filled in and cell no. 2 for the K_1 map has a 0 filled in.

Cell no. 3: The present state of the counter is 3 and the flip-flop changes from a 1 to a 0. For this to occur J is an X and K must be a 1. Thus cell no. 3 for the J_1 map has an X filled in and cell no. 3 for the K_1 map has a 1 filled in.

Step 5: excitation equations
The excitation equations obtained from the maps shown in Figure 6–26 are as follows: for Figure 6–26a and b, assuming that the don't cares are 1's,

Equation (6–3)
$$J_0 = 1$$

Equation (6–4)
$$K_0 = 1$$

and for Figure 6–26c and d,

Equation (6–5)
$$J_1 = Q_0$$

Equation (6–6)
$$K_1 = Q_0$$

Step 6: logic circuit
The logic circuit is implemented from Equations 6–3 through 6–6 as shown in Figure 6–27.

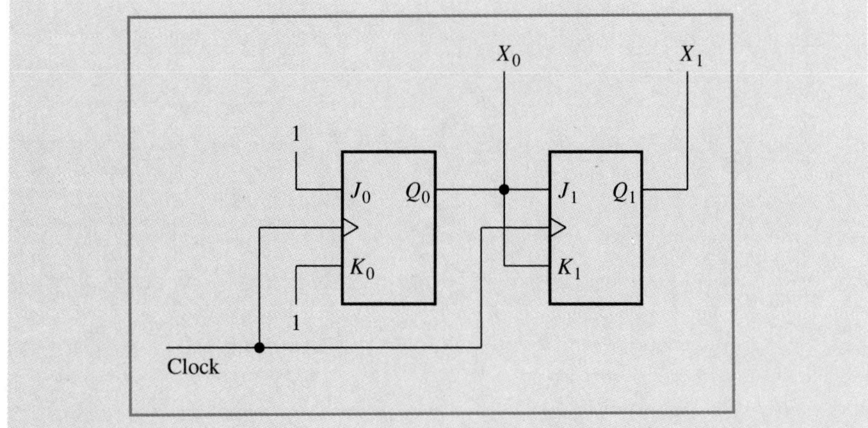

Figure 6–27 Logic circuit for the synchronous 2-bit binary up counter using *J-K* flip-flops.

It is apparent that the 2-bit up counter implemented with *J-K* flip-flops as shown in Figure 6–27 has a simpler logic circuit than the 2-bit up counter implemented with *D* flip-flops as shown in Figure 6–25 because of the additional Exclusive-*OR* gate. Generally, counters implemented with *J-K* flip-flops are considerably less complex than counters implemented with *D* flip-flops; this is due to the versatility and flexibility of the *J-K* flip-flop as discussed in Chapter 5. Notice that the only steps in the synthesis procedure that are dependent on the type of flip-flop to be used are the steps that involve the excitation maps and equations. If, however, the counting sequence is to be modified, the synthesis must be started from the beginning. The six steps that make up the synthesis procedure can be applied to the design of any counter. The following example shows the steps involved in the design of a 2-bit down counter.

EXAMPLE 6–5

Design a 2-bit synchronous binary down counter using *J-K* flip-flops.

SOLUTION

Step 1: state diagram

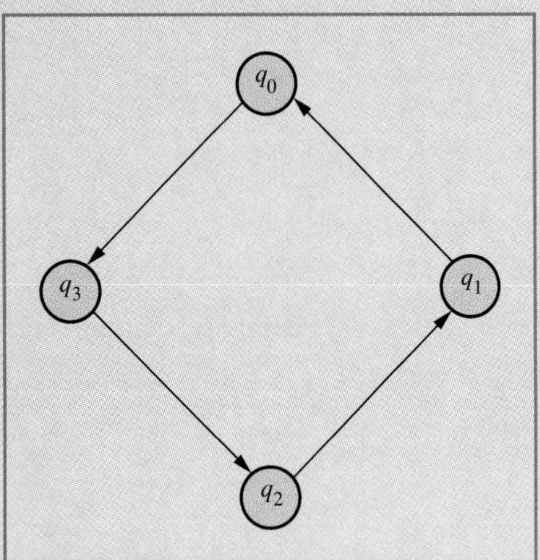

Figure 6–28 State diagram for a 2-bit synchronous binary down counter.

Step 2: state table

q^n	q^{n+1}
q_0	q_3
q_1	q_0
q_2	q_1
q_3	q_2

Step 3: transition table

$(Q_1\,Q_0)^n$	$(Q_1\,Q_0)^{n+1}$
0 0	1 1
0 1	0 0
1 0	0 1
1 1	1 0

Step 4: excitation maps

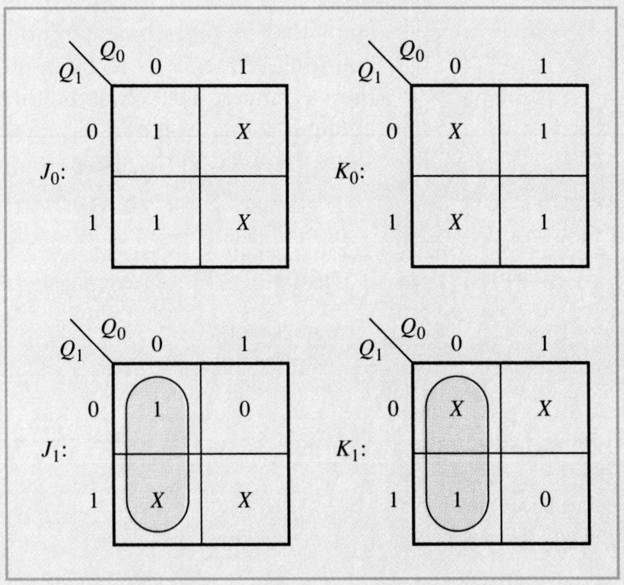

Figure 6–29
Excitation maps for the 2-bit synchronous binary down counter using *J-K* flip-flops.

J_0:

$Q_1 \backslash Q_0$	0	1
0	1	X
1	1	X

K_0:

$Q_1 \backslash Q_0$	0	1
0	X	1
1	X	1

J_1:

$Q_1 \backslash Q_0$	0	1
0	1	0
1	X	X

K_1:

$Q_1 \backslash Q_0$	0	1
0	X	X
1	1	0

Step 5: excitation equations

$$J_0 = 1$$
$$K_0 = 1$$
$$J_1 = \overline{Q}_0$$
$$K_1 = \overline{Q}_0$$

Step 6: logic circuit

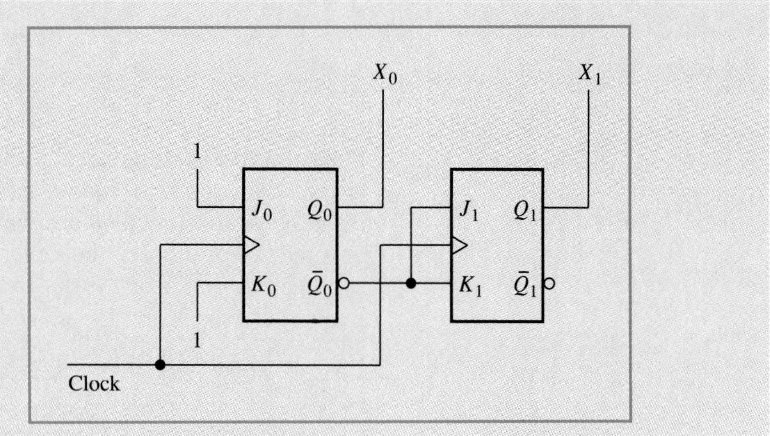

Figure 6–30 The 2-bit synchronous binary down counter.

The examples discussed so far have involved 2-bit binary counters that count in an ordered sequence. There are many applications that require counters that are larger in size and count in some predetermined sequence. The following example illustrates the use of the synthesis procedure to design a 3-bit counter that counts in a predetermined sequence.

EXAMPLE 6–6

Design a synchronous counter using D flip-flops that counts in the following sequence:

6, 3, 5, 0, 2, 6, 3, 5, 0, 2, 6, . . .

SOLUTION

Step 1: state diagram

Since the largest number in the count is 6, the size of the counter must be 3 bits in order to accommodate this number in its binary representation. A 3-bit counter has eight possible states. However, only five of the eight states (0, 2, 3, 5, and 6) will be used by the counter as shown in the state diagram of Figure 6–31.

Figure 6–31

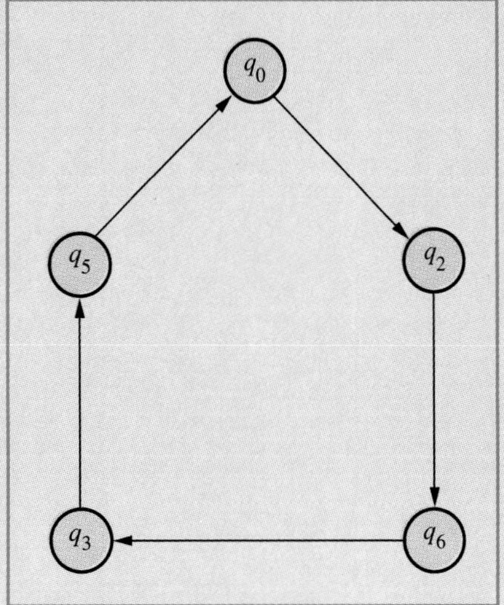

Step 2: state table

q^n	q^{n+1}
q_0	q_2
q_2	q_6
q_3	q_5
q_5	q_0
q_6	q_3

Note that the state entries in the present state (q^n) column are listed in increasing order. This is simply done for convenience when filling in the excitation maps later. The present state entries can be arranged in any order, however, it is recommended that they be arranged in increasing order in order to reduce the possibility of making errors during the construction of the excitation maps.

Step 3: transition table

$(Q_2\ Q_1\ Q_0)^n$	$(Q_2\ Q_1\ Q_0)^{n+1}$
0 0 0	0 1 0
0 1 0	1 1 0
0 1 1	1 0 1
1 0 1	0 0 0
1 1 0	0 1 1

Step 4: excitation maps

Figure 6–32

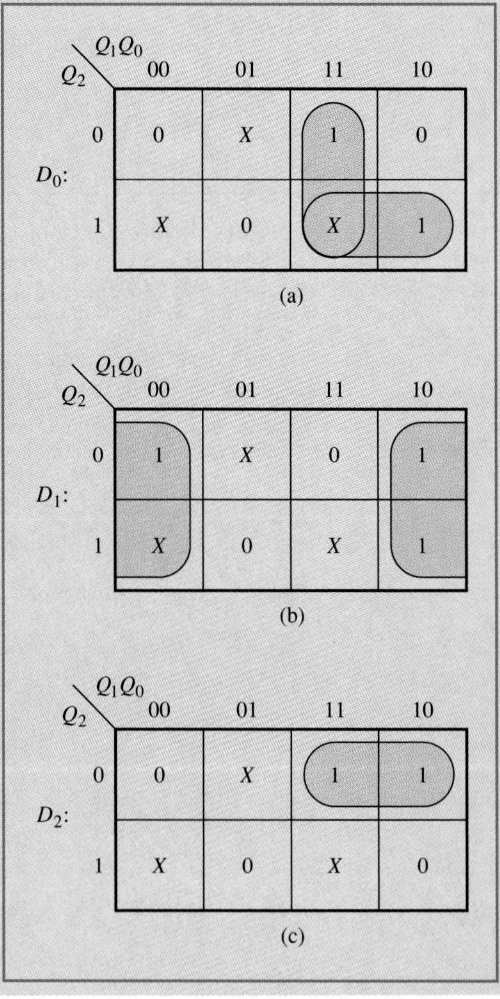

D_0:

$Q_2 \backslash Q_1Q_0$	00	01	11	10
0	0	X	1	0
1	X	0	X	1

(a)

D_1:

$Q_2 \backslash Q_1Q_0$	00	01	11	10
0	1	X	0	1
1	X	0	X	1

(b)

D_2:

$Q_2 \backslash Q_1Q_0$	00	01	11	10
0	0	X	1	1
1	X	0	X	0

(c)

Notice in the excitation maps that only cells 0, 2, 3, 5, and 6 have been filled in since these correspond to the present states. States 1, 4, and 7 have *x* (don't cares), since the counter is not designed to use these states and we don't care what the value of *D* is when the counter is in any of these states.

Step 5: excitation equations

$$D_0 = Q_1Q_0 + Q_2Q_1$$
$$D_1 = \overline{Q}_0$$
$$D_2 = \overline{Q}_2Q_1$$

Step 6: logic circuit

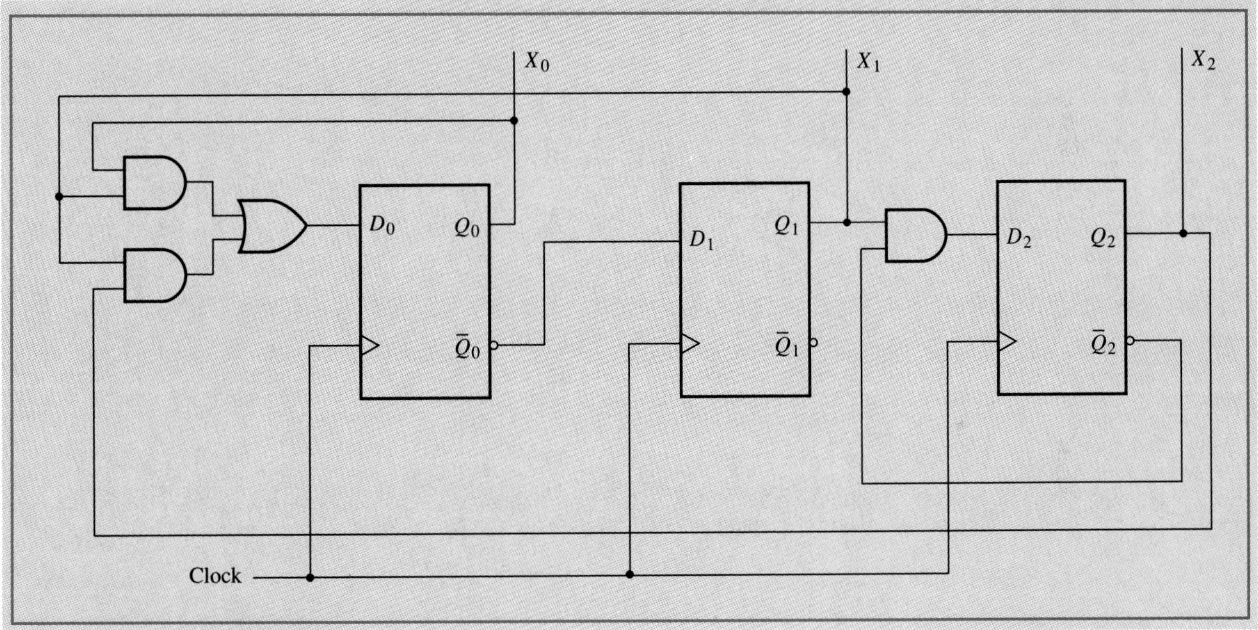

Figure 6–33
A synchronous counter that counts in a predetermined sequence.

To illustrate the design of a 4-bit synchronous counter let us consider the implementation of a asynchronous BCD counter similar to the one designed in Section 6–2 by using the synthesis procedure. Since the BCD code is a 4-bit code, the counter must be 4 bits in size but is modulus 10 and uses only the first 10 of the 16 possible states. The following example illustrates the design of the BCD up counter.

EXAMPLE 6–7

Design a synchronous BCD up counter using *J-K* flip-flops.

SOLUTION

Step 1: state diagram

Figure 6–34

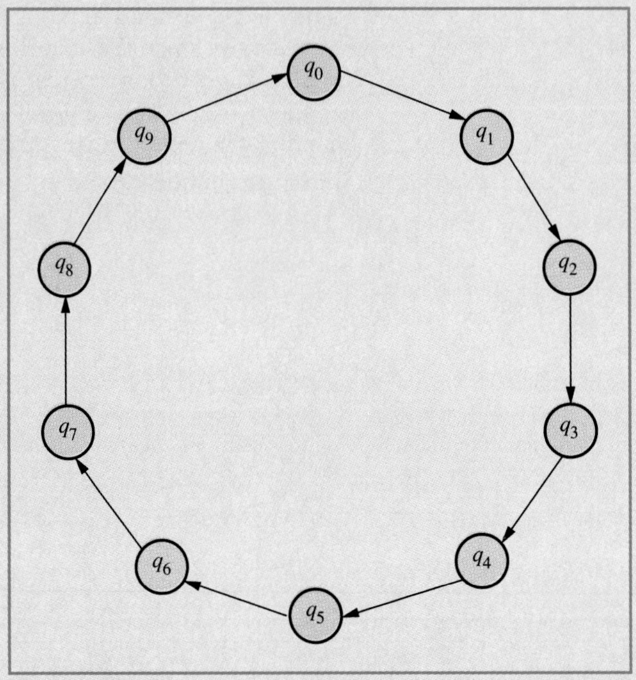

Step 2: state table

q^n	q^{n+1}
q_0	q_1
q_1	q_2
q_2	q_3
q_3	q_4
q_4	q_5
q_5	q_6
q_6	q_7
q_7	q_8
q_8	q_9
q_9	q_0

Step 3: transition table

$(Q_3\ Q_2\ Q_1\ Q_0)^n$	$(Q_3\ Q_2\ Q_1\ Q_0)^{n+1}$
0 0 0 0	0 0 0 1
0 0 0 1	0 0 1 0
0 0 1 0	0 0 1 1
0 0 1 1	0 1 0 0
0 1 0 0	0 1 0 1
0 1 0 1	0 1 1 0
0 1 1 0	0 1 1 1
0 1 1 1	1 0 0 0
1 0 0 0	1 0 0 1
1 0 0 1	0 0 0 0

Step 4: excitation maps

Figure 6–35

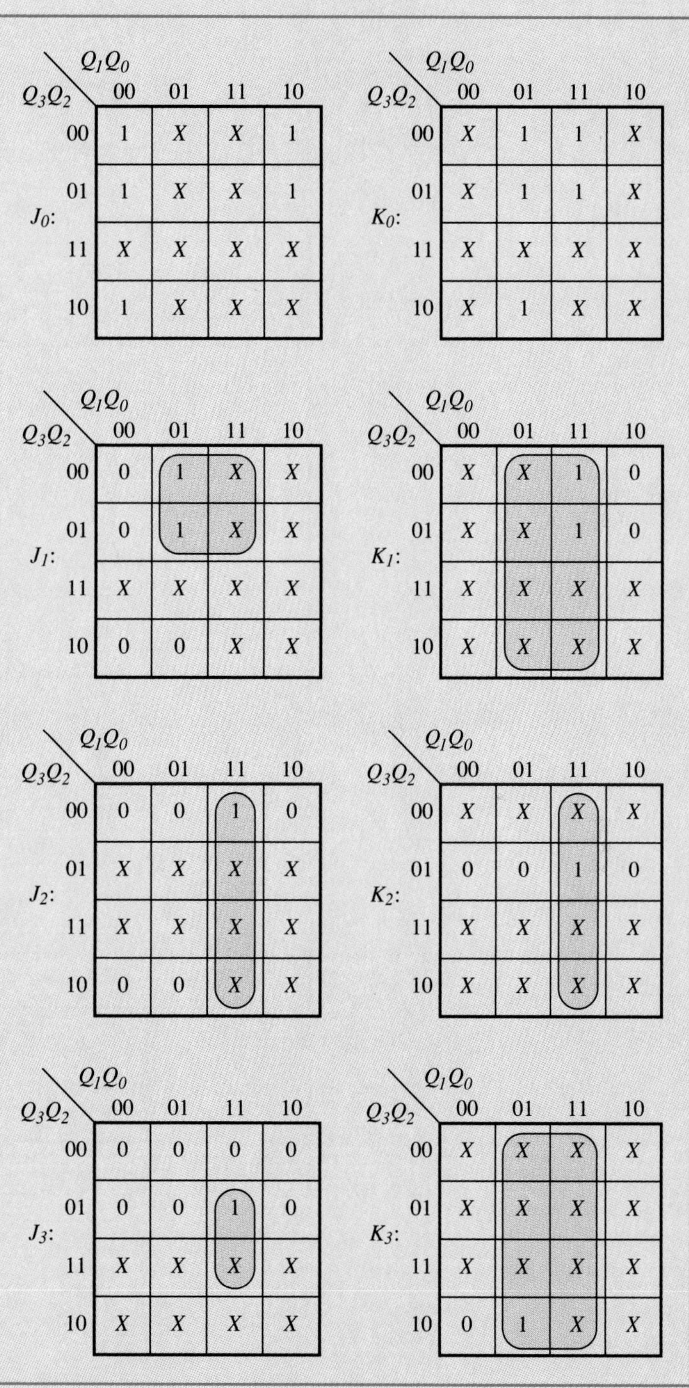

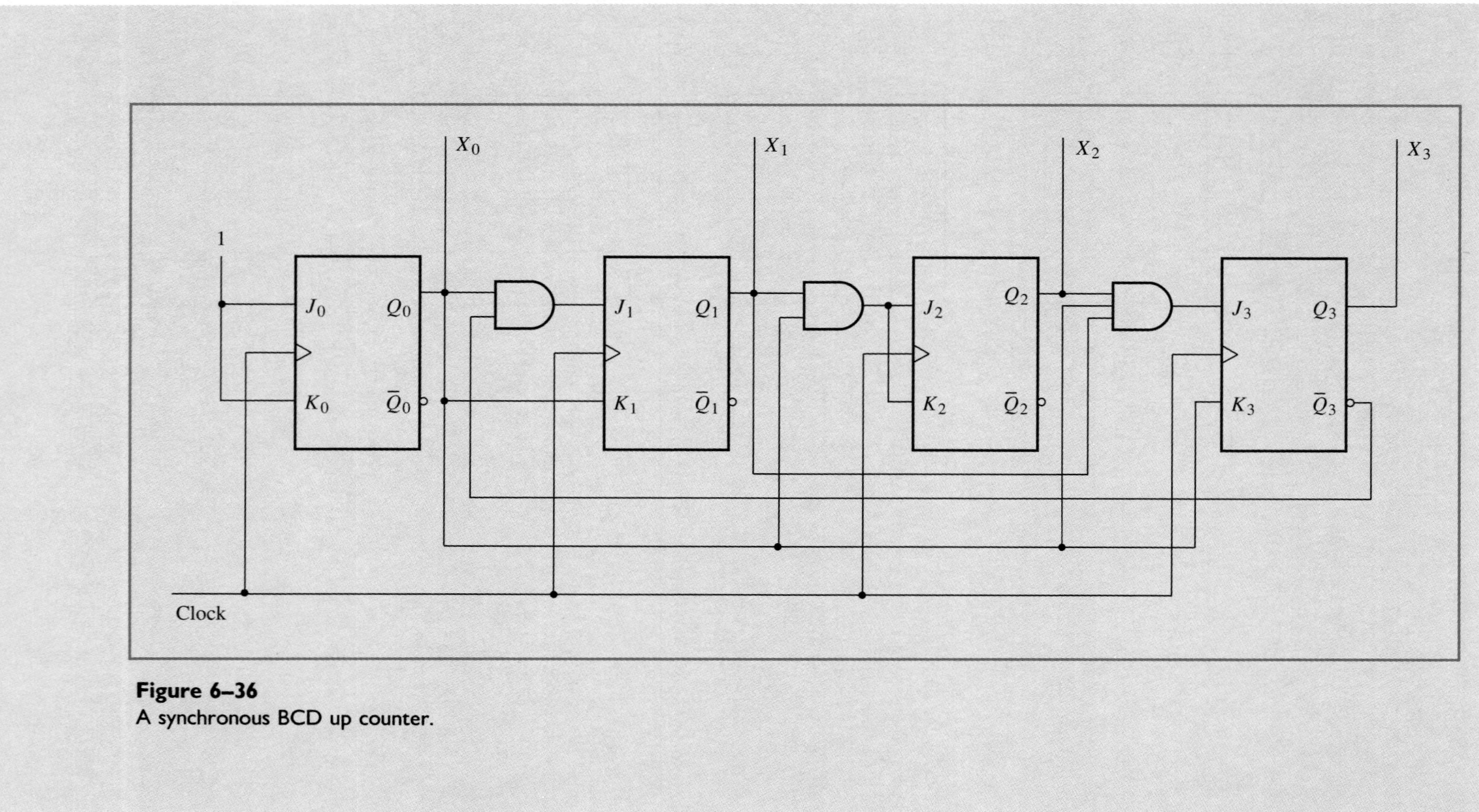

Figure 6–36
A synchronous BCD up counter.

Step 5: excitation equations

$$J_0 = 1$$
$$K_0 = 1$$
$$J_1 = Q_0\overline{Q}_3$$
$$K_1 = Q_0$$
$$J_2 = Q_1Q_0$$
$$K_2 = Q_1Q_0$$
$$J_3 = Q_2Q_1Q_0$$
$$K_3 = Q_0$$

Step 6: logic circuit. The logic circuit for the counter is shown in Figure 6–36.

The counters discussed so far were designed with a *dedicated* count. That is, each counter was designed for a specific counting sequence that could not be changed unless the entire circuit was redesigned. In Section 6–1 it was stated that a counter could have one or more *control* inputs that could be used to select the counting sequence of the counter. For example, consider a 2-bit synchronous binary counter that will count up or down on command. Such a counter is known as an *up/down counter* and its logic symbol is shown in Figure 6–37.

The counter has an additional input S (select) that will be used to select the counting sequence for the counter as follows:

If $S = 1$ then the counter counts *UP*
If $S = 0$ then the counter counts *DOWN*

With this information we can now go through the six-step synthesis procedure to design the logic circuit for the up/down counter. The counter will be implemented with D flip-flops (the design of the same counter using J-K flip-flops will be left as an exercise).

Figure 6–37
Logic symbol for a 2-bit up/down counter.

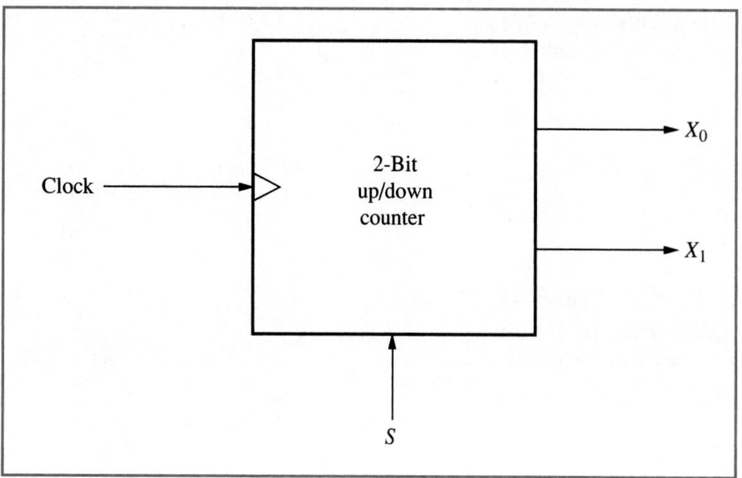

Figure 6–38
State diagram for the 2-bit up/down counter.

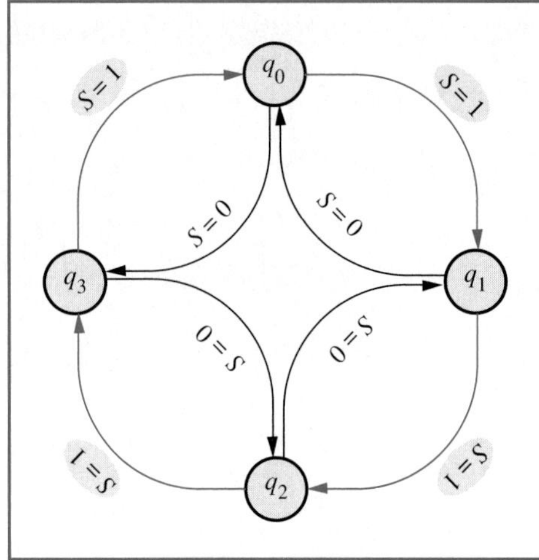

Step 1: state diagram

Notice that the state diagram (Figure 6–38) appears to be considerably different from the state diagrams seen so far. Since there are two possible next states for each present state based on the value of S (1 or 0), each state symbol has two arrows emanating from it, one representing a transition to the next state if $S = 1$, and the other representing a transition to the next state if $S = 0$. For example, if the present state is q_3, the next state is q_0 if $S = 1$ (up count) or q_2 if $S = 0$ (down count).

Step 2: state table

Since the up/down counter has two possible next states based on the value of S, the next state column of the state table will have two possible entries as shown in Table 6–3.

Step 3: transition table

Since the transition table is simply a translation of the state table to represent the outputs of the flip-flops, the transition table therefore has the same format as the state table as shown in Table 6–4.

Table 6–3
State table for a 2-bit up/down counter

	q^{n+1}	
q^n	$S = 0$	$S = 1$
q_0	q_3	q_1
q_1	q_0	q_2
q_2	q_1	q_3
q_3	q_2	q_0

Table 6–4
Transition table for a 2-bit up/down counter

	$(Q_1\ Q_0)^{n+1}$	
$(Q_1\ Q_0)^n$	$S = 0$	$S = 1$
0 0	1 1	0 1
0 1	0 0	1 0
1 0	0 1	1 1
1 1	1 0	0 0

Step 4: excitation maps

The excitation maps for each D flip-flop are shown in Figure 6–39. Each cell lists the value of D required for the flip-flop to make the transition to the next state, given the present state and the value of S. For example, consider the excitation map for D_0 (flip-flop no. 0) shown in Figure 6–39a.

Cell 0: The present state is 0 ($Q_1Q_0 = 00$) and $S = 0$. From the transition table, the next state is 3 ($Q_1Q_0 = 11$) and flip-flop no. 0 changes from a 0 to a 1, therefore D_0 has to be a 1.

Cell 2: The present state is 1 ($Q_1Q_0 = 01$) and $S = 0$. From the transition table, the next state is 0 ($Q_1Q_0 = 00$) and flip-flop no. 0 changes from a 1 to a 0, therefore D_0 has to be a 0.

Cell 4: The present state is 2 ($Q_1Q_0 = 10$) and $S = 0$. From the transition table, the next state is 1 ($Q_1Q_0 = 01$) and flip-flop no. 0 changes from a 0 to a 1, therefore D_0 has to be a 1.

Cell 6: The present state is 3 ($Q_1Q_0 = 11$) and $S = 0$. From the transition table, the next state is 2 ($Q_1Q_0 = 10$) and flip-flop no. 0 changes from a 1 to a 0, therefore D_0 has to be a 0.

Figure 6–39
Excitation maps for the 2-bit up/down counter.

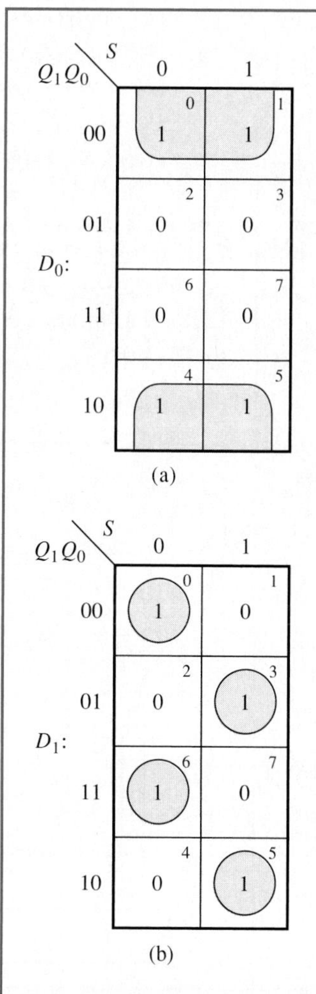

Cell 1: The present state is 0 ($Q_1Q_0 = 00$) and $S = 1$. From the transition table, the next state is 1 ($Q_1Q_0 = 01$) and flip-flop no. 0 changes from a 0 to a 1, therefore D_0 has to be a 1.

Cell 3: The present state is 1 ($Q_1Q_0 = 01$) and $S = 1$. From the transition table, the next state is 2 ($Q_1Q_0 = 10$) and flip-flop no. 0 changes from a 1 to a 0, therefore D_0 has to be a 0.

Cell 5: The present state is 2 ($Q_1Q_0 = 10$) and $S = 1$. From the transition table, the next state is 3 ($Q_1Q_0 = 11$) and flip-flop no. 0 changes from a 0 to a 1, therefore D_0 has to be a 1.

Cell 7: The present state is 3 ($Q_1Q_0 = 11$) and $S = 1$. From the transition table, the next state is 0 ($Q_1Q_0 = 00$) and flip-flop no. 0 changes from a 1 to a 0, therefore D_0 has to be a 0.

Similarly, we can fill-in each cell for D_1 (flip-flop no. 1) as shown in Figure 6–39b.

Step 5: excitation equations

The excitation equations for D_0 and D_1 can now be obtained from the excitation maps shown in Figure 6–39 as follows:

Equation (6–7)

$$D_0 = \overline{Q}_0$$

Equation (6–8)

$$D_1 = \overline{Q}_1\overline{Q}_0\overline{S} + \overline{Q}_1Q_0S + Q_1Q_0\overline{S} + Q_1\overline{Q}_0S$$

$$D_1 = \overline{Q}_1\overline{Q}_0\overline{S} + Q_1Q_0\overline{S} + \overline{Q}_1Q_0S + Q_1\overline{Q}_0S \quad \text{(commutative)}$$

$$D_1 = \overline{S}\,(\overline{Q}_1\overline{Q}_0 + Q_1Q_0) + S\,(\overline{Q}_1Q_0 + Q_1\overline{Q}_0) \quad \text{(distributive)}$$

$$D_1 = \overline{S}\,(\overline{Q_1 \oplus Q_0}) + S\,(Q_1 \oplus Q_0) \quad \text{(Exclusive-}OR\text{)}$$

$$D_1 = \overline{S \oplus (Q_1 \oplus Q_0)} \quad \text{(Exclusive-}OR\text{)}$$

Step 6: logic circuit

The logic circuit for the 2-bit up/down counter is implemented from Equations 6–7 and 6–8 and shown in Figure 6–40. Notice that flip-flop no. 0 is a toggle flip-flop and is not affected by S. This is because the LSB in a binary count always toggles regardless of whether the count is up or down.

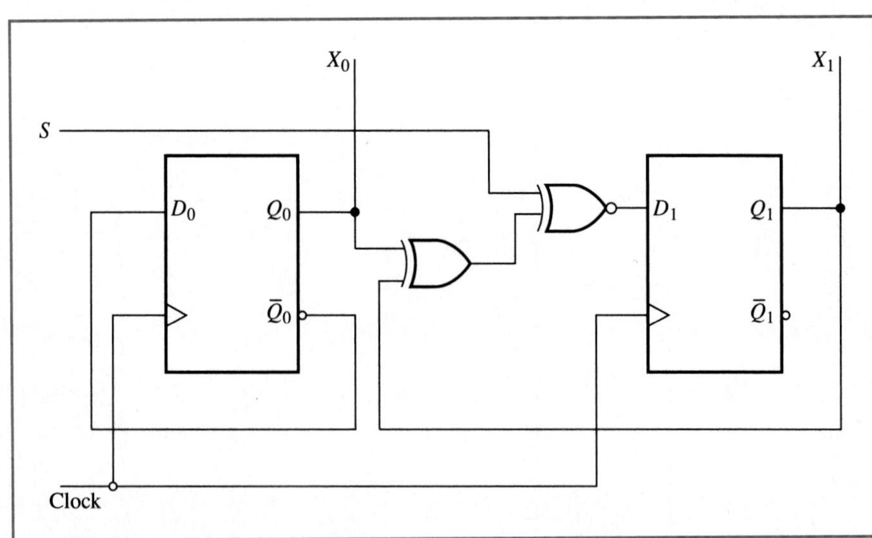

Figure 6–40 Logic circuit for the 2-bit synchronous up/down counter.

It is obvious that the logic circuits of the synchronous counters designed so far appear to be more complex in many instances than their asynchronous counterparts. This is because the synthesis procedure is generalized for the design of any type of counter and does not account for the various shortcuts that can be taken to make the circuit less complex. However, the synthesis procedure is flexible since it is applicable to any type or any size of counter. There is no formal procedure such as this for the design of asynchronous counters. Another advantage of the synthesis procedures discussed in this section is the ability to work through the steps backward and analyze a given sequential logic circuit without having to go through the tedious process of constructing a timing diagram. This procedure will be the subject of Section 6–5.

Review questions

1. Briefly outline the six-step procedure for the synthesis of synchronous counters.
2. What is the purpose of the state diagram?
3. What is the purpose of the state table?
4. What is the purpose of the transition table?
5. What is the purpose of the excitation maps?
6. What is the purpose of the excitation equations?

6–5 Analysis of synchronous counters

Section 6–2 introduced the timing diagram procedure for analyzing asynchronous counters and Section 6–3 applied the timing diagram to the analysis of synchronous counters. Timing diagrams can be a very efficient way to analyze a synchronous counter circuit as long as the circuit is relatively simple. The more complex the circuit, the more difficult it is to accurately plot a timing diagram that describes its operation. Since synchronous counters tend to be more complex than asynchronous counters, the timing diagram is not a very suitable means for analyzing synchronous counters, and therefore this section introduces a *formal procedure* of analysis that essentially reverses the six-step synthesis procedure discussed in the previous section.

The analysis of a synchronous counter involves all the steps required to obtain a state diagram from a given circuit. To illustrate the procedure for analyzing a synchronous counter, consider the analysis of the counter shown in Figure 6–25. We will assume that the function of this circuit is unknown since the purpose of analysis is to determine what the circuit does. The following steps are used in the analysis of the circuit.

Step 1: excitation equations

The excitation equations for the circuit are obtained by determining the logic equations for D_0 and D_1 from Figure 6–25 as follows:

Equation (6–9)
$$D_0 = \overline{Q}_0$$

Equation (6–10)
$$D_1 = Q_1 \oplus Q_0$$

Step 2: transition equations

The *transition equations* for each flip-flop allow us to determine the next state of each flip-flop, given the present conditions. Recall from Chapter 5 (Section 5–6) that the state transition equation for a D flip-flop is

Equation (6–11)
$$Q^{n+1} = D$$

Therefore for flip-flop no. 0

Equation (6–12)
$$Q_0^{n+1} = D_0$$

Substituting the excitation equation (Eq. 6–9) into Equation 6–12 we have

Equation (6–13)
$$Q_0^{n+1} = \overline{Q}_0$$

Similarly for flip-flop no. 1

Equation (6–14)
$$Q_1^{n+1} = D_1$$

Substituting the excitation equation (Eq. 6–10) into Equation 6–14 we have

Equation (6–15)
$$Q_1^{n+1} = Q_1 \oplus Q_0$$

Equations 6–13 and 6–15 are the transition equations for flip-flops no. 0 and 1, respectively. Equation 6–13 states that the next state of flip-flop no. 0 will be equal to the complement of its present state (toggle) and Equation 6–15 states that the next state of flip-flop no. 1 will be equal to the present state of flip-flop no. 1 exclusively-*OR*ed with the present state of flip-flop no. 0. Since the right-hand side of a transition equation always has variables that represent the present state, we choose to drop the superscript n for variables on the right-hand side.

Step 3: transition table

From the transition equations we can now construct the transition table for the counter since it is possible to determine the next state of each flip-flop given the present state. The transition table is constructed by first listing all possible combinations of present states

$(Q_1\ Q_0)^n$	
0	0
0	1
1	0
1	1

and then determining the next state of each flip-flop from its transition equation as shown in Table 6–5. For example, if the present state of flip-flop no. 0 is a 1, the next state is

$$Q_0^{n+1} = \overline{Q}_0$$
$$\therefore\ Q_0^{n+1} = \overline{1} = 0$$

Table 6–5
Transition table for circuit in Figure 6–25

$(Q_1\ Q_0)^n$	$(Q_1\ Q_0)^{n+1}$
0 0	0 1
0 1	1 0
1 0	1 1
1 1	0 0

and if the present state of flip-flop no. 0 is a 0, the next state is

$$Q_0^{n+1} = \overline{Q}_0$$
$$\therefore \quad Q_0^{n+1} = \overline{0} = 1$$

Thus the Q_0^{n+1} column in Table 6–5 is obtained by complementing the Q_0^n column. Similarly, for flip-flop no. 1, since the next state of this flip-flop is dependent on the present state of both flip-flops no. 0 and no. 1, if the present state of flip-flop no. 0 is 0 and the present state of flip-flop no. 1 is a 0, the next state of flip-flop no. 1 is

$$Q_1^{n+1} = Q_1 \oplus Q_0$$
$$Q_1^{n+1} = 0 \oplus 0 = 0$$

We can therefore fill in the Q_1^{n+1} column of Table 6–5 by simply exclusively-*OR*ing the Q_0^n and Q_1^n columns.

Step 4: state table

Since the state table is simply a translation of the transition table and relates the state of the flip-flops to the states of the counter, we can construct the state table shown in Table 6–6 by converting each binary combination in Table 6–5 to its appropriate state representation.

Step 5: state diagram

We can now represent the state table shown in Table 6–6 graphically in the form of a state diagram. To construct the state diagram we first draw the symbols that represent all possible states of the counter, and then referring to the state table we make the appropriate connections between the states as shown in Figure 6–41. Notice that the state diagram of Figure 6–41 is the same as the state diagram in Figure 6–23, thus verifying the analysis. The same procedure can be applied to circuits that

Table 6–6
State table for circuit in Figure 6–25

q^n	q^{n+1}
q_0	q_1
q_1	q_2
q_2	q_3
q_3	q_0

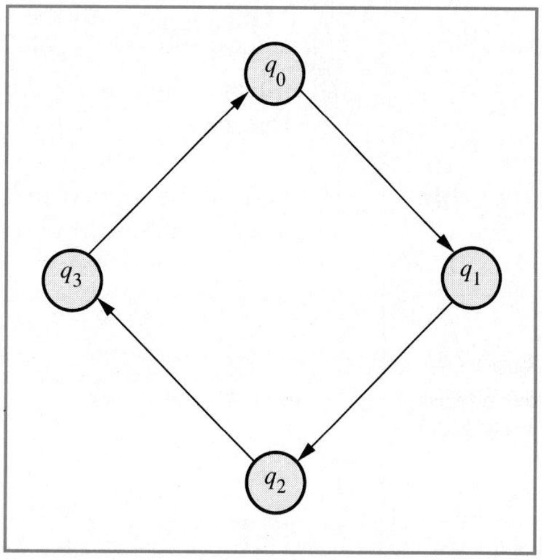

Figure 6–41
State diagram for the counter in Figure 6–25.

contain J-K flip-flops. For example, consider the analysis of the circuit shown in Figure 6–30.

Step 1: excitation equations

The excitation equations for the circuit are obtained by determining the logic equations for J_0, K_0, J_1, and K_1 as follows:

Equation (6–16)
$$J_0 = K_0 = 1$$

Equation (6–17)
$$J_1 = K_1 = \overline{Q}_0$$

Step 2: transition equations

Recall from Chapter 5 (Section 5–7) that the state transition equation for a J-K flip-flop is

$$Q^{n+1} = J\overline{Q}^n + \overline{K}Q^n$$

Therefore for flip-flop no. 0,

Equation (6–18)
$$Q_0{}^{n+1} = J_0\overline{Q}_0{}^n + \overline{K}_0Q_0{}^n$$

Substituting the excitation equation (Eq. 6–16) into Equation 6–18 (and dropping the superscript n on the right-hand side), we have

Equation (6–19)
$$Q_0{}^{n+1} = 1 \cdot \overline{Q}_0 + \overline{1} \cdot Q_0$$
$$Q_0{}^{n+1} = \overline{Q}_0 + 0 \cdot Q_0$$
$$Q_0{}^{n+1} = \overline{Q}_0$$

Similarly, for flip-flop no. 1

Equation (6–20)
$$Q_1{}^{n+1} = J_1\overline{Q}_1{}^n + \overline{K}_1Q_1{}^n$$

Substituting the excitation equation (Eq. 6–17) into Equation 6–20 (and dropping the superscript n) we have

Equation (6–21)
$$Q_1{}^{n+1} = \overline{Q}_0\overline{Q}_1 + \overline{\overline{Q}}_0Q_1$$
$$Q_1{}^{n+1} = \overline{Q}_0\overline{Q}_1 + Q_0Q_1$$
$$Q_1{}^{n+1} = \overline{Q_0 \oplus Q_1}$$

Equations 6–19 and 6–21 are the transition equations for flip-flops nos. 0 and 1, respectively.

Step 3: transition table

From the transition equations we can now construct the transition table for the counter as was done before. Notice that the next state of flip-flop no. 0 will toggle, and the next state of flip-flop no. 1 will be equal to the complement of the present state of flip-flop no. 0 Exclusively-*OR*ed with the present state of flip-flop no. 1. The resulting transition table is shown in Table 6–7.

Step 4: state table

The state table shown in Table 6–8 is constructed by translating the outputs of the flip-flops in the transition table to the states of the counter.

Table 6–7
Transition table for circuit in Figure 6–30

$(Q_1\ Q_0)^n$	$(Q_1\ Q_0)^{n+1}$
0 0	1 1
0 1	0 0
1 0	0 1
1 1	1 0

Table 6–8
State table for circuit in Figure 6–30

q^n	q^{n+1}
q_0	q_3
q_1	q_0
q_2	q_1
q_3	q_2

Step 5: state diagram

We can now represent the state table shown in Table 6–8 graphically in the form of a state diagram as shown in Figure 6–42. Notice that the state diagram of Figure 6–42 is the same as the state diagram in Figure 6–28, thus verifying the analysis.

One of the advantages of this procedure for analyzing synchronous counters is that it provides a thorough description of the circuit being analyzed. For example, in the previous section the 3-bit counter designed in Figure 6–33 counted through five of the eight possible states in the following sequence:

$$0, 2, 6, 3, 5, 0, 2, \ldots$$

Since the remaining states were not defined in the synthesis (hence the don't cares in the excitation maps) what would happen if the counter started counting in one of the states 1, 4, or 7? To determine this using the timing diagram approach we would have to draw three timing diagrams, each one starting off in one of the three undefined states. The timing diagram would then indicate what the next state(s) would be. This is a tedious process. Furthermore, since the objective of analysis is to determine the function of the circuit being analyzed, and if this were a circuit whose function was unknown, we would not know at which state to begin the timing diagram analysis and would have to use a trial-and-error approach. Using the formal five-step approach to analysis, one can obtain a complete state diagram that includes the main counting sequence of the counter as well as the connections among the undefined states. To illustrate this fact consider the analysis of the counter shown in Figure 6–33 in the following example.

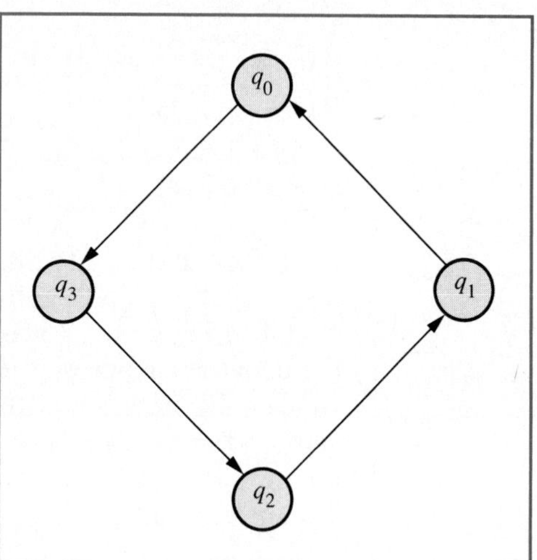

Figure 6–42
State diagram for the counter in Figure 6–30.

EXAMPLE 6–8

Analyze the logic circuit shown in Figure 6–33 by obtaining its state diagram.

SOLUTION

Step 1: excitation equations

$$D_0 = Q_1Q_0 + Q_2Q_1$$
$$D_1 = \overline{Q}_0$$
$$D_2 = \overline{Q}_2Q_1$$

Step 2: transition equations

$$\begin{aligned} Q_0{}^{n+1} &= D_0 \\ &= Q_1Q_0 + Q_2Q_1 \\ Q_1{}^{n+1} &= D_1 \\ &= \overline{Q}_0 \\ Q_2{}^{n+1} &= D_2 \\ &= \overline{Q}_2Q_1 \end{aligned}$$

Step 3: transition table

$(Q_2\ Q_1\ Q_0)^n$	$(Q_2\ Q_1\ Q_0)^{n+1}$
0 0 0	0 1 0
0 0 1	0 0 0
0 1 0	1 1 0
0 1 1	1 0 1
1 0 0	0 1 0
1 0 1	0 0 0
1 1 0	0 1 1
1 1 1	0 0 1

Step 4: state table

q^n	q^{n+1}
q_0	q_2
q_1	q_0
q_2	q_6
q_3	q_5
q_4	q_2
q_5	q_0
q_6	q_3
q_7	q_1

Step 5: state diagram. The state diagram for the counter is shown in Figure 6–43.

Even though the state diagram obtained in Example 6–8 (Figure 6–43) does not resemble the original state diagram (Figure 6–31) from which the circuit in Figure 6–33 was designed, by closer inspection of Figure 6–43 the count sequence 0, 2, 6, 3, 5, 0, . . . can be traced. From Figure 6–43 we can also determine what the next state of the counter will be if it starts off in states 1, 4, or 7. For example, if the counter begins at state 1, the next state is 0, and the count cycles through the main counting sequence. Similarly, if the counter begins at state 4, the next state is 2,

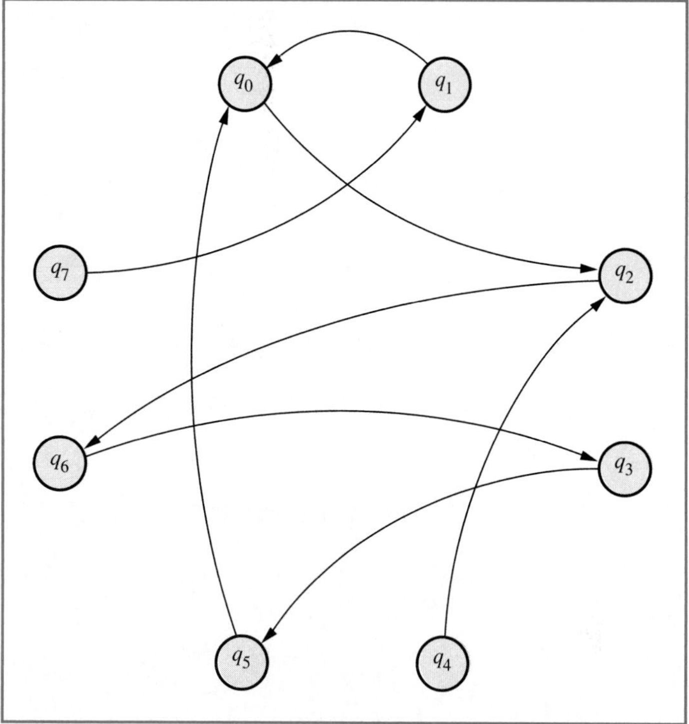

Figure 6–43
State diagram for the counter in Figure 6–33.

and count cycles through the main counting sequence. If, however, the counter begins at state 7, the next state is 1, which connects to the main counting sequence. Notice in Example 6–8 that we must list all possible present states in the transition table since we usually don't know the counting sequence of the counter to be analyzed.

The BCD counter designed in Example 6–7 is another example of a counter which when analyzed will produce a state diagram that is considerably different from the original state diagram from which the counter was designed. This is because the BCD counter uses 10 of the 16 possible states, and the state diagram obtained through analysis will also show the connections between the 6 undefined states besides the BCD counting sequence. The following example illustrates the analysis of the synchronous BCD counter designed in Figure 6–36.

EXAMPLE 6–9

Analyze the circuit shown in Figure 6–36 by obtaining a state diagram.

SOLUTION

Step 1: excitation equations

$$J_0 = K_0 = 1$$
$$J_1 = Q_0\overline{Q}_3$$
$$K_1 = Q_0$$
$$J_2 = K_2 = Q_1Q_0$$
$$J_3 = Q_2Q_1Q_0$$
$$K_3 = Q_0$$

Step 2: transition equations

$$
\begin{aligned}
Q_0^{n+1} &= J_0\bar{Q}_0 + \bar{K}_0 Q_0 \\
&= 1 \cdot \bar{Q}_0 + 0 \cdot Q_0 \\
&= \bar{Q}_0
\end{aligned}
$$

$$
\begin{aligned}
Q_1^{n+1} &= J_1\bar{Q}_1 + \bar{K}_1 Q_1 \\
&= Q_0\bar{Q}_3\bar{Q}_1 + \bar{Q}_0 Q_1
\end{aligned}
$$

$$
\begin{aligned}
Q_2^{n+1} &= J_2\bar{Q}_2 + \bar{K}_2 Q_2 \\
&= Q_1 Q_0\bar{Q}_2 + \overline{Q_1 Q_0}\, Q_2 \\
&= Q_1 Q_0\bar{Q}_2 + (\bar{Q}_1 + \bar{Q}_0)Q_2 \\
&= Q_1 Q_0\bar{Q}_2 + \bar{Q}_1 Q_2 + \bar{Q}_0 Q_2
\end{aligned}
$$

$$
\begin{aligned}
Q_3^{n+1} &= J_3\bar{Q}_3 + \bar{K}_3 Q_3 \\
&= Q_2 Q_1 Q_0\bar{Q}_3 + \bar{Q}_0 Q_3
\end{aligned}
$$

Step 3: transition table

$(Q_3\ Q_2\ Q_1\ Q_0)^n$	$(Q_3\ Q_2\ Q_1\ Q_0)^{n+1}$
0 0 0 0	0 0 0 1
0 0 0 1	0 0 1 0
0 0 1 0	0 0 1 1
0 0 1 1	0 1 0 0
0 1 0 0	0 1 0 1
0 1 0 1	0 1 1 0
0 1 1 0	0 1 1 1
0 1 1 1	1 0 0 0
1 0 0 0	1 0 0 1
1 0 0 1	0 0 0 0
1 0 1 0	1 0 1 1
1 0 1 1	0 1 0 0
1 1 0 0	1 1 0 1
1 1 0 1	0 1 0 0
1 1 1 0	1 1 1 1
1 1 1 1	0 0 0 0

Step 4: state table

q^n	q^{n+1}
q_0	q_1
q_1	q_2
q_2	q_3
q_3	q_4
q_4	q_5
q_5	q_6
q_6	q_7
q_7	q_8
q_8	q_9
q_9	q_0
q_A	q_B
q_B	q_4
q_C	q_D
q_D	q_4
q_E	q_F
q_F	q_0

Step 5: state diagram

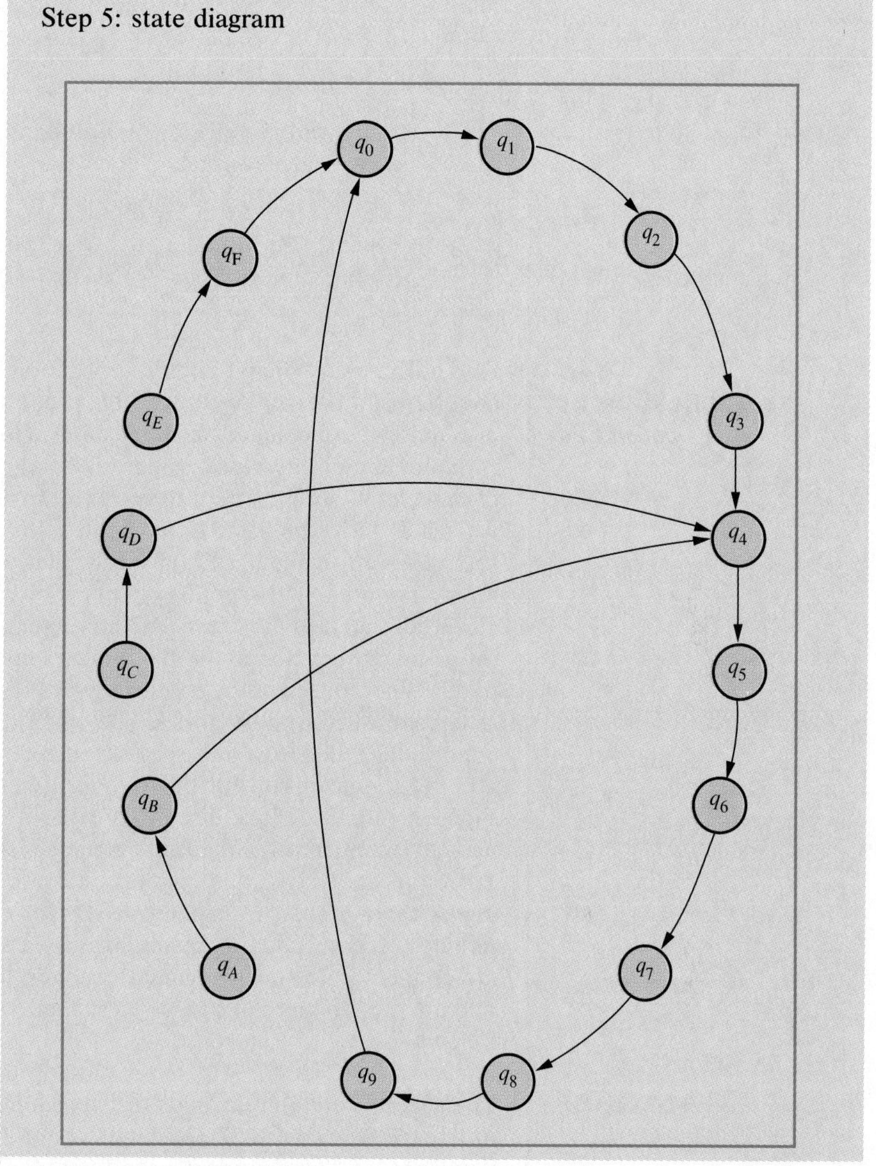

Figure 6–44
State diagram for the counter in Figure 6–36.

The state diagram in Figure 6–44 illustrates a BCD counting sequence and thus verifies the operation of the circuit as a BCD counter. Notice that the undefined states A, B, C, D, E, and F are all connected to the main BCD counting sequence. For example, if the counter starts off in state E, the next states are F and then 0. If the counter starts in state C, the next states are D, and then 4. Similarly, if the counter starts in state A, the next states are B and then 4.

It is apparent that the analysis technique discussed in this section has some advantages and disadvantages when compared with the timing diagram approach to analysis. Depending on the type of circuit, the formal approach to analysis can be the quickest path to a solution. However, some circuits are simple enough to be analyzed through a timing diagram. The formal approach is restricted to synchronous counters only while the timing diagram can be used to analyze both synchronous and asynchronous counters. Finally, the formal approach to analysis is the most complete and direct approach since the state diagram obtained from it accounts for all possible states of the circuit.

Review questions

1. Briefly outline the steps involved in the formal analysis of synchronous counters.
2. What are transition equations?
3. Compare the formal analysis procedure with the timing diagram procedure of analysis.

6–6 An application of counters

One of the many digital circuit applications that makes extensive use of counters is a *time-of-day clock.* This section applies many of the concepts of flip-flops and counters covered in previous sections to the design of a 24-hour time-of-day clock. Figure 6–45 shows the logic symbol for a time-of-day clock circuit. The circuit has a single clock input, *CLOCK,* which is used to keep the basic time of the counters, and outputs to drive three sets of seven-segment LED displays that will be used to display the hours, minutes, and seconds counts. For each (negative) transition of *CLOCK,* the seconds count will increment once from 00 to 59. Each time the seconds recycles to 00, the minutes count will increment once from 00 to 59. When the minutes count recycles to 00, the hours count will increment once from 00 to 23. The clock circuit can therefore be used to count hours, minutes, and seconds of real time.

The logic circuit for the 24-hour time-of-day clock is shown in Figure 6–46. The clock circuit consists of a series of counters that are used to count through the units of time—seconds, minutes, and hours. Since the hours, minutes, and seconds are displayed as two decimal digits, there are two 7-segment LEDs for each. Each unit of time (hours, minutes, and seconds) is maintained by a set of counters whose outputs are connected to seven-segment decoders to drive the LEDs. The operation of the circuit can be explained with reference to Figure 6–46 as follows.

Seconds counter

The seconds counter consists of two counters—one for each of the digits in the seconds count. The seconds count cycles from 00 to 59, and therefore a BCD counter is used to cycle through the 0 to 9 count of the

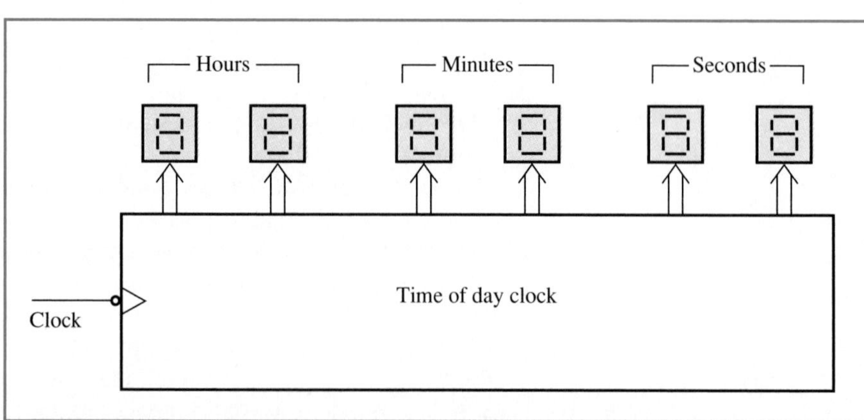

Figure 6–45 Logic symbol for a time-of-day clock.

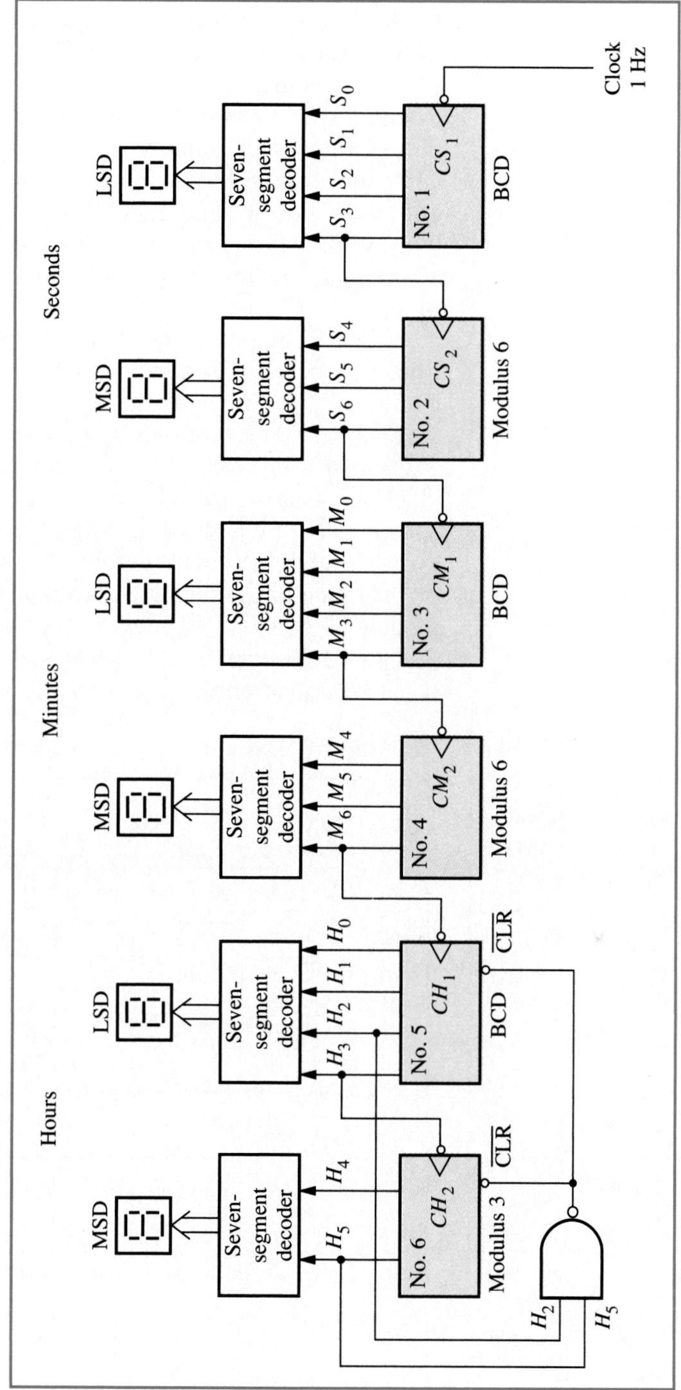

Figure 6–46
Logic circuit for a time-of-day clock.

LSD of the seconds count, and a modulus-6 counter is used to cycle through the 0 to 5 count of the MSD of the seconds count.

A periodic clock pulse with a frequency of 1 Hz provides the basic clock pulses for the BCD counter. Since the rate at which the clock pulses trigger the counter's clock (CS_1) is 1 pulse per second, the counter will count from 0 to 9 in 1-second increments. The timing diagram for the outputs of the BCD counter ($S_3S_2S_1S_0$) with reference to its clock input (CS_1) is shown in Figure 6–47. Notice that the output of the BCD counter, S_3, makes a negative transition when the counter recycles back to 0.

The S_3 output of the BCD counter therefore makes a negative transition every 10 seconds and is used as a clock for the modulus-6 counter (CS_2). The modulus-6 counter maintains the MSD count for seconds and therefore only requires 6 states, 0 through 5. When S_3 makes a negative transition every 10 seconds, since S_3 is connected to CS_2, the clock input of the modulus-6 counter, the counter will increment as shown in Figure 6–47. The outputs of the modulus-6 counter, $S_6S_5S_4$, will cycle through the counts 0 through 5.

Notice in the timing diagram of Figure 6–47 that the output of the modulus-6 counter, S_6, makes a negative transition when it recycles to 0. Since each time interval between CS_2 clock pulses is 10 seconds, this negative transition will occur every 60 seconds (1 minute).

Figure 6–47
Timing diagram for SECONDS count.

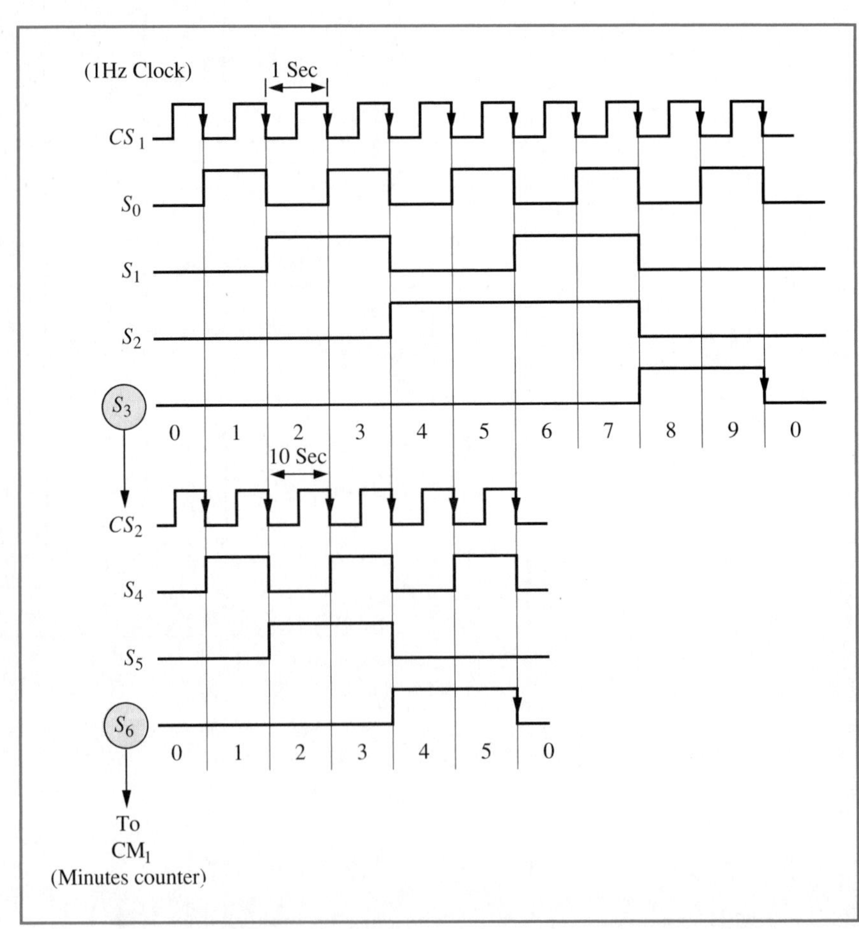

Minutes counter

Since the minutes count is exactly the same as the seconds count, the two counters that provide the minutes count are identical to the two counters that provide the seconds count. The minutes counter consists of two counters, one for each of the digits in the minutes count. The minutes count cycles from 00 to 59, and therefore a BCD counter is used to cycle through the 0 to 9 count of the LSD of the minutes count, and a modulus-6 counter is used to cycle through the 0 to 5 count of the MSD of the minutes count.

The clock input of the BCD counter, CM_1, is connected to S_6. Recall that S_6 will make a negative transition once every minute, and therefore the BCD counter will change states every minute. The BCD counter will therefore count from 0 to 9 in 1-minute increments. The timing diagram for the outputs of the BCD counter ($M_3M_2M_1M_0$) with reference to its clock input (CM_1) is shown in Figure 6–48. Notice that the output of the BCD counter, M_3, makes a negative transition when the counter recycles back to 0.

The M_3 output of the BCD counter therefore makes a negative transition every 10 minutes and is used as a clock for the modulus-6 counter. The modulus-6 counter maintains the MSD count for minutes and therefore only requires 6 states, 0 through 5. When M_3 makes a negative transition every 10 minutes, since M_3 is connected to CM_2 (the

Figure 6–48
Timing diagram for MINUTES count.

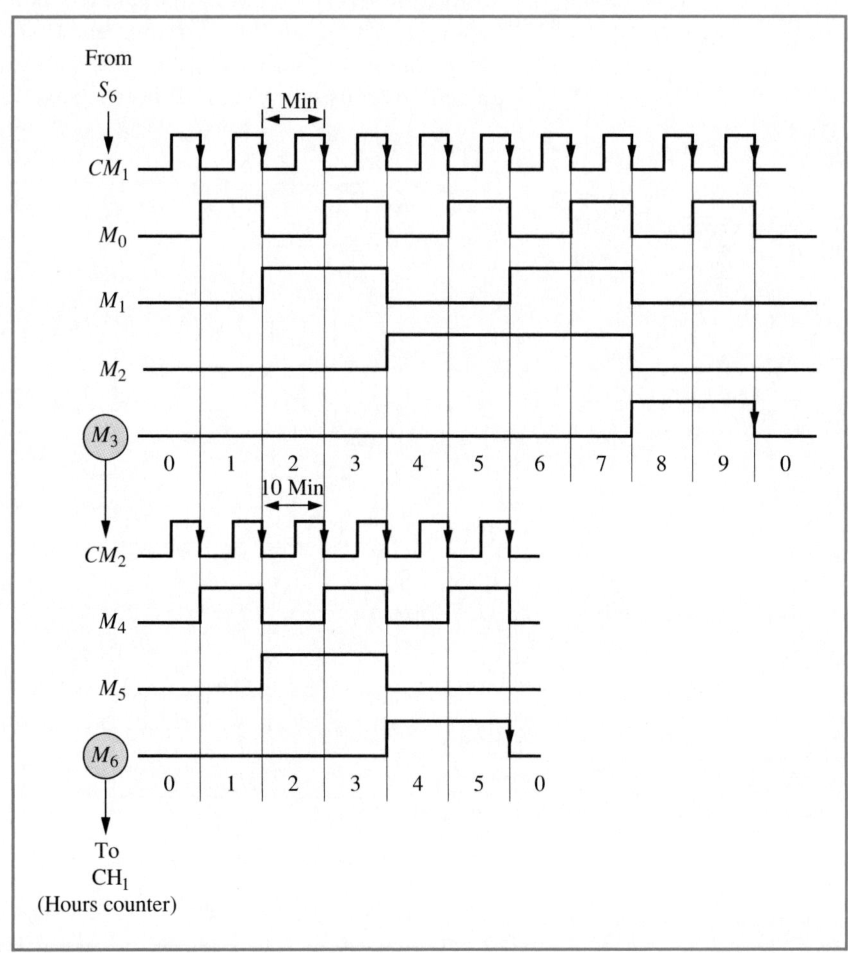

clock input of the modulus-6 counter), the counter will increment as shown in Figure 6–48. The outputs of the modulus-6 counter, $M_6M_5M_4$, will cycle through the counts 0 through 5.

Notice in the timing diagram of Figure 6–48 that the output of the modulus-6 counter, M_6, makes a negative transition when it recycles to 0. Since each time interval between CM_2 clock pulses is 10 minutes, this negative transition will occur every 60 minutes (1 hour).

Hours counter

The hours counter consists of two counters, one for each of the digits in the hours count. The hours count cycles from 00 to 23 and therefore a BCD counter is used to cycle through the 0 to 9 count of the LSD of the hours count, and a modulus-3 counter is used to cycle through the 0 to 2 count of the MSD of the hours count.

The clock input of the BCD counter, CH_1, is connected to M_6. Recall that M_6 will make a negative transition once every hour and therefore the BCD counter will change states every hour. The BCD counter will therefore count from 0 to 9 in 1-hour increments. The timing diagram for the outputs of the BCD counter ($H_3H_2H_1H_0$) with reference to its clock input (CH_1) is shown in Figure 6–49. Notice that the output of the BCD counter, H_3, makes a negative transition when the counter recycles back to 0.

The H_3 output of the BCD counter therefore makes a negative transition every 10 hours and is used as a clock for the modulus-3 counter. The modulus-3 counter maintains the MSD count for hours and therefore only requires three states, 0 through 2. When H_3 makes a negative transition every 10 hours, since H_3 is connected to CH_2 (the

Figure 6–49
Timing diagram for HOURS count.

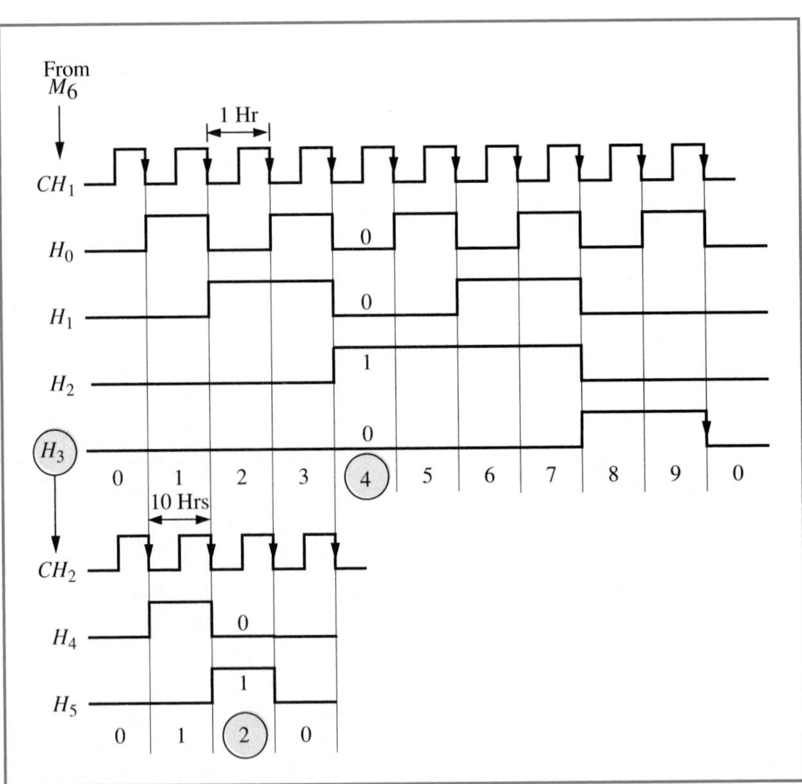

clock input of the modulus-3 counter), the counter will increment as shown in Figure 6–49. The outputs of the modulus-3 counter, H_5H_4, will cycle through the counts 0 through 2.

Notice in the timing diagram of Figure 6–49 that the output of the modulus-3 counter, H_5, makes a negative transition when it recycles to 0. Since each time interval between CH_2 clock pulses is 10 hours, theoretically this negative transition will occur every 30 hours. However, we want to recycle the counter *after* the hours count reaches 23. To do this we must reset the counter when the hours count reaches 24. Notice in the timing diagram of Figure 6–49 that when the hours count reaches 24

$$H_5H_4 = 10 \quad H_3H_2H_1H_0 = 0100$$

and this is the *only* time when output $H_5 = 1$ and $H_2 = 1$. Therefore when the two lines are at the logic 1 state we must reset both counters. The *NAND* gate shown in Figure 6–46 functions as a count decoder and will produce a logic 0 only when H_5 and H_2 are logic 1s. This active-low signal is used to directly reset (clear) the flip-flops that make up each counter through their direct $\overline{CLR}$ inputs. This scheme allows the hours counter to recycle after a count of 23.

Review questions

1. What type of counters could be used to implement the time-of-day clock—synchronous, asynchronous, or any type?
2. Could the entire time-of-day clock circuit be considered a synchronous or asynchronous (ripple) counter?
3. Why are the seven-segment decoders used in the circuit?

6–7 Frequency division

In Chapter 5 it was stated that a *T* flip-flop could be used to divide its input clock frequency by 2. If the output of the *T* flip-flop was connected to the clock input of another *T* flip-flop, the original clock frequency would then be divided by 4. Similarly, if a third *T* flip-flop was added in series the frequency division would be 8, and so on. Therefore each flip-flop in series divides its input clock by 2 so that the total frequency division is

$$2 \times 2 \times 2 = 8$$

The asynchronous counters discussed in Section 6–2 can be used as frequency division circuits since they are made up of *T* flip-flops connected in series. For example, notice in Figure 6–3 that the timing diagram for the 3-bit asynchronous down counter shown in Figure 6–2 divides the clock by 2 at output X_0, by 4 at output X_1, and by 8 at output X_2. Since the timing diagram for the asynchronous counter is the same as the timing diagram for a synchronous counter with the same counting sequence, we can conclude that the frequency division can also be accomplished by synchronous counters.

Since sequential binary counters can also be used for frequency division, they are also known as *divide-by-n* counters, where *n* is the modulus of the counter. For example, a 3-bit binary counter will have eight possible states (modulus 8) and therefore its most significant output

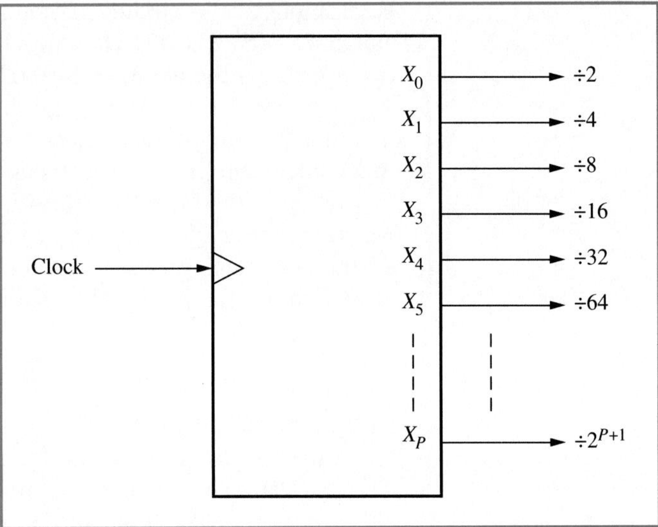

Figure 6–50
Divided outputs for binary counters.

will divide the clock frequency by 8. The 3-bit binary counter (modulus-8 counter) could also be referred to as a divide-by-8 counter. Similarly, a 4-bit binary counter would provide a maximum frequency division of 16 at its most significant output, while a 4-bit BCD counter (modulus-10 counter) would provide a maximum frequency division of 10 at its most significant output. The general frequency divisors for the outputs of binary counters is shown in Figure 6–50. Notice that the divisor by which the clock frequency is divided is a power of 2

$$2^{p+1}$$

where p is the bit position of the counter output. We can also use this relationship to determine the maximum frequency division that could be accomplished by an n-bit binary counter

$$2^n$$

where n is the size of the counter.

Counters can be *cascaded* to increase the frequency division. For example, Figure 6–51 shows how a divide-by-4 counter and a divide-by-8

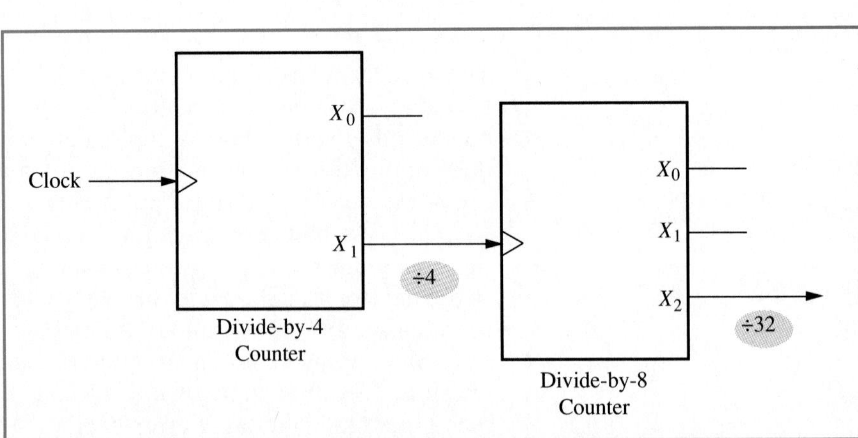

Figure 6–51
Divide-by-32 counter.

counter could be connected to produce a divide-by-32 counter. In general, to determine the total frequency division of two cascaded counters we must *multiply* the divisors of each counter.

EXAMPLE 6–10

Design a circuit to divide a periodic clock frequency by a factor of 160.

SOLUTION To accomplish the frequency division we can cascade a divide-by-16 counter with a divide-by-10 counter (BCD counter) to obtain a total frequency division of

$$16 \times 10 = 160$$

The circuit to do this is shown in Figure 6–52.

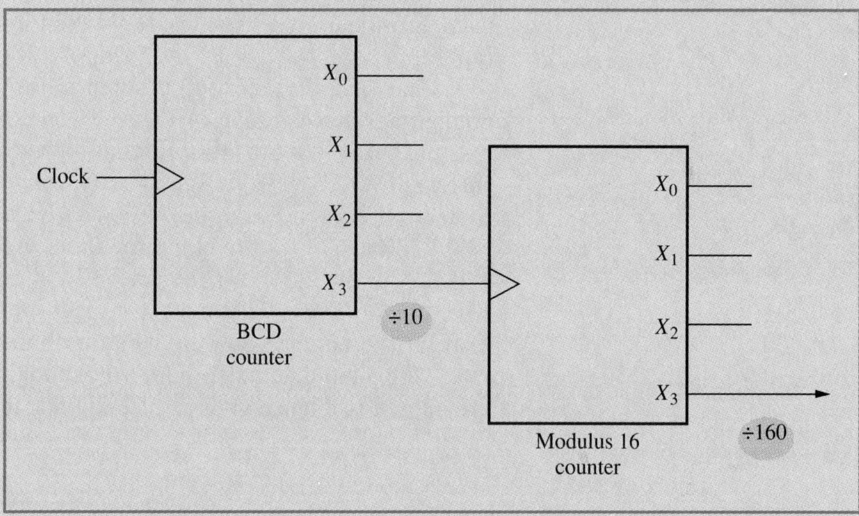

Figure 6–52
Divide-by-160 counter.

An example of frequency division was seen in the previous section for the design of the time-of-day clock. Notice in Figure 6–46 that the circuit effectively takes a 1 pulse per second (pps) input clock (at CS_1) and uses it to trigger a seconds counter. The seconds counter effectively divides the 1 pps clock first by 10 at S_3 and then by 60 at S_6 to produce a clock frequency of 1/60 pps or 1 pulse per minute (ppm). Similarly the 1-ppm clock triggers the minutes counters which also divide the 1-ppm clock by 60 and produces a clock frequency of 1/60 ppm or 1 pulse per hour (pph) at M_6. This clock is then used to trigger the hours counter.

Review questions

1. Why are binary counters known as divide-by-*n* counters?
2. Is frequency division limited to asynchronous counters? Why?
3. How can frequency division be increased?
4. What determines the maximum frequency division that can occur in a counter?

6–8 Integrated circuit logic

The 74xxx series of integrated circuits has several prepackaged counters available in the form of single ICs that can be used for a variety of counter applications. Both synchronous and asynchronous counters are available, many of which have additional control inputs for initializing the state of the counter and/or arrangements for reconfiguring the count.

The logic symbol and pin configuration of the 7493 divide-by-16 counter is shown in Figure 6–53. The 7493 is a four-stage (4-bit) modulus-16 binary ripple counter. The 4-bit output of the counter is made available at $Q_3Q_2Q_1Q_0$. Internally the 7493 consists of a divide-by-2 counter (single flip-flop) with clock input CP_0 and output Q_0, and a divide-by-8 counter with clock input CP_1 and outputs $Q_2Q_1Q_0$. The 7493 can therefore operate as two independent counters. By connecting the output of the divide-by-2 counter Q_0 with the clock input of the divide-by-8 counter CP_1, the 7493 can be configured as a divide-by-16 counter with a 4-bit output $Q_3Q_2Q_1Q_0$ and a clock input CP_0. The *master reset* lines MR_1 and MR_2 are used to asynchronously reset the counter when logic 1's are applied to *both* inputs.

The 7492 divide-by-12 counter whose logic symbol and pin configuration are shown in Figure 6–54 is similar in operation to the 7493 except that it is a modulus-12 counter with 4-bit outputs $Q_3Q_2Q_1Q_0$. Internally the 7492 is designed as two independent divide-by-2 and divide-by-6 counters that, like the 7493, can be connected to produce a divide-by-12 count. CP_0 is the clock for the divide-by-2 counter with output Q_0, and CP_1 is the clock for the divide-by-6 counter with outputs $Q_2Q_1Q_0$. Like the 7493, the 7492's master reset inputs MR_1 and MR_2 can be used to reset the counter outputs when both are at the logic 1 state.

The 7490 decade counter whose logic symbol and pin configuration are shown in Figure 6–55 is also similar in operation to the 7492 and 7493,

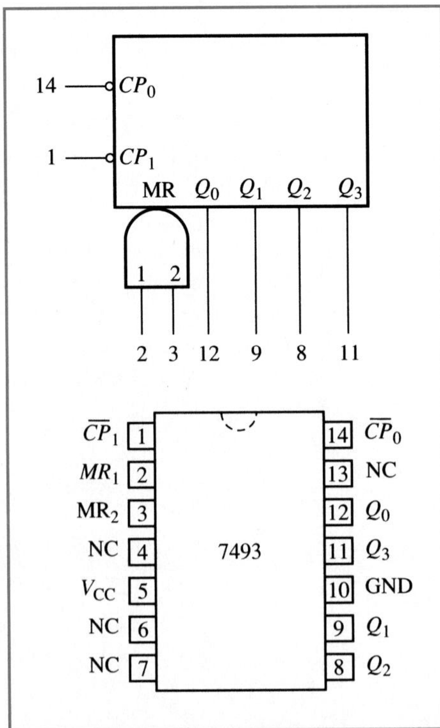

Figure 6–53 Logic symbol and pin configuration of the 7493 divide-by-16 counter.

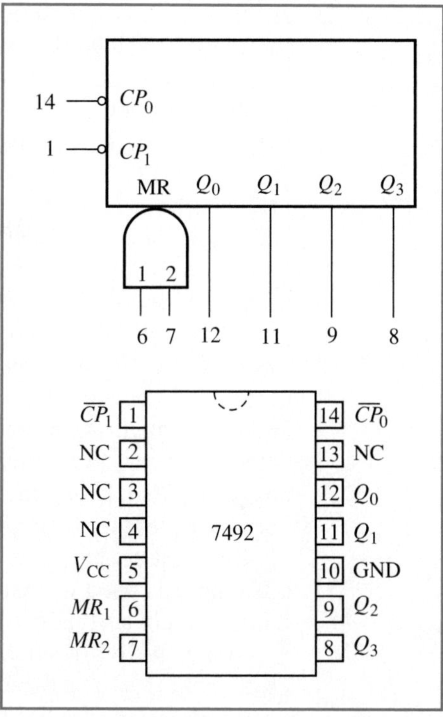

Figure 6–54 Logic symbol and pin configuration of the 7492 divide-by-12 counter.

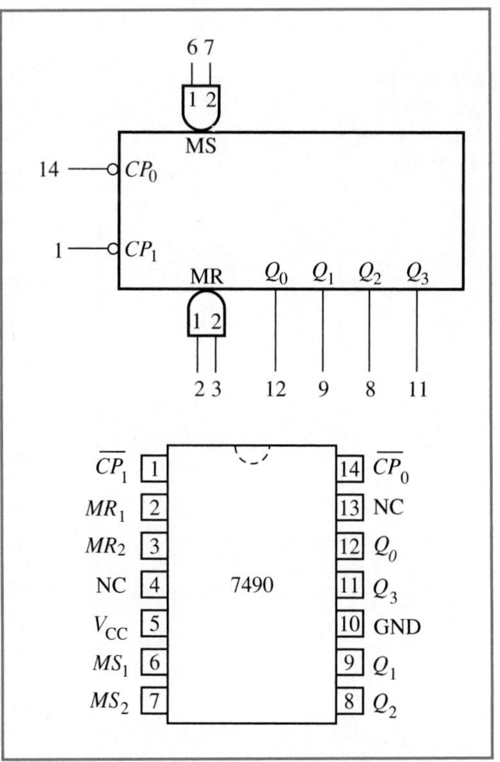

Figure 6–55 Logic symbol and pin configuration of the 7490 decade counter.

except that it is a modulus-10 BCD counter with 4-bit outputs $Q_3Q_2Q_1Q_0$. Internally the 7490 is designed as two independent divide-by-2 and divide-by-5 counters that, like the 7492 and 7493, can be connected to produce a divide-by-10 count. CP_0 is the clock for the divide-by-2 counter with output Q_0, and CP_1 is the clock for the divide-by-5 counter with outputs $Q_2Q_1Q_0$. Like the 7492 and 7493, the 7490's master reset inputs MR_1 and MR_2 can be used to reset the counter outputs when both are at the logic 1 state. The 7490 includes two additional *master set* inputs, MS_1 and MS_2, that are used to asynchronously set the outputs of the counter to the maximum count, 1001 (9) when *both* inputs are at the logic 1 state.

The 74177 presettable binary counter is similar in operation to the 7493. That is, it is a 4-bit modulus-16 ripple counter that is internally designed with a divide-by-2 and a divide-by-8 section. The logic symbol and pin configuration for the 74177 are shown in Figure 6–56. Unlike the 7493, however, the outputs of the 74177 can be asynchronously preset to any 4-bit binary combination so that on the next clock pulse the counter begins at a preset state. The data inputs $P_3P_2P_1P_0$ are used to enter the binary number that is to be preset in the counter and the Parallel Load ($\overline{PL}$) input is used to load (or latch) the data at the outputs $Q_3Q_2Q_1Q_0$. Only a single active-low master reset line $\overline{MR}$ is used to reset the counter.

The 74176 presettable decade counter is similar in operation to the 7490. That is, it is a 4-bit modulus-10 ripple counter that is internally designed with a divide-by-2 and a divide-by-5 section. The logic symbol and pin configuration for the 74176 are shown in Figure 6–57. Like the 74177, the outputs of the 74176 can be asynchronously preset to any 4-bit binary combination so that on the next clock pulse the counter begins at a preset state. The data inputs $P_3P_2P_1P_0$ are used to enter the binary number that is to be preset in the counter and the $\overline{PL}$ input is used to

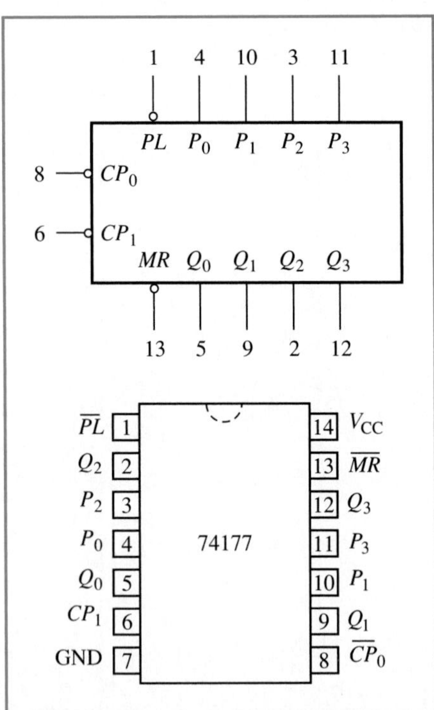

Figure 6–56
Logic symbol and pin configuration of the 74177 presettable binary counter.

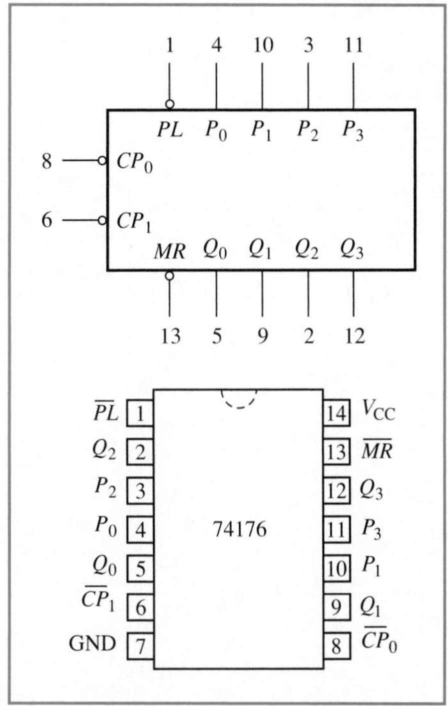

Figure 6–57 Logic symbol and pin configuration of the 74176 presettable decade counter.

load (or latch) the data at the outputs $Q_3Q_2Q_1Q_0$. Only a single active-low master reset line $\overline{MR}$ is used to reset the counter.

The 74176 and 74177 are asynchronous counters. For applications that require synchronous counters, the 74160 and 74161 provide the same functions, respectively, using synchronous counters instead.

The 74169 is a synchronous bidirectional modulus-16 binary counter that has data inputs for presetting the counter to any state, as well as a control input to make the counter count either *UP* or *DOWN*. The logic symbol and pin configuration for the 74169 is shown in Figure 6–58. The 74169 contains a *single* modulus-16 binary counter with outputs $Q_3Q_2Q_1Q_0$. The preset data inputs $P_3P_2P_1P_0$ are used to initialize the output of the counter to a predetermined state when the Parallel Enable ($\overline{PE}$) input is low *and* the clock (*CP*) makes a positive transition. The *UP*/*DOWN* ($U/\overline{D}$) input causes the counter to count *UP* (logic 1) or *DOWN* (logic 0). For counting to occur, both Count Enable Parallel ($\overline{\text{CEP}}$) and Count Enable Trickle ($\overline{\text{CET}}$) must be at the logic 0 states. The Terminal Count ($\overline{\text{TC}}$) output of the counter will be a logic 0 when the counter reaches 0000 in the *DOWN* mode or when the counter reaches 1111 in the *UP* mode.

The 74168 is a synchronous bidirectional BCD decade counter that, like the 74169, has data inputs for presetting the counter to any state, as well as a control input to make the counter count either *UP* or *DOWN*. The logic symbol and pin configuration for the 74168 are shown in Figure 6–59. The 74168 contains a *single* modulus-10 BCD counter with outputs $Q_3Q_2Q_1Q_0$. Like the 74169, the preset data inputs $P_3P_2P_1P_0$ are used to initialize the output of the counter to a predetermined state when $\overline{PE}$ is low *and* the clock (*CP*) makes a positive transition. The $U/\overline{D}$ input causes the counter to count *UP* (logic 1) or *DOWN* (logic 0). For counting to

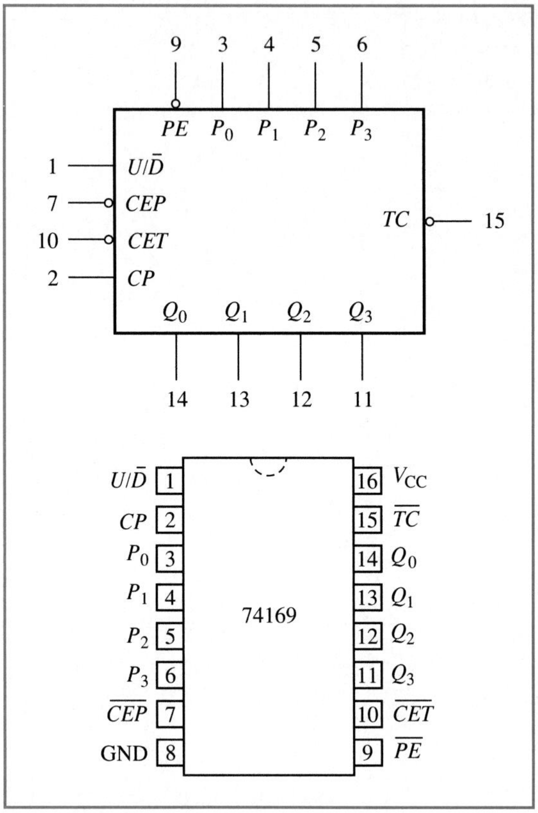

Figure 6–58
Logic symbol and pin configuration of the 74169 synchronous bidirectional modulus-16 binary counter.

Figure 6–59
Logic symbol and pin configuration of the 74168 synchronous bidirectional decade counter.

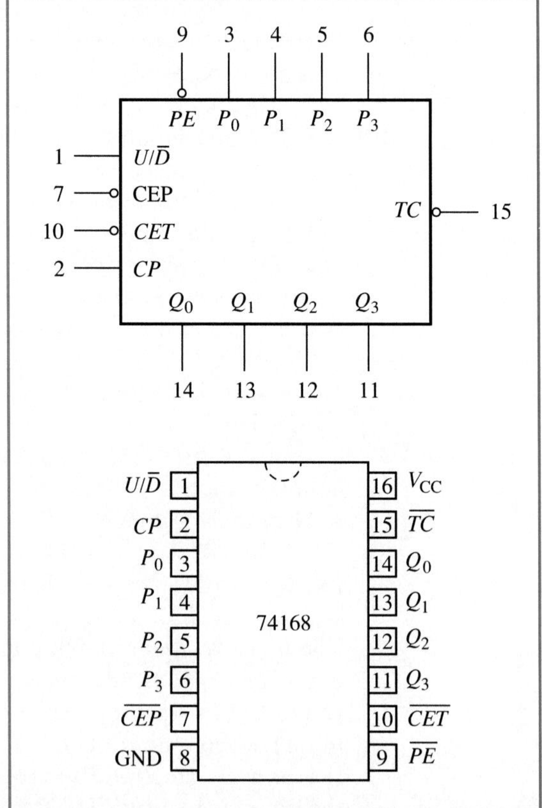

occur, both $\overline{CEP}$ and $\overline{CET}$ must be at the logic 0 states. The $\overline{TC}$ output of the counter will be a logic 0 when the counter reaches 0000 in the *DOWN* mode or when the counter reaches 1001 in the *UP* mode.

Review questions

1. Compare the functions of the 7493, 7492, and 7490 counters.
2. Compare the functions of the 74177 and 74176 counters.
3. Compare the functions of the 74169 and 74168 counters.

6–9 Troubleshooting counter circuits

Most problems and faults occur in counter circuits due to propagation delays. The effects of propagation delays are more common in asynchronous counters and at high frequencies. The effects of high-frequency operation on sequential logic circuits were seen in Chapter 5 where the propagation delays were more pronounced at higher frequencies and therefore did not comply with the recommended setup and hold times of the flip-flop. The effects of propagation delay on counters usually produce incorrect states in the counter.

For example, consider the 3-bit asynchronous binary down counter shown in Figure 6–2. The *ideal* timing diagram for the counter is shown in Figure 6–3. However, if the circuit is operated at very high clock rates, the *actual* timing diagram showing propagation delays is quite different as shown in Figure 6–60.

Note that the output X_0 (output of flip-flop $F0$) is delayed from the main clock transition by the propagation delay labeled t_{PD_0}. The output X_1 (output of flip-flop $F1$) is delayed from the transition of X_0 by the propagation delay labeled t_{PD_1}. Finally, the output X_2 (output of flip-flop $F2$) is delayed from the transition of X_1 by the propagation delay labeled t_{PD_2}. Therefore the output of the last flip-flop in series ($F2$) is delayed from the transition of the main clock by three propagation delays!

Notice in Figure 6–60 that the propagation delays cause the outputs of the flip-flops to change in the middle or close to the end of each interval thereby producing an unstable counter state for each transition of the main clock pulse. Also observe that there are many intervals (interval 1 for example) where the propagation delay causes the output X_2 to change after the next clock pulse has made its transition. This leads to an incorrect counting sequence for interval 1.

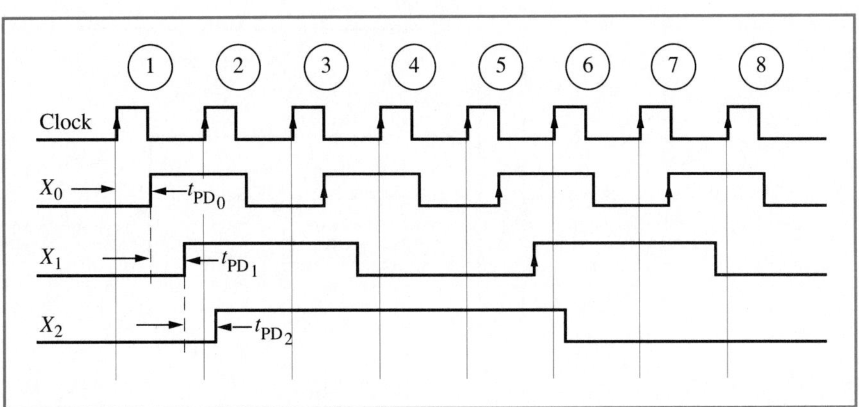

Figure 6–60
Timing diagram of a counter showing propagation delays.

Thus one of the possible causes for a fault in the counting sequence of a counter could be due to the flip-flop's propagation delay time, or the operating clock frequency of the counter. If a counter is to be operated at high clock frequencies, then flip-flops with very short propagation delays should be used. Counters that display wrong counts or missing counts should be checked for such problems by temporarily reducing the operating frequency of the circuit.

Synchronous counters are not as susceptible to the effects of propagation delay as asynchronous counters. This is because the flip-flops are triggered simultaneously, and therefore the propagation delay effects are not cumulative as in asynchronous counters. Therefore synchronous

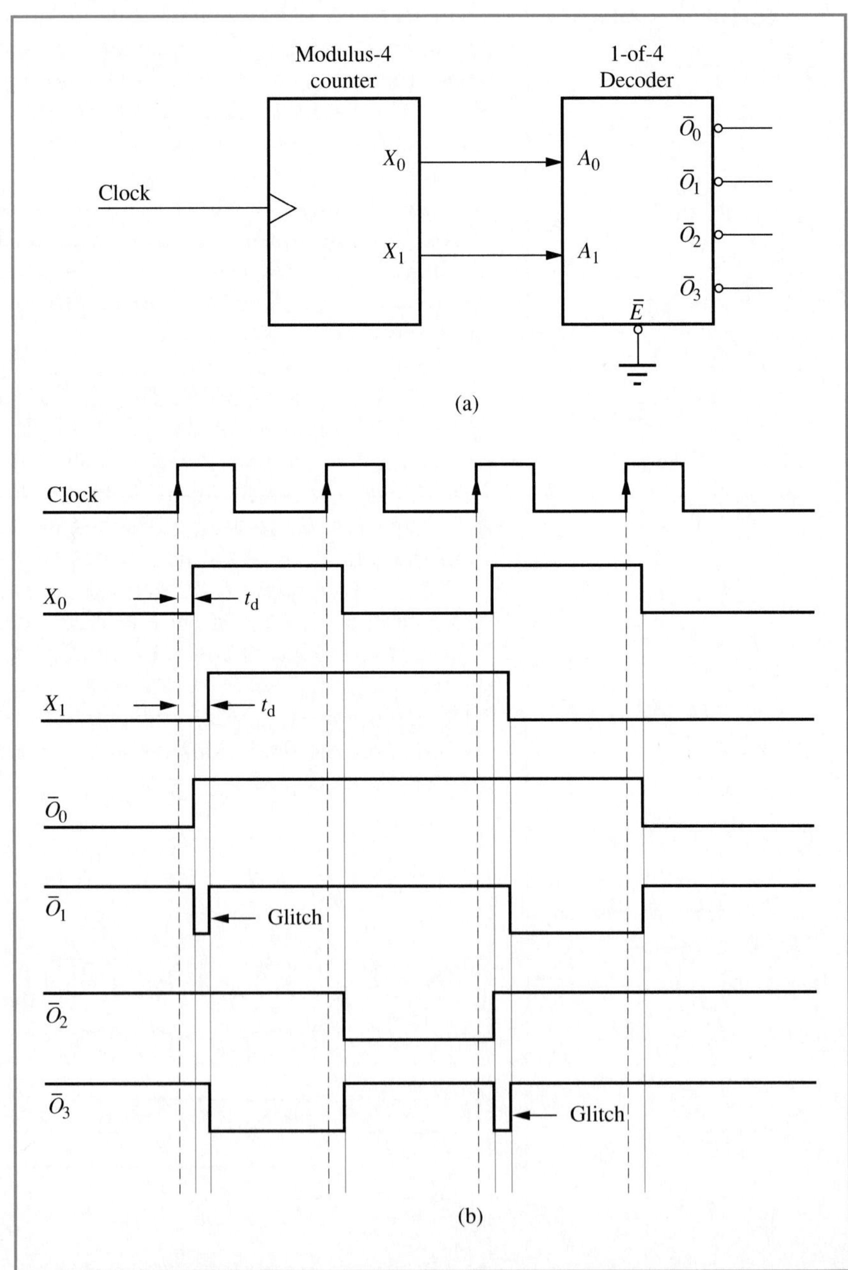

Figure 6–61 Detecting "glitches" in the output of a counter. (a) Circuit; (b) timing diagram.

counters, rather than asynchronous counters, should be used for high-frequency applications.

Another common fault that could cause problems in counter circuits is the effect of glitches in asynchronous counters that use counter decoding to recycle the count. If the counter is being cascaded with other counters, the glitch could cause false triggering of the other counters in the circuit. These glitches could also be caused by the propagation delays that exist in asynchronous counters. A technique that can be used to detect these glitches is to use a circuit similar to the one shown in Figure 6–61a. The circuit in Figure 6–61a shows an asynchronous modulus-4 counter whose outputs are connected to a 1-of-4 decoder. The decoder

Figure 6–62
Modification of Figure 6–61 to eliminate "glitches". (a) Circuit; (b) timing diagram.

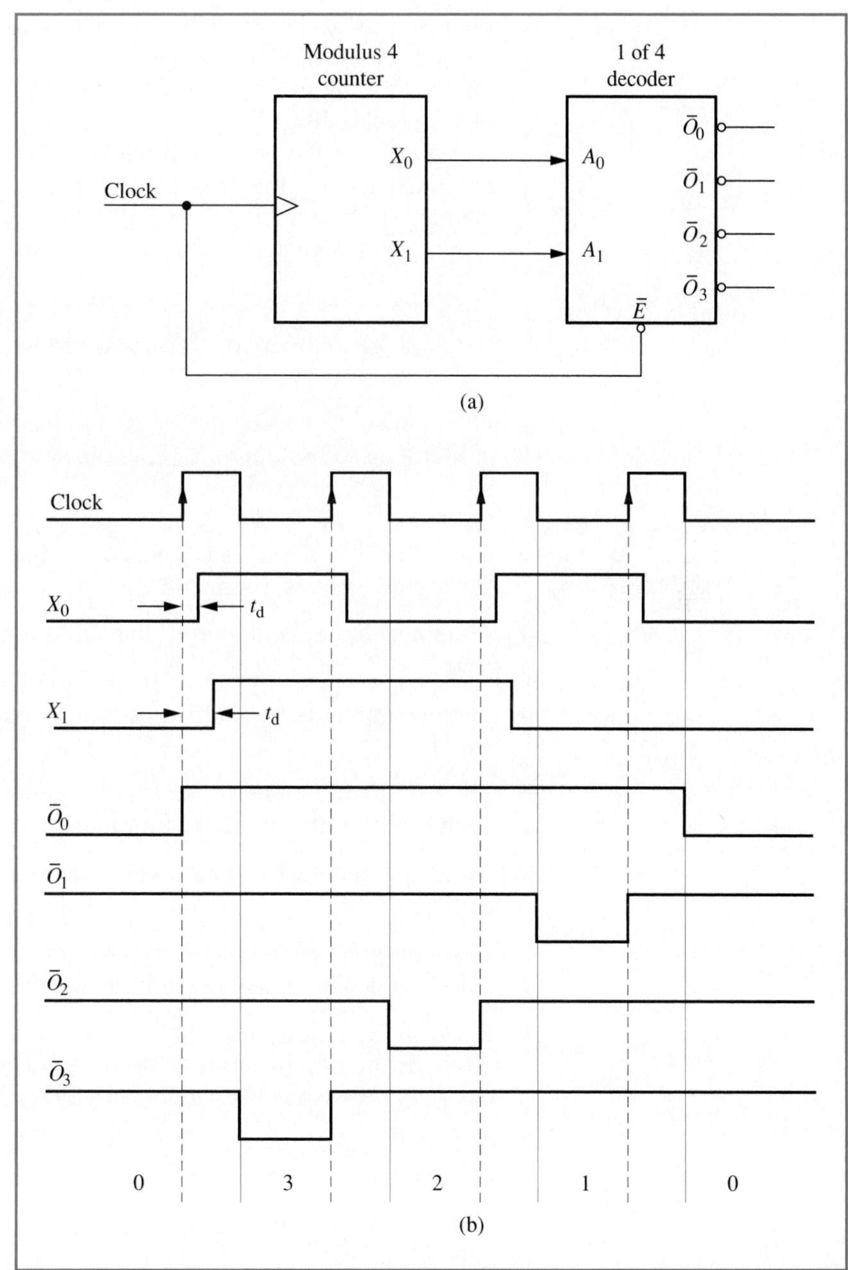

will activate one of its four outputs ($\overline{O}_0$ through $\overline{O}_3$) when it detects a particular count. Of course, a larger-size counter could also be used with a larger-size decoder. This type of circuit is used in many applications where it is necessary to determine if the counter is in one or more desired states.

The timing diagram of the circuit shown in Figure 6–61b shows how the small propagation delays generated by the output of each flip-flop in the counter can cause the decoder to falsely activate its output and produce glitches. Notice that the glitches on $\overline{O}_1$ and $\overline{O}_3$ are produced due to the cumulative effect of the propagation delays of each flip-flop. If these propagation delays do not exceed the high pulse width of the clock then we can use the low level of the clock to *strobe* the enable line of the decoder after the glitches disappear. The modifications to the circuit and the resulting timing diagram are shown in Figure 6–62. Now that the decoder is enabled only when the clock goes *LOW*, its output will not be affected by the propagation delays of X_0 and X_1 during the *HIGH* period of the clock pulse.

Again, if the problem of glitches in a circuit cannot be corrected by any other means, the only solution is to replace the asynchronous counters with synchronous counters that are less susceptible to propagation delays.

Review questions

1. What type of counter could produce incorrect counts at high frequencies?
2. What causes a counter to malfunction at high frequencies?
3. How is a glitch in the output of a counter produced?

6–10 SUMMARY

- A *counter* is a circuit that produces a numeric count each time an input clock pulse makes a transition.
- A counter has several output lines that can exist in several states called *counts*.
- The maximum number of states that a counter counts through is called its *modulus*.
- The maximum number of states of a counter depends on the relationship 2^n, where n is the number of outputs.
- All digital counters can be divided into two basic types—*asynchronous* and *synchronous*.
- Asynchronous counters are usually constructed with T flip-flops with their clock connected in a series arrangement.
- Because the clock pulses that trigger each flip-flop are generated by the previous flip-flop in series to produce a ripple effect, asynchronous counters are also called *ripple counters*.
- Once a counter *cycles* through all possible states it generally *recycles* through the same states and in the same sequence.
- A *BCD counter* counts from 0 to 9 and is known as a *modulus-10* or *decade counter*.

- A *synchronous counter* is made up of a set of flip-flops that are triggered simultaneously (in parallel) by a single *CLOCK* pulse.
- The term *synchronous* implies that all the flip-flops in the counter change states at the same time when the main clock pulse makes its transition.
- The counting sequence of a synchronous counter is not dependent on the type of flip-flop, negative or positive edge-triggered, used.
- The *state diagram* is a graphic representation of the states of the counter and the connections among its states.
- The *state table* is a translation of the state diagram and represents in tabular form the states of the counter and their connections.
- The *transition table* is a translation of the state table and relates each state of the counter to the individual outputs of the flip-flops.
- The *excitation maps* are used to determine what the control inputs of each flip-flop (*J* and *K*, or *D*) have to be for the flip-flop to make the transitions specified in the transition table.
- The *excitation equations* are obtained from the excitation maps and identify the connections for the control inputs of the counter's flip-flops.
- The *transition equations* allow us to determine what the next state of a flip-flop will be, given the present conditions.
- Since sequential binary counters can also be used for frequency division, they are also known as *divide-by-n* counters where *n* is the modulus of the counter.
- When two counters are cascaded to divide an input clock, the total frequency division obtained is the product of the divisors of each counter.
- The 74xxx series has both synchronous and asynchronous counters, many of which have additional control inputs for initializing the state of the counter and/or arrangements for reconfiguring the count.
- Asynchronous counters are more susceptible to the effects of propagation delay than synchronous counters.
- The effects of propagation delay at high frequencies could cause a counter to malfunction.

PROBLEMS

Section 6–2 Asynchronous counters

1. Design a 4-bit ripple *UP* counter using *J-K* flip-flops.
2. Design a 4-bit ripple *DOWN* counter using *D* flip-flops.
3. Analyze the circuit shown in Figure 6–63 by obtaining its timing diagram.
4. Analyze the circuit shown in Figure 6–64 by obtaining its timing diagram.
5. Design a modulus-6 ripple *UP* counter using positive edge-triggered *T* flip-flops.
6. Design an asynchronous count-down timer to count down from 7 to 0 and stop when the count reaches 0 (it should not recycle).

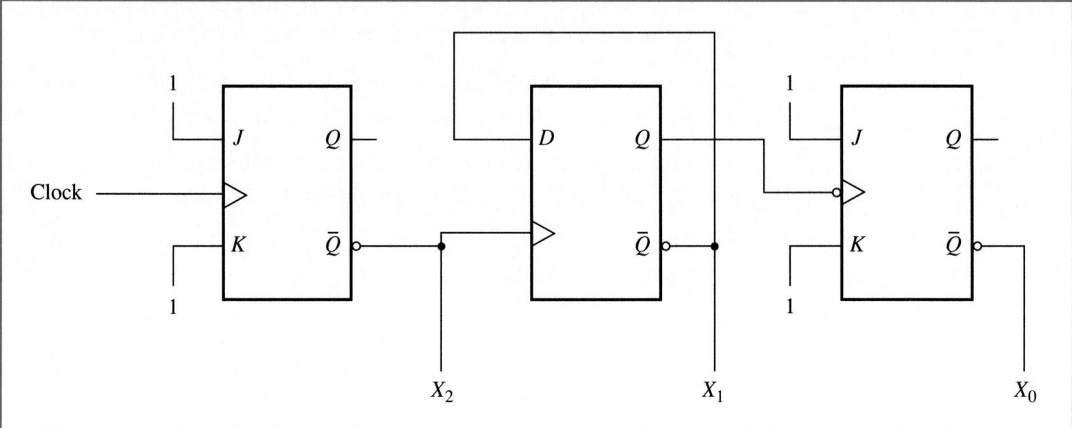

Figure 6–63

Section 6–3 Synchronous counters

7. Analyze the circuit shown in Figure 6–25 by obtaining its timing diagram.
8. Analyze the circuit shown in Figure 6–27 by obtaining its timing diagram.
9. Analyze the circuit shown in Figure 6–30 by obtaining its timing diagram.
10. Analyze the circuit shown in Figure 6–33 by obtaining its timing diagram. Start the timing diagram at state 1.
11. Analyze the circuit shown in Figure 6–33 by obtaining its timing diagram. Start the timing diagram at state 0.
12. Analyze the circuit shown in Figure 6–33 by obtaining its timing diagram. Start the timing diagram at state 4.
13. Analyze the circuit shown in Figure 6–33 by obtaining its timing diagram. Start the timing diagram at state 7.
14. Compare the results obtained from Problems 10 through 13 with the state diagram of Figure 6–43.

Section 6–4 Synthesis of synchronous counters

15. Design a 3-bit synchronous binary *UP* counter using *J-K* flip-flops.
16. Design a 3-bit synchronous binary *UP* counter using *D* flip-flops.
17. Design a 3-bit synchronous binary *DOWN* counter using *J-K* flip-flops.
18. Design a 3-bit synchronous binary *DOWN* counter using *D* flip-flops.
19. Design a synchronous counter using *D* flip-flops to count in the sequence

    9, F, 3, 0, 7, 9, F, . . .

20. Design a synchronous counter using *J-K* flip-flops to count in the sequence

    3, 7, 1, 0, 6, 5, 3, . . .

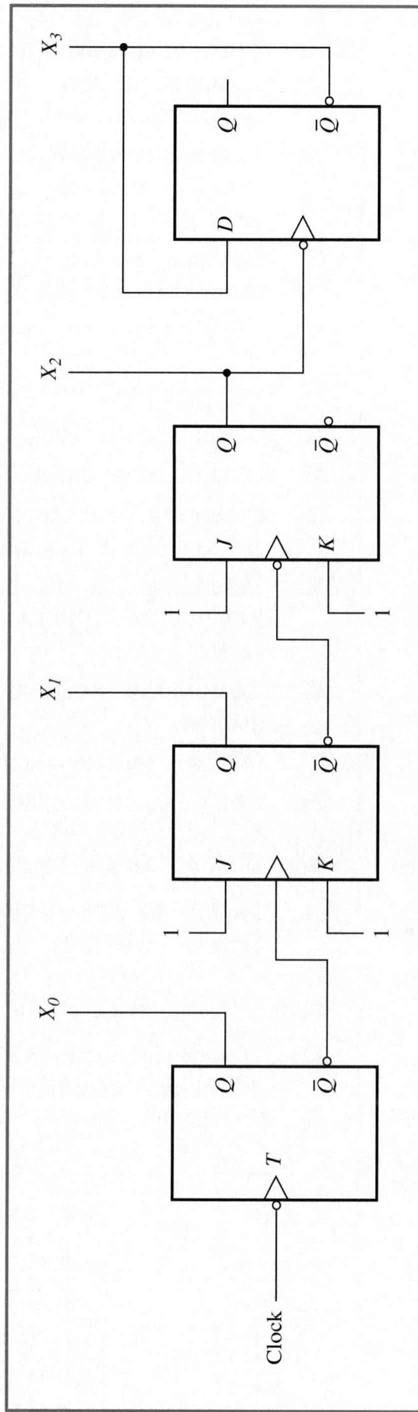

Figure 6–64

21. Design a 2-bit synchronous *UP/DOWN* counter that will count up or down on command. If the control input M is a logic 1 the counter should count down, and if the control input M is a logic 0 the counter should count up. Use *J-K* flip-flops in the design.
22. Design a synchronous counter that will count in the sequence: 2, 1, 3, 0, if control input $A = 1$, and in the sequence 0, 1, 3, 0, if control input $A = 0$. Use D flip-flops in the design.
23. Design a synchronous count-down counter that will count down from 7 to 0 and stop when the count reaches 0. Use D flip-flops in the design.

Section 6–5 Analysis of synchronous counters

24. Obtain a state diagram for the circuit shown in Figure 6–65.
25. Obtain a state diagram for the circuit shown in Figure 6–66.
26. Determine what the next state(s) of the counter designed in Problem 20 would be if the counter were to start off in states 2 and 4.
27. Determine what the next state(s) of the counter designed in Problem 19 would be if the counter were to start off in states 1, 2, 4, 5, 6, 8, *A*, *B*, *C*, *D*, *E*.
28. Analyze the circuit obtained for Problem 22 by obtaining its state diagram.
29. Obtain a state diagram for the circuit shown in Figure 7–26.
30. Obtain a state diagram for the circuit shown in Figure 7–28.

Section 6–6 An application of counters

31. Modify the 24-hour time-of-day clock in Figure 6–46 to function as a 12-hour time-of-day clock.

Section 6–7 Frequency division

32. Design a divide-by-80 counter circuit using a BCD counter and a 4-bit binary counter.

Figure 6–65

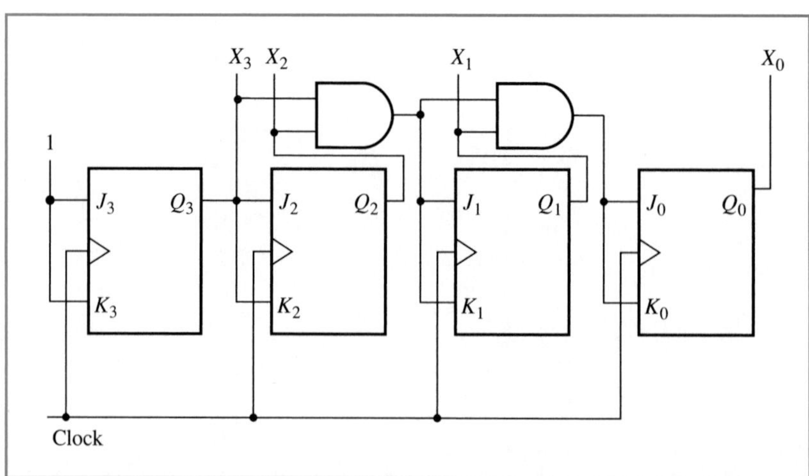

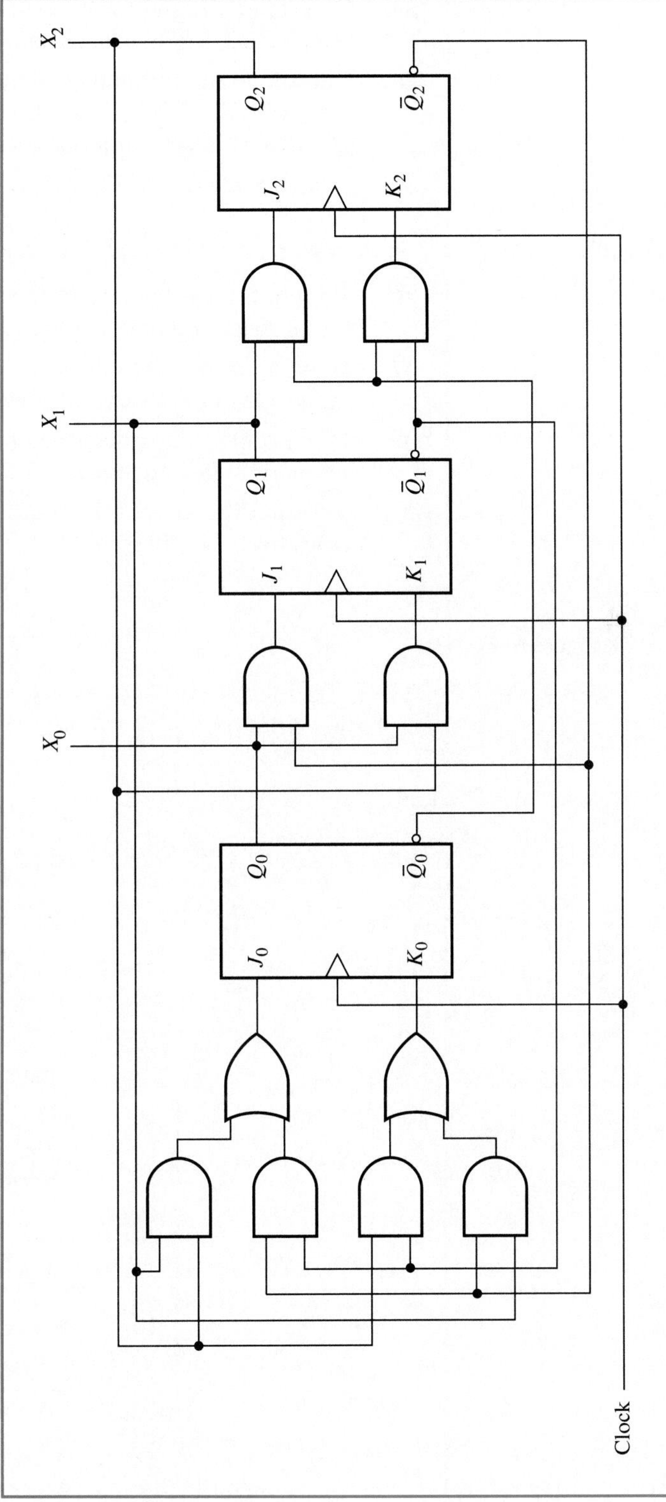

Figure 6–66

33. Determine the frequency of the *OUTPUT CLOCK* in Figure 6–67 if the *INPUT CLOCK* has a frequency of 4 kHz.

34. Determine the frequency of the *OUTPUT CLOCK* in Figure 6–67 if the *INPUT CLOCK* has a frequency of 200 Hz and the *OUTPUT CLOCK* is taken from output X_3 of the modulus-16 counter.

35. Design a divide-by-32 counter using two modulus-16 counters.

Troubleshooting

Section 6–9 Troubleshooting counter circuits

36. Draw a timing diagram of the counter shown in Figure 6–5 showing the propagation delays produced by each flip-flop.

37. Draw a timing diagram of the counter shown in Figure 6–14 showing the propagation delays produced by each flip-flop.

38. Discuss the timing diagrams produced in Problems 36 and 37 in terms of high-frequency operation of the two counters.

39. The circuit shown in Figure 6–68 produces spurious outputs when in operation. Troubleshoot the circuit by constructing a timing diagram

Figure 6–67

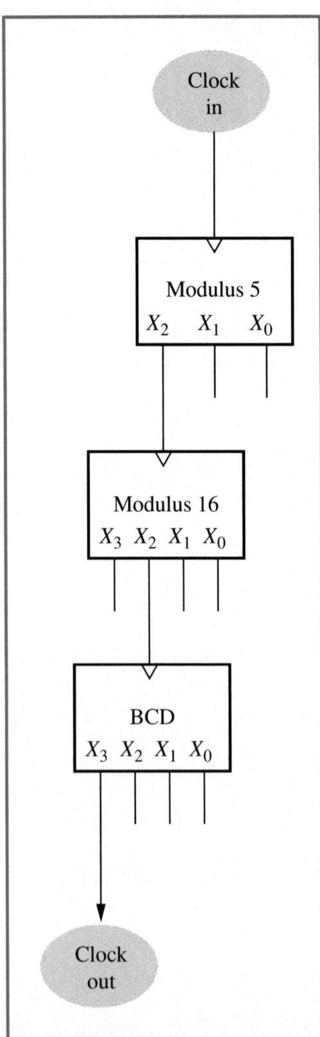

Figure 6–68

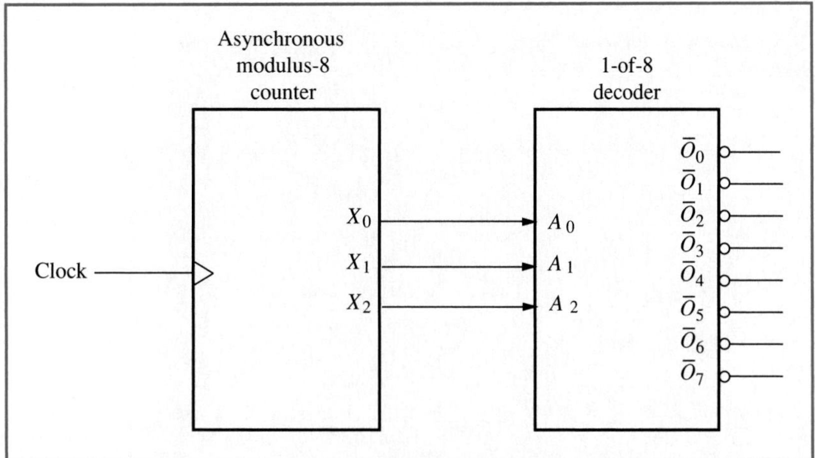

and determining where the glitches occur in the outputs. Suggest possible remedies.

40. Refer to the counter circuit shown in Figure 6–21. How would the circuit behave if one of the following faults existed in the circuit:
a. The output of the *AND* gate was shorted to GND.
b. The output of the *AND* gate was shorted to V_{CC}.

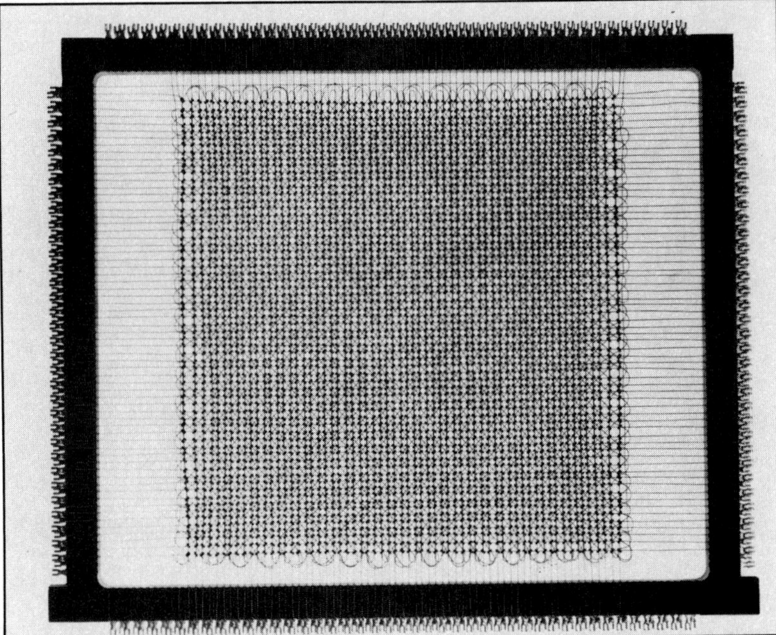

A computer must have the capability to store information in the form of binary numbers. The photograph shows a vintage magnetic memory plane that was used for this purpose. Thousands of these circuits can now be integrated on a single chip (Courtesy IBM).

OUTLINE

OBJECTIVES

The objectives of this chapter are to

- introduce the concepts of parallel data storage.
- study the operation of parallel registers and the manner in which data is loaded and retrieved.
- introduce the concepts of serial data storage.
- study the design and operation of shift registers and their timing characteristics.
- examine some simple applications of shift registers.
- introduce the concepts of register arrays (register files) and their applications.
- introduce the elementary concepts of memory circuits such as addressing, reading, and writing.
- apply addressing concepts to the expansion of register arrays.
- investigate the characteristics and operation of the various ICs that implement serial and parallel registers, and register files.

7 REGISTERS

7–1 INTRODUCTION

The *storage* of binary numbers is an important function of a digital computer and is also essential to several other types of digital circuit applications. *Registers* are logic circuits that are capable of storing single binary numbers of varying size while *memory systems* (to be examined in Chapters 8 and 9) can be viewed as arrays of registers that are capable of holding several binary numbers of varying size. Thus, flip-flops can store single bits, registers can store single binary numbers (several bits), and memories can store several binary numbers.

This chapter deals with the operation and design of various types of registers. Two broad categories of registers are covered here—parallel and serial registers. The two types of registers differ only in the manner in which the binary data (binary numbers) is loaded-in and retrieved from the register. Both types of registers are used to store binary data, but serial registers have other applications besides storage as will be seen in this chapter.

The concepts of memory architecture and design are also introduced in conjunction with the study of register arrays (register files). A more detailed study of memories is found in Chapters 8 and 9. The chapter concludes by examining the various hardware implementations of registers in the form of ICs.

7–2 Parallel registers

Parallel registers are used to store binary numbers of different sizes. The size of a parallel register is always equal to the number of bits in the largest binary number that it is capable of storing. For example, a 4-bit register is capable of storing all 4-bit numbers in the range 0000 through 1111. The word *parallel* is used to describe these registers because the bits that make up the number are loaded into the register in *parallel*. That is, they are loaded together, simultaneously. This is in contrast to the manner in which serial registers load data—1 bit at a time, or *serially*.

Figure 7–1 shows the logic symbol for a 4-bit parallel register. The register has four data inputs, $I_3I_2I_1I_0$ that allow a binary number to be stored into the register when the active-low *write line* ($\overline{WR}$) is activated (by a logic 0). The data stored in the register is available at all times at the output lines $O_3O_2O_1O_0$.

The logic circuit for the 4-bit parallel register is shown in Figure 7–2. The circuit uses a series of D flip-flops to store each bit that makes up the binary number. Since all the bits of the number are stored into the

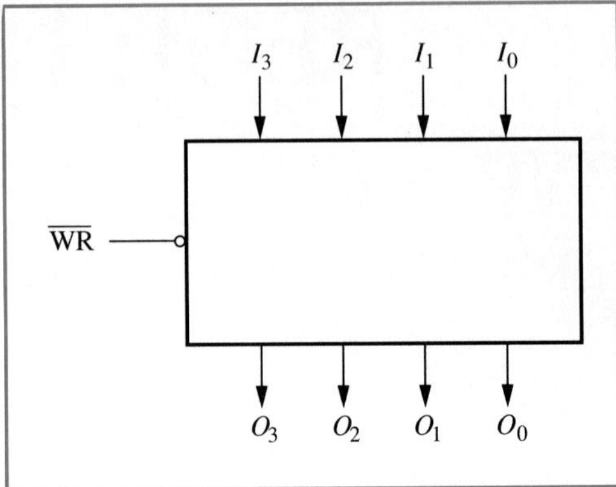

Figure 7–1
Logic symbol for a 4-bit parallel register.

flip-flops at one time, the clock inputs to all the flip-flops are connected together to make up the $\overline{WR}$ line. Since the flip-flops are negative edge-triggered, the flip-flops will trigger when the $\overline{WR}$ line changes to the logic 0 state. The input lines $I_3I_2I_1I_0$ are connected to the *D* inputs of the flip-flops to allow each flip-flop to latch a single bit of the input binary number. The *Q* outputs of the flip-flops hold the stored bits and therefore are connected to the register outputs $O_3O_2O_1O_0$.

A parallel register such as the one shown in Figure 7–2 therefore simply extends the concept of storage of a single bit using a *D* flip-flop to many bits (a binary number) using a series of *D* flip-flops. The size of the register can be expanded simply by adding more flip-flops to accommodate larger numbers. For example, Figure 7–3 shows the logic symbol and circuit for an 8-bit parallel register.

There are many applications that have the data outputs of several registers connected together on a common bus. In a scheme such as this, the outputs of all the registers must be isolated from the bus and should only put their data on the bus when selected. Figure 7–4a shows the logic symbol of a 4-bit parallel register with an active-low *read line* $\overline{RD}$. When the $\overline{RD}$ line is inactive (logic 1 state), the outputs of the register $O_3O_2O_1O_0$ are in a high-impedance state (*Z*). However, when the $\overline{RD}$ line

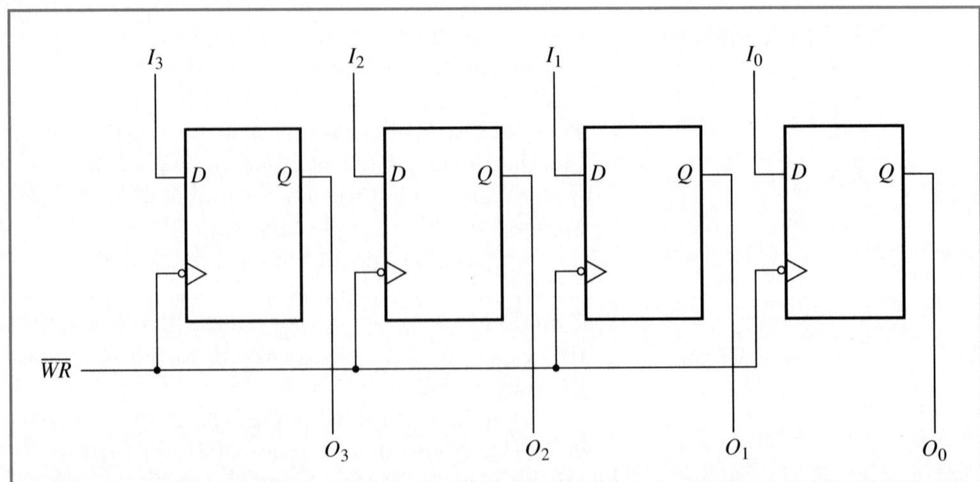

Figure 7–2
Logic circuit for a 4-bit parallel register.

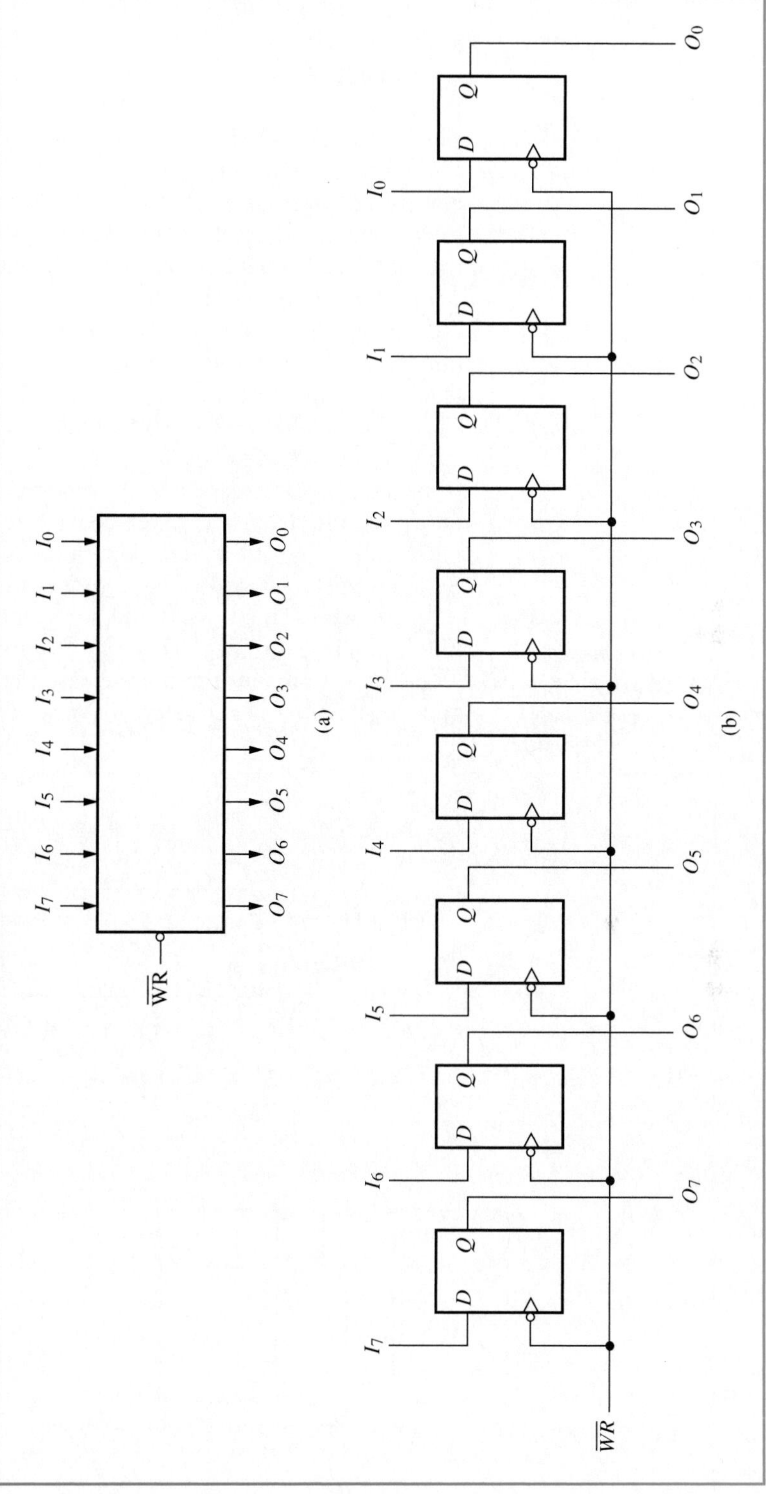

Figure 7–3 An 8-bit parallel register. (a) Logic symbol; (b) circuit.

is active (logic 0 state), the outputs of the register contain the data stored in the flip-flops.

Figure 7–4b shows the logic circuit for the 4-bit parallel register with tristate outputs. Notice that the circuit is similar to the circuit shown in Figure 7–2 except that tristate buffers have been added to the outputs of the flip-flops. The active-low enable lines of all the tristate buffers are connected together so that the $\overline{RD}$ line enables all the buffers when at the logic 0 state and disables all the buffers when at the logic 1 state. When the buffers are enabled ($\overline{RD}=0$), the outputs of the flip-flops, Q, are connected to the outputs of the register $O_3O_2O_1O_0$. However, when the buffers are disabled ($\overline{RD}=1$), the outputs of the register are in a high-impedance state and the outputs of the flip-flops are effectively cut off from the outputs of the register.

The parallel registers examined so far have had a separate set of input and output lines to load data into the register and to retrieve data from the register, respectively. Since at any given instant of time we are either reading from a register or writing to a register, but never both, it is more convenient to have a common set of lines to load and retrieve data. Such a register is said to have *bidirectional* data lines.

Figure 7–5a shows the logic symbol for a 4-bit parallel register with bidirectional data lines. To write (store) a 4-bit binary number, the logic levels that represent the number are placed on the data lines $D_3D_2D_1D_0$ and the $\overline{WR}$ line is activated. In order to read (retrieve) a 4-bit binary

Figure 7–4
A 4-bit parallel register with tristate outputs. (a) Logic symbol; (b) circuit.

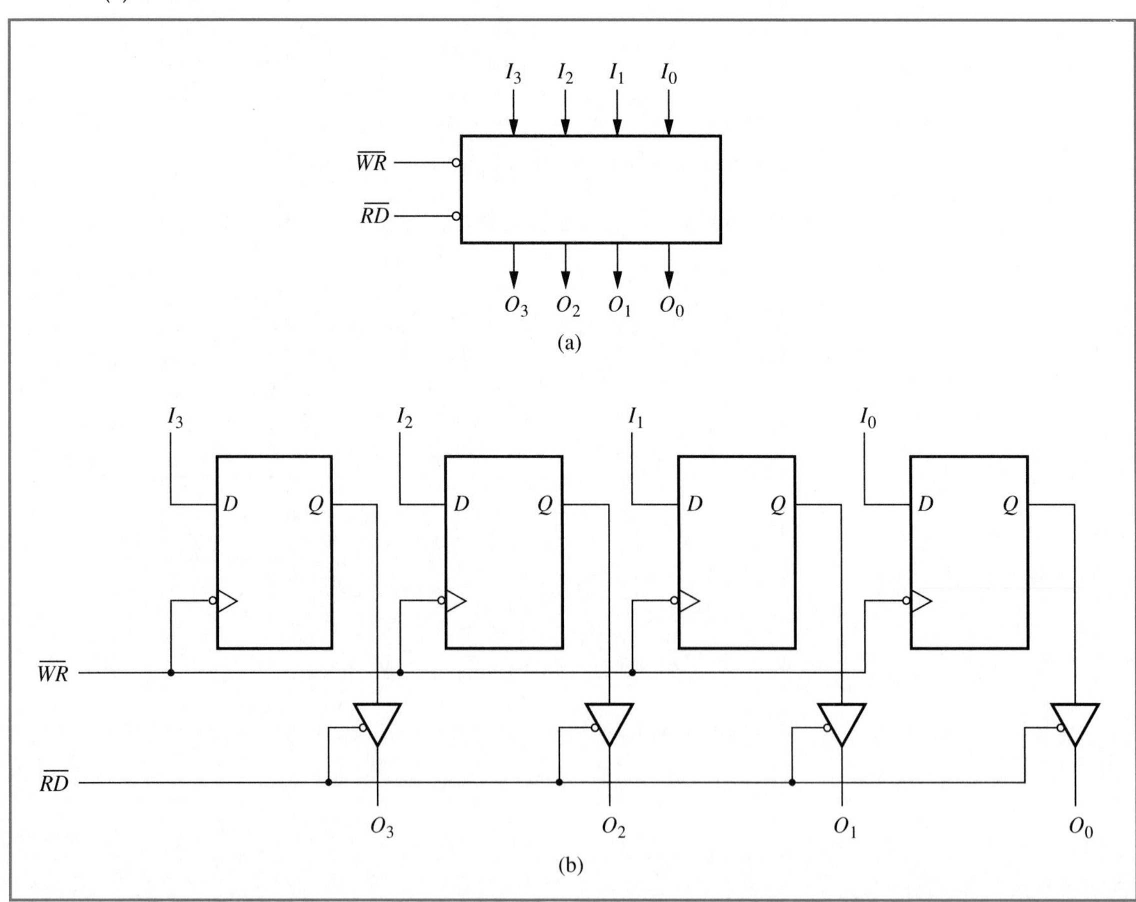

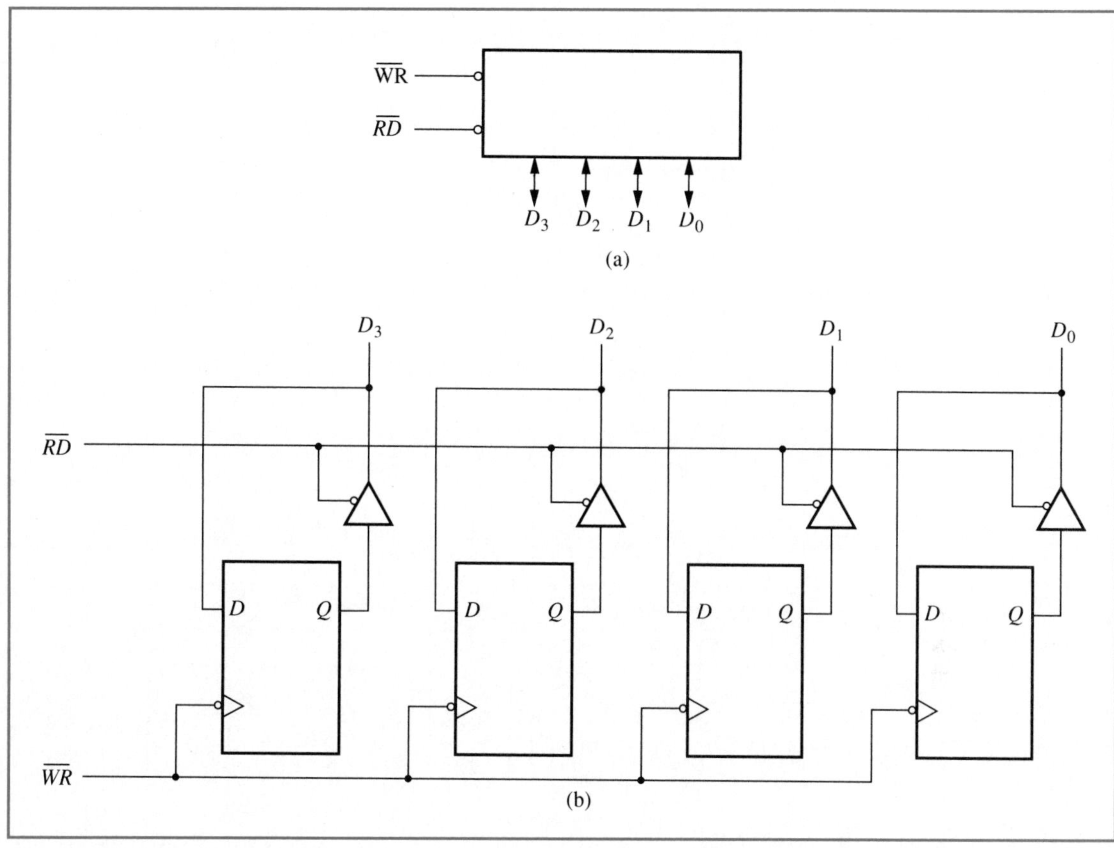

Figure 7–5
A 4-bit parallel register with bidirectional data lines. (a) Logic symbol; (b) circuit.

number from the register, the $\overline{RD}$ line is activated and the register puts out the data on the same set of lines $D_3D_2D_1D_0$.

Figure 7–5b shows the logic circuit for the 4-bit parallel register with bidirectional data lines. Notice the circuit is similar to Figure 7–4b except that we have connected the input and output lines together to produce a set of bidirectional data lines. At any given instant either the $\overline{RD}$ or the $\overline{WR}$ line may be activated (but never both). When data is to be stored into the register, the $\overline{RD}$ line is inactive, and therefore the tristate buffers isolate the outputs of the flip-flops from the data lines $D_3D_2D_1D_0$. The data on the D lines can then be latched into the flip-flops by activating the $\overline{WR}$ line. When data is to be retrieved, the $\overline{RD}$ line is activated which allows the outputs of the flip-flops to be connected to the data lines $D_3D_2D_1D_0$. The data retrieved from the outputs of the flip-flops is also fed back to the inputs. However, since the $\overline{WR}$ line is inactive, they do not have any effect on the circuit.

EXAMPLE 7–1

Figure 7–6 shows how the binary number 1011 is stored and retrieved using the parallel register designed in Figure 7–5.

In Figure 7–6a, since the $\overline{RD}$ line is inactive, the outputs of the tristate gates are open, thus isolating the outputs from the data lines. The number 1011 is applied to the D inputs and the $\overline{WR}$ line is pulsed to a logic 0. This latches the number 1011 into the flip-flops for storage.

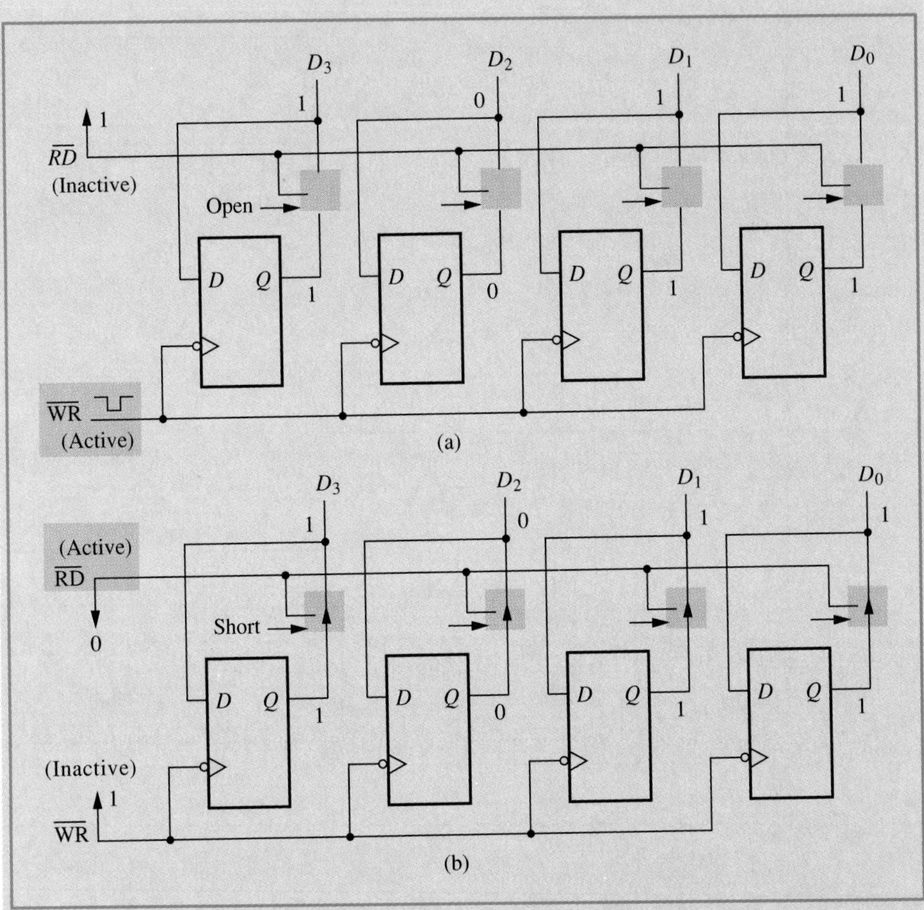

Figure 7–6
(a) Writing data into the register; (b) reading data from the register.

In Figure 7–6b, we have now activated the *RD* line while keeping the $\overline{WR}$ line inactive. This enables the tristate gates, thereby connecting the outputs of the flip-flops to the data lines. The data stored in the flip-flops 1011 is therefore retrieved and made available on the data lines *D*.

Parallel registers, or *parallel-load registers,* are therefore logic circuits that are used to hold (or store) single binary numbers. All the bits that make up the binary number are loaded and retrieved simultaneously. The process of storing a binary number is known as *writing* data while the process of retrieving a previously stored binary number is called *reading*. Not all registers use a parallel load/retrieve design. The next section examines a type of register in which data may be loaded into or retrieved from a register 1 bit at a time.

Review questions

1. What is the difference between a parallel register and a serial register?

2. What is the purpose of the $\overline{WR}$ line of a register?
3. What is the purpose of the $\overline{RD}$ line of a register?
4. Define the terms *reading* and *writing*.
5. What are bidirectional data lines?

7–3 Serial registers

The parallel registers examined in the previous section are sometimes referred to as *parallel-in parallel-out* registers since data is loaded into and retrieved from those registers in a parallel format. Serial registers load data serially, retrieve data serially, or both. When data is loaded serially into a register, each bit of the binary number is loaded individually—that is, sequentially, one bit at a time. Similarly, when data is retrieved serially from a register, each bit of the binary number stored in the register is retrieved individually.

Serial-in parallel-out registers

Figure 7–7a shows the logic symbol for a 4-bit *serial-in parallel-out* register. The register has a single input line I over which the data is loaded, and a 4-bit output $O_3O_2O_1O_0$ from which the stored data can be retrieved in parallel. The clock input is used to load into the register each bit of the binary number to be stored. A 4-bit binary number is loaded into the register by applying each bit that makes-up the number to the input line I. The bit is then loaded into the register by applying a clock pulse. It therefore takes 4 clock pulses to load a 4-bit binary number into the register. The logic circuit for the register is shown in Figure 7–7b. To understand the manner in which the register functions, consider the loading of the 4-bit binary number 1010. Since there are two 0-bits and

Figure 7–7 Serial-in parallel-out shift register. (a) Logic symbol; (b) circuit.

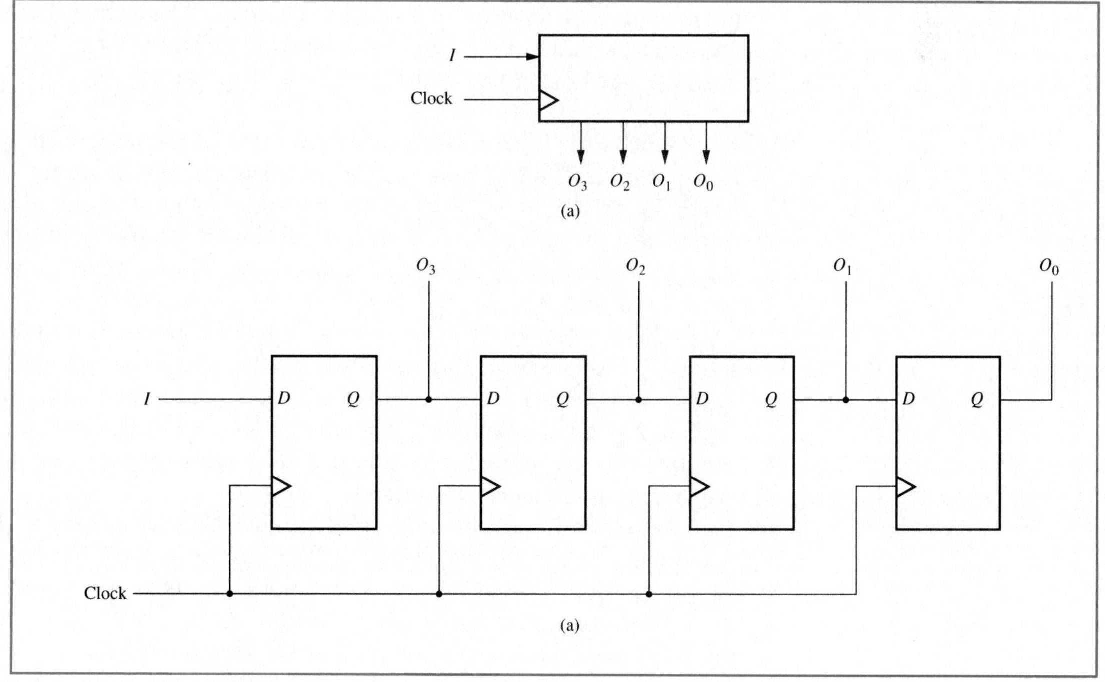

two 1-bits in this number, let us identify each bit by a subscript so that we can view the movement of a bit through the circuit

$$1_B 0_B 1_A 0_A$$

The LSB of the number is therefore 0_A and the MSB of the number is 1_B.

Figure 7–8a through 7–8e show a trace of the binary number for each pulse applied to the clock input. Each step in the trace can be explained as follows.

Clock ①: In Figure 7–8a before the first clock triggers the register, the LSB of the binary number 0_A, is applied to the I input. The outputs of the flip-flops, which are also the outputs of the register $O_3O_2O_1O_0$, are currently set to some logic levels that we are not concerned about and have identified by the don't care states X_A, X_B, X_C, and X_D. When the clock makes its transition, the output of each flip-flop will be equal to the logic level of its input before the clock pulse makes its transition. The result of the first clock transition is shown in Figure 7–8b. Notice that all the bits have *shifted* right one position so that the output O_3 is now equal to bit 0_A and bit X_D has been discarded.

Clock ②: Figure 7–8b shows the state of the register before the second clock pulse makes a transition. We have now applied the second significant bit, 1_A, of the binary number to the I line. When the clock pulse makes a transition, we again have a shift of all the bits to the right, and the new state of the register is shown in Figure 7–8c. Notice that X_C has now been discarded and $O_2 = 0_A$, $O_3 = 1_A$.

Clock ③: Figure 7–8c shows the state of the register before the third clock pulse makes a transition. We have now applied the third significant bit, 0_B, of the binary number to the I line. When the clock pulse makes a transition, we again have a shift of all the bits to the right, and the new state of the register is shown in Figure 7–8d. Notice that X_B has now been discarded and $O_1 = 0_A$, $O_2 = 1_A$, and $O_3 = 0_B$.

Clock ④: Figure 7–8d shows the state of the register before the fourth (and last) clock pulse makes a transition. We have now applied the MSB, 1_B, of the binary number to the I line. When the clock pulse makes a transition, we again have a shift of all the bits to the right and the new state of the register is shown in Figure 7–8e. Notice that X_A has now been discarded and $O_0 = 0_A$, $O_1 = 1_A$, $O_2 = 0_B$, and $O_3 = 1_B$.

Therefore at the end of four clock pulses, the entire 4-bit binary number 1010 has been *shifted in* the register since $O_3O_2O_1O_0 = 1010$. Notice that the previous number $X_AX_BX_CX_D$ that was stored in the register has been completely discarded. Because of the manner in which the data bits are shifted into (and out of) a serial register, serial registers are often known as *shift registers*.

Figure 7–9 shows the complete timing diagram for the shift register when the number 1010 is shifted in. We have assumed that the don't care states X_A, X_B, X_C, and X_D are zeros. Notice how each bit applied to the I input shifts right for each transition of the clock pulse and at the end of the fourth clock pulse, the outputs of the register $O_3O_2O_1O_0$ contain the complete number 1010. The following example illustrates the timing

Figure 7–8
Operation of the shift register.

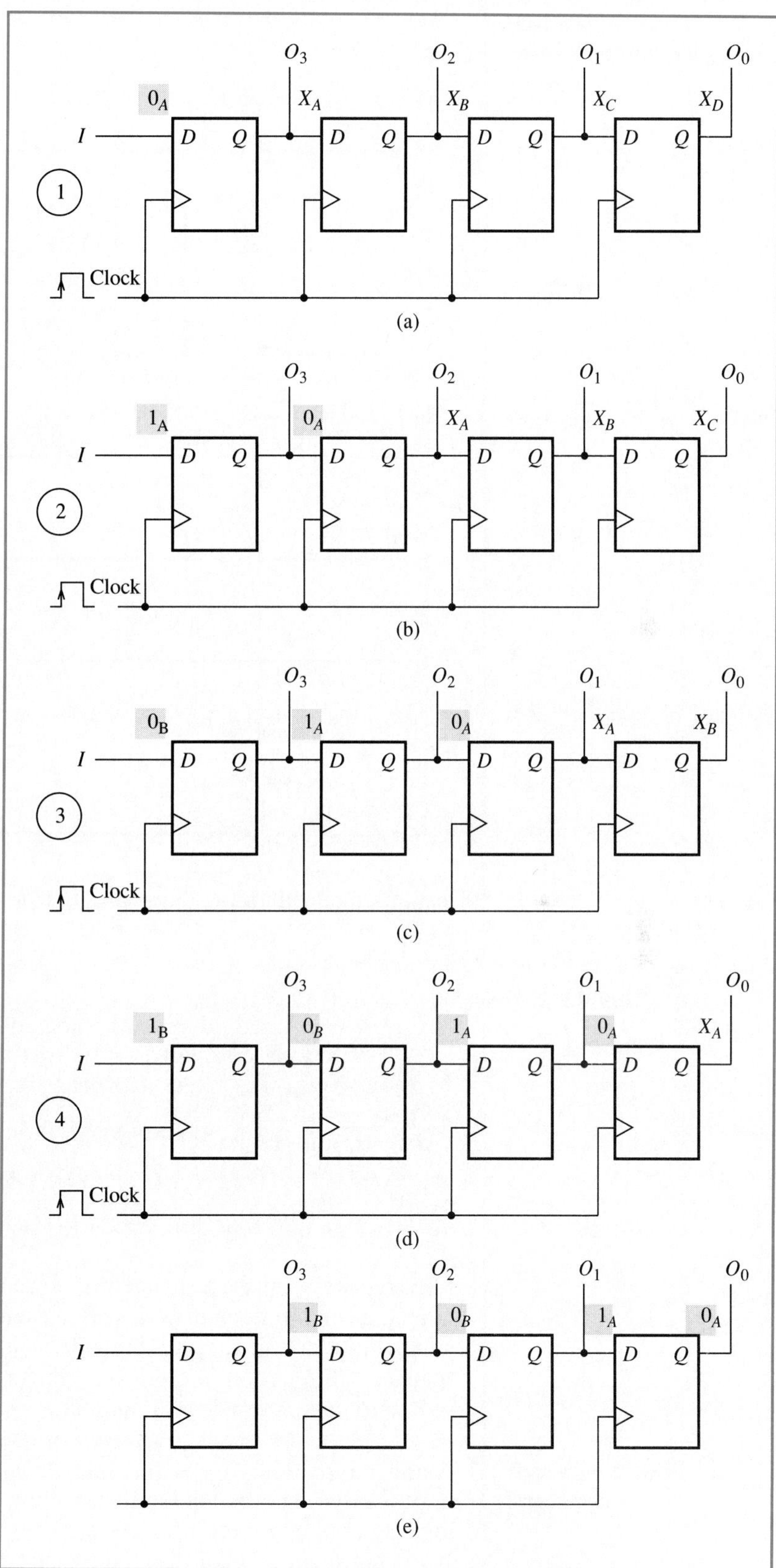

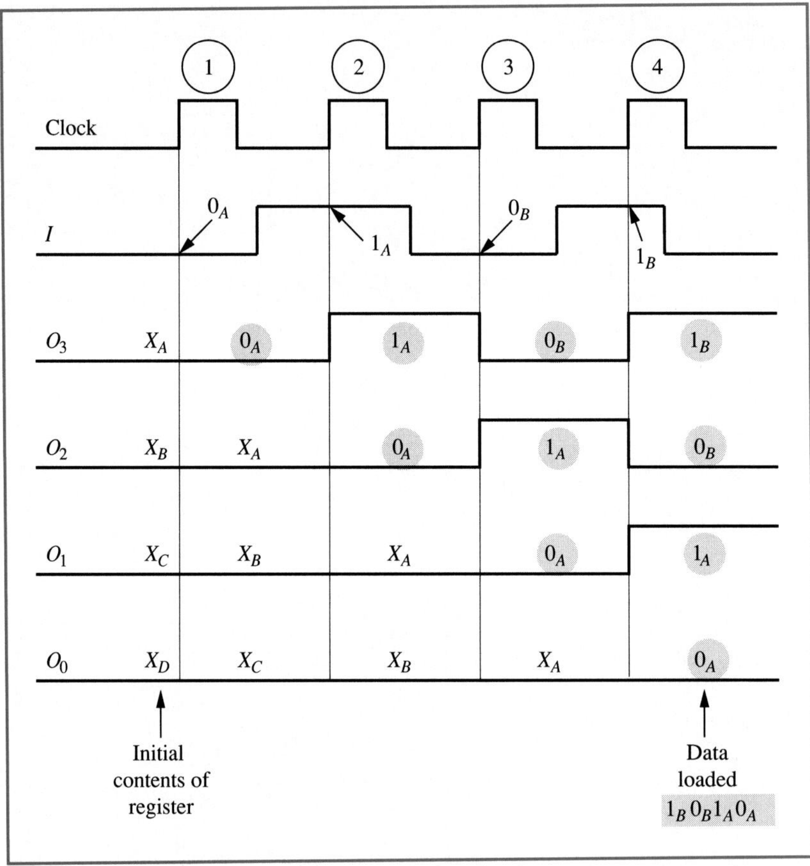

Figure 7–9 Timing diagram for the shift register.

diagram analysis of the shift register when loading another 4-bit binary number.

EXAMPLE 7–2

Construct a timing diagram for the shift register shown in Figure 7–7b when the 4-bit binary number 0110 is shifted in. Assume that the initial state of the register is 1111.

SOLUTION Let us identify each bit in the number 0110 as follows:

$$0_A 1_A 1_B 0_B$$

The timing diagram for the shift register is shown in Figure 7–10.

Shift registers can be expanded to store any size numbers simply by adding more flip-flops in series. For example, Figure 7–11 shows the logic symbol and circuit for an 8-bit shift register. Note that the 8-bit shift register will function like the 4-bit shift register but will require 8 clock pulses to completely load a number.

Serial-in serial-out registers

Another type of shift register, known as a *serial-in serial-out* register requires that data be loaded in serially as well as retrieved serially. The logic symbol and circuit for the shift register is shown in Figure 7–12. Notice in Figure 7–12a that the register has a single input *I* for loading the data serially, and a single output *O* for serially retrieving previously

Figure 7–10

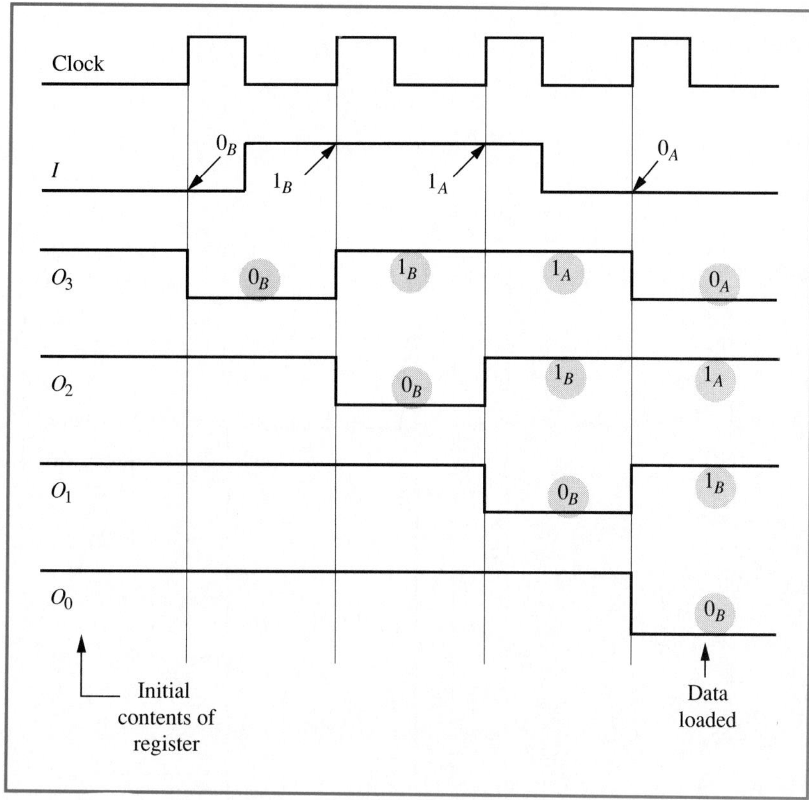

stored data. The clock input is used for both loading and retrieving data 1 bit at a time. The logic circuit for the shift register is a simple modification of the serial-in parallel-out shift register shown in Figure 7–7b. Since the data will not be retrieved in parallel, the individual outputs of the flip-flops are not accessed and therefore are not made available. Instead, however, the serial output of the shift register is the output of the last flip-flop in series. The loading of serial data is the same as for a serial-in parallel-out register. Recall that as data bits are shifted in on each clock transition, the existing bits in the register are discarded (shifted out) at the output of the last flip-flop. Thus to retrieve existing data in the register, the clock is triggered four times and the data bits appear at output O, LSB first (if the data was loaded-in LSB first).

EXAMPLE 7–3

Construct a timing diagram for the shift register shown in Figure 7–12 showing the storage and retrieval of the number 1011.

Figure 7–13a shows the storage of the number 1011 into the serial register of Figure 7–12, MSB first. Each bit is placed on the I line starting with the MSB. The bit is shifted into the register on the positive edge of the clock. After 4 clock pulses the entire number 1011 has been shifted into the register. The rightmost flip-flop in Figure 7–12 holds the MSB since it was shifted-in first.

Figure 7–13b shows the retrieval of the data 1011 that was previously stored in the register. On the first positive transition of the clock pulse, the MSB appears on the O line, since the MSB was

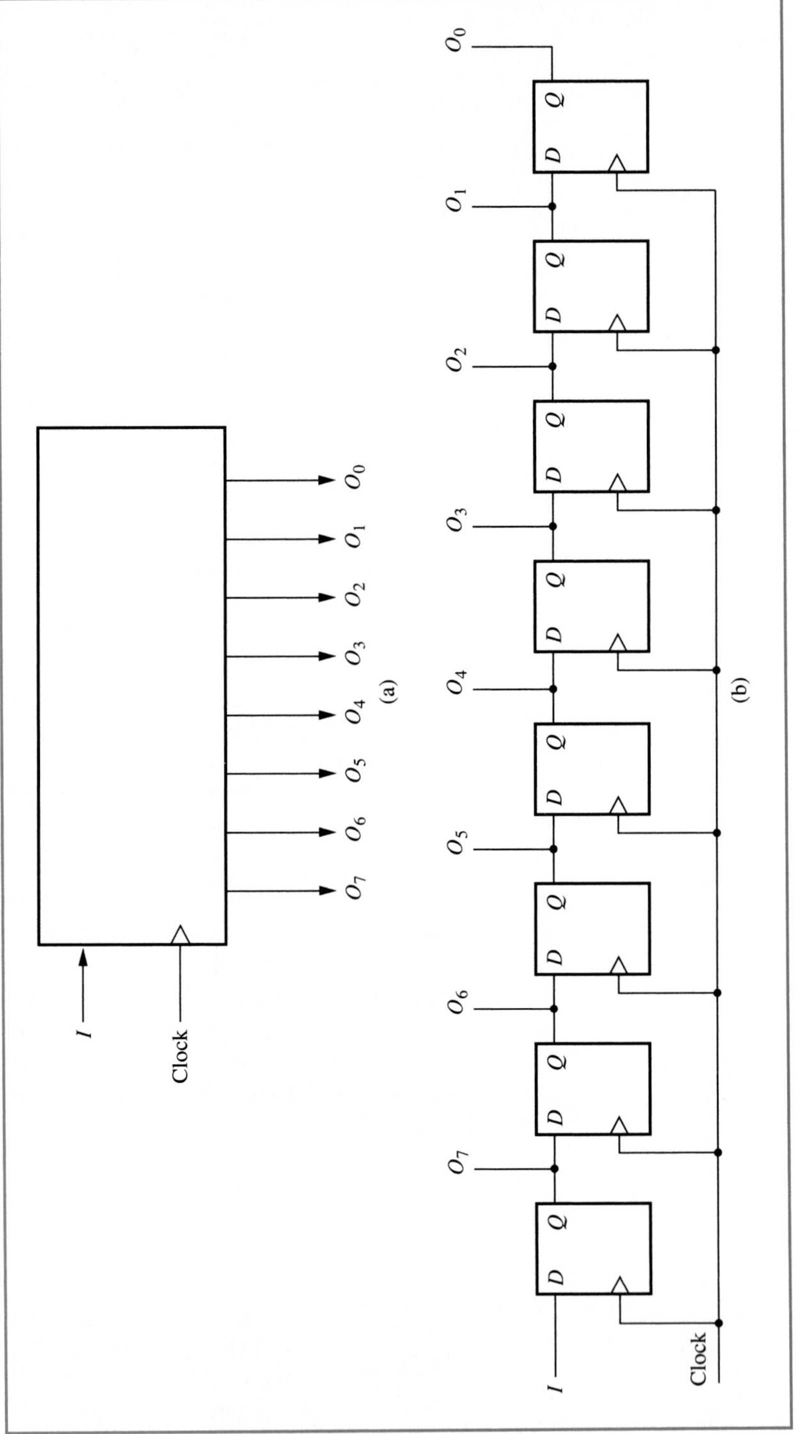

Figure 7–11 8-bit shift register.

Figure 7–12
Serial-in serial-out shift register.

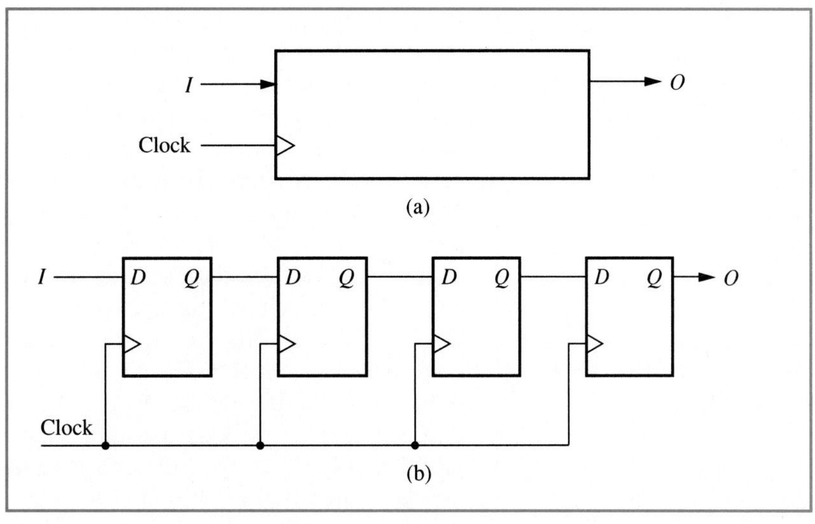

stored in the rightmost flip-flop. The other bits follow on each successive transition of the clock pulse, ending with the LSB. After 4 clock pulses the entire number 1011 has been retrieved serially.

Figure 7–13
Timing diagram for the serial-in serial-out shift register.

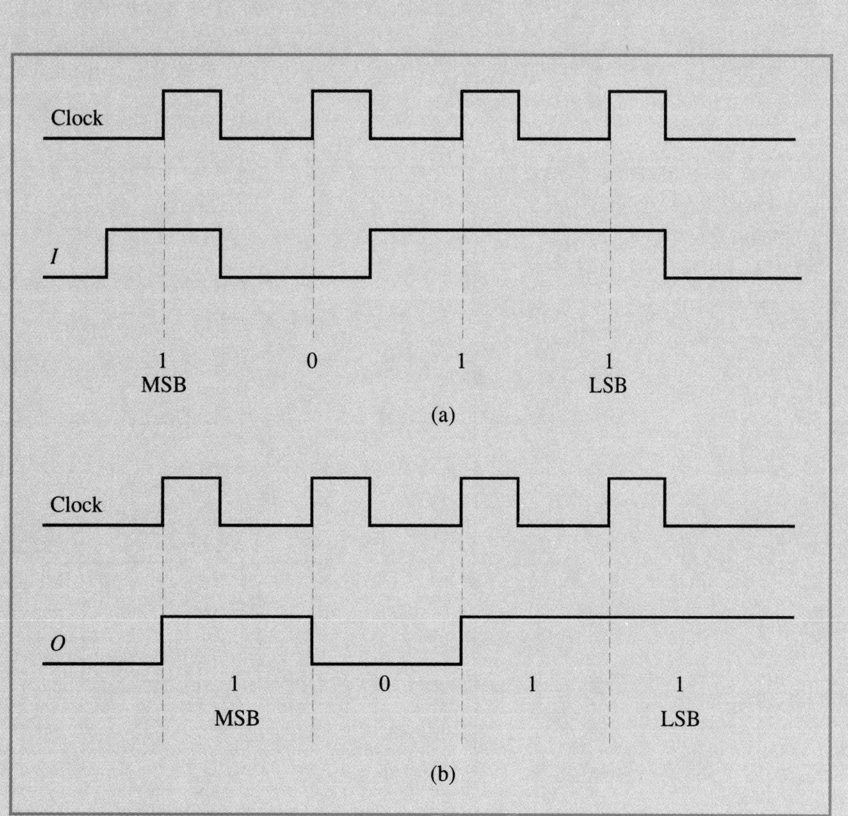

Parallel-in serial-out registers

The third type of shift register to be studied in this chapter is a *parallel-in serial-out* shift register. The logic symbol of the shift register is shown in Figure 7–14a. This type of shift register loads the data into the register in parallel, and retrieves the data serially. The data to be loaded into the register is placed on the input lines $I_3I_2I_1I_0$ and the *LOAD* line is activated by applying a logic 1 to it; this enables the parallel loading of data. A clock pulse is then applied to the clock input to actually load the data into the register. To retrieve data from the register serially, the *LOAD* line is made inactive by connecting it to a logic 0 (this disables the parallel loading of data) and 4 clock pulses are applied to the clock input to obtain the stored data serially at the output line O, LSB first.

The logic circuit for the parallel-in serial-out shift register is shown in Figure 7–14b. Notice that the D input of each flip-flop (except the first one) must be connected to one of two sources—the parallel input line I_n (where n is the bit number 0, 1, 2, or 3) or the output of the previous flip-flop in series Q. To load data into each flip-flop the data to be loaded in parallel, $I_3I_2I_1I_0$, has to be applied to the D inputs of the flip-flops and the flip-flops then triggered by a clock pulse. During loading, the Q outputs of the flip-flops must be disconnected from the D inputs of the flip-flops in series. During serial retrieval of data, the opposite must occur. That is, the inputs I_3, I_2, I_1 and I_0 must be disconnected from the D inputs of the flip-flops and the Q outputs of each flip-flop connected to the D input of the flip-flop in series so that the bits can be shifted through the circuit. To accomplish this task we use a two-channel single-bit multiplexer (such as the one designed in Chapter 4, Figure 4–58 or Figure 4–66). An expanded view of the multiplexer is shown in Figure 7–15. When the *LOAD* line is a logic 1 the *SELECT* input is a logic 1 and channel A is multiplexed to *OUT;* the I_n input is therefore connected to D and the Q input is cut-off

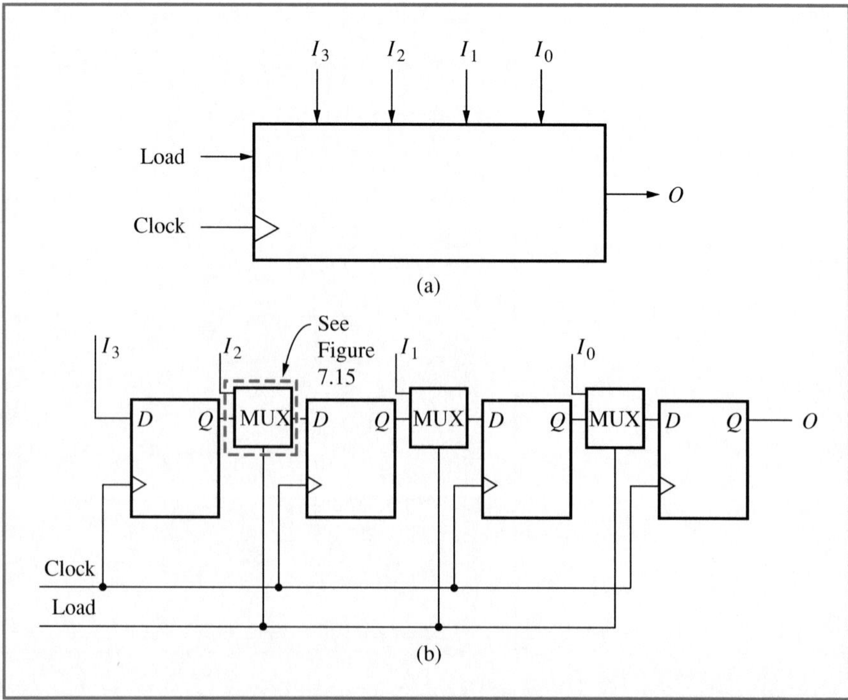

Figure 7–14
Parallel-in serial-out shift register. (a) Logic symbol; (b) circuit.

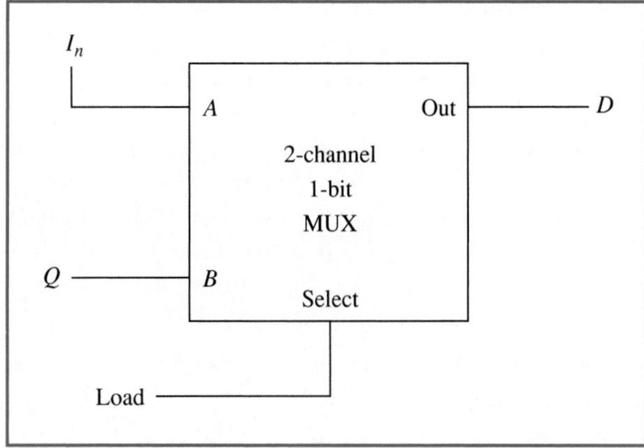

Figure 7–15
Multiplexer used in Figure 7–14.

from *D*. When the *LOAD* line is a logic 0 the *SELECT* input is a logic 0 and channel *B* is multiplexed to *OUT;* the *Q* input is therefore connected to *D* and the I_n input is cut off from *D*. Notice that no multiplexer is required for the first flip-flop since the only input to it is I_3.

Figure 7–16a shows the configuration of the circuit when $LOAD = 1$. Observe that the circuit resembles a parallel-load register and functions as such. Once the data has been loaded in, the data can be retrieved by setting *LOAD* to a logic 0. The circuit in Figure 7–16b shows the configuration of the circuit when $LOAD = 0$. Notice that the circuit now resembles a serial-out shift register and functions as such. Also note

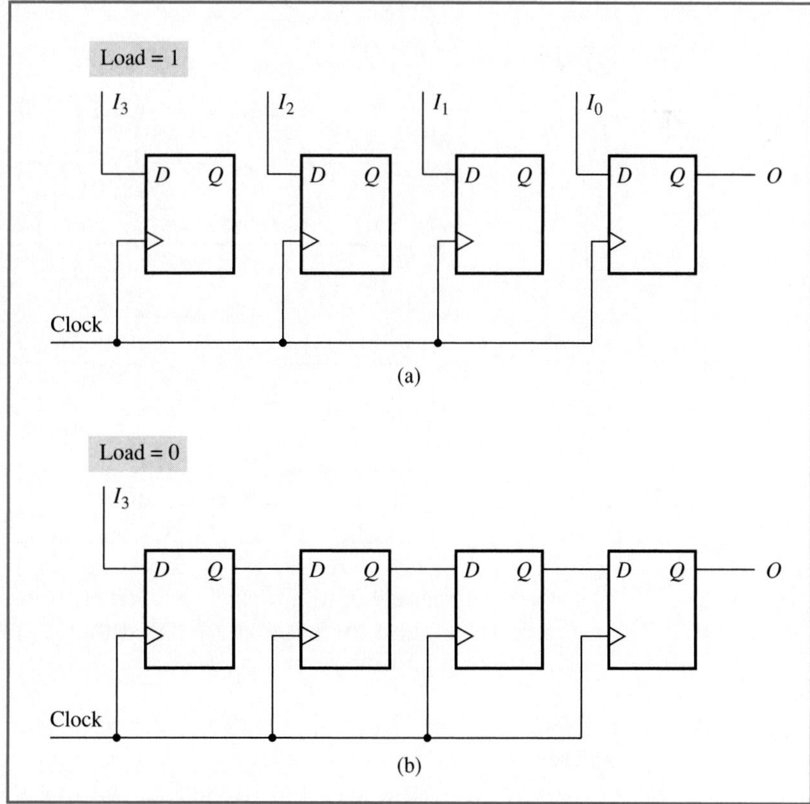

Figure 7–16
Operation of the parallel-in serial-out register. (a) Loading data into the register; (b) shifting data out.

that when $LOAD = 0$, the parallel-in serial-out shift register can also function as a serial-in serial-out shift register with serial input at I_3 and serial output at O.

EXAMPLE 7–4

Figure 7–17a shows the timing diagram to store the number 1011 into the register shown in Figure 7–14. The *LOAD* line is set to a logic 1 and the number 1011 is applied to the *I* inputs of the register. Since the data is loaded in parallel, only a single positive transition of the clock pulse is required to load the data into the register.

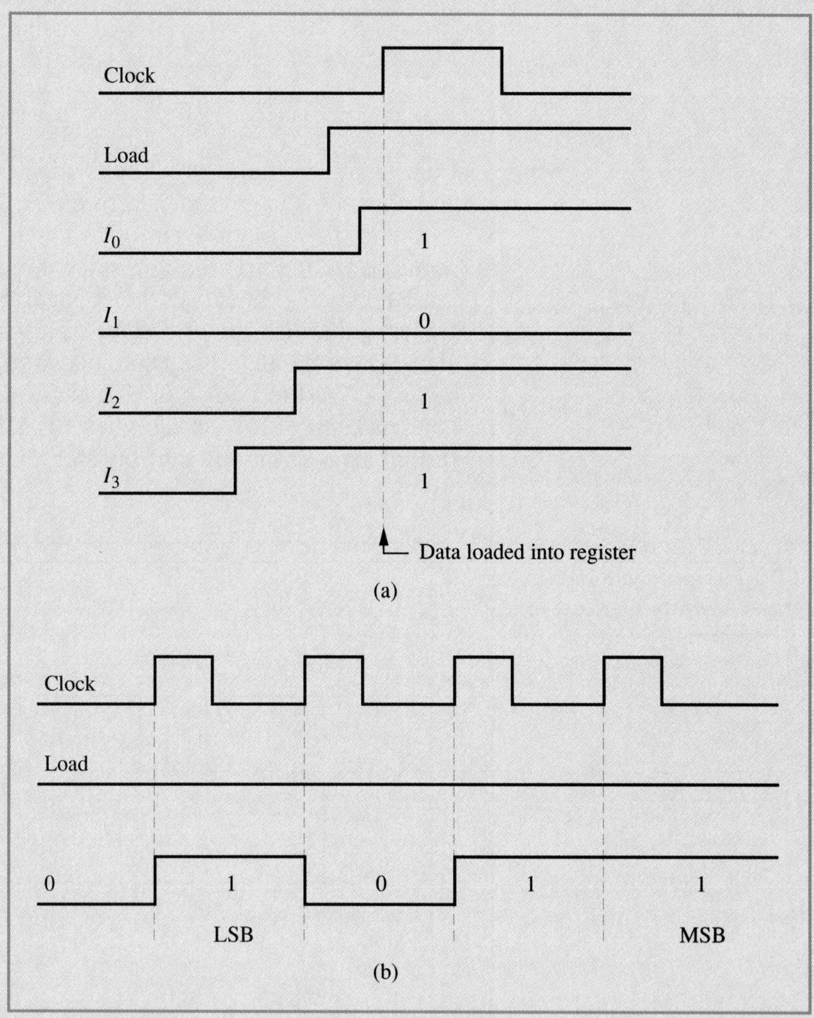

Figure 7–17 Timing diagram for the parallel-in serial-out register. (a) Loading data into the register; (b) shifting data out.

Figure 7–17b shows the timing diagram to retrieve the number 1011 that was previously stored in the register. The *LOAD* line is now set to a logic 0, and on each positive transition of the clock pulse the number 1011 is shifted out of the *O* line of the register, LSB first.

Bidirectional shift registers

The last type of shift register to be examined in this chapter is known as a *bidirectional shift register*. This type of register is very similar to a serial-in serial-out shift register except that the data can be shifted into and out of the register in both directions. The register also has parallel outputs for retrieving the data.

Figure 7–18a shows the logic symbol for a 4-bit bidirectional shift register. The register has a serial input line *SRI* to shift data into the register in the right direction (i.e., to the right), and a serial output line *SRO* to shift data out in the right direction. It also has another serial input, *SLI,* and another output line, *SLO,* to shift data into and out of the register in the left direction (i.e., to the left), respectively. The direction

Figure 7–18
A bidirectional shift register. (a) Logic symbol; (b) circuit.

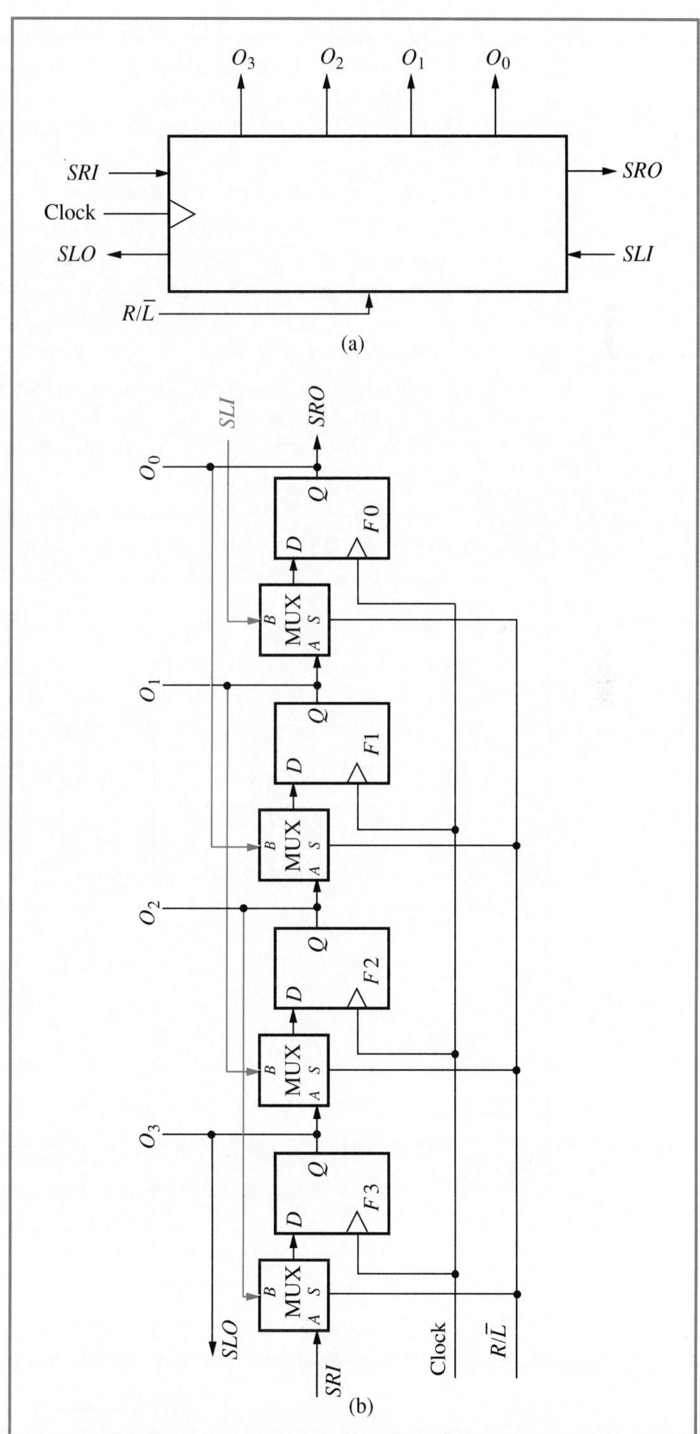

in which data is shifted (right or left) depends on the $R/\overline{L}$ (direction) line. If $R/\overline{L}=1$ then data is shifted in and out in the right direction, and if $R/\overline{L}=0$ then data is shifted in and out in the left direction. The *CLOCK* input is used to shift the bits in and out of the register. The content of the 4-bit register can be retrieved in parallel from the output lines $O_3O_2O_1O_0$.

The logic circuit for the bidirectional shift register is shown in Figure 7–18b. Notice that we have used a series of two-channel, single-bit multiplexers (similar to the one shown in Figure 7–15) to configure the "flow" of data to the right, that is, from *SRI* to flip-flops *F*3, *F*2, *F*1, and *F*0, to *SRO* if $R/\overline{L}=1$, or to the left, that is, from *SLI* to flip-flops *F*0, *F*1, *F*2, and *F*3, to *SLO* if $R/\overline{L}=0$. The parallel outputs $O_3O_2O_1O_0$ are obtained from the outputs of the flip-flops, and the *CLOCK* input synchronously loads each bit into the flip-flops.

Figure 7–19a shows the configuration of the circuit when the $R/\overline{L}$ line is set to a logic 1—right shift. Since the $R/\overline{L}$ line is connected to the select lines of the multiplexers, this effectively multiplexes channel *A* and connects *SRI* to the input of *F*3, the output of *F*3 to the input of *F*2, the output of *F*2 to the input of *F*1, the output of *F*1 to the input of *F*0, and the output of *F*0 to *SRO*. This accomplishes a right shift.

Figure 7–19b shows the configuration of the circuit when the $R/\overline{L}$ line is set to a logic 0—left shift. The multiplexers now multiplex the other input channel (*B*) so that this effectively connects *SLI* to the input of *F*0, the output of *F*0 to the input of *F*1, the output of *F*1 to the input of *F*2, the output of *F*2 to the input of *F*3, and the output of *F*3 to *SLO*. This accomplishes a left shift. Note that the order of the flip-flops drawn in Figure 7–19b have been rearranged for convenience.

Figure 7–19
Operation of the bidirectional shift register. (a) Right shift; (b) left shift.

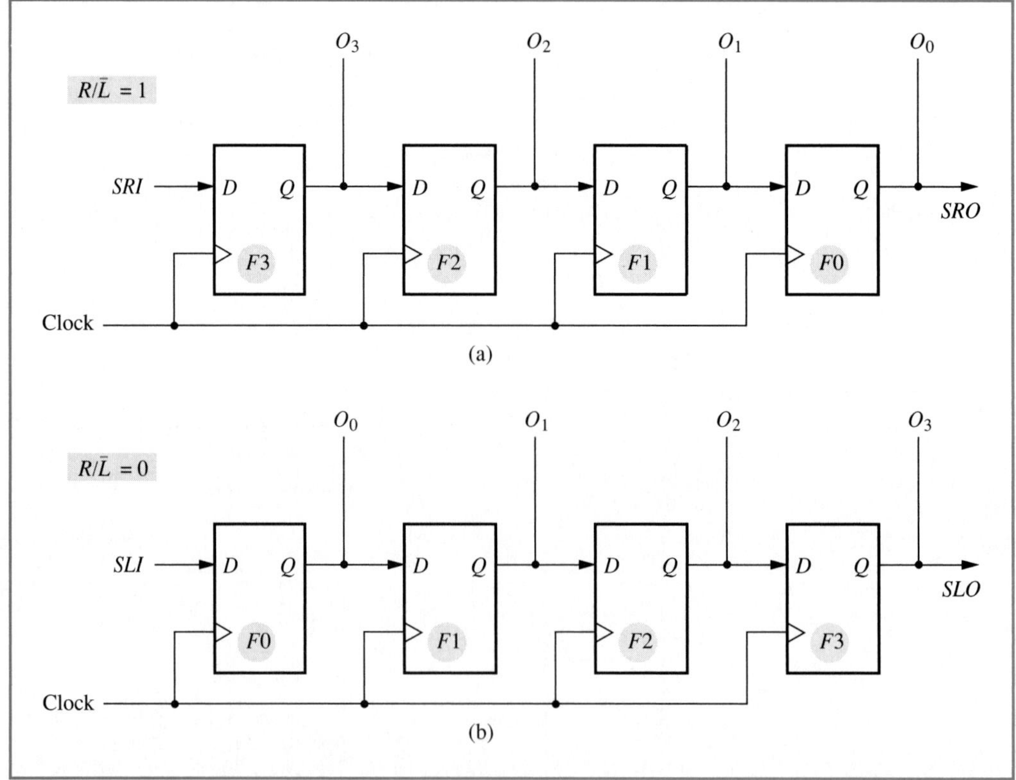

EXAMPLE 7–5

Assuming that the number 1011 has been shifted into the bidirectional shift register shown in Figure 7–18, draw a timing diagram of the parallel outputs when the $R/\overline{L}$ line is set to a logic 1, and when the $R/\overline{L}$ line is set to a logic 0.

SOLUTION The timing diagram of the parallel outputs $O_3O_2O_1O_0$ is shown in Figure 7–20a when $R/\overline{L} = 1$. Since we have assumed that the SRI line is tied to a logic 0, observe that after 4 clock pulses, the content of the register is 0000.

Figure 7–20 Timing diagram for the bidirectional shift register. (a) Right shift; (b) left shift.

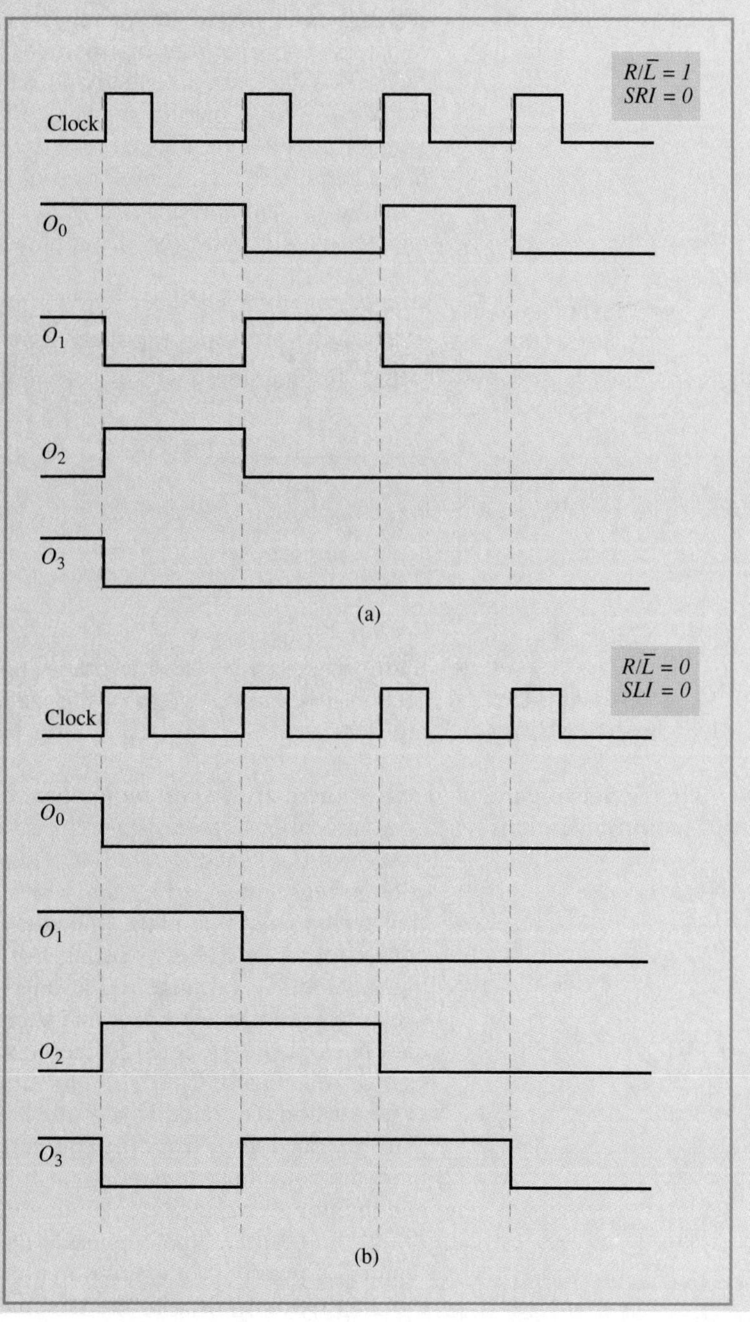

Figure 7–20b shows the timing diagram of the parallel outputs $O_3O_2O_1O_0$ when $R/\overline{L}=0$. Again, since we have assumed that the *SLI* line is tied to a logic 0, observe that after 4 clock pulses, the content of the register is 0000.

From the two timing diagrams shown in Figure 7–20 it is apparent that the bit pattern that appears on the output lines is different during the right and left shifts.

We can therefore see that both serial and parallel registers accomplish the same objective—storage of binary data. The manner in which data is loaded and/or retrieved is different in these two types of registers. It is apparent that parallel registers are faster than serial registers since data is loaded and retrieved on the transition of a single clock pulse for a parallel register but requires several clock pulses (the number depends on the size of the register) for a serial register. On the other hand, one advantage of serial registers is that the number of lines required to load and/or retrieve data is greatly reduced. The type of register used depends on the application.

Review questions

1. Why are serial registers known as shift registers?
2. Explain the operation of the four different types of shift registers.
3. What advantage does a parallel register have over a serial register?
4. What advantage does a serial register have over a parallel register?

7–4 Shift register applications

Shift registers have several other applications besides data storage. This section investigates a few of the more common applications of shift registers.

Serial data communications

One of the more important applications of shift registers is in digital data communication systems. One aspect of such a system is to transfer binary data from one place to another. The source and destination of the transfer could often be isolated by fairly long distances. Binary numbers can be transferred from one place (the source) to another (the destination) in one of two forms—parallel or serial. For parallel data transfers the entire binary number is transferred at one instant. Each bit in the number has a separate line dedicated to it and therefore the number of interconnecting data lines would be equal to the number of bits in the number being transferred. Figure 7–21 shows a simple parallel data communications system using two parallel registers. Assume that a 4-bit number $X_3X_2X_1X_0$ is to be transferred from the source *A* to the destination *B*. Notice that there are four interconnecting lines between *A* and *B* for the data, and one line for a common clock. The data is applied to the parallel register at *A* and the *CLOCK* line is pulsed. The *WR* line of the parallel register at *A* requires a positive transition to latch the data at its inputs, and therefore the data is transmitted to the destination at the positive transition of the

Figure 7–21
A parallel data communications circuit.

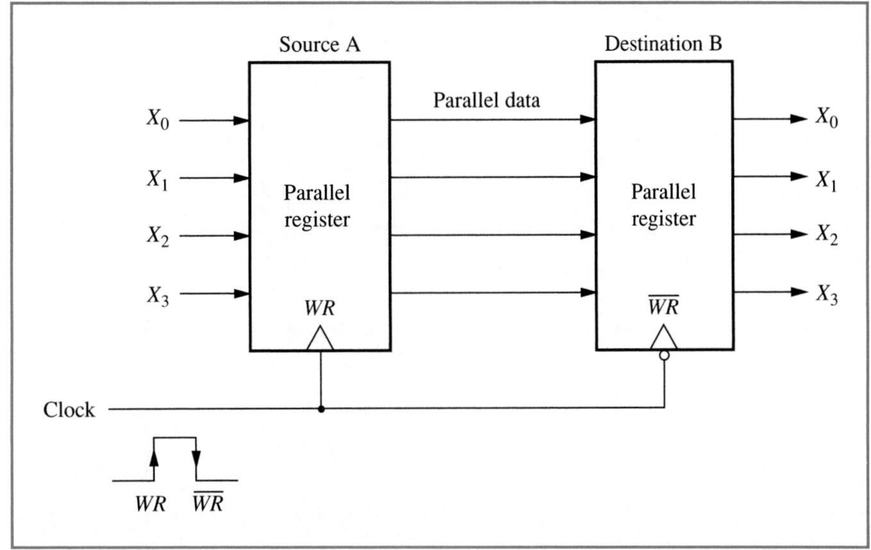

clock pulse. Since the same clock activates the $\overline{WR}$ line of the parallel register at B, the data transmitted by A is latched by this register on the negative transition of the clock pulse. The data $X_3X_2X_1X_0$ is therefore transmitted from A to B on 1 complete clock pulse.

In serial data communications systems, the source and destination transfer data over a single line. A binary number is transmitted serially over this line using a set of shift registers. Figure 7–22 shows a simplified circuit for transmitting a 4-bit binary number $X_3X_2X_1X_0$ serially from source A to destination B. Notice that this circuit has only two interconnecting lines between A and B—one for the serial data and one common clock. The shift register at A is a parallel-to-serial shift register (as shown in Figure 7–14). The binary number $X_3X_2X_1X_0$ is loaded into the register when the *LOAD* line is activated and then 4 clock pulses are required to shift the data out onto the serial data line. The serial data is shifted out on the positive transition of the clock pulse as shown in the timing diagram of Figure 7–23 which illustrates the transfer of the binary

Figure 7–22
A serial data communications circuit.

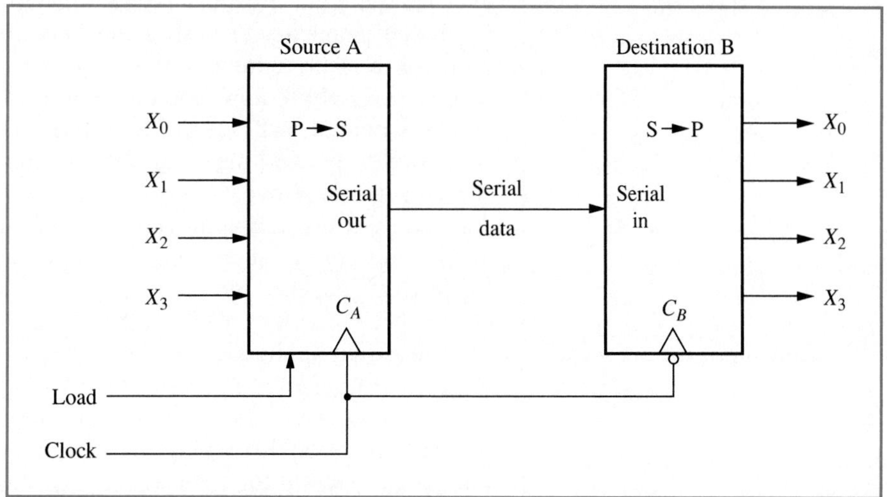

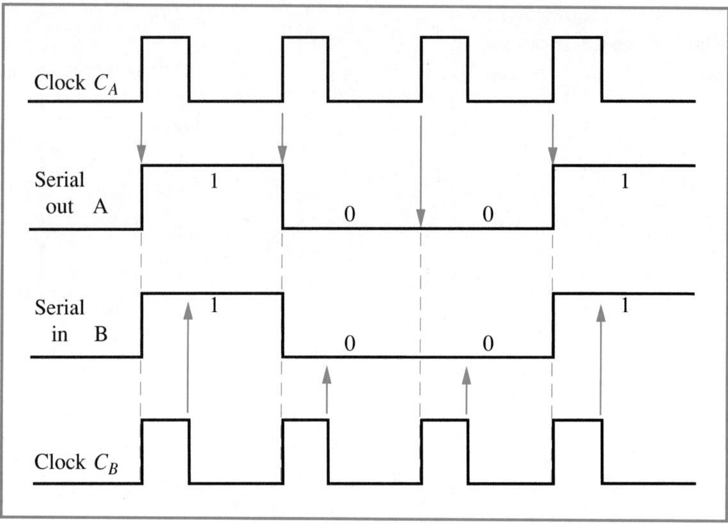

Figure 7–23 Timing diagram for serial data transfer.

number 1001. The same serial data line connects to a serial-to-parallel shift register (as shown in Figure 7–7) at destination *B*. This register is activated by the same clock that shifts the data out of the shift register at *A*, but the data on the serial data line is shifted in at *B* on the negative transition of the clock pulse. Shifting the data in on the negative transition allows the clock C_B to latch the data on the serial data line at approximately its midpoint thus allowing for minor variations in the timing of the two circuits. If the serial data in was sampled on the positive edge of the clock, any small difference in timing could lead to the wrong logic level(s) being shifted in. This type of serial data communications is known as *synchronous serial data communications* since a common clock is used to synchronize the transmitter *A* and the receiver *B*. The circuit in Figure 7–22 therefore transmits the data $X_3X_2X_1X_0$ from *A* to *B* on 4 complete clock pulses. The objectives of the two circuits shown in Figures 7–21 and 7–22 was to transfer a 4-bit number $X_3X_2X_1X_0$ from the source *A* to the destination *B*. The manner in which this was done was quite different, although the objective was met in both circuits. It is apparent that the advantage of serial data transfers over parallel data transfers is in a smaller number of interconnecting lines. A parallel data transfer system will require several data lines depending on the size of the numbers being transferred while a serial data transfer system will only require one data line regardless of the size of the numbers being transferred. The disadvantage of serial data transfers when compared to parallel data transfers is their slow speed. Each bit must be shifted separately over a single line, and therefore the time required to completely transfer a number serially will depend on the size of the number. In parallel data transfers the entire number is transferred on a single clock pulse.

Combination detector circuit

An interesting application of a shift register is in a commonly used circuit to detect a numerical code entered at a keypad. For example, a door could be unlocked on entering a four-digit code on a numeric keypad, or a six-digit code could be entered on a keypad to disable a security system. The block diagram for such a circuit is shown in Figure 7–24. A ten-key

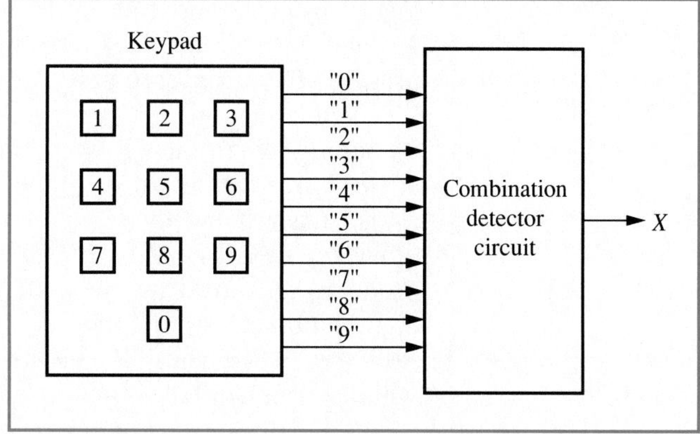

Figure 7–24
Block diagram for a combination detector circuit.

decimal keypad is connected to a circuit that we shall refer to as a *combination detector circuit*. Each time a key is pressed on the keypad, the keypad outputs a pulse (we will assume an active-high pulse) on the line corresponding to that key. For example, if the "8" key is pressed, a pulse appears on the line labeled "8," if the "4" key is pressed, a pulse appears on the line labeled "4," and so on. The combination detector circuit monitors the pulses on these ten lines, and if the right keys are pressed in the right sequence, the circuit activates its output line X to a logic 1 state. This output line could then be used to activate a device that unlocks a door, disables an alarm, or activates or deactivates any other device depending on the application.

Consider the design of a combination detector circuit to detect the combination "6284." That is, when this combination is entered at the keypad, the circuit must set X to a logic 1 and set X to a logic 0 if any other combination is entered. The circuit to implement this specification is shown in Figure 7–25. The circuit consists of a 4-bit shift register made up of D flip-flops with direct (asynchronous) active low $\overline{CLR}$ inputs. Recall from Chapter 5 that a logic 0 on the $\overline{CLR}$ input will directly reset the flip-flop without requiring a clock pulse. The first flip-flop in series, $F1$, has its D input connected to a logic 1. The objective is to shift this logic 1 to the output of the last flip-flop in series, $F4$. The output of $F4$ is X, the output of the combination detector circuit. In order to shift the

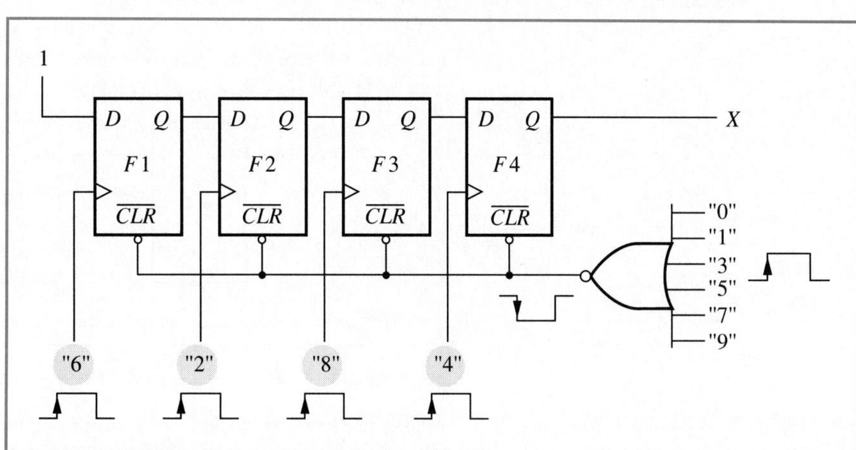

Figure 7–25
Logic circuit for a combination detector circuit.

logic 1 from the input of flip-flop $F1$ to the output of $F4$, flip-flops $F1$, $F2$, $F3$, and $F4$ must be clocked in sequence. That is, when the "6" key is pressed, $F1$ is triggered and the logic 1 is latched to its output. When the "2" key is pressed, $F2$ is triggered and the logic 1 (from $F1$) is latched to its output. When the "8" key is pressed, $F3$ is triggered and the logic 1 (from $F2$) is latched to its output. Finally, when the "4" key is pressed, $F4$ is triggered and the logic 1 (from $F3$) is latched to its output, X. If any other key besides "6," "2," "8," or "4" is pressed, the *NOR* gate produces an active-low pulse that clears all the flip-flops; the entire four-digit combination must then be reentered. Also notice that if the keys "6," "2," "8," and "4" are not pressed in the right sequence, the logic 1 will never be shifted to X.

Shift register counters

Another application of shift registers is in special-purpose counters known as *shift register counters.* Two of the more common types of shift register counters are the *ring counter* and the *Johnson counter.*

The logic circuit for a 3-bit ring counter is shown in Figure 7–26. The circuit consists of three D flip-flops configured as shift registers with the output of the last flip-flop $F0$ connected to the input of the first flip-flop $F2$. The outputs of the counter are $X_2X_1X_0$. When the counter is started, the $\overline{CLEAR}$ line is used to directly (asynchronously) clear all the flip-flops so that the initial state is 000. Flip-flop $F2$ is then asynchronously set by activating the preset ($\overline{PRE}$) line. Once the clock pulses are started, the logic 1 at the input of $F1$ is shifted to the next flip-flop, and the next, and so on, until it is fed back to the input of $F2$ again to recycle the count. The logic 1 thus appears to move (or rotate) through the counter in a *ring* or circular pattern and hence the name "ring counter."

The timing diagram for the counter is shown in Figure 7–27. Notice how the logic 1 at X_2 is shifted to X_1 and then to X_0, and then back again to X_2. The count for the 3-bit ring counter follows the sequence

$$4, 2, 1, 4, 2, 1, 4, \ldots$$

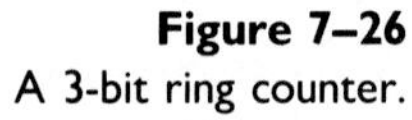

Figure 7–26
A 3-bit ring counter.

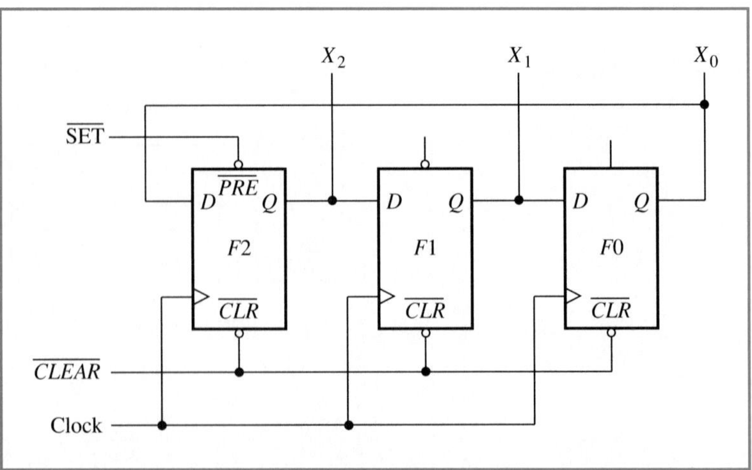

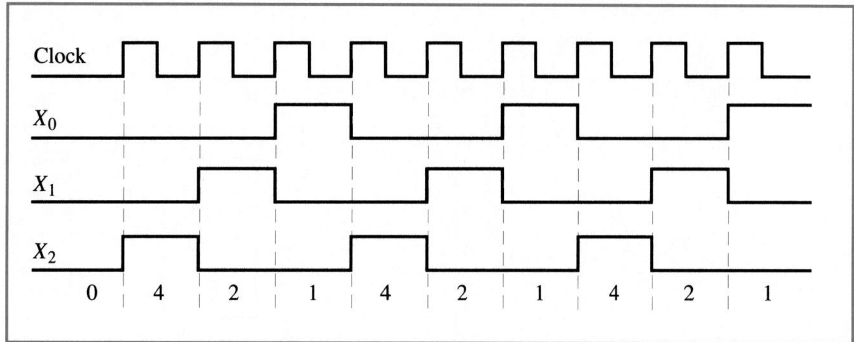

Figure 7–27
Timing diagram for a 3-bit ring counter.

or, in terms of the counter's outputs

1 0 0
0 1 0
0 0 1
1 0 0
0 1 0
0 0 1
1 0 0
⋮

Similarly, one can expect a 4-bit ring counter to follow the sequence:

8, 4, 2, 1, 8, 4, 2, 1, 8, . . .

or, in terms of the counter's outputs

1 0 0 0
0 1 0 0
0 0 1 0
0 0 0 1
1 0 0 0
0 1 0 0
0 0 1 0
⋮

The Johnson counter is basically a modification of the ring counter shown in Figure 7–26. A Johnson counter is constructed by connecting the *complement* of the output of the last flip-flop in series (*F*0) to the input of the first flip-flop in series (*F*2) as shown in Figure 7–28. This design produces a count sequence that appears to fill up the outputs with 1-bits, and then fill up the outputs with 0-bits, continuously repeating the cycle. The timing diagram for the Johnson counter is shown in Figure 7–29 and illustrates this effect.

Before the clock pulses to the counter are started, the counter is first cleared by activating its $\overline{CLEAR}$ line. The state of the counter's outputs is therefore 000 as shown in Figure 7–29. Since the output of *F*0 is a logic 0, a logic 1 is fed back to *F*2 and this logic 1 is shifted through the three flip-flops until it appears at the input of *F*0. When this occurs, on the next clock pulse, the $\overline{Q}$ output of *F*0 is a logic 0. This logic 0 is

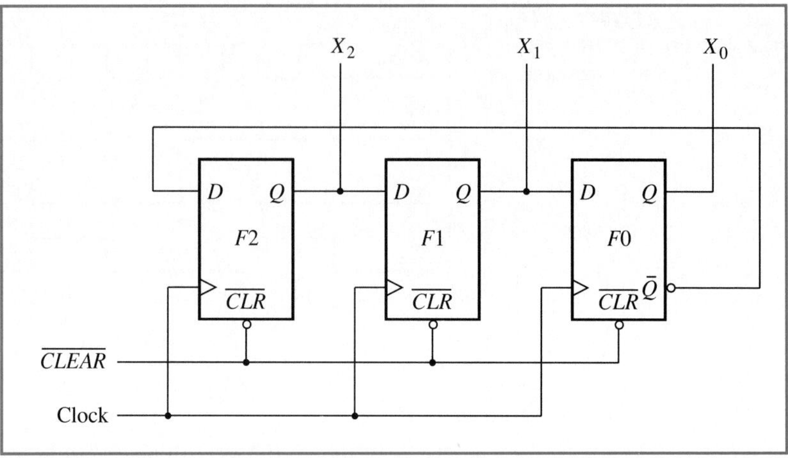

Figure 7–28
A 3-bit Johnson counter.

now shifted through the three flip-flops until it appears at the input of $F0$. The counter then recycles. The resulting count sequence for the Johnson counter is therefore

0, 4, 6, 7, 3, 1, 0, 4, 6, 7, 3, 1, 0, . . .

or, in terms of the counter's outputs

0 0 0
1 0 0
1 1 0
1 1 1
0 1 1
0 0 1
0 0 0
1 0 0
1 1 0
1 1 1
⋮

Similarly, one can expect a 4-bit Johnson counter to follow the sequence

0, 8, C, E, F, 7, 3, 1, 0, 8, C, . . .

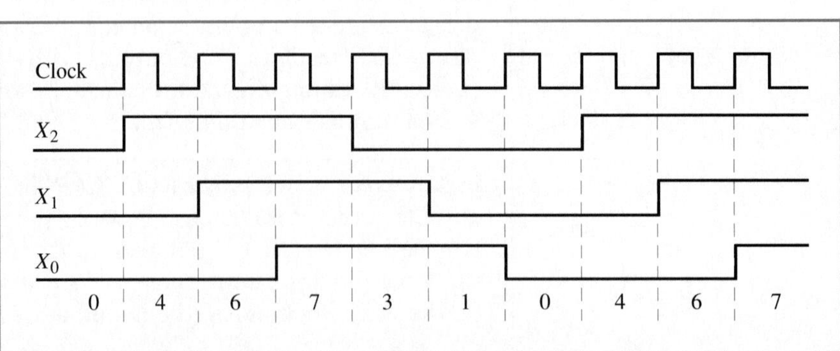

Figure 7–29
Timing diagram for a 3-bit Johnson counter.

or, in terms of the counter's outputs

0 0 0 0
1 0 0 0
1 1 0 0
1 1 1 0
1 1 1 1
0 1 1 1
0 0 1 1
0 0 0 1
0 0 0 0
1 0 0 0
1 1 0 0
⋮

Time delay circuit

A serial-in serial-out shift register can be used as a *time delay circuit* to produce a logic 1 at its output after a fixed amount of time. The circuit shown in Figure 7–30a uses the 4-bit shift register of Figure 7–12 to produce a time delay after 4 clock pulses have elapsed.

The timing diagram for the circuit shown in Figure 7–30b assumes that the *CLOCK* input is attached to a stream of clock pulses having a frequency of 1 kHz. It takes 4 clock pulses to shift the logic 1 from the input *I* of the shift register to the output *O*. Therefore, on the first clock pulse, the logic 1 is shifted into the register, and on the fourth clock pulse the logic 1 is shifted out. Since each clock period is 1 ms, the total time delay from the time the logic 1 was shifted in is 3 ms.

The amount of time delay that a serial-in serial-out shift register can produce can be modified by increasing or decreasing the size of the

Figure 7–30
A time delay circuit.

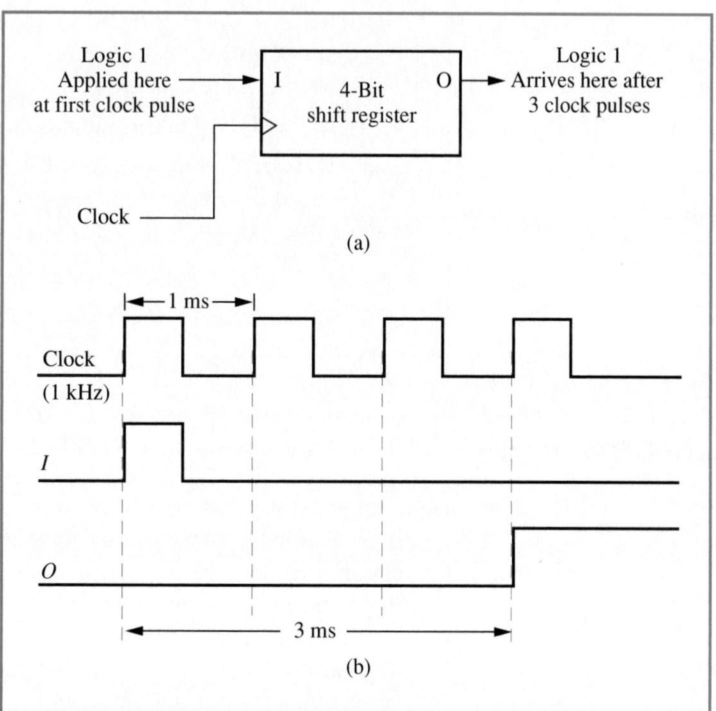

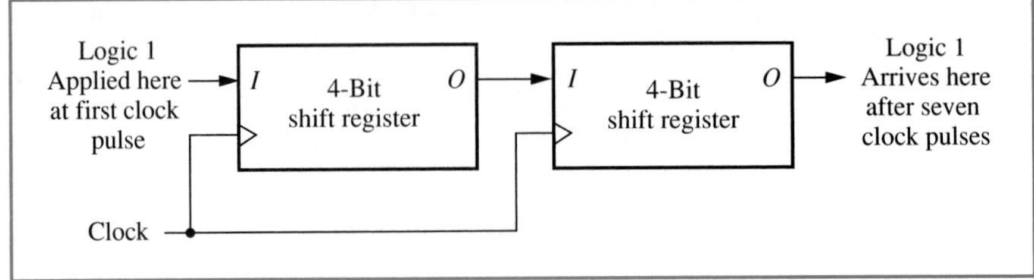

Figure 7–31
Using 2 shift registers to increase the time delay.

register. The amount of delay produced by a shift register (in clock pulses) will always be equal to 1 less than its size. For example, a 4-bit shift register will produce a delay of 3 clock pulses, an 8-bit shift register will produce a delay of 7 clock pulses, etc. The amount of time delay produced by this type of circuit can also be increased by cascading shift registers. For example, the circuit shown in Figure 7–31 uses two 4-bit shift registers to produce a delay of 7 clock pulses, since the two cascaded 4-bit shift registers function as a single 8-bit shift register. Again, if we assume that the clock frequency is 1 kHz, the total time delay will be 7 ms.

This section has investigated a few of the applications of shift registers. It is apparent that shift registers are used in very diverse applications ranging from storage circuits to counters. Parallel registers on the other hand are best suited for storage applications since speed is of primary importance in such applications. The next section examines such an application of parallel registers.

Review questions

1. State two ways in which data can be transferred from one point to another.
2. Describe the application of the shift register in a data communications system.
3. Describe an advantage of serial data communications over parallel data communications.
4. Describe an advantage of parallel data communications over serial data communications.
5. Describe the operation of the ring counter.
6. Describe the operation of the Johnson counter.
7. What type of shift register can be used to produce a time delay?

7–5 Register arrays

A *register array* or *register file* is a collection of registers incorporated into a single package. A register file is similar in concept and operation to a *memory system*. This section deals with the operation and design of register arrays and also introduces the elementary concepts of storage of binary numbers and memory systems. The concepts developed in this section will then be applied in Chapter 9 to the operation and design of larger and more practical memory circuits.

To illustrate the design of a register array, consider the set of four 4-bit parallel registers with bidirectional data lines shown in Figure 7–32. These registers are the same as the register shown in Figure 7–5 and

Figure 7–32
The building blocks of a register array.

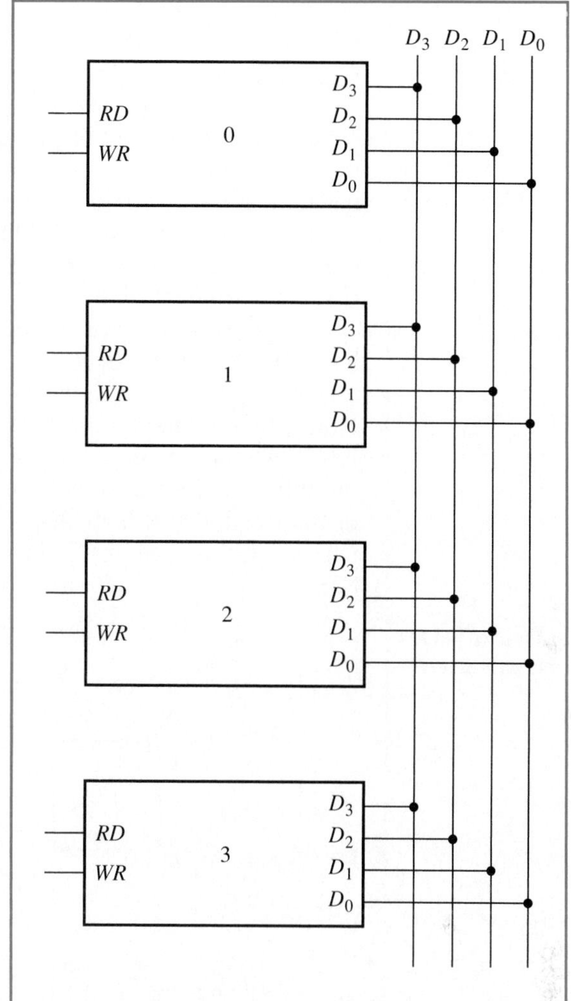

explained in Section 7–2, except that we have used active-high *RD* and *WR* lines. The bidirectional data lines of all the registers are connected together by means of a *data bus,* $D_3D_2D_1D_0$. Each register is numbered (from 0 to 3) for identification purposes. It we apply a binary number to the data bus, the number is applied to the data lines of *all* the registers. However, to write (store) the data into register 2, we must activate the *WR* line of register 2. Similarly, to write the data into register 0, we must activate the *WR* line of register 0. To read the data stored in a particular register we can activate the *RD* line of that register. The contents of the selected register will then be loaded onto the data bus from which it can be retrieved. Since the data outputs of all the registers are tristated and since the *RD* line enables the tristate gates at the outputs of each register, we must make sure that only one register's *RD* line is enabled at a time to prevent a situation known as *bus contention* from occurring. Bus contention is a condition that occurs when more than one register puts its stored data on the data bus. This generally causes a conflict in logic levels and the presence of invalid data on the bus. For example, if an attempt is made to read from register 1 and register 2 at the same time, and if register 1 has a 0000 stored in it while register 2 has a 1111 stored in it, what number will the data bus contain? Logically this is an undefined condition. We can therefore store or retrieve data from a particular register by activating its *WR* or *RD* line, respectively. However, this arrangement is quiet inconvenient if our circuit contained a much larger set of registers instead of just four. For example, if we had 32 registers in the circuit we would require 64 *RD* and *WR* lines to individually *address* (select) each register for reading or writing. It would be more efficient to have a common *RD* and *WR* line for reading and writing to all the registers and a single *SELECT* line to select a particular register for a read or write operation.

Figure 7–33 shows a modification of the circuit in Figure 7–32 to incorporate a single $R/\overline{W}$ line, and a separate *SELECT* line (S_0 through S_3) to select a particular register for either a read or write operation. Notice that we have now shown the data lines drawn in the form of a bus. Since at a given instant we will either be reading from a particular register or writing to a particular register, we can combine the functions of the *RD* and *WR* lines into a single line called $R/\overline{W}$ that will determine whether we are reading from a selected register ($R/\overline{W}=1$) or writing to a

Figure 7–33
The register array with select lines.

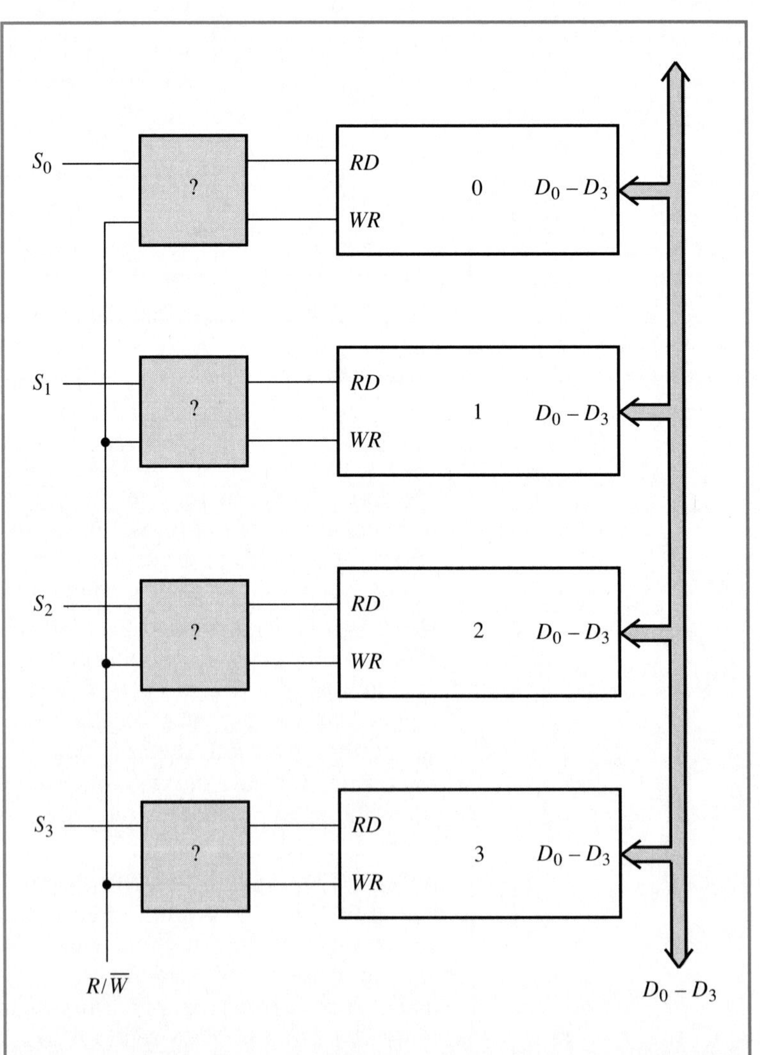

particular register ($R/\overline{W}=0$). For example, to read from register 0, the S_0 line is activated to a logic 1, and the $R/\overline{W}$ line is set to a logic 1. To write to register 0, the S_0 line is activated to a logic 1, and the $R/\overline{W}$ line is set to a logic 0. Similarly, to read from register 1, the S_1 line is activated to a logic 1, and the $R/\overline{W}$ line is set to a logic 1. To write to register 1, the S_1 line is activated to a logic 1, and the $R/\overline{W}$ line is set to a logic 0. We therefore need some combinational logic circuit (identified by the box with the ?) that will accept as input, the select line for a particular register S_n (where n is the register number) and the $R/\overline{W}$ line, and activate the *RD* and *WR* lines of the selected register appropriately for reading, writing, or no operation as shown in Table 7–1.

In Table 7–1, the first two combinations of $R/\overline{W}$ and S_n will not activate the *RD* or *WR* line of the register n since the register is not selected ($S_n=0$); notice that *RD* and *WR* are both logic 0's for the first two combinations. For the third combination $S_n=1$, and the register n is selected for a write operation since $R/\overline{W}=0$. Therefore $WR=1$ while *RD* is inactive at the 0 state. For the fourth combination $S_n=1$, and the register n is selected for a read operation since $R/\overline{W}=1$. Therefore $RD=1$ while *WR* is inactive at the 0 state. From the truth table shown in Table 7–1 we can now obtain the logic circuit for the "? boxes" shown in Figure 7–33 by first obtaining the logic equations for *RD* and *WR* as follows

Equation (7–1)
$$RD = R/\overline{W} \cdot S_n$$

Equation (7–2)
$$WR = \overline{R/\overline{W}} \cdot S_n$$

The logic circuit to implement Equations 7–1 and 7–2 is shown in Figure 7–34. This circuit is incorporated into each of the "? boxes" shown in Figure 7–33.

The circuit in Figure 7–33 now only uses a single select line, S_n, to select a particular register, n, for either a read or write operation which is controlled by a common $R/\overline{W}$ line. However, this circuit is still not practical for a large register set. We can reduce the number of lines required to select each register even further to produce a more efficient design that is practical for any number of registers.

At any given instant, only one of the four registers shown in Figure 7–33 is activated for a read or write operation. Therefore only one of the four select lines S_0 through S_3 will be active at a time. Thus, instead of selecting each register individually through the select lines we can select each register by means of a 2-bit binary code using a 1-of-4 decoder. Recall that a 1-of-4 decoder will accept a 2-bit binary code as input and activate one of 4 outputs to identify a particular code. The modified circuit is shown in Figure 7–35.

Observe in Figure 7–35 that we have replaced the "? boxes" with the circuit in Figure 7–34. The select lines are connected to a 1-of-4 decoder

Table 7–1

Operation	S_n	$R/\overline{W}$	RD	WR
No operation	0	0	0	0
No operation	0	1	0	0
Write	1	0	0	1
Read	1	1	1	0

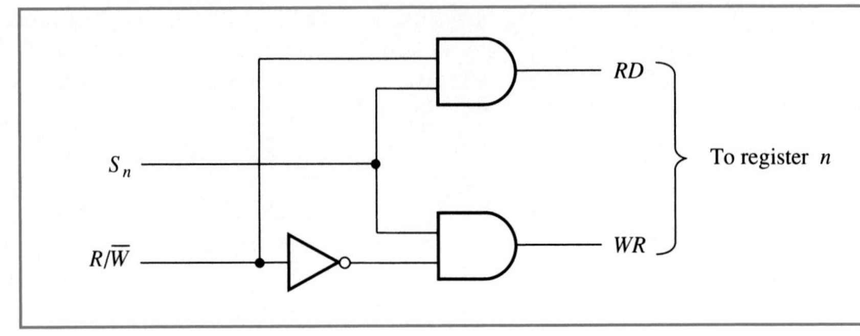

Figure 7–34
The read/write and select control circuit for each register.

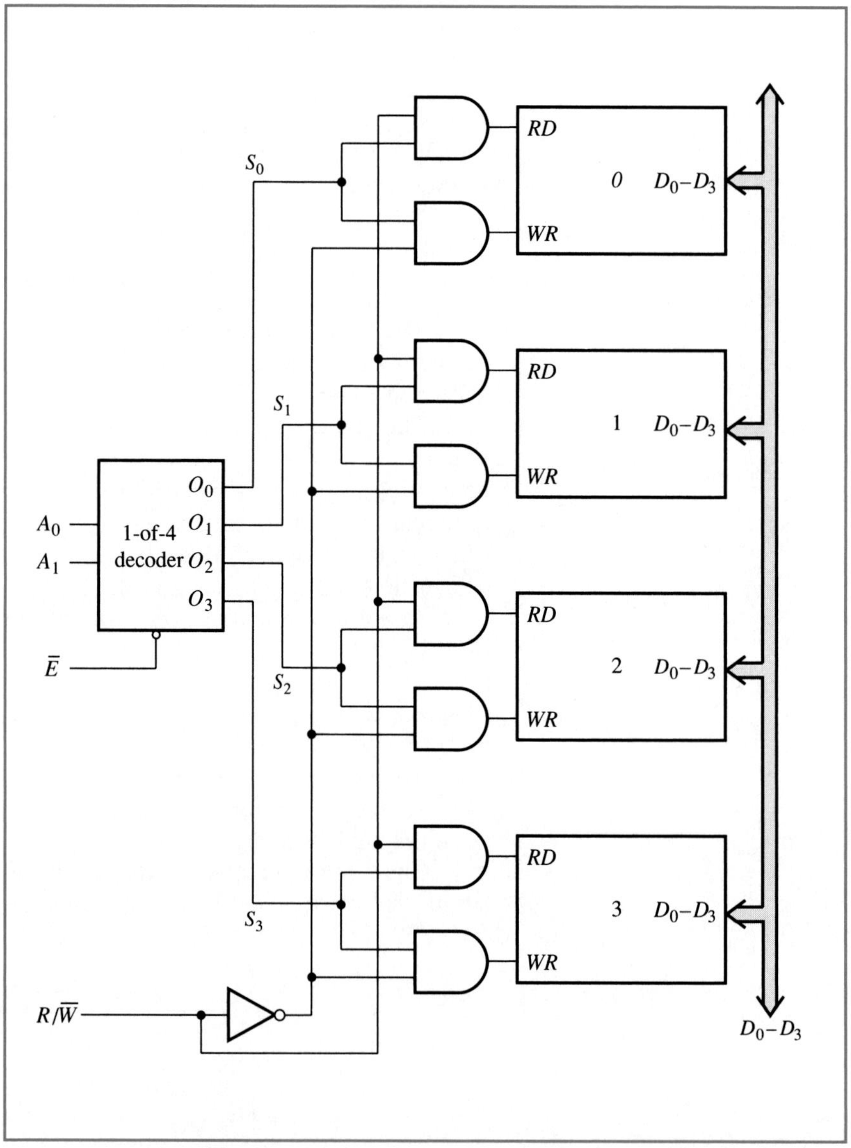

Figure 7–35
The complete 4 × 4 register array (file) circuit.

Table 7–2

Register	$\overline{E}$	A_1	A_0	S_0	S_1	S_2	S_3
Register 0	0	0	0	1	0	0	0
Register 1	0	0	1	0	1	0	0
Register 2	0	1	0	0	0	1	0
Register 3	0	1	1	0	0	0	1
None	1	X	X	0	0	0	0

such that each 2-bit code applied to its inputs A_1A_0 will select a particular register as shown in Table 7–2.

Since the input lines to the decoder A_1A_0 are used to *address* or select a register for a read or write operation, these lines are known as *address lines,* and the binary codes that are applied to these lines to select a particular register are known as *addresses*.

The active-low enable input of the decoder $\overline{E}$ is used to allow the decoder to function normally when at the 0 state and disables the decoder when at the 1 state. When the decoder is disabled all its outputs are inactive, and therefore if $\overline{E}$ is a logic 1 none of the registers can be selected and no data can be stored or retrieved from any register.

The circuit in Figure 7–35 is a register array also known as a 4 × 4 register file since it can store four binary numbers each being 4 bits (wide) in size. The logic symbol for the circuit is shown in Figure 7–36.

Register files can be designed for a variety of sizes and configurations. A register file is usually described in terms of the number of registers (R) and the size of each number stored (S). The product of the number of registers and the size of each register is the total capacity (in bits) of the register file. By convention, the number of registers (R) is always stated first and then the size (S) of each register—R × S. For example, Figure 7–37a shows the logic symbol for an 8 × 4 register file that is capable for storing eight 4-bit numbers for a total capacity of 32 bits. Notice that there are three address lines $A_2A_1A_0$ to address 2^3, or 8, registers and four data lines $D_3D_2D_1D_0$ to store and retrieve 4-bit numbers into and from each register. Figure 7–37b shows the logic symbol for a 4 × 8 register file that has the same storage capacity

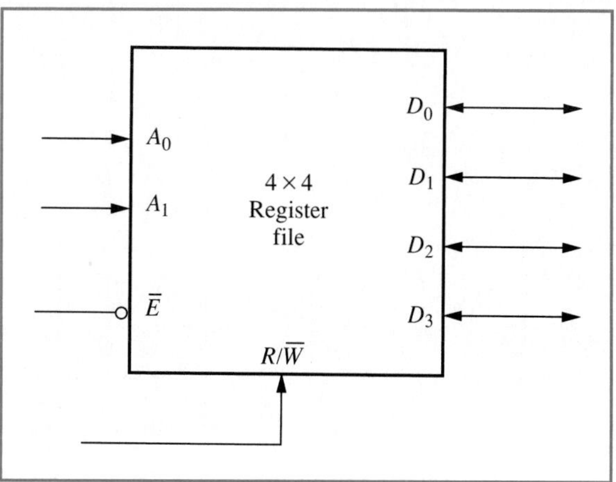

Figure 7–36
Logic symbol for the 4 × 4 register file.

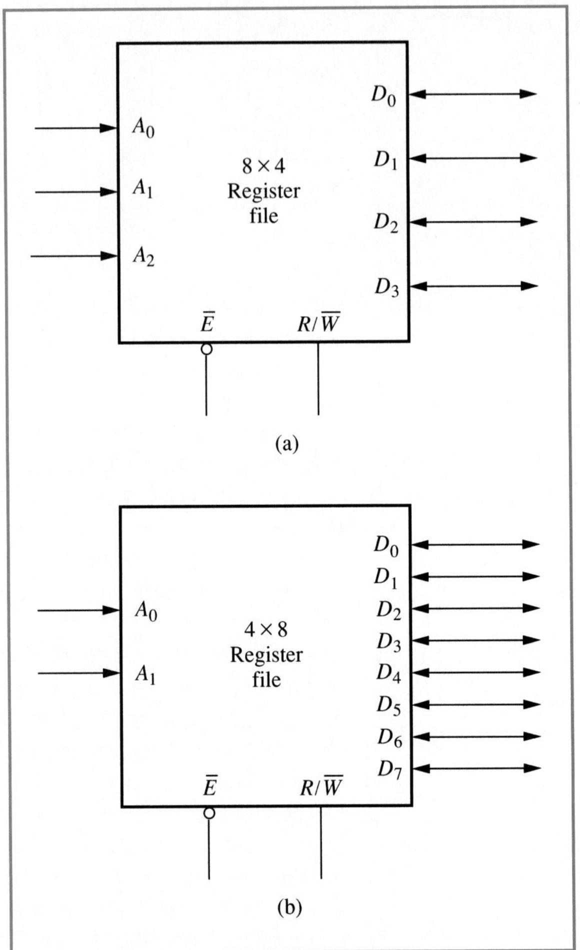

Figure 7–37
Logic symbols for (a) an 8 × 4 register file; (b) a 4 × 8 register file.

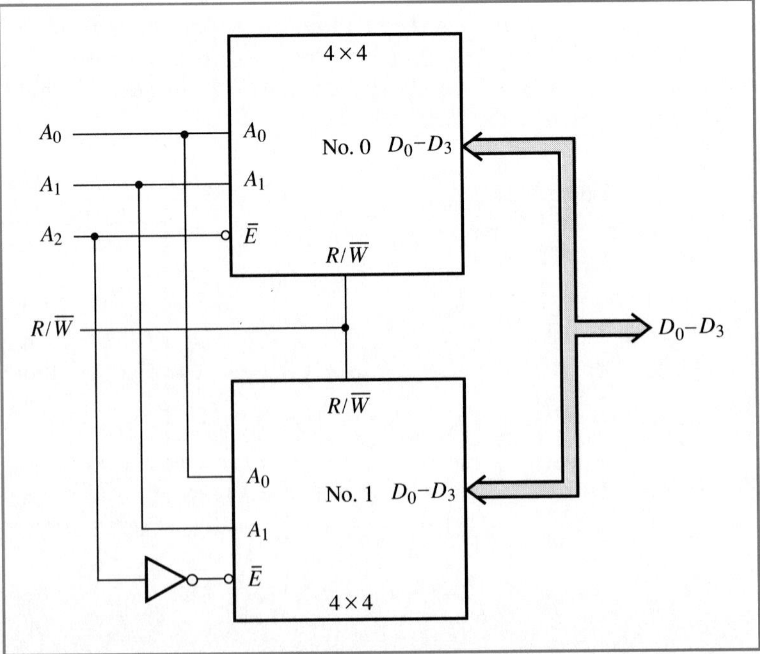

Figure 7–38
8 × 4 register file implemented with two 4 × 4 register files.

Table 7–3

	A_1	A_1	A_0	Register number
Register file no. 0	0	0	0	0
	0	0	1	1
	0	1	0	2
	0	1	1	3
Register file no. 1	1	0	0	0 (4)
	1	0	1	1 (5)
	1	1	0	2 (6)
	1	1	1	3 (7)

(32 bits) as the register file shown in Figure 7–37a. However, the configuration of the register file in Figure 7–37b is much different since it is organized as an array of four registers, each register capable of storing an 8-bit number. Address lines A_1A_0 select one of 4 registers while data lines $D_7D_6D_5D_4D_3D_2D_1D_0$ allow 8-bit numbers to be stored and retrieved into and from each selected register.

Register files can easily be expanded to store more numbers or store larger numbers. *Vertical expansion* involves increasing the number of registers in the file while *horizontal expansion* involves increasing the size of each register. For example, Figure 7–38 implements the 8 × 4 register file whose logic symbol is shown in Figure 7–37a by using two 4 × 4 register files.

The register file shown in Figure 7–38 will have a total of eight registers, four of which (0, 1, 2, and 3) will be located in register file no. 0 and the remaining four (4, 5, 6, and 7) in register file no. 1. The data lines $D_3D_2D_1D_0$ of both register files are connected together to form a common data bus to store and retrieve data to and from all eight registers. The $R/\overline{W}$ line controls the type of operation (read or write) performed on a selected register. Address lines A_1 and A_0 select the basic eight registers in each register file while address line A_2 selects either register file no. 0 (when at the logic 0 state) or register file no. 1 (when at the logic 1 state) as shown in Table 7–3.

Since the two register files in Figure 7–38 are treated as one register file circuit with a total of eight registers, we often refer to registers 0 through 3 of register file no. 1 as registers 4, 5, 6, and 7. Notice that the register numbers always correspond to the decimal or hexadecimal equivalents of the addresses.

EXAMPLE 7–6

Figure 7–39 shows a 16 × 4 register file constructed by using two of the 8 × 4 register files shown in Figure 7–37a. The addressing for the register file is shown in Table 7–4.

Notice in Figure 7–39 that address lines $A_2A_1A_0$ select the eight registers within each register file. However, when address line A_3 is a logic 0, register file no. 0 is enabled, and when address line A_3 is a logic 1, register file no. 1 is enabled. Therefore the registers in register file no. 0 have the addresses 0 through 7 and the registers in register file no. 1 have the addresses 8 through F. As before, the data lines connect to both register files but only the selected

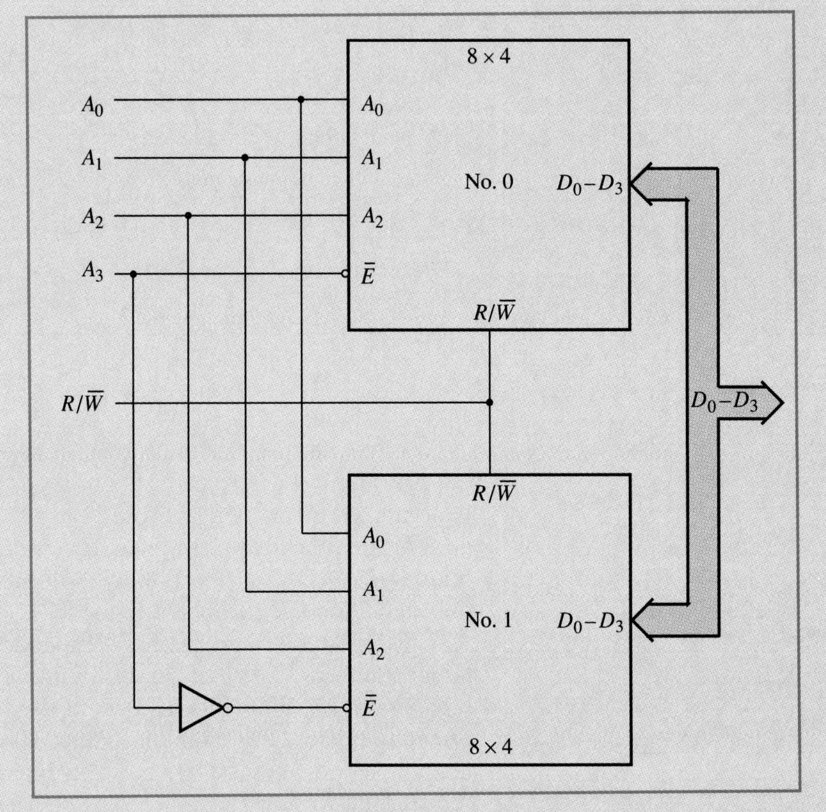

Figure 7–39
16 × 4 register file implemented with two 8 × 4 register files.

Table 7–4

	A_3	A_2	A_1	A_0	Register
Register file no. 0	0	0	0	0	0
	0	0	0	1	1
	0	0	1	0	2
	0	0	1	1	3
	0	1	0	0	4
	0	1	0	1	5
	0	1	1	0	6
	0	1	1	1	7
Register file no. 1	1	0	0	0	8
	1	0	0	1	9
	1	0	1	0	A
	1	0	1	1	B
	1	1	0	0	C
	1	1	0	1	D
	1	1	1	0	E
	1	1	1	1	F

(addressed) register in the selected (addressed) register file will be accessed for storage or retrieval of data. The $R/\overline{W}$ line establishes the direction of data flow into or out of the addressed register.

EXAMPLE 7–7

Figure 7–40 shows a 16 × 4 register file constructed by using four of the 4 × 4 register files shown in Figure 7–36. Observe that a 1-of-4 decoder is used to select one of four register files as the most significant address lines A_3A_2 cycle through all possible combinations of 2 bits as shown in Table 7–5.

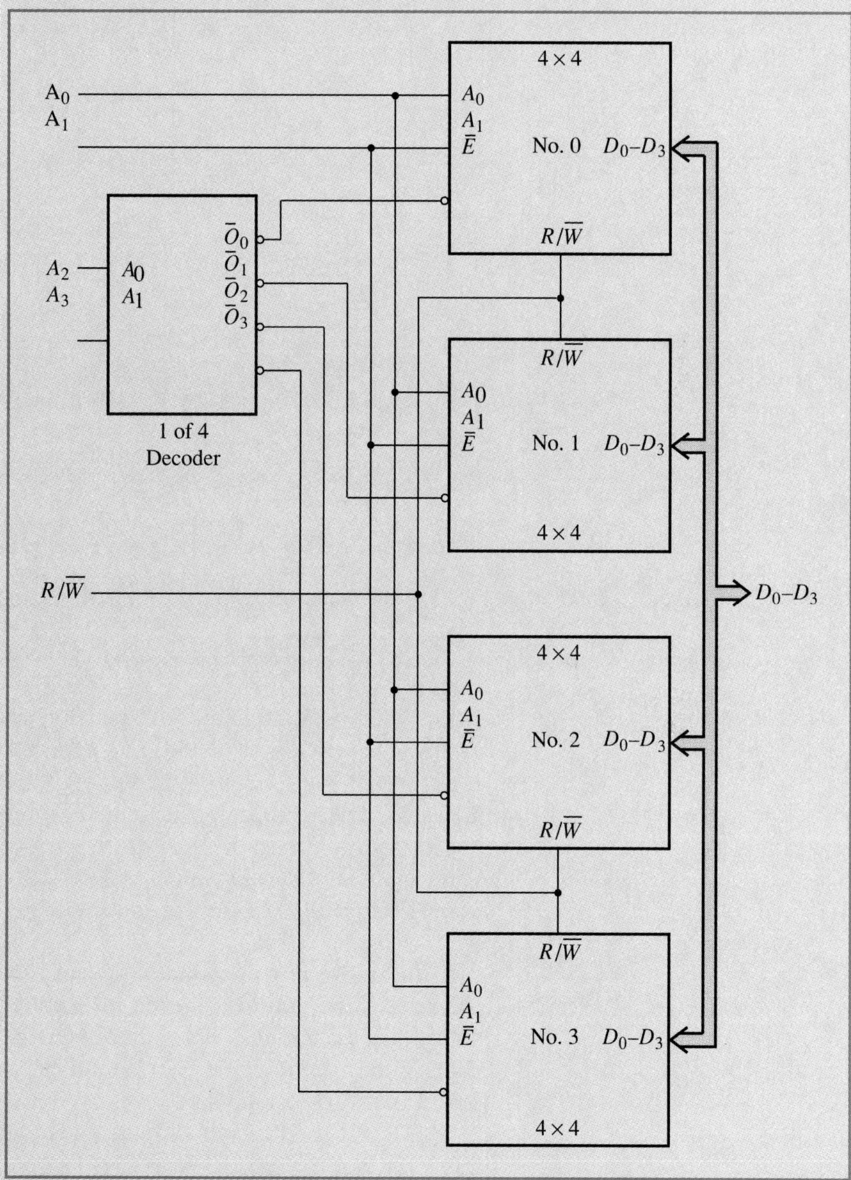

Figure 7–40 16 × 4 register file implemented with four 4 × 4 register files.

The addressing for the register file is shown in Table 7–5. Notice in Figure 7–40 that address lines A_1A_0 select the four registers within each selected register file. Address lines A_3A_2 are

Table 7–5

	A_3	A_2	A_1	A_0	Register
Register file no. 0	0	0	0	0	0
	0	0	0	1	1
	0	0	1	0	2
	0	0	1	1	3
Register file no. 1	0	1	0	0	4
	0	1	0	1	5
	0	1	1	0	6
	0	1	1	1	7
Register file no. 2	1	0	0	0	8
	1	0	0	1	9
	1	0	1	0	*A*
	1	0	1	1	*B*
Register file no. 3	1	1	0	0	*C*
	1	1	0	1	*D*
	1	1	1	0	*E*
	1	1	1	1	*F*

decoded and select one of the four register files as follows:

A_3A_2	Register File
0 0	0
0 1	1
1 0	2
1 1	3

Therefore the registers in register file no. 0 have the addresses 0 through 3, the registers in register file no. 1 have the addresses 4 through 7, the registers in register file no. 2 have the addresses 8 through *B*, and the registers in register file no. 3 have the addresses *C* through *F*. As before, the data lines (D_0–D_3) connect to all four register files, but only the selected (addressed) register in the selected (addressed) register file will be accessed for storage or retrieval of data. The $R/\overline{W}$ line establishes the direction of data flow into or out of the addressed register.

So far we have examined design procedures for vertical expansion. The size of the numbers stored in each register has remained the same, but we have combined several register files to increase the total number of registers available. For horizontal expansion, the number of address lines remains the same since we are not expanding the number of registers, but the number of data lines increases to accommodate larger numbers. For example, Figure 7–41 implements the 4 × 8 register file whose logic symbol is shown in Figure 7–37b by using two 4 × 4 register files.

Notice in Figure 7–41 that the address lines of both register files are connected together so that an address selects a register from *both* register files. However, note that the 8-bit number $D_7D_6D_5D_4D_3D_2D_1D_0$ is split between register file no. 0 ($D_3D_2D_1D_0$) and register file no. 1 ($D_7D_6D_5D_4$). Thus, when a particular address appears on the address lines, half of the 8-bit number is stored in the addressed register in register file no. 0 and

Figure 7–41
4 × 8 register file implemented with two 4 × 4 register files.

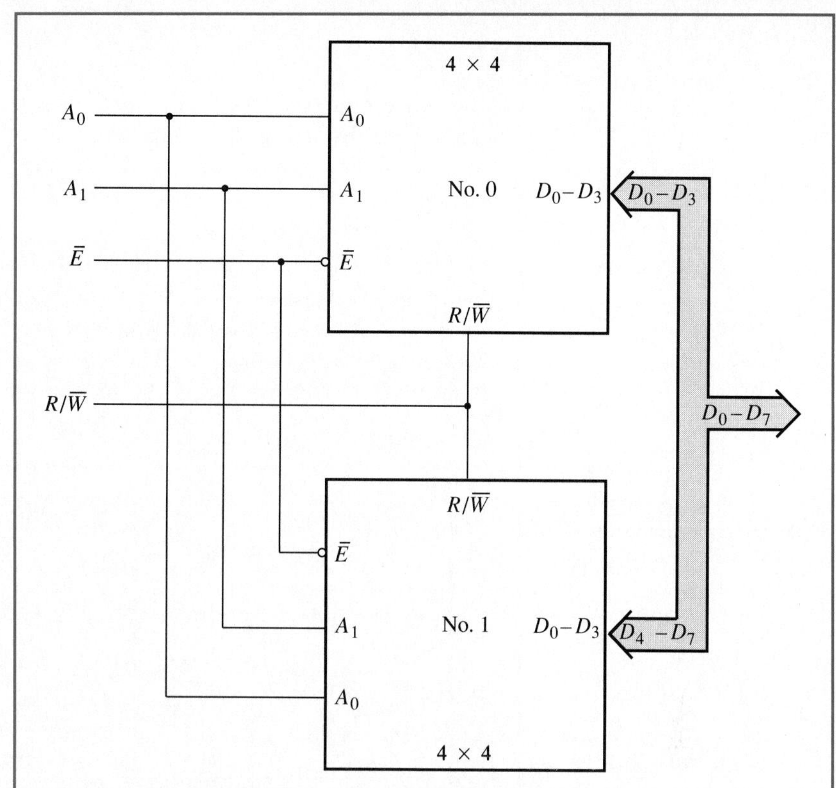

Figure 7–42
4 × 16 register file implemented with two 4 × 8 register files.

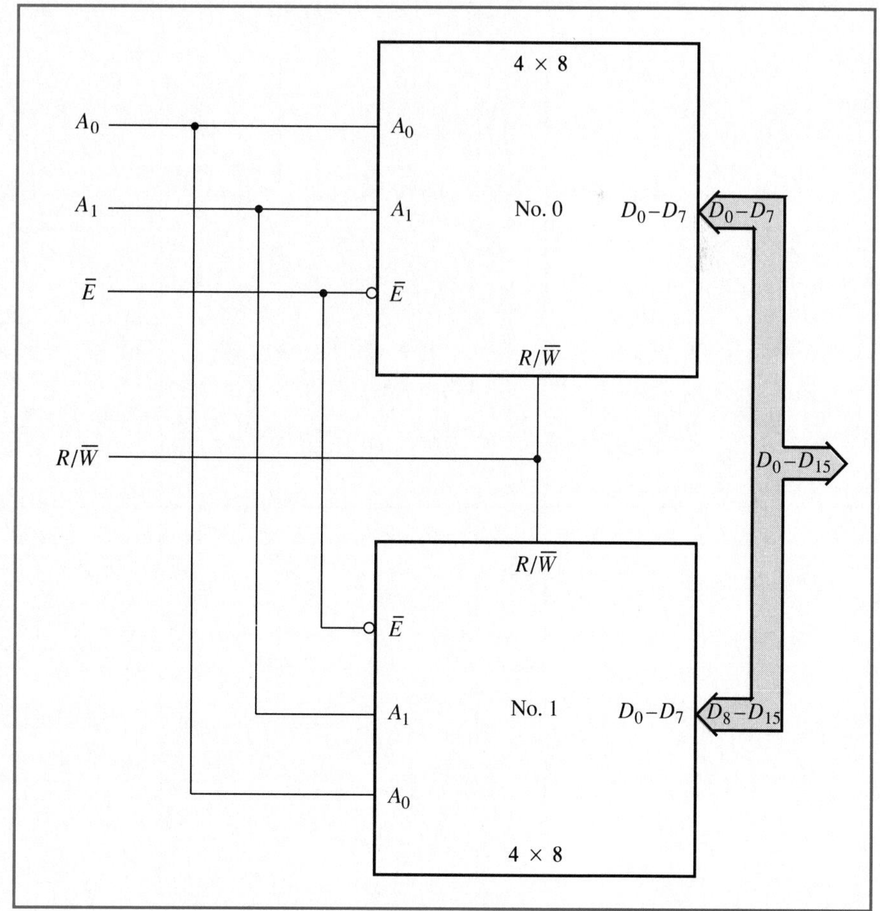

the other half is stored in the addressed register in register file no. 1. Therefore, by using this technique we can design a register file system to store a binary number of any size as illustrated by the following examples.

EXAMPLE 7–8

Figure 7–42 shows a 4 × 16 register file constructed by using two of the 4 × 8 register files shown in Figure 7–37b.

Notice that the 16-bit number to be stored in the register file is split up into two halves. The least significant 8 bits (D_0–D_7) is stored in register file no. 0, and the most significant 8 bits (D_8–D_{15}) is stored

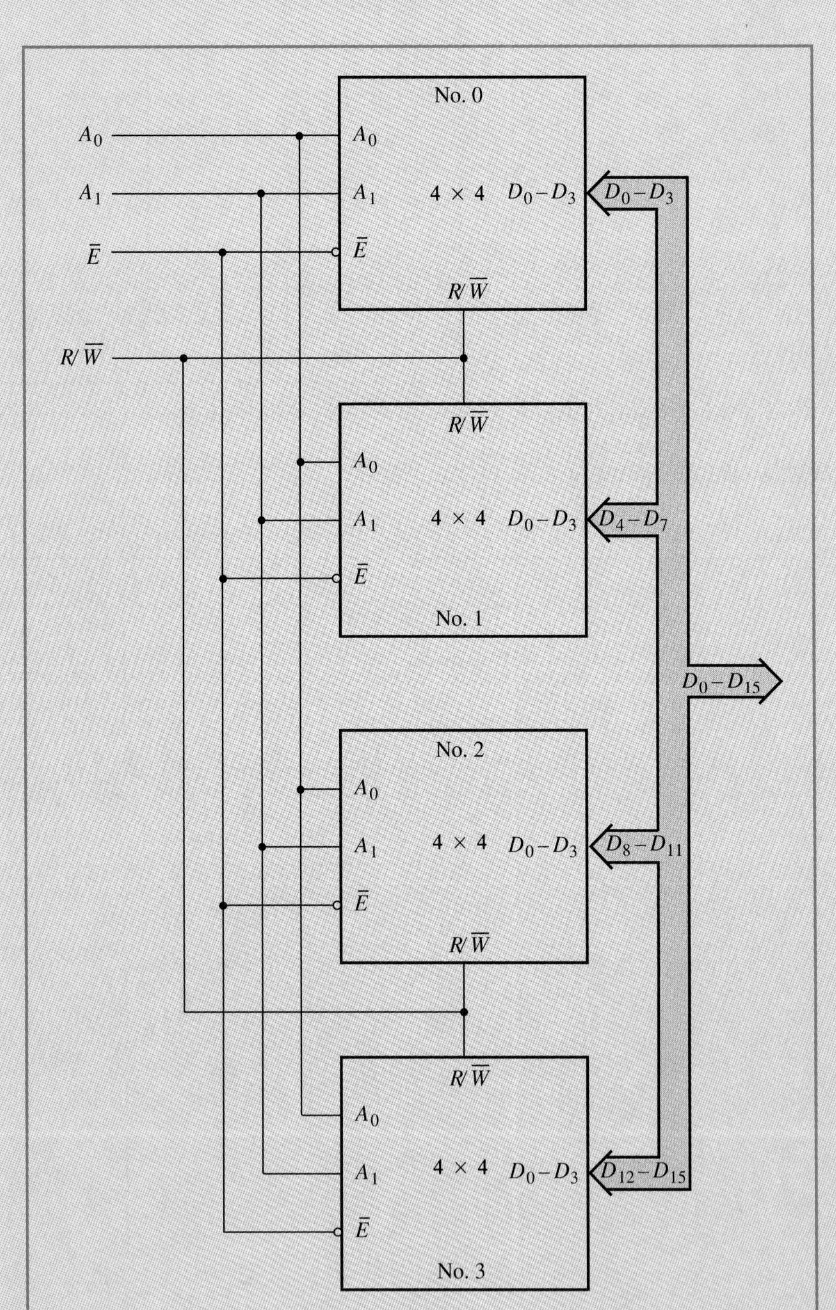

Figure 7–43 4 × 16 register file implemented with four 4 × 4 register files.

in register file no. 1. The same address is applied to both register files so that both halves of the 16-bit number are stored (or retrieved) at the same location in each register file. Since there are only two address lines we have four storage registers.

EXAMPLE 7–9

Figure 7–43 shows a 4 × 16 register file constructed by using four of the 4 × 4 register files shown in Figure 7–36.

In this circuit we are splitting the 16-bit number into four parts, and each part is stored in a 4 × 4 register file. D_0–D_3 is stored in register file no. 0, D_4–D_7 is stored in register file no. 1, D_8–D_{11} is stored in register file no. 2, and D_{12}–D_{15} is stored in register file no. 3. Again, we must address the same locations in all four register files in order to store the 16-bit number, and therefore address lines A_1A_0 select each of the four locations available for storage.

Expansion can also be done both horizontally and vertically as illustrated by the following example.

EXAMPLE 7–10

Figure 7–44 shows an 8 × 8 register file constructed by using four of the 4 × 4 register files shown in Figure 7–36. The addressing for the register file is shown in Table 7–6.

Table 7–6

	A_2	A_1	A_0	Register
Register files nos. 0 and 2	0	0	0	0
	0	0	1	1
	0	1	0	2
	0	1	1	3
Register files nos. 1 and 3	1	0	0	4
	1	0	1	5
	1	1	0	6
	1	1	1	7

The circuit in Figure 7–44 consists of four register files of the same size (4 × 4). Two of the register files are used for vertical expansion while the other two are used for horizontal expansion so that the total configuration becomes 8 × 8. Notice that register files 0 and 2 are enabled when address line A_2 is a logic 0 and register files 1, and 3 are enabled when address line A_2 is a logic 1. Each set contains four registers to give us a total of 8 registers. When registers 0 through 3 are addressed, the 8-bit number to be stored is split up between register files 0 and 2: D_0–D_3 is stored in register file no. 0 and D_4–D_7 is stored in register file no. 2. When registers 4 through 7 are addressed, the 8-bit number to be stored is split up between register files 1 and 3: D_0–D_3 is stored in register file no. 1 and D_4–D_7 is stored in register file no. 3. The $R/\overline{W}$ lines establish the direction of data flow either into or out of an addressed register and are therefore connected to a common $R/\overline{W}$ line for the entire system.

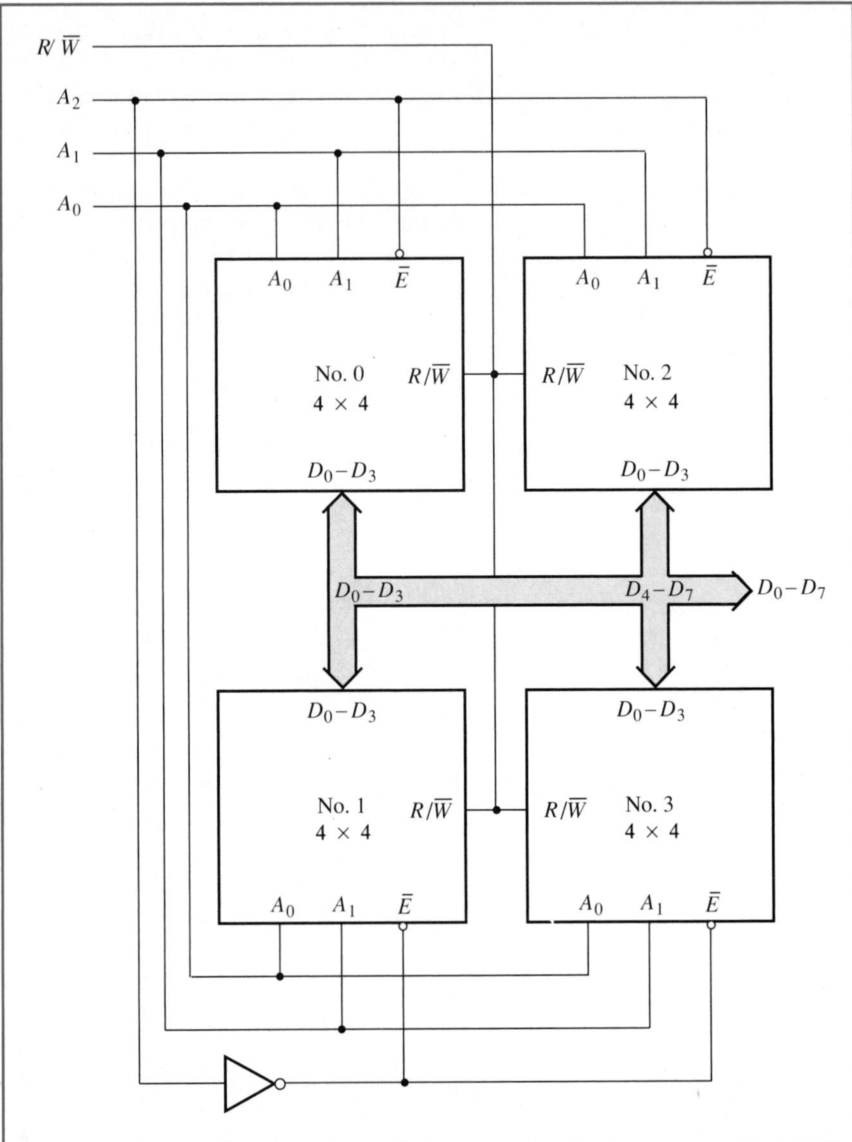

Figure 7–44
8 × 8 register file implemented with four 4 × 4 register files.

The concepts of storage of binary numbers developed in this section are important in the design and analysis of memory systems. From an external point of view, memory systems are identical to register files, and therefore all the concepts of addressing and expansion can be applied to memory design. The only difference between memory circuits and register files are architectural differences in the manner in which each bit is stored and retrieved. Also, memory systems are generally used as the primary storage circuits for digital computers and tend to be much larger in size. Chapter 9 investigates such memory systems and expands on the design techniques introduced here to include the design of larger memory systems used in digital computers.

Review questions

1. What is a register file?
2. How does one specify the configuration of a register file?

3. What is bus contention? How is it prevented?
4. What is the purpose of the address lines?
5. What is an address?
6. How is the size of a register file specified?
7. What is vertical expansion?
8. What is horizontal expansion?

7–6 Integrated circuit logic

There are several ICs that implement the various registers discussed in this section. Some ICs integrate the capabilities of both serial and parallel registers in one package while other ICs provide only the basic capabilities of serial or parallel registers. This section examines a few of the more common ICs available that implement serial and parallel registers as well as a register file.

The 74379 quad parallel register is a simple 4-bit parallel register that is similar in operation to the circuit examined in Figure 7–1. The logic circuit and pin configuration of the 74379 is shown in Figure 7–45. The 4-bit binary number to be stored is applied to the data inputs $D_3D_2D_1D_0$, and a pulse is applied to the *write* input, *CP*. On the positive edge of the clock pulse, the data is stored in the register. The stored data is available at all times at the output of the register $Q_3Q_2Q_1Q_0$ or in complementary form at $\overline{Q}_3\overline{Q}_2\overline{Q}_1\overline{Q}_0$. The 74379 also has an active-low enable line $\overline{E}$, which when at a logic 1 state will disable the register; that is, when $\overline{E}=1$ no data can be loaded into the register, but when $\overline{E}=0$, the register functions normally. The 74379 does not have tristated outputs (like the register in

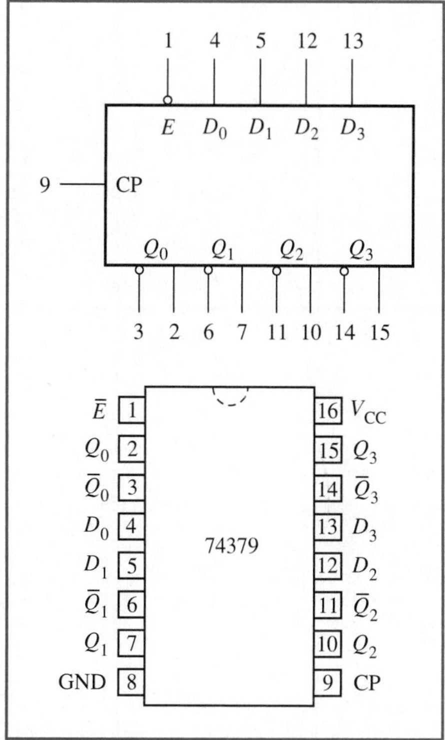

Figure 7–45 Logic symbol and pin configuration of the 74379 quad parallel register.

Figure 7–4), and therefore there is no *read* line necessary to retrieve the stored data.

The 74173 4-bit D-type register is also a parallel register that functions like the circuit shown in Figure 7–4. The logic symbol and pin configuration for the 74173 is shown in Figure 7–46. Like the 74379, the 74173 has a 4-bit data input $D_3D_2D_1D_0$ to load the data into the register when a pulse on the write line, CP, makes a positive transition. However, the outputs of the register $O_3O_2O_1O_0$ are tristated, and therefore to read data stored in the register, the active-low output enable lines $\overline{OE}_1$ *and* $\overline{OE}_2$ must both be at the logic 0 state. Note that if any one of the output enable lines is inactive at the logic 1 state, the output of the register will be (disabled) in a high-impedance state. The 74173 also has an asynchronous active-high master reset input MR which can be used to reset the register ($MR = 1$) independent of the clock input, CP. The active-low input enable lines $\overline{IE}_1$ and $\overline{IE}_2$ when active (at the logic 0 state) allow the data applied to $D_2D_2D_1D_0$ to be stored into the register. However, if any one of the input enable lines is inactive (at the logic 1 state), the data that was previously stored in the register will not be altered when an attempt is made to write new data into the register by activating CP.

Figure 7–47 shows the logic symbol and pin configuration of the 74178 4-bit shift register. The 74178 can function as a parallel-in parallel-out register, a parallel-in serial-out register, a serial-in parallel-out register, and a serial-in serial-out register.

For parallel-in parallel-out operation, the data to be stored into the register is applied to the inputs $P_3P_2P_1P_0$, and the active-high parallel

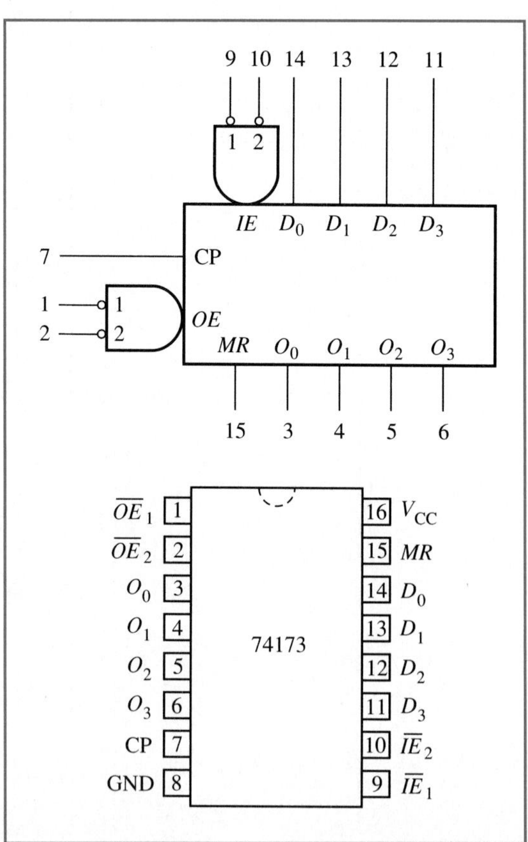

Figure 7–46 Logic symbol and pin configuration of the 74173 4-bit D type register.

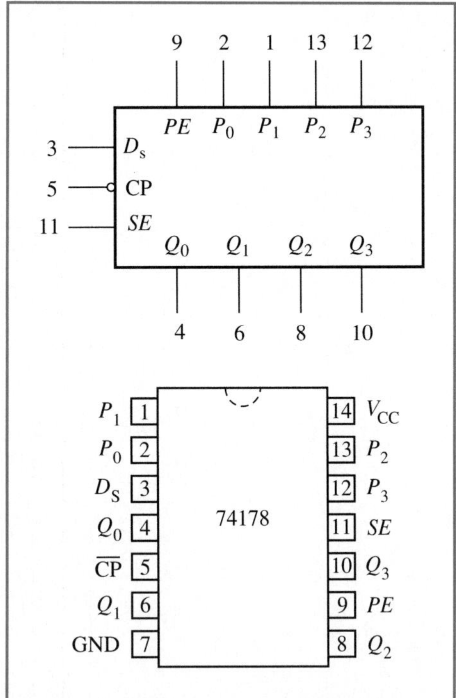

Figure 7–47 Logic symbol and pin configuration of the 74178 4-bit shift register.

enable input *PE* is set to a logic 1. The active-high serial enable input *SE* is set inactive at a logic 0. A clock pulse is applied to the $\overline{CP}$ input, and on the negative edge of the pulse the data at $P_3P_2P_1P_0$ is loaded into the register and is always available at the parallel register outputs $Q_3Q_2Q_1Q_0$.

For serial-in parallel-out operation, the serial data is applied to the serial data input line D_s, *most significant bit first*. The *SE* input is now activated at the logic 1 state, while the *PE* input is made inactive at the logic 0 state. The clock pulses are then applied to $\overline{CP}$, and on the negative transition of each clock pulse each bit is shifted into the register. After 4 clock pulses the entire 4-bit number is loaded into the register and is available at the parallel register outputs $Q_3Q_2Q_1Q_0$.

For serial-in serial-out operation, the serial data is loaded into the register as described for serial-in parallel-out operations. To retrieve the data from the register serially, the Q_3 output functions as a serial output, and on each negative transition of the clock pulse applied to $\overline{CP}$ the data is shifted out, *most significant bit first.*

For parallel-in serial-out operation, the parallel data is loaded into the register as described for parallel-in parallel-out operations. To retrieve the data in the register serially, *SE* is activated while *PE* is deactivated. As before, the Q_3 output functions as a serial output, and on each negative transition of the clock pulse applied to $\overline{CP}$ the data is shifted out, *most significant bit first.*

The 74194 is a 4-bit bidirectional universal shift register that incorporates the functions of all the different types of registers discussed in this chapter. It is capable of parallel-in parallel-out, serial-in parallel-out, parallel-in serial-out, serial-in serial-out, and bidirectional left-right shift functions. The logic symbol and pin configuration of the 74194 is shown in Figure 7–48.

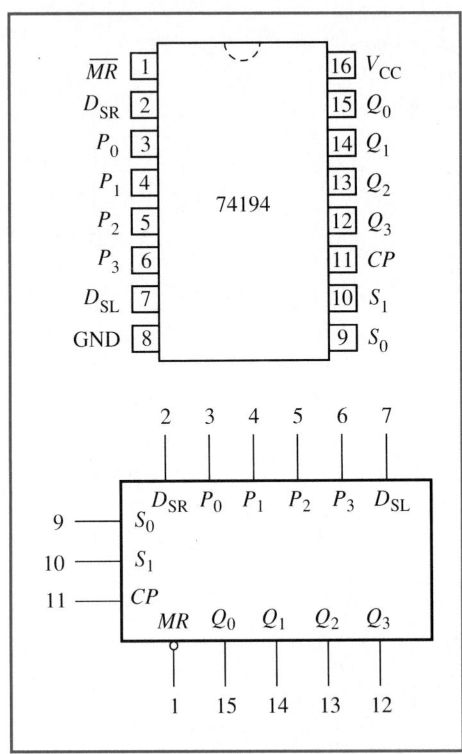

Figure 7–48
Logic symbol and pin configuration of the 74194 4-bit bidirectional universal shift register.

The 74194 has four parallel data inputs, $P_3P_2P_1P_0$, and four parallel data outputs, $Q_3Q_2Q_1Q_0$, that are used to store and retrieve data in parallel format. The Q_0 and Q_3 outputs can also be used to retrieve data serially. The master reset input $\overline{MR}$ is an active-low input that can be used to reset the contents of the register while the CP input is used to shift data in or out of the register on the positive edge of the clock pulse. The D_{SR} and D_{SL} inputs are used to shift data into the register. Since the 74194 is a bidirectional shift register, data is applied to the D_{SR} input when the shift direction is to the right and data is applied to the D_{SL} input when the shift direction is to the left. The two mode-control inputs S_1S_0 configure the 74194 to operate in one of the four following modes:

S_1	S_0	Function
L	L	Hold (do nothing)
L	H	Shift Right
H	L	Shift Left
H	H	Parallel Load

The 74670 is a 4 × 4 register file whose function is similar to the circuit shown in Figure 7–36. The logic symbol and pin configuration of the 74670 is shown in Figure 7–49. The 74670 register file has separate data inputs and data outputs, unlike the circuit of Figure 7–36. The data to be stored is applied to the data input lines $D_4D_3D_2D_1$, and the active-low write enable input $\overline{WE}$ is activated by applying a logic 0 to it. To retrieve data stored in a register, the active-low output enable line $\overline{OE}$ is activated by applying a logic 0 to it. This enables the tristated output lines $O_4O_3O_2O_1$ and makes the data available at these outputs.

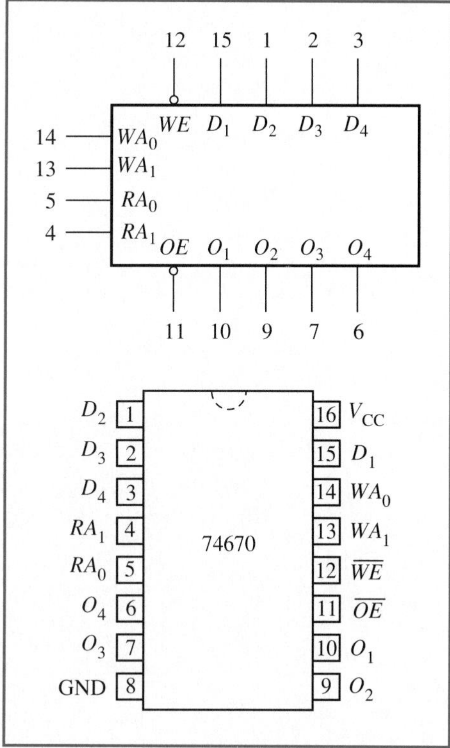

Figure 7–49 Logic symbol and pin configuration of the 74670 4 × 4 register file.

The D and O lines can be connected to produce a set of bidirectional data lines. To address one of four internal 4-bit registers the address must be applied to WA_1WA_0 *or* RA_1RA_0 depending on whether data is being *written* or *read*, respectively. These two sets of lines can be combined as a single set of address lines A_1A_0 to address the register file registers for read *and* write operations.

Review questions

1. Compare and contrast the operation of the 74379 with the circuits shown in Figures 7–2, 7–4, and 7–5.
2. Compare and contrast the operation of the 74173 with the circuits shown in Figures 7–2, 7–4, and 7–5.
3. Compare and contrast the operation of the 74178 with the circuits shown in Figures 7–7, 7–12, and 7–14.
4. Compare and contrast the operation of the 74194 with the circuit shown in Figure 7–18.
5. Compare and contrast the operation of the 74670 with the operation of the circuit shown in Figure 7–35.

7–7 Troubleshooting register circuits

Register circuits exhibit the same electrical faults as combinational logic circuits since their building blocks are flip-flops and flip-flops present the same faults as the basic logic gates. Therefore, register circuits that do not function properly can be tested for common problems such as shorted or open inputs or outputs using the basic test instruments—the logic probe, logic pulser, and current tracer. For example, consider the following troubleshooting case study.

EXAMPLE 7–11

The circuit in Figure 7–50 shows two 74178 ICs connected together to produce an 8-bit Johnson counter. On testing the circuit we find that following a reset the output of the counter produces the following sequences:

00000000
10000000
11000000
11100000
11110000
11110000
11110000
⋮

When a count of 1111000 is reached the counter stays in that state continuously until reset again. A functional counter would produce the following sequence after a count of 11110000:

11110000
11111000
11111100
11111110
11111111
01111111
⋮

From our observations of the counting sequence we can conclude that the fault lies in the connection between IC1 and IC2. The Q_3 output of IC1 appears to be functioning properly since the output of the counter, X_4, produces the correct levels during the valid counts. The fault therefore must be in the D_s input of IC2. D_s is probably internally open-circuited, since if it were shorted to GND or V_{CC}, X_4 would be held at a constant logic level.

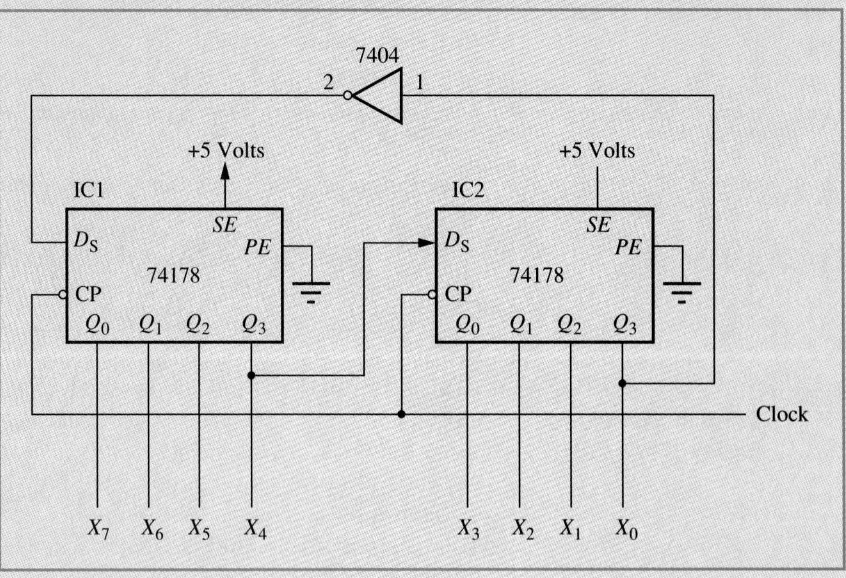

Figure 7–50
Register circuit containing a fault.

Registers, like counters, are also made up of flip-flops and are, therefore, also susceptible to faults at high operating frequencies, due to propagation delays. However, since the flip-flops of shift registers are synchronously clocked, they are not as affected by propagation delays as asynchronous counters. Most of the timing problems that occur in shift register circuits are primarily due to faults in the design of the circuit rather than faults due to propagation delays. For example, consider the timing diagram of the serial data communications system shown in Figure 7–23. Recall that the data was shifted out of the source on the positive edge of the clock and shifted into the destination on the negative edge of the clock. This was done to prevent any timing problems that could exist due to various factors such as delays due to line capacitance and setup times that are too short. Let us assume that the two clock inputs C_A and C_B shifted data in and out on the same edge of the clock pulse. Now if the rise times of the clocks C_A and C_B were different, this could cause the two clocks to be out of synchronization, which could lead to an error in data transmission. Also the setup time of the clock C_B would be too short since the serial data changes at almost the same time as the clock makes its transition. The remedy, of course, is to trigger the data out and the data in at different edges of the clock pulse as shown in Figure 7–23.

We can therefore apply many of the troubleshooting procedures developed in earlier chapters to isolate faults in register circuits. There are more advanced techniques of troubleshooting sequential logic circuits using more complex and expensive test equipment; applications of these techniques and equipment will be seen in Chapters 8 and 9 since they are designed to troubleshoot more complex circuits. For the simple circuits and their applications examined so far, most faults can be isolated and corrected with the basic test instruments such as the logic probe, pulser, and current tracer, and through the use of analytical reasoning.

Review questions

1. In what ways are the faults found in register circuits similar to the faults found in other logic circuits?
2. Why are register circuits not as susceptible to the effects of propagation delay as asynchronous counters?

7–8 SUMMARY

- *Registers* are logic circuits that are capable of storing single binary numbers of varying sizes.
- *Parallel registers* or *parallel-load registers* simultaneously load and retrieve all the bits that make up a binary number.
- *Serial registers* load and/or retrieve a binary number 1 bit at a time.
- The *write line* of a register is the line that is used to store (load), or *write,* data into the register.
- The *read line* of a register is the line that is used to retrieve, or *read,* data from the register.
- A register that stores and retrieves data over a common set of lines is said to have *bidirectional data lines.*
- Parallel registers are known as *parallel-in parallel-out* registers since data is stored and retrieved in parallel.

- Serial registers are often called *shift registers* because the bits that make up a binary number are shifted into or shifted out of the register 1 bit at a time.
- *Serial-in parallel-out* registers load data into the register serially but supply the data out in a parallel format.
- *Serial-in serial-out* registers load data into the register serially and supply the data out of the register serially.
- *Parallel-in serial-out* registers load data into the register in parallel but supply the data out of the register serially.
- *Bidirectional shift registers* can shift data into and out of a register in two directions—left and right.
- Parallel registers are characterized by fast loading and retrieving of data while serial registers are much slower.
- Serial registers require fewer lines to load and/or retrieve data while parallel registers require several lines to accomplish the same task.
- Shift registers are also used in special purpose counters known as *shift register counters*.
- Two of the more common types of shift register counters are the *ring counter* and the *Johnson counter*.
- A *register array* or *register file* is a collection of registers incorporated into a single package.
- The configuration of a register file is specified as $R \times S$ where R is the number of registers and S is the size of each register.
- The total storage capacity of a register file (in bits) is the product of the number of registers (R) and the size of each register (S).
- A register file is similar in concept and operation to a *memory system*.
- *Bus contention* is a conflict of logic levels that occurs when more than one register loads its data on a bus.
- *Address lines* are a set of lines that are used to select a particular register for reading or writing.
- An *address* is a binary number that uniquely identifies a particular register.
- Register files can easily be expanded to store larger numbers or to store more numbers.
- *Vertical expansion* involves increasing the number of registers in a register file.
- *Horizontal expansion* involves increasing the size of each register in a register file.
- Register circuits exhibit the same faults as other sequential and combinational logic circuits.
- Registers are not as susceptible to the effect of propagation delays at high frequencies as asynchronous counters because the flip-flops are synchronously clocked.

PROBLEMS

Section 7–2 Parallel registers

1. Design an 8-bit parallel register with bidirectional data lines and active-high *RD* and *WR* lines.
2. Implement the circuit in Problem 1 using *J-K* flip-flops instead of *D* flip-flops.
3. Design a 4-bit parallel register with bidirectional data lines and a single $R/\overline{W}$ line that will allow data to be loaded into the register when $R/\overline{W}=0$ and data to be retrieved from the register when $R/\overline{W}=1$.
4. Discuss what would happen if the $\overline{RD}$ and $\overline{WR}$ lines of Figure 7–5b were activated at the same time.

Section 7–3 Serial registers

5. Construct a timing diagram for the circuit shown in Figure 7–7b when the binary number 1100 is shifted in. Assume that the initial state of the register is 1010.
6. Construct a timing diagram for the circuit shown in Figure 7–7b when the binary number 0101 is shifted in. Assume that the initial state of the register is 1011.
7. Construct a timing diagram for the circuit shown in Figure 7–12b that shows the number 1010 shifted into the register and then shifted out.
8. Construct a timing diagram for the circuit shown in Figure 7–12b that shows the number 0101 shifted into the register and then shifted out.
9. Construct a timing diagram to *LOAD* the number 1010 into the register shown in Figure 7–14 and then retrieve the number from the register.
10. Construct a timing diagram to *LOAD* the number 0101 into the register shown in Figure 7–14 and then retrieve the number from the register.
11. Assuming that the number 0101 has been shifted into the bidirectional shift register shown in Figure 7–18, draw a timing diagram of the parallel outputs when the $R/\overline{L}$ line is set to a logic 1, and when the $R/\overline{L}$ line is set to a logic 0. Assume that *SLI* and *SRI* are both logic 0's.
12. Assuming that the number 1010 has been shifted into the bidirectional shift register shown in Figure 7–18, draw a timing diagram of the parallel outputs when the $R/\overline{L}$ line is set to a logic 1, and when the $R/\overline{L}$ line is set to a logic 0. Assume that *SLI* and *SRI* are both logic 0's.
13. Design a 4-bit shift register that is capable of loading and retrieving data in all possible formats—parallel-in parallel-out, serial-in serial-out, serial-in parallel-out, and parallel-in serial-out. Draw a logic symbol for this circuit and show the complete logic circuit for the shift register.

Section 7–4 Shift register applications

14. Construct a timing diagram for the circuit shown in Figure 7–22 when the binary number 1100 is transmitted from *A* to *B*.
15. Construct a timing diagram for the circuit shown in Figure 7–22 when the binary number 0101 is transmitted from *A* to *B*.
16. Expand the circuit in Figure 7–22 to transfer 8-bit numbers from *A* to *B*.
17. Construct a timing diagram for the circuit in Problem 16 when the binary number 10110010 is transmitted from *A* to *B*.
18. Modify the circuit in Figure 7–25 to detect the code "1936."
19. Expand the circuit in Figure 7–25 to detect the 5-bit combination "30217."
20. Design a 6-bit ring counter using *J-K* flip-flops.
21. Design a 6-bit Johnson counter using *J-K* flip-flops.
22. Draw a timing diagram for the circuit designed in Problem 20.
23. Draw a timing diagram for the circuit designed in Problem 21.
24. Determine the amount of delay that would occur in the circuit shown in Figure 7–30 if the clock had a frequency of 1 MHz.
25. Determine the amount of delay that would occur in the circuit shown in Figure 7–30 if the clock had a frequency of 5 MHz.
26. Design a time delay circuit that would produce a delay for exactly 15 clock periods. You should only use the registers discussed in this chapter.
27. Design a time delay circuit that would produce a delay for exactly 11 clock periods. You should only use the registers discussed in this chapter.

Section 7–5 Register arrays

28. Design the logic circuit for an 8 × 4 register file that has eight 4-bit registers.
29. Design the logic circuit for a 4 × 8 register file that has four 8-bit registers.
30. Design an expanded 16 × 4 register file system that has sixteen 4-bit registers using only 4 × 4 register files.
31. Design an expanded 16 × 8 register file system that has sixteen 8-bit registers using only 4 × 4 register files.
32. Design an expanded 64 × 8 register file system that has 64 8-bit registers using only 16 × 4 register files.

Section 7–6 Integrated circuit logic

33. Expand the 74178 4-bit shift register to accommodate 8 bits.
34. Design a 16-bit ring counter using 74178 ICs.
35. Design a 16-bit Johnson counter using 74178 ICs.
36. Construct the serial data communications circuit shown in Figure 7–22 using 74194 ICs.

37. Construct an 8-bit bidirectional shift register using two 74194 ICs.

38. Implement the 8 × 4 register file shown in Figure 7–37a using only 74670 ICs.

39. Implement the 4 × 8 register file shown in Figure 7–37b using only 74670 ICs.

Troubleshooting

Section 7–7 Troubleshooting register circuits

40. The 8-bit Johnson counter shown in Figure 7–50 produces the following count sequence when tested:

00000000
10000000
11000000
11100000
11110000
11111000
11111100
11111110
11111111
11111111
11111111
⋮

Discuss the possible fault(s) that could exist in the circuit.

41. The time delay circuit shown in Figure 7–51 uses a 74178 and a 74379 IC to produce a delay of five clock pulses, or 2.5 μs. When the circuit is tested the output pulse appears after only 2.25 μs. Troubleshoot the circuit and determine the possible fault. Suggest a solution for correcting the fault.

42. The ring counter shown in Figure 7–26 is to be operated at the maximum possible frequency that the flip-flops can handle. Determine if the circuit will work properly at that frequency by constructing a timing diagram showing the propagation delays. Use the flip-flop AC characteristics given in Chapter 5.

Figure 7–51

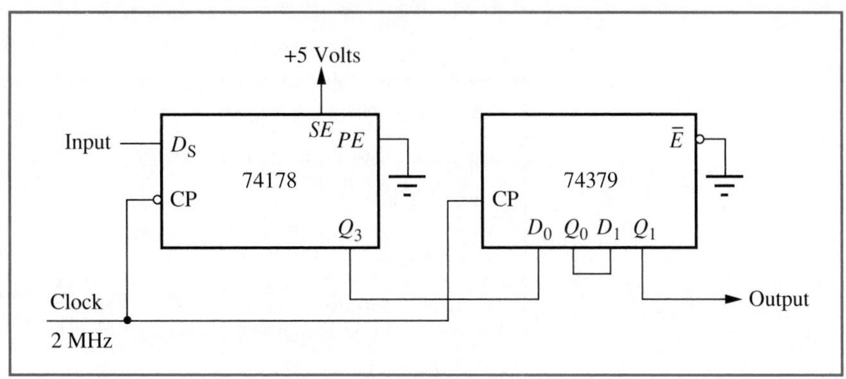

Magnetic tape is often used in large digital computers that have to backup large volumes of information in certain types of nonvolatile memory (Sperry Corporation).

OUTLINE

8–1 Introduction

8–2 Main memory and mass storage memory

8–3 Types of memory organizations

8–4 Memory addressing and access time

8–5 Alteration of memory contents: read/write and ROM devices

8–6 ROM classifications

8–7 Memory volatility

8–8 Memory cell data retention: static and dynamic memories

8–9 Addressing and read/write operation control

8–10 Semiconductor memory architecture

8–11 Introduction to the fundamental concepts of main memory design

8–12 Magnetic tape memory

8–13 Disk memory

8–14 Accessing I/O: isolated I/O and memory-mapped I/O

8–15 Fundamental concepts of troubleshooting

8–16 Summary

OBJECTIVES

The objectives of this chapter are to

- explain the operational concepts of various types of memories.
- introduce the concept of memory organization.
- explore methods of accessing memory locations and I/O devices.
- explain the method of read/write control when accessing a memory location or I/O device.
- study the hazards of operating on a memory location.
- categorize memory devices according to characteristics that are useful when implementing a memory design.
- explain fundamental concepts in the design of a memory system.
- develop procedures to troubleshoot faults in memory or I/O devices.

8 FUNDAMENTAL CONCEPTS OF MEMORY AND INPUT/OUTPUT

8–1 INTRODUCTION

Computers and products that are controlled by computers are instructed to do a task by their human master by means of a program. A *program* is a sequence of instructions that are written by a person (the programmer). The program is ultimately stored in a computer's memory as a sequence of binary numbers known as *machine instructions*. These programs are executed by the computer in an order prescribed by the programmer, and when executed in that order will achieve a desired task. Each instruction within a program directs the computer in a specific operation to be performed, such as adding data, *AND*ing data, and moving data from one place to another. These instructions are most often executed in the same order in which they are stored, which provides the computer with the correct sequence of operations needed to implement the task.

For a program to be executed, it must be stored in the computer system's memory, just as people store a set of instructions in memory when trained to do a task that they wish to remember. It is often necessary to store data as well as programs. The necessity to store programs and data requires that computers and other digital systems have memories.

Any computer memory is limited in its capacity to store programs and data. For this reason, only those programs and data currently being processed are stored in the computer's active memory, which is known as *main memory*. Those programs and data that are to be executed at another time are stored in an inactive back-up memory known as *mass storage*.

Main memory and mass storage memory are the primary memories in a computer system. Other, smaller memories may also be present in a computer or its peripheral support devices, such as a display monitor or printer. They function as temporary storage for program instructions or data in the course of program execution or as buffers between main memory and slower peripheral input or output devices.

Regardless of type, all memories are made up of memory locations that store a fixed number of binary bits. The bits stored at each memory location are the binary representation of either an instruction or data. (These concepts were introduced in earlier chapters.) These memory locations must be accessible to allow storage and retrieval of instructions or data.

In addition to memory a computer must have the means to communicate with external devices. It is through these peripheral devices that people and other machines communicate with a computer. Accessing these peripheral input and output devices (*I/O devices*) is very similar to accessing memory locations. For both, an address is used to identify the memory location or I/O device, and a control signal is used to determine storage or retrieval of instructions or data.

This chapter develops the fundamental concepts necessary to understand the various types of memories found in digital computers, their structures, and their methods of access. This material is essential to understanding the memory design techniques detailed in Chapter 9. Since accessing memory locations is similar in concept to accessing I/O devices, the latter sections of this chapter will apply the concepts of accessing memory locations to accessing I/O devices.

8–2 Main memory and mass storage memory

There are various types of memories and many ways to categorize them. One method has to do with whether a memory is used to store a program and data that are currently being processed or to store programs and data for later use. Memory that is used to store a program and data that are currently being operated on is known as the system's working memory, or *main memory*. *Mass storage memory* is used for programs and data that are to be executed or operated on at another time. This concept of working and mass storage memories is illustrated in Figure 8–1.

Figure 8–1 is an architectural representation of a basic computer system, in which the term "architecture" can be defined as a symbolic representation of a system showing the building blocks (function-blocks) of the system and their relationship to each other. Figure 8–1 will be studied in more detail later; for now we need only become familiar with its essence. The three main function-blocks in the figure are (1) the *central processing unit* (CPU), which is the "intelligence" of the computer system; (2) *memory*, which is the main memory (working memory) that stores both the program and data that are currently being operated on by the CPU; (3) *input/output* (I/O), a generic term for all

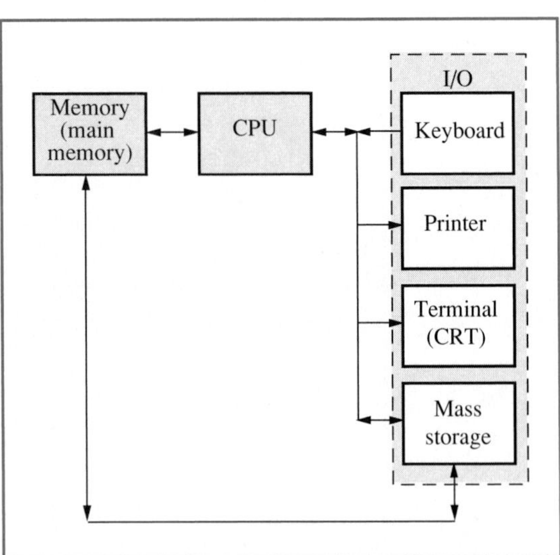

Figure 8–1 A computer architecture.

peripheral input and output devices that serve as communication links between the "outside world" and the CPU. Examples of such devices are the printer, keyboard, monitor, and mass storage devices (e.g., disk drives, tape drives).

We can see in Figure 8–1 that the connecting line between the CPU and main memory is bidirectional, which is necessary because the CPU must both retrieve (read) from and store (write) to this memory. Mass storage also has a bidirectional connection between it and main memory. This is to enable mass storage memory to load directly into main memory a program or data or both to be executed by the CPU, as well as to enable main memory to "free up" some of its space by placing programs and program results (data) that are not currently being executed into mass storage for later reference. Recall that for a program to be executed or for data to be accessible to the CPU, they must be stored in main memory.

It should be noted that main memory is essential to the computer system, since the CPU works from that memory, whereas mass storage memory is convenient, but, from a conceptual point of view, is not essential for basic operation of the system. However, as every user knows, mass storage soon becomes an essential item from a practical viewpoint because the main memory runs out of space as more programs and data are operated on by the CPU.

Why a memory technology may be utilized as either a main memory or mass storage memory has to do in part with its speed of operation and its memory capacity. The CPU operates at relatively fast speeds (microseconds to nanoseconds), and since it gets its instructions and data from main memory, it must operate at similar speeds to keep up with the CPU. To meet or exceed these timing requirements, *as established by the CPU*, main memory is implemented with semiconductor IC memory devices that are faster than the CPU. Unfortunately, semiconductor memory devices have relatively small memory capacity, which makes them unsuitable for mass storage applications. In contrast, for mass storage, storage capacity and portability are more important than speed. For these reasons mass storage devices use high-capacity media such as magnetic tape, floppy disk, hard disk, and magnetic bubble memories (MBM). At present, these mass storage devices have relatively slow operating speeds (milliseconds).

In this book we are interested primarily in main memory, since it is an integral part of a computer or other type of digital circuit that requires memory, rather than a peripheral device such as a mass storage memory system. However, so that the reader will become familiar with mass storage concepts, we will briefly describe mass storage devices and media.

The main memory of Figure 8–1 is composed of one or more semiconductor memory devices. To avoid any prerequisite study of electronics, we will discuss semiconductor memory devices from a systems point of view and defer the study of their internal electronic design to the logic technologies in Chapter 12. Readers who prefer to know the electronic principles before they study semiconductor memory devices may refer to Sections 12–1 to 12–3 of Chapter 12 for the fundamental concepts of electronic switches and logic gates. Section 8–3 should be studied before continuing with the remaining sections of Chapter 12.

Review questions

1. What are the three essential function-blocks of a computer?
2. Relative to the CPU, what is the difference between main memory and mass storage memory?
3. Why is there a bidirectional link between the main memory of Figure 8–1 and mass storage?
4. Why is main memory considered an essential part of a computer system whereas mass storage is not?
5. Why are main memories implemented with semiconductor devices?

8–3 Types of memory organizations

There are basically five categories that describe one or more characteristics or abilities of a memory. These five categories are (1) the organization of the memory and memory media, (2) the manner of accessing memory locations within the memory, (3) the ability to alter the contents of memory, (4) the ability of memory to retain its contents (volatility) when power is removed, and (5) the ability of memory to retain its contents even with power applied to the memory device, without continual refreshing of the contents. These characteristics are used by the system designer to determine whether or not a specific memory device can be used to implement a memory design. In this section we will investigate the organization of memories and memory media.

Memory internal organization

A memory is made up of many memory locations, and each location is made up of a fixed number of memory cells similar to register arrays examined in Chapter 7. Each cell can store a binary bit (1 or 0). The number of memory cells and their arrangement are collectively known as the *organization* of that memory. Some memories are organized such that they can store only 1 bit at each memory location, whereas others can store as many as 8 bits at each location; the size of the location depends on the number of memory cells per location. The more common memory cell organizations for semiconductor memories are 1, 4, or 8 bits per location. For example, a memory organized such that it has 256 locations and each location contains four memory cells is known as a 256 × 4 memory (256 by 4 bits), or a 1024-bit memory. Note that the × is treated as a dimensional indicator (by) rather than an arithmetic operator (multiply), even though the product of 256 and 4 is 1024, which is the total number of bits for that memory. If each of the 256 memory locations contained 8 memory cells, it would be a 256 × 8 (2048-bit) memory.

Figure 8–2 illustrates two simplified and condensed (only four locations) memory organizations, representing a 4 × 1-bit and a 4 × 4-bit memory. Each row of memory cells represents a memory location, and the number of memory cells per memory location represents the number of storage bits per location. Other examples of this type of memory organization are 1024 × 4 (4096 bits organized such that there are 1024 locations and 4 bits of storage per location), 2048 × 8 (16,384 bits), or 2048 × 1 (2048 bits).

Expanding on the concepts of memory organization, the various organizations of memory cells can be placed into one of four categories

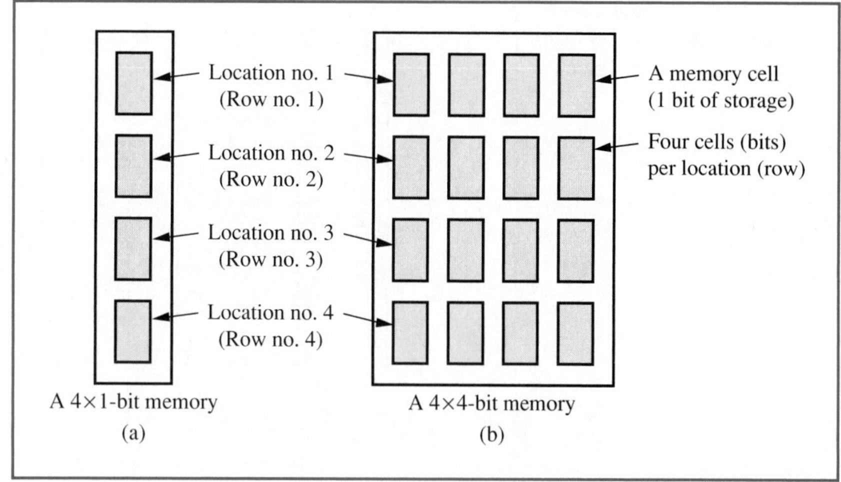

Figure 8–2
Examples of memory organization.

based on the dimensional characteristics of the memory. The four categories are:

1. One-dimensional array
2. Two-dimensional array
3. Three-dimensional array
4. Dynamic one-dimensional array

One-dimensional memory arrays

One-dimensional memory arrays are those memory cell organizations in which each memory location is represented as a row, each row having a fixed number of memory cells, such as the organizations illustrated in Figure 8–2.

Two-dimensional memory arrays

Two-dimensional memory arrays represent memory cell organization as a two-dimensional x-y coordinate system, such as those illustrated in Figure 8–3. Figure 8–3a shows 8 memory cells arranged in an 8 × 1-bit array (8 locations with 1 bit of storage per location), whereas Figure 8–3b shows a 16 × 1-bit organization. For the convenience of memory cell identification, each cell in the arrays of Figure 8–3 has been labeled with an identifying hexadecimal number: 0 through 7 for Figure 8–3a and 0 through F for Figure 8–3b. The location of any memory cell within either array of Figure 8–3 can be specified by stating the row (x coordinate) and the column (y coordinate) within which it appears. That is, memory cell 3 of Figure 8–3a is in row 01 and column 1. Because we can specify the location of a memory cell with a two-dimensional coordinate system (x and y), these types of arrays are called two-dimensional arrays.

Observe in Figure 8–3 that the rows and columns of the arrays are identified with binary numbers; these are specified in the index tables at the bottom and to the left of the array and are labeled as *Row Number* (x) and *Column Number* (y). The reason for using binary values to specify a memory location will be explained later. Note that the binary value created by combining the row and column binary numbers within which the cell is located, where the LSBs of that combination are formed by the column and the MSB positions are formed by the row, is numerically equivalent to the hexadecimal number used to identify the cell. For

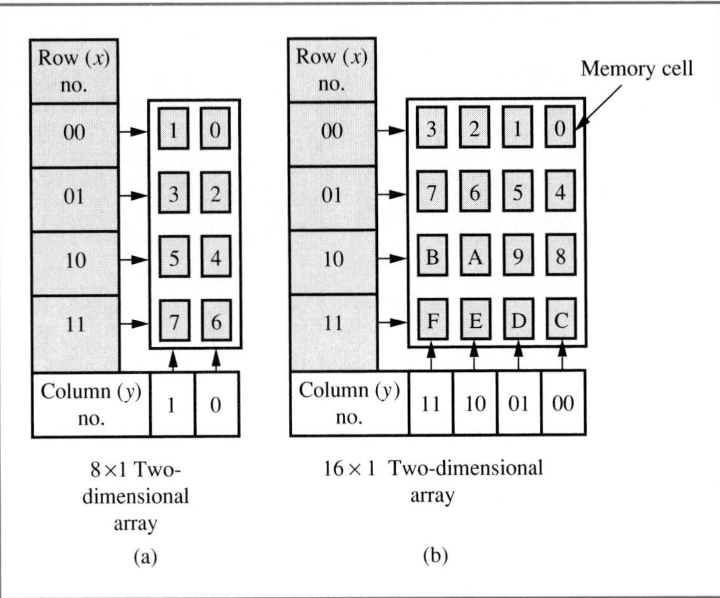

Figure 8–3 Two-dimensional memory cell organization.

example, using this *x-y* coordinate technique, cell 6 resides at location 110 (row 11 and column 0) in Figure 8–3a and at location 0110 (row 01 and column 10) in Figure 8–3b.

EXAMPLE 8–1

From Figure 8–3a and b, develop a table that specifies the location of each memory cell in binary, using the *x-y* coordinate system.

SOLUTION We know from the example above that the location is specified by combining the row and column binary values for each cell of Figure 8–3 with the *y*-coordinate (column) forming the LSBs and the *x*-coordinate (row) forming the MSBs. The resulting binary values are listed in the *Location* column of Table 8–1. The column of Table 8–1 labeled *Cell* contains the hexadecimal equivalent of the binary number in the *Location* column. That is, for Figure 8–3a, row 00 and column 0 locate cell 0, and when row and column binary numbers are combined the location number 000 is formed. When row 00 and column 00 of Figure 8–3b are combined, location number 0000 is formed, which is the binary location of memory cell 0. Likewise, 001 is the location of cell 1 in Figure 8–3a and 0001 is the location of cell 1 in Figure 8–3b.

Observe that Table 8–1 is basically a truth table in which the column labeled *Location* has all possible binary combinations present, 8 combinations for Figure 8–3a and 16 for Figure 8–3b, and (as previously stated) the cell identification number given in the column labeled *Cell* is the hexadecimal equivalent of the location number.

It is important that the reader understand the *x-y* coordinate system of identifying the location of a cell.

Three-dimensional memory arrays

Figure 8–4 illustrates the concept of a three-dimensional memory cell array. As before, we are using a small number of cells (64) in order to minimize the complexity of the illustration. As can be seen in Figure

Table 8–1
Memory location and memory cell per location for Figure 8–3a and b

Row	Column	Location	Cell
Figure 8–3a			
00	0	000	0
00	1	001	1
01	0	010	2
01	1	011	3
10	0	100	4
10	1	101	5
11	0	110	6
11	1	111	7
Figure 8–3b			
00	00	0000	0
00	01	0001	1
00	10	0010	2
00	11	0011	3
01	00	0100	4
01	01	0101	5
01	10	0110	6
01	11	0111	7
10	00	1000	8
10	01	1001	9
10	10	1010	A
10	11	1011	B
11	00	1100	C
11	01	1101	D
11	10	1110	E
11	11	1111	F

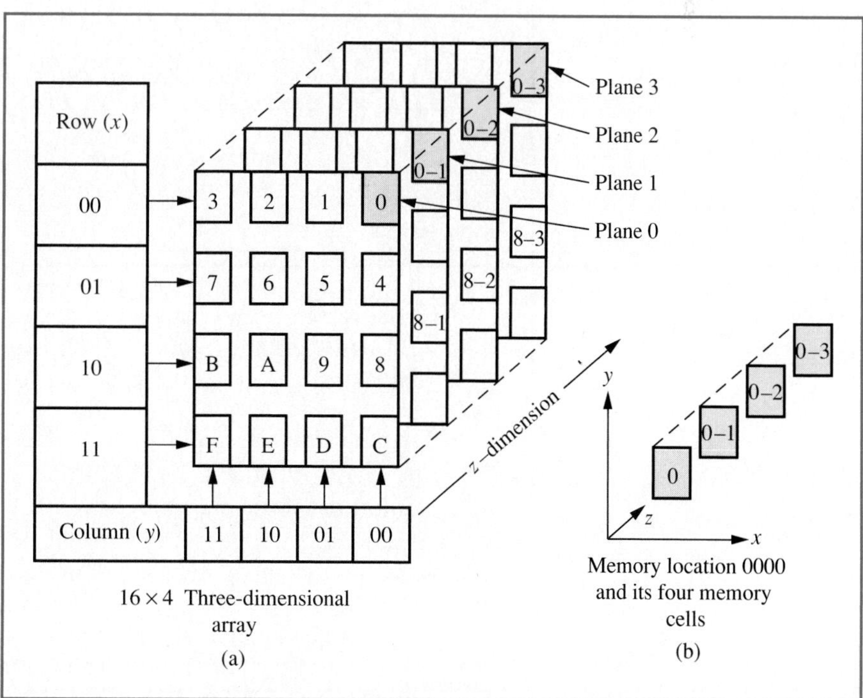

Figure 8–4
Three-dimensional memory cell organization.

8–4a, this organization has duplicate two-dimensional arrays that are "stacked" in a third dimension, which is called the *z*-coordinate. These stacked two-dimensional arrays are identified as planes (plane 0, plane 1, etc.), and the number of planes determines the number of memory cells for each memory location. That is, the bit size of each memory location for the memory of Figure 8–4a is four, since there are four planes of memory.

Memory locations of a three-dimensional memory are identified with the same *x-y* coordinate scheme as a two-dimensional array except that rather than identifying just one memory cell in plane 0, identically positioned memory cells in the other planes are also identified. For example, memory location 0000 of Figure 8–4a contains memory cells 0 of plane 0, 0-1 of plane 1, 0-2 of plane 2, and 0-3 of plane 3, as illustrated in Figure 8–4b. Likewise, memory location 1000 consists of memory cells 8, 8-1, 8-2, and 8-3. It can be concluded that *the x-y coordinates determine the number of memory locations*, which is 16 (0 through F) for the array of Figure 8–4a, and *the number of planes determines the storage bit size of each location.* Thus the memory of Figure 8–4 is a 16 × 4 memory. If one wished to devise a memory similar to that of Figure 8–4a but with eight memory cells per location (16 × 8 bit memory), then four more planes of memory cells would have to be added to the memory of Figure 8–4a.

In reality, when a three-dimensional array is manufactured using semiconductor technology, which uses a two-dimensional IC wafer, each plane of memory cells is integrated on the same IC wafer. The concept of a three-dimensional array, as illustrated in Figure 8–4a, is just a convenient way to visualize how multiple cells can reside at the same memory location.

EXAMPLE 8–2

In regard to the three-dimensional array of Figure 8–4(a)

1. What memory cells are at memory locations
   (a) 0111 (b) 1001 (c) 1111?
2. At what memory location are memory cells
   (a) 2, 2-1, 2-2, and 2-3 (b) A, A-1, A-2, and A-3?

SOLUTION The solution for part (1) is to decode the two LSBs of the memory location in order to obtain the column and the two MSBs to identify the row.

(a) Memory location 0111 selects column 11 and row 01, which, when cross-referenced using an *x-y* coordinate system, selects memory cells 7 of plane 0, 7-1 of plane 1, 7-2 of plane 2, and 7-3 of plane 3.
(b) 1001 selects column 01 and row 10, which selects memory cells 9, 9-1, 9-2, and 9-3.
(c) 1111 selects column 11 and row 11, which selects memory cells F, F-1, F-2, and F-3.

The solution for part (2) is the reverse of the solution for part (1)—the creation of a binary number that is equivalent to the hexadecimal memory location number.

(a) Memory cells 2, 2-1, 2-2, and 2-3 reside at memory location 2H(hexadecimal). The binary equivalent of 2H is 0010_2; hence 0010 is the binary location number for memory cells 2, 2-1, 2-2, and 2-3.

(b) Memory cells A, A-1, A-2, and A-3 reside at memory location AH. Since AH = 1010_2, 1010 is the correct memory location number.

A memory cell array can be configured in many different combinations, so long as both the number of rows and the number of columns is a power of two (2, 4, 8, 16, etc.). The eight memory cell array of Figure 8–3a could have been configured as two rows and four columns. With this organization the row numbers (x) would be 0 and 1 and the column numbers (y) would be 00, 01, 10, and 11 for determining the cell location number. The 16 memory cells of Figure 8–3b could have been organized with two rows and eight columns or with eight rows and two columns. For the former organization the row numbers (x) would be 0 and 1 and the column numbers (y) would be 000, 001, 010, 011, 100, 101, . . . 111 for cell location identification. Take note of binary patterns of the row and column numbers.

Dynamic one-dimensional memory arrays

Dynamic one-dimensional memory cell arrays are one-dimensional arrays in which the array is mobile (dynamic). Either these memory cells are on a movable memory medium, such as a magnetic tape, or the array itself is movable within the medium, such as is the case for a magnetic bubble memory (MBM).

Figure 8–5 illustrates dynamic one-dimensional memory cell arrays, which can be found in both semiconductor and magnetic memories. For convenience we shall use magnetic memories as the vehicle of explanation. Figure 8–5a illustrates a magnetic tape and Figure 8–5b a simplified MBM.

The read/write mechanism of Figure 8–5a is used to either read (retrieve) or write (store) binary bits from or to memory locations on the memory medium (magnetic tape). The read/write mechanism is stationary; therefore, the memory medium must move to align the memory location being accessed with the read/write mechanism. For example, if memory location N of Figure 8–5a is being accessed and then it is desired to access memory location $N + 3$, the tape must be moved so that location $N + 3$ aligns with the read/write mechanism. This dynamic movement to access a memory location accounts for the name of this type of memory.

As will be seen, mass storage devices using magnetic disks are similar in concept to those using magnetic tape. The major difference is that the memory medium is a disk that rotates to align memory locations with the read/write mechanism.

Figure 8–5b is another example of a dynamic one-dimensional memory array, which in concept is representative of an MBM. In this type of memory, the memory cells travel within a stationary magnetic medium in order to align with the read/write mechanism. The memory cells of an MBM are polarized magnetic domains that appear as bubbles when viewed under polarized light. They are polarized in one direction for a logic 1 and in the opposite direction for a logic 0. The memory cells are propagated through the magnetic medium in a loop so as to align them with the read/write mechanism. As a magnetic bubble is moved beneath the logic state detector, its polarization is detected and the logic voltage level corresponding to that polarization is output. To write a logic 0 into a memory cell that previously contained a logic 1 requires that the

Figure 8–5
A dynamic one-dimensional memory.

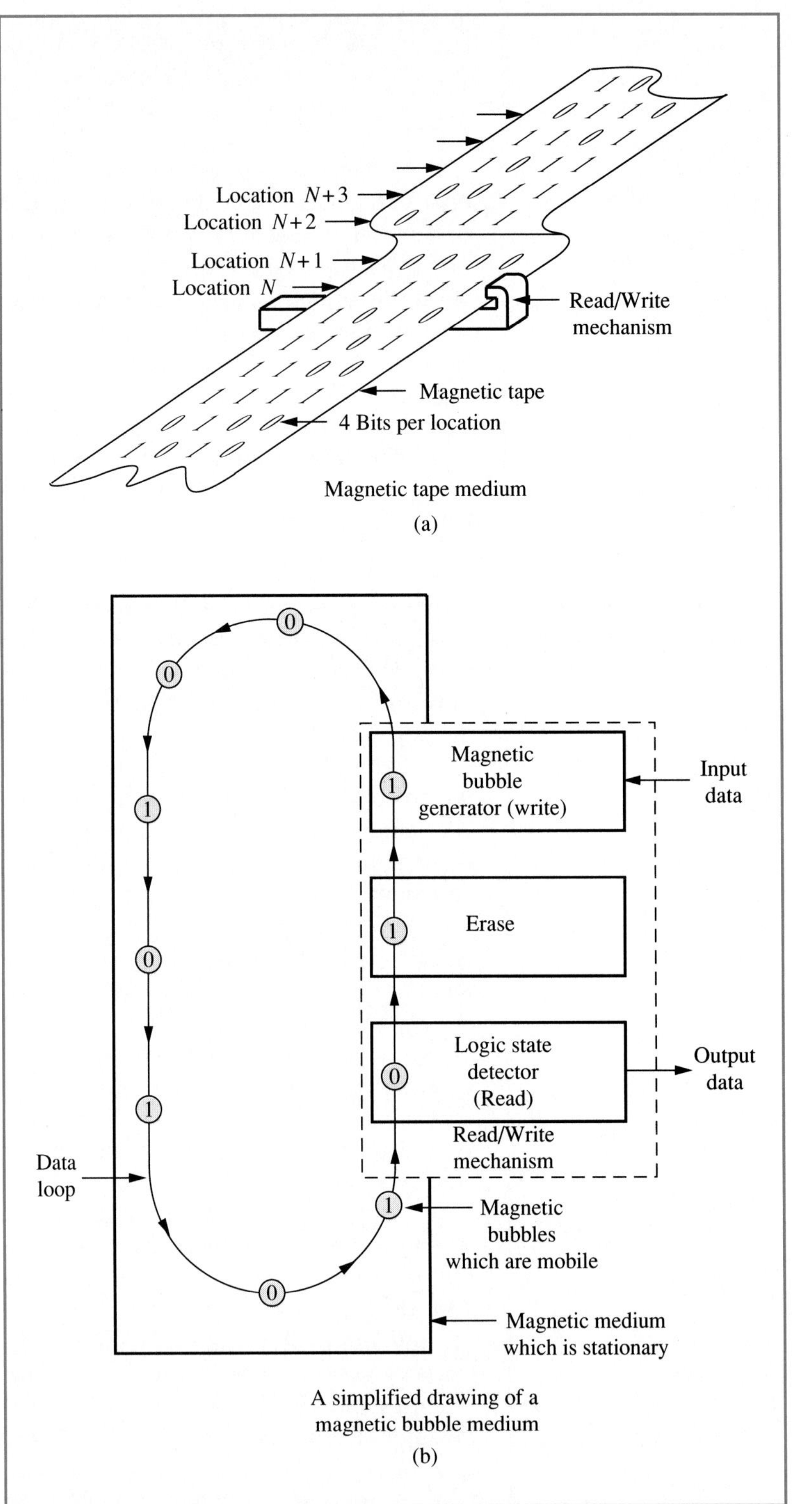

polarization of the logic 1 first be erased by the erase mechanism. Each magnetic bubble is erased after it is read, which frees that memory cell to have a new data bit written into it by the magnetic bubble generator. As the reader can see, the data of an MBM memory location is moved (i.e., dynamic) through the magnetic medium to be read. Also, a memory cell must be moved to the magnetic bubble generator in order to have data written into it.

Review questions

1. What is a memory cell?
2. How are the memory cells of an IC semiconductor memory organized?
3. How are the number of memory locations and the bit size of each location within a memory device specified?
4. What are the four types of dimensional arrays and how do they differ?
5. What is the essence of an *x-y* coordinate addressing scheme?
6. How is the third dimension formed in a three-dimensional array?
7. What characteristic is associated with dynamic one-dimensional arrays?
8. How do the memory cells of IC semiconductors, magnetic tapes, and magnetic bubbles store logic 1's and 0's?

8–4 Memory addressing and access time

Each memory location must be capable of being accessed so that binary bits representing either instructions or data may be stored in or retrieved from those locations. We can imagine each memory location as analogous to a person's residence; like a residence, each memory location has a unique *address*. Accessing a memory location is done by supplying the address of the location to the memory device containing that location, either within it or on its memory medium. The memory device possesses the necessary digital logic to find that location once it is given the address.

We have already developed the basic fundamental concepts for memory addresses, which is the *x-y* coordinate scheme for two-dimensional arrays and row identification for dynamic one-dimensional arrays. As an example, memory location addresses for the memory cell array of Figure 8–3 are given in Table 8–1 in the column labeled *Location*. They are identified by the row number in Figure 8–5a.

The parameter that indicates the speed with which a location within a memory device can be accessed is known as *access time*. Access time is defined as the time between application of a memory location address to a memory device (semiconductor memory, magnetic tape, bubble memory, floppy disk, etc.) and the accessing of the contents of that memory location.

There are conceptually four techniques for accessing memory locations. These techniques are known as *linear accessing*, *random accessing*, *first-in first-out* (FIFO), and *last-in first-out* (LIFO).

Linear accessing of memory

The linear accessing technique is used to access dynamic one-dimensional cell arrays. The access time for linear accessing varies with how far the medium or memory cells must travel. That is, if memory location 100 is being accessed and location 101 is to be accessed next, the access time is much shorter than if location 500 were the next location to be accessed.

EXAMPLE 8–3

Suppose that the illustration of Figure 8–5a represents a magnetic tape mass storage device with a speed of travel in either direction of 2 ms per memory location. What is the access time if (1) Memory location 0025 is being accessed and 0026 is to be accessed next? (2) Memory location 0026 is being accessed and 0500 is to be accessed next? (3) Memory location 0550 is being accessed and 0027 is to be accessed next?

SOLUTION

1. Since this requires that the tape be moved just one memory location, the access time is 2 ms.
2. This requires that the tape travel through 474 memory locations, which requires an access time of 948 ms.
3. This requires the tape to reverse its direction of travel, which could lengthen the access time, and then travel through 523 locations. At 2 ms per location, this is an access time of 1046 ms (plus the reversal time).

Random accessing of memory

The *random access technique* is one that accesses stationary memory locations on stationary media using the *x-y* coordinate system, such as the two- and three-dimensional arrays of Figures 8–3 and 8–4. The access time for this technique is constant; that is, any memory location within the array has the same access time regardless of the previous location being accessed. Access time for this technique is not position-dependent, as it is for linear accessing. To understand why this is so, recall that to locate a memory cell in a two-dimensional array or cells in a three-dimensional array, the row and column of the memory location are identified by the memory location address. The memory device must decode that address and then select the addressed row and column. The same amount of time is required to decode any row address and any column address. As a result, the access time of any combination of row and column is the same, regardless of the location. Because the access time of any memory location chosen at random is the same, a memory device with this characteristic is known as a *random access memory* (RAM). All modern main memories are composed of semiconductor RAM memory devices. As we shall see, the term "RAM" is often misused.

EXAMPLE 8–4

If 400 ns is required to select a column and row for a given memory location of Figures 8–3 and 8–4, what is the access time of those memory devices?

SOLUTION Since a memory location address simultaneously selects column and row for either array, then 400 ns is the access time for any memory location in those arrays. Hence, the access time of either memory device is 400 ns.

FIFO and LIFO accessed memories

FIFO and LIFO are acronyms for the method of accessing a certain type of memory. The data within a FIFO is accessed in a *first-in first-out* manner, and the data stored within a LIFO is accessed in a *last-in first-out* manner. That is, the first datum written into a FIFO memory is the first datum that can be read from it, whereas the last datum written into a LIFO is the first that can be read from it.

Conceptually there are similarities in the accessing techniques of FIFO and LIFO memories.

FIFO accessed memories

A 4 × 4 FIFO accessed memory is illustrated in Figure 8–6a. From Figure 8–6a we see that the FIFO is a dynamic one-dimensional array (i.e., mobile) that has been implemented with shift registers such as those studied in Chapter 7. Memory cells are the flip-flop of the shift registers. The arrows indicate the direction of the data shift between registers (cells). For simplicity, the clock and control logic are not shown. We note from Figure 8–6a that there is access to the FIFO only at those registers that make up location nos. 1 and 4. Location no. 1, known as the top of the FIFO, is used to input the datum, and location no. 4, known as the bottom of the FIFO, is used for outputting the datum. The rectangles identified as input data and output data represent a datum being written into the FIFO and a datum being read from it. The implementation of these rectangles is most often accomplished with registers.

Some FIFOs are designed such that a datum written into an empty FIFO (no data has been previously stored) via the top of the FIFO memory location is automatically shifted by control circuitry (not shown in Figure 8–6a) to the bottom of the FIFO, as indicated in Figure 8–6b. Figure 8–6b indicates that the datum written into the FIFO (1101) was written into the top of the FIFO and then automatically shifted to the bottom of the FIFO. This action fills the bottom location and makes that datum available to be read at the output (location no. 4); hence the datum is first-in first-out. The next datum written into the FIFO (0011) is automatically shifted from the top of the FIFO to the next available memory location (an empty one), which is the location preceding the bottom location, that is, location no. 3, as depicted in Figure 8–6c. As data is written into the FIFO this process continues until all locations of the FIFO are filled, as illustrated in Figure 8–6d if 1010 and 1110 were written into the FIFO.

Another method of filling a FIFO is to shift the entered datum just one location at a time. The datum written into this type of FIFO memory will push (shift) all subsequent data toward the bottom by one location. For a FIFO with four locations this would require that the four data entries be made before the first entry is available to be read at the bottom of the FIFO memory. For example, if the FIFO of Figure 8–6d had been filled using this technique, then the first datum written into the FIFO (1101) would have been written into the top of the FIFO but not shifted to the bottom of the FIFO. Hence, location no. 1 has 1101 written into it and the remaining locations are empty. Next, 0011 is written into the top of the FIFO (location no. 1), which shifts 1101 from location no. 1 to location no. 2. Then location no. 1 contains 0011, location no. 2 contains 1101, and location nos. 3 and 4 are empty. The next datum, 1010, is written into the top of the FIFO (location no. 1), which shifts the two previous data down by one location. As a result, location no. 1 contains 1010 (the new datum written into the FIFO), location no. 2 contains 0011 (shifted from location no. 1), location no. 3 contains 1101 (shifted from location no. 2), and location no. 4 is empty. Finally, 1110 is written into the top of the FIFO and all previous data is shifted down by one location, which results in the FIFO of Figure 8–6d.

Figure 8–6
Accessing a FIFO memory.

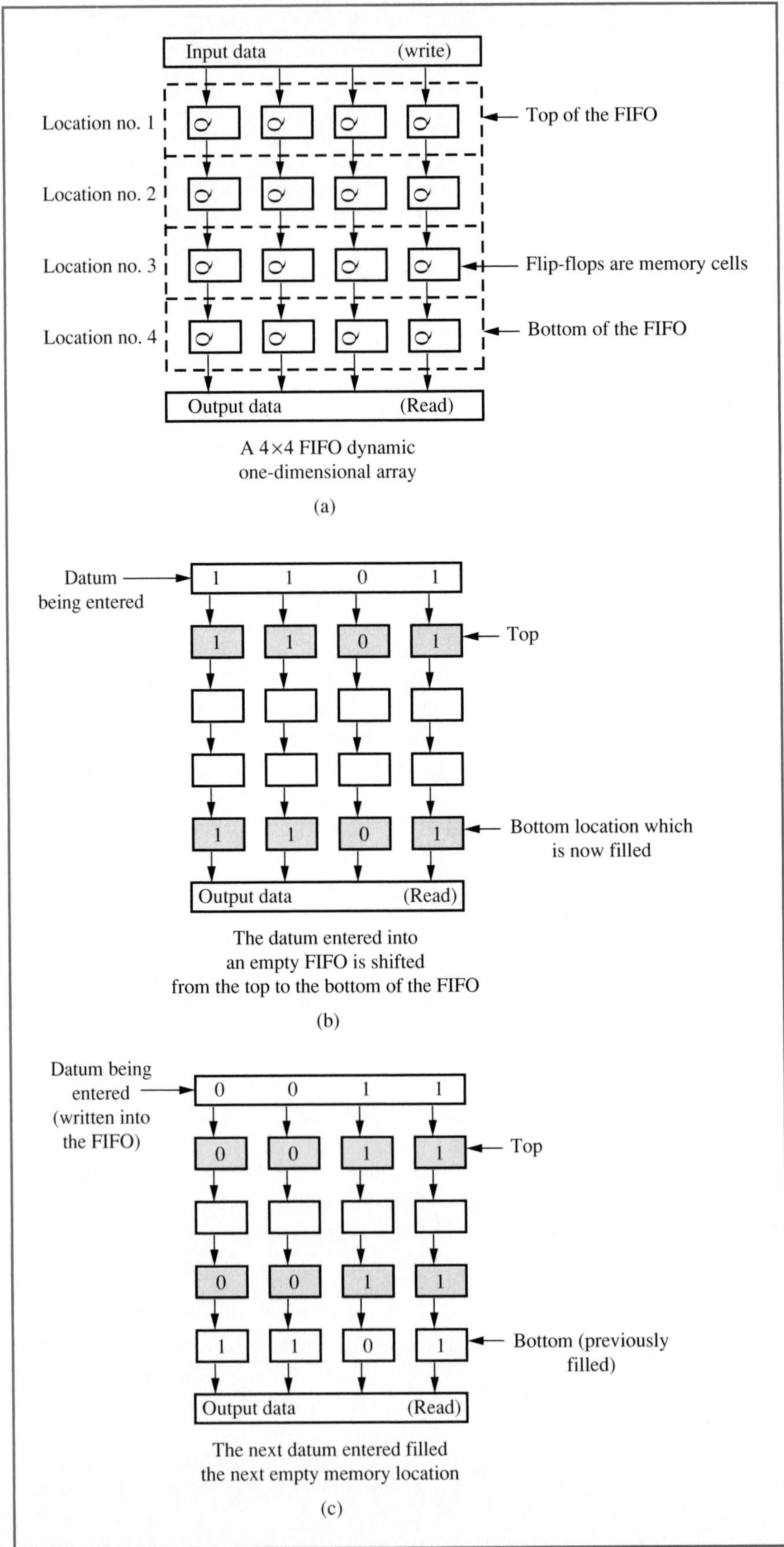

Figure 8–6
(continued)

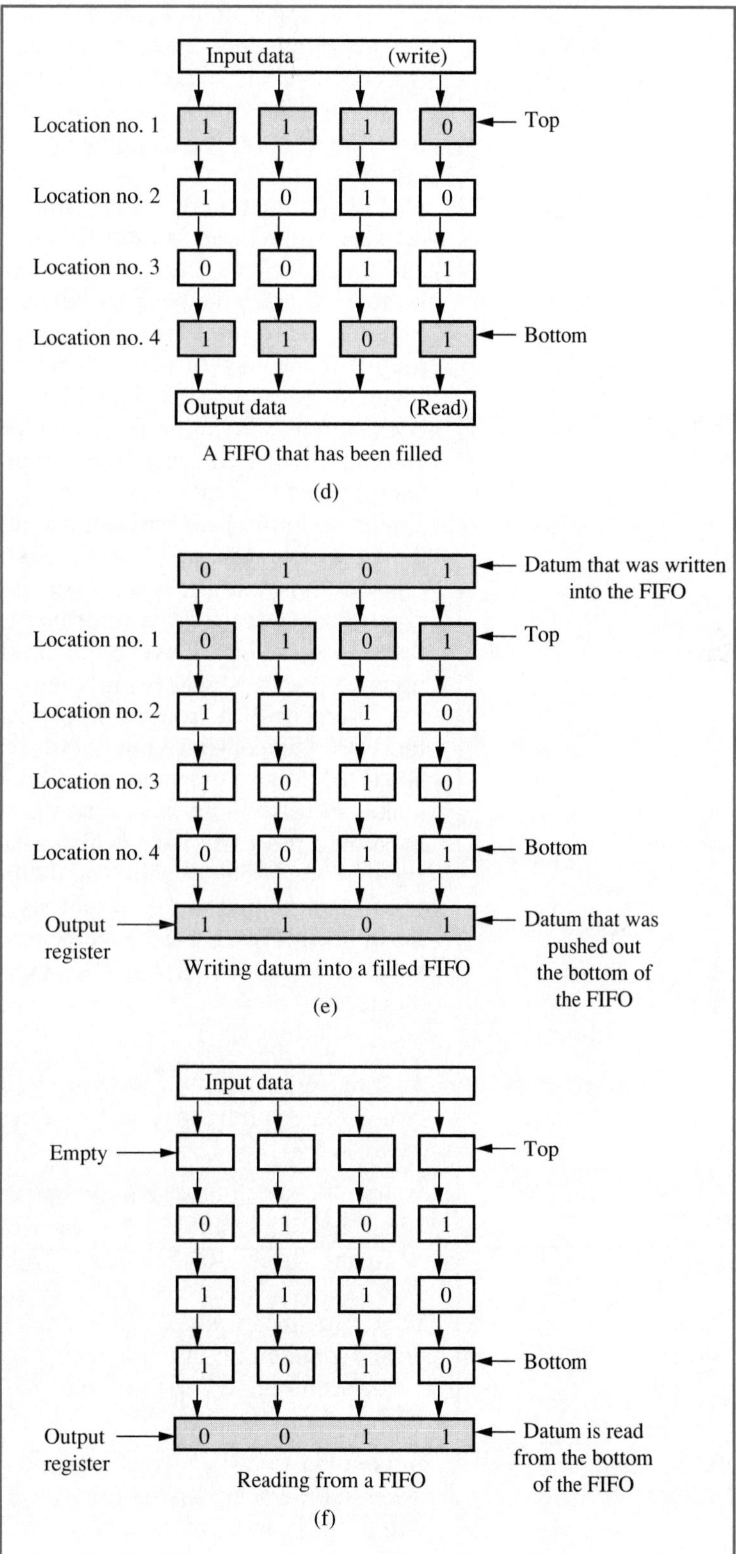

A FIFO that has been filled

(d)

Writing datum into a filled FIFO

(e)

Reading from a FIFO

(f)

The application dictates which FIFO is used; regardless of the technique, the most important point is that the first datum into the FIFO is the first datum that can be read from the FIFO.

Since the data is mobile in a FIFO memory and its organization is one-dimensional, it is classified as a dynamic one-dimensional organization.

When a datum is written into an already filled FIFO memory, the datum being written in the FIFO is, as usual, shifted into location no. 1 and the other data already in the FIFO is shifted one location toward the bottom. This can be seen by comparing the previously filled FIFO of Figure 8–6d to the FIFO memory of Figure 8–6e. Figure 8–6e represents the FIFO memory of Figure 8–6d after 0101 has been written into it. Figure 8–6e shows that the datum at location no. 1 in Figure 8–6d has been shifted to location no. 2 and location no. 1 contains the newly entered datum. Likewise, the datum at location no. 2 in Figure 8–6d has been shifted to location no. 3 in Figure 8–6e. This shifting of data continues in the FIFO memory until the datum that was at the bottom of the FIFO (location no. 4) is shifted out of the FIFO, which means that this datum is lost if it is not read. It is important that a user of a FIFO-type memory realize that improper use of a FIFO (entering more data than there are locations) results in lost data unless that data is read and processed as it is pushed out the bottom of the FIFO.

Reading data from a FIFO memory shifts the datum at the bottom of the FIFO memory into the output register. The data in the other locations is shifted down by one location, which empties the top location and makes it available to have a new datum written into it. This action is illustrated in Figure 8–6f, which assumes that the filled FIFO memory of Figure 8–6e is being read. The datum in location no. 4 of Figure 8–6e has been read (shifted to the output register), and the data at the other locations in Figure 8–6e have been shifted down one location in Figure 8–6f. The top location has been emptied as a result of reading this FIFO memory.

EXAMPLE 8–5

Suppose that 1100 were written into the FIFO memory of Figure 8–6d. Beginning with memory location no. 1, give the resulting data in each location.

SOLUTION 1100 is written into location no. 1, which is the top of the FIFO memory, and all previous data is shifted by one location. Hence, after 1100 has been written into the FIFO memory of Figure 8–6d, which is a filled FIFO memory, the content of location no. 1 is 1100, location no. 2 contains 1110, location no. 3 contains 1010, and location no. 4 contains 0011. The datum that was previously in location no. 4 (1101) was pushed out of the FIFO memory.

LIFO accessed memories

In contrast to a FIFO accessed memory is a *last-in first-out* (LIFO) accessed memory; that is, the last datum written in memory is the first datum that can be read from it. To illustrate the LIFO accessing technique, a representative simple dynamic one-dimensional 4 × 2-bit array is used, as shown in Figure 8–7. Figure 8–7a indicates that memory location no. 1 is the only accessible location in the array and therefore will be used to input and output the datum to and from this LIFO memory. The bidirectional arrows between the shift register flip-flops

Figure 8–7
Accessing a LIFO memory.

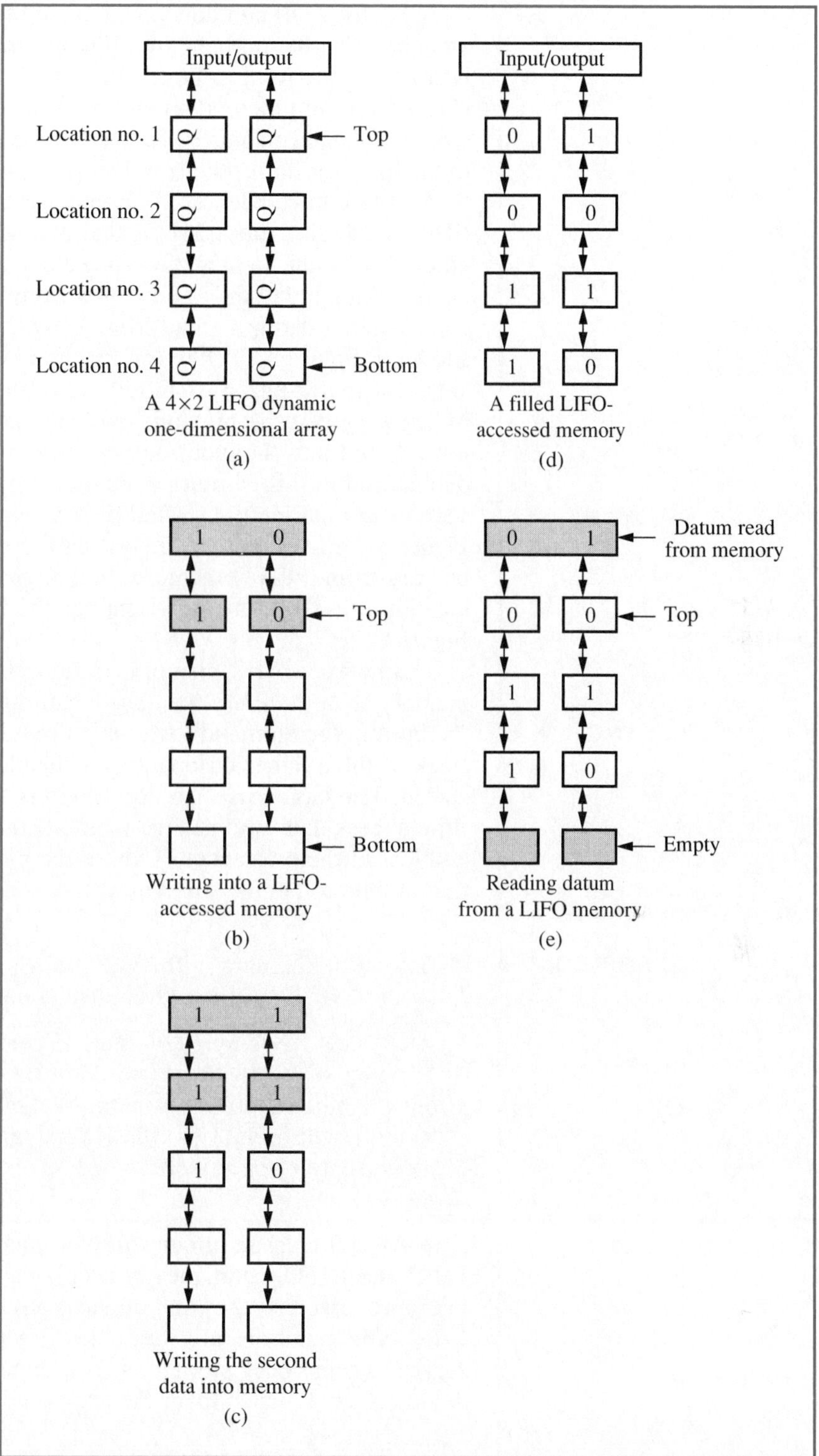

indicate that data can be shifted in either direction, ''down'' or ''up,'' where down is toward the bottom location and up is toward the top location. A datum is written (pushed) into the top location of a LIFO memory and all subsequent data is shifted one location down, just as in a FIFO accessed memory. The datum read from a LIFO is shifted out the top location and all subsequent data is shifted one location up.

Figure 8–7b contains an example of writing data into a LIFO-type memory. The datum being written into memory is 10, and as illustrated, it is stored in the top location. Figure 8–7c illustrates that the second datum (11) written into memory is stored at location no. 1 and the previously stored 10 has been shifted to location no. 2. Data is written into memory in this manner until the array is filled, which is illustrated in Figure 8–7d. So far there have been no differences in operation between a LIFO and a FIFO, excluding those FIFOs that automatically shift the datum being entered from the top location to the bottom location. The difference occurs when a datum is being read from a LIFO accessed memory.

When a datum is read from a LIFO memory, such as the memory shown in Figure 8–7d, the top location is the one that is read. The datum stored at that location is shifted out of memory, indicated in Figure 8–7e by showing that the previous contents of location no. 1 of Figure 8–7d are now shifted into the input/output register of Figure 8–7e. All subsequent data stored is shifted up one location when a read occurs, which can be seen by comparing the shifted data of Figure 8–7e with the original in Figure 8–7d. Notice that the last datum entered is the first datum that can be read from a LIFO memory, and also that reading data from the top location of a LIFO memory empties the bottom location, as indicated in Figure 8–7e.

As with a FIFO memory, if more data is written into a LIFO memory than there are memory locations, the excess data is shifted out the bottom location and lost, which requires the user or designer to keep track of the number of locations available and the amount of data to be stored. The jargon used to describe this situation is the statement that "the excess data was pushed (shifted) through the bottom (of memory) and lost in the bit-bucket," where the bit-bucket is imagined as a place that all bits to be discarded are "thrown."

EXAMPLE 8–6

Determine the data in the LIFO of Figure 8–7d if 10 is first pushed on the LIFO and then 11 is pushed on the LIFO.

SOLUTION Pushing 10 on the LIFO results in data 10, 01, 00, and 11 being stored, with 10 (stored at the bottom location) being pushed out the bottom and into the bit-bucket. Pushing 11 on the LIFO next would result in data 11, 10, 01, and 00 in the LIFO, with 11 being pushed into the bit-bucket.

As will be seen in our study of microprocessors, memories that use FIFO and LIFO techniques of accessing data are utilized to implement microprocessor designs and applications. We shall also discover that FIFO-type memories are well suited for some I/O devices. The FIFO serves as temporary storage of data that is to be processed by the I/O device, such as when main memory sends data to a printer.

Review questions

1. What is access time?
2. What is meant by RAM?
3. What is linear accessing?

4. How are FIFO and LIFO accessing similar and how do they differ?
5. How can data be lost in a FIFO and LIFO accessed memory?

8–5 Alteration of memory contents: read/write and ROM memory devices

In some memory devices, memory locations can be written to and read from by a function-block(s) within the system of Figure 8–1. A memory device with this characteristic is known as a *read/write memory* (R/W memory). When the contents of a memory cannot be altered by a function-block(s) within the system, that memory is known as a *read only memory* (ROM). The bits are written into a ROM by some manufacturing process or the action of an apparatus (a ROM programmer) that is not part of the system illustrated in Figure 8–1. The process of storing the bit patterns in each memory location of a ROM is known as *programming the ROM*. Since no function-block within the system of Figure 8–1 can write into a ROM, reading is the only operation that can be performed by the system—hence the term "read only memory."

There is an important fact to be remembered about R/W semiconductor memory devices. In the beginning of the evolution of the computer, main memories were often magnetic core memories, which were the only R/W memories that could be accessed as a RAM. For that reason these R/W memories were referred to as RAMs; hence the acronym RAM came to be synonymous with read/write memory. This is a misuse of the term "RAM," which actually indicates the accessing technique, whereas R/W indicates the alterability of memory contents. Nonetheless, reference to a RAM usually means R/W memory. We will follow convention and also use the acronym RAM to indicate an R/W memory, except when it would be clearer in meaning to use R/W.

Review questions

1. What is an R/W memory?
2. What is a ROM memory?
3. When would one use an R/W memory rather than a ROM?
4. Is it correct to refer to a memory as being a RAM when you mean it is an R/W memory? Why?
5. In industry is the acronym RAM misused, and if so when?

8–6 ROM classifications

There are four types of semiconductor ROMs, each requiring a different method to program it. The method of programming a ROM is directly related to two essential considerations in the design of a memory system—the time required to program the ROM and the cost of the programmed ROM. That is, in deciding which type of ROM to use, the designer must consider the cost of both the ROM itself and the programming of the ROM, as well as the time required for it to be programmed, which can cause delays in project construction. The four types of ROMs are (1) mask ROMs (MROM), (2) fuse link programmable ROMs (PROM), (3) ultraviolet light erasable ROMs (EPROM), also termed UV PROM, and (4) electrically erasable ROMs (EEPROM). Each has both positive and negative characteristics for a given application;

thus, the designer must weigh one characteristic against the others in order to make the optimal choice.

It should be noted that all modern ROM semiconductors are accessed as RAM; that is, all memory locations have the same access time even when locations are accessed at random.

MROM

A mask ROM (MROM) is programmed at the time of the manufacturing process using a series of photo masks; as a result, the data stored is permanent and cannot be changed. The process can take weeks, and the cost is very high for just a few ROMs; however, for large quantities, such as in high-volume production, the process becomes very inexpensive. MROMs are suited to high-volume applications that have sufficient turnaround time to include the manufacturing process. This excludes MROMs from a research and development environment as well as low-volume production applications.

PROM

Programmable ROMs (PROMs) are manufactured so that they can be programmed by the user by means of a special apparatus known as a PROM programmer. When they are manufactured each memory cell of a PROM has a logic 0 programmed into it. The PROM programmer supplies to those memory cells that are to be programmed in the logic 1 state a sufficiently large current to blow a fuse in the cell, resulting in an unalterable logic 1 state. Once programmed by the user, the content of PROM is permanent. The cost of a PROM is relatively low. The greatest disadvantages of PROMs are (1) the inability to alter its contents and (2) the time required to program a large number of PROMs. They are suited for applications with a medium-volume production. The trade-off between PROMs and MROMs is the time involved to program a PROM versus the high cost and long turnaround time for manufacturing a small number of MROMs. PROMs are not suited to research and development environments where programs are frequently changed.

EPROM (UV PROM)

Exposing the integrated circuit wafer of an erasable PROM (EPROM) to ultraviolet (UV) light for approximately 15 minutes erases the content of each memory cell (actually, it changes all memory cells to a logic 1). Once the contents of the EPROM have been erased, it can be reprogrammed using an EPROM programmer. EPROMs are easily recognizable by their quartz window, which exposes the integrated circuit wafer. Once the EPROM has been programmed, this window is covered with opaque tape to block out any ultraviolet light.

EPROMs are higher in cost than PROMs, but since their contents can be erased and programmed again this cost is offset in environments in which the contents of ROM are frequently changed, such as in research and development. EPROMs are also well suited for very low volume production, such as one or two items are being produced.

EEPROM

Electrically erasable PROMs (EEPROMs) are the only ROMs that can be programmed by a function-block within the computer system (usually the CPU). That is, the CPU can write data into an EEPROM, which erases the old data at the location(s) being written to and programs the new data into the EEPROM at the same time. One might question why a memory device with this characteristic is not classified as an R/W memory rather than a ROM. The reason is the time required for the CPU to write to an EEPROM memory location. It is relatively long (on the order of a millisecond); as a result, if an EEPROM were used to implement an R/W memory device in main memory, it would slow down the system considerably. Therefore, EEPROMs should be limited to applications that require very few write operations, such as back-up memory for power-sensitive R/W memory devices (see Section 8–7). For a back-up memory application, the contents of main memory are copied into an EEPROM or EEPROMs (depending on the number required for back-up) when an impending power failure is detected by the system.

EEPROMs are manufactured with two planes of memory within the integrated circuit: One plane is a typical R/W memory and the other duplicates the first plane in organization but is an EEPROM. Data can be copied from one plane to the other via command signals. For instance, when the CPU detects an impending power failure, it signals (via a pulse output to the EEPROM) that the memory contents in the R/W memory plane are to be written into the EEPROM plane. As a result, the data stored in the R/W memory plane will not be lost owing to a power failure. After power has been restored the process can be reversed; that is, the contents of the EEPROM can be copied into the R/W plane.

An EEPROM is actually an R/W memory with its own internal mass storage. Can EEPROMs therefore be used as mass storage devices? The answer is theoretically yes, but practically speaking no, since the number of memory locations available in an EEPROM is relatively small (remember that there are two memories in one IC) and an EEPROM is not very portable (one would have to remove the chip from the system, versus removing a disk or tape).

Since applications of EEPROMs are limited to read-mostly operations rather than R/W operations, they are sometimes referred to as *read-mostly memories*.

Review questions

1. How many types of ROMs are there?
2. What are the advantages and disadvantages of each type of ROM relative to their possible applications?
3. Why might one imagine an EEPROM as two memories in one?

8–7 Memory volatility

Another way of categorizing memories is by their memory content retention when power is removed from the memory device. If the memory contents are not lost or altered owing to the removal and reapplication of power to the memory device, then the memory device is known as a *nonvolatile memory*. A memory that does not retain its contents when power is removed is known as a *volatile memory*. Since the contents of ROMs are permanent, they are nonvolatile memories. In contrast, the majority of semiconductor R/W memory devices are volatile. Of course, this is why R/W memory devices within a main memory must unload their content into mass storage, or an EEPROM, if that content is to be saved when power is removed from the system. It should be noted that some computers have a battery back-up to retain the contents of their volatile memory.

It is of interest to note that another name for an EEPROM is *nonvolatile RAM*. The reason for this name is that it is an R/W (RAM) that can transfer its data to a nonvolatile memory plane.

Mass storage memory devices (disk, magnetic tape, and bubbles) are also nonvolatile memories. However, they are not referred to as nonvolatile RAMs; rather, that term is reserved for semiconductor nonvolatile R/W memories.

Review questions

1. What are the four types of ROMs?
2. Under what conditions would each of the four types be used?
3. Why are ROMs also RAMs?

8–8 Memory cell data retention: static and dynamic memories

The final category of memory is a characteristic associated with semiconductor R/W memory. It is characterized by the ability, or inability, of a memory cell to retain the stored logic level *even with power applied*. This characteristic is not to be confused with volatility—note the key words "with the power applied." If a semiconductor memory cell can retain its logic level as long as power is applied to the semiconductor memory device, it is known as a *static memory cell* and the semiconductor memory device is known as a *static R/W memory device*, or *static RAM*. However, if the logic level of a memory cell is lost over a period of time (usually less than 3 ms), even with power applied, and therefore must be continually refreshed (before 3 ms has elapsed) in order to retain the logic level, then the cell is known as a *dynamic memory cell*. A memory device made up of this type of cell is known as a *dynamic R/W memory* or *dynamic RAM* (DRAM). The important thing to remember is that once data is written into a static RAM it will retain (remember) those data so long as power is applied to the semiconductor memory device, or until the data is changed, whereas a dynamic RAM will not retain its data beyond approximately 3 ms and therefore requires that the data be continually refreshed. Insofar as serving as a memory, there is no difference between a static and a dynamic memory device. The difference is the required refreshing of the data within the dynamic memory.

The interval of time when refreshing of the data occurs is known as a *refresh cycle*, and refreshing is usually performed and controlled by a refresh cycle controller (dynamic RAM controller). Refresh cycle controllers are available as ICs and are discussed in Chapter 9.

The reason for the existence of dynamic memory devices is that they contain a larger number of memory cells than static memory devices. Of course, the disadvantage is the additional refresh cycle circuitry that is required and the longer access time.

The technology of a dynamic memory device basically involves the parasitic capacitance of a field effect transistor (FET) as the memory cell, greatly reducing the number of components needed to construct a memory cell. To write a logic 1 into a dynamic memory cell, the parasitic capacitor is charged to the logic 1 level; however, since parasitic capacitance has a large leakage current, the logic 1 level voltage quickly discharges (in approximately 3 ms), which is the reason for the required refresh cycle.

All memory devices discussed henceforth are assumed to be static unless stated otherwise.

Review questions

1. What is the difference between static and dynamic memories?
2. Why are dynamic RAMs sometimes preferred over static RAMs?
3. Why is it necessary to have a refresh cycle for dynamic RAMs but not for static RAMs?

8–9 Addressing and read/write operation control

To read or write from or to memory or I/O, two events are required and must occur in the following order: (1) Identifying the memory location or I/O device to be accessed. (2) Controlling when the read or write operation is to take place at the location or I/O device being accessed. The first of these two events occurs as a result of addressing the memory location or I/O device. The second occurs as a result of applying the proper control signal (a read or write control signal) to the memory or I/O device at the proper time. The control signal controls the actual read or write operation being performed at the addressed memory location or I/O device.

For example, if the CPU is to read a location in main memory, it must begin the process by applying the address of the memory location to main memory. Logic circuitry within main memory decodes that address and selects the appropriate memory location (refer to register arrays in Chapter 7). Once the addressed memory location has been selected, the CPU must tell that memory location when to output its contents so that the CPU can read it. Addressing memory and reading the contents of that address location are two separate events that occur at different times. To correlate these two events the CPU furnishes a read control signal (a write control signal when writing to memory) at the appropriate time after it has sent the address to memory.

In following sections we discuss how each of these events is implemented for various memory and I/O devices.

Review questions

1. How is memory or I/O accessed?
2. Once memory or I/O has been addressed, how does the CPU control the read or write operation?

8–10 Semiconductor memory architecture

Thus far we have investigated memories in a rather general manner. It is now appropriate to become more specific in our studies. We will study semiconductor memory device architecture in this section and its application to main memory in the next section. Last, we will investigate magnetic memories (tapes and disks).

As previously stated, to perform a read or write operation at a memory location, two events must take place. First, the address of the memory location to be accessed is sent by the CPU to memory. Second, the operation (read or write) that is to be performed must be initiated at the proper time. Let us investigate these two events.

Addressing semiconductor memories

From our study of two- and three-dimensional arrays we know that the address of a memory location within an array is the binary combination of the row and column. To access a memory location within a semiconductor memory, its address must be applied to the memory device and decoded by that device so that the appropriate row and column of the array are activated. The address decoder is an integral part of the memory device. The 8 × 2-bit memory device of Figure 8–8a will be used to illustrate this concept.

There are two decoders in Figure 8–8a, one to decode the two MSBs of the binary address (A_1 and A_2) and the other to decode the LSB

Figure 8–8 Address decoding of an 8 × 2 three-dimensional array.

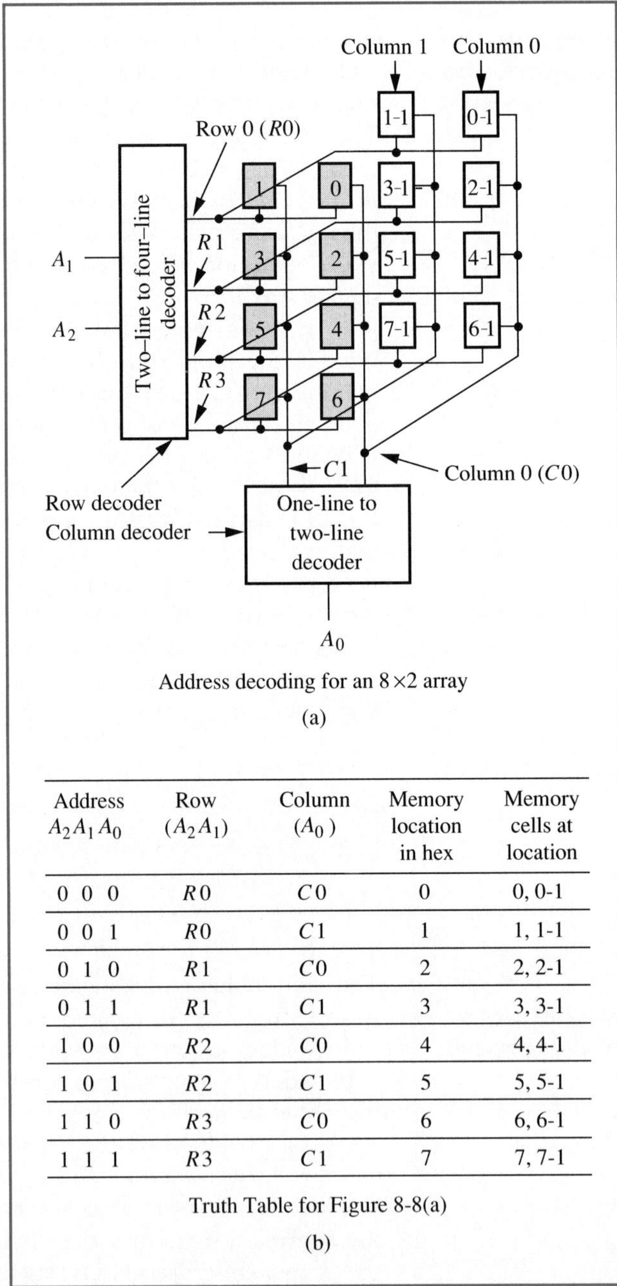

Address decoding for an 8×2 array

(a)

Address $A_2A_1A_0$	Row (A_2A_1)	Column (A_0)	Memory location in hex	Memory cells at location
0 0 0	$R0$	$C0$	0	0, 0-1
0 0 1	$R0$	$C1$	1	1, 1-1
0 1 0	$R1$	$C0$	2	2, 2-1
0 1 1	$R1$	$C1$	3	3, 3-1
1 0 0	$R2$	$C0$	4	4, 4-1
1 0 1	$R2$	$C1$	5	5, 5-1
1 1 0	$R3$	$C0$	6	6, 6-1
1 1 1	$R3$	$C1$	7	7, 7-1

Truth Table for Figure 8-8(a)

(b)

(A_0) of the address. Address bits A_1 and A_2 select the row ($R0$, $R1$, and so forth) and address bit A_0 selects the column ($C0$ or $C1$). From the truth table of Figure 8–8b, we see that address 000 will select row $R0$ and column $C0$. Using an x-y coordinate system, $R0$ and $C0$ will select memory location 0H (as before, an H is used to indicate base 16). Location 0H contains memory cells 0 and 0-1, as indicated in the truth table. Address 001 selects row $R0$ and column $C1$, which in turn selects memory location 1H. Memory location 1H contains memory cells 1 and 1-1, which can be verified from the truth table of Figure 8–8b. Using this same scheme, the remainder of the truth table is self-explanatory. It is important to realize that both the row and the column of the memory cell must be selected (activated) before any read/write operation can take place within that cell.

There is an equation that we have seen before which can be used to calculate the number of memory locations within a memory device:

Equation (8–1)

$$N = \text{Number of Binary Combinations} = 2^n$$

N is the number of binary combinations for n independent (n inputs) variables. Since the memory device of Figure 8–8a has three inputs (A_0, A_1, and A_2) which are relevant to addressing, then $n = 3$, and from Equation (8–1) we find that

$$N = 2^3 = 8$$

There are eight binary combinations for independent variables A_0, A_1, and A_2. These eight combinations can be determined in an organized method simply by generating eight consecutive binary numbers beginning with zero. To verify this we refer to the truth table of Figure 8–8b and find that indeed there are eight consecutive binary numbers, beginning with 000, in the address column.

EXAMPLE 8–7

From the truth table of Figure 8–8b and from what has been previously stated, we realize that the memory location hexadecimal number and the binary address are equivalent values.

1. What is the binary address for memory location
   (a) 2H (b) 5H (c) 7H?
2. What is the hexadecimal memory location for binary address
   (a) 001 (b) 101 (c) 110?

SOLUTION We convert hexadecimal values to binary values.
1. (a) 2H = 010_2 (b) 5H = 101_2 (c) 7H = 111_2
Now we convert binary values to hexadecimal values.
2. (a) 001_2 = 1H (b) 101_2 = 5H (c) 110_2 = 6H.

Verify these addresses and memory location values using the truth table of Figure 8–8b.

System bus architecture and buffering of the data bus

Now that we understand the concepts of addressing a memory location within a semiconductor memory device via its address, we must next learn the concepts of controlling the read/write operation to be performed on that accessed location. To reiterate, read/write operations must be synchronized with the device (CPU or mass storage device) that is performing the read/write operation. This synchronization is done under the control of the read and write control signals that are generated and output by the accessing device.

To develop the concepts associated with the control of read/write operations and broaden those addressing concepts already learned, let us take a systems approach to the workings of a computer. The system to be studied is that of Figure 8–9, which is a modified version of Figure 8–1 in which (1) the connecting lines between function-blocks have been expanded into busses, resulting in a *three-bus architecture* (address, data, and control busses); (2) an I/O selector is added, which is the I/O address decoder (it operates in much the same way as an address decoder within a memory device except that it decodes only I/O addresses); (3) mass storage is omitted, since it has little to do with the concepts we are developing here.

Since Figure 8–9 has three busses, the term "bus" needs definition and explanation. A bus is a line (a wire or other low-resistance medium) or lines used to electrically connect two or more devices in parallel. For example, the three devices of Figure 8–10 have four connections such that the three devices are in parallel. These four lines form a 4-bit bus.

If any two or more of the devices of Figure 8–10 were to use the bus at the same time by outputting data on it, there would be competition among the devices for use of the data bus, or more simply stated, *bus contention* would exist. Bus contention must be avoided. As stated in an earlier chapter, bus contention results in unintelligible data on a bus while it is in the state of contention (imagine the result if each device contending for the bus loaded a different binary bit pattern on it—a conflict would exist).

Read/write control

In the control of read/write operations, the data bus is used for transmitting data to and receiving data from different function-blocks within a system. The data bus of Figure 8–9 connects the CPU, main memory, and I/O devices. Consider the sequence of operations that occurs

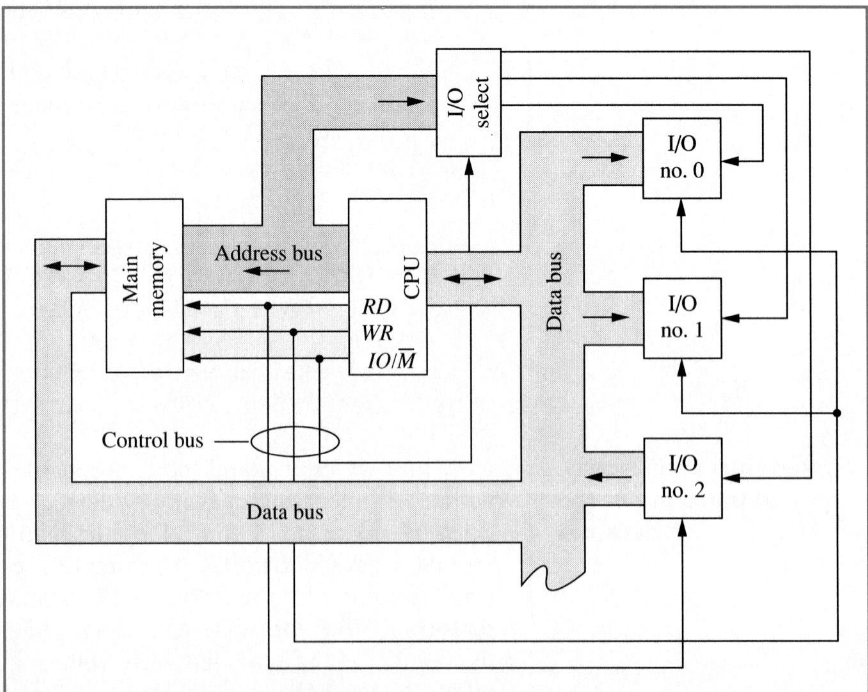

Figure 8–9
A simplified computer having a three-bus architecture.

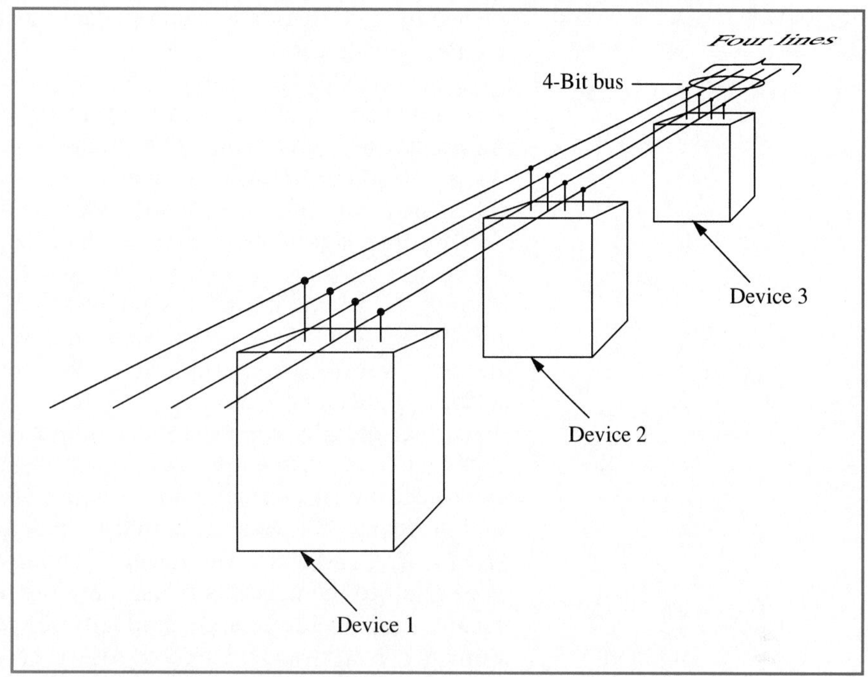

Figure 8–10
A 4-bit bus.

when the CPU fetches an instruction or data from main memory, as shown in Figure 8–9. The CPU beings the process by outputting the address of the memory location containing the instruction or data on the *address bus* and driving control signal $IO/\overline{M}$ to the appropriate logic level (low) for accessing memory. The $IO/\overline{M}$ line differentiates between addresses for I/O devices and memory by enabling either the main memory address decoder or the I/O address decoder (I/O selector). It is driven high by the CPU when the address put out by the CPU is an I/O address and low when the address is a memory address; hence, $IO/\overline{M}$ is driven low in this example. When $IO/\overline{M}$ is driven low, the semiconductor memory device of main memory (which contains that memory location) decodes the address on the address bus and thereby selects the addressed location within it. The CPU then sends a read control signal (*RD*) to the memory device via the *control bus*, causing the addressed location to transfer its contents to the data bus, from which the CPU can read it. The reading of that memory location is now complete. In summary, reading the contents of an addressed memory location simply means that the contents of an addressed memory location are copied (loaded) onto the data bus; since the CPU is in parallel with the memory device, that data can be latched from the data bus by the CPU. As we shall explain later in this chapter, reading an I/O device is similar to reading memory. The major difference is that control signal $IO/\overline{M}$ is driven high by the CPU, which disables the memory address decoders and enables the I/O address decoder (the I/O selector).

For the CPU to write to the main memory of Figure 8–9, the CPU again begins the operation by outputting the address of the memory location on the address bus and drives control signal $IO/\overline{M}$ low. The CPU then outputs the data to be stored on the data bus. The CPU next generates the write (*WR*) control signal and transmits it to main memory via the control bus. This causes the addressed semiconductor memory device to latch that data from the data bus and store it in the addressed memory location.

For the CPU to write to I/O, the process is similar to writing to memory except that the control signal $IO/\overline{M}$ is driven high by the CPU. As previously stated, with $IO/\overline{M}$ high the memory address decoders are disabled and the I/O address decoder is enabled, so that the address on the address bus selects an I/O device.

From Figure 8–9 we can see that the data bus is shared among many devices (main memory, CPU, and three I/O devices). Since it is shared, the potential for bus contention exists. *To prevent bus contention, all devices that have outputs connected to the data bus must have their outputs buffered with tristate logic*. With buffered outputs, only the addressed memory location or I/O device has its buffers on (enabled). This electronically connects those outputs, and only those outputs, to the data bus. The other output buffers are disabled [high-impedance (Hi-*Z*)], thus electronically isolating those outputs from the data bus. The enabling and disabling of a device's output buffers are done with control signals ($IO/\overline{M}$, *RD*, and *WR*). The result of buffering the outputs of devices connected to the data bus is that only the outputs of the addressed location or I/O device are electrically connected to the data bus; all other outputs are electrically isolated, thus preventing bus contention. We shall see shortly how this buffering is implemented.

Having reviewed the fundamental concepts of a data bus and bus contention, let us determine how R/W semiconductor memory devices achieve buffering from the data bus as well as the control of read/write operations. Figure 8–11 is a representative example of a 2-bit memory location that has memory cell *m* in plane 0 and memory cell *m*-1 in plane 1. It will be used to illustrate the concepts of buffering and the control of read/write operations. Figure 8–11 uses *D*-type flip-flops to implement the memory cells. There are actually a variety of semiconductor memory cells that could be used, but from a conceptual point of view, the *D*-type flip-flop can be used as a general representation of them all. Therefore, consider the input *D* as the means to write data into the memory cell and the output *Q* as the means to read the data stored in the memory cell. As a convenience the *D* inputs are not drawn in the conventional location in Figure 8–11. The memory location of Figure 8–11 is one representative location from an $N \times 2$-bit memory, where *N* is the number of memory locations within the array. For example, if the memory location of Figure 8–11 is to represent a memory location from the array of Figure 8–8a, then $N = 8$ and *m* is a number from 0 to 7. Specifically, if the memory location of Figure 8–11 is to represent memory location 010 of Figure 8–8a, then $m = 2$. However, if it is to represent memory location 101 in Figure 8–8a, then $m = 5$.

Figure 8–11 has two tristate buffers for each memory cell, which are labeled *R* and *W*. The purpose of these buffers is to provide buffering between the data lines (D_0 and D_1), output *Q*, and input *D* of the flip-flops. To see how the state (on or off) of these buffers is controlled, let us investigate those *AND* gates that drive their control lines. As we know, control signal $IO/\overline{M}$ is driven low when memory is accessed; and this enables both *AND* gates. Inputs *RD* and *WR* of those *AND* gates are controls that are driven by CPU control signals *RD* and *WR* and, via their logic level, determine whether a read or write operation is to occur.

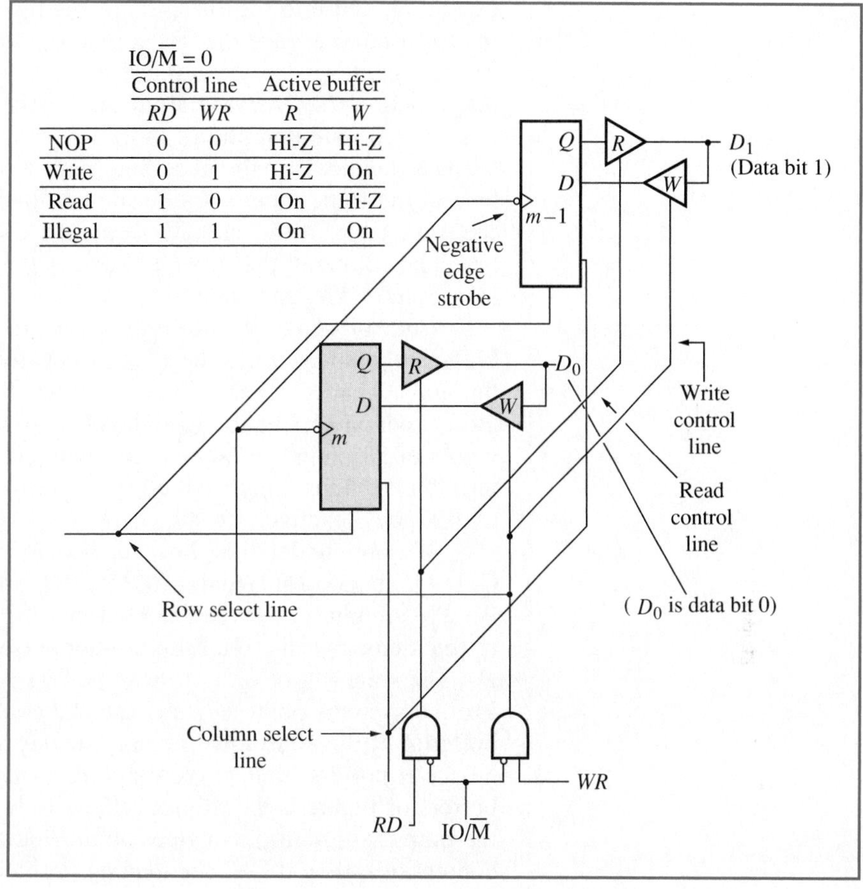

$IO/\overline{M} = 0$

	Control line		Active buffer	
	RD	*WR*	*R*	*W*
NOP	0	0	Hi-Z	Hi-Z
Write	0	1	Hi-Z	On
Read	1	0	On	Hi-Z
Illegal	1	1	On	On

Figure 8–11
Read/write control logic of an R/W memory location.

Inputs *RD* and *WR* are independent variables, and therefore the number of possible binary combinations for these two variables is

$$N = 2^2 = 4$$

These four binary combinations and their interpretation are given in the truth table of Figure 8–11. From this truth table we see that for the condition $RD = WR = 0$, *no-operation* (NOP) is to be performed, and therefore both buffers *R* and *W* are to be put in Hi-*Z*. This serves to buffer (isolate) the memory cell from data lines D_1 and D_0. With control signal $IO/\overline{M}$ being a logic low, both *AND* gates are enabled. However, since control signals *RD* and *WR* are also low, the output of each *AND* gate is low. With the outputs of the *AND* gates low, the buffers *R* and *W* are put in the Hi-*Z* state, which results in the desired effect and isolates the memory cells from the data bus lines. For the condition $RD = 0$ and $WR = 1$, the truth table of Figure 8–11 states that a write operation is to be performed. From the logic circuit we see that for this condition the *AND* gate with input *RD* puts out a low and the *R* buffer is off (in Hi-*Z* state), and the *AND* gate with input *WR* puts out a high, which turns the *W* buffers on. This condition causes the data to flow in the direction indicated by the arrow of the *W* buffer logic symbol—from data bus lines

D_0 and D_1 and into the D inputs of the flip-flops (a write operation). From the truth table we see that for a read operation, $RD = 1$ and $WR = 0$. These logic levels result in the R buffers being on and W buffers being off, which loads the data output by the Qs of the flip-flops onto the data bus lines (a read operation). Once the data is on the data bus lines it can be read (latched) by the accessing device (CPU). To have RD and WR high at the same time is an illegal and illogical operation, as indicated by the truth table. If this should occur, the CPU is malfunctioning and must be replaced (recall that the CPU drives the logic levels of the control signals RD, WR, and $IO/\overline{M}$).

Control signal WR not only serves to control the state of the W buffer but also provides the needed negative-edge strobe for triggering the flip-flops. That is, when WR goes active ($WR = 1$) it turns on the W buffer, which applies the logic levels on data lines D_0 and D_1 to the D inputs of flip-flops m and m-1, respectively. To strobe these logic levels into the flip-flops a negative-edge strobe pulse is required, which occurs as WR goes inactive (1 to 0).

We now understand how read/write operations are controlled by the CPU via its control signals—$IO/\overline{M}$, RD, and WR.

In summary, the read (R) buffers of Figure 8–11 buffer the outputs of the memory cells (Q of the flip-flops) from the data bus lines D_0 and D_1. The state (on or off) of these buffers is controlled by the signal RD. From a systems point of view, the RD control signal turns on the R buffer(s) of the addressed memory device and turns off all other memory device R buffers, thus preventing bus contention. While the write (W) buffers of Figure 8–11 are not related to bus contention (recall that only the outputs of multiple devices connected to a common bus can cause bus contention), they do reduce loading problems. A memory that is constructed with memory devices with buffered inputs reduces the loading of the data bus, since a data bus line sees the input loading of only a single W buffer rather than the D inputs of all the memory cells.

Let us duplicate the basic 2-bit single memory location of Figure 8–11 into two planes of eight identical locations, so as to produce an 8 × 2 R/W (RAM) memory device, as illustrated in Figure 8–12. Memory cells 0 to 7 are in plane 0 and cells 0-1 to 7-1 are in plane 1. Memory cells 0 and 0-1 form memory location 0, cells 1 and 1-1 form location 1, and so on. Data is input and output via the I/O pins I/O_0 and I/O_1, which are connected to the data bus lines D_0 and D_1, respectively. The memory locations of Figure 8–12 are accessed in exactly the same manner as the 8 × 2 array of Figure 8–8a. That is, *each memory cell has a row and column line going to it, and both must be active before the cell can be accessed for either a read or write operation* (the electronic principles are explained in Chapter 12). As a result, even though the outputs (Qs) of all flip-flops in a plane are connected to the R buffer for that plane only, the addressed memory cell serves as an active input to the R buffer. Likewise, the output data bit of the W buffer is received only by the addressed cell. Since the technique of addressing has been explained, it is not necessary to reproduce the row and column lines in the illustration of Figure 8–12, which results in a less complex drawing.

To see why we choose to use Figure 8–12 to represent the architecture of semiconductor R/W memories, let us briefly refer to Figure 9–4 of Chapter 9. Figure 9–4 is a manufacturer's data sheet for the

Figure 8–12
An 8 × 2 R/W memory device.

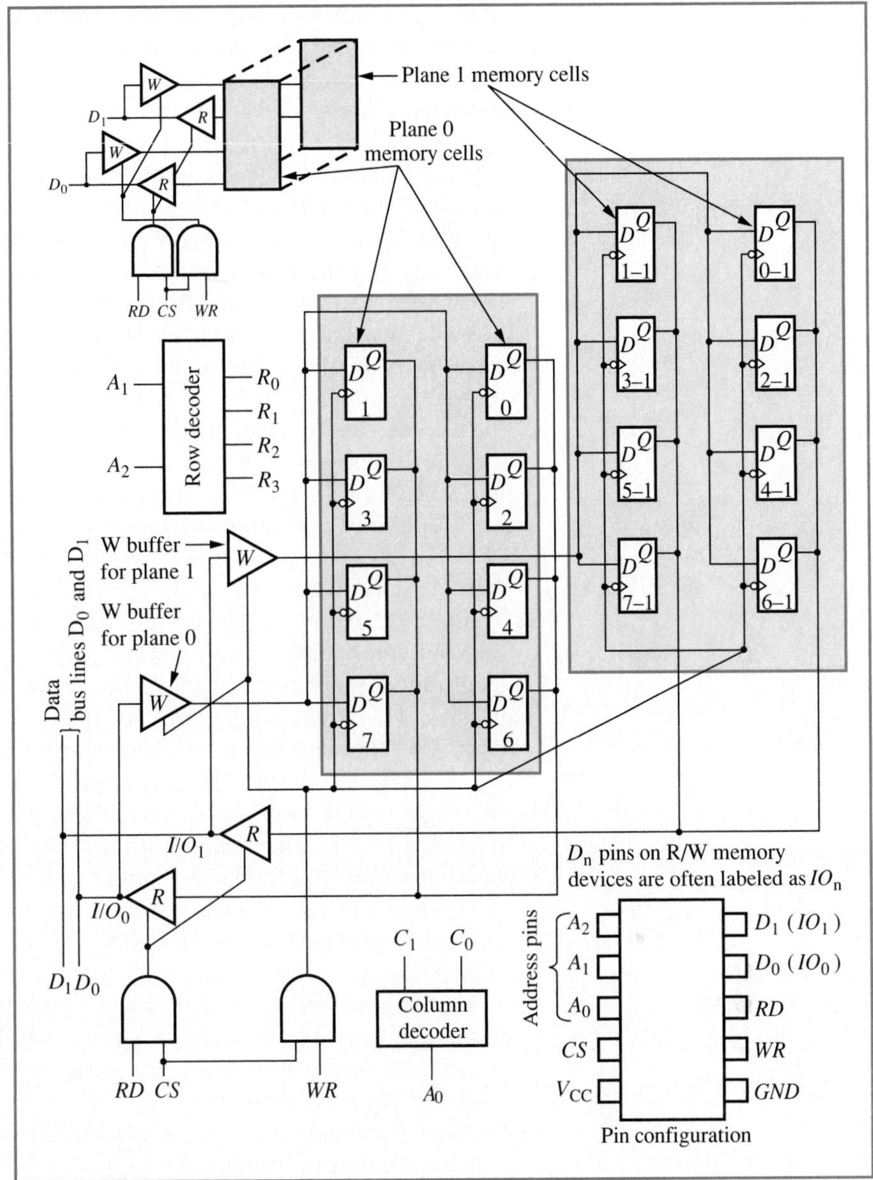

architecture of a 2148 semiconductor memory device. Comparing Figures 8–12 and 9–4, we see that on the data pins (I/O_1–I/O_4) of Figure 9–4 are the *R* and *W* buffers, but they are not labeled. Also not shown are the four planes of memory cells, which are contained within the function-block identified as the *memory array*. We know that there are four planes of memory cells because a plane of memory cells exists for each bit within a memory location, and, since this memory has 4 bits per location, there must be four planes of memory cells. The memory array function-block of Figure 9–4 is stated to have 64 rows and 64 columns. We can easily verify that there are 64 rows via Equation (8–1); that is, for six row address pins (A_4–A_9) there are 64 binary combinations and hence 64 rows. There are four column address pins (A_0 through A_3) and hence just 16 columns (2^4). However, since there are four planes of memory cells and each plane has four columns, there are indeed a total of 64 columns. It is to be hoped that this comparison has justified our conceptual representation of semiconductor memories, since industry has a similar representation.

In Figure 8–12, note the input *CS* (chip select). The purpose of this input is to allow the semiconductor memory device to be enabled and disabled. As we can see from the logic gates of Figure 8–12, if the level of input *CS* is a logic 0, then both *AND* gates are disabled (outputting logic 0's) and the *R* and *W* buffers are in the Hi-*Z* state. As a consequence, the memory device is electronically isolated from the data bus lines D_0 and D_1. However, if input *CS* is active (a logic 1), then the memory cells of the addressed memory location can be read (if $RD = 1$) or written into (if $WR = 1$). For the present, let it simply be stated that the address and control signal $IO/\overline{M}$ is instrumental in determining the logic level applied to input *CS*. Compare the *AND* gates of Figures 8–12 and 9–4 to verify their similarities in function.

The R/W operations of Figure 8–12 are basically the same as those of the memory location illustrated in Figure 8–11. That is, if $CS = 1$, $RD = 1$, and $WR = 0$ (a read operation), the *R* buffer of each plane of Figure 8–12 is on, and the output of each *addressed memory cell* (recall that only an addressed cell can be accessed) in both plane 0 and 1 is electronically connected to data bus lines D_0 and D_1 via their respective *R* buffers. This loads the data bits of those cells onto the data bus. Once the data is on the data bus it can be read by the CPU.

If a write operation is to be performed then $CS = 1$, $WR = 1$, and $RD = 0$. When input *WR* is driven high, it activates the *W* buffers of both planes, which electronically connect data bus lines D_0 and D_1 to the *D* inputs of each addressed memory cell in planes 0 and 1. However, notice that the data lines are loaded (load capacitance and any input current requirements) with the input of the *W* buffer only for each plane rather than the *D* inputs of every memory cell (8 in this case) in its plane. Even though the data bit is at the *D* input to every memory cell within the plane, *only one memory cell can be written into and it is the one being addressed* (both row and column are active). When *WR* creates a negative edge by going inactive (1 to 0), the data at the *D* inputs of the addressed memory location is strobed into those memory cells and the write operation is completed.

EXAMPLE 8–8

For the stated logic levels of the indicated input pins of Figure 8–12, determine the memory location being accessed and the operation to be performed.

1. $CS = 1$, $A_2 = 0$, $A_1 = 1$, $A_0 = 0$, $RD = 1$ and $WR = 0$
2. $CS = 1$, $A_2 = A_1 = A_0 = 1$, $RD = 0$ and $WR = 1$
3. $CS = 0$, $A_2 = A_1 = A_0 = 1$, $RD = 0$ and $WR = 1$
4. $CS = 1$, $A_2 = A_1 = A_0 = 1$, $RD = WR = 1$

SOLUTION

1. Since $CS = 1$, the memory device is enabled. The address input to the memory device is 2H, which accesses memory location 2H (memory cells 2 and 2-1). $RD = 1$ and $WR = 0$ drive the *W* buffers into the Hi-*Z* state and turn the *R* buffers on, which reads the data bits from memory cells 2 and 2-1 onto data bus lines D_0 and D_1, respectively.

2. The memory device is enabled (CS = 1) and the accessed memory location is 7H. Since RD = 0 and WR = 1, a write operation is to be performed on memory cells 7 and 7-1; hence, the data from data bus lines D_0 and D_1 is written into these cells when the negative edge of WR occurs (WR goes from 1 to 0).
3. Since CS = 0, the memory device is disabled (R and W buffers are off) and no read or write operation can be performed.
4. The memory device is enabled, but since RD = WR = 1, an illogical operation (read and write at the same time) is requested from the accessing device. The data will be read from the memory cell and then, on the negative edge of WR, that data will be written back into the cell. If the CPU drove RD = WR = 1, then the CPU would be malfunctioning, since it cannot read and write at the same time.

If the memory device of Figure 8–12 actually existed, the pin configuration of its dual in-line package (DIP) might appear as shown in the figure.

There are applications for which the designer or manufacturer of a memory device is not concerned with loading of the data bus lines for a write operation, such as if the array is small. Under this condition the designer or manufacturer would eliminate the W buffers.

A ROM has conceptually the same logic for addressing and reading memory locations as the R/W memory device of Figure 8–12. Obviously, for a ROM there is no write control circuitry. A memory location for an $N \times 2$ ROM might appear as illustrated in Figure 8–13. Notice that the memory cells are represented as flip-flops without the ability to have their logic levels changed (the D input and strobe input have been eliminated). From our previous discussion we know that the memory location of Figure 8–13 is first accessed via an address that selects the row and column of the memory cells within that memory location. This activates memory cells m and m-1 of that addressed memory location. After the memory cells have been activated, control signal RD (on a ROM this

Figure 8–13
Read control logic of a ROM memory location.

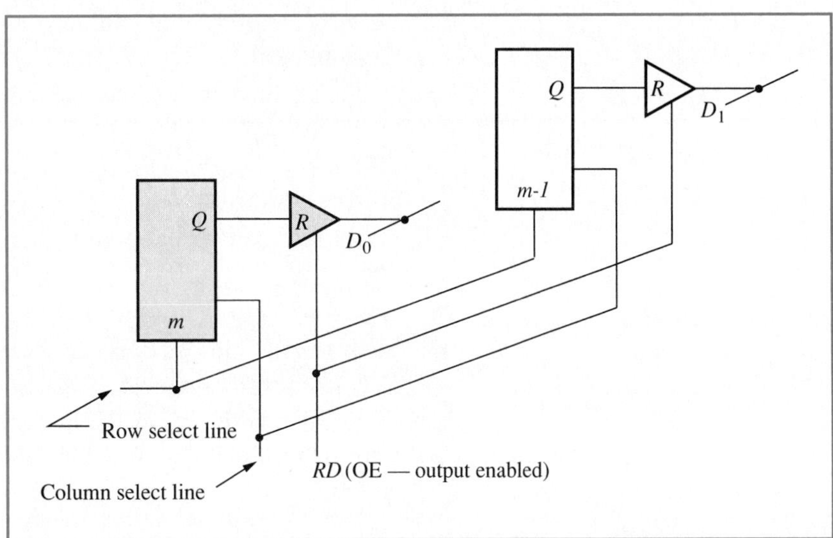

input is often labeled *OE*) is applied to the *R* buffers, which turn them on, resulting in an electronic connection between the *Q* outputs to data lines D_0 and D_1.

To incorporate the memory location of Figure 8–13 into an $N \times 2$ ROM, one would use reasoning similar to that used to develop the R/W array of Figure 8–12. In both cases the *CS* input is used to enable the ROM memory device.

The memory cell array organization of any memory device can be configured in a variety of ways, as was discussed in Section 8–3. From this discussion and Equation (8–1), we can reason that the address pins (A_0, A_1, A_2, etc.) can be distributed in any fashion relative to those used to address row or column, but their sum when applied to Equation 8–1 must equal the number of memory cells. If the memory device array of Figure 8–12 were organized such that there were two rows and four columns, then the row decoder would have a single input A_2 and the column decoder would have two inputs, A_1 and A_0. With three address pins there are eight memory locations, according to Equation (8–1). Manufacturers often organize the array of a semiconductor memory device so that the number of row-related address pins is as close as possible to the number of column-related address pins. This minimizes the access time of the semiconductor memory device, since the burden of address decoding is evenly distributed between row and column decoders.

Review questions

1. For all read/write operations two events must occur. Name the events and place them in the order in which they must occur.
2. What is the essence of an *x-y* coordinate accessing scheme?
3. How can one calculate the number of binary combinations for any number of inputs and also determine what those combinations are?
4. What purpose do the three busses of Figure 8–9 serve?
5. Why is it that only the CPU furnishes addresses, as indicated in Figure 8–9?
6. Why is it necessary to buffer all memory cell outputs? While it is not essential, why is it desirable to buffer the inputs of memory cells?
7. Before a memory cell can be either read from or written to, what condition must exist?
8. What purpose does the *CS* input serve for both ROM and R/W memory devices?
9. How are read/write operations performed on R/W memory devices?
10. How are read operations performed on ROMs?
11. How many address pins are there on a 4096×8 semiconductor memory IC?
12. For a five address pin IC, how many pins are used to address rows and how many to address columns of the memory cell array?

8–11 Introduction to the fundamental concepts of main memory design

We will focus on the design concepts of main memory systems in this section. As a review, recall that any time the CPU wishes to access main memory, it must first load the address of the memory location to be accessed onto the address bus and then drive control signal $IO/\overline{M}$ to a logic 0. That address input to the semiconductor memory device via its address pins is decoded by the device and as a result "locates" the desired memory address. Once the memory location has been selected, the CPU must generate the appropriate control signal (*RD* or *WR*) to perform the intended read or write operation.

Addressing techniques

There are two basic types of addressing, *fully decoded addressing* and *linear addressing*. A fully decoded addressing scheme is one in which the address lines of the address bus are decoded. This scheme maximizes the number of addresses available. For instance, if eight address lines were used to address memory locations and all eight were decoded, there would be $2^8 = 256$ memory addresses available, according to Equation (8–1). However, if two of those address bus lines are not decoded, leaving six that are decoded, the number of available addresses is reduced to $2^6 = 64$ (this is an important concept, especially when comparing fully decoded addressing to partial linear addressing).

Figure 8–14a illustrates the concept of a fully decoded addressing scheme. We see that address bus lines A_6 and A_7 are applied to the 1-of-4 decoder, and the outputs of the decoder are used to select a semiconductor memory IC via the IC's *CS* pin. The remaining address bus lines are decoded by the decoder (row and column) internal to each IC, and as a result the addressed memory location within the IC is selected via the *x-y* coordinate system previously discussed. Since each address line is decoded, the memory system of Figure 8–14a is a fully decoded addressing scheme with 2^8 (256) memory locations.

From Figure 8–14a we can conclude that for a fully decoded addressing scheme, those address lines serving as inputs to the decoder select an IC, whereas the address bus lines serving as inputs to the address pins of the ICs select the specific memory location within the selected IC. Being able to address any location within a memory IC, as well as addressing a specific IC, allows the addressing of any location in the memory system.

Linear addressing refers to the situation in which the address bus lines are not decoded, but instead each address line is used to select a semiconductor memory device. Figure 8–14b illustrates the concept of linear addressing. Each address line is connected to the *CS* of an IC. Hence, whenever the connecting address line has a logic 1 on it, the corresponding IC is selected. Notice that there is no linear method of selecting the memory location within the selected IC, since all semiconductor memory devices use a decoding process for internal memory location selection. As a result, linear addressing by itself cannot be used to address main memory locations. However, as will be shown, linear addressing can be used for addressing I/O.

Fully decoded and linear addressing can be combined to produce a usable scheme for addressing main memory, which we will refer to as *partial linear addressing*. A partial linear addressing scheme uses the necessary number of address lines as inputs to an IC memory device for addressing internal memory locations, just as was done in the fully

Figure 8–14
Main memory addressing schemes.

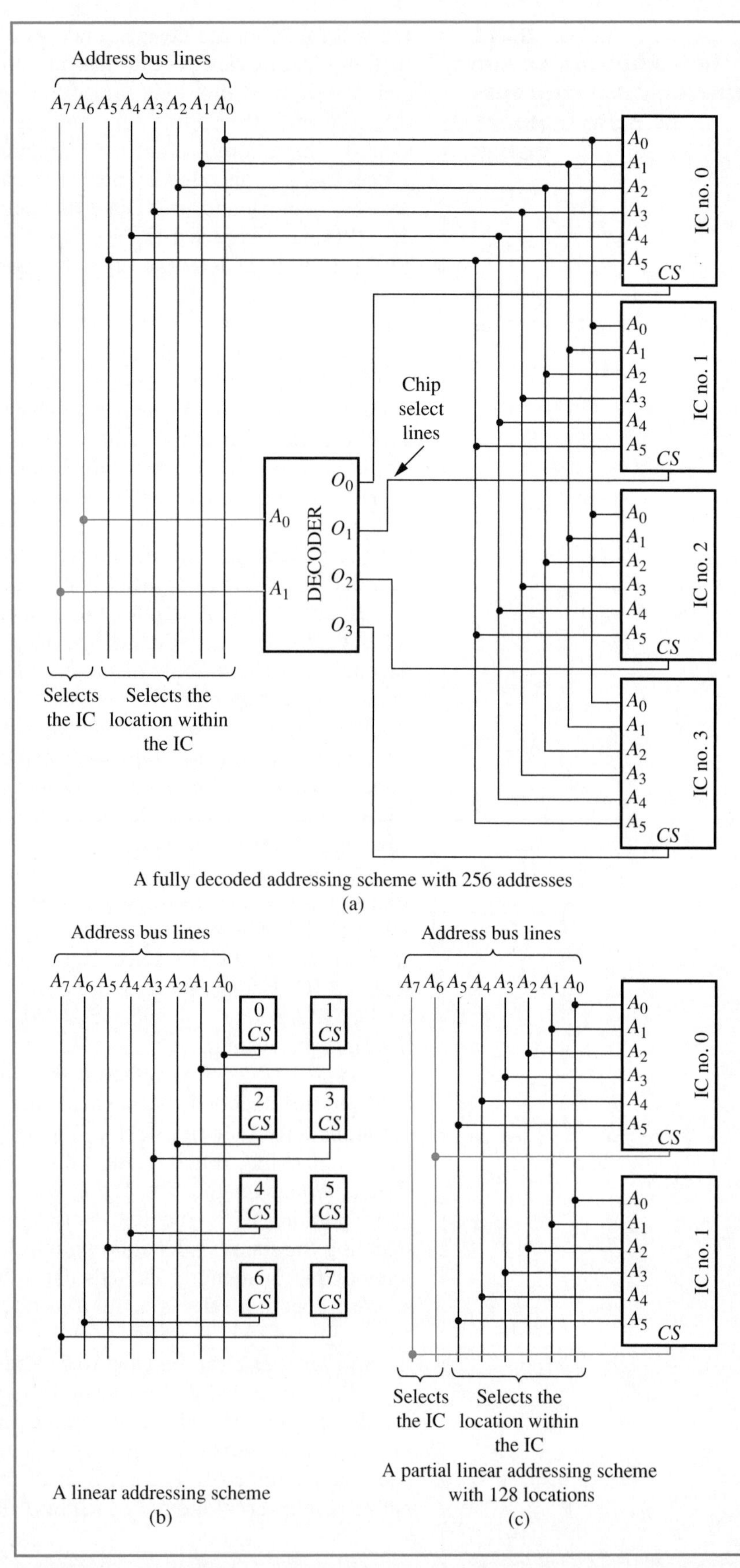

decoded addressing scheme. The remaining address lines are used to select the addressed IC memory device. Figure 8–14c illustrates a partial linear addressing scheme. Address bus lines A_0 through A_5 are used to select a memory location within an IC, and address lines A_6 and A_7 serve to select the IC that contains the location being addressed. With this addressing scheme there are 64 (2^6) locations within each IC; since there are two ICs, there are a total of 128 memory locations.

Comparing fully decoded addressing to partial linear addressing, we can see that a fully decoded scheme yields more addresses for a given number of address lines, whereas partial linear addressing simplifies the hardware (no decoder). The technique used in a design depends on the required memory capacity of the design and the ability to expand its capacity versus the desire to minimize design cost.

Memory organization

In a fully decoded addressing scheme, memory addresses are segmented according to the device within which they reside. As an example of this segmentation, let us consider the main memory of Figure 8–15. It is composed of four IC memory devices, with each memory device having four address pins (A_0–A_3), resulting in 16 memory locations per memory device ($2^4 = 16$). With four such memory devices there are a total of 64 memory locations in this memory system. The number of connections from the outputs of each memory device (O_n for ROM and IO_n for RAM, where RAM is to mean R/W memory) to the data bus lines indicates the bit size of each memory location, which in this case is eight. Therefore, this is a 64×8 memory.

The 64 memory locations of Figure 8–15 can be divided into four equal blocks, one block of memory per IC, with each block of memory locations being defined as a *page of memory*. It should be mentioned that often the contents of main memory are analogously viewed as made up of many pages, much like a book, and the contents of each IC memory device comprise one or more pages of that "book." This is the reason for the notation "page *n*" of Table 8–2 and "page select lines" of Figure 8–15.

If we were to determine the *hexadecimal* address for each of the 16 memory locations within each block of memory or memory device of Figure 8–15, we would find that the addresses would begin at 0 and end with F, just as we found for Figure 8–3b (see Table 8–1). To assign a unique address to each memory location within the memory system of Figure 8–15, we assign page numbers to each memory device, such as page 0 for ROM-0, page 1 for ROM-1, and so forth, and then prefix all memory locations within a memory device with the page number of that memory device. (*Note*: We use the suffix "H" to identify all hexadecimal numbers.) That is, memory address 0FH is memory location FH within page 0 (ROM-0), memory address 1FH is memory location FH within page 1 (ROM-1), memory address 2FH is memory location FH within page 2, and memory address 3FH is memory location FH within page 3. These addresses can be verified by referring to Table 8–2.

Table 8–2 can serve as both a wiring table and a truth table for the address bus lines of Figure 8–15. The wiring connections are shown in the first three rows of the rightmost six columns. The truth table is formed by the remaining rows under those six columns. The top row of Table 8–2 has the six address bus lines listed (A_0 through A_5). The next two rows

Table 8–2
Decoding the address bus of Figure 8–15

Address Bus Lines					A_5	A_4	A_3	A_2	A_1	A_0
Address Pins of the ROM and RAM ICs							A_3	A_2	A_1	A_0
Page Select Decoder				*Output On*	A_1	A_0				
ROM-0 Addresses	Page 0	0	0H	$O_0 = 1$	0	0	0	0	0	0
		0	1H		0	0	0	0	0	1
		0	2H		0	0	0	0	1	0
		0	3H		0	0	0	0	1	1
		⋮	⋮		⋮	⋮	⋮	⋮	⋮	⋮
		0	FH		0	0	1	1	1	1
ROM-1	Page 1	1	0H	$O_1 = 1$	0	1	0	0	0	0
		⋮	⋮		⋮	⋮	⋮	⋮	⋮	⋮
		1	FH		0	1	1	1	1	1
RAM-2	Page 2	2	0H	$O_2 = 1$	1	0	0	0	0	0
		⋮			⋮	⋮	⋮	⋮	⋮	⋮
		2	FH		1	0	1	1	1	1
RAM-3	Page 3	3	0H	$O_3 = 1$	1	1	0	0	0	0
		⋮	⋮		⋮	⋮	⋮	⋮	⋮	⋮
		3	FH		1	1	1	1	1	1
		Hexadecimal addresses			Page Select		Memory location address within the memory device			

are address pins for the ROMs and RAMs of Figure 8–15 as well as the page select decoder. These IC address pins are connected to address bus lines in their respective columns, as indicated in Table 8–2. That is, the second row of Table 8–2 states that address pins A_0 through A_3 of the ROMs and RAMs are connected to the corresponding address bus lines, whereas the third row of Table 8–2 states that address pins A_0 and A_1 of the decoder of Figure 8–15 are connected to address bus lines A_4 and A_5, respectively.

As previously stated, in addition to indicating address bus line connections to device pins, Table 8–2 also serves as a truth table for the address bus lines of Figure 8–15. That is, all possible binary combinations for the address bus lines are indicated in the rightmost six columns beginning in the fourth row. Of course, it is not necessary to show each combination to understand the concept; therefore, for the sake of convenience, many are not shown and are instead represented by dots. The binary address of each memory location is the binary combination of each of these rows. The corresponding hexadecimal value is shown in the column identified as the hexadecimal addresses. The page number being addressed is indicated for each block of memory addresses within that page of memory, as well as the specific IC (ROM-0, ROM-1, and RAM-2, and RAM-3). Note that the page number corresponds to the MSD of the hexadecimal address. Since there are four pages (0 to 3), each page contains 16 addresses (0H to FH), which can be verified by viewing the LSD position of the hexadecimal address. The column labeled as O_n is the page select column and shows which output of the page select decoder is active for any given address (logic 1—the other outputs are at a logic 0 state).

Figure 8–15
A fully decoded address main memory system.

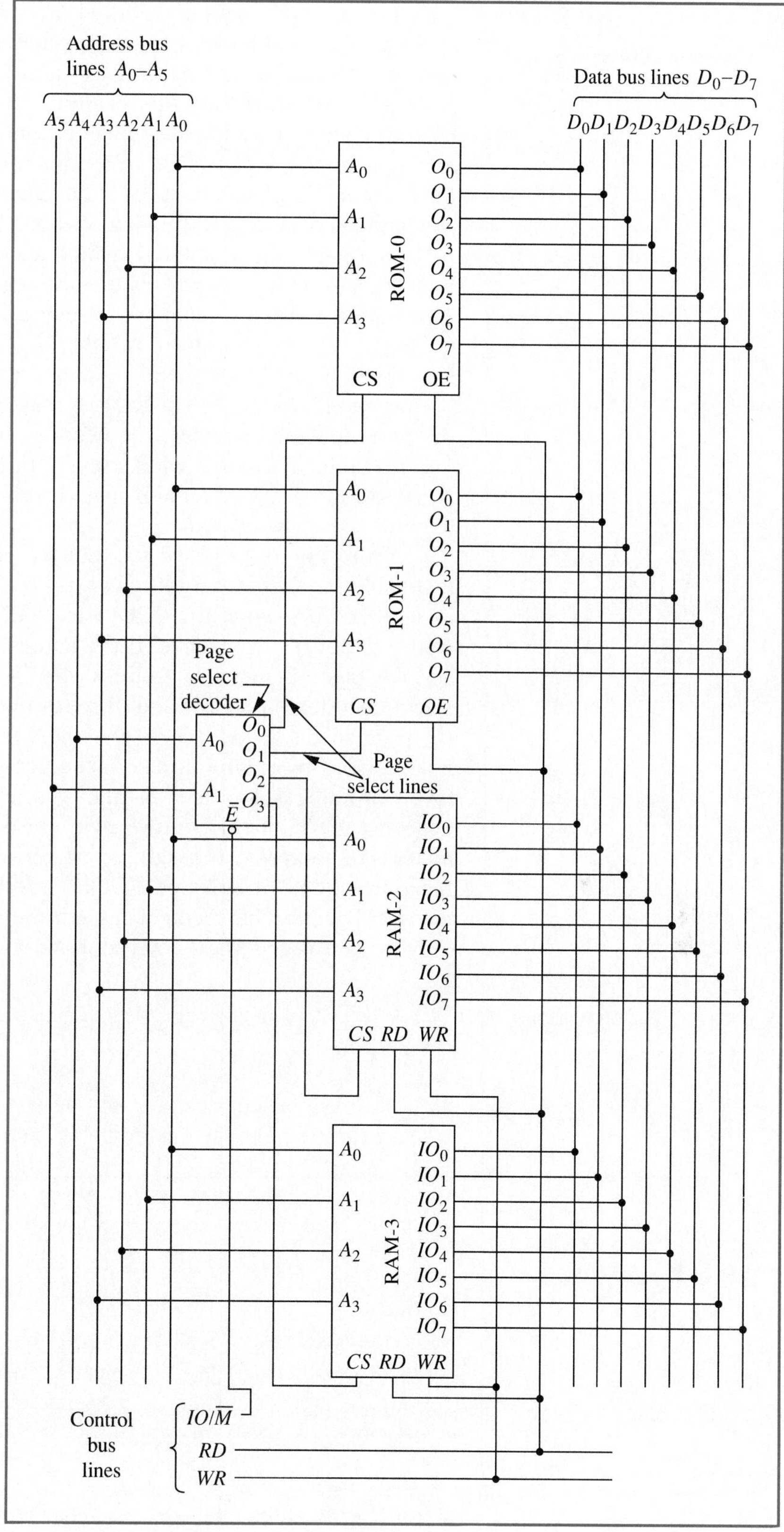

For any fully decoded addressing scheme, it is the page number prefix that is used to logically distinguish one page of addresses from another. The page number prefix of a memory address is decoded by the page select decoder, and the resulting active output of that decoder is used to enable the addressed page of memory via the *CS* input of the memory device. Refer again to the main memory system of Figure 8–15 for an example of this concept. The selection of memory pages is accomplished by connecting address bus lines A_4 and A_5 to inputs A_0 and A_1 of the page select decoder, which decodes the logic levels on address lines A_4 and A_5 and then activates the appropriate output (O_0 through O_3). This output is used to enable the selected memory device by driving its *CS* high. As can be seen from Table 8–2, if address bus lines A_5 and A_4 have logic levels 00 on them, the page select decoder activates its O_0 output, which would select ROM-0 (page 0), whereas if logic level 01 is on these address bus lines, the decoder activates its output O_1 and ROM-1 is selected. Similarly, if logic levels 10 are on address bus lines A_5 and A_4, RAM-2 is selected, and if logic levels 11 are on these address bus lines, RAM-3 is selected.

From Figure 8–15 we see that the page select decoder is enabled via its enable pin $\overline{E}$ driven by the control bus line $IO/\overline{M}$. Recall that the logic level of $IO/\overline{M}$ is what the CPU uses to distinguish memory addresses from I/O addresses. Hence, if the address output by the CPU onto the address bus is a memory address, the CPU drives $IO/\overline{M}$ low. The page select decoder is enabled and decodes that address; otherwise ($IO/\overline{M} = 1$), the page select decoder is disabled and the address on the address bus is not decoded by it (all outputs of the decoder are low). While the logic levels of address bus lines A_5 and A_4 are being decoded by the page select decoder, the IC memory devices are also decoding the logic levels of address bus lines A_4 through A_0, which select the specific memory location within the selected IC that is to be accessed. Of course, once a memory location has been accessed, the desired read or write operation may be performed via the appropriate control bus signal *RD* or *WR*.

EXAMPLE 8–9

For the bus logic levels indicated below, refer to Figure 8–15 and determine

1. the hexadecimal address of the memory location being accessed.
2. the page select decoder output that is acitve (high) and the resulting memory device being enabled via its *CS* input.
3. the specific memory location within the selected IC memory device being accessed.
4. the read or write operation being performed.

The bus logic levels are
(a) Address Bus Lines (A_5 through A_0) = 001001, $RD = 1$ and $WR = 0$.
(b) Address Bus Lines = 100010, $RD = 0$ and $WR = 1$.
(c) Address Bus Lines = 011110, $RD = 0$ and $WR = 1$.
(d) Address Bus Lines = 110101, $RD = 1$ and $WR = 0$.
For each of the above conditions $IO/\overline{M} = 0$.

SOLUTION

(a) The hexadecimal address for address bus logic levels 001001 is 09H ($001001_2 = 09_{16}$). The two most significant address bus lines are

decoded by the page select decoder, and for input logic levels 00 output O_0 is active, which enables ROM-0. The remaining bits (1001) are decoded by ROM-0, which selects location 9_{16}. Since $RD = 1$ and $WR = 0$, a read operation is to be performed at the addressed memory location 09H.
(b) The hexadecimal address is 22H ($100010_2 = 22_{16}$), which activates output O_2 of the page select decoder. This output enables RAM-2 and selects location 2_{16} (0010_2) within RAM-2. Since $WR = 1$, a write operation is performed at location 22H.
(c) The memory location being accessed is 1EH, which activates output O_1 of the page select decoder and, in turn enables ROM-1. Location EH (1110_2) within ROM-1 is being addressed. According to the logic levels of the control signal, a write operation is to take place. This is an illogical and illegal command (it cannot be done), since ROM-1 is a ROM. Therefore, there is an error in either the address or the control signal.
(d) Memory location 35H is being read. Output O_3 is active and enables RAM-3, and location 5H within RAM-3 is read.

Recall that the difference between fully decoded addressing and partial linear addressing is that there is no page select decoder in a partial linear addressing scheme, and hence no page select line coming from the page select decoder to enable the *CS* input of a memory device, as there is in a fully decoded addressing scheme. Rather, an address line from the address bus serves the function of selecting the addressed IC memory device. Figure 8–16 is the memory system of Figure 8–15 modified to use partial linear addressing. In comparison to Figure 8–15, the partial linear addressing scheme of Figure 8–16 requires two additional address bus lines (A_6 and A_7), since there is no page select decoder. That is, from two inputs the decoder of Figure 8–15 generates four outputs that are used to select IC memory devices. Therefore, without the decoder there must be four address bus lines to select the addressed IC memory device. Those address bus lines that are to serve as page select lines (A_4 through A_7) are *AND*ed with control signal $IO/\overline{M}$ (inverted) in order to distinguish between memory and I/O addresses. For the system of Figure 8–16, address bus line A_4 enables ROM-0 (page 0), A_5 selects ROM-1 (page 1), etc.

Table 8–3 provides a convenient address reference for the available addresses of Figure 8–16. The top six rows of Table 8–3 indicate which memory device pin is connected to each address bus line. For instance, the *CS* of ROM-0 is connected to address bus line A_4, as indicated by the fifth column from the right. When $A_4 = 1$, ROM-0 is selected. Likewise, the sixth column from the right states that address bus line A_5 is connected to *CS* of ROM-1, and when $A_5 = 1$ page 1 is selected. Address bus lines A_0 through A_3 are connected to address pins A_0 through A_3 of the RAMs and ROMs of Figure 8–16.

From Table 8–3 we can see that the biggest disadvantage to a partial linear addressing scheme is the decrease in the number of addresses available for the number of address lines used. Table 8–3 indicates that the first available address is 10H and that addresses 30H through 3FH, 50H through 5FH, 60H through 6FH, and 70H through 7FH are not available. That is, the available addresses are not continuous. Had we

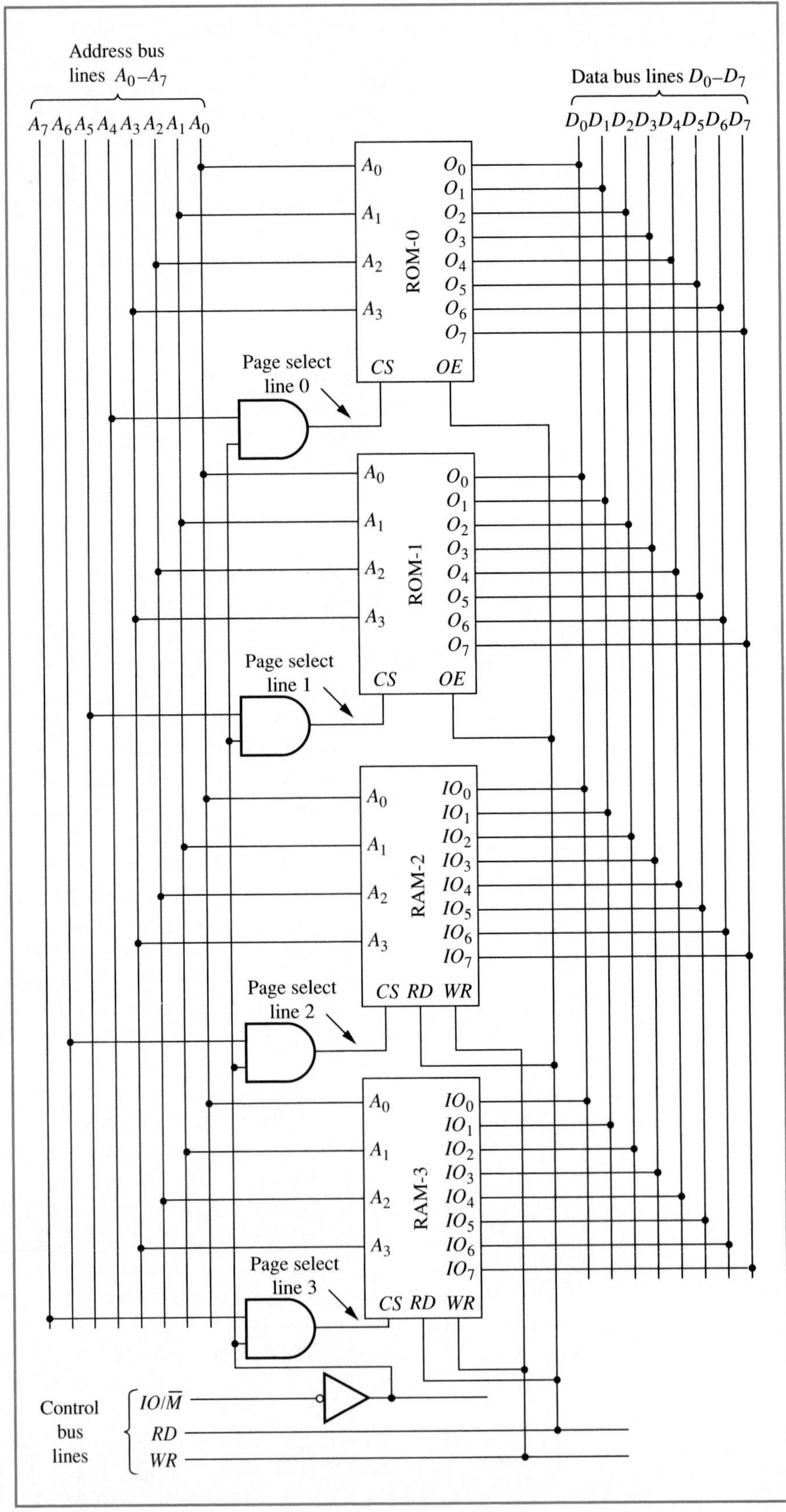

Figure 8–16
A partial linear addressed main memory system.

Table 8–3
"Decoding" the address bus of Figure 8–16

Address Bus Lines		A_7	A_6	A_5	A_4	A_3	A_2	A_1	A_0
ROMs and RAMs						A_3	A_2	A_1	A_0
ROM-0					CS				
ROM-1				CS					
RAM-2			CS						
RAM-3		CS							
ROM-0 addresses (Page 0)	10H ⋮ 1FH	0 ⋮ 0	0 ⋮ 0	0 ⋮ 0	1 ⋮ 1	0 ⋮ 1	0 ⋮ 1	0 ⋮ 1	0 ⋮ 1
ROM-1 addresses (Page 1)	20H ⋮ 2FH	0 ⋮ 0	0 ⋮ 0	1 ⋮ 1	0 ⋮ 0	0 ⋮ 1	0 ⋮ 1	0 ⋮ 1	0 ⋮ 1
RAM-2 addresses (Page 2)	40H ⋮ 4FH	0 ⋮ 0	1 ⋮ 1	0 ⋮ 0	0 ⋮ 0	0 ⋮ 1	0 ⋮ 1	0 ⋮ 1	0 ⋮ 1
RAM-3 addresses (Page 3)	80H ⋮ 8FH	1 ⋮ 1	0 ⋮ 0	0 ⋮ 0	0 ⋮ 0	0 ⋮ 1	0 ⋮ 1	0 ⋮ 1	0 ⋮ 1

fully decoded those eight address bus lines of Figure 8–20, there would have been $2^8 = 256$ available addresses rather than the 64 using partial linear addressing techniques. However, for designs requiring small numbers of memory locations, partial linear addressing is suitable, especially since linear addressing simplifies the hardware (no page select decoder).

EXAMPLE 8–10

For the addresses given, determine from Figure 8–16 the enabled memory device and the address of the accessed memory location within that memory device.
(1) 25H (2) 4AH (3) 65H

SOLUTION

1. 25H = 00100101_2. Since bit 5 is a logic 1, $A_5 = 1$, which enables ROM-1. Bits 3–0 have logic levels 0101, which is equivalent to 5H. Therefore, memory location 5H of ROM-1 is being accessed.
2. 4AH = 01001010_2. Since bit 6 is a logic 1, memory device RAM-2 is enabled. Therefore memory location AH within RAM-2 is being accessed.
3. 65H = 01100101_2. Two address lines have logic 1 level (bits 5 and 6), which would enable both ROM-1 and RAM-2. This is illegal (a nonexistent address); hence there is a fault with the address. Note that if $RD = 1$ and $WR = 0$ for this address, bus contention would exist (both ROM-1 and RAM-2 would be contending for the data bus).

It is the responsibility of the memory designer to identify valid addresses using a table similar to Tables 8–2 and 8–3. It is the responsibility of the user to use only valid addresses when writing programs.

Review questions

1. What is the principle of operation of a fully decoded addressing scheme?
2. What is the principle of operation of a linear addressing scheme?
3. How does partial linear addressing combine linear and fully decoded addressing?
4. How can one determine the memory capacity of a memory system by examining its address and data bus configuration?
5. How are read/write operations executed?
6. Why must read/write operations be synchronized with the CPU?

8–12 Magnetic tape memory

Magnetic tape is a medium that can be used to store data bits. When used for storage it serves as a memory, as illustrated in Figure 8–5a. Its use is limited to mass storage owing to its relatively long access time. Magnetic tape mass storage devices have memory capacities that greatly exceed the large megabyte capacity of hard disk memories found in many computers. As a result, magnetic tapes are often used as a means to back up the data stored on hard disk in case the system's hard disk memory "crashes" (losses its contents). Of course, magentic tapes can also be used to store programs and data for the main memory; however, the faster disk systems (floppy and hard disk) are most often used for that purpose.

The memory organization of a magentic tape is a dynamic one-dimensional array. To simplify the concepts of magnetic tape data formatting and accessing, the simplified representation of a magnetic tape mass storage device is illustrated in Figure 8–17. The tape is divided into two parts, the left side containing the binary address of the location and the right side the binary content. As the tape passes through the access mechanism, memory location addresses are detected and a read or write operation is performed on addressed locations using magentic coupling.

By means of magnetic coupling, any binary bit, whether from an address, data, or instruction, can be read from the magnetic tape or written onto it, with an R/W head similar in concept to the one illustrated in the inset of Figure 8–17. When current flows in the coil wound around the R/W head, a magnetic flux is produced in the head. Within the magnetic material of the head this flux travels unimpeded and is contained within the magnetic material that forms the R/W head. However, as the flux flows across the air gap of the R/W head, the flux spreads and couples with the magnetic tape. The coupling of the flux and magnetic tape creates a magnetic domain (magnetizes a spot on the tape) beneath the air gap of the R/W head, and this domain forms the memory cell. A magnetic domain is represented in the inset of Figure 8–17 as a dotted-line rectangle.

The direction of current flow in the coil of the R/W head determines the polarity of the magnetic domain and hence the logic state of the memory cell. To write a logic 1 into a magnetic domain, the coil current must flow in a specified direction; to write a logic 0 into the domain, the coil current must flow in the opposite direction. These magnetic domains are nonvolatile and therefore "remember" the logic state they were polarized to represent.

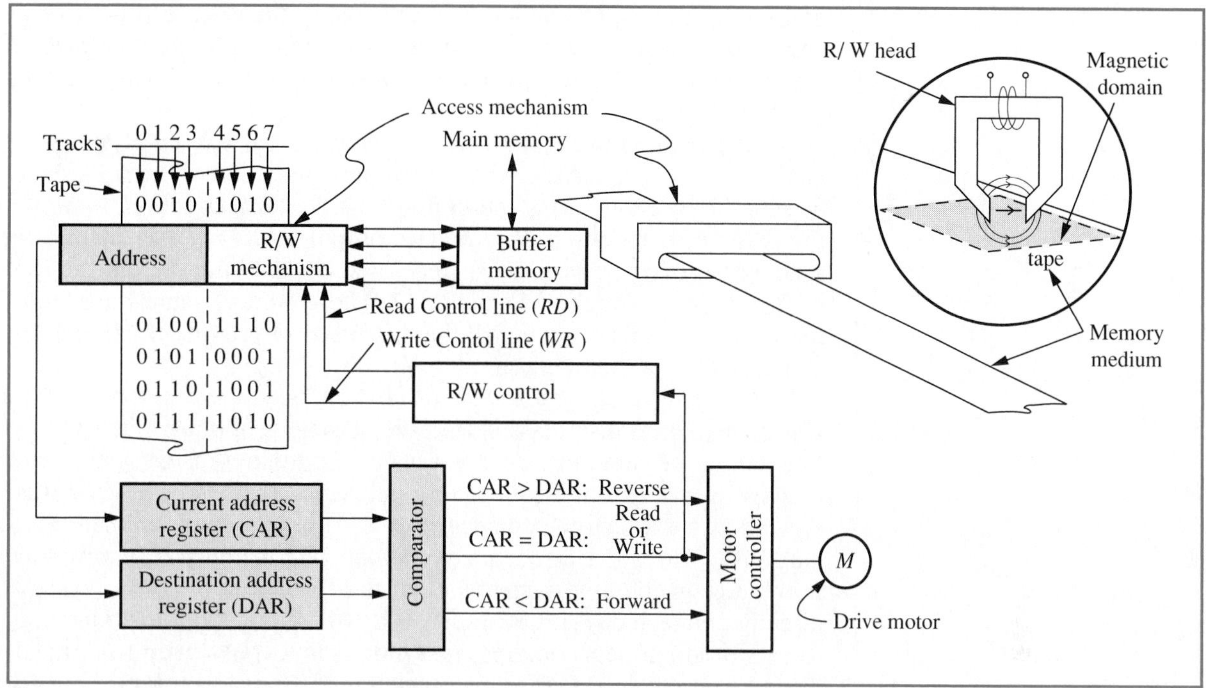

Figure 8–17
Architecture of an address locator for a magnetic tape mass storage device.

Reading from a tape is the reverse of the writing process. As the tape passes beneath the access mechanism, the flux created from the polarity of the domain passing beneath the R/W head couples with the R/W head via the air gap. This coupling results in a current flowing in the coil of the R/W head as long as movement occurs between the R/W head and the magnetic tape. The current direction is a function of the logic level (1 or 0) stored in the magnetic domain. The coil current direction is detected by electronic sensors within the mass storage device which detect whether a logic 1 or 0 was stored in that domain. There is an R/W head for each of the eight tracks of the memory medium in Figure 8–17. These R/W heads are housed within the access mechanism.

Let us now investigate how the accessing mechanism of Figure 8–17 operates. The four R/W heads in the address tracks (0 through 3) are in a read mode, whereas those in the data tracks (4 through 7) can be in either a read or a write mode. As the tape passes through the access mechanism, the R/W heads of tracks 0, 1, 2, and 3 read addresses sequentially from the tape. The address of the memory location currently passing through the access mechanism (location 0011 as shown) is known as the current address. This current address is loaded into the *current address register* (CAR). The address of the memory location to be accessed, known as the destination address, is loaded into the *destination address register* (DAR) by whatever device is to access that location on the tape. The contents of the CAR and DAR are input to a comparator, and a comparison is made between their magnitudes. If CAR > DAR (the current address is greater than the destination address), the tape has traveled too far and must be reversed; if CAR < DAR, the tape has not traveled far enough and must proceed further. When CAR = DAR the address of the memory location positioned beneath the access mechanism is also the address of the destination location. Under this condition, the desired operation (read or write) must be performed; once the operation is executed, the tape can either stop or move to the next location to be accessed. Because magnetic tapes are accessed in a sequential fashion, the acronym SAM (sequential access memory) is used to indicate their method of access.

As indicated in Figure 8–17, it is outputs CAR > DAR, CAR = DAR, and CAR < DAR of the comparator that control the rotational direction of the tape drive motor and thus the tape direction, as well as the read/write operation. Comparator output CAR = DAR enables the read/write control signals being generated by and output from the R/W control function-block of Figure 8–17. These control signals are input to the R/W mechanism, by which they are used to provide the timing for the actual read or write operation.

As we have already stated, magnetic tape storage devices are much slower than hard disk mass storage devices and, of course, the CPU. Since they are slow, one does not want to burden a faster system when reading or writing data from or to a magnetic tape, such as when it is serving as a mass storage device (Fig. 8–1) or as a back-up memory for a hard disk. Instead, a fast semiconductor FIFO memory is incorporated into the magnetic tape system so that the FIFO and the faster system are directly connected. With this approach, when data from main memory is to be stored on magnetic tape, that data is first transferred from main memory into the FIFO memory, which is identified as a buffer memory in Figure 8–17. Once the data is transferred to the buffer memory, the magentic tape mass storage device can write it to the tape at its own pace without CPU involvement. Similarly, when data is to be transferred to main memory from the magnetic tape system, the data to be transferred is read from the tape and written into the buffer memory. When all the data to be transferred has been loaded into the buffer (or when the buffer is filled), the data transfer takes place between the buffer memory and main memory. Since the buffer memory is constructed of semiconductors, the data transfer is fairly rapid and does not tie up the computer system. The buffer memory serves a similar purpose when connected to a hard disk system.

When mass storage has the ability to transfer data between itself and main memory without CPU involvement, it is said to have *direct memory access* (DMA) capability. DMA capability will be discussed again in the chapter investigating microprocessors—Chapter 10.

EXAMPLE 8–11

The mass storage device of Figure 8–17 has an access time of 2 ms for each address when accessed in sequence.

1. If CAR $= 4_{16} = 0100_2$ and DAR $= 9_{16} = 1001_2$, what are the logic levels of the comparator outputs and what is the access time for address 9_{16} when the magnetic tape is currently at location 4_{16}?
2. After address 9_{16} is accessed (CAR $= 9_{16}$), suppose the DAR is loaded with 3_{16}. What are the logic levels of the comparator outputs and what will be the access time for address 3_{16}?

SOLUTION

1. Since $4_{16} < 9_{16}$, the logic levels of the comparator outputs are (CAR > DAR) = 0, (CAR = DAR) = 0, and (CAR < DAR) =

1. Therefore, the tape coontinues in a forward direction. The tape must travel forward five locations, requiring an access time of (5 × 2 ms) 10 ms.
2. Comparator output logic levels are (CAR > DAR) = 1, (CAR = DAR) = 0, and (CAR < DAR) = 0. Therefore, the tape direction must be reversed. The tape must travel in reverse for six locations, resulting in an access time of 12 ms.

The data format of the tape in Figure 8–17 was presented in a simplified straightforward manner in order to focus on concepts and not mechanical details. However, it should be pointed out that the data format of Figure 8–17 is not as it actually appears on magnetic tape mass storage systems. Because magnetic tape is used to load and unload large blocks of data, including programs, between the memory medium and main memory (rather than addressing each datum individually), the data is grouped in blocks, and it is these blocks that are assigned addresses.

A representative format is illustrated in Figure 8–18a. With this type of formatting the access mechanism locates the block of data via its address, and the read/write operation is performed on the following memory locations. As we see from Figure 8–18a, there is other information on this tape, such as synchronizing data (mark), indications of how much data is in a block, and error-checking information. This information is necessary to synchronize the tape positioning with control of read/write operations via the R/W head. The control logic must be told the size of the data block so that it will know when to stop reading data from or writing data to the tape, and, of course, there must be some error checking. Each data block is divided into fields. Those fields identified as marks serve to identify the beginning of other fields, which are the address and data fields. The gaps allow the R/W head time to change modes (say from read to write). As illustrated in Figure 8–18a, there are usually nine tracks instead of the eight illustrated in Figure 8–17. The specific format is of interest, but more important to us are the concepts of addressing, reading, and writing.

There are many methods for coding 1's and 0's on magnetic tape; however, we shall discuss only three. The first is the *return-to-zero* (RZ). This method requires that a logic 1 or 0 be written in a specific time interval, identified as T in Figure 8–18b. The logic 1 pulse has a fixed duration T_d, which is well within the interval T, and as a result "returns to a zero" from logic level 1 within time T. We note that to write a logic 0 we simply do not write a logic 1 in the T interval.

The Manchester technique is illustrated in Figure 8–18c. To record a logic 1 it requires that a high be present during the entire time interval T. A logic 0 must traverse from a high to a low at the beginning of the interval T, and remain low for a specified time T_1, and then traverse from the low back to a high.

The third technique is the popular hobbyist-oriented method known as the Kansas City technique. It is named after the city in which it was devised. The Kansas City method used two tones to represent logic 1 and logic 0. A logic 1 is recorded as a 2.4-kHz tone and a logic 0 as a 1.2-kHz tone, as illustrated in Figure 8–18d. As in the other recording

Figure 8–18
A magnetic tape format and techniques for the storage of 1's and 0's.

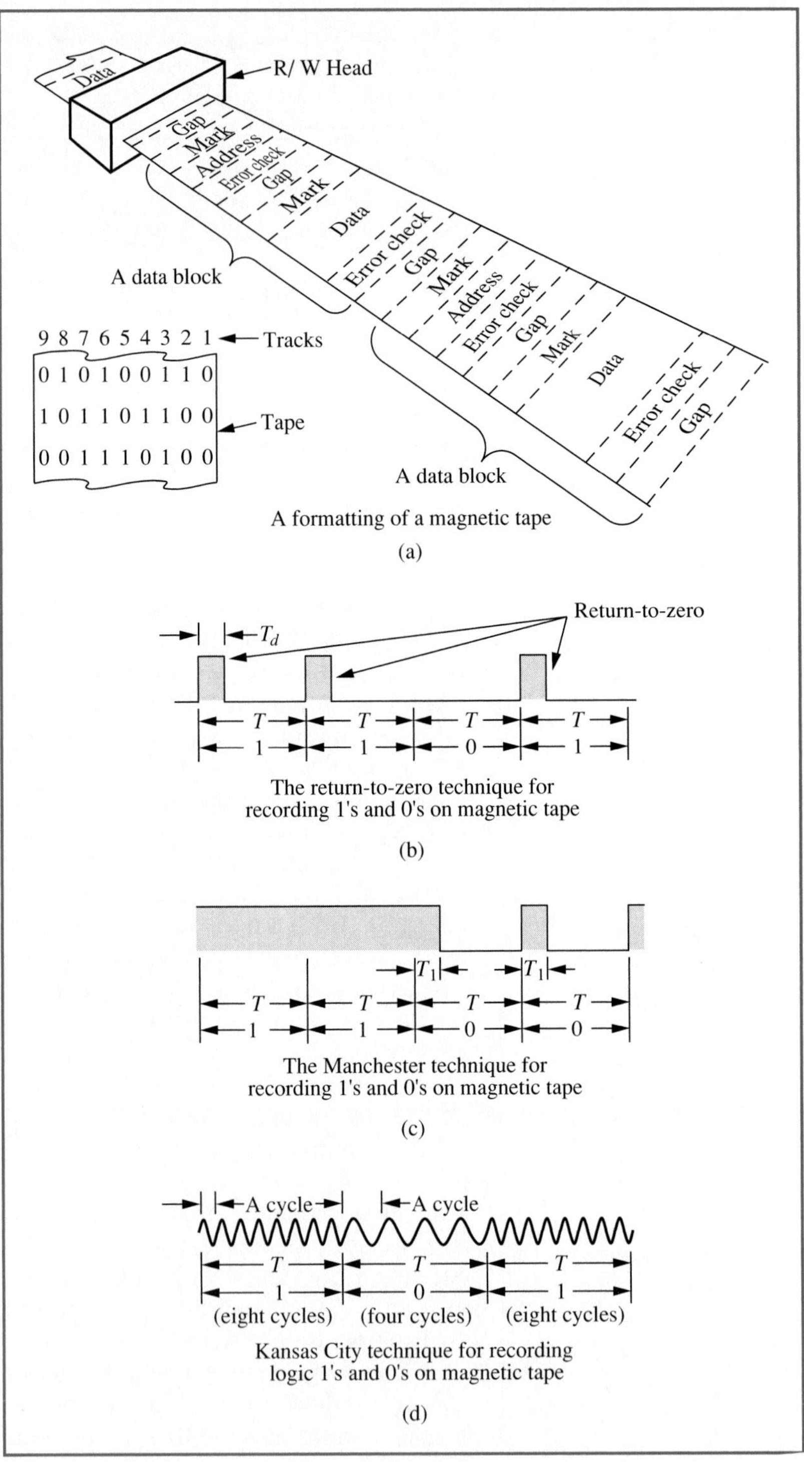

techniques, the interval within which the logic state is to be recorded (T) is fixed. The length of interval T is eight cycles of the logic 1 sinusoidal waveform or four cycles of a logic 0 waveform. Of course, this is because logic 1 has twice the frequency of logic 0.

Review questions

1. What functions do the CAR and DAR registers serve?
2. How does the access mechanism know when the location being accessed is aligned with the R/W heads?
3. What is the purpose of the FIFO memory?
4. Why is data stored in blocks on a magnetic tape, and how are these blocks located by the accessing mechanism?
5. What function does the address section serve in the format of Figure 8–18a?
6. What is the RZ technique to record a logic 1 or 0?
7. What is the Kansas City technique for recording a logic 1 or 0?

8–13 Disk memory

Magnetic disk storage media are also used by mass storage devices to store large blocks of data. Conceptually, magnetic tape and magnetic disk media are very similar. The tape is composed of magnetic material, whereas a disk is coated with a magnetic film. Floppy disks are made of very thin mylar and hard disks of rigid aluminum. The rigidity of the hard disk allows the disk to be rotated more rapidly and provides a more precise rotating surface, which enables more rapid access and more densely packed data. There are various sizes ($3\frac{1}{2}$ inch, $5\frac{1}{4}$ inch, and 8 inch) and types of disks, as well as various formats. Their memory has a dynamic one-dimensional organization very similar to that of magnetic tapes. We will base our conceptual discussion of disk memories on the floppy disk (using a standard 8-inch floppy disk as an example).

Floppy disks rotate in one direction at 360 rpm with an average access time of 200 ms. Typically an 8-inch disk can store 250 kilobytes per side. Like magnetic tapes, disks have tracks, but they are somewhat different. To visualize the format of a floppy disk, refer to Figure 8–19. The tracks of a magnetic disk are 77 concentric circles, as illustrated in Figure 8–19a. The disk is divided into 26 sectors, as shown in Figure 8–19b. Figure 8–19c shows one sector from Figure 8–19b. We define each track within a sector as a *sector track*. Figure 8–19d is a detailed illustration of the sector of Figure 8–19c showing the memory cells that are created by the R/W head. The R/W head reads or writes data (1 or 0) from or to memory cells within a sector track as the track is rotated beneath it in much the same manner as reading or writing from or to a magnetic tape memory cell. There are 128 memory cells per sector track. Digital control circuitry positions the R/W head over the addressed track.

From Figure 8–19a through d we can calculate the byte storage capacity of this type of formatted floppy disk. There are 26 sectors per disk, with each sector having 77 sector tracks, a total of 2002 sector tracks. With each sector track containing 128 bytes, there are 256,256 bytes of memory capacity available on one side of the disk.

The floppy disk is housed in a protective jacket, as indicated in Figure 8–19e. This jacket has an access window, which enables the R/W head to make contact with the disk. To provide a reference point on the disk, an index hole is made in the disk, as shown in Figure 8–19e. There is an access hole on the jacket that can be aligned with the index hole of the disk, allowing light to pass through when the two align. This light is detected by the floppy disk drive and is used to identify the first sector of

Figure 8–19
A floppy disk format.

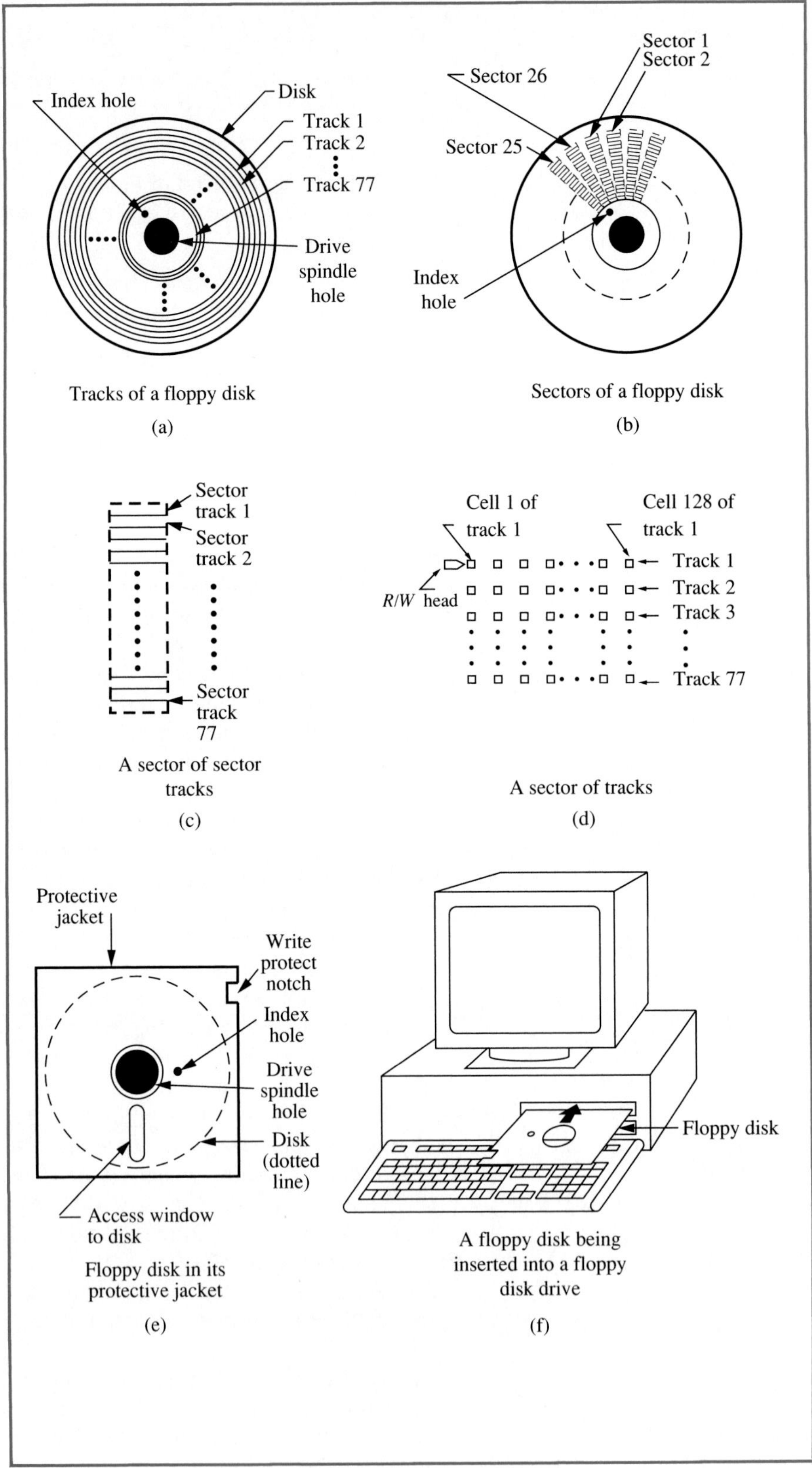

Figure 8–20 A simplified data format for a disk sector.

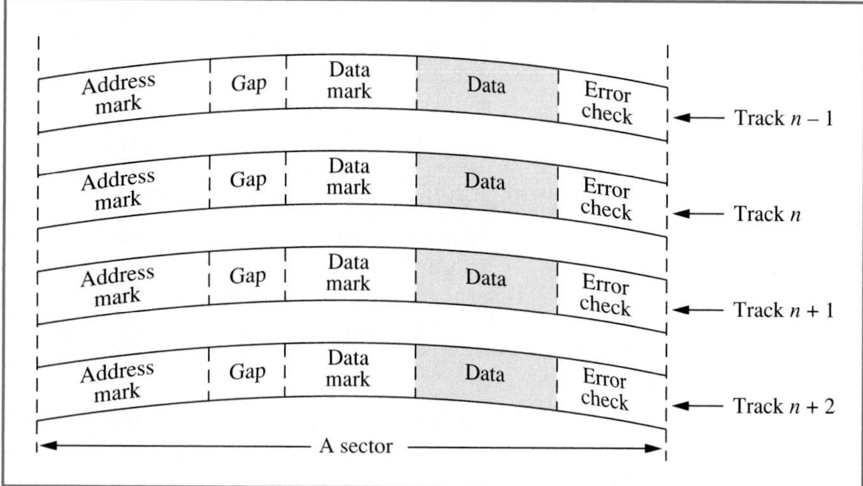

a track. The jacket has a write protect notch, which, when covered by the user, disables the write mode of the R/W head and thereby provides protection against unwanted writes on the disk. Figure 8–19f is an illustration of a floppy disk being inserted into a floppy disk drive.

There are various data formats for a floppy disk. A representative simplified format is shown in Figure 8–20. Using the same conceptual data format as for magnetic tape, we see that the data block address appears before the block of data, which allows the mass storage device to locate that data block. Once the portion of the track containing the data block has been located, the R/W head can perform the appropriate operation. To locate a data block on a disk, the destination address must contain sector number as well as track number. The access mechanism has logic circuits much like those of a magnetic tape (Fig. 8–17). There are registers with functions similar to those of CAR and DAR. That is, their content is input to a comparator, and the output of the comparator indicates whether the R/W head is in the correct sector and track. When the addressed sector is aligned, the R/W head and the control circuitry engage the R/W heads for the appropriate read or write operation.

Review questions

1. How does a track on a disk differ from a track on magnetic tape?
2. What is a disk sector?
3. How many sectors are there on a disk?
4. How many memory locations are there in each sector, and what is the word size (number of bits) for each location?
5. What is addressable on a disk?

8–14 Accessing I/O: isolated I/O and memory-mapped I/O

Note from the computer architecture of Figure 8–1 that if I/O peripheral devices were redefined such that I/O devices were simply "all devices external to the CPU," then main memory would also be considered an I/O. In reality this is the case; however, because of the very special nature and specific role of main memory, we tend to treat it as if it were not an I/O device. Because main memory is an I/O device (although it is never referred to as I/O), the techniques used to access it are similar to those for traditional I/O devices. Therefore, in our previous coverage of accessing memory locations, we have covered many of the essentials of accessing other I/O devices. That is, when the CPU is accessing an I/O device it must first output its address on the address bus and then drive control signal $IO/\overline{M}$ high, which enables the I/O selector so that it can decode that address (if decoding is used). The CPU then generates the appropriate read or write control signal. As you now realize, the process of accessing main memory and I/O is conceptually the same.

A computer system that distinguishes between memory and I/O addresses is known as an *isolated I/O system*. Isolated I/O systems utilize the control signal $IO/\overline{M}$ or a similar control signal to make such distinctions. The computer system of Figure 8–9 is an isolated I/O system, since it utilizes control signal $IO/\overline{M}$. The memory systems of Figures 8–15 and 8–16 are also isolated I/O memory systems, as indicated by the use of the control signal $IO/\overline{M}$.

Those computer systems that do not differentiate between memory and I/O addresses are referred to as *memory-mapped I/O systems*. These systems treat I/O devices simply as memory locations and assign addresses to both memory and I/O with no distinction between the two. This implies that the control signal $IO/\overline{M}$ is not required.

Converting the isolated I/O memory systems of Figures 8–15 and 8–16 to a memory-mapped I/O system requires the elimination of the control signal $IO/\overline{M}$ and an increase in the number of address bus lines to compensate for the additional I/O addresses. As a result, the system of Figure 8–15 would be modified so that the enable pin of the page select decoder would be tied low (strapped to ground), and the number of address bus lines would be increased. Adding an additional address bus line to the system of Figure 8–15 increases the number of addresses available from 64 to 128. Those additional 64 addresses could be assigned to I/O devices.

The computer system designer divides the available addresses into two parts—those assigned to memory and those assigned to I/O, a process known as mapping the address space. The addressing is designed so that there is no overlapping of the memory space and I/O space (i.e., there are no addresses common to both a memory location and an I/O device).

An isolated I/O system

Figure 8–21 is an isolated I/O memory system that uses a fully decoded addressing scheme. As previously shown, control signal $IO/\overline{M}$ partitions memory addresses from I/O addresses by enabling either the memory page select decoder or the I/O select decoder. As mentioned at the beginning of this section, accessing an I/O device is like accessing a memory location—the I/O device is selected via an address, which enables it, and then is read from or written to by activating the appropriate control signal. This select and read/write control of I/O devices is provided by the I/O select decoder and its control circuitry for read/write operations, as shown in Figure 8–21.

When an input I/O device (no. 0 or no. 3 of Figure 8–21) is being read by the CPU, the CPU outputs the address of the I/O device on the address bus and drives control line $IO/\overline{M}$ high. The I/O select decoder

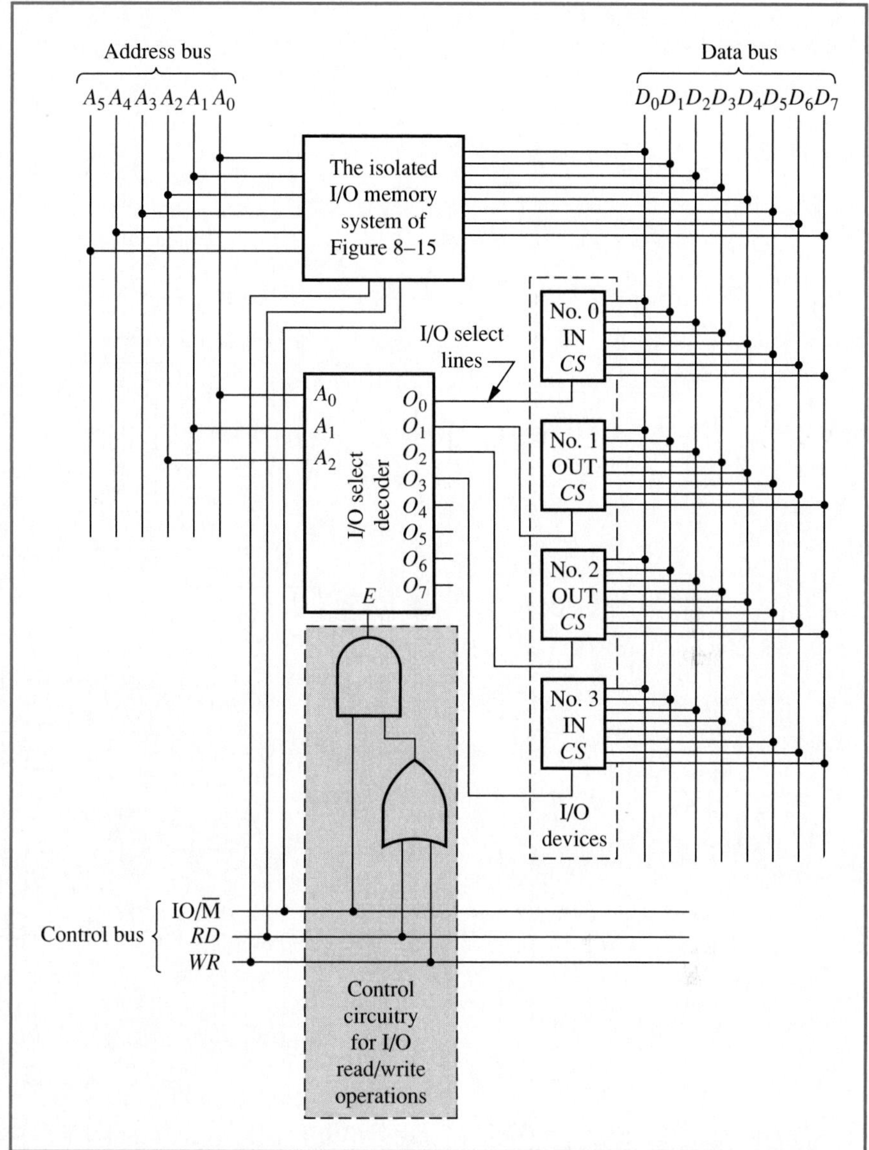

Figure 8–21
An isolated I/O memory system.

decodes that address and drives the appropriate output (O_0 or O_3) high when its enable pin E is driven high. As we see from the control circuitry for read/write operations of Figure 8–21, the enable pin E is not driven high until both $IO/\overline{M}$ and either RD or WR are also high. Hence, when $IO/\overline{M}$ and RD are driven high by the CPU, then the I/O address on the address bus is decoded and the appropriate output of the I/O select decoder enables the addressed I/O device. The addressed I/O then loads its datum (the datum that is to be read by the CPU) onto the data bus, from which the CPU can latch (read) it.

For the CPU to write to an I/O device (I/O devices nos. 1 and 2 in Figure 8–21), the CPU addresses the I/O device via the address bus. The CPU drives $IO/\overline{M}$ high, loads the datum to be written to the I/O device on the data bus, and drives WR high. With both $IO/\overline{M}$ and WR high, the addressed I/O device then latches the datum on the data bus (on the negative edge of WR), which constitutes an I/O-write operation.

Figure 8–22
A memory-mapped I/O system.

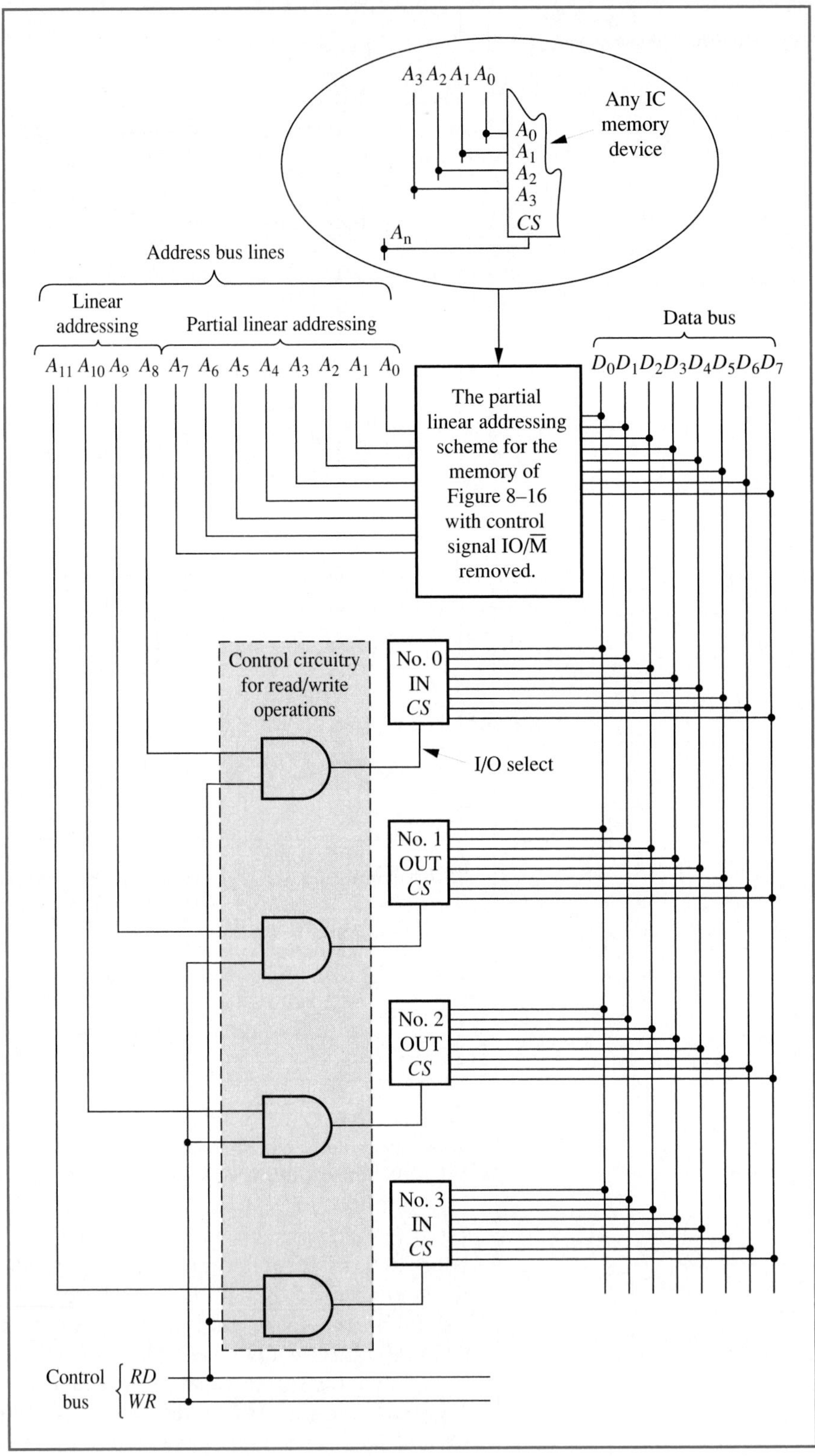

A memory-mapped I/O system

The system of Figure 8–22 is representative of a memory-mapped I/O system that uses linear addressing for accessing I/O devices. For the memory portion of this system, the memory of Figure 8–16 was modified so that it became a memory with memory-mapped I/O. To accomplish this modification the control signal $IO/\overline{M}$ and accompanying *AND* gates were eliminated, as shown in the inset of Figure 8–22, and four more address lines were added. To provide timing for the read/write operations, the *AND* gates of Figure 8–22 (identified as control circuitry) were incorporated. As a result, to access an I/O device the appropriate address line must be active (high) and the appropriate read (*RD*) or write (*WR*) control must also be active.

As an example of a write operation, if the CPU is to write to I/O device no. 1 of Figure 8–22, it must drive address line A_9 high and keep it high; when it has loaded the datum on the data bus, it then drives *WR* high. When these two inputs to the *AND* gate are high, the output of the *AND* gate is driven high, which enables I/O device no. 1, which then latches the datum that is on the data bus (on the negative edge of *WR*). As an example of a read operation, when the CPU is to read a datum from an I/O device, say I/O device no. 3, it first drives A_{11} high and then drives *RD* high at the appropriate time (when it is ready to latch the datum from the data bus). With both A_{11} and *RD* high, the output of the *AND* gate is high, which enables I/O device no. 3 via its *CS* pin. Enabling I/O device no. 3 causes it to load its datum onto the data bus, from which the CPU can latch it.

Review questions

1. What is the conceptual difference between a memory-mapped I/O system and an isolated I/O system?
2. How is an I/O read or write executed relative to CPU control signals *RD* and *WR*?
3. What signal determines whether a system is memory-mapped or isolated I/O? Explain how it works.
4. Why are the control signals *RD* and *WR* gated with an address line or I/O select line?
5. Why is the negative edge of *WR* of particular importance?
6. What are the conceptual differences (if any) between addressing memory and I/O?

8–15 Fundamental concepts of troubleshooting

Any fault in either a main memory or an I/O system can be detected by first observing the logic levels of the address and control signals and then detecting the resultant operation via the data bus. These events are illustrated in the timing diagram of Figure 8–23, where

1. An address is loaded on the address bus lines A_0 through A_n by the accessing device (CPU), which activates the *CS* of the addressed IC memory or I/O device, via decoders if an address decoding technique is used or via an address line if linear addressing is used.
2. If the system is an isolated I/O system, the control signal ($IO/\overline{M}$) that distinguishes memory addresses from I/O addresses is activated by the accessing device (CPU) at approximately the same time that

Figure 8–23
Timing diagrams for read and write operations.

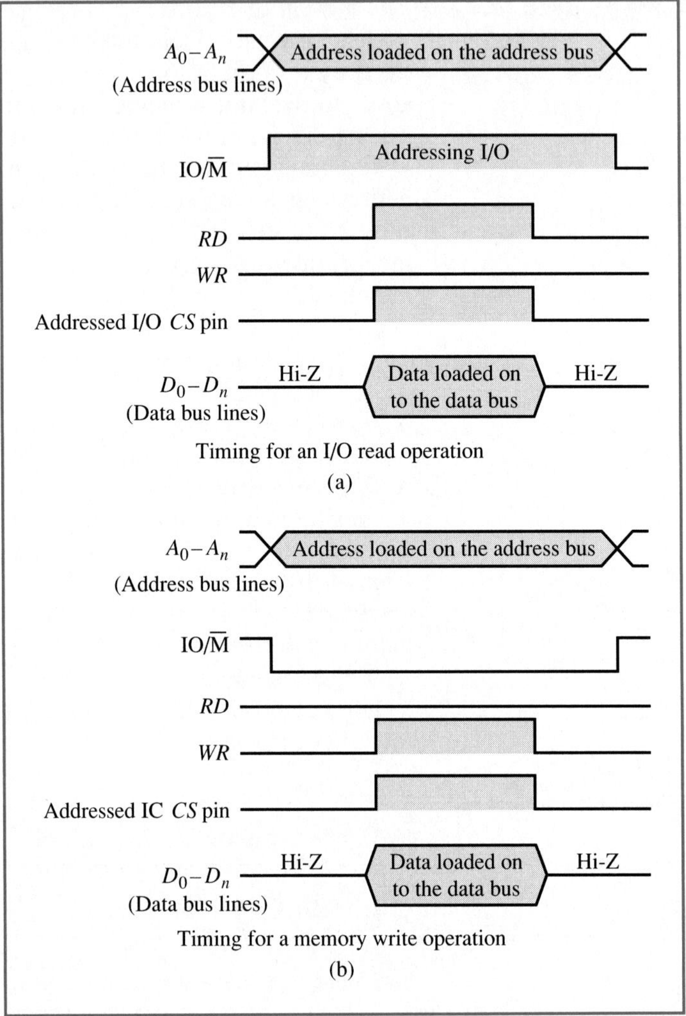

the address is loaded on the address bus (it is driven high for I/O addresses and low for memory addresses). This control signal is held at the appropriate logic for as long as the address is held on the address bus. From the logic level of *IO/$\overline{M}$*, Figure 8–23a shows an I/O access and Figure 8–23b a memory access. If the system is a memory-mapped system, there is no control signal (*IO/$\overline{M}$*) that distinguishes between memory and I/O addresses.

3. The accessing device (CPU) must activate the read (*RD*) or write (*WR*) control signal at the proper time—when the CPU is ready to read the datum from the data bus, as in the case of a read operation shown in Figure 8–23a, or when it has loaded the datum to be written to the addressed device onto the data bus for a write operation, shown in Figure 8–23b.

Note that when the data bus is not used for the reading or writing of data it is in a high-impedance (Hi-*Z*) state, meaning that all output buffers of both memory and I/O are turned off. Of course, when the data bus is in use, only the addressed device (memory IC or I/O) has its output buffers turned on, giving it exclusive use of the data bus during that time.

When troubleshooting, we must keep in mind the timing diagrams of Figure 8–23, which serve as a guide to troubleshooting diagnosis. From Figure 8–23 we will

1. apply and/or verify that there is a valid address on the address bus.
2. apply and/or verify that $IO/\overline{M}$ or a similar control signal is at the proper logic level for an isolated I/O system (there is no such control signal for memory-mapped I/O).
3. verify that the logic level on the CS pin of the addressed device is valid for the given address.
4. apply and/or verify that the proper read (RD) or write (WR) control signal is active for the intended operation.
5. verify that the datum on the data bus is correct and is present at the proper time. For a read operation this means that the datum in the addressed memory location or I/O device must be known so that it can be verified when it appears on the data bus. For a write operation the datum being written by the CPU must be known and also verified via the data bus.

"Apply and/or verify" means that if the memory and/or I/O system under test is integrated into a system, such as that in Figure 8–9, the CPU applies the logic levels for the address and control bus signals and it must be verified that they are correct. However, if the system under test has not been integrated into a completed system, these signals must be applied as part of the test environment.

There are two methods of troubleshooting. One involves static testing techniques, whereas the other is dynamic testing. In static testing, the logic 1's and 0's applied to the logic circuits under test use DC levels (static) to implement those logic levels, or logic levels change so slowly that their transition from one state to the other can be observed with the use of a volt-ohm meter (VOM) or logic probe. Static testing has until now been our major diagnostic tool. In dynamic testing, the logic 1's and 0's input to the logic circuitry under test change logic levels faster than can be observed with static-type test instruments. Of course, dynamic conditions are the conditions one experiences in the "real world." For example, when a modern CPU (microprocessor) is operating, it is outputting a new address on the address bus and reading or writing a datum from or to the data bus every few microseconds, as well as outputting control signals. To detect and observe dynamic logic levels, dynamic test equipment must be used.

Static troubleshooting

To statically troubleshoot memory and I/O systems, such as those in Figures 8–21 and 8–22, we refer to the troubleshooting guide previously listed. Let us consider the memory and I/O system of Figure 8–21 as the system to be statically tested; let it further be assumed that it has not been integrated into a complete system. Figure 8–24 illustrates the testing environment. Since there is no CPU to input addresses and control signals, the test environment must be configured to do so. The address bus lines A_0 through A_n, where $n = 5$ for the address bus of Figure 8–21, are being driven by the address static logic generator. The logic level of each output of this logic level generator, as well as all the other static logic level generators, is determined by the position of the slide switch of Figure 8–24b. If the slide switch is in the "down" position (to the left in

Figure 8–24a), a logic 0 is output; otherwise, a logic 1 is output. Applying the intended logic levels to control bus lines $IO/\overline{M}$, RD, and WR, we use the control signal static logic generator, which has the same structure as the address logic level generator. Once the address and appropriate $IO/\overline{M}$ logic level have been output, the logic level of the CS pin of the addressed device (memory IC device or I/O device) may be verified using a logic probe, as indicated in Figure 8–24a, or a logic level detector. If a datum is to be written into memory or I/O, the data static logic level generator serves that purpose via the data bus lines D_0 through D_m (m = 7). If a datum is being read from memory or I/O, the data static logic level detector is used. Each data line is connected to an input of the data static logic level detector, such as the one illustrated in Figure 8–24c. If a logic 1 is present at the input of the detector, the LED is on and a light is emitted; otherwise, the LED is off. Both the data static logic level detector and generator are buffered. These buffers have the opposite logic level for control of their output states (active or Hi-Z). Using these buffers allows one to isolate the read test equipment from the write test equipment and to eliminate bus contention in the case of the data static logic level generator. Connecting the buffer control line to RD controls which buffer is on.

To test the system of Figure 8–24a in a thorough and organized manner, the diagnostician needs to create a truth table that shows all binary combinations for the address bus lines and the resulting active output of the page select decoder or I/O select decoder, similar to the truth table of Table 8–2 for the memory system of Figure 8–15. The diagnostician then inputs those addresses, one at a time, via the test system of Figure 8–24a, and also sets the logic levels of the control bus lines. The CS pin of the addressed device is then checked. Last, a read and/or write on each memory location and I/O device is performed. That is, for R/W memory (RAM) a known binary pattern must first be written into memory before being read out for verification of storage. For ROM, the contents must be known so that a read verification for proper content and operation can be made. I/O devices must be similarly tested.

Static testing is limited to testing for faults resulting from an improperly wired system or the failure of an IC or other component.

Dynamic troubleshooting

Dynamic testing is conceptually the same as static testing, the difference being that the logic levels on the busses and the CS lines are rapidly changing—at the rate of micro- or nanoseconds. Under dynamic conditions, problems can arise that are not present during static testing. These types of faults may be timing problems due to propagation delay times or exceeding the operation speed of some IC or component in the system. Of course, actual working conditions are the best environment in which to test a system, which means dynamic testing. We therefore consider the memory and I/O systems under test to be integral parts of a total system, like the system of Figure 8–9. This means that if we are testing the memory and I/O systems of Figure 8–21, those logic levels that are produced by the CPU—that is, the address, control signal, and data—are rapidly changing. What is needed is test equipment capable of equally fast detection of logic levels and display of those logic levels in a form that can be interpreted. That is, since logic levels are changing in

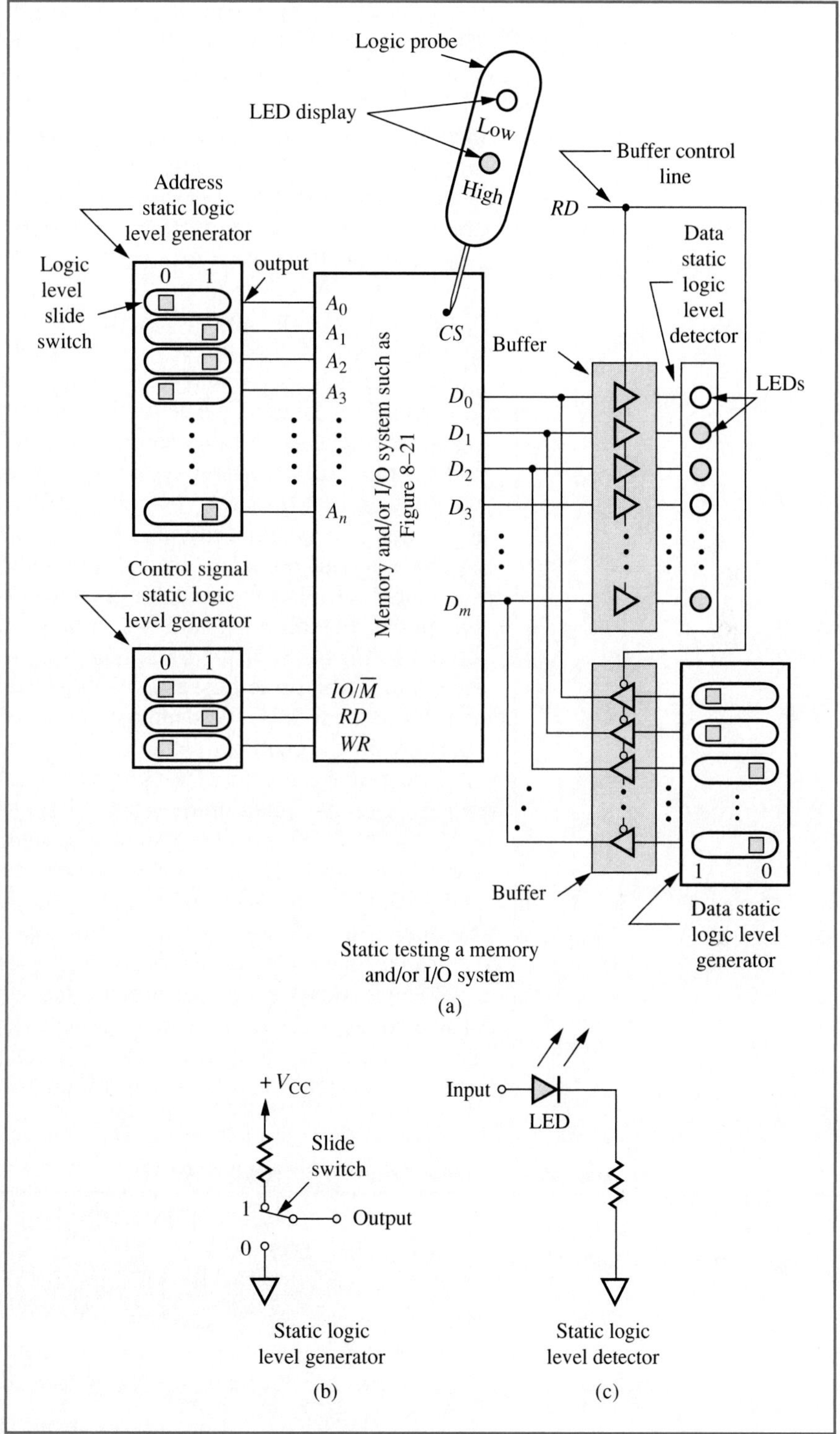

Figure 8–24
Static testing.

the micro- or nanosecond range, the results cannot be displayed at the rate at which they appear. Hence, they must be stored and then presented at a slower rate that will permit scrutiny by the observer.

Test equipment for dynamic logic level detection and display includes (1) *signature analyzers,* (2) *logic analyzers,* and (3) *microprocessor development systems.* We will limit our discussions to signature analyzers and logic analyzers, for their concepts of operation are also basic to microprocessor development systems.

The signature of an electrical point, known as a *node,* is determined by counting the logic 1's and the order in which they appear in an interval of time between two events (start and stop markers). From that count a four-digit code (0 through 9, A, C, F, H, P, U) is derived by the signature analyzer. This code is statistically unique to that particular string of data flow and therefore can serve as the signature of that node. Obviously, signature analysis is of diagnostic value only when the nodes (bus lines and *CS* pins) of a working system have been previously determined and recorded so that they may be compared to signatures of the same nodes of an identical system that has failed. Some manufacturers of electronic equipment print the signature of each node on the schematic of the equipment. If there were such a manufacturer of the memory and I/O systems of Figures 8–21 and 8–22 (integrated into a complete system such as that in Figure 8–9), that manufacturer would label the signatures of each line of the three busses as well as each *CS* pin on the schematic. When a system fails, the diagnostician can locate the failed node or nodes by comparing signatures of the failed system with those on the schematic. A more detailed analysis, possibly using a logic analyzer, may then be required, or the failed module may simply be replaced.

Figure 8–25 is a photograph of a signature analyzer manufactured by Hewlett-Packard. It has a 100% probability of detecting single-bit errors and a 99.98% probability of detecting multiple-bit errors. There is no maximum time between its start and stop limits. Up to 32 signatures can be stored for recall. It operates at a frequency of 25 MHz.

Logic analyzers perform much the same function as movie cameras. Movie cameras freeze a moment of an event by taking a snapshot of the event. Of course, they take many chronological snapshots of the event, and each snapshot is a static representation of a moment stored on film.

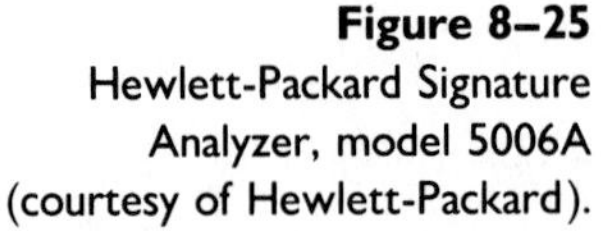
Figure 8–25
Hewlett-Packard Signature Analyzer, model 5006A (courtesy of Hewlett-Packard).

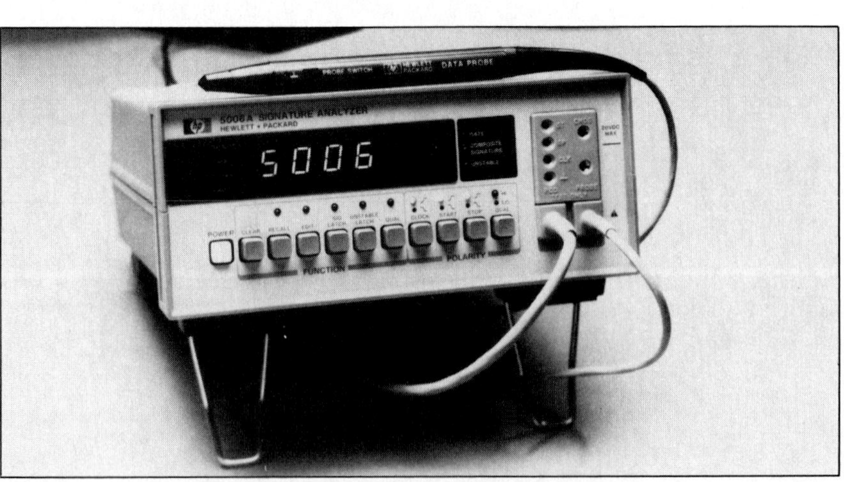

Logic analyzers take "snapshots" of the logic levels present on system busses and *CS* lines whenever commanded to do so by a trigger signal. After taking the snapshot the logic analyzer stores the logic levels in a FIFO memory in the same order in which they occurred. Just as importantly, each snapshot or series of snapshots of logic levels can be displayed at any rate desired, much as we can view a frame of movie film for as long as we wish.

A representative architecture of a logic analyzer is presented in Figure 8–26. We see that the architecture can be modularized into five function-blocks.

1. Input section, which consists of input channels, comparators, and voltage level detectors
2. Logic level storage unit, which is made up of an addressable FIFO and its control circuitry
3. Trigger source unit
4. Clock source
5. Display

The input section inputs data from the nodes being tested. Each system node under test has an input channel (C_0 through C_N) connected to it. To ensure that only "good clean logic levels" are being input to the logic level storage unit (the FIFO memory), comparators are used to "shape

Figure 8–26
The architecture of a logic analyzer.

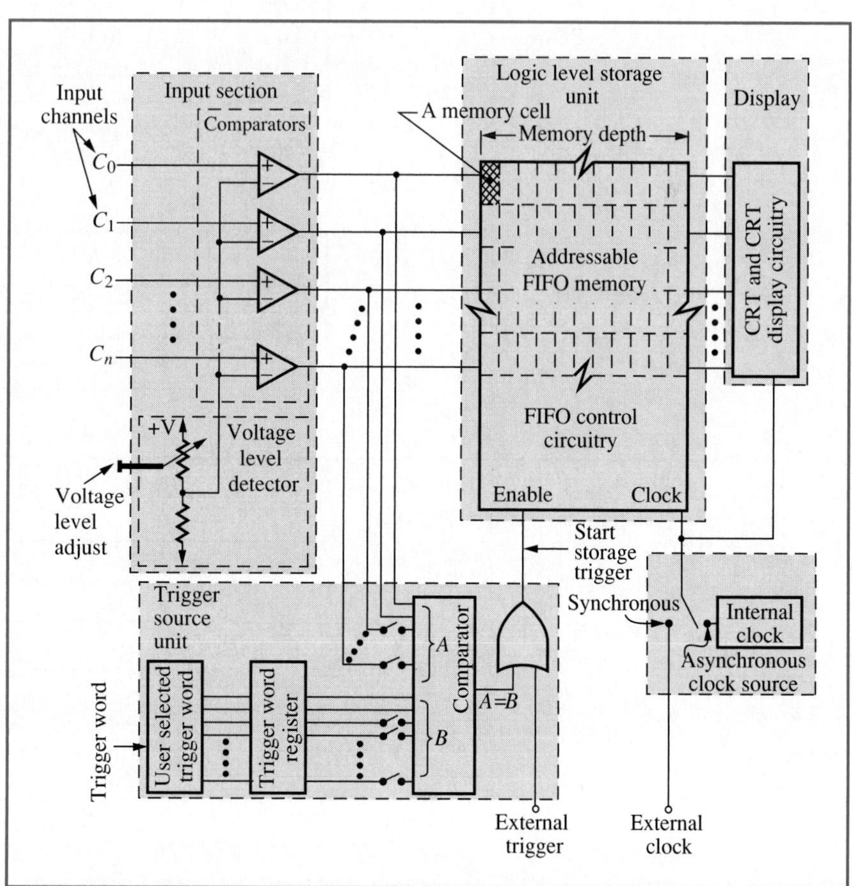

Figure 8–27
Filling the FIFO memory of the logic analyzer of Figure 8–26.

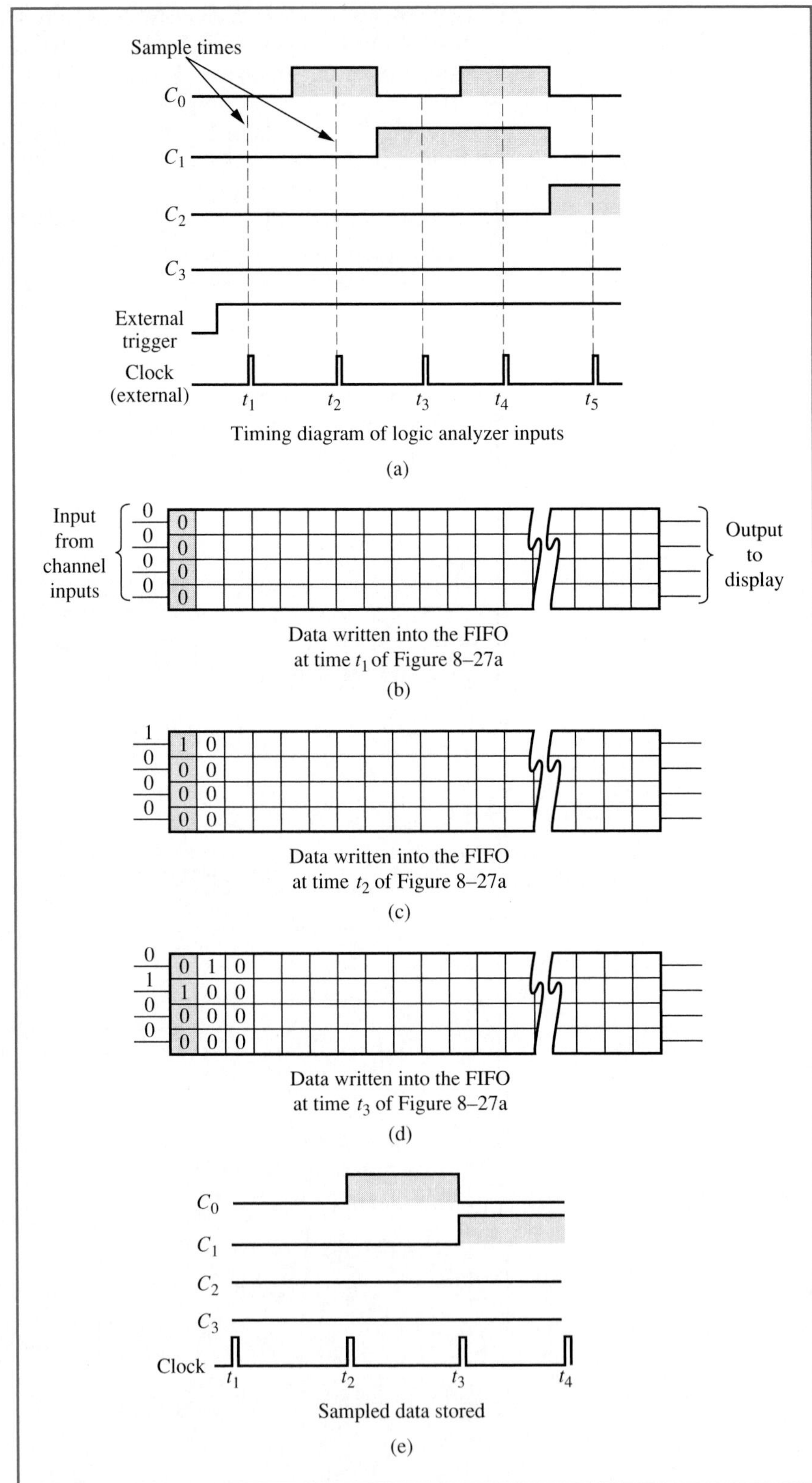

up'' the input waveforms. Also, the voltage level detector allows the logic analyzer to be adjusted for different logic families, which differ in the voltage levels that constitute a logic 1 or 0 (voltage requirements for various logic families are discussed in Chapter 12). Logic analyzers are available with a variety of channels. The least is 16 ($N = 16$) and the most is usually 80 ($N = 80$). For instance, if each node of the memory system of Figure 8–15 were to be tested, a minimum of 23 channels would be required (six address lines, eight data lines, three control bus lines, four *CS* lines, and one enable pin on the page select decoder). As can be easily seen, the number of input channels can rapidly increase as the system being tested becomes more complex.

As the nodes under test input data to the logic analyzer of Figure 8–26, that data is passed on to the logic level storage unit and also to input *A* of the comparator in the trigger source unit. The data is not stored in the FIFO unless its enable input is high, driven by the start storage trigger output by the trigger source unit. As we see by examining the trigger source unit function-block, the logic level of the start storage trigger is determined by the output of the comparator ($A = B$) or the external trigger input. If data storage in the FIFO is to be initiated when a certain binary bit pattern appears at the selected input channels (note that not all channel data need be input to the comparator—see the switches), that binary bit pattern, known as the *trigger word,* is loaded into the trigger word register via a keyboard or panel switches. When the trigger word appears at input *A* of the comparator, its output ($A = B$) will go high and hence the start storage trigger output will be driven high, which enables the FIFO for data storage. When the external trigger of the trigger source unit is used to drive the start storage trigger output high, it is connected to the node within the system that generates the signal with which data storage is to coincide. For example, if data collection and storage are to begin when *RD* goes high, then the external trigger input is connected to the *RD* line (node).

Enabling the FIFO to store data is determined by the logic level of the enable input of the FIFO control circuitry, as we have seen, but it is the clock input of the FIFO control circuitry that actually strobes the data into the FIFO. As we see from Figure 8–26, an external clock can be used to strobe data into the FIFO, or an internal clock can serve as the clock source. When an external clock source is used that is synchronized with the system being tested, the data collected is stored and displayed in synchrony with the tested system. For this reason, external timing of this type is known as the synchronous mode of operation. On the other hand, when an internal clock is used, no relationship exists between data collection, storage, and display timing and system timing; hence, the internal timing is known as an asynchronous mode of operation.

Note that the FIFO is identified as an addressable FIFO, which means that even though data is collected and shifted in a FIFO fashion, the data at any location can be addressed. This allows the user of a logic analyzer to define a reference point and then display the collected data before or after this reference point.

The collected data is displayed on a CRT. Because the data to be displayed is in FIFO memory and therefore in binary form, it can be displayed in any format desired, assuming that the logic analyzer has the

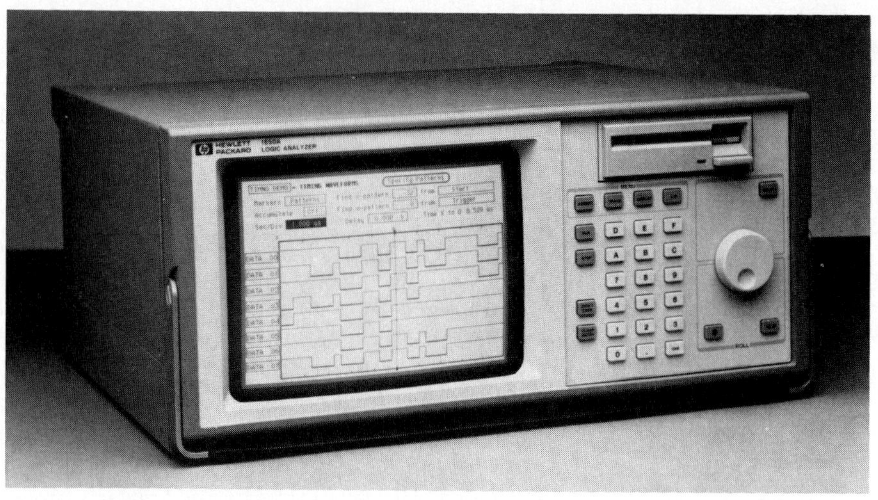

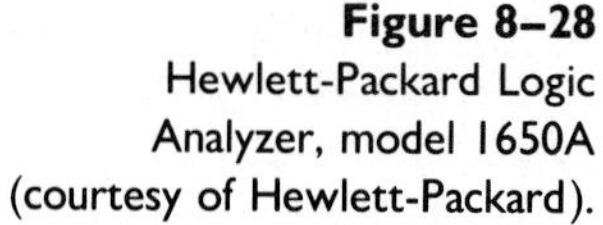

Figure 8–28
Hewlett-Packard Logic Analyzer, model 1650A (courtesy of Hewlett-Packard).

proper logic circuits. The only requirement is that the data be converted to the desired format before being displayed. For instance, if the address bus lines are being tested, it would be more convenient to display a hexadecimal value rather than a logic level diagram. Hence, the stored logic levels could be converted to hexadecimal characters for display on the CRT.

Figure 8–27 illustrates the storing of channel logic levels. Figure 8–27a represents the logic levels being input to channels C_0 through C_3, which could be any four nodes, such as address lines, data lines, or control bus lines. These channel inputs are shaped up and input to the FIFO via the comparators. At time t_1 the logic levels at the input of the FIFO, which is the output of the input section, are strobed into the FIFO, as illustrated in Figure 8–27b. At t_2 the channel datum has changed from 0000 to 0001 and is strobed into the FIFO, which also shifts the first datum by one location, as shown in Figure 8–27c. At t_3 the process is repeated, and the FIFO appears as in Figure 8–27d. Data is sampled and written into the FIFO until it is filled. On command by the user, the FIFO outputs the data to the display in the desired format. Note that if the sampled data is displayed as logic levels with the same timing at which it was collected, the displayed data is shifted to the right by one-half clock period, as illustrated in Figure 8–27e. This is why the output is really a logic level diagram rather than a timing diagram. If an accurate timing diagram is required, the logic analyzer must increase the sample time of the input logic levels by increasing its sample rate (increase the clock frequency).

The size of the FIFO is usually between 256 and 1028 words of storage (depth), and the word size (the number of bits stored per location) corresponds to the number of input channels.

Figure 8–28 is a photograph of the Hewlett-Packard Logic Analyzer model 1650A. Channel input connections are made via the five channel modules as shown, where only one module has the individual channel connectors inserted. The 1650A is an 80-channel logic analyzer that operates at 100 MHz. Among other features, the 1650A has five clock inputs and four clock qualifiers (a clock qualifier serves the same function as a trigger word), which enables the user to determine when the logic analyzer will begin storing the data at the nodes. It has methods to examine the stored data according to patterns or range of sample taken, and the displayed data can be formatted in a variety of ways, such as timing diagrams, state testings in hexadecimal, or inverse assembly language (assembly language will be discussed in Chapter 10). The test system performance can also be displayed as a bar graph.

We shall use both static and dynamic test equipment when troubleshooting memory and I/O designs in the chapters to follow.

Review questions

1. When troubleshooting a read or write operation, whether statically or dynamically, what nodes would be tested in order to make a complete diagnosis of the system and why should those nodes be monitored?
2. What are the five troubleshooting steps to be used as a guide?
3. What is the essential difference between static and dynamic testing?
4. When is static testing the preferred choice?
5. When is dynamic testing necessary?
6. How does a signature analyzer function?
7. How does a logic analyzer function?
8. Under what test conditions can a signature analyzer and logic analyzer be used?

8–16 SUMMARY

- Memories can be categorized in a variety of ways. A classification based on CPU access to the data stored has two dominant memories: main memory and mass storage memory.
- *Main memory* is the working memory of the system, and its contents are programs and data that are currently being operated upon by the CPU.
- *Mass storage memory* stores mass amounts of data from program execution or programs that are to be executed at a later time.
- The portability of the mass storage medium allows convenient storage of the medium (tape or disk) as well as transport of programs and data between systems.
- Semiconductor memory devices used in the design of main memory have memory cell organizations that are either two- or three-dimensional arrays. If the width of the memory location is 1-bit, it is a two-dimensional array; if the width is greater than 1-bit, it is a three-dimensional array.
- Memory media such as tapes and disks use dynamic linear one-dimensional memory cell arrays.
- There are four different techniques of accessing a memory location—dynamic linear, RAM, FIFO, and LIFO.

- The access time for dynamic linear accessing is a variable, since it is a position-dependent access time.
- The access time for RAM devices is the same for all memory locations within the device.
- The access time for FIFOs and LIFOs is a function of how many shifts must be made in order to access the desired data.
- A FIFO memory has accessible input and output memory locations, but those locations between the two are usually not accessible (the FIFO of a logic analyzer is an exception).
- A LIFO memory has only one accessible memory location, which is defined as the top of those memory locations comprising the LIFO.
- The ability of the CPU or a mass storage device to alter the contents of a semiconductor memory device categorizes semiconductor memory devices into one of two types—either a ROM or an R/W memory.
- The CPU can only read the contents of a ROM, whereas it can both read from and write to an R/W memory device. The acronym RAM is erroneously used as a synonym for R/W.
- There are many types of ROMs, each having its advantages and disadvantages.
- EEPROMs are read-mostly R/W memories and are sometimes referred to as nonvolatile RAMs.
- Magnetic tape media are accessed in a serial fashion (SAM). Their formatting is such that the address locates a block of data. Also present in the format is information providing error checking and indicating the size of the data block.
- Disks have data stored on concentric circles known as tracks. On an 8-inch disk there are 77 tracks that are sectioned into 26 sectors.
- There are two types of disks—floppy and hard disks. Floppy disks are constructed from a flexible material whereas hard disks are rigid. Hard disks have larger memory capacities and are more expensive.
- A memory device that cannot retain the contents of its memory locations on the removal of power is known as a *volatile memory device*.
- Those memory devices that do retain the contents of their memory locations, even without power, are known as *nonvolatile memory devices*.
- *Static memories* are memory devices that retain their contents so long as power is applied to the memory device.
- *Dynamic memories* cannot retain their contents without being refreshed.
- The CPU-generated signals *RD* and *WR* control read and write operations.
- There are three basic techniques for addressing memory locations and I/O devices—fully decoded addressing, linear, and partial linear addressing.
- The *fully decoded addressing* technique decodes the logic level of the address bus. These address bus lines are decoded by the memory device

and the page select decoder, or the I/O select decoder. The decoder internal to the memory device selects the addressed location within the memory device, and the select decoder selects the memory or I/O device.

- *Partial linear and linear addressing* is different from fully decoded addressing in that there are no select decoders; instead, address bus lines are used to select memory and I/O devices.
- There are two techniques for allocating addresses to memory and I/O—isolated and memory-mapped I/O. *Isolated I/O systems* partition memory addresses from I/O addresses via the $IO/\overline{M}$ control signal. *Memory-mapped I/O systems* do not distinguish between memory and I/O addresses.
- Memory-mapped I/O systems require less hardware to implement (no I/O select decoder), but they have fewer addresses available.
- Accessing I/O is very similar to accessing memory. The major difference is that I/O has no internal memory that is addressed via the system address bus; hence, I/O access requires only that its *CS* be driven to the active level.
- I/O outputs that are connected to the data bus are buffered in order to prevent bus contention.
- In static troubleshooting, DC voltages are input to the circuit under test, and static test equipment, such as a logic probe, is used to detect the resultant logic levels.
- Static testing is acceptable for troubleshooting catastrophic failures such as IC failure and wiring problems.
- Dynamic troubleshooting tests for catastrophic failures and dynamic-related problems such as timing and coupling due to stray capacitance.
- Signature analysis is a dynamic test but requires that the signatures of the system nodes be known.
- Logic analyzers are also dynamic test equipment. They sample the data on nodes according to a trigger signal and store that data in a FIFO in the order sampled. When the collected data is displayed there are a variety of formats from which the user can choose. The user can compare the displayed data with that of a truth table for diagnostic purposes.

PROBLEMS

Section 8–1 Introduction

1. What are program instructions and what purpose do they serve?
2. What are programs and what purpose do they serve?
3. Why must a computer system have a memory?

Section 8–2 Main memory and mass storage memory

4. What does the term "architecture" mean?
5. How does the function of main memory differ from that of mass storage memory?
6. What are the criteria of main memory and mass storage memory? As part of your answer, provide a rationale for your statements.

7. Why does the CPU work from main memory rather than mass storage memory?
8. Why is main memory implemented with semiconductor memory devices?

Section 8–3 Types of memory organization

9. What are the five basic categories that can be used to classify memories?
10. For the given memory organizations determine the number of memory locations and memory cells per location.
    a. 1024 × 8 bits
    b. 2048 × 8 bits
    c. 4096 × 1 bit
11. For a 64 × *N*-bit memory device, which is either a two- or three-dimensional array, determine the possible number of row and column configurations of the array using an *x-y* coordinate system, such as was used in Figures 8–3 and 8–4.
12. Repeat Problem 8–11 for a 32 × *N*-bit memory device.
13. From the evolving patterns of Problems 11 and 12, develop a table that indicates the combinations of columns and rows for *N* memory locations.
14. If it is desired to have the number of rows and columns equal, or as nearly equal as possible, for a memory cell organization, how many columns and rows would exist in Problems 11 and 12?
15. Using the same *x-y* coordinate scheme used to identify the memory locations of Figures 8–2 and 8–4, develop a memory cell organization for a 16 × 1-bit memory device that has eight rows and two columns for memory location identification.
16. For the resultant memory cell array of Problem 15, develop a table similar to Table 8–1 giving the location and cell identification numbers for every possible binary combination of the rows and column. Then compare this table with Table 8–1 and note that the results are the same.

Section 8–4 Memory addressing and access time

17. If a magnetic tape mass storage device has an access time of 2 ms per location, what is the access time if the current address is 10_{10} and the next location to be accessed is
    a. 256_{10}
    b. 500_{10}
    c. 05_{10}
18. Repeat Problem 17 but add an additional 4 ms each time the direction of travel must be reversed.
19. If the access time of a RAM is 2 μs, what is the access time of any location within that RAM?
20. Suppose the FIFO of Figure 8–6a requires 0.20 μs to shift data from one location to the next. If the FIFO were empty and a datum was written into the FIFO, how long would it take for the datum to be available at the output location?

21. If the FIFO of Problem 20 had another 4 bits of data written into it, what would be the delay time before this datum appeared in the next available location?

22. Suppose that the LIFO of Figure 8–7 requires 0.2 μs to shift data from one location to the next. How much time is required to read a datum that has had two other data entries pushed on top of it?

23. Suppose that the LIFO of Problem 22 had three 2-bit datum entries pushed on it. How much time is required to push this data down and return the original datum to the top location?

Section 8–5 Alteration of memory contents: read/write and ROM devices

24. When would a ROM rather than an R/W memory be used in an application?

25. Can a ROM also be a RAM? Why?

26. Why must all mass storage devices have read/write capability?

Section 8–6 ROM classifications

27. In a manufacturing environment (large-volume production), what type of ROMs might be used?

28. In a research and development environment, what type of ROMs might be used?

29. What type of ROM might be used by an engineering firm that produces one to three copies of a product?

30. If you were designing a computer that processed data so important that you did not dare lose it as a result of a power failure to the system, what type of ROM could you use as a nonvolatile R/W memory back-up?

Section 8–7 Memory volatility

31. What is volatility?

32. What is a nonvolatile RAM?

Section 8–8 Memory cell data retention: static and dynamic memories

33. When would a dynamic RAM be chosen over a static RAM?

34. What is the disadvantage of using dynamic RAMs?

35. Are both static and dynamic RAMs volatile?

Section 8–9 Addressing and read/write operation control

36. When accessing memory or I/O, what two events are required and in what order?

37. How does the CPU control read and write operations?

Section 8–10 Semiconductor memory architecture

38. The numbers of address pins are listed for four different memory devices. Determine the number of memory locations within each device.
    a. 8
    b. 10
    c. 11
    d. 12
39. For the number of address pins listed in Problem 38, what combination of row and column address pins might you expect? Refer to Problem 14 and Figure 8.
40. The numbers of address pins of three memory devices are listed. In each case, determine the type of row and column address pin decoder, such as specified in Figure 8–8a.
    a. 10
    b. 11
    c. 13
41. For a 13 address pin memory device, how many columns and rows are there?
42. Does the number of bits at a memory location affect its address?
43. Why is it logical that the CPU of a system, such as the computer illustrated in Figure 8–9, be the function-block that supplies addresses (both memory and I/O)?
44. The illustration of Figure 8–9 shows the CPU as the origin of the control signals $IO/\overline{M}$, RD and WR. Explain why it is logical that the CPU is the source for control signals.
45. To access the memory cells of Figure 8–11 for either a read or a write operation, what is the first event that must occur? Explain why it must occur first.
46. Explain the purpose of the R and W buffers of the memory cells of Figure 8–12.
47. In Figure 8–12 we see that the outputs (Q) of all memory cells within a plane are connected. Why doesn't this cause bus contention on that connecting line?
48. Suppose you were asked to describe the internal design of a memory

Figure 8–29
The pin configuration of a memory device.

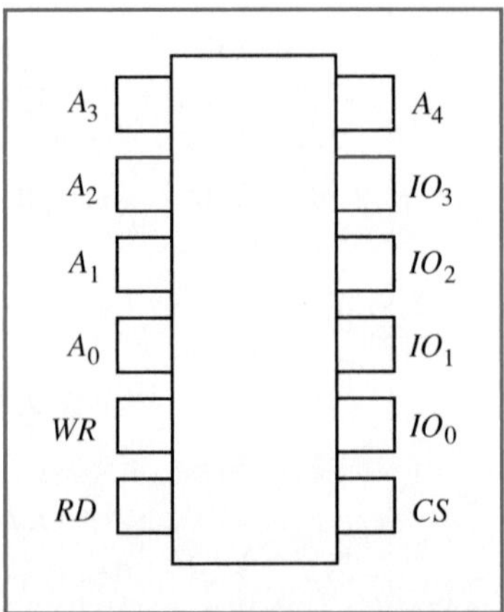

device that has the pin configuration of Figure 8–29. Using Figure 8–8 as a reference, give the requested description.

49. Using the ROM memory cell of Figure 8–13, design an 8 × 2 ROM.

Section 8–11 Introduction to the fundamental concepts of main memory design

50. In a fully decoded address scheme, what function does the page select decoder serve?

51. What is meant by the expression "a page of memory"?

52. Let us define a page of memory to be equal to the least number of memory locations within any memory device used to implement a main memory system. Using this definition, how many memory locations would there be in a page of memory if there were 12 address pins on the system's ROMs and 10 address pins on the system's RAMs?

53. If you designed a main memory system using ROMs and RAMs that had 11 address pins per memory device
    a. How many memory locations would each memory device contain?
    b. How many memory locations would a page of memory contain if the definition of Problem 52 is used?
    c. For a fully address-decoded 8K memory (8192 memory locations) how many address bus lines are required?
    d. Using partial linear addressing, how many address bus lines are required for an 8K memory?

54. For part (c) of Problem 53, what address bus lines would serve as inputs to a page select decoder?

55. For part (d) of Problem 53, what address bus lines would be used to activate the *CS* inputs of the memory devices?

56. Design an 8K fully decoded main memory using the memory devices of Figure 8–30, such that there are 2K of ROM and 6K of RAM. ROM is to begin at address zero. Draw a schematic of your design similar to Figure 8–15. Create a table that shows the connections of component pins to the address bus lines and also the beginning and ending address of each page of memory, similar to Table 8–2.

57. Repeat Problem 56 but use partial linear addressing.

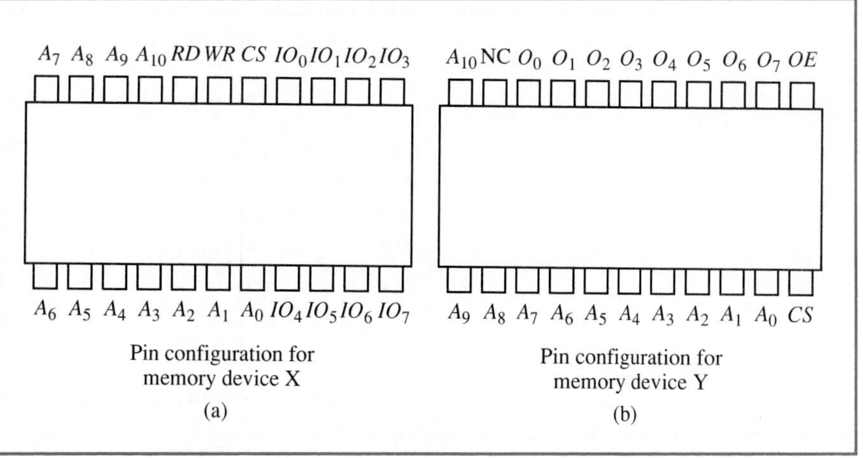

Figure 8–30
Pin configurations for memory device *X* and *Y*.

58. What is the essential difference between the static memories of Problems 56 and 57 and similar designs using dynamic memory devices?

Section 8–12 Magnetic tape memory

59. If the CAR of the magnetic tape mass storage device of Figure 8–17 contained 5H and the device were told to access location BH next, what would be the contents of the DAR and the logic level of the comparator outputs at the moment it was told to access location BH next?

60. Assume that the magnetic tape mass storage device of Figure 8–17 has a travel access time of 2 ms per location. What would be the access time for location BH if CAR = 5H?

61. The contents of the CAR of Figure 8–17 is EH, and its DAR is loaded with 2H. If the tape direction was forward when 2H was loaded into the DAR, what would be the access time if we must allow 4 ms for tape reversal and 2 ms per location for travel and access?

62. How does a magnetic memory cell differ from a semiconductor memory cell?

63. When a block of data is being read from a disk, how do the R/W heads know when to read the block of data from the disk?

64. Explain the difference between the RZ and Manchester techniques for recording 1's and 0's.

65. Why would the Kansas City technique for recording logic 1's and 0's be advantageous for telephone transmission?

Section 8–13 Disk memory

66. How many seconds are required for a disk rotating at 360 rpm to complete a revolution?

67. Based on the answer to Problem 66, what is the maximum access time for a disk?

68. Why is the access time of a floppy disk so much shorter than that of a magnetic tape mass storage device?

69. Explain how data storage is partitioned on a disk.

70. Why can a hard disk be accessed faster and have a larger storage capacity than a floppy disk?

Section 8–14 Accessing I/O: isolated I/O and memory-mapped I/O

71. Into the memory design of Figure 8–15 incorporate five I/O devices: three input and two output devices. It is to be a fully decoded system. Construct a table similar to Table 8–2 showing the addresses of memory and I/O.

72. Repeat Problem 71, but modify the resultant design to be a memory-mapped I/O system.

73. To the design of Figure 8–21 add another 16 bytes of ROM.

74. Using 128 × 8 memory ICs for RAM and ROM, design a 768 × 8

memory. There are to be 256 bytes of ROM and these 256 bytes are to occupy the first 256 addresses. To this memory add three I/O devices, one of which is an output device. This is to be a partial linear addressed and memory-mapped I/O system. Construct a table similar to Table 8–2.

75. Repeat Problem 74 but design the system to be a fully decoded, isolated I/O system.

Troubleshooting

Section 8–15 Fundamental concepts of troubleshooting

76. For the design of Problem 71 devise a static test. Make a drawing similar to Figure 8–24.

77. Devise a dynamic test for the design of Problem 71. Create a drawing similar in concept to Figure 8–24, but using a logic analyzer.

78. For the dynamic test of Problem 71 create a timing diagram for reading memory locations 0000H, 0001H, and 0002H. At location 0000H data 21H is stored, at 0001H data 31H is stored, and at 0002H data 44H is stored. For these three addresses assume that samples were made and stored in a logic analyzer.

79. Can a signature analyzer be used to diagnose a fault with the newly constructed design of Problem 77?

Technology now allows us to design highly complex digital circuits and fabricate them in a tenth of the space required a few years ago. The photograph shows a complete digital computer circuit on a single printed circuit board (Courtesy Chips & Technology, Inc.).

OUTLINE

OBJECTIVES

The objectives of this chapter are to

- introduce various types of memory systems and applications.
- explain the characteristics of operation of some specific semiconductor memory devices.
- explain memory device timing parameters and how they relate to system requirements.
- apply the fundamental concepts set forth in this and the previous chapter to the design of main memory systems.
- apply troubleshooting techniques to the memory designs developed in this chapter.

9 DESIGN AND APPLICATIONS OF SEMICONDUCTOR MEMORIES

9–1 INTRODUCTION

In Chapter 8 we developed fundamental concepts that were concerned primarily with memory cell organization, various accessing techniques, and the control of read/write operations. In this chapter we will apply that information to various memory applications and to the design of main memory systems.

When designing a semiconductor memory system, choices must be made concerning the specific memory devices to be used in the implementation of the design, which often requires choosing among various technologies, such as transistor-transistor logic (TTL), metal oxide semiconductor (MOS), and complementary MOS (CMOS). In this chapter, the chosen technologies and the reasons for choosing those technologies will be a matter of stated fact. When we study the logic technologies in Chapter 12 we will explain the reasons for those choices.

Since this is a broad-based text on digital fundamentals, rather than one limited to the design of memory systems, a presentation on the design of memory systems must be somewhat selective. The authors believe that the focus of memory design should be on main memory systems, since main memory is an essential part of a computer system. In addition, the principles of main memory design also apply to the design of other memory systems. In fact, often the only differences between a main memory and other types of memory systems are the name and its application; the principles of design are the same. To complete our study of memories we shall take a quick look at some other commonly used semiconductor memories and their applications in computers.

9–2 Applications of semiconductor memories

When we refer to a specific type of memory system, we are classifying that system according to its application. Each application has operational characteristics that determine the type of memory system and the semiconductor memory devices to be used in implementation of that design. For instance, if we are designing the portion of main memory from which the CPU is fetching and executing a *bootstrap program* (a program that gives the CPU its beginning instructions), a ROM must be used to store that program. Of course this ROM must have a faster access time than the CPU with which it is intended to work.

In Chapter 8 various types of memories were discussed but two applications were emphasized—main memory and mass storage memory.

We will now expand that list to include five additional memories, all of which are implemented with semiconductor technology:

1. Scratch pad memory
2. Buffer memory
3. Stack memory
4. Queue memory
5. Look-up table memory

In the following material we will discuss the primary application of each of these five memory types.

Scratch pad memory

Scratch pad memory serves as a temporary memory for intermediate data calculations, although the final data is stored in main memory. This is analogous to a scratch pad used by a person calculating his or her federal income tax. The intermediate values are calculated and saved on a scratch pad so as to be readily accessible for other required calculations. Once these intermediate values have served their purpose in determining the final value, which is to be recorded on the proper "permanent" government form, the scratch pad data can be discarded. Recording the final value on the proper form is analogous to storing data in main memory or mass storage memory, depending on how permanent you mean "permanent" to be.

There are several assumptions behind the concept of scratch pad memory: (1) The data stored is considered temporary, meaning that it should quickly serve its purpose in the execution of a program so that it can be discarded to make room for other scratch pad data. (2) Scratch pad data is readily available. Therefore, data retrieval time from scratch pad memory should be faster than from main memory; otherwise, why not store the data in the main memory? (3) The capacity of a scratch pad memory is small (perhaps only eight memory locations), since the data stored is brief and temporary.

Scratch pad memories are often implemented with registers and are manufactured as an internal part of the CPU. Registers A, B, C, and D of Figure 9–1 are representative of scratch pad memory within a CPU. The ALU, which is also an integral part of the CPU but is not shown, performs all calculations and therefore requires a scratch pad memory for storage and retrieval of intermediate data. Manufacturing the scratch pad memory as part of the CPU allows the read/write operations required with intermediate calculations to be performed much faster than if the CPU had to go to an external memory, such as main memory (recall that to access main memory the CPU must generate addresses and control signals). We can imagine scratch pad memory as a one-dimensional memory array. Chapter 10 deals with the design that might be used to implement a scratch pad memory.

Sometimes programmers set aside a portion of main memory to act as a scratch pad memory. This type of scratch pad memory serves only to "partition" scratch pad data from other data (from a programming organization point of view) and does not offer the "readily available" aspect of a true scratch pad memory. Notice that a programmer's view of memory can be different from that of a designer.

Buffer memory

Buffer memories serve as a buffer between a fast operating device, such as a computer or CPU, and a slow device, such as an I/O. We saw an example of this in the magnetic tape mass storage of Figure 8–17. Another example is a computer sending data to a printer. The computer can send out data much faster than the printer can print it, which would result in an ineffective use of the computer if it had to wait for one byte of data to be printed before sending the next byte. A solution is for the printer to have a large-capacity buffer memory to which the computer can write its data at a high rate of speed. The printer can fetch the data to be printed from the buffer memory at its own rate, releasing the computer for other duties during printing.

A FIFO memory can serve as a buffer memory, as illustrated in Figure 8–6. Of course a memory with a larger capacity than that in the figure is required.

The memory device used to implement a buffer memory must have a write cycle time equal to or faster than the CPU. As stated in Chapter 8, shift registers can be used to construct a FIFO-style memory. A memory much like main memory could also be used, with an added controller (to furnish addresses and control signals) that would access its memory location as a FIFO.

EXAMPLE 9–1

Suppose a computer were sending a 1-K (1024) block of ASCII data to a printer that does not have a buffer memory. If the printer can print 100 ASCII characters per second, how long will the computer be "tied up" while this 1-K block of ASCII data is being printed?

SOLUTION At a rate of 100 characters per second, it takes 1/100 second (10 ms) for the printer to print one character. To print 1024 characters therefore takes 10.24 seconds (1024×0.01), which is the amount of time the computer will be tied up.

EXAMPLE 9–2

To make the printer of Example 9–1 more efficient, let us add a FIFO buffer memory, which has a capacity of at least 1 K. If the computer can transmit an ASCII character to the buffer memory every 5 μs and the write cycle time of the buffer memory is sufficiently fast, how long will the computer be tied up printing this 1-K block of ASCII characters?

SOLUTION The computer can write this 1-K block of data in 5.12 ms ($1024 \times 5\ \mu$s). Once the computer has transmitted the 1-K block of data to the buffer memory, it can perform other functions. Once the block of data is in the buffer memory, the printer can print it at its own pace, which from Example 9–1 is 10.24 seconds. Obviously, 5.12 ms of computer time is more efficient than 10.24 seconds.

Stack memory

A *stack memory* is used to store data temporarily as the result of an *interrupt request* to the CPU. To understand the function of a stack, imagine that a CPU is executing a program, known as the main program. It is requested by some I/O device to interrupt its execution of the main program to service that I/O device. To do so, the CPU must leave the main program and jump to another program, referred to as the *I/O service subroutine*, which instructs the CPU on how to service the requesting I/O device. When the CPU jumps to the service routine, it stores (pushes) the address of the next main program instruction to be executed, referred to as the *return address*, on the stack. Also pushed on the stack are CPU register contents, such as its scratch pad memory, which might be lost as a result of servicing the interrupt. When the CPU has finished executing the I/O subroutine, it pops (reads) the data that was pushed on the stack, which loads the CPU registers with the same data and address (the return address) it had at the time of the interrupt. As a result, the next address output by the CPU for an instruction fetch, after it has completed the interrupt, is the return address of the main program. This pushing and popping of data and return address on and off stack allows the CPU to be interrupted (jump to a subroutine) and then return and resume execution of the main program without loss of data or place of execution within the main program.

Figure 9–1 illustrates the concepts of a stack. Figure 9–1a is a symbolic representation of both the CPU and its stack memory. It shows the stack as a portion of the main memory; as indicated by the bidirectional arrow of the data bus, it is an R/W memory. There are two points of access to this main memory, one via the top of the stack when accessed as a LIFO, which is its normal mode of access; and any location within the stack when accessed as any standard RAM (we mean RAM in its true sense here).

To get a feel of how the stack and the CPU interrelate, let us investigate Figure 9–1a. Figure 9–1a shows two data busses, one internal to the CPU and the other external to the CPU, which is known as the *system data bus*. Between the two data busses is a buffer to prevent bus contention between the internal and system data busses.

Memory addresses supplied by the CPU can be furnished from one of two sources—the program counter (PC) or the stack pointer (SP) of Figure 9–1a. If the CPU is fetching a program instruction from memory, the PC (PCH and PCL; see definitions below) supplies the address; however, if the CPU is doing a stack operation, the SP furnishes the address. The content of the source address register (PC or SP) is latched by the address latch and then loaded on the address bus, which will access the appropriate memory location. The address decoders, shown as part of main memory, symbolically represent all address decoders involved in decoding the address bus lines. We are assuming that main memory is 1 byte wide, as are all the CPU registers except the PC and SP, which are 16 bits (16-bit addresses). PCH represents the high-order byte of the PC, and PCL represents the low-order byte.

As an example of the CPU using the stack, imagine the CPU being interrupted by an I/O device. Suppose that just before the interrupt the content of the PC of Figure 9–1a is 0500H and that of the SP is FFFFH. This means that the CPU is in the process of fetching and executing an instruction from the main program at memory location 0500H. The stack pointer is pointing to the top of the stack, which from the contents of the SP is address FFFFH. While the CPU is executing the instruction at location 0500H, the I/O device requests an interrupt via the interrupt request input pin. When the CPU finishes execution of that instruction, it acknowledges the interrupt request by transferring control (jumps) to the I/O service subroutine. Before the CPU leaves the main program it automatically increments and pushes the contents of its PC (plus one),

Figure 9–1
A LIFO stack memory.

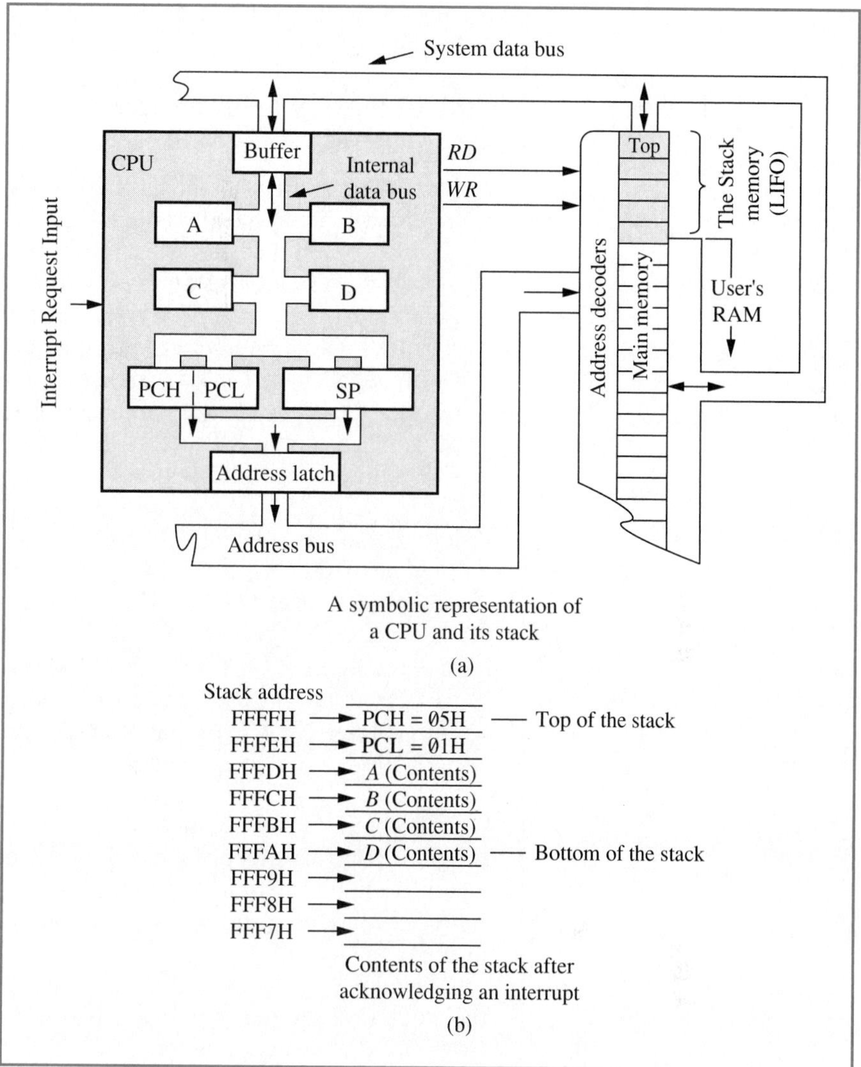

which is the return address 0501H, on the stack. It then transfers jumps to the subroutine, where the first instructions tell the CPU to push the contents of its scratch pad memory (registers A, B, C, and D) on the stack. To accomplish this requires the following steps: (1) The contents of the SP (SP = FFFFH) are latched by the address latch and output on the address bus, which accesses memory location FFFFH (top of the stack). (2) The contents of the PC (PC = 0501H) are pushed on the stack by first writing the contents of PCH (PCH = 05H) into memory location FFFFH, via the internal data bus and system data bus. This is illustrated in Figure 9–1b. (3) The SP is then decremented by one by the CPU (SP = FFFEH), and the content of PCL (PCL = 01H) is pushed on the stack. (4) The CPU jumps to the subroutine, where it is instructed to push the contents of register A on the stack. The SP is decremented again (SP = FFFDH), and the content of register A is pushed on the stack (see Figure 9–1b). This pushing of register contents continues until all scratch pad memory contents are saved on the stack. To pop data off the stack the process is reversed; that is, the stack is read and the SP is incremented by one. Note that the last data in is the first data out, which is LIFO-type accessing.

EXAMPLE 9–3

Using Figure 9–1b as a model, pop data from the stack and explain how the stack is accessed as well as where the data is to go.

SOLUTION When the stack of Figure 9–1b was filled, the content of the SP was FFFAH (the address of the bottom location of the stack). To pop data from the stack, the content of the SP is latched by the address latch, from which it is output on the address bus. This accesses location FFFAH. The CPU then activates control signal *RD*, and the content of location FFFAH (contents of register D) is loaded on the system data bus and also on the internal data bus (the buffer is active), from which it is latched by register D. The SP is incremented (SP = FFFBH) and the content of stack location FFFBH is loaded into register C. The SP is again incremented (SP = FFFCH) and the content of stack location FFFCH is loaded into register B. The SP is incremented (SP = FFFDH), and the content of stack location FFFDH is loaded in register A. The SP is again incremented (SP = FFFEH), and the content of stack location FFFEH, which is 01H, is loaded in PCL. The SP is again incremented (SP = FFFFH), and the content of stack location FFFFH, which is 05H, is loaded in PCH. The stack has been emptied and the CPU is back to its original state at the time of the interrupt.

It should again be noted that these pushes and pops are under program control, except for pushing the content of the PC on the stack, which is an automatic consequence of an interrupt acknowledge. A representative program to push and pop data on and off a stack is shown in Figure 9–2.

Figure 9–2 shows that the data of scratch pad memory is pushed on the stack (PUSH and POP are actual 8085 microprocessor instructions). After the PUSH instructions are carried out, the service routine is executed. After execution of the subroutine, the scratch pad data is POPped off the stack. Last, a return instruction (RET) causes the return

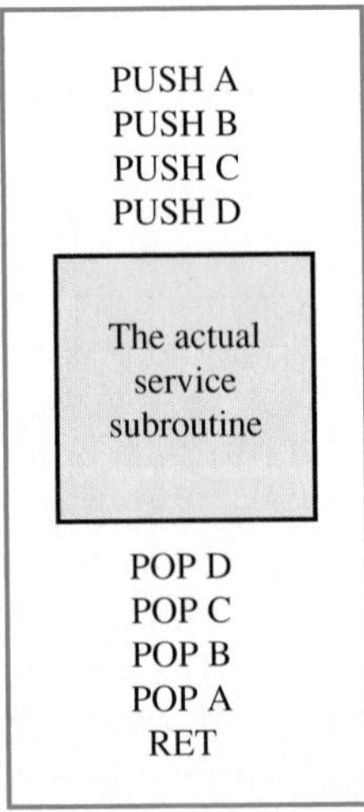

Figure 9–2
A representative I/O service subroutine format.

address to be popped from the stack and loaded into the PC (recall that the return address is automatically pushed on the stack when the interrupt request is acknowledged).

To reiterate, the stack memory is just a designated portion of main memory. What makes it different from main memory is not the hardware, for it is physically the same as main memory, but rather the addressing sequence executed by the CPU. The CPU addresses this portion of main memory so that it functions as a LIFO-type memory. It could be said that in this case the CPU functions as a LIFO controller for accessing the stack.

Queue memory

A *queue memory* is used in some microprocessor designs to speed up the instruction fetch and execute cycle, as well as to make more efficient use of the system data bus. To accomplish this, the CPU fetches instructions from main memory and stores them in proper sequence for execution in the queue, which is much like a small-capacity buffer memory (shift register) with FIFO-type accessing. To make better use of the system data bus, the CPU fills its queue at the same time it is executing instructions that do not require use of the system data bus. As a result, when the CPU finishes the execution of one instruction and is ready to fetch and execute the next, it fetches the next instruction from its queue rather than from main memory, which speeds up the instruction-fetching process, since internal fetches are much faster than external ones. This process is known as *pipelining*. Six- and 4-byte queues are common in some microprocessors.

Figure 9–3 is a simplified illustration of a microprocessor queue memory. Instructions are loaded in the queue as they are fetched from main memory. Since the queue is accessed as a FIFO, the first instruction fetched is the first one executed. The CPU fetches an instruction for execution from the queue and loads it into its instruction register (IR), where it is decoded for execution. As an instruction is fetched from the queue, the above procedure empties the entry location, allowing the CPU to fetch another instruction for the queue at a convenient time, that is, when the CPU is not using the data bus to execute an instruction.

Figure 9–3
An instruction queue.

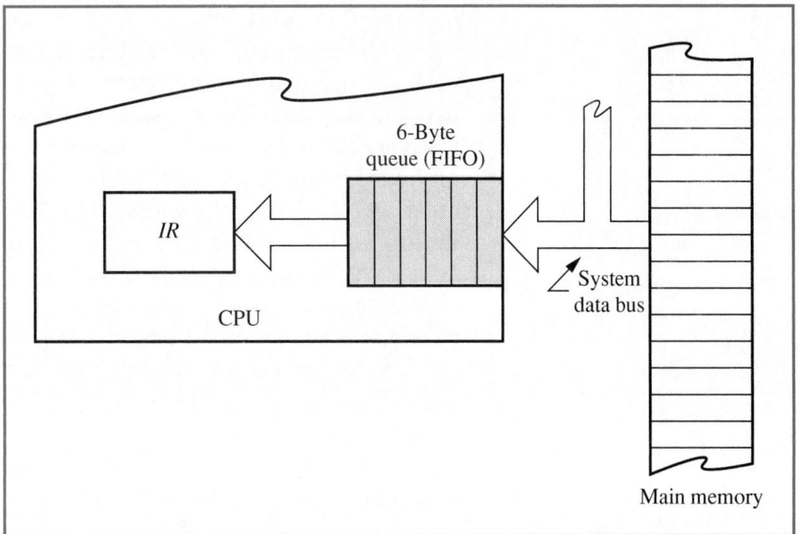

EXAMPLE 9–4

Of the instructions listed below, which ones could the CPU execute while at the same time fetching another instruction from main memory to fill its queue?

(a) ADD A,B. This instruction adds the contents of the CPU's registers A and B and then stores the sum in register A.
(b) MOV D,C. This instruction moves (copies) the data in register C into register D.
(c) MOV B,DATA_1. This instruction moves the datum at memory location DATA_1, which is symbolic representation of an address, into register B.

SOLUTION

(a) Executing ADD A,B does not require use of the system data bus, since both registers are internal to the CPU. Therefore, the CPU could execute this instruction and at the same time fetch another instruction for its queue.
(b) The CPU could execute MOV D,C and also fetch another instruction for the queue for the same reason as in Problem a.
(c) To execute MOV B,DATA_1, the CPU must use the system data bus to fetch the datum from memory location DATA_1; therefore, it cannot fetch another instruction at the same time.

Although the CPU cannot simultaneously execute and fetch instructions 100% of the time, two times out of three isn't a bad average.

Look-up table (translation table)

There are many situations in which data must be converted (translated) from one format to another, as done by the encoders described in Chapter 4. One method of data conversion for a computer is for a program to manipulate and perform calculations on the data in order to transform it into a usable form. Another method is to use a *look-up table* to translate the data from one form to another, such as converting a person's name into that person's telephone number. For this conversion, the look-up table is the telephone book. Another example is use of a table of sine values to find the sine of 30 degrees.

To manipulate or perform calculations on data via a program requires CPU time, which for some applications may be more time than the application can tolerate. For instance, suppose a spacecraft had an on-board navigational computer. If it took this computer 10 seconds to perform the calculations necessary to convert coordinate parameters into a navigational decision that had to be made in less than 3 seconds, that on-board computer would not be an appropriate device for such an application. However, if this computer could be relieved of having to perform those calculations by having the CPU use a look-up table approach for the answer, which required just 5 μs, the on-board computer would be an appropriate device. Also, there are applications in which data conversions must be made, but there is no "intelligence" (CPU) in the system to perform calculations, as in a system composed of combinational logic. Look-up tables work very well for these types of applications.

For look-up table applications, we need a device that can be programmed so that for a given input, which is the binary number to be converted, the converted value will appear as output. A memory device is well-suited for this application.

Let us determine some other characteristics of this look-up table so that we may decide on the type of memory device to use. (1) Is there a CPU in this system that can write the look-up table values in the memory device? (2) Must the look-up table be present from a "cold start"? Let us suppose the answer to the first question is no, which means a ROM must be used to implement the look-up table. If the answer to the second question is yes, again a ROM must be used. Only if the answer to the first question is yes and the second question is no might the system designer use a RAM. Even under these conditions a ROM might still be a better choice, for the designer may not want the look-up table taking up RAM memory, which is usually at a premium.

EXAMPLE 9–5

As an example of a ROM look-up table application, program a ROM so that it will convert (encode) 4-bit binary numerical values into their corresponding binary ASCII values. Recall from Chapter 1 that ASCII is a coding scheme for representing characters on a keyboard.

SOLUTION We begin by constructing a table that shows the logic states of the independent variables (the inputs) and the desired logic states of the dependent variables (the outputs). The results are given in Table 9–1. The leftmost column of Table 9–1 lists the ASCII characters that are to be translated into their equivalent binary ASCII values. The next four columns comprise the inputs to the ROM, which are address pins (A_3 through A_0). Under those four columns are listed all possible addresses in binary, each address

Table 9–1
ROM look-up table to convert binary to ASCII

	ROM												
ASCII character	*Input* A_3	A_2	A_1	A_0	*Contents* m_7	m_6	m_5	m_4	m_3	m_2	m_1	m_0	*ASCII hex value*
0	0	0	0	0	0	0	1	1	0	0	0	0	3 0
1	0	0	0	1	0	0	1	1	0	0	0	1	3 1
2	0	0	1	0	0	0	1	1	0	0	1	0	3 2
3	0	0	1	1	0	0	1	1	0	0	1	1	3 3
4	0	1	0	0	0	0	1	1	0	1	0	0	3 4
5	0	1	0	1	0	0	1	1	0	1	0	1	3 5
6	0	1	1	0	0	0	1	1	0	1	1	0	3 6
7	0	1	1	1	0	0	1	1	0	1	1	1	3 7
8	1	0	0	0	0	0	1	1	1	0	0	0	3 8
9	1	0	0	1	0	0	1	1	1	0	0	1	3 9
A	1	0	1	0	0	1	0	0	0	0	0	1	4 1
B	1	0	1	1	0	1	0	0	0	0	1	0	4 2
C	1	1	0	0	0	1	0	0	0	0	1	1	4 3
D	1	1	0	1	0	1	0	0	0	1	0	0	4 4
E	1	1	1	0	0	1	0	0	0	1	0	1	4 5
F	1	1	1	1	0	1	0	0	0	1	1	0	4 6

corresponding to the hexadecimal equivalent of the ASCII character given in the leftmost column. The next eight columns are the contents of each memory cell (m_7 through m_0) at the corresponding address. The binary patterns listed beneath them are the codes representing each ASCII character. The rightmost column gives the hexadecimal equivalent of the ASCII binary code. Hence, at memory location 0000_2 the ROM will have been programmed with 00110000_2, which is equivalent to hexadecimal value 30. Hexadecimal 30 is the code for ASCII character 0. At location 0001_2 the binary value 00110001_2 is programmed, and so on, until at the last location, 1111_2, the binary value 01000110_2 is programmed, the binary value for ASCII character F.

This look-up table application may be called a binary-to-ASCII encoder.

It is important that the reader realize the relationship between the addresses and the data stored at those addresses, since this relationship is often the key to encoder solutions.

EXAMPLE 9–6

Suppose a 16 × 8-bit ROM is programmed according to Table 9–1. What is the output of this ROM if the input is (a) 5H (b) 7H (c) AH (d) EH?

SOLUTION Referring to Table 9–1 we find that (a) 5H is address 0101, which will output 00110101 (35H) (b) 37H (c) 41H (d) 45H.

Review questions

1. What are the five memory applications discussed in this section?
2. What are the essential characteristics of each of the five memory applications?
3. Why is a scratch pad memory often manufactured as an integral part of the CPU?
4. When would a scratch pad memory be a portion of main memory?
5. When a memory is referred to as a buffer memory, what is implied about its application?
6. What type of accessing is best for a buffer memory? Why?
7. What is the major function of a stack memory?
8. The stack memory of a computer system is a designated portion of main memory. How is a stack memory different from a main memory?
9. What function does a queue memory serve, and what type of accessing is used?
10. What is a look-up table and how is it implemented?

9–3 Memory device architecture and timing characteristics

Before we can design a memory, we must choose the semiconductor memory devices to be used in the design. These choices are based on the performance requirements of the memory system, such as memory capacity and organization, speed of operation (access time and write cycle time), and power consumption. Of course, cost is also a major consideration. Selection of a memory device is often based on the technology used in its fabrication. The technologies from which semiconductors are manufactured [TTL, emitter coupled logic (ECL), N-channel metal-oxide semiconductor (NMOS), and complementary MOS (CMOS)] have characteristics that make one better suited than another for a particular application. When we study the electronics of these technologies we will understand why they differ, but such knowledge is not necessary at this time. Table 9–2 lists characteristics of semiconductor technologies.

Summarizing the information of Table 9–2, the bipolar technologies (bipolar transistors are the active devices) TTL and ECL are the fastest, but they have less memory capacity than the others. NMOS and CMOS [metal-oxide-semiconductor field-effect transistors (MOSFETS) are the active devices] have larger memory capacities but are slower. CMOS devices consume less power.

There is a large selection of semiconductor memory devices. The priority of the characteristics listed in Table 9–2 depends on the application. The following memory systems are representative of the relative importance of the three parameters.

1. **Mainframe computer.** Mainframe computers are powerful and fast, with a large memory capacity. The trade-off is between speed and memory capacity. If speed is the primary requirement, ECL technology must be used, but additional ECL memory devices are necessary to provide the desired memory capacity. However, if slower operating speed and therefore fewer memory devices are acceptable, then NMOS technology can be used.
2. **Personal computer** (PC). A PC trades operating speed for memory capacity, since by its very nature it is to be small enough to fit on a desktop. Also, the CPU of a PC is a microprocessor (to be studied in Chapter 10) that is usually fabricated from one of the MOS technologies, which lessens the need for an exceptionally fast main memory. As a result, main memories of PCs are also fabricated from either NMOS or CMOS. If a PC has a battery back-up power supply, power consumption can be critical.

Table 9–2
Characteristics of semiconductor technologies

	Technology	Bit capacity	Speed of operation	Power
RAM's	TTL	576	45 ns	1W
	ECL	1–16 K	20 ns	0.7–1W
	NMOS (dynamic)	16–64 K	150 ns	0.2–0.3W
	NMOS (static)	4–16 K	50 ns	0.6–0.8W
	CMOS	16 K	150 ns	0.3W
PROM's	NMOS EPROM	16–64 K	200–450 ns	0.5–0.8W
	CMOS EPROM	16–64 K	250–300 ns	40mW/MHz
	TTL PROM	1–32 K	35–65 ns	0.8–1W

3. **Portable PCs.** Portable PCs have the same basic requirements as a desktop PC except that power consumption must also be considered. Hence, CMOS should be a primary consideration for implementing main memory.
4. **Microcontrollers.** Microcontrollers are microprocessor-based "computers" that are dedicated to the control of some function or process. Such applications are often referred to as an embedded application of the microcontroller and the microcontroller is referred to as an embedded controller. Because they usually interface with other machines, they generally do not have a keyboard or terminal. Owing to their limited abilities and narrowness of purpose, the programs they execute require relatively little memory capacity. Microcontrollers must often be physically small so that they can be integrated into the system they are controlling, such as a robotic mechanism. As a result of the physical constraints and microprocessor operating speed, a microcontroller memory is most often of the MOS (NMOS or CMOS) technology.

Designing a main memory with any of the technologies listed in Table 9–2 is conceptually the same; it is basically a matter of choosing the technology that meets the system's memory capacity and speed of operation requirements. Of course, cost is also a major factor, especially in a high-production environment.

To become familiar with semiconductor memory devices, we will study some representative archtectures and timing characteristics of dominant memory device technologies, which are NMOS and CMOS. Data sheets and excerpts from data sheets of the semiconductor memory

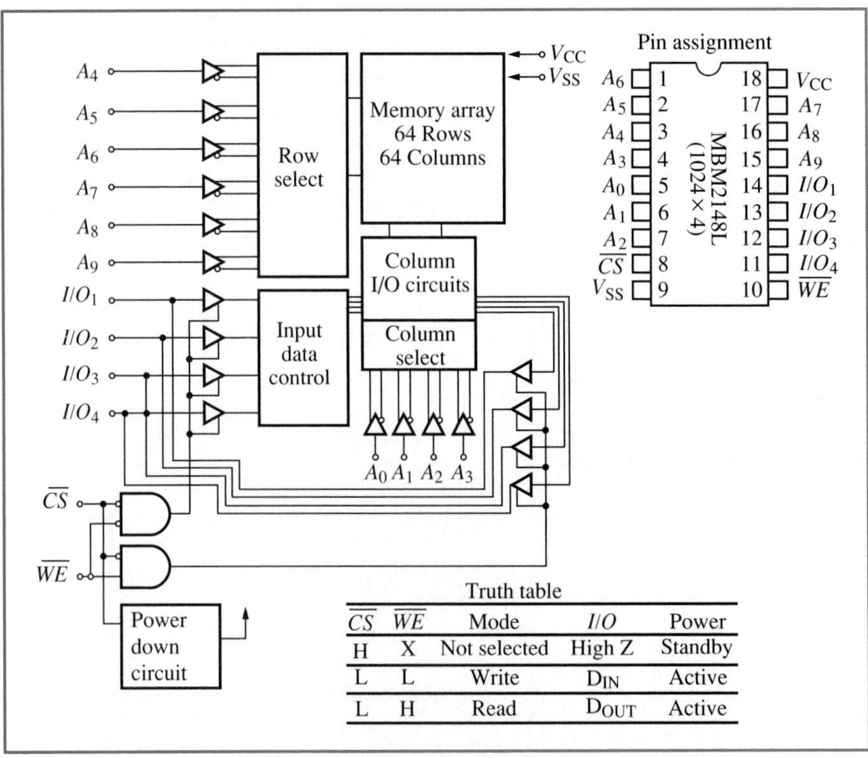

$\overline{CS}$	$\overline{WE}$	Mode	I/O	Power
H	X	Not selected	High Z	Standby
L	L	Write	DIN	Active
L	H	Read	DOUT	Active

Figure 9–4 The architecture and pin configuration of a 2148 NMOS 1024 × 4-bit static RAM. (Courtesy of FUJITSU Microelectronics, Inc.)

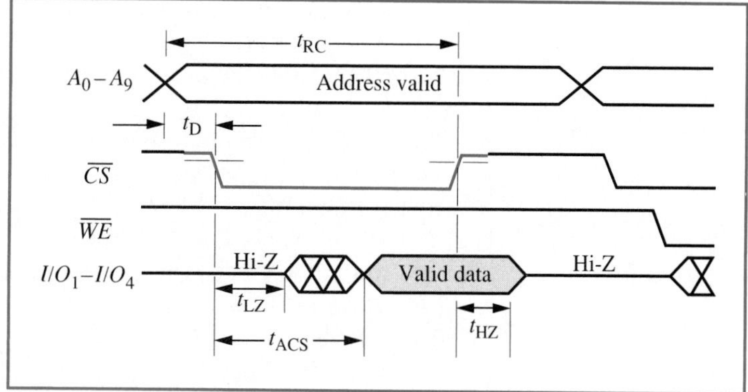

Figure 9–5 Basic read timing diagram for a 2148.

devices studied are found in Appendix C. Figure 9–4 shows the architecture and pin configuration and Figures 9–5 and 9–6 show the read/write timing diagrams for the 2148 static RAM. The 2148 has ten address pins (A_0 to A_9) and therefore 1024 (2^{10}) memory locations. The four pins labeled I/O_1 to I/O_4, which are similar in function to the I/O pins of Figure 8–12, are for reading and writing of data. From the address and data pins we can reason that the 2148 is a 1024 × 4 (4096)-bit memory. Notice that the memory array block states that there are 64 rows, which for six address pins (A_4 through A_9) agrees with the figure 2^6, but that same block also states that there are 64 columns, for which there are only four address pins (A_0 through A_3). As we pointed out in Chapter 8, the manufacturer has included in this number the columns formed by the *z*-coordinate planes (see Figure 8–12). Knowing this, we find that for four address pins and four planes of bits we have $2^4 \times 4 = 64$ columns, which agrees with the manufacturer.

The $\overline{CS}$ pin of Figure 9–4 serves to activate the 2148, just as pin $\overline{CS}$ did in Figure 8–12; however, active logic is a low. If $\overline{CS} = 1$, the chip is disabled (not selected) and all read/write buffers are in the Hi-*Z* state. The chip is in the *standby mode*, and the power consummation is reduced by as much as 60%. When pin $\overline{CS}$ of the 2148 is a logic 0, the *AND* gates to which it inputs are enabled. The other input to those *AND* gates, $\overline{WE}$ (write enable), controls whether the operation to be performed at the addressed location is a read or a write.

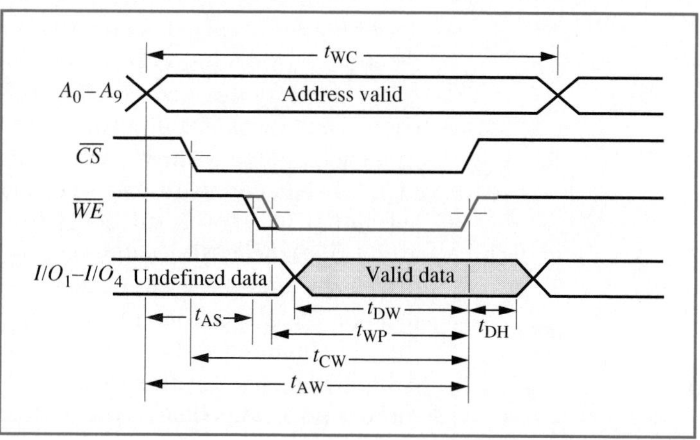

Figure 9–6 Basic write timing diagram for a 2148.

If $\overline{WE}$ is a logic 0, which is the active logic level for this pin, a write is to be performed; if $\overline{WE}$ is a logic 1, a read is to be done. We can reason this from the architecture of Figure 9–4. If $\overline{WE} = 0$ and $\overline{CS} = 0$, the "top" *AND* gate to which these signals are input will output a logic 1, which activates the data (I/O_n) buffers that allow data to flow from the data pins into the memory cell array via the input data control function block. When $\overline{WE} = 1$ and $\overline{CS} = 0$, the bottom *AND* gate outputs a logic 1, which activates the read buffers.

You should recognize these concepts from Figure 8–12; the application of these concepts aids in understanding the architecture of Figure 9–4.

Figure 9–5 is a basic read timing diagram for many static RAMs but has been "personalized," via pin identification, to the 2148. Industry has standardized timing specification parameters in a generic manner, but they become personalized when parameter values for a specific memory device are supplied by the manufacturer in the data specification sheets of that device. These timing specifications usually appear in table form and are labeled *AC characteristics*.

The timing diagram of Figure 9–5 shows the logic levels of the address pins A_0 through A_9, the chip select pin $\overline{CS}$, the write enable pin $\overline{WE}$, and the data pins I/O_1 through I/O_4. To interpret the meanings of these symbolic representations, let us first consider the address pins. When an address is input to a memory device, each address pin has either a logic 0 or a logic 1 applied to it. To simplify and generalize the timing diagram of those multiple address pins, they are represented as a group (A_0 through A_9) with both logic levels present. The crossover; or "*X*," is the point of possible transition from one logic level to the other, such as when a change of address is being applied. When the address is stable (and usable), it is identified as *address valid* for that interval of time. The same symbolic representation applies to the data pins I/O_1 through I/O_4 as to the address pins. Each datum being read is present on those data pins in the interval labeled *valid data*. The "hash-marked" interval just before the valid data interval is a period of transition. That is, the memory cells being read are transferring their respective logic levels to the data pins via the read buffers, but since all four read buffers do not activate at the same instant (there is turn-on time that is inherently different for each buffer), there is a period of transition. The interval labeled Hi-*Z* occurs when the I/O buffers are in the off state, which isolates those pins.

To interpret the timing parameters of Figure 9–5, recall that the accessing device, such as a CPU, first outputs the address of the memory location on the address bus, and this address is input to the address pins of the memory device as well as to the page selector decoder. The active page select decoder output then drives the $\overline{CS}$ pin of the memory device to its active level, which enables the chip. There is some delay between input of the address to the page select decoder and output of the page selector, which for this decoder is a logic low ($\overline{CS}$). The authors have arbitrarily labeled this delay time t_D in Figure 9–5. Once the $\overline{CS}$ pin is driven low and the write enable pin is high, the addressed memory location makes its content available at pins I/O_4 through I/O_1. Owing to the time required to turn on the output buffers of the 2148, there is a delay between activation of the $\overline{CS}$ pin and availability of the datum at the I/O pins. This delay time is the access time of the 2148 measured relative to the $\overline{CS}$ pin being enabled. As indicated in Figure 9–5, the *chip select access time* is identified as t_{ACS}. Since the data pins of the 2148 are connected to the system data bus, the datum being read from the 2148 is also available on the system data bus after t_{ACS} seconds have elapsed. Parameter t_{HZ} indicates the time from deselection of the chip ($\overline{CS} = 1$) until the read buffers actually go into the Hi-Z state. The reading device (the CPU) must read the datum while it is available and stable on the system data bus. As we see from Figure 9–5, this occurs at the positive edge of $\overline{CS}$. The *read cycle time* t_{RC} measures the minimum time required to complete a read operation.

The primary parameters of Figure 9–5 are the chip select access time t_{ACS} and the read cycle time t_{RC}. These parameters are used as a gauge for the read operation speed of a 2148 RAM.

Note from Figure 9–5 that the address was input to the device before the $\overline{CS}$ pin was driven low. This timing diagram would also be valid if the $\overline{CS}$ pin were driven low at the same instant ($t_D = 0$) the address was input to the address pins, which is the case when partial linear addressing is used. The timing of Figure 9–5 would not be valid if the $\overline{CS}$ pin of the 2148 were driven low before a valid address was applied to the address pins.

The write timing parameters of Figure 9–6 are interpreted in the following manner: The minimum time before the write enable pin can be driven active ($\overline{WE} = 0$), as measured from the time the address is input to the address pins, is t_{AS} (the address set-up time). Note that since timing parameter t_{AS} is a timing requirement of the 2148, this ensures that address decoding will be completed by the 2148 before the write enable is activated. Timing parameter t_{WP} is the minimum pulse width allowable for a write operation to occur. This specifies a minimum requirement that must be met by the device driving the write enable pin ($\overline{WE}$), which is the CPU. Timing parameter t_{CW} specifies a minimum time for the $\overline{CS}$ pin to be active. Timing parameter t_{DW} specifies the minimum time that the datum being written into memory must be present before the write enable pin goes inactive ($\overline{WE} = 1$). Timing parameter t_{DH} is the minimum time that the datum must be held on the data pins after the positive edge of $\overline{WE}$. Parameter t_{AW} is the minimum time that the positive edge of $\overline{WE}$ can appear after the address has been input to the address pins. The *write cycle time* t_{WC} is the minimum time required to complete a write operation. Timing parameters t_{DW}, t_{AH}, and t_{WC} are the most critical, for they basically determine the speed at which the memory device can be written to.

There is a situation that is potentially damaging when a write operation is to be done. The condition occurs when the chip select is active before the write enable (as shown in Figure 9–6), which is the logic state for a memory read. Under this condition the read buffers of the 2148 are active, resulting in output of the datum on the data pins (labeled as undefined data). Also, if during this period the device writing to the 2148 loads the datum being written on the data bus, bus contention exists until the write enable pin goes active and puts the read buffers in the Hi-Z state. Whether damage results depends on the current capabilities (to be studied later) of the 2148 and the device writing to it. A way to avoid this problem is to activate the chip select at the same time as or after the write enable, or to make sure that the writing device (the CPU) does not

Table 9–3
Representative AC parameters of a 2148

Read parameters		Write parameters	
t_{ACS} =	55 ns max	t_{DW} =	20 ns min
t_{LZ} =	20 ns min	t_{WP} =	40 ns min
t_{HZ} =	20 ns max	t_{AW} =	50 ns min
t_{RC} =	55 ns min	t_{WC} =	55 ns min

write on the data bus except during interval t_{WP}. The latter criterion is the one that we will use in our designs.

The parameters listed in Table 9–3 are representative timing values for one variety of 2148 (there are different versions of the 2148).

From Table 9–3 we see that the read access time (t_{ACS}) is 55 ns, as is t_{RC}, for a 2148. The write parameters are as follows: t_{DW} is 20 ns, t_{WP} is 40 ns, and t_{WC} is 55 ns. The fastest access time of this variety of 2148 for a read or write operation is 55 ns. The values of t_{RC} and t_{WC} are used as an overall gauge of speed for 2148 RAM, but t_{ACS}, t_{AW}, t_{WP}, and t_{DW} are essential for detailed read/write timing coordination. It is interesting to note that the 2148 timing parameter t_{DH} has a value of zero, as does t_{AS} (not shown in Table 9–3).

It is important for the later use that the significance of timing parameters t_{ACS}, t_{DW}, t_{WP}, and t_{AW} be well understood.

Figure 9–7
Architecture and pin configuration of an 8128 NMOS 2048 × 8-bit static RAM. (Courtesy of FUJITSU Microelectronics, Inc.)

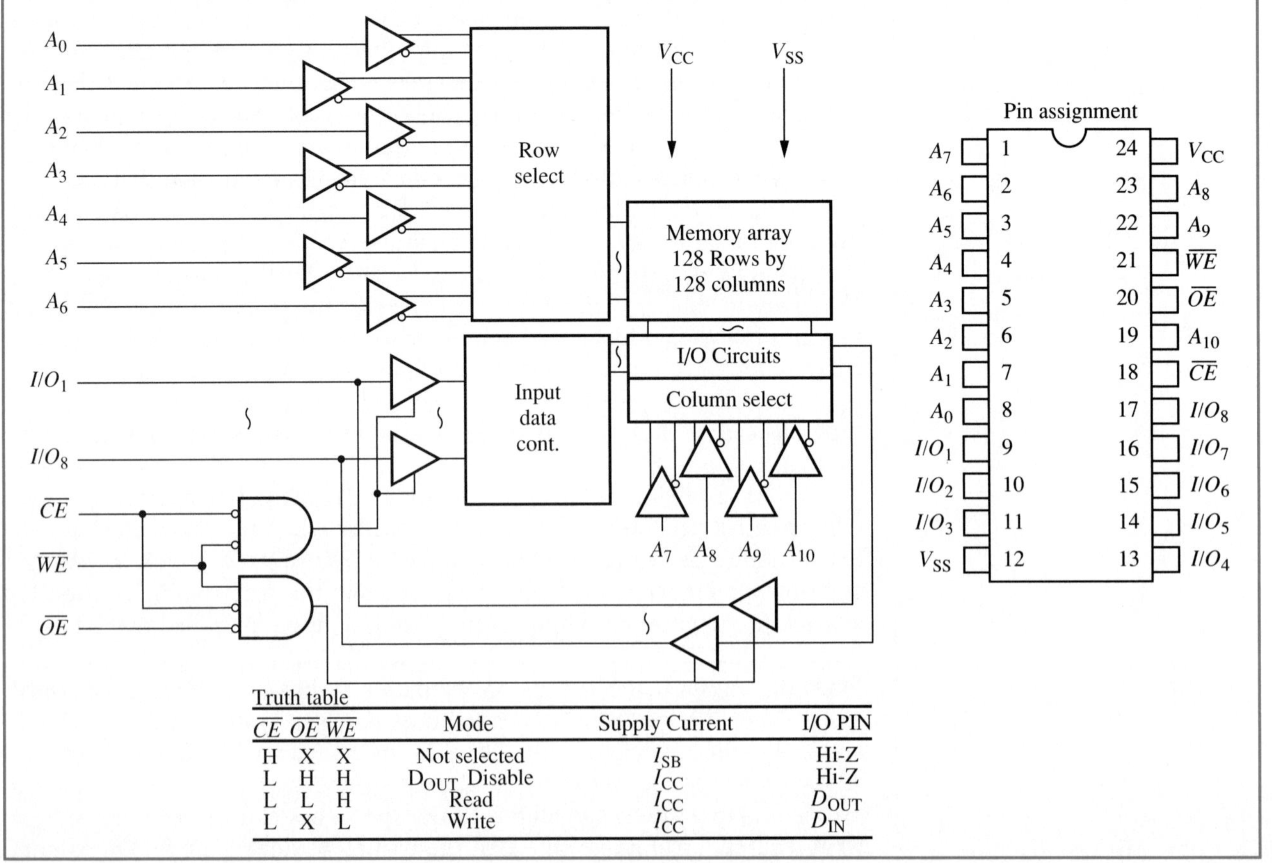

Truth table

$\overline{CE}$	$\overline{OE}$	$\overline{WE}$	Mode	Supply Current	I/O PIN
H	X	X	Not selected	I_{SB}	Hi-Z
L	H	H	D_{OUT} Disable	I_{CC}	Hi-Z
L	L	H	Read	I_{CC}	D_{OUT}
L	X	L	Write	I_{CC}	D_{IN}

Figure 9–7 shows the architecture and pin configuration of an 8128 NMOS static RAM, which has a memory capacity of 16,384 (2048 × 8) bits. There are 11 address pins (A_0 through A_{10}), which is equivalent to 2048 memory locations, and eight data pins (I/O_1 through I/O_8), which indicates that each location is 8 bits wide. We note that the 8128 has an input pin with a name different from any pin on the 2148—the output enable in $\overline{OE}$. From the architecture we can reason that in order to activate the read buffers, the output of the three-input *NAND* gate must be high. This requires that the chip enable input ($\overline{CE}$) be low, the write enable ($\overline{WE}$) high, and the output enable pin ($\overline{OE}$) low. The output enable pin $\overline{OE}$ provides more control over read operations than does the 2148. That is, a read for a 2148 is controlled by the chip select (chip select and chip enable serve the same function; they just have different names) and the write enable, whereas a read is controlled by the chip select, write enable, and output enable for a 8128.

Figure 9–8 is a basic read timing diagram for an 8128. Comparing the timing diagrams of Figures 9–5 and 9–8, we see that the addition of the output enable ($\overline{OE}$) is the major difference. The timing parameters t_{RC}, t_{ACE}, and t_{OE} are of most interest, for they are measures of how fast a read operation can be performed. Time parameter t_{ACE}, the chip enable access time, provides the maximum elapsed time between activation of the chip enable and output of valid data (datum) at the I/O pins, assuming that the output enable pin has been activated for at least t_{OE} seconds. Time parameter t_{OE} is the maximum time that the output enable pin $\overline{OE}$

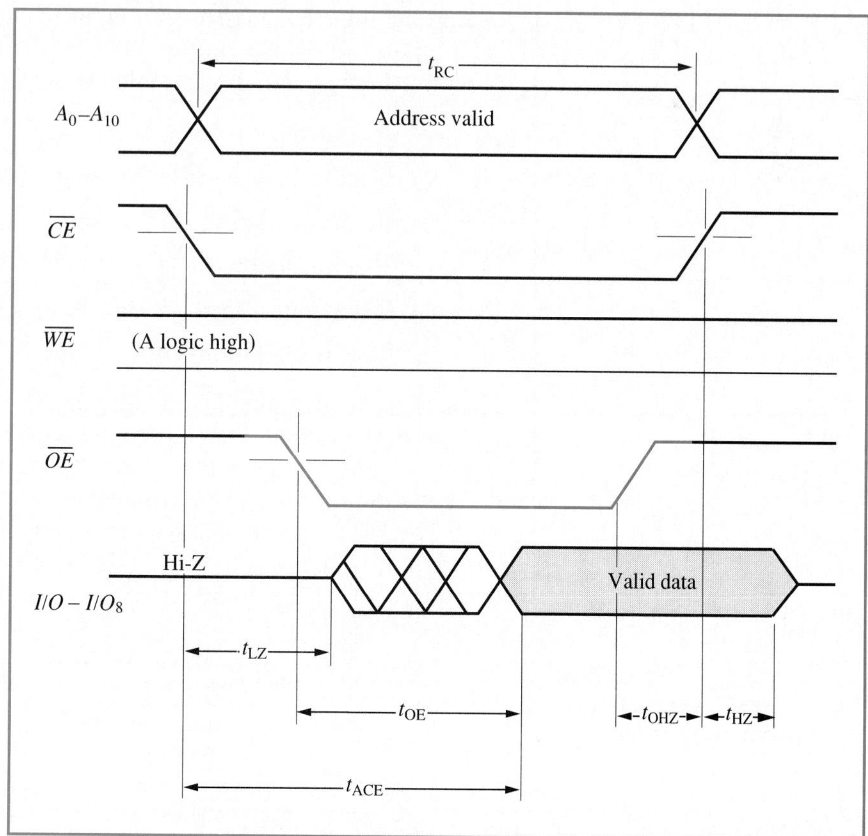

Figure 9–8
Basic read timing diagram for an 8128.

must be low before valid data (datum) is output at the I/O pins, assuming that the chip enable pin $\overline{CE}$ has been active for at least t_{ACE} seconds.

The write timing diagram for the 8128 is basically the same as Figure 9–6. The major difference is the output enable $\overline{OE}$, which for a write operation requires $\overline{OE}$ to be driven high (verify this from Figure 9–7). For one type of 8128, some representative AC parameter values are given in Table 9–4.

Figure 9–9 shows the architecture of a CMOS static RAM that has 16,384 bits of storage capacity. Since there are 11 address pins and eight I/O pins, the memory cell organization is 2048 × 8 bits. It appears to be drawn differently from those previously studied; that is, most logic gates, decoders, etc. are represented as function-blocks rather than discrete logic gates and function-blocks. The former approach entails simplifying the architectural drawing by omitting some details; however, it has the same conceptual meaning.

We see that this memory device has three input pins labeled $\overline{WE}$, $\overline{CS}$, and $\overline{CE}$. By this time the reader should be able to decipher the function of these inputs. $\overline{WE}$ is the write enable pin and $\overline{CE}$ is a chip enable pin, and both are active-low. $\overline{CS}$ is a chip select pin. The $\overline{CS}$ input activates those buffers by which it is directly input as well as those with inputs labeled *CSB* (chip select buffer). Likewise, input $\overline{CE}$ enables all buffers with *CEB* (chip enable buffer) as an input. Notice that the I/O pins require $\overline{CE}$ (*CEB*) and $\overline{WE}$ to be active. Both the row and column access pins buffers are enabled by *CEB* ($\overline{CE}$), and the write buffers are activated by $\overline{WE}$ and $\overline{CS}$ (*CSB*). From the architecture we can reason that both the $\overline{CE}$ and $\overline{CS}$ pins must be active in order to perform a read operation, since *CEB* ($\overline{CE}$) is an input to the I/O pin buffers and *CSB* ($\overline{CS}$) is an input to the write buffers, as well as $\overline{WE}$. In fact, there are six different timing diagrams possible for a read or write operation. These are

1. A $\overline{WE}$-controlled read ($\overline{CE}$ = low, $\overline{CS}$ = low)
2. A $\overline{CE}$-controlled read ($\overline{WE}$ = high, $\overline{CS}$ = low)
3. A $\overline{CS}$-controlled read ($\overline{WE}$ = high, $\overline{CE}$ = low)
4. A $\overline{WE}$-controlled write ($\overline{CE}$ = low, $\overline{CS}$ = low)
5. A $\overline{CE}$-controlled write ($\overline{WE}$ = low, $\overline{CS}$ = low)
6. A $\overline{CS}$-controlled write ($\overline{WE}$ = low, $\overline{CE}$ = low)

Each of these diagrams is unique, yet they are conceptually similar. We will study the timing diagram for the read condition in number three and the write condition in number four. From these two representative timing diagrams the reader can determine the others.

The 8417 read timing diagram of Figure 9–10a is chip select controlled, which means that $\overline{CS}$ is the controlling pin for a read operation. From the architecture we reasoned that $\overline{WE}$, $\overline{CS}$, and $\overline{CE}$ are

Table 9–4
Representative AC parameters of an 8128

Read parameters	Write parameters
t_{ACE} = 100 ns max	t_{DW} = 40 ns min
t_{OE} = 50 ns max	t_{WP} = 85 ns min
t_{LZ} = 0 ns min	t_{AW} = 95 ns min
t_{RC} = 100 ns min	t_{WC} = 100 ns min

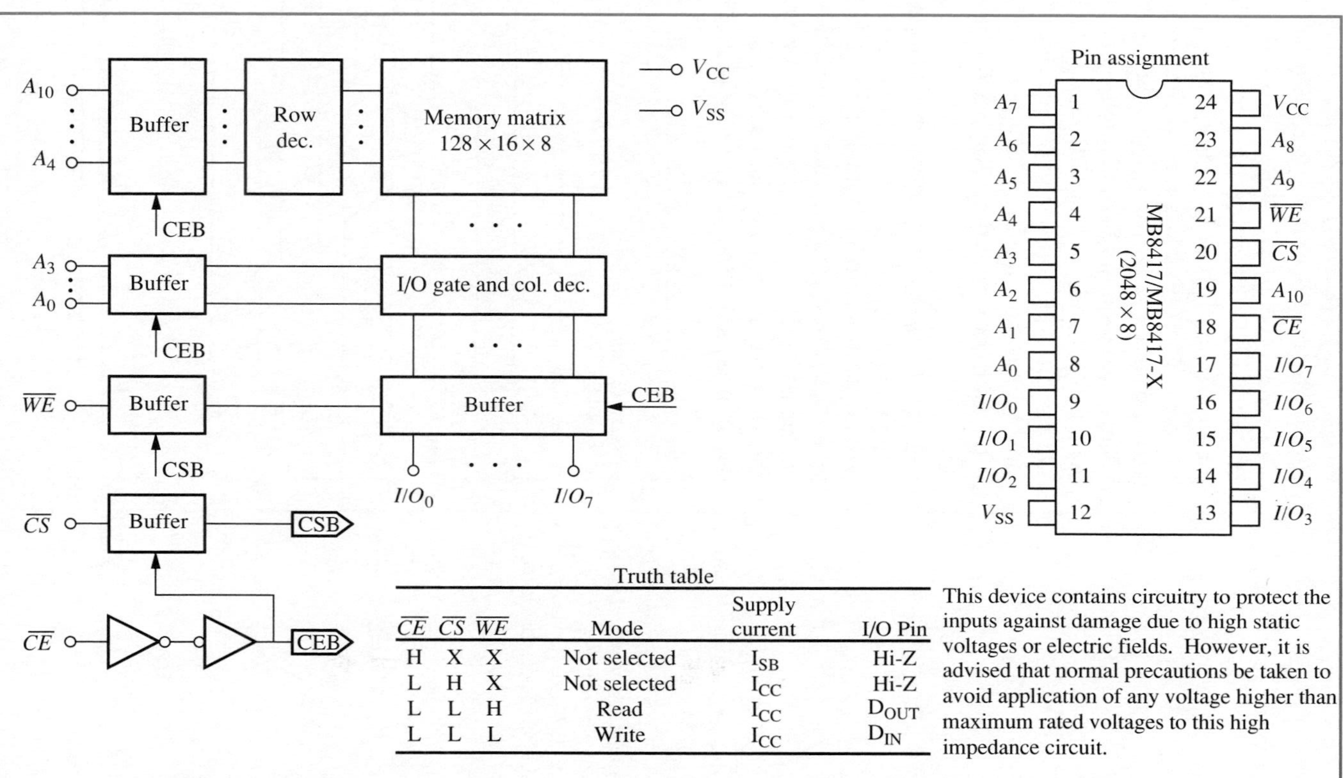

Truth table

$\overline{CE}$	$\overline{CS}$	$\overline{WE}$	Mode	Supply current	I/O Pin
H	X	X	Not selected	I_{SB}	Hi-Z
L	H	X	Not selected	I_{CC}	Hi-Z
L	L	H	Read	I_{CC}	D_{OUT}
L	L	L	Write	I_{CC}	D_{IN}

This device contains circuitry to protect the inputs against damage due to high static voltages or electric fields. However, it is advised that normal precautions be taken to avoid application of any voltage higher than maximum rated voltages to this high impedance circuit.

Figure 9–9 Architecture and pin configuration of an 8417 CMOS static RAM. (Courtesy of FUJITSU Microelectronics, Inc.)

involved in all read/write operations and that in order for input pin $\overline{CS}$ to control the read operation, the write enable ($\overline{WE}$) must be high and the chip enable ($\overline{CE}$) must be low before the chip select input becomes active. From Figure 9–10a we see that the $\overline{CS}$ access time t_{ACS} is measured from the time that $\overline{CE}$ goes low until the datum is output (D_{OUT} valid) at the I/O pins. The $\overline{CS}$ access time is the maximum time required to read a datum from an 8417 using $\overline{CS}$ as the read control pin. Parameter t_{AA} is the address access time, the maximum time required to read the content of a memory location, measured from when the address is applied to the address pins. Of course the $\overline{CS}$ pin must have been active t_{ACS} seconds, and $\overline{WE}$ must be at a logic high level and $\overline{CE}$ at a logic low. Parameter t_{AS} is the address set-up time; for the 8417 the minimum required time is 0 ns.

Figure 9–10b is a timing diagram for a write operation in which the write enable pin ($\overline{WE}$) is the write control pin. For the $\overline{WE}$ to control the write operation, the chip enable pin ($\overline{CE}$) must be active prior to $\overline{WE}$ being active. Figures 9–6 and 9–10b are very similar, except that the chip enable and select pins of Figure 9–10b are assumed to be low and are not shown. Parameter t_{DS} is the data set-up time, the minimum time that the datum must be at the data pins prior to the rising edge of the write enable

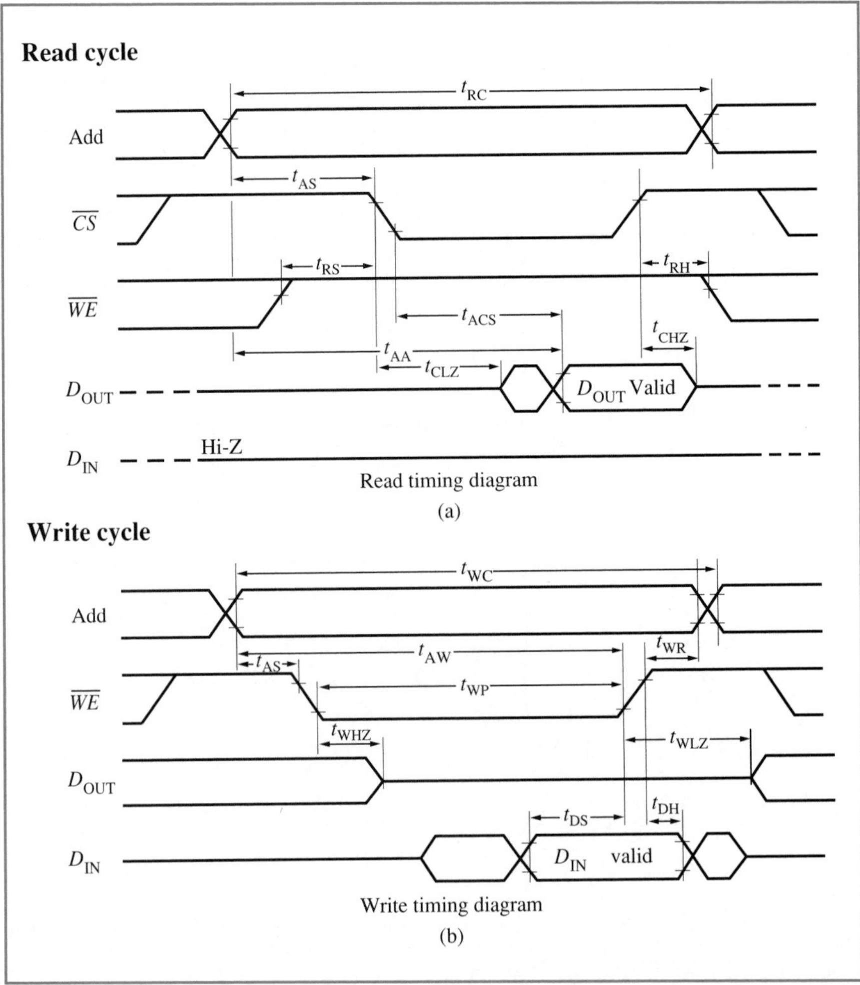

Figure 9–10 Representative read and write timing diagrams for an 8417 CMOS RAM. (Courtesy of FUJITSU Microelectronics, Inc.)

Table 9–5
Representative AC parameters of an 8417

Read parameters	Write parameters
t_{ACS} = 100 ns max	t_{DS} = 60 ns min
t_{AA} = 200 ns max	t_{WP} = 140 ns min
t_{AS} = 0 ns min	t_{AW} = 160 ns min
t_{RC} = 200 ns min	t_{WC} = 200 ns min

signal $\overline{WE}$ (the datum is actually written into the 8417 on the positive edge of signal $\overline{WE}$). The $\overline{WE}$ pin must be active for a minimum of t_{WP} seconds. AC parameter t_{AS} is the address set-up time and is 0 ms for the 8417. Table 9–5 is a listing of some 8417 AC parameters.

The 2764 of Figure 9–11 is an NMOS UV EPROM device with 65,536 bits of memory, which are arranged as 8192 × 8 bits. Note the window in the chip, which permits UV light to strike the silicon die for erasures. The reader should confirm the memory organization via the number of address and data pins (the data pins are output pins labeled O_n). Pins V_{PP} and $\overline{PGM}$ are programming pins and are not discussed, since the EPROM programmer supplies the proper voltages and timing for programming the EPROM. When the 2764 is in the memory system for normal operations, the $\overline{PGM}$ pin is tied high. The chip enable pin $\overline{CE}$ enables the chip for a read operation, whereas the output enable pin $\overline{OE}$ enables the read buffers, as illustrated in Figure 8–13.

The read timing diagram is illustrated in Figure 9–12. The 2764 chip enable pin $\overline{CE}$ must be low t_{CE} seconds before the datum can be output

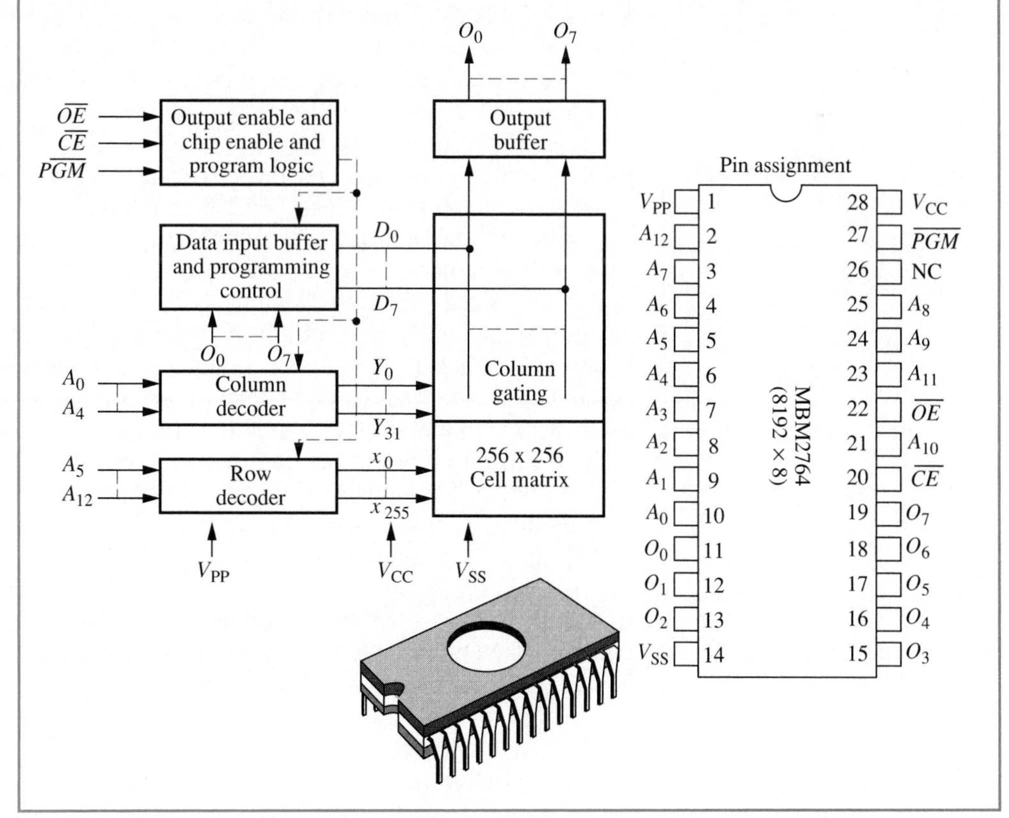

Figure 9–11
Architecture and pin configuration of a 2764 8192 × 8-bit NMOS UV EPROM. (Courtesy of FUJITSU Microelectronics, Inc.)

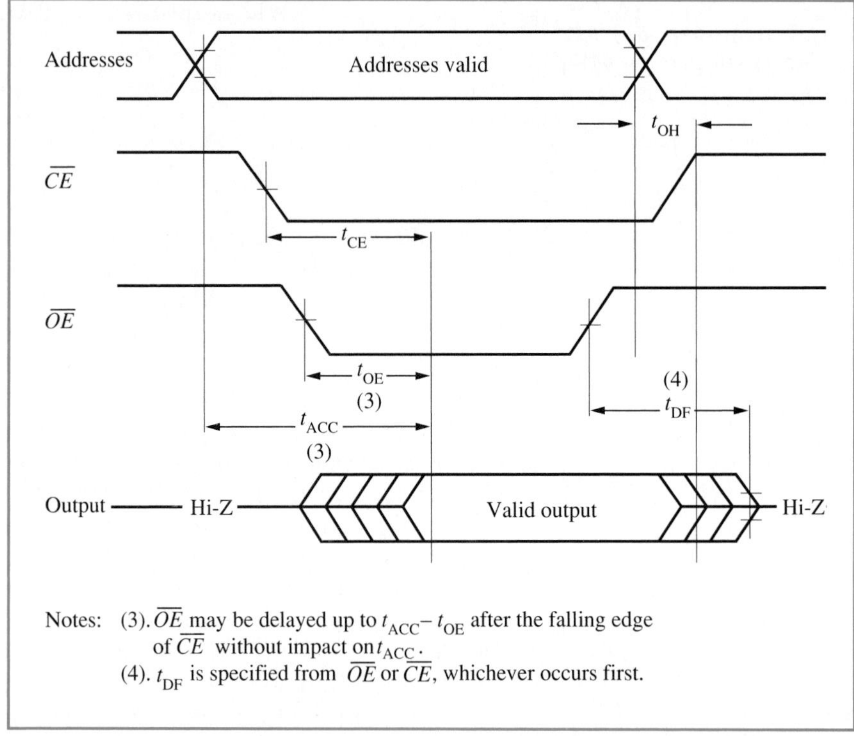

Figure 9–12 Timing diagram for a 2764. (Courtesy of FUJITSU Microelectronics, Inc.)

on the data pins O_0 through O_7. When the output enable pin $\overline{OE}$ is driven low, t_{OE} seconds elapse before the datum appears at the output pins, and from the time $\overline{OE}$ is deactivated (0 to 1) the datum may remain at the data pins for t_{DF} seconds. An overall measure of how fast a read operation can be performed is the access time t_{ACC}, but this measurement requires that the timing specification of $\overline{CE}$ and $\overline{OE}$ be met. t_{ACC} will be used as the primary indicator of how fast the 2764 can be read. For one variety of 2764, $t_{ACC} = 200$ ns(max), $t_{CE} = 200$ ns(max), and $t_{OE} = 70$ ns(max).

The architecture and pin configuration of Figure 9–13 are for an 8116 NMOS dynamic RAM, or *DRAM*. The 8116 is a 16,384-bit memory that is organized as 16,384 × 1 bit. To address 16,384 memory locations, 14 address pins (2^{14}) are required, yet the 8116 has only 7 address pins (A_0 through A_6). These seven address pins are multiplexed for row and column addresses, which can be seen from the architecture of Figure 9–13. Multiplexing pin function allows the pin size to be minimized. The address pins are input in parallel to both the row and column address buffers, which are latches in this case, and decoders. From the architecture we can reason that the signal input to the row address strobe ($\overline{RAS}$) pin acts as a strobe to latch the address on the address pins into the row address buffer, and the signal input on the column address strobe ($\overline{CAS}$) pin provides the strobe for latching the bit levels of the address pins into the column address buffer. Obviously, then, the full 14-bit address is divided into two 7-bit parts.

From the read and write timing diagram of Figure 9–14, we see that the 7-bit row address must be applied to the address pins first, followed by the 7-bit column address. From both timing diagrams we also note the negative edge (1 to 0, where these levels are identified as V_{IHC} and V_{IL}) of

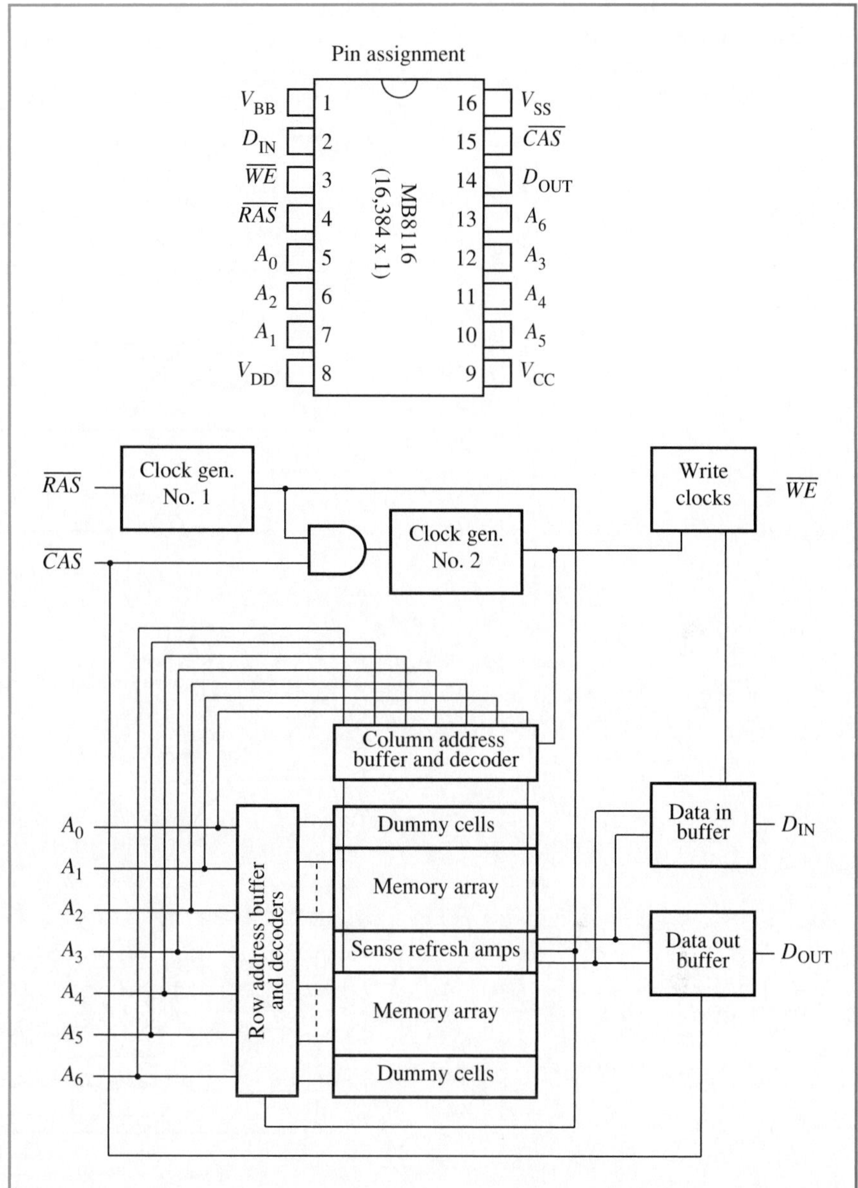

Figure 9–13 Architecture and pin configuration of an 8116 NMOS 16,384-bit DRAM. (Courtesy of FUJITSU Microelectronics, Inc.)

$\overline{RAS}$ strobes row address and the negative edge of $\overline{CAS}$ strobes column address. From both the architecture and pin configuration of Figure 9–13 we see that there are two data pins, D_{IN} and D_{OUT}. With proper timing of the signals input to $\overline{WE}$ and $\overline{CAS}$, the two data pins can be connected to form just one I/O data pin, which is preferred. The timing is oriented to first identifying the type of access to be made (read or write) and therefore requires that $\overline{WE}$ go active before the negative edge of $\overline{CAS}$ occurs.

In Figure 9–14 we note that either a read or a write operation begins with the negative edge of $\overline{RAS}$. The row address must be on the address pins for a minimum of t_{ASR} seconds (the row address set-up time) prior to the negative edge of $\overline{RAS}$. For the 8116E, t_{ASR} is 0 ns, meaning that the address may be applied at the same time $\overline{RAS}$ goes low. This row address

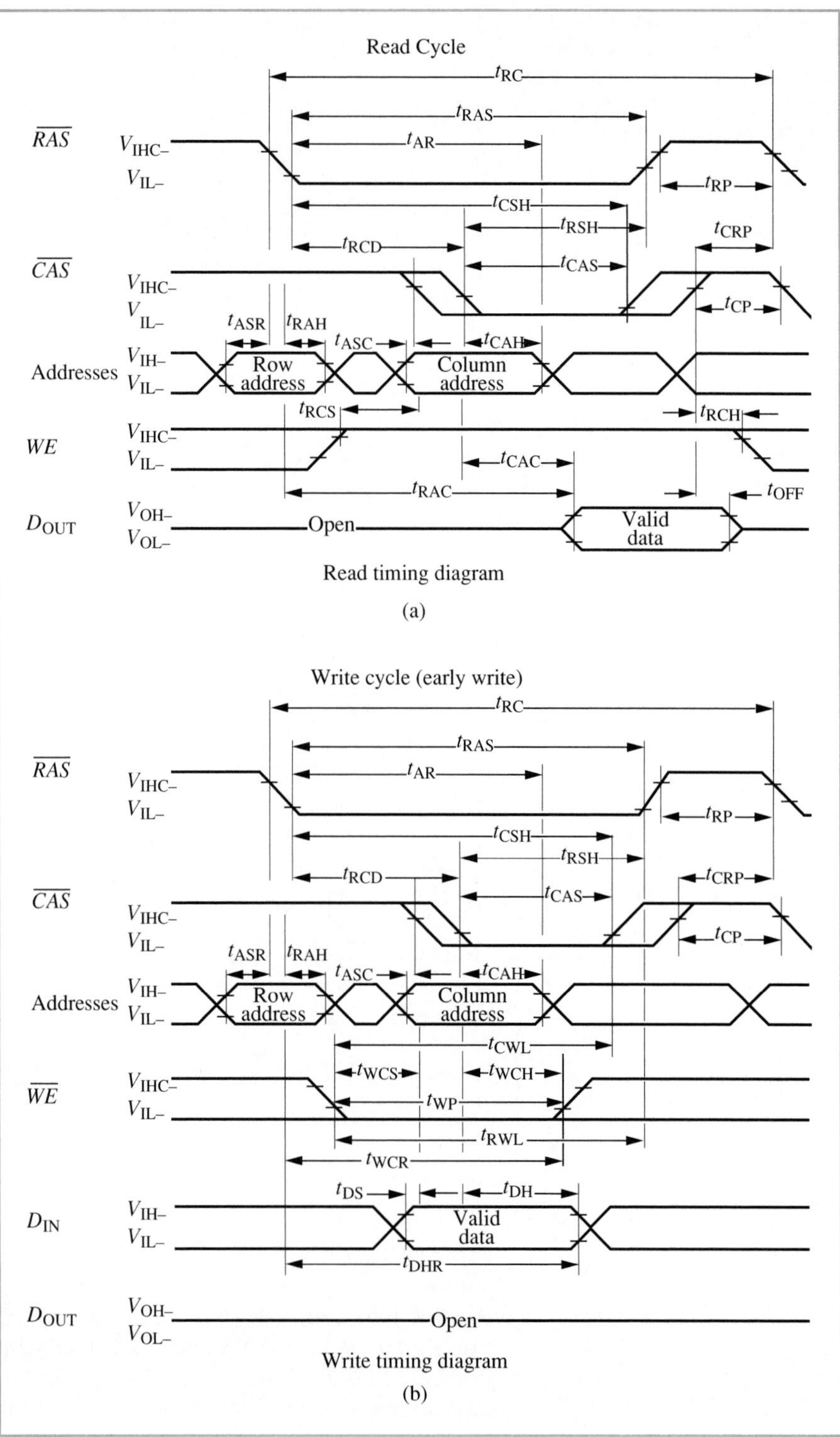

Figure 9–14
Read and write timing diagrams for an 8116 DRAM. (Courtesy of FUJITSU Microelectronics, Inc.)

must be held on the address pins for t_{RAH} seconds (the row address hold time), which for the 8116E is a minimum of 25 ns. After t_{RAH} seconds the address pins function as column address pins. The column address is input to the address pins, and on the negative edge of $\overline{CAS}$ this address will be latched by the column address buffer. From either timing diagram we note that the column address hold time, t_{CAH}, is measured with reference to the negative edge of $\overline{CAS}$, which for the 8116E is 55 ns minimum. This completes latching of the row and column address for either a read or a write operation.

The timing diagram of Figure 9–14a indicates that to perform a read operation the write enable pin $\overline{WE}$ must be a logic high and the $\overline{RAS}$ and $\overline{CAS}$ pins must be low. We note that there are two read access times, one for $\overline{RAS}$ and the other for $\overline{CAS}$. The read row access time t_{RAC} is measured from the negative edge of $\overline{RAS}$, and for the 8116E is a maximum of 200 ns. The read column access time t_{CAC} is measured from the negative edge of CAS, and for the 8116E is a maximum of 135 ns. Since both $\overline{RAS}$ and $\overline{CAS}$ are required for a read operation, and t_{RAC} is the longer of the two access times, t_{RAC} is the limiting access time and is therefore considered the access time of the 8116E.

The write timing diagram of Figure 9–14b indicates that, in this case, the write operation is controlled by the negative edge of $\overline{CAS}$. The reasoning behind this statement is based on the timing reference line in the "middle" of "valid data." The write strobe should be timed so that the datum being written into memory is stable on the data pin(s) and not in transition. Since the datum should be stable near the middle of the time period during which it is input to the data pin(s), we look there for the expected strobe pulse, which is referenced to the negative edge of $\overline{CAS}$. According to Figure 9–14b there is a data set-up time t_{DS}, which is 0 ns for the 8116E. The data hold time, t_{DH}, is 55 ns minimum, which means that the datum must be held on the data pins 55 ns after the negative edge of $\overline{CAS}$.

There are other timing diagrams associated with DRAMs. For instance, the refresh cycle, which must occur every 2 to 3 ms to refresh the data in the DRAM, is an important timing requirement. Fortunately, there are chips (DRAM controllers) available that take care of all read/write signal generation, divide the address into row and column, and provide refresh cycles as well. For the intended level of this text, there is little need for any further detailed timing diagrams of a DRAM, since a DRAM controller would be used to provide the proper signals and timing to a DRAM. A brief explanation of a DRAM controller will be given in the next section.

Review questions

1. Which is the fastest technology—TTL, NMOS, or CMOS?
2. Which technology has the highest density?
3. For what type of applications is power consumption critical?
4. How to microcontroller memories differ from microcomputer memories?
5. What is the essence of the access time of a memory device?
6. How is t_{ACS} of Figure 9–5 similar to t_{ACE} of Figure 9–8?
7. What is the relationship between t_{ACE} and t_{OE} in Figure 9–8, relative to performing a read operation?
8. The input pin CS of Figure 8–12 is similar in function to which input pin of Figure 9–7?

9. How could the inputs *RD* and *WR* of Figure 8–12 be modified to serve the same function as $\overline{WE}$ of Figure 9–7?
10. What is meant by stand-by mode, and why does this mode consume less power?
11. Why does the write strobe occur "in the middle" of the period during which the datum is on the data pins?
12. What is the major conceptual difference between the architectural drawings of Figures 9–7 and 9–9?
13. Why is the $\overline{CS}$ access time t_{ACS} used to gauge the access time of the 8417?
14. From the write timing diagram of Figure 9–10b, explain why, on the positive edge of $\overline{WE}$, the datum on the data pins(s) is written into the addressed memory location?
15. What parameter of Figure 9–12 determines the access time of 2764?
16. What is a dynamic RAM, and how does it differ from a static RAM?
17. How do designers and manufacturers of DRAMs economize on the number of pins required?
18. Why are DRAMs attractive for large memory applications?
19. What is the purpose of the inputs $\overline{RAS}$ and $\overline{CAS}$ of Figure 9–14?
20. What parameter of Figure 9–14a determines the access time of the 8116 DRAM?

9–4 AC compatibility

In this section we shall apply the concepts of Section 8–5 to the design of main memory systems. In designing a memory system one of many criteria that must be met is operational speed compatibility between the device reading memory and the main memory. We will assume that the memory systems we are designing are for microprocessor-based systems; that is, a microprocessor is used to implement the CPU. Since main memory is supportive to the microprocessor, the microprocessor sets the criteria that are to be met by main memory. Figure 9–15 shows two representative timing diagrams for read (a) and write (b) operations of a microprocessor unit (MPU). For either operation the MPU first outputs the address of the memory location on its address pins. Recalling those concepts developed from Figure 8–9 and relating them to the timing diagrams of Figure 9–15, we understand that the address is applied to both the address pins of the memory device and the page select decoder, via the address bus. Next the MPU activates its read control pin $\overline{RD}$, which differs from the read control *RD* of Figure 8–9 only in its active logic level. This control signal is used as an input to memory to enable the read circuitry of memory, such as the memory in Figure 8–15. Memory responds to this control signal by outputting the datum of the addressed memory location to its data pins and the data pins of the MPU via the data bus. The read timing diagram of Figure 9–15a indicates that

Figure 9–15
A representative timing diagram for a microprocessor.

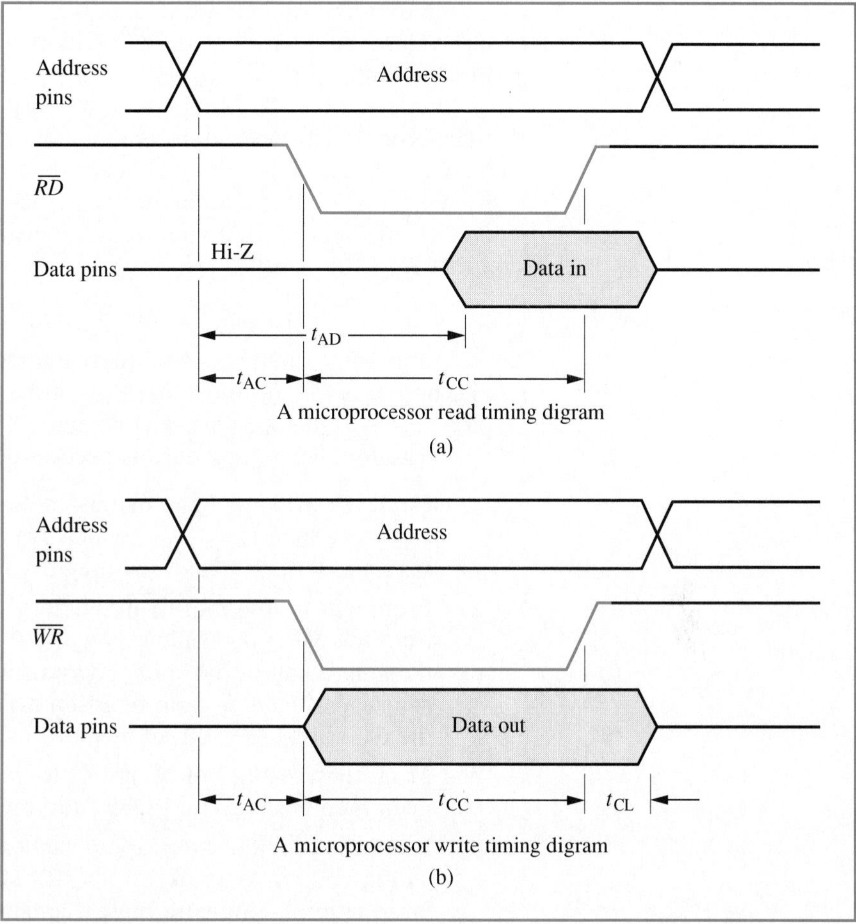

the datum being read from a memory location, labeled *Data in*, must be at the MPU's data pins on or before t_{AD} seconds. *We can conclude that for an MPU and a memory device to be time-compatible, the access time t_{ACC} of the memory device must be less than (faster than) the read time t_{AD} of the MPU*, where t_{ACC} is meant as a generic symbol. Mathematically stated, this is expressed as

Equation (9–1)

$$t_{AD} > t_{ACC}.$$

The positive edge of $\overline{RD}$ serves as a strobe for the MPU to read (latch) the datum from the data bus, which brings the datum internal to the MPU via the MPU buffer of Figure 9–1. Notice that the positive edge of $\overline{RD}$ occurs in the "middle" of *Data in*, which means that the actual read occurs at t_{AC} plus t_{CC} after the address is output by MPU on the address bus.

From the write timing diagram of Figure 9–15b we note that the MPU outputs on the data bus the datum it is writing to memory t_{AC} seconds after it outputs the address of the memory location on the address bus. The MPU signals that it has written this datum on the data bus by pulling its write control pin $\overline{WR}$ low. The MPU holds $\overline{WR}$ low for as long as it holds the datum on the data bus, which is t_{CC} plus t_{CL} seconds. We note that since the positive edge of $\overline{WR}$ occurs while the output datum is stable on the data bus, it can be used as a write strobe pulse to the addressed memory location so as to latch that datum from the data bus.

In order for the MPU and a memory device to be compatible in timing for a write operation, the memory device must be capable of having a datum written into it faster than the MPU can write a datum to it. We can state this matematically by using write control parameters t_{AC} and t_{CC} of Figure 9–15b and the write enable parameter t_{AW} of Figures 9–6 and 9–10b.

Equation (9–2)

$$t_{AC} + t_{CC} > t_{AW}$$

Table 9–6 contains some representative AC parameter values for a microprocessor. These parameters will be used as criteria for the AC parameters of memory and I/O devices.

The following procedure is used in designing a memory system:

1. From the MPU AC parameters, determine the read access time t_{AD} and write time $t_{AC} + t_{CC}$, which are used to determine the minimum read and write times of the memory devices to be used.
2. From a description of main memory use, determine its capacity and from the MPU determine its width (8 or 16 bits). That is, by knowing the length and purpose of programs and data to be stored the capacity of memory can be estimated, and by knowing the size of the data bus the width of memory can be determined.
3. From the volatility of programs to be stored, determine how much of main memory is to be ROM and how much is to be RAM.
4. From the previous steps the designer knows the speed, capacity, width, and make-up (ROM and RAM) of main memory. With this information, along with power consumption, cost, and availability considerations, the designer can decide on the optimal semiconductor memory devices to use in implementing a design.

We will take a more convenient approach, allowing our focus to be on concepts and design of relatively small main memories, which in no way limits the design concepts being developed. So much of memory design is repetitive that the design of a small memory is conceptually the same as that of a large memory; dealing with a small memory lessens the bookkeeping and the size of the resulting schematic.

As examples of implementing the first step of the design procedures outlined above, let us compare the MPU AC parameter values of Table 9–6 with the semiconductor memory device AC parameters of Tables 9–3, 9–4, and 9–5 to determine which of those devices are AC compatible with the MPU. As a first approximation of compatibility, let us compare the sum of t_{AC} and t_{CC} of the MPU with the read and write cycle time of the memory devices, which must be faster than the read and write time of the MPU. That is

Equation (9–3)

$$t_{AC} + t_{CC} > \text{Read/write cycle time of memory}$$

Table 9–6
Representative MPU AC parameters

t_{AD} = 600 ns max	t_{AC} = 250 ns min
t_{CC} = 450 ns min	t_{CL} = 50 ns min

From Table 9–6 we find for this MPU

$$t_{AC} + t_{CC} = (250 + 450)\text{ns} = 700 \text{ ns}$$

Returning to Figure 9–5, we see that t_{RC} is used to reference the read cycle time of the 2148, and from Table 9–3 we find that this parameter is 55 ns. Since 55 ns is less than 700 ns, Equation 9–3 is satisfied and the 2148 meets the first criterion for timing compatibility for a read operation. For a first approximation of write compatibility, we see from Figure 9–6 that the 2148 uses t_{WC} to specify the write cycle time; according to Table 9–3, parameter t_{WC} is also 55 ns. Equation 9–3 is satisfied, and the 2148 has passed the first test for write timing compatibility.

Comparing the read/write timing parameters for the 8417, as represented in Figure 9–10, we see that the read and write cycle times are also represented by t_{RC} and t_{WC}. From Table 9–4 we find that for this CMOS device the times are 100 ns, which is almost twice as long as for the 2148, but these times still easily satisfy Equation 9–3. Hence, on a first approximation basis, the 8128 qualifies as AC compatible with this MPU.

The read/write cycle times for the 8116 DRAM are specified by t_{RC}, which is 375 ns. The 8116 also has sufficient speed to be compatible with this MPU.

The 2764 is an EPROM and therefore we need to be concerned only with read timing. From Figure 9–12 we see that t_{RC} is not a specified AC parameter; therefore we must reason which 2764 parameter(s) can be used as a measure of AC compatibility. From the MPU read timing diagram of Figure 9–15a, we know that the MPU latches the datum at its data pins on the positive edge of its read signal $\overline{RD}$, which is $t_{AC} + t_{CC}$. Therefore the memory device being read—the 2764 in this case—must have the datum being read from it on the data bus, via its data pins, in that amount of time. In fact, Figure 9–15a indicates that the datum must be on the data pins t_{AD} seconds after the MPU outputs the address on the data bus, and the datum must be held on the data bus by the memory device until $t_{AC} + t_{CC}$. If at that point the AC parameter t_{ACC} of the 2764 is less than t_{AD} of the MPU, the two are AC compatible. As a note of interest for later use, in our design we use the read control signal $\overline{RD}$ of the MPU to hold the datum being read from the 2764 on the data bus. As stated earlier, t_{ACC} for the 2716 is 200 ns and from Table 9–6 we find that t_{AD} is 600 ns; hence as a first approximation we find that the 2764 and the MPU are AC compatible. Since we have already started detailed analysis of the 2764's timing diagram, it will be convenient to continue and finish our AC compatibility test. From Figure 9–15a we see that the MPU requires that the datum being read be held on the data bus for a minimum of

Equation (9–4)

$$\text{Time data held} = (t_{CC} + t_{AC}) - t_{AD}$$

If we use the MPU read control signal $\overline{RD}$ as an input to the output enable pin $\overline{OE}$ of the 2764 (see Figure 9–17), then t_{OE} of Figure 9–12 must be less than the Time data held of Equation 9–4. Applying the appropriate data of Table 9–6 to Equation 9–4 we find that the Time data held is equal to 100 ns. It was stated earlier that t_{OE} for the 2764 is 70 ns (maximum), so the 2764 and this MPU are AC compatible and no further

AC test need be made. It is important to understand the concepts of AC parameter comparison for the 2764 and MPU, for they will serve as a basis for timing analysis of the other memory devices.

For a more detailed comparison of AC read parameters, we must determine if the memory devices are quick enough to meet the MPU specification of t_{AD}, as we have already done for the 2764. From Figure 9–5 we can estimate that by using the MPU read control signal $\overline{RD}$ to activate the chip select pin $\overline{CS}$ of the 2148 (the write enable pin $\overline{WE}$ must be high), t_{ACS} must be less than t_{AD}. Comparing those values of Tables 9–3 and 9–6 (55 ns and 600 ns, respectively), we see that the 2148 is much faster than the MPU and is therefore AC compatible for a read operation. It is important that the reader understand how we analyzed Figure 9–5 to use MPU control signal $\overline{RD}$ to drive $\overline{CS}$ of the 2148.

From the read timing of Figure 9–8 we conclude that we can use the output enable pin $\overline{OE}$ of the 8128 to control a read operation ($\overline{CE}$ could also be used). If we use $\overline{OE}$, then t_{OE} specifies the read time relative to the negative edge of $\overline{OE}$. This implies that, in a memory design, the MPU read control signal $\overline{RD}$ can be used to drive $\overline{OE}$ of the 8128 (see Figure 9–17). Since t_{OE} is 50 ns and t_{AD} is 600 ns, the 8128 is more than fast enough to be AC compatible with this MPU for read operations.

From Figure 9–10a we can again conclude that MPU control signal $\overline{RD}$ can be used to drive chip select pin $\overline{CS}$ of the 8417 and, if so, t_{ACS} must be compared to t_{AD}. From Table 9–5 we see that t_{ACS} is 100 ns, which is much faster than 600 ns. Hence, the 8417 is AC compatible for read operations.

Now let us determine write timing compatibility of the memory devices using reasoning similar to that for read operations. Reviewing the MPU write timing diagram of Figure 9–15b, the MPU writes the datum that is to be stored in memory on the data bus, via its data pins, t_{AC} seconds after the MPU outputs the address on the address bus. The MPU holds that datum on the data bus for t_{CC} plus t_{CL} seconds and signals that the datum is on the data bus by driving its write control pin $\overline{WR}$ low during that time. It can be seen that the positive edge of $\overline{WR}$ occurs while the datum is still present (and stable) on the data bus; therefore, control signal $\overline{WR}$ can serve as a strobe pulse. As a result, we shall use the write control signal $\overline{WR}$ of the MPU to drive the write enable pin $\overline{WE}$ of the RAM memory devices.

From the timing diagram of the 2148, as illustrated in Figure 9–6, we can reason that if the write control signal of the MPU is used to drive the write enable pin $\overline{WE}$ of the 2148 (see Figure 9–23), the datum must be on the data pins of the 2148, via the data bus, a minimum of t_{DW} seconds before the positive edge of $\overline{WE}$. Of course $\overline{WE}$ has the same timing as $\overline{WR}$ if $\overline{WR}$ and $\overline{WE}$ are connected. Since memory must respond to the write control of the MPU, t_{DW} of the 2148 must be less than t_{CC} of the MPU. From Table 9–3 we find that t_{DW} is 20 ns and from Table 9–6 we see that t_{CC} is 450 ns, which indicates write timing compatibility. Lastly, Figure 9–6 indicates that $\overline{WE}$ must be active for a minimum of t_{WP} seconds. Since the MPU write control signal will be driving the $\overline{WE}$ pin and (from Figure 9–15b) will be active for t_{CC} seconds (450 ns) and t_{WP} is 40 ns for the 2148, this requirement is also satisfied. Hence, the 2148 is AC compatible with this MPU for both read and write operations.

Likewise, from Table 9–4 we find that t_{DW} is 40 ns, which indicates that the 8128 is also AC compatible for write operations. From our previous analysis we can conclude that the 8128 is AC compatible with this MPU.

From the 8417 timing diagram of Figure 9–10b we can reason, by using the MPU control signal $\overline{WR}$ to drive the 8417 write enable pin $\overline{WE}$, that the MPU must write the datum to the 8417 data pins, via the data bus, t_{DS} seconds before the positive edge of $\overline{WE}$ (MPU control signal $\overline{WR}$) occurs. Since the MPU writes the datum on the data bus at the negative edge of $\overline{WR}$ and holds it there for at least t_{CC} seconds, at which time the positive edge of $\overline{WR}$ ($\overline{WE}$) occurs, the 8417 AC parameter t_{DS} must be less than the MPU AC parameter t_{CC}. From Table 9–5 we find that t_{DS} is 60 ns and from Table 9–6 we see that t_{CC} is 450 ns, which agrees with the criterion. Also, Figure 9–10b indicates that $\overline{WE}$ must be active for a minimum t_{WP} seconds, or 140 ns; since t_{CC} is 450 ns, this criterion is also met.

It is evident that modern memory devices are certainly fast enough to be compatible with an MPU that operates at speeds indicated in Table 9–6. The specifications of Table 9–6 are representative of an NMOS 8-bit microprocessor.

There are other timing diagrams that can be applied to the sampling of memory devices we have studied, as well as timing diagrams for devices we have not studied. The principal concepts of timing diagrams have been sufficiently explained to enable you to interpret other timing diagrams with little assistance.

To demonstrate the remaining three steps of our design procedure, we will consider some design examples in the next section. You may wish to refer to and review the design concepts established in Chapter 8.

Review questions

1. Why do MPU AC parameters determine the timing criterion for memory?
2. From the read timing diagram of Figure 9–15a, at what time does the MPU actually latch (read) the datum from its data pins?
3. From the timing diagram of Figure 9–15b, when does the MPU write its output datum to its data pins?
4. When the MPU writes data to a device (memory) via the data bus, how does the MPU signal the device that the datum is available on the data bus?
5. Generally speaking, what is the critical timing between the MPU and memory for a read or write operation?
6. For RAMs without an output enable pin $\overline{OE}$ and instead a write enable pin $\overline{WE}$ to control write operations, how can read operations be controlled?
7. For RAMs with both chip enable and output enable pins, such as the 8128, could a read operation be controlled by the chip enable pin rather than the output enable pin, as illustrated in Figure 9–8?
8. Why does the 8116 DRAM require input signals $\overline{RAS}$ and $\overline{CAS}$?
9. Why must a DRAM have a refresh cycle? How often must a DRAM be refreshed?

9–5 Designing static memories

From previous analysis we know that all the memory devices of the previous section are AC compatible with the MPU AC specifications of Table 9–6. Therefore, in order to have AC compatibility for the design examples of this section we will assume MPU specifications from Table 9–6 and will limit our selection of memory devices to those covered in the previous section. To learn the process of memory design we shall consider an example.

EXAMPLE 9–7

An 8-bit microprocessor–based system requires a fully decoded memory with memory-mapped I/O with 5 K of memory for its bootstrap loader and 3.5 K of *user memory space*, memory in which the user can store programs and data. Note the implication that the bootstrap loader (recall that the bootstrap loader is a program that instructs the MPU from a cold start) is not a user program; rather, it is a permanently installed program that cannot be altered by the MPU.

SOLUTION From the description of memory, 5 K of ROM is required, since the bootstrap loader must be stored in a nonvolatile memory, and 3.5 K of RAM is specified, since the user memory space must be capable of read and write operations. Since the MPU is 8 bits, the data bus will be 8 bits wide; hence, we want 8-bit memory locations. Having only the 2764 EPROM available, we implement our design with a single 2764, which results in an 8 K × 8 of ROM rather than 5 K. For the 3.5 K of user memory we can use the 2148 (1 K × 4), the 8128 (2 K × 8), or the 8417 CMOS (2 K × 8). To use the 2148 would require that two 2148s be paralleled to achieve 8-bit memory locations and to get 3.5 K locations would require four pairs of 2148 memory devices. We have better options available. The 8128 and 8417 are our choices. We shall use an NMOS device, which is the 8128. Using the 8128 means that the user memory space is 4 K, since we must use two 8128s to get at least 3.5 K of memory.

Now that we know the memory devices we will be working with, we can assign memory device address pins to system address bus lines, just as was done in Table 8–2. We need to know the size of the address bus, which is determined by the number of address pins of the MPU. Most 8-bit microprocessors have 16 address pins; hence, we shall assume that the address bus is composed of 16 address lines, which are identified as A_0 through A_{15}. Table 9–7 shows these address bus lines as the top row. In the next row we list the address pins of the memory device whose addresses must begin at 0000H (on a cold start this MPU goes to address 0000H for the first instruction), which is the 2764. The 2764 has 13 address pins, identified as A_0 through A_{12}, and they form the second row of Table 9–7. We aligned address pins A_0 through A_{12} of the 2764 to address bus lines of the same subscript. As a result, the lower 13 bits on the address bus pinpoint one of 8192 (2^{13}) locations within the 2764 memory device when the 2764 is chip enabled. The columns formed by these pin and bus line alignments in Table 9–7 are interpreted as bus connects when implemented in hardware. The third row of Table 9–7 is the memory device whose first address is

Table 9–7
Decoding the address bus of Example 9–7

System address bus lines		A_{15}	A_{14}	A_{13}	A_{12}	A_{11}	A_{10}	A_9	A_8	A_7	A_6	A_5	A_4	A_3	A_2	A_1	A_0
2764 Address pins					A_{12}	A_{11}	A_{10}	A_9	A_8	A_7	A_6	A_5	A_4	A_3	A_2	A_1	A_0
8128 Address pins (no. 1 and no. 2)							A_{10}	A_9	A_8	A_7	A_6	A_5	A_4	A_3	A_2	A_1	A_0
74LS138 pins (page selector)		A_2	A_1	A_0													
2764-0 Address range	0000H to	0	0	0	0	0	0	0	0	0	0	0	0	0	0	0	0
	1FFFH	0	0	0	1	1	1	1	1	1	1	1	1	1	1	1	1
8128-1 Address range	2000H to	0	0	1	X	X	0	0	0	0	0	0	0	0	0	0	0
	3FFFH	0	0	1	X	X	1	1	1	1	1	1	1	1	1	1	1
8128-2 Address range	4000H to	0	1	0	X	X	0	0	0	0	0	0	0	0	0	0	0
	5FFFH	0	1	0	X	X	1	1	1	1	1	1	1	1	1	1	1

the one following the last address in the 2764, which will be the beginning address of the user's memory space. Then the third row of Table 9–7 lists the address pins of the 8128's, which have 11 pins each.

From Table 9–7 we see that we do not have the same number of address pins for the 2764 and the 8128's, which was not the case for the simplified design of Figure 8–14a. Let us follow procedures similar to those used to design Figure 8–14a and use address bus lines A_{13} through A_{15} as inputs to a page select decoder. We will use a 74LS138 decoder, whose logic symbol and truth table are shown in Figure 9–16. The fourth row of Table 9–7 represents the input pins A_0 through A_2 of the 74LS138, which are connected to address bus

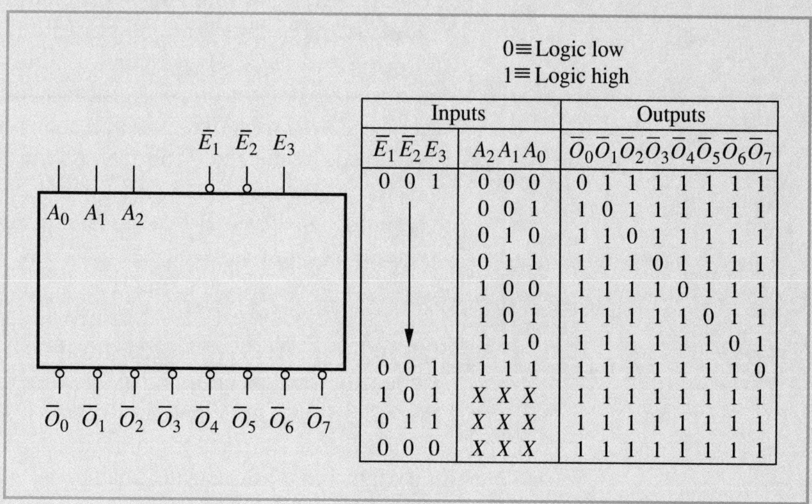

Inputs		Outputs
$\bar{E}_1 \bar{E}_2 E_3$	$A_2 A_1 A_0$	$\bar{O}_0\bar{O}_1\bar{O}_2\bar{O}_3\bar{O}_4\bar{O}_5\bar{O}_6\bar{O}_7$
0 0 1	0 0 0	0 1 1 1 1 1 1 1
	0 0 1	1 0 1 1 1 1 1 1
	0 1 0	1 1 0 1 1 1 1 1
	0 1 1	1 1 1 0 1 1 1 1
	1 0 0	1 1 1 1 0 1 1 1
	1 0 1	1 1 1 1 1 0 1 1
	1 1 0	1 1 1 1 1 1 0 1
0 0 1	1 1 1	1 1 1 1 1 1 1 0
1 0 1	X X X	1 1 1 1 1 1 1 1
0 1 1	X X X	1 1 1 1 1 1 1 1
0 0 0	X X X	1 1 1 1 1 1 1 1

Figure 9–16
Logic symbol and truth table for a 74LS138.

lines A_{13} through A_{15}. These first four rows serve as a wiring table for connecting device pins to address bus lines.

We use Table 9–7 to determine the addresses within each memory device. All we really need is the first and last address within each memory device, since we will then know the range of addresses. To determine the first address, just apply 0's to the address pins of the memory device being studied; to determine the last address, apply all 1's to those address pins. The page selector pins A_0 through A_2 have a binary value that corresponds to the assigned page number of the device. Since 2764-0 is the first page (page 0), the 74LS138 address pins will have 0's applied to them. The beginning address for the 2764, as indicated by the first address row of Table 9–7, has the 2764 address pins equal to 0's, while the second address row has logic 1's input to those address pins, representing the last address within the 2764. The 2764 has an address range of 0000H to 1FFFH.

The next range of addresses shown in Table 9–7 is for the 8128-1, where the suffix distinguishes between 8128's. To deselect the 2764 and instead select 8128-1, the logic levels input to address pins A_2 through A_0 of the 74LS138 must change to 001, as is indicated in Table 9–7. Since the 8128-1 has only 11 address pins, rather than 13 as the 2764 has, when the 8128-1 is selected address bus lines A_{11} and A_{12} have no function. As a result, the logic levels on those two address bus lines do not matter. When the logic level of a bus line or pin has no meaning, the logic level is known as a *don't care* and is symbolized by an X, as indicated in Table 9–7. In that table, 0's were used for the X's in the beginning address and 1's were used for the X's in the last address of the 8128-1. For this condition, the address range of the 8128-1 is 2000H to 3FFFH. Within this range of addresses there are 8192 memory locations ($3FFF_{16} - 2000_{16} + 1 = 8192_{10}$). However, since the 8128 has only 11 address pins, it has only 2048 memory locations, which is four times fewer than the 8196 addresses allocated to it. This means that every 8128-1 memory location has four addresses. For instance, any one of the following addresses accesses the same location: 2000H, 2800H, 3000H, and 3800H. You should verify this by beginning with address 2000H and then substituting all binary combinations for the two don't care (XX) address bus lines A_{11} and A_{12} of Table 9–7. Of course, the reason there are four addresses for each location is that there are two don't care address bus lines ($2^2 = 4$). The terminology used to describe the condition of multiple addresses for a location is *address foldback*. Note that to have address foldback is to have "wasted" addresses. Foldback may be acceptable in small memory systems but not in large ones. We will see how to avoid address foldback later in this chapter.

To select the 8128-2, the logic levels of address bus lines A_{15} through A_{13} must be changed to 010, as indicated in Table 9–7. Repeating the procedures used to determine the address range for the 8128-1, we find that the 8128-2 has an address range of 4000H to 5FFFH. For the reasons already discussed for the 8128-1, the 8128-2 will also have address foldback.

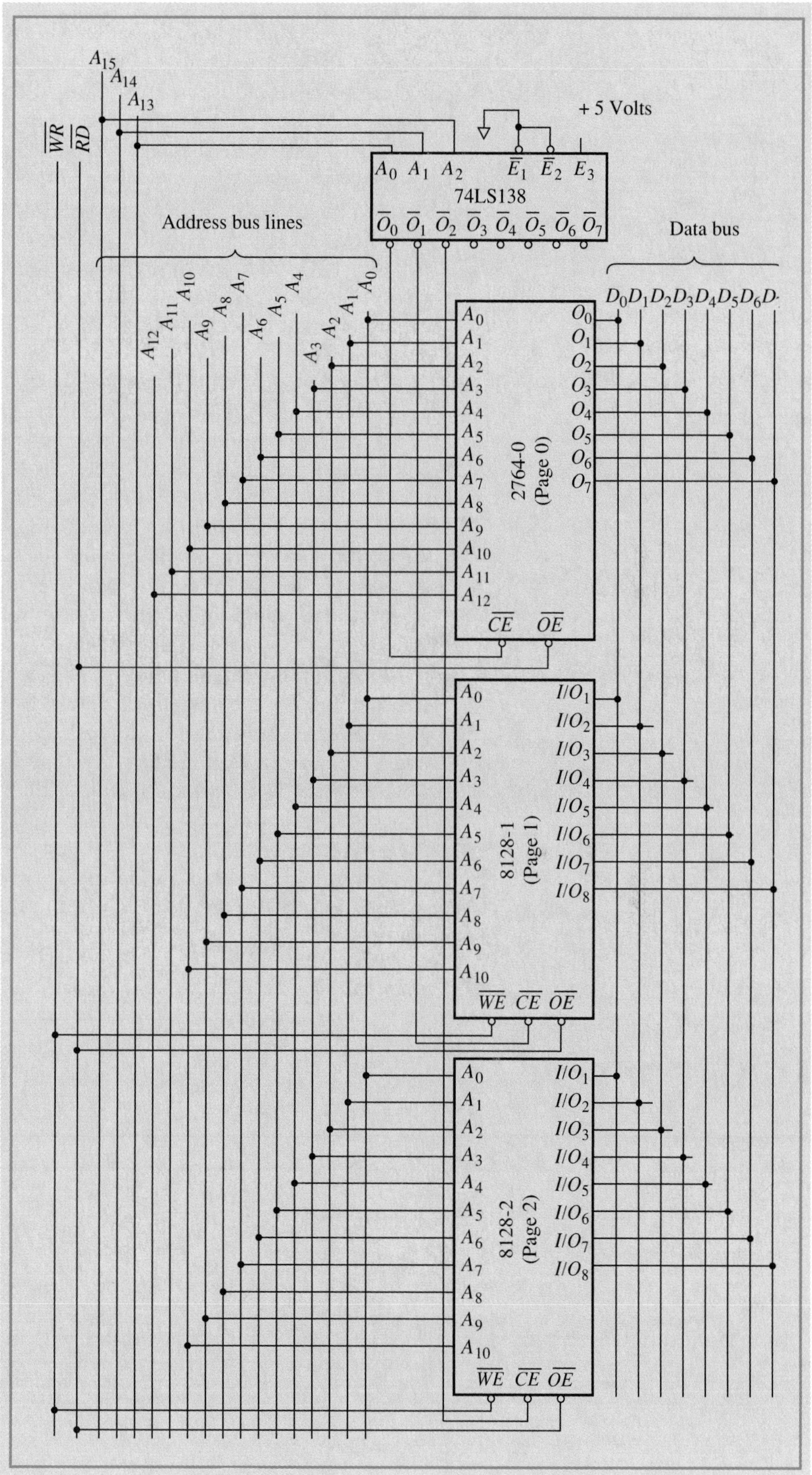

Figure 9–17
Memory design of Example 9–7.

As previously stated, Table 9–7 indicates how we are going to connect the address pins of each chip to the address bus lines of the system. That is, those address pins in the same column are connected to the address bus line in that column. For instance, the A_0 pins of both 8128's and the 2764 are connected to address bus line A_0, and address pin A_0 of the 74LS138 is connected to A_{13} of the address bus. Figure 9–17 illustrates the implementation of the wiring indicated in Table 9–7. Note that the outputs of the page select decoder (74LS138) go to the chip enable ($\overline{CE}$) of the corresponding page number; for example, output $\overline{O}_0$ is connected to the chip enable of page 0. Outputs $\overline{O}_3$ to $\overline{O}_7$ of the 74LS138 are available for memory expansion.

An address locates and selects the memory location to be accessed; however, as we know, there must also be proper control of the read or write operation that is to be performed. MPU-generated control signals $\overline{RD}$ and $\overline{WR}$ are used to control memory read and write operations. From our analysis of timing diagrams, we have already reasoned that we will connect the MPU write control signal line $\overline{WR}$ to the write enable pin $\overline{WE}$ of the 8128's to control write operations. To control read operations we connect the MPU read control signal line $\overline{RD}$ to the output enable $\overline{OE}$ of each memory device, as is shown in Figure 9–17.

From the address ranges of Table 9–7 a *memory map* can be constructed for the system of Figure 9–17. A memory map is a geometrical representation (a rectangle) of assigned memory addresses for a given memory system and is a convenient reference for both hardware and software designers. The hardware designer may refer to it when expanding or otherwise modifying memory, and the programmer refers to it to obtain available addresses for programming. The illustration of Figure 9–18 is a memory map for the memory of Figure 9–17. Note the foldback due to the don't cares.

Figure 9–18
Memory map of the memory in Figure 9–17.

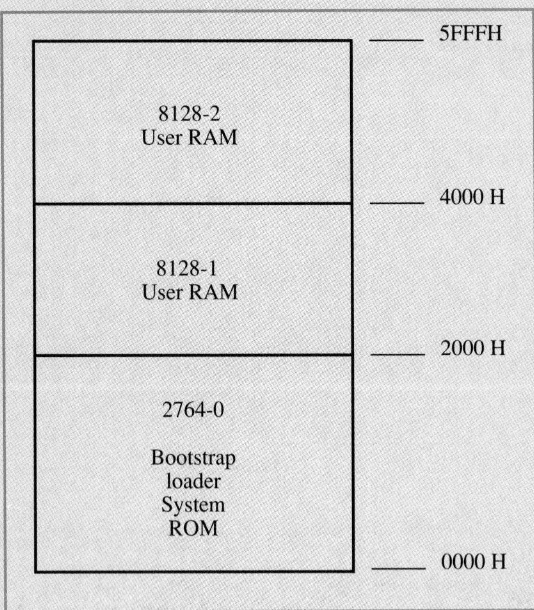

Before designing the next memory, let us outline the design procedures of Example 9–7.

1. Based on memory specifications, memory allocation is divided into ROM and RAM.
2. By knowing the required memory capacity and width of ROM and RAM, initial choices of memory devices can be made to implement the design.
3. These initial choices are checked for AC compatibility with the MPU.
4. An address bus decoding table (Table 9–7) is constructed, which indicates the wiring of the memory devices and page selector to the address bus as well as supplying memory addresses for a memory map.
5. The MPU must exercise control over all read and write operations, which is accomplished by having the MPU read control signal $\overline{RD}$ drive the out enable pin $\overline{OE}$ of the memory device and the MPU write control signal drive the write enable pin $\overline{WE}$ of the memory device. Of course an output from the page selector drives the chip enable pin $\overline{CE}$ (or chip select) of the memory device.

The address foldback of Example 9–7 is due in part to the different number of address pins for each memory device. From Table 9–7 we see that address bus lines A_0 through A_{12} are dedicated to memory location selection within a device, the memory device with the greater number of address pins determining which address bus lines are to be used for this purpose. Of course it is address bus lines A_{13} through A_{15} that select the memory device. Now, since the 2764 has two more address pins than the 8128's, there are 2^2 times as many addresses in the 2764 as there are in either 8128, which is why there were four addresses for every memory location within each 8128. To eliminate address foldback due to a mismatch of the number of address pins, the memory device with the least capacity determines the address bus lines to be used for addressing a memory location within a memory device. A page of memory of a system is defined as the number of memory locations within the memory device with the least capacity. Hence, a page of memory in Example 9–7 would be 2^{11} (2048) locations, as determined by the number of address pins on an 8128. Those memory devices with a greater capacity are "divided" into pages of the same size, as defined by the device with the smallest capacity. That is, the number of pages assigned to a memory device will be

Equation (9–5)

$$\text{Number of page assignments (NPA)} = 2^m/2^n$$

where m is equal to the number of address pins of the memory device in question and n is equal to the number of address pins of the memory device within the system with the least memory capacity. Thus, for a memory system composed of 2764's and 8128's, the n is 11 and m is 13 for the 2764. According to Equation 9–5, the NPA for the 2764 is

$$\text{NPA2764} = 2^{13}/2^{11} = 4$$

Figure 9–19
Examples of page select line assignments to memory devices.

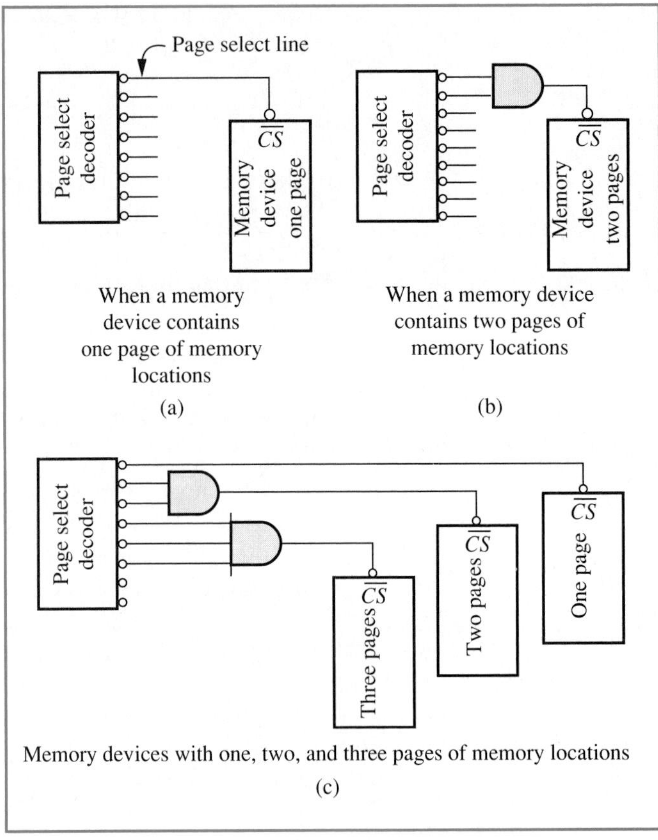

Thus, when 2764's and 8128's are used in the same system, a 2764 is imagined to be divided into four pages (2 K per page).

When any page of memory locations within a memory device is addressed, the page selector will select that memory device. Hence, if a memory device contains one page of memory locations, one page select line (an output of the page select decoded) will be dedicated to enabling its chip enable, or chip select, pin, as illustrated in Figure 9–19a. If a memory device contains two pages of memory locations, two page select lines are dedicated to selecting the memory device, as illustrated in Figure 9–19b. The *AND* gate of Figure 9–19b serves to enable the memory device if either of its two pages is addressed. Figure 9–19c shows three memory devices, each having a different number of memory pages, but in each case there is a one-to-one correspondence between the number of dedicated page select lines and the number of pages.

EXAMPLE 9–8

Redesign the memory system of Example 9–7 such that address foldback is eliminated.

SOLUTION We can skip many design procedure steps and begin with the creation of the address bus decoding table (similar to Table 9–7). The difference between the address bus pin assignments for this table and that of Table 9–7 is assignment of the page select decoder address pins A_0 through A_2. Since the page select decoder is used to select pages of memory location rather than just memory devices, its address pins are used to decode those address bus lines that have the page address. These are the address bus lines that are numerically to the left of the most significant address pin of the memory device with the least number of address pins, which you will recall is the memory device that defines the number of memory locations in a page of memory. As we see from Table 9–7, address bus line A_{11} is the first numerically more significant address bus line to the left of the most significant address pin of the 8128's, which defines a page of memory to be 2^{11} memory locations. The next three address bus lines, A_{11} to A_{13}, are decoded by the page select decoder (74LS138) to select the addressed page of memory, as indicated in Table 9–8.

We must next determine the number of memory pages in each memory device using Equation 9–5. As we have already determined for this system, $n = 11$, since the 8128 has the least number of address pins. The NPA for the 8128's is

$$\text{NPA}(8128) = 2^{11}/2^{11} = 1 \text{ page}$$

and for the 2764

$$\text{NPA}(2764) = 2^{13}/2^{11} = 4 \text{ pages.}$$

Table 9–8
Fully decoded address bus of Example 9–8

System address bus lines		A_{15}	A_{14}	A_{13}	A_{12}	A_{11}	A_{10}	A_9	A_8	A_7	A_6	A_5	A_4	A_3	A_2	A_1	A_0
2764 Address pins					A_{12}	A_{11}	A_{10}	A_9	A_8	A_7	A_6	A_5	A_4	A_3	A_2	A_1	A_0
8128 Address pins (no.1 and no.2)							A_{10}	A_9	A_8	A_7	A_6	A_5	A_4	A_3	A_2	A_1	A_0
74LS138 Page selector				A_2	A_1	A_0											
Page	Address																
2764-0	0000H	0	0	0	0	0	0	0	0	0	0	0	0	0	0	0	0
Page 0	to																
	07FFH	0	0	0	0	0	1	1	1	1	1	1	1	1	1	1	1
2764-0	0800H	0	0	0	0	1	0	0	0	0	0	0	0	0	0	0	0
Page 1	to																
	0FFFH	0	0	0	0	1	1	1	1	1	1	1	1	1	1	1	1
2764-0	1000H	0	0	0	1	0	0	0	0	0	0	0	0	0	0	0	0
Page 2	17FFH	0	0	0	1	0	1	1	1	1	1	1	1	1	1	1	1
2764-0	1800H	0	0	0	1	1	0	0	0	0	0	0	0	0	0	0	0
Page 3	1FFFH	0	0	0	1	1	1	1	1	1	1	1	1	1	1	1	1
8128-1	2000H	0	0	1	0	0	0	0	0	0	0	0	0	0	0	0	0
Page 4	27FFH	0	0	1	0	0	1	1	1	1	1	1	1	1	1	1	1
8128-2	2800H	0	0	1	0	1	0	0	0	0	0	0	0	0	0	0	0
Page 5	2FFFH	0	0	1	0	1	1	1	1	1	1	1	1	1	1	1	1

Hence, each of the two 8128's is assigned one page of addresses and the 2764 is assigned four pages of addresses. For this memory system there is a total of six pages of memory, as shown in Table 9–8.

Table 9–8 indicates that the 2764-0 has four pages of memory that are identified with page address 0, 1, 2, and 3. The 8128-1 has one page of memory, and that page address is 4. The 8128-2 also has one page of memory, which is assigned page address 5. Notice that the numerical suffix after each memory device no longer can be associated with a page number, as in Example 9–7, but rather is used as chip identification. The reader should also notice that the binary patterns of the page selector in Table 9–8 correspond to the page addresses. That is, throughout the entire range of memory addresses (0000H to 07FFH) for page 0, the 74LS138 address pins A_2 through A_0 have the binary bit pattern 000, for page 1 (0800H to 0FFFH) the bit pattern of these three address pins is 001, for page 2 it is 010, for page 3 it is 011, and so on. Thus, it can be stated that for this memory system address bus lines A_{13} through A_{11} are used to address memory pages.

Address bus lines A_{14} and A_{15} are not used for addressing and therefore are not decoded. As a result, they could be considered don't cares. However, this would allow address foldback due to unused address bus lines and, as in Example 9–7, there would be 2^2 addresses for each location (verify that addresses 0000H, 4000H, 8000H, and C000H address the same location if A_{14} and A_{15} are considered don't cares). To prevent foldback due to unused address bus lines, we incorporate those unused lines in the addressing scheme. They are used to enable one of the page selectors' enable pins ($\overline{E}_1$, $\overline{E}_2$, or E_3). If we assume that logic levels of A_{14} and A_{15} are 0 for a valid address, as is assumed in Table 9–8, we can determine from Table 9–9 that we must use a *NOR* gate if E_3 is used (E_3 requires a logic 1 input) or an *OR* gate if either $\overline{E}_1$ or $\overline{E}_2$ (they require a logic 0 input) is used to prevent foldback. Let us somewhat arbitrarily choose to enable E_3. Then a two-input *NOR* gate is used to detect when both address bus lines are low, and its output is used to enable the page selector via its E_3 pin, as indicated in Figure 9–20.

The design of this example is basically the same as that of Example 9–7, except for the connections of the 74LS138 address pins and enable pin E_3. For that reason, to simplify the wiring diagram of this memory design, we symbolically abbreviate bus connections and detail just the 74LS138 connections. For detailed bus connections for the memory devices, refer to Figure 9–17. Figure 9–20 illustrates the design.

Table 9–9
Truth table for A_{14} and A_{15}

		Output			
A_{15}	A_{14}	*AND*	*NAND*	*OR*	*NOR*
0	0	0	1	0	1
0	1	0	1	1	0
1	0	0	1	1	0
1	1	1	0	1	0

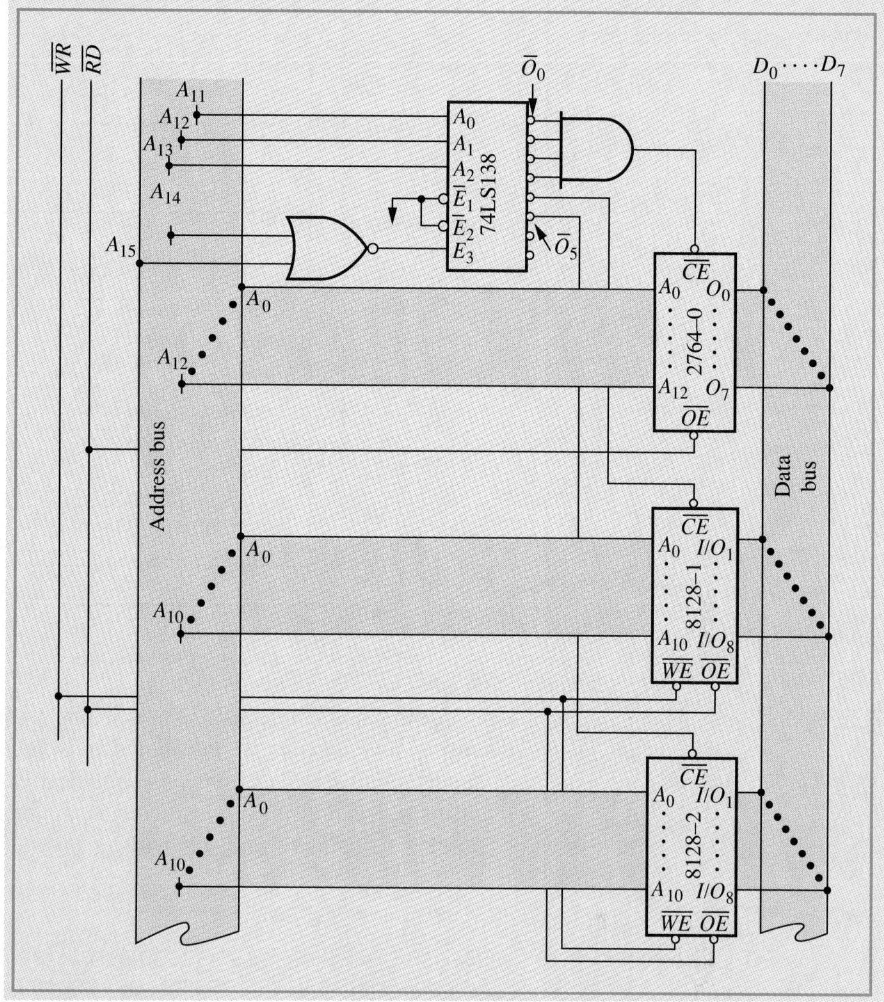

Figure 9–20
Memory design of Example 9–8 (no address foldback).

The memory map for the design of Figure 9–20 can be obtained from the address ranges listed under the *Address* heading in Table 9–8. This memory map is shown in Figure 9–21.

Expanded page addressing

The 74LS138 has eight outputs, which limits the number of pages that can be addressed by a single 74LS138 to eight. To expand the number of page addresses available, up to eight 74LS138's can be paralleled, which yields 64 pages of memory. To parallel 74LS138's, their address pins A_0 through A_2 are connected and then tied to the address bus lines that are used to address memory pages. Other numerically more significant address lines are used to address (enable) the desired page selector. Figure 9–22 shows four 74LS138s that have been paralleled. Address bus lines A_{11} through A_{15} were arbitrarily chosen to represent those address bus lines that are to furnish page addresses. Using five address bus lines means that 32 page addresses are available. Address bus lines A_{14} and A_{15} are used to enable the page selectors according to the logic levels of the truth table shown. Any time a logic 0 is to enable a pin, either $\overline{E}_1$ or $\overline{E}_2$ can be used, otherwise enable pin E_3 is used.

There are many semiconductor memory devices that do not have memory widths of 8 bits, yet they are used in designs with 8-bit widths. A

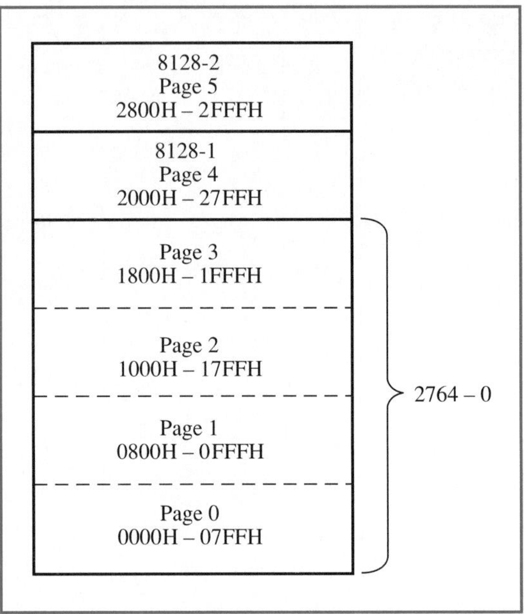

Figure 9–21 Memory map for the memory of Figure 9–20.

very common width in DRAMs is 1 bit, as for the 8116 (see Fig. 9–13). Memory devices may be paralleled to achieve the desired width. For instance, eight 8116's could be paralleled in order for each memory location to have an 8-bit width, or a pair of 2148's (Figure 9–4) could be paralleled to achieve an 8-bit width.

EXAMPLE 9–9

Design a fully decoded 9-K memory system with isolated I/O that does not have address foldback. The system is to have 4 K of ROM and 5 K of RAM. Use a 2732 for the ROM and 2148's for RAM. A 2732 has pin functions similar to a 2764 ($\overline{OE}$, $\overline{CE}$, etc.), but it has 12 address pins.

SOLUTION Using 2148's requires pairing to achieve 8-bit memory locations, which is indicated in Figure 9–23, where the suffix A identifies the lower nibble (D_0 through D_3) and suffix B identifies the upper nibble D_4 through D_7). Table 9–10 shows decoding of the address bus. From Table 9–10 we see that address bus lines A_{10} through A_{15} could be used for page addressing (A_9 is the most significant bus line common to all memory devices). To determine the number of memory pages required, we calculate the number of memory pages per device using Equation 9–5. Since the number of address pins of the 2148 is the least, n is equal to 10 and the 2148's define the number of memory locations to a page of memory (1024). Thus,

$$\text{NPA } 2732 = 2^{12}/2^{10} = 4 \text{ pages}$$

and of course

$$\text{NPA } 2148 = 1 \text{ page}$$

Figure 9–22
Expanded page addressing.

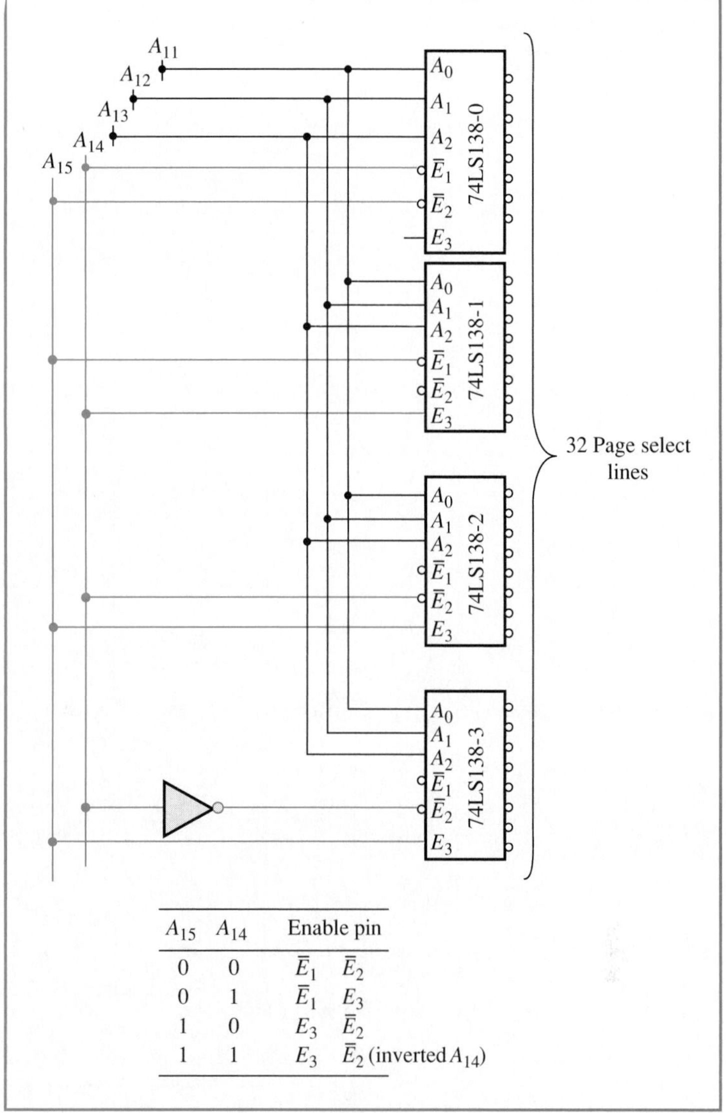

A_{15}	A_{14}	Enable pin
0	0	$\overline{E}_1$ $\overline{E}_2$
0	1	$\overline{E}_1$ E_3
1	0	E_3 $\overline{E}_2$
1	1	E_3 $\overline{E}_2$ (inverted A_{14})

Since the system is to have 9 K of memory, and a page is 1 K, nine pages of memory are required. The 2732 has four pages and five 2148 pairs have five pages. To address nine pages of memory, four address bus lines ($2^4 = 16$) are required. From Table 9–10 we see that these four address bus lines are A_{10} through A_{13}, where A_{13} is used to enable (E), the appropriate page selector enable pin. With just one address bus line used to enable the appropriate page selector, there can be just two page selectors ($2^1 = 2$). When A_{13} is a logic 0, 74LS138-6 of Figure 9–23 is enabled via its $\overline{E}_1$ pin, and when it is a logic 1, 74LS138-7 is enabled via its E_3 pin. Address bus lines A_{14} and A_{15} are again used to prevent address foldback, as indicated in Figure 9–23. From Table 9–9 we again reason that if the logic levels of the address lines are to be 0's, a *NOR* gate must be

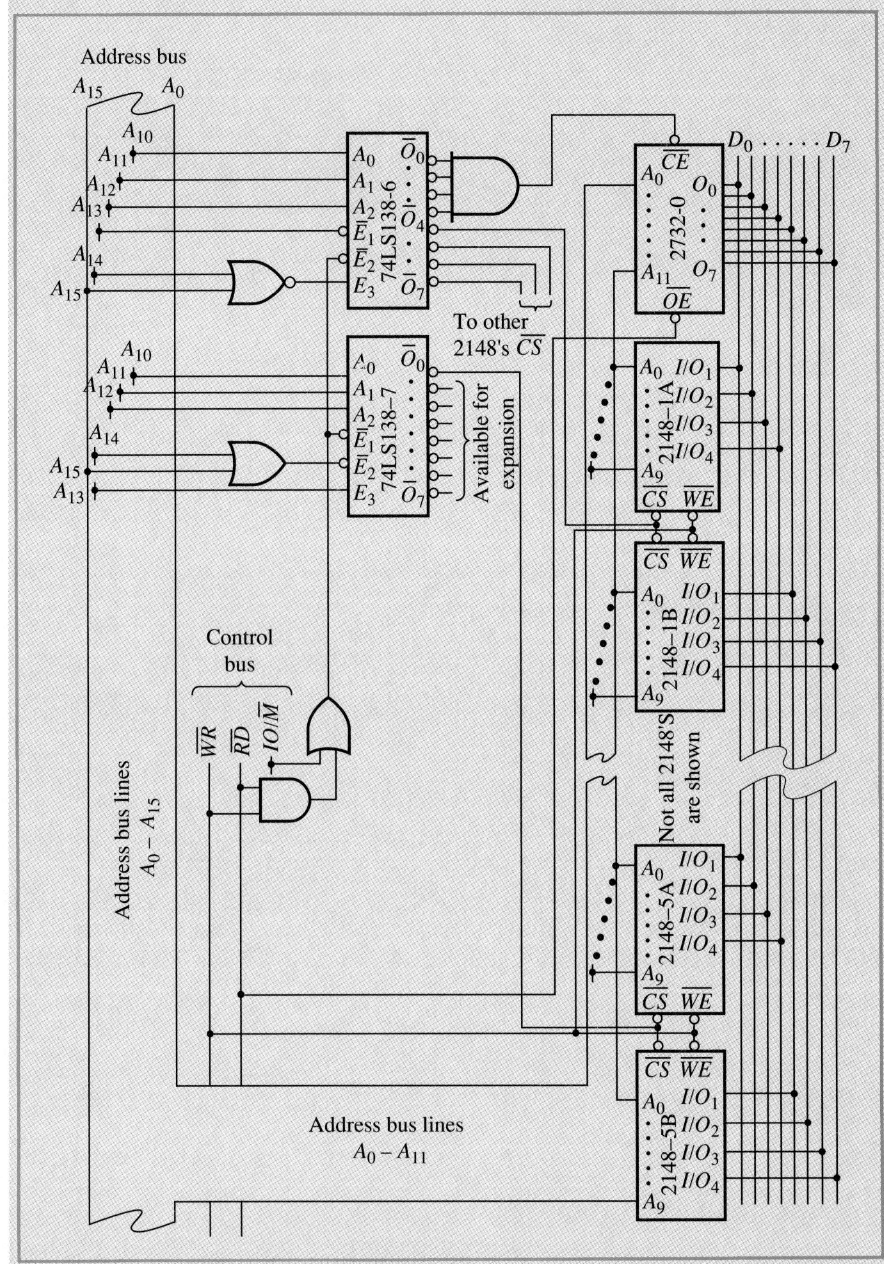

Figure 9–23
Memory design of Example 9–9.

used when a logic 1 output is required and an *OR* gate when a logic 0 output must occur. As a point of interest, note that the binary bit patterns of address bus lines A_{13} through A_{10} in Table 9–10 are a truth table sequence, which is equivalent to the decimal value that identifies the page number.

As we know, to control the read and write operations MPU control signals $\overline{RD}$ and $\overline{WE}$ must be used. However, the 2148's do not have an output enable pin with which to control read operations as the 8128's do. To control read operations we use the $\overline{RD}$ control signal to enable any page selector that is addressed to select a page

Table 9–10
Fully decoded address bus of Example 9–9

System address bus lines		A_{15}	A_{14}	A_{13}	A_{12}	A_{11}	A_{10}	A_9	A_8	A_7	A_6	A_5	A_4	A_3	A_2	A_1	A_0
2732 Address pins						A_{11}	A_{10}	A_9	A_8	A_7	A_6	A_5	A_4	A_3	A_2	A_1	A_0
2148 Address pins								A_9	A_8	A_7	A_6	A_5	A_4	A_3	A_2	A_1	A_0
74LS138 pins				E	A_2	A_1	A_0										
2732-0	0000H	0	0	0	0	0	0	0	0	0	0	0	0	0	0	0	0
Page 0	to																
	03FFH	0	0	0	0	0	0	1	1	1	1	1	1	1	1	1	1
2732-0	0400H	0	0	0	0	0	1	0	0	0	0	0	0	0	0	0	0
Page 1	to																
	07FFH	0	0	0	0	0	1	1	1	1	1	1	1	1	1	1	1
2732-0	0800H	0	0	0	0	1	0	0	0	0	0	0	0	0	0	0	0
Page 2	to																
	0BFFH	0	0	0	0	1	0	1	1	1	1	1	1	1	1	1	1
2732-0	0C00H	0	0	0	0	1	1	0	0	0	0	0	0	0	0	0	0
Page 3	to																
	0FFFH	0	0	0	0	1	1	1	1	1	1	1	1	1	1	1	1
2148-1	1000H	0	0	0	1	0	0	0	0	0	0	0	0	0	0	0	0
Page 4	to																
	13FFH	0	0	0	1	0	0	1	1	1	1	1	1	1	1	1	1
2148-2	1400H	0	0	0	1	0	1	0	0	0	0	0	0	0	0	0	0
Page 5	to																
	17FFH	0	0	0	1	0	1	1	1	1	1	1	1	1	1	1	1
2148-3	1800H	0	0	0	1	1	0	0	0	0	0	0	0	0	0	0	0
Page 6	to																
	1BFFH	0	0	0	1	1	0	1	1	1	1	1	1	1	1	1	1
2148-4	1C00H	0	0	0	1	1	1	0	0	0	0	0	0	0	0	0	0
Page 7	1FFFH	0	0	0	1	1	1	1	1	1	1	1	1	1	1	1	1
2148-5	2000H	0	0	1	0	0	0	0	0	0	0	0	0	0	0	0	0
Page 8	to																
	23FFH	0	0	1	0	0	0	1	1	1	1	1	1	1	1	1	1

of memory from a 2148. In this case, that is both 74LS138's. At first it would seem logical to connect $\overline{RD}$ to either $\overline{E}_1$ or $\overline{E}_2$ of the 74LS138's. However, the 74LS138's could not be enabled for write operations. However, inputting $\overline{RD}$ and $\overline{WR}$ into an *AND* gate and having the output of that gate drive either $\overline{E}_1$ or $\overline{E}_2$ enables the 74LS138 for a read or a write operation. To isolate memory addresses from I/O addresses, the output of this *AND* gate is *OR*ed with control signal $IO/\overline{M}$. This logic circuit is illustrated in Figure 9–23.

We must verify that the timing is compatible, since now the timing will be referenced to the negative edge of the 2148's signal

$\overline{CS}$. From the timing diagrams of Figures 9–5, 9–6, and 9–15 we can reason that for the timing to be compatible,

$$t_{ACS} < t_{AD}$$

and

$$t_{WP} < t_{CC}$$

From Tables 9–3 and 9–6 we see that these conditions are satisfied.

The reader can easily construct a memory map from Table 9–10.

To convert the fully decoded address memory systems of Figures 9–17, 9–20, and 9–23 to partial linear address memory systems, one must eliminate the page selectors and enable the memory devices using address bus lines. Using an address bus line to enable a memory device means that it must be dedicated solely to enabling a memory device and cannot be used for any other purpose. As we have seen, this often leads to "wasted" addresses, in that addresses exist for nonexistent memory locations.

EXAMPLE 9–10

Redesign the memory system of Figure 9–20, using partial linear addressing instead of fully decoded addressing. Construct a memory map and compare it with the memory map for Figure 9–20.

SOLUTION As in previous designs, we begin constructing the address bus by developing an address bus table, which is shown in Table 9–11. This table indicates that there are three address bus lines available for selecting (enabling) memory devices. These address bus lines are A_{13}, A_{14}, and A_{15} and will be used to enable memory devices 2764-0, 8128-1, and 8128-2, as shown in Figure 9–25. Address bus lines A_{11} and A_{12} are don't cares (X) when an 8128 is addressed. Therefore, foldback exists. The authors have chosen to represent the don't cares as 0's in the address ranges of Table 9–11. By comparing Figures 9–20 and 9–25 it is easy to see

Table 9–11
Partial linear addressing of Example 9–10

System address bus lines		A_{15}	A_{14}	A_{13}	A_{12}	A_{11}	A_{10}	A_9	A_8	A_7	A_6	A_5	A_4	A_3	A_2	A_1	A_0
2764 Address pins					A_{12}	A_{11}	A_{10}	A_9	A_8	A_7	A_6	A_5	A_4	A_3	A_2	A_1	A_0
8128 Address pins							A_{10}	A_9	A_8	A_7	A_6	A_5	A_4	A_3	A_2	A_1	A_0
2764-0	C000H	1	1	0	0	0	0	0	0	0	0	0	0	0	0	0	0
	to																
	DFFFH	1	1	0	1	1	1	1	1	1	1	1	1	1	1	1	1
8128-1	B000H	1	0	1	X	X	0	0	0	0	0	0	0	0	0	0	0
	B7FFH	1	0	1	X	X	1	1	1	1	1	1	1	1	1	1	1
8128-2	6000H	0	1	1	X	X	0	0	0	0	0	0	0	0	0	0	0
	67FFH	0	1	1	X	X	1	1	1	1	1	1	1	1	1	1	1

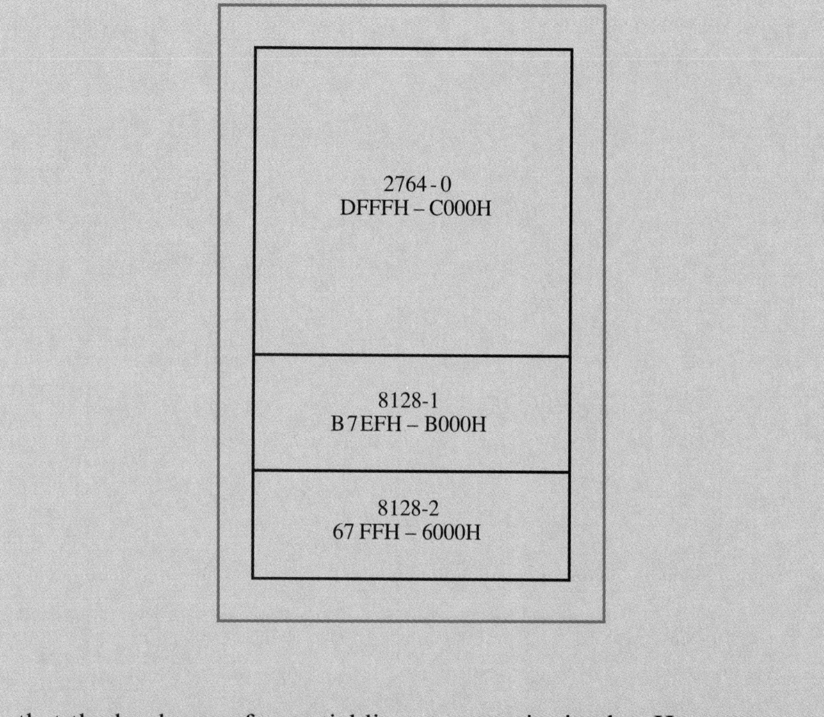

Figure 9–24
Memory map for the memory of Figure 9–25.

that the hardware of a partial linear system is simpler. However, comparing the memory maps of Figures 9–21 and 9–24 shows the disadvantages of partial linear addressing—the wasted memory addresses due to foldback and the number of addresses for which no memory exists (0000H to 6000H, for example).

Partial linear addressing is often used in systems that require smaller memory capacities, as these systems do not require utilization of every possible address. The cost and space savings resulting from the reduced hardware are a real benefit. The user would be furnished a memory map to identify which memory addresses exist. When a system designer must make the most of the available memory space, a fully decoded system is the solution.

There are microprocessors that require that ROM be placed at the beginning address of memory (0000H) and others that require that ROM be placed at the top end of memory ($A_{15} = 1$). This difference is due to the design of MPUs that causes them to automatically go (*vector*) to a specific memory location to fetch the first instruction of the bootstrap loader as a result of either a reset or a cold start. Under these conditions, some MPUs are designed to vector to location zero, whereas others vector to the top of memory.

A designer must be able to manipulate the placement of memory pages within the memory map. In partial linear addressing this simply means that the designer must chose the address line and its logic level to enable a memory device. For instance, to place 2764-0 at memory location zero, address lines A_{15} and A_{14} could be inverted before being input to the chip enable pins of Figure 9–25. Under this condition, to enable 2764-0 the logic levels of the address lines are such that $A_{13} = 0$ and $A_{14} = A_{15} = 0$, which places 2764-0 at address zero. To enable 8128-

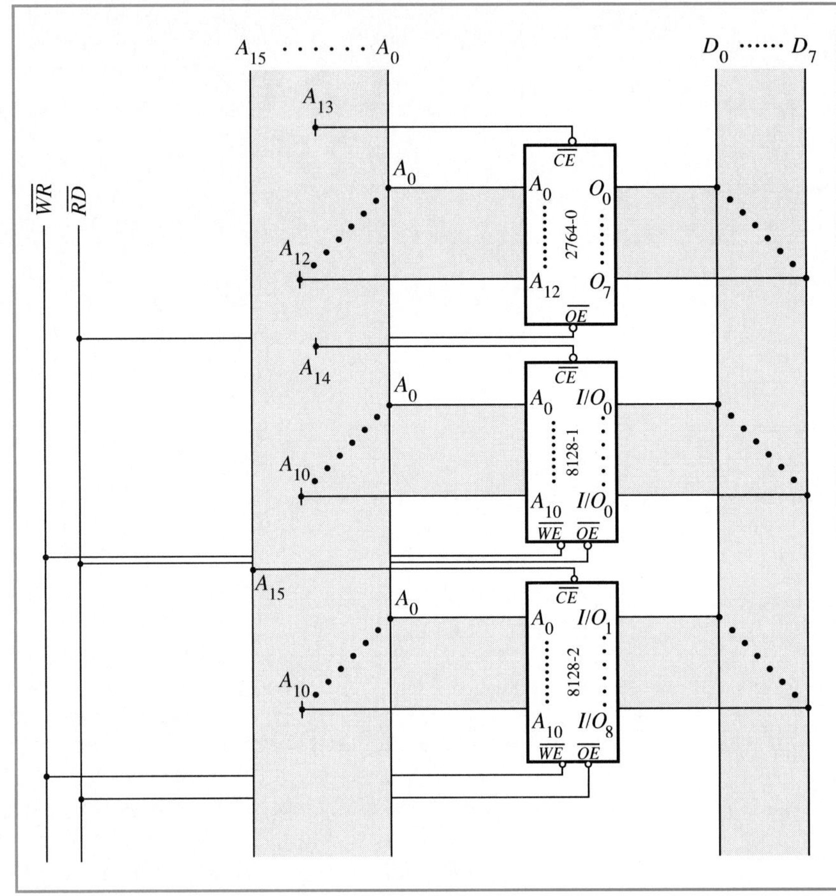

Figure 9–25 Partial linear addressing for Example 9–10.

1 requires that $A_{13} = 1$, $A_{14} = 1$, and $A_{15} = 0$, and to enable 8128-2 requires that $A_{13} = 1$, $A_{14} = 0$, and $A_{15} = 1$.

To manipulate which ICs contain what memory pages within a fully address decoded memory, the designer simply selects which decoder output is used to select each IC. As you will realize, there are various combinations that exist for the placement of memory pages.

Review questions

1. What is the rationale of a fully decoded address design?
2. What is the rationale of a partial linear address memory design?
3. How is the page selector(s) enabled?
4. How are the timing of RAM read/write operations controlled by an MPU?
5. How are ROM read operations controlled by an MPU?
6. What is foldback?
7. How can foldback be prevented?
8. What is a memory map, and of what value is it?

9. What is the significance of Table 9–9 to the 74LS138's of Figure 9–23?
10. Explain the reason for the read/write *AND* gate of Figure 9–23?
11. Without going into the details of specific AC parameters, what timing criteria must be met for MPU-initiated read/write operations, relative to memory devices?
12. What is paging?
13. How is paging implemented?
14. How is expanded page addressing implemented?

9–6 Dynamic memories

Thus far the memories we have designed have had static RAMs (SRAMs). However, many memory systems are better suited to DRAMs because of a DRAM's low power consumption and high-density memory capacity. The disadvantage is the DRAM's required refresh cycle, which means additional circuitry and timing considerations. We shall take a survey approach to studying and designing dynamic memory systems.

Recall from our study of the 8116 DRAM that memory address pins are multiplexed via address pins A_0 through A_6. These seven pins first serve as seven row address pins and then as seven column address pins, which results in a 14-bit address (16,384 locations). The timing diagrams of Figure 9–14 indicate that a row strobe ($\overline{RAS}$) and column strobe ($\overline{CAS}$) must be input to the 8116 so that it latches the 7-bit row address applied to its address pins and then the 7-bit column address. To the MPU there is no difference between a static RAM address and a DRAM address. As a result, the MPU outputs this address as any standard 16-bit binary number on the system's 16-line address bus, just as it did for memories with static RAMs. It is necessary to take the 16-bit address and divide it in half, a low byte and a high byte, and apply those bytes to 8116 address pins A_0 through A_6. When these address bytes are being input to those address pins, strobe pulses must be generated and applied to 8116 input pins $\overline{RAS}$ and $\overline{CAS}$ at the appropriate time. Rather than waste engineering time designing the logic to do this, as well as engineering the logic to execute a memory refresh cycle, it is better to purchase a chip manufactured for this specific task. Such a device is known as a *refresh controller* or a *dynamic RAM controller*. Figure 9–26 illustrates some of the essential addressing functions of a dynamic RAM controller. What has been omitted are the memory refresh functions, which require a better knowledge of microprocessor technology. The DRAM controller of Figure 9–26 has 16 address pins that are connected to the system address bus lines. If the 16/$\overline{64}$-K pin is strapped high, the DRAM controller treats the 16-bit address as a 14-bit address (16 K); if strapped low, all 16 address pins are valid (64 K). The DRAM controller divides the 16-bit address into two 7-bit addresses, for this case, which are output at its output pins O_0 through O_6 (notice that O_7 is not connected). The timing signals $\overline{RAS}$ and $\overline{CAS}$ are generated by the DRAM controller and are input to the 8116's so that the row and column addresses are latched by the 8116's, as previously discussed. The DRAM controller has its own crystal, since its refresh cycles are independent of the MPU's clock. The DRAM has MPU

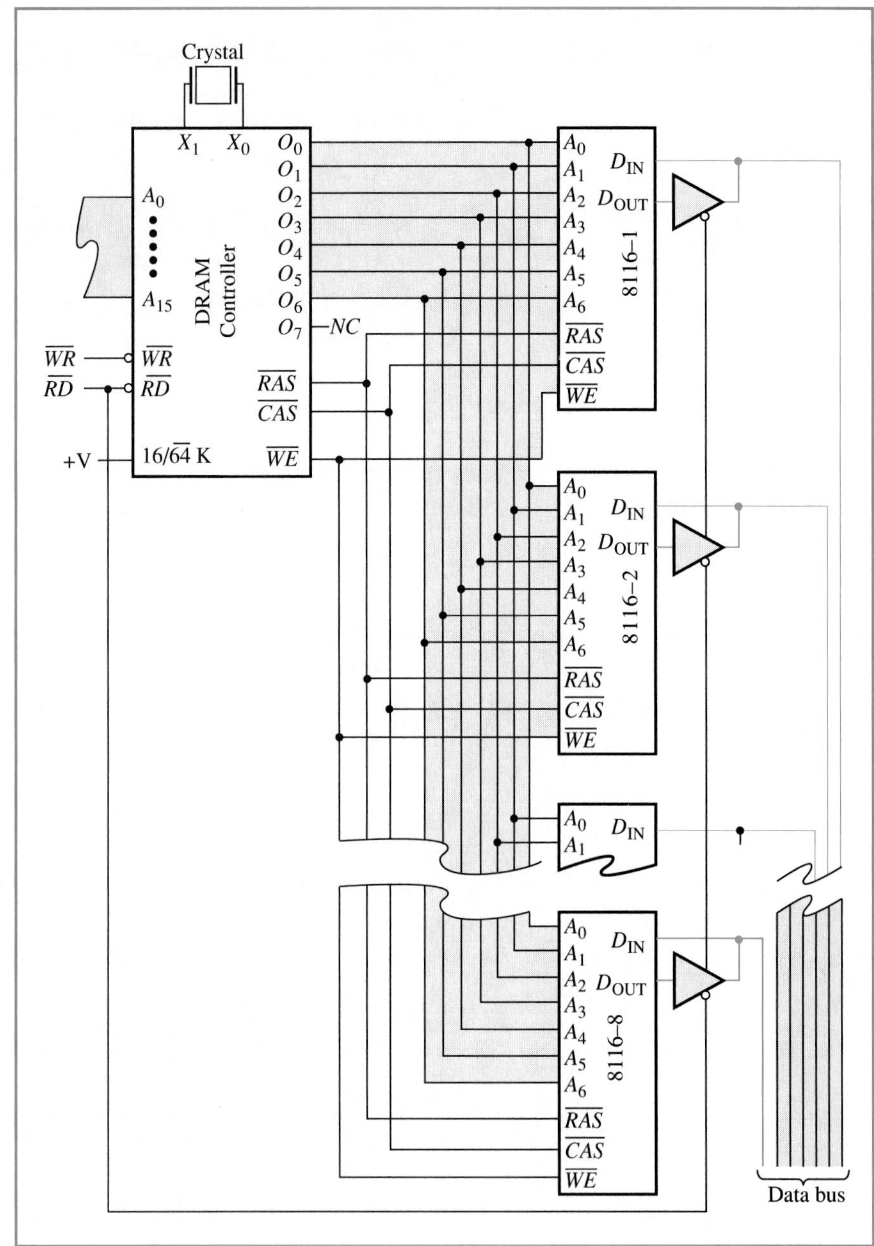

Figure 9–26
DRAM controller for a 48-K dynamic memory.

control signals $\overline{RD}$ and $\overline{WR}$ as inputs so that DRAM read and write operations are synchronized with the MPU.

Review questions

1. What are the advantages and disadvantages of dynamic memories?
2. Why must dynamic memories be refreshed?
3. What is the essence of a DRAM controller?

9–7 Troubleshooting

The concepts and procedures for troubleshooting memory systems were developed and outlined in Chapter 8. In this section we will apply those concepts and procedures to testing memory designs with both static and dynamic tests.

To decide whether static or dynamic testing is appropriate, the diagnostician must recall that static testing is limited to diagnosing circuit continuity problems and catastrophic failure of components. Even then the failed design must be such that static logic levels can be applied, such as in the case of a newly constructed memory not yet integrated into the system. All other cases require dynamic testing.

We shall first devise a static testing environment for the memory of Figure 9–17. We begin by determining the number of test nodes and then classifying them as either input nodes, output nodes, or bidirectional nodes. As illustrated in Figure 8–24, input nodes require static logic generators to input the desired logic levels to the system, whereas the output nodes require static logic level detectors to detect the logic levels resulting from the logic levels applied to the input nodes. The bidirectional nodes require both static logic level generators and detectors.

The memory system of Figure 9–23 has 19 primary input nodes, which are composed of 16 address bus lines and three control lines. The system has 14 secondary output nodes, a secondary output node being one that must remain intact in order for the system to function properly. Lastly, there are eight bidirectional nodes, which are the data bus line. Figure 9–27 is an illustration of the static test configuration for the memory of Figure 9–23.

We shall apply static logic level generators to the primary input nodes, as illustrated in Figure 9–27. We connect both static logic level generators and detectors, which are buffered, to the bidirectional nodes (data bus lines), just as we did in Figure 8–24. Figure 9–27 shows a read/write switch to control these buffers (turn them on and off). The secondary outputs are identified in Figure 9–27 as circles with numbers.

Using the truth table of Table 9–10 as a map of the memory space, we can select addresses of various memory locations and then either read the contents of those locations for ROMs or write in a known data byte and read it back in the case of a RAM. For each read or read/write operation we can use a logic probe to test the corresponding logic levels at the secondary outputs. For instance, to test location 0510H, we would put the static logic levels generator slide switches for the address bus lines to output logic levels 0000010100010000. From Table 9–10 we can determine that page 1 of the 2732-0 is being addressed. Since this is a ROM and can only be read, then we drive control signal $\overline{RD}$ low via its static logic level generator. Of course $\overline{WR}$ must be driven high and $IO/\overline{M}$ low via their static logic level generators. With the address and control signal logic levels properly set, the secondary output number 14 can be checked for its resultant logic level, which must be a logic 0 in order to enable 2732-0. If this test point is low and the logic level on $\overline{OE}$ is also low, then the content of the memory location (which must be known in advance of the test) should be present on the data bus and detected by the static logic level detectors.

If the correct datum does not appear on the data bus with the proper address and control signals applied, we must begin to diagnose the

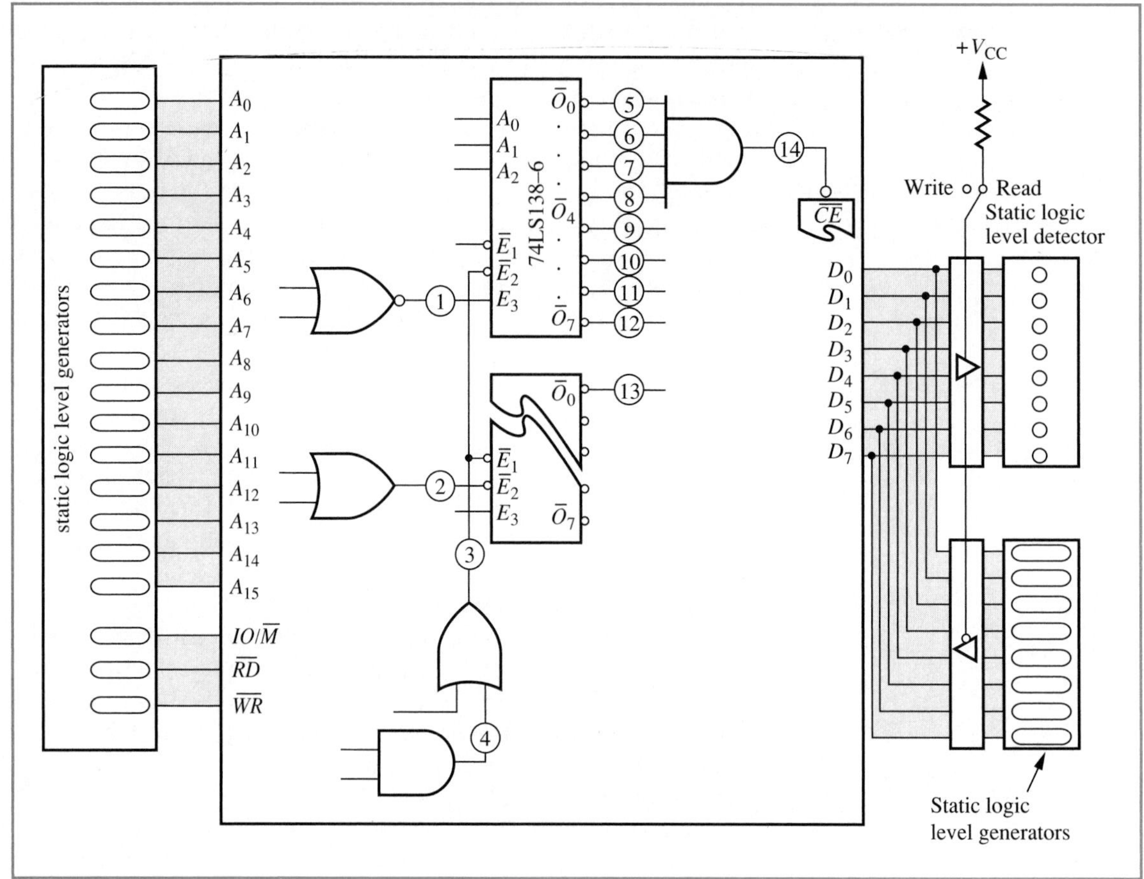

Figure 9–27
Static testing of the memory of Figure 9–23.

problem, which means starting at the point of the detected fault and working back to its source. Suppose the content of location 0510H of Figure 9–27 is not what is on the data bus, but rather the bus is floating (it is in a Hi-*Z* state). This means that the 2732-0 is not being enabled. We must verify the logic level of secondary output 14 and $\overline{OE}$. The output enable is connected to $\overline{RD}$, and we find its logic level to be low. However, we tested secondary output 14 with a logic probe and it was high. We now know that the problem exists with the *AND* gate or output $\overline{O}_0$ of the page select decoder 74LS138-6. If the output of $\overline{O}_0$ was a logic 1, the decoder has failed or its enable circuitry is not functioning properly. To test the enable circuitry we would use a logic probe to verify the logic levels on secondary outputs 1, 2, and 3, that is, a logic 1 at E_3 and logic 0's at $\overline{E}_1$ and $\overline{E}_2$. If these are correct, the decoder has failed and must be replaced. If one or more of these secondary nodes has the wrong logic level on it, we must continue to backtrack through the logic. Let us suppose that secondary node 3 had a logic high; this means that either the *OR* gate or *AND* gate has failed. If secondary output 4 has a logic 0 on it, either the *OR* gate or the *AND* gate has failed.

To summarize and formalize an approach to troubleshooting memory systems or any other digital system, we must do the following:

1. Configure the test environment by deciding what is to be tested. Once the test objectives are decided, such as reading the content of a memory location, the diagnostician can reason from a truth table(s) and a system schematic(s) what nodes must be under control of the test as well as those to be monitored.

2. When the controlled and monitored nodes have been determined, the diagnostician must decide whether the test nodes are primary or secondary. Primary input nodes are driven by logic level generators, which are not part of the system, and primary output nodes are to be monitored with logic level detectors. *Secondary input nodes* are considered part of the test and are driven by circuitry within the system, such as the address bus lines (nodes) driven by the CPU. Secondary output nodes are nodes that must be monitored without being modified for testing. Secondary output nodes require that their logic levels be monitored with test equipment such as a logic probe or logic analyzer. By classifying the test nodes as primary or secondary, the diagnostician has also determined how the nodes are to be tested.

3. Decide what nodes are to be tested for correct logic levels. The logic levels of the input nodes are checked first, and then the output nodes that respond to those input nodes are tested for their logic levels. Once an incorrect response is found, the diagnostician can trace the fault back through the logic circuits until its source is identified.

4. Trace back through the logic looking for the source of the fault. The diagnostic techniques for tracing through logic circuits were discussed in the early chapters, beginning with Chapter 2. The above four steps apply to both static and dynamic testing. You may want to review our previous analysis of the static test of Figure 9–27 and relate the above four steps to our diagnosis.

EXAMPLE 9–11

Suppose that for the test configuration of Figure 9–27 it was determined that the datum read from address 1250H was not the datum previously written into that location. All input nodes were tested and had the correct logic levels. All secondary output nodes were tested and these logic levels were detected using a logic probe: 1 = 1, 2 = 0, 3 = 0, 4 = 0, 5 through 8 = 1, 9 = 0, 10 through 12 = 1, 13 = 0, and 14 = 1. What could be causing the fault?

SOLUTION Before we begin the diagnosis let us understand what is implied by the statement, "the datum read . . . was not the datum previously written into that location." Address 1250H is in page 4, as can be determined from Table 9–10. Page 4 is made up of two 2148's, which are RAMs. Since we have no way of knowing for certain the content of a RAM memory location, we must test it by first writing a known data byte into that location and then reading the content of that location to verify that the written datum was indeed stored. To really be certain of a memory location, AAH should be written into it and then read from it and then the write/read process repeated for the complement of AAH(55H). This ensures that each memory cell can store both logic levels (a cell could have failed such that its logic level is always 1 or 0).

To begin the diagnosis, we would refer to Table 9–10 and the test configuration of Figure 9–27. From Table 9–10 we determine that address 1250H is in page 4 of the memory space and that page 4 is contained in the 2148-1. From Figure 9–27 (and Figure 9–23) we find that page 4 is enabled by page select line $\overline{O}_4$ of 74LS138-6, which is secondary output 9. This is the start of our backward trace. By knowing that secondary output 9 enabled (chip select) the 2148-1 pair, we also know that the correct logic level is a low, which was the logic level measured, and that the other outputs of the page selectors must be high. Logic 1's were measured at the other decoder secondary test points except for 13, which was measured as low. This identifies a fault and we can now focus our diagnostic efforts on 74LS138-7 and from there continue our fault tracing. To determine if the decoder is at fault, we check the logic levels at its enable pins. Test point 2 should be high and 3 low. Test points 2 and 3 were measured to be logic 0's. This leads us to conclude that the *OR* gate has failed, unless the $\overline{E}_2$ pin of 74LS138-7 has failed by shorting to ground. This can be checked when the *OR* gate is removed.

As we stated in Chapter 8, dynamic troubleshooting is conceptually similar to static troubleshooting. The differences are the logic level generators and detectors, that is, the test equipment. Both generators and detectors must be dynamic, as is obvious. If the memory is an integral part of a system, the CPU acts as the logic level generator (it outputs addresses and control signals, and when writing it also outputs the datum). If the memory is isolated from a system (no CPU present), there must be a dynamic logic level generator that generates addresses and control signals. A counter and logic gates could be utilized for this task. In either case, the logic level detectors are implemented with a signature analyzer or logic analyzer.

To dynamically test the memory of Figure 9–23 we must use a logic analyzer, since no correct signatures are given (none is on the schematic). As outlined by the four steps to be followed for troubleshooting, we must identify the objective of our test. If it is to verify the content of a memory location, we know that the principal nodes of logic level verification are the data bus lines. In addition to the principal nodes we must also monitor input nodes such as enable pins and control lines for proper logic levels. Then the test configuration of Figure 9–27 is valid if the static generators and detectors are removed. There is no replacement required for generators if a CPU is present; otherwise there must be dynamic generators instead of static ones. We are assuming that a CPU is present. To verify the logic levels on the input nodes, an input channel of the logic analyzer must be connected to each input node, which requires 16 such nodes for the address bus and 3 for the control bus. In addition, we need eight input channels of the logic analyzer to detect the logic levels on the data bus lines. So far we are requiring that 27 nodes be continuously monitored, which requires that the input anayzer have 27 input channels. To monitor test points 1 through 14 of Figure 9–27 requires 14 more channels. If the logic analyzer has 80 channels, an input channel of the logic analyzer may be connected to each of these nodes, a total of 41 channels. If the logic analyzer does not have enough channels, a channel would have to be shared for the remaining nodes and moved from one node to another for each sample interval. We will assume an 80-channel logic analyzer.

Recall that there must be some event (either a word or a trigger pulse) with which to start the data sampling. We must determine what event is to act as the reference point. For Example 9–11 we could use the address 1250H as the trigger word. Under this condition, when address 1250H was output on the address bus, the logic analyzer would sample the nodes and display their logic levels on a CRT. This display is then analyzed for the various logic levels, just as if it were a static test.

EXAMPLE 9–12

Suppose that when memory location 2000H of the memory system of Figure 9–23 is addressed and read, the data bus is found to be in the Hi-*Z* state. Using the nodes of Figure 9–27, diagnose the problem.

SOLUTION A logic analyzer is used to detect the logic levels on the nodes for the various sample times. Figure 9–28 is a representation of the logic analyzer's display, where the channels connected to the address and data bus lines are not displayed individually, as they actually are, but rather are grouped together symbolically. Also for simplicity and convenience, test points 6 through 12 are not shown.

Since address 2000H is the memory location in question, we use 2000H as the trigger word to start the logic analyzer reading the logic levels at those nodes connected to its channels and storing that sampled data in its FIFO memory. Recall that, once triggered, the logic analyzer continues to collect data until the logic analyzer buffer memory (the FIFO) is full. The sampled data can then be displayed in a variety of formats. We will choose to display the data as a timing diagram, as illustrated in Figure 9–28.

From Figure 9–28 we see that outputting the address 2000H on the address bus triggered the logic analyzer, which then began to collect the nodal data at the rate indicated by the sample clock. Once the FIFO was filled, the sample data was displayed in the chosen format and from that display we have Figure 9–28, which can be used much like a truth table. That is, we divide the timing diagram into intervals of time (t_1, t_2, t_3, etc.), with the boundaries of these intervals defined by a change in logic state of any of the input nodes that are responsible for a corresponding change in an output node. The logic changes of $\overline{RD}$ define the time interval boundaries (nos. 3 and 4 are outputs). Having defined the intervals, we can analyze the logic levels within the intervals for correctness.

From the statement of the detected fault (data bus in Hi-*Z*) we know that the addressed memory IC is not being selected. From Table 9–10 we find that address 2000H should select and enable page 8 of memory, which means the 2148-5 should be enabled. From Figures 9–23 and 9–27 this means that output 13 is to be low for a read (when $\overline{RD}$ is low). Therefore, it is when $\overline{RD}$ is low that is of most interest. During interval t_2 of Figure 9–28 test point 13 is high, as it is also in t_4. Thus the page selector 74LS138-7 is not being enabled, or the decoder has failed. This leads us to examine the logic levels of test nodes 2 and 3, each of which must be low during

Figure 9–28
The display of a logic analyzer for Example 9–12.

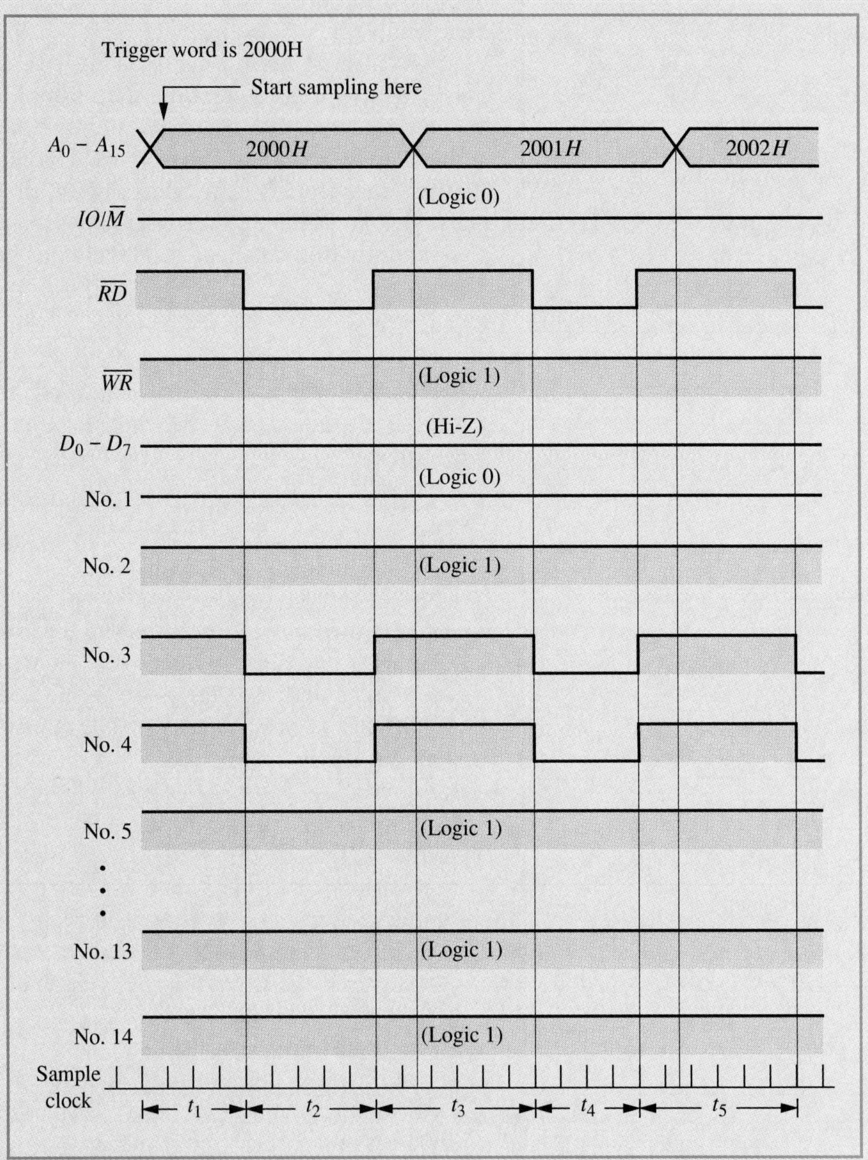

these intervals of t_2 and t_4. In both intervals test node 2 is high, whereas test node 3 is correct. Hence, it is the *OR* gate of test node 2 that has failed.

In our examples we used trigger words to signal the logic analyzer to begin sampling and storage of nodal data. If our diagnosis were to be focused on a read or write operation, we could use control signal $\overline{RD}$ or $\overline{WR}$ to trigger the logic analyzer. Often a microprocessor has a signal that triggers the beginning of a fetch and execute cycle of an instruction (*ALE* for an 8085 and *VMA* for a 6800). These signals can be used to sample data based on the instruction fetch and execute cycle.

9–8 SUMMARY

There are basically seven types of memories:

1. Main memory
2. Mass storage memory
3. Scratch pad memory
4. Buffer memory
5. Stack memory
6. Queue memory
7. Look-up table memory

Scratch pad memory stores intermediate and other temporary-type data. Scratch pad memory should be easily accessed so as to minimize the CPU's data retrieval time, which often means that it is internal to the microprocessor.

Buffer memories serve as a buffer between two devices, such as a computer and a printer. A buffer memory can be used to free up the computer from the printer.

Stack memory also contains temporary data, such as CPU register contents and the return address as the result of an interrupt, but it differs from scratch pad memory in the manner of access and its physical location. Stack memories are usually a portion of main memory and are accessed in a LIFO fashion.

Queue memory is internal to a CPU and holds instructions that were prefetched from main memory. As the CPU is ready to fetch an instruction, it does the fetching from its queue rather than from main memory, which speeds up the instruction fetch time and does not tie up the system data bus for an instruction fetch.

Look-up table memories have been programmed with values that translate an input (address) to a desired output, which is the content of the addressed memory location.

Manufacturers furnish data sheets for their semiconductor memory devices, and among other information these data sheets contain an architectural drawing and timing diagram. A memory designer can get a feel for how the memory device functions from its architecture, and this also helps the designer determine its organization. *Timing diagrams* indicate timing specifications of the memory device.

Timing specifications for the microprocessor dictate timing parameters that must be met by memory devices (including I/O devices). The microprocessor requires that the datum being read from memory be on the data bus in t_{AD} seconds from the time it outputs an address and be held there until the positive edge of $\overline{RD}$. For write operations the microprocessor outputs the datum it is writing to memory t_{AC} seconds from the time it outputs the address on the address bus. It holds that datum on the data bus for t_{CC} seconds.

- All memory devices must be checked for AC compatibility by determining if their read/write response time is faster than that of the microprocessor.

- In designing a memory, the designer chooses memory devices based on characteristics such as capacity, organization, speed, volatility, and cost.

- There are two addressing schemes, fully decoded and linear. *Fully decoded addressing* decodes all address bus lines, whereas partial linear does not; rather, *partial linear addressing* assigns some address bus lines to enable memory devices via the $\overline{CS}$ pin.

- Memory devices of the same memory system with differing numbers of address pins can produce nonexistent addresses unless paging is used.

- *Address foldback* exists unless all address bus lines are used in determining memory addresses. Foldback results in multiple addresses for the same location.

- Dynamic memories have address pins that are shared among functions (*multiplexed*). They first serve as row address input pins and then as column address input pins. Dynamic memories also require refresh cycles in order for the device to remember the data stored. A DRAM controller can be used to demultiplex the address pins as well as to provide the required refresh cycle.

- To static test a memory, static level generators are used to input addresses, control signals, and data when a write operation is performed.

- *Static testing* is limited to testing for component failure and wiring problems.

- *Static logic level detectors* are used to detect the logic levels of the nodes under static testing.

- As with any digital troubleshooting procedure, a truth table can be used to assist in determining the correct logic level for an output for any given input. The truth table enables the diagnostician to identify an incorrect response. Once the error has been located, the diagnostician can trace back through the logic circuits until the source of the error is located.

Dynamic testing is conceptually the same as static testing. The difference in implementation is that dynamic testing requires logic level generators that have changing logic levels. This implies test equipment that can detect changing logic levels and display this data so that it can be observed and interpreted by the user. The signature analyzer and logic analyzer are two pieces of dynamic test equipment.

Signature analysis is well suited for troubleshooting memory failures when nodal signatures are known.

Logic analyzers can present data in several formats, of which a *timing diagram* is one. The diagnostician can use this timing diagram much like a truth table and from it identify correct and incorrect logic levels on the test system's nodes.

PROBLEMS

Section 9–2 Applications of semiconductor memories

1. What is the primary function of a scratch pad memory?
2. Why is a scratch pad memory usually an integral part of a microprocessor rather than a designated portion of main memory, as is a stack memory?
3. Using registers, design a 4 × 8 scratch pad memory, to be composed of registers A, B, C, and D of Figure 9–1a.
4. Why are buffer memories accesssed using FIFO techniques?
5. Design a 4 × 8 buffer memory.
6. Since a stack memory is often a portion of main memory, how do the two differ?
7. Could a stack memory be accessed using FIFO techniques?
8. With reference to Figure 9–1 and the explanation given of a stack memory, what determines how a stack memory is accessed?
9. Explain how using a queue memory can enable more efficient use of the system data bus insofar as the MPU is concerned.
10. Can a queue memory be likened to a small-capacity buffer memory?
11. When is a look-up table memory a viable solution to a programming problem?
12. Explain how look-up table memories are implemented.
13. Design an encoder (translation table) to convert hexadecimal values to seven-segment LED code, using the two codes given in Table 9–12.
14. Design a look-up table that will square an integer with values of 0 to 9 and give the answer as the binary equivalent.

Section 9–3 Memory device architecture and timing characteristics

15. Which of the technologies described in this section is the fastest?
16. Which technology has the highest RAM bit density?

Table 9–12
Translation table for hex to seven segment

Hexadecimal value	Binary code	Seven-segment code
0	0000	00111111
1	0001	00000110
2	0010	01011011
3	0011	01001111
4	0100	01100110
5	0101	01101101
6	0110	01111101
7	0111	00000111
8	1000	01111111
9	1001	01101111
A	1010	01110111
B	1011	01111100
C	1100	00101001
D	1101	01011110
E	1110	01111001
F	1111	01110001

17. Why is MOS technology so popular?

18. What characterizes the main memory of a PC?

19. How would the main memory of a PC differ from the main memory of a microcontroller?

20. If a semiconductor memory device has an organization of 4096 × 8 bits, how many address and data pins does it have?

21. From the row and column address pin assignments of the 8128 architecture of Figure 9–7, determine how many rows and columns there are. Explain the discrepancy between the number of columns stated in the architecture (128) and the number calculated from the number of address pins.

22. Relative to Figure 9–5, when will a 2148 output the contents of an accessed memory location?

23. Why are memory device AC parameters often referenced with respect to an address being input to the memory device via the address pins?

24. From the write timing diagram of Figure 9–6, explain the AC parameter t_{DW}. When the MPU is writing to a 2148, how does t_{DW} relate to MPU timing?

25. What is the functional difference between $\overline{CE}$ and $\overline{OE}$ of an 8128?

26. What is the functional difference between t_{ACE} and t_{OE} of Figure 9–8?

27. Notice in the memory matrix function-block of Figure 9–9 that the memory organization is listed as 128 × 16 × 8. Relate this method of notation to Figure 8–12 and the answer to Problem 21.

28. Explain the significance of the time interval specified by AC parameters $t_{AS} + t_{ACS}$ in Figure 9–10a.

29. Explain the meaning of data output, as indicated by D_{OUT}, in Figure 9–10b.

30. Explain the purpose of the output enable pin and chip enable pin on the 2764.

31. What is the significance of AC parameter t_{ACC} in Figure 9–12? As part of your answer relate the meaning of this AC specification to a memory designer.

32. Name the major advantages and disadvantages of using DRAMs in a memory design.

Section 9–4 AC compatibility

33. Whenever a microprocessor executes a read or write operation there are three events that must take place. Referring to Figure 9–15, determine these three events and briefly explain the purpose of each event. You may also wish to refer to Figure 8–9.

34. Explain the significance of AC parameters t_{AD} and $t_{AC} + t_{CC}$ in Figure 9–15a. Relate your explanation to an MPU-reading main memory.

35. Which memory devices with the following AC specification are or are not AC compatible with an MPU with the AC specifications of Table 9–6? (Explain your answer.)
(a) t_{ACS} of Figure 9–5 is 700 ns.
(b) t_{DW} of Figure 9–6 is 500 ns.
(c) t_{OE} of Figure 9–8 is 470 ns.
(d) t_{DS} of Figure 9–10b is 470 ns.

36. Explain the implications of the time when $t_{AC} + t_{CC}$ occurs in Figure 9–15b, relative to the MPU writing to memory.

Section 9–5 Designing static memories

37. How can the enable pins of a 74LS138 be utilized when expanding the page addresses beyond eight pages?

38. Explain the purpose of creating an address bus table such as Tables 9–7, 9–8, 9–10, and 9–11.

39. Explain the purpose of a page selector in a fully decoded addressing scheme.

40. Address foldback creates multiple addresses for a memory location, while using memory devices with an unequal number of address pins can result in addresses for nonexistent memory locations. How can each of these situations be prevented?

41. Can foldback and nonexistent memory locations, as discussed in Problem 40, always be prevented? Explain your answer.

42. What defines a page of memory?

43. How are multiple memory page assignments to a memory device implemented in hardware?

44. How does partial linear addressing differ from fully decoded addressing?
45. When might a memory designer choose to use partial linear addressing versus fully decoded addressing?
46. Design a fully decoded main memory that is without address foldback or nonexistent addresses in the memory map for a 16-bit address bus. This memory is to have 1 K of nonvolatile memory for the bootstrap loader and 4 K of user RAM. The beginning address of the bootstrap loader is to be 0000H. Use 2716 and 8417 memory devices. The pin configuration of a 2716 is basically the same as that of a 2764 or 2732 except that it has just ten address pins (it is AC compatible with this MPU). As part of your design solution, develop an address bus table and memory map.
47. Redesign the memory of Problem 46 using partial linear addressing. Do not concern yourself with the starting address of the bootstrap loader.
48. Some MPUs fetch the first instruction to be executed after a restart or cold start from address zero. This would require the bootstrap loader program to begin at location zero, as specified in Problem 46. Redesign the memory of Problem 47 so that the beginning address of the 2716 is 0000H.
49. There are MPUs that fetch the first instruction to be executed after a cold start or reset from the top portion of memory (the highest possible address). Redesign the memory of Problem 46 for this type of MPU.

Section 9–6 Dynamic memories

50. Redesign the memory of Problem 46, but use 8128's instead of 8417's.
51. From a designer's point of view, what is the difference between using 8417's and 8128's?

Troubleshooting

Section 9–7 Troubleshooting

52. If we were to verify correct operation for each memory location of Figure 9–27, using static testing, how many addresses must be generated?
53. Suppose that the decoder of the memory-mapped I/O memory of Figure 9–17 was suspected of failure. How would you test the decoder using static testing?
54. Devise a static test to verify proper operation of the design in Figure 9–17. Illustrate the test configuration in a manner similar to Figure 9–27.
55. Suppose that the chip enable of the 2764-0 in Figure 9–17 was "stuck" high. Devise a static test to locate the problem.

56. Repeat Problem 54 using dynamic testing. Identify what nodal signal might be used for the trigger source.

57. Repeat Problem 55 but use a logic analyzer. Assume that a CPU is present.

58. Devise a dynamic test to verify proper operation of page 5 of the design in Figure 9–20. Use a nodal signal as the trigger source.

The personal computer is finding its way into almost all aspects of our everyday lives. Here a personal computer is being used with graphics software to display a map of Europe (Courtesy IBM).

OUTLINE

OBJECTIVES

The objectives of this chapter are to

- describe the architecture of a generic microprocessor.
- explain the function of each function-block within the MPU architecture and how it is controlled by instructions of the instructions set.
- explain the fundamental concepts of MPU AC characteristics.
- introduce the concepts of interfacing an MPU to system hardware, that is, memory and I/O.
- present the fundamentals of instruction coding.
- introduce programming concepts.
- survey some real microprocessors.

10 INTRODUCTION TO COMPUTER ARCHITECTURE AND MICROPROCESSORS

10–1 INTRODUCTION

Microprocessor technology is a major application of digital circuits. A *microprocessor* is a semiconductor IC that as a minimum is capable of implementing a CPU, such as the one depicted in Figure 8–9. Hence, the microprocessor unit (MPU) provides the pseudointelligence of a system, because it manipulates data, for example, by performing arithmetic and logic operations and by reading data from and writing data to memory and I/O. Most significantly, based on data computational results, the MPU is capable of making decisions. We say that the MPU has no real intelligence, but rather a pseudointelligence, because at present it is the human intellect that provides the intelligence of an MPU. The MPU's pseudointelligence is provided by programs that instruct the MPU about the task to be performed by providing a step-by-step procedure.

The IC wafer on which an MPU is manufactured is very small, often slightly less than the size of a thumbnail, and contains thousands of MOSFETs (there are also bipolar MPUs). Because of the small size of an MPU and its pseudointellect—that is, its ability to process data and make decisions, it can be used in most applications requiring data processing and/or mechanism control. MPUs are found in medical equipment, automobiles, home appliances, PCs, electronic games, etc. In fact, one could say that an MPU can be used for any application requiring decision capability and/or process control.

Microprocessor technology can be divided into three areas of study: (1) MPU architecture, (2) hardware, and (3) software. MPU architecture provides a function-block diagram representation of the MPU, which serves to illustrate its internal structure. By knowing and understanding the internal structure of an MPU, a user (system designer and/or programmer, who can be the same person) can quickly refer to an architectural diagram in order to be reminded of the capabilities and limitations of an MPU as well as use it as an aid in understanding program instructions.

Hardware considerations are the physical aspects of the system, such as the pin configuration and system connections to those pins. From the memory designs of Chapter 9 the reader should already have a feel for hardware considerations such as assignments and connections of address pins, data pins, and the control bus signal pins for read and write operations. As stated above, the architecture diagram aids in system hardware design and implementation by serving as a quick reference.

Software is a term used to describe programs and MPU instructions that make up a program. *The software controls the hardware via the MPU*. To write a microprocessor program, the person writing the program (the *programmer*) has a specific set of instructions the MPU is designed to recognize and execute, which is the *instruction set* for that MPU type. The programmer must understand and have a general feel for the repertoire of instructions in an instruction set. As we have already stated, the MPU architecture diagram can serve as a visual aid to the programmer in helping him or her remember the software capabilities of an MPU as well as the general characteristics of the instruction set.

It is clear that MPU architecture is a major focal point for understanding a specific type of microprocessor and the workings of the instructions within its instruction set. For this reason we will devote much of this chapter to developing the concepts of MPU architecture. There are various types of MPUs available; however, their principal concepts of operation are very similar. To understand the concepts of one MPU is to understand them all. We shall take a somewhat generic approach, but our generic MPU is oriented to a specific microprocessor; the 8085. At the end of the chapter we will survey three specific MPUs, the 8085, 8086, and 6800, to apply our generic architecture to specific MPUs.

Our investigation and use of software will be limited, since comprehensive coverage of programming is beyond the scope of this book. We encourage the reader to continue digital studies by studying microprocessors and their applications.

10–2 Computer architecture

To logically develop MPU architecture we must first understand computer architecture. Fortunately, we have already studied most of the essential concepts of computers in Chapters 2 through 9. Therefore, much of this section will review, collect, and organize our knowledge of computer systems.

Figure 10–1 is an illustration of a computer architecture, much like Figure 8–9 but expanded. This architecture has three-bus structure and also function-blocks MPU, main memory, and I/O. The three busses are the address bus, data bus, and control bus; the control bus is formed by those lines over which MPU-generated signals $\overline{RD}$, $\overline{WR}$, *DMA ACK*, $IO/\overline{M}$, and *INT ACK* are transmitted, as well as I/O-generated signals *DMA REQ* and *INT REQ*. The computer architecture of Figure 10–1 shows MPU address pins A_0 through A_n connected to the address bus and the MPU data pins D_0 to D_m connected to the data bus. This MPU can generate 2^n addresses, as indicated by the number of MPU address pins, and can read or write m data bits in parallel for each read or write operation, since there are m data pins. The process of accessing a main memory location was explained in Chapter 9.

To review some of the concepts developed in Chapter 9, recall that the MPU outputs the address of the memory location to be accessed on the address bus. The page selector enables the memory device within main memory that contains that location. The accessed memory location has the datum either read from it or written into it, depending on the program instruction being executed. If the datum is being read from memory, the accessed memory location loads its contents on the data bus when MPU-generated read control signal $\overline{RD}$ is driven active. The MPU

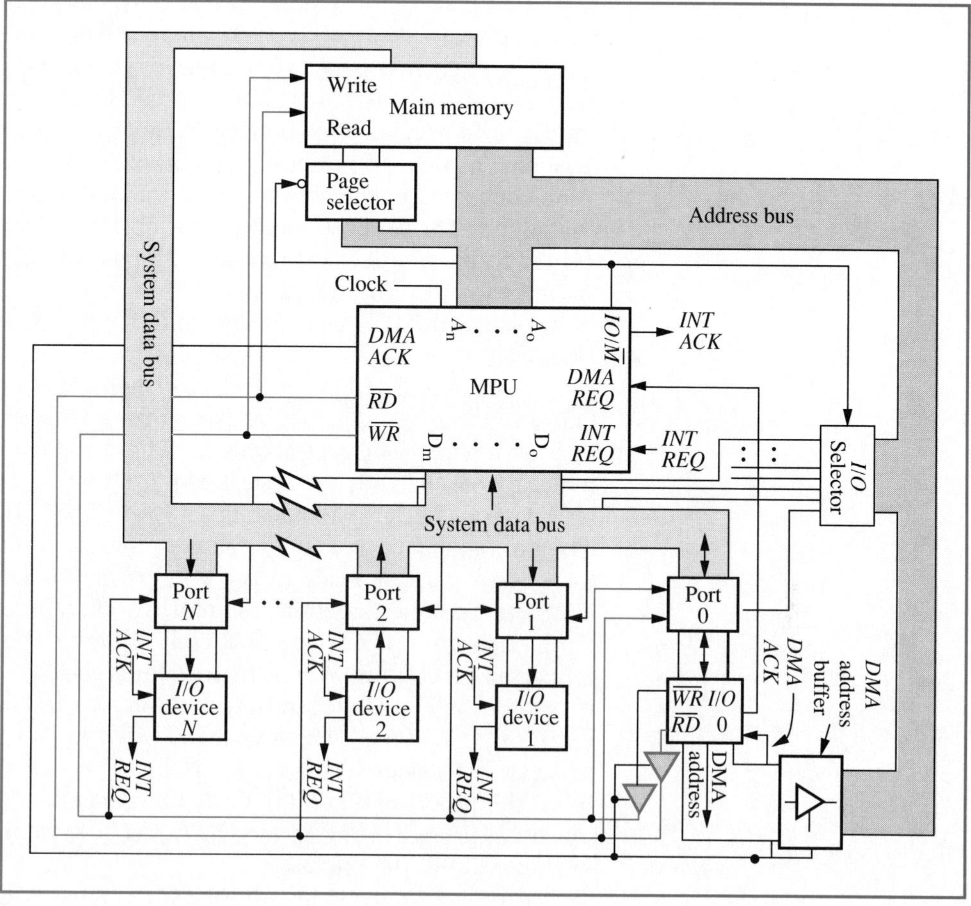

Figure 10–1
A computer architecture.

latches that datum off the data bus on the positive edge of that read control signal. If the MPU is writing to the accessed location, the MPU outputs the datum on the data bus and drives the write control signal $\overline{WR}$ low at the same time. The positive edge of $\overline{WR}$ serves as a strobe for memory; as a result, memory latches the datum on the data bus into the accessed memory location.

Reading or writing to an I/O device is done in a similar fashion to reading or writing to a memory location. As we already know from Chapters 8 and 9, the difference between memory and I/O operations has to do with the logic state of MPU pin $IO/\overline{M}$.

Note that the system of Figure 10–1 employs a fully decoded addressing scheme, as suggested by the presence of the page selector. Also, as in Figure 8–9, the architecture of Figure 10–1 indicates that addressed I/O ports (I/O ports are interfaces for I/O devices) are enabled via the I/O selector, which functions much like the page selector except for the method whereby it is enabled. That is, it is enabled when the $IO/\overline{M}$ control signal is a logic high.

The architecture of Figure 10–1 has $N + 1$ I/O devices that are buffered from the data bus by a port with the corresponding identification number. A *port* is the interfacing device that connects an I/O device to the computer system via the data bus. The hardware implementation of a port often consists of tristate buffers for input ports (port 2) and latches for output ports (ports 1 and *N*), where the direction of the arrow specifies whether the port is an input or output port. Port 0 is bidirectional and therefore requires a combination of both buffers and latches. The necessity for buffering input ports follows the same reasoning as that for buffering memory cells; that is, all outputs that are to be connected to the data bus must be buffered to prevent bus contention. The purpose of an output port is to latch the datum being written to the I/O device from the data bus. In many instances the port is an integral part of the I/O device. I/O port read and write operations are controlled with MPU control signals $\overline{RD}$ and $\overline{WR}$, just as they are for memory.

I/O device 0 of Figure 10–1 has DMA capability, which could mean that it is a mass storage device (recall from Chapter 8 that DMA means that the device has direct access to main memory). I/O device 0 can request from the MPU, via the *DMA REQ* pin, a direct access to memory (a DMA); that is, it is requesting to bypass the MPU and read or write a block of data from or to main memory directly. The reason the request for a DMA must be made to the MPU is that under normal conditions the MPU controls the three busses (address, data, and control busses); but during a DMA, I/O device 0 will control them. Thus, the MPU must relinquish its control. When I/O device 0 requests a DMA, and the MPU is ready to relinquish those busses, it notifies I/O device 0 by activating its *DMA ACK* signal, which enables the device to execute the DMA process. I/O device 0 will keep *DMA REQ* in the active logic state during the DMA process. When I/O device 0 drives the *DMA REQ* pin low (inactive) the MPU knows the DMA has been completed and then takes back control of the busses.

To better understand why the MPU must relinquish control of the busses, it is I/O device 0 that must generate the memory addresses where the block of data is to be stored or from which it is to be read, whichever is the case. Since the MPU normally generates addresses, for a DMA it must relinquish control and use of the address bus and I/O device 0 must take control. To implement this process with the hardware, the address pins of the MPU are internally tristated and are put in the Hi-*Z* state during a DMA. The *DMA ACK* signal is also used to turn on the DMA address buffers of Figure 10–1, which allows I/O device 0 to gain control of the address bus for as long as the *DMA ACK* signal is active.

As we have stated, not only must I/O device 0 control the address bus during a DMA, but it must also control the control bus, for I/O device 0 must generate and transmit, via the control bus, the appropriate read or write control signal. As with its address and data pins, during a DMA the MPU puts its control signal pins in the Hi-*Z* state and activates *DMA ACK*, which turns on the two read and write control tristate buffers of I/O device 0. This action enables the device to control the $\overline{RD}$ and $\overline{WR}$ lines of the control bus by becoming the source of those control signals. By controlling the control signals $\overline{RD}$ and $\overline{WR}$, it also controls the data bus.

When I/O device 0 is finished with the DMA, it drives its *DMA REQ* pin to the inactive logic state, which is detected by the MPU. In response, the MPU drives its *DMA ACK* pin to the inactive logic state, which puts all of I/O device 0's DMA-related buffers in the Hi-*Z* state, and the MPU again takes control of the busses.

An exchange of request and acknowledge between two devices is known as a *handshake*. In the case just described for a DMA, the handshake is the request by I/O device 0 for a DMA of the MPU by activation of its *DMA REQ* pin and the acknowledgment of that request by the MPU by activation of its *DMA ACK* pin. Specifically, these events are called a *DMA handshake*. Handshaking between two devices is necessary when the two devices share control of one or more busses or when one device needs the attention of the other.

Some I/O devices are able to make a request to the MPU that it interrupt the program it is currently executing and service the I/O by going to (vectoring) and executing the program that services its needs, such as reading a data byte that it must input to the MPU. When the MPU acknowledges the interrupt request, which often is as soon as it finishes the instruction execution that was underway at the time of the request (it will not stop in the middle of executing an instruction), it activates its *INT ACK* pin. The *INT ACK* signal is input to the requesting I/O device, acknowledging that its request for service by the MPU has been granted. For illustrative purposes, all I/O devices except 0 were given interrupt capability. The interrupt request and acknowledge exchange between MPU and I/O is an *interrupt handshake*.

From the computer architecture of Figure 10–1 and related explanations, you should have an understanding of how a computer system operates. It is to be hoped that you have also begun to relate these concepts to the memory designs seen in Chapter 9.

Review questions

1. What function does the MPU signal $IO/\overline{M}$ serve?
2. What do the page and I/O selectors have in common and how do they differ?
3. What is the function of an I/O port?
4. Explain how the MPU controls read and write operations.
5. What is meant by the term "handshaking"?
6. Why is handshaking essential?
7. Why does the MPU electronically isolate itself during a DMA?
8. Why does the DMA-requesting I/O device take control of the system busses during a DMA?

10–3 Microprocessor architecture

Our approach to developing and understanding MPU architecture is to divide it into sections according to function(s). We will partition the architecture into five sections:

1. Addressing section
2. Instruction section
3. Data storage section
4. Arithmetic, logic, and decision section
5. DMA and interrupt section

These five sections should be sufficient for describing the essence of all MPU architectures, yet we will find that it does not specifically match any. Fortunately, from an educational point of view, understanding the architecture and operation of any MPU provides insight into the architecture and operation of all of them.

Addressing section

There are three different registers that can be used to address memory. These are the program counter (PC), data counter (DC), and stack pointer (SP). The PC and SP were discussed in Chapter 9 (see Figure 9–1) but are briefly reviewed here. Each of these three registers addresses memory for a specific type of memory content. The contents of the PC are output on the address bus by the MPU when the MPU is fetching a program instruction from memory. The contents of the SP are output on the address bus by the MPU when the MPU is accessing the stack for the purpose of either reading or writing the datum from or to it as the result of a *POP, RET*, or *PUSH* instruction. Recall that when it acknowledges an interrupt request, the MPU automatically pushes the return address on the stack, which is popped off the stack using the *RET* instruction. You may wish to review Figures 9–1 and 9–2 as well as Example 9–3 if this material has been forgotten, that is, if your memory has become dynamic and you must therefore enter a refresh cycle. The DC is used as the addressing register when the MPU is accessing memory for the purpose of storing or retrieving data.

The only method for addressing I/O is via the I/O pointer (I/OP) register. Any time I/O is addressed, this register supplies the address of the I/O port being accessed.

Figure 10–2 is an illustration of the address section. The address latch, which has buffered outputs, serves as a buffer between the internal

Figure 10–2
Address section of an MPU architecture.

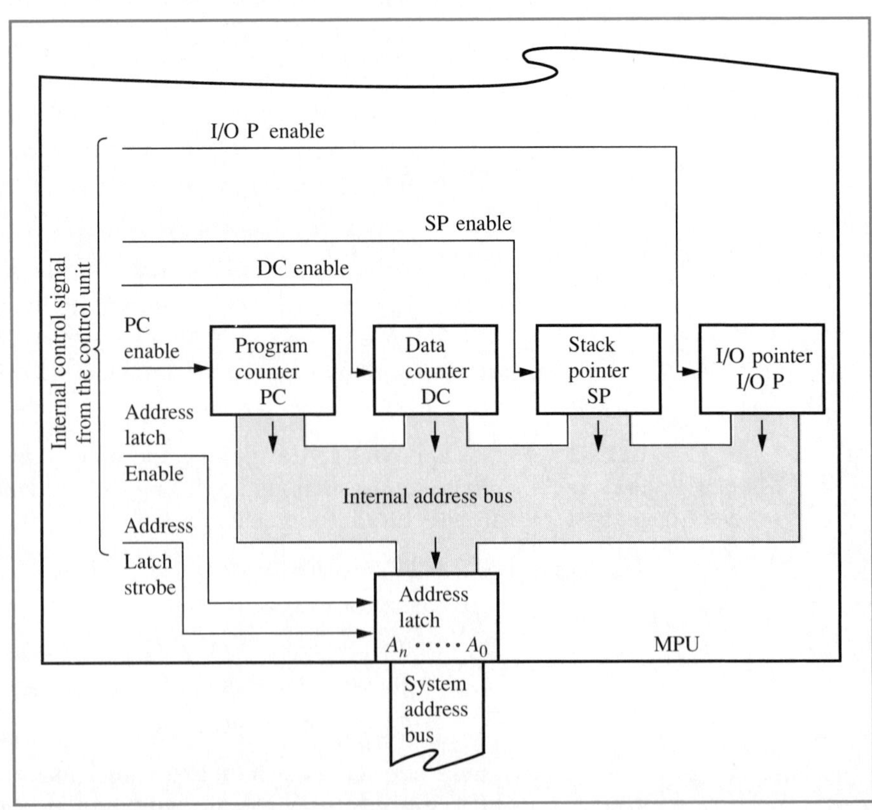

address bus of the MPU and the external system address bus. The output pins of the address latch are A_0 through A_n, which are represented in Figure 10–1. The address latch also serves as a latch in which the desired address can be stored by the MPU via its control unit. That is, the outputs of the PC, DC, SP, and I/OP are tristated, and the MPU, via its internal control unit, can enable the output buffers of that register whose contents are to be latched by the address latch. When the appropriate enable line is driven active by the control unit, the output tristated buffers of that address register (PC, DC, SP, or I/OP) output its contents on the MPU's internal address bus. While that address is on the internal address bus, the control unit generates and transmits a strobe signal to the address latch so that it latches the address off the internal data bus. The control unit then activates the address latch enable line, which turns on the output buffers of the address latch, causing the content of the address latch to be loaded on the address bus.

As the control unit (CU) outputs the memory or I/O address on the address bus, it also drives its $IO/\overline{M}$ pin either high (for I/O) or low (for memory) to indicate whether the address being output is an I/O or memory address. As we saw in earlier chapters and in Figure 10–1, this signal is used to enable the appropriate selector.

Instruction section

When an instruction is fetched from memory it is stored in an MPU register so that the binary code representing the instruction can be decoded. Decoding the instruction's binary code tells the MPU which specific instruction of its instruction set it has fetched from memory. When it identifies the instruction to be executed, the MPU "knows" what steps to take in order to execute that instruction. As discussed in Chapter 9 and illustrated in Figure 9–3, to make a more efficient use of the data bus, some MPUs have an instruction queue within which instructions can be prefetched and stored while awaiting execution. Our model MPU will have an instruction queue.

Figure 10–3 illustrates the instruction section. The first observation to be made from Figure 10–3 is that the last function-block in the instruction fetch and execute stream is the Control Unit (CU). Also note that the CU controls the operation of all other function-blocks. In fact, the CU also controls the function-blocks of Figure 10–2. Indeed, the CU controls the operations of all function-blocks within the MPU. The CU may be likened to a special-purpose computer that is dedicated to the execution of instructions. The purpose of the instruction fetch and execute stream is to instruct the CU as to which instruction to execute. From previous studies we know that if the CU is truly a computer, it must have memory, I/O, and busses over which it can control and communicate with its memory and I/O (refer to Figures 8–9 and 10–1). From Figures 10–2 and 10–3 we can see that the CU does indeed have three busses and I/O, which are the various internal registers. The CU's memory is internal to the CU and is used to hold the program that contains the procedures and steps necessary to execute the instructions in the MPU's instruction set. This program is known as the *microprogram*. Microprograms should not be confused with *macroprograms*, which are stored in main memory and executed by the MPU. A microprogram is written by the microprocessor designer and stored in the CU's internal memory (MROM) at the time of manufacture. A microprogram is

Figure 10–3
The instruction section of an MPU architecture.

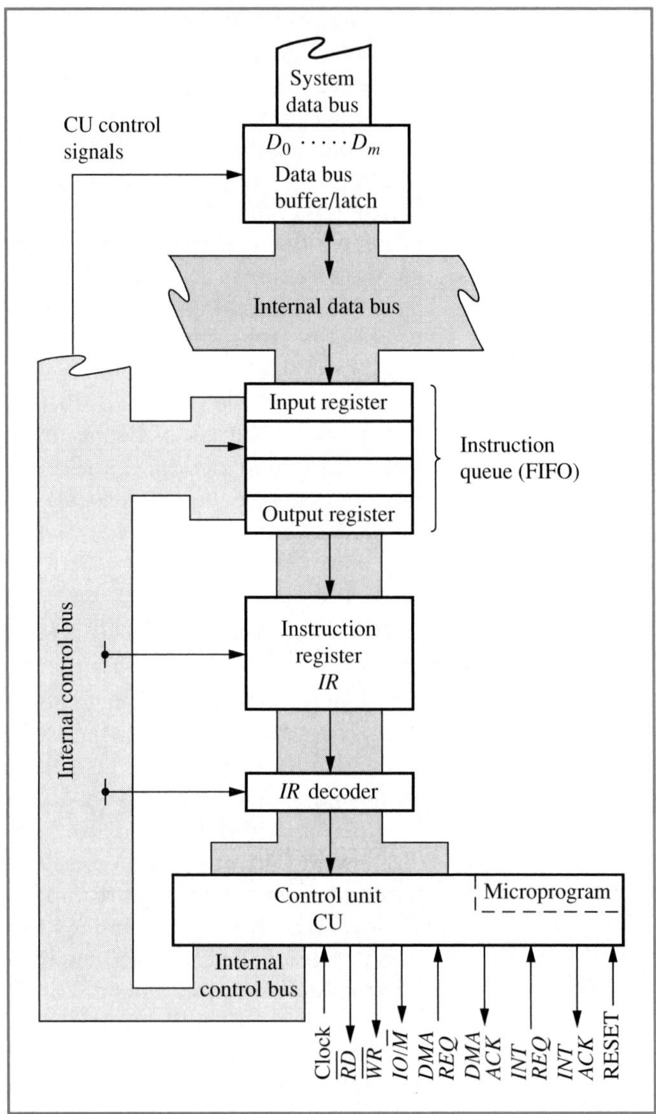

executed by the CU. For most MPUs the microprogram cannot be altered by the user.

In some discussions we will make reference to instructions from both the microprogram and a macroprogram. To distinguish between the two instructions, we will use *microinstruction* to mean an instruction from the microprogram and *instruction,* or *macroinstruction* for clarity, to identify one from the macroprogram.

Now that we understand the purpose of the instruction fetch and execute stream as well as the CU, let us start at the beginning and discuss each of its function-blocks. The data bus buffer/latch of Figure 10–3 is composed of tristate buffers and latches. It serves to buffer the internal data bus from the system data bus, as well as to latch the datum on the system data bus for MPU read operations and load the datum on the data bus for MPU write operations. The CU controls these read/write operations and therefore controls the data bus buffer/latch. The data bus buffer/latch has data pins D_0 through D_m, which are connected to the system's data bus, as is indicated in Figure 10–1. The CU drives the tristate buffers in the Hi-*Z* state when the MPU is not using the data bus, such as during a DMA.

When an instruction is being fetched from memory, the CU begins by instructing the MPU to output the memory address on the address bus via the address latch of Figure 10–2 (refer to Figure 9–15 for MPU timing). The CU then drives the read control pin $\overline{RD}$ active, which (we recall from Chapters 8 and 9) causes the addressed memory location to load its content (the instruction) onto the system data bus. While the instruction is on the data bus, the CU strobes the data bus buffer/latch of Figure 10–3, causing it to latch the instruction off the system data bus. The CU loads the instruction on the internal data bus and then strobes the instruction queue (a FIFO memory), which causes it to latch the instruction off the internal data bus. This instruction is shifted in the queue to the first empty location. This fetching of instructions continues until the queue is filled. While filling the queue, the CU may also be executing instructions. From the queue the CU loads the instruction to be executed into the instruction register, where it is decoded and identified for CU execution.

As pointed out in Chapter 9, the presence of a queue allows more efficient use of the data bus and faster instruction fetching and execution. The use of a queue is known as *pipelining*.

As the CU completes the execution of an instruction, it fetches the next instruction from the queue. It does so by shifting the contents of the queue and strobing the *instruction register* (IR), causing the IR to latch the instruction as it is shifted out of the queue's output register. The instruction now resides in the IR and can be decoded by the *IR decoder*. Decoding results in activating outputs at the IR decoder, which are inputs to the CU. These active CU inputs impact on the CU in two ways. They cause the CU to vector to that portion of the microprogram that pertains to the execution of the instruction in the IR, and they enable the appropriate CU circuitry required for the execution of that instruction. When the CU completes execution of the last microinstruction of the microprogram, pertaining to the execution of the instruction in the IR, it fetches the next instruction to be executed from the queue, loads it into the IR, and repeats the instruction execution cycle. This fetch and execute cycle of instruction continues until the macroprogram has been executed.

There are other ways for the CU to be vectored to a specified portion of the microprogram than by decoding an instruction in the IR. Activating CU external inputs *DMA REQ, INT REQ,* and *RESET* also causes this vectoring to occur. For instance, if input *DMA REQ* is driven active, the CU is vectored to the portion of the microprogram that contains the microinstructions for executing a DMA. Similar events occur if either input *INT REQ* or *RESET* is driven active.

Timing for CU operations is provided by the clock. The clock signal is a repetitive series of pulses of fixed rate and duty cycle. It provides not only CU timing but also system read and write timing, except during a DMA, via CU control signals $\overline{RD}$ and $\overline{WR}$.

Because many MPUs do not have an instruction queue, let us consider the instruction section without one. Without the instruction queue the CU must fetch the instructions from main memory. The CU proceeds as it did with the queue; that is, it increments the contents of the PC (the address of the next instruction) and loads its contents on the address bus. The CU then activates the read control signal $\overline{RD}$, which enables the output pin of the memory device being addressed. The addressed memory location outputs its content (the instruction) on the data bus, from which the CU causes the data bus buffer/latch to latch it. The CU then loads the instruction from the data bus buffer/latch to the IR via the internal data bus (recall that the queue is absent). Once the macroinstruction is in the IR, it is decoded and executed via microinstructions, just as if an instruction queue were present. After execution of the instruction is completed by the CU, the CU increments the PC and fetches and executes the next instruction.

It is interesting to speculate how an MPU designer might implement an IR decoder. He or she might design the IR decoder from logic gates, as with any combinational logic design. Another approach is to use a ROM that functions as a look-up table or encoder. The input to the IR decoder (ROM), which is the binary code of the macroinstruction, serves as an address. The content of that addressed location would have been programmed with the binary bit size and pattern required to enable the CU logic circuitry to execute the instruction and vector the CU to the appropriate portion of the microprogram, as discussed previously. If an MPUs instructions are 8 bits, there could be as many as 256 (2^8) instructions in its instruction set if there is an instruction for each possible binary code. Some MPU instruction sets use almost every binary code possible to create an instruction set, whereas others do not. This means that MPUs that use all or most binary codes have a relatively large instruction set compared to those that have the same bit size IR but do not have an instruction for many of the binary codes. MPUs intended to function in a wide range of applications need an instruction set that is general and large, and hence utilize all or most of the binary patterns. An MPU intended for a limited variety of applications would not need as many instructions and therefore would use a smaller number of the binary codes for its instruction set. An example of an MPU with a limited instruction set is a microcontroller.

Data storage section

The data section of an MPU was first discussed in Chapter 9 and is illustrated in a portion of Figure 9–1. Figure 10–4 is an illustration of the data storage section. Registers A, B, C, and D constitute the internal data storage for the MPU, and hence these registers function as a scratch pad memory. A datum is written into or read from one of these registers via the internal data bus and the read and write signals originating from the CU. The CU read and write signals of Figure 10–4 are not to be confused with the external read and write control signals $\overline{RD}$ and $\overline{WR}$. These are for internal use only.

As an example of how the data section functions, suppose that in the process of executing an instruction, the MPU is to read a data byte from memory and store it in register A. The CU will have addressed the memory location and loaded its content on the data bus by driving the read control signal $\overline{RD}$ active. On the positive edge of $\overline{RD}$ the CU latches that data off the data bus and into the data bus buffer/latch, as previously described. The CU then loads that data byte onto the internal data bus, from which register A can latch it by means of a strobe signal from the CU to the read input of register A.

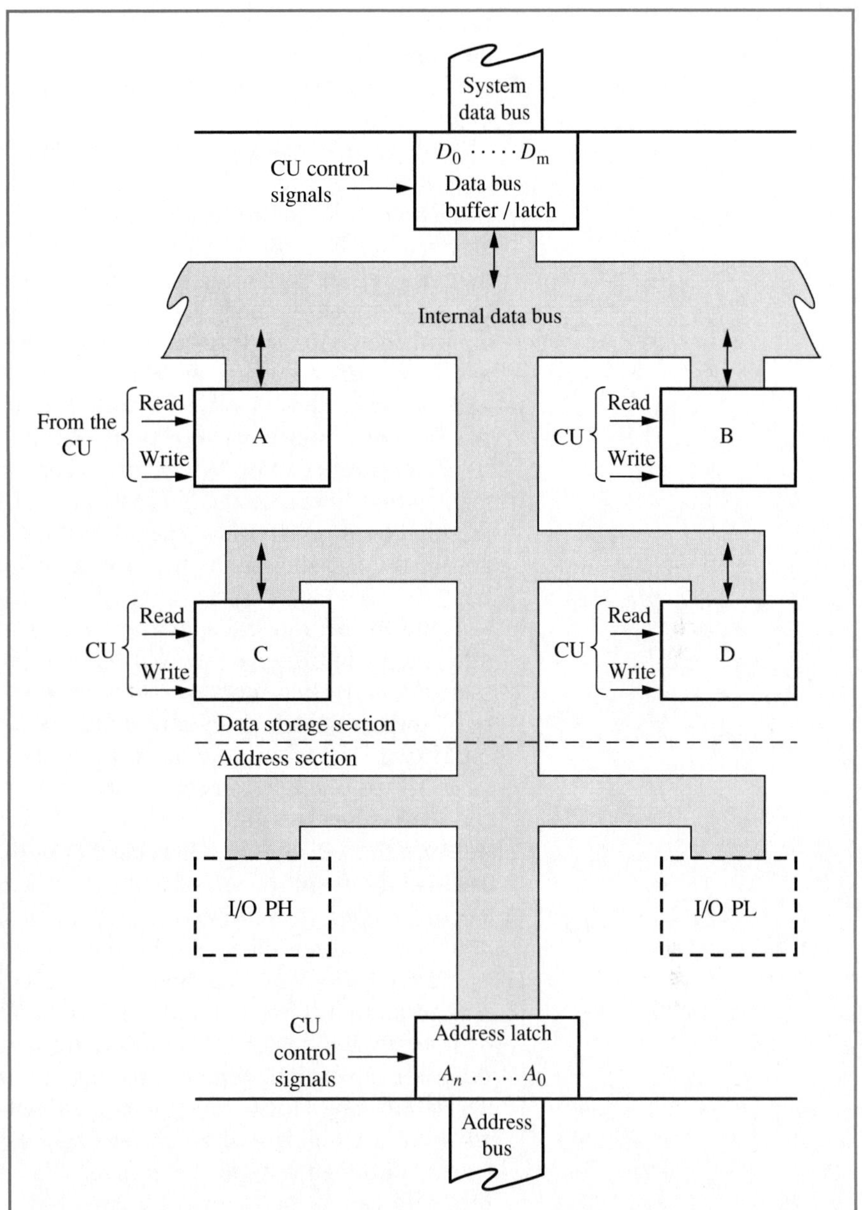

Figure 10–4
The data section of an MPU architecture.

For the MPU to write the content of one of its scratch pad registers to memory, the CU drives the write input of that register active, which loads the content of that register onto the internal data bus by turning on its output buffers. Recall that all outputs connected to a bus must be buffered to prevent bus contention. While the CU is holding that datum on the internal data bus (by keeping the write input of the register active), it also activates the buffers of the data bus buffer/latch, which electronically connects the internal data bus to the system data bus and thereby loads the datum on the internal data bus onto the system data bus. While these actions are occurring internally, the CU has also output the address of the memory location on the address bus and t_{AC} seconds later (see Figure 9–15) activates its external write control signal $\overline{WR}$. With the address of the memory location on the address bus and the datum to be stored in memory on the system data bus, the CU deactivates $\overline{WR}$, creating the positive edge needed to strobe the datum on the system data bus into the addressed memory location.

If we refer back to our discussion of the addressing section and Figure 10–2, we recall that the data counter (DC) is used by the CU to furnish memory addresses for data storage or retrieval. The DC in this architecture is not a specific register but rather the pairing of scratch pad memory registers (pairing is required, since memory addresses are 16 bits but scratch pad memory registers are 8 bits). Under macroprogram control the CU is instructed to create a DC by pairing either registers A and B or registers C and D. Registers B and D form the low byte of an address and registers A and C form the high byte. When the address of the DC (a register pair) is to be output on the address bus, the CU loads the content of the low-byte register pair on the internal data bus and latches that byte into the address latch. The CU then loads the content of the high-byte register into the address latch via the internal data bus. Of course the CU then loads the content of the address latch on the address bus.

From the above discussion we can state that in order to make maximum use of registers A, B, C, and D, the MPU has been designed so that they have two functions—to serve as scratch pad memory registers, or as the low and high byte of a DC. Which function they serve depends on instructions within the macroprogram. A register pair could be acting as a DC for one macroinstruction and as a scratch pad memory register for another instruction.

So far we have not discussed how I/O addresses are formed. They are also the result of registers much like the scratch pad register pairs. The difference is that these registers are used exclusively to store I/O addresses and not data; that is, they are not programmable by the user (there are no instructions that can be used to program their content). To remind us that these I/O address function-blocks (I/OPH and I/OPL, for high and low bytes of the address) are part of the architecture but are not programmable, their function-blocks are phantomed in, using dotted lines. These function-blocks have been included in the address section of Figure 10–4 but are configured much the same as scratch pad registers to remind us that although they function much like a DC, they differ from a DC in that data cannot be stored in them.

Arithmetic, logic, and decision section

Figure 10–5 is an illustration of the arithemetic, logic, and decision sections of an MPU architecture. The ALU is the function-block that performs all arithmetic (add, substract, and so forth) and logic (*AND*, *OR*, and so forth) operations on data. The data to be operated on is input to the ALU via the two registers that are identified as *ACC*1 and *ACC*2, ACC being an abbreviation for accumulator. These registers are called accumulators because they collect, or accumulate, ALU results.

The ALU flags are flip-flops whose logic state depends on the result of an ALU operation. There can be any number of ALU flags, but we will imagine that there are four for our generic MPU. These four flags are the sign flag (*S*-flag), zero-flag (*Z*-flag), carry-flag (*C*-flag), and the parity flag (*P*-flag). If an ALU-related instruction is executed whose result could be either positive or negative, such as when a subtraction is performed

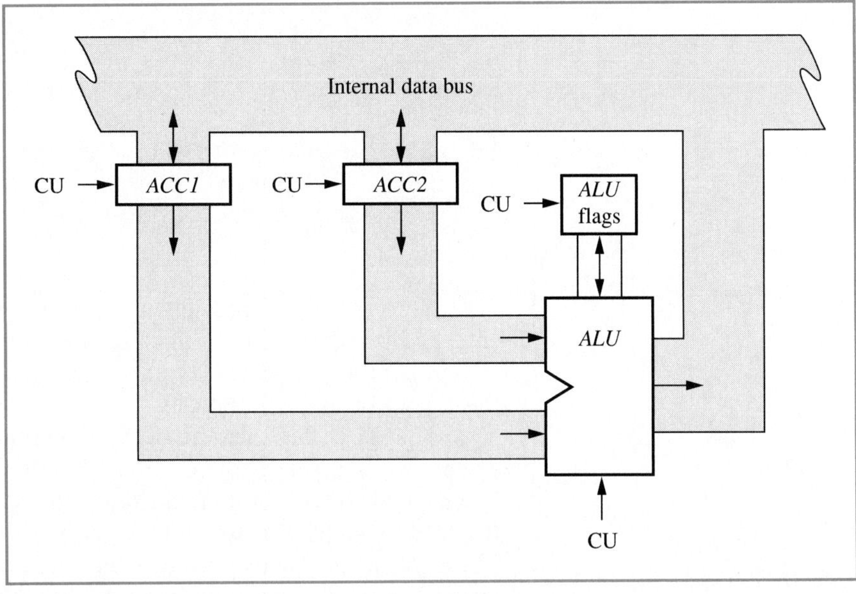

Figure 10–5 Arithmetic, logic, and decision section of an MPU architecture.

between the two accumulators, the logic state of the *S*-flag is determined accordingly. If the result of the ALU operation is negative, the *S*-flag is set (1); if the result is positive, the *S*-flag is reset (0). The logic state of the *S*-flag will be the same as the logic state of the MSB of the ALU result, which corresponds to the mechanics of complement arithmetic. The logic state of the *Z*-flag is set if the result of the ALU operation is a zero; otherwise it is reset. The *C*-flag is set if a carry or borrow resulted from the MSB of the ALU operation; otherwise it is reset. The *P*-flag is set if the ALU result is even parity; otherwise it is reset.

EXAMPLE 10–1

For the indicated ALU operations, determine the logic states of the *S*-, *Z*-, and *C*-flags. All data is 4 bits in length.

(a) 1000 + 0001
(b) 1000 + 1001
(c) 0100 − 1000
(d) 1101 *AND* 0010
(e) 1101 *AND* 0001
(f) 0100 *OR* 0111

SOLUTION

(a)
```
  1000
  0001
  ----
  1001
```
S-flag is reset
Z-flag is reset
C-flag is reset (no carry)

(b)
```
   1000
   1001
   ----
 1 0001
```
S-flag is reset
Z-flag is reset
C-flag is set

(c)
```
    0100
   −1000
   -----
 1  1100
```
S-flag is set
Z-flag is reset
C-flag is set

(d)
```
     1101
 AND 0010
     ----
     0000
```
S-flag is reset
Z-flag is set
C-flag is not affected for an *AND* operation

(e) 1101
AND 0001
0001

S-flag is reset
Z-flag is reset
C-flag is not affected

(f) 0100
OR 0111
0111

S-flag is reset
Z-flag is reset
C-flag is not affected

Note that the *C*-flag has no meaning in logic operations (*AND, OR*, etc.) and therefore is not affected. Sometimes flags that are affected by an ALU operation seem to have no logical reason for being affected, such as the *S*-flag for logic operations.

The ALU flags are the decision-making mechanisms of an MPU. There are MPU macroinstructions that enable the logic state of a flag to be checked, based on its logic state, the programmer can choose whether the MPU should execute an instruction or not—hence a decision is made. These are known as *conditional instructions*. Usually conditional instructions are conditional branch instructions; that is, if the conditional branch instruction is executed, the MPU branches from one part of the program to another part, allowing selective parts of a program to be executed.

EXAMPLE 10–2

Suppose that an instruction from an MPU's instruction set is a conditional jump (*J*) instruction, which is dependent on the logic state of the *Z*-flag. This instruction has two forms, one based on the *Z*-flag being set (*Z*) and the other based on its being reset (*NZ*). The instruction for execution of the instruction for the *Z*-flag being set is written as *JZ ADDR* (jump to *ADDR* if ALU result is zero); it is written as *JNZ ADDR* (jump to *ADDR* if ALU result is nonzero) if the branch instruction is to be executed if the *Z*-flag is reset. To determine whether it is to execute the instruction *JZ ADDR*, the MPU first checks the logic state of the *Z*-flag. If the *Z*-flag is set (a logic 1) the MPU jumps to the 16-bit memory address following *JZ*, which is symbolized by *ADDR*; otherwise it continues to the next instruction. The execution of *JNZ ADDR* is similar to that of the *JZ* instruction except that it jumps to memory location *ADDR* if the *Z*-flag is a logic 0.

If the contents of *ACC*1 and *ACC*2 are as given, what is the result when the MPU executes the three-instruction program shown?

```
AND ACC1, ACC2     (AND ACC1 and ACC2)
JNZ    0200H
JZ     0500H
END    (end of program)
```

(a) *ACC*1 = 00001000
*ACC*2 = 00001110
(b) *ACC*1 = 00011111
*ACC*2 = 10000000

SOLUTION The datum of registers $ACC1$ and $ACC2$ is ANDed.

(a) 00001000
AND <u>00001110</u>
00001000
Z-flag is reset

Therefore, when instruction JNZ is loaded into the IR as the next instruction, it is executed since $Z = 0$, and the MPU jumps to memory location 0200H.

(b) 00011111
AND <u>10000000</u>
00000000
Z-flag $= 1$

The JNZ instruction is not executed. The MPU fetches the next instruction, which is JZ. It is executed and the MPU branches to memory location 0500H.

An application of computer decision making via the ALU flags is the writing of letters that are personalized to the recipient (Example 10–3). Suppose a company wishes to mail out marketing promotional letters that are tailored to the recipient. The company would have a file, which is a collection of data stored on disk, on each individual to whom it intends to send a letter. A file can be very short and simple, or it can be very long and complex; in either case the file contains a marketing profile of the recipient. Many of these characteristics can be specified with the logic state of a single bit, such as gender. If the binary code that contains these marketing characteristics comes in, for example, a byte, there are eight marketing characteristics. To interrogate the data byte for a market characteristic, the programmer must write a program that isolates the sought-after information from other information in that byte of data. The technique often used is called *masking*. Masking uses a binary bit pattern, called the *mask*, which either ANDs or ORs, the data and the mask. When the mask is ANDed with the data byte being interrogated, all bits are masked out except the bit of interest. That is, all bits being masked are forced to a known logic state, zero. Hence, those bits to be masked out are zeros in the mask. We could mask those unwanted bits to a logic one by ORing the data byte with the mask-byte, so that the mask-byte would have logic 1's in all unwanted bit positions. Either logic operation affects the ALU flags, which can be detected and affect the execution (or not) of a conditional instruction.

EXAMPLE 10–3

Using an approach similar to the "program" of Example 10–2, write a program that vectors to memory location 0200H if the recipient is female and to address 0500H if the recipient is male. The market profile of this recipient was stored at main memory location 0800H when it was written in from mass storage via a DMA. This market profile data byte is structured so that the recipient's gender is indicated by bit position D_3, where the LSB position is identified as D_0. If the recipient is female, D_3 is a logic 1; otherwise D_3 is a logic

0. The instruction *MOV ACC*1,[*ADDR*] can be used to move the data byte (the marketing profile) stored at location *ADDR* into register *ACC*1, and instruction *MVI ACC*2 can be used to move the binary value of the mask-byte into *ACC*2.

SOLUTION

```
MOV ACC1,[0800H]
MVI ACC2, 08H

AND ACC1,ACC2
JNZ 0200H
JZ 0500H

END
```

The first instruction moves (*MOV*) the data at memory location 0800H into *ACC*1. If the recipient is female bit $D_3=1$, and bit $D_3=0$ if the recipient is male. The other seven bits are of no interest at this time. The instruction *MVI ACC*2,08H moves the immediate data (*MVI*), which is the mask-byte, into *ACC*2. Mask-byte 08H (00001000_2) will mask the *AND*ed result bits to a logic 0 except for D_3, which is the bit that indicates gender. After *ACC*1 and *ACC*2 are *AND*-ed, and if the recipient is female, the *Z*-flag is reset to 0 (nonzero result), and if the recipient is male the *Z*-flag is set to a logic 1. Hence if the recipient is female ($Z=0$) the MPU vectors to location 0200H, and if male ($Z=1$) the MPU vectors to 0500H. Memory location 0200H is the beginning of the program that personalizes the letters to women, and beginning at memory location 0500H personalizes those to men.

It is to be hoped after Examples 10–2 and 10–3 that the reader understands why the ALU flags are the function-block on which programming decisions are made.

DMA and interrupt section

As illustrated in Figure 10–3, the DMA and interrupt section has been included as a part of the CU. We could represent them with separate architectural function-blocks, as was done for the other sections, but we would find each of them to be no more than a single function-block with external handshaking pins (request and acknowledge) and internal output and input going to the CU to provide DMA and interrupt control.

The purpose of the DMA and interrupt handshaking pins was thoroughly discussed in Section 10–2, so we will only briefly review them here. The *DMA REQ* input pin is the request pin on which an I/O device with DMA capability can request the MPU to release control and use of the system data bus so that it may take control and use the data bus for a DMA. When the MPU is ready to release the data bus, it notifies the I/O device by activating its *DMA ACK* output pin, which signals the I/O device to take control. The I/O device holds the *DMA REQ* pin active until the DMA is completed. The MPU monitors the logic state of that pin, and during the time that the I/O device holds it active, the MPU holds the *DMA ACK* pin active and puts all of its bus-related pins in the Hi-*Z* state, which electronically isolates the MPU from the busses, thereby allowing the I/O device to control the system busses. When the

I/O device has completed the DMA, it drives the *DMA REQ* pin inactive, which is sensed by the MPU. The MPU then drives its *DMA ACK* pin inactive and takes its bus-related pins out of the Hi-*Z* state. At that point the MPU takes back control and use of the data bus for the normal operation of fetching and executing instructions.

Interrupt handshaking involves an I/O device requesting an interrupt of the MPU via the *INT REQ* input pin. When an interrupt request has been made of the MPU and it completes the execution of the current instruction, it notifies the I/O device that it is ready to service its request by activating its *INT ACK* pin and vectoring to a prespecified memory location, which contains the first instruction of the program subroutine (service program) that services the device. Prior to vectoring to this subroutine, the MPU stores the return address on the stack. The first instructions of the subroutine are *PUSH* instructions, which save the contents of scratch pad memory by pushing (storing) their contents on the stack. At the end of the subroutine *POP* instructions restore the original scratch pad data pushed on the stack by writing it to the scratch pad registers. The last instruction of the subroutine is return (*RET*), which pops the return address off the stack and writes it into the PC, enabling the MPU to return to that address.

The partial MPU architecture of Figure 10–3 shows just one interrupt request pin; however, some MPUs have multiple interrupt request pins. If there is just one interrupt request pin serving more than a single I/O device, there must be external logic to establish interrupt request identities and priorities. This I/O identification and interrupt priority logic is not shown in Figure 10–1 and is beyond the scope of this book.

Figure 10–6 is a composite of Figures 10–2 through 10–5. For convenience it shows two CU control busses when in fact there is just one. Also, the CU control bus signals going to registers B, D, I/OPL, and SP are not shown connected to the CU control bus although they actually are. The CU control bus signals symbolize the control that the CU has over all other function-blocks in carrying out the process of instruction fetch and execution. Although just one line is shown coming from the CU control bus and going to a function-block, there may be more than one.

ALU results are output on the internal data bus, as the ALU arrow direction of Figure 10–6 indicates. Once this resultant datum is on the internal data bus, the CU can cause any programmable register (there is an instruction that allows a register to be written to) to latch that datum off the data bus. Registers *ACC*1, *ADD*2, A, B, C, and D are programmable. Under this condition, the resultant ALU output can be stored under program control in any scratch pad register or in register *ACC*2 or *ACC*1. However, there are MPUs that do not allow a choice of location of data storage. These MPUs automatically store the results in *ACC*1. Notice that for this type of MPU the original datum in *ACC*1 is destroyed (it is written over by the ALU results). For this type of MPU, if the datum in *ACC*1 is not to be lost as the result of an ALU operation, it must be stored in scratch pad memory before execution of the ALU-related instruction.

Rather than having a single clock input to the CU, as illustrated in Figure 10–3, an MPU would probably have two input pins for connecting

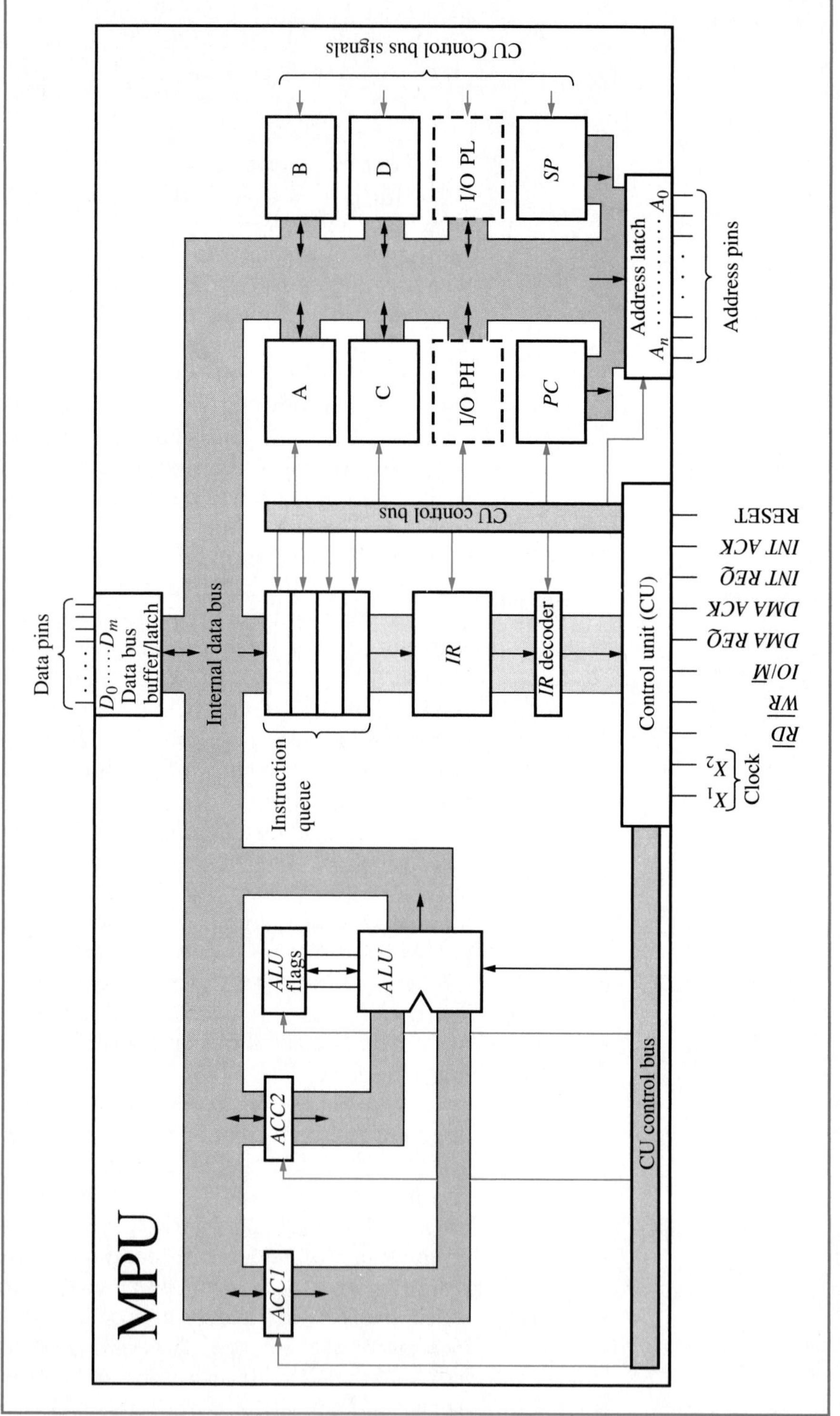

Figure 10–6 An MPU architecture.

a crystal (X_1 and X_2) as shown in Figure 10–6. The crystal is the source of MPU clock pulses, but the output signal of the crystal must be wave-shaped into pulses by the internal circuitry known as the clock (not shown) before it can be used for MPU timing. Some MPUs do not have an on-board clock. If this is the case, there must be external circuitry to support this function. As one can imagine, the amount of MPU support hardware can be reduced by selecting the proper MPU for an application.

Review questions

1. How many different registers can be used by the MPU architecture of Figure 10–6 to address memory, and when is each of these used?
2. How many registers can be used to address I/O?
3. What is the function of the address latch?
4. What is an instruction queue and what purpose does it serve?
5. Explain the purpose of the IR and IR decoder.
6. In general terms, define the CU and explain its function.
7. What is a microprogram, and how does it differ from a macroprogram?
8. Explain handshaking and why it is necessary.
9. What is the dual function of scratch pad memory registers A, B, C, and D?
10. How are the logic states of the ALU flags altered?
11. What purpose does the ALU serve?
12. Under what condition would the contents of *ACC*1 be destroyed as a result of an ALU operation?

10–4 MPU timing

The basic external timing of an MPU was introduced in Chapter 9 and is illustrated in Figure 10–7. We know from Chapter 9 that any time the MPU is to access either memory or I/O, the read or write cycle begins with the MPU outputting an address on the address bus of Figure 10–1, which is indicated by the timing diagram of Figure 10–7. During the time that the address is being output, the MPU also drives its $IO/\overline{M}$ pin to logic 1 if the address is an I/O address and to logic 0 if it is a memory address, enabling the appropriate selector in Figure 10–1. The MPU then activates the appropriate control bus pin—$\overline{RD}$ for a read and $\overline{WR}$ for a write. If an MPU read is being performed, the read control signal enables the addressed memory location or I/O port to load its datum on the system data bus, making that datum available to the MPU data pins via the system data bus of Figure 10–1. As indicated in Figure 10–7, the MPU latches that data off the data bus on the positive edge of $\overline{RD}$, which completes the read cycle. If the MPU had been performing a write operation, it would activate its write control pin $\overline{WR}$ at the same time it outputs the datum on the data bus. The write control pin is held active as long as the MPU holds the datum on the data bus. Just before the MPU removes that datum from the data bus (t_{CL} seconds), it drives $\overline{WR}$ inactive, from which a positive edge is formed. As indicated in Figure 10–1, this positive edge serves as a strobe to the addressed memory location or I/O port.

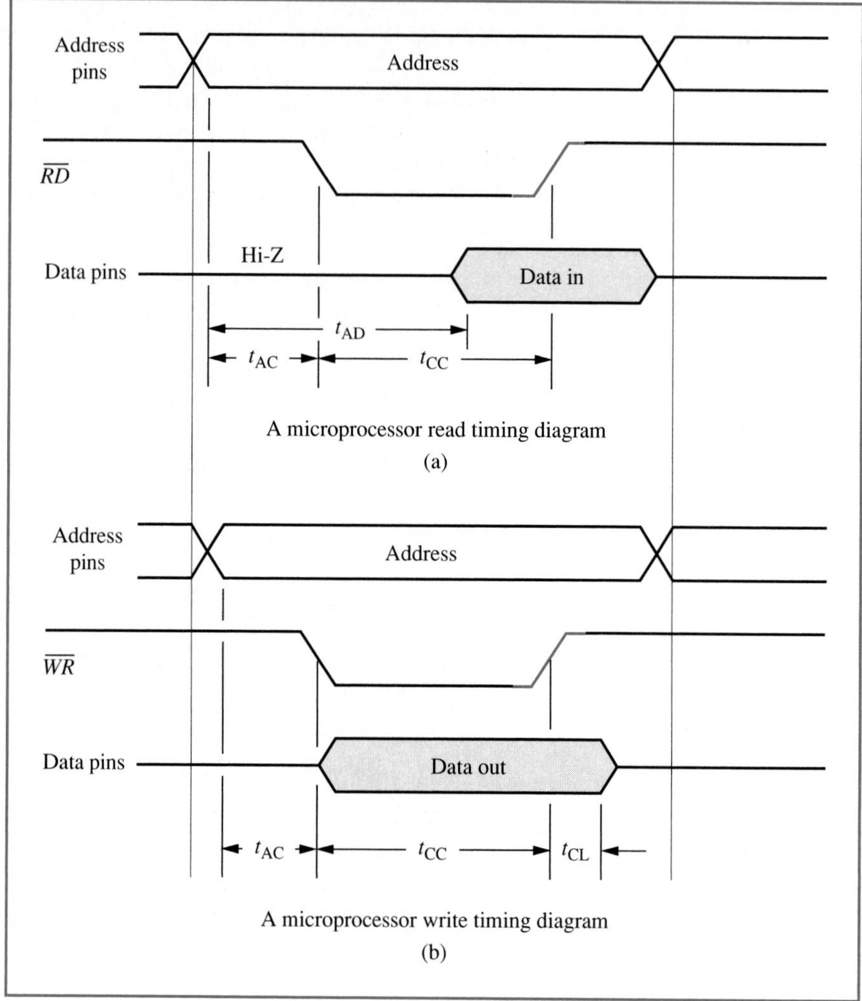

Figure 10–7
A representative timing diagram for an MPU.

The timing diagram for an interrupt or reset is basically that of a read. Recall that for either of these cases the MPU vectors to a specified memory location and fetches the first instruction of the subroutine program that is to be executed, which is just a normal memory read cycle. The only thing that sets an interrupt and reset apart from a normal memory read is how the vector address is obtained. For an interrupt or reset, the CU loads the vector address into the PC as a result of the interrupt request or reset. For a normal instruction fetch, the memory address is established by the CU incrementing the PC.

Review questions

1. What two events must occur for every MPU read and write operation?
2. What purpose does the MPU read control signal serve to the system?
3. What purpose does the MPU write control signal serve to the system?

4. Rationalize the timing of the $IO/\overline{M}$ pin.
5. When does the MPU actually read the datum from the data bus?
6. When does memory or I/O normally latch the datum being written by the MPU?

10–5 MPU instruction set

We will develop a hypothetical instruction set for the MPU architecture of Figure 10–6. The first thing to note is that this instruction set is valid for a specific architecture, the one of Figure 10–6. Just as the programmer must know the instruction set of an MPU, he or she must know the architecture of that MPU to properly apply those instructions.

Before we can develop an instruction set, we must either know or decide upon the capabilities of the architecture. That is, we must know which registers are to be programmable and between which registers data transfers can occur. We must also know the capabilities of the ALU. Since this hypothetical instruction set's only purpose is to serve as a vehicle for explaining concepts, let us keep it simple. Suppose that the ALU can perform just four arithmetic and logic functions, which are addition, subtraction, *AND*ing, and *OR*ing. We allow register data manipulation but restrict it to simply moving data between the programmable registers A, B, C, D, *ACC*1, *ACC*2, SP, and memory. The SP must be programmable, since it points to the stack, which is a defined portion of main memory and therefore may have different locations in different systems. To summarize the capabilities of the architecture thus far, it must be capable of ALU operations and the transfer of data between registers, memory, and the SP.

From our discussions on decision-making and vectoring, we know that this architecture must also be capable of branching from one place in memory to another. The branch instructions we have already looked at are *JZ* and *JNZ* (jump on zero and jump on nonzero, respectively), which are conditional branch instructions. We also require some unconditional branch instructions whose execution is independent of the ALU flags. Adding the ability of branching to this architecture expands its capabilities to include ALU operations, data transfer, and branching.

Last, we must allow this architecture to transfer data between register *ACC*1, and I/O. We have limited the exchange of I/O data to *ACC*1 and I/O to simplify coding of the instruction set. The reader might wonder why we do not just group I/O data transfers under the general heading of data transfers, which has already been listed. The reason has to do with the $IO/\overline{M}$ pin. Since this pin partitions the hardware into I/O and memory data transfers (recall the I/O and page selectors of Figure 10–1), there must also be a similar partitioning of instructions in the instruction set. When those memory-related data transfer instructions are executed the CU drives the $IO/\overline{M}$ pin low, and when an I/O data transfer instruction is executed the CU drives that pin high.

This completes the capabilities of this simple instruction set. Any instruction within the instruction set can be grouped into one of the following four categories: (1) ALU, (2) branching, (3) memory data transfer, or (4) I/O data transfer.

Our next task is to create a detailed listing of the instructions that comprise the instruction set, along with an explanation of the MPU operation to be performed. A sampling of that listing is given in Table 10–1. These samplings of instructions have been grouped according to the four MPU operations—ALU, branching, memory data transfer, and I/O data transfer.

In Table 10–1, the term *mnemonic* means "memory aid." The mnemonic used to represent an instruction is an abbreviation that symbolizes the MPU operation to be performed. Mnemonics are a shorthand method for writing instructions, but they are for the programmer's convenience, since the MPU understands only binary code, which it receives via the IR and IR decoder. Thus, for each instruction there must also be a corresponding binary code.

To design binary codes for instructions of the instruction set, a template for each grouping is devised to enable binary encoding of an instruction. To create these instruction templates, we must decipher from

Table 10–1
The instruction set for the MPU architecture of Figure 10–6

	Mnemonic	Explanation of instruction
ALU	*ADD R1,R2*	Add contents of registers *R*1 and *R*2 and deposit the sum in register *R*1, where *R*1 is *ACC*1 but *R*2 can be register *ACC*2, A, B, C, or D. Set ALU flags accordingly.
	*AND R*1,*R*2	*AND* registers *R*1 and *R*2. Deposit the results in *R*1. *R*1 and *R*2 as above. Set ALU flags.
Branch	*JMP ADDR*	Jump to memory location whose address is specified by *ADDR*.
	CALL ADDR	Jump to the memory location specified by *ADDR* and push the return address on the stack.
	JNZ ADDR	Jump to *ADDR* if $Z=0$; otherwise do not execute.
	JZ ADDR	Jump to *ADDR* if $Z=1$; otherwise do not execute.
	CNZ ADDR	Jump to *ADDR* if $Z=0$ and push return address on the stack; otherwise do not execute.
	CZ ADDR	Jump to *ADDR* if $Z=1$ and push return address on the stack.
Data transfer	*MOV R*1,*R*2	Move the data in register *R*2 into register *R*1. *R*1 and *R*2 can be registers *ACC*1, *ACC*2, A, B, C, or D.
	*MOV R*2,*M*	Move the data from memory location *M*, whose address is in register pair C and D, and load that data into register *R*1.
	MOV M,*R*2	Move the data from register *R*1 into memory location *M*. *R*2 can be either register *ACC*1, *ACC*2, A, B, C, or D. The address of location *M* is the contents of register pair C and D.
	MVI R,I	Move the immediate data *I* into register *R*. *R* can be *ACC*1, *ACC*2, A, B, C, or D.
I/O	*IN ADDR*	Read the port specified by *ADDR* and load the data read into *ACC*1
	OUT ADDR	Write the content of *ACC*1 to port *ADDR*.

Table 10–2
Binary code for encoding MPU operation types

Binary code	MPU operation
00	ALU
01	Branch
10	Data transfer
11	I/O

the created instruction set of Table 10–1 what must be encoded into an instruction. From Table 10–1 we can determine, among other things, that we must encode (1) the type of MPU operation, (2) the specific instruction within the group of MPU operations, and (3) the registers (if any) that are involved in the execution of the instruction.

Any MPU operation is one of our possible operations and therefore requires two binary bits to specify it. Table 10–2 shows the 2-bit binary codes chosen to encode the MPU operation that the instruction will expedite when executed. Table 10–3 is made up of binary codes for some specific operations for two of those MPU operation categories. Table 10–4 provides the binary codes for encoding registers *R*1 and *R*2 of the instruction set of Table 10–1.

Let us assume that the MPU of Figure 10–6 is an 8-bit microprocessor. Its IR is 8 bits and therefore requires MPU operation codes (*op-codes*) that consist of 8 bits. Figure 10–8 provides a sampling of templates for some instructions of the instruction set of Table 10–1.

Bit positions T_7 and T_6 of the op-code templates, which is the first byte of multiple-byte instructions, of Figure 10–8 are for encoding the type of MPU operation, via the codes of Table 10–2. The template for *ADD ACC*1, *R*2 of Figure 10–8a is for instruction *ADD R*1, *R*2 of Table 10–1. From the template we see that bits T_7 and T_6 are 00, corresponding to the ALU code of Table 10–2. Bits T_5 and T_4 of that template are for encoding the specific ALU operation to be performed, according to the code of Table 10–3. Since this is an *ADD* instruction, 00 is placed in those two bit positions. Had the instruction been a subtract (*SUB*), 01 would be placed in bit positions T_5 and T_4 according to Table 10–3. Bit position T_3 is not required to code *ADD ACC*1, *R*1 and therefore becomes a don't care. Of course, at coding time a 1 or 0 will appear in that bit position, and we will use 0's for all X's. The next 3 bits are for coding which register content will be added to *ACC*1. If the programmer wishes to add the content of register B to *ACC*1, the mnemonic will appear as

Table 10–3
Codes for encoding operations

Code	ALU operation
00	*ADD*
01	*SUB* (subtract)
10	*AND*
11	*OR*

Code	Branch
00	Conditional jump
01	Conditional call
10	Jump
11	Call

Table 10–4
Codes for encoding registers

Code	Register
000	Immediate data
001	*ACC*1
010	*ACC*2
011	A
100	B
101	C
110	D
111	*M* (C and D)

*ADD ACC*1, B and the op-code is 00000100 ($X=0$), since according to Table 10–4 the code for register B is 100. However, if the mnemonic is *ADD ACC*1, D, the binary code is 00000110.

The *AND* template of Figure 10–8b also shows bit positions T_7 and T_6 as 00, which is in agreement with the code of Table 10–2 for an

T_7	T_6	T_5	T_4	T_3	T_2	T_1	T_0
0	0	0	0	*X*	*R*2	*R*2	*R*2

ADD *ACC*1,*R*2
(a)

0	0	1	0	*X*	*R*2	*R*2	*R*2

AND *ACC*1,*R*2
(b)

0	1	1	0	*X*	*X*	*X*	*X*

JMP ADDR

ADDRL (low-byte)

ADDRH (high-byte)
(c)

1	0	*R*1	*R*1	*R*1	*R*2	*R*2	*R*2

MOV R1, R2
(d)

1	0	*R*	*R*	*R*	0	0	0

MVI R, I

Immediate data
(e)

Figure 10–8
Example encoding templates for the instruction set of Table 10–1.

ALU-type operation. Bits T_5 and T_4 are 10, which according to Table 10–3 identifies the ALU operation as an *AND*. Bits T_2 to T_0 are used to identify the register, just as for an *ADD* instruction.

The template of Figure 10–8c states that jump instructions require three bytes. The first is the op-code, which specifies that a jump is to be executed. The next two bytes are known as *operands*; these are the low byte (*ADDRL*) and high byte (*ADDRH*) of the address that the MPU is to jump to. The template states that bits T_7 and T_6 are 01, which agrees with Table 10–2 for a branch instruction. Bits T_5 and T_4 specify the branch operation, which is jump according to the code of Table 10–3. If the MPU were to jump to address 0506H, the mnemonic would be written as *JMP* 0506H and the 3-byte binary code would be 01100000 ($X=0$) for the op-code, 00000110 for *ADDRL*, and 00000101 for *ADDRH*.

The remaining two templates of Figure 10–8 are data transfers; hence T_7 and T_6 are 10 in both cases, as specified in Table 10–2. For the template of Figure 10–8d the remaining bits specify the registers using the codes of Table 10–4. For instance, the binary code of instruction *MOV A,B* is 10011100, since according to Table 10–4 the binary code is 011 for register A and 100 for register B.

The *MVI R,I* instruction of Figure 10–8e is a 2-byte instruction. The first byte is the op-code and specifies the instruction to be executed. The second byte is an operand and is the (immediate) data to be operated on, which in this case will be copied into the register specified in the op-code. For the mnemonic *MVI C*,08H, from its template the binary code is 10101000 for the op-code *MVI C* and 00001000 for the immediate data 08H. Note that the binary code 000 from Table 10–4 is used to identify this instruction as having immediate data, meaning that the data is encoded as part of the instruction.

EXAMPLE 10–4

From the architecture of Figure 10–6 we reasoned that there must be an instruction with which to program the SP. Can an *MVI SP,I* instruction be properly coded using the template of Figure 10–8e? Justify your answer. If the answer is no, what might be some possible solutions?

SOLUTION No, since there is no code for SP in Table 10–4.

Some solutions are the following: (1) Assign the 000 code of Table 10–4 to identify SP and devise another method to code immediate data. This would also mean increasing the *MVI SP,I* instruction to three bytes, since the SP is a 16-bit register. (2) The best solution is to create another subset of data transfer instructions in addition to those shown in Table 10–1. The mnemonic might appear as *LD SP,ID*, which when executed would load (*LD*) the immediate 16 bits of data (*ID*) into register SP. We know from Table 10–2 that bits T_7 and T_6 must be 10, but the remainder of the template is yet to be defined.

The purpose of this example is to demonstrate to the reader options that the designer of an MPU encounters.

EXAMPLE 10–5

For the following mnemonics encode their corresponding binary codes using the templates of Figure 10–8 and the instruction definitions of Table 10–1.

(a) *ADD ACC*1,D (b) *AND ACC*1,C
(c) *JMP* 12ABH (d) *MOV* C,D
(e) *MVI ACC*2,50H

SOLUTION We arbitrarily let $X=0$ for all templates.

(a) Using the template of Figure 10–8a

$$ADD\ ACC1,D=00000110$$

since D = 110 from Table 10–4.

(b) *AND ACC*1, C = 00100101

(c) *JMP* 12ABH
JMP = 01100000
ABH = 10101011
12H = 00010010

(d) *MOV* C,D = 10101110
since C = 101 and D = 110 from Table 10–4.

(e) *MVI ACC*2,50H
*MVI ACC*2 = 10010000
50H = 01010000

The reader should understand that with an 8-bit IR there are 256 binary combinations, and therefore there could be as many mnemonics. If the architecture of an MPU has the capability and versatility to require 256 instructions, there would of course be an instruction for each 8-bit binary combination. Most microprocessors do not use all binary combinations. Of those binary combinations used, almost no manufacturer would put don't cares in the templates, but rather would put in either a 1 or a 0.

Information concerning the instruction set of an MPU is supplied by its designer and/or manufacturer. Fortunately the user does not have to write programs using the binary code format of the instruction set but instead uses the more convenient mnemonic form. However, the mnemonic program must be translated into the binary code, since this is the only form the MPU "understands." The program can then be stored in main memory and executed by the MPU.

The process of translating a program from mnemonics to binary code is known as *assembly*. If a person does the assembly, the translation process is known as *hand assembly*. Computer programs that assemble mnemonics into binary code are called *assemblers*.

A program in mnemonic form is identified as a *source program* or a *source file*, since it is the source file of the assembler. The file created by the assembly process is known as the *object file*, or *machine code* file, since the object of translation (assembly) is to create a binary file that can be executed by an MPU (sometimes referred to as a machine).

EXAMPLE 10–6

Using the mnemonics of Table 10–1, write a program that will read a data byte from ports 3H and 4H and determine their sum. This sum is to be written to port 0006H and stored at memory location 2134H.

SOLUTION

MVI D,34H
MVI C,21H

IN 0003H
*MOV B,ACC*1

IN 0004H
*ADD ACC*1,*B*

OUT 0006H
*MOV M,ACC*1

END

The first two instructions initialize registers D and C with the address of memory location 2134H (see *M* of Table 10–4). The *IN* instruction reads the data byte of port 3 into *ACC*1. This data byte must be stored in scratch pad memory, since port 4 is also to be read. The *MOV B,ACC*1 saves the data byte read from port 3 into register B. Next the data byte of port 4 is read and stored in *ACC*1. The two data bytes can be added using *ADD ACC*1,*B*. The sum of this addition is automatically written into register *ACC*1 and written out to port 6 using the instruction *OUT* 0006H. Since the *OUT* instruction is a nondestructive instruction (it did not alter the content of *ACC*1), *MOV M,ACC*1 can be used to write the content of *ACC*1 to memory, where register pair D and C serves as the data counter.

As we have already stated, before an MPU can execute the program of Example 10–6 it must first be assembled and stored in main memory. To execute a program the MPU would repeat the fetch and execute cycle discussed earlier. An important part of the execute cycle is the decoding of the binary code in the IR, which tells the CU which instruction is to be executed and vectors the CU to the appropriate location within the microprogram to execute the instruction.

Review questions

1. What determines the capability of an instruction set?
2. What determines the maximum number of instructions within an instruction set?
3. What are some considerations in determining the instruction set templates?
4. When might the machine code of an instruction exceed a single byte?
5. What are mnemonics, and what purpose do they serve?
6. Can an MPU execute a mnemonic?
7. What is assembly?
8. What is an assembler?

10–6 Memory and I/O interfacing: the creation of a microcomputer

To *interface* two or more devices means to join them together with what is required to make the joining compatible. The "required hardware" is referred to as *glue* and includes chips, resistors, and capacitors. Often the glue is buffers and latches, such as the I/O ports of Figure 10–1 that interface I/O devices to the MPU via the data bus.

Once memory and I/O have been interfaced to an MPU, a microcomputer has been created, as suggested by the computer architecture of Figure 10–1. In Chapter 9 we designed memories that are compatible with the MPU of Figure 10–6.

Figure 10–9 shows the glue for interacing I/O devices *N* and *M* to the system data bus. I/O device *N* requires an input port, and therefore a buffer will serve as its interface. When the MPU is instructed to read data from I/O device *N*, the MPU accesses that device by the usual accessing

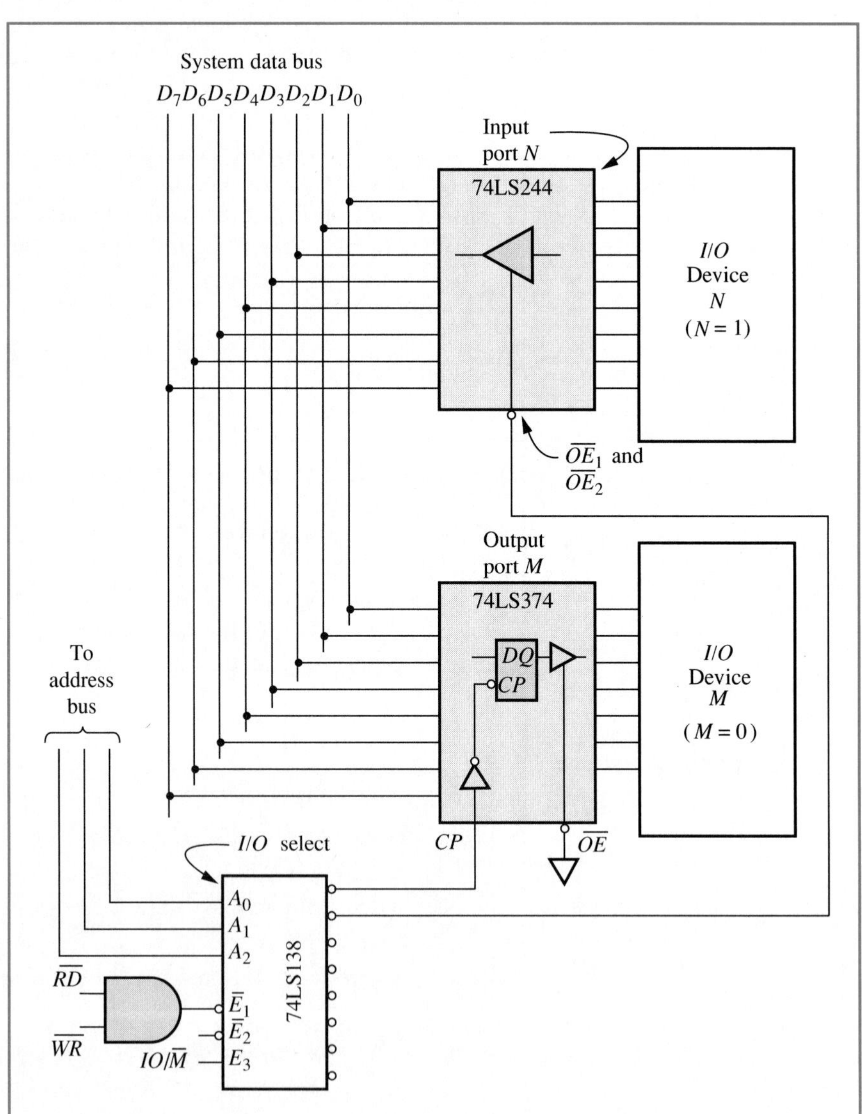

Figure 10–9 Input and output ports.

method by first outputting the address on the address bus and then activating the read control signal $\overline{RD}$. Since the buffer (74LS244) does not have a *CS* pin for address selection and an output enable for read timing, the selection and timing must be done via the output enable pins ($\overline{OE}$) of the 74LS244. To accomplish this the I/O selector is enabled by the MPU read and write control signals by *AND*ing them and having the output of that *AND* gate drive either of the active-low enable pins ($\overline{E}_1$ or $\overline{E}_2$). The other active-low enable pin can be used to prevent address foldback, if

Figure 10–10
A microprocessor-based microcomputer.

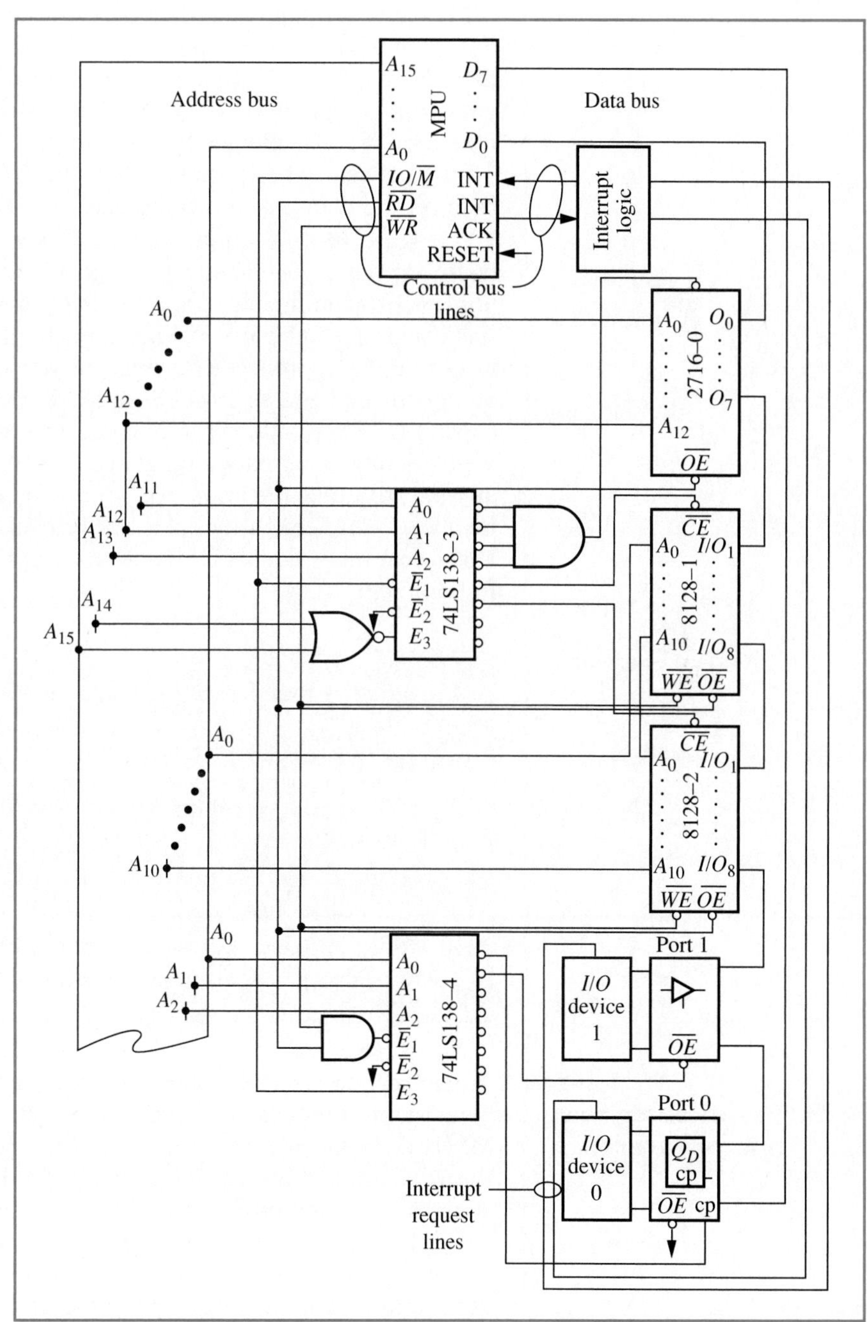

that is desired. For isolated I/O systems, $IO/\overline{M}$ is used to enable the I/O address decoder by being connected to E_3, since $IO/\overline{M} = 1$ if the address is an I/O address. Address pins A_0 to A_2 of the 74LS138 go to corresponding address bus lines so that it can decode the I/O port address on the address bus. If more than eight I/O ports are to be interfaced, the I/O selector must be expanded, as was done for the page selector of Figure 9–22.

The output port of Figure 10–9 is a latch. When the MPU writes a data byte to this port, the 74LS138 decodes the address and activates the addressed port (0 in this case) when the write enable is driven active. The MPU loads and holds the data byte on the data bus and during that period of time it also holds $\overline{WR}$ low. Just before removing the data byte from the data bus it drives $\overline{WR}$ inactive, which forms a positive-edge strobe as $\overline{WR}$ goes from a logic 0 to a logic 1. This positive edge is duplicated at the output of the I/O selector by disabling the 74LS138 when $\overline{WR} = 1$. This positive-edge strobe is input to the 74LS374 latch, which latches the data byte off the data bus and completes the write operation.

Figure 10–10 is an illustration of a microprocessor-based microcomputer. The memory design of Figure 9–20 was used to implement main memory, and the I/O ports of Figure 10–9 are used to implement the I/O ports of Figure 10–10. Take note of how $IO/\overline{M}$ is used to enable the appropriate decoder, as depicted in the computer architecture of Figure 10–1. Both I/O devices are given interrupt request capability. These interrupt requests are input to interrupt request logic, which is implemented with logic gates and establishes interrupt priority and interrupt identification. There are commercial chips available that can be used to implement the interrupt request logic. Neither I/O device has DMA capability, so MPU pins *DMA REQ* and *DMA ACK* do not appear in Figure 10–10.

Review questions

1. What is required to interface memory and I/O to an MPU?
2. What logic is usually used as interface glue for an input I/O device?
3. What logic is usually used as interface glue for an output I/O device?
4. What is the role of the MPU's $IO/\overline{M}$ pin relative to memory and I/O addressing?
5. Why must the I/O selector not only select the addressed port but also provide timing for read and write operations?

10–7 A survey of some real microprocessors

In this section we will make a cursory study of the architectures of three popular microprocessors—the 8085, 6800, and 8086. The 8085 and 6800 MPUs are 8-bit microprocessors and the 8086 is a 16-bit microprocessor. We will compare the architecture of each to that of Figure 10–6, looking at similarities and differences.

The architecture of the 8085 is illustrated in Figure 10–11a. Comparing it to the architecture of Figure 10–6, we see that the two are very similar. The number in parentheses in the function-block of the 8085

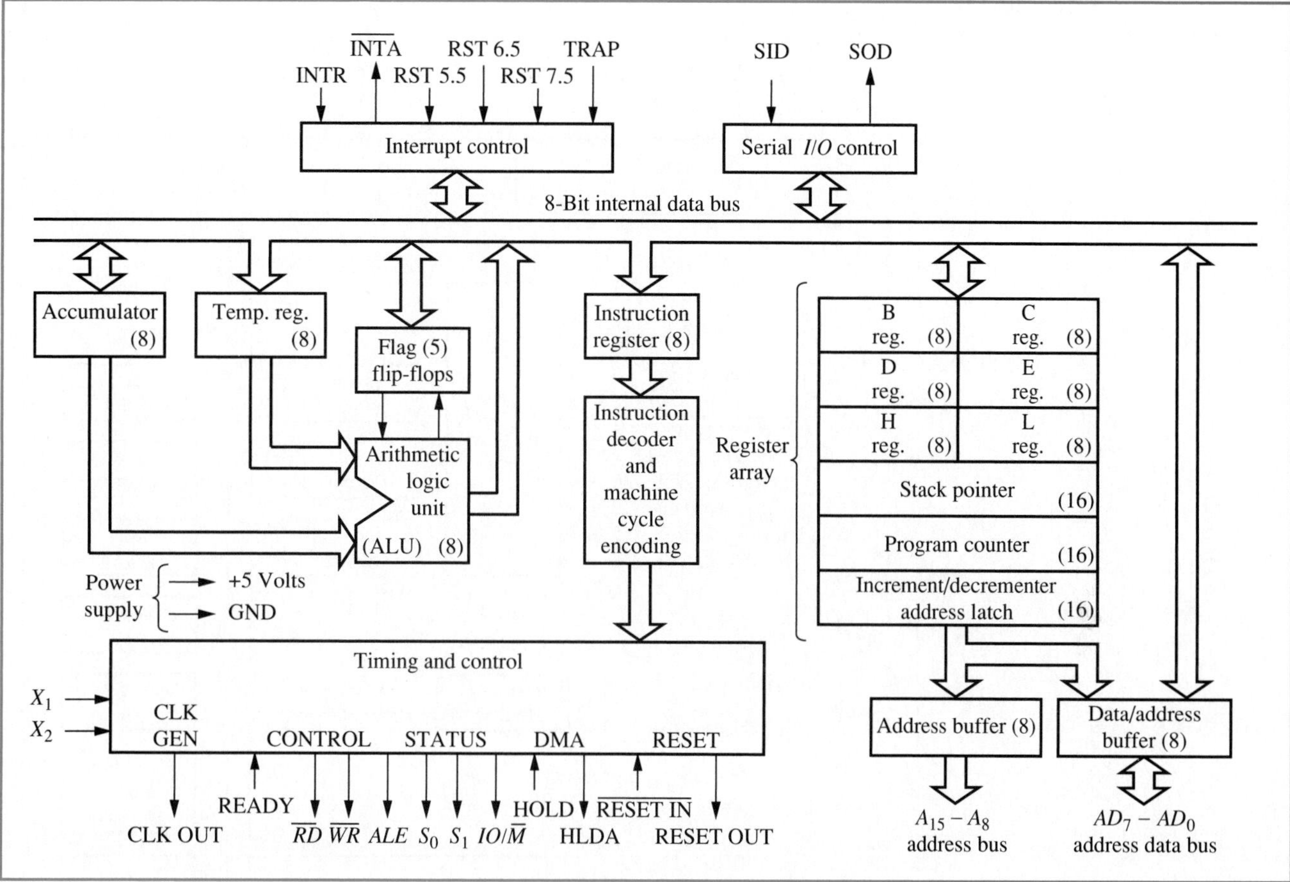

Figure 10–11a
The architectures of the 8085 microprocessor.

architecture indicates the bit size of that function-block. The internal data bus is 8 bits. Register *ACC*1 of Figure 10–6 is labeled as accumulator in the 8085 architecture, and *ACC*2 of Figure 10–6 is identified as being a temporary register in Figure 10–11a. The term *temporary register* means that the register is not programmable by the user but is used by the CU to temporarily hold the datum that is to be input to the ALU. Thus, there cannot be an instruction of the mnemonic form *ADD ACC*1,*ACC*2, since *ACC*2 is not programmable. 8085 add instructions appears as *ADD* B, *ADD* C, *ADD* D, etc.; it is not necessary to specify accumulator in the mnemonic since all ALU operations must have the contents of the accumulator as one of its inputs. The second input to the ALU (the temporary register) will be a copy of the register contents specified in the mnemonic (B, C, D, etc.). There are six scratch pad memory/data counter registers (B, C, D, E, H, and L). To conserve on the number of pins required, the low-byte address pins are multiplexed (time shared) with the data pins. These multiplexed pins are identified as AD_0 through AD_7. The 8085 architecture has multiple interrupt request inputs (*INTR, RST* 5.5, *RST* 6.5, *RST* 7.5, and *TRAP*). The interrupt acknowledge output is identified as $\overline{INTA}$. The DMA handshaking pins are *HOLD* for the request and *HLDA* for the acknowledge. The 8085 has two serial I/O

Figure 10–11b
The architecture of the 6800 microprocessor.

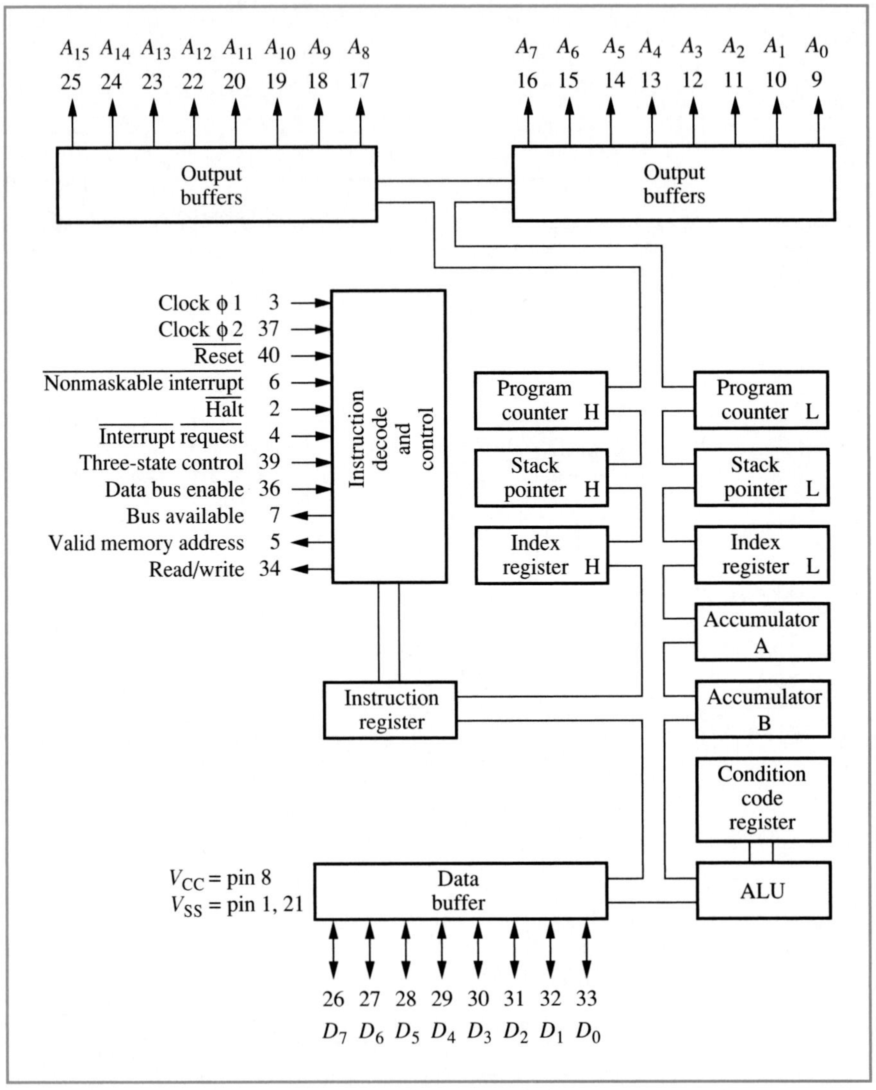

ports (a single-bit port), one of which is an input port (SID) and the other an output port (SOD). The architecture of Figure 10–6 does not have serial ports. The 8085 does not have an instruction queue. It has two I/O addressing registers, which are temporary registers, much like I/OPH and I/OPL in Figure 10–6; however, they are not shown in its architecture. The generic microprocessor architecture of Figure 10–6 was modeled after the 8085.

Figure 10–11b is an illustration of the 6800 microprocessor. It has a PC and SP. The counter (DC) is the index register. There is no $IO/\overline{M}$ pin to distinguish between memory and I/O addresses because the 6800 addresses I/O ports as if they were memory locations, which is known as memory-mapped I/O addressing (see Chapter 8). There are two programmable accumulators identified as A and B. We see that there are 8 data bus pins and 16 address pins. Note that the 6800 architecture lumps the IR decoder and CU into one function-block. There are no internal scratch pad registers or instruction queue. Rather than having

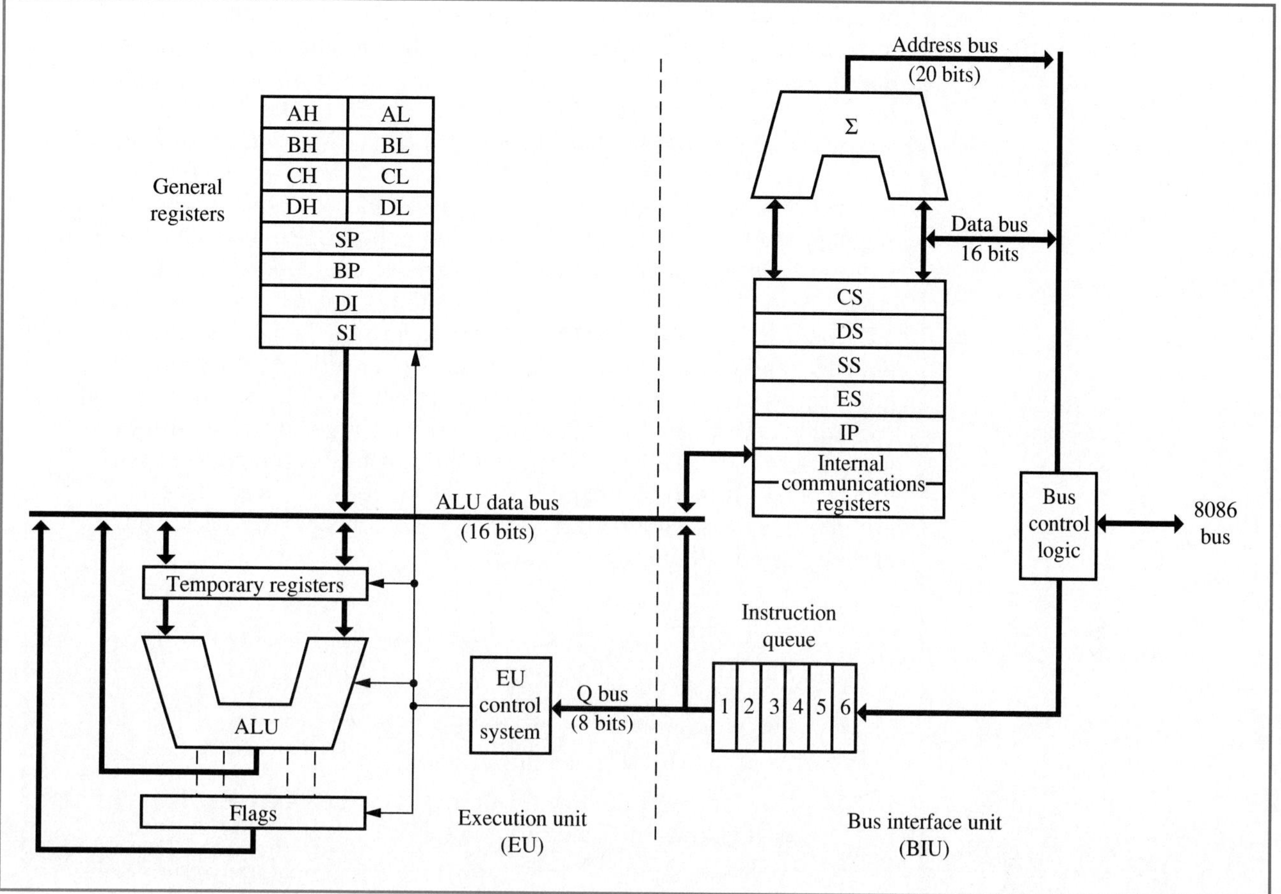

Figure 10–11c
The architecture of the 8086 microprocessor.

separate read and write control pins, the 6800 combines them in a single pin which is identified as read/write. When reading memory or I/O the 6800 drives that pin high, and when writing it drives the pin low. The pin identified as three-state control is the DMA request pin, and the bus-available pin is the DMA acknowledge pin. The data bus enable pin is used to put the 6800's data pins in the Hi-*Z* state during a DMA. The remainder of the pins have labels that are either self-explanatory or require more detailed explanation than is necessary here.

The 8086 is a 16-bit microprocessor whose architecture is illustrated in Figure 10–11c. Its architecture is very different from that of Figure 10–6, which is one of the reasons it was chosen for the comparison. However, despite the differences, the concepts of Figure 10–6 are valid and applicable to the 8086 or any other unknown architecture.

The 8086 architecture is divided into two parts, the bus interface unit (BIU) and execution unit (EU). These two units divide the responsibility of fetching and executing an instruction. The BIU fetches all instructions and data, which means it must access (address) memory or I/O and read or write the accessed memory location or I/O port. If an instruction is being fetched, the BIU writes it into the instruction queue (now you know the other reason the 8086 was chosen). The EU is responsible for the execution of instructions and the manipulation of data as a result of executing those instructions. The EU fetches those instructions to be executed from the instruction queue, as previously discussed for the instruction queue of Figure 10–6. Basically, addresses for accessing data memory locations are formed by the addition of a segment register's contents (DS, SS, or ES) and the contents of one of the EU general registers (SP, BP, SI, or DI). Addresses for fetching instructions from memory are formed by adding the contents of the code segment register (CS) and the instruction pointer (IP). The adder (ALU) shown in the BIU is for adding these register contents to form addresses while the ALU of the EU is for operating on data. Of course having an EU and a BIU allows the 8086 to implement pipelining. The IR, IR decoder, and CU are now shown in the EU in order to keep the architectural drawing simple and uncluttered. The EU has scratch pad registers A, B, C, and D, which are two 8-bit registers divided into a high (H) and low (L) byte. The remainder of the architecture is self-explanatory. It is to be hoped that this brief look at these three architectures has demonstrated that certain concepts are basic to the architecture of all microprocessors.

Review questions

1. What registers of the 8085 form its scratch pad memory?
2. The CU is labeled as what 8085 function-block?
3. What are the 8085 serial I/O ports?
4. How does the 8085 differ in interrupt capability from the MPU architecture of Figure 10–6?
5. What is meant by "the low-byte address pins are multiplexed with the data pins on the 8085"?
6. Does the 6800 have an internal scratch pad memory?
7. Does the 6800 have two programmable accumulators?
8. What purpose does the 6800 index register serve?
9. What are the EU and BIU of the 8086?
10. How can the 8086 make more efficient use of the data bus than the 8085 or 6800?
11. Why does the 8086 show an ALU symbol in both the EU and BIU?

10–8 TROUBLESHOOTING

Since microprocessors are on a single integrated circuit wafer, if any single MPU function fails, the entire MPU fails catastrophically and must be replaced.

To devise a test for an MPU we must recall that it is the primary source of all dynamic logic levels. That is, the system MPU generates the system's addresses, control signals, and data on the data bus when the MPU is performing a write operation. Measuring dynamic logic levels requires that a logic analyzer be used to monitor those logic levels. The test configuration is nearly the same as those dynamic tests discussed in Chapter 9. There are two differences: (1) The instruction being executed at the time the data samples were taken affects the proper timing diagram for an MPU. (2) The MPU has input pins that, when activated, can modify the operation of the MPU, such as an interrupt or reset. Hence, to investigate thoroughly the many complex faults that can arise requires a more in-depth knowlege of an MPU's instruction set and the function of some input pins than is within the scope of this book. Therefore, we discuss here only some of the more straightforward aspects of troubleshooting an MPU.

Our test configurations dynamically monitor (1) address bus nodes, (2) control signal bus nodes, (3) data bus nodes, and (4) MPU input nodes that alter the sequence of program execution. Using the MPU of Figure 10–6, we will statically (or dynamically) activate MPU inputs, DMA request, interrupt request, and reset and observe whether the proper MPU response occurs. In addition to the above, we should check for the obvious, such as proper power supply voltages, grounds, and whether an unused input has been left disconnected (this should never be done).

EXAMPLE 10–7

Suppose that the MPU of Figure 10–10 is to be tested for correctness for each memory location and I/O device. Devise the test.

SOLUTION To monitor each MPU node requires 29 logic analyzer channels—16 channels for the address bus, 3 for the control bus, 8 for the data bus, and 2 for the interrupt request and acknowledge. We must have the MPU execute a program that accesses each memory location and I/O device; that is, each ROM location and input I/O device is read and each RAM location is written into and then read, while output devices are written to. This test is programmed in a 2716 UV PROM, which is then placed in the IC socket for 2716-0.

Once the programmed 2716 is in the system, the MPU's reset pin is activated. Resetting this MPU causes it to vector to address 0000H, from which it fetches and executes the first instruction. Using 0000H as the trigger word, the logic analyzer samples and fills its FIFO memory with nodal data. If the FIFO stores 128 words (128×29 bits for 29 channels), then every 128 addresses a new trigger word must be updated and input to the logic analyzer. This updated trigger word is the first address of the next 128 samples to be displayed. Updating the trigger word and resetting the MPU continues until the MPU has accessed every memory and I/O address. The display of the logic analyzer is then checked for correctness.

Seldom would one want, or need, to go through the task of testing every memory location and I/O device via a logic analyzer, since there are self-diagnostic programs that can accomplish the task faster and better. Most often the diagnostician is able to narrow the diagnosis to an event or events by means of the symptoms of the failure, as was the case in troubleshooting the system of Chapter 9.

EXAMPLE 10–8

Suppose that locations within the 8128-1 of Figure 10–10 failed the test. By this we mean that a program executed by the MPU wrote a known byte of data into the 8128-1's locations, but the data read back was not what had been written. Devise a test to locate the fault.

SOLUTION We must verify that the $\overline{CE}$, $\overline{WE}$, and $\overline{OE}$ have the proper logic levels for the MPU read or write operation. From the $\overline{CE}$ of the 8128-1 of Figure 10–10 we find that the $\overline{O}_4$ output of the page selector 74LS138-3 must be low when 8128-1 is addressed. From the truth table of a 74LS138 we also know that the inputs A_2, A_1, and A_0 must have logic levels 1, 0, and 0, respectively, in order to activate output $\overline{O}_4$. In addition to the correct logic levels on the address pins of the 74LS138-3, the enable pins must have the proper logic levels.

As a result of the above analysis we will connect the channels of the logic analyzer to $\overline{CE}$, $\overline{WE}$, and $\overline{OE}$ of the 8128-1 and to A_0, A_1, A_2, $\overline{E}_1$, and E_3 of the 74LS138-3 ($\overline{E}_2$ is strapped to ground; therefore, it could be tested with a logic probe or VOM).

If the $\overline{\text{CE}}$ node of the 8128-1 is high when the 8128-1 is addressed, either the page select *AND* gate has failed or the active output of the 74LS138-3 is at fault. Checking the 74LS138-3 output for proper operation requires that its address and enable pins be checked for proper logic levels. Enable pin $\overline{E}_1$ is directly connected to $IO/\overline{M}$, and E_3 is driven by address lines A_{13} and A_{14}, so these pins can be easily checked for correct logic levels. If E_3 is a logic low and the address bus lines A_{14} and A_{15} have the correct logic level, the *NOR* gate has failed or enable pin E_3 of the 74LS138-3 has shorted internally to ground if the *NOR* gate is operating properly. By disconnecting the *NOR* gate from the 74LS138-3 we can determine which is the case and replace either the *NOR* gate or the 74LS138-3.

If the 74LS138-3 checks out, either $\overline{WE}$ or $\overline{OE}$ of the 8128-1 has failed. Their logic levels can also be verified via the display.

If the interrput logic of the MPU in Figure 10–10 is to be tested, either I/O device 0 or 1 must be forced to request an interrupt. With the logic analyzer monitoring the address bus and MPU inputs *INT* and *INT ACK*, the logic level on *INT* is used as the logic analyzer trigger source. When the interrupt request is made, the logic analyzer begins to sample and store the nodal data. This data can be displayed by the logic analyzer, where it is checked for correctness. If the interrupt request is working properly, the address on the address bus should be the vector address which vectors the MPU to the I/O service routine, and the logic level on interrupt acknowledge (*INT ACK*) should be high.

Although troubleshooting MPUs is involved and is highly dependent on the program being executed, the foregoing discussion should provide basic techniques and insight about methodology.

Review questions

1. Why can the MPU be termed the system's logic level generator?
2. What MPU output nodes can be monitored for verification of MPU operation?
3. How can the input nodes of an MPU be tested (excluding data pins)?
4. What is the essential difference between testing a memory/IO system and testing an MPU that is part of a total system (i.e., one that contains MPU, memory, and I/O?

10–9 SUMMARY

- The computer architecture of Figure 10–1 shows the MPU communicating with memory and I/O via the system data bus. The selection of the memory location or I/O device is made by the MPU over the address bus.

- The control bus is used by the MPU to transmit read and write control signals for controlling use of the data bus and to provide read and write operation timing. Control signal $IO/\overline{M}$ differentiates between I/O and memory addresses and is used to enable the appropriate selector (I/O or page).

- An I/O device, such as mass storage, may have the capability of requesting direct access to memory (DMA). When an I/O device has DMA capability, it must control all busses because it must supply the memory location address, control signals, and data.

- An I/O port interfaces an I/O device to the system data bus and is accessed by the MPU.

- The MPU architecture of Figure 10–6 has three registers with which to address memory—the PC for fetching instructions, the SP for accessing the stack, and the DC (a register pair) for accessing data.

- There is just one register pair for addressing I/O—the nonprogrammable register pair *I/OPH* and *I/OPL* of Figure 10–6.

- The instruction stream of this architecture is the instruction queue, IR, IR decoder, and CU.

- The CU is a special-purpose computer that is dedicated to the execution of instructions from the instruction set.

- The ALU performs the actual arithmetic and logic operations, and the logic states of the ALU flags are determined by the results of ALU operations. These flags are used as program decision makers by testing the logic levels of these flags.

- The instruction set of an MPU is determined by the capabilities of the MPU's architecture.

- Instructions have two forms—*mnemonics*, which are designed to be readable by the programmer, and *machine code*, which is readable by the MPU and is the binary code equivalent of a mnemonic instruction.

- The MPU can execute only machine code. To translate a program written in mnemonics into machine code, the programmer can hand-assemble the mnemonics or use an assembler.

- The 8085 has an architecture very similar to that of Figure 10–6, but without an instruction queue and with serial I/O ports.

- The 6800 differs in architecture from Figure 10–6 by not having scratch pad memory and an $IO/\overline{M}$ pin (or one that is similar in function).

- The 8086 is different in many ways from the MPU architecture in Figure 10–6, but the most obvious difference is the existence of the BIU and EU.

- Having an architecture which contains a BIU and an EU allows pipelining to be implemented.

- Troubleshooting an MPU is similar to troubleshooting memory or I/O. The major difference is its dependence on program execution and the MPU's ability to have its program sequence altered via instructions and external requests, such as interrupts and DMAs.

PROBLEMS

Section 10–2 Computer architecture

1. Explain the purpose of each bus in Figure 10–1.
2. When we refer to bus control signals, what is it that is being controlled?
3. What is the purpose of the two selectors in Figure 10–1, and how are they enabled?
4. What is the necessity of handshaking?
5. What are the events that take place for
   (a) DMA handshake
   (b) interrupt handshake
6. Explain why I/O device 0 of Figure 10–1 is the only I/O device directly connected to the address and control bus via buffers.

Section 10–3 Microprocessor architecture

7. What are the various methods that the MPU architecture of Figure 10–6 has to address
   (a) memory
   (b) I/O ports
8. What function-block controls the content that is latched by the address latch of Figure 10–6?
9. How are the contents of locations addressed by the PC and SP different?
10. How does the method for determining the next sequential address differ between the PC and SP? Explain the purpose of this difference.
11. When an MPU fetches an instruction, where does it store it and why?
12. What purpose do the IR and IR decoder serve?
13. Define the CU of the architecture of Figure 10–6.
14. What are the three essential function-blocks that a computer must have, and how does the CU meet these criteria?
15. What is a microprogram, and how does it differ from a macroprogram?
16. Why does the architecture of Figure 10–6 show control lines going from the CU to every function-block?
17. All ALU operations must be done with two data at a time. How does the architecture of Figure 10–6 indicate this?
18. What are the ALU flags, and what purpose do they serve?
19. Why can an MPU reset be categorized in the same way as an interrupt and what effect on the MPU do both events have in common?
20. What function does the $IO/\overline{M}$ pin serve?
21. The control signals of some microprocessor-based systems require that the control bus signals be made up of $\overline{I/O\ R}$, $\overline{I/O\ W}$, $\overline{MEM\ R}$, and $\overline{MEM\ W}$, which are separate I/O and memory read and write control signals. Using the control signals of Figure 10–6, design the necessary logic.
22. What is the function of the data bus buffer/latch of Figure 10–6?

Section 10–4 MPU timing

23. Suppose that the machine code for instruction *MVI A*,50H begins at memory location 102CH, followed by the instruction *ADD ACC*1,A. Using the timing diagram of Figure 10–7 and the architecture of Figures 10–1 and 10–6, explain in detail the events that are occurring on the various busses and within the MPU architecture to fetch and execute these two instructions. Imagine the instruction queue to be empty, just as if the MPU had vectored to location 101CH as the result of an interrupt.

24. When does the MPU latch the datum being read from the data bus?

25. How does the MPU signal an addressed memory location or I/O device that it has the datum being written on the data bus?

Section 10–5 MPU instruction set

26. Explain how the designer and/or manufacturer of a microprocessor determines what type of instructions will make up an instruction set.

27. What is a mnemonic and what is the significance of mnemonics?

28. What is a source file and an object file? How are object files different from machine code files?

29. What is an assembler and what does it do?

30. What is an instruction template?

31. How are the configurations of instruction templates arrived at by microprocessor designers?

32. An IR decoder can be a decoder or a look-up table. Explain each of these decoding techniques relative to this application.

33. Hand assemble the following mnemonic instructions using Table 10–1, the templates of Figiure 10–8, and the binary codes of Tables 10–2, 10–3, and 10–4.

a. *ADD ACC*1,D b. *ADD ACC*1,A
c. *AND ACC*1,D d. *JMP* 0500H
e. *MOV* A,C f. *MOV* C,A
g. *MOV M*,B h. *MOV* B,*M*
i. *MVI* D,7AH j. *MVI* B,3BH

34. Using your knowledge of the instruction templates of Figure 10–8a and b, hand assemble *SUB ACC*1,C and *OR ACC*1,D.

35. Using the templates of Figure 10–8, why is it that instructions such as *ADD* B,C, or *AND* D,C cannot exist?

36. Is the architecture a limiting factor in creating instructions *ADD* B,C and *AND* D,C? Explain your answer.

Section 10–6 Memory and I/O interfacing: the creation of a microcomputer

37. What is an I/O port, and what purpose does it serve?

38. What type of digital circuit is often used to implement I/O ports? Explain your answer.

39. Redesign the system of Figure 10–10 so that its memory will be a memory-mapped I/O.

40. Redesign the memory of Figure 10–10 so that partial linear addressing is used.

41. Repeat Problem 40 using linear addressing for I/O.

Section 10–7 A survey of some real microprocessors

42. Compare the architectures of the 8085 and 6800.

43. Compare the architecture of the 8086 to that of the 8085 and the 6800.

Troubleshooting

Section 10–8 Troubleshooting

44. Suppose you want to test the DMA operation of a system. How would you configure the test?
45. If the MPU of Figure 10–10 failed when reading I/O device 1, what test procedures would you use?
46. Repeat Problem 45 but with I/O device 0.
47. How would you configure a test for the 2716-0 of Figure 10–10?

The erasable programmable read-only memory (EPROM) is an IC that can be programmed with a sequence of binary numbers to instruct a microcomputer or to replace complex logic circuits (Reprinted by permission of Intel Corporation. © Intel Corporation 1989).

OUTLINE

OBJECTIVES

The objectives of this chapter are to

- introduce the concept of the programmable fuse-link.
- explain PLD notation.
- introduce programmable logic elements (PLE).
- introduce programmable array logic (PAL).
- introduce programmable logic arrays (PLA).
- investigate some combinational and sequential logic applications for PLEs, PALs, and PLAs.

PROGRAMMABLE LOGIC DEVICES

11–1 INTRODUCTION

In the introductory chapters of this book we developed design techniques that enabled us to seek combinational and sequential logic circuit solutions to design problems. We then introduced the microprocessor as a possible alternate solution. The microprocessor has the advantage of being programmable, so that changes in design circumstances can be accommodated by programming changes. A design change for combinational and sequential logic designs often requires scrapping much of the old design and its hardware and beginning anew. This suggests that microprocessor technology has made combinational and sequential logic circuits obsolete; however, this is not the case. In fact, most microprocessor-based systems require some combinational logic circuits to interconnect the various parts of the system, that is, it is the combinational logic that ''glues'' a microprocessor system together. Logic circuit designs that are tailored for a particular application are often faster than a microprocessor solution, but they are less flexible and often occupy more printed circuit board area because of the many ICs required to implement them.

This chapter introduces a third possible solution for a design problem—programmable logic devices (PLD). PLDs are ICs that contain many logic gates, and in some cases registers also, that can be configured (programmed) by the user into various logic combinations. This approach offers flexibility in the specific combinational or sequential logic design to be implemented by the PLD, with the additional benefit that the logic design is contained in one 20- or 24-pin IC rather than as many as six ICs. We shall investigate three types of PLDs: (1) PLE, (2) PAL, and (3) PLA.

11–2 The programmable fuse-link

Regardless of the PLD, the essence of its programmability is the *fuse-link*, which links two points electrically. If a connection is desired between the two points, the fuse is left intact; if no connection is needed between the two points, the fuse is removed by ''blowing'' it. The ability of the user to select which fuses remain and which are removed implements the programmability of a PLD. A programmer (a PROM programmer or similar apparatus) is used to ''burn'' those fuses to be blown.

Figure 11–1 illustrates the concept of the fuse-link. The logic gate of Figure 11–1a is a four-input *AND* gate that has fuse-links at each input. As stated, the user can selectively blow these fuse-links, implementing

Figure 11–1
Fuse-link examples.

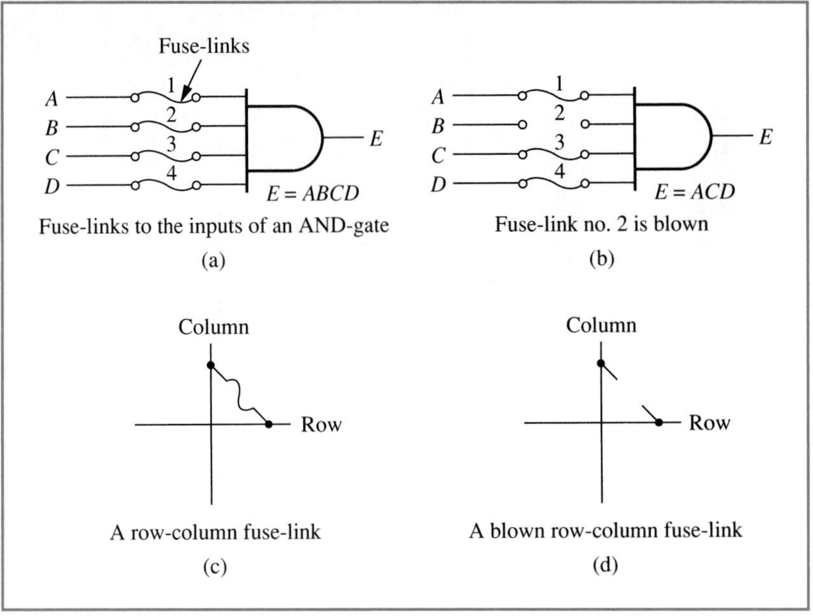

the programmability of the *AND* gate. The *AND* gate of Figure 11–1b has the no. 2 fuse-link blown ($B = 1$) and as a result the Boolean expression for output E is

$$E = A \cdot 1 \cdot C \cdot D = ACD$$

Observe that an open input is equivalent to a logic 1 for an *AND* gate. The reason for this will be explained in Chapter 12.

The row and column lines of Figure 11–1c have the same logic level, whereas blowing the connecting fuse-link makes the row and column lines independent of each other. Figure 11–1d illustrates a blown fuse-link for a row and column connection.

Figure 11–2a is an illustration of how fuse-links might be utilized in a PLD. The Boolean independent variables are the inputs at pins I_0 to I_2. The user may select (program) which of those variables will act as inputs to *AND* gates A, B, and C via the row-column fuse-links. The *OR* gate fuse-links are used to program which *AND* gate outputs will be *OR*ed together.

The logic circuit of Figure 11–2b is the PLD of Figure 11–2a that has been programmed. The only fuse-link of *AND* gate A that has been left intact is the Z input; hence, the output of *AND* gate A is Z (recall that an open input is the same as a logic 1 applied at that input). *AND* gate B has all fuse-links intact, so its output is equal to the Boolean expression XYZ. The output of *AND* gate C can be expressed by the Boolean expression XZ. *OR* gate 0 will *OR* just the output of *AND* gate A, since that output is the only one with its fuse-link intact. Note that an open input of an *OR* gate is equivalent to a logic 0. Similarly, *OR* gate 1 will *OR* the output of *AND* gate B, and *OR* gate 2 will *OR* the output of *AND* gates B and C. The Boolean expressions for outputs O_0 through O_2 are

$$O_0 = Z$$
$$O_1 = XYZ$$
$$O_2 = XYZ + ZX$$

Figure 11–2
The fuse-links of a PLD.

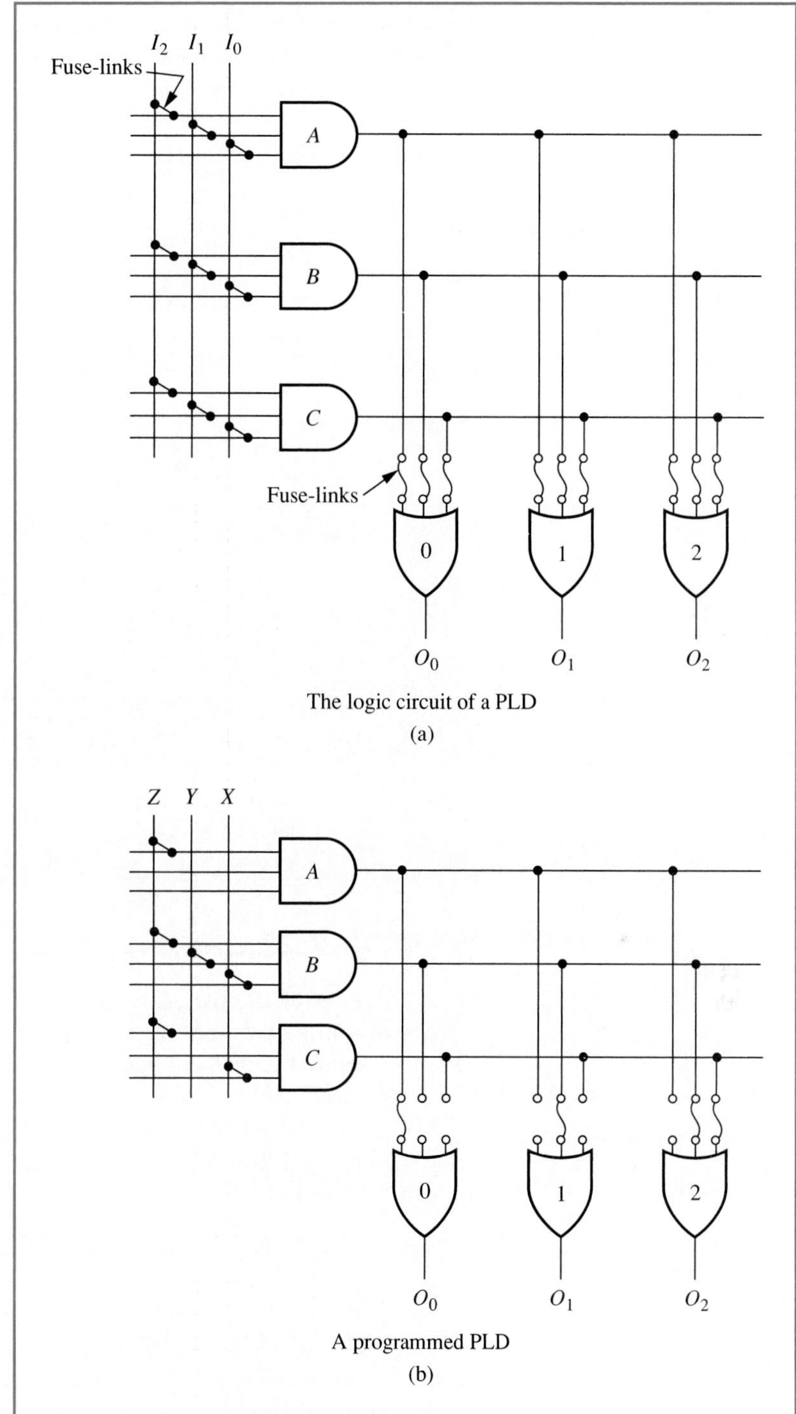

EXAMPLE 11–1

Program the PLD of Figure 11–2a to yield the Boolean expression

$$W = XY + YZ + XZ$$

SOLUTION As in Figure 11–2b, in Figure 11–3 we will arbitrarily input variable X at input I_0, Y at I_1, and Z at I_2. Only one output is required, so we use the output of *OR* gate 0; however, notice that if outputs O_1 and O_2 are not required for other Boolean expressions and the *OR* gate fuses are left intact, any one of the outputs will yield the desired Boolean expression.

AND gate *A* is programmed to yield the product *XY*, *AND* gate *B* is programmed for *YZ*, and *AND* gate *C* is programmed for *XZ*.

Figure 11–3
A PLD programmed for the Boolean expression of Example 11–1.

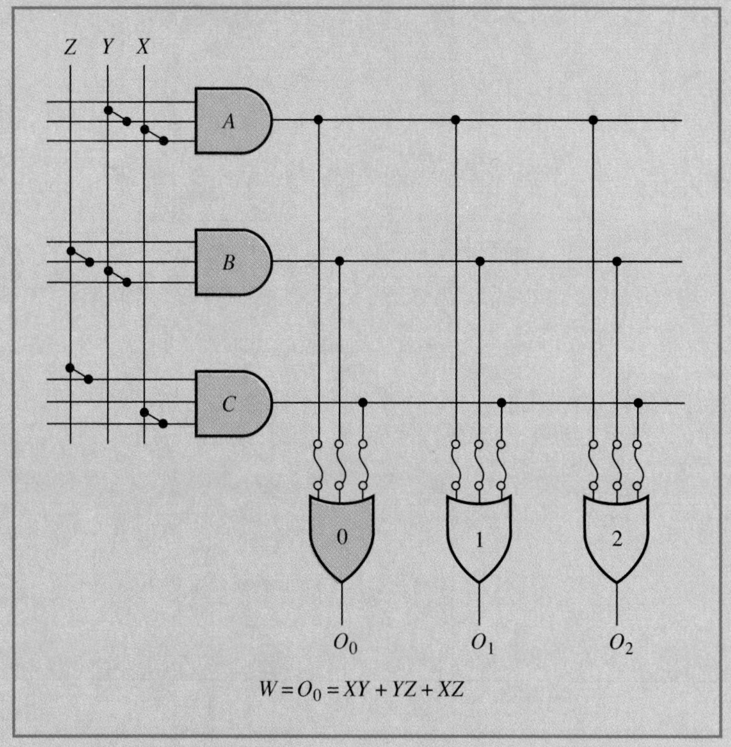

PLDs offer the user the capability to implement logic circuit designs and to minimize the number of ICs, since many of the logic circuits can be programmed within a single 20- or 24-pin PLD.

Review questions

1. What relationship exists between programmability and fuse-links?
2. What is the resultant logic state for a blown fuse at the input of an *AND* gate?
3. What is the resultant logic state for a blown fuse at the input of an *OR* gate?
4. What is a PLD?
5. How can PLDs be programmed to implement combinational logic?

11–3 Programmable logic elements (PLE) and PLD symbolic notation

To develop PLD notation, it is best to have a model that can be referenced for examples. Because PLEs are the most fundamental PLDs, they will be used to develop some basic concepts and notation.

As stated in the Introduction, there are basically three types of PLDs—PLE, PAL, and PLA. Except for the placement of the fuse-link, all three types of PLDs have essentially the same conceptual logic-gate architecture. It is their fuse-link placement and hence their programming flexibility that differentiate one type from another. A PLE has a fixed *AND* gate array (they are not programmable) whose outputs serve as inputs to programmable (fused) *OR* gates. Figure 11–4 is the logic architecture of a simplified two-input, four-output PLE. A PLE is very similar to a PROM but differs in that it has a specific architecture that lends itself to implementing combinational logic designs. When a PLD is used to implement combinational logic, rather than as a PROM, it is referred to as a programmable logic element (PLE).

The two inputs of the PLE of Figure 11–4 are identified as I_0 and I_1. If this PLD were to function as a PROM, these inputs would be used as address pins. However, in a PLE we can think of these input variables in a more generic sense; that is, these inputs are just pins to which Boolean independent variables are input to the PLE. The two input buffers have as outputs the input variables and their complements. Hence, the output of *AND* gate number 0 (m_0) is minterm $\bar{I}_1\bar{I}_0$, m_1 is minterm $\bar{I}_1 I_0$, and so on. The output of each *AND* gate is a minterm of a sum-of-the-products (SOP) Boolean equation. Table 11–1 shows the Boolean expression for each minterm as a function of the generic input variables I_1 and I_0 and their complements.

In using a PLD as a PLE, its flexibility is the ability to program which minterm(s) will serve as inputs to the *OR* gate of Figure 11–4. If a fuse-link is left intact, the respective minterm is input to that *OR* gate. In contrast, when a fuse-link is blown, the minterm served by that fuse-link is not input to that *OR* gate; instead an equivalent logic 0 is applied. The

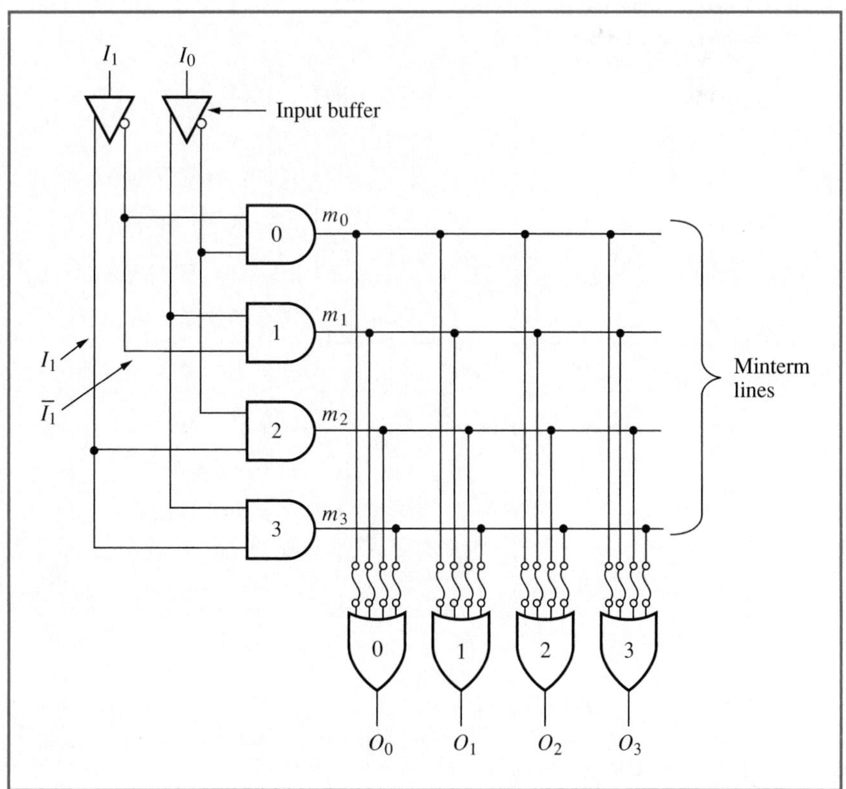

Figure 11–4
The logic architecture of a PLE.

Table 11–1
Boolean expressions for the minterms of the PLD of Figure 11–4

Minterm	Boolean expression	Input logic states
m_0	$\overline{I}_1\overline{I}_0$	00
m_1	$\overline{I}_1 I_0$	01
m_2	$I_1\overline{I}_0$	10
m_3	$I_1 I_0$	11

Boolean expression for the output of each *OR* gate has the potential of being equal to

Equation (11–1)

$$O_n = m_0 + m_1 + m_2 + m_3 = \sum_{n=0}^{3} m_n$$

The reader should recognize that Equation 11–1 is in the form of an SOP Boolean expression.

EXAMPLE 11–2

It is desired to use a PLE, similar to Figure 11–4, to implement the given Boolean expression, where *A* and *B* are the independent (input) variables and the dependent (outputs) variables are *X*, *Y*, *Z*, and *W*.

$$X = \sum_m 1, 2$$

$$Y = 3$$

$$Z = \sum_m 0, 3$$

$$W = \sum_m 0, 1, 2$$

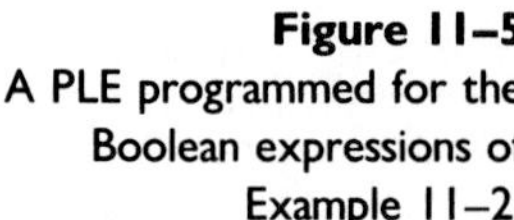

Figure 11–5
A PLE programmed for the Boolean expressions of Example 11–2.

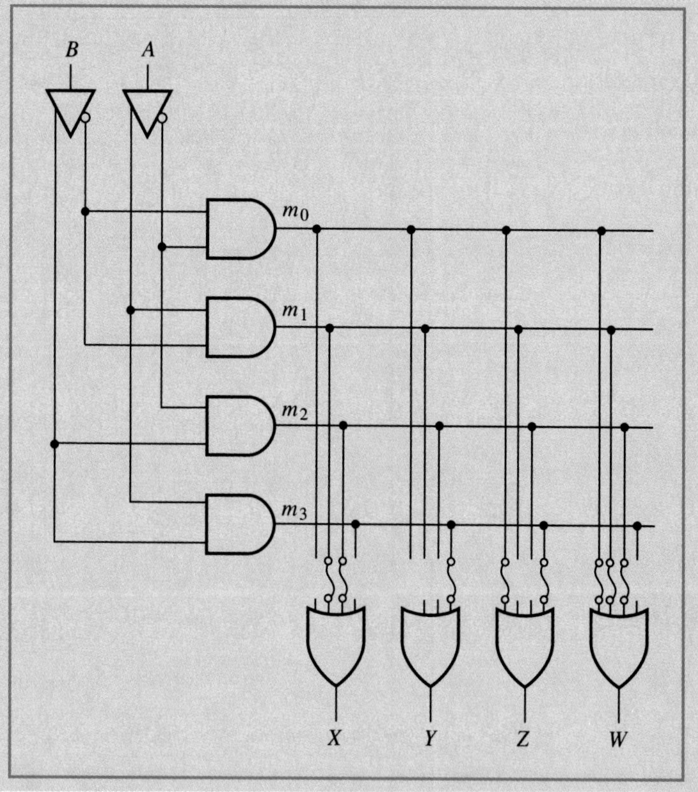

SOLUTION The independent variables (the input variables) A and B are defined as $A = I_0$ and $B = I_1$. The dependent variables (the output variables) X, Y, Z, and W are defined as $O_0 = X$, $O_1 = Y$, $O_2 = Z$, and $O_3 = W$. Implementation of these definitions is illustrated in Figure 11–5.

From Figure 11–5 we observe that the *OR* gate for output X has the fuses for the minterms m_0 and m_3 blown and fuses for minterms m_1 and m_2 still intact. This yields the desired output of

$$X = \sum_m 1, 2$$

The authors believe the logic for the dependent variables Y, Z, and W is now self-explanatory.

Note that a PLE is limited to standard unsimplified minterms. For instance, had dependent variable X equaled the sum of minterms 1 and 3, which is equivalent to A (A is the LSB) when simplified, it could not be implemented in simplified form. PLAs and PALs do not have this restriction.

The number of minterms of a PLE and, of course, the number of *AND* gates are directly related to the number of input variables. From previous chapters we know that for n independent variables there are 2^n binary combinations of those independent variables. Since there is an *AND* gate for each combination (a minterm), such as is shown in Table 11–1, there are also 2^n *AND* gates. The outputs of these *AND* gates are the minterms of an SOP-type Boolean equation. To reiterate, these same *AND* gates would be labeled address decoders if the PLD were being used as a PROM rather than a PLE.

The number of *OR* gates of a PLE can vary, just as can the bit size (width) of a PROM. In fact, it is the number of *OR* gates that determines the bit size of a PROM. For instance, if the PLD of Figure 11–4 were being used as a PROM, each memory location would contain 4 bits, 1 bit per *OR* gate. The *OR* gate fuse-links would be programmed for each minterm line (m_0, m_1, m_2, and m_3) so as to store the desired bit pattern for each memory location. For example, suppose that the PLD of Figure 11–5 were programmed to be used as a PROM rather than a PLE. The programmed bit pattern would be that of Table 11–2. The technique of deciphering the programmed bit patterns of a PLD, which is to function as a PROM, is to realize that each minterm line is a memory location and that each *OR* gate input is a programmed data bit for that location. If the fuse-link of an *OR* gate input is intact, a logic 1 is output at the *OR* gate when that minterm line is addressed; if the fuse-link is blown, a logic 0 is output. As an example of how this works, consider address 00 input (via

Table 11–2
The PLE of Figure 11–5 used as a PROM

Address		Active address line (product-line)	Programmed bit stored			
B	A		X	Y	Z	W
0	0	m_0 (P_0)	0	0	1	1
0	1	m_1 (P_1)	1	0	0	1
1	0	m_2 (P_2)	1	0	0	1
1	1	m_3 (P_3)	0	1	1	0

inputs A and B) to the PROM of Figure 11–5. Minterm line m_0 goes high and the other minterm lines are low. The *OR* gate of output X has its fuse-link to m_0 blown; therefore, the X output is a logic 0, as indicated by the first row of Table 11–2. The connection from minterm m_0 to the input of the *OR* gate with output Y is also blown, so Y is also a logic 0, which is also indicated in Table 11–2. The connections from minterm line m_0 to *OR* gate outputs Z and W are intact; as a result, these outputs go high when the minterm m_0 goes high, which is also indicated by the first row of Table 11–2.

The number of *OR* gate outputs for a PLE varies according to the number of dependent variables (outputs) to be implemented. If a PLE is used to implement a design with two dependent variables, two outputs are all that is required. If there are to be five dependent variables, there must be five *OR* gates, and so on. There is no strong relationship between the number of *OR* gates and the number of independent variables (inputs) for a PLE, just as there is no direct relationship between the number of input variables (the number of address bits) of a PROM and its bit size. However, if for a PLE the number of outputs becomes too large, there is a redundancy in minterms available for each output. What can be said with certainty is for n input variables there are 2^n *AND* gates, where the output of each *AND* gate is a minterm of an SOP Boolean expression. The number of *OR* gate outputs can be any number that a designer and/or manufacturer wishes to include as part of the IC. Of course the size of the silicon wafer and the number of pins on the IC have a limiting effect.

A practical upper limit on the number of storage bits within a PROM is 64 K (65,536) bits (recall the 2764 PROMs of Chapter 9). Because microprocessors have data bus structures that are multiples of 8 bits, these 64-K PROMs are often configured as an 8 K $\times$ 8 memory. With 8 K locations, there are 13 inputs (address pins). However, when a PLD is being fabricated for use as a PLE, rather than as a PROM, the number of inputs is decreased, but the number of outputs is increased. A PLE may have six inputs and as many as 16 outputs. This redistribution of input variables versus output variables seems reasonable, since seldom would a logic design have more than six independent variables, but there could be many dependent variables that are to be generated from those input variables.

EXAMPLE 11–3

For a PLE with six inputs and 16 outputs, what is the maximum number of minterms that an SOP Boolean expression can have for any one of those outputs?

SOLUTION For six inputs there are 2^6 (64) minterms. If the connections between all minterm lines (m_0, m_1, . . . , jm_{63}) and the fuse-links to an *OR* gate are left intact, that *OR* gate would *OR* each minterm of the *AND* gate array. The resulting Boolean expression for this output would be

$$O_n = \Sigma_m\ 0, 1, 2, 3, 4, 5, 6, 7, 8, 9, \ldots, 63$$

Imagine having to make an architectural drawing, similar to Figure 11–4, for the PLE of Example 11–3. There would be six inputs (I_0 through I_5) and input buffers, 64 *AND* gates, and 16 *OR* gates. Each of the 64 *AND* gates would have six inputs, these inputs being combinations of the input variables and their complements, which form the minterms of the resulting SOP equations. Each of the 16 *OR* gates would have 64 programmable (fuse-linked) inputs. (If you were not discouraged from making this drawing by the 64 *AND* gates, surely these 16 *OR* gates with 64 inputs each would do the job.)

PDL symbolic notation

The large number of logic gate inputs and fuse-links that, in part, make up a PLD architectural drawing practially requires that a simpler symbolic representation be adopted. The following PLD notation has been adopted by industry:

1. Show a single input line for all logic gates. For the *AND* gates these lines are known as *product lines*.
2. Put an *X* at any programmable (fuse-linked) connection where the fuse is intact.
3. Use a dot, as is standard practice in electronic schematics, to indicate a fixed (nonprogrammable) electrical connection.

Figure 11–6 is a logic architectural drawing of the PLE of Figure 11–5 using this PLD symbolic notation. The *AND* gates in Figure 11–6 have a single input that is identified as a product line, as stated above. Those inputs (*A* and *B*) to the *AND* gate are identified as fixed (nonprogrammable) connections by the use of a dot. The connections of those programmable *OR* gate inputs that are fuse-linked (the fuse-link is intact) to a minterm line (m_0, m_1, m_2, and m_3) are indicated by an *X* at the intersection of the *OR* gate input and the minterm line. Fuse-links that are blown make no connection to the minterm line and are so indicated by the absence of a × at the intersection of minterm line and the *OR* gate input.

Figure 11–6
The PLE of Figure 11–5 using PLD symbolic notation.

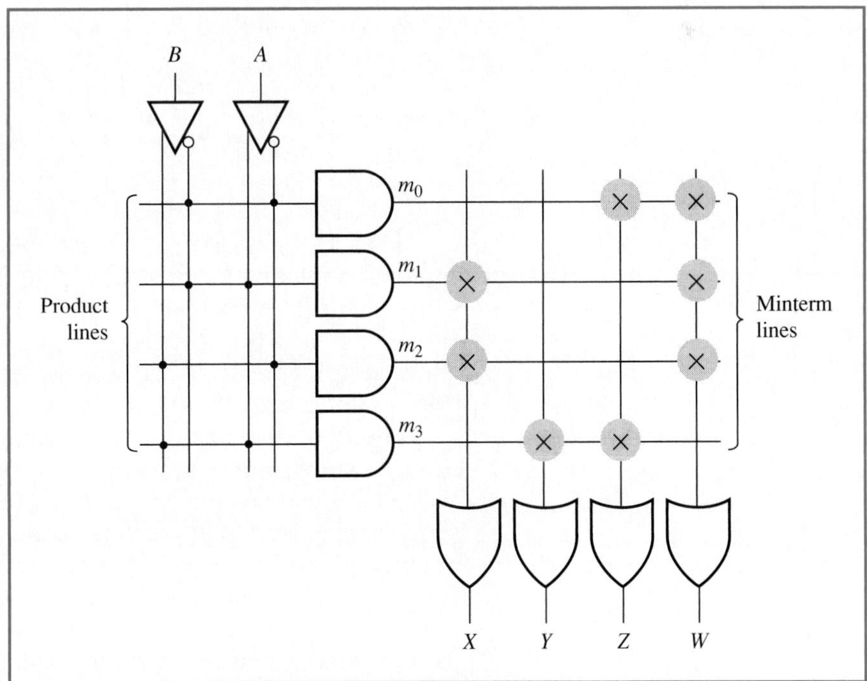

EXAMPLE 11–4

Using a PLE with four inputs and four outputs, similar in architecture to Figure 11–4, design the 2-bit parallel binary adder of Figure 4–15. Show the programmed logic architecture using PLD symbolic notation.

SOLUTION As in Chapter 4, we first construct a truth table, which relates the logic states of the independent variables (inputs) and the dependent variables (outputs). From this truth table we determine the SOP Boolean expression for each of the outputs. These equations are simplified and then implemented. We follow the same steps except that we do not simplify the Boolean equations, since PLE devices implement only minterms.

Minterm	B_1	A_1	B_0	A_0	S_1	S_0	C
0	0	0	0	0	0	0	0
1	0	0	0	1	0	1	0
2	0	0	1	0	0	1	0
3	0	0	1	1	1	0	0
4	0	1	0	0	1	0	0
5	0	1	0	1	1	1	0
6	0	1	1	0	1	1	0
7	0	1	1	1	0	0	1
8	1	0	0	0	1	0	0
9	1	0	0	1	1	1	0
10	1	0	1	0	1	1	0
11	1	0	1	1	0	0	1
12	1	1	0	0	0	0	1
13	1	1	0	1	0	1	1
14	1	1	1	0	0	1	1
15	1	1	1	1	1	0	1

The design equations are

$$S_0 = \Sigma_m\, 1, 2, 5, 6, 9, 10, 13, 14$$

$$S_1 = \Sigma_m\, 3, 4, 5, 6, 8, 9, 10, 15$$

$$C = \Sigma_m\, 7, 11, 12, 13, 14, 15$$

The PLE must have at least 2^4 (16) *AND* gates, product lines, and minterm lines, since there are four independent variables (A_0, B_0, A_1, and B_1). Since there are three dependent variables (S_0, S_1, and C), the chosen PLE must have a minimum of three outputs. As indicated in Figure 11–7, the selected PLE has four outputs, one of which is not used.

The Boolean equations are programmed into the PLE of Figure 11–7. To verify proper programmed connections, we refer to the Boolean design equation and find that the "input line" of *OR* gate S_0 must be connected to minterm lines m_1, m_2, m_5, etc. The *OR* gate for output S_0 in Figure 11–7 has been properly programmed. We repeat program verification for *OR* gates S_1 and C and find them also to be programmed according to the design equations. The *OR* gate that is not used indicates that all fuse-links were left intact. Since it

Figure 11–7
A 2-bit parallel binary adder implemented with a PLE.

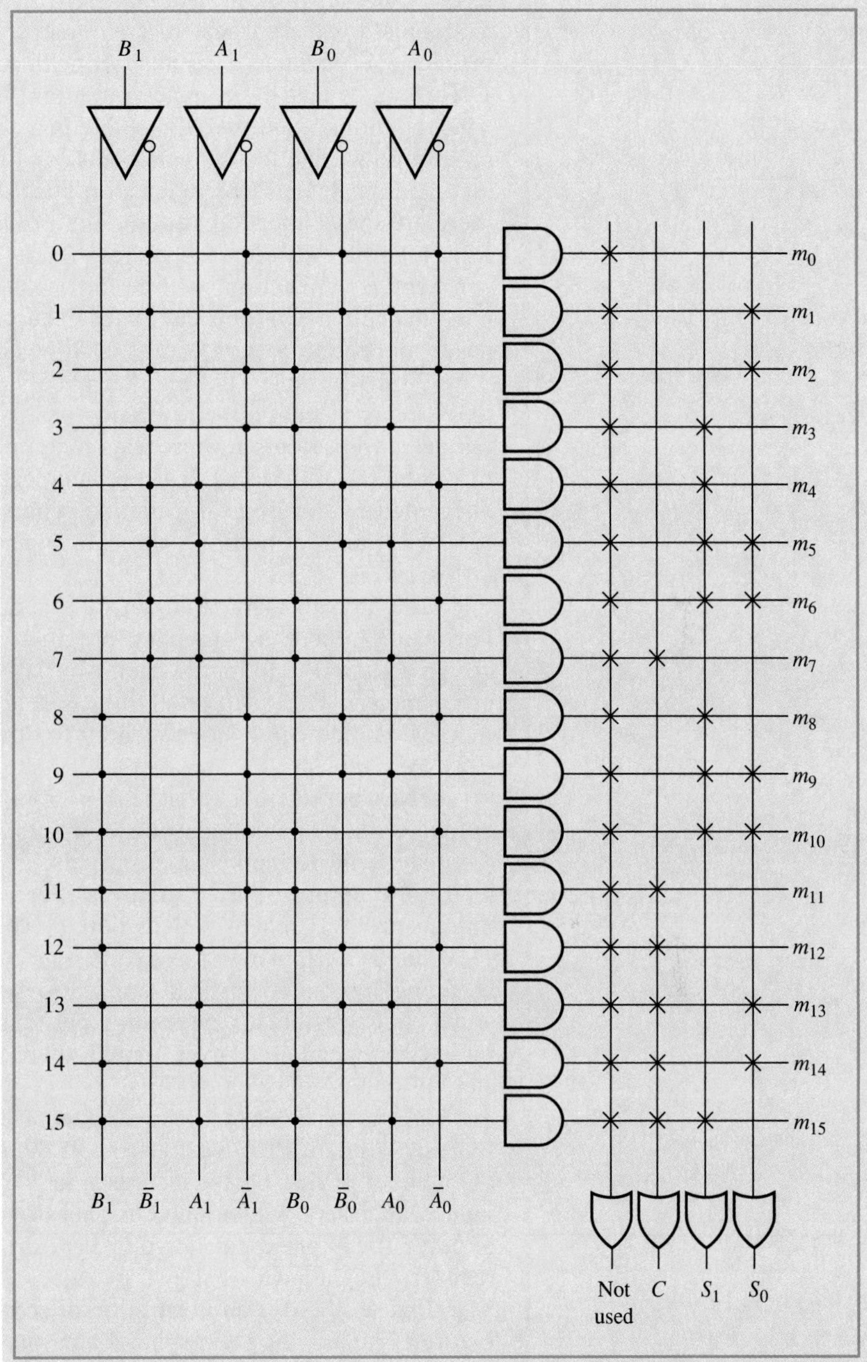

is not used, and by leaving the fuses intact, that *OR* gate can be considered a spare that may be programmed at a later date if needed.

The reader may want to refer to the input variable vertical lines (A_0, $\overline{A}_0$, B_0, etc.) and notice the pattern of connections with the product lines (*AND* gate inputs). This pattern is the same as counting in binary; that is, the connections appear in a pattern equivalent to the bit position assigned to the variable. If we arbitrarily assign A_0 to be the LSB and B_1 to be the MSB (just for the purpose of creating a truth table in an organized manner), A_0 occupies a bit position with a weighting factor of 2^0, which is 1. This states that the connection pattern is one (just as is the bit pattern when counting), and therefore there is a connection made with every other product line for the variable A_0 and its complement. The variable B_0 (and its complement) occupies a bit position with a weighting factor of 2^1; therefore the connection pattern is in product line pairs. As we can see from Figure 11–7, the vertical lines for B_0 and its complement have two consecutive connections with product lines and then two consecutive non-connections with product lines. The vertical line for input variable A_1 has a connection pattern of 2^2 and B_1 has a pattern of 2^3.

From Example 11–4 another observation is made: to program an SOP Boolean equation, one simply locates the vertical input line to the *OR* gate, whose output represents the dependent variable being programmed, and put an X at the intersection of that line and the minterm line stated by the Boolean equation. The variable C has minterms 7, 11, 12, 13, 14, and 15, and *OR* gate C of Figure 11–7 has X's at those intersections.

The actual programming of PLE is done with a PROM programmer. Fortunately, there are computer programs available that take Boolean design equations and format them so that they are compatible with PROM programmers. One such program was developed by Monolithic Memories, Inc. (MMI) and can be identified by the trademark PLEASM (PLE Assembler).

PLEs are manufactured with various numbers of inputs and outputs—5 inputs and 8 outputs, 6 inputs and 16 outputs, 9 inputs and 4 outputs, and 12 inputs and 8 outputs. In addition to various numbers of inputs and outputs, they also have different output capability. Buffered outputs are available as well as outputs with buffered registers. Figure 11–8 shows a sampling of available outputs. As we already know, buffering is required for PLE applications whose outputs are connected to a bus, and registers are necessary any time the output logic state must be retained (remembered) even though there has been a change in the logic state of one or more input variables.

Let us consider the logic architecture of Figure 11–9 to be representative of PLE logic arrays. If the number of inputs is increased, the number of *AND* gates increases to 2^n, where n is the number of inputs, and there will be one *OR* gate per output. If the outputs are buffered they appear as shown in Figure 11–8a, and if they have buffered registers they appear as shown in Figure 11–8b.

Figure 11–10 is an illustration of some commercially available PLEs. Figure 11–10a is the logic symbol and pin configuration of a PLE5P16. We see that the *AND/OR* logic array symbol is greatly simplified by being represented by a rectangle. There are five inputs that are identified as *I*0, *I*1, *I*2, *I*3, and *I*4 (pins 7 through 11), which means there are 32 (2^5) *AND* gates. There are 16 buffered outputs, which are identified as *O*1, *O*2, *O*3, . . ., *O*16. The buffer enable input is identified as $\overline{E}$ and is pin 12. The power supply V_{CC} is 5 V DC and is pin 6, and the ground is pin 18.

Figure 11–10b is another PLE with buffered outputs; however, there are two enable inputs, pins 15 and 16, as can be reasoned from the *AND* gate that controls the state of the three-state buffers. Since the inputs to

Figure 11–8
Output variations of a PLE.

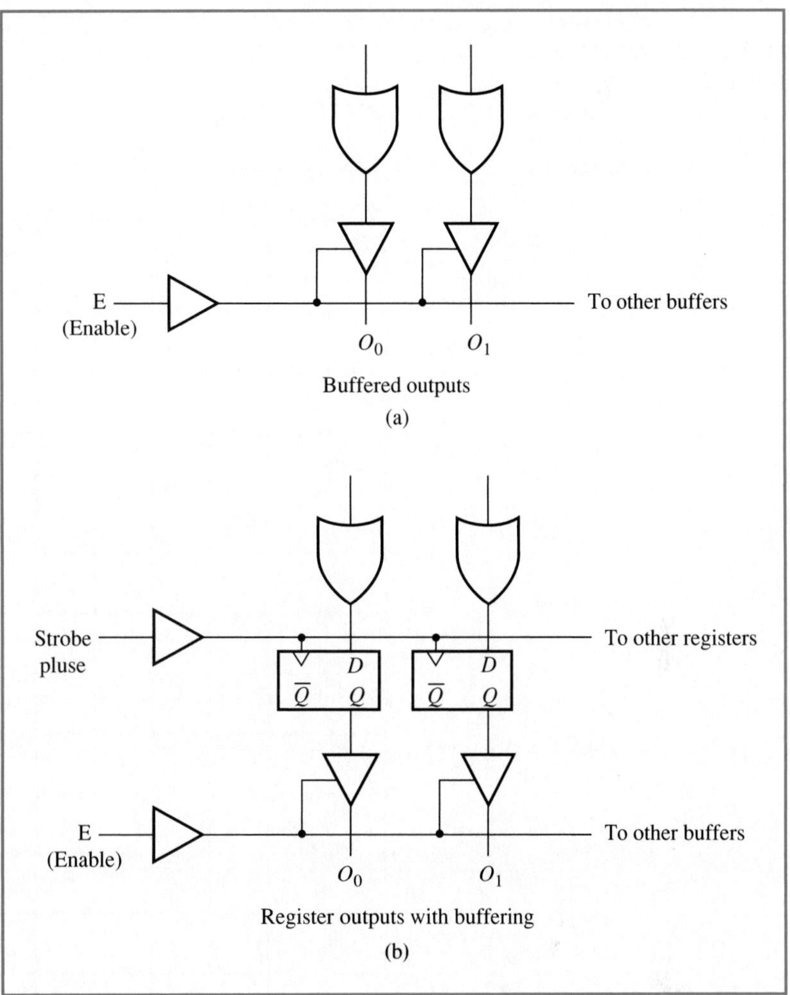

this *AND* gate are active lows, the enable signals are also active lows. The inputs are not labeled but are pins 1, 2, 3, 4, 5, 17, 18, and 19. The outputs are not labeled but are obvious because of the buffers.

Figure 11–10c is very similar to the PLE of Figure 11–10b except there are 10 inputs rather than eight, and four enable inputs are required. Two of those inputs are an active-low (pins 20 and 21) and two are active-high (pins 18 and 19).

Figure 11–10d shows a PLE10R8; it has eight buffered register outputs, which are easily recognized, and there are 10 inputs. Output buffers are enabled via an active-low at pin 21 and the *D*-type flip-flop, which is the synchronous enable flip-flop, set at pin 19. The strobe pulse for the flip-flops is input at pin 18. Pin 20 is an input pin to which an initialization input signal can be applied. When low, this allows one of 16-bit patterns to be set into the output registers, where the desired pattern is determined by the logic levels of input pins 5, 6, 7, and 8. This feature allows the flip-flops to be set to a desired binary combination, which can be useful in some sequential or feedback designs. If this feature is not to be used, pin 20 is connected to a logic high.

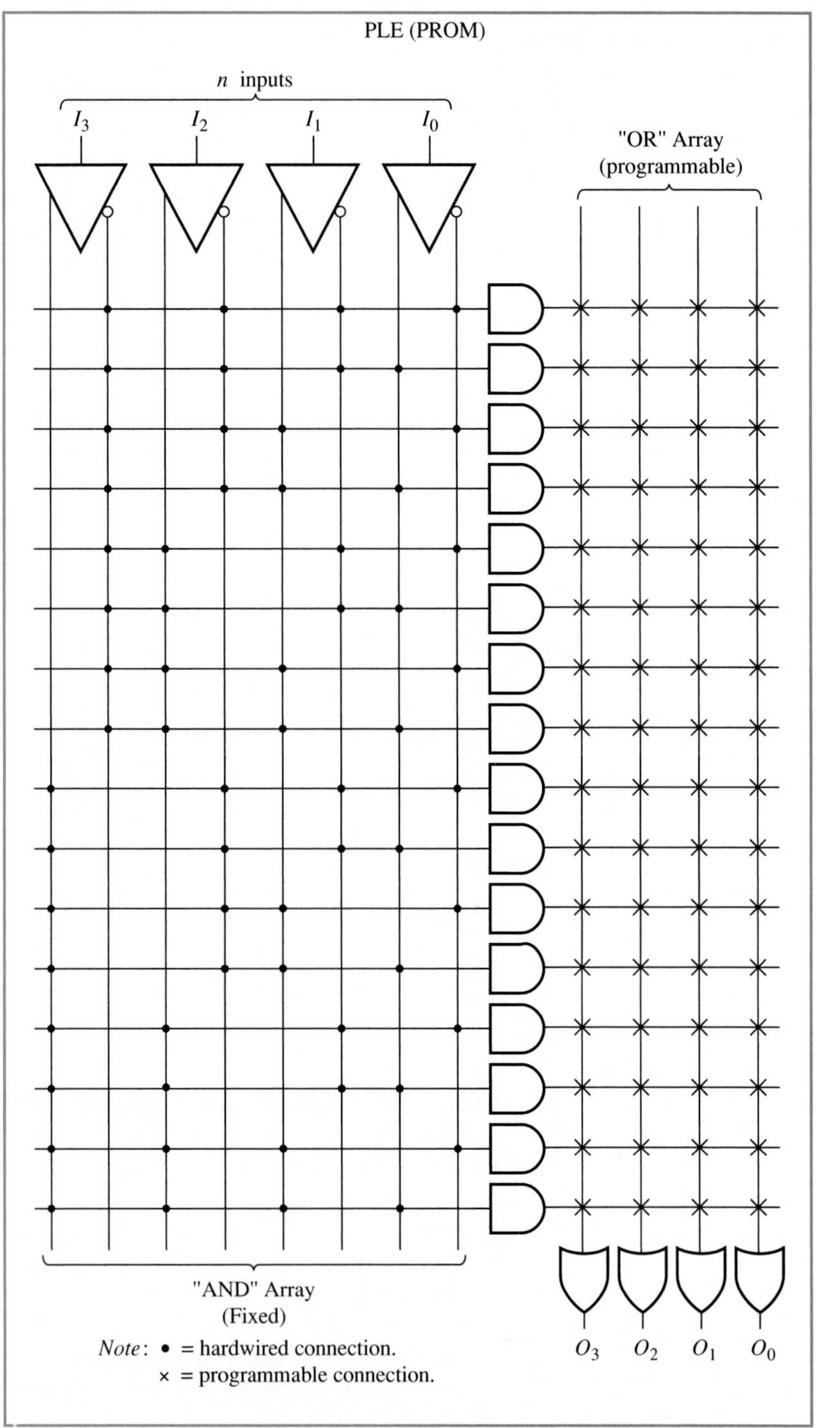

Figure 11–9
Representative *AND/OR* logic array of a PLE. (Copyright © Advance Micro Devices, Inc., 1990. Reprinted with permission of copyright owner. All rights reserved.)

There are a variety of PLEs available from a variety of sources. Regardless of the source, the identifying nomenclature for a PLE is similar. Figure 11–11 illustrates a common approach to PLD identification. The left-most symbols identify the PLD type (PLE, PLA, or PAL). PLE is a registered trademark of MMI; therefore, they use PLE for their product line of programmable logic elements. For a five-input

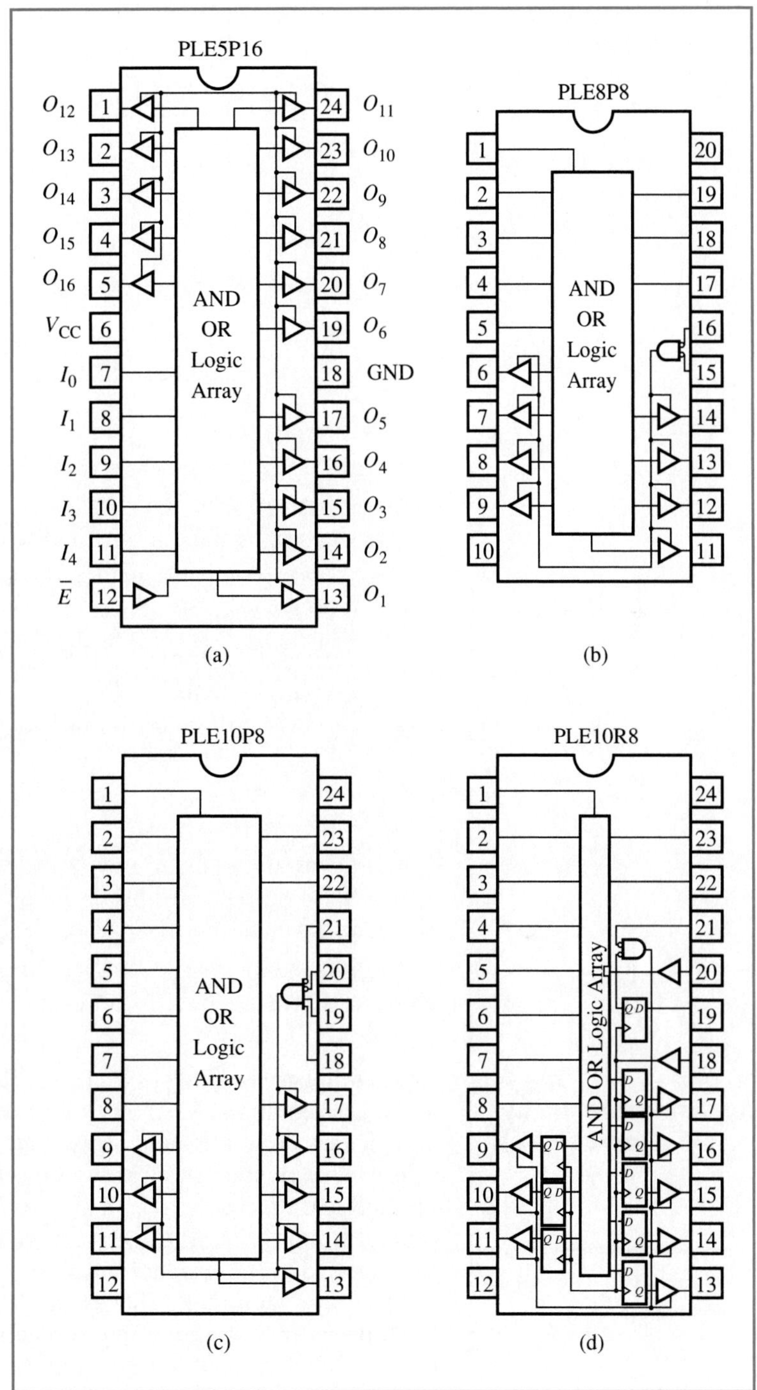

Figure 11–10
Samples of PLE devices. (Copyright © Advanced Micro Devices, Inc., 1990. Reprinted with permission of copyright owner. All rights reserved.)

PLE the number 5 is in the II position of the product identification symbol. The *TOP* symbol indicates whether the outputs are registers (R), nonregistered (P), and so on. The number of outputs is designated by *OO*. The PLE5P16 in Figure 11–10a has five inputs and 16 outputs, which are nonregistered (P). The other PLEs of Figure 11–10 are identified by the same notation.

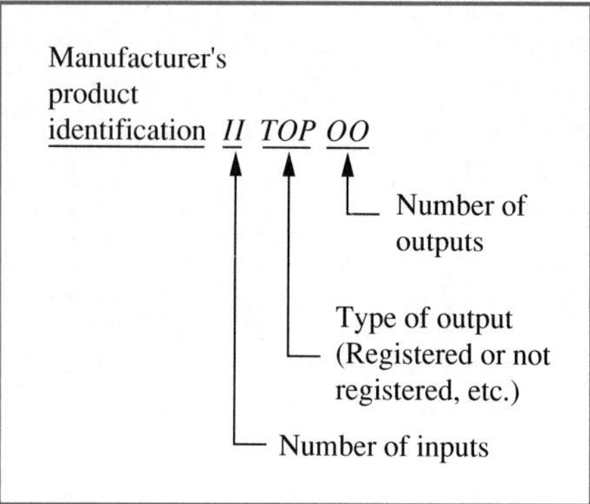

Figure 11–11 PLD product identification nomenclature.

Review questions

1. Relative to the *AND/OR* logic array of a PLE or PROM, which gates have fixed inputs and which have programmable inputs?
2. What relationship is there between the number of inputs and the number of *AND* gates?
3. What relationship is there between the number of inputs and the number of outputs?
4. What part of an SOP Boolean equation does the *AND* gate yield?
5. What part of an SOP Boolean equation does the *OR* gate implement?
6. Given the logic architecture of a PLE and the SOP equation to be programmed, how can a person methodically program (on paper) a PLE?
7. What is the nomenclature for a PLE?

11–4 Programmable array logic (PAL)

A PLD that is similar to a PLE or PROM, but differs from a PLE in that it has a programmable *AND* gate array with a fixed *OR* gate array is known as a programmable array logic (PAL) device. The acronym PAL is a registered trademark of Monolithic Memories, Inc. Figures 11–1a and 11–2a illustrate the programmable *AND* gate. Figure 11–12 illustrates the general format of a PAL using conventional PLD notation. The reader can verify that the product lines are programmable but the inputs to the *OR* gate are fixed. Note that each *OR* gate can *OR* four product terms. Until programmed, the output of each *AND* gate, the product term, is the product of each input variable and its complement, which is a logic 0. Hence, if a product line is not programmed (all of its fuse-links are left intact), the output of that *AND* gate is a permanent logic 0. The symbolic notation used to indicate a product line with all independent variable (input variables) fuse-links intact is a × within the *AND* gate logic symbol of that product line (see Figure 11–13).

EXAMPLE 11–5

Use the PAL of Figure 11–12 to implement the dependent variables *Y* and *Z* if the independent variables are *A*, *B*, and *C* and the

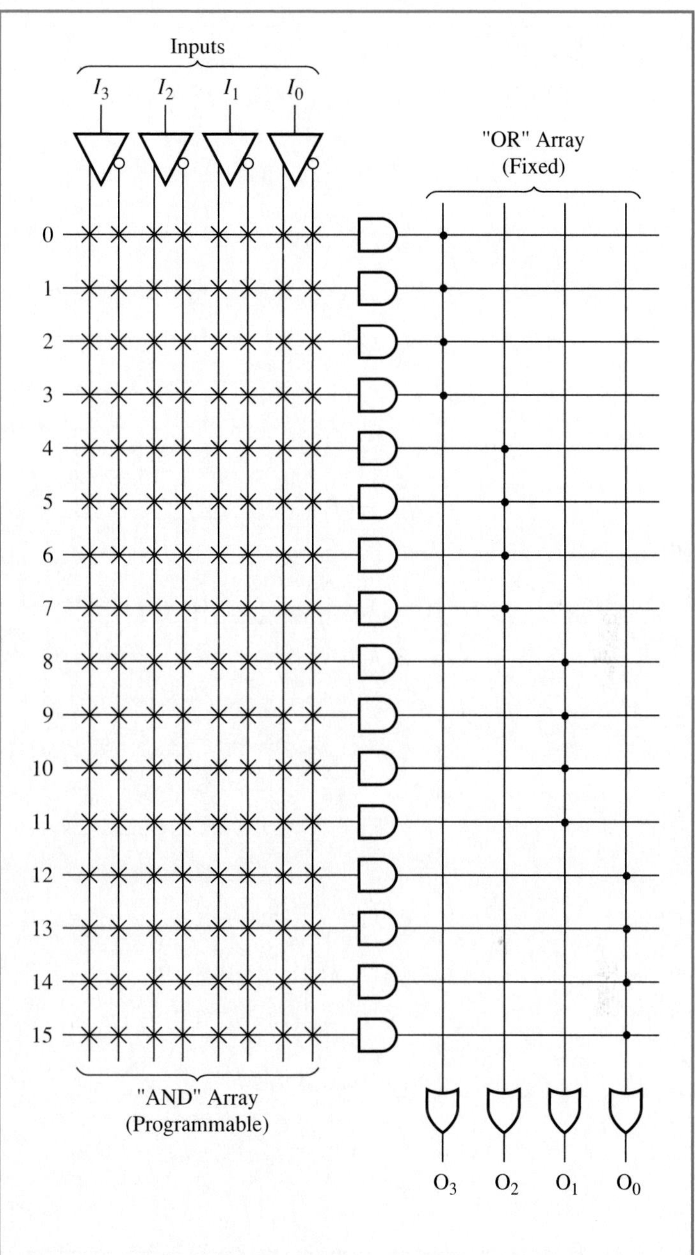

Figure 11–12
The logic architecture of a PAL. Copyright © Advanced Micro Devices, Inc., 1990. Reprinted with permission of copyright owner. All rights reserved.)

Boolean design equations are

$$Y = BC + AC + \overline{A}B\overline{C}$$
$$Z = \Sigma_m\ 0, 3, 6 \qquad (C \text{ is the LSB position})$$

SOLUTION Only two outputs are being used. We choose output O_0 for the dependent variable Y and output O_1 for dependent variable Z. We do not use outputs O_2 and O_3; as a result, the *AND* gates that these two *OR* gates *OR* together do not need to be programmed. Product lines 0 through 7 of Figure 11–13 show an *X* within the

Figure 11–13
The programmed PAL for Example 11–5.

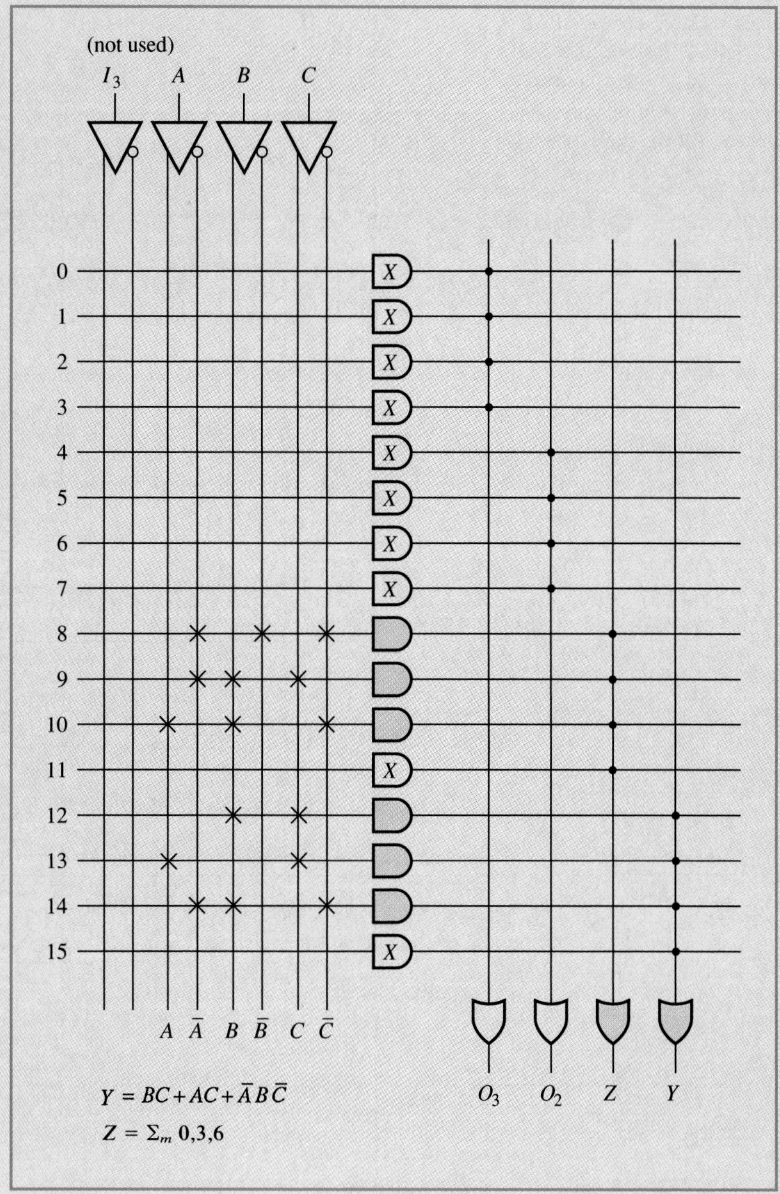

respective *AND* gate symbol to indicate this. These outputs can serve as spares for later circuit revisions if needed.

From Figure 11–3 we can reason that to program a PAL we pick a product term from the Boolean equation and implement it with one of the *AND* gates being *OR*ed into the *OR* gate that was chosen to represent the dependent variable. For dependent variable *Y* we used product line 12 to implement the product term BC, product line 13 to implement AC, and product line 14 to implement the product term $\overline{A}B\overline{C}$. Product line 15 is not used, so all its fuse-links are left intact. The procedure is repeated for dependent variable *Z*.

To reiterate, the unused input has its fuse-links intact for product lines 0 to 7, 11, and 15.

Figure 11–14
Variations of the basic PAL logic array.

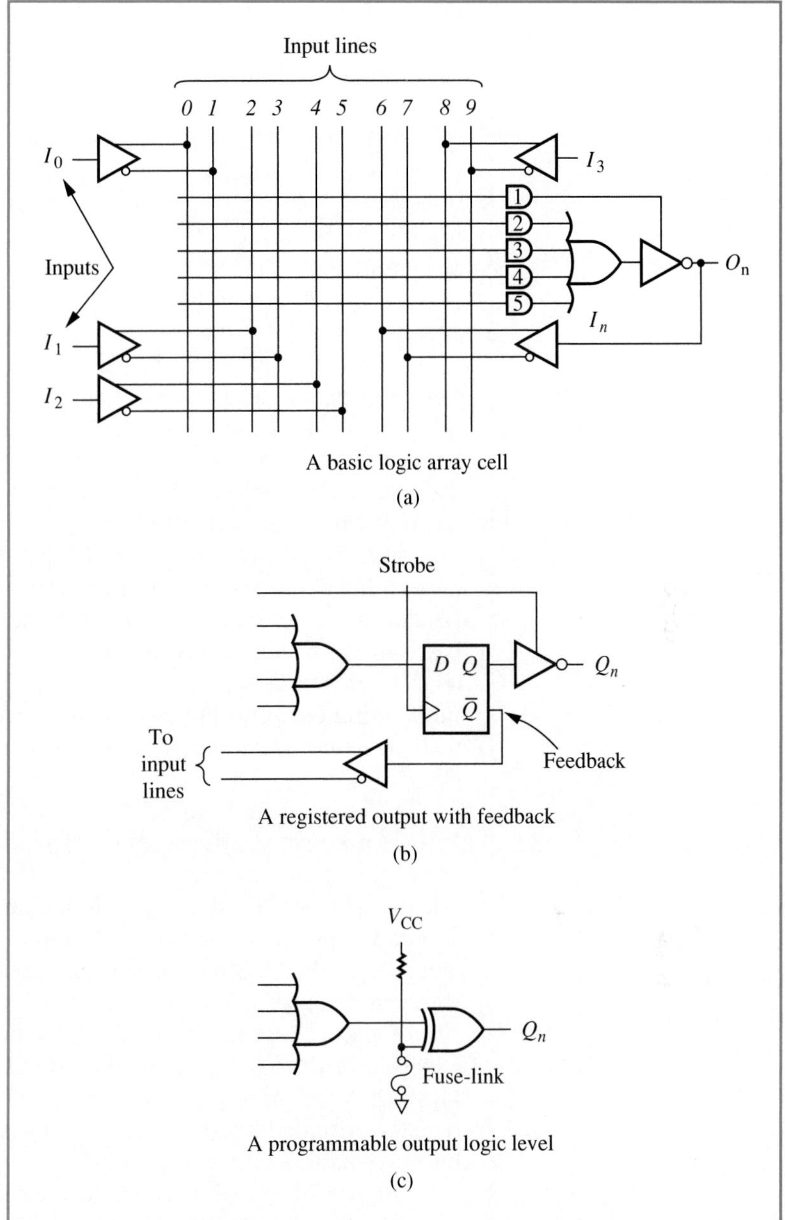

Just as there are a variety of PLEs, there are also various types of PALs. The basic nomenclature of Figure 11–11 also applies to PALs. There are PALs that have inverted buffered outputs (the *TOP* designation is an L) as well as noninverted buffered outputs (the *TOP* is H), buffered outputs that can be programmed to be either inverted or noninverted output (the *TOP* is P), registered outputs (the *TOP* is R), and programmable I/O pins (some manufacturers use a V, for versatile, to indicate the *TOP*). Figure 11–14 shows a sampling of these variations of the basic PAL.

Figure 11–14a is a logic array cell of a PAL, which is conceptually the same as the logic array of Figure 11–12. The major difference between the two is that they are drawn slightly differently and a buffered output and its control *AND* gate have been added in Figure 11–14a (*AND* gate 1). *AND* gates 2, 3, 4, and 5 of Figure 11–14a serve the same function as any four *AND* gates connected to the same *OR* gate input line in Figure 11–12.

Figure 11–14a uses a method that visually individualizes the *OR* gate inputs, which is similar to the method used to indicate the *OR* gate inputs in Figure 11–4; regardless of the technique used to represent the *OR* gate inputs, the meaning is the same. *AND* gate 1 allows the user to program an output that may be used as an input or output pin. That is, if an output is not required but additional inputs are needed, the fuse-links to *AND* gate 1 can be left intact to drive its output to a permanent logic 0, which puts the output buffer in the Hi-*Z* state. As a result of this action, the output buffer is disabled and that pin can be used as input pin I_n. On the other hand, if all the input fuse-links of *AND* gate 1 are blown, it permanently outputs a logic 1, which enables the buffer, and the pin is then an output pin. *AND* gate 1 can also be programmed so that one or more of the input variables can serve as an input to it, thereby allowing the inputs to control the output buffer. For instance, if *AND* gate 1 was programmed so that only input I_0 was an input to it, the logic level of I_0 would control the output buffer of output O_n. For *AND* gate 1 to have been programmed in this manner, input I_0 must be used as an output enable for output O_n. Also note from Figure 11–14a that when a pin is functioning as an output pin (O_n), the attached input (I_n) becomes a feedback line ($O_n = I_n$), which works out nicely for sequential logic designs.

Observe from Figure 11–14a that the number of *AND* gates for a PAL is not necessarily equal to 2^n as it is for PLEs. A PAL has a number of product terms available to the user which may or may not be minterms, since the *AND* gate inputs are programmable. Also, the reader should be aware of the inverted output for this logic array cell, which means that the DeMorgan theorem must be used to properly format the Boolean design equations.

Figure 11–14b is basically the same logic array as Figure 11–14a except that the logic state of the output is latched by the register. The logic array cell of Figure 11–14b has a register output similar to the PLE register outputs already discussed. Feedback is provided via the feedback buffer, which is connected to $\overline{Q}$.

Figure 11–14c illustrates, in concept, how the active state of an output is programmable. The Boolean equation for an *X-OR* gate can be expressed as

$$Q_n = A\overline{B} + \overline{A}B$$

If we define input variable *B* to be the variable that is the grounded input, and the fuse is left intact, then

$$Q_n = A\overline{0} + A0 = A$$

and the output is not inverted. However, if the fuse-link is blown, $B = 1$ and

$$Q_n = A\overline{1} + \overline{A}1 = \overline{A}$$

which states that the output of the *OR* gate is *A* inverted.

Figures 11–15 and 11–16 are samples of PAL devices. Figure 11–15 is a PAL20L8, and from the nomenclature of Figure 11–11 we know that

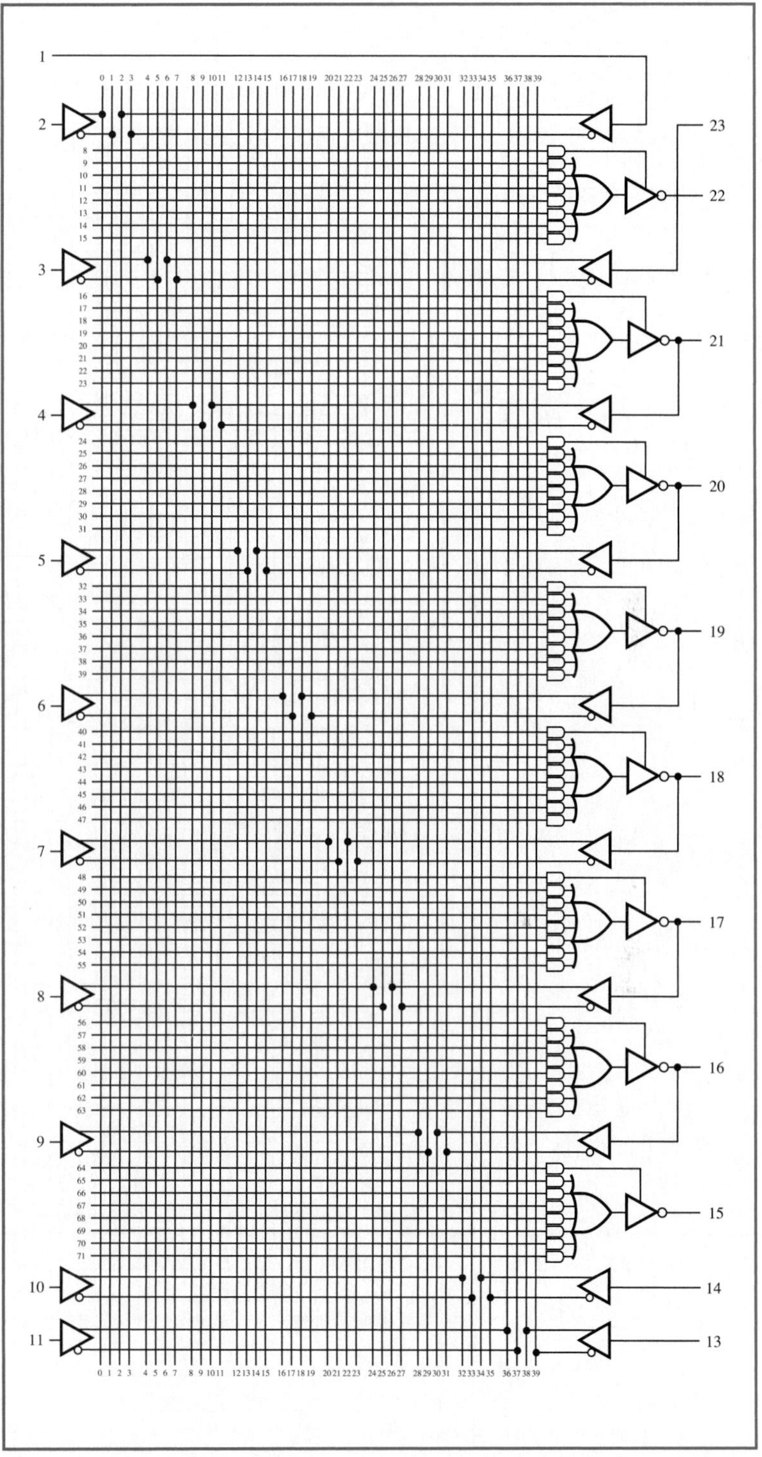

Figure 11–15
Logic architecture of the PAL20L8. (Copyright © Advanced Micro Devices, Inc., 1990. Reprinted with permission of copyright owner. All rights reserved.)

it is a PAL with 20 inputs and 8 outputs, which are buffered with inverting buffers (L). The inputs and outputs are identified with pin numbers rather than symbols. There are 14 input pins—1 through 11, 13, 14, and 23, and two output pins—15 and 22. In addition to these I/O pins there are six others that are user programmable as either input or output pins by means of the technique discussed for Figure 11–14a. Pins 16 to 21

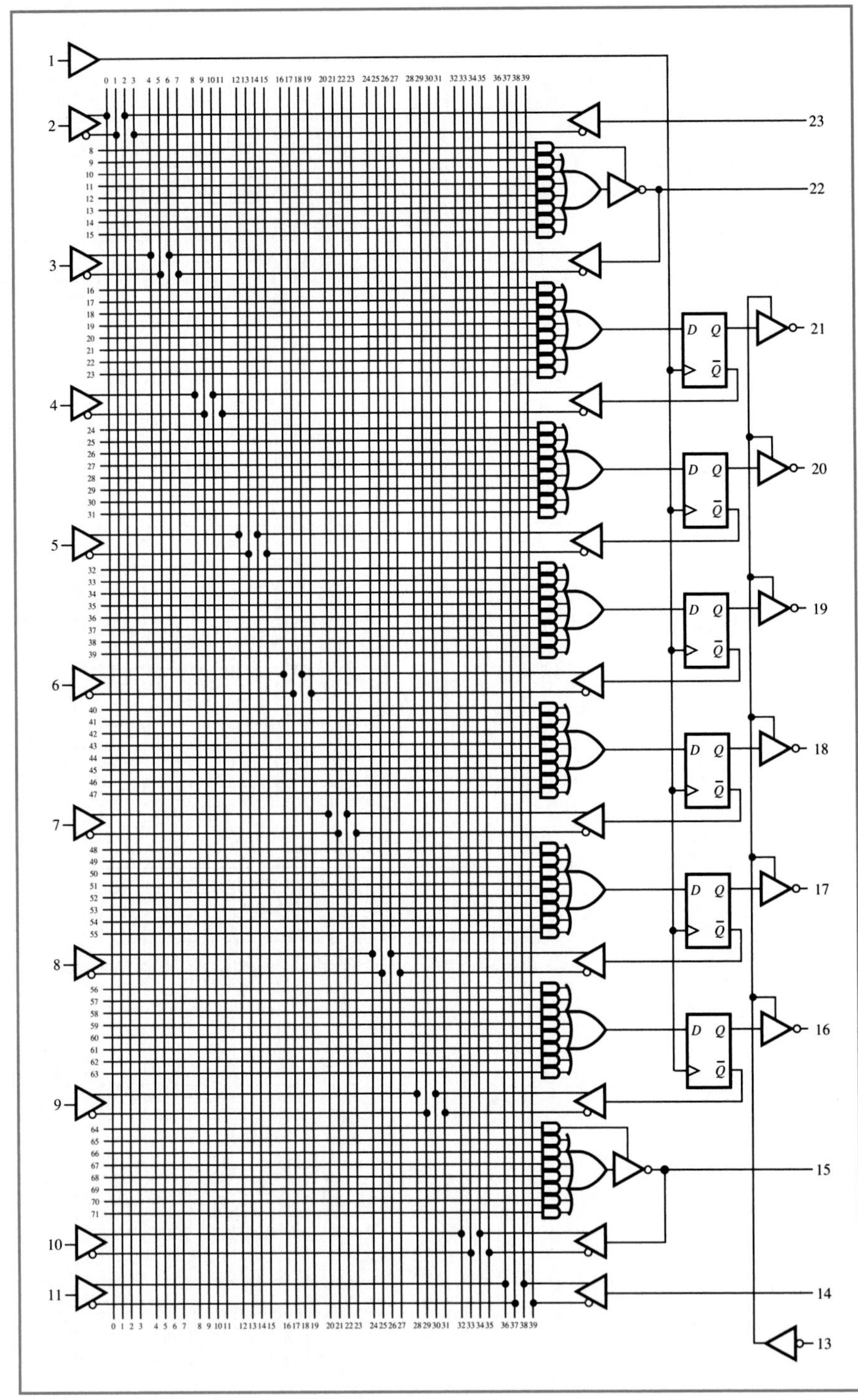

Figure 11–16
Logic architecture of the PAL20R6. (Copyright © Advanced Micro Devices, Inc., 1990. Reprinted with persmission of copyright owner. All rights reserved.)

are the programmable pins. The PAL20L8 is factory erasable, via UV light, just as an EPROM is, and a version of the PAL20L8 is available with a window for user erasability. The vertical lines 0 to 39 are the independent variables and their complements, and the horizontal lines are product lines. As can be observed from Figure 11–16, the PAL20R6 is much like the PAL20L8 except that it has six registered L-type outputs.

If a designer uses inverted output (L) type PALs POS-type Boolean design equations are better to work with. This can be verified from the basic logic array cell of Figure 11–14a. The Boolean equation for output O_n is

$$O_n = \overline{m_2 + m_3 + m_4 + m_5}$$

Applying DeMorgan's theorem we find that

$$O_n = \overline{m}_2 \cdot \overline{m}_3 \cdot \overline{m}_4 \cdot \overline{m}_5 = \mathrm{M}_2 \cdot \mathrm{M}_3 \cdot \mathrm{M}_4 \cdot \mathrm{M}_5$$

which is a POS-type Boolean expression.

EXAMPLE 11–6

Use a PAL20L8 to implement the following Boolean design equations:

$$X = \Sigma_m\, 1, 2, 3, 7 \text{ where } X(A, B, C)$$

$$Y = \Sigma_m\, 0, 3, 5, 6 \text{ where } Y(D, E, F)$$

$$Z = \Sigma_m\, 2, 3, 4, 5, 6 \text{ where } Z(G, H, J)$$

SOLUTION Since the PAL20L8 has inverted outputs, we first convert the SOP Boolean equations to equivalent POS Boolean equations. We then apply DeMorgan's theorem to properly format the Boolean equations for implementation. To convert from SOP to POS, we need only form products of those maxterms that are not present in the SOP Boolean equation. Therefore

$$X = \Sigma_m\, 1, 2, 3, 7 = \Pi_M\, 0, 4, 5, 6 \text{ where } X(A, B, C)$$

$$Y = \Sigma_m\, 0, 3, 5, 6 = \Pi_M\, 1, 2, 4, 7 \text{ where } Y(D, E, F)$$

$$Z = \Sigma_m\, 2, 3, 4, 5, 6 = \Pi_M\, 0, 1, 7 \text{ where } Z(G, H, J)$$

After minimization, where possible, we find that

$$X = (B + C)(\overline{A} + B)(\overline{A} + C)$$

$$\begin{aligned} Y &= \Pi_M\, 1, 2, 4, 7 \\ &= (D + E + \overline{F})(D + \overline{E} + F) \\ &\quad (\overline{D} + E + F)(\overline{D} + \overline{E} + \overline{F}) \end{aligned}$$

$$Z = (G + H)(\overline{G} + \overline{H} + \overline{J})$$

Applying DeMorgan's theorem we have

$$\begin{aligned} \overline{X} &= \overline{(B + C)(\overline{A} + B)(\overline{A} + C)} \\ &= (\overline{B + C}) + (\overline{\overline{A} + B}) + (\overline{\overline{A} + C}) = \overline{B}\overline{C} + A\overline{B} + A\overline{C} \end{aligned}$$

Negating the equation one more time, so as to return the $\overline{X}$ variable back to its original logic state of X, we find that

$$\overline{\overline{X}} = X = \overline{\overline{B}\overline{C} + A\overline{B} + A\overline{C}}$$

Using the same conversion principles for Y and Z, we have

$$Y = \overline{\overline{D}\overline{E}F + \overline{D}E\overline{F} + D\overline{E}\overline{F} + DEF}$$

$$Z = \overline{\overline{G}\overline{H} + GHJ}$$

Figure 11–17 shows the PAL20L8 of Figure 11–15 programmed to implement X, Y, and Z. For example, the independent variables, A, B, C, and D were input at pins 1, 2, 3, and 4, respectively, as indicated in the table below. Due to these pin assignments, the

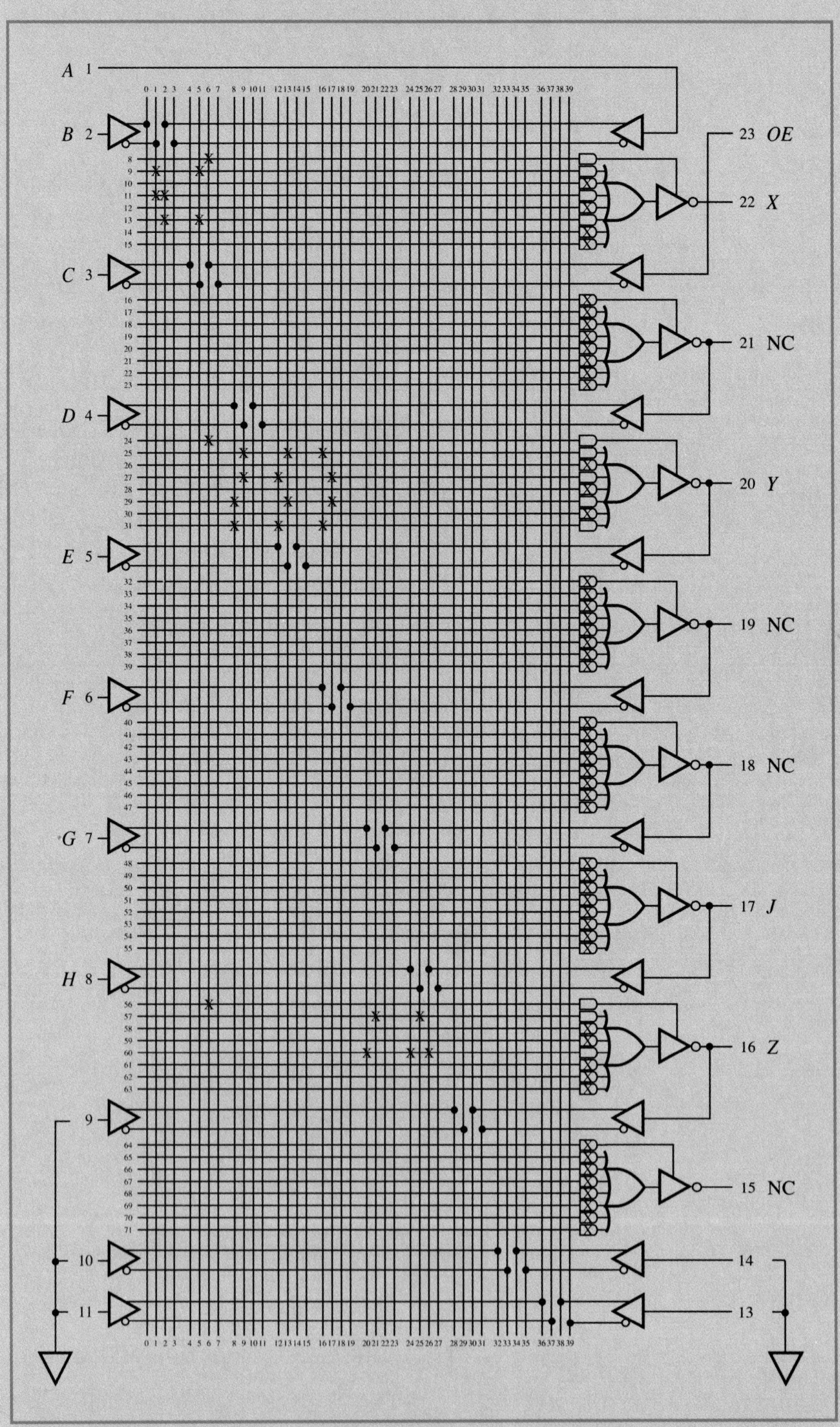

Figure 11–17
A PAL20L8 programmed for the Boolean equations of Example 11–6.

variables and their complements are available on the vertical lines indicated in the table.

Independent variable	Vertical input line
A(pin 1)	2
$\overline{A}$	3
B (pin 2)	0
$\overline{B}$	1
C (pin 3)	4
$\overline{C}$	5
D (pin 4)	8
$\overline{D}$	9
and so on.	

As an academic exercise we decided to program pin 17 to function as an input pin for the independent variable J rather than use a standard input pin. This was accomplished by leaving all fuse-links of the tristate buffer control line intact (as indicated by the X within the control *AND* gate), which causes that *AND* gate to output a fixed logic 0, and as a result, drive that output buffer in the Hi-Z state.

Since just three outputs are required (plus the one that is programmed as an input pin), the four outputs not used were not connected (NC) and all fuse-links were left intact. The outputs that were used did not require the use of all of their *AND* gate arrays; as a result, for convenience the authors left at least every other *AND* gate fully fused (making programming a little easier).

To program the PAL we simply follow the product line that is to be used for a product term and put an X at each intersection of the product line and the vertical input lines that are to be connected to that product line. For instance, since output pin 22 was chosen for the dependent variable X, horizontal product lines 8 to 15 must be programmed. The second *AND* gate from the top (product line 9) was chosen to implement the product $\overline{B}\overline{C}$ of the Boolean equation $\overline{B}\overline{C} + A\overline{B} + A\overline{C}$. Then product line 9 is connected to vertical input lines 1, which is $\overline{B}$, and 5, which is $\overline{C}$. Product line 11 is used to implement the product $A\overline{B}$, so product line 11 is connected to vertical input lines 2 and 1. Lastly, product line 13 is used to implement the product $A\overline{C}$, which requires that vertical input lines 2 and 5 be connected to that product line.

Input pin 23 is used as an output enable pin; therefore, the product line of each control *AND* gate (product lines 8, 24, and 56) is connected to the vertical input line 6. Notice that if an active-low output enable were required, the product lines of those control *AND* gates would be connected to the vertical input line 7, which is $O\overline{E}$.

To reduce the effects of electrical noise, all input pins must be connected to an inactive logic state. For that reason we have grounded input pins 9, 10, 11, 13, and 14.

EXAMPLE 11–7

Implement a 2-bit binary down-counter using a PAL20R6 (see Figure 11–16). All register outputs of a PAL20R6 are inverted; that is, rather than have Q as the output variable it is the complement $\overline{Q}$. As a result, outputs $\overline{Q}_2$ and $\overline{Q}_1$ of those flip-flops used to construct the up-counter must serve to output the count. The following truth table can be used to extract the design equations.

Q_2	Q_1	$\bar{Q}_2$	$\bar{Q}_1$	D_2	D_1
1	1	0	0	0	1
1	0	0	1	1	0
0	1	1	0	1	1
0	0	1	1	0	0

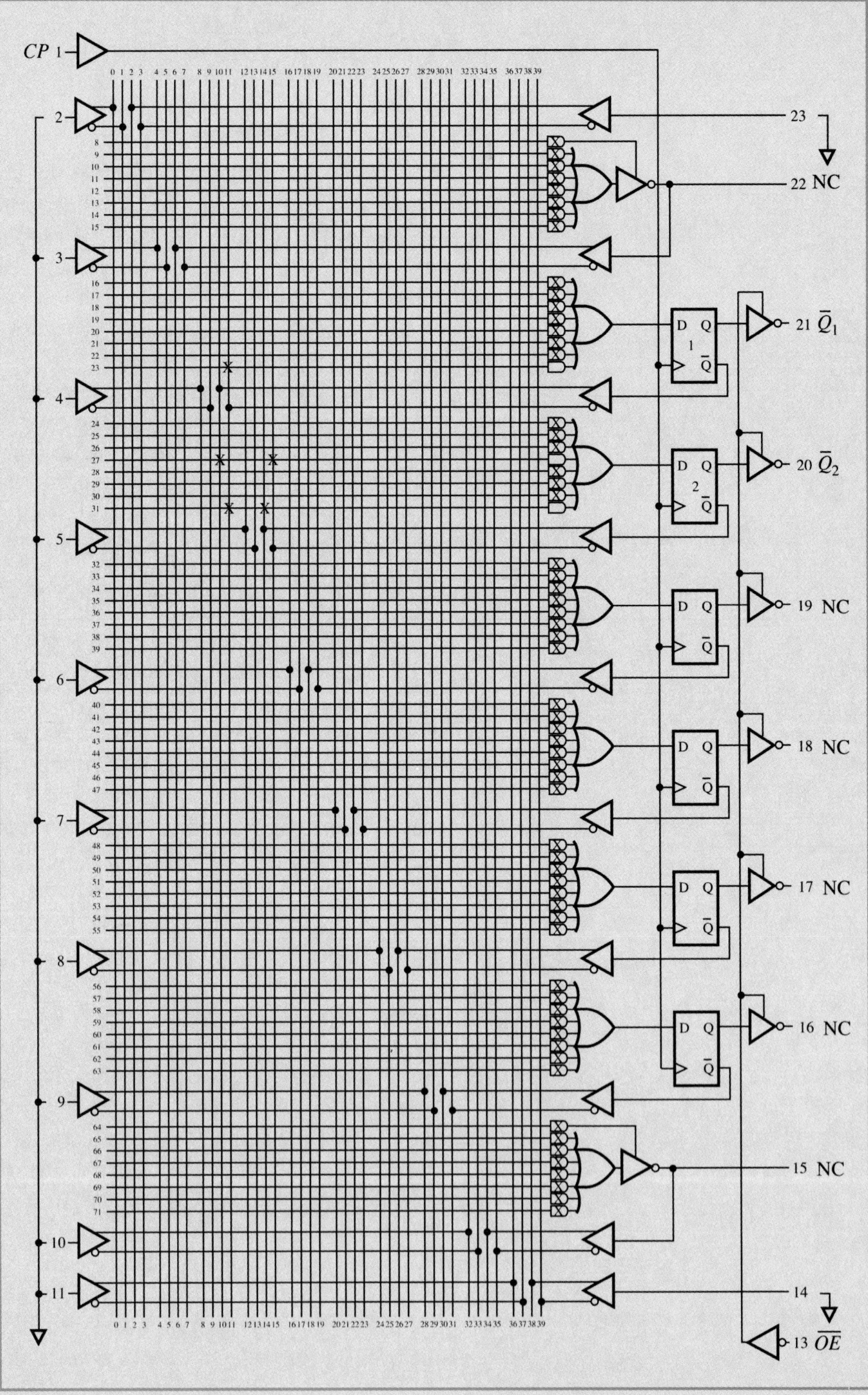

Figure 11–18
A PAL20R6 programmed as a 2-bit down-counter.

The design equations are

$$D_2 = \overline{\overline{Q}}_2\overline{Q}_1 + \overline{Q}_2\overline{\overline{Q}}_1 = Q_2\overline{Q}_1 + \overline{Q}_2 Q_1$$

$$D_1 = \sum_m 0, 2 = \overline{\overline{Q}}_1 = Q_1$$

We assign the D-type flip-flops of output pins 21 and 20 to implement $\overline{Q}_1$ and $\overline{Q}_2$, respectively, as indicated in Figure 11–18. Q_1 is available at vertical line 11 ($\overline{Q}_1$ is available at vertical line 10) and Q_2 is available at vertical line 15 ($\overline{Q}_2$ is available at vertical line 14).

To implement the design equations for the D input of flip-flop 1, horizontal product line 23 is connected to the vertical input line 11, which is Q_1. The D input for flip-flop 2 (D_2) uses the horizontal product line 27 for the product $Q_2\overline{Q}_1$ and product line 31 for the product $\overline{Q}_2 Q_1$. Hence, product line 27 is connected to vertical input lines 10 and 15, and product line 31 is connected to vertical input lines 11 and 14.

Review questions

1. What is the essential difference between a PLE and a PAL?
2. What function does the *AND* gate array of a PLE serve?
3. What is the symbolic meaning of a product line?
4. Is the quantity of *AND* gates in the *AND* gate array of a PAL necessarily equal to 2^n, where n is the number of input variables?
5. What is the significance of the *OR* gate inputs being fixed?
6. How can a PAL's programmable I/O pin be programmed to be either an input or an output?
7. From PLD product identification nomenclature, how are the number of inputs and outputs designated as well as the output type?
8. When is it that SOP-type Boolean equations rather than POS equations are best used to implement a logic design, and how does that relate to a PAL?
9. Why is a PALIIROO-type PAL required for sequential logic design?
10. What is the essence of how a programmable *X-OR* gate implements the active logic state (L or H) of an output?
11. What techniques are used to program a PAL by hand, as was done for the PAL of Figures 11–3, 11–17, and 11–18?

11–5 Programmable logic array (PLA)

A programmable logic array (PLA) is similar to a PAL, except that for a PLA both the *AND* gate and *OR* gate arrays are programmable. The PLD of Figure 11–2a is a PLA. Having the ability to program both arrays yields PLDs with the greatest flexibility. The price paid for this increased flexibility is a slight loss in performance speed (an increase in propagation delay time—see Chapter 12) and a decrease in the number of logic gates on the silicon wafer owing to the increased number of fuse-links. Figure 11–19 shows a logic architectural diagram of a PLA.

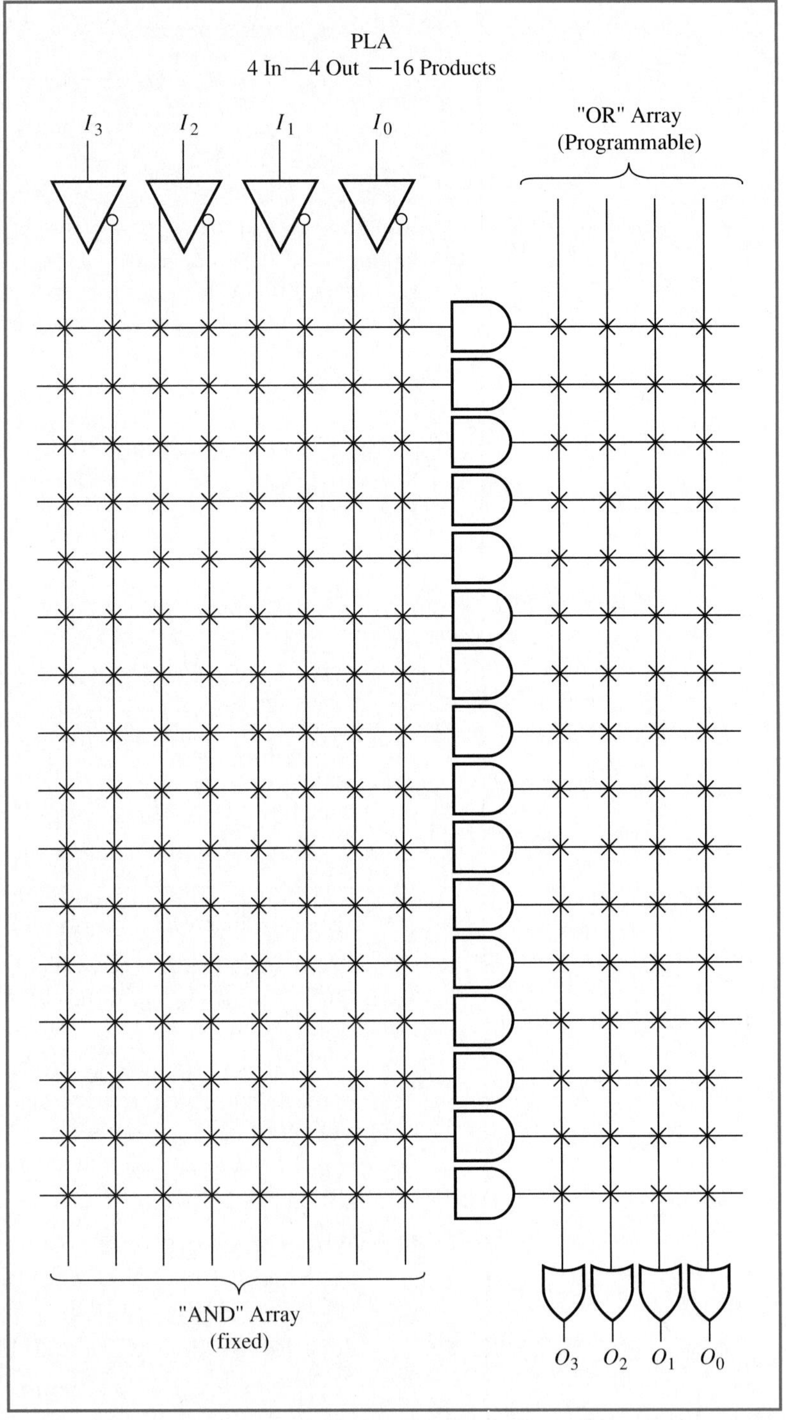

Figure 11–19
The logic architecture of a PLA. (Copyright © Advanced Micro Devices, Inc., 1990. Reprinted with permission of copyright owner. All rights reserved.)

Comparing the PLA of Figure 11–19 with that of Figure 11–12, one can see that a PLA is not limited to four product-terms per output, but rather can be programmed to have as many product terms as there are *AND* gates in the *AND* gate array. Comparison of the PLE of Figure 11–9 and the PAL of Figure 11–12 with the PLA of Figure 11–19 shows that a PLA has the programming capability of both PLE and PAL logic architectures. As with PAL, the number of *AND* gates in the *AND* gate array does not have to be equal to 2^n (where n equals the number of inputs) unless the designer of the PLA intends that the user have these *AND* gates available for *OR*ing all minterms.

EXAMPLE 11–8

Using the PLA of Figure 11–19, implement the Boolean design equations

$$Y = \overline{A}\overline{C} + \overline{A}BD + \overline{B}C + A\overline{B}C + AC\overline{D}$$
$$W = \overline{A}\overline{C} + AB\overline{C} + ABD + \overline{B}C$$

Figure 11–20
A programmed PLA for Example 11–8.

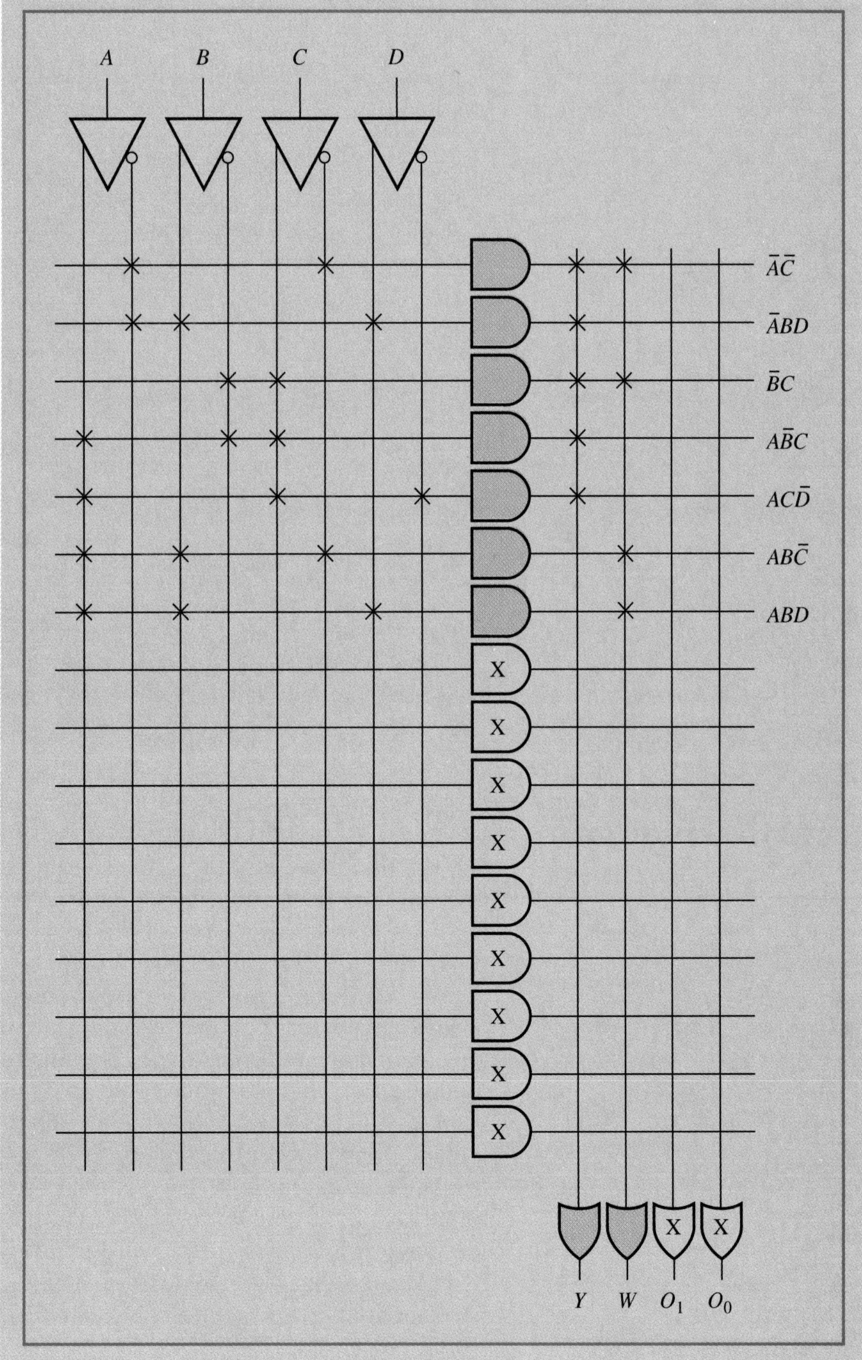

SOLUTION We first program the product lines by taking each nonrepeated product term in the Boolean design equation and connecting the indicated variables to that product line, just as was done for the PAL. The first product line of Figure 11–20 is connected to variables $\bar{A}$ and $\bar{C}$; its product $\bar{A}\bar{C}$ is the first product term of the equation for Y. For convenience, when programming the *OR* gate array, the authors identified the product $\bar{A}\bar{C}$ at the product-term line (the output of the *AND* gate). The next product line was programmed for the second product ($\bar{A}BD$) of the equation for Y. The product-term line for that *AND* gate was identified as $\bar{A}BD$. This procedure is repeated for all products in the Boolean equations for Y and W. Those *AND* gates that are not needed are not programmed. Notice that equations Y and W have two products in common.

Once the *AND* gate array has been programmed, the *OR* gate can be programmed by putting a $\times$ at the intersection of each horizontal product-term line in the Boolean equation and the vertical input line for the *OR* gate being programmed. The vertical input line for *OR* gate Y is connected to the five products indicated by the Boolean design equation. The *OR* gate with outputs O_1 and O_0 are not being used.

The nomenclature for a PLA is indicated by Figure 11–11. Hence, a PLA with 16 inputs and eight registered outputs would be identified as a PLA16R8.

Review questions

1. How are PLAs and PLEs alike, and how do they differ?
2. How are PLAs and PALs alike, and how do they differ?
3. What are the major advantages and disadvantages of PLAs compared to PLEs and PALs?

11–6 TROUBLESHOOTING

PLDs, whether PLEs, PLAs, or PALs, are programmable combinational logic circuits contained within a single semiconductor integrated circuit and are therefore viewed as a system function-block rather than individual logic circuits for troubleshooting purposes. To troubleshoot a PLD for a fault is a very straightforward process. By means of a schematic of the programmed PLD and/or a truth table of the input and output variables, the diagnostician knows the correct output responses for any combination of logic levels for the input variables. With this information the diagnostician can devise a test to evaluate the PLD either statically or dynamically. That is, either static or dynamic logic level generators are used to generate logic levels for the input variables and either static or dynamic logic level detectors are used to detect the resultant output logic levels. If an output logic level is not correct, the PLD has failed and must be replaced. Of course, to verify failure this output must be isolated from the input(s) it is driving to ensure that it is not the input of the circuit being driven that has failed, for example, an input that has become a permanent short to ground due to an electronic failure.

Review questions

1. Why must PLDs be viewed as function-block diagrams relative to troubleshooting?
2. If a fault is found with a PLD, whether the fault is with an input or output, what is the remedy?
3. What is the essence of any PLD static or dynamic test?

11–7 SUMMARY

- Programmable logic devices are structured primarily of *AND* and *OR* gate arrays. The inputs to either the *AND* gates, the *OR* gates, or both are programmable via a fuse-link. These programmable fuse-links afford the PLD a flexibility that ordinary combinational logic circuits do not have. A 20- or 24-pin PLD can often replace up to six ICs containing typical logic gates.
- The type of PLD is determined by which logic gate array is programmable. If the logic of the *AND* gate array is fixed (not programmable) and the *OR* gate array is programmable, the PLD is a programmable logic element (PLE), or PROM. If the logic of the *AND* gate array is programmable and that of the *OR* gate array is fixed, the PLD is a programmable array logic (PAL) device. When the logic of both arrays is programmable, the PLD is a programmable logic array (PLA). PLEs are the least flexible of the PLDs, PLAs are the most flexible, and PAL is in the middle.
- Programmability extracts a price of increased propagation delay time and reduced logic density on the silicon wafer. PAL is a convenient compromise between flexibility and performance. Using a PAL is seldom a sacrifice, since most logic designs do not require the program flexibility of a PLA.
- PLDs have a variety of output combinations. There are buffered outputs that are noninverting [active-high (H)] or inverting [active-low (L)]. Some PLDs have registers as their outputs (*D*-type flip-flops), with or without feedback. Lastly, some PLDs have programmable I/O pins. The user must decide which PLD best accommodates the design.
- PLD nomenclature readily identifies a PLD type. The first symbols usually identify the type of device (PLE, PAL, or PLA), and in some instances the manufacturer as well. The number following the type identification indicates the number of inputs, the next symbol specifies the output configuration, and the next number indicates the number of outputs available.
- To symbolically program a PLD, an *x-y* coordinate system is used. A $\times$ is placed at any horizontal and vertical intersection that is to be connected to indicate that the fuse-link is to remain intact.
- In troubleshooting, the PLD is viewed as an IC that is either properly functioning or has failed and must be replaced, rather than as individual logic circuits.
- To troubleshoot a PLD, the logic levels of the input variables are applied and the output variables are tested for correct responses. An incorrect response constitutes a failure of the PLD.

PROBLEMS

Section 11–2 The programmable fuse-link

1. Explain the concept of PLD programmability and explain what a fuse-link is.
2. Use the PLD of Figure 11–2a to implement the Boolean design equations

$$X = AB + BC$$
$$Y = AC$$

Show the fuse-links that are intact and those that are blown in the logic diagram of Figure 11–2a.

3. Use the PLD of Figure 11–2a to program the Boolean design equation $X = \overline{A}B + AB$. Make a drawing of the logic circuit.
4. Modify the PLD of Figure 11–2a so that the Boolean design equation

$$Y = A\overline{B} + \overline{A}B$$

can be implemented. Show the logic diagram of your modified circuit as it appears when programmed.

5. Show the logic circuit, similar to the logic circuit of Figure 11–2a, that is capable of implementing a design with three independent variables (inputs) and four dependent variables (outputs). The logic circuit is to accommodate the independent variables and their complements.
6. Use the logic circuit of Problem 5 to program the Boolean equations.

$$X = \sum_{m} 0, 3, 5, 6$$
$$Y = \sum_{m} 1, 3, 5, 6, 7$$
$$Z = \sum_{m} 0, 6, 7$$

The independent variables are A, B, and C. For the truth table assign A to be the MSB and C to be the LSB. Show the resultant programmed logic circuit for the simplified Boolean equations.

7. In order to accommodate all possible minterms, how many *AND* gates must there be in a PLD?

Section 11–3 Programmable logic elements (PLE) and PLD symbolic notation

8. What are the conceptual differences between the PLE of Figure 11–4 and the PLD of Figure 11–2a?
9. If a PLE has six input pins for inputting independent variables how many *AND* gates are there?
10. What relationship is there between the number of *AND* gates and *OR* gates for a PLE? Explain your answer.
11. How many inputs does each *OR* gate of a PLE have? Explain your answer.

12. Use the PLE of Figure 11–4 to implement the Boolean design equations

$$A = \sum_m 0, 3$$

$$B = \sum_m 1, 2$$

$$D = 0$$

$$E = \sum_m 0, 2, 3$$

Show the resultant programmed logic diagram, which should be similar to Figure 11–5.

13. Repeat Problem 2 using the PLD symbolic notation for indicating a fuse-link that is to remain intact and single lines to indicate the inputs to the *AND* and *OR* gate arrays, as in Figures 11–6 and 11–7.

14. Use a PLE as a look-up table to convert binary hexadecimal values 0 through F into their equivalent ASCII binary codes. You are to design the PLE required. Show the programmed version of your PLE, which will be similar to the programmed PLE of Figure 11–7.

15. Modify the outputs of Problem 14 so that they may be connected to the data bus of a microprocessor-based system.

16. Modify the outputs of Problem 14 so that their output logic states can be latched (remembered) and then read at some later time.

17. What is the major restriction to the flexibility of a PLE?

18. Can the PLE of Figure 11–9 be used to implement the simplified version of the Boolean equation

$$A = \sum_m 1, 3, 5, 7, 10, 11, 12, 13$$

where $A(X, Y, W, Z)$?

Explain your answer.

19. What would be the PLD product identification for
    a. the PLE of Problem 15?
    b. the PLE of Problem 16?

Section 11–4 Programmable array logic (PAL)

20. How does a PAL differ from a PLE?

21. Why is a PAL more flexible than a PLE?

22. What relationship exists between the number of input variables and logic gates in the *AND* and *OR* gate array of a PAL?

23. Use the PAL of Figure 11–12 to implement the simplified Boolean equation of Problem 18.

24. How many inputs and outputs are required for a PAL to be used to implement the Boolean design equations given below?

    a. $Y(A, B, C)$
       $Z(A, B, C)$

    b. $A(X, Y, W, Z)$ $D(X, Y, W, Z)$
       $B(X, Y, W, Z)$ $E(X, Y, W, Z)$
       $C(X, Y, W, Z)$ $H(X, Y, W, Z)$

25. Construct a PAL that will accommodate the implementation of the Boolean design equations given below.

$$W = \sum_m 0, 1, 2, 7$$

$$X = \sum_m 3, 5, 6, 7$$

$$Y = \sum_m 0, 3, 4, 6$$

$$Z = \sum_m 0, 3$$

where the independent variables are A, B, and C with A = MSB and C = LSB of the truth table.

26. How would you modify the PAL of Problem 25 if it were to be interfaced with a microprocessor-based system?

27. Explain how an output pin can be programmed to function as an input pin. Use the basic logic array cell of Figure 11–14a as a model for your explanation.

28. Explain how an *X-OR* gate can be programmed to determine if a logic state is inverted or not. Use Figure 11–14c as a model for your explanation.

29. Repeat Problem 25, but use the PAL of Figure 11–15 to implement the design.

30. Repeat Problem 25, but use the PAL of Figure 11–16 to implement the design. Pin 22 is to be programmed as an input pin.

31. Design a three-stage up-counter and implement the design using a PAL20R6 (see Figure 11–16).

Section 11–5 Programmable logic array (PLA)

32. What are the advantages and disadvantages of a PLA over a PLE or PAL?

33. What is the relationship between the number of independent variables and the number of *AND* and *OR* gates in the logic gate arrays of a PLA?

34. Use the PLA of Figure 11–19 to implement the following Boolean design equations:

$$G = \sum_m 0, 5, 10, 11, 14, 15$$

$$H = \sum_m 4, 7, 10, 11, 14, 15$$

$$J = \sum_m 1, 3, 9, 10, 11, 14, 15$$

$$K = \sum_m 1, 3, 5, 10, 11, 14, 15$$

where the independent variables are A, B, C, and D (A = MSB).

Troubleshooting

Section 11–6 Troubleshooting

35. Devise a static test to troubleshoot the PLE of Figure 11–7.

36. How would you dynamically test the PLE of Figure 11–7?

37. Devise a dynamic test for the PAL of Figure 11–13 and state those input logic levels for which there should be a logic 1 at output Y.

38. Devise a static test for the PAL of Figure 11–17.

39. What logic levels would be input to test output Z of Figure 11–17?

40. Devise a test configuration for the PAL of Figure 11–18 and show the expected output for $\overline{Q}_1$ and $\overline{Q}_2$.

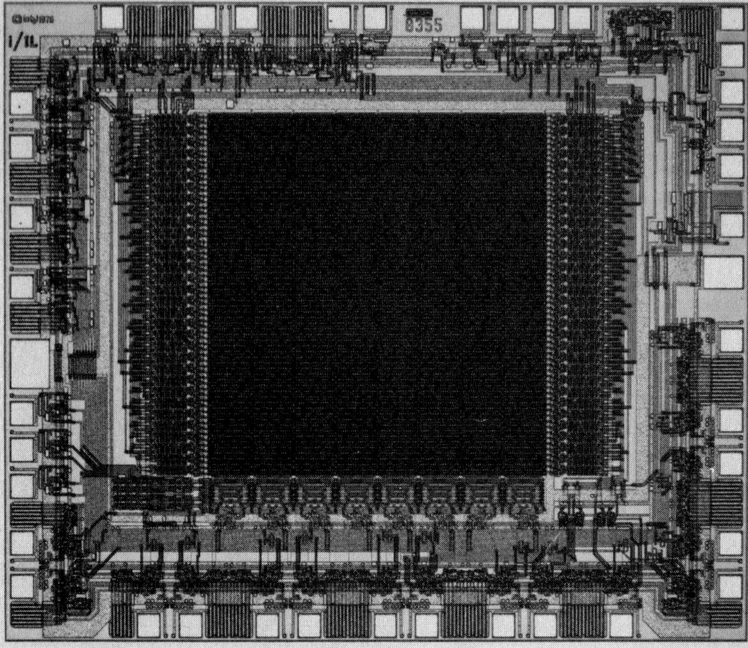

Various designs or technologies are used to fabricate highly complex digital circuits onto a piece of silicon inside an integrated circuit. The photograph shows a typical memory chip capable of storing approximately 16,000 bits of information (Reprinted by permission of Intel Corporation. © Intel Corporation 1989).

OUTLINE

OBJECTIVES

The objectives of this chapter are to:

- explain the operating characteristics of bipolar and MOSFET electronic switches.
- describe the structuring of electronic switches that form logic gates.
- analyze the voltage and current requirements of logic families.
- explain loading rules for various logic families.
- investigate the propagation delay time characteristic of various logic families.
- examine the memory cell for different technologies.
- investigate some parameters that are specified in manufacturers' data sheets.
- apply the knowledge learned in this chapter to some representative designs.

12 THE TECHNOLOGY OF LOGIC FAMILIES

12–1 INTRODUCTION

Many different logic families may be used to implement a digital design. Logic families include transistor-transistor logic (TTL), resistor-transistor logic (RTL), diode-transistor logic (DTL), emitter-coupled logic (ECL), N-channel MOS (NMOS), and complementary MOS (CMOS). We will study the more popular of these logic families, which are TTL, NMOS, and CMOS. As we have already seen, each of these logic families has characteristics that are better suited for a particular application than are other families. By studying the electronics of these logic families we shall understand why logic families have specific characteristics and have more insight into the loading characteristics introduced in Chapter 2. Among the characteristics of particular interest are power consumption, voltage levels, current requirements, loading rules, and propagation delay time.

We will investigate which logic families are directly compatible and which need interfacing circuitry. This will enable a "mix and match" of logic families when necessary or desirable.

The electronic principles of memory cells will also be examined and will be related to the design concepts developed in Chapters 8 and 9. In addition to understanding the "whys" of memory access time, we will be able to improve the memory designs and I/O port interfacing dealt with in Chapter 9 by considering loading criteria.

12–2 The electronic switch

Regardless of which logic family is being studied, the electronic switch is the key to the technology in question. An *electronic switch* is a transistor (bipolar or MOS) that is always in one of two states—on or off. All logic families are constructed of either bipolar switches (TTL, RTL, DTL, or ECL) or MOSFET switches (PMOS, NMOS, or CMOS).

Figure 12–1a is an illustration of a bipolar junction transistor (BJT) configured as an electronic switch. Figure 12–1b and c are equivalent circuits of that electronic switch for the on and off states. Figure 12–1a shows the transistor voltage parameters V_{BE} and V_{CE} as well as the electronic switch voltage V_I, which is the input voltage, and the output voltage V_O. Note that the transistor voltage parameter V_{CE} is the same as the output voltage V_O of the electronic switch. The purpose in labeling the same voltage with two different symbols has to do with our prior knowledge and our present interest. Those who have studied BJTs in other courses, such as an electronic device course, are accustomed to using the collector-to-emitter voltage V_{CE} and are familiar with the

Figure 12–1
An NPN bipolar electronic switch.

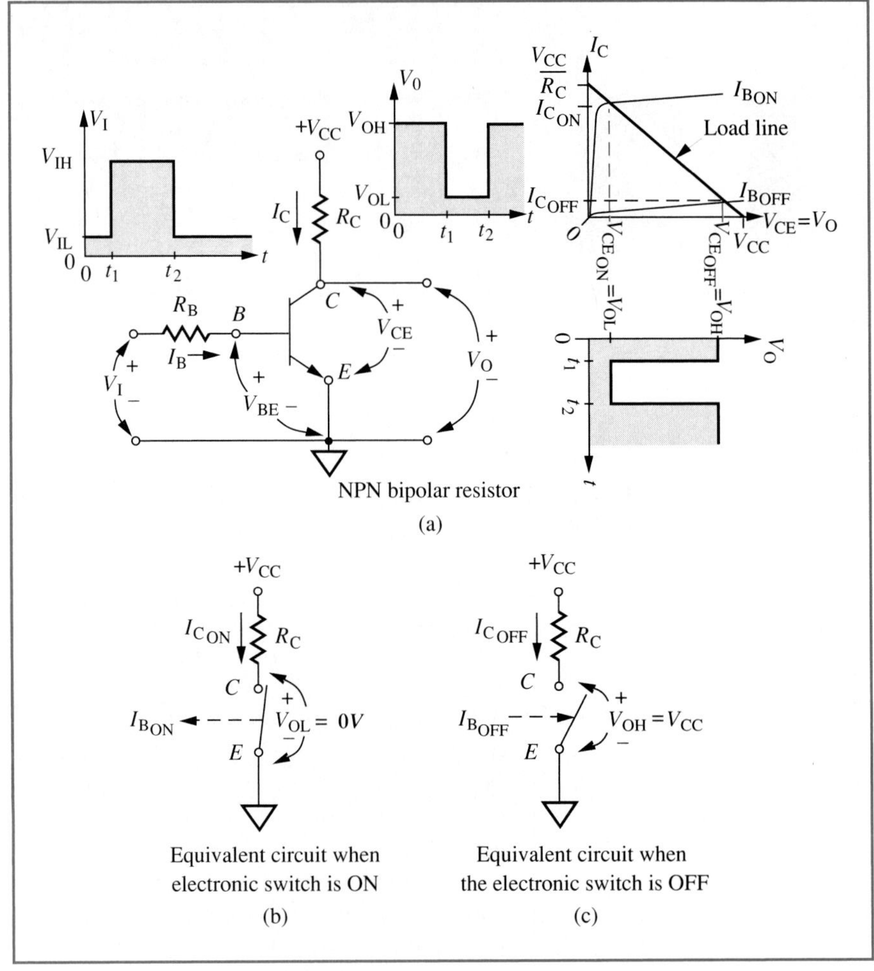

characteristic curve of Figure 12–1a. However, in digital electronics the interest focuses on the electronic switch parameters V_I and V_O. We will use both notations as we study the electronic switch.

To understand the BJT electronic switch we must know the cause-and-effect relationship of its parameters V_{BE}, I_B, I_C, and V_{CE}. To turn the NPN transistor of Figure 12–1a on, the base-to-emitter voltage V_{BE} must have the polarity shown and be greater than approximately 0.6 volt. When this condition occurs, the base-to-emitter junction of the transistor is *forward biased* (forward biasing turns on the base-to-emitter junction like a switch), and base current I_B flows through the base-to-emitter junction to ground. The electronic switch is designed so that R_B has a magnitude that allows a relatively large amount of base current to flow when the base-to-emitter junction is turned on. In the characteristic curves of the BJT this base current has been labeled $I_{B\ ON}$. Whenever any magnitude of base current flows, it causes collector-to-emitter current to flow, which is similar to turning on the collector-to-emitter junction. If the base current is large enough, it turns the collector-to-emitter junction on "hard" so that its equivalent circuit is a closed switch, as depicted in Figure 12–1b. When the electronic switch is on, the output voltage is zero—the voltage level for a logic 0. As a result, when the electronic switch of Figure 12–1a is turned on, its output is driven to the logic low (0) level, symbolically designated V_{OL} (V for voltage, O for output, and L for logic low), as we recall from Section 2–9.

To relate the events and magnitudes of the electronic switch for the on condition to the characteristic curve of the BJT, refer to Figure 12–1a. We see that when base current $I_{B\ ON}$ is flowing through the base-to-emitter junction, the point of intersection with the load line when projected vertically to the V_{CE} axis indicates that an output voltage of $V_{CE\ ON}$ is present at V_{CE}. The magnitude of $V_{CE\ ON}$ is the same as V_{OL}. We also note that the collector current is at a maximum and is identified as $I_{C\ ON}$, which also is shown in Figure 12–1b.

If the base-to-emitter junction of Figure 12–1a is not forward biased, the BJT is turned off and its collector-to-emitter equivalent circuit is that of an open circuit, which is illustrated in Figure 12–1c. If the BJT is not forward biased, V_{BE} must be less than approximately 0.6 volt. Under this condition the base-to-emitter junction is an equivalent open circuit, which means that there is no base current. Without base current there is no collector current I_C, so there must be an equivalent open circuit between the collector and the emitter, which is the case (Fig. 12–1c). With the collector-to-emitter circuit open, the collector-to-emitter voltage of Figure 12–1c is at voltage level V_{CC}. V_{CE} is the output voltage V_O, and since its voltage level is V_{CC}, which is a logic high (1), the output voltage is a logic high. The output voltage of V_O is designated V_{OH} for the logic high state (V for voltage, O for output and H for logic high).

To relate the off condition to the BJT's characteristic curve of Figure 12–1a, the condition of no base current (or almost none—there is always leakage current) is labeled $I_{B\ OFF}$. The intersection of $I_{B\ OFF}$ with the load line results in a voltage magnitude of $V_{CE\ OFF}$ for V_{CE}, which is a logic high level designated V_{OH}.

Note in each equivalent circuit of Figure 12–1b and c that base current I_B is symbolically shown to control the position of the toggle. If the base current is large enough, it closes the switch; otherwise, the switch is open.

In Figure 12–1a there is an input (V_I) and output (V_O) voltage waveform. Let us review the events of the electronic switch using these voltage waveforms. During the interval $t_1 < t < t_2$, $V_I = V_{IH}$ (the input voltage is a logic high) and its magnitude and polarity are correct for forward biasing the base-to-emitter junction of the BJT. This turns on the collector-to-emitter junction, producing output voltage V_{OL}. However, when the input voltage is at the logic 0 level (V_{IL}) the magnitude is not sufficient to turn the base-to-emitter junction on and there is little to no base current, which is indicated by $I_{B\ OFF}$ (only a little leakage current flows). Without base current there is no collector current, and the collector-to-emitter voltage is at a logic high level (V_{OH}), as indicated in the output waveform of Figure 12–1a. Note from these two waveforms that the electronic switch is an inverter.

It is important to understand that bipolar technologies are current-driven; that is, in addition to a base-to-emitter voltage of sufficient magnitude and proper polarity, a *base-to-emitter current is required to turn a BJT switch on.*

There are a variety of MOSFETs—the junction FET, the channel depletion mode, and the channel enhancement mode—and each of the modes can be either a negative (N) or a positive (P) channel. They are all voltage-controlled devices. The more popular mode is the N-channel enhancement mode, which will be used to discuss the workings of a MOSFET electronic switch.

The N-channel enhancement mode MOSFET of Figure 12–2a has areas at each end that are heavily doped with negative-(N) yielding–type material. This doping results in either end, the drain (D) and the source (S), being N-type semiconductor material. Between the source and the drain is the channel, which has been doped with a material that yields a P-type semiconductor. Before current can flow from source to drain, the channel must be changed to a pseudo–N-type material. That is, while the

Figure 12–2 An N-channel enhancement mode MOSFET electronic switch.

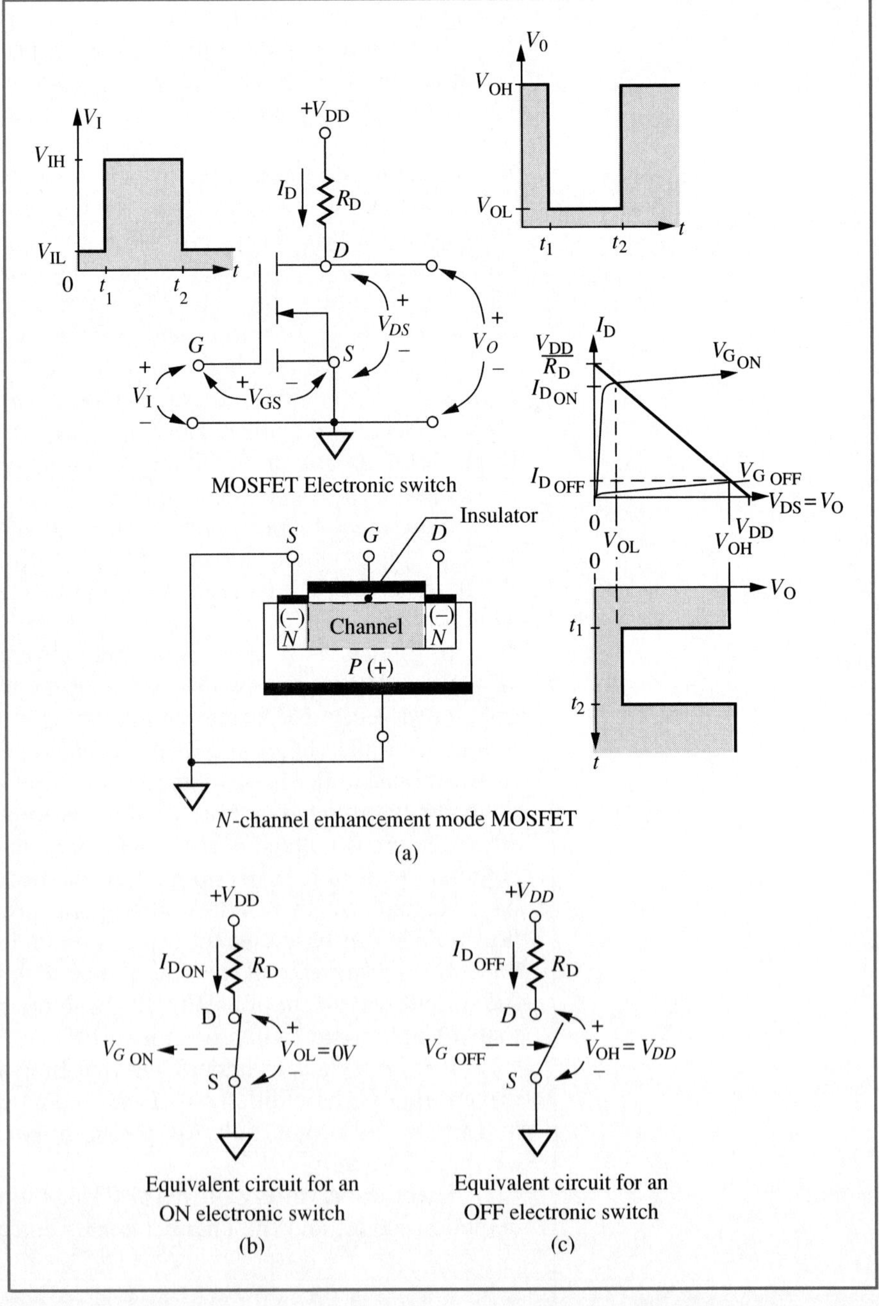

channel doping is not physically changed, we must find a method that has the same net result. To change the channel so that it appears to be an N-type channel, a positive voltage is placed on the gate (G). As illustrated in Figure 12–2a, the gate is physically placed over the channel but is insulated from it by a thin layer of metal oxide. The schematic symbol for a MOSFET shows the gate insulated from the drain and source by a space as well as the space existing between source, channel, and drain. This positive gate voltage attracts electrons (opposites attract) throughout the channel material and concentrates them beneath the gate, which runs the length of the channel. At the same time this positive gate voltage repels positive charges (holes) from the channel. As a result, the concentration of N-type and P-type charges within the channel changes in favor of the N-type charges (the channel has been *enhanced* with current carrier-electrons), which has the effect of changing the channel to an N-type material, causing electron current to flow between the source and drain. As the positive gate voltage is increased, it attracts more electrons and repels more holes in the channel, which increases the current flow between the source and drain (I_D).

The characteristic curve of Figure 12–2a supports what has been said. At a gate voltage of $V_{G\ OFF}$, which is insufficient to attract enough electrons within the channel for current flow, there is no drain current. However, from the characteristic curve we see that even with a gate voltage of $V_{G\ OFF}$ there is a small value of gate leakage current, $I_{D\ OFF}$. Ideally $I_{D\ OFF}$ is zero, which means the electronic switch is open-circuited, as illustrated in Figure 12–2c. With the electronic switch open, the output voltage V_O of the switch is V_{DD}, which is a logic high, and the output voltage for this condition is identified as V_{OH}.

As indicated by the characteristic curve, as the gate voltage is increased to $V_{G\ ON}$, the drain current increases to $I_{D\ ON}$. When the drain current has a value of $I_{D\ ON}$, the electronic switch is considered closed, as illustrated in Figure 12–2b. From the characteristic curve we can also see that the output voltage of the MOSFET is nearly at zero volts and, consistent with our notation, is designated V_{OL}.

Like the bipolar electronic switch, the MOSFET electronic switch is an inverter. The difference between the two is that the logic state of a MOSFET is voltage-controlled (gate voltage), whereas that of the bipolar switch is current-controlled (base current).

Review questions

1. For a logic high input voltage, what is the logic level of the output of an electronic switch?
2. What input parameter of a bipolar switch controls the logic level of its output?
3. What input parameter of a MOSFET switch controls the logic level of its output?
4. Both bipolar and MOSFET characteristic curves indicate that for the off condition (open circuit) there is a small collector ($I_{C\ OFF}$), or drain ($I_{D\ OFF}$) current. Why are these currents present?
5. What do the symbols V_{OL} and V_{OH} represent? Explain the meaning of each letter of the symbols.

12–3 Fundamental electronic principles of logic gates

As we saw in Chapter 2, electronic switches can be used to construct logic gates. Figure 12–3 shows how two electronic switches form an *AND* gate when connected in series and an *OR* gate when connected in parallel. The position of the toggle of each switch is controlled by base currents I_{B1} and I_{B2} for bipolar electronic switches and gate voltages V_{G1} and V_{G2} for MOSFET switches. The accompanying truth tables verify the logic of each circuit, where the inputs I_B and V_G are simply labeled no. 1 and no. 2. An input logic 0 means no base ($I_{B\ OFF}$) for the bipolar switch and no gate voltage ($V_{G\ OFF}$) for the MOSFET switches, whereas a logic 1 input means $I_{B\ ON}$ or $V_{G\ ON}$. Although actual logic gates are more complicated than those of Figure 12–3, they do function in a similar fashion.

Transistor-transistor logic (TTL)

The circuits of Figure 12–4 are a TTL inverter and a tristate buffer with inverted output. We will discuss Figure 12–4a now but defer the discussion of Figure 12–4d until multiple emitters (Q_1) have been discussed. They are both included in the same figure because of their similarities in structure and operation.

Let us begin by taking a quick overview of this inverter. The diodes D_1 and D_2 are not part of the electronic switches. D_1 serves to protect Q_1 from a reverse polarity input voltage. D_2 ensures that Q_3 is off when Q_2 is on. Transistor Q_1 controls the base current of Q_2 and hence determines whether Q_2 is on or off. Transistor Q_2 in turn determines whether transistor Q_3 or Q_4 is on or off. If Q_2 is off, Q_3 is on and Q_4 is off;

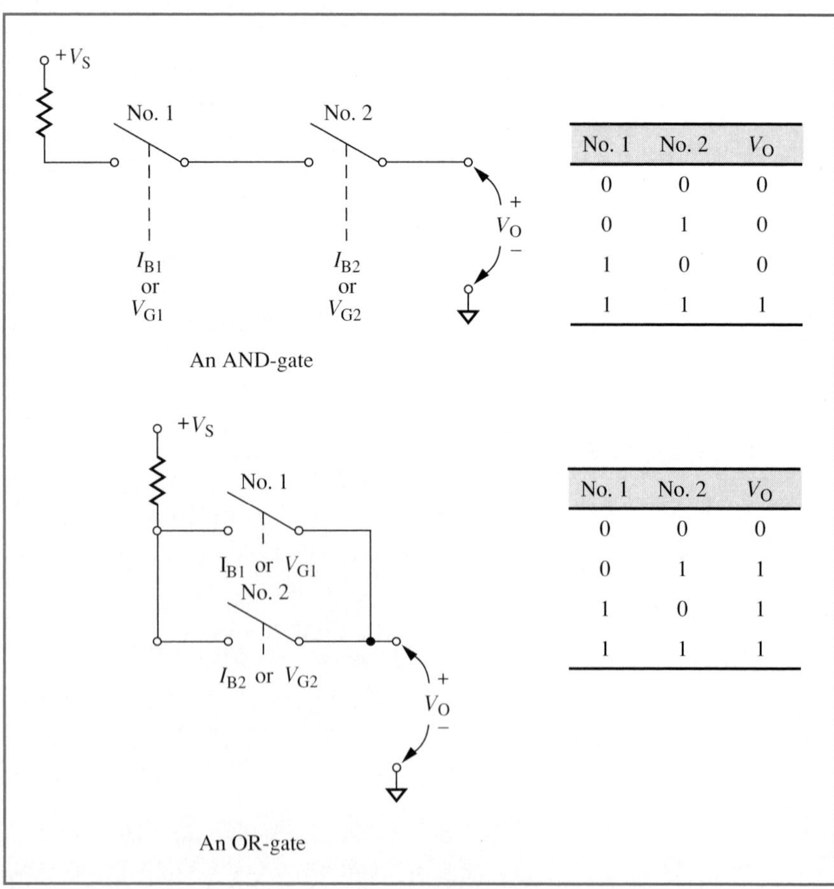

No. 1	No. 2	V_O
0	0	0
0	1	0
1	0	0
1	1	1

No. 1	No. 2	V_O
0	0	0
0	1	1
1	0	1
1	1	1

Figure 12–3 Logic gates constructed from electronic switches.

Figure 12–4
Basic TTL logic circuits.

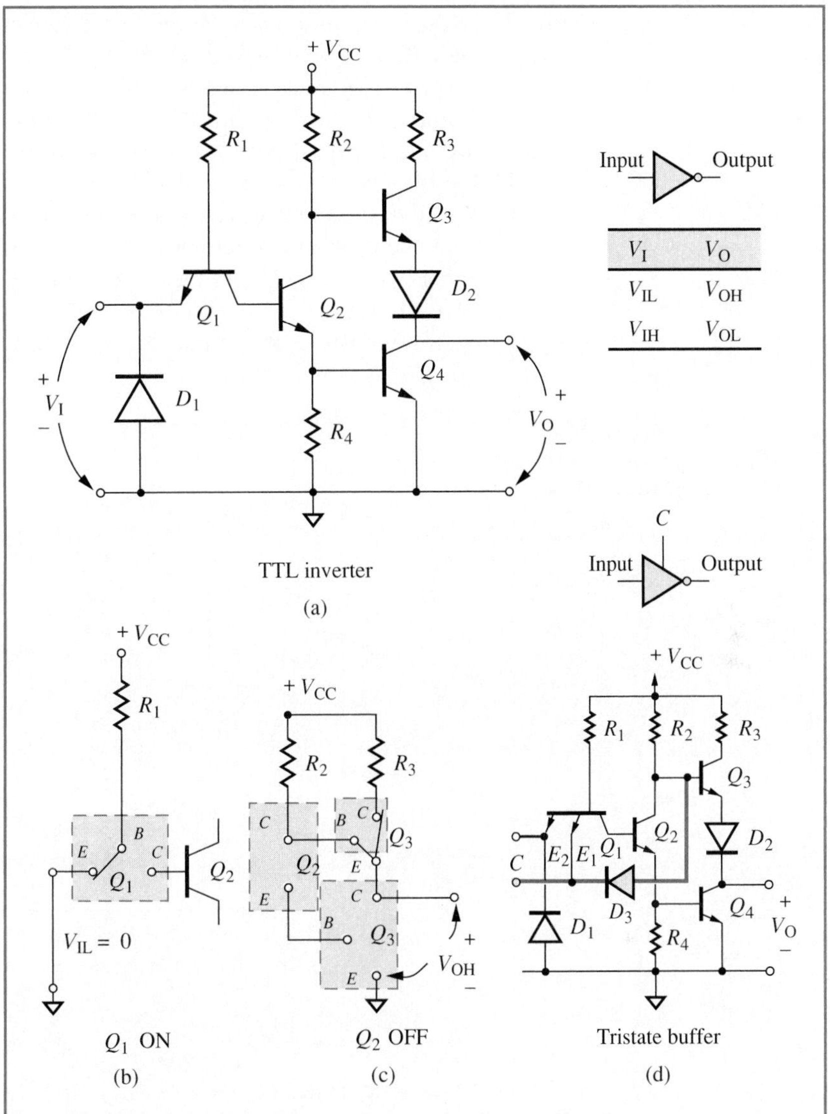

however, if Q_2 is on, Q_3 is off and Q_4 is on. The series arrangement of Q_3 and Q_4 is known as a *totem pole*. When Q_4 is off the output voltage is a logic high, and when Q_4 is on the output voltage is a logic low.

Before analyzing the inverter of Figure 12–4a, let us introduce the equivalent electronic switch circuit for Q_1. This equivalent circuit is a double-pole–single-throw switch, as indicated in Figure 12–4b. The reason for this type of equivalent circuit is that current will either be flowing through the base-to-emitter junction or the base-to-collector junction.

Let us apply a logic 0 and then a logic 1 to the input V_I of the TTL inverter of Figure 12–4a and determine the resultant output logic level. Applying a logic 0 to the input, which is designated V_{IL} in the truth table of Figure 12–4a, pulls the emitter of Q_1 to ground (or close to it). With the emitter at approximately zero volts (ground) and the base pulled to V_{CC} (+5 volts), via R_1, the base-to-emitter junction is forward biased and

Q_1 turns on and base-to-emitter current flows, as shown in Figures 12–4b and 12–5a. Even though Q_1 is on, there is effectively no collect-to-emitter current (I_C), since there is no source for that current. That is, the current would have to flow through the base-collector junction of Q_2 in order to provide a closed path from ground to the power supply V_{CC}. As shown in Figure 12–5a, the base-to-collector of Q_2 is an equivalent open circuit. Without sufficient base-to-emitter current to Q_2 it is off, and it has an equivalent open collector-to-emitter circuit, as shown in Figures 12–4c and 12–5a. With Q_2 acting as an equivalent open circuit there can be no base-to-emitter current to Q_4, but R_2 provides a direct path through which base-to-emitter current can be supplied by the power supply to Q_3. R_2

Figure 12–5
Equivalent circuits for the TTL inverter of Figure 12–4.

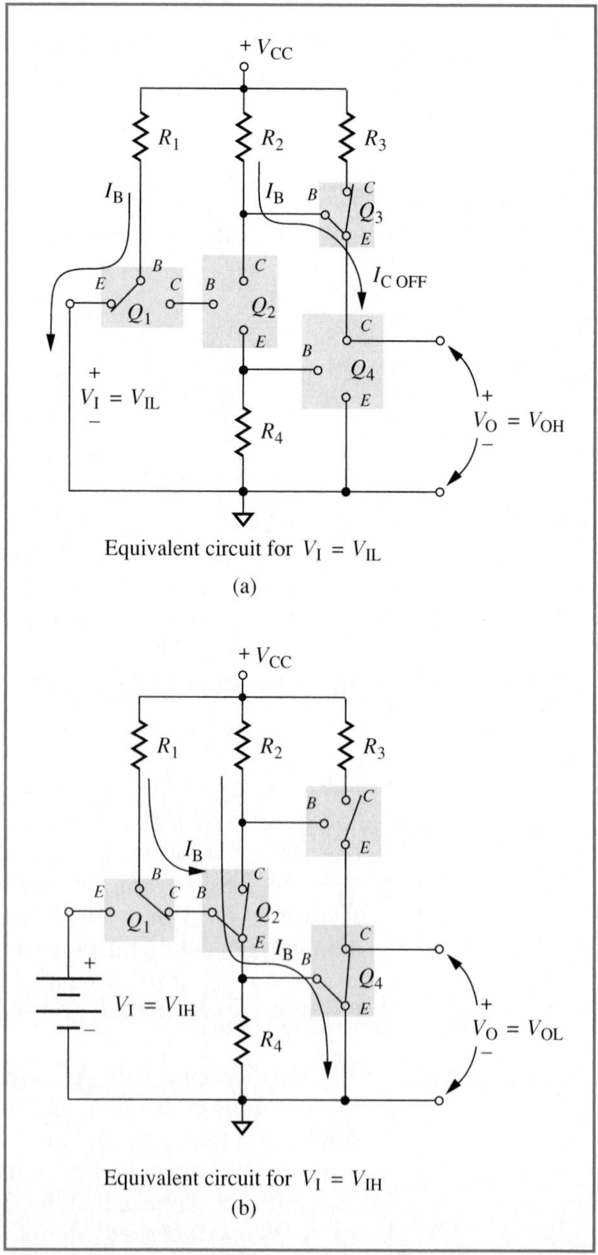

also "pulls" the base potential of Q_3 up toward V_{CC} volts. With approximately V_{CC} volts on the base of Q_3 and a source for a base-to-emitter current, Q_3 turns on. It should be noted that I_B of Q_3 is small, since Q_4 is off, but even with Q_4 off $I_{C\ OFF}$ amperes can flow through Q_4, and Q_3 is designed so that this much base-to-emitter current will turn Q_3 on, which is illustrated in Figure 12–5a. When Q_3 is on, its collector-emitter is an effective short circuit; it pulls the collector of Q_4 toward V_{CC}, which is the logic 1 state that is designated V_{OH} in the truth table of Figure 12–4a. Therefore $V_O = V_{OH}$, as shown in Figure 12–5a.

Inputting V_{IH} (a logic 1) volts at the emitter of Q_1 does not forward bias Q_1's base-to-emitter junction. As a result, that junction is an equivalent open circuit, as shown in Figure 12–5b. The emitter of Q_2 is referenced to ground via R_4. This has the effect of referencing the collector of Q_1 to ground. With the collector of Q_1 pulled toward ground potential and its base near V_{CC} volts, the base-to-collector junction of Q_1 is forward biased, acting as a forward biased diode that is designed to turn on and act as an equivalent short circuit, as shown in Figure 12–5b. This provides a source of base-to-emitter current for Q_2, which turns Q_2 on, as indicated in Figure 12–5b. There is now a base-to-emitter current path for Q_4, which turns Q_4 on and pulls the collector of Q_4 to approximately zero volts. Pulling V_O to near ground potential results in a logic level of V_{OL}.

Compare the equivalent circuits of Figure 12–5a and b and note that Q_2 controls whether Q_3 or Q_4 is on (both cannot be on at the same time), and Q_1 controls the state of Q_2.

Why not use the simple inverter of Figure 12–1a rather than the more complex one of Figure 12–4? The answer is the reduced current drain at the output. The inverter of Figure 12–4 is designed so that there is always a high impedance between V_{CC} and ground, since either Q_3 or Q_4 is off.

The inverter of Figure 12–4a can be modified so that it becomes the *NAND* gate of Figure 12–7. To make this modification, Q_1 is designed and manufactured with multiple emitters, one emitter for each input. Figure 12–6a shows the schematic symbol of a multiemitter transistor, and Figure 12–6b is its diode equivalent. The base-to-emitter diodes are connected so that if one or more of these diodes are forward biased, base-to-emitter current can flow and turn Q_1 on; otherwise, the base-to-collector diode is forward biased and a base-to-collector current will flow. Since Q_1 of Figure 12–6a has either a base-to-emitter current flow or a base-to-collector current flow, the equivalent circuit of a multiemitter transistor is also a double-pole–single-throw switch. As a result, the equivalent circuits of Figure 12–5 are also valid for analyzing the *NAND* gate of Figure 12–7.

When all inputs to Q_1 of Figure 12–7 are a logic high, the base-to-emitter junction of each equivalent diode is not forward biased, resulting in no base-emitter current for Q_1. With this condition, the equivalent circuit of Figure 12–5b is valid and the output voltage V_O is a logic 0, which agrees with the truth table of Figure 12–7. Anytime a logic low (V_{IL}) is input to one or more emitters of Q_1, a path exists for base-to-emitter current to flow, and Q_1 is turned on. With Q_1 on, the equivalent circuit of Figure 12–5a is valid, and the output voltage is a logic high ($V_O = V_{OH}$). From the truth table of Figure 12–7 we see that indeed any

Figure 12–6
Multiemitter transistor Q_1.

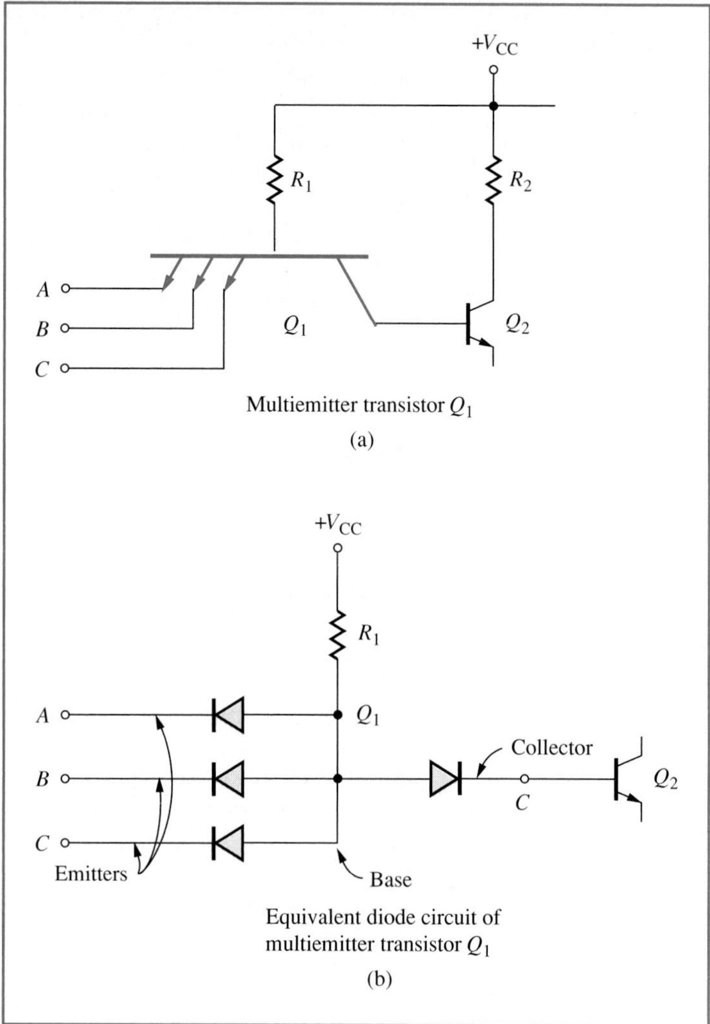

time an input is at a logic low the equivalent circuit of Figure 12–5a is valid and V_O is a logic 1 (V_{OH}).

Return now to the tristate buffer of Figure 12–4d. If the control line C is a logic high, D_3 and emitter E_1 are reverse biased and the circuit functions as the inverter of Figure 12–4a. However, if control C is pulled to a logic low, diode D_3 is forward biased, which pulls the base of Q_3 low, turning Q_3 off. Emitter E_1 of transistor Q_1 is also forward biased, which turns Q_1 on and turns Q_2 off. As a result of diode D_3 and transistor Q_1 being on, both transistors Q_3 and Q_4 are off. Hence, the output impedance is very high (Hi-Z) and very little current can be sinked or sourced (the output has no current drive capability). As a result, the Hi-Z output can be easily driven to either a logic 0 or 1 by any other output connected to it.

Normally the outputs of logic gates cannot be connected because this may result in damage to the output transistor Q_4 as well as bus contention. One type of TTL technology is designed so that outputs can be connected: *open-collector* logic. The top half of the output totem pole

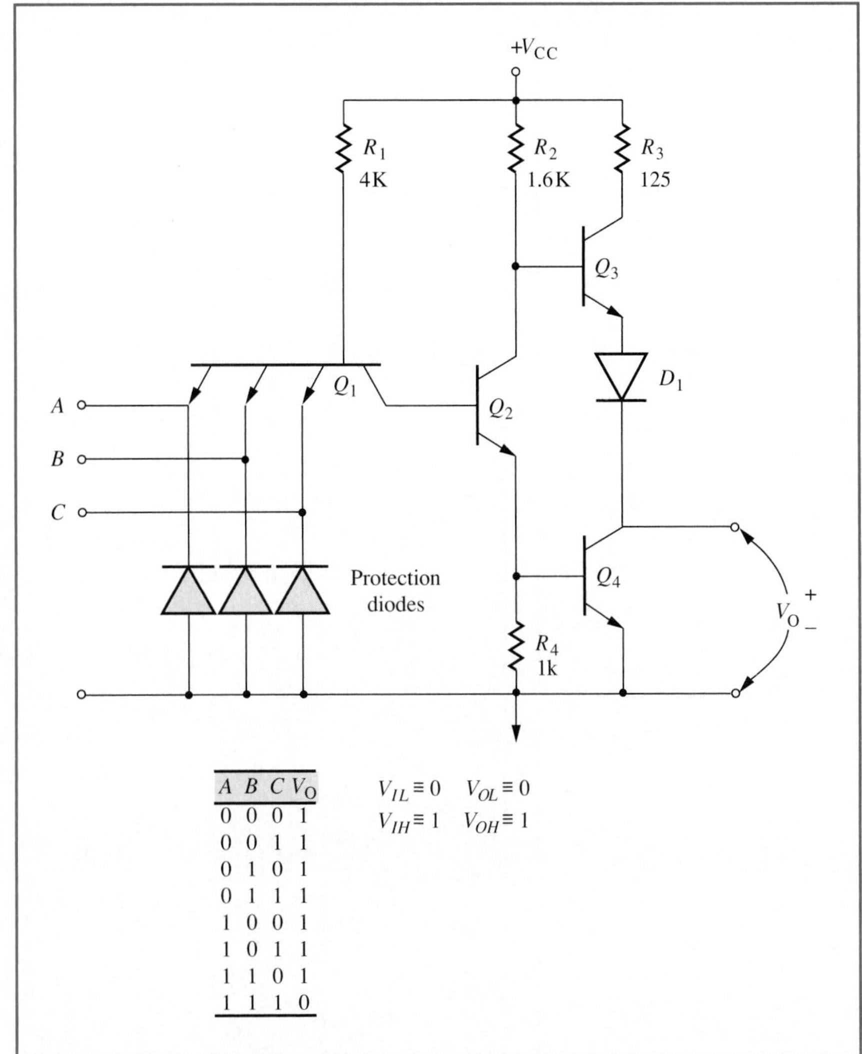

A	B	C	V_O
0	0	0	1
0	0	1	1
0	1	0	1
0	1	1	1
1	0	0	1
1	0	1	1
1	1	0	1
1	1	1	0

Figure 12–7
TTL three-input *NAND* gate.

is removed, as illustrated for the *NAND* gate of Figure 12–8a. In place of the removed portion of the totem pole is an external *pull-up resistor R* (Figure 12–8b), which configures Q_4 as an electronic switch similar to Figure 12–1a. From Figure 12–8b we see that by connecting the outputs of open-collector logic gates, *NAND* gates in this case, we have paralleled the output transistors (Q_4) of each *NAND* gate. As we know, paralleling electronic switches implements *OR* logic, but owing to negation the resultant logic is an *AND*. Thus the logic at the connection is the *AND*ing of the three *NAND* gates whose outputs are connected, which is why the resultant logic from wiring the outputs of open-collected logic is known as *wired-AND logic*. Figure 12–8c shows the symbolic representation for wired-*AND* logic.

The pull-up resistor derives its name from the fact that it ''pulls'' the collector(s) of Q_4 to V_{CC} when Q_4 is off. It also serves to limit the current from the power supply when Q_4 is on. The magnitude of this current must be less than the magnitude specified by the manufacturer as damaging to Q_4.

Figure 12–8
Open-collector *NAND* gate.

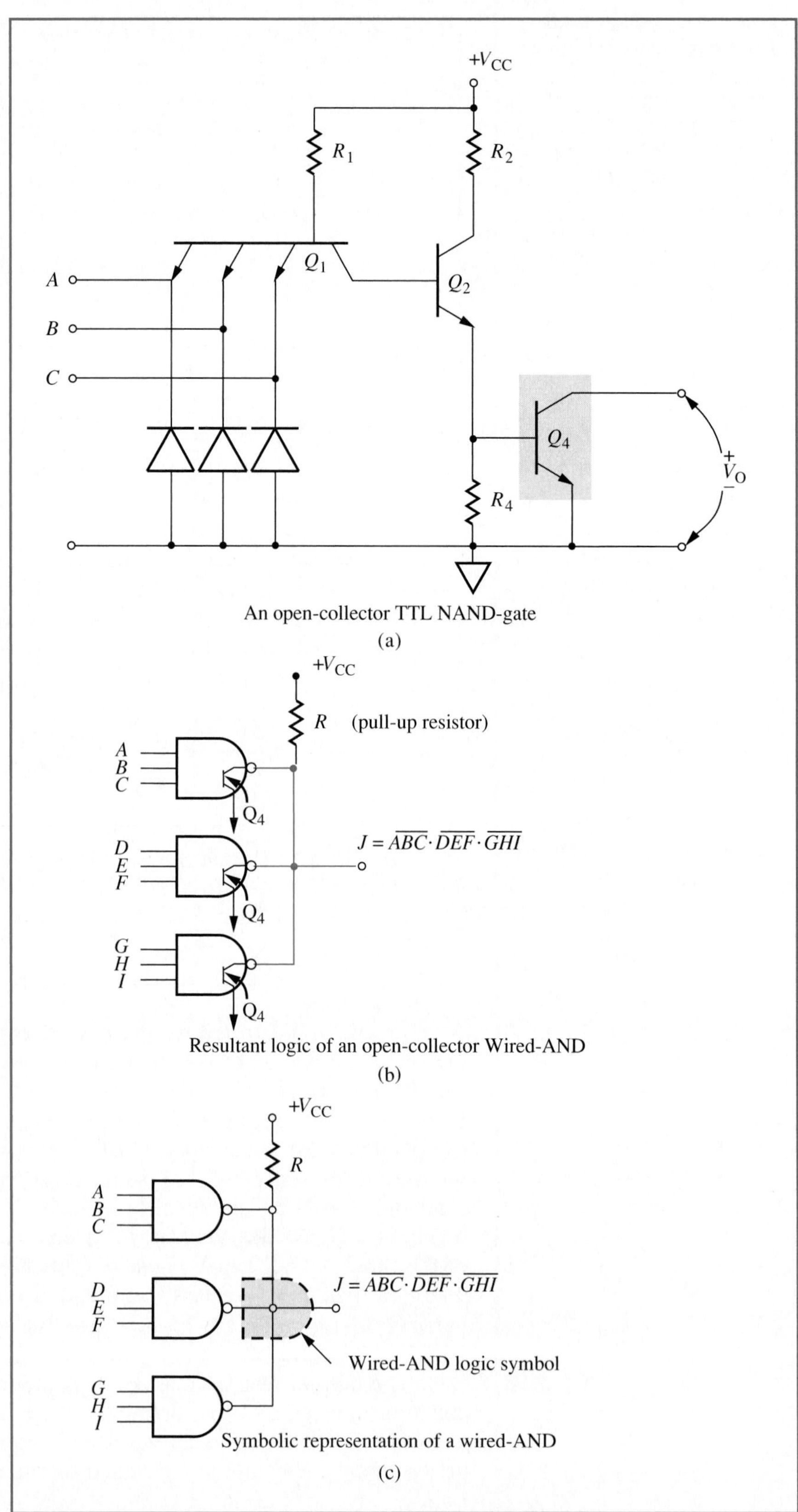

There are various open-collector TTL logic functions available, such as *NAND* gates, *NOR* gates, *AND* gates, *OR* gates, multiplexers, registers, and memory chips. Regardless of the logic function, its output operates as described for the *NAND* gate of Figure 12–8.

The on-to-off switching time of the bipolar transistors can be shortened by not allowing those electronic switch transistors to be turned on so "hard," that is, by not letting them be driven into saturation. To accomplish this the base current must be limited as the transistor approaches $I_{B\ ON}$, which is shown on the characteristic curve of Figure 12–1a. To limit the base current a specially designed bypass diode, known as a Schottky diode, is placed from the base to the collector of a bipolar transistor, as shown in Figure 12–9a. Note that the cathode of a Schottky diode is drawn as an ʃ in the schematic. This diode has a very low forward biased voltage drop of approximately 0.2 volt. As a result, with the base-to-emitter junction of the transistor at approximately 0.6 volt (forward biased voltage drop) and the collector voltage V_{CE} for an on transistor at about 0.4 volt ($V_{CE\ ON}$), the Schottky diode is forward biased and turns on, which bypasses any excessive base current to the collector. This action limits the base current to the amount needed to turn on the transistor and keeps it from being driven into saturation. When the transistor is to be turned off using this technique, there are fewer excessive charges in the base-to-emitter region of the transistor (NP junctions act like capacitors in that they collect charges) that must first be drained off. A Schottky transistor is created when the Schottky diode of Figure 12–9a is manufactured as an integral part of the transistor, and its logic symbol is that of Figure 12–9b. Schottky transistors are of the bipolar TTL family, and therefore logic gates that utilize Schottky transistors are still TTL devices. Schottky-based logic gates are designated with an ʃ to differentiate them from others of the same family.

Figure 12–9c is a Schottky *NAND* gate. Although there are some circuit changes in the *NAND* gate of Figure 12–9c, when compared to the *NAND* gate of Figure 12–7, these changes are not related to the use of Schottky transistors but rather are for the purpose of improving performance. To electronically analyze the *NAND* gate of Figure 12–9c we would proceed exactly as we did for the standard TTL *NAND* gate of Figure 12–7.

N-channel MOS logic (NMOS)

A disadvantage to manufacturing TTL logic is the implementation of resistors. Resistors require a lot of integrated circuit area, which means relatively few logic gates can be put on a TTL chip. As a result, TTL is restricted to *small-scale integration* (SSI), which means that there are a maximum of 10 logic gates on a single chip, *medium-scale integration* (MSI), which means 10 to 100 gates, or *large-scale integration* (LSI), which means 100 to 1000 gates. MOS technology integrated circuits use MOSFETs as biasing resistors; that is, these MOSFETs are substituted for R_D of Figure 12–2a. This technique allows the number of logic gates per chip to be greatly increased. At present, the upper limit is approximately 200,000 logic gates per chip, which is designated as *very-large-scale integration* (VLSI). MOS technology is used primarily for VLSI applications, such as microprocessors and memory chips.

Figure 12–9
Schottky electronics and logic gate.

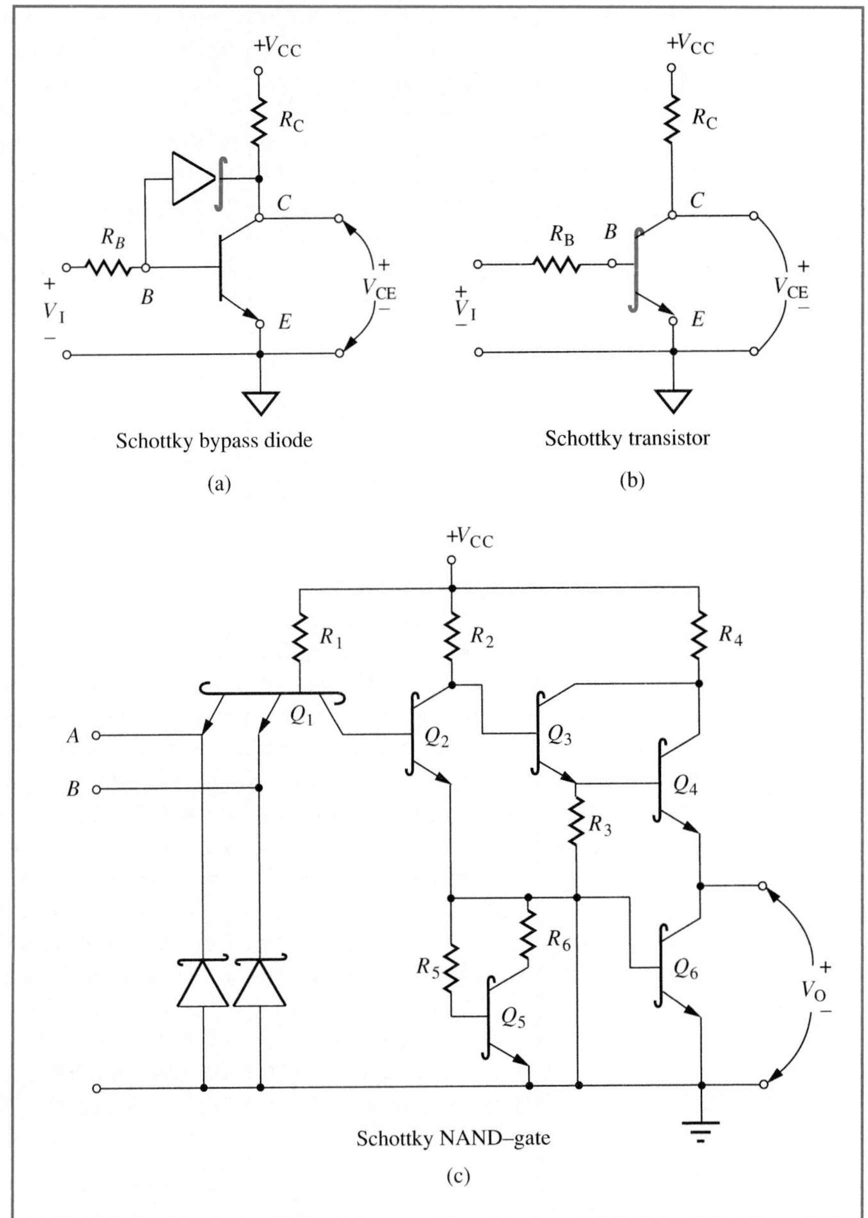

The MOSFET identified as Q_R in Figure 12–10a and b serves as bias resistor R_D. The gate of Q_R is tied to V_{DD}, which biases it to conduct (partially turns it on) so as to offer an impedance of R_D ohms. The two MOSFET electronic switches of Figure 12–10a, Q_A and Q_B, are connected in parallel. From Figure 12–3 we know that paralleling of switches is an *OR* configuration. However, since electronic switches invert the logic (negative logic), the paralleling of electronic switches Q_A and Q_B must result in a negated *OR*, which is a *NOR* gate.

To electronically verify the type of logic gate of Figure 12–10a we will input the logic levels as indicated in the truth table. In keeping with our notation, V_{IL} and V_{OL} are logic 0's, and V_{IH} and V_{OH} are logic 1's. Let us input a logic 0 to inputs *A* and *B*. Both Q_A and Q_B will be off,

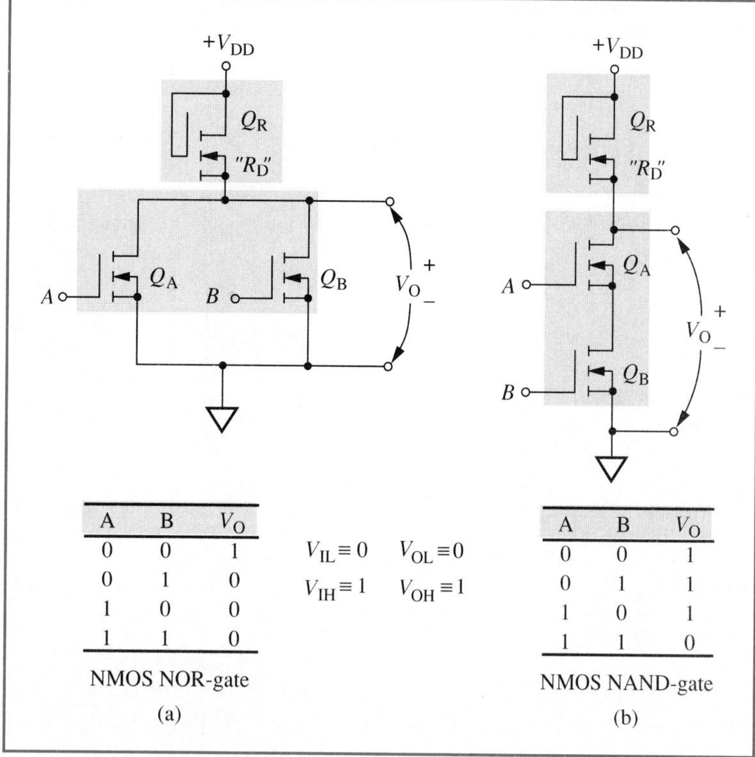

A	B	V_O
0	0	1
0	1	0
1	0	0
1	1	0

NMOS NOR-gate
(a)

A	B	V_O
0	0	1
0	1	1
1	0	1
1	1	0

NMOS NAND-gate
(b)

Figure 12–10 NMOS logic gates.

thereby pulling the output up to a logic 1, which agrees with the first row of the truth table. Applying a logic 0 to input *A* and a logic 1 to input *B* results in Q_A being turned off and Q_B being turned on. With Q_B on, the output is pulled to ground potential, which is a logic 0. This agrees with the second row of the truth table. Because Q_A and Q_B are in parallel, any time either or both electronic switches of Figure 12–10a are on, the output is a logic 0, which is also stated by the truth table. This completes the verification.

When electronic switches are connected in series, they configure a logic *AND* according to Figure 12–3. Since electronic switches Q_A and Q_B of Figure 12–10b are in series and electronic switches invert, the resultant logic will be a negated *AND* gate, or *NAND* gate. This can be verified electronically by inputting a logic 0 to both switches at inputs *A* and *B*. Under this condition both switches are off and the output is pulled toward V_{DD}, which is a logic 1. In fact, owing to the series connection of the switches, any time either or both switches are off, the output is pulled to V_{DD}. The output is other than a logic 1 only when both switches are on—that is, both inputs *A* and *B* are a logic 1—since then the output is pulled to ground (a logic 0). This is the truth table of a *NAND* gate.

Complementary MOS logic (CMOS)

To reduce power consumption the resistance of Q_R of Figure 12–10 could be increased. In fact, if the resistance value of Q_R were in the megaohm range, the current drawn from the power supply V_{DD} would be greatly reduced. A MOS technology that accomplishes this goal is referred to as complementary MOS (CMOS). CMOS uses the complement channel type of electronic switch to structure Q_R. As shown in Figure 12–11, the bias resistors Q_{RA} and Q_{RB} are P-channel MOSFETs, and the electronic

Figure 12–11
CMOS logic gates.

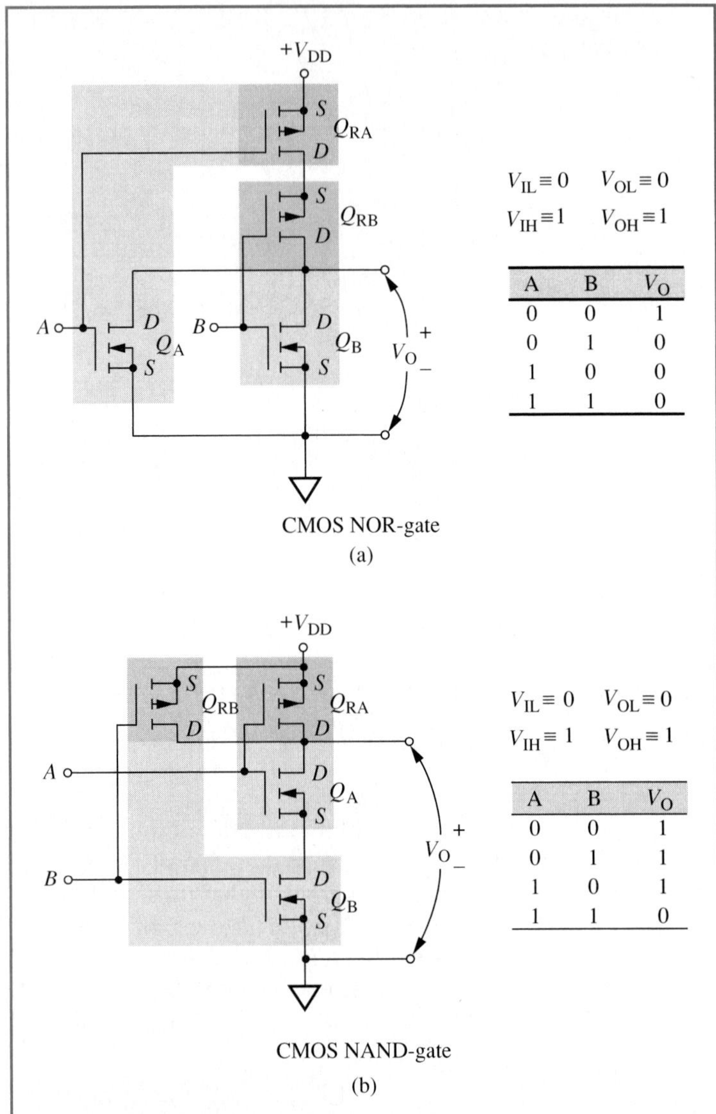

switches Q_A and Q_B are N-channel MOSFETs. With complementary channels connected so that their gates are in parallel, as are MOSFETs Q_A and Q_{RA} and MOSFETs Q_B and Q_{RB}, turning on an electronic switch (Q_A or Q_B) turns off the corresponding bias resistor (Q_{RA} or Q_{RB}), which means there is always a large impedance from the power supply to ground.

Figure 12–11a is configured as a *NOR* gate and Figure 12–11b as a *NAND* gate, as is evident from the electronic switch arrangements. That is, electronic switches Q_A and Q_B are in parallel in Figure 12–11a, which is an *OR* arrangement; however, since electronic switches negate the logic, this is really a *NOR* gate. The two electronic switches of Figure 12–11b are in series, which is an *AND* configuration, but the resultant logic is a *NAND* configuration due to negation. To make the electronic principles clearer, let us apply a logic 0 (which is a few tenths of a volt) to both inputs *A* and *B* of Figure 12–11a. The logic 0 on the gates of Q_A and Q_B is not large enough to enhance their channels; therefore, both electronic switches are off. However, those logic 0's are also applied to the gates of the P-channel MOSFETs Q_{RA} and Q_{RB}. With the source of Q_{RA} at a positive V_{DD} potential, by writing a loop equation it can be determined that the magnitude and polarity of the gate-to-source voltage is negative V_{DD} volts (for simplicity assume a logic 0 to be 0 volts). A negative potential of magnitude V_{DD} is sufficient to enhance the P-channel of Q_{RA}, which turns it on. Turning Q_{RA} on pulls the source of Q_{RB} to $+V_{DD}$, which then biases the gate of Q_{RB} with $-V_{DD}$ volts and also turns on Q_{RB}. With Q_{RA} and Q_{RB} on and electronic switches Q_A and Q_B off, the output is pulled up to potential V_{DD}, which is a logic 1. This agrees with the logic states shown in the truth table.

Since the electronic switches of Figure 12–11a are connected in parallel, if either or both switches are turned on the output is shorted to ground and the output is a logic 0, which agrees with the truth table. To "walk through" the electronics for an example of a logic 1 input, let us apply a logic 1 to input *B* and a logic 0 to input *A*. From previous discussion we know that for the logic 0 input to *A*, Q_A is off and Q_{RA} is on, which pulls the source of Q_{RB} to V_{DD}. The logic 1 (near $+V_{DD}$ volts) on the gate of Q_B sufficiently enhances its N-channel so that it turns on; hence Q_B will be on. Since that same logic 1 ($+V_{DD}$) is on the gate of P-channel Q_{RB}, its gate-to-source potential is zero volts (recall that Q_{RA} is on), which does not enhance the channel of Q_{RB} and it therefore is turned off. With Q_B on, the output is a logic 0, which agrees with the truth table, but little current is drawn from the power supply V_{DD} since Q_{RB} is off. The reader can verify the remaining logic conditions of the truth table.

To trace through an electronic state of Figure 12–11b, let us input logic 0's to both *A* and *B*. The channels of both Q_A and Q_B are not sufficiently biased to enhance them for source-to-drain current conduction. Thus, both Q_A and Q_B are off and there is an equivalent open circuit from the drain of Q_A to the source of Q_B. The logic 0's input at *A* and *B* result in a bias potential of $-V_{DD}$ volts between the source-to-gate of Q_{RA} and Q_{RB}, and this situation is sufficient biasing to turn them on. With either Q_{RA} or Q_{RB} turned on, and with Q_A and Q_B off, the drain of Q_A is pulled to $+V_{DD}$ volts, which is a logic 1. This agrees with the first row of the truth table of Figure 12–11b.

From the series configuration of Figure 12–11b we know that the only way to get a logic 0 at V_O is to turn both Q_A and Q_B on. This requires that logic 1's be input to *A* and *B*, which results in both Q_{RA} and Q_{RB} being turned off and again greatly reduces any current flow from the power supply V_{DD} to ground. With inputs *A* and *B* at a logic 1 level, Q_A and Q_B are on and the output voltage is V_{OL}, which agrees with the truth table, and there is very little current drain from V_{DD} since Q_{RA} and Q_{RB} are off.

We have not investigated each and every logic gate configuration for the technologies discussed (standard TTL, open-collector TTL, Schottky TTL, NMOS, and CMOS), nor have we discussed all the technologies available (RTL, DTL, ECL, I^2L, PMOS, or HMOS). We have examined only the essential concepts of the most prominent technologies to establish a solid background for reasoning our way through an analysis of those technologies and logic gates not studied. More importantly, from the fundamental knowledge we have developed, we will have a better understanding of manufacturers' data sheet specifications so as to interpret their significance relative to design requirements.

Review questions

1. What is required to turn on a bipolar electronic switch?
2. What controls the toggle position of a MOSFET electronic switch?
3. When either a bipolar or MOSFET electronic switch is off, what is the logic level of the output voltage?
4. When the output voltage of an electronic switch is V_{OL}, what is the input logic level of that switch, whether it is a bipolar or MOSFET electronic switch?
5. What must be the gate-to-source voltage polarity of an NMOS and PMOS electronic switch in order to turn them on?
6. How are enhancement-type MOSFETs turned on?
7. What is significant about the totem pole output of TTL technology?
8. What purpose does Q_2 serve in Figures 12–4 and 12–7?
9. How does a multiemitter bipolar transistor function?
10. Is input current I_{IL} required of a TTL gate for the logic condition V_{IL}?
11. What is a wired-*AND* logic gate?
12. What is a Schottky transistor, and how does it improve the performance of the standard TTL technology?
13. How does MOS technology improve the logic gate density of an integrated circuit?
14. For the logic condition of V_{IL}, does an NMOS device require current I_{IL} in order to function?
15. In what circuit configuration are the electronic switches of a *NOR* gate connected?
16. In what circuit configuration are the electronic switches of an *AND* gate connected?
17. What is meant by complementary channel?
18. How does CMOS reduce power consumption?

12–4 Electronic memory cells

Figure 12–12 is the schematic of a static bipolar R/W memory cell, which is a flip-flop. There are two multiemitter transistors—Q_1 and Q_2. The generic Boolean variable Q was arbitrarily chosen to be represented by the output of Q_1, and the output of Q_2 represents the complement of the variable, which is the negation of Q ($\overline{Q}$). As we know, a static memory cell can have a logic 1 written into it ($Q=1$ and $\overline{Q}=0$), or a logic 0 ($Q=0$ and $\overline{Q}=1$), whichever logic state is to be remembered by the memory cell. As explained in Chapter 9, a static memory cell retains the logic of the bit stored in it so long as power is applied to the cell (it is volatile).

Transistors Q_1 and Q_2 of Figure 12–12 are electronic switches that are dependent on each other for the needed forward biased base-to-emitter potential in order to be turned on. That is, the base of Q_1 is connected to the collector of Q_2, and the base of Q_2 is connected to the collector of Q_1; therefore, if Q_1 is off, its collector is pulled up to V_{CC} by

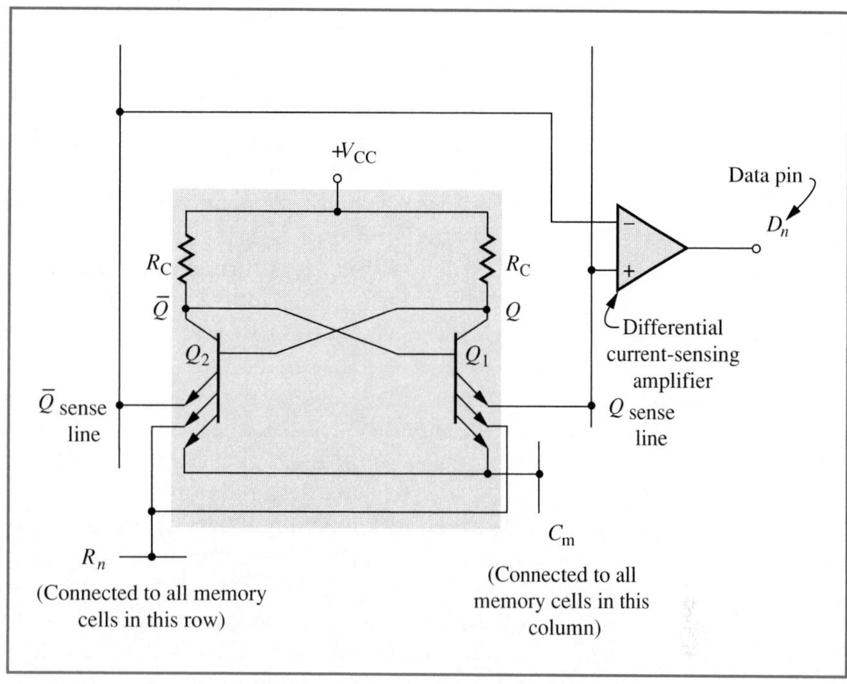

Figure 12–12
Static bipolar R/W memory cell.

R_C, and this voltage is applied to the base of Q_2, which will forward bias Q_2 and turn it on. With Q_2 on, its collector voltage is pulled low and as a result the base voltage is not of sufficient magnitude to forward bias Q_1 and it remains off. The memory cell remains in that state ($Q = 1$ and $\overline{Q} = 0$) as long as power is applied to the cell, or until Q_1 is turned on or Q_2 is turned off.

When the memory cell of Figure 12–12 is not being accessed, the row (R_n) and/or column (C_m) lines are inactive, which is a logic low. With either or both R_n and C_m low, the base-to-emitter current of the on transistor has a low-resistance path to ground, which keeps that transistor on. The off transistor has no base current and therefore does not use the row or column line as a low-resistance path to ground.

When a memory cell is to be read, the row and column lines are first driven to a logic high by decoding the address being input to the memory chip (see Fig. 8–8a). When both row and column lines go high, this reduces the base-to-emitter forward biased voltage so that the four emitters connected to R_n and C_m are not forward biased, even the on transistor. However, the sense lines are at a logic low, which means that there is a base-to-emitter forward biased voltage present to keep the on transistor forward biased and conducting. This base-to-emitter current is conducted to the sense line. The off transistor is not forward biased and therefore does not conduct base-to-emitter current to its sense line. The sense line with base current is detected by a current-sensing differential amplifier, which outputs the proper logic voltage level V_{OL} at the data pin. The sense line connected to the nonconducting transistor would not have a base-to-emitter current flowing in it, which of course would be sensed by the current-sensing amplifier and as a result its output would be V_{OH}.

To write to the memory cell of Figure 12–12, again the R_n and C_m lines are driven high via row and column address decoders within the memory chip. As we know from the timing diagram of an MPU, the datum being written to memory is loaded on the data bus and then latched by the memory chip on the positive edge of $\overline{WR}$. The memory chip then applies the latched logic level to the Q sense line (the emitter of Q_1), and the complement of that logic level is applied to the sense line of $\overline{Q}$ (the emitter of Q_2). If the logic level to be stored is a logic 1, Q_1 is turned off but Q_2 is turned on, since a logic 0 would be applied to the emitter of Q_2. The address is then removed from the address pins by the MPU, causing the row and column lines to go to a logic low and thereby provide a path for the base current of the on transistor to continue to flow. This keeps Q_2 on and Q_1 off after the logic level 1 being written into the memory cell has been removed from the sense lines.

In summary, to read the data bit stored in the memory cell of Figure 12–12, or to write a data bit into it, the row and column lines must be driven to a logic high state. If the memory cell is being read, this action results in a base-to-emitter current flowing in the sense line that is connected to the emitter of the on transistor. The sense line connected to the off transistor does not have a current flowing in it. A current-sensing differential amplifier detects the presence or absence of current in these sense lines and outputs to the data pin the appropriate logic level. When writing to the memory cell, the logic level being stored is applied to Q, and the complement of that logic level is applied to $\overline{Q}$ by way of their respective sense lines. This turns on the transistor whose emitter is connected to the sense line with the logic low, and the other transistor is turned off. The row and column lines then go inactive (low), and the logic state written into the cell remains.

The configuration of the static R/W NMOS memory cell of Figure 12–13 is similar to that of the static bipolar memory cell of Figure 12–12. The MOSFETs Q_1 and Q_2 are cross-coupled so that each enhancement gate voltage is dependent on the other's drain voltage. Whenever electronic switch Q_1 or Q_2 is on, its drain-to-source voltage is pulled to a logic low and is therefore inadequate to enhance the channel of the other to support conduction, and as a result it is turned off. The MOSFETs identified as Q_B are the biasing resistors, just as R_D is in Figure 12–10. When the cell is to be read or written to, its row and column lines are activated, as they were in Figure 12–12. This action turns on the MOSFETs identified as Q_X (for row) and Q_Y (for column). If a read is to occur, the gate of Q_R is driven high (it can be designed for an active low or high), which turns Q_R on. With Q_X, Q_Y, and Q_R on, the logic state of Q_1 is transferred to the data pin $\overline{D}_n$. If the cell is to be written into, the data bit to be stored is input at data pin D_n, and the write input is driven high (again, it could be designed to be low), which turns Q_W on. Since Q_Y and Q_X are already on, this loads the data bit on the write line and applies it to the gate of Q_1. If the data bit is a logic 1 it turns on Q_1 and pulls the drain of Q_1 low, which turns off Q_2. If the data bit is a logic 0, Q_1 is turned off and Q_2 is turned on. Notice that the logic level at the drain of Q_1 is the inverse of the logic level input at pin D_n. This is the reason for the negation symbol of the read output pin $\overline{D}_n$. Of course, the read output can be inverted, and then the write data pin and read data pin could be made to be one pin.

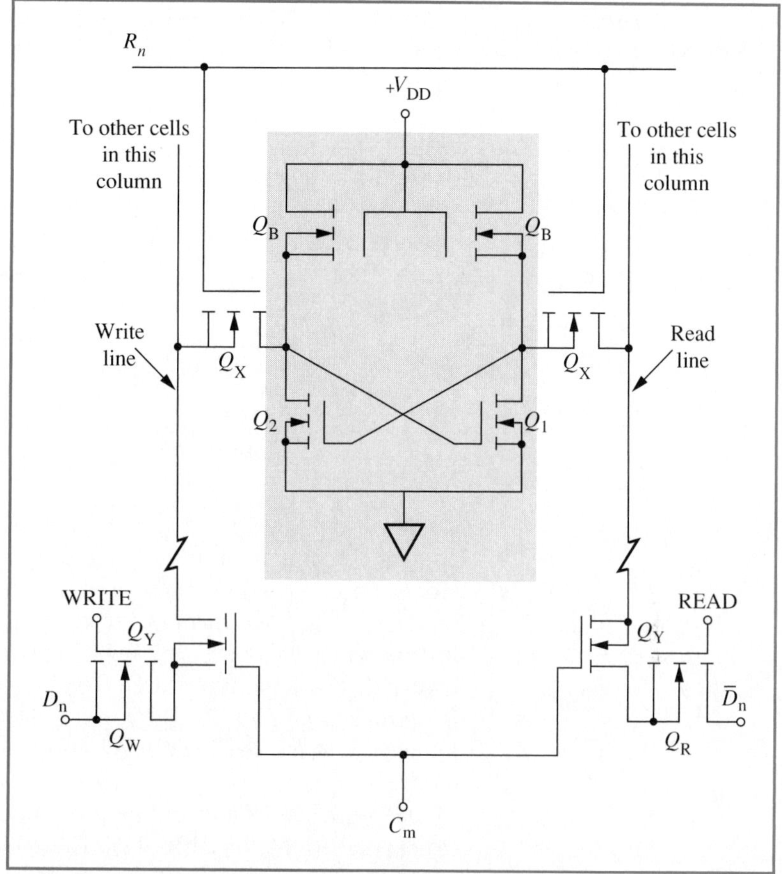

Figure 12–13 An NMOS static R/W memory cell.

The sampling of R/W memory cells should give the reader a feeling for static R/W memory cells, regardless of the logic family technology. The only memory cell that is fundamentally different is that of a dynamic memory, which we shall examine next.

There are a variety of dynamic memory cell arrangements, but the one illustrated in Figure 12–14 is representative of the principle of operation of all of them. The MOSFETs are standard MOSFETs but are drawn in a more conventional and convenient fashion. The storage mechanism is the capacitor C, which is a parasitic capacitance of the MOSFET Q_2. The logic level being read from the cell appears at C_O and is the inverse logic level stored on C. Because C is a parasitic capacitance, it has a relatively large amount of leakage current. This leakage is large enough to discharge C within 2 to 3 ms, so it must be refreshed, as discussed in Chapter 9. Because of the reduced number of components per cell, dynamic memories have the largest bit capacity.

When a data bit is to be written into the dynamic memory cell of Figure 12–14, it is input at D_{IN} and the gate of Q_1 is activated so that Q_1 turns on. With Q_1 on, the capacitor C is charged (or discharged) to the logic level of the data bit. Once the data bit has been stored on C, the write control signal is driven inactive, Q_1 is turned off and C is buffered (no current is drained from it) between Q_1 and the gate of Q_2.

When a data bit is to be read from the dynamic memory cell of Figure 12–14, the output capacitor C_O is precharged to a logic 1. The read

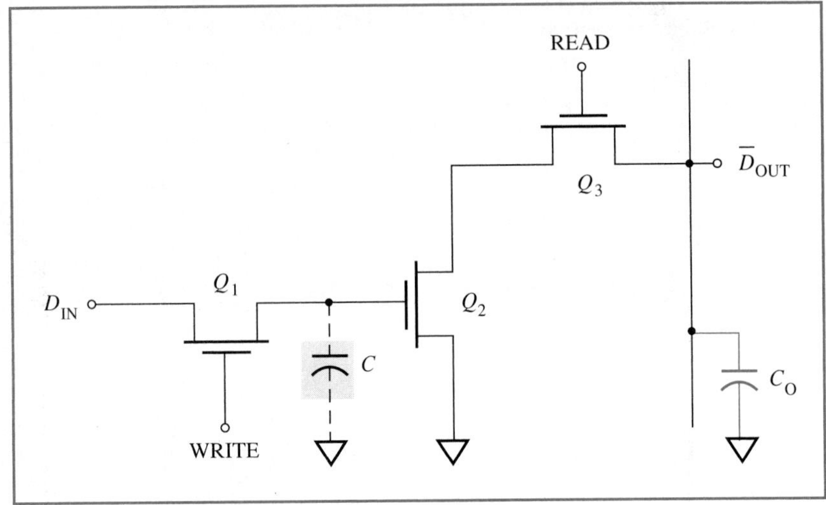

Figure 12–14 A dynamic memory cell.

control signal is then driven active, which turns Q_3 on. If a logic 1 is stored on C, Q_2 is turned on and the logic 1 stored on C_O is discharged to a logic 0. The logic level of the data bit output at $\overline{D}_{OUT}$, via C_O, is the inverse of that stored, which means that the output logic bit must be inverted. If a logic 0 is stored on C, Q_2 is off, and even with Q_3 on there is no discharge path for the logic 1 stored on C_O and the output at $\overline{D}_{OUT}$ is a logic 1. Again, the logic level output is the inverse of the logic bit stored.

Figure 12–15 conceptually represents a typical ROM of any type (MROM, PROM, or EPROM). The ROM of Figure 12–15 is an $N \times$ 4-bit ROM; that is, it has N rows of memory locations and 4 bits per location. The boxes represent the memory cells, which are implemented with

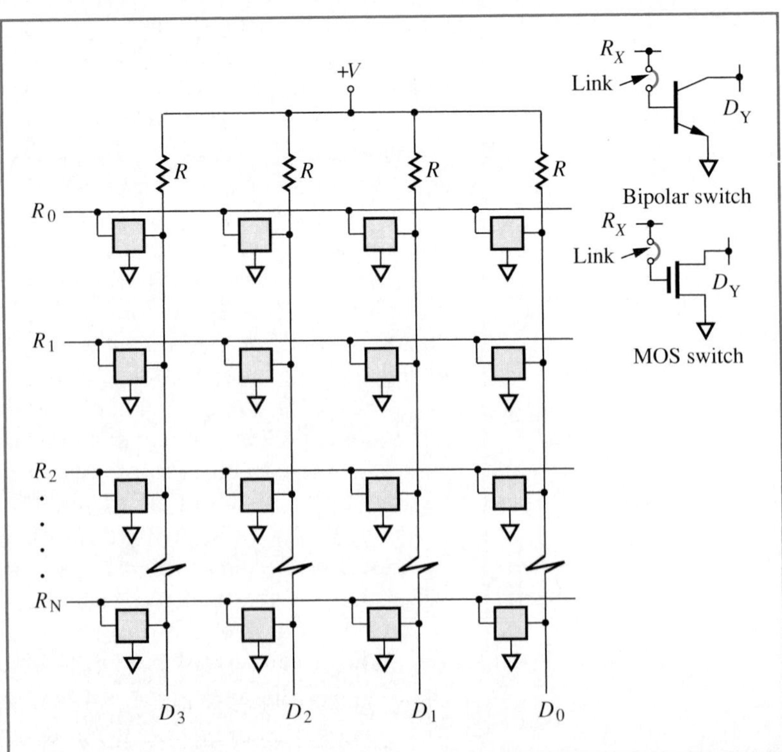

Figure 12–15 A typical ROM.

electronic switches. Representatives of those switches are shown for the bipolar and MOS family. Bipolar switches have a fuse-link in the base, whereas MOS switches have a fuse-link in the gate. If a logic 1 is to be programmed into a memory cell, this link is removed; otherwise it remains. If the ROM is a mask ROM (MROM), these links are put in or removed as part of the manufacturing process. If the ROM is a PROM, all links are manufactured in the electronic switches but are removed at the discretion of the user. A PROM programmer is used to cause an excessive base current to flow in those bipolar transistors for which a logic 1 is to be programmed, thereby blowing the *fuse-link* in the base, or applying the gate fuse-link with excessive gate voltage for a MOS switch. For either family, the link is removed for those memory cells that are to be programmed with a logic 1 and left intact for those to be programmed with a 0.

When a memory location is to be read, the appropriate row is activated (driven high) via the address decoder of that ROM. When that address row goes high it turns on all electronic switches connected to it that have their links still intact. When these switches are turned on, they pull the data line (D_n) connected to them to ground, which results in a logic 0. Those electronic switches whose links are removed and therefore are always off do not short their data line to ground when the row line goes high, and as a result these links are pulled to $+V$ volts, which is a logic 1. The pull-up "resistors" identified as R are resistors if the family is TTL but are MOSFETs if the ROM is from the MOS family.

Review questions

1. What do all static memory cells have in common?
2. How is a bipolar static R/W memory cell addressed, and how does that affect the electronic switches of that cell?
3. How is a bit written into a bipolar memory cell?
4. What electronic switches are turned on as the result of an address in an NMOS static R/W memory cell?
5. What function do the electronic switches Q_W and Q_R of Figure 12–13 serve?
6. What makes dynamic memory cells dynamic?
7. What is a refresh cycle, and why is it necessary for DRAMs?

12–5 Designations of logic families

The TTL logic family has numerous variations, whereas the MOS logic family is composed of basically the three we have already introduced—NMOS, PMOS, and CMOS. TTL was the first major logic family, and as a result established many industry standards. Because of the importance and variety of TTL logic available, the authors believe that the major variations of TTL and their designations warrant discussion as well as the dominant types of MOS logic—NMOS and CMOS.

There are five major classifications of TTL logic:

1. Standard
2. Schottky (S)
3. Low current drain, or low power (L)
4. High speed, or high current drain (H)
5. Low power using Schottky transistors (LS)

The first TTL logic available is now known as *standard TTL*. Its designation is either 54XX or 74XX, the numbers indicating the operating temperature range of the integrated circuit. Those ICs with designations of 74 operate from 0 to 70°C, which meets most commercial requirements and those designated 54 operate in a temperature range of −55 to 125°C, which meets military specifications. The two or three digits following the number specify the logic. For instance, 7400 and 5400 indicate that the IC contents are quad two-input *NAND* gates (four 2-input *NAND* gates per chip), and 54181 and 74181 are 4-bit ALUs. Standard TTL logic ICs are designated the 54/74 family. Figure 12–7 shows a standard TTL three-input *NAND* gate, and an integrated circuit that contains three of these *NAND* gates is designated either 5410 or 7410, depending on the operating temperature range.

Variations of the 54/74 series were implemented to improve performance, for example, by increased operating speed, reduced power consumption, or both. Many of these performance improvements were accomplished by simply changing the values of resistors, such as R_2, R_3, and R_4 of Figure 12–7, whereas others required circuit modifications such as replacing R_4 of Figure 12–7 with circuitry composed of R_5, R_6, and Q_5 of Figure 12–9. Some speed improvements were made if Schottky transistors were used instead of the standard bipolar transistor.

The S (Schottky) series is faster than the standard 54/74 and uses Schottky transistors to achieve the improved performance. Its designation is 54S/74S. That same triple three-input *NAND* gate that is designated 5410 or 7410 for standard TTL is designated 54S10 or 74S10 for the Schottky version (they are pin-for-pin compatible).

In the low-power (L) series the values of the resistors are increased. For example, in Figure 12–7, R_2 would be increased to 20 K ohms, R_3 to 500 ohms, and R_4 to 12 K ohms. With these increased resistances less current is drained from the power supply. The disadvantage is that the operating time also increases. This low-power series is designated 54L/74L. Hence, the previously mentioned triple three-input *NAND* gate IC would be designated 54L10 or 74L10.

In the high-power (H) series, the biasing resistance values are reduced, which increases the current drain, in order to achieve greater output current as well as a reduction in operating speed. For instance, R_3 of Figure 12–7 would be reduced to 50 ohms. This high-power and higher-speed series is designated 54H/74H. Therefore, the triple three-input *NAND* gate IC would be designated either 54H10 or 74H10.

The last circuit modification designation we shall consider is the low-power Schottky (LS) variety. This series uses a combination of Schottky transistors and circuit modifications, as in Figure 12–9. In fact, Figure 12–9 is a two-input *NAND* gate of the LS variety. If there are four of these *NAND* gates per IC, the designation is 54LS00 or 74LS00.

If a CMOS IC is pin compatible with that of a 54/74 series, it is designated with a prefix 74C; otherwise it has a 4000 (4001, 4002, etc.) designation. Our interest is restricted to the TTL pin-compatible 54/74 series.

There are five different types of 74 CMOS series logic devices: 74C, 74HC, 74HCT, 74AC, and 74ACT. The 74C was the first TTL pin-compatible CMOS. Its major disadvantage is its relatively low value for I_{OL} (0.36 mA), which means it could not sink the input current I_{IL} of a single 74 series TTL load (−1.6 mA). If 74C CMOS were to drive a TTL logic device, the latter would have to be from the 74LS series, and even then it would be incompatible by 0.04 mA.

The 74HC series is a high-voltage CMOS (it can be operated up to +15 volts). It is a counterpart to TTL open-collector logic devices. 74HC logic devices are open-drain devices and are used in wired-*AND* configurations. A disadvantage to the 74HC series is its minimum input voltage level for a logic 1 (V_{IH}), which is 3.5 volts; this is also true for the 74C series. Since TTL has a V_{OH} of 2.4 volts, interface circuitry must be used if 74 series TTL logic devices are to drive 74HC CMOS logic devices.

The 74HCT CMOS was designed to correct the incompatibility of the 74HC series with TTL logic devices. For the 74HCT series V_{IH} is 2.0 volts.

The 74AC series is similar to the 74HC series—that is, it has open-drain technology—but differs in the amount of current it can sink in the logic low state (I_{OL} = 24 mA). The AC series is also faster than the HC series.

The 74ACT series is the AC series but is directly compatible with TTL logic because it has an input voltage requirement for a logic high (V_{IH}) of 2.0 volts.

After load currents have been examined, Table 12–5 will be presented, which allows easy reference for the voltage level specifications of the various logic families discussed. At that time examples will also be given.

Review questions

1. What designation is used to signify standard TTL logic ICs?
2. What is the difference between 54 and 74 logic ICs?
3. What performance characteristic is improved with use of the Schottky transistor?
4. Over the 54/74 series, what performance improvements are noted with (a) 54L/74L series? (b) 54H/74H series? (c) 54LS/74LS series?
5. How is a CMOS IC designated if there is a 54/74 IC with which it is pin compatible?
6. What do the designations given below indicate?
   (a) 74C (b) 74HC (c) 74HCT (d) 74AC (e) 74ACT

12–6 Definition of logic voltage levels and loading specifications

Recall from Section 2–9 that *logic voltage levels* are simply the specified magnitudes of input and output voltages that define the maximum voltage of a logic 0 and the minimum voltage of a logic 1. *Loading* is the act of connecting outputs to inputs. Of primary interest are the resultant currents and output capacitance due to loading. *Load currents* are the result of connecting an output of a digital device to one or more inputs. *Load capacitance* is the capacitance of the output due to the input capacitance of the driven circuits. Various consequences can result from loading an output. For example, the output voltage levels may not meet specifications for a logic 1 or 0; there could be damage to the output circuitry due to excessive current demands; or propagation time delays

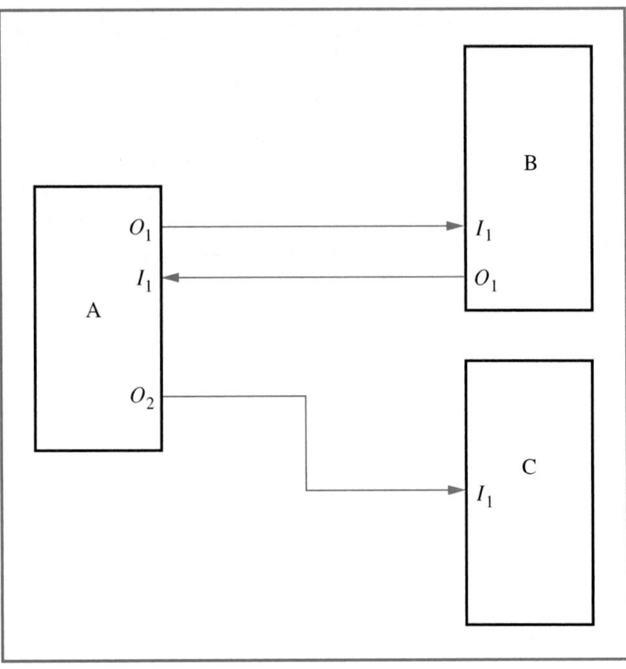

Figure 12–16
Illustration of driver and driven.

may become unacceptable owing to the loading capacitance (the adding of parasitic capacitance). It is these effects of loading that interest us.

Two terms that will be used to develop concepts that have evolved from loading considerations: *driver* and *driven*. A *driver* is an output that must drive the inputs connected to it to the appropriate logic level, which is either a logic 0 or 1. Figure 12–16 illustrates this concept. Device A has two outputs, O_1 and O_2, and device B has one output, O_1. Device A has one input, as do devices B and C, which are labeled I_1 for each device. These inputs and outputs are accessible via pins on the IC. The outputs of Figure 12–16 are the driver pins and the inputs are the driven pins. A device may have both driver and driven pins. The memory device of Figure 9–23 provides confirmation of this, since $\overline{CS}$, $\overline{RD}$, $\overline{WE}$, and the address inputs A_0 through A_9 are examples of driven pins, whereas IO pins function as both driver and driven pins. The IO pins are driven during write operations and are driver pins during read operations.

EXAMPLE 12–1

Identify whether the pins of the page select decoder and the ROMs of Figure 9–20 are drivers, driven, or both.

SOLUTION Input pins are driven pins, output pins are driver pins, and those that serve as both input and output are both driven and driver, depending on the operation being executed.

The page selector (74LS138) has driver pins $\overline{O}_0$ through $\overline{O}_5$ and driven pins A_0 to A_2 and E_3. The address pins of the 2764-0 are driven pins, as are pins $\overline{CE}$ and $\overline{OE}$. The output pins O_0 through O_7 of the 2764-0 are driver pins. The address pins of both 8128's are driven pins, as are pins $\overline{CE}$, $\overline{OE}$, $\overline{WE}$, and the I/O data pins for write operations. The driver pins for the 8128's are the I/O data pins in a read mode of operation. The input pins of the *NOR* gate are driven, and its output pin is a driver.

Inputs of semiconductor digital-type devices have specifications that must be met by their drivers in order for the driver and driven to be compatible. If driver and driven devices are incompatible, some circuitry must be placed between them to make them compatible. Recall that this circuitry is known as an interface, or glue, since it lies between the two devices. Interface circuitry can be as simple as a resistor or fairly complex, with multiple transistors.

To determine if two digital devices are electronically compatible, one uses the manufacturer's data specifications for the input and output pins of these devices. The driver of every driven pin must be checked for compatibility of both voltage levels and loading current. One need know very little of the internal electronics to make these comparisons. Usually just knowing which families of technology (TTL, MOS, CMOS, etc.) are involved is sufficient.

Manufacturers categorize the data specification of a digital device into two basic groups, which are listed as *DC* or *AC characteristics*. Direct current (DC) characteristics are concerned with voltage and current specifications, and alternating current (AC) characteristics provide timing specifications, as we already know from Chapter 9. The reader should note that the meanings of DC and AC have been stretched beyond their original meanings when applied to circuit analysis.

To review the symbolic notation used: The data sheet of a digital device, whether it is a logic gate or a memory device, provides output and input DC parameters using an O subscript to denote an output parameter and an I subscript to denote an input parameter. If the parameter is a voltage, a V is used to identify it, whereas an I is used to denote a current. For example, if a manufacturer wishes to identify an output voltage, the notation V_O is used, and V_I is used to denote an input voltage. I_I is used to represent an input current and I_O to indicate an output current. In addition, manufacturers use the subscripts L and H to indicate the specified logic level, where L (low) designates a logic 0 and H (high) indicates a logic 1. Thus V_{IH} signifies an input (I) voltage (V) which is a logic 1 (H), and I_{OL} signifies an output (O) current (I) in the 0 logic state (L). Table 12–1 shows all possible symbolic representations of input and output voltages and currents in either logic state.

Figure 12–17 is an illustration of two devices that have both input and output pins, the input pins represented by an I and the output pins by an O. As we have already noted, the outputs are drivers and the inputs are driven by these drivers. Figure 12–17 also contains a table that indicates which parameters of devices A and B are to be compared for compatibility. Recalling that the driver output parameters must be

Table 12–1
Symbolic representations of input and output DC parameters

DC parameter	Explanation of parameter
V_{OH}	Output voltage in the logic 1 state
V_{OL}	Output voltage in the logic 0 state
V_{IH}	Input voltage in the logic 1 state
V_{IL}	Input voltage in the logic 0 state
I_{OH}	Output current in the logic 1 state
I_{OL}	Output current in the logic 0 state
I_{IH}	Input current in the logic 1 state
I_{IL}	Input current in the logic 0 state

Figure 12–17
Parameters of comparison.

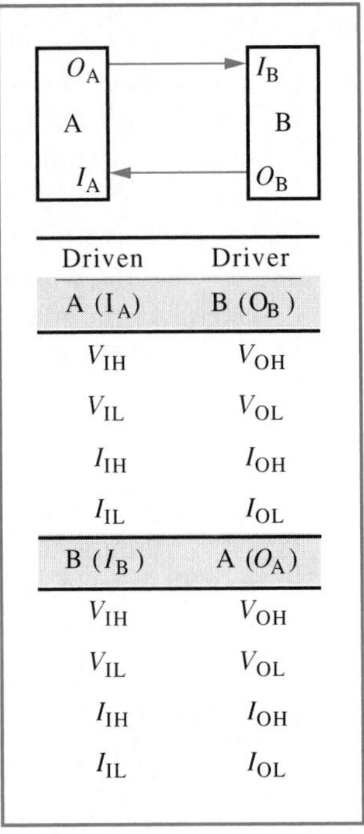

Driven	Driver
A (I_A)	B (O_B)
V_{IH}	V_{OH}
V_{IL}	V_{OL}
I_{IH}	I_{OH}
I_{IL}	I_{OL}
B (I_B)	A (O_A)
V_{IH}	V_{OH}
V_{IL}	V_{OL}
I_{IH}	I_{OH}
I_{IL}	I_{OL}

compatible with the input parameters of the driven, a comparison must be made of both voltage and current parameters for the logic 0 and logic 1 states. As shown, O_A is the driver and I_B is the driven. In the first row of that table, V_{IH} and V_{OH} are compared for voltage magnitude and polarity compatibility, as are V_{IL} and V_{OL} in the second row. The third and fourth rows contain current parameters I_{IH}, I_{OH}, I_{IL}, and I_{OL}, which must be compatible in magnitude and direction. A comparison must also be made for O_B as the driver pin and I_A as the driven pin, which is done in the next four rows.

Voltage compatibility

To understand how to compare the input and output voltage parameters of Figure 12–17, we must first understand that the driven input parameters dictate criteria for the driver (output). That is, if an input pin has a minimum voltage level for a logic 1 (V_{IH}), the driver (V_{OH}) must meet or exceed this level if the input and output are to be voltage compatible for the logic 1 state. Also, the driven pin has a maximum acceptable voltage for a logic 0 (V_{IL}), and V_{OL} must be equal to or less than this maximum value in order for V_{IL} and V_{OL} to be compatible. Polarities must be the same, which for most modern semiconductors is not a problem, since TTL logic has set a pseudostandard of +5 volts DC for the power supply voltage (V_{CC}). The required relationships between the input and output voltages are given in Table 12–2.

To visualize the relationships of input voltages V_I and output voltages V_O of Table 12–2, refer to Figure 12–18. As previously stated,

Table 12–2
V_O and V_I criteria

$$V_{IH}\text{ (min)} < V_{OH}\text{ (min)}$$
$$V_{IL}\text{ (max)} > V_{OL}\text{ (max)}$$

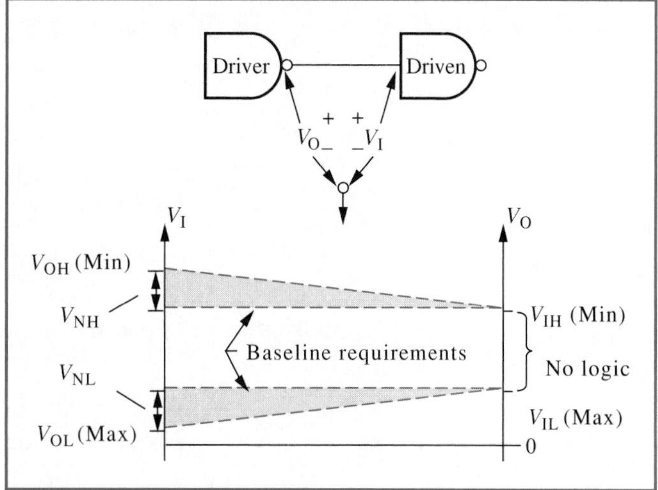

Figure 12–18
V_I and V_O voltage level specifications.

voltage level criteria are determined by the input voltage levels of the driven device; hence Figure 12–18 shows V_{IH} and V_{IL} as establishing baseline requirements that the driver must meet or even exceed. These input and output voltage levels are guaranteed specifications that are supplied by the manufacturer. If the logic 1 level of the output voltage V_{OH} is theoretically equal to or greater than V_{IH}, the two devices are voltage compatible for the logic 1 level; otherwise they are not compatible and interfacing must be done. The equal-to condition is too "close" for a good design and is ignored as acceptable. Therefore, Figure 12–18 represents V_{OH} as exceeding V_{IH} by

Equation (12–1)

$$V_{OH} - V_{IH} = V_{NH}$$

As illustrated in Figure 12–18, V_{NH} is a voltage safety factor, which is the amount of voltage variation that can exist between output and input under worst-case conditions (minimum voltage levels). This voltage safety factor is known as the *noise margin* for the logic 1 state. Similarly, V_{IL} is a guaranteed maximum voltage that can be input to the device and be interpreted by the device as a logic 0, which establishes a baseline maximum voltage requirement for V_{OL}, as indicated in Figure 12–18. V_{NL} is the noise margin for the logic 0 state and is equal to

Equation (12–2)

$$V_{IL} - V_{OL} = V_{NL}$$

The reader should understand the reasoning of Table 12–2 to the extent that he or she can duplicate it without reliance on memory.

Voltage specifications for standard TTL (54/74 series) are given in Table 12–3.

Table 12–3
TTL voltage specifications

V_{IH}	=	2.0 volts min
V_{IL}	=	0.8 volt max
V_{OH}	=	2.4 volts min
V_{OL}	=	0.4 volt max

EXAMPLE 12–2

Suppose that the voltage specifications for devices *A* and *B* of Figure 12–17 are as follows.

1. Devices *A* and *B* are TTL devices; therefore, refer to Table 12–3.
2. $V_{IH} = 3.2$ volts, $V_{IL} = 0.2$ volt
   $V_{OH} = 2.7$ volts and $V_{OL} = 0.4$ volt

Are these devices voltage compatible for conditions 1 and 2?

SOLUTION Applying the inequalities of Table 12–2 for the specifications of condition 1, we have

1.

$$V_{IH} = 2.0 \text{ volts} < 2.4 \text{ volts} = V_{OH}$$
$$V_{IL} = 0.8 \text{ volt} > 0.4 \text{ volt} = V_{OL}$$

Since all of the inequalities of Table 12–2 agree with the specifications of condition 1, devices *A* and *B* are voltage compatible. The current specifications for devices *A* and *B* must also be compared for compatibility.

2. Filling in the values of Table 12–2 with the specified values for condition 2, we find

$$V_{IH} = 3.2 \text{ volts} < 2.7 \text{ volts} = V_{OH}$$

Since this inequality is not true, devices *A* and *B* are not compatible in the logic 1 state. Testing for the logic 0 state we find

$$V_{IL} = 0.2 \text{ volt} > 0.4 \text{ volt} = V_{OL}$$

which is also not true. As a result of these comparisons, we find that devices *A* and *B* are not voltage compatible in either logic state. Therefore, if the two devices are to be used together, an interface must be used to make the two devices voltage compatible.

Current compatibility

To understand the meaning of current specifications, it is necessary to first understand the meanings of the terms *current sink* and *current source*. The *NAND* gates of Figure 12–19 (*NAND* gates are the dominant logic gate used in industry and for that reason are often used in a generic sense) are used to illustrate the meaning of these two terms. *Current*

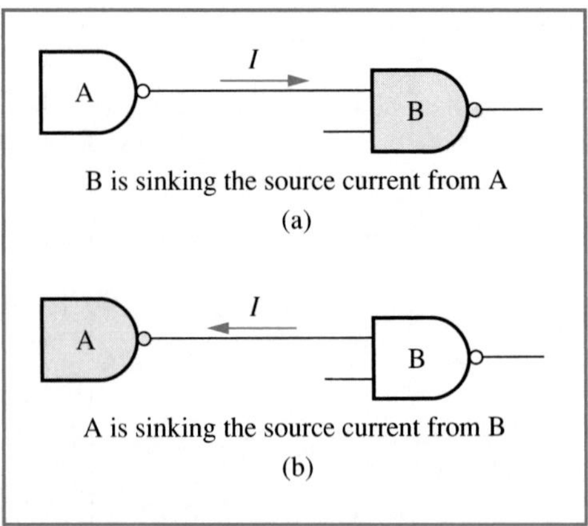

Figure 12–19
Examples of sourcing and sinking of current.

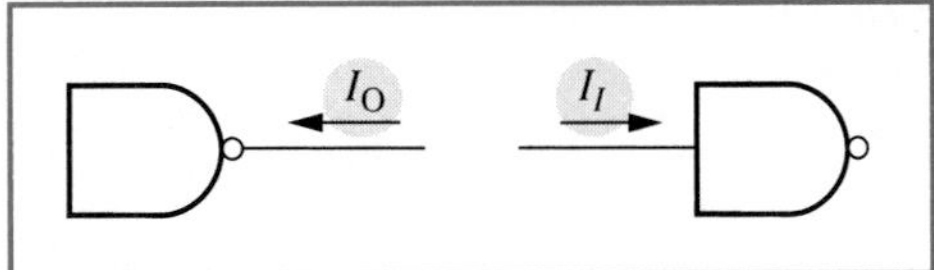

Figure 12–20
The assumed current directions for output current I_O and input current I_I.

source describes an input or output that acts as a source of current, such as the output of *NAND* gate *A* in Figure 12–19a, or the input of *NAND* gate *B* in Figure 12–19b. A *current sink* is an input or output that acts as receiver (a sink) of a source current, such as input *B* in Figure 12–19a and the output of device *A* in Figure 12–19b. The direction and magnitude of a current depend on the logic level and the technology of the device (TTL, MOS, etc.).

Since the pin of a device (either input or output) may be a sink for one logic level and a source for the other, industry has established a common direction for all input and output currents, regardless of the logic level. As indicated in Figure 12–20, all currents, I_O or I_I, are assumed to flow into the device, which also says that industry assumes all pins to be current sinks. Intuitively, if not in fact, we know this cannot be, since there must be one or more current sources and current sinks whenever two or more pins are connected (if there is at least one current source, there must be at least one current sink, since what goes in must come out). Figure 12–21 illustrates the physical logic of this statement; in the figure there are many current sinks but not one source of current for those sink currents, which is illogical. When a current is actually a source current (i.e., one that flows in the opposite direction from that assumed in Figure 12–20), industry assigns a negative sign to its magnitude in the DC specifications. For instance, a statement in the DC specification table of a device that $I_{IL} = -1.6$ mA means that for this input, the logic 0 level current (I_{IL}) is a source current (the negative sign states it is flowing out of the pin) with a magnitude of 1.6 mA. Of course, it is implied that there must be one or more current-sinking outputs (I_{OL}) connected to this input pin that are capable of handling (sinking) this source current of 1.6 mA.

Current considerations are most important if the technology (bipolar or MOSFET) used to implement a design requires source and/or sink current to operate. As we know, bipolar (TTL, RTL, ECL, etc.) technology is a current-controlled technology and therefore requires

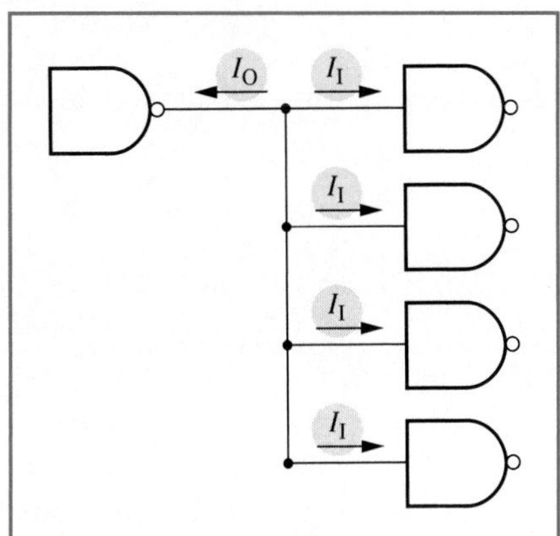

Figure 12–21
The assumed current direction for multiple device connection.

Table 12–4
Current specifications for TTL technology

I_{IH}	=	40 μA max
I_{IL}	=	−1.6 mA max
I_{OH}	=	−400 μA max
I_{OL}	=	16 mA max

certain currents to operate, whereas MOSFET technology (NMOS, PMOS, or CMOS) is a voltage-controlled technology and does not require either a sink or a source current. For instance, recall that TTL technology requires an input source current (to turn on Q_1) for the logic 0 state (I_{IL}) of an input. This implies that there must also be an output, or outputs, connected to this input that can sink that source current to ground; otherwise, where would the source current go? Hence, the specified magnitude of the sink current (I_{OL}) of the output must be larger in magnitude than that of I_{IL}, or the sum of I_{IL}'s if the output is driving more than one input, and the sink and source currents must be of opposite sign. In contrast, I_{IH} for TTL is an input sink current that is not required for operation but flows if it is provided with a path to ground. This path is provided via the driver output. There is a problem for this condition if the output has to drive too many inputs, such that the sum of the individual input sink currents exceeds that of the source current (I_{OH}) specification. If this occurs, the voltage drops internal to the driver device are large enough that the output voltage cannot reach the specified minimum voltage level to satisfy the guaranteed minimum voltage (V_{OH}) for a logic 1. In summary, for TTL technology we must be concerned with both the sinking and the sourcing of current for both logic states. For the logic 0 condition we must ensure that the I_{OL} sink current exceeds the source current I_{IL}. For the logic 1 state we must ensure that the source current I_{OH} is greater than the sink current I_{IH} so that the output is at least minimum voltage V_{OH}. Table 12–4 gives the current load specifications for standard TTL technology.

EXAMPLE 12–3

If the logic circuit of Figure 12–22 is implemented with standard TTL *NAND* gates (7400), is there current compatibility?

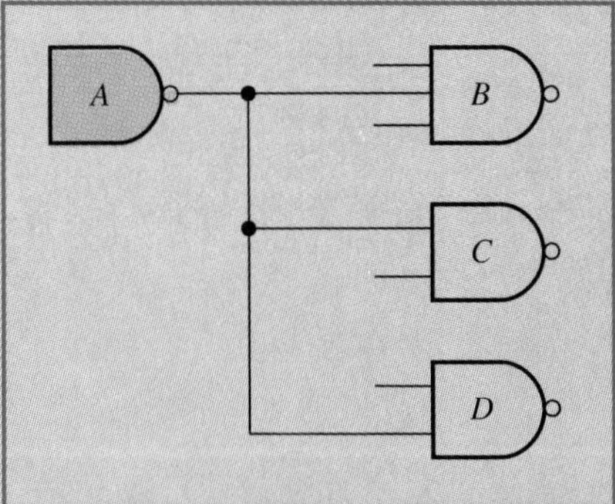

Figure 12–22
NAND gate *A* driving *NAND* gates *B*, *C*, and *D*.

SOLUTION To test for current compatibility we must consider both the logic 0 and logic 1 states; if compatibility fails for either logic state, the logic gates are not compatible. Testing the logic 1 state for current compatibility, we compare I_{IH} of *NAND* gates *B*, *C*, and *D* of Figure 12–22 with I_{OH} of *NAND* gate *A*. The current specifications of Table 12–4 state that I_{IH} of each TTL input is a sink current (positive sign) and its magnitude is guaranteed not to exceed 40 μA. Since there are three inputs (gates *B*, *C*, and *D*), 120 μA (3 × 40 μA) of sink current is drawn from the source current I_{OH} of gate *A*. $I_{OH} = -400\ \mu$A, which means it can reliably source up to 400 μA (note that I_{IH} and I_{OH} are opposite in sign) and therefore is capable of sourcing the 120 μA. These gates are compatible for the logic 1 state.

Testing the logic 0 state, we find from the current specifications of Table 12–4 that $I_{IL} = -1.6$ mA, which means that in the logic 0 state a TTL input sources 1.6 mA of current. As previously stated, this source current must be present in order for the *NAND* gate to operate. Since there are three of these inputs connected to the output of gate *A* in Figure 12–22, I_{OL} of gate *A* must exceed 3 I_{IL} (3 × 1.6 mA = 4.8 mA). For current direction compatibility, the source and sink currents must be of opposite sign. From the current specifications, we find that $I_{OL} = 16$ mA, which exceeds the source current magnitude requirement of 4.8 mA and is also sign compatible with I_{IL}. Since there is current compatibility for both logic states, the TTL-implemented logic design of Figure 12–22 is current compatible.

EXAMPLE 12–4 Can a TTL output drive 12 TTL inputs relative to current requirements?

SOLUTION With 12 inputs the output must sink 12 × I_{IL} [12 × (−1.6 mA) = −19.2 mA] of source current for the logic 0 state. Since $I_{OL} = 16$ mA, one TTL output cannot sink 19.2 mA and the logic design would not be compatible in the logic 0 state, which means the choice of technology must be altered or the design changed.

Even though current compatibility has failed based on the logic 0 state analysis, for practice let us also determine current compatibility (or noncompatibility) for the logic 1 state. $I_{IH} = 40\ \mu$A maximum; therefore, under worst-case conditions (all 12 inputs sinking the maximum current) I_{OH} must be capable of sourcing 480 μA (12 × 40 μA) of current without degradation of the voltage level required for V_{OH}. From Table 12–4 we see that the manufacturer guarantees just 400 μA ($I_{OH} = -400\ \mu$A) of source current, which is not sufficient. Hence, the logic circuit also fails current compatibility for the logic 1 state.

To provide a more convenient method of accessing load currents, TTL circuit manufacturers have devised the *unit load* (UL). A unit load is a standardized load that is representative of maximum TTL input loading on a TTL output. The *NAND* gate was chosen to be representative of TTL logic. The load currents of Table 12–4 are for a TTL *NAND* gate, and therefore the unit load is based on these values. Since the input parameters set the criteria to be met by the driver, it is logical that the input load currents be used to define the UL. Because there are two logic states and the input load current is different for each, the UL must be defined for both logic states. The UL is defined to be

Equation (12–3)

$$\text{UL (logic 1 state)} = 40\ \mu\text{A}$$

Equation (12–4)

$$\text{UL (logic 0 state)} = 1.6\ \text{mA}$$

Note that the UL definitions are based on magnitudes, and therefore the designer must be conscious of sign (current direction) when mixing logic technologies, such as TTL and CMOS.

EXAMPLE 12–5

The current DC characteristics of a TTL logic device are listed in the manufacturer's data specifications as

$$I_{IH} = 1.25\ \text{UL and } I_{IL} = 2\ \text{UL}$$

1. How much sink and source current does this represent, and what minimum requirements do these specifications impose on the driver?
2. How many of these types of input can a standard TTL output drive?

SOLUTION

1. To determine the amount of load current, we simply multiply the number of ULs by the magnitude of the current standard for the input logic level in question. Then

$$I_{IH} = (\text{UL}) \times 40\ \mu\text{A} = 1.25(40)\ \mu\text{A} = 50\ \mu\text{A}$$

and

$$I_{IL} = -(\text{UL}) \times 1.6\ \text{mA} = -2 \times 1.6\ \text{mA} = -3.2\ \text{mA}$$

2. $I_{IH} = 50\ \mu$A states that this input could sink as much as 50 μA from the driver. Therefore, the driver's I_{OH} specification must not exceed 50 μA of source current in order that degradation of V_{OH} does not occur. For a standard TTL output $I_{OH} = -400\ \mu$A; since $I_{IH} = 50\ \mu$A, eight (400 μA/50 μA) such inputs could be driven for the logic 1 state. For the logic 0 state $I_{IL} = -3.2$ mA, meaning that this input will source 3.2 mA, which the driver output must be capable of sinking. Determining the number of inputs that can be driven by a standard TTL output for the logic 0 state, we find that $I_{OL}/I_{IL} = 16$ mA/3.2 mA = five inputs. A standard TTL output can drive eight inputs based on logic 1 load current specifications and five inputs based on logic 0 load current specifications. Obviously it cannot do both, and therefore we must use the worst-case situation, which is five inputs.

The terminology used to describe the number of inputs an output of the same technology can drive is termed *fan out*. The fan out of an output is determined by dividing the output current for both logic states by the input current of the same logic state. The integer of that quotient is the fan out for that output. For instance, the fan out for a standard TTL output driving a standard TTL input is

$$\underset{\text{for } H}{\text{Fan out}} = \frac{I_{OH}}{I_{IH}} = \frac{400\ \mu A}{40\ \mu A} = 10$$

and

$$\underset{\text{for } L}{\text{Fan out}} = \frac{I_{OL}}{I_{IL}} = \frac{16\ mA}{1.6\ mA} = 10$$

For this case the fan out is the same, regardless of the logic state.

EXAMPLE 12–6

Determine the approximate value of I_{OL} for a TTL output that has a fan out of 15.

SOLUTION This output can drive 15 UL inputs and since 1 UL = 1.6 mA, $I_{OL} = 15 \times I_{IL} = 15 \times 1.6\ mA = 24\ mA$.

Now that we have studied logic voltage levels and load currents and understand their relevance to compatibility, let us summarize them for various TTL and CMOS families in table form. Table 12–5 is a listing of the logic voltage levels and load currents for some of the logic families previously mentioned.

The *NAND* gate has become the pseudostandard, which is the reason for using the *NAND* gate (the last two digits are 00) as the gate of comparison. Many of the logic devices (*OR, AND, NOR*, etc.) have these same specifications, so they are representative of the entire series. We use the specifications of Table 12–5 to represent the series rather than just the *NAND* gate of that series.

Table 12–5 clearly shows those technologies that are current controlled and those that are voltage controlled. That is, the input current I_{IL} for the current-dependent TTL family is in the milliampere range,

Table 12–5
Voltage levels and load currents for various logic families

	Series	V_{IL} max (volts)	V_{IH} min (volts)	V_{OL} max (volts)	V_{OH} min (volts)	I_{IL} max	I_{IH} max (μA)	I_{OL} max (mA)	I_{OH} max (mA)
TTL Family	7400	0.8	2.0	0.4	2.4	−1.6 mA	40	16	−0.4
	74L00	0.7	2.0	0.4	2.4	−0.18 mA	10	3.6	−0.2
	74H00	0.8	2.0	0.4	2.4	−2.0 mA	50	20	−0.5
	74S00	0.8	2.0	0.5	2.7	−2.0 mA	50	20	−1.0
	74LS00	0.8	2.0	0.5	2.7	−0.4 mA	20	8.0	−0.4
CMOS Family	74C00	1.5	3.5	0.33	4.5	−1.0 μA	1.0	0.36	−0.36
	74HC00	1.2	3.5	0.33	3.84	−1.0 μA	1.0	4.0	−4.0
	74HCT00	0.8	2.0	0.33	3.84	−1.0 μA	1.0	4.0	−4.0
	74AC00	0.8	3.5	0.33	3.84	−1.0 μA	1.0	24	−4.0
	74ACT00	0.8	2.0	0.33	3.84	−1.0 μA	1.0	24	−4.0

whereas that for the CMOS family is in the microampere range. Input current for a CMOS device is leakage current. Also notice those series that have the greatest noise margins V_{NH} and V_{NL}, according to Equations 12–1 and 12–2, and the largest output current sinking capability for a logic low (I_{OL}).

EXAMPLE 12–7

From the specifications of Table 12–5 and Equations 12–1 and 12–2, determine the noise margins for the following: (a) standard TTL (b) 74LS (c) 74C (d) 74ACT

SOLUTION

(a)
$$V_{NH} = V_{OH} - V_{IH} = 2.4 - 2.0 = 0.4 \text{ volt}$$
$$V_{NL} = V_{IL} - V_{OL} = 0.8 - 0.4 = 0.4 \text{ volt}$$

(b)
$$V_{NH} = 2.7 - 2.0 = 0.7 \text{ volt}$$
$$V_{NL} = 0.8 - 0.5 = 0.3 \text{ volt}$$

(c)
$$V_{NH} = 4.5 - 3.5 = 1.0 \text{ volt}$$
$$V_{NL} = 1.5 - 0.33 = 1.17 \text{ volts}$$

(d)
$$V_{NH} = 3.84 - 2.0 = 1.84 \text{ volts}$$
$$V_{NL} = 0.8 - 0.33 = 0.47 \text{ volt}$$

Recall that noise margin is the measure of acceptable voltage variation within a logic level. It can then be said that noise margin is a voltage tolerance. If this variation in voltage were due to noise, such as an unwanted voltage spike, the noise margin for a logic level (*H* or *L*) would specify how large that spike could be before the input voltage level of the driven device would be out of the specified voltage limits for a logic 0 (V_{IL}) or logic 1 (V_{IH}). This tolerance for noise is referred to as *noise immunity*. From Example 12–7 we see that standard TTL has the least immunity to noise for a logic high (0.4 volt) and the 74ACT series has the largest immunity to noise (1.84 volts). The 74LS series has the least immunity to logic 0 level noise, and the 74C series has the largest. High values of noise immunity are a most desirable characteristic, especially if the logic circuits are to be placed in a noisy environment (for example, in close proximity to electric motors that create power surges).

EXAMPLE 12–8

Suppose you had to design a logic circuit that was to be placed in an environment where 0.8-volt spikes often appeared on the connecting lines (printed circuit lines) of the logic circuit. From the logic series of Example 12–7, which would you use?

SOLUTION Since this 0.8-volt spike could appear in either logic level, V_{NL} and V_{NH} must be equal to or greater than 0.8 volt. Only the 74C series meets this design criterion.

As we have previously stated, it is sometimes desirable to mix logic families within a design. To do this the voltage levels and load currents must be checked for compatibility using the same methods that we have used previously.

EXAMPLE 12–9

Suppose that an older 74 series logic circuit design is to be interfaced with low-power Schottky (LS) and CMOS 74AC logic circuits. Check for both input and output compatibility and determine whether the 74LS and 74AC logic can be interfaced directly or interfacing circuitry must be used.

SOLUTIONS From the table of Figure 12–17 we see which voltage levels and current levels are to be compared for compatibility, and Table 12–5 states the criteria for voltage level compatibility. The criteria for load current compatibility are the following.

1. Are input currents required?
2. If input currents are required, are they necessary in the logic low state, the logic high, or both?
3. Are the required sink and source currents compatible in both magnitude and sign (direction)?

We shall create a table similar to the one in Figure 12–17. The column to the left contains driver specifications and columns to the right contain specifications for those inputs that are to be driven.

Comparing voltage logic levels of the 74 and 74LS series, we find

$$V_{OL} = 0.4 \text{ volt} < V_{IL} = 0.8 \text{ volt}$$

which agrees with the criteria of Table 12–2, and

$$V_{OH} = 2.4 \text{ volts} > V_{IH} = 2.0 \text{ volts}$$

which also agrees with the criteria of Table 12–2. Therefore, the 74 and 74LS logic circuits are voltage compatible for 74 logic driving 74LS logic.

Let us now compare the 74 and 74LS series for current compatibility. From the stated criteria the answer to the first question, which asks if input current is required, is yes, since both series are from the TTL family. Now that we know that an input current is required, we must determine for which logic state. From previous discussion and the analysis of Figure 12–5, we know that input current is necessary for just the logic low state and in this logic state the driven input acts as a current source and the driver output must be capable of sinking this source current. Then the condition $I_{OL} > I_{IL}$ must be met. From Table 12–6 we find for the condition when the 74 series is driving 74LS

$$I_{OL} = 16 \text{ mA} > I_{IL} = -0.4 \text{ mA}$$

which means that the 74 series can sink up to 16 mA and the 74LS series requires a source (the negative sign) current of just 0.4 mA. Hence, the two series are compatible for the logic low state.

From Figure 12–5b we know that input current is not required for the logic high state. However, the specifications of Table 12–6

Table 12–6
Voltage logic levels and load current comparison

Driver 7400		Driven 74LS00		Driven 74AC00	
V_{OL} =	0.4 volt	V_{IL} =	0.8 volt	V_{IL} =	0.8 volt
V_{OH} =	2.4 volts	V_{IH} =	2.0 volts	V_{IH} =	3.5 volts
I_{OL} =	16.0 mA	I_{IL} =	−0.4 mA	I_{IL} =	−1.0 μA
I_{OH} =	−0.4 mA	I_{IH} =	20.0 μA	I_{IH} =	1.0 μA
74LS00		7400			
V_{OL} =	0.5 volt	V_{IL} =	0.8 volt		
V_{OH} =	2.7 volts	V_{IH} =	2.0 volts		
I_{OL} =	8.0 mA	I_{IL} =	−1.6 mA		
I_{OH} =	−0.4 mA	I_{IH} =	40.0 μA		
74AC00		7400			
V_{OL} =	0.33 volt	V_{IL} =	0.8 volt		
V_{OH} =	3.84 volts	V_{IH} =	2.0 volts		
I_{OL} =	24.0 mA	I_{IL} =	−1.6 mA		
I_{OH} =	−4.0 mA	I_{IH} =	40.0 μA		

state that the 74 can source up to a magnitude of 0.4 mA in the logic high state if required ($I_{OH} = -0.4$ mA) and specifications for the 74LS series state that it will drain (sink) a maximum of 20 μA (leakage) if the source current is available. Therefore, the 74LS series would draw 20 μA from the 74 series, but it is not necessary for operation. As a result, the logic low state specifications must be met for load current compatibility. Since the 74 and 74LS series are compatible for both logic voltage levels and load currents, the two series can be directly connected when a 74 series device drives a 74LS device, if they have the specifications of Table 12–5 (check actual data sheets of logic devices in question).

Checking compatibility for the 74 series driving the 74AC series, we find from the first and second rows of Table 12–6 that

$$V_{OL} = 0.4 \text{ volt} < V_{IL} = 0.8 \text{ volt}$$

and

$$V_{OH} = 2.4 \text{ volts} < V_{IH} = 3.5 \text{ volts}$$

From Table 12–2 we find that logic voltage level compatibility fails for the high state, and hence the two series are not directly compatible. To interface the two series, where a 74 series is driving a 74AC series, there must be an interface circuit that corrects the logic-high voltage level.

Checking for compatibility when 74LS series is driving 74 series logic, we find

$$V_{OL} = 0.5 \text{ volt} < V_{IL} = 0.8 \text{ volt}$$

and

$$V_{OH} = 2.7 \text{ volts} > V_{IH} = 2.0 \text{ volts}$$

which agrees with the inequalities of Table 12–2. Thus the two series are logic voltage level compatible. Current specifications state that a 74 series logic device must source 1.6 mA of current in the logic low state and the 74LS can sink up to 8.0 mA. Hence they are compatible for the logic low state, and, since current is not required for a logic high, and the 7400 sinks (40 μA) less than the 74LS can source (0.4 mA) without degradation of the output voltage specified for V_{OH}, the two series are also current compatible.

When 74AC drives 74 series logic we find the two are voltage and current compatible.

In Example 12–9 we checked for load current compatibility, but we did not determine the fan out for each configuration and we know that an output may have to drive many inputs. Load current specifications limit the number of inputs that can be connected to an output. For instance, when the 74 series logic is driving 74LS logic, the fan out is 40 (16 mA/0.4 mA). Any design that has multiple inputs connected to an output must be checked for current loading to ensure that load current specifications have not been exceeded. If they are, a driver must be used as an interfacing circuit between the output and inputs. Drivers are simply buffers without the control for Hi-*Z* state and with large output load current specifications.

EXAMPLE 12–10

Suppose that the NMOS MPU of Figure 10–10 has DC load current specifications

$$I_{OL} = 2 \text{ mA}, I_{OH} = -400 \ \mu\text{A}$$

and the input current for both logic levels is just leakage current (recall its NMOS technology) with values of $\pm$ 10 μA. Is this MPU load current compatible with its memory?

The logic level input currents for the 8128's (NMOS) are leakage currents I_{LI} and I_{LO}, with values of $\pm$ 10 μA. Output load current specifications are I_{OL} = 2.1 mA and I_{OH} = -1 mA. The 2716 has input leakage currents that are $\pm$ 10 μA, and its output load currents are I_{OL} = 2 mA and I_{OH} = -400 μA.

SOLUTION When a 8128 is being read by the MPU the 8128 data pins are driving the MPU data pins. We compare the output load current of the 8128 to the input load current of the MPU for both logic levels. The specifications indicate that the 8128 can sink 2.1 mA in the logic low state (I_{OL} = 2.1 mA), but the MPU requires no input current, just leakage values of $\pm$ 10 μA. This is a fan out of 210 (2.1 mA/10 μA). For the logic high state, the 8128 can source up to 1 mA (I_{OH} = -1 mA) and the MPU again requires no input current but draws a leakage current of 10 μA. Hence the fan out of the 8128 for a logic high is 100. Then the worst case is a fan out of 100. Hence, one hundred 8128 inputs can be connected to this MPU.

For reading the 2716, its data pins drive the data pins of the MPU. The 2716 can sink 2 mA of current in the logic low state and source 400 μA in the logic high state. The MPU requires no input current, but 10 μA of leakage current will flow. Hence, the worst-case fan out is 40.

The MPU is writing to the 8128 when the MPU outputs an address to its address pins, data to its data pins, or control signals to its enable pins. Checking the fan out of the MPU relative to its writing to an 8128, we find that the MPU can sink 2 mA of current in the logic low state and source 400 μA in the high state. The input current to the 8128 is a leakage current of $\pm$ 10 μA values; hence the worst-case fan out is 40. The 2716 has the same write conditions as the 8128 (except the MPU writes only to the address and enable pins) and therefore also has a worst-case fan out of 40.

Considering all worst-case fan outs, this MPU can have a total of forty 2716's and/or 8128's connected to it, which is well within the three memory chips of this design. However, the I/O 74LS138 *NOR* gate loading must be also considered.

The reader should make certain that he or she understands the criteria for DC compatibility and is able to evaluate the compatibility of various logic families. In addition to understanding compatibility, the reader must be able to determine the fan out of an output relative to the inputs it is driving.

Review questions

1. What are the symbolic designations for the voltage logic levels and load currents? Explain the reasoning of those symbols used.
2. What are the requirements for compatibility relative to logic voltage levels and load currents?
3. What is fan out, and what role does it play in the design of logic circuits?
4. What is a unit load?
5. Why is it that MOS family devices do not require current to operate?
6. Why do bipolar family devices require current to operate? Are both input and output currents required, and for which logic state?

12–7 Propagation delay and load capacitance

No voltage or current can change its magnitude, polarity, or direction instantaneously, as is the case of all things with mass. There is a delay from the time the input of a logic device theoretically causes a change in the output logic voltage level to the time when it actually occurs. Figure 12–23 illustrates this concept.

Figure 12–23 shows the input logic level going from a logic 0 to a logic 1 and the input of the logic circuit (a *NAND* gate is shown for the purpose of demonstration) detects the logic 1 level at the time t_1. The *NAND* gate begins to drive its output to the logic 0 level, but there is a delay from time t_1 to t_2. This delay is due to internal parasitic capacitance of the electronic switches requiring time to charge or discharge. Once they are charged or discharged to the proper voltage level, the electronic switches begin to make the transition to the opposite state, which occurs

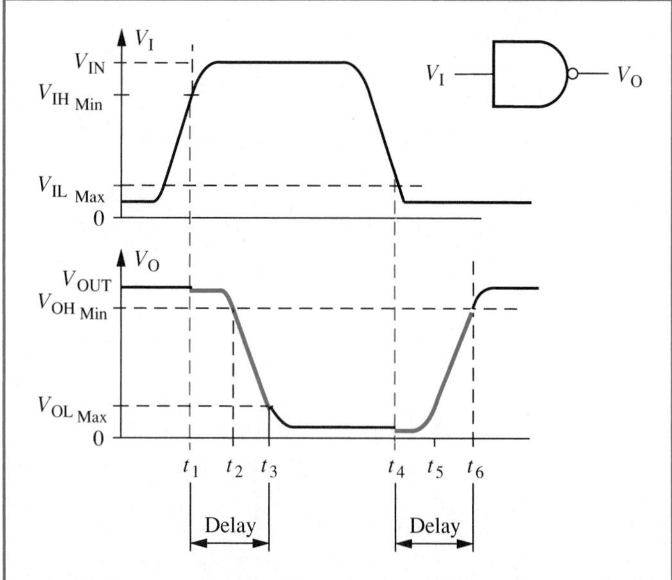

Figure 12–23
Logic level transition delay.

in the interval t_2 to t_3. At the time t_3 the output voltage is at a logic 0 voltage level and the transition from a logic 1 to a logic 0 has been completed. Hence, the total delay time, as measured from when the input detected a logic 1 to when the output reached a logic 0, is $t_3 - t_1$.

As the input voltage level of Figure 12–23 traverses from a logic 1 to a logic 0, there is also a delay in the output voltage going from a logic 0 to a logic 1 due to the same parasitic capacitors. When the input voltage reaches the logic 0 level, at time t_4, the electronic switch begins to change states. At time t_5 the output voltage actually begins to change from a logic 0 to a logic 1. At time t_6 the output voltage has reached the logic 1 state and the transition is complete. The total delay time is $t_6 - t_4$.

These delay times are known as *propagation delay time* and are symbolically represented as t_p. If the propagation delay time is the delay time in going from a low (0) to a high (1), it is designated t_{pLH}, and t_{pHL} is used to signify the propagation delay time in going from a high to a low.

Propagation delays, both t_{pLH} and t_{pHL}, are measured with respect to the 50% level between a logic 1 (V_{OH}) and a logic 0 (V_{OL}). Using the same waveforms as Figure 12–23, propagation delay parameters are defined as illustrated in Figure 12–24.

In addition to parasitic capacitance, propagation delay times are affected by capacitance that appears at the output of a logic device. As a result, every output pin must not have its specified load capacitance exceeded in order that the propagation delay time remain within the manufacturer's specifications. That is, each input pin has an input capacitance (C_{IN}) associated with it, and when connected in parallel these input capacitances are additive. Hence, as inputs are connected to an output, these input capacitances add and increase the output load capacitance. However, their sum must remain below the specified load capacitance (C_L) of the output pin. Most often manufacturers provide a graph of propagation delay versus load capacitance. From this graph the user can determine the allowable load capacitance for an acceptable propagation delay time.

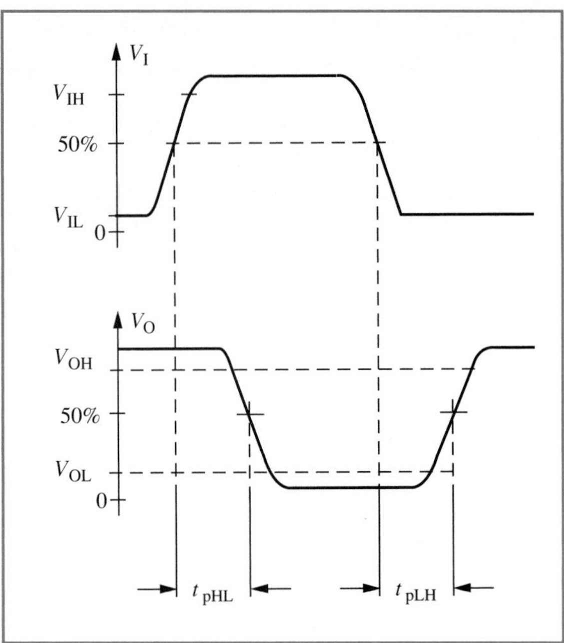

Figure 12–24 Propagation delay times.

EXAMPLE 12–11 If the load capacitance of an 8128 output pin is 100 pF and the input capacitance of a 74C00 is 6 pF, how many 74C00 inputs can be connected to an 8128?

SOLUTION Considering just capacitive loading, sixteen (100/6) 74C00 inputs can be connected to an 8128 output.

As a pulse is propagated through logic circuits, the *total propagation delay* of the pulse is the sum of propagation delay time for each logic circuit it passes through.

EXAMPLE 12–12 If a pulse passes through five 74C00 *NAND* gates, what is the resultant propagation delay time? Assume a load capacitance of 60 pF, which results in a propagation time of 58 ns.

SOLUTION The total delay time is 290 ns (5×58 ns).

Propagation delay times are most critical when two events must occur simultaneously, such as when two logic levels are *AND*ed. If these two events begin from their source with proper synchronization but each must pass through different logic circuits, they can arrive at their destination out of synchronization. This misalignment is due to their different propagation delay times.

EXAMPLE 12–13 From the MPU read timing diagram of Figure 10–7a we know that the address and read control signals originate from the MPU according to the timing indicated. When this timing diagram is applied to the system of Figure 10–10, how is it altered?

SOLUTION Some address bits are applied instantaneously to the 2716-0 (A_0 through A_{12}) and 8128's (A_0 through A_{10}), whereas others are delayed owing to the propagation delay of the 74LS138-3 and the *AND* gate. But even the enabling of the 74LS138-3 is delayed by the propagation delay of the *NOR* gate (approximately 5 ns). Thus the total delay of the effects of address lines A_{11} through A_{15} is the sum of the propagation delay time of the *NOR* gate, the 74LS138-3, and the *AND* gate. If the propagation delay time for both gates is 5 ns and the 74LS138 has a propagation delay of 27 ns, the total is 37 ns. This is acceptable if $t_{AC} + t_{CC} - t_{AD}$ of Figure 9–15 is greater than the delay time. From Table 9–6 we find that 250 ns + 450 ns − 600 ns = 100 ns is greater than 37 ns. Therefore the delay is acceptable.

Review questions

1. What is propagation delay, and why does it exist?
2. Why can propagation delay time be a problem?
3. Why does increasing the load capacitance increase the propagation delay?
4. Why does increasing the number of inputs an output must drive also increase the propagation delay time?
5. When logic circuits are cascaded, what effect does that have on the overall propagation delay time, and why is it affected in this manner?

12–8 TROUBLESHOOTING

In this chapter we have investigated the ''whys'' of logic level voltage specification, current, and capacitance loading. In Chapter 2 and this chapter we used these specifications to ensure that our designs were within specification limits for fan out and interfacing various technologies. To troubleshoot on the basis of fan out and the mixing of technologies, we simply drive the output under test to the two logic states and measure them to verify that V_{OH} and V_{OL} are within specifications. If they are, there is no fault with that output. If an output fails to meet the output voltage specifications, it may be that the output is overloaded and is either sinking or sourcing too much current, as discussed in this chapter. The input voltage must also be verified as being within specifications.

To test for capacitance load we compare the input pulse that gates the logic circuit on and off and measure the resultant propagation delay times. If these are greater than allowed, the amount of capacitance load must be reduced. To reduce the capacitance loading we can attempt to replace the driven logic circuit with logic circuits that have less input capacitance. Alternatively, we could choose to buffer the output of the driver, therefore seeing just a single input.

The troubleshooting techniques relative to the material of this chapter are similar in concept to those techniques previously developed; that is, input a known logic state and measure its resultant output state. From this, determine if input and output parameters are within specifications. If they are not in specification, check for incompatibility and loading requirements via the manufacturer's data sheets. If compatibility and loading are within specifications, then this specific device is out of specification due to an internal fault and must be replaced.

Review questions

1. What are the results of loading problems?
2. How are loading problems detected?
3. How can loading problems be corrected?

12–9 SUMMARY

- Digital circuitry contains electronic switches that operate in one of two states, on and off. There are bipolar and MOSFET electronic switches.
- The state of a bipolar electronic switch is controlled by the presence or absence of base current.
- The state of a MOSFET electronic switch is controlled by the presence or absence of a gate voltage.
- There are various bipolar-based logic circuits, but the dominant family is TTL. Also, there are various types of TTL logic, such as 54/74, 74S, and 74LS. These variations are the result of changing biasing resistance values and/or using the faster Schottky transistors.
- MOS family logic circuits also come in various alterations of the basic MOS circuit. There are MOS logic circuits available in NMOS, PMOS, and CMOS. NMOS and CMOS are the dominant MOS technologies.
- MOS technology enabled VLSI integrated circuits to become a reality.
- Logic voltage levels and load current symbols are standardized, and manufacturers specify these voltage levels and load currents in data sheets. In order for two logic families to be compatible the logic voltage levels at the output of the driver must be of proper magnitude and polarity, so that these levels are properly interpreted as 1's and 0's at the input of the driven logic circuit.
- For load current compatibility, if load currents are required for either logic level there must be a current source and sink. The current sink must be capable of sinking the required source current.
- *Noise margin* is a measure of voltage tolerance that exists between output and input logic voltage levels for either logic level.
- When noise margin is applied to logic families' tolerance of noise, it is known as *noise immunity*.
- A high noise immunity is desired.
- Owing to the parasitic capacitance of electronic switches, they do not turn on or off instantaneously. This results in a propagation delay time.
- As a pulse propagates through a number of cascaded logic circuits, its *propagation delay time* is the sum of propagation delays of each individual logic circuit it passes through.
- Propagation delay time is a potential problem when synchronization must occur between two or more pulses, such as in MPU read and write operations.
- To test a node for loading and/or mismatched technologies, the output voltage parameters are measured.

When a fault due to loading is detected (including too much input capacitance), a buffer (driver) could be the solution.

PROBLEMS

Section 12–2 The electronic switch

1. Explain why the voltage logic level V_{IL} cannot turn on the electronic switch of Figure 12–1.
2. Why does the gate voltage of the electronic switch of Figure 12–2 have to be of magnitude $V_{G\ ON}$, which is the same as V_{IH}, in order to turn it on?
3. What logic gate is formed by connecting two or more switches in series?
4. When switches are paralleled, what logic gate is formed?

Section 12–3 Fundamental electronic principles of logic gates

5. What is the primary function of Q_1 of the TTL inverter that is shown in Figure 12–4?
6. What is the function of Q_2 in the schematic of Figure 12–4?
7. TTL logic has a totem pole output. Explain what is meant by this term and what its advantage is over the output of the electronic switch of Figure 12–1.
8. What is the essential difference in operation between the circuits of Figures 12–4 and 12–7?
9. When any emitter input (A, B, or C) of Figure 12–7 is at ground potential, what is the approximate magnitude of that emitter current if V_{CC} is 5 volts?
10. If the inputs of the *NAND* gate of Figure 12–7 were not connected—that is, they were left ''floating''—what would be the resultant logic level at the output? This illustrates why inputs should always be connected—never leave an input unconnected.
11. What is the advantage of open-collector logic?
12. Explain why connecting open-collector outputs together results in *AND*ing the output logic of those gates.
13. Explain how the diodes at the inputs of TTL logic protect the circuit.
14. What is the essential difference between a standard bipolar transistor and a Schottky bipolar transistor?
15. Why does a Schottky transistor have a faster transition time than a standard bipolar transistor?
16. Why do the NMOS logic circuits drain less current from their power supply than does a TTL gate?
17. Why is MOS logic used to produce VLSI ICs and not TTL logic?
18. Explain why by just a quick examination of the circuits of Figure 12–10 the logic gates are easily identified as *NOR* and *NAND* gates.
19. Why do the circuits of Figure 12–10 justify the statement that the input currents of those circuits are only leakage currents?

20. Using switches, create equivalent circuits that can be used to verify the truth tables for the circuits of Figure 12–10.
21. Repeat Problem 20 for the circuits of Figure 12–11.
22. Why does CMOS drain less current than either TTL or NMOS?
23. What is significant in the wiring connections of Q_A and Q_B of Figure 12–11 relative to determining the resultant type of logic gate?
24. Using circuit analysis, prove that when Q_A is on, Q_{RA} is off and vice versa.

Section 12–4 Electronic memory cells

25. How does the static memory cell ensure that when one of its electronic switches is on the other is off?
26. Why must both row and column of both Figures 12–12 and 12–13 be high in order to access the cell?
27. Explain how read/write operations are performed on the memory cell of Figure 12–12.
28. Repeat Problem 27 for the memory cell of Figure 12–13.
29. What is the purpose of Q_W and Q_R in the memory cell of Figure 12–13?
30. Explain how a dynamic memory cell functions.
31. How is a memory cell programmed in the ROM memory cells of Figure 12–15?
32. What constitutes a memory location in the ROM of Figure 12–15, and how is it read?

Section 12–5 Designations of logic families

33. List designations used to identify the TTL logic families, and explain the difference in electronics from the TTL standard.
34. List designations used to identify the various types within the MOS family.

Section 12–6 Definition of logic voltage levels and loading specifications

35. What is meant by driver and driven?
36. When referring to a driver logic circuit, are input or output parameters of interest?
37. Relative to compatibility, what is the significance of the table in Figure 12–17?
38. Explain the significance of Table 12–2. Use Figure 12–18 as the basis of your explanation.
39. How does a manufacturer indicate a source current in the DC specifications?
40. What is fan out and how is it determined?
41. Check these logic families for compatibility.
    a. 74L driven by 74C
    b. 74S driven by 74HC
    c. 74H driven by 74AC
    d. 74HC driven by 74S

42. Check the I/O section of Figure 10–10 for compatibility if the I/O ports of Figure 10–9 are used.
43. Check the memory of Figure 9–17 for load current compatibility if the 2764 has the same DC specifications as a 2716 (see Example 12–10).
44. What would be the resultant propagation delay time of any one of the 74LS138 decoders of Figure 9–22? (Do not give a numerical answer.)

Section 12–7 Propagation delay and load capacitance

45. What causes a logic gate to have a propagation delay?
46. Explain what would constitute the resultant propagation delay time of the chip enable input signal ($\overline{CE}$) of the 2732 of Figure 9–23?
47. Explain what would be the resultant propagation delay time of the chip enable and output enable of any of the memory ICs of Figure 9–25.
48. When can propagation delay be a problem?

Troubleshooting

Section 12–8 Troubleshooting

49. If the fan out of a TTL *NAND* gate is exceeded, what is the expected output response?
50. How would you know that the capacitance being driven by an *OR* gate has been exceeded?
51. If exceeding fan out produces logic levels out of specification, what are the possible problems and how would you test for them?
52. What is a remedy for overloading an output?
53. If the pull-up resistor of an open-collector *NAND* gate were too large, what would be the symptom detected?
54. How could you detect a failed tristate buffer? State your reasoning.

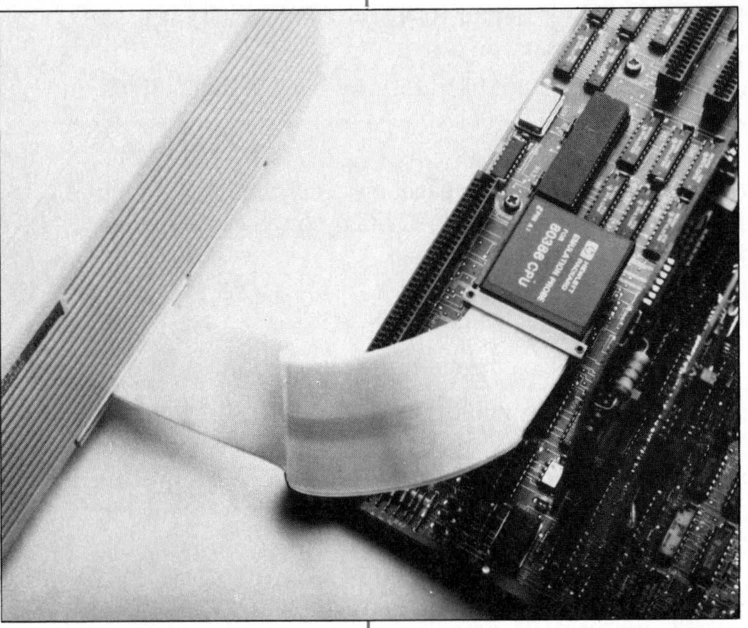

A digital computer is designed to interface with various components. The photograph shows an instrument known as an emulator/analyzer that assists in the design and development of the components that make-up and interact with the digital computer (Photo courtesy Hewlett-Packard Company).

OUTLINE

13–1 Introduction

13–2 Interfacing parallel I/O devices

13–3 Interfacing serial I/O devices

13–4 The operational amplifier

13–5 Digital-to-analog conversion

13–6 Analog-to-digital conversion

13–7 Troubleshooting

13–8 Summary

OBJECTIVES

The objectives of this chapter are to

- explain how to interface parallel and serial I/O devices.
- study some devices that are specifically for interfacing parallel and serial I/O devices.
- introduce operational amplifiers as they apply to DACs and ADCs.
- examine the essential concepts of digital-to-analog conversion.
- explain the electronic principles involved in the operation of a DAC.
- examine the essential concepts of analog-to-digital conversion.
- explain the electronic principles involved in the operation of an ADC.
- implement the concepts of DACs and ADCs as interfacing between digital and analog systems.

13 INTERFACING PARALLEL, SERIAL, AND ANALOG I/O DEVICES

13–1 INTRODUCTION

As we are already aware, various types of I/O devices must be connected to the data bus of the system. Sometimes an I/O device can be directly connected to the data bus, whereas other I/O devices require additional circuitry, such as tristate buffers or latches. Whatever the case, the process of connecting I/O devices to the data bus is known as *interfacing the I/O device to the system*. The interface circuitry may range from a very simple circuit (made up of wires) to one that is quite complex.

Each I/O interfacing is a path over which the CPU can communicate with an I/O device, and the interface is known as an *I/O port*. An I/O port is a place, or circuitry, that provides access to the data bus.

Input/output devices can basically be classified into one of three groups: (1) parallel, (2) serial, or (3) analog. *Parallel I/O devices* are those that transfer or receive data bits in parallel. An 8-bit parallel I/O device reads to or writes from the 8-bit data bus lines in parallel. Parallel I/O devices are the type shown in the memory and I/O designs of Chapter 8. *Serial I/O devices* can transfer, or receive, just 1 bit at a time. *Analog I/O devices* do not digitize data (they do not put the data in a format of 1's and 0's) but rather process data in analog form. The data is a continuous waveform and therefore has no sudden jumps (like going from a 1 to a 0). An audio amplifier is an example of an analog device.

There are many applications in which a digital computer is used to control a process, such as adjusting the fuel/air mixture ratio of an automobile engine or the speed of an electric motor. Controlling a process digitally may require that the digital output, which is composed of bits, be converted to a single analog like output voltage whose magnitude is proportional to the digital value being converted. This type of conversion is known as a digital-to-analog conversion. The circuitry that performs the conversion is known as a *digital-to-analog converter* (DAC). There are also applications in which the reverse conversion must occur, known as analog-to-digital conversion; in this case, the conversion is performed by an *analog-to-digital converter* (ADC).

We will study parallel, serial, and analog interfacing. Digital-to-analog conversion and analog-to-digital conversion will be studied from both a conceptual and an electronic point of view. The electronics of DACs and ADCs will be approached analytically; however, those not interested in a detailed analysis of the electronics may omit that material without any loss of essential concepts.

13–2 Interfacing parallel I/O devices

In Chapter 8 we discussed the need for buffering input devices (to prevent bus contention) and output devices (to reduce loading). To interface a parallel I/O device, tristate buffers and latches are used, as illustrated in Figure 13–1, which shows input device N being interfaced with the data bus via tristate buffers. When I/O device N is addressed, its I/O select goes to the active state (high) and the buffers turn on, via the CS pin, which in turn loads the datum of input device N on the data bus. On the other hand, when output device M is written to by the CPU the datum written on the data bus is latched by the latch when its I/O select line goes from a 0 to a 1 (positive edge). Both the tristate buffers and the latches provide access to the data bus and are therefore I/O ports. For specific I/O ports, refer to devices such as the 74125, 74126, and 74LS374 in Appendix C. Buffers and latches were also examined in Chapter 2.

There is a large demand for devices that can serve as parallel I/O ports. Industry has provided a somewhat universal parallel I/O interface that is applicable to almost any parallel I/O port. The configuration of this interfacing device is programmable, which enables it to be tailored to a particular application. To program the interfacing device, the CPU writes a control-word to it on start up of the system. The control-word (often an 8-bit code) is read by the device and stored in its control register. The

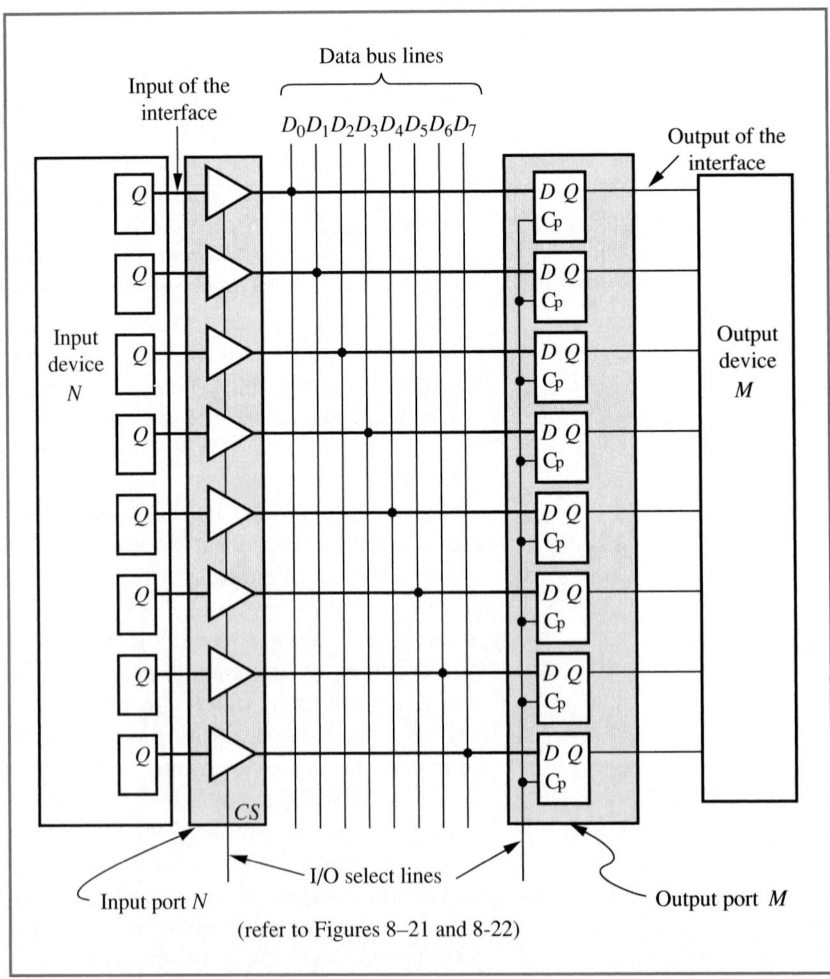

Figure 13–1
I/O port interfacing.

control-word is then decoded by the interfacing device. It then configures itself according to the dictates of the control-word and remains in that configuration so long as power is applied to the interfacing device.

Figure 13–2 is representative of a programmable interface. It has three I/O ports (A, B, and C) to which I/O devices 0, 1, and 2 are connected. We can see from the directions of the arrows that, in this instance, port A has been programmed to be bidirectional, port B has been programmed to function as an output port, and port C is an input port. The address pins A_0 and A_1 permit each port to be assessed, as well as the control register. That is, logic level 00 accesses port A, logic level 01 accesses port B, logic level 10 accesses port C, and logic level 11 accesses the control register. The control register is the internal register that contains the control-word that configures the interfacing device. As already stated, the CPU writes the control-word to the control register. The write ($\overline{WR}$) and read ($\overline{RD}$) pins control read/write operations to or from the ports, as well as to or from the control register of the interfacing device. The chip select pin $\overline{CS}$ enables the interfacing device and is used in conjunction with addressing the interfacing device via an I/O select line from the I/O selector decoder.

One specific programmable I/O interfacing device is an 8255. It is identified by the acronym PPI, which stands for *programmable peripheral interface*. Figure 13–3 illustrates its architecture and pin configuration. There are three 8-bit ports. Port A pins are identified as PA_7 through PA_0 and port B pins as PB_7 through PB_0. Port C pins have been divided into two groups of four and are identified as PC_7 through PC_4 and PC_3 through PC_0. Port C pins are divided so that they can provide either of two functions: an 8-bit port, as illustrated in Figure 13–2, or handshaking capability for ports A and B. Since handshaking is a request and acknowledge action, two pins per port are required, or a total of four pins. The address, write, read, and $\overline{CS}$ pins function like the interfacing device of Figure 13–2, except that their active logic level is a low.

Figure 13–2
Programmable I/O interfacing.

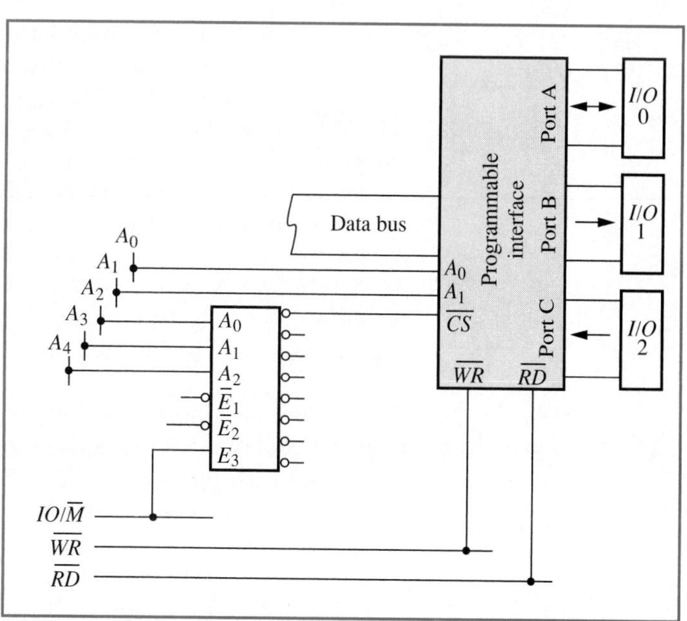

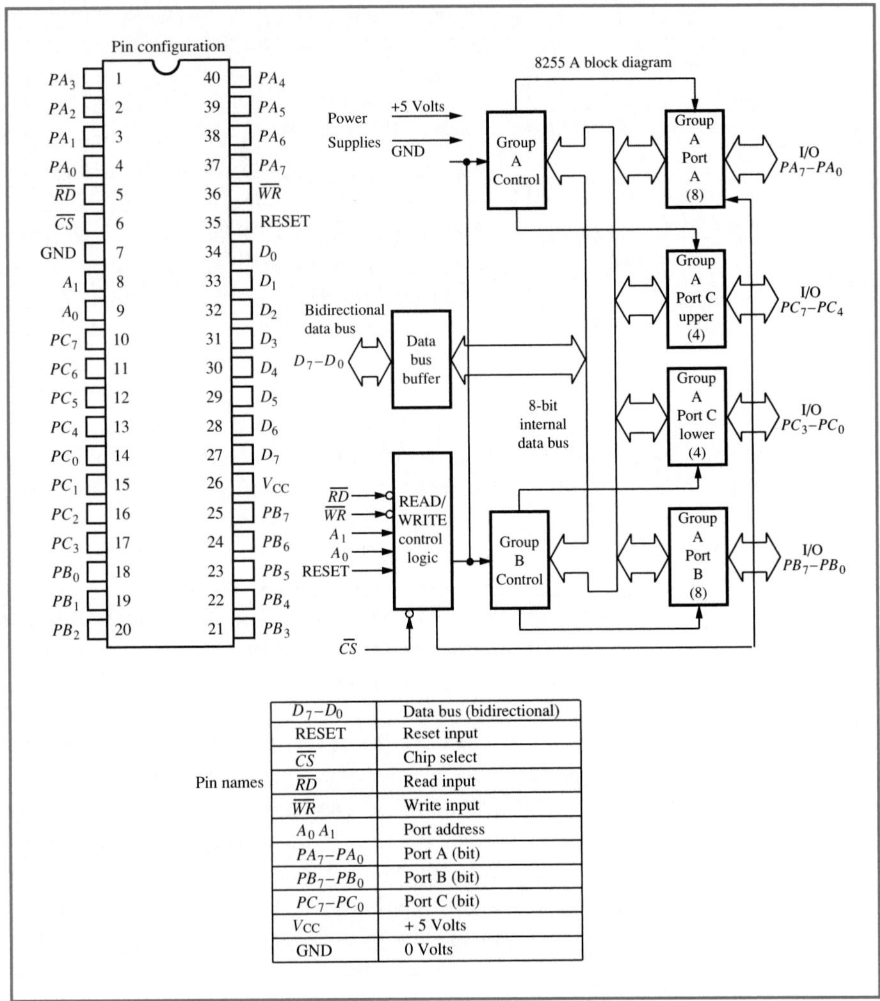

D_7–D_0	Data bus (bidirectional)
RESET	Reset input
$\overline{CS}$	Chip select
$\overline{RD}$	Read input
$\overline{WR}$	Write input
$A_0\ A_1$	Port address
PA_7–PA_0	Port A (bit)
PB_7–PB_0	Port B (bit)
PC_7–PC_0	Port C (bit)
V_{CC}	+ 5 Volts
GND	0 Volts

Figure 13–3 The architecture and pin configuration of an 8255 PPI. (Courtesy of INTEL Corp.)

To program the 8255 the CPU outputs the address of the 8255 on the address bus and the control-word on the data bus a ''short'' time later in accordance with the MPU timing diagram of Figure 9–15. The address selects the 8255 via its $\overline{CS}$ pin and also addresses the control register within the 8255 (which is similar to addressing a memory location within a semiconductor memory). To access the control register both A_1 and A_0 must be a logic high. The CPU then activates the write control signal of the control bus, to which $\overline{WR}$ of the 8255 is connected, while it is holding the datum (the control-word) on the data bus. The control signal causes the control-word to be latched off the data bus by the control register. The control-word is decoded and the 8255 then configures itself accordingly.

To access a port, the address must enable the $\overline{CS}$ pin and input the proper logic levels to the address pins of the 8255. To address port A requires that A_1 and A_0 be a logic 0. To address port B, A_1 must be low and A_0 high, and port C requires $A_1 = 1$ and $A_0 = 0$. Whether a read or write operation is performed on the port is a function of the logic level on the read and write pins.

There are many more features to the 8255, but these functions and basic operations are sufficient for our purposes. There are other parallel interfacing devices besides the 8255, but conceptually its purpose and operation are representative of all of them.

Review questions

1. What is parallel interfacing?
2. What two types of logic circuits are used to interface parallel I/O devices?
3. What is an I/O port, and how does it relate to the system and I/O device?
4. What is a PPI and how does it work?

13–3 Interfacing serial I/O devices

I/O devices that output, or input, 1 bit of data at a time are termed *serial I/O devices*. When a serial I/O device is interfaced to a data bus (data bus lines are read or written to in parallel), there must be an interfacing device between the serial I/O device and the parallel data bus. The interfacing device must assemble serial data into a parallel format before loading it on the data bus when the serial I/O device is an input I/O device. When an output serial I/O device is interfaced to the data bus, the interface must latch the parallel data being written to the serial I/O device and then disassemble that data and transfer it to the serial I/O device 1 bit at a time. Shift registers can be used to implement both types of interfaces.

Figure 13–4 illustrates shift registers from Chapter 7 used as serial-to-parallel and parallel-to-serial interfaces. The serial-to-parallel interface shifts data into the interfacing unit 1 bit at a time. The shift rate is controlled by the clock, which may be synchronous or asynchronous with the I/O device. After the 8 bits of data have been shifted to the interface,

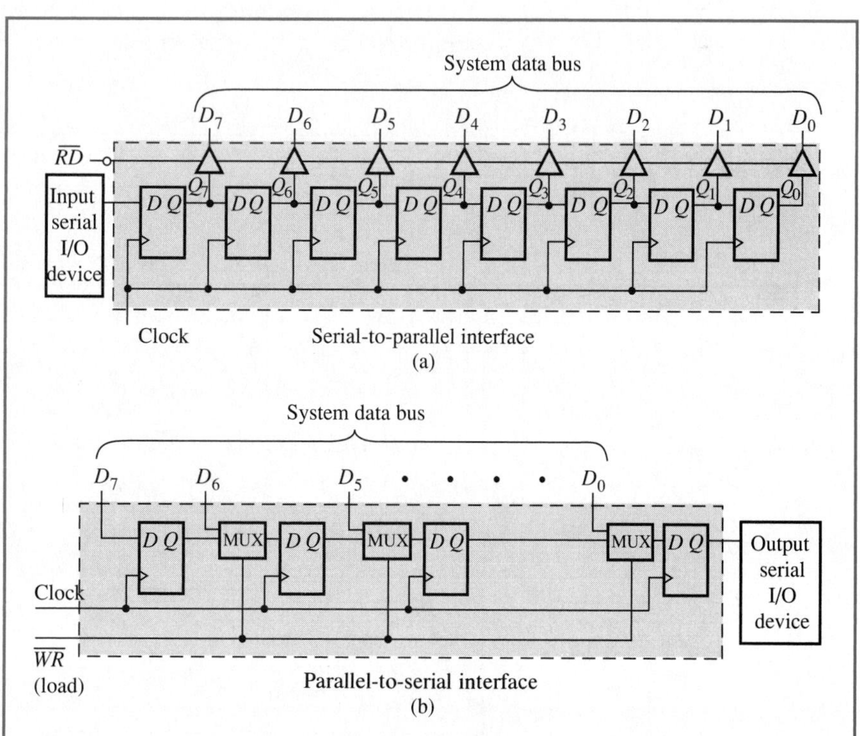

Figure 13–4
Serial I/O interfacing.

the CPU activates control signal $\overline{RD}$, which turns on the buffers and loads the datum on the 8-bit data bus in parallel.

The parallel-to-serial interface of Figure 13–4b functions opposite to the serial-to-parallel circuit of Figure 13–4a. The parallel datum on the data bus is written into the parallel-to-serial interface device when its write pin is driven low. The datum is then shifted to the output serial I/O device 1 bit at a time, where the rate of shift is determined by the clock rate.

As with parallel-to-parallel interfacing devices, there is also demand for serial interfacing devices. One such device is a *universal synchronous/asynchronous receiver/transmitter* (USART). Within a single IC both serial-to-parallel and parallel-to-serial circuits exist. Figure 13–5 is conceptually representative of a USART. The system data bus is connected to pins D_7 through D_0 of the USART. An input serial I/O device sends its data over the receive line, which is input at the *RXD* (receive data) pin. The USART transmits serial data to an output I/O device over the transmit line, which is connected to output *TXD* (transmit data) of the USART. The rate of serial data transmission is determined by the clock rate input at *TXC*, and the clock rate input at *RXC* determines the rate of serial data reception by the USART. The read and write operations discussed in Figure 13–4a and b are controlled by the logic levels on pins $\overline{RD}$ and $\overline{WR}$. The $\overline{CS}$ pin is activated as a result of the USART being addressed. The $C/\overline{D}$ pin determines whether the USART is reading or writing data ($C/\overline{D} = 0$) or a control-word ($C/\overline{D} = 1$). The USART can be programmed for certain data formats.

The 8251 is a very popular USART. Its architecture and pin configuration are shown in Figure 13–6. From the additional pins of the 8251 the reader can deduce that it is much more complex than this discussion has indicated; however, we have set forth the essence of an 8251 and of USARTs in general.

Before closing our discussion of serial data interfacing devices we should mention a standard known as *RS-232C*, one that many serial I/O

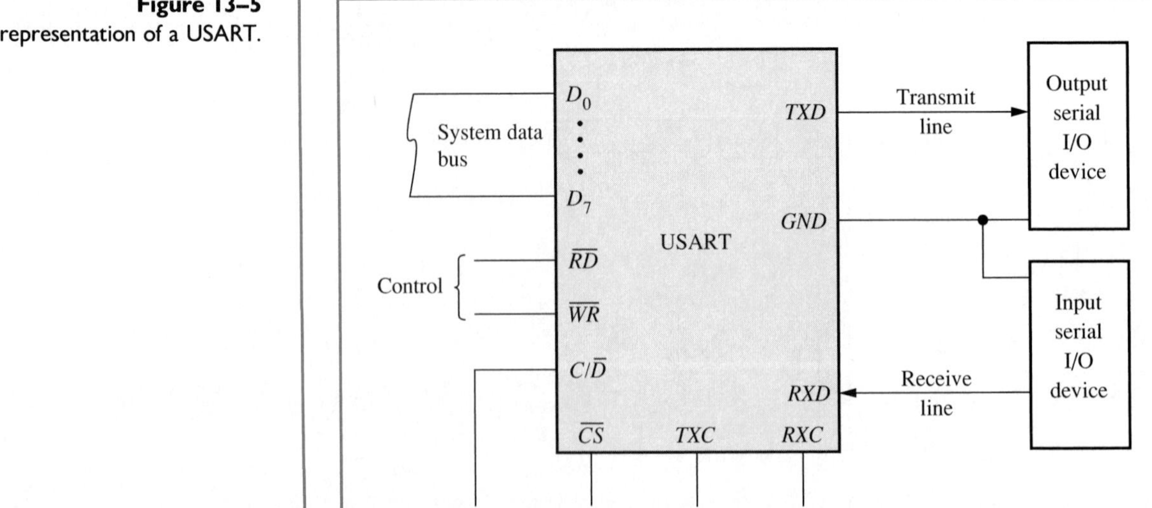

Figure 13–5
A representation of a USART.

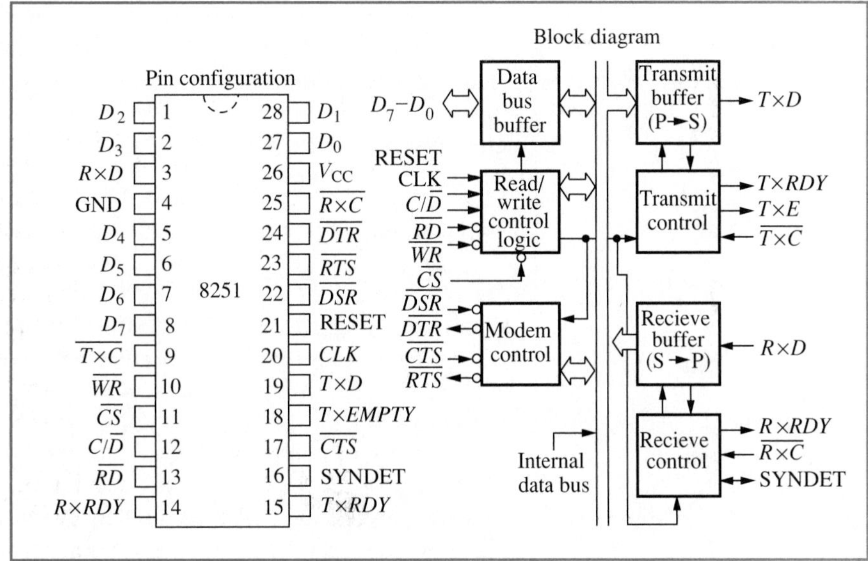

Figure 13–6 The architecture and pin configuration of the 8251 USART. (Courtesy of INTEL Corp.)

devices adhere to. The formal title of this standard is "Interface Between Data Terminal Equipment and Data Communication Equipment Employing Serial Binary Data Interchange." Among other things, it defines the voltage levels necessary to represent a logic 1 and 0, which are from +3 to +25 volts for a logic 0 and from −3 to −25 volts for a logic 1. The reason for such large voltage levels is that an RS-232C line is often used for the exchange of data over long distances; with large voltage levels noise is less of a problem. Most modern ICs operate with a 5-volt DC power supply, so there must be an interface between a data bus operating over a range of 0 to 5 volts and an RS-232C serial I/O. These interfaces are available.

Review questions

1. How can a shift register function as a parallel-to-serial interface?
2. How can a shift register function as a serial-to-parallel interface?
3. What is a USART, and how do such devices function?
4. What is an RS-232C standard, and why does it require interfacing?

13–4 The operational amplifier

The operational amplifier (op-amp) is an essential component of many DACs and ADCs. Therefore, we will examine the principles of op-amps necessary for understanding the operation of DACs and ADCs.

As with any active electronic component, such as a transistor, op-amps can be operated in either a nonlinear or a linear mode. In the nonlinear mode they operate in either extreme of their operating range; that is, they are either on or off. An amplifier operated as a nonlinear device is essentially an electronic switch, which is another way of saying it is operated as a digital device. The electronic switches of Figures 12–1a and 12–2a are examples of amplifiers that were designed to operate in a nonlinear mode.

Linear amplifiers are designed to operate within the range bounded by the extremes defined by "fully turned on" and "fully turned off." This range of operation is known as its *active region* or *linear range of operation*. As examples, the linear range of operation for the characteristic curve of Figure 12–1a can be defined by the inequality

$$V_{CE\,\mathrm{ON}} < V_{CE} < V_{CE\,\mathrm{OFF}}$$

The inequality

$$V_{\mathrm{OL}} < V_{\mathrm{DS}} < V_{\mathrm{OH}}$$

defines the linear range of operation for the characteristic curve of Figure 12–2a. Linear amplifiers are what we normally think of as an amplifier; that is, the output has the same waveform appearance as the input, but amplified.

An op-amp is a differential amplifier in that it amplifies the difference of the voltage signals applied at its two inputs. Depending on its external circuit configuration, the op-amp amplifies the difference signal e_d in either a linear or a nonlinear fashion.

The symbol for an op-amp is shown in Figure 13–7a, and Figure 13–7b shows the characteristic curve of an op-amp. The symbol shows two power supplies (*V*), one with positive and the other with negative polarity. The magnitude and polarity of the output voltage v_o are bounded by the magnitudes and polarities of the power supplies. The maximum magnitude of the output voltage, positive or negative, is approximately 0.5 volt less than the magnitude of the power supply with the corresponding polarity. These maximum magnitudes are identified as saturation voltages, $V^+{}_{\mathrm{SAT}}$ for the magnitude of the positive saturation voltage and $V^-{}_{\mathrm{SAT}}$ for the magnitude of the negative output saturation voltage, which is illustrated in Figure 13–7b.

The schematic of Figure 13–7a has two inputs, one positive and the other negative. The positive sign indicates that any voltage input at this input will be in phase with the amplified voltage at the output. In contrast, the amplified output voltage will be 180 degrees out of phase with the voltage applied at the negative input. The gain of the op-amp is designated A_{OL} (open-loop gain), and any voltage at either input is amplified by this gain. To determine the resultant output voltage from voltages being applied at both inputs, we use the discussion above and the superposition theorem. The output voltage resulting from input voltage e_A of Figure 13–7a is

$$v'_o = e_A A_{\mathrm{OL}}$$

The output voltage resulting from e_B is

$$v''_o = e_B(-A_{\mathrm{OL}}) = -e_B A_{\mathrm{OL}}$$

Using the superposition theorem to determine the resultant output, we find

$$v_o = v_o' + v''_o$$

or

Equation (13–1)

$$v_o = (e_A - e_B)A_{\mathrm{OL}}$$

Now, the difference voltage at the inputs can be expressed as

Equation (13–2)

$$e_d = e_A - e_B$$

Figure 13–7
The symbol and characteristic curves of an op-amp.

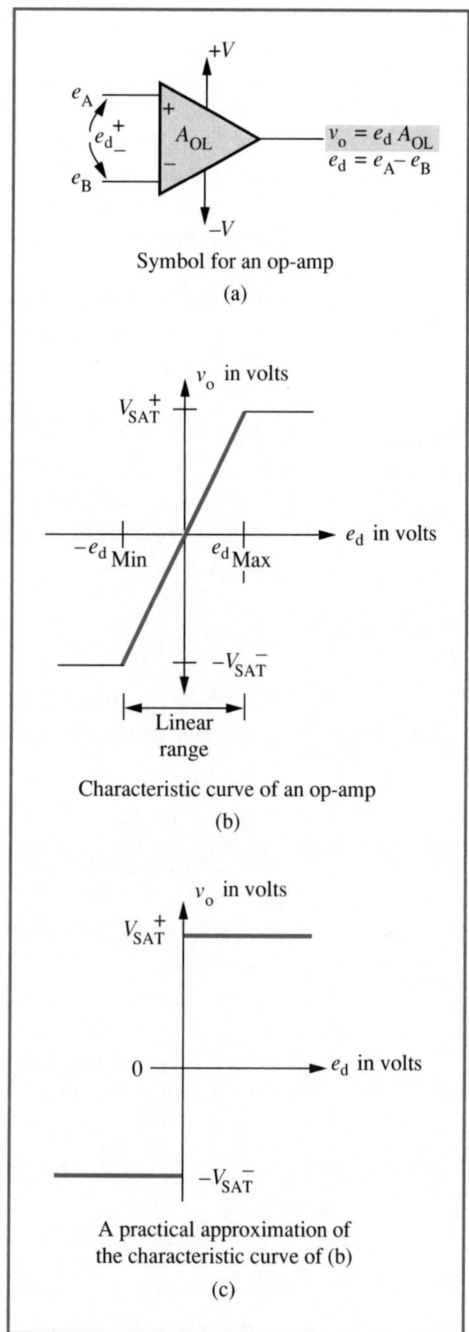

Thus Equation 13–1 can be written as

Equation (13–3)
$$v_o = e_d A_{OL}$$

Equation 13–3 confirms that this is a difference amplifier, since it amplifies the difference of two input voltages. The polarity of e_d shown in Figure 13–7a is an assumed polarity and is valid if $e_A > e_B$.

The characteristic curve of Figure 13–7b shows that an op-amp is operating in a linear mode so long as its input voltage is within the range of

Equation (13–4)

$$-e_{d\,\min} < e_d < e_{d\,\max}$$

If e_d is outside this range, the output voltage is driven into saturation and the op-amp is functioning in a nonlinear mode. Notice that if the negative power supply is removed, and its place of connection is grounded, then $-V^-{}_{\mathrm{SAT}}$ is 0 volts, which is a logic 0. In addition, if we use a +5-volt power supply for the positive power supply, then $V^+{}_{\mathrm{SAT}}$ is a logic 1 ($V_{\mathrm{OH}} = 5 - 0.5 = 4.5$ volts). Hence, by grounding the negative power supply connection and using a +5-volt power supply to implement $+V$, the output is voltage compatible for all TTL and CMOS families, as can be verified from Table 12–5.

Nonlinear mode of operation

As stated by Equation 13–4 and illustrated in Figure 13–7b, if the difference voltage is greater than $e_{d\,\max}$ or less than $-e_{d\,\min}$, the op-amp is operating in a nonlinear mode. Using a realistic value of 10^5 for A_{OL}, let us "get a feel" for the value of $e_{d\,\max}$. From Equation 13–3 we can solve for e_d; hence

Equation (13–5)

$$e_d = \frac{v_o}{A_{\mathrm{OL}}}$$

Since we are solving for maximum conditions, let the output voltage be at $V^+{}_{\mathrm{SAT}}$. Then Equation 13–5 becomes

Equation (13–6)

$$e_{d\,\max} = \frac{V^+{}_{\mathrm{SAT}}}{A_{\mathrm{OL}}}$$

If the positive polarity power supply is +5 volts DC (V_{CC}), $V^+{}_{\mathrm{SAT}}$ is equal to +4.5 volts, as discussed previously. Substituting values into Equation 13–6, we find that

$$e_{d\,\max} = \frac{4.5}{10^5} = 45\ \mu\mathrm{V}$$

If we use the same reasoning for the negative voltage range ($V^- = -5$ volts DC), we find that

$$e_{d\,\min} = \frac{-4.5}{10^5} = -45\ \mu\mathrm{V}$$

The linear range for e_d is thus $\pm 45\ \mu$V (a 90-μV swing). This voltage range is so small that we may consider it to be zero, which yields the characteristic curve of Figure 13–7c. From this figure we reason that the magnitude and polarity of the output voltage is determined by the polarity of voltage e_d. From Figure 13–7c we can mathematically state that if

Equation (13–7)

$$e_d > 0 \text{ volts},\ v_o = V^+{}_{\mathrm{SAT}}$$

and if

Equation (13–8)

$$e_d < 0 \text{ volts},\ v_o = -V^-{}_{\mathrm{SAT}}$$

The characteristic curve of Figure 13–7c is that of a binary digital device; that is, it operates in one of two states.

Nonlinear applications

The practical characteristic curve of Figure 13–7c is that of a nonlinear device and represents the output voltage as a function of the input voltage. The most common application of an op-amp operated in a nonlinear mode is a comparator.

EXAMPLE 13–1

For the circuit of Figure 13–8a derive equations that can be used to predict the op-amp output voltage as a function of the input voltage $e(t)$. Then apply those equations to the input voltage waveform of Figure 13–8b and plot the resultant output voltage.

SOLUTION From Equations 13–7 and 13–8 we know that to predict the output voltage, all we need to know is the polarity of e_d as a function of $e(t)$. To find this relationship, write a loop equation at the input circuit of Figure 13–8a. Hence,

$$e(t) - e_d = 0.$$

Figure 13–8
The op-amp comparator.

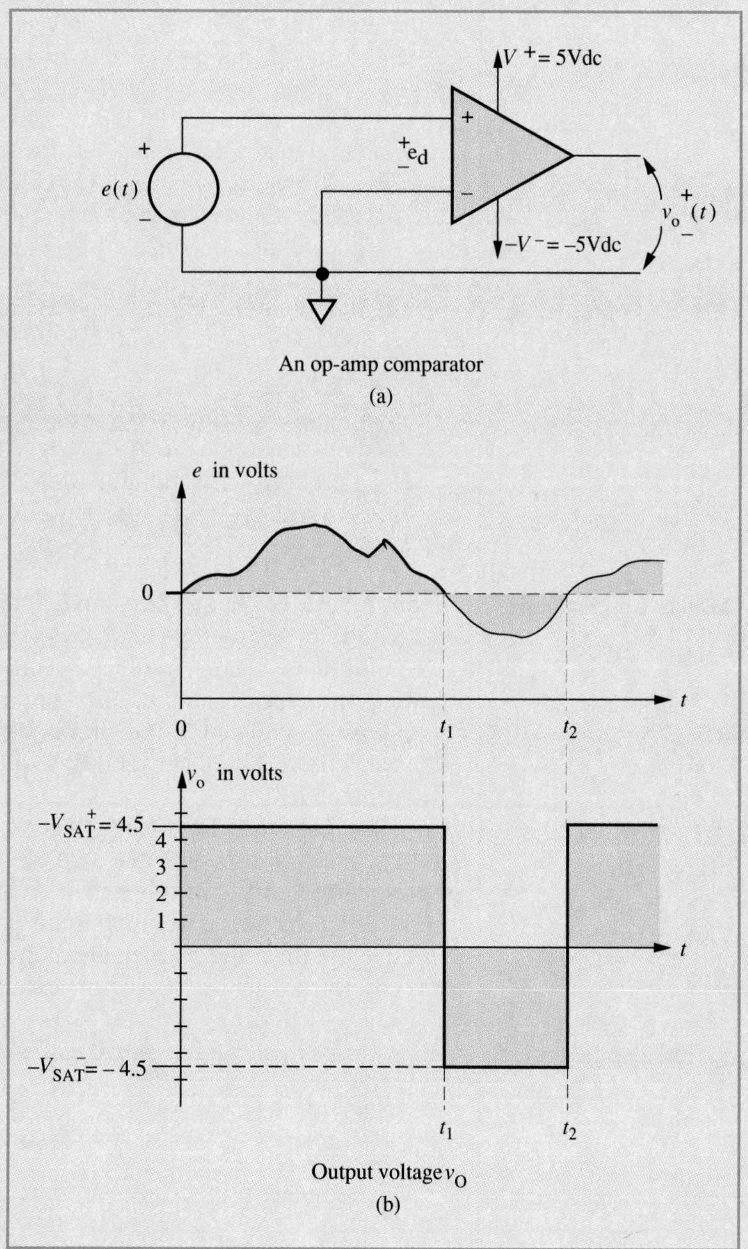

Solving for dependent variable e_d (the polarity and magnitude of e_d depends on e), we have

Equation (13–9)

$$e_d = e(t)$$

By substituting Equation 13–9 into Equation 13–7, we find that

$$e(t) > 0 \text{ volts}, \ v_o = V^+{}_{SAT}$$

From the plot of $e(t)$ in Figure 13–8b we see that this inequality is true for the time range 0 to t_1 and as a result $v_o = V^+{}_{SAT}$ over this same interval of time.

When $e(t)$ is negative, Equation 13–8 applies and the output is -4.5 volts, as shown in the interval of time from t_1 to t_2 in Figure 13–8b.

From the mathematical analysis and the plot of v_o in Figure 13–8b, it can be stated that *an op-amp operating in an open-loop (OL) mode, that is, without negative feedback, is a comparator*. In the case of the circuit of Figure 13–8a the input voltage $e(t)$ is compared with the reference voltage (the voltage at the negative input), which is zero volts. As a result, when $e(t)$ becomes greater than the reference voltage (0 volts) the output is $+4.5$ volts, and when $e(t)$ drops below the reference voltage the output voltage becomes -4.5 volts.

EXAMPLE 13–2

For the circuit of Figure 13–9 apply the input voltage of Figure 13–8b and plot the output voltage.

SOLUTION The circuit of Figure 13–9a has the negative power supply pin connected to ground. As a result, when $e_d < 0$, the output voltage goes to zero volts rather than $-V^-{}_{SAT}$. The solution of Example 13–1 is valid for this circuit, except that $-V^-{}_{SAT}$ is equal to zero volts, as indicated by the output voltage plot of Figure 13–9b.

Linear applications

From Figure 13–7b and Equation 13–5 we realize that it is the large magnitude of open-loop gain (A_{OL}) that causes an open-loop op-amp to be a nonlinear device, as graphically indicated in Figure 13–7c. To use an op-amp in a linear mode of operation requires that *the effects of open-loop gain* be reduced in the circuit. To accomplish this, negative feedback is used. Figure 13–10 shows negative feedback being implemented with the resistor R_f.

The circuit of Figure 13–10 is an analog adder; that is, the output voltage is proportional to the sums of the input voltages e_0, e_1, e_2, etc. To analytically verify that this is true, we begin the analysis by writing a nodal equation at the negative input of the op-amp. Those not interested in the electronic and mathematical details of the analysis may progress directly to the answer (Equation 13–12), which is printed in color.

Equation (13–10)

$$\frac{e_o - (-e_d)}{R_0} + \frac{e_1 - (-e_d)}{R_1} + \frac{e_2 - (-e_d)}{R_2} + \ldots + \frac{e_N - (-e_d)}{R_N} + \frac{v_o - (-e_d)}{R_f} = 0$$

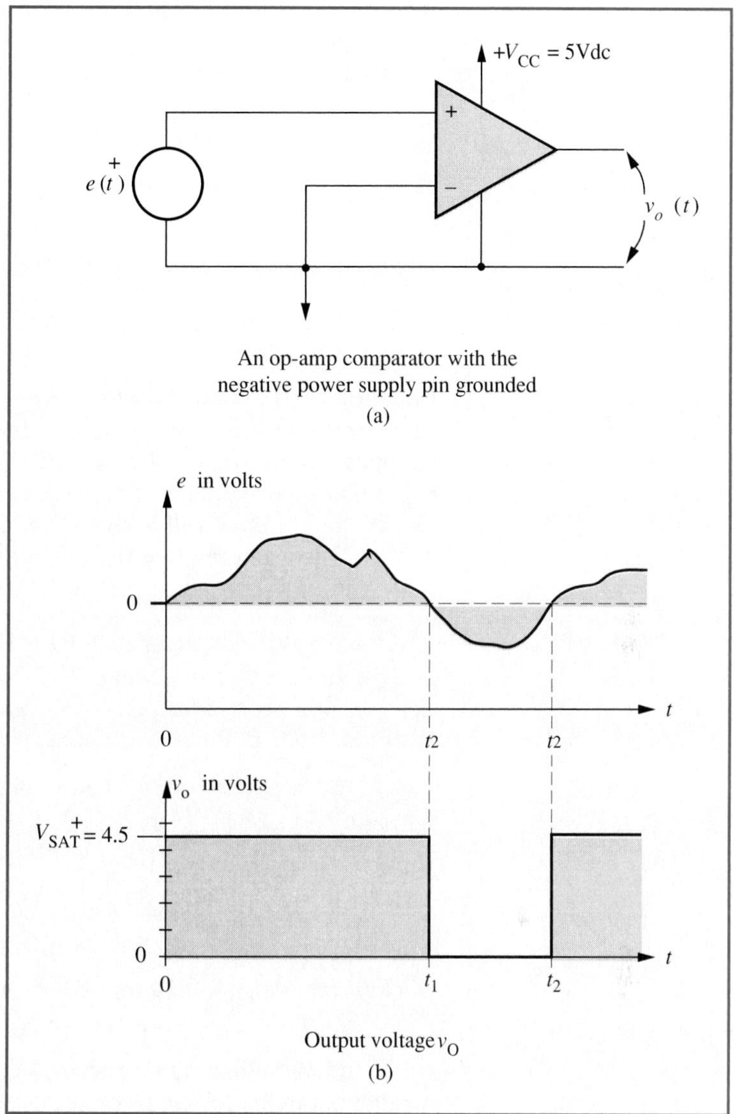

Figure 13–9
An op-amp comparator with a reference of 0 volts.

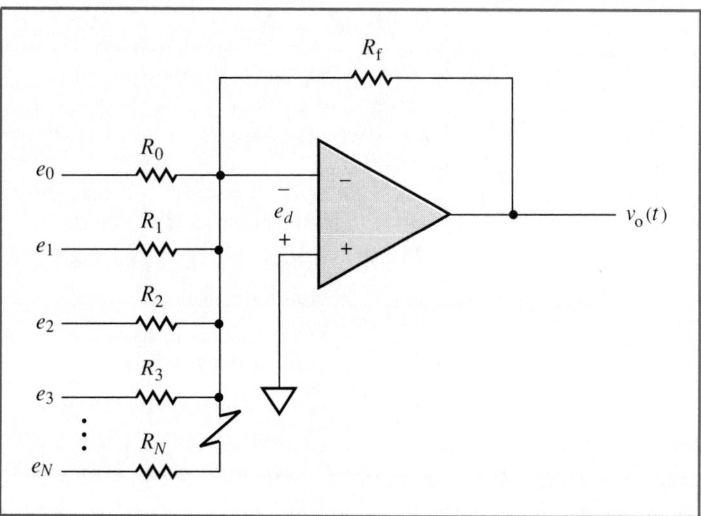

Figure 13–10
An adder.

We know that for all practical purposes $e_d = 0$. Equation 13–10 can then be written as

Equation (13–11)

$$\frac{e_0}{R_0} + \frac{e_1}{R_1} + \frac{e_2}{R_2} + \ldots + \frac{e_N}{R_N} + \frac{v_o}{R_f} = 0$$

Solving Equation 13–11 v_o gives the equation we were seeking, which is

Equation (13–12)

$$v_o = -\left[\left(\frac{R_f}{R_0}\right)e_0 + \left(\frac{R_f}{R_1}\right)e_1 + \ldots + \left(\frac{R_f}{R_N}\right)e_N\right]$$

Equation 13–12 states that the output voltage is 180 degrees out of phase (the negative sign) with the sum of the products of coefficients and input voltages, where the coefficients (R_f/R_0, R_f/R_1, etc.) are gains of the individual input voltages. We shall use these gains as *weighting factors*; that is, their values will assign a "weight of influence" to each input voltage when determining the magnitude of the output voltage.

EXAMPLE 13–3

Suppose that the adder of Figure 13–10 had three input voltages with values of $e_0 = 2$ volts, $e_1 = -3$ volts, and $e_2 = 1.5$ volts. If $R_f = 100$ K, $R_0 = 50$ K, $R_1 = 30$ K, and $R_2 = 25$ K, what is the output voltage for power supplies equal to + and − 25 volts DC?

SOLUTION Substituting these values into Equation 13–12 we find that

$$v_o = -\left[\left(\frac{100}{50}\right)2 + \left(\frac{100}{30}\right)(-3) + \left(\frac{100}{25}\right)1.5\right]$$

$$= -(4 - 10 + 6) = 0 \text{ volts}$$

Note the weighting factor of the coefficients.

Understanding op-amp–based comparators and adders is sufficient for understanding op-amps as they relate to DACs and ADCs.

Review questions

1. What is the meaning of the positive and negative sign at the input of an op-amp?
2. Why can an op-amp be classified as a difference amplifier?
3. In what two modes of operation can an op-amp operate?
4. Why is the open-loop gain of an op-amp responsible for its nonlinear characteristic?
5. Why does negative feedback enable an op-amp to function in a linear mode?
6. When an op-amp is used without negative feedback, why is it a voltage comparator?
7. What is meant by "weighting factor" as it applies to Equation 13–12?

13–5 Digital-to-analog conversion

An *analog signal* is a signal that is continuous; that is, it has no points of discontinuity. The input voltage $e(t)$ of Figure 13–8b is an analog signal. All digital signals are discontinued signals. The output voltage v_o of that same figure is a *digital signal*. This waveform is discontinuous since it changes abruptly at t_1 and t_2.

The "real world" functions in an analog mode. However, people have created a digital world through digital devices and computers. Often it is convenient, or even necessary to cross back and forth between these two worlds, which is the reason for DACs and ADCs.

To develop the concepts of DACs we will study two basic techniques of converting a digital signal to an analog signal. These two techniques are (1) a resistor ladder network known as an *R*-2*R* ladder, (2) an *R*-2*R* ladder that is buffered with an op-amp circuit that is either a voltage-follower, an adder, or an inverting amplifier.

R-2*R* ladder network

Figure 13–11 is a schematic of a three-input *R*-2*R* ladder network. If additional inputs are required, the pattern of Figure 13–11 is repeated for each additional input. The inputs e_0, e_1, and e_2 are digital inputs, so their voltage level is either V_{OL} (0) or V_{OH} (1). To analyze the circuit of Figure 13–11 we write nodal equations at each node. The nodal voltages are identified by v_1, v_2, and v_o. We solve for v_o, which is the converted analog signal. If the reader wishes to skip the analysis of Figure 13–11 he or she may advance to the answer of Equation 13–13. As with the analysis of the adder, the answer is printed in color for those wishing to omit the analysis. This format will be used for all circuit analyses.

Writing a nodal equation at node 1 of the *R*-2*R* ladder network, we have

$$0 = -\frac{v_1}{2R} + \frac{e_0 - v_1}{2R} + \frac{v_2 - v_1}{R}$$

At node 2 we find that

$$0 = \frac{v_1 - v_2}{R} + \frac{e_1 - v_2}{2R} + \frac{v_o - v_2}{R}$$

At node 3, which is the output node

$$0 = \frac{v_2 - v_o}{R} + \frac{e_2 - v_o}{2R}$$

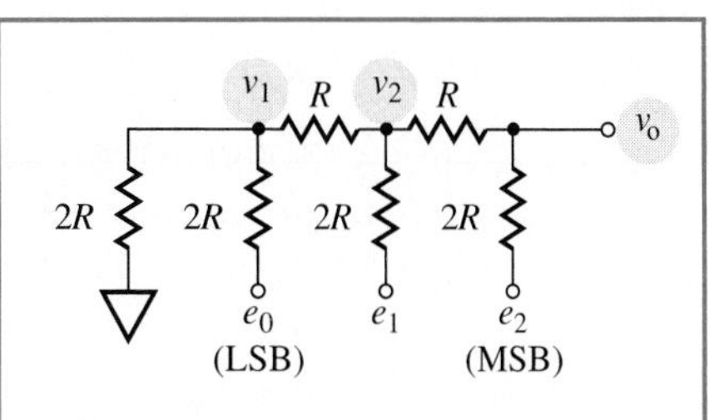

Figure 13–11
R-2*R* resistor ladder network.

Dividing the Rs out of the three nodal equations and setting the knowns (e_1, e_2, and e_3) equal to the unknowns (v_1, v_2, and v_3) we have

$$\begin{aligned}
\frac{e_0}{2} &= 2v_1 \quad - v_2 \\
\frac{e_1}{2} &= -v_1 + \left(\frac{5}{2}\right) v_2 \quad - v_o \\
\frac{e_2}{2} &= \quad - v_2 + \left(\frac{3}{2}\right) v_o
\end{aligned}$$

From these three equations we structure the determinate D

$$D = \begin{vmatrix} 2 & -1 & 0 \\ -1 & \frac{5}{2} & -1 \\ 0 & -1 & \frac{3}{2} \end{vmatrix}$$

$$D = 2\left[\left(\frac{5}{2}\right)\left(\frac{3}{2}\right) - 1\right] + 1\left[-1\left(\frac{3}{2}\right)\right] = 4$$

Now solving for Dv_o we have

$$Dv_o = \begin{vmatrix} 2 & -1 & \frac{e_0}{2} \\ -1 & \frac{5}{2} & \frac{e_1}{2} \\ 0 & -1 & \frac{e_2}{2} \end{vmatrix}$$

$$= 2\left[\frac{(5/2)e_2}{2} + \frac{e_1}{2}\right] + \left[\frac{-e_2}{2} + \frac{e_0}{2}\right]$$

$$= 2e_2 + e_1 + \frac{e_0}{2}$$

Dividing Dv_o by the determinate ($D = 4$)

$$v_o = \frac{e_2}{2} + \frac{e_1}{4} + \frac{e_0}{8}$$

Putting these quantities under common denominator

Equation (13–13)

$$v_o = \frac{1}{8}(4e_2 + 2e_1 + e_0)$$

Let us generalize Equation 13–13 so that it is valid for any number of inputs by recognizing a pattern of the coefficients. Using n to represent the nth input and N to represent the total number of inputs, Equation 13–13 can be written as

Equation (13–14)

$$v_o = \frac{1}{2^N}\left(\sum_{n=0}^{N-1} 2^n e_n\right)$$

Note from Equations 13–13 and 13–14 that each input voltage has a weighting function, which is the coefficient of each input voltage and is equal to 2^{n-N}; as a result; input voltage e_2 is given more weight than e_1 or e_0. This means that the MSB of, say, a 3-bit binary input would be connected to input e_2 and the LSB would be connected to e_0, since the MSB should have more weight when determining the output voltage than any of the other bits and e_0 should have the least weight.

EXAMPLE 13–4

For the number of inputs specified, draw the appropriate schematic for an R-$2R$ ladder network and from Equation 13–14 determine the equation for the output voltage v_o.
(a) $N = 4$ (b) $N = 8$

SOLUTION Figure 13–12 shows the schematics for parts (a) and (b).
(a) For $N = 4$, Equation 13–14 becomes

$$v_o = \frac{1}{2^4}\left(\sum_{n=0}^{4-1} 2^n e_n\right)$$

$$= \frac{1}{16}(2^0 e_0 + 2^1 e_1 + 2^2 e_2 + 2^3 e_3)$$

$$v_o = \frac{1}{16}(e_0 + 2e_1 + 4e_2 + 8e_3)$$

Figure 13–12
4- and 8-bit R-$2R$ ladder networks.

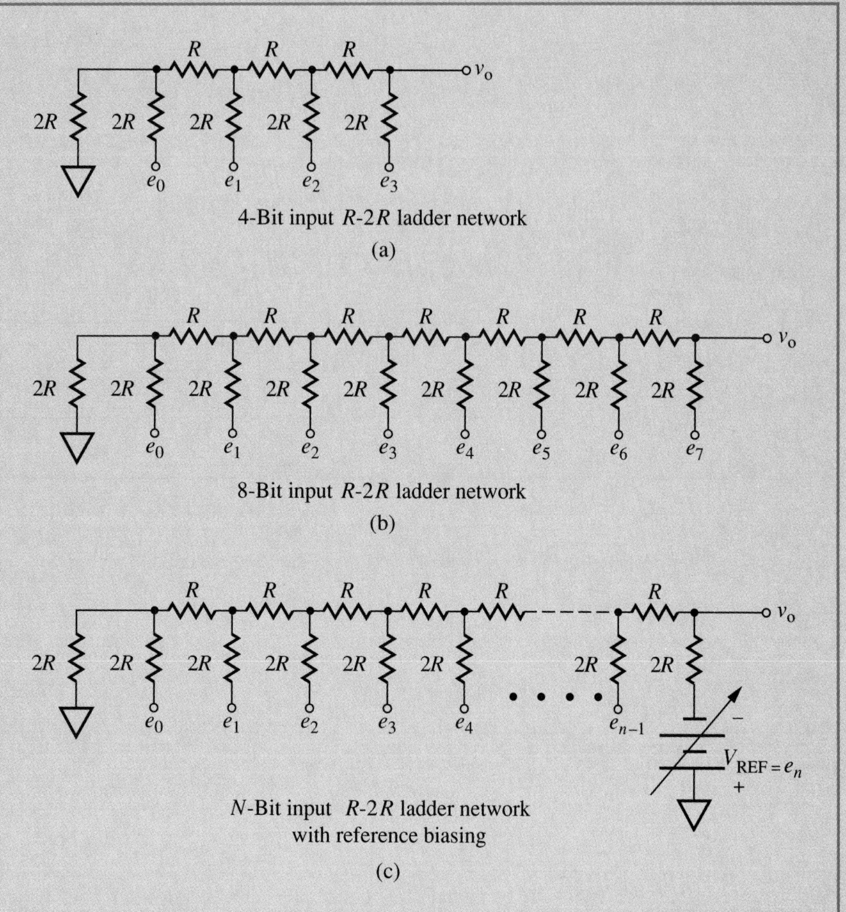

(b) For $N = 8$

$$v_o = \frac{1}{256}(e_0 + 2e_1 + 4e_2 + 8e_3 + 16e_4 + 32e_5 + 64e_6 + 128e_7)$$

EXAMPLE 13–5

Suppose that the output of a 74LS series 4-bit up-counter is to be converted to an analog signal. Use an R-$2R$ ladder network as shown in Figure 13–13a to make the conversion. Graph the analog output signal as the digital counter progresses from count 0000 to 1111.

SOLUTION Using the equation from Example 13–4a, which is

$$v_o = \frac{1}{16}(e_0 + 2e_1 + 4e_2 + 8e_3)$$

Figure 13–13 Analog output voltage for a 4-bit R-$2R$ ladder DAC.

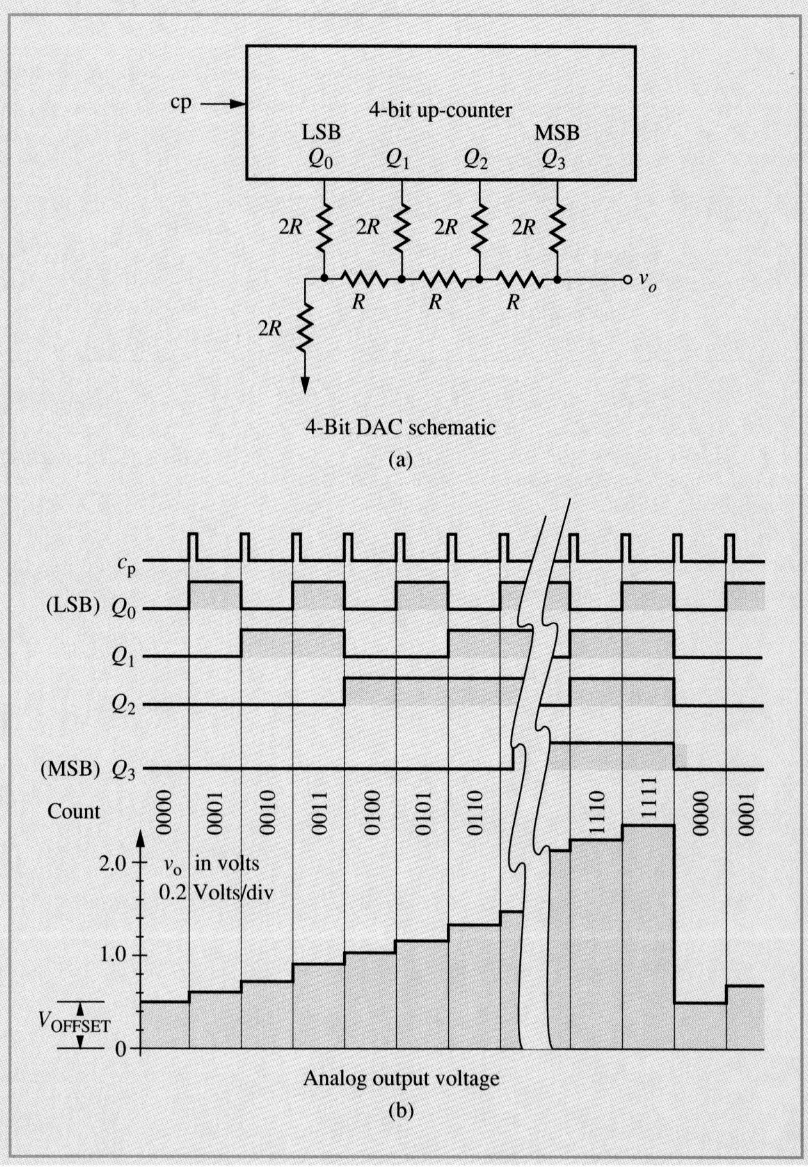

we can calculate the analog output voltage from the 16 binary numbers that make up the count. From Table 12–5 we know that a logic 0 output voltage (V_{OL}) is 0.5 volt and a logic 1 (V_{OH}) is 2.7 volts.

For the first count of 0000 all the input voltages are $V_{OL} = 0.5$ volt. Then

$$v_o = \frac{1}{16}[0.5 + 2(0.5) + 4(0.5) + 8(0.5)] = 0.469 \text{ volt}$$

When the count is 0001

$$v_o = \frac{1}{16}[2.7 + 2(0.5) + 4(0.5) + 8(0.5)] = 0.606 \text{ volt}$$

Count = 0010

$$v_o = \frac{1}{16}[0.5 + 2(2.7) + 4(0.5) + 8(0.5)] = 0.744 \text{ volt}$$

Count = 0011

$$v_o = \frac{1}{16}[2.7 + 2(2.7) + 4(0.5) + 8(0.5)] = 0.881 \text{ volt}$$

Count = 0100

$$v_o = \frac{1}{16}[0.5 + 2(0.5) + 4(2.7) + 8(0.5)] = 1.018 \text{ volts}$$

Count = 0101

$$v_o = \frac{1}{16}[2.7 + 2(0.5) + 4(2.7) + 8(0.5)] = 1.156 \text{ volts}$$

Count = 0110

$$v_o = \frac{1}{16}[0.5 + 2(2.7) + 4(2.7) + 8(0.5)] = 1.294 \text{ volts}$$

Count = 1110

$$v_o = \frac{1}{16}[0.5 + 2(2.7) + 4(2.7) + 8(2.7)] = 2.394 \text{ volts}$$

Count = 1111

$$v_o = \frac{2.7}{16}[1 + 2 + 4 + 8] = 2.531 \text{ volts}$$

The output voltages for each corresponding count are plotted in Figure 13–13b.

From Figure 13–13b the reader can observe that the analog output voltage has an offset voltage of V_{OFFSET} when the binary input is 0000 and that the output voltage is not a true analog signal in that it is not a smooth, continuous curve. However, it can be considered close enough to be an analog signal; if a smoother (more continuous) output voltage is desired, a capacitive load can be put at the output. To get rid of the offset voltage, a reference biasing voltage source V_{REF} can be added to the R-$2R$

ladder network, as illustrated in Figure 13–12c. Using this technique the negative reference voltage can be viewed as the n input voltage; that is, $e_n = -V_{REF}$. Making this addition to the ladder of Figure 13–11 would modify Equation 13–13 to appear as

$$v_o = \frac{1}{8}(-4V_{REF} + 2e_1 + e_0)$$

since $e_2 = -V_{REF}$. To incorporate this change into Equation 13–14,

Equation (13–15)

$$v_o = \frac{1}{2^N}\left(\sum_{n=0}^{N-2} 2^n e_n - 2^{N-1}V_{REF}\right)$$

To cancel the offset voltage V_{OFFSET} of Figure 13–13b would require that

$$\sum_{n=0}^{N-2} 2^n e_n - 2^{N-1}V_{REF} = 0$$

when $e_n = V_{OL}$, which is the first count. Hence, solving for V_{REF}

$$V_{REF} = \frac{1}{2^{N-1}}\left(\sum_{n=0}^{N-2} 2^n e_n\right)$$

where $e_n = V_{OL}$. This equation states that V_{REF} must be adjusted to a magnitude of V_{OFFSET} volts, which is equal to

$$\frac{1}{2^{N-1}}\left(\sum_{n=0}^{N-2} 2^n e_n\right) \text{ volts}$$

where $e_n = V_{OL}$.

More importantly, the DACs of Figure 13–13 will convert a digital signal into a single output analog voltage whose voltage level is directly proportional to the binary value of the input.

Buffered *R*-2*R* ladder network

The *R*-2*R* ladder network has one major disadvantage—the input resistance of the load must be at least ten times larger than the value *R* of the network. This requirement is necessary so as not to "load down" the ladder network so that Equation 13–14 and/or 13–15 is no longer valid. There are three op-amp solutions to this problem. One is to use the adder circuit of Figure 13–10 when it is properly designed to yield an output voltage equal to the magnitude of Equation 13–14 or 13–15. To take care of the sign difference we can use an inverting amplifier at the output of the adder.

Using the same technique to express the adder op-amp output voltage of Equation 13–12 in a general fashion, as was used to express the *R*-2*R* ladder output voltage of Equation 13–14, we find that Equation 13–12 can be written as

Equation (13–16)

$$v_o = -\sum_{n=0}^{N-1}\left(\frac{R_f}{R_n}\right)e_n$$

Since Equation 13–14 can be written as

Equation (13–17)

$$v_o = \sum_{n=0}^{N-1} 2^{n-N}e_n$$

equating weighting functions (coefficients) we have

$$\frac{R_f}{R_n} = 2^{n-N}$$

or

Equation (13–18)

$$R_n = 2^{-(N-n)} R_f$$

EXAMPLE 13–6

If the feedback resistor R_f of Figure 13–10 has a value of 100 K, what values should the remaining resistors have, according to Equation 13–18, for a 4-bit digital input?

SOLUTION Evaluating Equation 13–18 for $N = 4$

$$R_0 = 2^{-(4-0)} \times 100\text{ K} = \frac{1}{16}(100\text{ K}) = 6.25\text{ K}$$

$$R_1 = 2^{-(4-1)} \times 100\text{ K} = \frac{1}{8}(100\text{ K}) = 12.50\text{ K}$$

$$R_2 = 2^{-2} \times 100\text{ K} = 25\text{ K}$$

$$R_3 = 50\text{ K}$$

Notice that in addition to having an inverted output (the negative sign of Equation 13–16), the solution of Example 13–6 requires the use of five different values of resistors in the manufacture of this design. Because of the price break given for quantity purchases, it would be more desirable if the number of different values of resistors could be reduced. The *R*-2*R* ladder is desirable because it requires only two values of resistors. Incorporating the *R*-2*R* ladder network with an op-amp that has an input resistance in the megaohm range, so that the op-amp buffers the *R*-2*R* ladder from the load, would be a suitable solution.

The *voltage-follower* of Figure 13–14 can serve as buffer between the *R*-2*R* ladder network and the load, which is represented by resistor R_L. Voltage-followers have a voltage gain of one, so that Equation 13–14 or 13–15 would be valid to describe the output voltage of the voltage-follower.

We can modify the adder of Figure 13–10 so that it is an inverting amplifier and then use it as a buffer for an *R*-2*R* ladder network. The inverting amplifier can be designed to duplicate the output voltage as described by Equations 13–14 and 13–15, or it can be designed so that a scaling factor may be multiplied by those equations, depending on the gain designed into the amplifier.

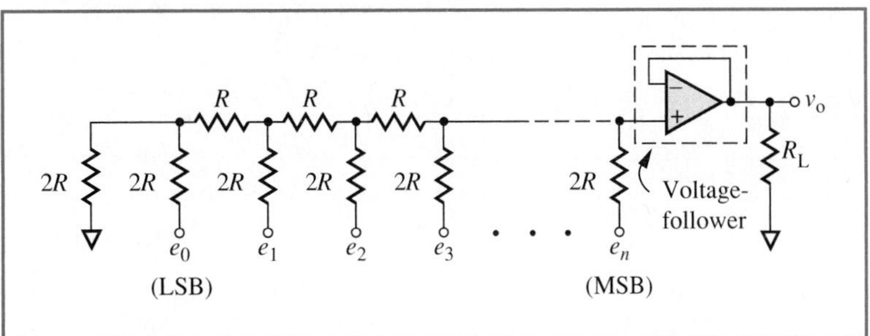

Figure 13–14
An *R*-2*R* ladder buffered with an op-amp voltage follower.

To determine how to modify Figure 13–10 so that it is an inverting amplifier, notice from Equation 13–12 that if all inputs but one, say e_0, are forced to zero volts (shorted to ground) then

$$v_o = -\left(\frac{R_f}{R_0}\right) e_0$$

Shorting these inputs to ground allows us to remove them and their associated resistor R_n. The resulting circuit is the inverting amplifier of Figure 13–15a. Figure 13–15b is the inverting amplifier of (a) with the output of an R-$2R$ ladder serving as the input voltage, where the Ds (digital data) are used to indicate inputs.

To derive the equation for the output voltage of Figure 13–15b, we use Thevenin's theorem to determine the equivalent circuit for the R-$2R$ ladder, which is the same as Figure 13–12c if the reference voltage is considered as input voltage e_n. A Thevenin's equivalent circuit is a voltage source and resistor in series, like e_0 and R_0 of Figure 13–15a. Hence, e_0 can be viewed as Thevenin's open-circuit voltage and R_0 as Thevenin's equivalent resistance. Applying Thevenin's theorem to the ladder of Figure 13–12c, we find that the open-circuit voltage is described by Equation 13–15, that is, v_o, and that Thevenin's equivalent resistance is R. Since the inverting amplifier of Figure 13–15a is the Thevenin's equivalent of Figure 13–15b, the output voltage equation v_o of Figure 13–15b is

Equation (13–19)

$$v_o = -\left(\frac{R_f}{R}\right)\frac{1}{2^N}\left[\sum_{n=0}^{N-1} 2^n e_n + 2^{N-1}V_{REF}\right]$$

Figure 13–15
An inverting amplifier buffering an R-$2R$ ladder.

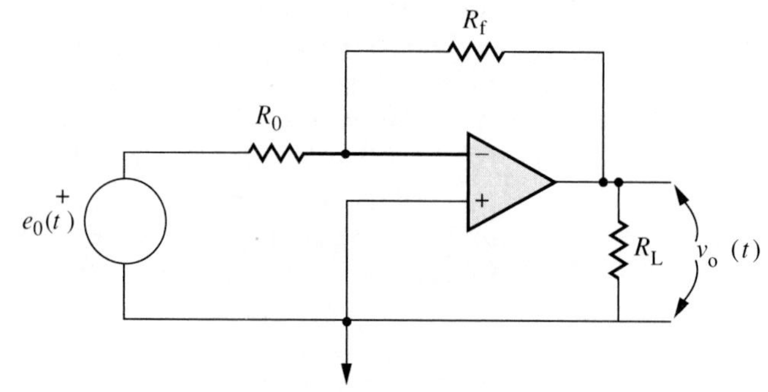

An op-amp inverting amplifier
(a)

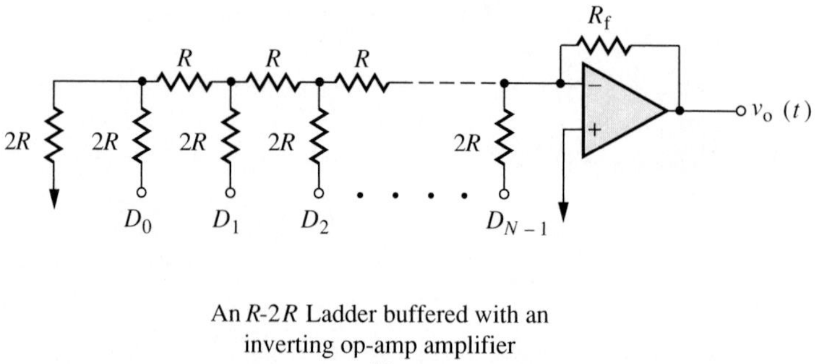

An R-$2R$ Ladder buffered with an inverting op-amp amplifier
(b)

EXAMPLE 13–7

If resistor R of Figure 13–15b has a value of 20 K
(a) Design the circuit of Figure 13–15b so that Equation 13–19 is equal to Equation 13–15.
(b) Design the circuit so that Equation 13–15 is scaled (amplified) by a factor of 2.
(c) Adjust V_{REF} to null out V_{OFFSET} for the design of part (a) if the digital input is an 8-bit CMOS counter.

SOLUTION

(a) From Equation 13–19 we know that R_f/R must equal 1; since $R = 20$ K, R_f must also equal 20 K.
(b) For a gain of 2, Equation 13–19 states that $R_f/R = 2$, or $R_f = 2R$. Then $R_f = 40$ K.
(c) From the equation

$$V_{REF} = \frac{1}{2^{N-1}}\left(\sum_{n=0}^{N-1} 2^n e_n\right)$$

CMOS inputs ($V_{OL} = 0.33$ volt) when $N = 9$ (V_{REF} is an input), we have

$$V_{REF} = \frac{1}{2^8}(e_0 + 2e_1 + 4e_2 + \ldots + 128e_7)$$

$$= \frac{1}{256} V_{OL}\,(1 + 2 + 4 + 8 + 16 + 32 + 64 + 128)$$

$$= \left(\frac{1}{256}\right)(0.33)\,(255) = 0.329 \text{ volt}$$

Other techniques are available to implement a DAC, but most are just variations of the concepts presented.

Figure 13–16a shows the schematic symbol for a DAC and Figure 13–16b illustrates a representative application of a DAC in which the converted digital signal is driving an analog load, such as an electrical motor. From the symbol of Figure 13–16a we see that the inputs are on the left and the outputs are on the right, with the power supply and reference voltage at the top. The output labeled R_{FB} is an external pin for access to the feedback resistor labeled R_f in Figure 13–15b. There is a feedback resistor manufactured within the DAC, but since its value may not be large enough the manufacturer has made it possible to increase the feedback resistance by adding external resistor R_2 in series with internal resistor R_{FB}. If the DAC user does not want an inverting amplifier, such as that in Figure 13–16, then the pin for R_{FB} is not connected and the voltage-follower of Figure 13–14 is used instead.

Figure 13–16a also shows that for this DAC there are 8 bits of digital data that can be input via data pins DB_0 through DB_7. From Figure 13–16b we see that these pins are connected to the source of the digital data to be converted, which in this case is an MPU via the data bus (the MPU writes the data byte to be converted to the DAC). The data byte on the data bus is latched by the DAC under control of the write control signal, which is input at the $\overline{WR}$ pin. The $\overline{CS}$ pin serves the same function as the chip selects of memory ICs, which is to enable the DAC chip. Figure 13–16b shows the $\overline{CS}$ pin being activated via the I/O selector,

Figure 13–16
A DAC system application.

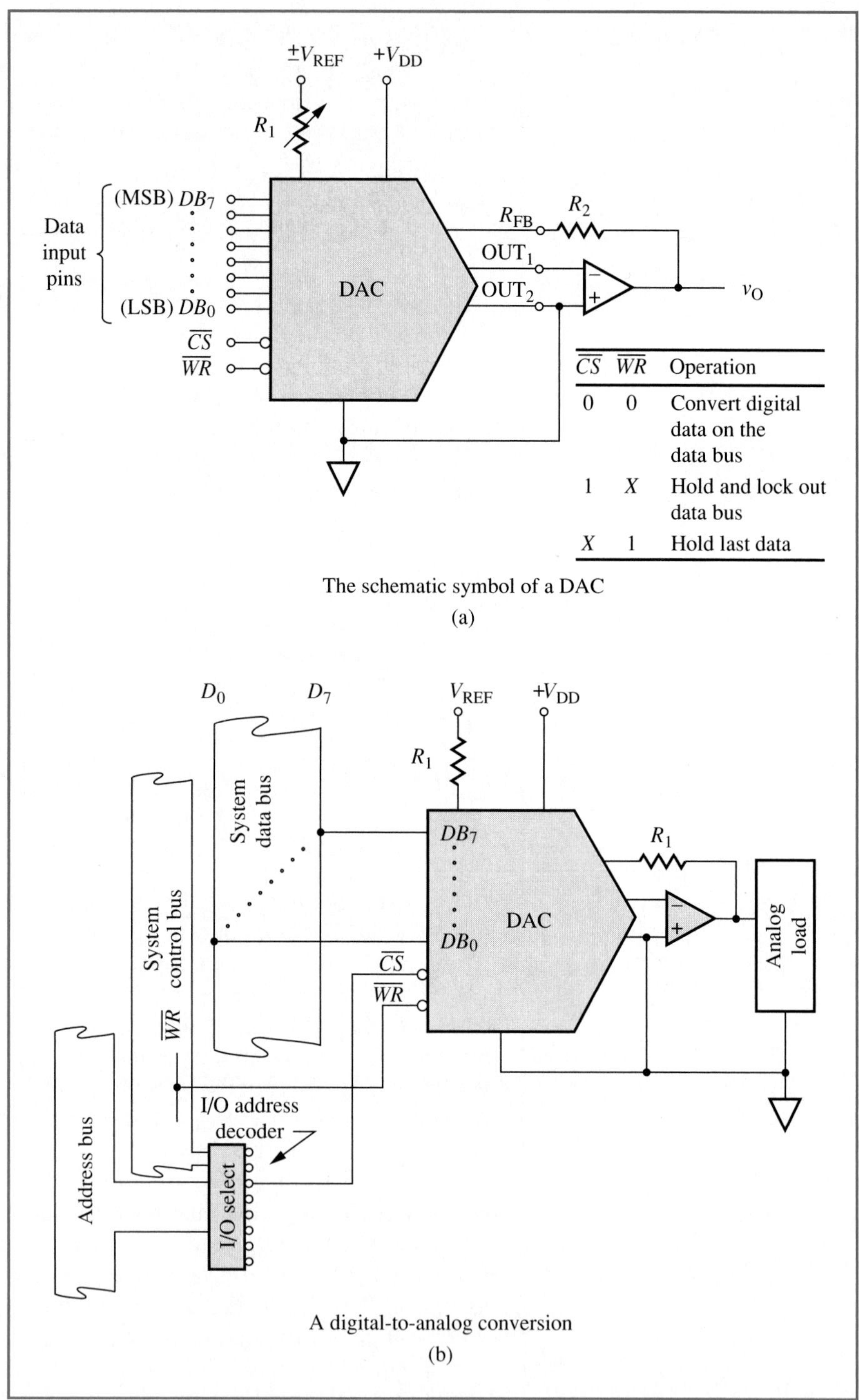

The schematic symbol of a DAC
(a)

A digital-to-analog conversion
(b)

which employs the same design concepts that select the I/O ports of Figure 10–9. We note from the truth table of Figure 13–16a that the $\overline{CS}$ pin must be at a logic low in order for the DAC to convert data. When the write enable pin is activated, that is, is driven low by the MPU when it has written a data byte on the data bus, and if $\overline{CS}$ is also low at this time, the DAC converts the data byte on the data bus and outputs the converted analog signal at output OUT_1. As shown in Figure 13–16b, the write enable pin of the DAC is connected to the write control signal of the control bus. We realize from the connections of Figure 13–16b that a DAC is just another I/O port to the MPU.

Some DACs are designed so that the reference voltage V_{REF} serves as a multiplier of the output OUT_1. For this type of DAC the output voltage of the op-amp of Figure 13–16a would be a modification of Equation 13–14 such that V_{REF} is a multiplier. Hence,

Equation (13–20)

$$v_o = -V_{\mathrm{REF}}\left(\frac{R_{FB} + R_2}{R}\right)\frac{1}{2^N}\sum_{n=0}^{N-1} 2^n e_n$$

Note from Equation 13–20 that if V_{REF} is made a negative voltage, the output voltage of the op-amp is no longer negative.

EXAMPLE 13–8

Suppose that the DAC of Figure 13–16b has an $R_{FB} = R$ (of the internal R-$2R$ ladder) $= 50$ K and $R_2 = 25$ K. What polarity and magnitude must V_{REF} have for the output of the op-amp to be equal to

$$v_o = 15\left(\frac{1}{2^N}\right)\sum_{n=0}^{N-1} 2^n e_n$$

SOLUTION Equating the above equation to Equation 13–20 and using the stated resistance values, this would require that

$$-V_{\mathrm{REF}}\left(\frac{75 \times 10^3}{50 \times 10^3}\right) = 15$$

or

$$V_{\mathrm{REF}} = -\frac{2}{3}(15) = -10 \text{ volts}$$

There are some definitions associated with DACs with which we should be familiar. These are

1. *Resolution*. The resolution of a magnitude is the smallest change that can be detected or resolved. For a DAC it is a change in output voltage that must be resolved. Since any change in output voltage is the result of a corresponding change in the digital input, the smallest change in output voltage is due to the weight assigned to the LSB. As an example, from Figure 13–14 we are reminded that $n = 0$ for the LSB, and from Equation 13–14 we can calculate that the change in the analog output voltage *for this equation* is

Equation (13–21)

$$\textit{Resolution} = \frac{1}{2^n} e_0 \text{ volts}$$

where $e_0 = V_{\mathrm{OH}}$ and the gain of the voltage-follower is unity.

2. *Quantization error*. As with any quantity that is expressed with a finite number (a limited number of digit positions), there is an inherent error. That is, if it is stated that an object is 0.42 cm long, the error exists in the unstated third position to the right of the decimal point. The length may be closer to 0.41723 cm, if the magnitude was truncated. However, now the 3 is in question, for it may also be the result of truncation. Of course a digital number can be truncated just like an analog number. Then

the weight of the LSB of a digital number is the bit in question and it may be positive or negative by one-half of the weight of that LSB. For example, the quantization error (QE) of Equation 13–14 can be expressed as

Equation (13–22)

$$QE = \pm\left[\frac{1}{2}\left(\frac{1}{2^N}\right)e_0\right] = \pm\left[\left(\frac{1}{2^{N+1}}\right)e_0\right]$$

3. *Accuracy*. The accuracy is the difference between the calculated analog output voltage and the measured value. These differences can be attributed to noise, drift, difference in calculated and actual gain, offset potential, etc.

EXAMPLE 13–9

What are the resolution, QE, and accuracy of the DAC of Example 13–8 for a 74LS 8-bit digital input if the measured output voltage was 15.03 volts for an input of 00111010?

SOLUTION For this DAC $N = 8$

1. Resolution is determined from the analog output voltage equation for $n = 0$. From Example 13–8 the weight of e_0 is
Resolution $= (15/2^8)\ (2^0 e_0) = 15/256(2.7)$
$= 0.158$ volt (truncated) where $V_{OH} = 2.7$ volts.
Then 0.158 volt is the smallest detectable change in the analog signal.
2. QE $= \pm\ \frac{1}{2}(0.158)$ volt $= \pm\ -0.079$ volt, since the weight of e_0 is as stated above.
3. The calculated value of the output voltage is

$$\begin{aligned} v_o &= \frac{15}{256}(e_0 + 2e_1 + 4e_2 + 8e_3 + 16e_4 + 32e_5 + 64e_6 + 128e_7) \\ &= 0.0586\,[0.5 + 2(2.7) + 4(0.5) + 8(2.7) + 16(2.7) + 32(2.7) \\ &\quad + 64(0.5) + 128(0.5)] \\ &= 0.0586(255.10) = 14.95 \text{ volts (truncated)} \end{aligned}$$

Since the measured voltage is 15.03 volts, the difference between the calculated and measured value is −0.08 volts. As a result,

$$\text{Accuracy} = -0.08/14.95 \times 100 = -0.54\% \text{ (truncated)}$$

Review questions

1. What function does a weighting factor serve?
2. Which DAC input should have the largest weighting factor and why?
3. Why is the output of an *R*-2*R* ladder buffered?
4. When would a voltage-follower be used to buffer an *R*-2*R* ladder network?
5. When would an amplifier be used to buffer an *R*-2*R* ladder?
6. What function does the write enable pin of a DAC serve?
7. Why is it often desirable that a DAC have latching capability?
8. What function does the chip enable of a DAC pin serve, and how is it utilized in a system design?

9. What is the function of the R_{FB} pin of a DAC, and under what circumstance is it used?
10. What signal is at the output OUT_1 of a DAC?

13–6 Analog-to-digital conversion

There are many applications that require that an analog signal be converted to a digital signal. Such a conversion is necessary when the amplitude of an analog signal is to be detected by a digital computer or some other digital device and then the amplitude of that analog signal is to be processed by the computer. For example, computers control the environment of many residences and workplaces. To control such parameters as temperature, humidity, and air flow, the computer must "know" the temperature and humidity of its source for fresh air (the air outside) as well as the required volume of fresh air (how many people are in the room). Since parameters such as temperature, humidity, and the air flow of a room are analog signals, they must be converted to digital signals before a digital computer can process them. Of course, the magnitude of the converted digital signal must be directly proportional to the amplitude of the analog signal. In other words, it is the reverse of a digital-to-analog conversion.

As with a digital-to-analog conversion, there are various methods to achieve analog-to-digital conversion. Regardless of the technique, the device that performs the conversion is known as an analog-to-digital converter (ADC). There are three techniques of conversion that we shall investigate to demonstrate the basic concepts of all ADCs. These techniques are simultaneous ADC, tracking ADC, and successive approximation ADC.

Simultaneous ADC

Figure 13–17 is a schematic of a 3-bit simultaneous ADC. As an overview, the analog signal e_a is applied to the positive inputs of the seven op-amps, which are serving as comparators, since they are being used in a nonlinear mode of operation (open-loop). The comparison voltage (the reference voltage) for each comparator is the voltage input at the negative input, which is identified as V_1, V_2, V_3, etc. These reference voltages are the result of the voltage divider formed by the DC voltage V_{REF} and the series resistors identified as R (recall that an op-amp has a very large input impedance—consider it to be an open circuit). As the analog signal is input to the comparators, their output reflects whether the analog signal is greater than or less than their reference voltage by outputting either a logic 1 or 0. These logic levels are input to the priority-encoder, which encodes the appropriate binary number for the magnitude of analog signal being input.

To analyze the ADC of Figure 13–17, let us determine how the comparators function. Figure 13–17 has an inset that is a representative model of the comparators, which is identified as the nth op-amp ($n = 1, 2, 3$, etc.). From earlier work we know that if $e_d > 0$ volts the output voltage of O_n is a logic 1 (V_{SAT}); otherwise it is a logic 0 (the negative power supply input of the op-amp is connected to ground). Since it is the polarity of the input voltage e_d that controls the output logic level of the comparator, let us write a loop equation at the input and solve for e_d.

Figure 13–17
A simultaneous ADC.

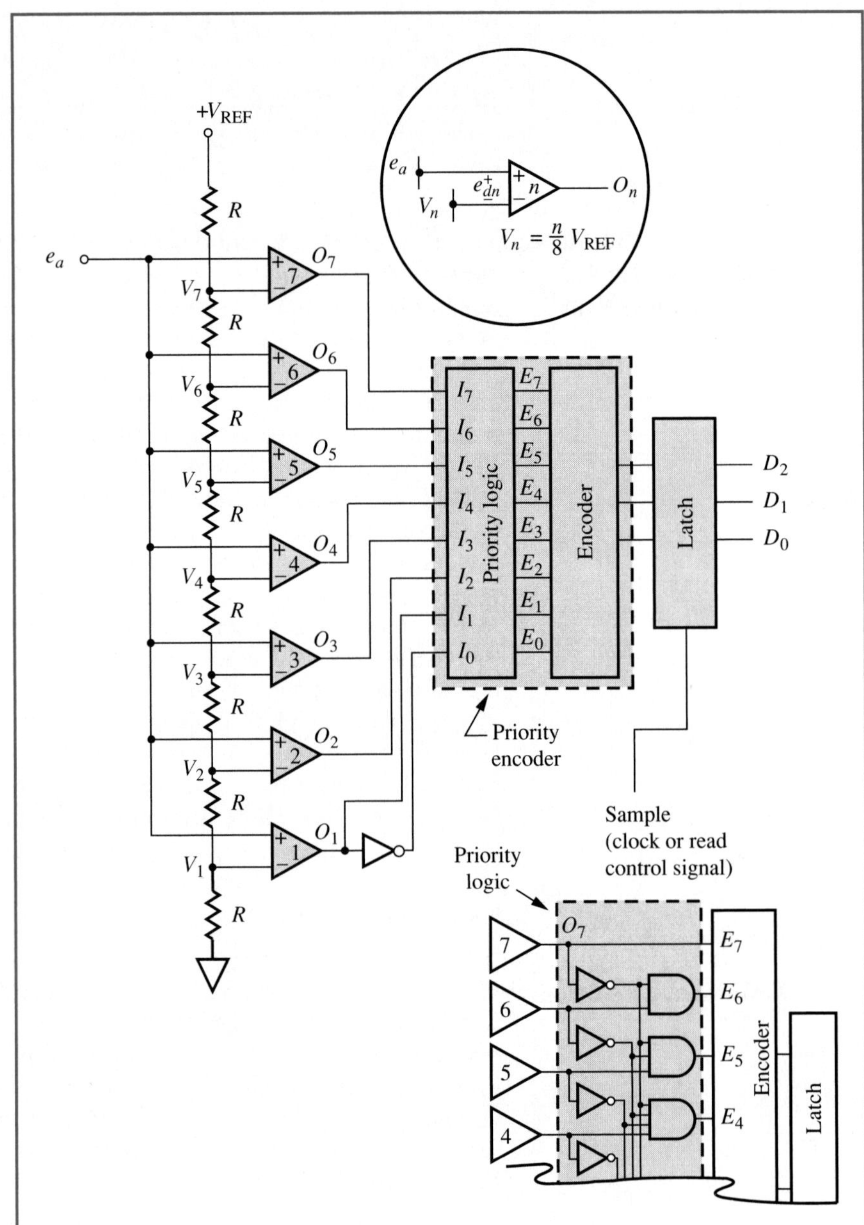

Hence

$$V_n + e_d - e_a = 0$$

from which we find that

Equation (13–23)

$$e_d = e_a - V_n$$

Solving for the condition $e_d > 0$ we have

$$e_d = e_a - V_n > 0 \text{ and if}$$

Equation (13–24)

$$e_a > V_n$$

then $O_n = 1$; otherwise $O_n = 0$.

From Equation 13–24 we can reason that for various voltage ranges of e_a

$e_a < V_1$	then the outputs of all comparators are 0.
$V_1 < e_a < V_2$	then $O_1 = 1$ and the rest are 0.
$V_2 < e_a < V_3$	then $O_1 = O_2 = 1$ and the rest are 0.
$V_3 < e_a < V_4$	then $O_1 = O_2 = O_3 = 1$ and the rest are 0, etc.

These ranges of e_a and the resultant logic level of the comparators are given in Table 13–1. From Figure 13–17 note that the outputs of the comparators are connected to the corresponding inputs of the priority encoder, or $O_n = I_n$.

As we can observe from Table 13–1, as e_a increases above a reference voltage V_n (V_1, V_2, V_3, etc.), the output of the corresponding comparator is a logic 1 as well as those comparator outputs with preceding reference voltages. That is, for the first row of Table 13–1 $e_a > V_1$, the magnitude of e_a is less than all reference voltages and as a result the outputs of all comparators are a logic 0. For the second row e_a increases above V_1 but remains below V_2, and as a result output $O_1 = 1$ ($I_1 = 1$). For the third row e_a increases above V_2 but remains below V_3, and both outputs O_1 and O_2 are a logic 1. This continues until e_a increases above v_7 and the output of all the comparators is a logic 1.

To derive a general equation for the reference voltages, we apply the voltage divider equation, which is

Equation (13–25)

$$V_n = \left(\frac{n}{N}\right) V_{\text{REF}} = \left(\frac{n}{8}\right) V_{\text{REF}}$$

where N is the number of resistors R ($N = 8$ for the ADC of Figure 13–17).

After the analog signal has been digitized via the comparator outputs, the digitized output must be converted to a binary number. We use an encoder to accomplish this portion of the conversion. Encoders require that one, and only one, input be active (a logic 1 in this case) at any given time; hence we must ensure that this occurs for all ranges of e_a. As we can see from Table 13–1, many outputs are active at the same time once e_a is greater than V_2 volts. A solution is to input these outputs to priority logic, which is illustrated as an inset in Figure 13–17. The priority logic outputs a logic 1 for the comparator with the highest priority whose

Table 13–1
Truth table for the priority encoder of Figure 13–17

Voltage range of e_a	**Priority logic**								**Encoder**								**ADC output**		
	I_7	I_6	I_5	I_4	I_3	I_2	I_1	I_0	E_7	E_6	E_5	E_4	E_3	E_2	E_1	E_0	D_2	D_1	D_0
$e_a < V_1$	0	0	0	0	0	0	0	1	0	0	0	0	0	0	0	1	0	0	0
$V_1 < e_a < V_2$	0	0	0	0	0	0	1	0	0	0	0	0	0	0	1	0	0	0	1
$V_2 < e_a < V_3$	0	0	0	0	0	1	1	0	0	0	0	0	0	1	0	0	0	1	0
$V_3 < e_a < V_4$	0	0	0	0	1	1	1	0	0	0	0	0	1	0	0	0	0	1	1
$V_4 < e_a < V_5$	0	0	0	1	1	1	1	0	0	0	0	1	0	0	0	0	1	0	0
$V_5 < e_a < V_6$	0	0	1	1	1	1	1	0	0	0	1	0	0	0	0	0	1	0	1
$V_6 < e_a < V_7$	0	1	1	1	1	1	1	0	0	1	0	0	0	0	0	0	1	1	0
$e_a > V_7$	1	1	1	1	1	1	1	0	1	0	0	0	0	0	0	0	1	1	1

$V_n = (n/N)\, V_{\text{REF}} = (n/8)\, V_{\text{REF}}$.

output is a logic 1, and all other comparator outputs are disabled via *AND* gates in the priority logic (see inset). V_7 has the highest priority and V_1 has the lowest. Table 13–1 confirms the function of the priority logic, since the outputs of the priority logic, which are the inputs of the encoder, have only one active output at any given time. The encoder encodes the binary inputs, which are listed as E_0, E_1, E_2, etc., in Table 13–1, to the ADC output binary number shown in columns D_2, D_1, and D_0 of Table 13–1.

EXAMPLE 13–10

In Figure 13–17 the analog signal shown in Figure 13–18 is applied to the input e_a of the ADC. If V_{REF} is 16 volts DC, determine the corresponding converted digital output of e_a and superimpose this digital plot on the analog plot of e_a.

SOLUTION From Equation 13–25 the reference voltages of each comparator can be determined.

$$V_n = \left(\frac{n}{8}\right) 16 \text{ volts}$$

so

$$V_1 = \left(\frac{1}{8}\right) 16 = 2 \text{ volts}$$

$$V_2 = \left(\frac{2}{8}\right) 16 = 4 \text{ volts}$$

and so on. As a guide for determining the various ranges of e_a, a vertical axis for V_n was created beside the axis for e_a in Figure 13–18.

Applying the values for V_1, V_2, V_3, etc., to Table 13–1, we can determine the various voltage ranges for e_a and the corresponding ADC output. For instance, $V_1 = 2$ volts and $V_2 = 4$ volts; from Table 13–1 we find that the range 2 volts $< e_a <$ 4 volts is in the second row from the top, which results in ADC output 001. From the plot of e_a in Figure 13–18 this range of e_a is valid over the time

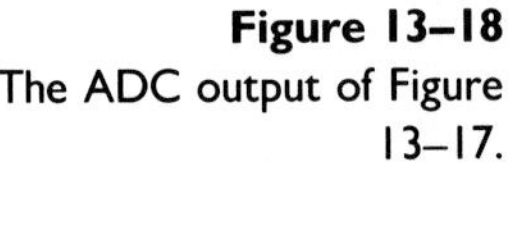
Figure 13–18
The ADC output of Figure 13–17.

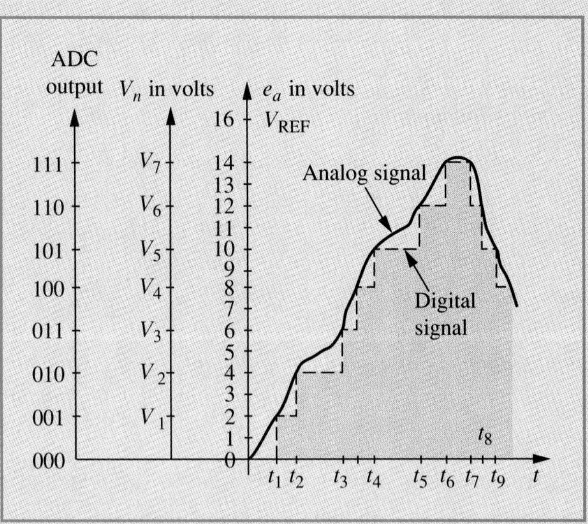

interval $t_2 - t_1$. When e_a increases to 4 volts, then 4 volts $= V_2 < e_a < V_3 = 6$ volts. The ADC output is 010 according to Table 13–1 and is in the time interval of $t_3 - t_2$ according to Figure 13–18. The digital magnitude of the dotted line of Figure 13–18 corresponds to the vertical scale of the ADC output and is the superimposed digital plot of e_a.

As can be seen from the digital output shown in Figure 13–18, the digital ADC output is an approximation of the analog signal. To improve the approximation would require more comparators, which would increase the number of comparison voltage ranges of the analog signal. It is the number of comparators required that is the major disadvantage to this type of ADC. That is, since most MPUs have a minimum data bus size of 8 bits, to encode the analog signal into bytes would require 255 (256 − 1) op-amps and associated priority logic. Two hundred fifty-five comparators would yield 256 analog signal ranges, just as the seven comparators of Figure 13–17 yield eight analog ranges. This increase in the number of analog ranges and ADC output bit size would reduce the quantization error (QE) and improve the resolution.

EXAMPLE 13–11

Apply the concepts of resolution and quantization error to Example 13–10.

SOLUTION By definition, the resolution is equal to the weight assigned to the LSB. For this application the LSB changes for every 2-volt change of the analog signal; hence

$$\text{Resolution} = 2 \text{ volts}$$

The QE is plus or minus one-half the weight of the LSB, which for this application is one-half of 2 volts. Then

$$\text{QE} = \pm 1 \text{ volt}$$

EXAMPLE 13–12

What would be the resolution and QE for the ADC output of Example 13–10 if the ADC contained 255 comparators?

SOLUTION As we have already discussed, with 255 comparators there would be 256 analog ranges. To determine the size of each range we can use Equation 13–25. Thus

$$V_n = \left(\frac{n}{256}\right)16 = n62.5 \text{ mV}$$

So each analog range is 62.5 mV; therefore the LSB will represent 62.5 mV. Then

$$\text{Resolution} = 62.5 \text{ mV}$$

$$\text{and the QE} = \pm\left(\tfrac{1}{2}\right) 62.5 \text{ mV} = \pm 31.25 \text{ mV}.$$

An ADC with less than 8 bits can be used with an 8-bit system. It is the required resolution of the output that determines the bit size of the ADC.

Tracking ADC

The digital output of a tracking ADC simply follows the magnitude of the analog input signal. Figure 13–19 is a block diagram of a tracking ADC. The up-down counter is the heart of this type of ADC, for its output is used as the voltage of comparison to the analog input voltage, where equality is being sought. As shown, a *clock pulse* (a continuous stream of pulses with a fixed repetition rate) is input to the up-down counter, causing the counter to either increase or decrease its count based on the logic level of its up/$\overline{\text{down}}$ input. If the logic level is a high at input up/$\overline{\text{down}}$, it will count up and increase its count; if it is a logic low, it counts down and decreases its count. It is the output of the comparator that determines the logic level of this input. That is, when $e_d > 0$ volts, the output of the comparator is a high; when $e_d < 0$ volts, the output of the comparator is low.

The output of the up-down counter is converted to an analog voltage e'_a, which is proportional to the digital count. Identifying the signal at the negative input of the comparator is e'_a and the input signal at the positive input as e_{aH}, then by writing a loop equation we find that

$$e_d = e_{aH} - e'_a$$

From this equation we can reason that if

Equation (13–26)

$$e_{aH} > e'_a$$

the counter counts up, and if

Equation (13–27)

$$e_{aH} < e'_a$$

the counter counts down.

The analog signal e_a is sampled at a fixed rate and its amplitude is held at the value sampled by the sample/hold (S/H) circuitry of Figure 13–19. The output of the S/H circuit is the sampled-and-held value of the analog signal and is identified in Figure 13–19 as e_{aH}, which is input to the positive input of the comparator. The counter digital output is

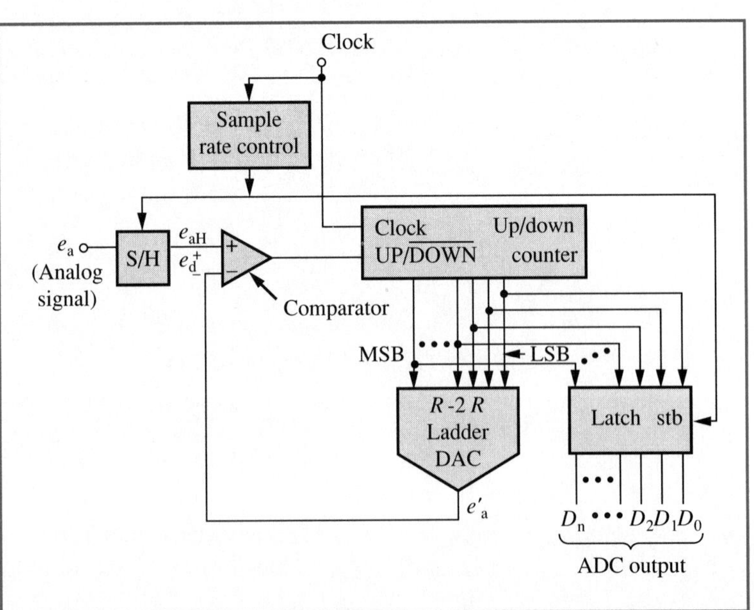

Figure 13–19
A tracking ADC block diagram.

converted to an analog signal e'_a. As we have stated, the comparator compares these two voltages. When Equation 13–26 is valid, the digital number of the counter is less than the sampled analog signal and therefore the digital count must be increased, which is why the counter acts as an up-counter. In contrast, when Equation 13–27 is valid, the digital value of the counter is greater than the sampled value of the analog signal and therefore must be decreased, which is why the counter must act as a down-counter for this condition. The digital number of the counter never equals the amplitude of the sampled analog signal but rather follows it. The accuracy of the conversion depends on how fast the counter increases or decreases its count versus how fast a new sample is taken of the analog signal e_a. The faster the count, relative to the sample time, the better the accuracy. It is hoped that the digital value can track the analog signal with no more than a difference (error) of one count. This being true, the QE would be plus or minus the weight of the LSB.

As each new sample is taken, the counter contains the binary equivalent of the analog amplitude of the previous sample. This binary number is to be latched by the latch of Figure 13–19, usually at the same sample rate as that of the analog signal.

Figure 13–20 illustrates the concepts that we have discussed. At t_1 a sample of e_a was made and held by the S/H circuitry. The amplitude of the analog signal at that sample time was e_{aH1}. The output of the DAC (e'_a) is less than e_{aH1}; hence the counter up-counts until time t_2. At t_2 the voltage e'_a is greater by one count than e_{aH1}, so the counter down-counts. From t_2 to t_3 the counter increases and decreases by one count about the e_{aH1} value, and it is this digital value that would be considered equivalent to e_{aH1}. Notice that the equivalent digital value was essentially reached for the analog signal sampled at t_1 at time t_2 and oscillated about that value until the new sample was taken at t_3. The latch would latch this value and output it at pins D_0 through D_n. A new sample e_{aH2} is taken at t_3 and the counter increases until t_4, at which time the count is too large by 1. From t_4 to t_5 the counter again oscillates about the sampled value e_{aH2}. A new sample is taken at t_5. We notice that the digital equivalent has not enough time (clock pulses) in the time interval t_5 to t_6 for the

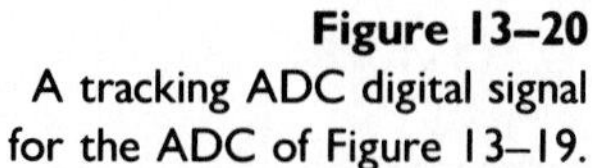
Figure 13–20
A tracking ADC digital signal for the ADC of Figure 13–19.

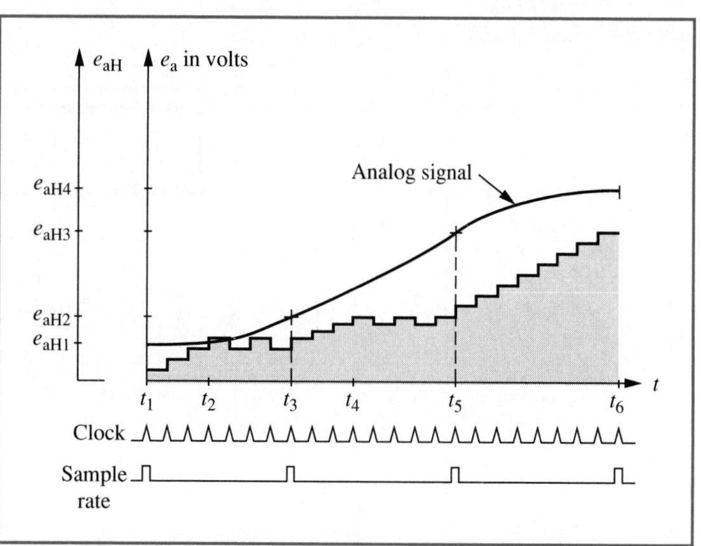

counter to reach a value equivalent to e_{aH3}; hence this design would fail. There are several solutions.

1. Increase the weight of the LSB so that for each count, e'_a will increase faster. This solution sacrifices resolution and causes a higher QE.
2. Lengthen the sample time. This means we could miss fluctuations in e_a that appear between samples, which is not a good idea and implies that to determine the sample we must have some idea of how rapidly the analog signal may change.
3. Increase the repetition rate of the clock pulses per sample. This is probably the best solution.

Successive approximation ADC

Figure 13–21 is an illustration of the essential function-blocks of a successive approximation ADC. The reader should compare Figures 13–21 and 13–19 and note that a successive approximation ADC utilizes the content of a successive approximation register (SAR) for comparison with the sampled-and-held analog signal e_{aH} via the DAC output, rather than the content of an up-down counter. Other than that, the two are essentially the same.

The SAR is a register (the series of rectangles within the SAR of Figure 13–21 represents flip-flops) whose contents are successively adjusted and compared to the sampled-and-held value of the analog signal e_{aH}. To successively adjust and compare the content of the SAR and e_{aH}, each bit position of the SAR is successively set to a logic 1, beginning with the MSB, and then this new "adjusted" digital value is converted to e'_a, via the *DAC*, and compared to e_{aH}. If the most recent bit that was set to a logic 1 causes the content of the SAR to be larger than e_{aH} ($e'_a > e_{aH}$) that bit is reset to a logic 0, since the content of the SAR is too large; otherwise it remains a logic 1. Specifically, when the start input of the SAR is activated, the SAR sets its MSB to a logic 1 and all other bits to a logic 0, which is the mid-value of the SAR, as the first approximation to e_{aH}. As we have stated, the content of the SAR is converted to an

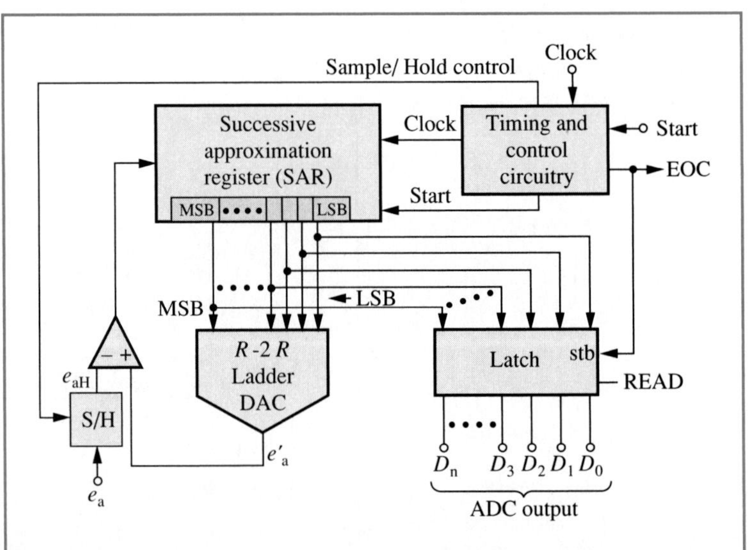

Figure 13–21
Successive approximation ADC.

analog signal e'_a and compared to e_{aH}. As can be determined by writing a loop equation at the input of the comparator, if $e'_a > e_{aH}$ then the output voltage of the comparator will be a logic 1, and if $e'_a < e_{aH}$ the comparator output voltage will be a logic 0. When $e'_a > e_{aH}$ the content of the SAR is too large and must be reduced, whereas if $e'_a < e_{aH}$ the content of the SAR is too small and therefore must be increased. It is the logic level of the comparator output that makes the adjustments to the content of the SAR. If the output of the comparator is a logic 1 ($e'_a > e_{aH}$) the MSB is reset to a logic 0, whereas if the output of the comparator is a logic 0 ($e'_a < e_{aH}$) the MSB logic level is left as a logic 1. After the MSB has been set to a logic 1 and then tested and adjusted, the SAR sets its second MSB to a logic 1 and repeats the test-and-adjust procedure for this bit position. The process is then repeated for the third, fourth, etc. positions until the LSB has been tested and adjusted, at which time the successive approximation is completed and the content of the SAR is the digital approximation of the sampled analog voltage e_{aH}. This digital value is latched and output via pins D_0 through D_n and at the same time an end-of-conversion (EOC) is output. The EOC signal can be used to request an interrupt of an MPU, since the ADC would have digital data for the MPU at that time.

EXAMPLE 13–13

Suppose that an 8-bit ADC, such as that shown in Figure 13–21, is to be used to convert the analog signal of Figure 13–22. Following Figure 13–20, superimpose the resulting ADC output on the same graph as the analog signal. Assume that the SAR is 74ACT technology.

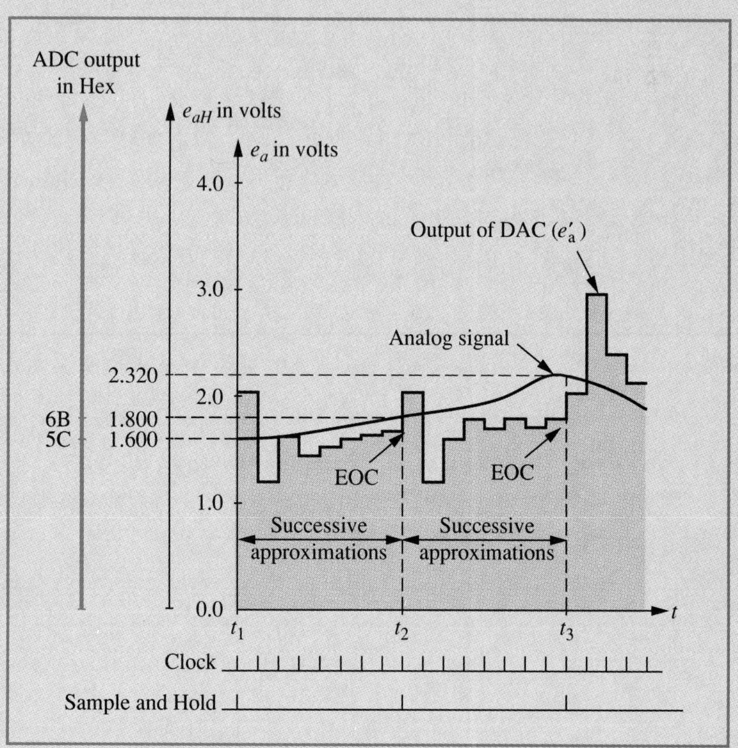

Figure 13–22
Successive approximation of an analog signal.

SOLUTION Since there are eight flip-flops within the SAR, there are eight comparisons and adjustments before EOC occurs. Hence, a new sample of the analog is taken every eight clock pulses, as indicated in Figure 13–22.

Assuming that the DAC does not have amplification, as could be the case if the DAC of Figure 13–15 were used, Equation 13–14 can be used to determine the magnitude of e'_a. Hence for an 8-bit DAC

$$e'_a = \frac{1}{256}(128e_7 + 64e_6 + 32e_5 + 16e_4 + 8e_3 + 4e_2 + 2e_1 + e_0)$$

At t_1 the first sample of e_a is made and held. Let us suppose that the sampled-and-held value is 1.600 volts. While the sample of e_a is being taken, the start command is issued to the SAR, which causes the MSB of the SAR (e_7) to be set to a logic 1 and the other seven bits to be reset to a logic 0. Since for 74ACT logic V_{OH} = 3.84 volts and V_{OL} = 0.33 volt, when SAR = 10000000 we find that

$$e'_a = \frac{1}{256}[128(3.84) + 127(0.33)] = 2.083 \text{ volts}$$

Since e'_a = 2.083 volts > 1.600 volts = e_{aH}, the MSB is reset and then the second most significant bit is set. For SAR = 01000000 the DAC output voltage is

$$e'_a = \frac{1}{256}[128(0.33) + 64(3.84) + 63(0.33)] = 1.206 \text{ volts}$$

For this adjustment e'_a < 1.600 volts; hence the logic 1 remains and the next bit is set. When SAR = 01100000

$$e'_a = (1/256)\,[128(0.33) + 64(3.84) + 32(3.84) + 31(0.33)] = 1.630$$

volts, which is greater than 1.600 volts so the bit is reset and the next most significant bit is set. When SAR = 01010000, e'_a = 1.426 volts which is less than 1.600 volts; therefore the bit remains a 1 and the next most significant bit is set.

SAR = 01011000
e'_a = 1.535 volts < e_{aH}; set the next bit.
SAR = 01011100
e'_a = 1.590 volts < e_{aH}; set the next bit.
SAR = 01011110
e'_a = 1.618 volts > e_{aH}; reset the bit and set the next bit.
SAR = 01011101

e'_a = 1.604 volts > 1.600 volts. Therefore, the LSB is reset and SAR = 01011100 = 5CH and e'_a = 1.590 volts, which differs by 0.010 volts from the actual sampled-and-held value of 1.600 volts.

Since all registers of the SAR have been adjusted (set or reset), the successive approximation has been completed. The content of the SAR is latched and output via the EOC signal. Figure 13–22 shows the digital output (ADC output) in hexadecimal values. Note that the resolution is 15 mV and the QE is ±7.5 mV. For this sample the accuracy is −0.625%.

At t_2 another sample of e_a is made and held. For this sample let us suppose that the sample-and-held value is e_{aH} = 1.800 volts. Again the MSB of the SAR is set and the process is again repeated.

$$
\begin{aligned}
\text{SAR} &= 10000000 \\
e'_a &= 2.083 \text{ volts} > e_{aH} \\
\text{SAR} &= 01000000 \\
e'_a &= 1.206 \text{ volts} < e_{aH} \\
\text{SAR} &= 01100000 \\
e'_a &= 1.630 \text{ volts} < e_{aH} \\
\text{SAR} &= 01110000 \\
e'_a &= 1.864 \text{ volts} > e_{aH} \\
\text{SAR} &= 01101000 \\
e'_a &= 1.755 \text{ volts} < e_{aH} \\
\text{SAR} &= 01101100 \\
e'_a &= 1.809 \text{ volts} > e_{aH} \\
\text{SAR} &= 01101010 \\
e'_a &= 1.782 \text{ volts} < e_{aH} \\
\text{SAR} &= 01101011 \\
e'_a &= 1.796 \text{ volts} < e_{aH}
\end{aligned}
$$

The conversion is completed; the EOC output is driven high (requesting and interrupt of the MPU) and the content of the SAR (SAR = 01101011 = 6BH) is latched and output to the data pins. At t_3 another sample is taken and the process is repeated. As before, for each EOC the content of the SAR is latched and output at pins D_0 through D_7. The output of the ADC can be read by activating its read pin.

Note from Figure 13–22 that there is a delay time that is eight clock pulses from the time a sample of the analog signal is made until the EOC occurs and the digital output is displayed. As in the case of the tracking ADC, how much delay time exists is directly related to the repetition rate of the clock and, of course, the sampling rate.

Review questions

1. What function do the comparators of a simultaneous ADC serve?
2. Why must a priority encoder be employed in a simultaneous ADC?
3. What is the difficulty in fabricating a simultaneous ADC?
4. How do a tracking ADC and a successive approximation ADC function?
5. What function does the up-down counter serve in a tracking ADC?
6. What function does the successive approximation register serve in a successive approximation ADC?
7. What are resolution and quantization error?

13–7 TROUBLESHOOTING

To troubleshoot parallel I/O device interfaces, we may use either static or dynamic testing. For either case we would simply address the I/O port under test and then either read data from it or write to it. If the I/O port is an input port, such as input port *N* of Figure 13–1, a static logic level generator or the input device itself may be used to input the datum to be read. When the read control signal is activated (by the CPU or a logic level generator) the *CS* pin of the interface buffer (I/O port) also goes active and in turn activates those buffers, which loads the datum of input I/O device *N* on the data bus. Monitoring the address, control, and data busses dynamically, the logic analyzer samples the logic levels on these busses. To capture the data samples of interest, the logic analyzer can be triggered on the address of I/O device *N*. The captured data can then be displayed and checked to see that the datum being input to port *N* is also on the data bus while the read control signal is active.

To test an output port dynamically, such as port *M* in Figure 13–1, we again use a logic analyzer to monitor the address, control, data busses, and the output of the interface (I/O port). Using the address of output device *M* as the trigger-word, when it appears, the logic analyzer captures the data on the busses as well as the output datum of port *M*. Examining the display of this sampled data, we can check to see that while the write control signal is active the datum on the data bus also appears at the output of port *M*. A static test of an output port is conducted in a similar fashion.

To troubleshoot a PPI dynamically, we again use a logic analyzer to monitor the busses as well as I/O ports' nodal connections to the I/O devices. Using the I/O port address as the trigger-word, the datum captured from the data bus would be that being read or written to the port. Once the captured data is displayed, verification of proper operation can be made using the same methods for the I/O ports previously discussed. That is, if an input port is being tested, such as port C of Figure 13–2, then, with its address as the trigger-word, the displayed datum from the data bus can be checked to see that the datum being input to port C is the same as that on the data bus during the interval that the read pin $\overline{RD}$ is active. For an output port, such as port B of Figure 13–2, the datum being output by the port must be the same as that on the data bus when port B of the PPI is addressed and the write pin $\overline{WR}$ is activated. The bidirectional port A would be tested for a read and then a write.

Testing serial I/O interfaces requires that for a serial-to-parallel interface, the assembled serial input datum be the same as the parallel datum output and the reverse for a parallel-to-serial interface. To test the serial-to-parallel interface of Figure 13–4a, we would use a logic analyzer that is triggered on the address of that serial I/O port. Monitoring the input node to the interface and the data bus would enable us to verify assembly of the serial datum. For a parallel-to-serial interface we would monitor the data bus and the output of the interface.

To dynamically test the USART of Figure 13–5, we use its address as the trigger-word and monitor nodes $\overline{CS}$, $\overline{RD}$, $\overline{WE}$, *TXD*, and *RXD*. Similar to the test of those serial interfaces in Figure 13–4, we verify that for a read the serial bits input to the USART via *RXD* (and assembled) contain the same datum as on the data bus while the read pin is active. For a write, the datum on the data bus and the serial data output at *TXD* are monitored to verify that their bit compositions are the same.

To test a DAC or ADC statically we simply input the static data and measure the resultant output data. For example, to statically test the *R*-2*R* buffered ladder of Figure 13–14, we connect static logic level generators to inputs e_0 through e_n and measure output voltage v_o with a volt-ohm-meter (VOM). To statistically test the simultaneous ADC of Figure 13–17 we use a DC supply to simulate the input analog voltage e_a. This input voltage is varied from zero volts to the maximum allowed for the design. Static logic level detectors or a logic probe are used to detect the digital output at D_2 to D_0.

To test DACs and ADCs dynamically, we need test instruments that can display both digital and analog signals. For instance, for a 4-bit *R*-2*R* ladder DAC we would use the display of Figure 13–13. If a particular logic analyzer has the ability to display both digital and analog values, our testing problems are solved; otherwise we must improvise by using a standard dual-trace oscilloscope.

To test a DAC dynamically with a logic analyzer, we monitor the digital inputs and analog output and use the address of the DAC as the trigger-word. Using the DAC of Figure 13–16b as our test model, we monitor the $\overline{CS}$ node as well as the write node ($\overline{WR}$). We would have a display of the sampled data similar to Figure 13–13b, but for eight digital inputs instead of three, during the time that the write node was active. To produce the output voltage of Figure 13–13 requires that the digital input data begin at zero and increase to the maximum count. To input the desired digital data the inputs of the DAC could be interfaced to the system data bus and the CPU programmed to perform this count. Otherwise an up-counter could be used.

If a dual-trace oscilloscope is used to monitor the output voltage of a DAC, one input channel of the oscilloscope should be connected to the LSB of the digital input of the DAC and the other channel to the DAC output. Monitoring the LSB of the input provides us with a reference for each change in the output voltage, as can be imagined by viewing Figure 13–13 without the presence of Q_1, Q_2, and Q_3.

To test the ADC of Figure 13–17, the amplitude of the analog input signal must be monitored as well as the digital output D_2 through D_0. When the read control signal is active, the amplitude of the analog input voltage e_a can be compared to the digital output for correctness. The address of the ADC or the read control signal can be used to trigger the logic analyzer. The ADCs of Figures 13–19 and 13–21 are tested in much the same manner.

Review questions

1. What is the procedure for testing a parallel I/O port?
2. What is the procedure for testing a serial I/O port?
3. What is the procedure for testing a DAC?
4. What is the procedure for testing an ADC?

13–8 SUMMARY

- The data written to or read from a system bus is written in parallel fashion.
- There are two parallel I/O interfaces: tristate buffers and latches. *Tristate buffers* are used to implement input ports, and *latches* are used to implement output ports.

- A *PPI* is a parallel interface that can be programmed so that its ports are either input or output. Port A can be programmed to be a bidirectional port.
- Serial I/O devices require interfacing to a parallel data bus. For a serial input port the serial data is assembled by the interface before being loaded in parallel on the data bus. A serial output port requires that the interface disassemble parallel data bus data into a stream of serial data. Shift registers are used to implement both types of interfaces.
- It is often necessary to interface digital systems with analog systems. For a digital system to output to an analog system, the digital data must be converted to an equivalent analog signal. For an analog system to input its data to a digital system, the analog signal must be converted to digital data. The interface for a digital system to output data to an analog system is a *digital-to-analog converter* (DAC). The interface for the reverse data flow is an *analog-to-digital converter* (ADC).
- There are various DACs. We investigated several, but all of them are based on either the *R*-2*R* ladder network or an op-amp adder. All operate by assigning a weight to each bit position of the digital number to be converted. The MSB is given the greatest weight and the LSB is assigned the least.
- There are also various ADCs. Those we studied either compare the analog signal directly, as does the simultaneous ADC, or make a comparison with the content of a digital circuit (either a counter or a register). As these comparisons are made, the resulting digital signal is adjusted to be within one count (the resolution) of equivalent value to the analog signal. In the case of the simultaneous ADC, the adjustment is done via a priority encoder. The adjustments for the tracking ADC are implemented with an up-down counter, and the adjustment is accomplished with a successive approximation register for the successive approximation ADC.
- Regardless of the conversion being performed, there is a quantization error, and by the very nature of quantization the resolution of the answer is limited by the number of bits used to represent the signal. It is the LSB that determines the resolution and the quantization error.
- To test a DAC the digital input is compared to the analog output voltage while the write control signal is active.
- To test an ADC the analog input amplitude is compared to the digital output while the read control signal is active.

PROBLEMS

Section 13–2 Interfacing parallel I/O devices

1. How would an input port differ when implemented with a 74125 rather than a 74126?
2. How could an output port be implemented with *J-K* flip-flops?
3. Design an I/O system that has six I/O ports. Use programmable interfaces such as illustrated in Figure 13–2 to implement the six ports.

Section 13–3 Interfacing serial I/O devices

4. Using the serial-to-parallel interface of Figure 13–4a, design a serial-to-parallel interface and show it integrated into a system similar to the programmable interface of Figure 13–2.
5. Repeat Problem 4 for the parallel-to-serial interface of Figure 13–4b.
6. What is a USART and essentially how does it work?
7. Why are clocks necessary for a USART?

Section 13–4 The operational amplifier

8. Why is an open-loop op-amp considered a nonlinear device?
9. Configure the op-amp comparator of Figure 13–8a so that $e(t)$ is input at its inverting input $(-)$ and the positive input is grounded. Then analytically determine the output voltage of the op-amp for the signal $e(t)$.
10. Configure the comparator of Figure 13–9a so that the positive power supply pin is grounded and apply -5 volts DC to the negative power supply pin. Analytically determine the resultant output voltage for the input signal e of Figure 13–9b.
11. What is the maximum output voltage of an op-amp?
12. For linear applications of an op-amp, why must negative feedback be added?
13. Analytically show that the difference voltage e_d is approximately equal to zero voltage for linear applications of an op-amp.
14. Design the adder of Figure 13–10 so that its output voltage is equal to

$$v_o = -(10e_2 + 5e_1 + 2.5e_0)$$

Let $R_f = 100$ K.

15. Design the adder of Figure 13–10 for the output voltage stated in Problem 14 with the restriction that the maximum feedback current I_f will be approximately 0.1 mA. For ease of analysis, assume that the output voltage is V_{SAT}, with the input voltages approximately equal to zero volts. The power supply voltages are + and -19.5 volts DC.
16. For the adder of Problem 15, what is the output voltage if

a. $e_0 = 2.0$ volts, $e_1 = -1.6$ volts, $e_2 = 0.3$ volt

b. $e_0 = 0.1$ volt, $e_1 = 0.5$ volt, $e_2 = 3.0$ volts

Section 13–5 Digital-to-analog conversion

17. Determine the output voltage of a voltage-follower–buffered R-$2R$ ladder network if the digital input has standard TTL technology and is equal to

a. 1001

b. 1010

c. 10000001

d. 10000010

18. Determine the value of V_{REF} of Figure 13–12 if the digital input has 74C technology and has
    (a) 4 bits.
    (b) 8 bits.
    (c) 16 bits.
19. Plot the analog output of the DAC of Figure 13–13a if the input digital sequence is 0000, 1000, 1100, 1110, and 1111 and is from the 74C logic family. Take note of the decrease in voltage change as the bit position decreases.
20. If R_f of Figure 13–15 is 100 K and it is desired that the voltage output of the R-$2R$ ladder network be amplified by 10, what must be the value of R?
21. Design the R-$2R$ amplifier of Figure 13–15b such that a gain of 10 is achieved and the maximum feedback current I_f (the current flowing through R_f) is approximately 0.1 mA. For simplicity of analysis, assume that the output voltage is V_{SAT} and the input voltages are zero. The negative power supply is 15 volts DC, and the positive power supply input is grounded.
22. Repeat Problem 21 but use resistor values in stock, which are 10 K, 12 K, 13 K, 15 K, 100 K, 120 K, and 150 K. What is the value of I_f using these standard values?
23. For R = 13 K and R_f = 130 K of Figure 13–15b, what is the magnitude of the actual worst-case feedback current I_f?
24. What is the largest digital number that can be converted by the DAC of Figure 13–15b?
25. If the DAC of Figure 13–15b is designed for a gain of −5, what is the maximum output voltage of the op-amp if the digital technology is 74LS and the number of bits is
    (a) $N = 4$
    (b) $N = 8$
26. For the 8-bit DAC of Figure 13–15b, with 74LS logic driving it, what is the maximum gain that the amplifier can have? Assume that the negative power supply is 20 volts DC.
27. What are the resolution and QE for Problem 25?
28. What is the accuracy of the DAC of Problem 25a and b if the measured output voltage for part (a) is −12.45 volts and that for (b) is −13.21 volts, with all inputs a logic 1?

Section 13–6 Analog-to-digital conversion

29. For a 4-bit simultaneous ADC, how many op-amps and resistors R are there?
30. For a 4-bit simultaneous ADC, how many inputs are there to the priority encoder?
31. For a 4-bit simultaneous ADC with V_{REF} = 20 volts DC, what magnitude is the maximum comparator reference voltage?
32. For the ADC of Problem 31, what is the maximum amplitude of an analog signal that can be converted?

33. Derive an equation for the ADC of Figure 13–17 that can serve as a guide in selecting a value for V_{REF} for a given analog signal.

34. For an analog voltage that is a ramp, which peaks to 5 volts in 5 ms, plot the digital output similar to Figure 13–18. The ADC is to be a 4-bit simultaneous ADC.

35. If the DAC of Figure 13–19 is a 4-bit DAC, how many clock pulses must there be before a sample and hold of the analog can be made?

36. For the 4-bit tracking ADC of Figure 13–19, what is the approximate upper limit of an analog signal amplitude that can be converted? Assume that the DAC does not have an amplified output and the DAC has 74C technology.

37. The 4-bit tracking ADC of Problem 36 is to convert a 5-volt ramp that peaks in 5 ms. Plot the results using Figure 13–20 as a guide.

38. Repeat Problem 37 for an 8-bit successive approximation ADC. The ADC is constructed from 74C technology.

39. Repeat Problem 37 for a 4-bit successive approximation ADC, using Figure 13–22 as a guide to describe the resultant plot. The DAC has 74C technology.

Troubleshooting

Section 13–7 Troubleshooting

40. Devise a static test for the parallel ports of Figure 13–1.

41. Devise a static test for the programmable I/O interface of Figure 13–2.

42. Devise a dynamic test for the programmable interface of Figure 13–2.

43. Devise a dynamic test for the USART of Figure 13–5.

44. Devise a static test for the *R*-2*R* buffered ladder of Figure 13–15b.

45. Devise a dynamic test that uses an up-counter to generate the digital input to a three-input DAC (similar to that of Figure 13–14) and a dual-trace oscilloscope to detect the resultant output voltage.

46. If the up-counter of Problem 45 has 74C technology, produce a drawing of the expected output voltage and LSB of the input voltage.

47. Changing the digital input to 3 bits, make a timing diagram for the DAC of Figure 13–16b similar to the one displayed by a logic analyzer.

48. Devise a dynamic test for the ADC of Figure 13–19.

Appendix A Powers of 2

Powers of Two

2^n	n	2^{-n}
1	0	1 0
2	1	0 5
4	2	0 25
8	3	0 125
16	4	0 062 5
32	5	0 031 25
64	6	0 015 625
128	7	0 007 812 5
256	8	0 003 906 25
512	9	0 001 953 125
1 024	10	0 000 976 562 5
2 048	11	0 000 488 281 25
4 096	12	0 000 244 140 625
8 192	13	0 000 122 070 312 5
16 384	14	0 000 061 035 156 25
32 768	15	0 000 030 517 578 125
65 536	16	0 000 015 258 789 062 5
131 072	17	0 000 007 629 394 531 25
262 144	18	0 000 003 814 697 265 625
524 288	19	0 000 001 907 348 632 812 5
1 048 576	20	0 000 000 953 674 316 406 25
2 097 152	21	0 000 000 476 837 158 203 125
4 194 304	22	0 000 000 238 418 579 101 562 5
8 388 608	23	0 000 000 119 209 289 550 781 25
16 777 216	24	0 000 000 059 604 644 775 390 625
33 554 432	25	0 000 000 029 802 322 387 695 312 5
67 108 864	26	0 000 000 014 901 161 193 847 656 25
134 217 728	27	0 000 000 007 450 580 596 923 828 125
268 435 456	28	0 000 000 003 725 290 298 461 914 062 5
536 870 912	29	0 000 000 001 862 645 149 230 957 031 25
1 073 741 824	30	0 000 000 000 931 322 574 615 478 515 625
2 147 483 648	31	0 000 000 000 465 661 287 307 739 257 812 5
4 294 967 296	32	0 000 000 000 232 830 643 653 869 628 906 25
8 589 934 592	33	0 000 000 000 116 415 321 826 934 814 453 125
17 179 869 184	34	0 000 000 000 058 207 660 913 467 407 226 562 5
34 359 738 368	35	0 000 000 000 029 103 830 456 733 703 613 281 25
68 719 476 736	36	0 000 000 000 014 551 915 228 366 851 806 640 625
137 438 953 472	37	0 000 000 000 007 275 957 614 183 425 903 320 312 5
274 877 906 944	38	0 000 000 000 003 637 978 807 091 712 951 660 156 25
549 755 813 888	39	0 000 000 000 001 818 989 403 545 856 475 830 078 125
1 099 511 627 776	40	0 000 000 000 000 909 494 701 772 928 237 915 039 062 5
2 199 023 255 552	41	0 000 000 000 000 454 747 350 886 464 118 957 519 531 25
4 398 046 511 104	42	0 000 000 000 000 227 373 675 443 232 059 478 759 765 625
8 796 093 022 208	43	0 000 000 000 000 113 686 837 721 616 029 739 379 882 812 5
17 592 186 044 416	44	0 000 000 000 000 056 843 418 860 808 014 869 689 941 406 25
35 184 372 088 832	45	0 000 000 000 000 028 421 709 430 404 007 434 844 970 703 125
70 368 744 177 664	46	0 000 000 000 000 014 210 854 715 202 003 717 422 485 351 562 5
140 737 488 355 328	47	0 000 000 000 000 007 105 427 357 601 001 858 711 242 675 781 25
281 474 976 710 656	48	0 000 000 000 000 003 552 713 678 800 500 929 355 621 337 890 625
562 949 953 421 312	49	0 000 000 000 000 001 776 356 839 400 250 464 677 810 668 945 312 5
1 125 899 906 842 624	50	0 000 000 000 000 000 888 178 419 700 125 232 338 905 334 472 656 25
2 251 799 813 685 248	51	0 000 000 000 000 000 444 089 209 850 062 616 169 452 667 236 328 125
4 503 599 627 370 496	52	0 000 000 000 000 000 222 044 604 925 031 308 084 726 333 618 164 062 5
9 007 199 254 740 992	53	0 000 000 000 000 000 111 022 302 462 515 654 042 363 166 809 082 031 25
18 014 398 509 481 984	54	0 000 000 000 000 000 055 511 151 231 257 827 021 181 583 404 541 015 625
36 028 797 018 963 968	55	0 000 000 000 000 000 027 755 575 615 628 913 510 590 791 702 270 507 812 5
72 057 594 037 927 936	56	0 000 000 000 000 000 013 877 787 807 814 456 755 295 395 851 135 253 906 25
144 115 188 075 855 872	57	0 000 000 000 000 000 006 938 893 903 907 228 377 647 697 925 567 676 950 125
288 230 376 151 711 744	58	0 000 000 000 000 000 003 469 446 951 953 614 188 823 848 962 783 813 476 562 5
576 460 752 303 423 488	59	0 000 000 000 000 000 001 734 723 475 976 807 094 411 924 481 391 906 738 281 25

Appendix B* Explanation of the ANSI/IEEE 91-1984 logic symbols

1.0 INTRODUCTION

The International Electrotechnical Commission (IEC) has been developing a very powerful symbolic language that can show the relationship of each input of a digital logic circuit to each output without showing explicitly the internal logic. At the heart of the system is dependency notation, which will be explained in Section 4.

The system was introduced in the USA in a rudimentary form in IEEE/ANSI Standard Y32.14-1973. Lacking at that time a complete development of dependency notation, it offered little more than a substitution of rectangular shapes for the familiar distinctive shapes for representing the basic functions of AND, OR, negation, etc. This is no longer the case.

Internationally, Working Group 2 of IEC Technical Committee TC-3 has prepared a new document (Publication 617-12) that consolidates the original work started in the mid 1960's and published in 1972 (Publication 117-15) and the amendments and supplements that have followed. Similarly for the USA, IEEE Committee SCC 11.9 has revised the publication IEEE Std 91/ANSI Y32.14. Now numbered simply IEEE Std 91-1984, the IEEE standard contains all of the IEC work that has been approved, and also a small amount of material still under international consideration. Texas Instruments is participating in the work of both organizations and this document introduces new logic symbols in accordance with the new standards. When changes are made as the standards develop, future editions will take those changes into account.

The following explanation of the new symbolic language is necessarily brief and greatly condensed from what the standards publications will contain. This is not intended to be sufficient for those people who will be developing symbols for new devices. It is primarily intended to make possible the understanding of the symbols used in this data book and is somewhat briefer than the explanation that appears in several of TI's data books on digital logic. However, it includes a new section (6.0) that explains several symbols for actual devices in detail. This has proven to be a powerful learning aid.

*The following material has been reprinted by permission of Texas Instruments. It has been reset to conform to the style of this text.

Figure 1
Symbol composition

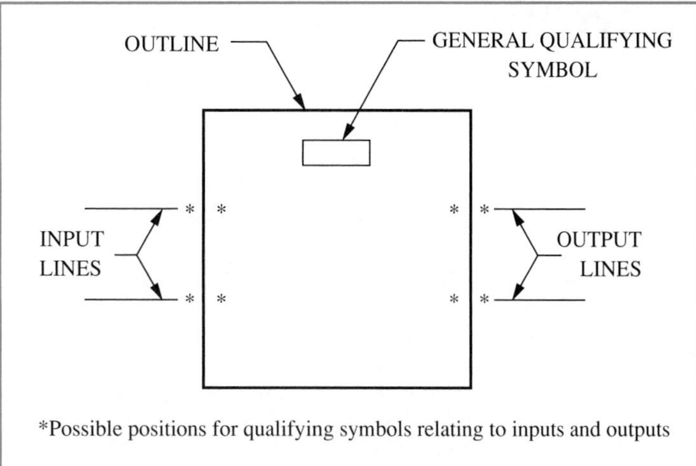

2.0 Symbol composition

A symbol comprises an outline or a combination of outlines together with one or more qualifying symbols. The shape of the symbols is not significant. As shown in Figure 1, general qualifying symbols are used to tell exactly what logical operation is performed by the elements. Table 1 shows general qualifying symbols defined in the new standards. Input lines are placed on the left and output lines are placed on the right. When an exception is made to that convention, the direction of signal flow is indicated by an arrow as shown in Figure 9.

3.0 Qualifying symbols

3.1 General qualifying symbols

Table 1 shows general qualifying symbols defined by IEEE Standard 91. These characters are placed near the top center or the geometric center of a symbol or symbol element to define the basic function of the device represented by the symbol or of the element.

X/Y is the general qualifying symbol for identifying coders, code converters, and level converters. X and Y may be used in their own right to stand for some code or either or both may be replaced by some other indication of the code or level such as BCD or TTL. As might be expected, interface circuits make frequent use of this set of qualifying symbols.

3.2 Qualifying symbols for inputs and outputs

Qualifying symbols for inputs and outputs are shown in Table 2 and will be familiar to most users with the possible exception of the logic polarity and analog signal indicators. The older logic negation indicator means that the external 0 state produces the internal 1 state. The internal 1 state means the active state. Logic negation may be used in pure logic diagrams; in order to tie the external 1 and 0 logic states to the levels H (high) and L (low), a statement of whether positive logic (1 = H, 0 = L) or negative logic (1 = L, 0 = H) is being used is required or must be assumed. Logic polarity indicators eliminate the need for calling out the logic convention and are used in this data book in the symbology for actual devices. The presence of the triangle polarity indicator indicates that the L logic level will produce the internal 1 state (the active state) or that, in the case of an output, the internal 1 state will produce the external L level. Note how the active direction of transition for a dynamic input is indicated in positive logic, negative logic, and with polarity indication.

Table 1
General qualifying symbols

Symbol	Description	
&	AND gate or function.	
≥ 1	OR gate or function. The symbol was chosen to indicate that at least one active input is needed to activate the output.	
= 1	Exclusive-OR. One and only one input must be active to activate the output.	
1	The one input must be active.	
▷ or ◁	A buffer or element with more than usual output capability (symbol is oriented in the direction of signal flow).	
⎍	Schmitt trigger; element with hysteresis.	
X/Y	Coder, code converter, level converter.	
	The following are examples of subsets of this general class of qualifying symbol used in this book:	
	BCD/7-SEG	BCD to 7-segment display driver.
	TTL/MOS	TTL to MOS level converter.
	CMOS/PLASMA DISP	Plasma-display driver with CMOS-compatible inputs.
	MOS/LED	Light-emitting-diode driver with MOS-compatible inputs.
	CMOS/VAC FLUOR DISP	Vacuum-fluorescent display driver with CMOS-compatible inputs.
	CMOS/EL DISP	Electroluminescent display driver with CMOS-compatible inputs.
	TTL/GAS DISCH DISPLAY	Gas-discharge display driver with TTL-compatible inputs.
SRGm	Shift register. m is the number of bits.	

When nonstandardized information is shown inside an outline, it is usually enclosed in square brackets [like these]. The square brackets are omitted when associated with a nonlogic input, which is indicated by an X superimposed on the connection line outside the symbol.

3.3 Symbols inside the outline

Table 3 shows some symbols used inside the outline. Note particularly that open-collector (open-drain), open-emitter (open-source), and three-state outputs have distinctive symbols. Also note that an EN input affects all of the outputs of the element and has no effect on inputs. An EN input affects all the external outputs of the element in which it is placed, plus the external outputs of any elements shown to be influenced by that element. It has no effect on inputs. When an enable input affects only certain outputs, affects outputs located outside the indicated influence of the element in which the enable input is placed, and/or affects one or more inputs, a form of dependency notation will indicate this (see Section 4.9). The effects of the EN input on the various types of outputs are shown.

It is particularly important to note that a D input is always the data input of a storage element. At its internal 1 state, the D input sets the storage element to its 1 state, and at its internal 0 state it resets the storage element to its 0 state.

Table 2
Qualifying symbols for inputs and outputs

Logic negation at input. External 0 produces internal 1.

Logic negation at output. Internal 1 produces external 0.

Active-low input. Equivalent to in positive logic

Active-low output. Equivalent to in positive logic

Active-low input in the case of right-to-left signal flow

Active-low output in the case of right-to-left signal flow

Signal flow from right to left. If not otherwise indicated, signal flow is from left to right.

Bidirectional signal flow

Dynamic inputs active on indicated transition

POSITIVE LOGIC	NEGATIVE LOGIC	POLARITY INDICATION
1 → 0	1 → 0	not used
not used	not used	H → L
0 → 1	0 → 1	L → H

Nonlogic connection. A label inside the symbol will usually define the nature of this pin.

Input for analog signals (on a digital symbol) (see Figure 11).

Input for digital signals (on an analog symbol) (see Figure 11).

3.4 Combinations of outlines and internal connections

When a circuit has one or more inputs that are common to more than one element of the circuit, the common-control block may be used. This is the only distinctively shaped outline used in the IEC system. Figure 2 shows that unless otherwise qualified by dependency notation, an input to the common-control block is an input to each of the elements below the common-control block.

Table 3
Symbols inside the outline

Symbol	Description
	Bi-threshold input (input with hysteresis).
	N-P-N open-collector or similar output that can supply a relatively low-impedance L level when not turned off. Requires external pull-up. Capable of positive-logic wired-AND connection.
	Passive-pull-up output is similar to N-P-N open-collector output but is supplemented with a built-in passive pull-up.
	N-P-N open-emitter or similar output that can supply a relatively low-impedance H level when not turned off. Requires external pull-down. Capable of positive-logic wired-OR connection.
	Passive pull-down output is similar to N-P-N open-emitter output but is supplemented with a built-in passive pull-down.
	3-state output.
	Output with more than usual output capability (symbol is oriented in the direction of signal flow).
EN	Enable input When at its internal 1-state, all outputs are enabled. When at its internal 0-state, open-collector, open-emitter, and three-state outputs are at external high-impedance state, and all other outputs (i.e., totem-poles) are at the internal 0-state.
J, K, R, S, T	Usual meanings associated with flip-flops (e.g., R = reset, T = toggle)
D	Data input to a storage element equivalent to: S R
→m ←m	Shift right (left) inputs, m = 1, 2, 3, etc. If m = 1, it is usually not shown.
o … m	Binary grouping. m is highest power of 2. Produces a number equal to the sum of the weights of the active inputs.
	Input line grouping . . . indicates two or more terminals used to implement a single logic input, e.g., differential inputs.

Figure 2
Common-control block

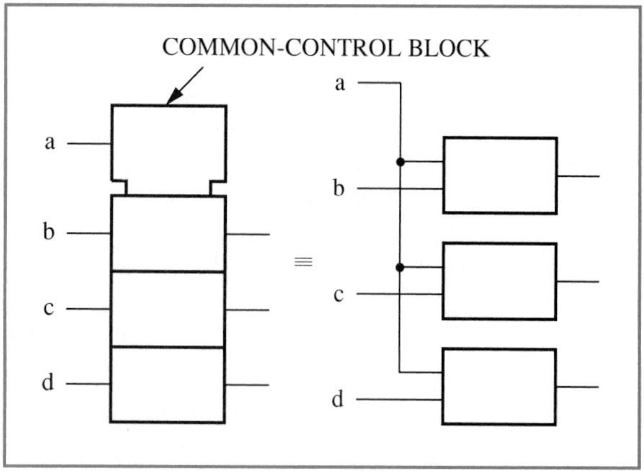

The outlines of elements may be embedded within one another or abutted to form complex elements, in which case the following rules apply. There is no logic connection between elements when the line common to their outlines is in the direction of signal flow. There is at least one logic connection when the line common to two outlines is perpendicular to the direction of signal flow. If no indications are shown on either side of the common line, it is assumed that there is only one logic connection. If more than one internal connection exists between adjacent elements, the number of connections will be clarified by the use of one or more of the internal connection symbols from Table 4 and/or appropriate qualifying symbols or dependency notation.

Table 4 shows symbols that are used to represent internal connection with specific characteristics. The first is a simple noninverting connection, the second is inverting, the third is dynamic. As with this symbol and an external input line, the transition from 0 to 1 on the left produces a momentary 1-state on the right. The fourth symbol is similar except that the active transition on the left is from 1 to 0.

Only logic states, not levels, exist inside symbols. The negation symbol (○) is used internally even when direct polarity indication (▷) is used externally.

The binary grouping symbol will be explained more fully in Section 6.11. Binary-weighted inputs are arranged in order and the binary weights of the least-significant and the most-significant lines are indicated by numbers. In this document weights of input and output lines will be

Table 4
Symbols for internal connections

Symbol	Description
(internal connection symbol)	Internal connection. 1 state on left produces 1 state on right.
(negated internal connection symbol)	Negated internal connection. 1 state on left produces 0 state on right.
(dynamic internal connection symbol)	Dynamic internal connection. Transition from 0 to 1 on left produces transitory 1 state on right.
(negated dynamic internal connection symbol)	Dynamic internal connection. Transition from 1 to 0 on left produces transitory 1 state on right.

represented by powers of two usually only when the binary grouping symbol is used, otherwise decimal numbers will be used. The grouped inputs generate an internal number on which a mathematical function can be performed or that can be an identifying number for dependency notation. This number is the sum of the weights (1, 2, 4 . . . 2^n) of those input standing at their 1 states. A frequent use is in addresses for memories.

Reversed in direction, the binary grouping symbol can be used with outputs. The concept is analogous to that for the inputs and the weighted outputs will indicate the internal number assumed to be developed within the circuit.

In an array of elements, if the same general qualifying symbol and the same qualifying symbols associated with inputs and outputs would appear inside each of the elements of the array, these qualifying symbols are usually shown only in the first element. This is done to reduce clutter and to save time in recognition. Similarly, large identical elements that are subdivided into smaller elements may each be represented by an unsubdivided outline. The MC3446 symbol illustrates this principle.

4.0 Dependency notation

Some readers will find it more to their liking to skip this section and proceed to the explanation of the symbols for a few actual devices in 6.0. Reference will be made there to various parts of this section as it is needed. If this procedure is followed, it is recommended that 5.0 be read after 6.0 and then all of 4.0 be reread.

4.1 General explanation

Dependency notation is the powerful tool that sets the IEC symbols apart from previous systems and makes compact, meaningful symbols possible. It provides the means of denoting the relationship between inputs, outputs, or inputs and outputs without actually showing all the elements and interconnections involved. The information provided by dependency notation supplements that provided by the qualifying symbols for an element's function.

In the convention for the dependency notation, use will be made of the terms "affecting" and "affected." In cases where it is not evident which inputs must be considered as being the affecting or the affected ones (e.g., if they stand in an AND relationship), the choice may be made in any convenient way.

So far, eleven types of dependency have been defined but only the eight used in this book are explained. They are listed below in the order in which they are presented and are summarized in Table 5 following 4.10.2.

Section	Dependency type or other subject
4.2	G, AND
4.3	General Rules for Dependency Notation
4.4	V, OR
4.5	N, Negate (Exclusive-OR)
4.6	Z, Interconnection
4.7	X, Transmission
4.8	C, Control
4.9	EN, Enable
4.10	M, Mode

4.2 G (AND) dependency

A common relationship between two signals is to have them ANDed together. This has traditionally been shown by explicitly drawing an AND gate with the signals connected to the inputs of the gate. The 1972 IEC publication and the 1973 IEEE/ANSI standard showed several ways to show this AND relationship using dependency notation. While ten other forms of dependency have since been defined, the ways to invoke AND dependency are now reduced to one.

In Figure 3 input **b** is ANDed with input **a** and the complement of **b** is ANDed with **c**. The letter G has been chosen to indicate AND relationships and is placed at input **b**, inside the symbol. A number considered appropriate by the symbol designer (1 has been used here) is placed after the letter G and also at each affected input. Note the bar over the 1 at input **c**.

Figure 3
G dependency between inputs

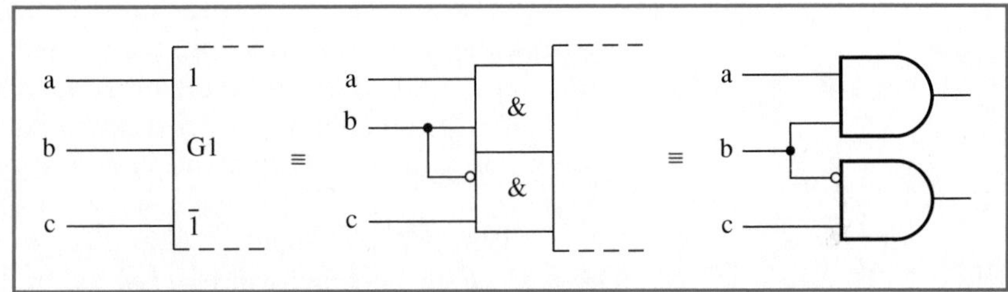

In Figure 4, output **b** affects input **a** with an ANDed relationship. The lower example shows that it is the internal logic state of **b**, unaffected by the negation sign, that is ANDed. Figure 5 shows input **a** to be ANDed with a dynamic input **b**.

The rules for G dependency can be summarized thus:

When a G*m* input or output (*m* is a number) stands at its internal 1 state, all inputs and outputs affected by G*m* stand at their normally

Figure 4
G dependency between outputs and inputs

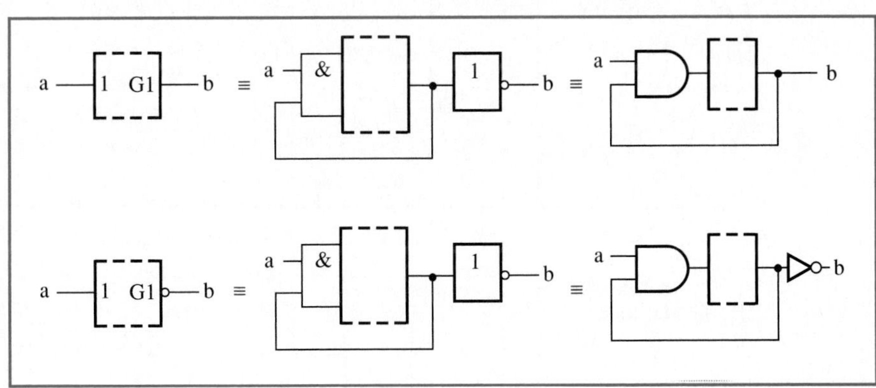

Figure 5
G dependency with a dynamic input

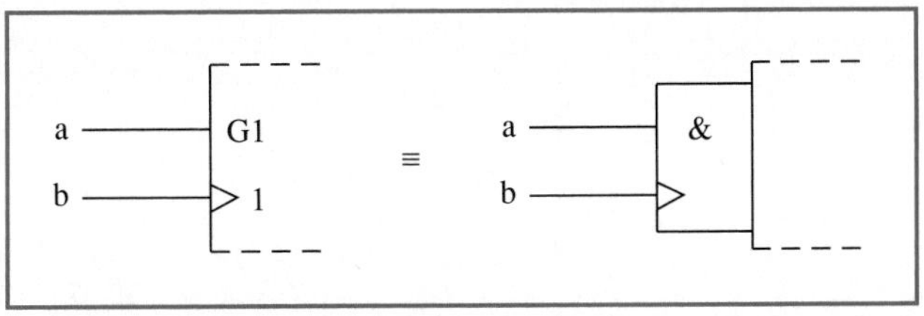

defined internal logic states. When the Gm input or output stands at its 0 state, all inputs and outputs affected by Gm stand at their internal 0 states.

4.3 Conventions for the application of dependency notation in general

The rules for applying dependency relationships in general follow the same pattern as was illustrated for G dependency.

Application of dependency notation is accomplished by:

1. labeling the input (or output) *affecting* other inputs or outputs with the letter symbol indicating the relationship involved (e.g., G for AND) followed by an identifying number, appropriately chosen, and
2. labeling each input or output *affected* by that affecting input (or output) with that same number.

If it is the complement of the internal logic state of the affecting input or output that does the affecting, then a bar is placed over the identifying numbers at the affected inputs or outputs (Figure 3).

If two affecting inputs or outputs have the same letter and same identifying number, they stand in an OR relationship to each other (Figure 6).

If the affected input or output requires a label to denote its function (e.g., "D"), this label will be *prefixed* by the identifying number of the affecting input (Figure 12).

If an input or output is affected by more than one affecting input, the identifying numbers of each of the affecting inputs will appear in the label of the affected one, separated by commas. The normal reading order of these numbers is the same as the sequence of the affecting relationships (Figure 12).

Figure 6
ORed affecting inputs

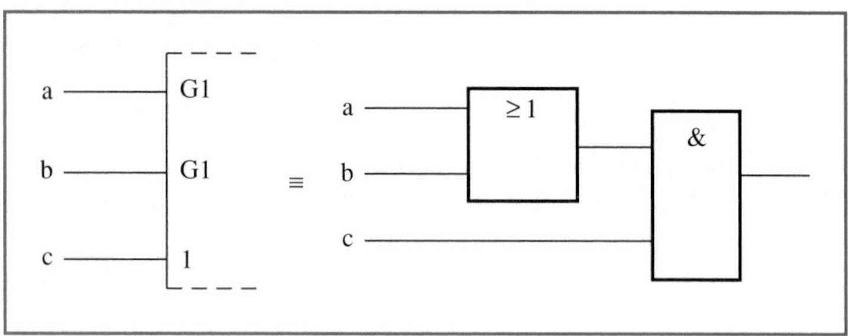

Figure 7
V (OR) dependency

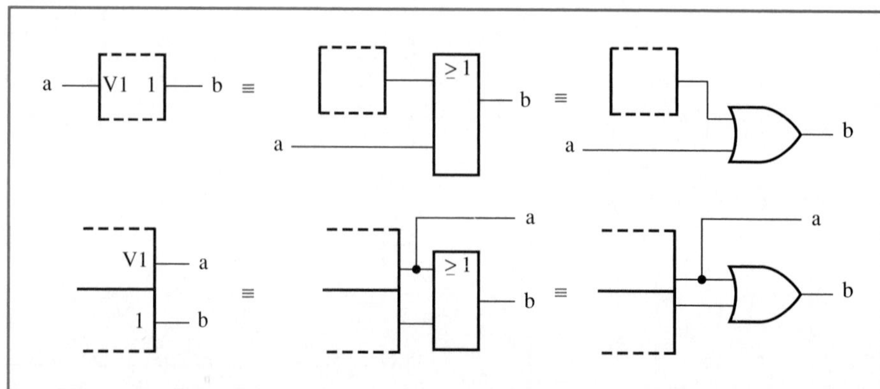

4.4 V (OR) dependency

The symbol denoting OR dependency is the letter V (Figure 7).

When a Vm input or output stands at its internal 1 state, all inputs and outputs affected by Vm stand at their internal 1 states. When the Vm input or output stands at its internal 0 state, all inputs and outputs affected by Vm stand at their normally defined internal logic states.

4.5 N (negate) (Exclusive-OR) dependency

The symbol denoting negate dependency is the letter N (Figure 8). Each input or output affected by an Nm input or output stands in an Exclusive-OR relationship with the Nm input or output.

Figure 8
N (Negate) (Exclusive-OR) dependency

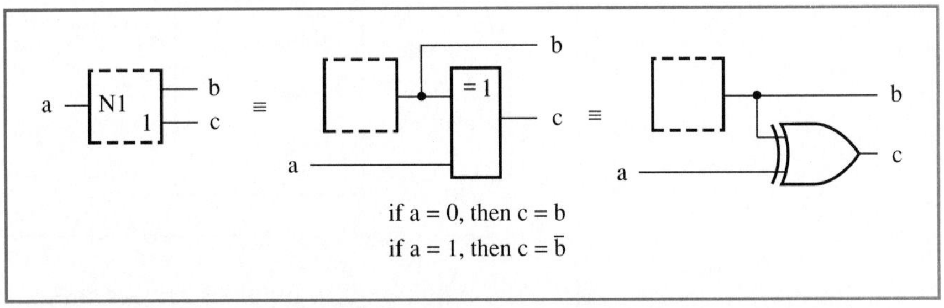

When an Nm input or output stands at its internal 1 state, the internal logic state of each input and each output affected by Nm is the complement of what it would otherwise be. When an Nm input or output stands at its internal 0 state, all inputs and outputs affected by Nm stand at their normally defined internal logic states.

4.6 Z (interconnection) dependency

The symbol denoting interconnection dependency is the letter Z.

Interconnection dependency is used to indicate the existence of internal logic connections between inputs, outputs, internal inputs, and/or internal outputs.

Figure 9
Z (interconnection) dependency

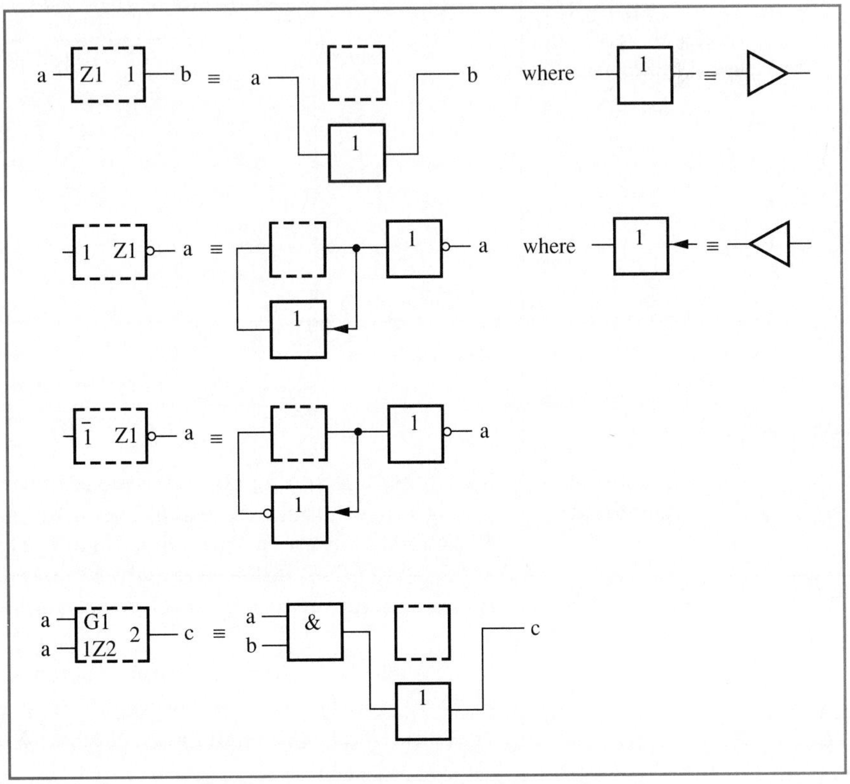

The internal logic state of an input or output affected by a Z*m* input or output will be the same as the internal logic state of the Z*m* input or output, unless modified by additional dependency notation (Figure 9).

4.7 X (transmission) dependency

The symbol denoting transmission dependency is the letter X.

Transmission dependency is used to indicate controlled bidirectional connections between affected input/output ports (Figure 10).

Figure 10
X (transmission) dependency

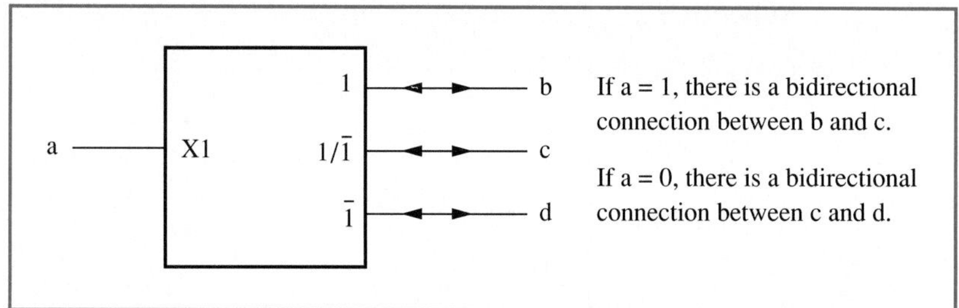

When an X*m* input or output stands at its internal 1 state, all input-output ports affected by this X*m* input or output are bidirectionally connected together and stand at the same internal logic state or analog signal level. When an X*m* input or output stands at its internal 0 state, the connection associated with this set of dependency notation does not exist.

Although the transmission paths represented by X dependency are inherently bidirectional, use is not always made of this property. This is analogous to a piece of wire, which may be constrained to carry current in only one direction. If this is the case in a particular application, then the directional arrows shown in Figures 10 and 11 would be omitted.

Figure 11
Analog data selector (multiplexer/demultiplexer)

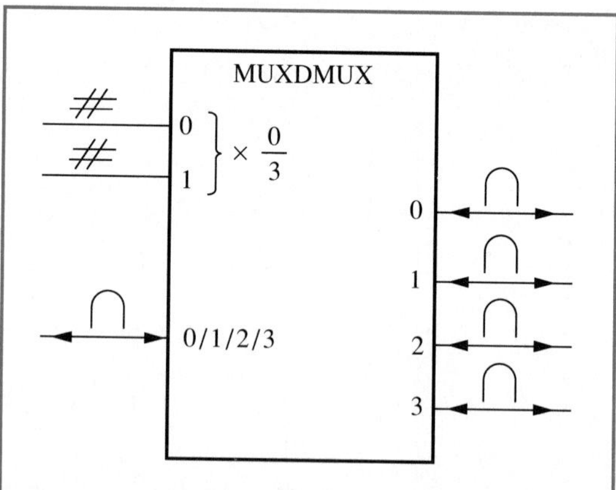

4.8 C (control) dependency

The symbol denoting control dependency is the letter C.

Control inputs are usually used to enable or disable the data (D, J, K, R, or S) inputs of storage elements. They may take on their internal 1 states (be active) either statically or dynamically. In the latter case the dynamic input symbol is used as shown in the second example of Figure 12.

When a C*m* input or output stands at its internal 1 state, the inputs affected by C*m* have their normally defined effect on the function of the element, i.e., these inputs are enabled. When a C*m* input or output stands

Figure 12
C (control) dependency

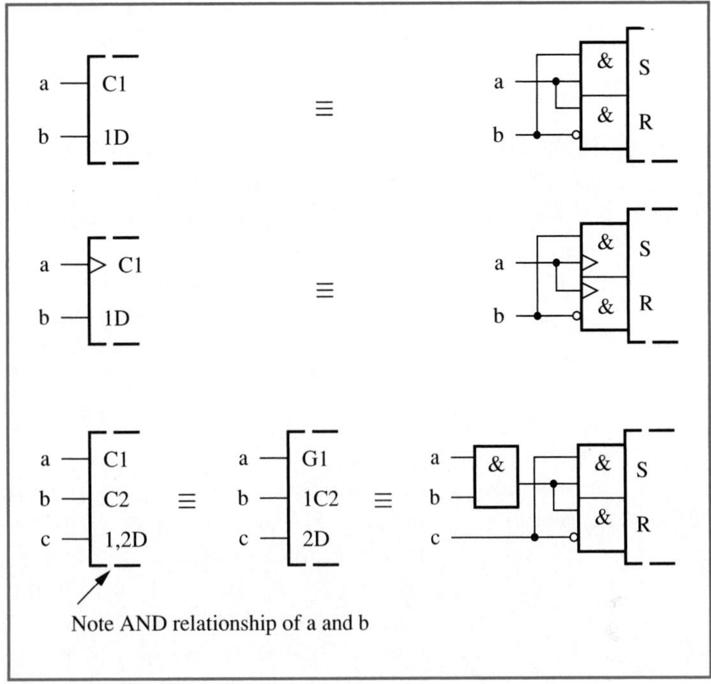

at its internal 0 state, the inputs affected by C*m* are disabled and have no effect on the function of the element.

4.9 EN (enable) dependency

The symbol denoting enable dependency is the combination of letters EN.

An EN*m* input has the same effect on outputs as an EN input, see 3.3, but it affects only those outputs labeled with the identifying number *m*. It also affects those inputs labeled with the identifying number *m*. By contrast, an EN input affects all outputs and no inputs. The effect of an EN*m* input on an affected input is identical to that of a C*m* input (Figure 13).

When an EN*m* input stands at its internal 1 state, the inputs affected by EN*m* have their normally defined effects on the function of the element and the outputs affected by this input stand at their normally defined internal logic states, i.e., these inputs and outputs are enabled.

When an EN*m* input stands at its internal 0 state, the inputs affected by EN*m* are disabled and have no effect on the function of the element, and the outputs affected by EN*m* are also disabled. Open-collector

Figure 13
EN (enable) dependency

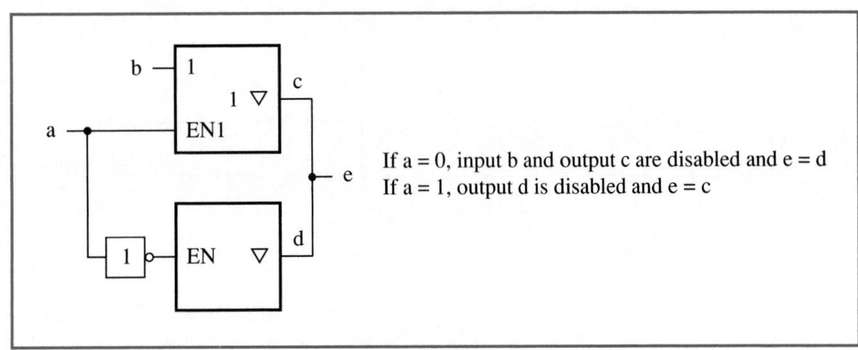

outputs are turned off, three-state outputs stand at their high-impedance state, and all other outputs (e.g., totem-pole outputs) stand at their internal 0 states.

4.10 M (mode) dependency

The symbol denoting mode dependency is the letter M.

Mode dependency is used to indicate that the effects of particular inputs and outputs of an element depend on the mode in which the element is operating.

If an input or output has the same effect in different modes of operation, the identifying numbers of the relevant affecting M*m* inputs will appear in the label of that affected input or output between parentheses and separated by solidi, e.g., (1/2)CT = 0 ≡ 1CT = 0/2CT = 0 where 1 and 2 refer to M1 and M2.

4.10.1 M dependency affecting inputs

M dependency affects inputs the same as C dependency. When an M*m* input or M*m* output stands at its internal 1 state, the inputs affected by this M*m* input or M*m* output have their normally defined effect on the function of the element, i.e., the inputs are enabled.

When an M*m* input or M*m* output stands at its internal 0 state, the inputs affected by this M*m* input or M*m* output have no effect on the function of the element. When an affected input has several sets of labels separated by solidi (e.g., C4/2 → /3+), any set in which the identifying number of the M*m* input or M*m* output appears has no effect and is to be ignored. This represents disabling of some of the functions of a multifunction input.

The circuit in Figure 14 has two inputs, **b** and **c**, that control which one of four modes (0, 1, 2, or 3) will exist at any time. Inputs **d**, **e**, and **f** are D inputs subject to dynamic control (clocking) by the **a** input. The numbers 1 and 2 are in the series chosen to indicate the modes so inputs **e** and **f** are only enabled in mode 1 (for parallel loading) and input **d** is only enabled in mode 2 (for serial loading). Note that input **a** has three functions. It is the clock for entering data. In mode 2, it causes right shifting of data, which means a shift away from the control block. In mode 3, it causes the contents of the register to be incremented by one count.

Figure 14
M (mode) dependency affecting inputs

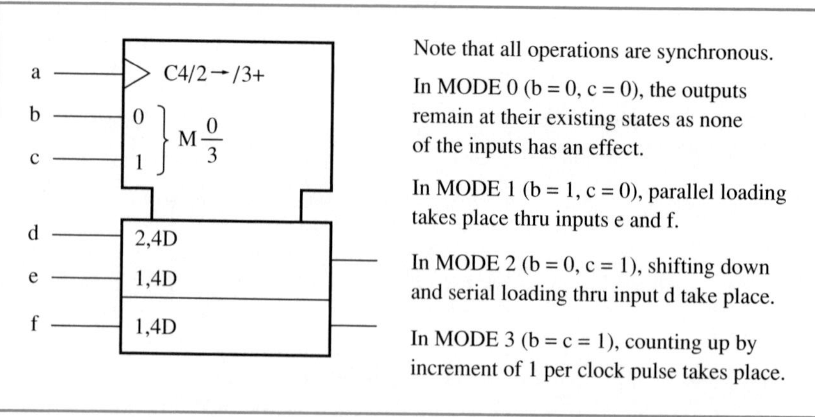

4.10.2 M dependency affecting outputs

When an M*m* input or M*m* output stands at its internal 1 state, the affected outputs stand at their normally defined internal logic states, i.e., the outputs are enabled.

When an M*m* input or M*m* output stands at its internal 0 state, at each affected output any set of labels containing the identifying number of that M*m* input or M*m* output has no effect and is to be ignored. When an output has several different sets of labels separated by solidi (e.g., 2,4/3,5), only those sets in which the identifying number of this M*m* input or M*m* output appears are to be ignored.

Figure 15
Type of output determined by mode

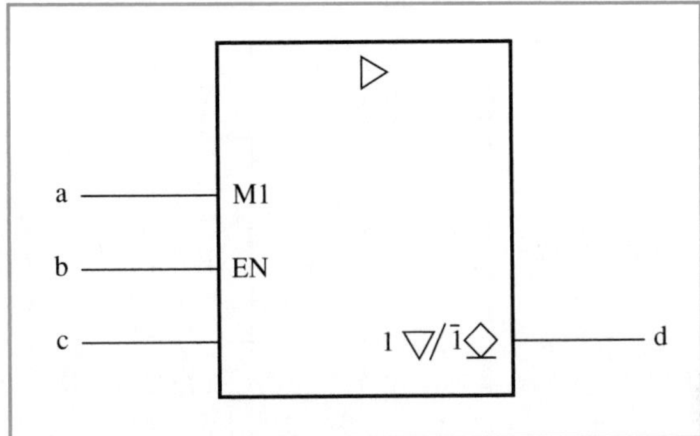

Figure 15 shows a symbol for a device whose output can behave like either a 3-state output or an open-collector output depending on the signal applied to input **a**. Mode 1 exists when input **a** stands at its internal 1 state and, in that case, the three-state symbol applies and the open-element symbol has no effect. When **a** = 0, mode 1 does not exist so the three-state symbol has no effect and the open-element symbol applies.

Table 5
Summary of dependency notation

Type of dependency	Letter Symbol*	Affecting input at its 1-state	Affecting input at its 0-state
Control	C	Permits action	Prevents action
Enable	EN	Permits action	Prevents action of inputs ◇ outputs turned off ▽ outputs at external high impedance Other outputs at internal 0 state
AND	G	Permits action	Imposes 0 state
Mode	M	Permits action (mode selected)	Prevents action (mode not selected)
Negate (Ex-NOR)	N	Complements state	No effect
OR	V	Imposes 1 state	Permits action
Transmission	X	Bidirectional connection exists	Bidirectional connection does not exist
Interconnection	Z	Imposes 1 state	Imposes 0 state

*These letter symbols appear at the AFFECTING input (or output) and are followed by a number. Each input (or output) AFFECTED by that input is labeled with that same number.

5.0 Bistable elements

The dynamic input symbol and dependency notation provide the tools to identify different types of bistable elements and make synchronous and asynchronous inputs easily recognizable (Figure 16).

Figure 16
Latches and flip-flops

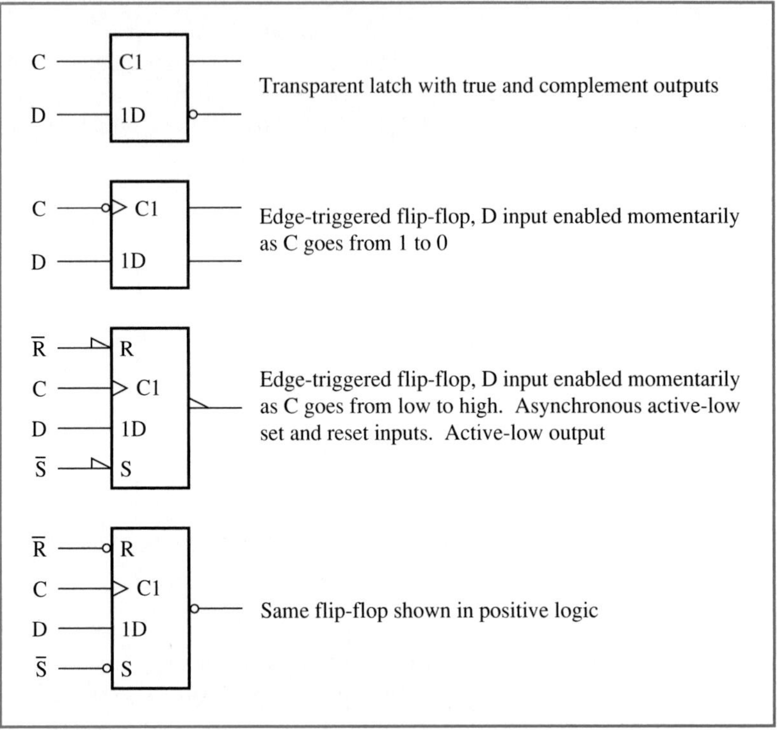

Transparent latches have a level-operated control input. The D input is active as long as the C input is at its internal 1 state. The outputs respond immediately. Edge-triggered elements accept data from D, J, K, R, or S inputs on the active transition of C.

Notice that synchronous inputs can be readily recognized by their dependency labels (a number preceding the functional label, 1D in these examples) compared to the asynchronous inputs (S and R), which are not dependent on the C inputs. Of course if the set and reset inputs were dependent on the C inputs, their labels would be similarly modified (e.g., 1S, 1R).

6.0 Examples of actual device symbols

The symbols explained in this section include some of the most complex in this book. These were chosen, not to discourage the reader, but to illustrate the amount of information that can be conveyed. It is likely that if one reads these explanations and follows them reasonably well, most of the other symbols will seem simple indeed. The explanations are intended to be independent of each other so they may seem somewhat repetitious. However each illustrates new principles. They are arranged more or less in the order of complexity.

6.1 SN75430 dual peripheral positive-AND driver

There are two identical sections. The symbology is complete for the first element; the absence of any symbology for the second element indicates it is identical. The two elements share pin 1 as an input. Had there been more than two elements this would have been indicated more conveniently with a common input block.

Each of the two elements is indicated by the & to be an AND gate. The output, pin 3, is active low so to this point the device is what would

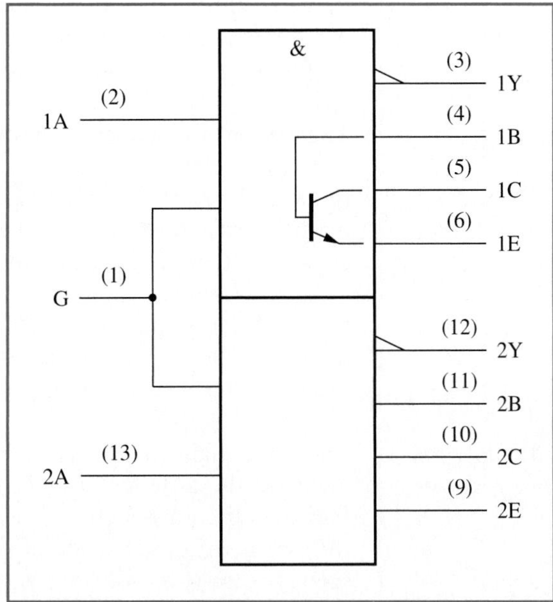

commonly be called a NAND gate. An extension of symbology used for analog devices has been used to show a floating transistor. Its emitter, base, and collector are shown lined up with the terminals to which they are connected; they are not connected internally to anything else. The device is usually used with pin 3 connected to pin 4 providing an inverting driver output and converting the NAND to AND.

6.2 SN75437 quadruple peripheral driver

There are four identical sections. The symbology is complete for the first element; the absence of any symbology for the other elements indicates

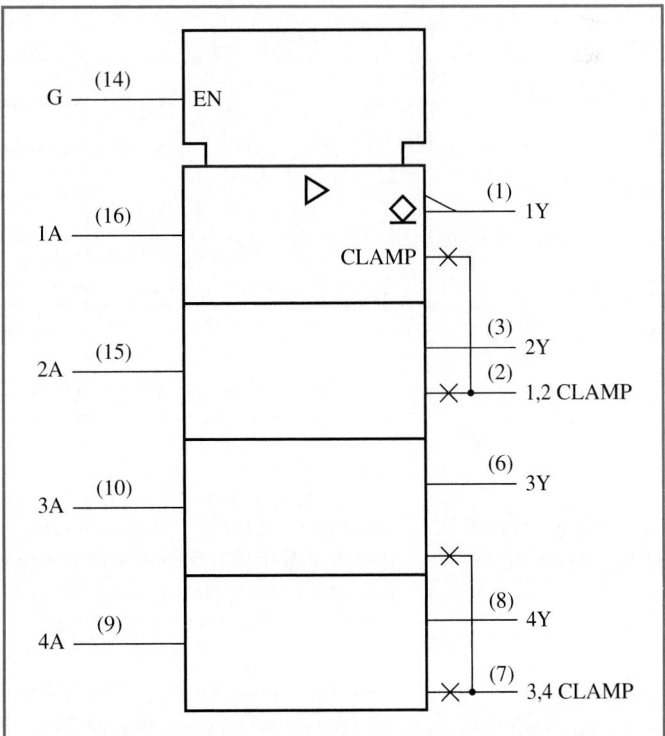

they are identical. The top two elements share a common output clamp, pin 2. This is shown to be a nonlogic connection by the superimposed X on the line. The function for this type of connection is indicated briefly and not necessarily exactly by a small amount of text within the symbol. The bottom two elements likewise share a common clamp.

Each element is shown to be an inverter with amplification (indicated by ▷). Taking TTL as a reference, this means that either the input is sensitive to lower level signals, or the output has greater drive capability than usual. The latter applies in this case. The output is shown by ⎐ to be open collector.

All the outputs share a common EN input, pin 14. See Figure 2 for an explanation of the common control block. When EN = 0 (pin 14 is low), the outputs, being open-collector types, are turned off and go high.

6.3 SN75128 8-channel line receiver

There are eight identical sections. The symbology is complete for the first element; the absence of any symbology for the next three elements indicates they are identical. Likewise the symbology is complete for the fifth element; the absence of any symbology for the next three elements indicates they are identical to the fifth.

Each element is shown to be an inverter with amplification (indicated by ▷). Taking TTL as a reference, this means that either the input is sensitive to lower level signals, or the output has greater drive capability than usual. The former applies in this case. Since neither the symbol for open-collector (⎐) or 3-state (▽) outputs is shown, the outputs are of the totem-pole type.

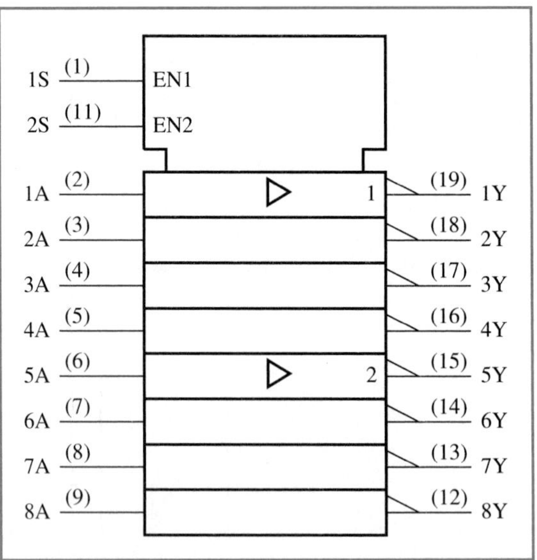

The top four outputs are shown to be affected by affecting input number 1, which is EN1, meaning they will be enabled if EN1 = 1 (pin 1 is high). See 4.9 for an explanation of EN dependency. If pin 1 is low, EN1 = 0 and the affected outputs will go to their inactive (high) levels. Similarly, the lower four outputs are controlled by pin 11.

6.4 SN75122 triple line drivers

There are two identical sections. The symbology is complete for the first section; the absence of any symbology for the next section indicates it is

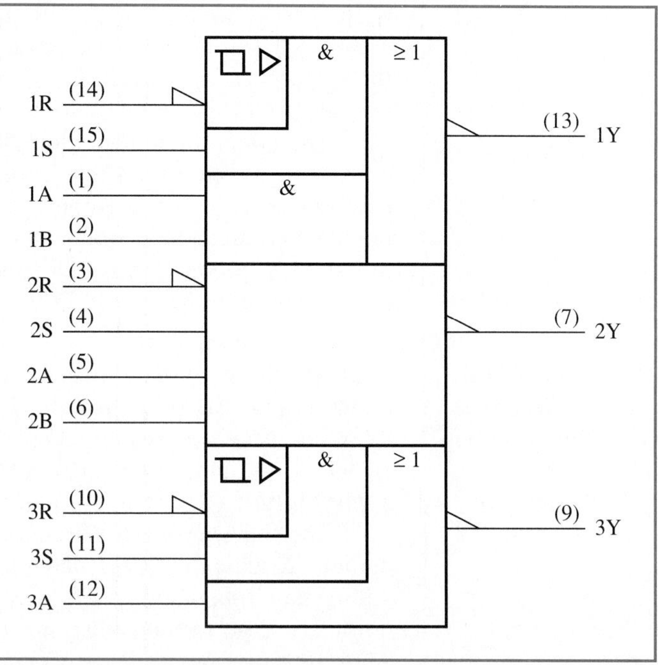

identical. Likewise the symbology is complete for the third section, which is similar, but not identical, to the first and second.

The top section may be considered to be an OR element (≥1) with two embedded ANDs (&), one of which has an active-low amplified input (▷) with hysteresis (⎍), pin 14. This is ANDed with pin 15 and the result is ORed with the AND of pins 1 and 2. The output of the OR, pin 13, is active-low.

The third section is identical to the first except that pin 12 has no input ANDed with it. Since neither the symbol for open-collector (◇) or 3-state (▽) outputs is shown, the outputs are of the totem-pole type.

6.5 SN75142 dual line receivers

There are two identical sections. The symbology is complete for the first section; the absence of any symbology for the second section indicates it is identical.

Each section may be considered to be a 3-input OR element (≥1). The first input, pin 2, is common to both OR elements. See Figure 2 for an explanation of the common-control block. The second input in the case

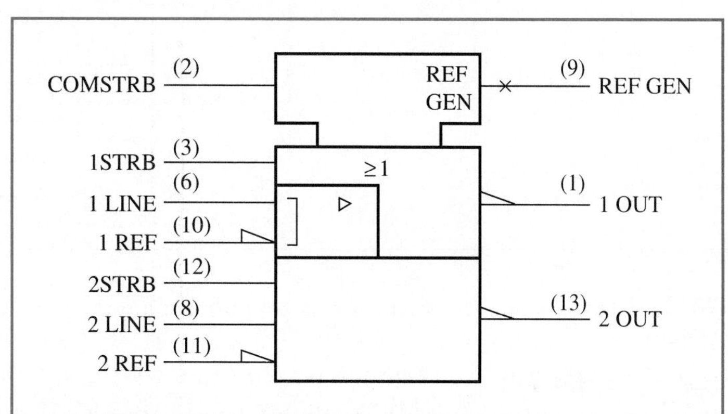

of the first section is pin 3. The third input is a differential pair shown coming into an embedded element with amplification (▷). Since neither the symbol for open-collector (◇̲) or 3-state (▽) outputs is shown, the outputs are of the totem-pole type. The outputs are active-low.

The common control block is sometimes used as a point of placement for an output that originates with either more than one or, as in this case, none of the elements in the array. Pin 9 is shown to be a nonlogic connection by the superimposed X on the line. The function for this type of connection is indicated briefly and not necessarily exactly by a small amount of text within the symbol.

6.6 DS8831 quad single-ended or dual differential line drivers

There are four similar elements in the array. Each element is shown to be noninverting with amplification (indicated by ▷). Taking TTL as a reference, this means that either the input is sensitive to lower level signals or the output has greater drive capability than usual. The latter applies in this case. The outputs are shown by ▽ to be of the 3-state type.

The top two outputs are shown to be affected by affecting input number 2, which is EN2, meaning they will be enabled if EN2 = 1. See 4.9 for an explanation of EN dependency. If EN2 = 0, the affected outputs will go to their high-impedance (off) states. EN2 is the output of an AND gate (indicated by &) whose active-low inputs are pins 1 and 2. Both pins 1 and 2 must be low to enable pins 3 and 5. Likewise both pins 14 and 15 must be low to enable pins 11 and 13 through EN3.

Input pins 6 and 10 are shown to be affected by affecting input number 1, which is N1, meaning they will be negated if N1 = 1. See 4.5 for an explanation of N (negate or exclusive-OR) dependency. If N1 = 0, the input signals are not negated. N1 is the output of an OR gate (indicated by ≥1) whose active-high inputs are pins 7 and 9. Thus if either of these pins are high, then the second and third elements become inverters.

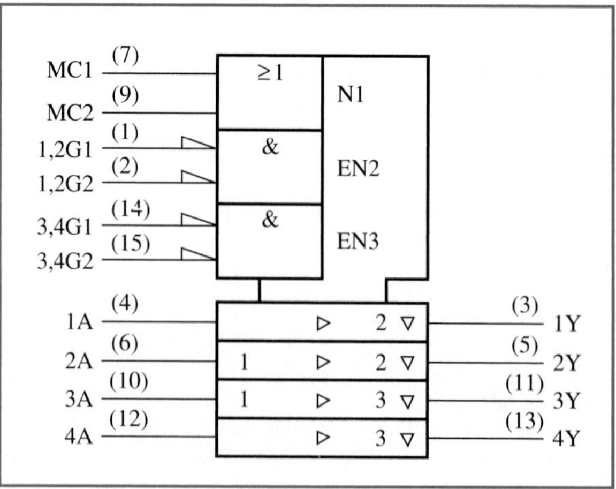

6.7 SN75113 differential line drivers with split 3-state outputs

There are two similar elements in the array. The first is a 2-input AND element (indicated by &); the second has only a single input. Both elements are shown to have special amplification (indicated by ▷). Taking TTL as a reference, this means that either the input is sensitive to lower level signals, or the output has greater drive capability than usual. The latter applies in this case.

Each element has four outputs. Pins 4 and 3 are a pair consisting of one open-emitter output (⩣) and one open-collector output (⩢). Relative to the AND function, both are active high. Pins 1 and 2 are a similar pair but relative to the AND function, both are active low. All outputs of a single, unsubdivided element always have identical internal logic states determined by the function of the element except when otherwise indicated by an associated symbol or label inside the element. Here there is no such contrary indication. All four outputs are shown to be affected by affecting input number 1, which is EN1, meaning they will all be enabled if EN1 = 1. See 4.9 for an explanation of EN dependency. If EN1 = 0, all the affected outputs will be turned off. EN1 is the output of an AND gate (indicated by &) whose active-high inputs are pins 7 and 9. Both pins 7 and 9 must be high to enable the outputs of the top element. Assuming they are enabled and that pins 5 and 6 are both high, the internal state of all four outputs will be a 1. Pins 4 and 3 will both be high, pins 1 and 2 will both be low. The part is designed so that pins 3 and 4 may be connected together creating an active-high 3-state output. Likewise pins 1 and 2 may be connected together to create an active-low 3-state output.

All that has been said about the first element regarding its outputs and their enable inputs also applies to the second element. Pins 9 and 10 are the enable inputs in this case.

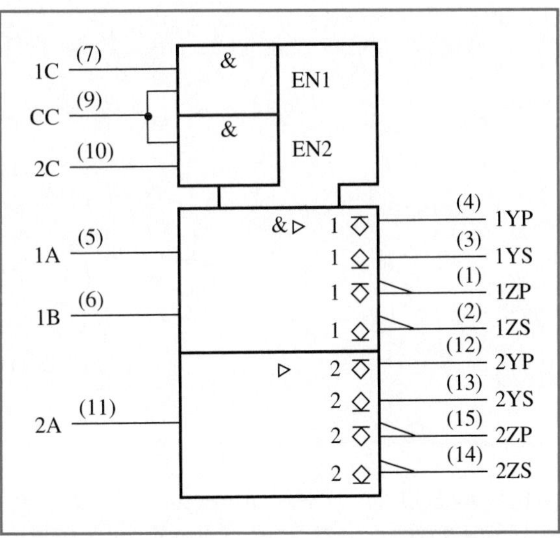

6.8 SN75163A octal general-purpose interface bus transceiver

There are eight I/O ports on each side, pins 2 through 9 and 12 through 19. There are eight identical channels. The symbology is complete for the first channel; the absence of any symbology for the other channels indicates they are identical. The eight bidirectional channels each have amplification from left to right, that is, the outputs on the right have increased drive capability (indicated by ▷), and the inputs on the right all have hysteresis (indicated by ⎍).

The outputs on the left are shown to be 3-state outputs by the ▽. They are also shown to be affected by affecting input number 4, which is

EN4, meaning they will be enabled if EN4 = 1 (pin 1 is low). See 4.9 for an explanation of EN dependency. If EN4 = 0 (pin 1 is high), the affected outputs will go to their high-impedance (off) states.

The labeling at pin 2, which applies to all the outputs on the right, is unusual because the outputs themselves have an unusual feature. The label includes both the symbol for a 3-state output (▽) and for an open-collector output (⎐), separated by a slash indicating that these are alternatives.

The symbol for the 3-state output is shown to be affected by affecting input number 1, which is M1, meaning the ▽ label is valid when M1 = 1 (pin 11 is high), but is to be ignored when M1 = 0 (pin 11 is low). See 4.10 for an explanation of M (mode) dependency. Likewise the symbol for the open-collector output is shown to be affected by affecting input number 2, which is M2, meaning the ⎐ label is valid when M2 = 1 (pin 11 is low), but is to be ignored when M2 = 0 (pin 11 is high). These labels are enclosed in parentheses (used as in algebra); the numeral 3 indicates that in either case the output is affected by EN3. Thus the right-hand outputs will be off if pin 1 is low. It can now be seen that pin 1 is the direction control and pin 11 is used to determine whether the outputs are of the 3-state or open-collector variety.

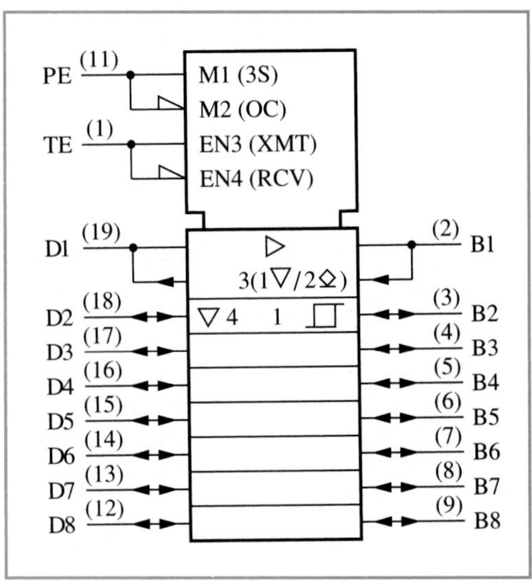

6.9 SN75161 octal general-purpose interface bus transceiver

There are eight I/O ports on each side, pins 2 through 9 and 12 through 19. Pin 13 is not only an I/O port; the lines running into the common-control block (see Figure 2) indicate that it also has control functions. Pins 1 and 11 are also controls. The eight bidirectional channels each have amplification from left to right, that is, the outputs on the right have increased drive capability (indicated by ▷), and the inputs on the right all have hysteresis (indicated by ⎍). All of the outputs are shown to be of the 3-state type by the ▽ symbol except for the outputs at pins 9, 4, and 5, which are shown to have passive pullups by the ⎐ symbol.

Starting with a typical I/O port, pin 18, the output portion is identified by an arrow indicating right-to-left signal flow and the three-state output symbol (▽). This output is shown to be affected by affecting input number 1, which is EN1, meaning it will be enabled as an output if EN 1 = 1 (pin 11 is high). See 4.9 for an explanation of EN dependency. If pin 11 is low, EN1 = 0 and the output at pin 18 will be in its high-impedance (off) state. This also applies to the 3-state outputs at pins 13 and 19 and to the passive-pullup output at pin 9. On the other hand, the outputs at pins 8, 2, 3, and 12 all are affected by the complement of EN1. This is indicated by the bar over the 1 at each of those outputs. They are enabled only when pin 11 is low. Thus one function of pin 11 is to serve as direction control for the first, third, fourth, and fifth channels.

Similarly it can be seen that pin 1 serves as direction control for the sixth, seventh, and eighth channels. If pin 1 is high, transmission will be from left to right in the sixth channel, right to left in the seventh and eighth. These transmissions are reversed if pin 1 is low.

The direction control for the second channel, EN3, is more complex. EN3 is the output of an OR (≥1) function. One of the inputs to this OR is the active-high signal on pin 13. This signal is shown to be affected at the input to the OR gate by affecting input number 5, which is G5, meaning that pin 13 is ANDed with pin 1 before entering the OR gate. See 4.2 for an explanation of G (AND) dependency. The other input to the OR is the active-low signal on pin 13. This signal is ANDed with the complement of pin 11 before entering the OR gate. This is indicated by the G4 at pin 1 and the 4 with a bar over it at pin 13. Thus for EN3 to stand at the 1 state, which would enable transmission from pin 14 to pin 7, both pins 13 and 1 must be high or both pins 13 and 11 must be low.

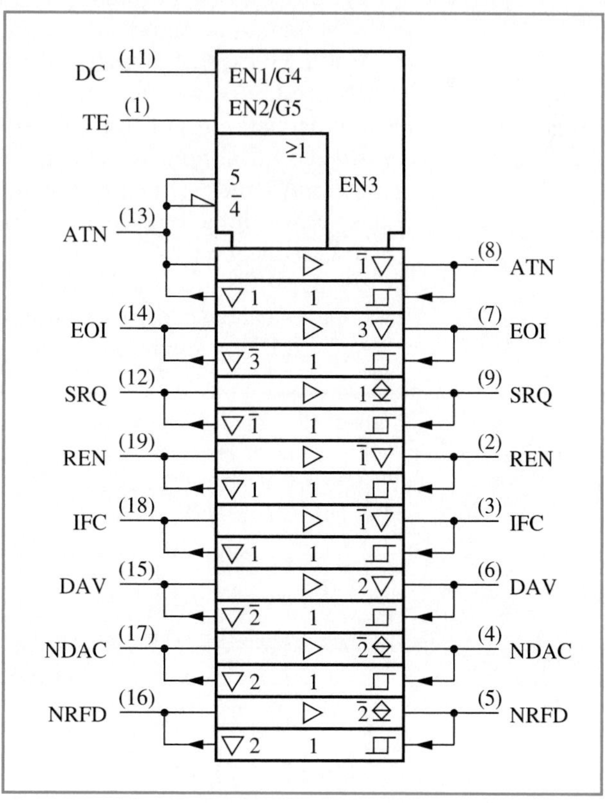

6.10 SN75584 7-segment cathode driver with latched inputs

The heart of this device and its symbol is a BCD to 7-segment decoder. Supplying the inputs to the decoder are four elements with D inputs. A D input indicates a storage element. All four D inputs are shown to be affected by affecting input number 10, which is C10, meaning they will all be enabled if C10 = 1. See 4.8 for an explanation of C dependency and 5.0 for a discussion of bistable elements. Since the C input is not dynamic, the storage element is a transparent latch. While C = 1, meaning in this case while pin 6 is high, the outputs of the latches (and hence the inputs of the decoder) will follow the input pins 8, 2, 3, and 7. When C = 0 (pin 6 goes low), those inputs are latched.

The BCD inputs to the decoder are labeled **1**, **2**, **4**, and **8** corresponding to the weights of the inputs. The outputs are labeled **a** through **g** corresponding to the accepted segment designations for 7-segment displays. When the decoder is in operation, an internal number is produced that is equal to the sum of the weights of the BCD inputs that stand at their 1 states. This causes those outputs corresponding to the segments needed to display that number to take on their 1 states. For example, if pins 8 (weight 1) and 7 (weight 8) were high and pins 2 and 3 were low while pin 6 was high, the internal number would be the sum of 8 and 1. All the segment outputs except **e** would be active (low).

The remaining input to the decoder, pin 5, is an EN input. An EN input affects all the outputs of the element in which it is placed. When EN = 0, all the segment outputs take on their 0 states. Being active low, that means they are forced high.

Located below the decoder is another transparent latch. In this case its active-low output is brought out to a terminal. This latch is also under the control of C10 (pin 6). The output, pin 17, is shown to be affected by affecting input number 11, which is EN11, meaning it will be enabled if EN11 = 1 (pin 5 is high). See 4.9 for an explanation of EN dependency. Notice that while the effect of pin 5 is the same for the latch output as for the decoder outputs, it is necessary to use EN dependency for the latter since an EN input affects all outputs of the element in which it is placed and any other elements shown to be affected by that element. The latch is shown to have no logic connection to the decoder making it necessary to use dependency notation to show that its output is also controlled by pin 5.

Located below that latch is another element whose function is defined by its single input, pin 4. This is shown to be a nonlogic

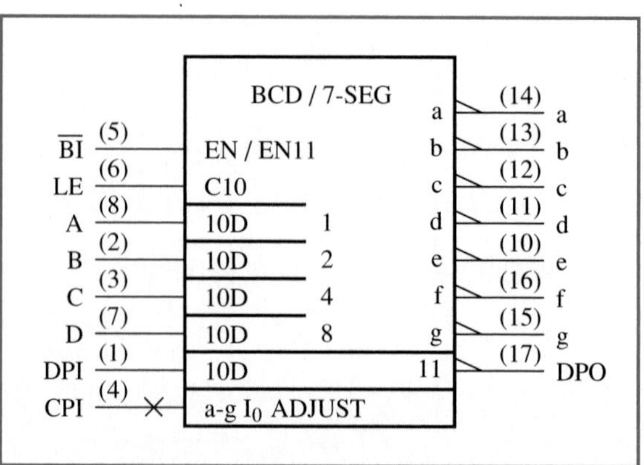

connection by the superimposed X on the line. The function for this type of connection is indicated briefly and not necessarily exactly by a small amount of text within the symbol. In this case the function of the element is to adjust the output current of the decoder, but not that of the latch.

6.11 SN75500A AC plasma display driver with CMOS-compatible inputs

The heart of this device and its symbol is an 8-bit shift register. It has a single D input, pin 2, which is shown to be affected by affecting input number 9, which is C9, meaning it will be enabled if C9 = 1. See 4.8 for an explanation of C dependency and 5.0 for a discussion of bistable elements. Since the C input is dynamic, the storage elements are edge-triggered flip-flops. While C = 1, which in this case will occur on the transition of pin 3 from low to high, the state of the D input will be stored. Pin 2 is shown to be active low so to store a 1, pin 2 must be low.

In addition to controlling the D input, pin 3 is shown by /→ to have an additional function. As pin 3 goes from low to high, data stored in the shift register is shifted one position. The right-pointing arrow means that the data is shifted away from the control block (down).

On the right side of the symbol an abbreviation technique has been used that is practical only when the internal labels and the pin numbers are both consecutive. Thus it should be clear that the input of the element whose output is pin 5 is affected by affecting input number 2, just as the input of the element whose output is pin 4 is affected by affecting input number 1. Affecting inputs 1 through 8 are Z inputs (Z1 through Z8), which means their signals are transferred directly to the output elements. See 4.6 for an explanation of Z dependency.

The inputs of the 32 implicitly shown output elements are also shown to be affected by affecting inputs numbers 11, 12, 13, and 14 in

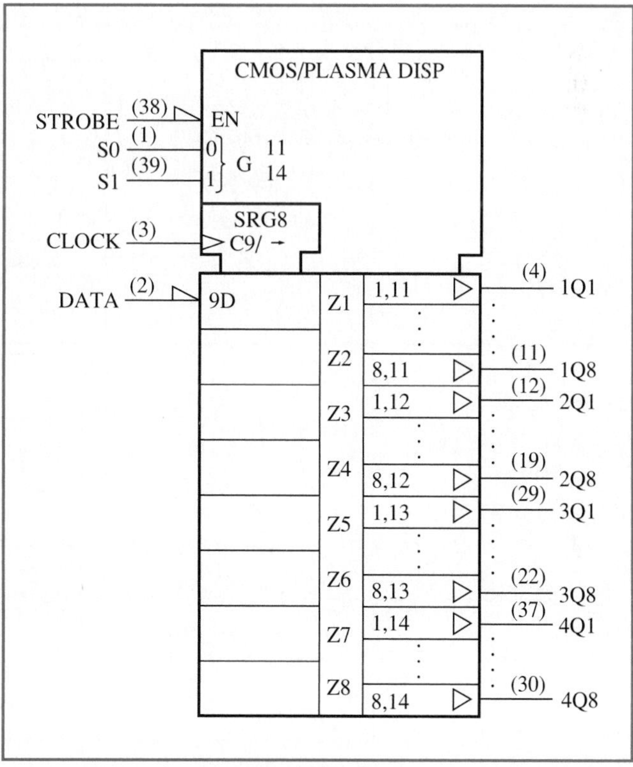

four blocks of eight each. These inputs will be found in the common control block preceded by a letter G and a brace. The brace is called the binary grouping symbol. It is equivalent to a decoder with outputs in this case driving four G inputs (G11, G12, G13, and G14). The weights of the inputs to the coder are shown to be 2^0 and 2^1 for pins 1 and 39, respectively. The decoder has four outputs corresponding to the four possible sums of the weights of the activated decoder inputs. If pins 1 and 39 are both low, the sum of the weights = 0 and G11 = 1. If pin 1 is low while pin 39 is high, the sum = 2 and G13 = 1 and so forth. G indicates AND dependency, see 4.2. Only one of the four affecting G inputs at a time can take on the 1 state. The block of eight output elements affected by that G input are enabled; the 0 state is imposed on the other 24 output elements and externally those output pins are low.

Because of their high-current, high-voltage characteristics, the outputs are labeled with the amplification symbol ▷. All the outputs share a common EN input, pin 38. See Figure 2 for an explanation of the common control block. When EN = 0 (pin 38 is high), the outputs take on their internal 0 states. Being active high, that means they are forced low.

6.12 SN75551 electroluminescent row driver with CMOS-compatible inputs

The heart of this device and its symbol is a 32-bit shift register. It has a single D input, pin 24, which is shown to be affected by affecting input number 40, which is C40, meaning it will be enabled if C40 = 1. See 4.8 for an explanation of C dependency and 5.0 for a discussion of bistable elements. Since the C input is dynamic, the storage elements are edge-triggered flip-flops. While C = 1, which in this case will occur on the transition of pin 20 from high to low, the state of the D input will be

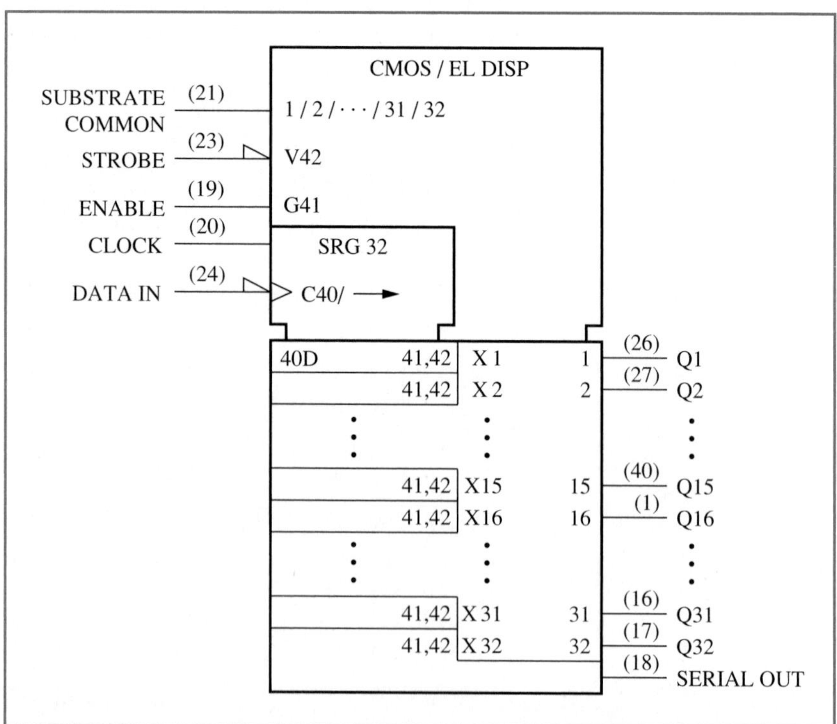

stored. Pin 24 is shown to be active high so to store a 1, pin 24 must be high.

In addition to controlling the D input, pin 20 is shown by /→ to have an additional function. As pin 20 goes from high to low, data stored in the shift register is shifted one position. The right-pointing arrow means that the data is shifted away from the control block (down). The internal outputs of the shift register are all shown to be affected by affecting inputs 41 and 42. Affecting input 41 is G41, meaning that pin 19 is ANDed with each of the internal register outputs. If pin 19 is high, the affected outputs are enabled. If pin 19 is low, the 0 state is imposed on the affected outputs. See 4.2 for an explanation of G (AND) dependency. Affecting input 42 if V42, meaning that pin 23 (active low) is ORed with each of the internal register outputs. If pin 23 is high, V42 = 0 and the affected outputs are enabled. If pin 23 is low, V42 = 1 and the 1 state is imposed on the affected outputs. See 4.4 for an explanation of V (OR) dependency. The affect of V42 is taken into account after that of V41 because of the order in which the labels appear. This means that the imposition of the 1 state by pin 23 would take precedance over the imposition of the 0 state by pin 19 in case both inputs were active. Pin 18 is shown to be an output directly from the thirty-second stage of the shift register. The dependency label 41,42 does not apply to this output, so pins 19 and 23 do not affect it.

An abbreviation technique has been used for the shift register elements, the output lines, and the associated dependency notation. This technique is practical only when the internal labels and the pin numbers are both consecutive. Thus it should be clear that the output at pin 28 is affected by affecting input number 3, just as the output at pin 27 is affected by affecting input number 2. Affecting inputs 1 through 32 are X inputs (X1 through X32). If one of these X inputs stands at the 1 state, there is a connection established between the ports labeled with the number of the X input. In the case of X2, there would be a connection between pin 27 and pin 21. Pin 21 (labeled 1/2 . . . /31/32) is the common point for all the connections indicated by X dependency in this symbol. See 4.7 for an explanation of X dependency.

Appendix C*

Manufacturer's integrated-circuit data sheets

Numerical index to Data Sheets

*The material on pages 820–913 has been reprinted courtesy of Signetics Company, a division of North American Philips Corporation.

Signetics

Logic Products

7400, LS00, S00
Gates

Quad Two-Input NAND Gate
Product Specification

TYPE	TYPICAL PROPAGATION DELAY	TYPICAL SUPPLY CURRENT (TOTAL)
7400	9ns	8mA
74LS00	9.5ns	1.6mA
74S00	3ns	15mA

ORDERING CODE

PACKAGES	COMMERCIAL RANGE V_{CC} = 5V ±5%; T_A = 0°C to +70°C
Plastic DIP	N7400N, N74LS00N, N74S00N
Plastic SO	N74LS00D, N74S00D

NOTE:
For information regarding devices processed to Military Specifications, see the Signetics Military Products Data Manual.

FUNCTION TABLE

INPUTS		OUTPUT
A	B	Y
L	L	H
L	H	H
H	L	H
H	H	L

H = HIGH voltage level
L = LOW voltage level

INPUT AND OUTPUT LOADING AND FAN-OUT TABLE

PINS	DESCRIPTION	74	74S	74LS
A, B	Inputs	1ul	1Sul	1LSul
Y	Output	10ul	10Sul	10LSul

NOTE:
Where a 74 unit load (ul) is understood to be 40μA I_{IH} and −1.6mA I_{IL}, a 74S unit load (Sul) is 50μA I_{IH} and −2.0mA I_{IL}, and 74LS unit load (LSul) is 20μA I_{IH} and −0.4mA I_{IL}.

PIN CONFIGURATION

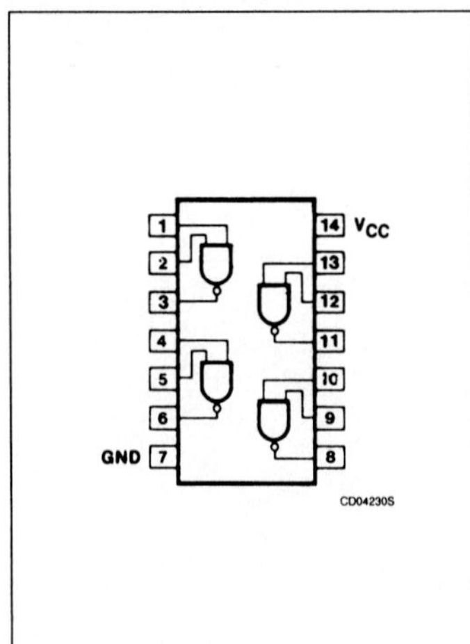

LOGIC SYMBOL

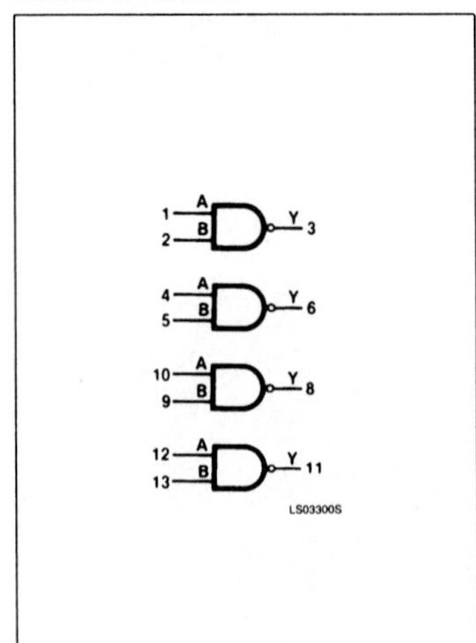

LOGIC SYMBOL (IEEE/IEC)

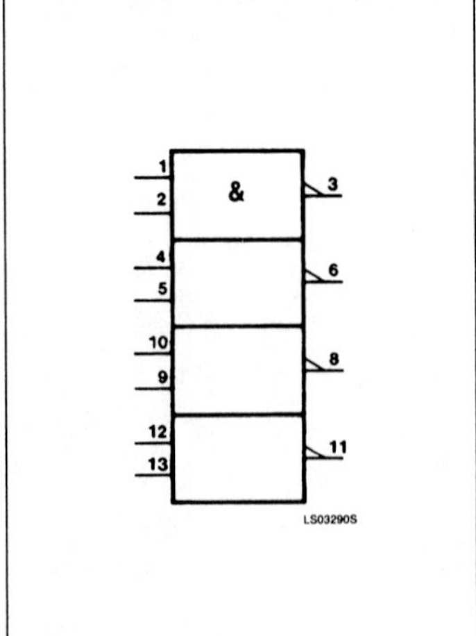

ABSOLUTE MAXIMUM RATINGS (Over operating free-air temperature range unless otherwise noted.)

PARAMETER		74	74LS	74S	UNIT
V_{CC}	Supply voltage	7.0	7.0	7.0	V
V_{IN}	Input voltage	−0.5 to +5.5	−0.5 to +7.0	−0.5 to +5.5	V
I_{IN}	Input current	−30 to +5	−30 to +1	−30 to +5	mA
V_{OUT}	Voltage applied to output in HIGH output state	−0.5 to $+V_{CC}$	−0.5 to $+V_{CC}$	−0.5 to $+V_{CC}$	V
T_A	Operating free-air temperature range	0 to 70			°C

RECOMMENDED OPERATING CONDITIONS

PARAMETER		74			74LS			74S			UNIT
		Min	Nom	Max	Min	Nom	Max	Min	Nom	Max	
V_{CC}	Supply voltage	4.75	5.0	5.25	4.75	5.0	5.25	4.75	5.0	5.25	V
V_{IH}	HIGH-level input voltage	2.0			2.0			2.0			V
V_{IL}	LOW-level input voltage			+0.8			+0.8			+0.8	V
I_{IK}	Input clamp current			−12			−18			−18	mA
I_{OH}	HIGH-level output current			−400			−400			−1000	μA
I_{OL}	LOW-level output current			16			8			20	mA
T_A	Operating free-air temperature	0		70	0		70	0		70	°C

TEST CIRCUITS AND WAVEFORMS

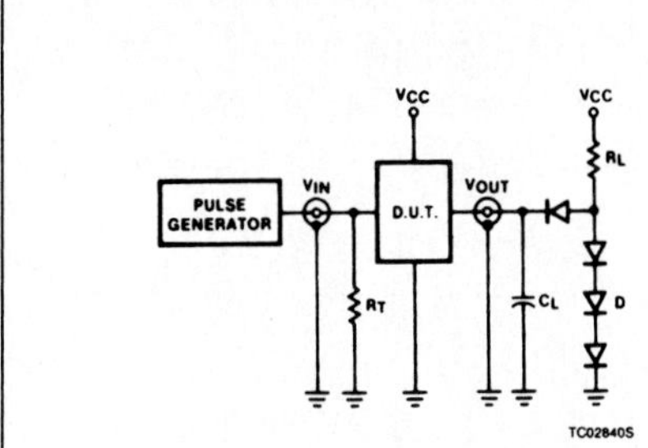

Test Circuit For 74 Totem-Pole Outputs

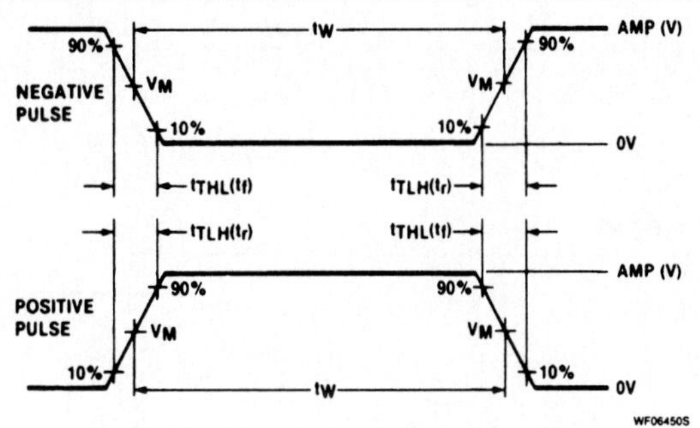

V_M = 1.3V for 74LS; V_M = 1.5V for all other TTL families.

Input Pulse Definition

DEFINITIONS
R_L = Load resistor to V_{CC}; see AC CHARACTERISTICS for value.
C_L = Load capacitance includes jig and probe capacitance; see AC CHARACTERISTICS for value.
R_T = Termination resistance should be equal to Z_{OUT} of Pulse Generators.
D = Diodes are 1N916, 1N3064, or equivalent.
t_{TLH}, t_{THL} Values should be less than or equal to the table entries.

FAMILY	INPUT PULSE REQUIREMENTS				
	Amplitude	Rep. Rate	Pulse Width	t_{TLH}	t_{THL}
74	3.0V	1MHz	500ns	7ns	7ns
74LS	3.0V	1MHz	500ns	15ns	6ns
74S	3.0V	1MHz	500ns	2.5ns	2.5ns

Gates 7400, LS00, S00

DC ELECTRICAL CHARACTERISTICS (Over recommended operating free-air temperature range unless otherwise noted.)

PARAMETER		TEST CONDITIONS[1]		7400			74LS00			74S00			UNIT
				Min	Typ[2]	Max	Min	Typ[2]	Max	Min	Typ[2]	Max	
V_{OH}	HIGH-level output voltage	V_{CC} = MIN, V_{IH} = MIN, V_{IL} = MAX, I_{OH} = MAX		2.4	3.4		2.7	3.4		2.7	3.4		V
V_{OL}	LOW-level output voltage	V_{CC} = MIN, V_{IH} = MIN	I_{OL} = MAX		0.2	0.4		0.35	0.5			0.5	V
			I_{OL} = 4mA (74LS)					0.25	0.4				V
V_{IK}	Input clamp voltage	V_{CC} = MIN, $I_I = I_{IK}$				−1.5			−1.5			−1.2	V
I_I	Input current at maximum input voltage	V_{CC} = MAX	V_I = 5.5V			1.0						1.0	mA
			V_I = 7.0V						0.1				mA
I_{IH}	HIGH-level input current	V_{CC} = MAX	V_I = 2.4V			40							μA
			V_I = 2.7V						20			50	μA
I_{IL}	LOW-level input current	V_{CC} = MAX	V_I = 0.4V			−1.6			−0.4				mA
			V_I = 0.5V									−2.0	mA
I_{OS}	Short-circuit output current[3]	V_{CC} = MAX		−18		−55	−20		−100	−40		−100	mA
I_{CC}	Supply current (total)	V_{CC} = MAX	I_{CCH} Outputs HIGH		4	8		0.8	1.6		10	16	mA
			I_{CCL} Outputs LOW		12	22		2.4	4.4		20	36	mA

NOTES:

1. For conditions shown as MIN or MAX, use the appropriate value specified under recommended operating conditions for the applicable type.
2. All typical values are at V_{CC} = 5V, T_A = 25°C.
3. I_{OS} is tested with V_{OUT} = +0.5V and V_{CC} = V_{CC} MAX + 0.5V. Not more than one output should be shorted at a time and duration of the short circuit should not exceed one second.

AC WAVEFORM

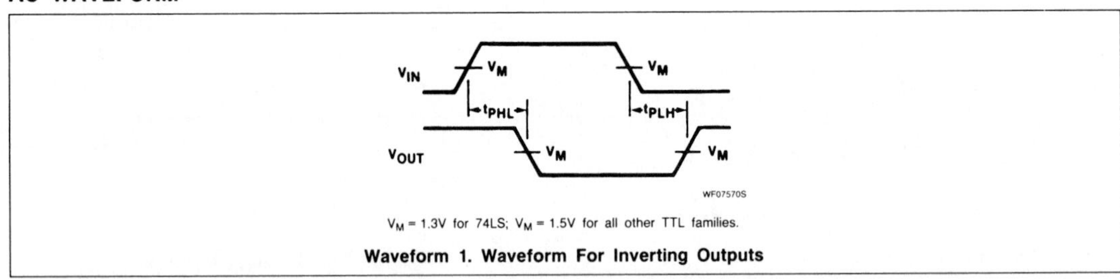

V_M = 1.3V for 74LS; V_M = 1.5V for all other TTL families.

Waveform 1. Waveform For Inverting Outputs

AC ELECTRICAL CHARACTERISTICS T_A = 25°C, V_{CC} = 5.0V

PARAMETER		TEST CONDITIONS	74		74LS		74S		UNIT
			C_L = 15pF, R_L = 400Ω		C_L = 15pF, R_L = 2kΩ		C_L = 15pF, R_L = 280Ω		
			Min	Max	Min	Max	Min	Max	
t_{PLH} t_{PHL}	Propagation delay	Waveform 1		22 15		15 15		4.5 5.0	ns

Signetics

7402, LS02, S02
Gates

Quad Two-Input NOR Gate
Product Specification

Logic Products

TYPE	TYPICAL PROPAGATION DELAY	TYPICAL SUPPLY CURRENT (TOTAL)
7402	10ns	11mA
74LS02	10ns	2.2mA
74S02	3.5ns	22mA

ORDERING CODE

PACKAGES	COMMERCIAL RANGE $V_{CC} = 5V \pm 5\%$; $T_A = 0°C$ to $+70°C$
Plastic DIP	N7402N, N74LS02N, N74S02N
Plastic SO	N74LS02D, N74S02D

NOTE:
For information regarding devices processed to Military Specifications, see the Signetics Military Products Data Manual.

FUNCTION TABLE

INPUTS		OUTPUT
A	B	Y
L	L	H
L	H	L
H	L	L
H	H	L

H = HIGH voltage level
L = LOW voltage level

INPUT AND OUTPUT LOADING AND FAN-OUT TABLE

PINS	DESCRIPTION	74	74S	74LS
A, B	Inputs	1ul	1Sul	1LSul
Y	Output	10ul	10Sul	10LSul

NOTE:
Where a 74 unit load (ul) is understood to be 40μA I_{IH} and −1.6mA I_{IL}, a 74S unit load (Sul) is 50μA I_{IH} and −2.0mA I_{IL}, and 74LS unit load (LSul) is 20μA I_{IH} and −0.4mA I_{IL}.

PIN CONFIGURATION

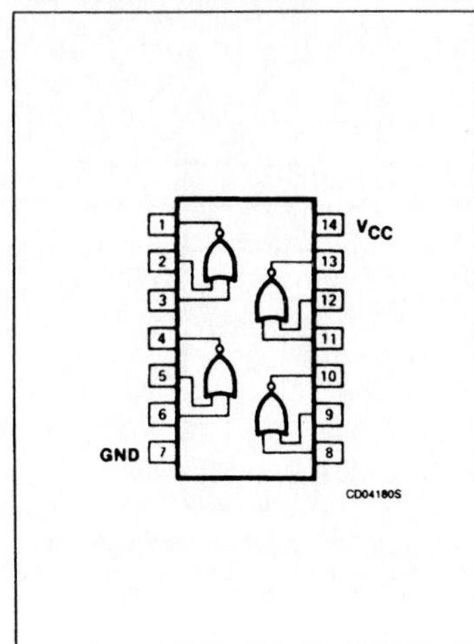

LOGIC SYMBOL

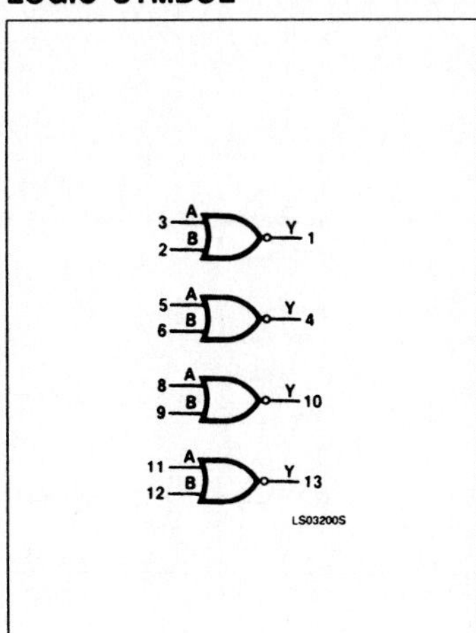

LOGIC SYMBOL (IEEE/IEC)

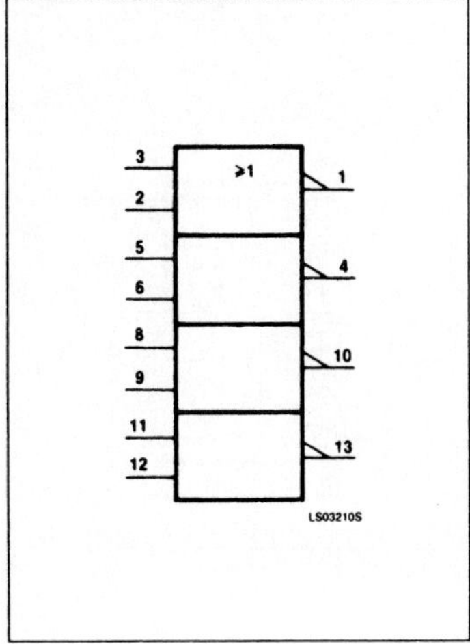

Signetics

7403, S03
Gates

Quad Two-Input NAND Gate (Open Collector)
Product Specification

Logic Products

TYPE	TYPICAL PROPAGATION DELAY	TYPICAL SUPPLY CURRENT (TOTAL)
7403	35ns (t_{PLH}) 8ns (t_{PHL})	8mA
74S03	5ns (t_{PLH}) 4.5ns (t_{PHL})	13mA

ORDERING CODE

PACKAGES	COMMERCIAL RANGE V_{CC} = 5V ±5%; T_A = 0°C to +70°C
Plastic DIP	N7403N, N74S03N
Plastic SO	N74S03D

NOTE:
For information regarding devices processed to Military Specifications, see the Signetics Military Products Data Manual.

FUNCTION TABLE

INPUTS		OUTPUT
A	B	Y
L	L	H
L	H	H
H	L	H
H	H	L

H = HIGH voltage level
L = LOW voltage level

INPUT AND OUTPUT LOADING AND FAN-OUT TABLE

PINS	DESCRIPTION	74	74S
A, B	Inputs	1ul	1Sul
Y	Output	10ul	10Sul

NOTE:
Where a 74 unit load (ul) is understood to be 40μA I_{IH} and −1.6mA I_{IL}, a 74S unit load (Sul) is 50μA I_{IH} and −2.0mA I_{IL}.

PIN CONFIGURATION

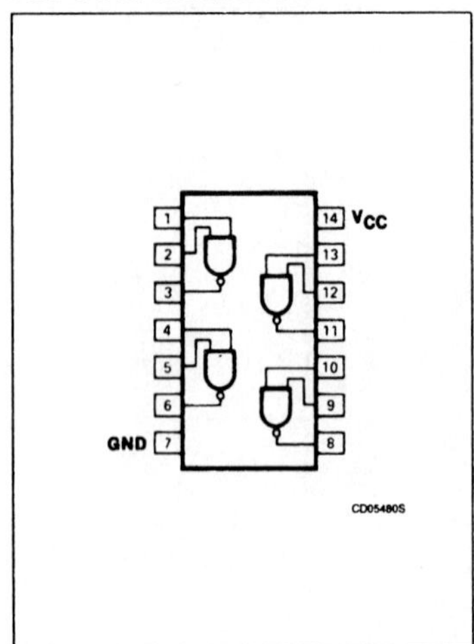

LOGIC SYMBOL

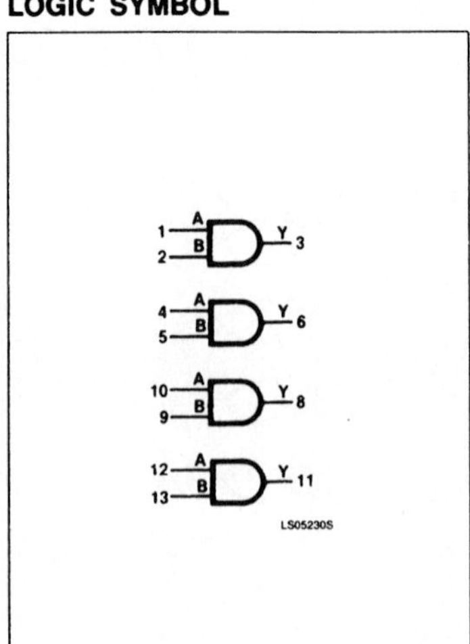

LOGIC SYMBOL (IEEE/IEC)

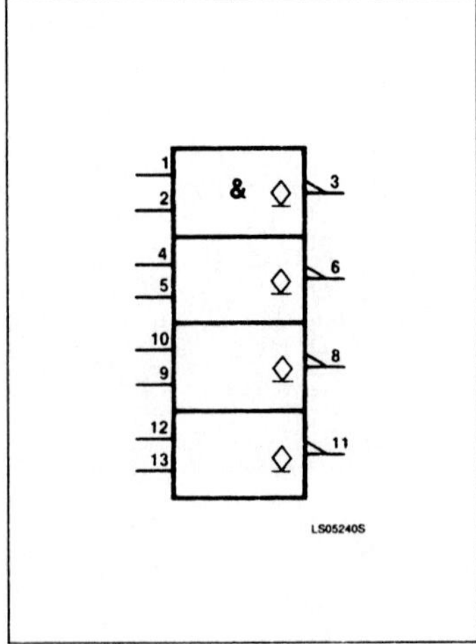

Signetics

7404, LS04, S04
Inverters

Hex Inverter
Product Specification

Logic Products

TYPE	TYPICAL PROPAGATION DELAY	TYPICAL SUPPLY CURRENT (TOTAL)
7404	10ns	12mA
74LS04	9.5ns	2.4mA
74S04	3ns	22mA

ORDERING CODE

PACKAGES	COMMERCIAL RANGE $V_{CC} = 5V \pm 5\%$; $T_A = 0°C$ to $+70°C$
Plastic DIP	N7404N, N74LS04N, N74S04N
Plastic SO	N74LS04D, N74S04D

NOTE:
For information regarding devices processed to Military Specifications, see the Signetics Military Products Data Manual.

FUNCTION TABLE

INPUT	OUTPUT
A	Y
L	H
H	L

H = HIGH voltage level
L = LOW voltage level

INPUT AND OUTPUT LOADING AND FAN-OUT TABLE

PINS	DESCRIPTION	74	74S	74LS
A	Input	1ul	1Sul	1LSul
Y	Output	10ul	10Sul	10LSul

NOTE:
Where a 74 unit load (ul) is understood to be 40μA I_{IH} and −1.6mA I_{IL}, a 74S unit load (Sul) is 50μA I_{IH} and −2.0mA I_{IL}, and 74LS unit load (LSul) is 20μA I_{IH} and −0.4mA I_{IL}.

PIN CONFIGURATION

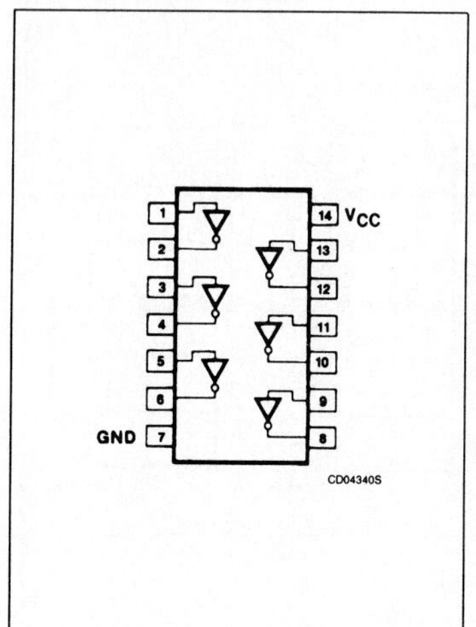

LOGIC SYMBOL

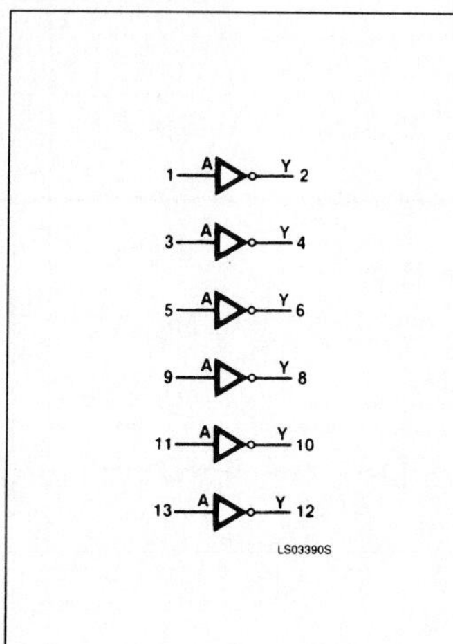

LOGIC SYMBOL (IEEE/IEC)

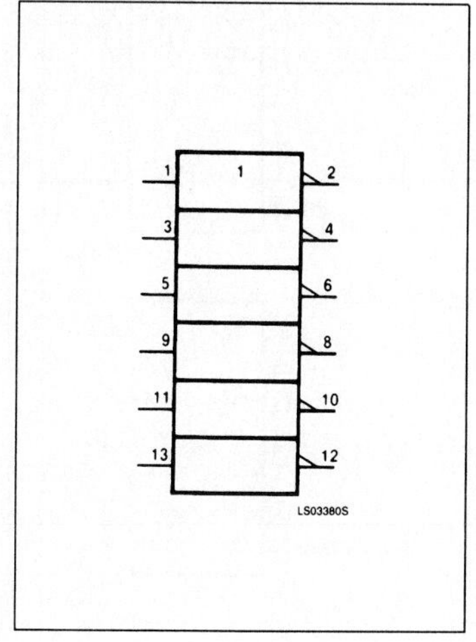

Signetics

Logic Products

7406, 07
Inverter/Buffer/Drivers

'06 Hex Inverter Buffer/Driver (Open Collector)
'07 Hex Buffer/Driver (Open Collector)
Product Specification

TYPE	TYPICAL PROPAGATION DELAY	TYPICAL SUPPLY CURRENT (TOTAL)
7406	10ns (t_{PLH}) 15ns (t_{PHL})	31mA
7407	6ns (t_{PLH}) 20ns (t_{PHL})	25mA

ORDERING CODE

PACKAGES	COMMERCIAL RANGE $V_{CC} = 5V \pm 5\%$; $T_A = 0°C$ to $+70°C$
Plastic DIP	N7406N, N7407N
Plastic SO	N7406D, N7407D

NOTE:
For information regarding devices processed to Military Specifications, see the Signetics Military Products Data Manual.

FUNCTION TABLE

'06		'07	
INPUT	OUTPUT	INPUT	OUTPUT
A	Y	A	Y
H	L	H	H
L	H	L	L

H = HIGH voltage level
L = LOW voltage level

INPUT AND OUTPUT LOADING AND FAN-OUT TABLE

PINS	DESCRIPTION	74
A	Input	1ul
Y	Output	10ul

NOTE:
Where a 74 unit load (ul) is understood to be 40μA I_{IH} and −1.6mA I_{IL}.

PIN CONFIGURATION

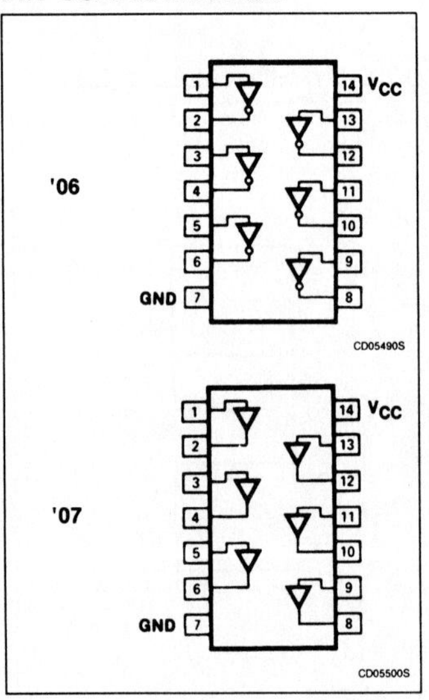

LOGIC SYMBOL

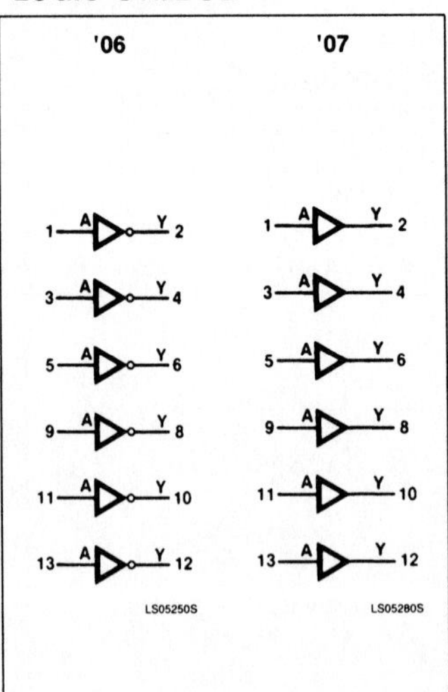

LOGIC SYMBOL (IEEE/IEC)

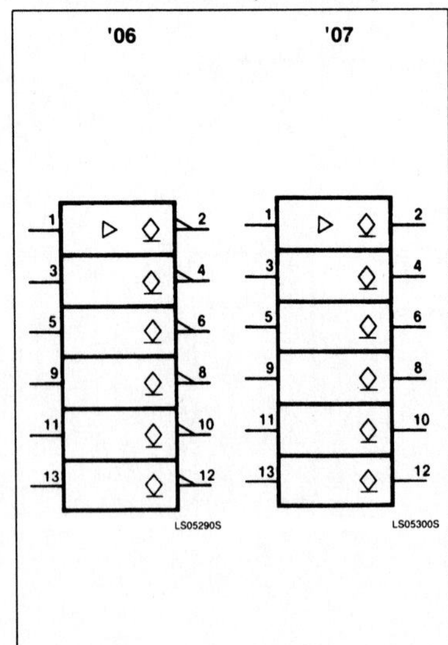

Inverter/Buffer/Drivers — 7406, 07

ABSOLUTE MAXIMUM RATINGS (Over operating free-air temperature range unless otherwise noted.)

PARAMETER		74	UNIT
V_{CC}	Supply voltage	7.0	V
V_{IN}	Input voltage	−0.5 to +5.5	V
I_{IN}	Input current	−30 to +5	mA
V_{OUT}	Voltage applied to output in HIGH output state	−0.5 to +30	V
T_A	Operating free-air temperature range	0 to 70	°C

RECOMMENDED OPERATING CONDITIONS

PARAMETER		74			UNIT
		Min	Nom	Max	
V_{CC}	Supply voltage	4.75	5.0	5.25	V
V_{IH}	HIGH-level input voltage	2.0			V
V_{IL}	LOW-level input voltage			+0.8	V
I_{IK}	Input clamp current			−12	mA
V_{OH}	HIGH-level output voltage			30	V
I_{OL}	LOW-level output current			40	mA
T_A	Operating free-air temperature	0		70	°C

TEST CIRCUITS AND WAVEFORMS

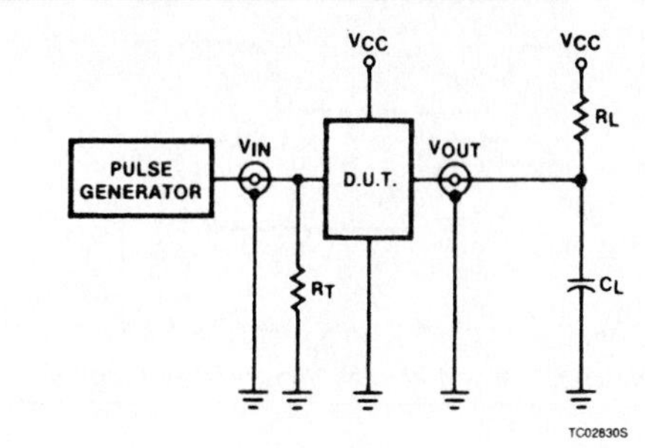

Test Circuit For 74 Open Collectors Outputs

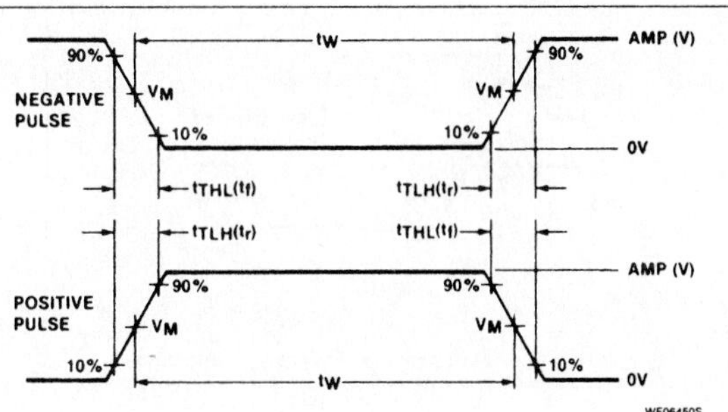

V_M = 1.3V for 74LS; V_M = 1.5V for all other TTL families.

Input Pulse Definition

DEFINITIONS

R_L = Load resistor to V_{CC}; see AC CHARACTERISTICS for value.
C_L = Load capacitance includes jig and probe capacitance; see AC CHARACTERISTICS for value.
R_T = Termination resistance should be equal to Z_{OUT} of Pulse Generators.
D = Diodes are 1N916, 1N3064, or equivalent.
t_{TLH}, t_{THL} Values should be less than or equal to the table entries.

FAMILY	INPUT PULSE REQUIREMENTS				
	Amplitude	Rep. Rate	Pulse Width	t_{TLH}	t_{THL}
74	3.0V	1MHz	500ns	7ns	7ns
74LS	3.0V	1MHz	500ns	15ns	6ns
74S	3.0V	1MHz	500ns	2.5ns	2.5ns

Inverter/Buffer/Drivers 7406, 07

DC ELECTRICAL CHARACTERISTICS (Over recommended operating free-air temperature range unless otherwise noted.)

	PARAMETER	TEST CONDITIONS[1]			7406, 7407 Min	Typ[2]	Max	UNIT
I_{OH}	HIGH-level output current	V_{CC} = MIN, V_{IH} = MIN, V_{IL} = MAX, V_{OH} = 30V					250	μA
V_{OL}	LOW-level output voltage	V_{CC} = MIN, V_{IH} = MIN, V_{IL} = MAX	I_{OL} = 16mA				0.4	V
			I_{OL} = 30mA				0.7	V
			I_{OL} = 40mA				0.7	V
V_{IK}	Input clamp voltage	V_{CC} = MIN, I_I = I_{IK}					−1.5	V
I_I	Input current at maximum input voltage	V_{CC} = MAX, V_I = 5.5V					1.0	mA
I_{IH}	HIGH-level input current	V_{CC} = MAX, V_I = 2.4V					40	μA
I_{IL}	LOW-level input current	V_{CC} = MAX, V_I = 0.4V					−1.6	mA
I_{CC}	Supply current (total)	V_{CC} = MAX	I_{CCH} Outputs HIGH	'06		30	48	mA
			I_{CCL} Outputs LOW			32	51	mA
			I_{CCH} Outputs HIGH	'07		29	41	mA
			I_{CCL} Outputs LOW			21	30	mA

NOTES:
1. For conditions shown as MIN or MAX, use the appropriate value specified under recommended operating conditions for the applicable type.
2. All typical values are at V_{CC} = 5V, T_A = 25°C.

AC WAVEFORMS

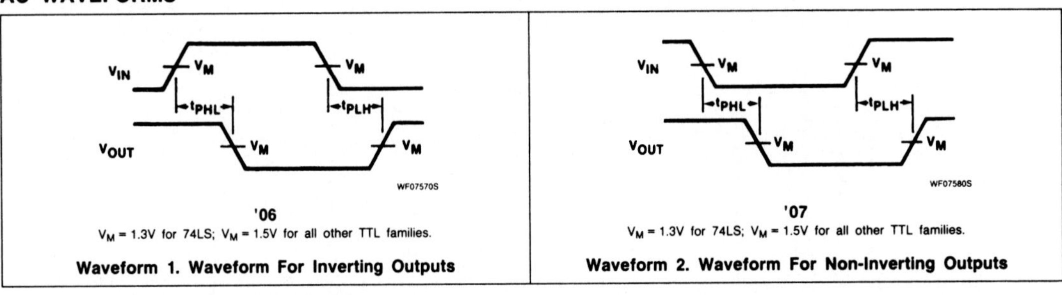

'06
V_M = 1.3V for 74LS; V_M = 1.5V for all other TTL families.

Waveform 1. Waveform For Inverting Outputs

'07
V_M = 1.3V for 74LS; V_M = 1.5V for all other TTL families.

Waveform 2. Waveform For Non-Inverting Outputs

AC ELECTRICAL CHARACTERISTICS T_A = 25°C, V_{CC} = 5.0V

	PARAMETER	TEST CONDITIONS	7406 C_L = 15pF, R_L = 110Ω Min	Max	7407 C_L = 15pF, R_L = 110Ω Min	Max	UNIT
t_{PLH} t_{PHL}	Propagation delay	Waveform 1, '06 Waveform 2, '07		15 23		10 30	ns

Signetics

7408, LS08, S08
Gates

Quad Two-Input AND Gate
Product Specification

Logic Products

TYPE	TYPICAL PROPAGATION DELAY	TYPICAL SUPPLY CURRENT (TOTAL)
7408	15ns	16mA
74LS08	9ns	3.4mA
74S08	5ns	25mA

ORDERING CODE

PACKAGES	COMMERCIAL RANGE V_{CC} = 5V ±5%; T_A = 0°C to +70°C
Plastic DIP	N7408N, N74LS08N, N74S08N
Plastic SO	N74LS08N, N74S08N

NOTE:
For information regarding devices processed to Military Specifications, see the Signetics Military Products Data Manual.

FUNCTION TABLE

INPUTS		OUTPUT
A	B	Y
L	L	L
L	H	L
H	L	L
H	H	H

H = HIGH voltage level
L = LOW voltage level

INPUT AND OUTPUT LOADING AND FAN-OUT TABLE

PINS	DESCRIPTION	74	74S	74LS
A, B	Inputs	1ul	1Sul	1LSul
Y	Output	10ul	10Sul	10LSul

NOTE:
Where a 74 unit load (ul) is understood to be 40μA I_{IH} and −1.6mA I_{IL}, a 74S unit load (Sul) is 50μA I_{IH} and −2.0mA I_{IL}, and 74LS unit load (LSul) is 20μA I_{IH} and −0.4mA I_{IL}.

PIN CONFIGURATION

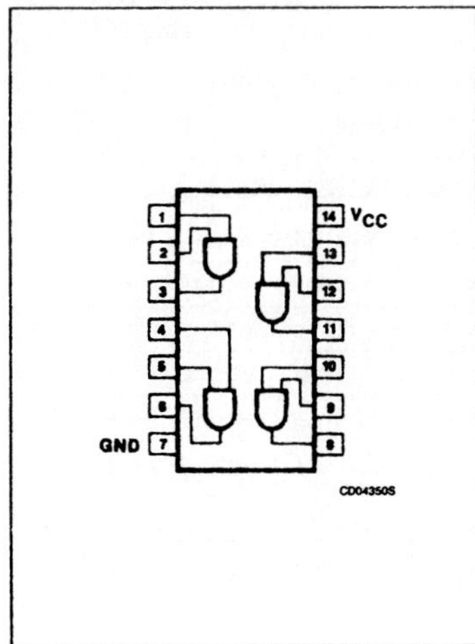

LOGIC SYMBOL

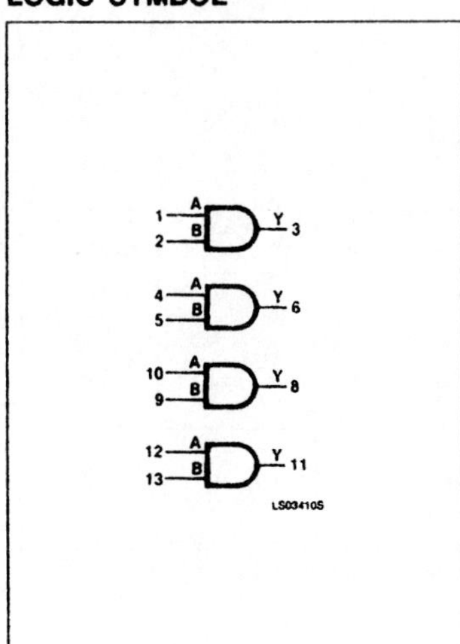

LOGIC SYMBOL (IEEE/IEC)

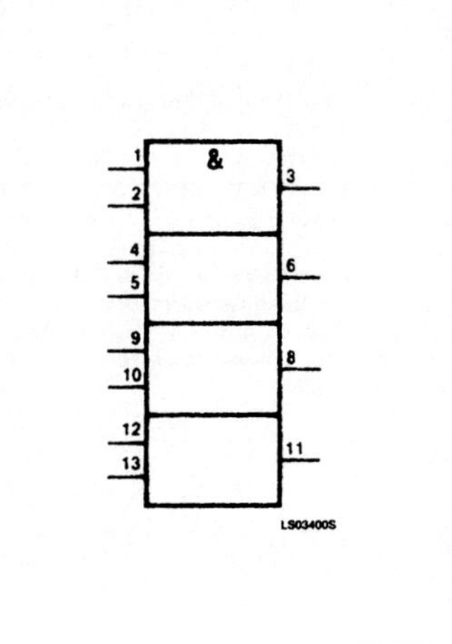

Gates 7408, LS08, S08

ABSOLUTE MAXIMUM RATINGS (Over operating free-air temperature range unless otherwise noted.)

PARAMETER		74	74LS	74S	UNIT
V_{CC}	Supply voltage	7.0	7.0	7.0	V
V_{IN}	Input voltage	−0.5 to +5.5	−0.5 to +7.0	−0.5 to +5.5	V
I_{IN}	Input current	−30 to +5	−30 to +1	−30 to +5	mA
V_{OUT}	Voltage applied to output in HIGH output state	−0.5 to $+V_{CC}$	−0.5 to $+V_{CC}$	−0.5 to $+V_{CC}$	V
T_A	Operating free-air temperature range	0 to 70			°C

RECOMMENDED OPERATING CONDITIONS

PARAMETER		74			74LS			74S			UNIT
		Min	Nom	Max	Min	Nom	Max	Min	Nom	Max	
V_{CC}	Supply voltage	4.75	5.0	5.25	4.75	5.0	5.25	4.75	5.0	5.25	V
V_{IH}	HIGH-level input voltage	2.0			2.0			2.0			V
V_{IL}	LOW-level input voltage			+0.8			+0.8			+0.8	V
I_{IK}	Input clamp current			−12			−18			−18	mA
I_{OH}	HIGH-level output current			−800			−400			−1000	μA
I_{OL}	LOW-level output current			16			8			20	mA
T_A	Operating free-air temperature	0		70	0		70	0		70	°C

TEST CIRCUITS AND WAVEFORMS

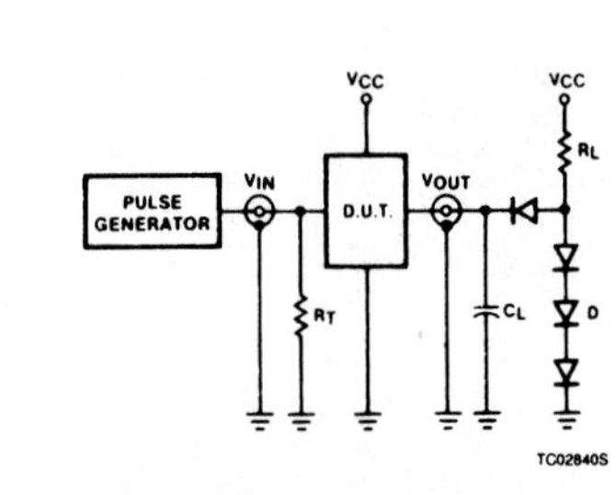

Test Circuit For 74 Totem-Pole Outputs

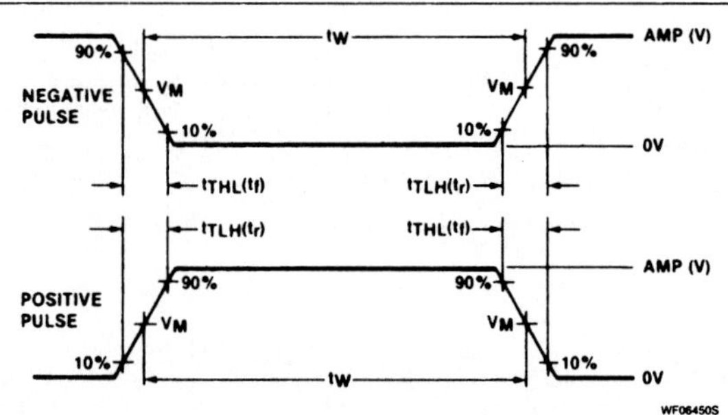

V_M = 1.3V for 74LS; V_M = 1.5V for all other TTL families.

Input Pulse Definition

DEFINITIONS
R_L = Load resistor to V_{CC}; see AC CHARACTERISTICS for value.
C_L = Load capacitance includes jig and probe capacitance; see AC CHARACTERISTICS for value.
R_T = Termination resistance should be equal to Z_{OUT} of Pulse Generators.
D = Diodes are 1N916, 1N3064, or equivalent.
t_{TLH}, t_{THL} Values should be less than or equal to the table entries.

FAMILY	INPUT PULSE REQUIREMENTS				
	Amplitude	Rep. Rate	Pulse Width	t_{TLH}	t_{THL}
74	3.0V	1MHz	500ns	7ns	7ns
74LS	3.0V	1MHz	500ns	15ns	6ns
74S	3.0V	1MHz	500ns	2.5ns	2.5ns

DC ELECTRICAL CHARACTERISTICS (Over recommended operating free-air temperature range unless otherwise noted.)

PARAMETER		TEST CONDITIONS[1]		7408			74LS08			74S08			UNIT
				Min	Typ[2]	Max	Min	Typ[2]	Max	Min	Typ[2]	Max	
V_{OH}	HIGH-level output voltage	V_{CC} = MIN, V_{IH} = MIN, I_{OH} = MAX		2.4	3.4		2.7	3.4		2.7	3.4		V
V_{OL}	LOW-level output voltage	V_{CC} = MIN, V_{IL} = MAX	I_{OL} = MAX		0.2	0.4		0.35	0.5			0.5	V
			I_{OL} = 4mA (74LS)					0.25	0.4				V
V_{IK}	Input clamp voltage	V_{CC} = MIN, $I_I = I_{IK}$				−1.5			−1.5			−1.2	V
I_I	Input current at maximum input voltage	V_{CC} = MAX	V_I = 5.5V			1.0						1.0	mA
			V_I = 7.0V						0.1				mA
I_{IH}	HIGH-level input current	V_{CC} = MAX	V_I = 2.4V			40							μA
			V_I = 2.7V						20			50	μA
I_{IL}	LOW-level input current	V_{CC} = MAX	V_I = 0.4V			−1.6			−0.4				mA
			V_I = 0.5V									−2.0	mA
I_{OS}	Short-circuit output current[3]	V_{CC} = MAX		−18		−55	−20		−100	−40		−100	mA
I_{CC}	Supply current (total)	V_{CC} = MAX	I_{CCH} Outputs HIGH		11	21		2.4	4.8		18	32	mA
			I_{CCL} Outputs LOW		20	33		4.4	8.8		32	57	mA

NOTES:
1. For conditions shown as MIN or MAX, use the appropriate value specified under recommended operating conditions for the applicable type.
2. All typical values are at V_{CC} = 5V, T_A = 25°C.
3. I_{OS} is tested with V_{OUT} = +0.5V and $V_{CC} = V_{CC}$ MAX + 0.5V. Not more than one output should be shorted at a time and duration of the short circuit should not exceed one second.

AC WAVEFORM

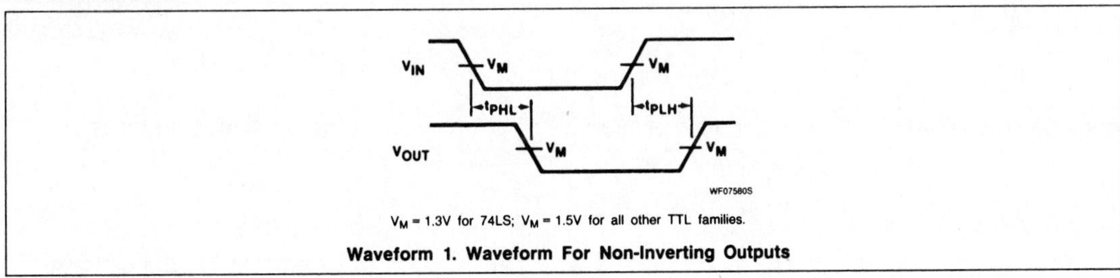

V_M = 1.3V for 74LS; V_M = 1.5V for all other TTL families.

Waveform 1. Waveform For Non-Inverting Outputs

AC ELECTRICAL CHARACTERISTICS T_A = 25°C, V_{CC} = 5.0V

PARAMETER		TEST CONDITIONS	74		74LS		74S		UNIT
			C_L = 15pF, R_L = 400Ω		C_L = 15pF, R_L = 2kΩ		C_L = 15pF, R_L = 280Ω		
			Min	Max	Min	Max	Min	Max	
t_{PLH} t_{PHL}	Propagation delay	Waveform 1		27 19		15 20		7.0 7.5	ns

Signetics

Logic Products

7410, 7411, LS10, LS11, S10, S11 Gates

Triple Three-Input NAND ('10), AND ('11) Gates
Product Specification

TYPE	TYPICAL PROPAGATION DELAY	TYPICAL SUPPLY CURRENT (TOTAL)
7410	9ns	6mA
74LS10	10ns	1.2mA
74S10	3ns	12mA
7411	10ns	11mA
74LS11	9ns	2.6mA
74S11	5ns	19mA

ORDERING CODE

PACKAGES	COMMERCIAL RANGE V_{CC} = 5V ±5%; T_A = 0°C to +70°C
Plastic DIP '10	N7410N, N74LS10N, N74S10N
'11	N7411N, N74LS11N, N74S11N
Plastic SO '10	N74LS10D, N74S10D
Plastic SO '11	N74LS11D, N74S11D

NOTE:
For information regarding devices processed to Military Specifications, see the Signetics Military Products Data Manual.

FUNCTION TABLE

INPUTS			OUTPUTS	
A	B	C	Y('10)	Y('11)
L	L	L	H	L
L	L	H	H	L
L	H	L	H	L
L	H	H	H	L
H	L	L	H	L
H	L	H	H	L
H	H	L	H	L
H	H	H	L	H

H = HIGH voltage level
L = LOW voltage level

INPUT AND OUTPUT LOADING AND FAN-OUT TABLE

PINS	DESCRIPTION	74	74S	74LS
A – C	Inputs	1ul	1Sul	1LSul
Y	Output	10ul	10Sul	10LSul

NOTE:
Where a 74 unit load (ul) is understood to be 40μA I_{IH} and −1.6mA I_{IL}, a 74S unit load (Sul) is 50μA I_{IH} and −2.0mA I_{IL}, and 74LS unit load (LSul) is 20μA I_{IH} and −0.4mA I_{IL}.

PIN CONFIGURATION

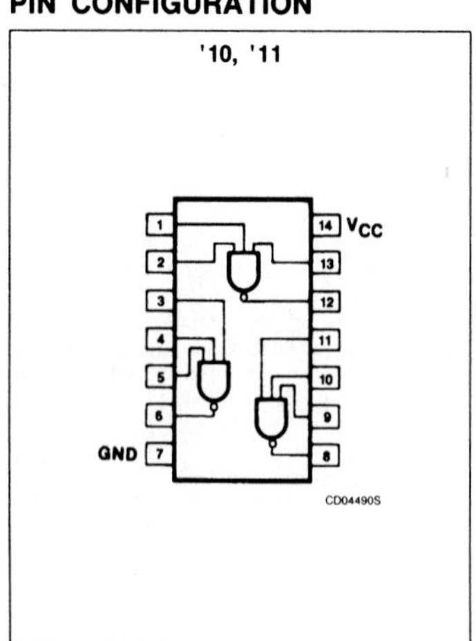

LOGIC SYMBOL

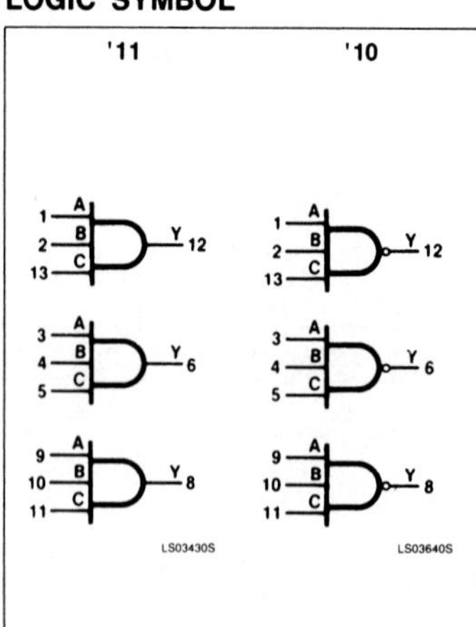

LOGIC SYMBOL (IEEE/IEC)

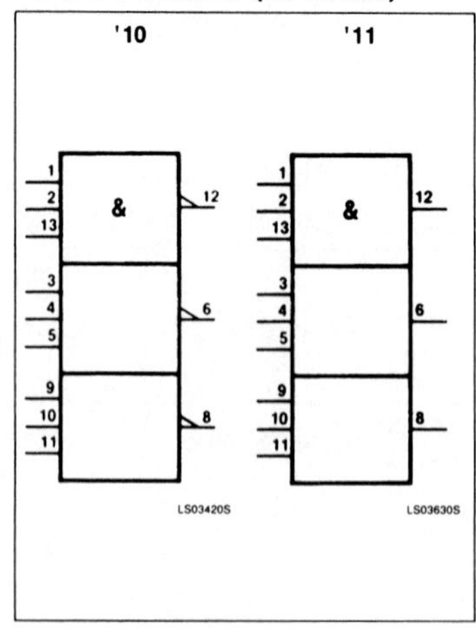

Signetics

7414, LS14
Schmitt Triggers

Hex Inverter Schmitt Trigger
Product Specification

Logic Products

DESCRIPTION

The '14 contains six logic inverters which accept standard TTL input signals and provide standard TTL output levels. They are capable of transforming slowly changing input signals into sharply defined, jitter-free output signals. In addition, they have greater noise margin than conventional inverters.

Each circuit contains a Schmitt trigger followed by a Darlington level shifter and a phase splitter driving a TTL totem-pole output. The Schmitt trigger uses positive feedback to effectively speed-up slow input transition, and provide different input threshold voltages for positive and negative-going transitions. This hysteresis between the positive-going and negative-going input thresholds (typically 800mV) is determined internally by resistor ratios and is essentially insensitive to temperature and supply voltage variations.

TYPE	TYPICAL PROPAGATION DELAY	TYPICAL SUPPLY CURRENT (TOTAL)
7414	15ns	31mA
74LS14	15ns	10mA

ORDERING CODE

PACKAGES	COMMERCIAL RANGE $V_{CC} = 5V \pm 5\%$; $T_A = 0°C$ to $+70°C$
Plastic DIP	N7414N, N74LS14N
Plastic SO	N74LS14D

NOTE:
For information regarding devices processed to Military Specifications, see the Signetics Military Products Data Manual.

INPUT AND OUTPUT LOADING AND FAN-OUT TABLE

PINS	DESCRIPTION	74	74LS
A	Inputs	1ul	1LSul
Y	Output	10ul	10LSul

NOTE:
Where a 74 unit load (ul) is understood to be 40μA I_{IH} and −1.6mA I_{IL}, and 74LS unit load (LSul) is 20μA I_{IH} and −0.4mA I_{IL}.

PIN CONFIGURATION

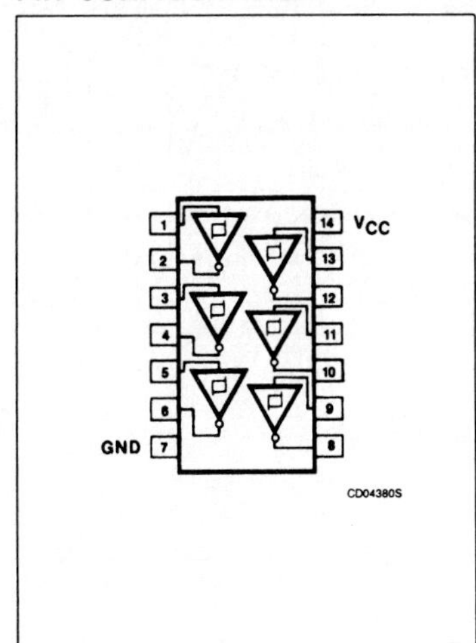

LOGIC SYMBOL

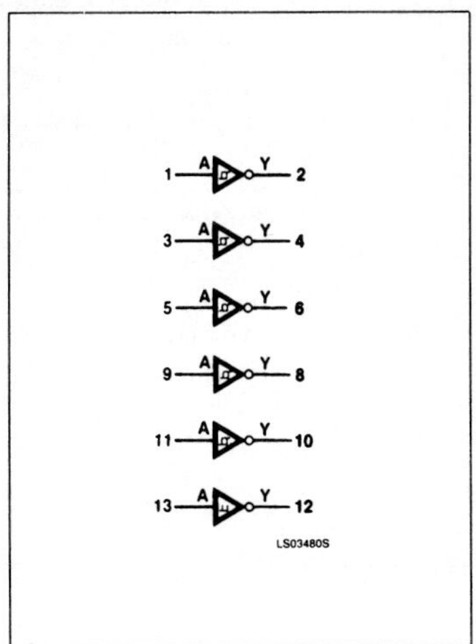

LOGIC SYMBOL (IEEE/IEC)

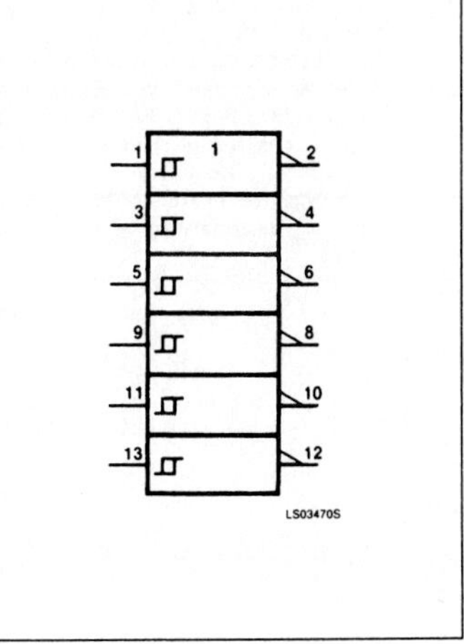

Schmitt Triggers — 7414, LS14

ABSOLUTE MAXIMUM RATINGS (Over operating free-air temperature range unless otherwise noted.)

PARAMETER		74	74LS	UNIT
V_{CC}	Supply voltage	7.0	7.0	V
V_{IN}	Input voltage	−0.5 to +5.5	−0.5 to +7.0	V
I_{IN}	Input current	−30 to +5	−30 to +1	mA
V_{OUT}	Voltage applied to output in HIGH output state	−0.5 to $+V_{CC}$	−0.5 to $+V_{CC}$	V
T_A	Operating free-air temperature range	0 to 70		°C

RECOMMENDED OPERATING CONDITIONS

PARAMETER		74			74LS			UNIT
		Min	Nom	Max	Min	Nom	Max	
V_{CC}	Supply voltage	4.75	5.0	5.25	4.75	5.0	5.25	V
I_{IK}	Input clamp current			−12			−18	mA
I_{OH}	HIGH-level output current			−800			−400	μA
I_{OL}	LOW-level output current			16			8	mA
T_A	Operating free-air temperature	0		70	0		70	°C

TEST CIRCUITS AND WAVEFORMS

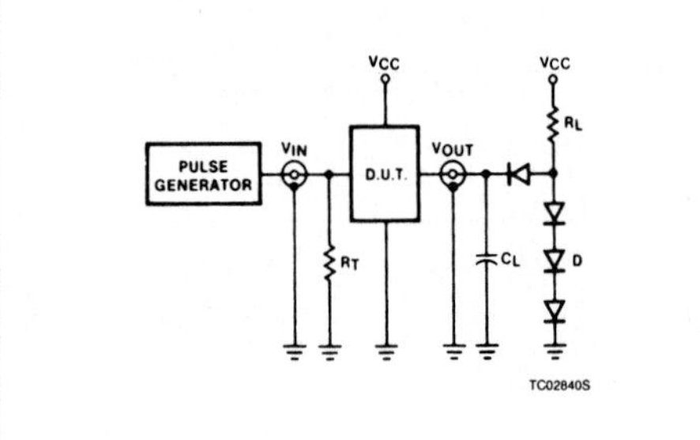

Test Circuit For 74 Totem-Pole Outputs

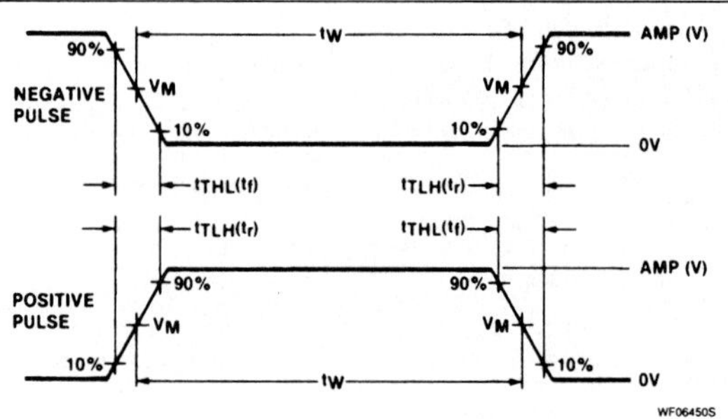

V_M = 1.3V for 74LS; V_M = 1.5V for all other TTL families.

Input Pulse Definition

DEFINITIONS
R_L = Load resistor to V_{CC}; see AC CHARACTERISTICS for value.
C_L = Load capacitance includes jig and probe capacitance; see AC CHARACTERISTICS for value.
R_T = Termination resistance should be equal to Z_{OUT} of Pulse Generators.
D = Diodes are 1N916, 1N3064, or equivalent.
t_{TLH}, t_{THL} Values should be less than or equal to the table entries.

FAMILY	INPUT PULSE REQUIREMENTS				
	Amplitude	Rep. Rate	Pulse Width	t_{TLH}	t_{THL}
74	3.0V	1MHz	500ns	7ns	7ns
74LS	3.0V	1MHz	500ns	15ns	6ns
74S	3.0V	1MHz	500ns	2.5ns	2.5ns

DC ELECTRICAL CHARACTERISTICS (Over recommended operating free-air temperature range unless otherwise noted.)

PARAMETER		TEST CONDITIONS[1]	7414 Min	7414 Typ[2]	7414 Max	74LS14 Min	74LS14 Typ[2]	74LS14 Max	UNIT
V_{T+}	Positive-going threshold	$V_{CC} = 5.0V$	1.5	1.7	2.0	1.4	1.6	1.9	V
V_{T-}	Negative-going threshold	$V_{CC} = 5.0V$	0.6	0.9	1.1	0.5	0.8	1.0	V
ΔV_T	Hysteresis ($V_{T+} - V_{T-}$)	$V_{CC} = 5.0V$	0.4	0.8		0.4	0.8		V
V_{OH}	HIGH-level output voltage	$V_{CC} = MIN$, $V_I = V_{T-MIN}$, $I_{OH} = MAX$	2.4	3.4		2.7	3.4		V
V_{OL}	LOW-level output voltage	$V_{CC} = MIN$, $V_I = V_{T+MAX}$; $I_{OL} = MAX$		0.2	0.4		0.35	0.5	V
		$V_{CC} = MIN$, $V_I = V_{T+MAX}$; $I_{OL} = 4mA$ (74LS)					0.25	0.4	V
V_{IK}	Input clamp voltage	$V_{CC} = MIN$, $I_I = I_{IK}$			−1.5			−1.5	V
I_{T+}	Input current at positive-going threshold	$V_{CC} = 5.0V$, $V_I = V_{T+}$		−0.43			−0.14		mA
I_{T-}	Input current at negative-going threshold	$V_{CC} = 5.0V$, $V_I = V_{T-}$		−0.56			−0.18		mA
I_I	Input current at maximum input voltage	$V_{CC} = MAX$; $V_I = 5.5V$			1.0				mA
		$V_{CC} = MAX$; $V_I = 7.0V$						0.1	mA
I_{IH}	HIGH-level input current	$V_{CC} = MAX$; $V_I = 2.4V$			40				µA
		$V_{CC} = MAX$; $V_I = 2.7V$						20	µA
I_{IL}	LOW-level input current	$V_{CC} = MAX$, $V_I = 0.4V$			−1.2			−0.4	mA
I_{OS}	Short-circuit output current[3]	$V_{CC} = MAX$	−18		−55	−20		−100	mA
I_{CC}	Supply current (total)	$V_{CC} = MAX$; I_{CCH} Outputs HIGH		22	36		8.6	16	mA
		$V_{CC} = MAX$; I_{CCL} Outputs LOW		39	60		12	21	mA

NOTES:

1. For conditions shown as MIN or MAX, use the appropriate value specified under recommended operating conditions for the applicable type.
2. All typical values are at $V_{CC} = 5V$, $T_A = 25°C$.
3. I_{OS} is tested with $V_{OUT} = +0.5V$ and $V_{CC} = V_{CC}$ MAX + 0.5V. Not more than one output should be shorted at a time and duration of the short circuit should not exceed one second.

FUNCTION TABLE

INPUT	OUTPUT
A	Y
0	1
1	0

AC WAVEFORM

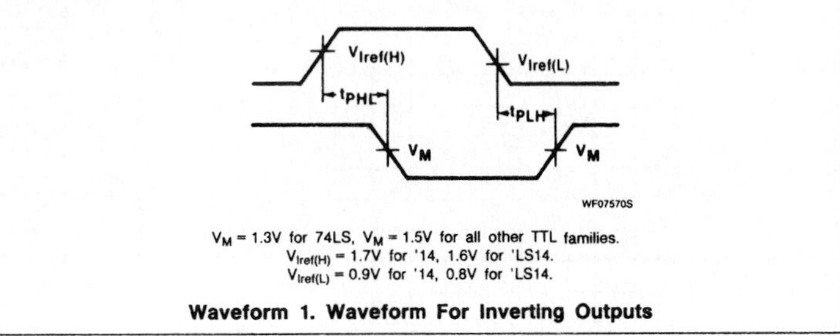

$V_M = 1.3V$ for 74LS, $V_M = 1.5V$ for all other TTL families.
$V_{Iref(H)} = 1.7V$ for '14, 1.6V for 'LS14.
$V_{Iref(L)} = 0.9V$ for '14, 0.8V for 'LS14.

Waveform 1. Waveform For Inverting Outputs

Schmitt Triggers 7414, LS14

AC ELECTRICAL CHARACTERISTICS $T_A = 25°C$, $V_{CC} = 5.0V$

PARAMETER		TEST CONDITIONS	74		74LS		UNIT
			$C_L = 15pF$, $R_L = 400\Omega$		$C_L = 15pF$, $R_L = 2k\Omega$		
			Min	Max	Min	Max	
t_{PLH} t_{PHL}	Propagation delay	Waveform 1		22 22		22 22	ns

TYPICAL PERFORMANCE CHARACTERISTICS

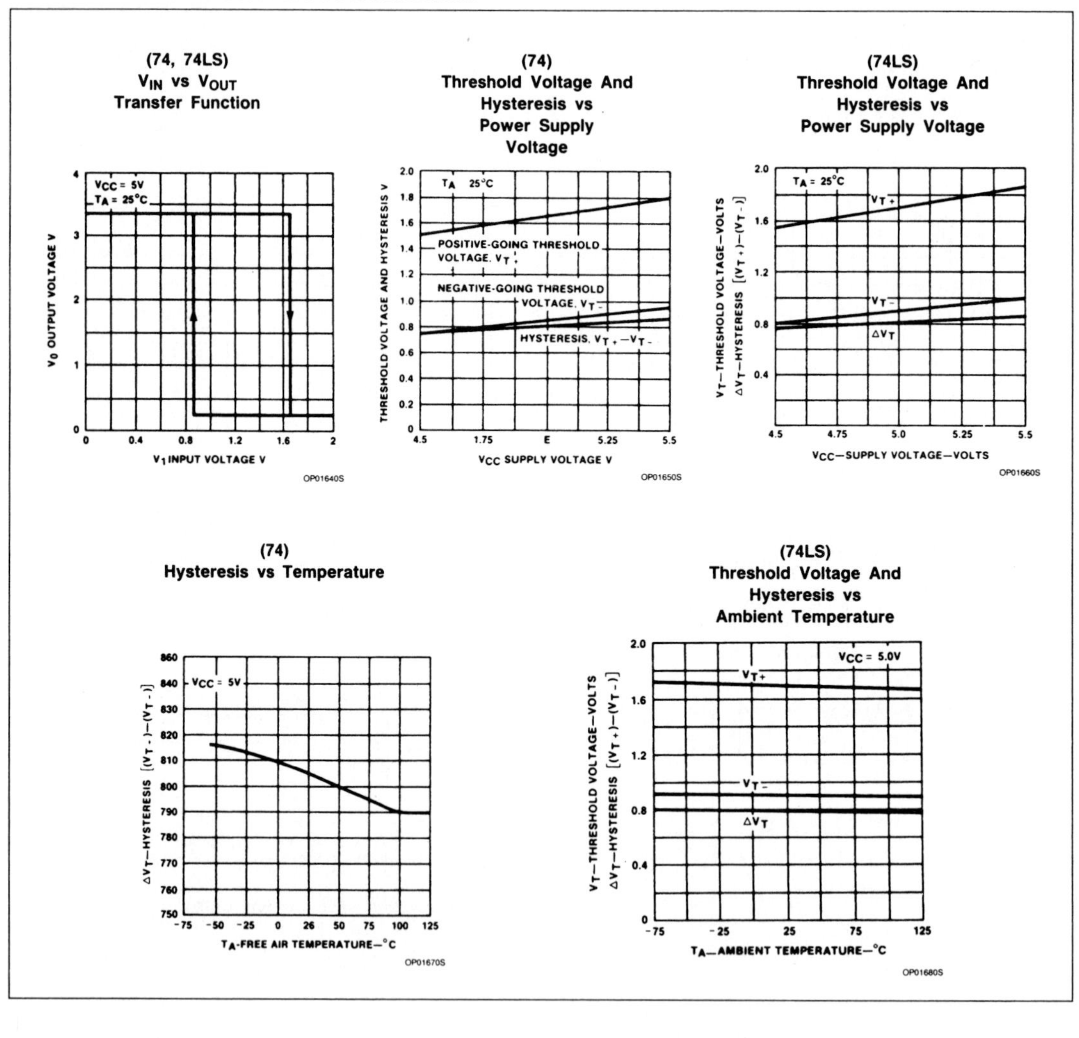

Signetics

Logic Products

7420, 7421, LS20, LS21, S20
Gates

Dual Four-Input NAND ('20) AND ('21) Gate
Product Specification

TYPE	TYPICAL PROPAGATION DELAY	TYPICAL SUPPLY CURRENT (TOTAL)
7420	10ns	8mA
74LS20	10ns	0.8mA
74S20	3ns	8mA
7421	12ns	8mA
74LS21	9ns	1.7mA

ORDERING CODE

PACKAGES	COMMERCIAL RANGE $V_{CC} = 5V \pm 5\%$; $T_A = 0°C$ to $+70°C$
Plastic DIP '20	N7420N, N74LS20N, N74S20N
'21	N7421N, N74LS21N
Plastic SO	N74LS20D, N74S20D, N74LS21D

NOTE:
For information regarding devices processed to Military Specifications, see the Signetics Military Products Data Manual.

FUNCTION TABLE

INPUTS				OUTPUTS	
A	B	C	D	Y('20)	Y('21)
L	X	X	X	H	L
X	L	X	X	H	L
X	X	L	X	H	L
X	X	X	L	H	L
H	H	H	H	L	H

H = HIGH voltage level
L = LOW voltage level
X = Don't care

INPUT AND OUTPUT LOADING AND FAN-OUT TABLE

PINS	DESCRIPTION	74	74S	74LS
A – D	Inputs	1ul	1Sul	1LSul
Y	Output	10ul	10Sul	10LSul

NOTE:
Where a 74 unit load (ul) is understood to be 40μA I_{IH} and −1.6mA I_{IL}, a 74S unit load (Sul) is 50μA I_{IH} and −2.0mA I_{IL}, and 74LS unit load (LSul) is 20μA I_{IH} and −0.4mA I_{IL}.

PIN CONFIGURATION

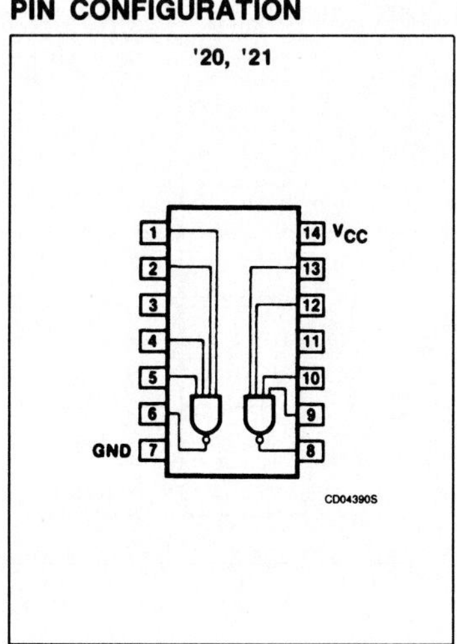

LOGIC SYMBOL

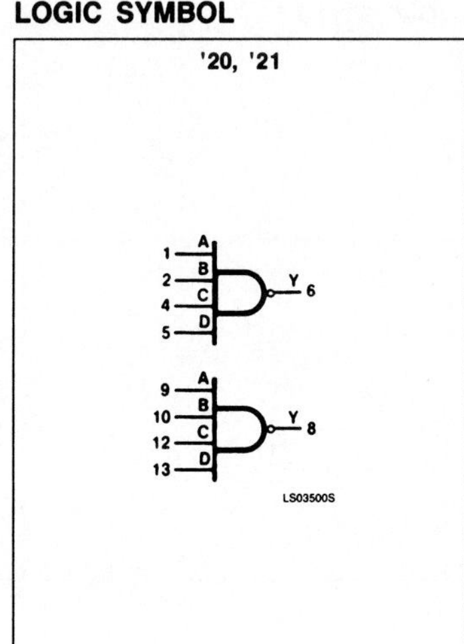

LOGIC SYMBOL (IEEE/IEC)

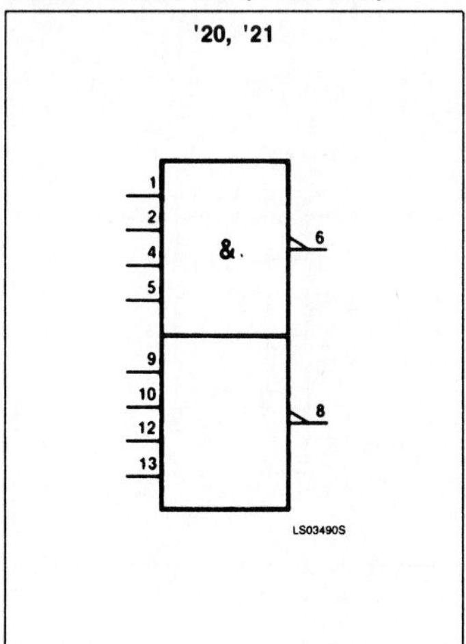

Signetics

7427, LS27
Gates

Triple Three-Input NOR Gate
Product Specification

Logic Products

TYPE	TYPICAL PROPAGATION DELAY	TYPICAL SUPPLY CURRENT (TOTAL)
7427	9ns	13mA
74LS27	10ns	2.7mA

ORDERING CODE

PACKAGES	COMMERCIAL RANGE $V_{CC} = 5V \pm 5\%$; $T_A = 0°C$ to $+70°C$
Plastic DIP	N7427N, N74LS27N
Plastic SO	N74LS27D

NOTE:
For information regarding devices processed to Military Specifications, see the Signetics Military Products Data Manual.

FUNCTION TABLE

INPUTS			OUTPUT
A	B	C	Y
L	L	L	H
X	X	H	L
X	H	X	L
H	X	X	L

H = HIGH voltage level
L = LOW voltage level
X = Don't care

INPUT AND OUTPUT LOADING AND FAN-OUT TABLE

PINS	DESCRIPTION	74	74LS
A – C	Inputs	1ul	1LSul
Y	Output	10ul	10LSul

NOTE:
Where a 74 unit load (ul) is understood to be 40μA I_{IH} and −1.6mA I_{IL}, a 74LS unit load (LSul) is 20μA I_{IH} and −0.4mA I_{IL}.

PIN CONFIGURATION

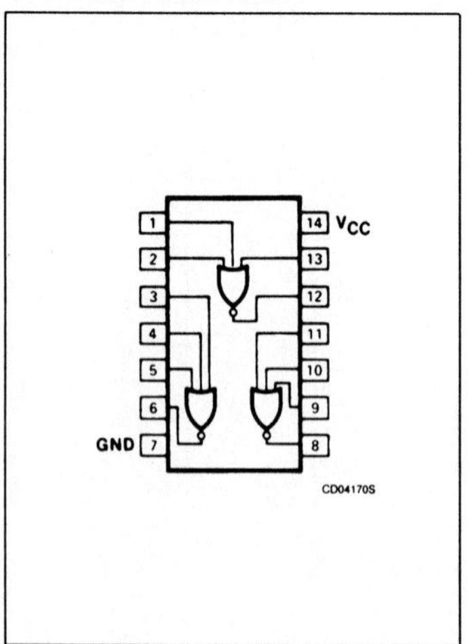

LOGIC SYMBOL

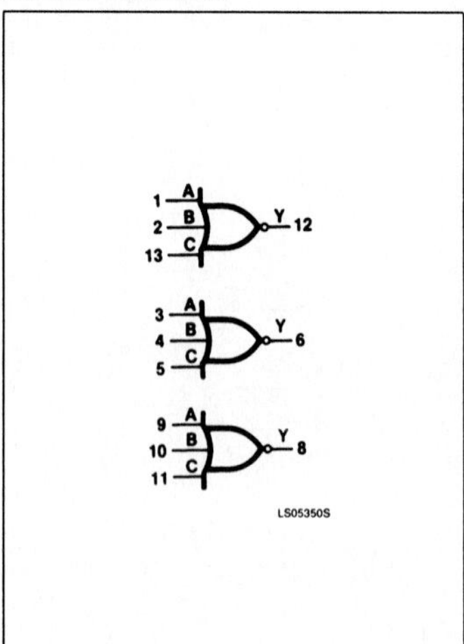

LOGIC SYMBOL (IEEE/IEC)

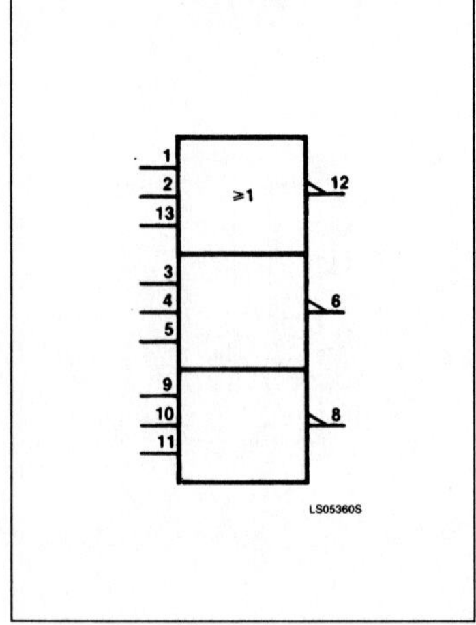

Signetics

Logic Products

7430, LS30
Gates

Eight-Input NAND Gate
Product Specification

TYPE	TYPICAL PROPAGATION DELAY	TYPICAL SUPPLY CURRENT (TOTAL)
7430	11ns	2mA
74LS30	11ns	0.5mA

ORDERING CODE

PACKAGES	COMMERCIAL RANGE $V_{CC} = 5V \pm 5\%$; $T_A = 0°C$ to $+70°C$
Plastic DIP	N7430N, N74LS30N
Plastic SO	N74LS30D

NOTE:
For information regarding devices processed to Military Specifications, see the Signetics Military Products Data Manual.

FUNCTION TABLE

INPUTS								OUTPUT
A	B	C	D	E	F	G	H	Y
L	X	X	X	X	X	X	X	H
X	L	X	X	X	X	X	X	H
X	X	L	X	X	X	X	X	H
X	X	X	L	X	X	X	X	H
X	X	X	X	L	X	X	X	H
X	X	X	X	X	L	X	X	H
X	X	X	X	X	X	L	X	H
X	X	X	X	X	X	X	L	H
H	H	H	H	H	H	H	H	L

H = HIGH voltage level
L = LOW voltage level
X = Don't care

INPUT AND OUTPUT LOADING AND FAN-OUT TABLE

PINS	DESCRIPTION	74	74LS
A – H	Inputs	1ul	1LSul
Y	Output	10ul	10LSul

NOTE:
Where a 74 unit load (ul) is understood to be 40μA I_{IH} and −1.6mA I_{IL}, and a 74LS unit load (LSul) is 20μA I_{IH} and −0.4mA I_{IL}.

PIN CONFIGURATION

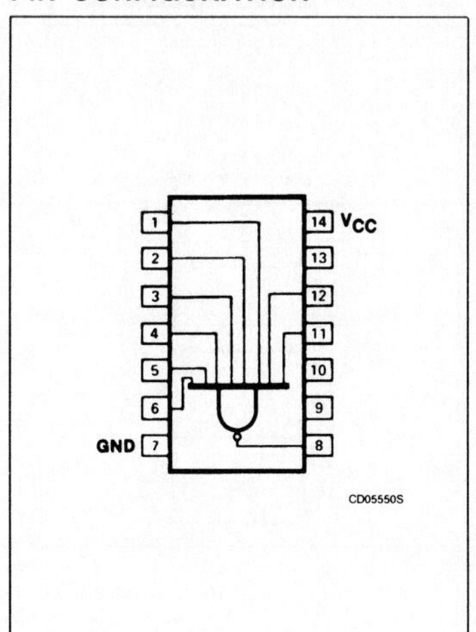

LOGIC SYMBOL

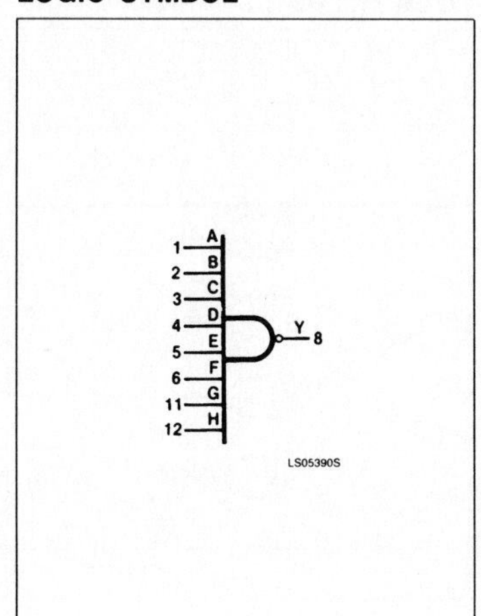

LOGIC SYMBOL (IEEE/IEC)

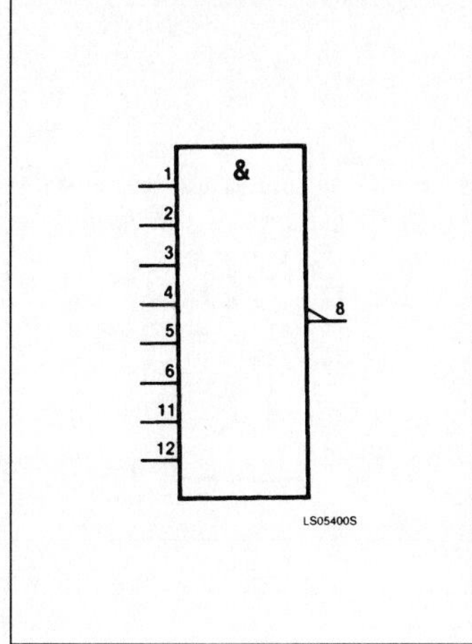

Signetics

7432, LS32, S32
Gates

Quad Two-Input OR Gate
Product Specification

Logic Products

TYPE	TYPICAL PROPAGATION DELAY	TYPICAL SUPPLY CURRENT (TOTAL)
7432	12ns	19mA
74LS32	14ns	4.0mA
74S32	4ns	28mA

ORDERING CODE

PACKAGES	COMMERCIAL RANGE $V_{CC} = 5V \pm 5\%$; $T_A = 0°C$ to $+70°C$
Plastic DIP	N7432N, N74LS32N, N74S32N
Plastic SO – 14	N74LS32D, N74S32D

NOTE:
For information regarding devices processed to Military Specifications, see the Signetics Military Products Data Manual.

FUNCTION TABLE

INPUTS		OUTPUT
A	B	Y
L	L	L
L	H	H
H	L	H
H	H	H

H = HIGH voltage level
L = LOW voltage level

INPUT AND OUTPUT LOADING AND FAN-OUT TABLE

PINS	DESCRIPTION	74	74S	74LS
A, B	Inputs	1ul	1Sul	1LSul
Y	Output	10ul	10Sul	10LSul

NOTE:
Where a 74 unit load (ul) is understood to be $40\mu A$ I_{IH} and $-1.6mA$ I_{IL}, and a 74S unit load (Sul) is $50\mu A$ I_{IH} and $-2.0mA$ I_{IL}, and a 74LS unit load (LSul) is $20\mu A$ I_{IH} and $-0.4mA$ I_{IL}

PIN CONFIGURATION

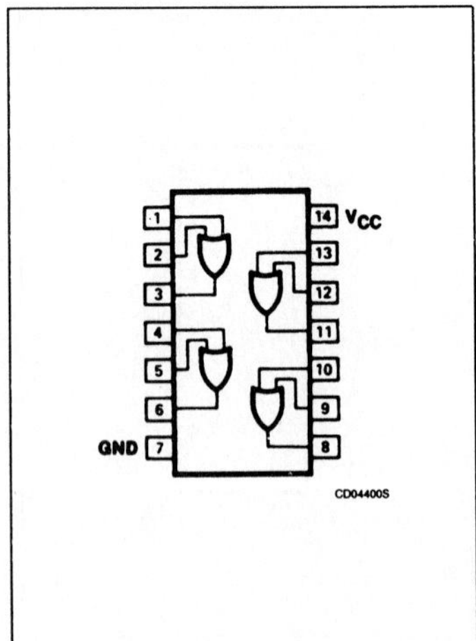

LOGIC SYMBOL

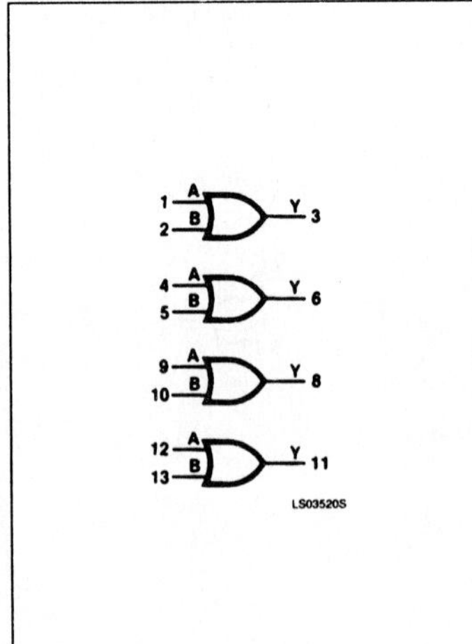

LOGIC SYMBOL (IEEE/IEC)

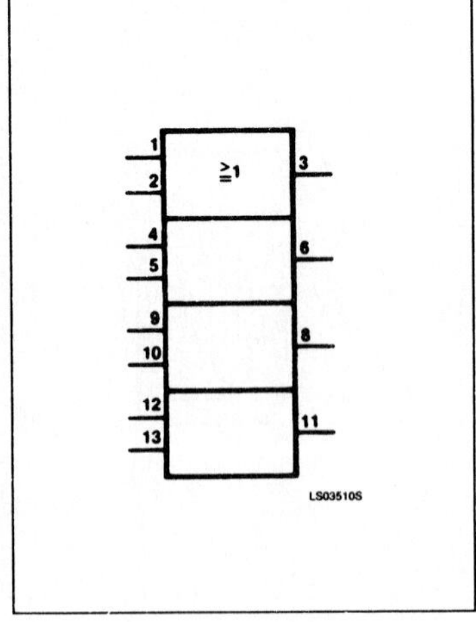

Signetics

7474, LS74A, S74
Flip-Flops

Dual D-Type Flip-Flop
Product Specification

Logic Products

DESCRIPTION

The '74 is a dual positive edge-triggered D-type flip-flop featuring individual Data, Clock, Set and Reset inputs; also complementary Q and $\overline{Q}$ outputs.

Set ($\overline{S}_D$) and Reset ($\overline{R}_D$) are asynchronous active-LOW inputs and operate independently of the Clock input. Information on the Data (D) input is transferred to the Q output on the LOW-to-HIGH transition of the clock pulse. The D inputs must be stable one set-up time prior to the LOW-to-HIGH clock transition for predictable operation. Although the Clock input is level-sensitive, the positive transition of the clock pulse between the 0.8V and 2.0V levels should be equal to or less than the clock-to-output delay time for reliable operation.

TYPE	TYPICAL f_{MAX}	TYPICAL SUPPLY CURRENT (TOTAL)
7474	25MHz	17mA
74LS74A	33MHz	4mA
74S74	100MHz	30mA

NOTE:
For information regarding devices processed to Military Specifications, see the Signetics Military Products Data Manual.

ORDERING CODE

PACKAGES	COMMERCIAL RANGE $V_{CC} = 5V \pm 5\%$; $T_A = 0°C$ to $+70°C$
Plastic DIP	N7474N, N74LS74AN, N74S74N
Plastic SO	N741S74A, N74S74D

NOTE:
For information regarding devices processed to Military Specifications, see the Signetics Military Products Data Manual.

INPUT AND OUTPUT LOADING AND FAN-OUT TABLE

PINS	DESCRIPTION	74	74S	74LS
D	Input	1ul	1Sul	1LSul
$\overline{R}_D$	Input	2ul	3Sul	2LSul
$\overline{S}_D$	Input	1ul	2Sul	2LSul
CP	Input	2ul	2Sul	1LSul
Q,$\overline{Q}$	Outputs	10ul	10Sul	10LSul

NOTE:
Where a 74 unit load (ul) is understood to be 40μA I_{IH} and −1.6mA I_{IL}, a 74S unit load (Sul) is 50μA I_{IH} and −2.0mA I_{IL}, and 74LS unit load (LSul) is 20μA I_{IH} and −0.4mA I_{IL}.

PIN CONFIGURATION

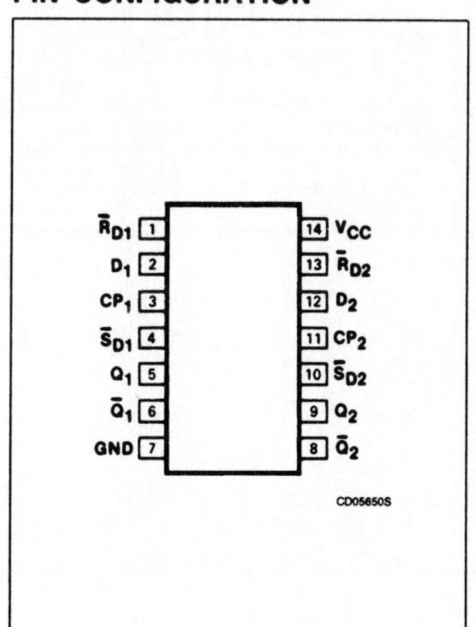

LOGIC SYMBOL

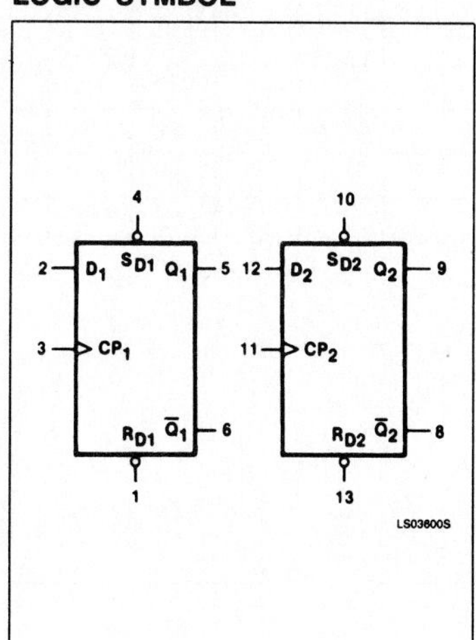

LOGIC SYMBOL (IEEE/IEC)

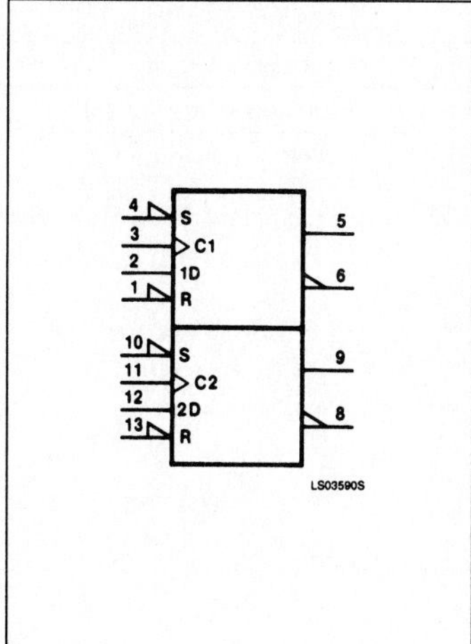

Flip-Flops 7474, LS74A, S74

LOGIC DIAGRAM

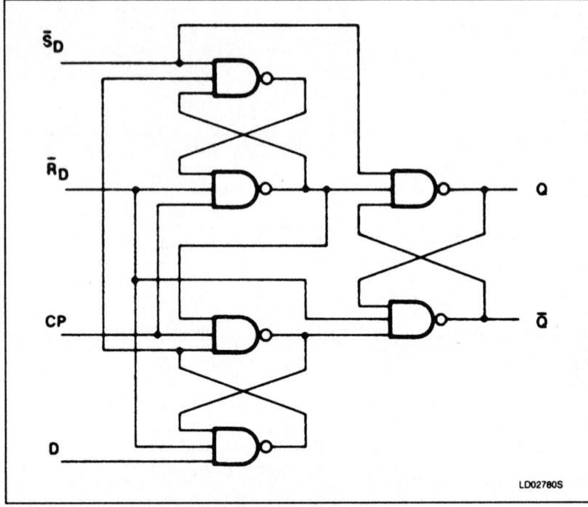

MODE SELECT — FUNCTION TABLE

OPERATING MODE	INPUTS				OUTPUTS	
	$\bar{S}_D$	$\bar{R}_D$	CP	D	Q	$\bar{Q}$
Asynchronous Set	L	H	X	X	H	L
Asynchronous Reset (Clear)	H	L	X	X	L	H
Undetermined(1)	L	L	X	X	H	H
Load "1" (Set)	H	H	↑	h	H	L
Load "0" (Reset)	H	H	↑	l	L	H

H = HIGH voltage level steady state.
h = HIGH voltage level one set-up time prior to the LOW-to-HIGH clock transition.
L = LOW voltage level steady state.
l = LOW voltage level one set-up time prior to the LOW-to-HIGH clock transition.
X = Don't care.
↑ = LOW-to-HIGH clock transition.

NOTE:
(1) Both outputs will be HIGH while both $\bar{S}_D$ and $\bar{R}_D$ are LOW, but the output states are unpredictable if $\bar{S}_D$ and $\bar{R}_D$ go HIGH simultaneously.

ABSOLUTE MAXIMUM RATINGS (Over operating free-air temperature range unless otherwise noted.)

PARAMETER		74	74LS	74S	UNIT
V_{CC}	Supply voltage	7.0	7.0	7.0	V
V_{IN}	Input voltage	−0.5 to +5.5	−0.5 to +7.0	−0.5 to +5.5	V
I_{IN}	Input current	−30 to +5	−30 to +1	−30 to +5	mA
V_{OUT}	Voltage applied to output in HIGH output state	−0.5 to $+V_{CC}$	−0.5 to $+V_{CC}$	−0.5 to $+V_{CC}$	V
T_A	Operating free-air temperature range	0 to 70			°C

RECOMMENDED OPERATING CONDITIONS

PARAMETER		74			74LS			74S			UNIT
		Min	Nom	Max	Min	Nom	Max	Min	Nom	Max	
V_{CC}	Supply voltage	4.75	5.0	5.25	4.75	5.0	5.25	4.75	5.0	5.25	V
V_{IH}	HIGH-level input voltage	2.0			2.0			2.0			V
V_{IL}	LOW-level input voltage			+0.8			+0.8			+0.8	V
I_{IK}	Input clamp current			−12			−18			−18	mA
I_{OH}	HIGH-level output current			−400			−400			−1000	μA
I_{OL}	LOW-level output current			16			8			20	mA
T_A	Operating free-air temperature	0		70	0		70	0		70	°C

DC ELECTRICAL CHARACTERISTICS (Over recommended operating free-air temperature range unless otherwise noted.)

PARAMETER		TEST CONDITIONS[1]			7474			74LS74A			74S74			UNIT
					Min	Typ[2]	Max	Min	Typ[2]	Max	Min	Typ[2]	Max	
V_{OH}	HIGH-level output voltage	V_{CC} = MIN, V_{IH} = MIN, V_{IL} = MAX, I_{OH} = MAX			2.4	3.4		2.7	3.4		2.7	3.4		V
V_{OL}	LOW-level output voltage	V_{CC} = MIN, V_{IH} = MIN, V_{IL} = MAX	I_{OL} = MAX			0.2	0.4		0.35	0.5			0.5	V
			I_{OL} = 4mA (74LS)						0.25	0.4				V
V_{IK}	Input clamp voltage	V_{CC} = MIN, I_I = I_{IK}					−1.5			−1.5			−1.2	V
I_I	Input current at maximum input voltage	V_{CC} = MAX	V_I = 5.5V				1.0						1.0	mA
			V_I = 7.0V	D input						0.1				mA
				$\overline{R}_D$ input						0.2				mA
				$\overline{S}_D$ input						0.2				mA
				CP input						0.1				mA
I_{IH}	HIGH-level input current	V_{CC} = MAX	V_I = 2.4V	D input			40							μA
				$\overline{R}_D$ input			120							μA
				$\overline{S}_D$ input			80							μA
				CP input			80							μA
			V_I = 2.7V	D input						20			50	μA
				$\overline{R}_D$ input						40			150	μA
				$\overline{S}_D$ input						40			100	μA
				CP input						20			100	μA
I_{IL}	LOW-level input current[5]	V_{CC} = MAX	V_I = 0.4V	D input			−1.6			−0.4				mA
				$\overline{R}_D$ input			−3.2			−0.8				mA
				$\overline{S}_D$ input			−1.6			−0.8				mA
				CP input			−3.2			−0.4				mA
			V_I = 0.5V	D input									−2	mA
				$\overline{R}_D$ input									−6	mA
				$\overline{S}_D$ input									−4	mA
				CP input									−4	mA
I_{OS}	Short-circuit output current[3]	V_{CC} = MAX			−18		−57	−20		−100	−40		−100	mA
I_{CC}	Supply current[4] (total)	V_{CC} = MAX				17	30		4	8		30	50	mA

NOTES:

1. For conditions shown as MIN or MAX, use the appropriate value specified under recommended operating conditions for the applicable type.
2. All typical values are at V_{CC} = 5V, T_A = 25°C.
3. I_{OS} is tested with V_{OUT} = +0.5V and V_{CC} = V_{CC} MAX + 0.5V. Not more than one output should be shorted at a time and duration of the short circuit should not exceed one second.
4. Measure I_{CC} with the Clock inputs grounded and all outputs open, with the Q and $\overline{Q}$ outputs HIGH in turn.
5. Set is tested with reset HIGH and reset is tested with set HIGH.

Flip-Flops — 7474, LS74A, S74

AC ELECTRICAL CHARACTERISTICS $T_A = 25°C$, $V_{CC} = 5.0V$

PARAMETER		TEST CONDITIONS	74		74LS		74S		UNIT
			$C_L = 15pF$, $R_L = 400\Omega$		$C_L = 15pF$, $R_L = 2k\Omega$		$C_L = 15pF$, $R_L = 280\Omega$		
			Min	Max	Min	Max	Min	Max	
f_{MAX}	Maximum clock frequency	Waveform 1	15		25		75		MHz
t_{PLH} t_{PHL}	Propagation delay Clock to output	Waveform 1		25 40		25 40		9 9	ns
t_{PLH} t_{PHL}	Propagation delay Set or Reset to output	Waveform 2 Waveform 2 CP = HIGH		25 40		25 40		6 13.5	ns
t_{PHL}	Set or Reset to output	Waveform 2 CP = LOW		40		40		8	ns

NOTE:
Per industry convention, f_{MAX} is the worst case value of the maximum device operating frequency with no constraints on t_r, t_f, pulse width or duty cycle.

AC SET-UP REQUIREMENTS $T_A = 25°C$, $V_{CC} = 5.0V$

PARAMETER		TEST CONDITIONS	74		74LS		74S		UNIT
			Min	Max	Min	Max	Min	Max	
$t_W(H)$	Clock pulse width (HIGH)	Waveform 1	30		25		6		ns
$t_W(L)$	Clock pulse width (LOW)	Waveform 1	37				7.3		ns
$t_W(L)$	Set or reset pulse width (LOW)	Waveform 2	30		25		7		ns
$t_s(H)$	Set-up time (HIGH) data to clock	Waveform 1	20		20		3		ns
$t_s(L)$	Set-up time (LOW) data to clock	Waveform 1	20		20		3		ns
t_h	Hold time data to clock	Waveform 1	5		5		2		ns

AC WAVEFORMS

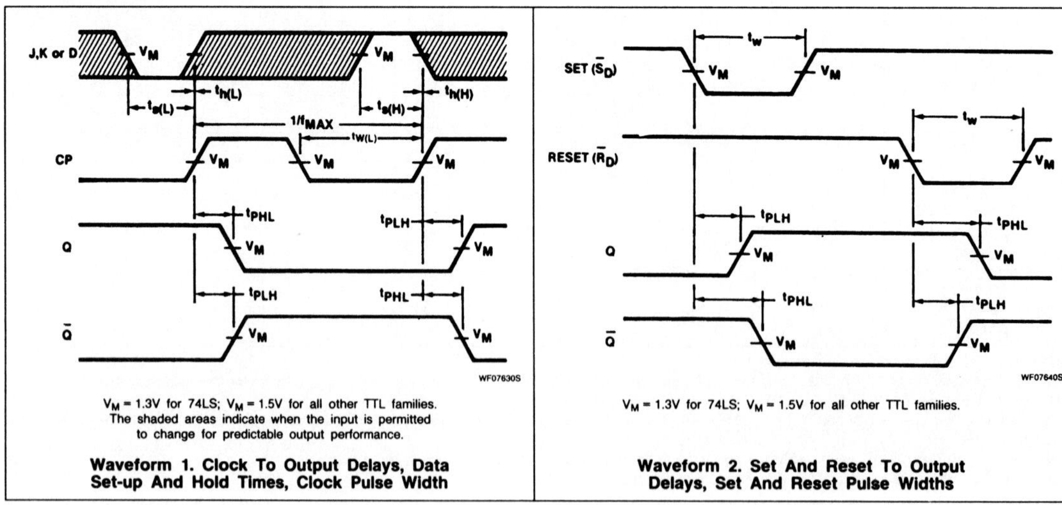

V_M = 1.3V for 74LS; V_M = 1.5V for all other TTL families.
The shaded areas indicate when the input is permitted to change for predictable output performance.

Waveform 1. Clock To Output Delays, Data Set-up And Hold Times, Clock Pulse Width

V_M = 1.3V for 74LS; V_M = 1.5V for all other TTL families.

Waveform 2. Set And Reset To Output Delays, Set And Reset Pulse Widths

TEST CIRCUITS AND WAVEFORMS

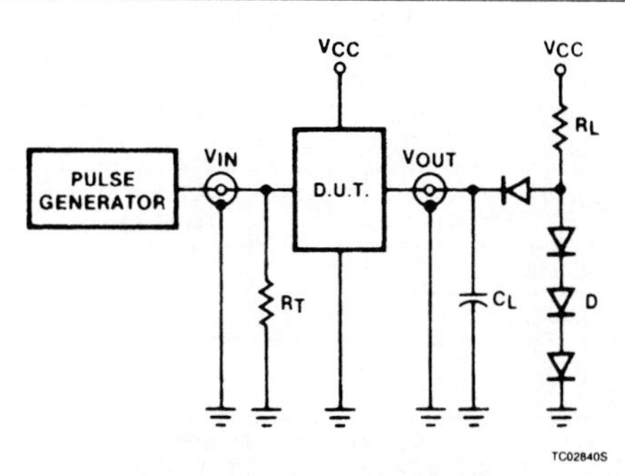

Test Circuit For 74 Totem-Pole Outputs

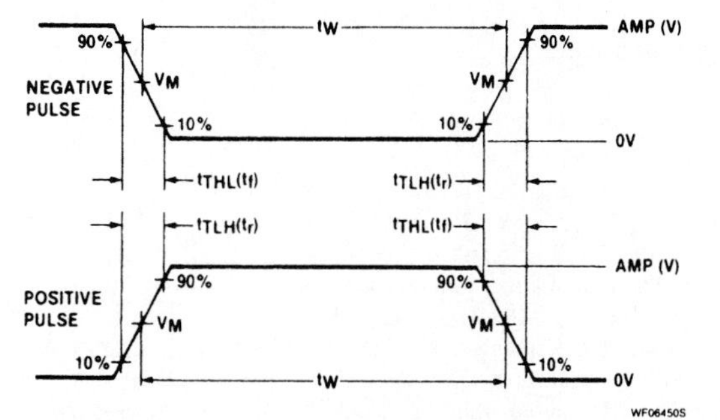

V_M = 1.3V for 74LS; V_M = 1.5V for all other TTL families.

Input Pulse Definition

DEFINITIONS

R_L = Load resistor to V_{CC}; see AC CHARACTERISTICS for value.
C_L = Load capacitance includes jig and probe capacitance; see AC CHARACTERISTICS for value.
R_T = Termination resistance should be equal to Z_{OUT} of Pulse Generators.
D = Diodes are 1N916, 1N3064, or equivalent.
t_{TLH}, t_{THL} Values should be less than or equal to the table entries.

FAMILY	INPUT PULSE REQUIREMENTS				
	Amplitude	Rep. Rate	Pulse Width	t_{TLH}	t_{THL}
74	3.0V	1MHz	500ns	7ns	7ns
74LS	3.0V	1MHz	500ns	15ns	6ns
74S	3.0V	1MHz	500ns	2.5ns	2.5ns

Signetics

7476, LS76
Flip-Flops

Dual J-K Flip-Flop
Product Specification

Logic Products

DESCRIPTION

The '76 is a dual J-K flip-flop with individual J, K, Clock, Set and Reset inputs. The 7476 is positive pulse-triggered. JK information is loaded into the master while the Clock is HIGH and transferred to the slave on the HIGH-to-LOW Clock transition. The J and K inputs must be stable while the Clock is HIGH for conventional operation.

The 74LS76 is a negative edge-triggered flip-flop. The J and K inputs must be stable only one set-up time prior to the HIGH-to-LOW Clock transition.

The Set ($\overline{S}_D$) and Reset ($\overline{R}_D$) are asynchronous active LOW inputs. When LOW, they override the Clock and Data inputs, forcing the outputs to the steady state levels as shown in the Function Table.

TYPE	TYPICAL f_{MAX}	TYPICAL SUPPLY CURRENT (TOTAL)
7476	20MHz	10mA
74LS76	45MHz	4mA

ORDERING CODE

PACKAGES	COMMERCIAL RANGE V_{CC} = 5V ±5%; T_A = 0°C to +70°C
Plastic DIP	N7476N, N74LS76N

NOTE:
For information regarding devices processed to Military Specifications, see the Signetics Military Products Data Manual.

INPUT AND OUTPUT LOADING AND FAN-OUT TABLE

PINS	DESCRIPTION	74	74LS
$\overline{CP}$	Clock input	2ul	2LSul
$\overline{R}_D$, $\overline{S}_D$	Reset and Set inputs	2ul	2LSul
J, K	Data inputs	1ul	1LSul
Q, $\overline{Q}$	Outputs	10ul	10LSul

NOTE:
Where a 74 unit load (ul) is understood to be 40µA I_{IH} and −1.6mA I_{IL}, and a 74LS unit load (LSul) is 20µA I_{IH} and −0.4mA I_{IL}.

PIN CONFIGURATION

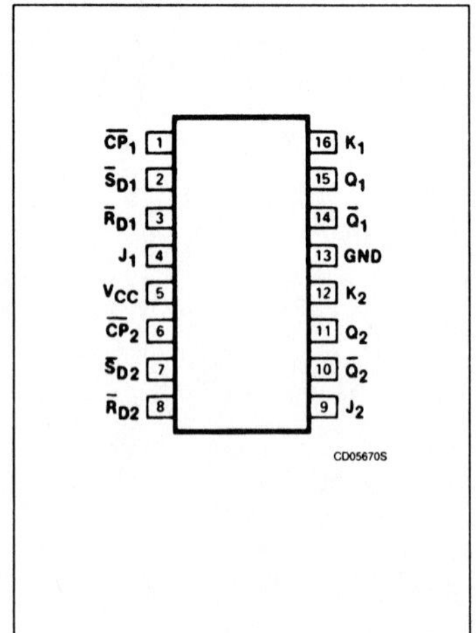

LOGIC SYMBOL

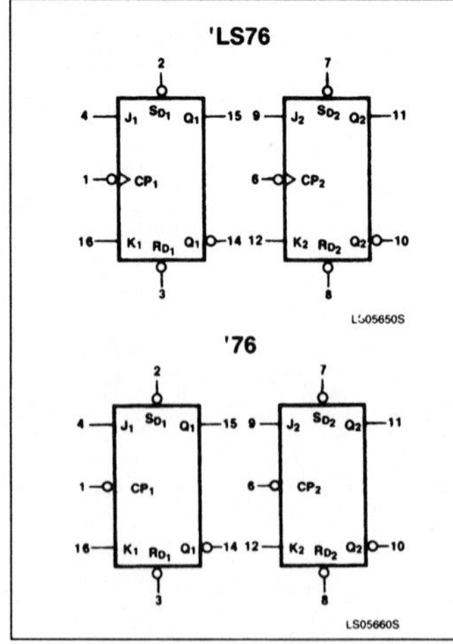

LOGIC SYMBOL (IEE/IEC)

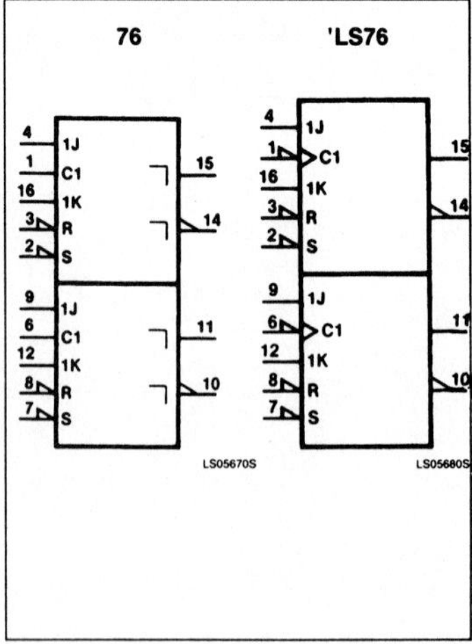

LOGIC DIAGRAM

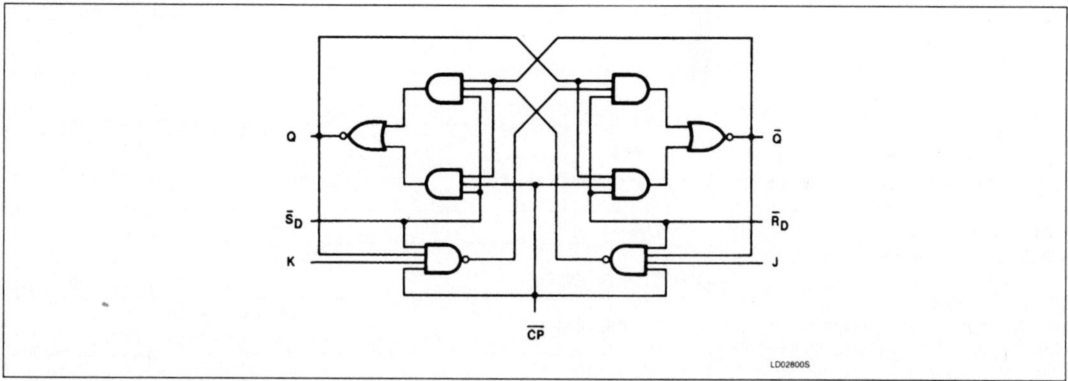

FUNCTION TABLE

OPERATING MODE	INPUTS					OUTPUTS	
	$\bar{S}_D$	$\bar{R}_D$	$\overline{CP}$(2)	J	K	Q	$\bar{Q}$
Asynchronous set	L	H	X	X	X	H	L
Asynchronous reset (Clear)	H	L	X	X	X	L	H
Undetermined(1)	L	L	X	X	X	H	H
Toggle	H	H	⎍	h	h	$\bar{q}$	q
Load "0" (Reset)	H	H	⎍	l	h	L	H
Load "1" (Set)	H	H	⎍	h	l	H	L
Hold "no change"	H	H	⎍	l	l	q	$\bar{q}$

H = HIGH voltage level steady state.
h = HIGH voltage level one set-up time prior to the HIGH-to-LOW Clock transition.(3)
L - LOW voltage level steady state.
l = LOW voltage level one set-up time prior to the HIGH-to-LOW Clock transition.(3)
q = Lower case letters indicate the state of the referenced output prior to the HIGH-to-LOW Clock transition.
X = Don't care.
⎍ = Positive Clock pulse.

NOTES:
1. Both outputs will be HIGH while both $\bar{S}_D$ and $\bar{R}_D$ are LOW, but the output states are unpredictable if $\bar{S}_D$ and $\bar{R}_D$ go HIGH simultaneously.
2. The 74LS76 is edge triggered. Data must be stable one set-up time prior to the negative edge of the Clock for predictable operation.
3. The J and K inputs of the 7476 must be stable while the Clock is HIGH for conventional operation.

Signetics

7483, LS83A
Adders

4-Bit Full Adder
Product Specification

Logic Products

FEATURES

- **High speed 4-bit binary addition**
- **Cascadeable in 4-bit increments**
- **LS83A has fast internal carry lookahead**
- **See '283 for corner power pin version**

DESCRIPTION

The '83 adds two 4-bit binary words (A_n plus B_n) plus the incoming carry. The binary sum appears on the Sum outputs ($\Sigma_1 - \Sigma_4$) and the outgoing carry (C_{OUT}) according to the equation:

$$C_{IN} + (A_1 + B_1) + 2(A_2 + B_2) + 4(A_3 + B_3) + 8(A_4 + B_4) = \Sigma_1 + 2\Sigma_2 + 4\Sigma_3 + 8\Sigma_4 + 16C_{OUT}$$

Where (+) = plus.

Due to the symmetry of the binary add function, the '83 can be used with either all active-HIGH operands (positive logic) or with all active-LOW operands (negative logic). See Function Table. With active-HIGH inputs, C_{IN} cannot be left open; it must be held LOW when no "carry in" is intended. Interchanging inputs of equal weight does not affect the operation, thus C_{IN}, A_1, B_1, can arbitrarily be assigned to pins 10, 11, 13, etc.

TYPE	TYPICAL ADD TIMES (TWO 8-BIT WORDS)	TYPICAL SUPPLY CURRENT (TOTAL)
7483	23ns	66mA
74LS83A	25ns	19mA

ORDERING CODE

PACKAGES	COMMERCIAL RANGE $V_{CC} = 5V \pm 5\%$; $T_A = 0°C$ to $+70°C$
Plastic DIP	N7483N, N74LS83AN
Plastic SO	N74LS83AD

NOTE:
For information regarding devices processed to Military Specifications, see the Signetics Military Products Data Manual.

INPUT AND OUTPUT LOADING AND FAN-OUT TABLE

PINS	DESCRIPTION	74	74LS
A_1, B_1, A_3, B_3, C_{IN}	Inputs	2ul	
A_2, B_2, A_4, B_4	Inputs	1ul	
A, B	Inputs		2LSul
C_{IN}	Input		1LSul
Sum	Outputs	10ul	10LSul
Carry	Output	5ul	10LSul

NOTE:
Where a 74 unit load (ul) is understood to be 40μA I_{IH} and −1.6mA I_{IL}, and a 74LS unit load (LSul) is 20μA I_{IH} and −0.4mA I_{IL}.

PIN CONFIGURATION

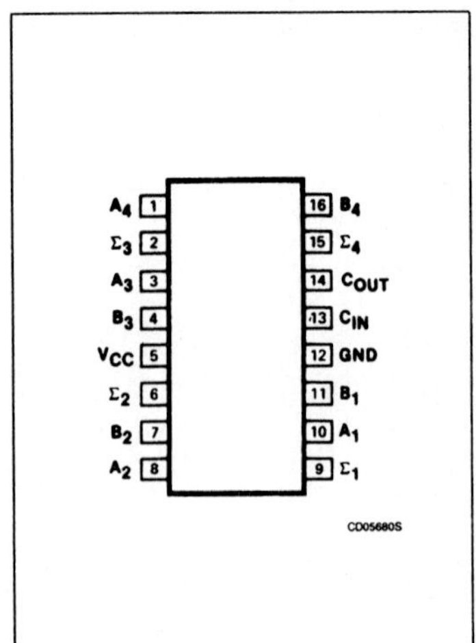

LOGIC SYMBOL

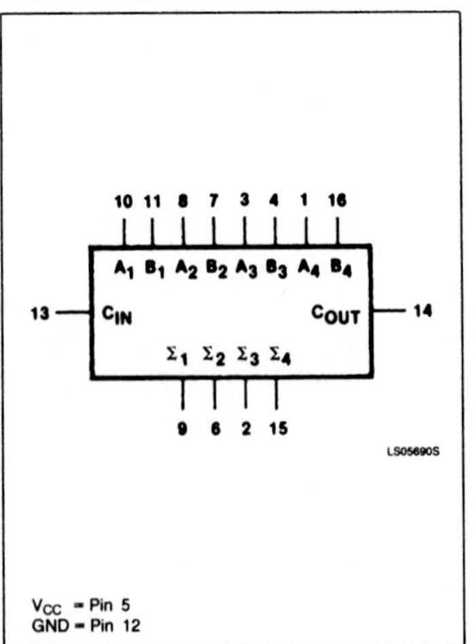

LOGIC SYMBOL (IEEE/IEC)

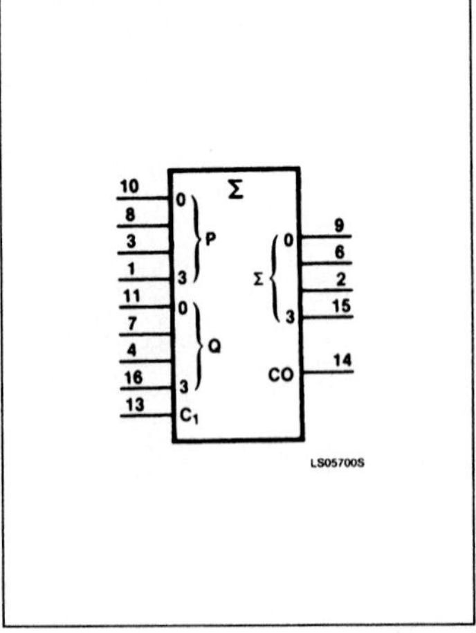

LOGIC DIAGRAM

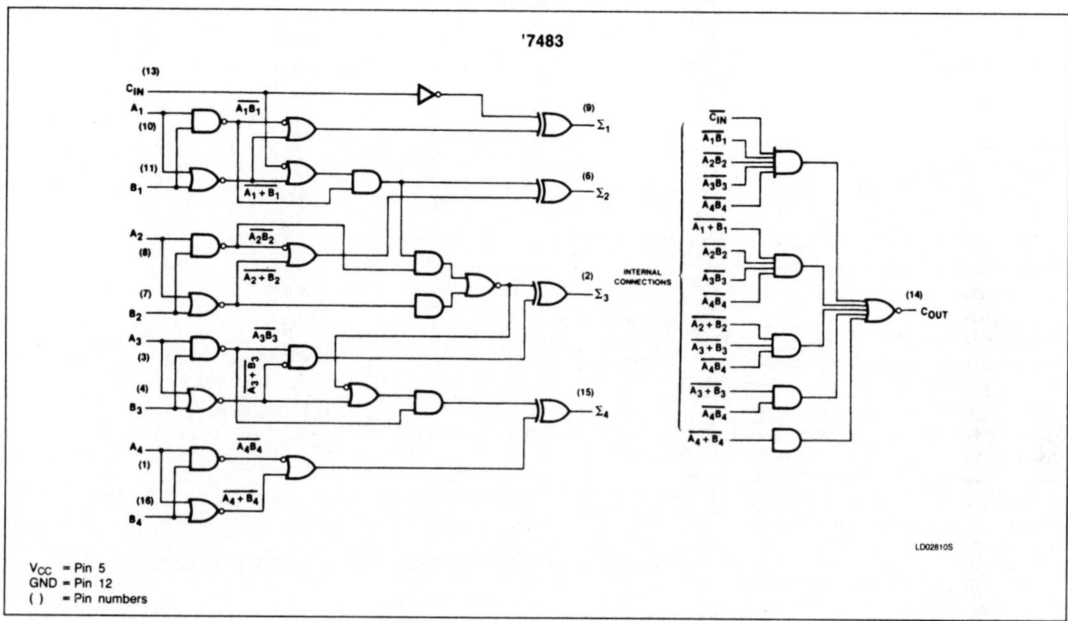

LOGIC DIAGRAM

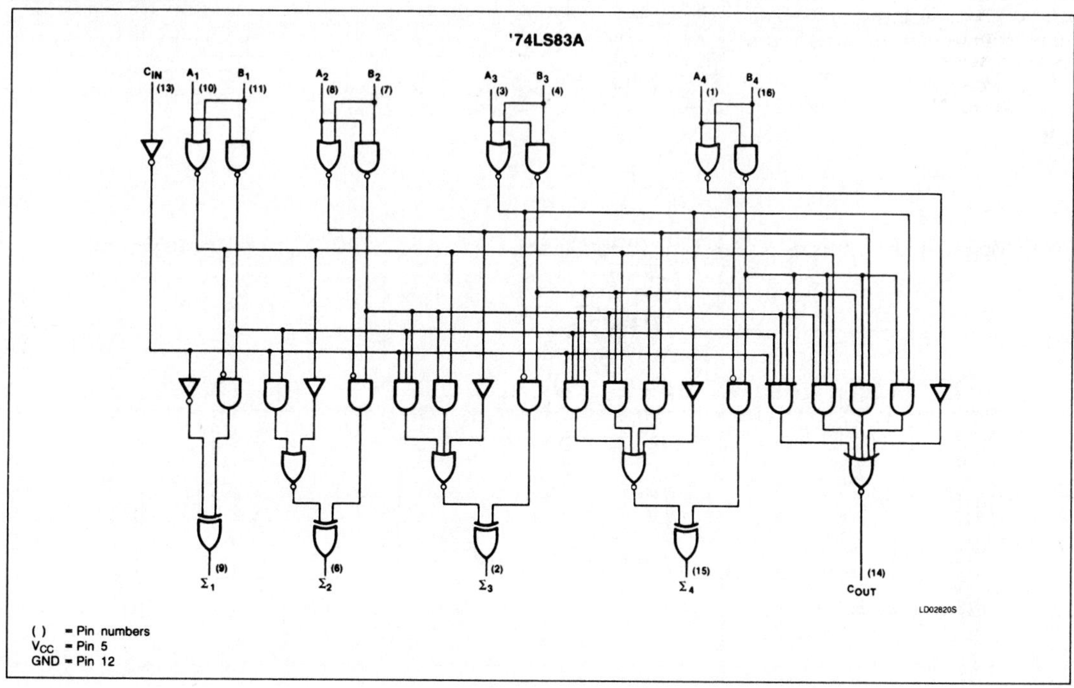

Signetics

Logic Products

7485, LS85, S85
Comparators

4-Bit Magnitude Comparator
Product Specification

FEATURES

- **Magnitude comparison of any binary words**
- **Serial or parallel expansion without extra gating**
- **Use 74S85 for very high speed comparisons**

DESCRIPTION

The '85 is a 4-bit magnitude comparator that can be expanded to almost any length. It compares two 4-bit binary, BCD, or other monotonic codes and presents the three possible magnitude results at the outputs. The 4-bit inputs are weighted ($A_0 - A_3$) and ($B_0 - B_3$), where A_3 and B_3 are the most significant bits.

The operation of the '85 is described in the Function Table, showing all possible logic conditions. The upper part of the table describes the normal operation under all conditions that will occur in a single device or in a series expansion scheme.

TYPE	TYPICAL PROPAGATION DELAY	TYPICAL SUPPLY CURRENT (TOTAL)
7485	23ns	55mA
74LS85	23ns	10mA
74S85	12ns	73mA

ORDERING CODE

PACKAGES	COMMERCIAL RANGE $V_{CC} = 5V \pm 5\%$; $T_A = 0°C$ to $+70°C$
Plastic DIP	N7485N, N74LS85N, N74S85N
Plastic SO	N74LS85D, N74S85D

NOTE:
For information regarding devices processed to Military Specifications, see the Signetics Military Products Data Manual.

INPUT AND OUTPUT LOADING AND FAN-OUT TABLE

PINS	DESCRIPTION	74	74S	74LS
$A_0 - A_3$, $B_0 - B_3$, $I_{A=B}$	Inputs	3ul	3Sul	3LSul
$I_{A<B}$, $I_{A>B}$	Inputs	1ul	1Sul	1LSul
A = B, A < B, A > B	Outputs	10ul	10Sul	10LSul

NOTE:
Where a 74 unit load (ul) is understood to be 40μA I_{IH} and −1.6mA I_{IL}, a 74S unit load (Sul) is 50μA I_{IH} and −2.0mA I_{IL}, and 74LS unit load (LSul) is 20μA I_{IH} and −0.4mA I_{IL}.

PIN CONFIGURATION

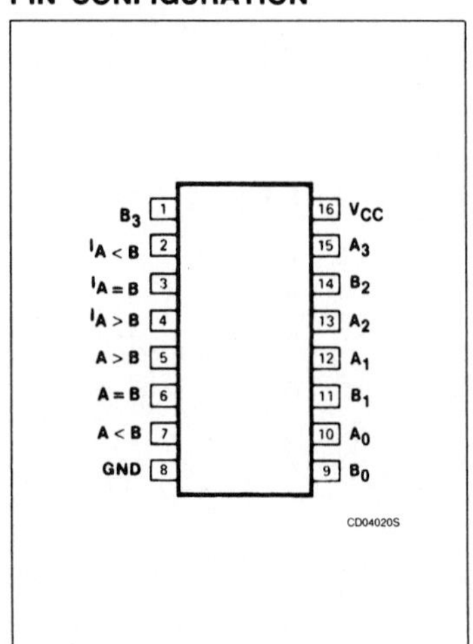

LOGIC SYMBOL

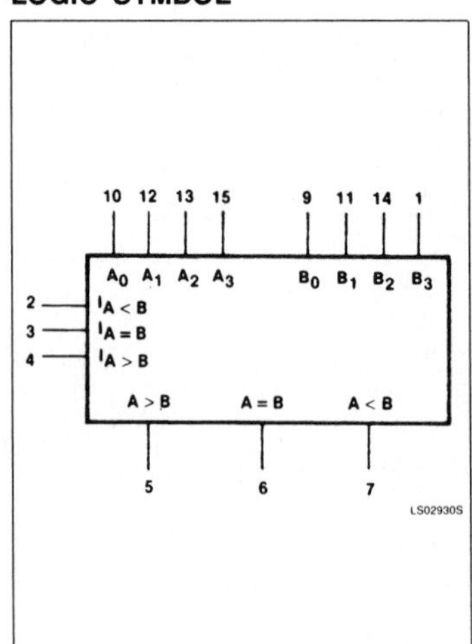

LOGIC SYMBOL (IEEE/IEC)

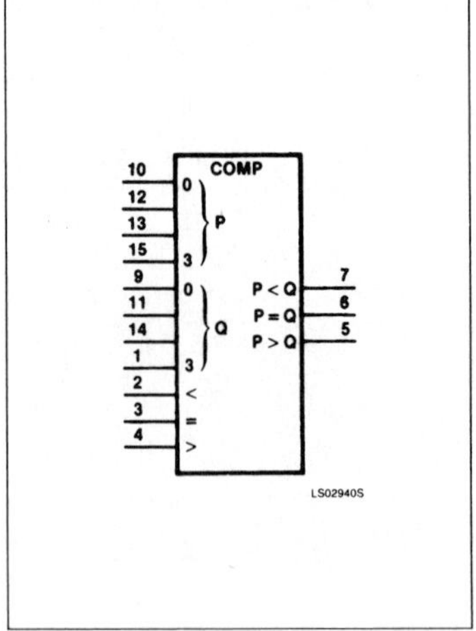

Comparators

7485, LS85, S85

LOGIC DIAGRAM

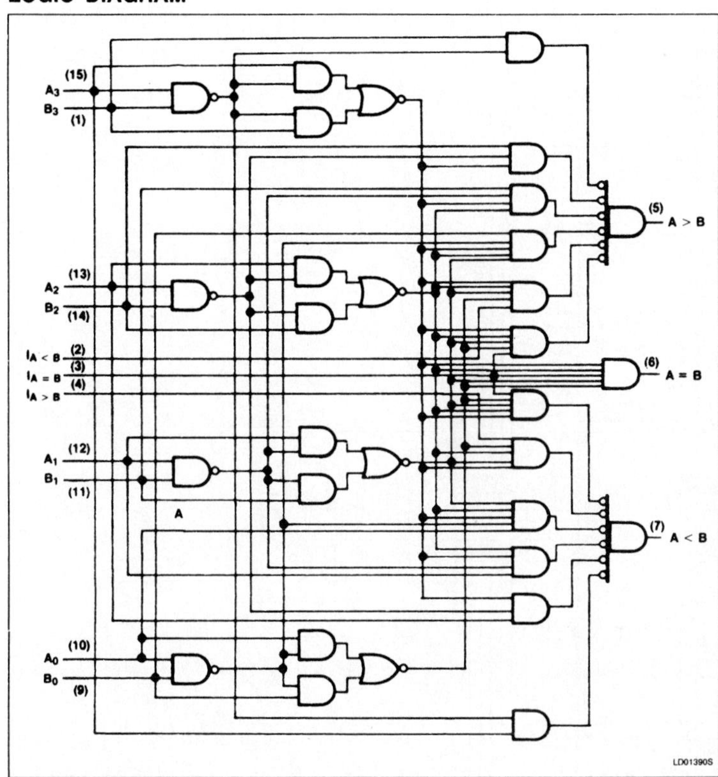

In the upper part of the table the three outputs are mutually exclusive. In the lower part of the table, the outputs reflect the feed-forward conditions that exist in the parallel expansion scheme.

The expansion inputs $I_{A>B}$, $I_{A=B}$, and $I_{A<B}$ are the least significant bit positions. When used for series expansion, the A > B, A = B and A < B outputs of the least significant word are connected to the corresponding $I_{A>B}$, $I_{A=B}$, and $I_{A<B}$ inputs of the next higher stage. Stages can be added in this manner to any length, but a propagation delay penalty of about 15ns is added with each additional stage. For proper operation the expansion inputs of the least significant word should be tied as follows: $I_{A>B}$ = LOW, $I_{A=B}$ = HIGH, and $I_{A<B}$ = LOW.

The parallel expansion scheme shown in Figure 1 demonstrates the most efficient general use of these comparators. In the parallel expansion scheme, the expansion inputs can be used as a fifth input bit position except on the least significant device which must be connected as in the serial scheme. The expansion inputs are used by labeling $I_{A>B}$ as an "A" input, $I_{A<B}$ as a "B" input and setting $I_{A=B}$ LOW. The '85 can be used as a 5-bit comparator only when the outputs are used to drive the ($A_0 - A_3$) and ($B_0 - B_3$) inputs of another '85 device. The parallel technique can be expanded to any number of bits as shown in Table 1.

FUNCTION TABLE

COMPARING INPUTS				CASCADING INPUTS			OUTPUTS		
A_3, B_3	A_2, B_2	A_1, B_1	A_0, B_0	$I_{A>B}$	$I_{A<B}$	$I_{A=B}$	A > B	A < B	A = B
$A_3 > B_3$	X	X	X	X	X	X	H	L	L
$A_3 < B_3$	X	X	X	X	X	X	L	H	L
$A_3 = B_3$	$A_2 > B_2$	X	X	X	X	X	H	L	L
$A_3 = B_3$	$A_2 < B_2$	X	X	X	X	X	L	H	L
$A_3 = B_3$	$A_2 = B_2$	$A_1 > B_1$	X	X	X	X	H	L	L
$A_3 = B_3$	$A_2 = B_2$	$A_1 < B_1$	X	X	X	X	L	H	L
$A_3 = B_3$	$A_2 = B_2$	$A_1 = B_1$	$A_0 > B_0$	X	X	X	H	L	L
$A_3 = B_3$	$A_2 = B_2$	$A_1 = B_1$	$A_0 < B_0$	X	X	X	L	H	L
$A_3 = B_3$	$A_2 = B_2$	$A_1 = B_1$	$A_0 = B_0$	H	L	L	H	L	L
$A_3 = B_3$	$A_2 = B_2$	$A_1 = B_1$	$A_0 = B_0$	L	H	L	L	H	L
$A_3 = B_3$	$A_2 = B_2$	$A_1 = B_1$	$A_0 = B_0$	L	L	H	L	L	H
$A_3 = B_3$	$A_2 = B_2$	$A_1 = B_1$	$A_0 = B_0$	X	X	H	L	L	H
$A_3 = B_3$	$A_2 = B_2$	$A_1 = B_1$	$A_0 = B_0$	H	H	L	L	L	L
$A_3 = B_3$	$A_2 = B_2$	$A_1 = B_1$	$A_0 = B_0$	L	L	L	H	H	L

H = HIGH voltage level
L = LOW voltage level
X = Don't care

Comparators

7485, LS85, S85

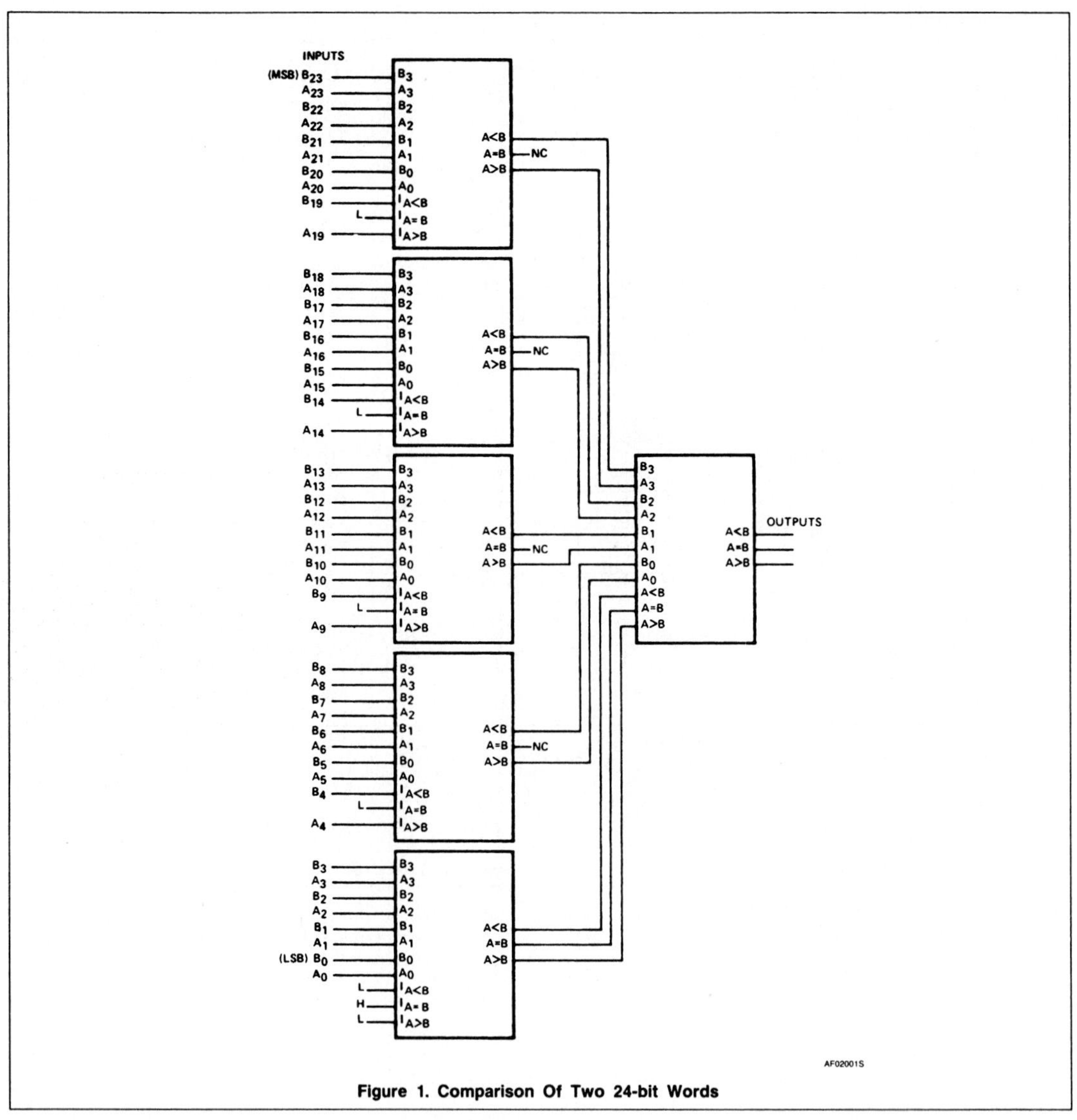

Figure 1. Comparison Of Two 24-bit Words

Table 1

WORD LENGTH	NUMBER OF PACKAGES	TYPICAL SPEEDS		
		74	74S	74LS
1 – 4 Bits	1	23ns	12ns	23ns
5 – 25 Bits	2 – 6	40ns	22ns	46ns
25 – 120 Bits	8 – 31	63ns	34ns	69ns

Signetics

7486, LS86, S86 Gates

Quad Two-Input Exclusive-OR Gate
Product Specification

Logic Products

TYPE	TYPICAL PROPAGATION DELAY	TYPICAL SUPPLY CURRENT (TOTAL)
7486	14ns	30mA
74LS86	10ns	6.1mA
74S86	7ns	50mA

ORDERING CODE

PACKAGES	COMMERCIAL RANGE $V_{CC} = 5V \pm 5\%$; $T_A = 0°C$ to $+70°C$
Plastic DIP	N7486N, N74LS86N, N74S86N
Plastic SO	N74LS86D, N74S86D

NOTE:
For information regarding devices processed to Military Specifications, see the Signetics Military Products Data Manual.

FUNCTION TABLE

INPUTS		OUTPUT
A	B	Y
L	L	L
L	H	H
H	L	H
H	H	L

H = HIGH voltage level
L = LOW voltage level

INPUT AND OUTPUT LOADING AND FAN-OUT TABLE

PINS	DESCRIPTION	74	74S	74LS
A, B	Inputs	1ul	1Sul	1LSul
Y	Output	10ul	10Sul	10LSul

NOTE:
Where a 74 unit load (ul) is understood to be 40μA I_{IH} and −1.6mA I_{IL}, a 74S unit load (Sul) is 50μA I_{IH} and −2.0mA I_{IL}, and a 74LS unit load (LSul) is 20μA I_{IH} and −0.4mA I_{IL}.

PIN CONFIGURATION

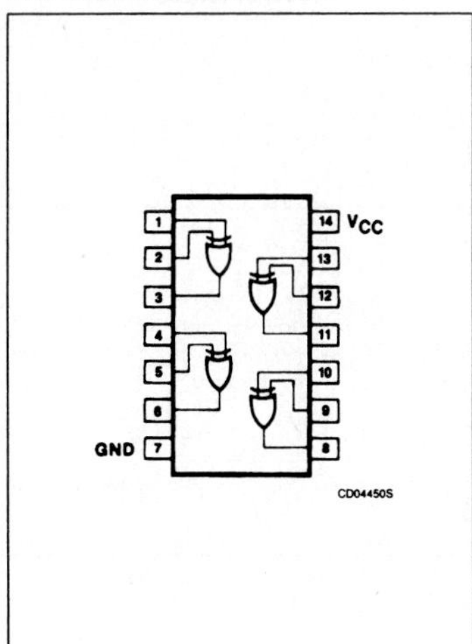

LOGIC SYMBOL

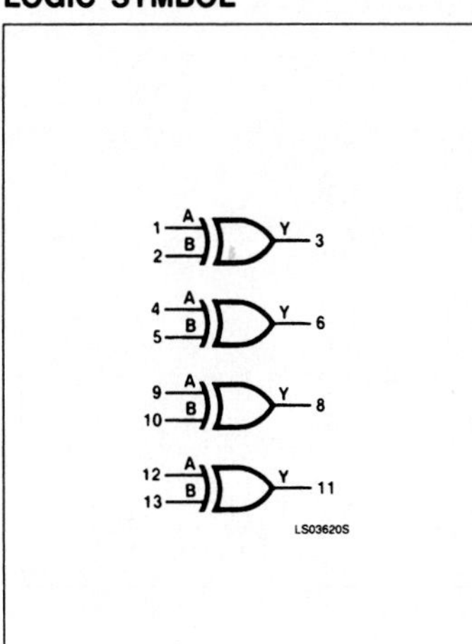

LOGIC SYMBOL (IEEE/IEC)

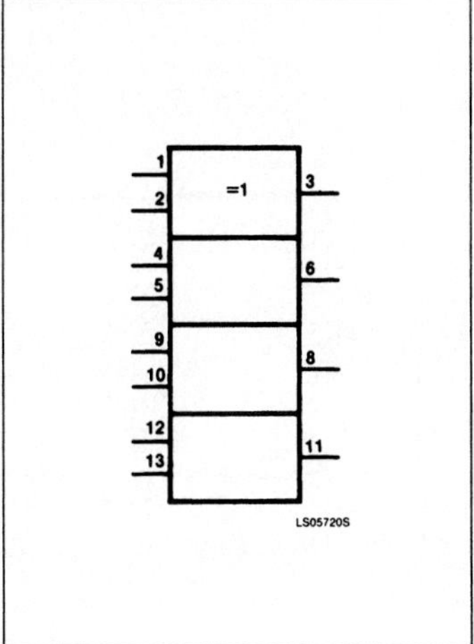

Signetics

7490, LS90
Counters

Decade Counter
Product Specification

Logic Products

DESCRIPTION

The '90 is a 4-bit, ripple-type Decade Counter. The device consists of four master-slave flip-flops internally connected to provide a divide-by-two section and a divide-by-five section. Each section has a separate Clock input to initiate state changes of the counter on the HIGH-to-LOW clock transition. State changes of the Q outputs do not occur simultaneously because of internal ripple delays. Therefore, decoded output signals are subject to decoding spikes and should not be used for clocks or strobes.

A gated AND asynchronous Master Reset ($MR_1 \cdot MR_2$) is provided which overrides both clocks and resets (clears) all the flip-flops. Also provided is a gated AND asynchronous Master Set ($MS_1 \cdot MS_2$) which overrides the clocks and the MR inputs, setting the outputs to nine (HLLH).

Since the output from the divide-by-two section is not internally connected to the succeeding stages, the device may be operated in various counting modes. In a BCD (8421) counter the $\overline{CP}_1$ input must be externally connected to the Q_0 output. The $\overline{CP}_0$ input receives the incoming count producing a BCD count sequence. In a symmetrical Bi-quinary divide-by-ten counter the Q_3 output must be connected externally to the $\overline{CP}_0$ input. The input count is then applied to the CP_1 input and a divide-by-ten square wave is obtained at output Q_0. To operate as a divide-by-two and a divide-by-five counter no external interconnections are required. The first flip-flop is used as a binary element for the divide-by-two function ($\overline{CP}_0$ as the input and Q_0 as the output). The $\overline{CP}_1$ input is used to obtain a divide-by-five operation at the Q_3 output.

TYPE	TYPICAL f_{MAX}	TYPICAL SUPPLY CURRENT
7490	30MHz	30mA
74LS90	42MHz	9mA

ORDERING CODE

PACKAGES	COMMERCIAL RANGE $V_{CC} = 5V \pm 5\%$; $T_A = 0°C$ to $+70°C$
Plastic DIP	N7490N, N74LS90N

NOTE:
For information regarding devices processed to Military Specifications, see the Signetics Military Products Data Manual.

INPUT AND OUTPUT LOADING AND FAN-OUT TABLE

PINS	DESCRIPTION	74	74LS
$\overline{CP}_0$	Input	2ul	6LSul
$\overline{CP}_1$	Input	4ul	8LSul
MR, MS	Inputs		1ul
$Q_0 - Q_3$	Outputs	10ul	10LSul

NOTE:
Where a 74 unit load (ul) is understood to be $40\mu A$ I_{IH} and $-1.6mA$ I_{IL}, and a 74LS unit load (LSul) is $20\mu A$ I_{IH} and $-0.4mA$ I_{IL}.

PIN CONFIGURATION

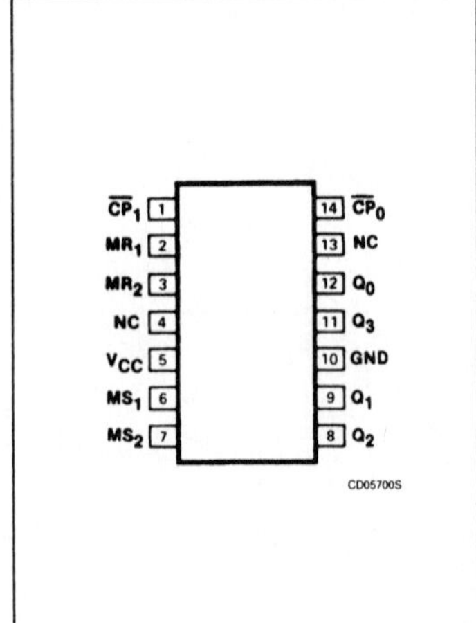

LOGIC SYMBOL

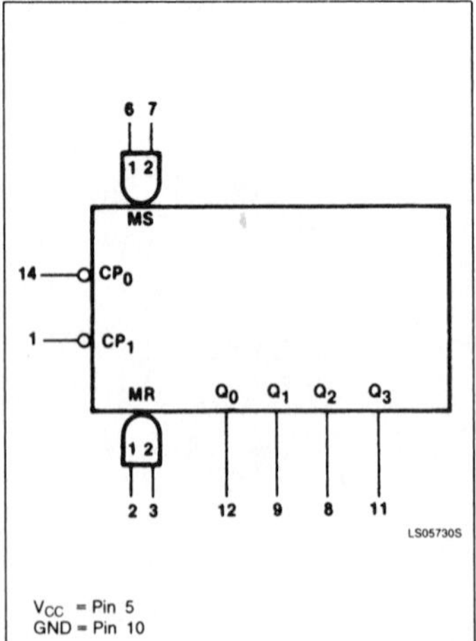

LOGIC SYMBOL (IEEE/IEC)

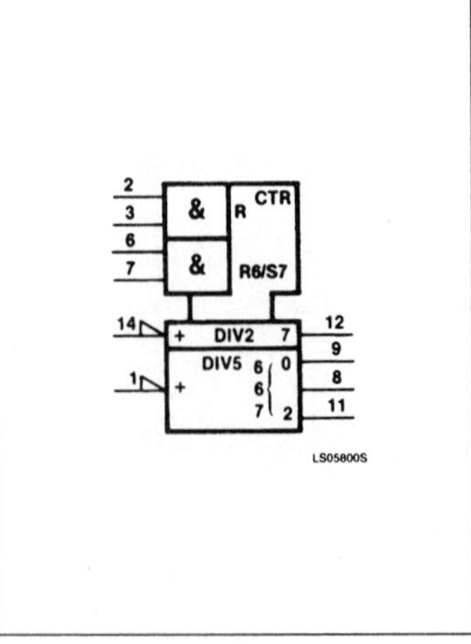

LOGIC DIAGRAM

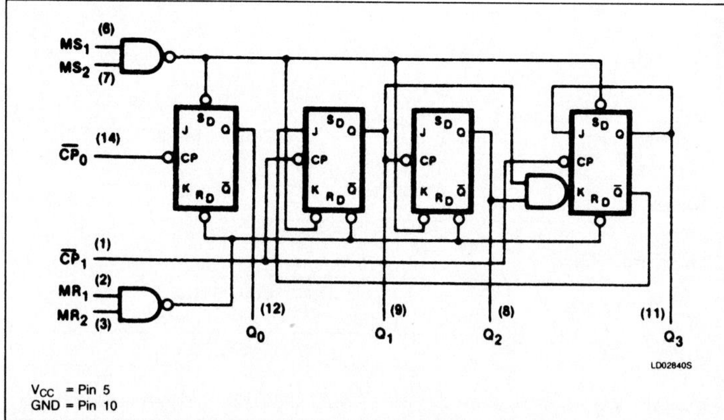

MODE SELECTION — FUNCTION TABLE

RESET/SET INPUTS				OUTPUTS			
MR_1	MR_2	MS_1	MS_2	Q_0	Q_1	Q_2	Q_3
H	H	L	X	L	L	L	L
H	H	X	L	L	L	L	L
X	X	H	H	H	L	L	H
L	X	L	X	Count			
X	L	X	L	Count			
L	X	X	L	Count			
H	L	L	X	Count			

H = HIGH voltage level
L = LOW voltage level
X = Don't care

BCD COUNT SEQUENCE — FUNCTION TABLE

COUNT	OUTPUTS			
	Q_0	Q_1	Q_2	Q_3
0	L	L	L	L
1	H	L	L	L
2	L	H	L	L
3	H	H	L	L
4	L	L	H	L
5	H	L	H	L
6	L	H	H	L
7	H	H	H	L
8	L	L	L	H
9	H	L	L	H

NOTE:
Output Q_0 connected to input $\overline{CP}_1$.

ABSOLUTE MAXIMUM RATINGS (Over operating free-air temperature range unless otherwise noted.)

	PARAMETER	74	74LS	UNIT
V_{CC}	Supply voltage	7.0	7.0	V
V_{IN}	Input voltage	−0.5 to +5.5	−0.5 to +7.0	V
I_{IN}	Input current	−30 to +5	−30 to +1	mA
V_{OUT}	Voltage applied to output in HIGH output state	−0.5 to $+V_{CC}$	−0.5 to $+V_{CC}$	V
T_A	Operating free-air temperature range	0 to 70		°C

NOTE:
V_{IN} is limited to +5.5V on $\overline{CP}_0$ and $\overline{CP}_1$ inputs on the 74LS90 only.

RECOMMENDED OPERATING CONDITIONS

	PARAMETER	74			74LS			UNIT
		Min	Nom	Max	Min	Nom	Max	
V_{CC}	Supply voltage	4.75	5.0	5.25	4.75	5.0	5.25	V
V_{IH}	HIGH-level input voltage	2.0			2.0			V
V_{IL}	LOW-level input voltage			+0.8			+0.8	V
I_{IK}	Input clamp current			−12			−18	mA
I_{OH}	HIGH-level output current			−800			−400	μA
I_{OL}	LOW-level output current			16			8	mA
T_A	Operating free-air temperature	0		70	0		70	°C

Signetics

7492, LS92
Counters

Divide-By-Twelve Counter
Product Specification

Logic Products

DESCRIPTION

The '92 is a 4-bit, ripple-type Divide-by-12 Counter. The device consists of four master-slave flip-flops internally connected to provide a divide-by-two section and a divide-by-six section. Each section has a separate Clock input to initiate state changes of the counter on the HIGH-to-LOW clock transition. State changes of the Q outputs do not occur simultaneously because of internal ripple delays. Therefore, decoded output signals are subject to decoding spikes and should not be used for clocks or strobes.

A gated AND asynchronous Master Reset ($MR_1 \cdot MR_2$) is provided which overrides both clocks and resets (clears) all the flip-flops.

TYPE	TYPICAL f_{MAX}	TYPICAL SUPPLY CURRENT
7492	28MHz	28mA
74LS92	42MHz	9mA

ORDERING CODE

PACKAGES	COMMERCIAL RANGE $V_{CC} = 5V \pm 5\%$; $T_A = 0°C$ to $+70°C$
Plastic DIP	N7492N, N74LS92N

NOTE:
For information regarding devices processed to Military Specifications, see the Signetics Military Products Data Manual.

INPUT AND OUTPUT LOADING AND FAN-OUT TABLE

PINS	DESCRIPTION	74	74LS
MR	Master reset inputs	1ul	1LSul
CP_0	Input	2ul	6LSul
CP_1	Input	4ul	8LSul
$Q_0 - Q_3$	Outputs	10ul	10LSul

NOTE:
Where a 74 unit load (ul) is understood to be 40μA I_{IH} and −1.6mA I_{IL}, and a 74LS unit load (LSul) is 20μA I_{IH} and −0.4mA I_{IL}.

PIN CONFIGURATION

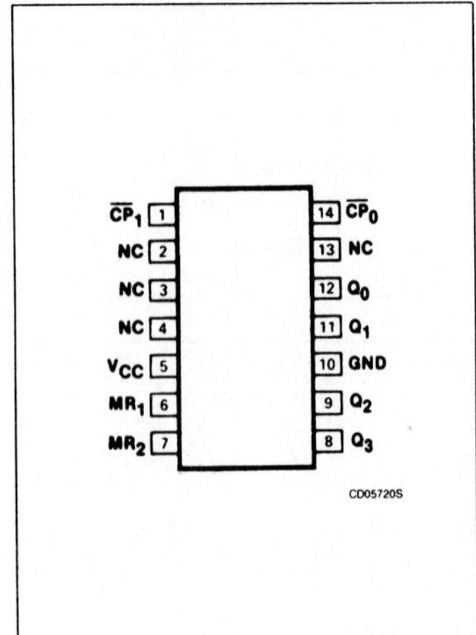

LOGIC SYMBOL

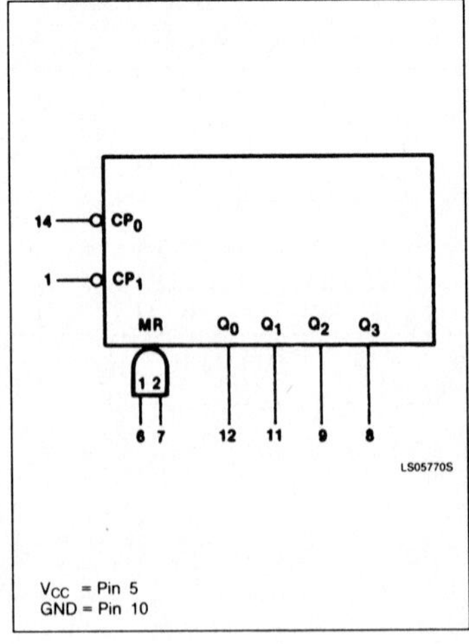

LOGIC SYMBOL (IEEE/IEC)

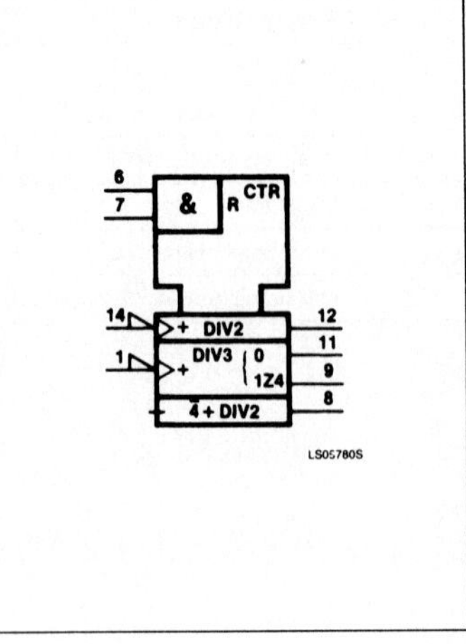

Counters

7492, LS92

LOGIC DIAGRAM

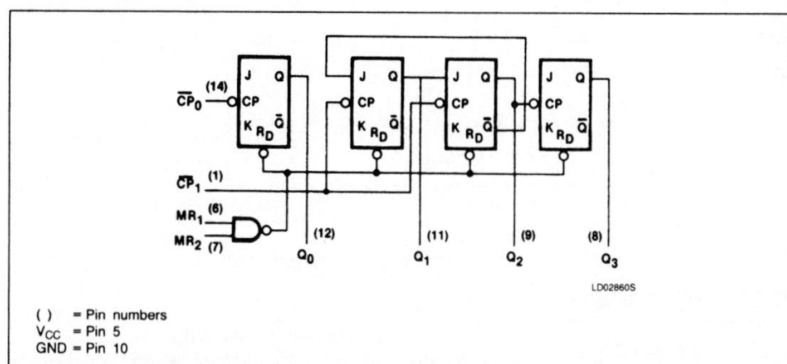

Since the output from the divide-by-two section is not internally connected to the succeeding stages, the device may be operated in various counting modes. In a Modulo-12, Divide-by-12 Counter the $\overline{CP}_1$ input must be externally connected to the Q_0 output. The $\overline{CP}_0$ input receives the incoming count and Q_3 produces a symmetrical divide-by-12 square wave output. In a divide-by-six counter no external connections are required. The first flip-flop is used as a binary element for the divide-by-two function. The $\overline{CP}_1$ input is used to obtain divide-by-three operation at the Q_1 and Q_2 outputs and divide-by-six operation at the Q_3 output.

FUNCTION TABLE

COUNT	OUTPUTS			
	Q_0	Q_1	Q_2	Q_3
0	L	L	L	L
1	H	L	L	L
2	L	H	L	L
3	H	H	L	L
4	L	L	H	L
5	H	L	H	L
6	L	L	L	H
7	H	L	L	H
8	L	H	L	H
9	H	H	L	H
10	L	L	H	H
11	H	L	H	H

NOTE:
Output Q_0 connected to input $\overline{CP}_1$.

MODE SELECTION

RESET INPUTS		OUTPUTS			
MR_1	MR_2	Q_0	Q_1	Q_2	Q_3
H	H	L	L	L	L
L	H	Count			
H	L	Count			
L	L	Count			

H = HIGH voltage level
L = LOW voltage level
X = Don't care

ABSOLUTE MAXIMUM RATINGS (Over operating free-air temperature range unless otherwise noted.)

PARAMETER		74	74LS	UNIT
V_{CC}	Supply voltage	7.0	7.0	V
V_{IN}	Input voltage	−0.5 to +5.5	−0.5 to +7.0	V
I_{IN}	Input current	−30 to +5	−30 to +1	mA
V_{OUT}	Voltage applied to output in HIGH output state	−0.5 to $+V_{CC}$	−0.5 to $+V_{CC}$	V
T_A	Operating free-air temperature range	0 to 70		°C

NOTE:
V_{IN} is limited to 5.5V on $\overline{CP}_0$ and $\overline{CP}_1$ inputs only on the 74LS92.

RECOMMENDED OPERATING CONDITIONS

PARAMETER		74			74LS			UNIT
		Min	Nom	Max	Min	Nom	Max	
V_{CC}	Supply voltage	4.75	5.0	5.25	4.75	5.0	5.25	V
V_{IH}	HIGH-level input voltage	2.0			2.0			V
V_{IL}	LOW-level input voltage			+0.8			+0.8	V
I_{IK}	Input clamp current			−12			−18	mA
I_{OH}	HIGH-level output current			−800			−400	μA
I_{OL}	LOW-level output current			16			8	mA
T_A	Operating free-air temperature	0		70	0		70	°C

Signetics

7493, LS93 Counters

4-Bit Binary Ripple Counter
Product Specification

Logic Products

DESCRIPTION

The '93 is a 4-bit, ripple-type Binary Counter. The device consists of four master-slave flip-flops internally connected to provide a divide-by-two section and a divide-by-eight section. Each section has a separate Clock input to initiate state changes of the counter on the HIGH-to-LOW clock transition. State changes of the Q outputs do not occur simultaneously because of internal ripple delays. Therefore, decoded output signals are subject to decoding spikes and should not be used for clocks or strobes.

A gated AND asynchronous Master Reset ($MR_1 \cdot MR_2$) is provided which overrides both clocks and resets (clears) all the flip-flops.

Since the output from the divide-by-two section is not internally connected to the succeeding stages, the device may be operated in various counting modes. In a 4-bit ripple counter the output Q_0 must be connected externally to input $\overline{CP}_1$.

TYPE	TYPICAL f_{MAX}	TYPICAL SUPPLY CURRENT (TOTAL)
7493	40MHz	28mA
74LS93	42MHz	9mA

ORDERING CODE

PACKAGES	COMMERCIAL RANGE $V_{CC} = 5V \pm 5\%$; $T_A = 0°C$ to $+70°C$
Plastic DIP	N7493N, N74LS93N
Plastic SO	N74LS93D

NOTE:
For information regarding devices processed to Military Specifications, see the Signetics Military Products Data Manual.

INPUT AND OUTPUT LOADING AND FAN-OUT TABLE

PINS	DESCRIPTION	74	74LS
MR	Master reset inputs	1ul	1LSul
$\overline{CP}_0$	Input	2ul	6LSul
$\overline{CP}_1$	Input	2ul	4LSul
$Q_0 - Q_3$	Outputs	10ul	10LSul

NOTE:
Where a 74 unit load (ul) is understood to be 40μA I_{IH} and −1.6mA I_{IL}, and a 74LS unit load (LSul) is 20μA I_{IH} and −0.4mA I_{IL}.

PIN CONFIGURATION

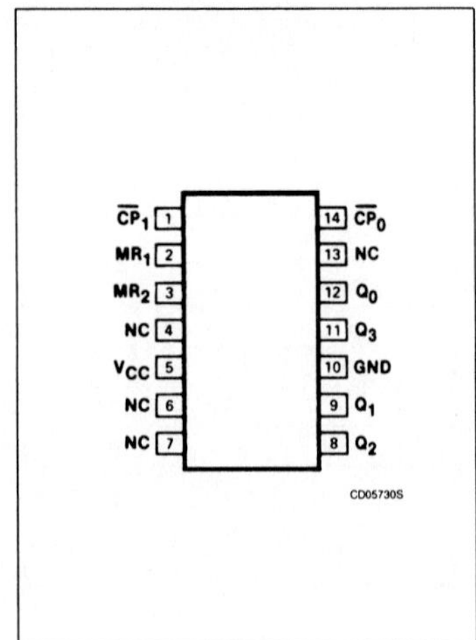

LOGIC SYMBOL

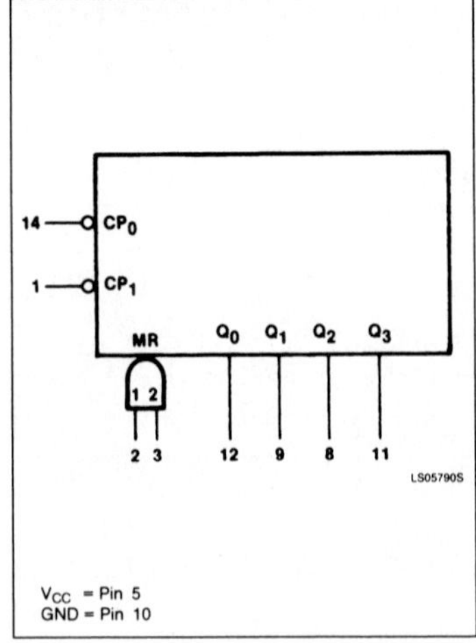

LOGIC SYMBOL (IEEE/IEC)

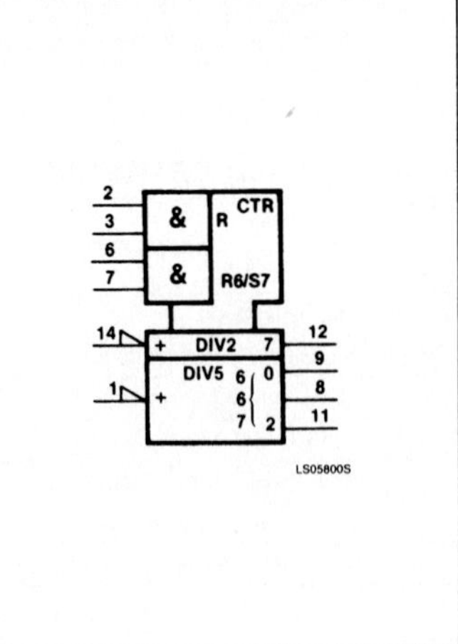

Counters

7493, LS93

LOGIC DIAGRAM

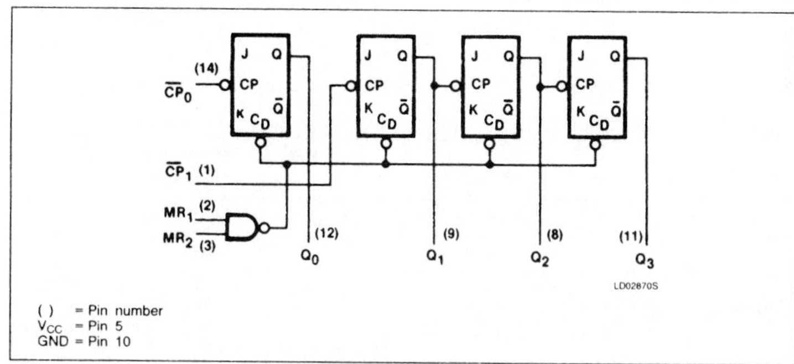

The input count pulses are applied to input $\overline{CP}_0$. Simultaneous divisions of 2, 4, 8 and 16 are performed at the Q_0, Q_1, Q_2 and Q_3 outputs as shown in the Function Table.

As a 3-bit ripple counter the input count pulses are applied to input $\overline{CP}_1$. Simultaneous frequency divisions of 2, 4 and 8 are available at the Q_1, Q_2 and Q_3 outputs. Independent use of the first flip-flop is available if the reset function coincides with reset of the 3-bit ripple-through counter.

FUNCTION TABLE

COUNT	OUTPUTS			
	Q_0	Q_1	Q_2	Q_3
0	L	L	L	L
1	H	L	L	L
2	L	H	L	L
3	H	H	L	L
4	L	L	H	L
5	H	L	H	L
6	L	H	H	L
7	H	H	H	L
8	L	L	L	H
9	H	L	L	H
10	L	H	L	H
11	H	H	L	H
12	L	L	H	H
13	H	L	H	H
14	L	H	H	H
15	H	H	H	H

NOTE:
Output Q_0 connected to input $\overline{CP}_1$.

MODE SELECTION

RESET INPUTS		OUTPUTS			
MR_1	MR_2	Q_0	Q_1	Q_2	Q_3
H	H	L	L	L	L
L	H	Count			
H	L	Count			
L	L	Count			

H = HIGH voltage level
L = LOW voltage level
X = Don't care

ABSOLUTE MAXIMUM RATINGS (Over operating free-air temperature range unless otherwise noted.)

PARAMETER		74	74LS	UNIT
V_{CC}	Supply voltage	7.0	7.0	V
V_{IN}	Input voltage	−0.5 to +5.5	−0.5 to +7.0	V
I_{IN}	Input current	−30 to +5	−30 to +1	mA
V_{OUT}	Voltage applied to output in HIGH output state	−0.5 to $+V_{CC}$	−0.5 to $+V_{CC}$	V
T_A	Operating free-air temperature range	0 to 70		°C

RECOMMENDED OPERATING CONDITIONS

PARAMETER		74			74LS			UNIT
		Min	Nom	Max	Min	Nom	Max	
V_{CC}	Supply voltage	4.75	5.0	5.25	4.75	5.0	5.25	V
V_{IH}	HIGH-level input voltage	2.0			2.0			V
V_{IL}	LOW-level input voltage			+0.8			+0.8	V
I_{IK}	Input clamp current			−12			−18	mA
I_{OH}	HIGH-level output current			−800			−400	μA
I_{OL}	LOW-level output current			16			8	mA
T_A	Operating free-air temperature	0		70	0		70	°C

Signetics

74LS112, S112
Flip-Flops

Dual J-K Edge-Triggered Flip-Flop
Product Specification

Logic Products

DESCRIPTION

The '112 is a dual J-K negative edge-triggered flip-flop featuring individual J, K, Clock, Set and Reset inputs. The Set ($\overline{S}_D$) and Reset ($\overline{R}_D$) inputs, when LOW, set or reset the outputs as shown in the Function Table regardless of the levels at the other inputs.

A HIGH level on the Clock ($\overline{CP}$) input enables the J and K inputs and data will be accepted. The logic levels at the J and K inputs may be allowed to change while the $\overline{CP}$ is HIGH and the flip-flop will perform according to the Function Table as long as minimum setup and hold times are observed. Output state changes are initiated by the HIGH-to-LOW transition of $\overline{CP}$.

TYPE	TYPICAL f_{MAX}	TYPICAL SUPPLY CURRENT (TOTAL)
74LS112	45MHz	4mA
74S112	125MHz	15mA

ORDERING CODE

PACKAGES	COMMERCIAL RANGE $V_{CC} = 5V \pm 5\%$; $T_A = 0°C$ to $+70°C$
Plastic DIP	N74S112N, N74LS112N
Plastic SO	N74LS112D, N74S112D

NOTE:
For information regarding devices processed to Military Specifications, see the Signetics Military Products Data Manual.

INPUT AND OUTPUT LOADING AND FAN-OUT TABLE

PINS	DESCRIPTION	74S	74LS
$\overline{CP}$	Clock input	2Sul	4LSul
$\overline{R}_D$, $\overline{S}_D$	Reset and set inputs	3.5Sul	3LSul
J, K	Data inputs	1Sul	1LSul
Q, $\overline{Q}$	Outputs	10Sul	10LSul

NOTE:
A 74 unit load (ul) is 50μA I_{IH} and −2.0mA I_{IL}, and a 74LS unit load (LSul) is 20μA I_{IH} and −0.4mA I_{IL}.

PIN CONFIGURATION

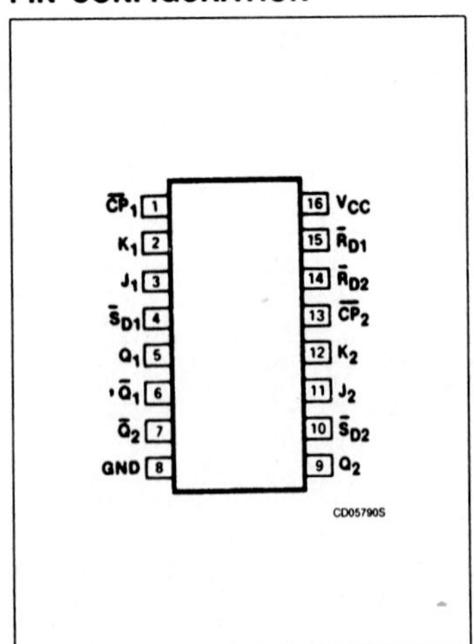

LOGIC SYMBOL

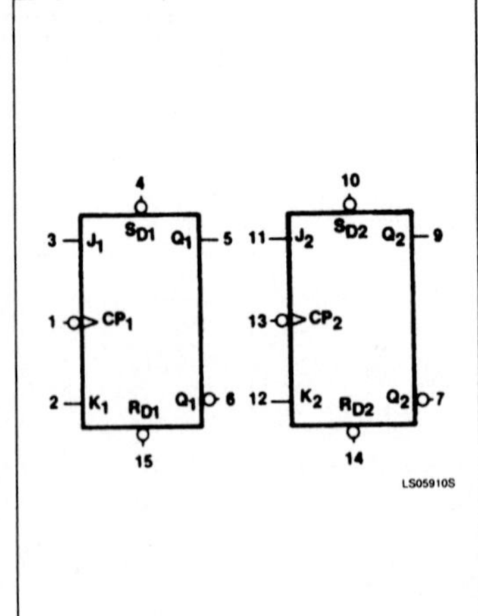

LOGIC SYMBOL (IEEE/IEC)

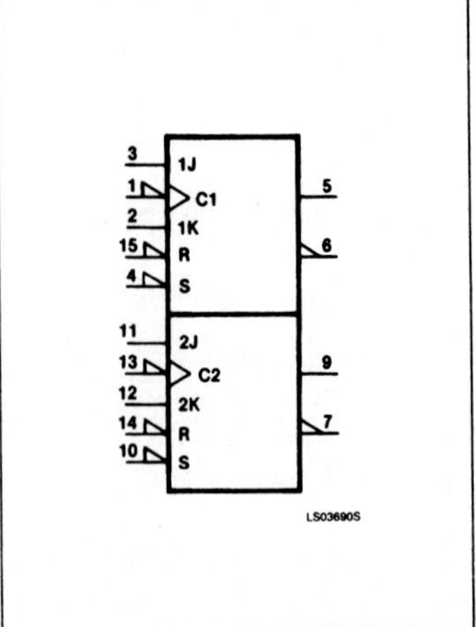

LOGIC DIAGRAM

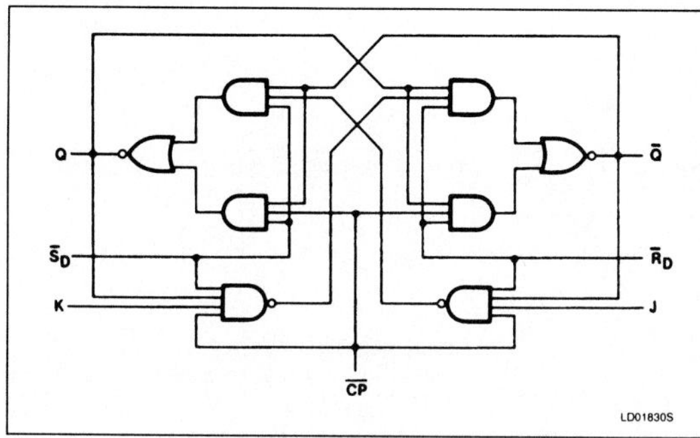

FUNCTION TABLE

OPERATING MODE	INPUTS					OUTPUTS	
	$\overline{S}_D$	$\overline{R}_D$	$\overline{CP}$	J	K	Q	$\overline{Q}$
Asynchronous set	L	H	X	X	X	H	L
Asynchronous reset (clear)	H	L	X	X	X	L	H
Undetermined	L	L	X	X	X	H	H
Toggle	H	H	↓	h	h	$\overline{q}$	q
Load "0" (reset)	H	H	↓	l	h	L	H
Load "1" (set)	H	H	↓	h	l	H	L
Hold "no change"	H	H	↓	l	l	q	$\overline{q}$

H = HIGH voltage level steady state.
h = HIGH voltage level one set-up time prior to the HIGH-to-LOW Clock transition.
L = LOW voltage level steady state.
l = LOW voltage level one set-up time prior to the HIGH-to-LOW Clock transition.
q = Lower case letters indicate the state of the referenced output one set-up time prior to the HIGH-to-LOW Clock transition.
X = Don't care.
↓ = HIGH-to-LOW Clock transition.

NOTE:
Both outputs will be HIGH while both $\overline{S}_D$ and $\overline{R}_D$ are LOW, but the output states are unpredictable if $\overline{S}_D$ and $\overline{R}_D$ go HIGH simultaneously.

ABSOLUTE MAXIMUM RATINGS (Over operating free-air temperature range unless otherwise noted.)

PARAMETER		74LS	74S	UNIT
V_{CC}	Supply voltage	7.0	7.0	V
V_{IN}	Input voltage	−0.5 to −7.0	−0.5 to +5.5	V
I_{IN}	Input current	−30 to +1	−30 to +5	mA
V_{OUT}	Voltage applied to output in HIGH output state	−0.5 to $+V_{CC}$	−0.5 to $+V_{CC}$	V
T_A	Operating free-air temperature range	0 to 70		°C

RECOMMENDED OPERATING CONDITIONS

PARAMETER		74LS			74S			UNIT
		Min	Nom	Max	Min	Nom	Max	
V_{CC}	Supply voltage	4.75	5.0	5.25	4.75	5.0	5.25	V
V_{IH}	HIGH-level input voltage	2.0			2.0			V
V_{IL}	LOW-level input voltage			+0.8			+0.8	V
I_{IK}	Input clamp current			−18			−18	mA
I_{OH}	HIGH-level output current			−400			−1000	μA
I_{OL}	LOW-level output current			8			20	mA
T_A	Operating free-air temperature	0		70	0		70	°C

Signetics

74121 Multivibrator

Monostable Multivibrator
Product Specification

Logic Products

FEATURES

- **Very good pulse width stability**
- **Virtually immune to temperature and voltage variations**
- **Schmitt trigger input for slow input transitions**
- **Internal timing resistor provided**

DESCRIPTION

These multivibrators feature dual active LOW going edge inputs and a single active HIGH going edge input which can be used as an active HIGH enable input. Complementary output pulses are provided.

Pulse triggering occurs at a particular voltage level and is not directly related to the transition time of the input pulse. Schmitt-trigger input circuitry (TTL hysteresis) for the B input allows jitter-free triggering from inputs with transition rates as slow as 1 volt/second, providing the circuit with an excellent noise immunity of typically 1.2 volts. A high immunity to V_{CC} noise of typically 1.5 volts is also provided by internal latching circuitry. Once fired, the outputs are independent of further transitions of the inputs and are a function only of the timing components. Input pulses may be of any duration relative to the output pulse. Output pulse length may be varied from 20 nanoseconds to 28 seconds by choosing appropriate timing components. With no external timing components (i.e., R_{int} connected to V_{CC}, C_{ext} and R_{ext}/C_{ext} open), an output pulse of typically 30 or 35 nanoseconds is achieved which may be used as a dc triggered reset signal. Output rise and fall times are TTL compatible and independent of pulse length.

Pulse width stability is achieved through internal compensation and is virtually independent of V_{CC} and temperature. In most applications, pulse stability will only be limited by the accuracy of external timing components.

Jitter-free operation is maintained over the full temperature and V_{CC} ranges for more than six decades of timing capacitance (10pF to 10μF) and more than one decade of timing resistance (2kΩ to 30kΩ for the 54121 and 2KΩ to 40kΩ for the 74121). Throughout these ranges, pulse width is defined by the relationship: (see Figure 1)

$$t_W(out) = C_{ext} R_{ext} \ln 2$$

$$t_W(out) \cong 0.7\, C_{ext} R_{ext}$$

TYPE	TYPICAL PROPAGATION DELAY	TYPICAL SUPPLY CURRENT (TOTAL)
74121	43ns	18mA

ORDERING CODE

PACKAGES	COMMERCIAL RANGE $V_{CC} = 5V \pm 5\%$; $T_A = 0°C$ to $+70°C$
Plastic DIP	N74121 N
Plastic SO	N74121 D

NOTE:
For information regarding devices processed to Military Specifications, see the Signetics Military Products Data Manual.

PIN CONFIGURATION

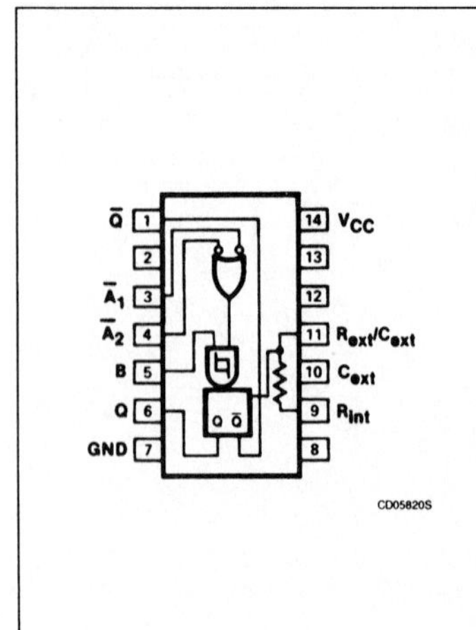

LOGIC SYMBOL

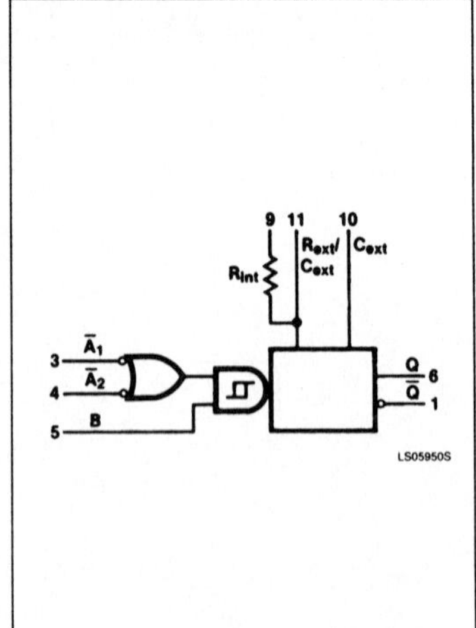

LOGIC SYMBOL (IEEE/IEC)

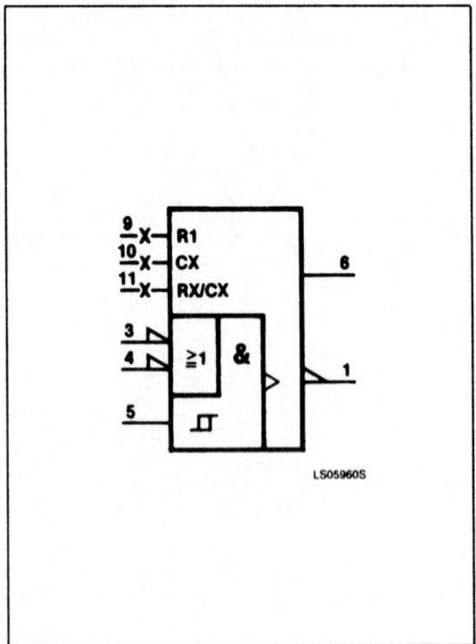

Multivibrator 74121

FUNCTION TABLE

INPUTS			OUTPUTS	
$\overline{A}_1$	$\overline{A}_2$	B	Q	$\overline{Q}$
L	X	H	L	H
X	L	H	L	H
X	X	L	L	H
H	H	X	L	H
H	↓	H	⎍	⊔
↓	H	H	⎍	⊔
↓	↓	H	⎍	⊔
L	X	↑	⎍	⊔
X	L	↑	⎍	⊔

H = HIGH voltage level
L = LOW voltage level
X = Don't care
↑ = LOW-to-HIGH transition
↓ = HIGH-to-LOW transition

INPUT AND OUTPUT LOADING AND FAN-OUT TABLE

PINS	DESCRIPTION	74
$\overline{A}_1$, $\overline{A}_2$	Inputs	1ul
B	Input	2ul
Q, $\overline{Q}$	Outputs	10ul

NOTE:
A 74 unit load (ul) is understood to be 40μA I_{IH} and −1.6mA I_{IL}.

In circuits where pulse cutoff is not critical, timing capacitance up to 1000μF and timing resistance as low as 1.4kΩ may be used.

ABSOLUTE MAXIMUM RATINGS (Over operating free-air temperature range unless otherwise noted.)

PARAMETER		74	UNIT
V_{CC}	Supply voltage	7.0	V
V_{IN}	Input voltage	−0.5 to +5.5	V
I_{IN}	Input current	−30 to +5	mA
V_{OUT}	Voltage applied to output in HIGH output state	−0.5 to +V_{CC}	V
T_A	Operating free-air temperature range	0 to 70	°C

RECOMMENDED OPERATING CONDITIONS

PARAMETER			74			UNIT
			Min	Nom	Max	
V_{CC}	Supply voltage		4.75	5.0	5.25	V
I_{IK}	Input clamp current				−12	mA
I_{OH}	HIGH-level output current				−400	μA
I_{OL}	LOW-level output current				16	mA
dv/dt	Rate of rise or fall of input pulse	B input	1			V/s
		$\overline{A}_1$, $\overline{A}_2$ inputs	1			V/μs
T_A	Operating free-air temperature		0		70	°C

Multivibrator 74121

DC ELECTRICAL CHARACTERISTICS (Over recommended operating free-air temperature range unless otherwise noted.)

PARAMETER		TEST CONDITIONS[1]		74121 Min	Typ[2]	Max	UNIT
V_{T+}	Positive-going threshold at $\overline{A}$ and B	V_{CC} = MIN				2.0	V
V_{T-}	Negative-going threshold at $\overline{A}$ and B	V_{CC} = MIN		0.8			V
V_{OH}	HIGH-level output voltage	V_{CC} = MIN, V_{IH} = MIN, V_{IL} = MAX, I_{OH} = MAX		2.4	3.4		V
V_{OL}	LOW-level output voltage	V_{CC} = MIN, V_{IH} = MIN, V_{IL} = MAX, I_{OL} = MAX			0.2	0.4	V
V_{IK}	Input clamp voltage	V_{CC} = MIN, $I_I = I_{IK}$				−1.5	V
I_I	Input current at maximum input voltage	V_{CC} = MAX, V_I = 5.5V				1.0	mA
I_{IH}	HIGH-level input current	V_{CC} = MAX, V_I = 2.4V	$\overline{A}_1$, $\overline{A}_2$ inputs			40	μA
			B input			80	μA
I_{IL}	LOW-level input current	V_{CC} = MAX, V_I = 0.4V	$\overline{A}_1$, $\overline{A}_2$ inputs			−1.6	mA
			B input			−3.2	mA
I_{OS}	Short-circuit output current[3]	V_{CC} = MAX		−18		−55	mA
I_{CC}	Supply current (total)	V_{CC} = MAX	Quiescent		13	25	mA
			Triggered		23	40	mA

NOTES:
1. For conditions shown as MIN or MAX, use the appropriate value specified under recommended operating conditions for the applicable type.
2. All typical values are at V_{CC} = 5V, T_A = 25°C.
3. I_{OS} is tested with V_{OUT} = +0.5V and V_{CC} = V_{CC} MAX + 0.5V. Not more than one output should be shorted at a time and duration of the short circuit should not exceed one second.

AC ELECTRICAL CHARACTERISTICS T_A = 25°C, V_{CC} = 5.0V

PARAMETER		TEST CONDITIONS	74 (C_L = 15pF, R_L = 400Ω) Min	Max	UNIT
t_{PLH} t_{PHL}	Propagation delay $\overline{A}$ input to Q & $\overline{Q}$ output	Waveform 1 C_{ext} = 80pF, R_{int} to V_{CC}		70 80	ns
t_{PLH} t_{PHL}	Propagation delay B input to Q & $\overline{Q}$ output	Waveform 2 C_{ext} = 80pF, R_{int} to V_{CC}		55 65	ns
t_W	Minimum output pulse width	C_{ext} = 0pF, R_{int} to V_{CC}	20	50	ns
t_W	Output pulse width	C_{ext} = 80pF, R_{int} to V_{CC}	70	150	ns
		C_{ext} = 100pF, R_{ext} = 10kΩ	600	800	ns
		C_{ext} = 1μF, R_{ext} = 10kΩ	6.0	8.0	ms

AC SET-UP REQUIREMENTS $T_A = 25°C$, $V_{CC} = 5.0V$

PARAMETER		TEST CONDITIONS	74		UNIT
			Min	Max	
t_W	Minimum input pulse width to trigger	Waveforms 1 & 2	50		ns
R_{ext}	External timing resistor range		1.4	40	kΩ
C_{ext}	External timing capacitance range		0	1000	μF
	Output duty cycle	$R_{ext} = 2kΩ$		67	%
		$R_{ext} = R_{ext}(Max)$		90	%

AC WAVEFORMS

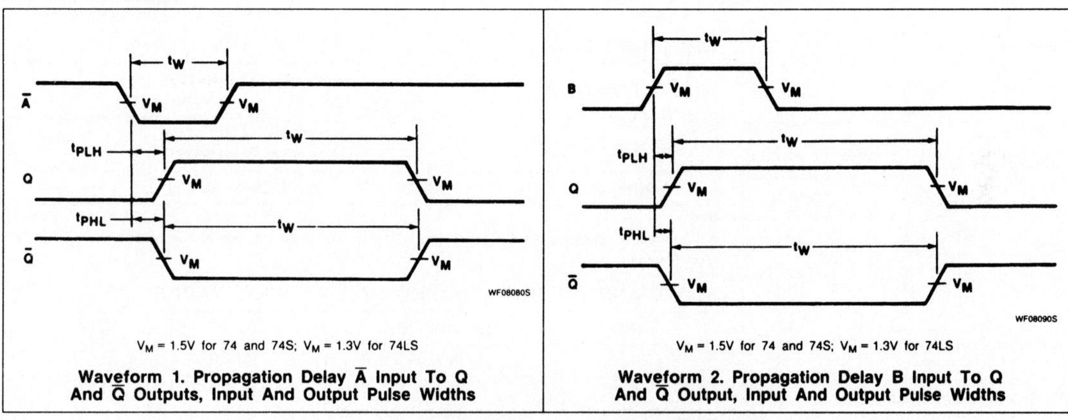

V_M = 1.5V for 74 and 74S; V_M = 1.3V for 74LS

Waveform 1. Propagation Delay A̅ Input To Q And Q̅ Outputs, Input And Output Pulse Widths

V_M = 1.5V for 74 and 74S; V_M = 1.3V for 74LS

Waveform 2. Propagation Delay B Input To Q And Q̅ Output, Input And Output Pulse Widths

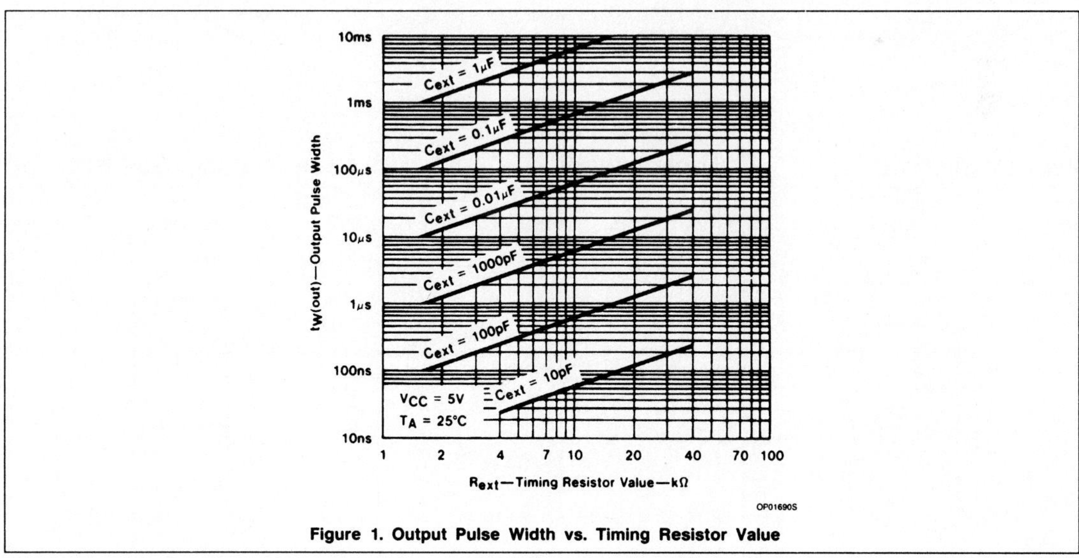

Figure 1. Output Pulse Width vs. Timing Resistor Value

Signetics

74125, 74126, LS125A, LS126A
Buffers

Logic Products

Quad 3-State Buffer
Product Specification

TYPE	TYPICAL PROPAGATION DELAY	TYPICAL SUPPLY CURRENT (TOTAL)
74125	10ns	32mA
74LS125A	8ns	11mA
74126	10ns	36mA
74LS126A	9ns	12mA

FUNCTION TABLE '125, 'LS125A

INPUTS		OUTPUT
C	A	Y
L	L	L
L	H	H
H	X	(Z)

FUNCTION TABLE '126, 'LS126A

INPUTS		OUTPUT
C	A	Y
H	L	L
H	H	H
L	X	(Z)

H = HIGH voltage level
L = LOW voltage level
X = Don't care
(Z) = HIGH impedance (off)

ORDERING CODE

PACKAGES	COMMERCIAL RANGE V_{CC} = 5V ±5%; T_A = 0°C to +70°C
Plastic DIP	N74125N, N74LS125N N74126N, N74LS126N
Plastic SO	N74LS125AD

NOTE:
For information regarding devices processed to Military Specifications, see the Signetics Military Products Data Manual.

INPUT AND OUTPUT LOADING AND FAN-OUT TABLE

PINS	DESCRIPTION	74	74LS
All	Inputs	1ul	1LSul
All	Outputs	10ul	30LSul*

NOTE:
Where a 74 unit load (ul) is understood to be 40μA I_{IH} and −1.6mA I_{IL}, and a 74LS unit load (LSul) is 20μA I_{IH} and −0.4mA I_{IL}.

PIN CONFIGURATION

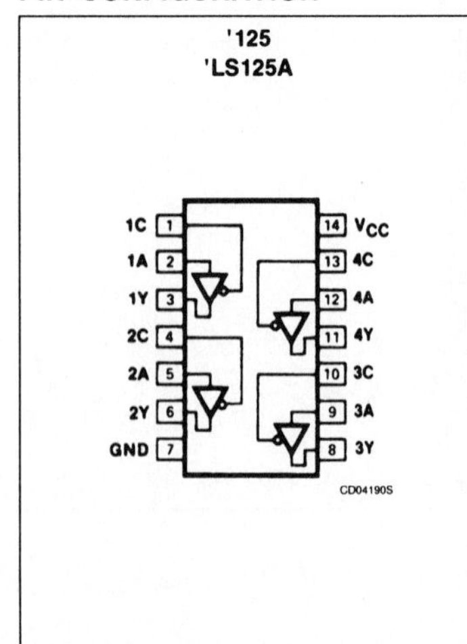

LOGIC SYMBOL

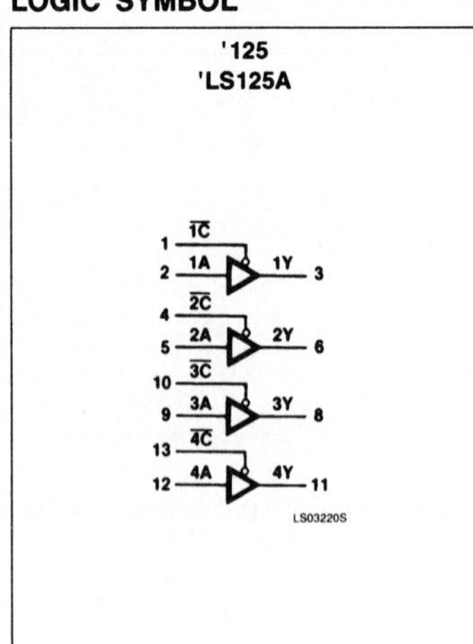

LOGIC SYMBOL (IEEE/IEC)

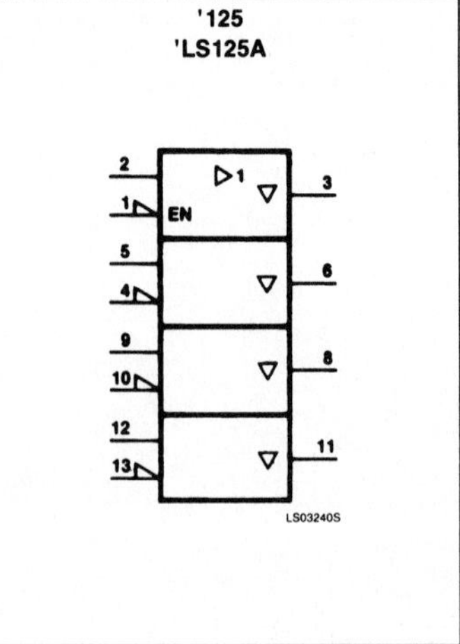

Buffers 74125, 74126, LS125A, LS126A

PIN CONFIGURATION

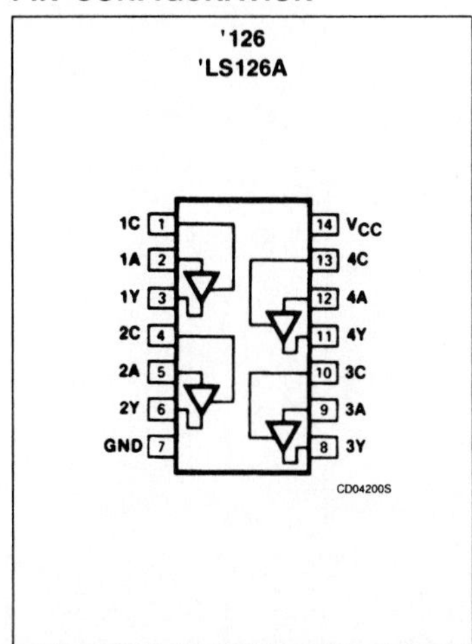

LOGIC SYMBOL

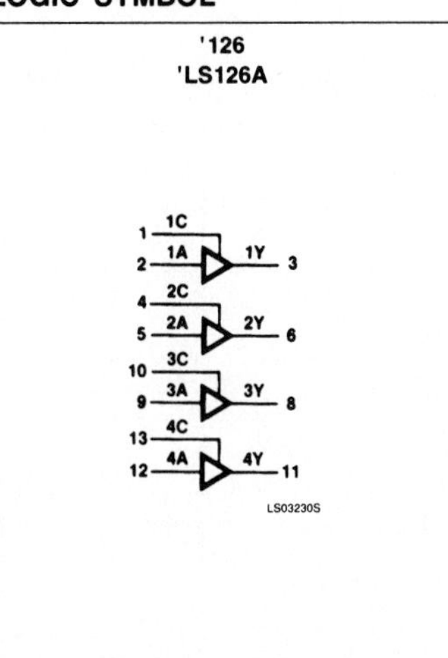

LOGIC SYMBOL (IEEE/IEC)

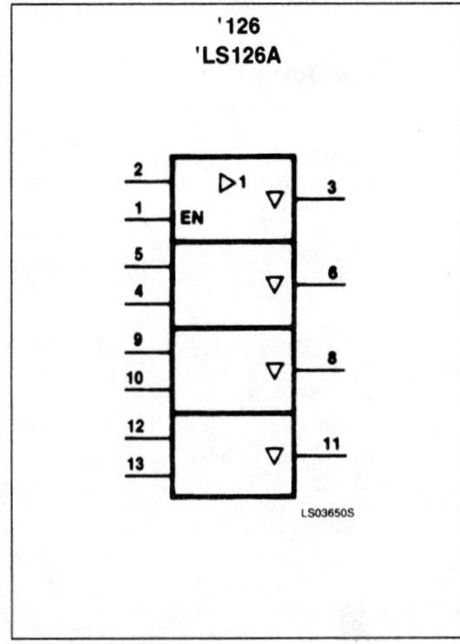

ABSOLUTE MAXIMUM RATINGS (Over operating free-air temperature range unless otherwise noted.)

PARAMETER		74	74LS	UNIT
V_{CC}	Supply voltage	7.0	7.0	V
V_{IN}	Input voltage	−0.5 to +5.5	−0.5 to +7.0	V
I_{IN}	Input current	−30 to +5	−30 to +1	mA
V_{OUT}	Voltage applied to output in HIGH output state	−0.5 to $+V_{CC}$	−0.5 to $+V_{CC}$	V
T_A	Operating free-air temperature range	0 to 70		°C

RECOMMENDED OPERATING CONDITIONS

PARAMETER		74			74LS			UNIT
		Min	Nom	Max	Min	Nom	Max	
V_{CC}	Supply voltage	4.75	5.0	5.25	4.75	5.0	5.25	V
V_{IH}	HIGH-level input voltage	2.0			2.0			V
V_{IL}	LOW-level input voltage			+0.8			+0.8	V
I_{IK}	Input clamp current			−12			−18	mA
I_{OH}	HIGH-level output current			−5.2			−2.6	mA
I_{OL}	LOW-level output current			16			24	mA
T_A	Operating free-air temperature	0		70	0		70	°C

Buffers 74125, 74126, LS125A, LS126A

DC ELECTRICAL CHARACTERISTICS (Over recommended operating free-air temperature range unless otherwise noted.)

PARAMETER		TEST CONDITIONS[1]		74125 74126			74LS125A 74LS126A			UNIT
				Min	Typ[2]	Max	Min	Typ[2]	Max	
V_{OH}	HIGH-level output voltage	V_{CC} = MIN, V_{IH} = MIN, V_{IL} = MAX, I_{OH} = MAX		2.4	3.1		2.4			V
V_{OL}	LOW-level output voltage	V_{CC} = MIN, V_{IH} = MIN, V_{IL} = MAX	I_{OL} = MAX			0.4		0.35	0.5	V
			I_{OL} = 12mA (74LS)					0.25	0.4	V
V_{IK}	Input clamp voltage	V_{CC} = MIN, $I_I = I_{IK}$				−1.5			−1.5	V
I_{OZH}	Off-state output current, HIGH-level voltage applied	V_{CC} = MAX, V_{IH} = MIN, V_{IL} = MAX, V_O = 2.4V				40			20	μA
I_{OZL}	Off-state output current, LOW-level voltage applied	V_{CC} = MAX, V_{IH} = MIN, V_{IL} = MAX, V_O = 0.4V				−40			−20	μA
I_I	Input current at maximum input voltage	V_{CC} = MAX	V_I = 5.5V			1.0				mA
			V_I = 7.0V						0.1	mA
I_{IH}	HIGH-level input current	V_{CC} = MAX	V_I = 2.4V			40				μA
			V_I = 2.7V						20	μA
I_{IL}	LOW-level input current	V_{CC} = MAX, V_I = 0.4V				−1.6			−0.4	mA
I_{OS}	Short-circuit output current[3]	V_{CC} = MAX		−28		−70	−40		−130	mA
I_{CC}	Supply current (total)	V_{CC} = MAX	'125		32	54		11	20	mA
			'126		36	62		12	22	mA

NOTES:

1. For conditions shown as MIN or MAX, use the appropriate value specified under recommended operating conditions for the applicable type.
2. All typical values are at V_{CC} = 5V, T_A = 25°C.
3. I_{OS} is tested with V_{OUT} = +0.5V and V_{CC} = V_{CC} MAX + 0.5V. Not more than one output should be shorted at a time and duration of the short circuit should not exceed one second.

AC WAVEFORMS

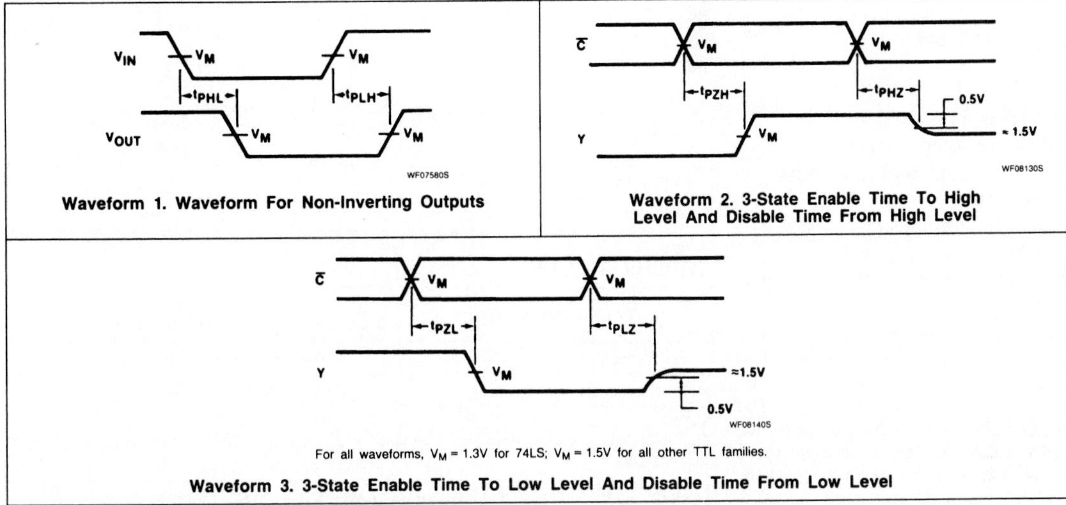

Waveform 1. Waveform For Non-Inverting Outputs

Waveform 2. 3-State Enable Time To High Level And Disable Time From High Level

For all waveforms, V_M = 1.3V for 74LS; V_M = 1.5V for all other TTL families.

Waveform 3. 3-State Enable Time To Low Level And Disable Time From Low Level

AC ELECTRICAL CHARACTERISTICS T_A = 25°C, V_{CC} = 5.0V

PARAMETER		TEST CONDITIONS	74125		74LS125A		74126		74LS126A		UNIT
			C_L = 50pF, R_L = 400Ω		C_L = 45pF, R_L = 667Ω		C_L = 50pF, R_L = 400Ω		C_L = 45pF, R_L = 667Ω		
			Min	Max	Min	Max	Min	Max	Min	Max	
t_{PLH} t_{PHL}	Propagation delay Data to output	Waveform 1		13 18		15 18		13 18		15 18	ns
t_{PZH}	Enable to HIGH	Waveform 2		17		20		18		25	ns
t_{PZL}	Enable to LOW	Waveform 3		25		25		25		35	ns
t_{PHZ}	Disable from HIGH	Waveform 2, C_L = 5pF		8.0		20		16		25	ns
t_{PLZ}	Disable from LOW	Waveform 3, C_L = 5pF		12		20		18		25	ns

Signetics

74LS138, S138
Decoders/Demultiplexers

1-Of-8 Decoder/Demultiplexer
Product Specification

Logic Products

FEATURES

- **Demultiplexing capability**
- **Multiple input enable for easy expansion**
- **Ideal for memory chip select decoding**
- **Direct replacement for Intel 3205**

DESCRIPTION

The '138 decoder accepts three binary weighted inputs (A_0, A_1, A_2) and when enabled, provides eight mutually exclusive, active LOW outputs ($\overline{0}-\overline{7}$). The device features three Enable Inputs: two active LOW ($\overline{E}_1$, $\overline{E}_2$) and one active HIGH (E_3). Every output will be HIGH unless $\overline{E}_1$ and $\overline{E}_2$ are LOW and E_3 is HIGH. This multiple enable function allows easy parallel expansion of the device to a 1-of-32 (5 lines to 32 lines) decoder with just four '138s and one inverter.

The device can be used as an eight output demultiplexer by using one of the active LOW Enable inputs as the Data input and the remaining Enable inputs as strobes. Enable inputs not used must be permanently tied to their appropriate active HIGH or active LOW state.

TYPE	TYPICAL PROPAGATION DELAY	TYPICAL SUPPLY CURRENT (TOTAL)
74LS138	20ns	6.3mA
74S138	7ns	49mA

ORDERING CODE

PACKAGES	COMMERCIAL RANGE V_{CC} = 5V ±5%; T_A = 0°C to +70°C
Plastic DIP	N74S138N, N74LS138N
Plastic SO	N74LS138D, N74S138D

NOTE:
For information regarding devices processed to Military Specifications see the Signetics Military Products Data Manual.

INPUT AND OUTPUT LOADING AND FAN-OUT TABLE

PINS	DESCRIPTION	74S	74LS
All	Inputs	1Sul	1LSul
All	Outputs	10Sul	10LSul

NOTE:
Where a 74S unit load (Sul) is 50μA I_{IH} and −2.0mA I_{IL}, and a 74LS unit load (LSul) is 20μA I_{IH} and −0.4mA I_{IL}.

PIN CONFIGURATION

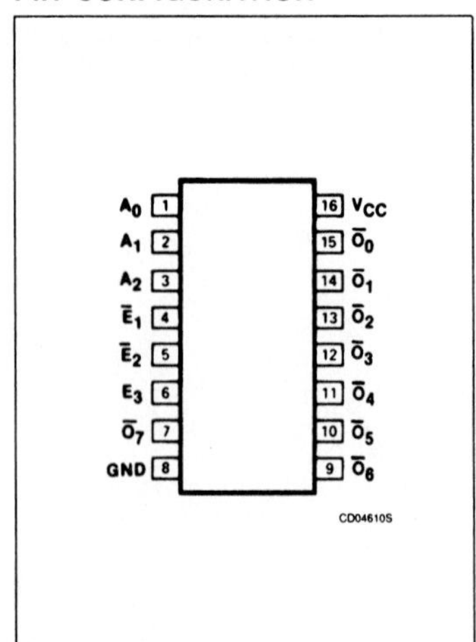

LOGIC SYMBOL

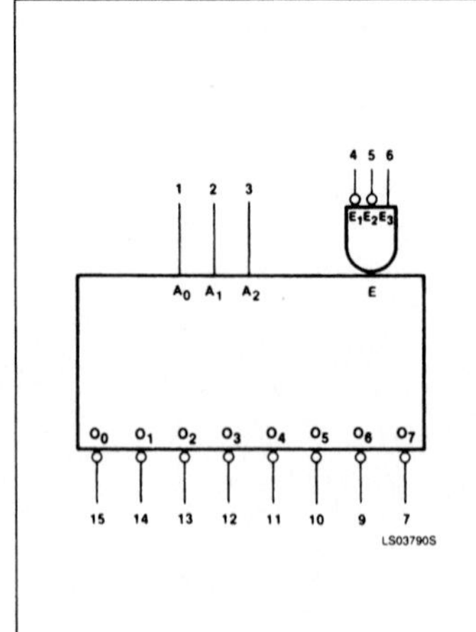

LOGIC SYMBOL (IEEE/IEC)

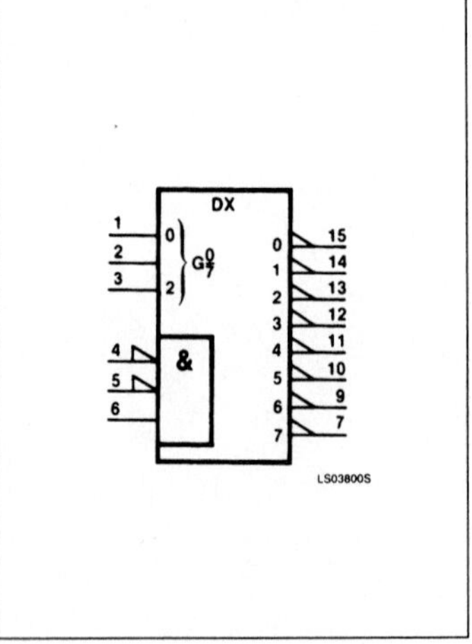

LOGIC DIAGRAM

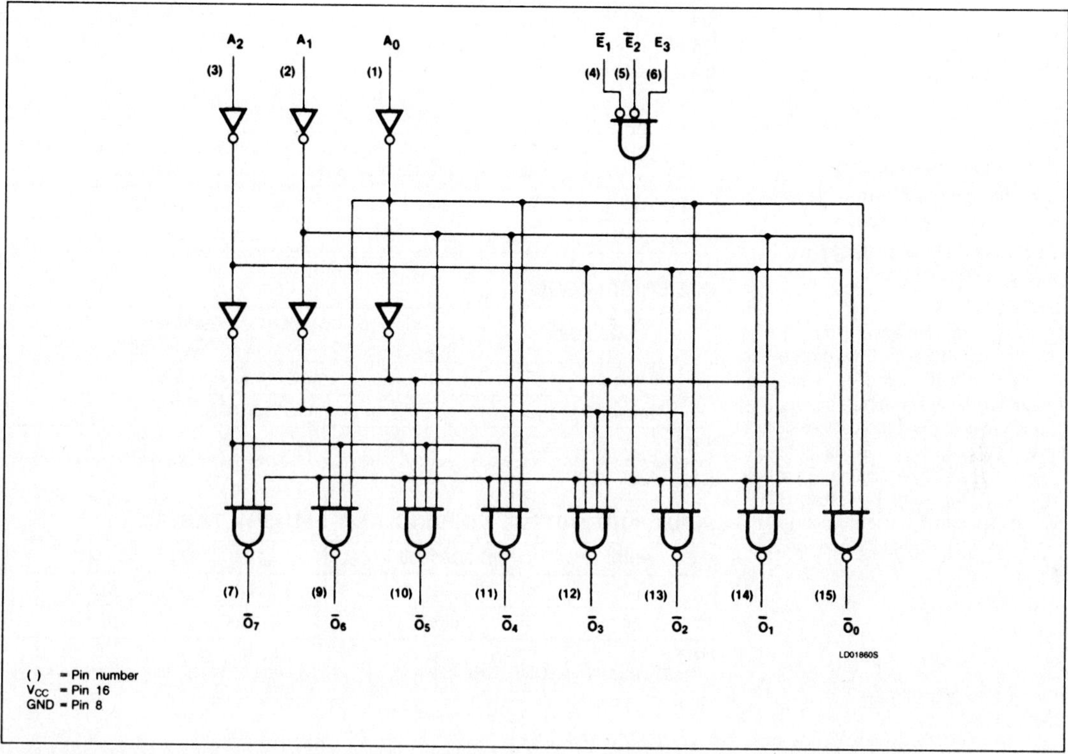

FUNCTION TABLE

INPUTS						OUTPUTS							
$\overline{E}_1$	$\overline{E}_2$	E_3	A_0	A_1	A_2	$\overline{0}$	$\overline{1}$	$\overline{2}$	$\overline{3}$	$\overline{4}$	$\overline{5}$	$\overline{6}$	$\overline{7}$
H	X	X	X	X	X	H	H	H	H	H	H	H	H
X	H	X	X	X	X	H	H	H	H	H	H	H	H
X	X	L	X	X	X	H	H	H	H	H	H	H	H
L	L	H	L	L	L	L	H	H	H	H	H	H	H
L	L	H	H	L	L	H	L	H	H	H	H	H	H
L	L	H	L	H	L	H	H	L	H	H	H	H	H
L	L	H	H	H	L	H	H	H	L	H	H	H	H
L	L	H	L	L	H	H	H	H	H	L	H	H	H
L	L	H	H	L	H	H	H	H	H	H	L	H	H
L	L	H	L	H	H	H	H	H	H	H	H	L	H
L	L	H	H	H	H	H	H	H	H	H	H	H	L

H = HIGH voltage level
L = LOW voltage level
X = Don't care

Signetics

74LS139, S139

Decoders/Demultiplexers

Dual 1-of-4 Decoder/Demultiplexer
Product Specification

Logic Products

FEATURES

- **Demultiplexing capability**
- **Two independent 1-of-4 decoders**
- **Multifunction capability**
- **Replaces 9321 and 93L21 for higher performance**

DESCRIPTION

The '139 is a high-speed, dual 1-of-4 decoder/demultiplexer. This device has two independent decoders, each accepting two binary weighted inputs (A_0, A_1) and providing four mutually exclusive active LOW outputs ($\overline{0}-\overline{3}$). Each decoder has an active LOW Enable ($\overline{E}$). When $\overline{E}$ is HIGH, every output is forced HIGH. The Enable can be used as the Data input for a 1-of-4 demultiplexer application.

TYPE	TYPICAL PROPAGATION DELAY (ENABLE AT 2 LOGIC LEVELS)	TYPICAL SUPPLY CURRENT (TOTAL)
74LS139	19ns	6.8mA
74S139	6ns	60mA

ORDERING CODE

PACKAGES	COMMERCIAL RANGE $V_{CC} = 5V \pm 5\%$; $T_A = 0°C$ to $+70°C$
Plastic DIP	N74S139N, N74LS139N
Plastic SO	N74LS139D, N74S139D

NOTE:
For information regarding devices processed to Military Specifications, see the Signetics Military Products Data Manual.

INPUT AND OUTPUT LOADING AND FAN-OUT TABLE

PINS	DESCRIPTION	74S	74LS
All	Inputs	1Sul	1LSul
All	Outputs	10Sul	10LSul

NOTE:
A 74S unit load (Sul) is 50μA I_{IH} and −2.0mA I_{IL}, and a 74LS unit load (LSul) is 20μA I_{IH} and −0.4mA I_{IL}.

PIN CONFIGURATION

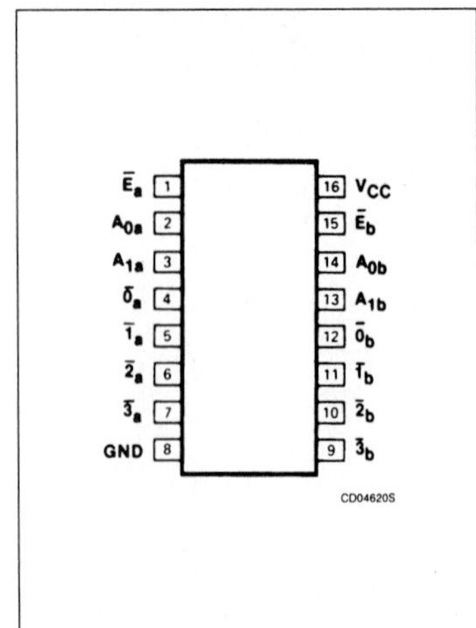

LOGIC SYMBOL

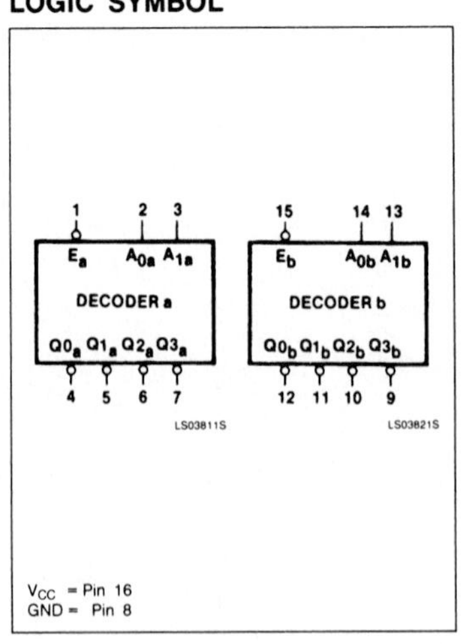

LOGIC SYMBOL (EEE/IEC)

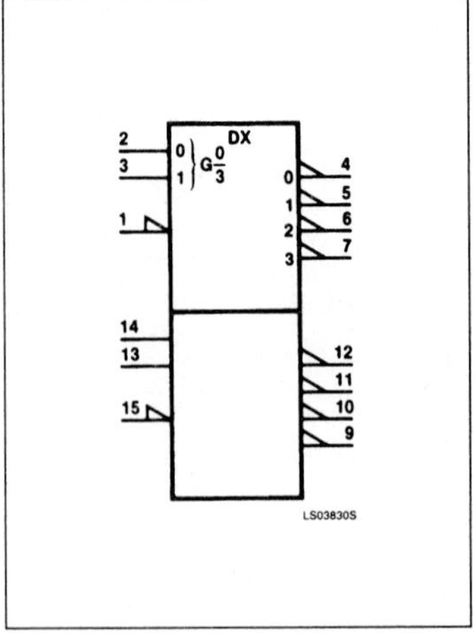

LOGIC DIAGRAM

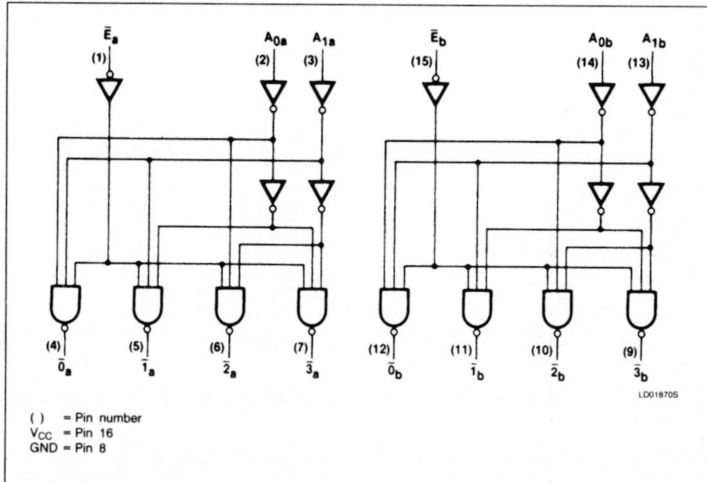

FUNCTION TABLE

INPUTS			OUTPUTS			
$\bar{E}$	A_0	A_1	$\bar{0}$	$\bar{1}$	$\bar{2}$	$\bar{3}$
H	X	X	H	H	H	H
L	L	L	L	H	H	H
L	H	L	H	L	H	H
L	L	H	H	H	L	H
L	H	H	H	H	H	L

H = HIGH voltage level
L = LOW voltage level

ABSOLUTE MAXIMUM RATINGS (Over operating free-air temperature range unless otherwise noted.)

PARAMETER		74LS	74S	UNIT
V_{CC}	Supply voltage	7.0	7.0	V
V_{IN}	Input voltage	−0.5 to +7.0	−0.5 to +5.5	V
I_{IN}	Input current	−30 to +1	−30 to +5	mA
V_{OUT}	Voltage applied to output in HIGH output state	−0.5 to $+V_{CC}$	−0.5 to $+V_{CC}$	V
T_A	Operating free-air temperature range	0 to 70		°C

RECOMMENDED OPERATING CONDITIONS

PARAMETER		74LS			74S			UNIT
		Min	Nom	Max	Min	Nom	Max	
V_{CC}	Supply voltage	4.75	5.0	5.25	4.75	5.0	5.25	V
V_{IH}	HIGH-level input voltage	2.0			2.0			V
V_{IL}	LOW-level input voltage			+0.8			+0.8	V
I_{IK}	Input clamp current			−18			−18	mA
I_{OH}	HIGH-level output current			−400			−1000	μA
I_{OL}	LOW-level output current			8			20	mA
T_A	Operating free-air temperature	0		70	0		70	°C

Signetics

74147
Encoder

10-Line-To-4-Line Priority Encoder
Product Specification

Logic Products

FEATURES

- **Encodes 10-line decimal to 4-line BCD**
- **Useful for 10-position switch encoding**
- **Used in code converters and generators**

DESCRIPTION

The '147 9-input priority encoder accepts data from nine active-LOW inputs ($\bar{I}_1 - \bar{I}_9$) and provides a binary representation on the four active-LOW outputs ($A_0 - A_3$). A priority is assigned to each input so that when two or more inputs are simultaneously active, the input with the highest priority is represented on the output, with input line $\bar{I}_9$ having the highest priority.

The device provides the 10-line-to-4-line priority encoding function by use of the implied decimal "zero." The "zero" is encoded when all nine data inputs are HIGH, forcing all four outputs HIGH.

TYPE	TYPICAL PROPAGATION DELAY	TYPICAL SUPPLY CURRENT (TOTAL)
74147	10ns	46mA

ORDERING CODE

PACKAGES	COMMERCIAL RANGE $V_{CC} = 5V \pm 5\%$; $T_A = 0°C$ to $+70°C$
Plastic DIP	N74147N

NOTE:
For information regarding devices processed to Military Specifications see the Signetics Military Products Data Manual.

INPUT AND OUTPUT LOADING AND FAN-OUT TABLE

PINS	DESCRIPTION	74
All	Inputs	1ul
All	Outputs	10ul

NOTE:
A 74 unit load (ul) is understood to be $40\mu A$ I_{IH} and $-1.6mA$ I_{IL}.

PIN CONFIGURATION

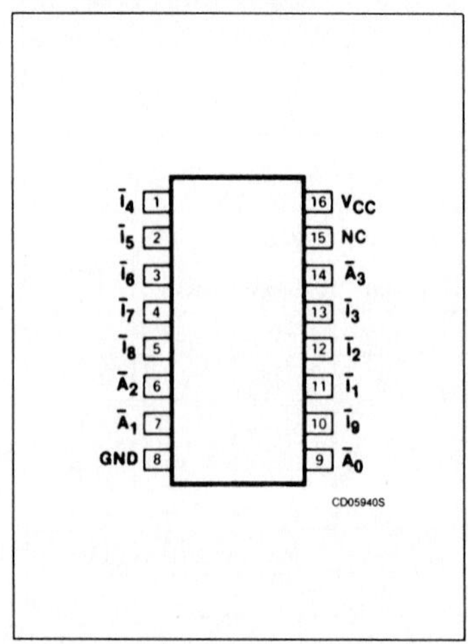

LOGIC SYMBOL

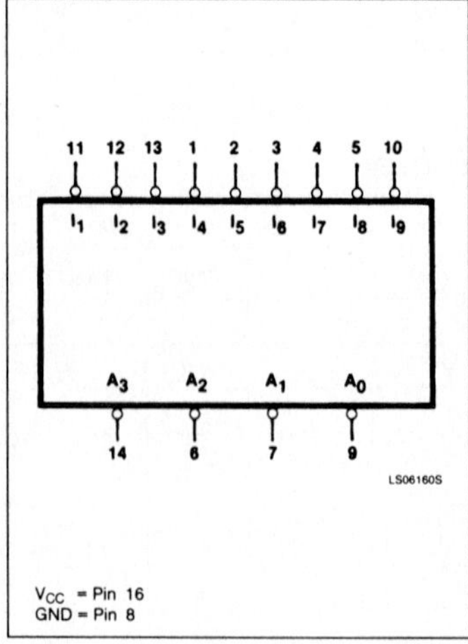

LOGIC SYMBOL (IEEE/IEC)

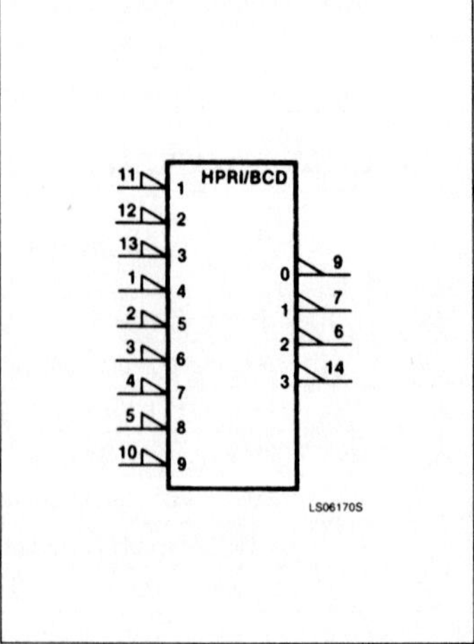

Encoder 74147

LOGIC DIAGRAM

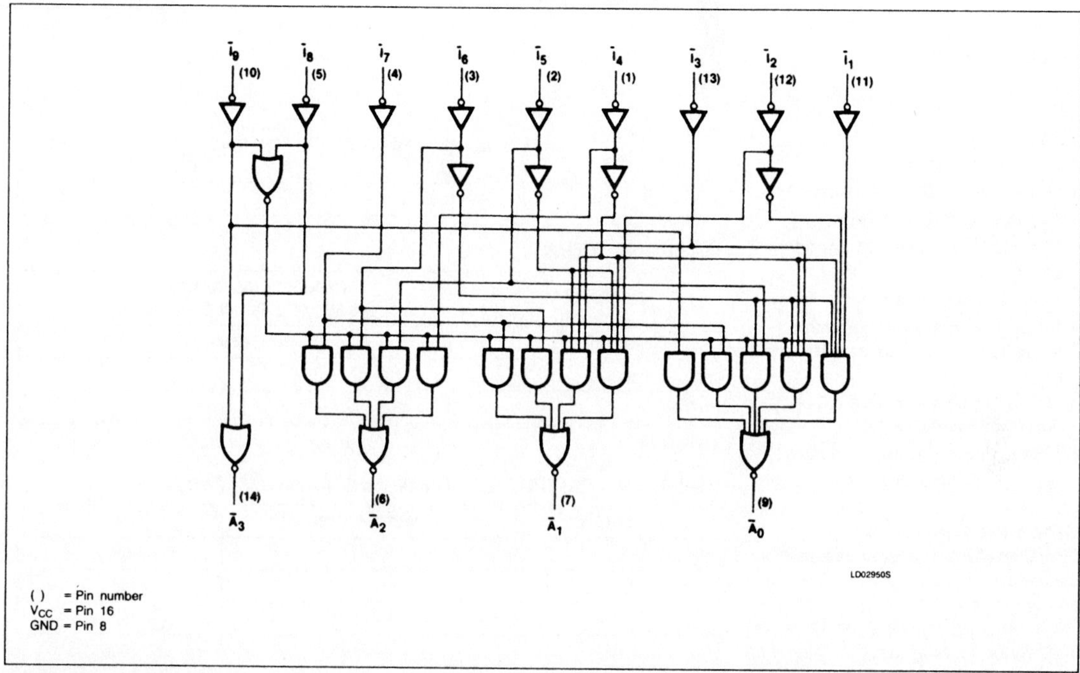

FUNCTION TABLE

INPUTS									OUTPUTS			
$\bar{I}_1$	$\bar{I}_2$	$\bar{I}_3$	$\bar{I}_4$	$\bar{I}_5$	$\bar{I}_6$	$\bar{I}_7$	$\bar{I}_8$	$\bar{I}_9$	$\bar{A}_3$	$\bar{A}_2$	$\bar{A}_1$	$\bar{A}_0$
H	H	H	H	H	H	H	H	H	H	H	H	H
X	X	X	X	X	X	X	X	L	L	H	H	L
X	X	X	X	X	X	X	L	H	L	H	H	H
X	X	X	X	X	X	L	H	H	H	L	L	L
X	X	X	X	X	L	H	H	H	H	L	L	H
X	X	X	X	L	H	H	H	H	H	L	H	L
X	X	X	L	H	H	H	H	H	H	L	H	H
X	X	L	H	H	H	H	H	H	H	H	L	L
X	L	H	H	H	H	H	H	H	H	H	L	H
L	H	H	H	H	H	H	H	H	H	H	H	L

H = HIGH voltage level
L = LOW voltage level
X = Don't care

Signetics

74148
Encoder

8-Input Priority Encoder
Product Specification

Logic Products

FEATURES

- Code conversions
- Multi-channel D/A converter
- Decimal-to-BCD converter
- Cascading for priority encoding of "N" bits
- Input Enable capability
- Priority encoding — automatic selection of highest priority input line
- Output Enable — active LOW when all inputs HIGH
- Group Signal output — active when any input is LOW

DESCRIPTION

The '148 8-input priority encoder accepts data from eight active-LOW inputs and provides a binary representation on the three active-LOW outputs. A priority is assigned to each input so that when two or more inputs are simultaneously active, the input with the highest priority is represented on the output, with input line $\bar{I}_7$ having the highest priority.

TYPE	TYPICAL PROPAGATION DELAY	TYPICAL SUPPLY CURRENT (TOTAL)
74148	10ns	38mA

ORDERING CODE

PACKAGES	COMMERCIAL RANGE $V_{CC} = 5V \pm 5\%$; $T_A = 0°C$ to $+70°C$
Plastic DIP	N74148N
Plastic SO	

NOTES:
For information regarding devices processed to Military Specifications, see the Signetics Military Products Data Manual.

INPUT AND OUTPUT LOADING AND FAN-OUT TABLE

PINS	DESCRIPTION	74
$\bar{I}_0$	Input	1ul
$\bar{I}_1 - \bar{I}_7$	Inputs	2ul
$\overline{EI}$	Input	2ul
All	Outputs	10ul

NOTE:
A 74 unit load (ul) is understood to be $40\mu A$ I_{IH} and $-1.6mA$ I_{IL}.

PIN CONFIGURATION

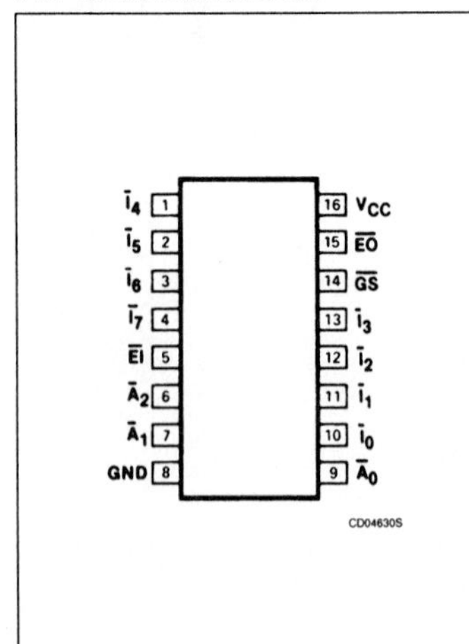

LOGIC SYMBOL

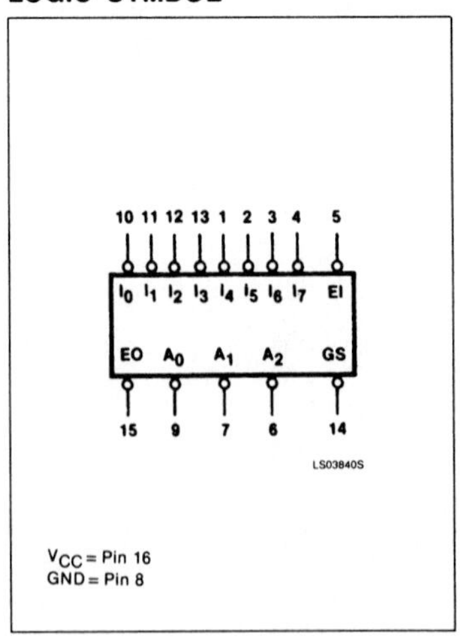

LOGIC SYMBOL (IEEE/IEC)

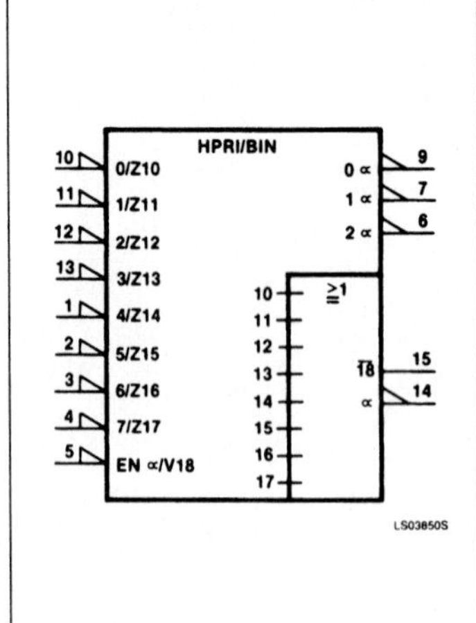

Encoder

74148

A HIGH on the Enable Input ($\overline{EI}$) will force all outputs to the inactive (HIGH) state and allow new data to settle without producing erroneous information at the outputs.

A Group Signal ($\overline{GS}$) output and an Enable Output ($\overline{EO}$) are provided with the three data outputs. The $\overline{GS}$ is active-LOW when any input is LOW; this indicates when any input is active. The $\overline{EO}$ is active-LOW when all inputs are HIGH. Using the Enable Output along with the Enable Input allows priority encoding of N input signals. Both $\overline{EO}$ and $\overline{GS}$ are active-HIGH when the Enable input is HIGH.

LOGIC DIAGRAM

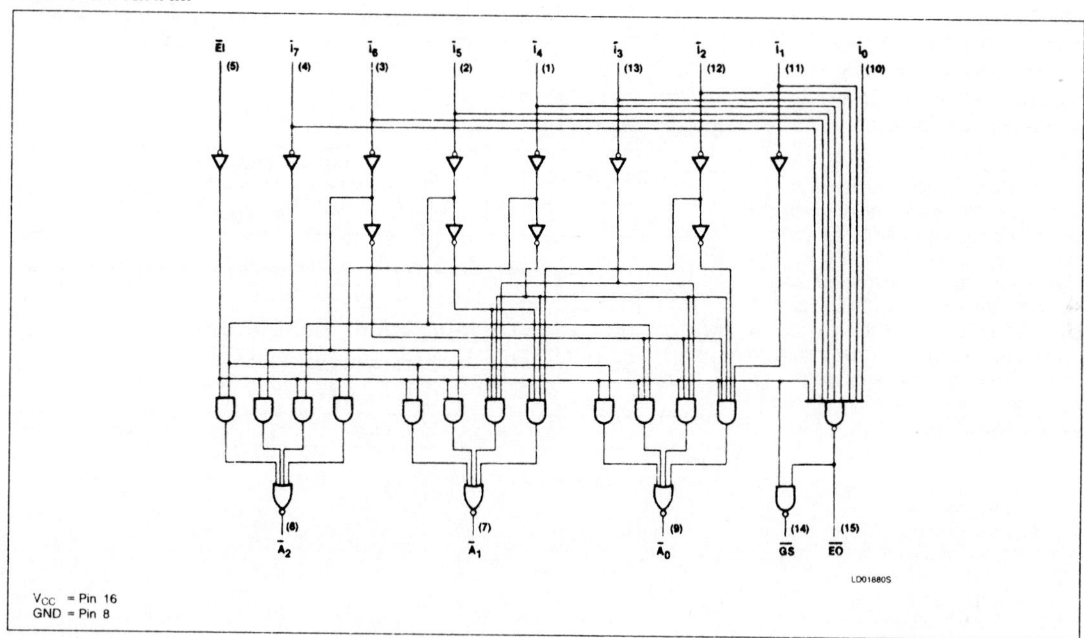

FUNCTION TABLE

INPUTS									OUTPUTS				
$\overline{EI}$	$\bar{I}_0$	$\bar{I}_1$	$\bar{I}_2$	$\bar{I}_3$	$\bar{I}_4$	$\bar{I}_5$	$\bar{I}_6$	$\bar{I}_7$	$\overline{GS}$	$\bar{A}_0$	$\bar{A}_1$	$\bar{A}_2$	$\overline{EO}$
H	X	X	X	X	X	X	X	X	H	H	H	H	H
L	H	H	H	H	H	H	H	H	H	H	H	H	L
L	X	X	X	X	X	X	X	L	L	L	L	L	H
L	X	X	X	X	X	X	L	H	L	H	L	L	H
L	X	X	X	X	X	L	H	H	L	L	H	L	H
L	X	X	X	X	L	H	H	H	L	H	H	L	H
L	X	X	X	L	H	H	H	H	L	L	L	H	H
L	X	X	L	H	H	H	H	H	L	H	L	H	H
L	X	L	H	H	H	H	H	H	L	L	H	H	H
L	L	H	H	H	H	H	H	H	L	H	H	H	H

H = HIGH voltage level
L = LOW voltage level
X = Don't care

Signetics

74150
Multiplexer

16-Input Multiplexer
Product Specification

Logic Products

FEATURES
- **Select data from 16 sources**
- **Demultiplexing capability**
- **Active-LOW enable or strobe**
- **Inverting data output**

DESCRIPTION
The '150 is a logical implementation of a single-pole, 16-position switch with the switch position controlled by the state of four Select inputs. S_0, S_1, S_2, S_3. The Multiplexer output ($\overline{Y}$) inverts the selected data. The Enable input ($\overline{E}$) is active-LOW. When $\overline{E}$ is HIGH the $\overline{Y}$ output is HIGH regardless of all other inputs. In one package the '150 provides the ability to select from 16 sources of data or control information.

TYPE	TYPICAL PROPAGATION DELAY	TYPICAL SUPPLY CURRENT (TOTAL)
74150	17ns	40mA

ORDERING CODE

PACKAGES	COMMERCIAL RANGE $V_{CC} = 5V \pm 5\%$; $T_A = 0°C$ to $+70°C$
Plastic DIP	N74150N

NOTE:
For information regarding devices processed to Military Specifications, see the Signetics Military Products Data Manual.

INPUT AND OUTPUT LOADING AND FAN-OUT TABLE

PINS	DESCRIPTION	74
All	Inputs	1ul
$\overline{Y}$	Output	10ul

NOTE:
A 74 unit load (ul) is understood to be 40μA I_{IH} and −1.6mA I_{IL}.

PIN CONFIGURATION

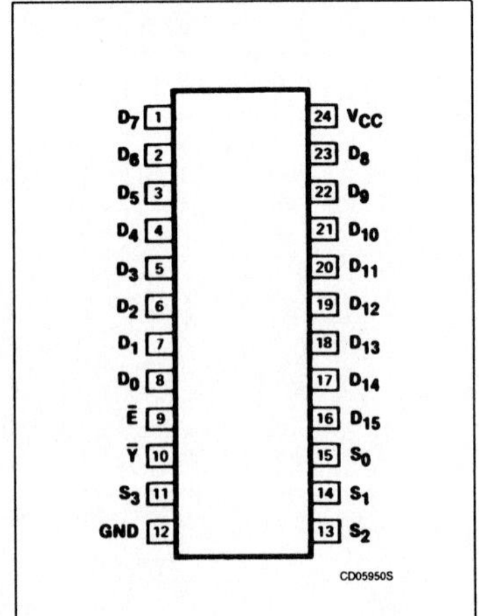

LOGIC SYMBOL

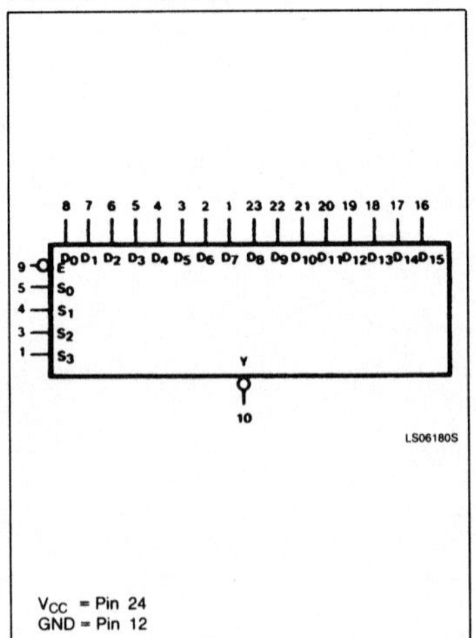

LOGIC SYMBOL (IEEE/IEC)

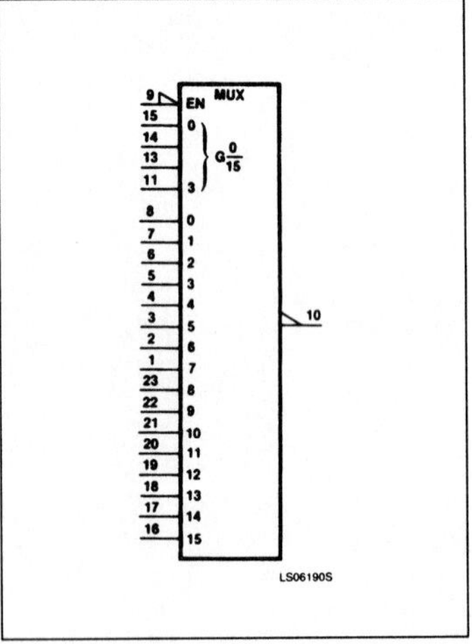

Multiplexer 74150

LOGIC DIAGRAM

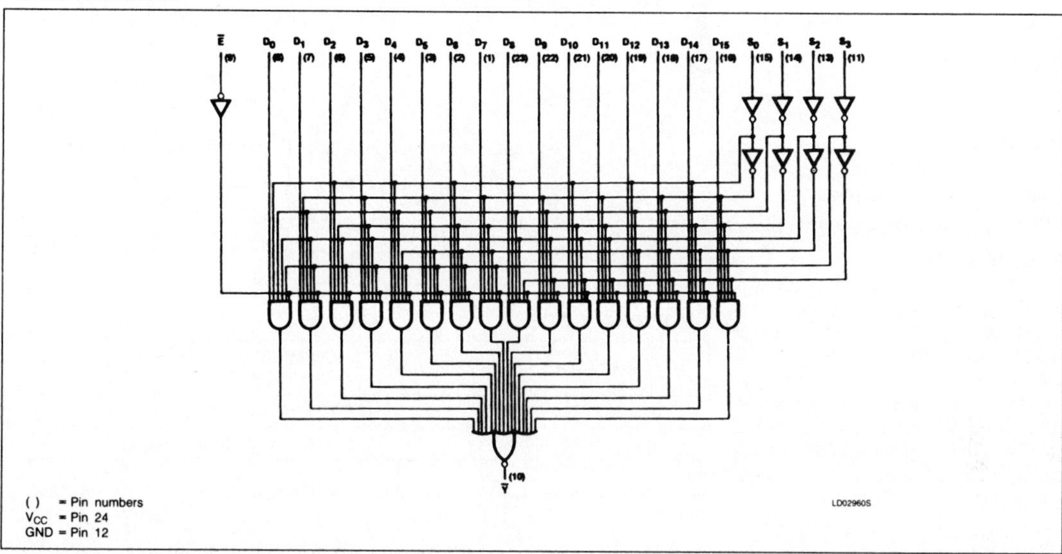

FUNCTION TABLE

INPUTS																					OUTPUT
S_3	S_2	S_1	S_0	$\overline{E}$	D_0	D_1	D_2	D_3	D_4	D_5	D_6	D_7	D_8	D_9	D_{10}	D_{11}	D_{12}	D_{13}	D_{14}	D_{15}	$\overline{Y}$
X	X	X	X	H	X	X	X	X	X	X	X	X	X	X	X	X	X	X	X	X	H
L	L	L	L	L	L	X	X	X	X	X	X	X	X	X	X	X	X	X	X	X	H
L	L	L	L	L	H	X	X	X	X	X	X	X	X	X	X	X	X	X	X	X	L
L	L	L	H	L	X	L	X	X	X	X	X	X	X	X	X	X	X	X	X	X	H
L	L	L	H	L	X	H	X	X	X	X	X	X	X	X	X	X	X	X	X	X	L
L	L	H	L	L	X	X	L	X	X	X	X	X	X	X	X	X	X	X	X	X	H
L	L	H	L	L	X	X	H	X	X	X	X	X	X	X	X	X	X	X	X	X	L
L	L	H	H	L	X	X	X	L	X	X	X	X	X	X	X	X	X	X	X	X	H
L	L	H	H	L	X	X	X	H	X	X	X	X	X	X	X	X	X	X	X	X	L
L	H	L	L	L	X	X	X	X	L	X	X	X	X	X	X	X	X	X	X	X	H
L	H	L	L	L	X	X	X	X	H	X	X	X	X	X	X	X	X	X	X	X	L
L	H	L	H	L	X	X	X	X	X	L	X	X	X	X	X	X	X	X	X	X	H
L	H	L	H	L	X	X	X	X	X	H	X	X	X	X	X	X	X	X	X	X	L
L	H	H	L	L	X	X	X	X	X	X	L	X	X	X	X	X	X	X	X	X	H
L	H	H	L	L	X	X	X	X	X	X	H	X	X	X	X	X	X	X	X	X	L
L	H	H	H	L	X	X	X	X	X	X	X	L	X	X	X	X	X	X	X	X	H
L	H	H	H	L	X	X	X	X	X	X	X	H	X	X	X	X	X	X	X	X	L
H	L	L	L	L	X	X	X	X	X	X	X	X	L	X	X	X	X	X	X	X	H
H	L	L	L	L	X	X	X	X	X	X	X	X	H	X	X	X	X	X	X	X	L
H	L	L	H	L	X	X	X	X	X	X	X	X	X	L	X	X	X	X	X	X	H
H	L	L	H	L	X	X	X	X	X	X	X	X	X	H	X	X	X	X	X	X	L
H	L	H	L	L	X	X	X	X	X	X	X	X	X	X	L	X	X	X	X	X	H
H	L	H	L	L	X	X	X	X	X	X	X	X	X	X	H	X	X	X	X	X	L
H	L	H	H	L	X	X	X	X	X	X	X	X	X	X	X	L	X	X	X	X	H
H	L	H	H	L	X	X	X	X	X	X	X	X	X	X	X	H	X	X	X	X	L
H	H	L	L	L	X	X	X	X	X	X	X	X	X	X	X	X	L	X	X	X	H
H	H	L	L	L	X	X	X	X	X	X	X	X	X	X	X	X	H	X	X	X	L
H	H	L	H	L	X	X	X	X	X	X	X	X	X	X	X	X	X	L	X	X	H
H	H	L	H	L	X	X	X	X	X	X	X	X	X	X	X	X	X	H	X	X	L
H	H	H	L	L	X	X	X	X	X	X	X	X	X	X	X	X	X	X	L	X	H
H	H	H	L	L	X	X	X	X	X	X	X	X	X	X	X	X	X	X	H	X	L
H	H	H	H	L	X	X	X	X	X	X	X	X	X	X	X	X	X	X	X	L	H
H	H	H	H	L	X	X	X	X	X	X	X	X	X	X	X	X	X	X	X	H	L

H = HIGH voltage level
L = LOW voltage level
X = Don't care

Signetics

74151, LS151, S151
Multiplexers

8-Input Multiplexer
Product Specification

Logic Products

FEATURES
- **Multifunction capability**
- **Complementary outputs**
- **See '251 for 3-state version**

DESCRIPTION
The '151 is a logical implementation of a single-pole, 8-position switch with the switch position controlled by the state of three Select inputs, S_0, S_1, S_2. True (Y) and Complement ($\overline{Y}$) outputs are both provided. The Enable input ($\overline{E}$) is active LOW. When $\overline{E}$ is HIGH, the $\overline{Y}$ output is HIGH and the Y output is LOW, regardless of all other inputs. The logic function provided at the output is:

$$Y = \overline{E} \bullet (I_0 \bullet \overline{S}_0 \bullet \overline{S}_1 \bullet \overline{S}_2 + I_1 \bullet S_0 \bullet \overline{S}_1 \bullet \overline{S}_2 + I_2 \bullet \overline{S}_0 \bullet S_1 \bullet \overline{S}_2 + I_3 \bullet S_0 \bullet S_1 \bullet \overline{S}_2 + I_4 \bullet \overline{S}_0 \bullet \overline{S}_1 \bullet S_2 + I_5 \bullet S_0 \bullet \overline{S}_1 \bullet S_2 + I_6 \bullet \overline{S}_0 \bullet S_1 \bullet S_2 + I_7 \bullet S_0 \bullet S_1 \bullet S_2$$

In one package the '151 provides the ability to select from eight sources of data or control information. The device can provide any logic function of four variables and its negation with correct manipulation.

TYPE	TYPICAL PROPAGATION DELAY (ENABLE TO $\overline{Y}$)	TYPICAL SUPPLY CURRENT (TOTAL)
74151	18ns	29mA
74LS151	12ns	6mA
74S151	9ns	45mA

ORDERING CODE

PACKAGES	COMMERCIAL RANGE $V_{CC} = 5V \pm 5\%$; $T_A = 0°C$ to $+70°C$
Plastic DIP	N74151N, N74LS151N, N74S151N
Plastic SO	N74LS151D, N74S151D

NOTE:
For information regarding devices processed to Military Specifications, see the Signetics Military Products Data Manual.

INPUT AND OUTPUT LOADING AND FAN-OUT TABLE

PINS	DESCRIPTION	74	74S	74LS
All	Inputs	1ul	1Sul	1LSul
All	Outputs	10ul	10Sul	10LSul

NOTE:
Where a 74 unit load (ul) is understood to be $40\mu A$ I_{IH} and $-1.6mA$ I_{IL}, a 74S unit load (Sul) is $50\mu A$ I_{IH} and $-2.0mA$ I_{IL}, and 74LS unit load (LSul) is $20\mu A$ I_{IH} and $-0.4mA$ I_{IL}.

PIN CONFIGURATION

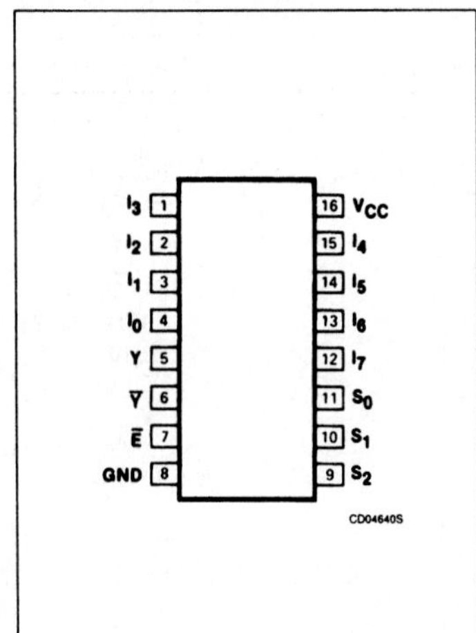

LOGIC SYMBOL

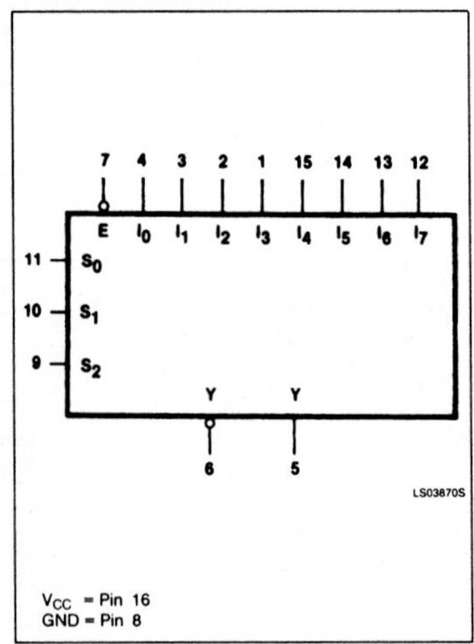

LOGIC SYMBOL (IEEE/IEC)

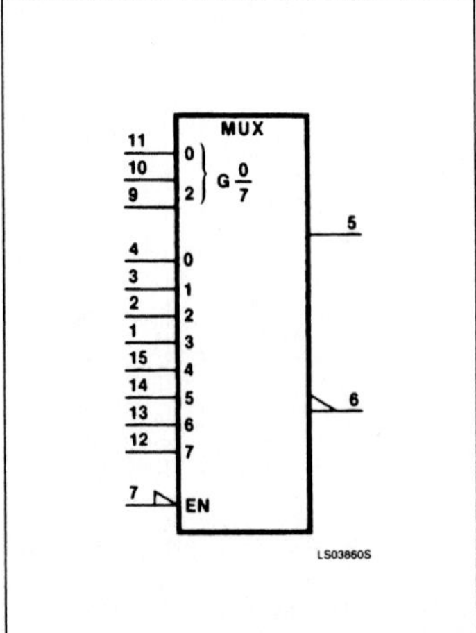

LOGIC DIAGRAM

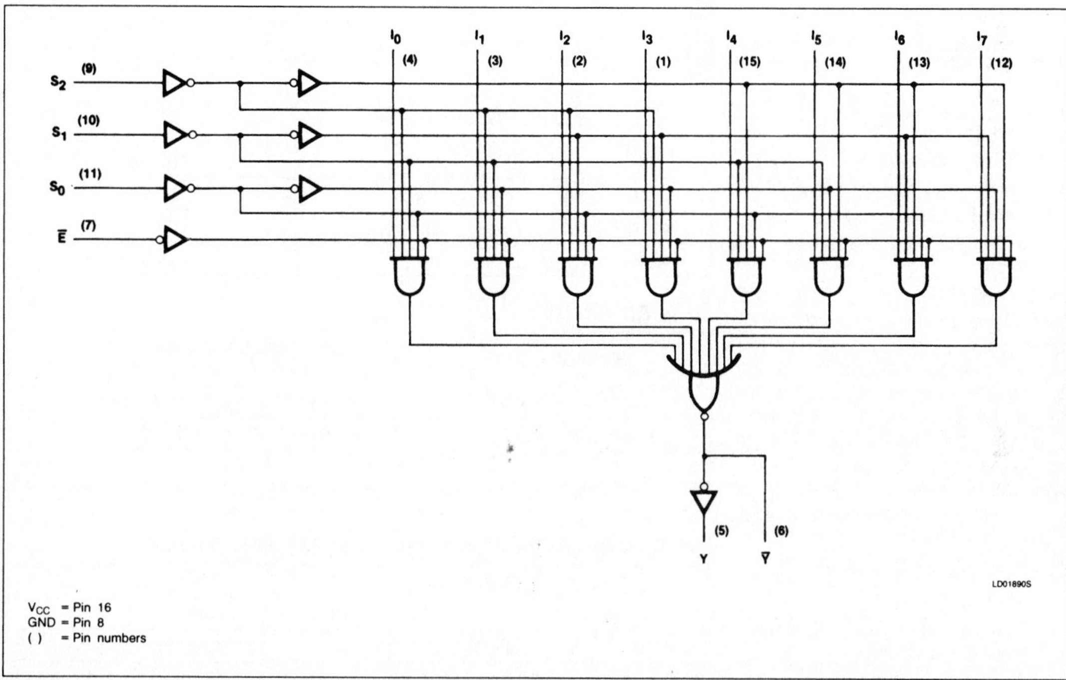

FUNCTION TABLE

INPUTS												OUTPUTS	
$\overline{E}$	S_2	S_1	S_0	I_0	I_1	I_2	I_3	I_4	I_5	I_6	I_7	$\overline{Y}$	Y
H	X	X	X	X	X	X	X	X	X	X	X	H	L
L	L	L	L	L	X	X	X	X	X	X	X	H	L
L	L	L	L	H	X	X	X	X	X	X	X	L	H
L	L	L	H	X	L	X	X	X	X	X	X	H	L
L	L	L	H	X	H	X	X	X	X	X	X	L	H
L	L	H	L	X	X	L	X	X	X	X	X	H	L
L	L	H	L	X	X	H	X	X	X	X	X	L	H
L	L	H	H	X	X	X	L	X	X	X	X	H	L
L	L	H	H	X	X	X	H	X	X	X	X	L	H
L	H	L	L	X	X	X	X	L	X	X	X	H	L
L	H	L	L	X	X	X	X	H	X	X	X	L	H
L	H	L	H	X	X	X	X	X	L	X	X	H	L
L	H	L	H	X	X	X	X	X	H	X	X	L	H
L	H	H	L	X	X	X	X	X	X	L	X	H	L
L	H	H	L	X	X	X	X	X	X	H	X	L	H
L	H	H	H	X	X	X	X	X	X	X	L	H	L
L	H	H	H	X	X	X	X	X	X	X	H	L	H

H = HIGH voltage level
L = LOW voltage level
X = Don't care

Signetics

74153, LS153, S153
Multiplexers

Dual 4-Line To 1-Line Multiplexer
Product Specification

Logic Products

FEATURES

- **Non-inverting outputs**
- **Separate enable for each section**
- **Common select inputs**
- **See '253 for 3-state version**

DESCRIPTION

The '153 is a dual 4-input multiplexer that can select 2 bits of data from up to eight (8) sources under control of the common Select inputs (S_0, S_1). The two 4-input multiplexer circuits have individual active LOW Enables ($\overline{E}_a$, $\overline{E}_b$) which can be used to strobe the outputs independently. Outputs (Y_a, Y_b) are forced LOW when the corresponding Enables ($\overline{E}_a$, $\overline{E}_b$) are HIGH.

$Y_a = \overline{E}_a \bullet (I_{0a} \bullet \overline{S}_1 \bullet \overline{S}_0 + I_{1a} \bullet \overline{S}_1 \bullet S_0 + I_{2a} \bullet S_1 \bullet \overline{S}_0 + I_{3a} \bullet S_1 \bullet S_2)$

$Y_b = \overline{E}_b \bullet (I_{0b} \bullet \overline{S}_1 \bullet \overline{S}_0 + I_{1b} \bullet \overline{S}_1 \bullet S_0 + I_{2b} \bullet S_1 \bullet \overline{S}_0 + I_{3b} \bullet S_1 \bullet S_2)$

TYPE	TYPICAL PROPAGATION DELAY	TYPICAL SUPPLY CURRENT (TOTAL)
74153	18ns	36mA
74LS153	18ns	6.2mA
74S153	9ns	45mA

ORDERING CODE

PACKAGES	COMMERCIAL RANGE $V_{CC} = 5V \pm 5\%$; $T_A = 0°C$ to $+70°C$
Plastic DIP	N74153N, N74LS153N, N74S153N
Plastic SO	N74LS153D, N74S153D

NOTE:
For information regarding devices processed to Military Specifications, see the Signetics Military Products Data Manual.

INPUT AND OUTPUT LOADING AND FAN-OUT TABLE

PINS	DESCRIPTION	74	74S	74LS
All	Inputs	1ul	1Sul	1LSul
All	Outputs	10ul	10Sul	10LSul

NOTE:
Where a 74 unit load (ul) is understood to be $40\mu A$ I_{IH} and $-1.6mA$ I_{IL}, a 74S unit load (Sul) is $50\mu A$ I_{IH} and $-2.0mA$ I_{IL}, and 74LS unit load (LSul) is $20\mu A$ I_{IH} and $-0.4mA$ I_{IL}.

PIN CONFIGURATION

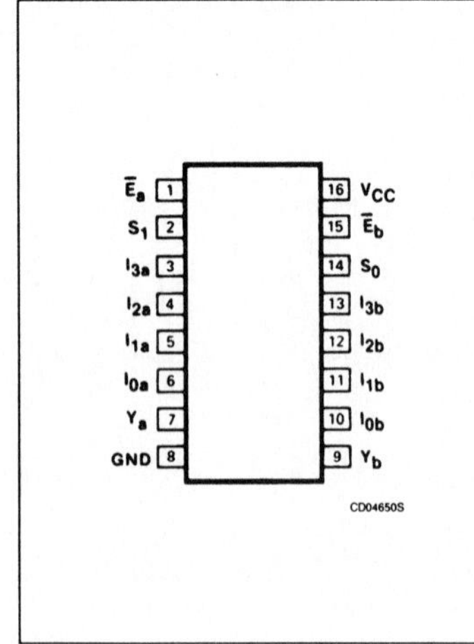

LOGIC SYMBOL

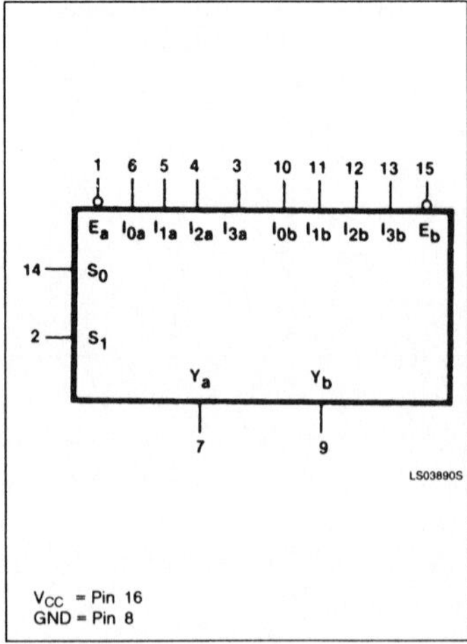

LOGIC SYMBOL (IEEE/IEC)

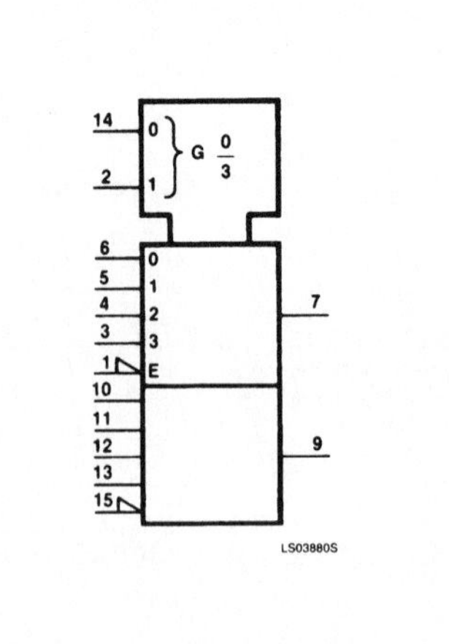

Multiplexers 74153, LS153, S153

The '153 can be used to move data to a common output bus from a group of registers. The state of the Select inputs would determine the particular register from which the data came. An alternative application is as a function generator. The device can generate two functions or three variables. This is useful for implementing highly irregular random logic.

LOGIC DIAGRAM

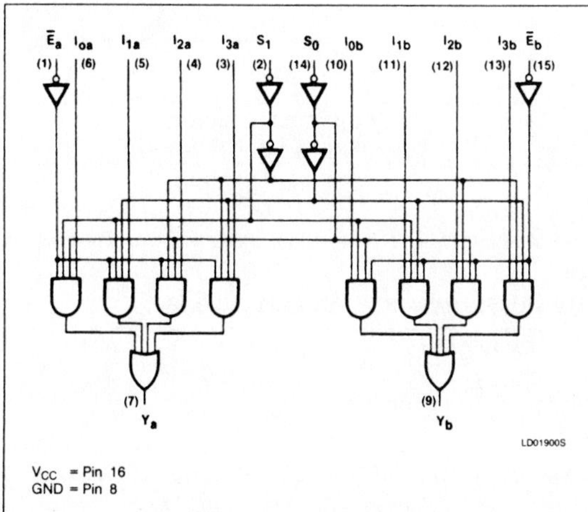

V_{CC} = Pin 16
GND = Pin 8

FUNCTION TABLE

SELECT INPUTS		INPUTS (a or b)					OUTPUT
S_0	S_1	$\bar{E}$	I_0	I_1	I_2	I_3	Y
X	X	H	X	X	X	X	L
L	L	L	L	X	X	X	L
L	L	L	H	X	X	X	H
H	L	L	X	L	X	X	L
H	L	L	X	H	X	X	H
L	H	L	X	X	L	X	L
L	H	L	X	X	H	X	H
H	H	L	X	X	X	L	L
H	H	L	X	X	X	H	H

H = HIGH voltage level
L = LOW voltage level
X = Don't care

ABSOLUTE MAXIMUM RATINGS (Over operating free-air temperature range unless otherwise noted.)

PARAMETER		74	74LS	74S	UNIT
V_{CC}	Supply voltage	7.0	7.0	7.0	V
V_{IN}	Input voltage	−0.5 to +5.5	−0.5 to +7.0	−0.5 to +5.5	V
I_{IN}	Input current	−30 to +5	−30 to +1	−30 to +5	mA
V_{OUT}	Voltage applied to output in HIGH output state	−0.5 to $+V_{CC}$	−0.5 to $+V_{CC}$	−0.5 to $+V_{CC}$	V
T_A	Operating free-air temperature range	0 to 70			°C

RECOMMENDED OPERATING CONDITIONS

PARAMETER		74			74LS			74S			UNIT
		Min	Nom	Max	Min	Nom	Max	Min	Nom	Max	
V_{CC}	Supply voltage	4.75	5.0	5.25	4.75	5.0	5.25	4.75	5.0	5.25	V
V_{IH}	HIGH-level input voltage	2.0			2.0			2.0			V
V_{IL}	LOW-level input voltage			+0.8			+0.8			+0.8	V
I_{IK}	Input clamp current			−12			−18			−18	mA
I_{OH}	HIGH-level output current			−800			−400			−1000	μA
I_{OL}	LOW-level output current			16			8			20	mA
T_A	Operating free-air temperature	0		70	0		70	0		70	°C

Signetics

74154, LS154
Decoder/Demultiplexers

1-of-16 Decoder/Demultiplexer
Product Specification

Logic Products

FEATURES
- **16-line demultiplexing capability**
- **Mutually exclusive outputs**
- **2-input enable gate for strobing or expansion**

DESCRIPTION
The '154 decoder accepts four active HIGH binary address inputs and provides 16 mutually exclusive active LOW outputs. The 2-input enable gate can be used to strobe the decoder to eliminate the normal decoding "glitches" on the outputs, or it can be used for expansion of the decoder. The enable gate has two AND'ed inputs which must be LOW to enable the outputs.

The '154 can be used as a 1-of-16 demultiplexer by using one of the enable inputs as the multiplexed data input. When the other enable is LOW, the addressed output will follow the state of the applied data.

TYPE	TYPICAL PROPAGATION DELAY	TYPICAL SUPPLY CURRENT (TOTAL)
74154	21ns	34mA
74LS154	15ns	9mA

ORDERING CODE

PACKAGES	COMMERCIAL RANGE $V_{CC} = 5V \pm 5\%$; $T_A = 0°C$ to $+70°C$
Plastic DIP	N74154N, N74LS154N

NOTE:
For information regarding devices processed to Military Specifications, see the Signetics Military Products Data Manual.

INPUT AND OUTPUT LOADING AND FAN-OUT TABLE

PINS	DESCRIPTION	74	74LS
All	Inputs	1ul	1LSul
All	Outputs	10ul	10LSul

NOTE:
Where a 74 unit load (ul) is understood to be 40μA I_{IH} and −1.6mA I_{IL}, and a 74LS unit load (LSul) is 20μA I_{IH} and −0.4mA I_{IL}.

PIN CONFIGURATION

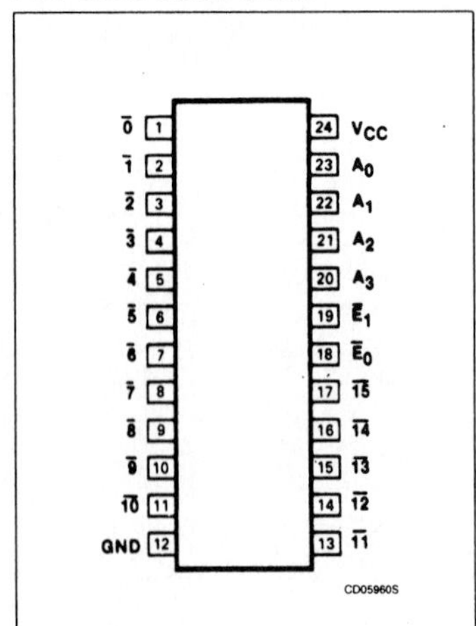

LOGIC SYMBOL

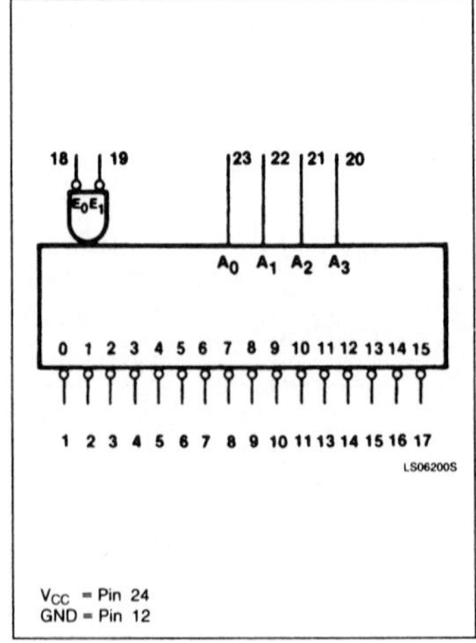

LOGIC SYMBOL (IEEE/IEC)

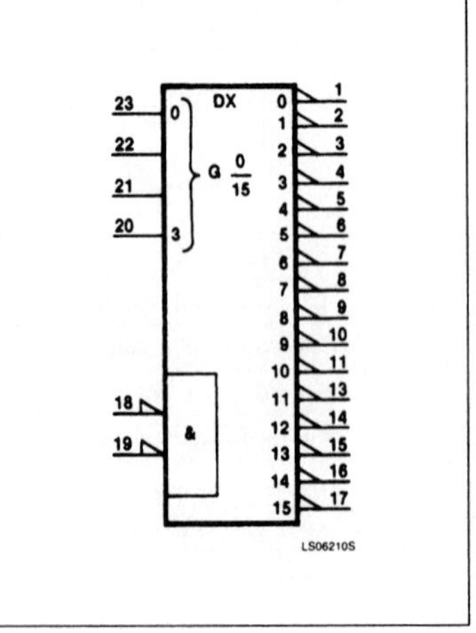

Decoder/Demultiplexers 74154, LS154

LOGIC DIAGRAM

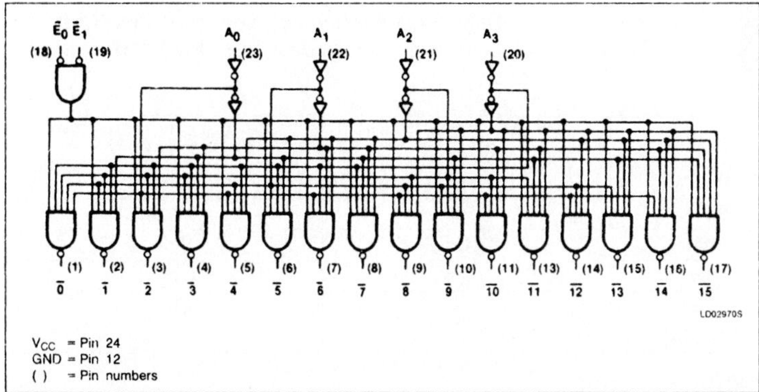

FUNCTION TABLE

INPUTS						OUTPUT															
E_0	E_1	A_3	A_2	A_1	A_0	0	1	2	3	4	5	6	7	8	9	10	11	12	13	14	15
L	H	X	X	X	X	H	H	H	H	H	H	H	H	H	H	H	H	H	H	H	H
H	L	X	X	X	X	H	H	H	H	H	H	H	H	H	H	H	H	H	H	H	H
H	H	X	X	X	X	H	H	H	H	H	H	H	H	H	H	H	H	H	H	H	H
L	L	L	L	L	L	L	H	H	H	H	H	H	H	H	H	H	H	H	H	H	H
L	L	L	L	L	H	H	L	H	H	H	H	H	H	H	H	H	H	H	H	H	H
L	L	L	L	H	L	H	H	L	H	H	H	H	H	H	H	H	H	H	H	H	H
L	L	L	L	H	H	H	H	H	L	H	H	H	H	H	H	H	H	H	H	H	H
L	L	L	H	L	L	H	H	H	H	L	H	H	H	H	H	H	H	H	H	H	H
L	L	L	H	L	H	H	H	H	H	H	L	H	H	H	H	H	H	H	H	H	H
L	L	L	H	H	L	H	H	H	H	H	H	L	H	H	H	H	H	H	H	H	H
L	L	L	H	H	H	H	H	H	H	H	H	H	L	H	H	H	H	H	H	H	H
L	L	H	L	L	L	H	H	H	H	H	H	H	H	L	H	H	H	H	H	H	H
L	L	H	L	L	H	H	H	H	H	H	H	H	H	H	L	H	H	H	H	H	H
L	L	H	L	H	L	H	H	H	H	H	H	H	H	H	H	L	H	H	H	H	H
L	L	H	L	H	H	H	H	H	H	H	H	H	H	H	H	H	L	H	H	H	H
L	L	H	H	L	L	H	H	H	H	H	H	H	H	H	H	H	H	L	H	H	H
L	L	H	H	L	H	H	H	H	H	H	H	H	H	H	H	H	H	H	L	H	H
L	L	H	H	H	L	H	H	H	H	H	H	H	H	H	H	H	H	H	H	L	H
L	L	H	H	H	H	H	H	H	H	H	H	H	H	H	H	H	H	H	H	H	L

H = HIGH voltage level
L = LOW voltage level
X = Don't care

ABSOLUTE MAXIMUM RATINGS (Over operating free-air temperature range unless otherwise noted.)

PARAMETER		74	74LS	UNIT
V_{CC}	Supply voltage	7.0	7.0	V
V_{IN}	Input voltage	-0.5 to $+5.5$	-0.5 to $+7.0$	V
I_{IN}	Input current	-30 to $+5$	-30 to $+1$	mA
V_{OUT}	Voltage applied to output in HIGH output state	-0.5 to $+V_{CC}$	-0.5 to $+V_{CC}$	V
T_A	Operating free-air temperature range	0 to 70		°C

Signetics

74157, 74158, LS157, LS158, S157, S158

Data Selectors/Multiplexers

Logic Products

'157 Quad 2-Input Data Selector/Multiplexer (Non-Inverted)
'158 Quad 2-Input Data Selector/Multiplexer (Inverted)
Product Specification

DESCRIPTION

The '157 is a quad 2-input multiplexer which selects four bits of data from two sources under the control of a common Select input (S). The Enable input ($\overline{E}$) is active LOW. When $\overline{E}$ is HIGH, all of the outputs (Y) are forced LOW regardless of all other input conditions.

Moving data from two groups of registers to four common output busses is a common use of the '157. The state of the Select input determines the particular register from which the data comes. It can also be used as a function generator. The device is useful for implementing highly irregular logic by generating any four of the 16 different functions of two variables with one variable common.

TYPE	TYPICAL PROPAGATION DELAY	TYPICAL SUPPLY CURRENT (TOTAL)
74157	13ns	30mA
74LS157	13ns	9.7mA
74S157	7.4ns	50mA
74158	13ns	30mA
74LS158	13ns	4.8mA
74S158	6ns	40mA

ORDERING CODE

PACKAGES	COMMERCIAL RANGES $V_{CC} = 5V \pm 5\%$; $T_A = 0°C$ to $+70°C$
Plastic DIP	N74157N, N74LS158N, N74S157N N74LS157N, N74S158N, N74LS158N
Plastic SO	N74LS157D, N74S158D

NOTE:
For information regarding devices processed to Military Specifications, see the Signetics Military Products Data Manual.

INPUT AND OUTPUT LOADING AND FAN-OUT TABLE

PINS	DESCRIPTION	74	74S	74LS
S, $\overline{E}$	Inputs	1ul	2Sul	2LSul
Data	Inputs	1ul	1Sul	1LSul
All	Outputs	10ul	10Sul	10LSul

NOTE:
Where a 74 unit load (ul) is understood to be 40μA I_{IH} and −1.6mA I_{IL}, a 74S unit load (Sul) is 50μA I_{IH} and −2.0mA I_{IL}, and a74LS unit load (LSul) is 20μA I_{IH} and −0.4mA I_{IL}.

PIN CONFIGURATION

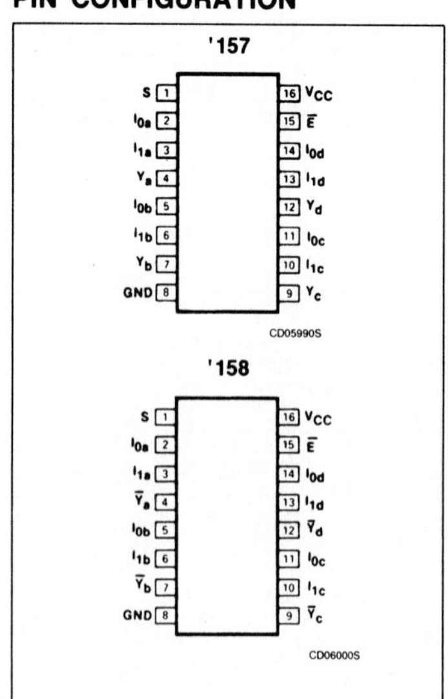

LOGIC SYMBOL

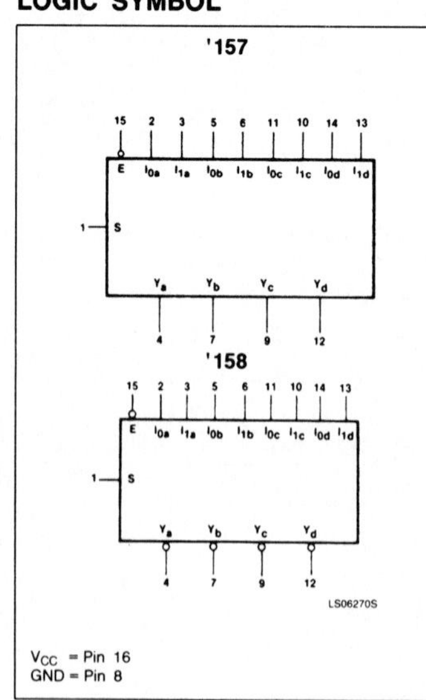

LOGIC SYMBOL (IEEE/IEC)

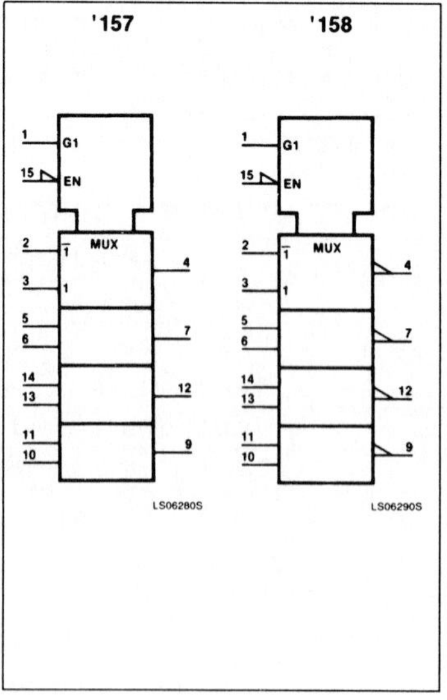

Data Selectors/Multiplexers 74157, 74158, LS157, LS158, S157, S158

LOGIC DIAGRAM, '157

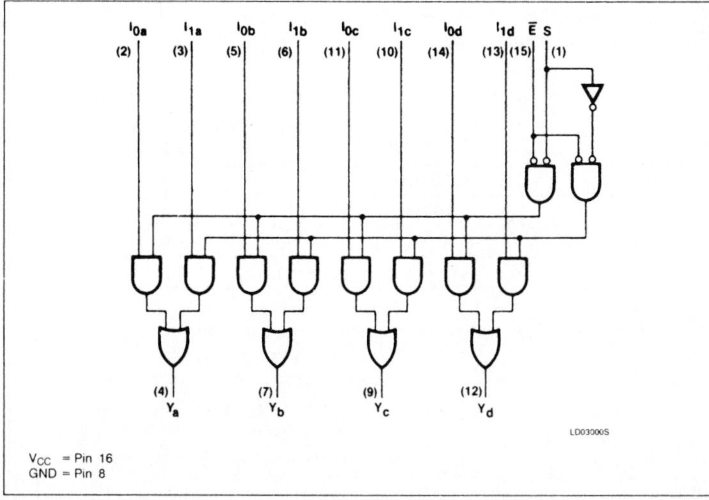

LOGIC DIAGRAM, '158

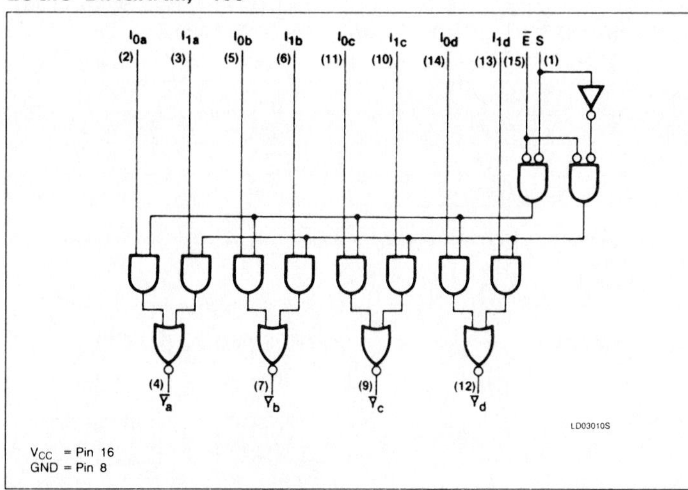

The device is the logic implementation of a 4-pole, 2-position switch where the position of the switch is determined by the logic levels supplied to the Select input. Logic equations for the outputs are shown below:

$Y_a = \overline{E} \cdot (I_{1a} \cdot S + I_{0a} \cdot \overline{S})$
$Y_b = \overline{E} \cdot (I_{1b} \cdot S + I_{0b} \cdot \overline{S})$
$Y_c = \overline{E} \cdot (I_{1c} \cdot S + I_{0c} \cdot \overline{S})$
$Y_d = \overline{E} \cdot (I_{1d} \cdot S + I_{0d} \cdot \overline{S})$

The '158 is similar but has inverting outputs:

$\overline{Y}_a = \overline{E} \cdot (I_{1a} \cdot S + I_{0a} \cdot \overline{S})$
$\overline{Y}_b = \overline{E} \cdot (I_{1b} \cdot S + I_{0b} \cdot \overline{S})$
$\overline{Y}_c = \overline{E} \cdot (I_{1c} \cdot S + I_{0c} \cdot \overline{S})$
$\overline{Y}_d = \overline{E} \cdot (I_{1d} \cdot S + I_{0d} \cdot \overline{S})$

FUNCTION TABLE, '157

ENABLE	SELECT INPUT	DATA INPUTS		OUTPUT
$\overline{E}$	S	I_0	I_1	Y
H	X	X	X	L
L	H	X	L	L
L	H	X	H	H
L	L	L	X	L
L	L	H	X	H

H = HIGH voltage level
L = LOW voltage level
X = Don't care

FUNCTION TABLE, '158

ENABLE	SELECT INPUT	DATA INPUTS		OUTPUT
$\overline{E}$	S	I_0	I_1	$\overline{Y}$
H	X	X	X	H
L	L	L	X	H
L	L	H	X	L
L	H	X	L	H
L	H	X	H	L

H = HIGH voltage level
L = LOW voltage level
X = Don't care

ABSOLUTE MAXIMUM RATINGS (Over operating free-air temperature range unless otherwise noted.)

PARAMETER		74	74LS	74S	UNIT
V_{CC}	Supply voltage	7.0	7.0	7.0	V
V_{IN}	Input voltage	−0.5 to +5.5	−0.5 to +7.0	−0.5 to +5.5	V
I_{IN}	Input current	−30 to +5	−30 to +1	−30 to +5	mA
V_{OUT}	Voltage applied to output in HIGH output state	−0.5 to $+V_{CC}$	−0.5 to $+V_{CC}$	−0.5 to $+V_{CC}$	V
T_A	Operating free-air temperature range	0 to 70			°C

Signetics

74LS168A, 74LS169A, S168A, S169A
4-Bit Bidirectional Counters

Logic Products

4-Bit Up/Down Synchronous Counter
Product Specification

FEATURES
- **Synchronous counting and loading**
- **Up/down counting**
- **Modulo 16 binary counter — '169A**
- **BCD decade counter — '168A**
- **Two Count Enable inputs for n-bit cascading**
- **Positive edge-triggered clock**

DESCRIPTION
The '168A is a synchronous, presettable BCD decade up/down counter featuring an internal carry look-ahead for applications in high-speed counting designs. Synchronous operation is provided by having all flip-flops clocked simultaneously so that the outputs change coincident with each other when so instructed by the Count Enable inputs and internal gating. This mode of operation eliminates the output spikes which are normally associated with asynchronous (ripple clock) counters. A buffered Clock input triggers the flip-flops on the LOW-to-HIGH transition of the clock.

TYPE	TYPICAL f_{MAX}	TYPICAL SUPPLY CURRENT (TOTAL)
74LS168A	32MHz	20mA
74S168A	70MHz	100mA
74LS169A	32MHz	20mA
74S169A	70MHz	100mA

ORDERING CODE

PACKAGES	COMMERCIAL RANGE V_{CC} = 5V ±5%; T_A = 0°C to +70°C
Plastic DIP	N74LS168AN, N74S168AN N74LS169AN, N74S169AN
Plastic SO	N74LS169AD, N74LS169AD, N74S169AD

NOTE:
For information regarding devices processed to Military Specifications, see the Signetics Military Products Data Manual.

INPUT AND OUTPUT LOADING AND FAN-OUT TABLE

PINS	DESCRIPTION	74S	74LS
$\overline{PE}$	Input	1Sul	2LSul
$\overline{CET}$	Input	2Sul	1LSul
Other	Inputs	1Sul	1LSul
All	Outputs	10Sul	10LSul

NOTE:
Where a 74S unit load (Sul) is understood to be 50μA I_{IH} and −2.0mA I_{IL} and a 74LS unit load (LSul) is 20μA I_{IH} and −0.4mA I_{IL}.

PIN CONFIGURATION

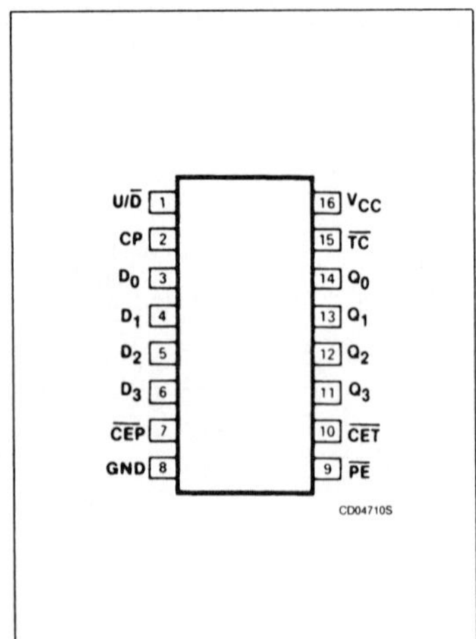

LOGIC SYMBOL

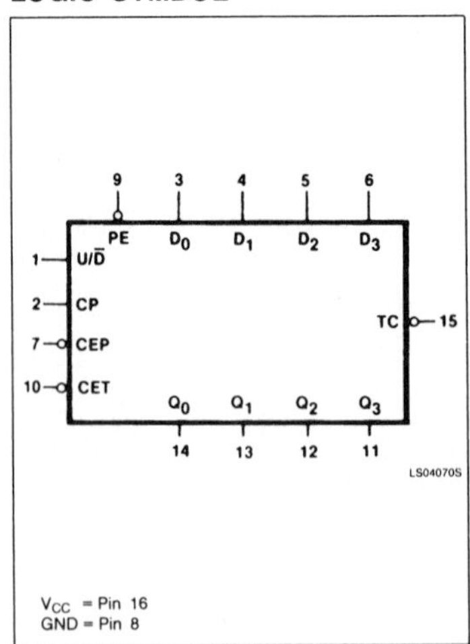

LOGIC SYMBOL (IEEE/IEC)

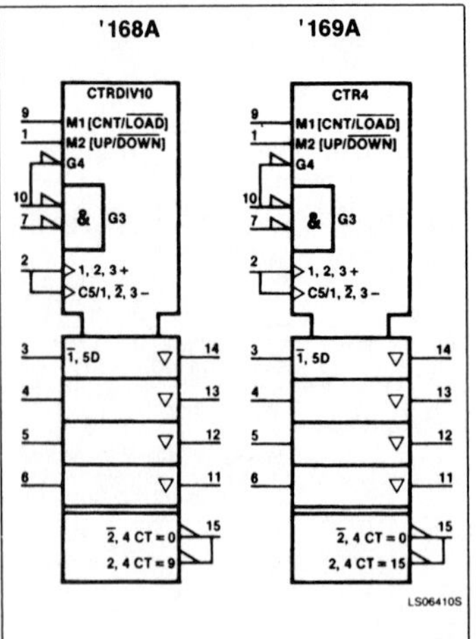

4-Bit Bidirectional Counters 74LS168A, 74LS169A, S168A, S169A

The counter is fully programmable; that is, the outputs may be preset to either level. Presetting is synchronous with the clock and takes place regardless of the levels of the Count Enable inputs. A LOW level on the Parallel Enable ($\overline{PE}$) input disables the counter and causes the data at the D_n input to be loaded into the counter on the next LOW-to-HIGH transition of the clock.

The direction of counting is controlled by the Up/Down (U/$\overline{D}$) input; a HIGH will cause the count to increase, a LOW will cause the count to decrease.

The carry look-ahead circuitry provides for cascading counters for n-bit synchronous applications without additional gating. Instrumental in accomplishing this function are two Count Enable inputs ($\overline{CET} \cdot \overline{CEP}$) and a Terminal Count ($\overline{TC}$) output. Both Count Enable inputs must be LOW to count. The $\overline{CET}$ input is fed forward to enable the $\overline{TC}$ output. The $\overline{TC}$ output thus enabled will produce a LOW output pulse with a duration approximately equal to the HIGH level portion of the Q_0 output. This LOW level $\overline{TC}$ pulse is used to enable successive cascaded stages. See Figure A for the fast synchronous multistage counting connections.

The '169A is identical except that it is a Modulo 16 counter.

LOGIC DIAGRAM, '168A

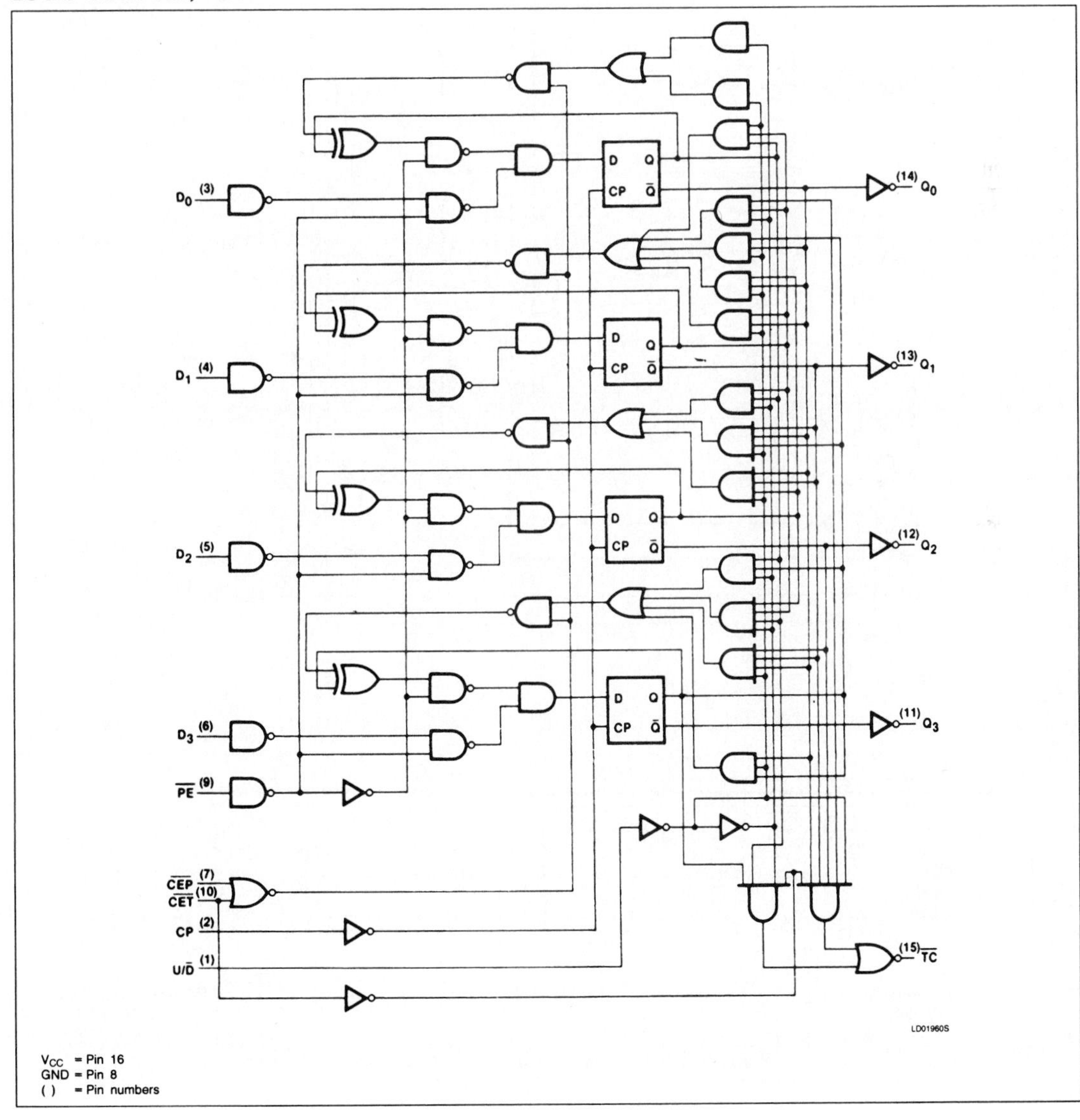

V_{CC} = Pin 16
GND = Pin 8
( ) = Pin numbers

4-Bit Bidirectional Counters 74LS168A, 74LS169A, S168A, S169A

LOGIC DIAGRAM, '169A

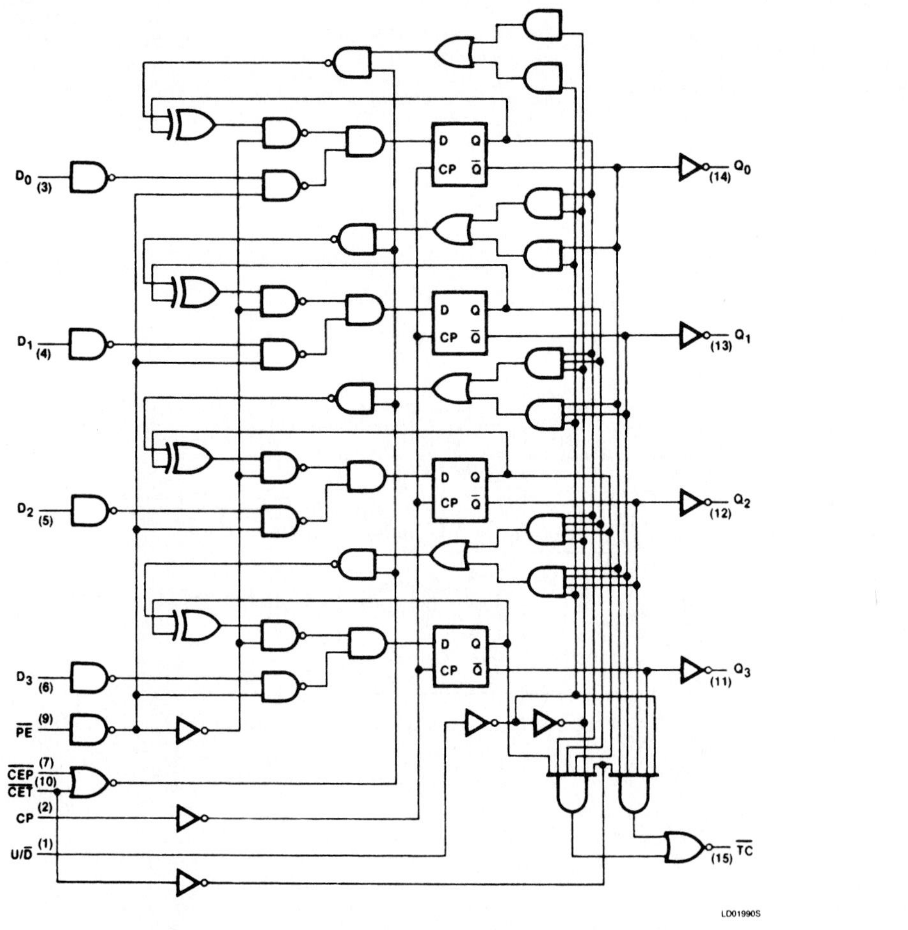

V_{CC} = Pin 16
GND = Pin 8
( ) = Pin numbers

LD01990S

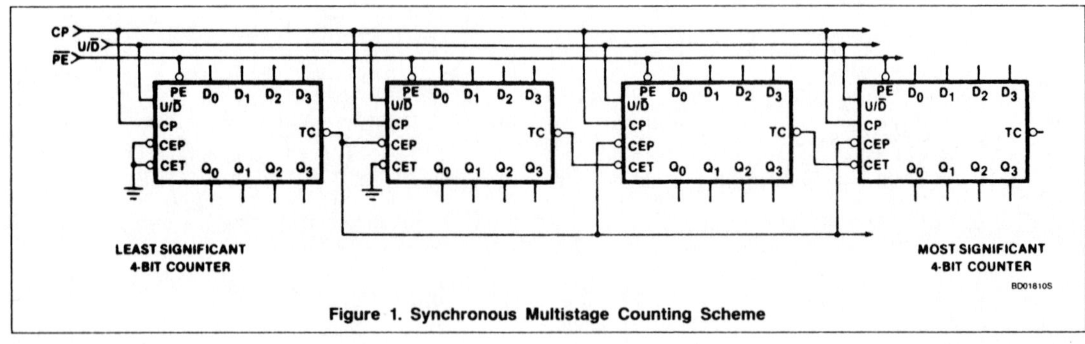

Figure 1. Synchronous Multistage Counting Scheme

4-Bit Bidirectional Counters 74LS168A, 74LS169A, S168A, S169A

MODE SELECT — FUNCTION TABLE

OPERATING MODE	INPUTS						OUTPUTS	
	CP	U/$\overline{D}$	$\overline{CEP}$	$\overline{CET}$	$\overline{PE}$	D_n	Q_n	$\overline{TC}$
Parallel Load	↑ ↑	X X	X X	X X	l l	l h	L H	(1) (1)
Count Up	↑	h	l	l	h	X	Count Up	(1)
Count Down	↑	l	l	l	h	X	Count Down	(1)
Hold (do nothing)	↑ ↑	X X	h X	X h	h h	X X	q_n q_n	(1) H

H = HIGH voltage level steady state
h = HIGH voltage level one setup time prior to the LOW-to-HIGH clock transition
L = LOW voltage level steady state
l = LOW voltage level one setup time prior to the LOW-to-HIGH clock transition
X = Don't care
q = Lower case letters indicate the state of the referenced output prior to the LOW-to-HIGH clock transition
↑ = LOW-to-HIGH clock transition

NOTE:
1. The $\overline{TC}$ is LOW when $\overline{CET}$ is LOW and the counter is at Terminal Count. Terminal Count Up is (HHHH) and Terminal Count Down is (LLLL) for '169A. The $\overline{TC}$ is LOW when $\overline{CET}$ is LOW and the counter is at Terminal Count. Terminal Count Up is (HLLH) and Terminal Count Down is (LLLL) for '168A.

WAVEFORM (Typical Load, Count, and Inhibit Sequences)

Illustrated below is the following sequence for the '168A. The operation of the '169A is similar.

1. Load (preset) to BCD seven.
2. Count up to eight, nine (maximum), zero, one, and two.
3. Inhibit.
4. Count down to one, zero (minimum), nine, eight, and seven.

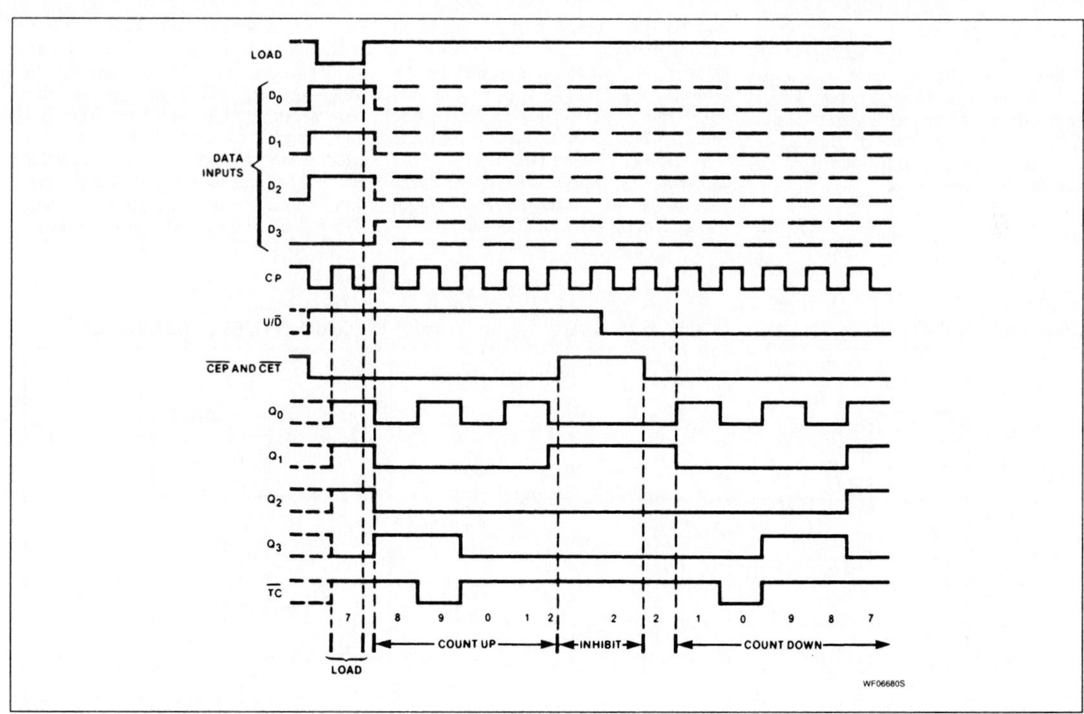

Signetics

74S172
Register File

16-Bit Multiple Port Register File (3-State)
Product Specification

Logic Products

FEATURES
- **Simultaneous and independent Read and Write operations**
- **Expandable to 1024 words on n-bits**
- **3-State outputs**

DESCRIPTION
The '172 is a high-performance, 16-bit multiport register file with 3-State outputs organized as eight words of two bits each. Multiple address decoding circuitry is used so that the read and write operation can be performed independently on up to three word locations. Data can be written into two word locations through input Port "A" or input Port "C" while data is simultaneously read from both output Port "B" and output Port "C".

Port "A" is an input port which can be used to write two bits of data (D_{A0}, D_{A1}) into one of eight register locations selected by the Address inputs (A_{A0}, A_{A1}, A_{A2}). When the Write Enable ($\overline{WE}_A$) input is LOW one set-up time prior to the LOW-to-HIGH transition of the Clock (CP) input, the data is written into the selected location.

TYPE	TYPICAL f_{MAX}	TYPICAL SUPPLY CURRENT (TOTAL)
74S172	40MHz	160mA

ORDERING CODE

PACKAGES	COMMERCIAL RANGE V_{CC} = 5V ±5%; T_A = 0°C to +70°C
Plastic DIP	N74S172N

NOTE:
For information regarding devices processed to Military Specifications, see the Signetics Military Products Data Manual.

INPUT AND OUTPUT LOADING AND FAN-OUT TABLE

PINS	DESCRIPTION	74S
All	Inputs	1Sul
All	Outputs	8Sul

NOTE:
A 74S unit load (Sul) is 50μA and I_{IH} −2.0mA I_{IL}.

Port "B" is an output port which can be used to read two bits of data from one of eight register locations selected by the Address inputs (A_{B0}, A_{B1}, A_{B2}). When the Read Enable ($\overline{RE}_B$) is LOW, the selected 2-bit word appears on outputs Q_{B0} and Q_{B1}. When $\overline{RE}_B$ is HIGH, the Q_{B0} and Q_{B1} outputs are in the HIGH impedance "off" state. The read operation is independent of the clock.

Port "C" is a read/write port that has separate Data input and Data output sections, but common Address inputs (A_{C0}, A_{C1}, A_{C2}). Data can be simultaneously written into and read from the same register location. Port "C" can be used to write data into one location while Port "A" is writing into a different location, but data cannot be written reliably into the same location simultaneously.

PIN CONFIGURATION

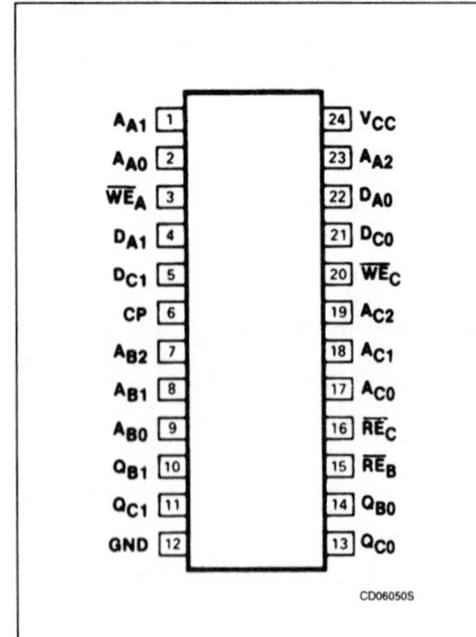

LOGIC SYMBOL

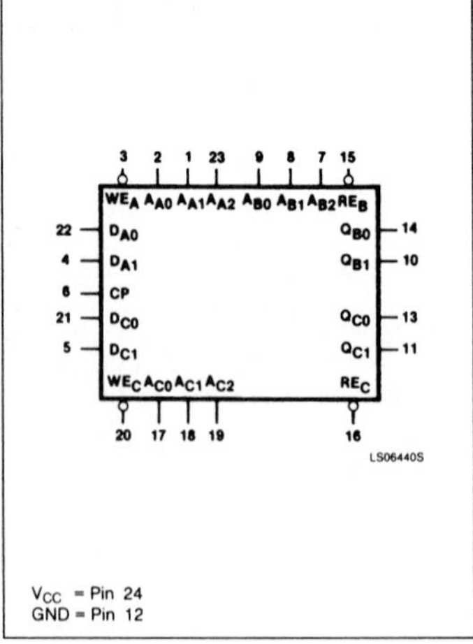

LOGIC SYMBOL (IEEE/IEC)

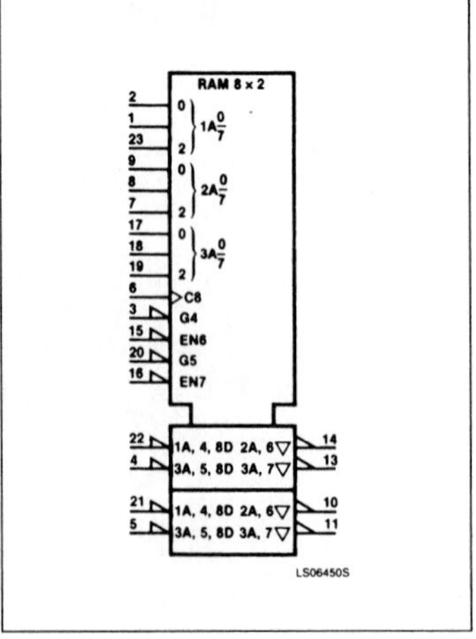

Register File 74S172

If both Ports "A" and "C" are enabled for writing into the same location during the same clock cycle, the LOW data will predominate if there is a conflict.

The register operation is essentially a master-slave flip-flop. Each master acts as a transparent D latch when selected by the "A" or "C" address and the clock and applicable write enable are LOW. The data in the master is transferred to the slave (or output section) following the LOW-to-HIGH transition of the Clock (CP). The Address inputs must be stable while the Clock and Write Enable inputs are LOW to ensure retention of data previously written into the other locations. Any number of masters can be altered while the clock and write enable are LOW, but the new data will not be loaded into the slaves, or be available at the outputs, until the clock goes HIGH.

BLOCK DIAGRAM

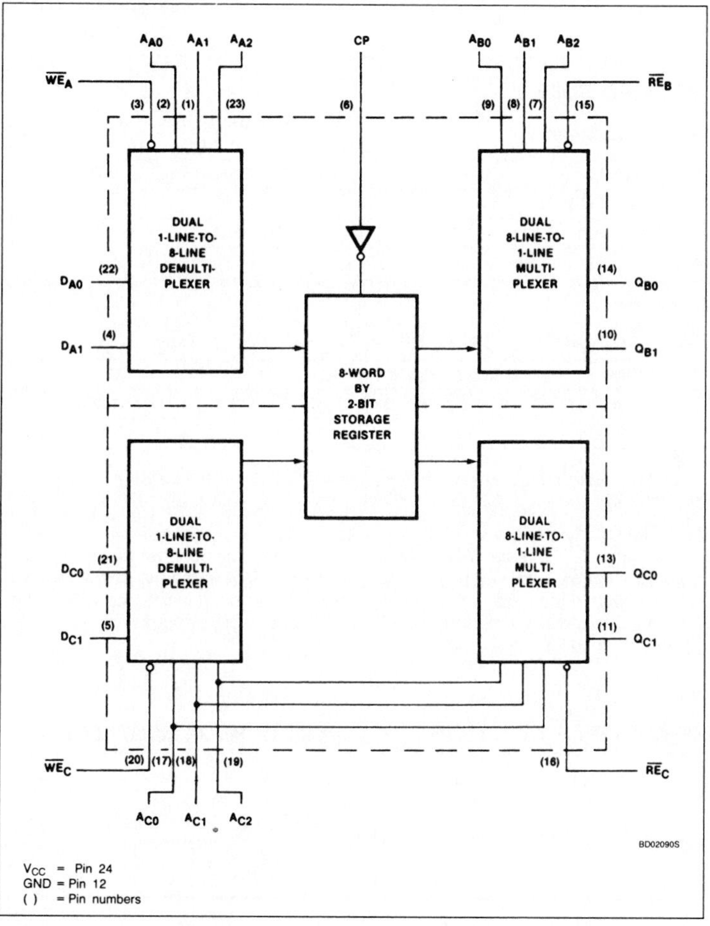

V_{CC} = Pin 24
GND = Pin 12
( ) = Pin numbers

WRITE MODE SELECT TABLE

OPERATING MODE	INPUTS			ADDRESSED REGISTER
	$\overline{CP}$	$\overline{WE}$	D_n	
Write data[a]	↑ ↑	l l	l h	L H
Hold[b]	↓	h	X	no change

READ MODE SELECT TABLE

OPERATING MODE	INPUTS		OUTPUTS
	$\overline{RE}$	Addressed Register	Q_n
Read	L L	L H	L H
Disabled	H	X	(Z)

H = HIGH voltage level steady state.
h = HIGH voltage level one set-up time prior to the LOW-to-HIGH or HIGH-to-LOW clock transition.
L = LOW voltage level steady state.
l = LOW voltage level one set-up time prior to the LOW-to-HIGH clock transition.
X = Don't care.
(Z) = HIGH impedance (off) state.
↑ = LOW-to-HIGH clock transition.
↓ = HIGH-to-LOW clock transition.

NOTES:

a. The Write Address (A_A and A_C) to the "internal register" must be stable while $\overline{WE}$ and CP are LOW for conventional operation.
b. The Write Enable must be HIGH before the HIGH-to-LOW clock transition to ensure that the data in the register is not changed.

ABSOLUTE MAXIMUM RATINGS (Over operating free-air temperature range unless otherwise noted.)

PARAMETER		74S	UNIT
V_{CC}	Supply voltage	7.0	V
V_{IN}	Input voltage	−0.5 to +5.5	V
I_{IN}	Input current	−30 to +5	mA
V_{OUT}	Voltage applied to output in HIGH output state	−0.5 to $+V_{CC}$	V
T_A	Operating free-air temperature range	0 to 70	°C

Signetics

74173, LS173

Flip-Flops

Quad D-Type Flip-Flop With 3-State Outputs
Product Specification

Logic Products

FEATURES

- **Edge-triggered D-type register**
- **Gated Input enable for hold "do nothing" mode**
- **3-State output buffers**
- **Gated output enable control**
- **Pin compatible with the 8T10 and DM8551**

DESCRIPTION

The '173 is a 4-bit parallel load register with clock enable control, 3-State buffered outputs and master reset. When the two Clock Enable ($\overline{E}_1$ and $\overline{E}_2$) inputs are LOW, the data on the D inputs is loaded into the register synchronously with the LOW-to-HIGH Clock (CP) transition. When one or both $\overline{E}$ inputs are HIGH one set-up time before the LOW-to-HIGH clock transition, the register will retain the previous data. Data inputs and Clock Enable inputs are fully edge triggered and must be stable only one set-up time before the LOW-to-HIGH clock transition.

The Master Reset (MR) is an active HIGH asynchronous input. When the MR is HIGH, all four flip-flops are reset (cleared) independently of any other input condition.

TYPE	TYPICAL f_{MAX}	TYPICAL SUPPLY CURRENT (TOTAL)
74173	35MHz	50mA
74LS173	50MHz	20mA

ORDERING CODE

PACKAGES	COMMERCIAL RANGE $V_{CC} = 5V \pm 5\%$; $T_A = 0°C$ to $+70°C$
Plastic DIP	N74173N, N74LS173N
Plastic SO-16	N74LS173D
Plastic SOL-16	CD7186D

NOTE:
For information regarding devices processed to Military Specifications, see the Signetics Military Products Data Manual.

INPUT AND OUTPUT LOADING AND FAN-OUT TABLE

PINS	DESCRIPTION	74	74LS
All	Inputs	1ul	1LSul
All	Outputs	10ul	30LSul

NOTE:
Where a 74 unit load (ul) is understood to be 40μA I_{IH} and −1.6mA I_{IL} and a 74LS unit load (LSul) is 20μA I_{IH} and −0.4mA I_{IL}.

The 3-State output buffers are controlled by a 2-input NOR gate. When both Output Enable ($\overline{OE}_1$ and $\overline{OE}_2$) inputs are LOW, the data in the register is presented at the Q outputs. When one or both $\overline{OE}$ inputs is HIGH, the outputs are forced to a HIGH impedance "off" state. The 3-State output buffers are completely independent of the register operation; the $\overline{OE}$ transition does not affect the clock and reset operations.

PIN CONFIGURATION

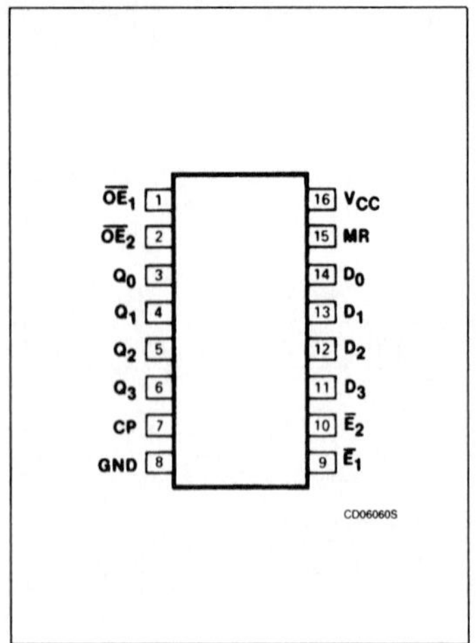

LOGIC SYMBOL

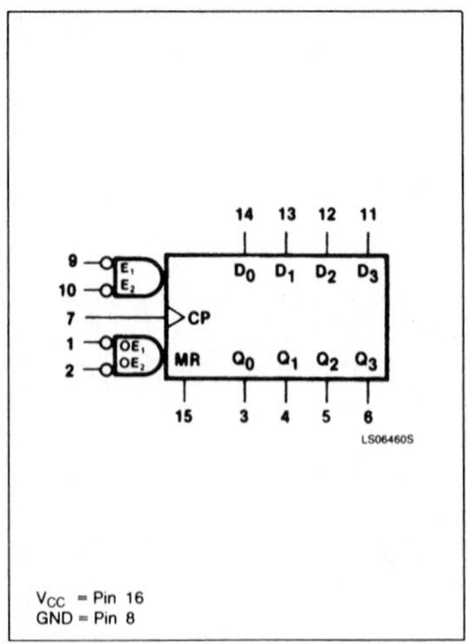

LOGIC SYMBOL (IEEE/IEC)

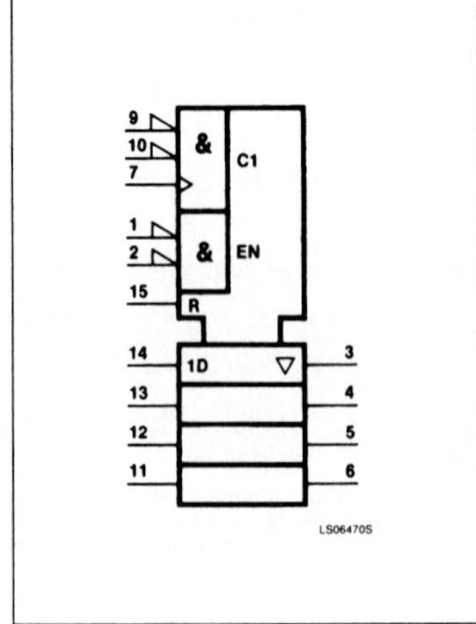

LOGIC DIAGRAM

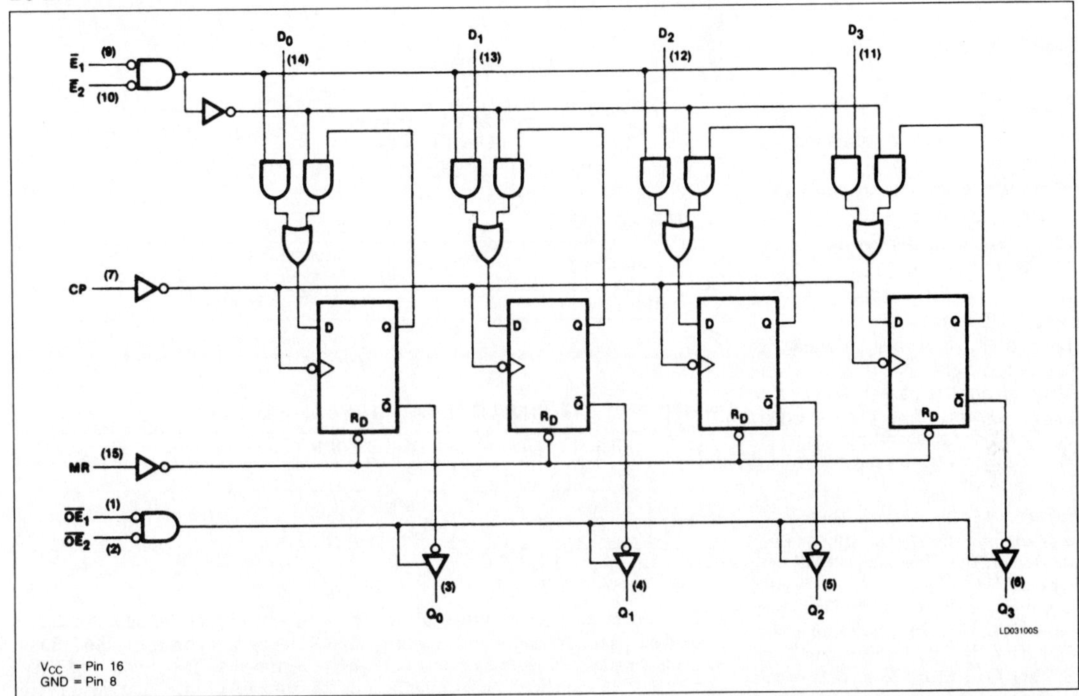

MODE SELECT — FUNCTION TABLE

REGISTER OPERATING MODES	INPUTS					OUTPUTS
	MR	CP	$\overline{E}_1$	$\overline{E}_2$	D_n	Q_n (Register)
Reset (clear)	H	X	X	X	X	L
Parallel load	L L	↑ ↑	l l	l l	l h	L H
Hold (no change)	L L	X X	h X	X h	X X	q_n q_n

3-STATE BUFFER OPERATING MODES	INPUTS			OUTPUTS
	Q_n (Register)	$\overline{OE}_1$	$\overline{OE}_2$	Q_0, Q_1, Q_2, Q_3
Read	L H	L L	L L	L H
Disabled	X X	H X	X H	(Z) (Z)

H = HIGH voltage level.
h = HIGH voltage level one set-up time prior to the LOW-to-HIGH clock transition.
L = LOW voltage level.
l = LOW voltage level one set-up time prior to the LOW-to-HIGH clock transition.
q_n = Lower case letters indicate the state of the referenced input (or output) on set-up time prior to the LOW-to-HIGH clock transition.
X = Don't care.
(Z) = HIGH impedance "off" state.
↑ = LOW-to-HIGH clock transition.

Signetics

Logic Products

74180

Parity Generator/Checker

9-Bit Odd/Even Parity Generator/Checker
Product Specification

FEATURES

- **Word length easily expanded by cascading**
- **Generate even or odd parity**
- **Checks for parity errors**
- **See '280 for faster parity checker**

DESCRIPTION

The '180 is a 9-bit parity generator or checker commonly used to detect errors in high speed data transmission or data retrieval systems. Both Even and Odd parity enable inputs and parity outputs are available for generating or checking parity on 8-bits.

True active-HIGH or true active-LOW parity can be generated at both the Even and Odd outputs. True active-HIGH parity is established with Even Parity enable input (P_E) set HIGH and the Odd Parity enable input (P_O) set LOW. True active-LOW parity is established when P_E is LOW and P_O is HIGH. When both enable inputs are at the same logic level, both outputs will be forced to the opposite logic level.

Parity checking of a 9-bit word (8 bits plus parity) is possible by using the two enable inputs plus an inverter as the ninth data input. To check for true active-HIGH parity, the ninth data input is tied to the P_O input and an inverter is connected between the P_O and P_E inputs. To check for true active-LOW parity, the ninth data input is tied to the P_E input and an inverter is connected between the P_E and P_O inputs.

Expansion to larger word sizes is accomplished by serially cascading the '180 in 8-bit increments. The Even and Odd parity outputs of the first stage are connected to the corresponding P_E and P_O inputs, respectively, of the succeeding stage.

TYPE	TYPICAL PROPAGATION DELAY, $P_O = 0V$	TYPICAL SUPPLY CURRENT
74180	36ns	34mA

ORDERING CODE

PACKAGES	COMMERCIAL RANGE $V_{CC} = 5V \pm 5\%$; $T_A = 0°C$ to $+70°C$
Plastic DIP	N74180N

NOTE:
For information regarding devices processed to Military Specifications, see the Signetics Military Products Data Manual.

INPUT AND OUTPUT LOADING AND FAN-OUT TABLE

PINS	DESCRIPTION	74
$I_0 - I_7$	Data inputs	1ul
P_E, P_O	Parity inputs	2ul
Σ_E, Σ_O	Parity outputs	10ul

NOTE:
A 74 unit load (ul) is understood to be 40µA I_{IH} and −1.6mA I_{IL}.

PIN CONFIGURATION

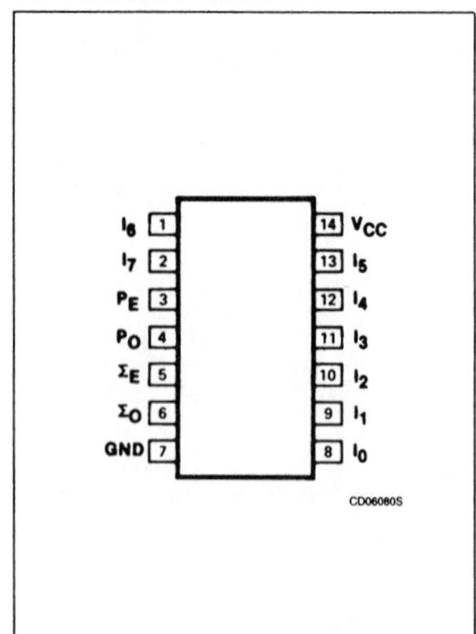

LOGIC SYMBOL

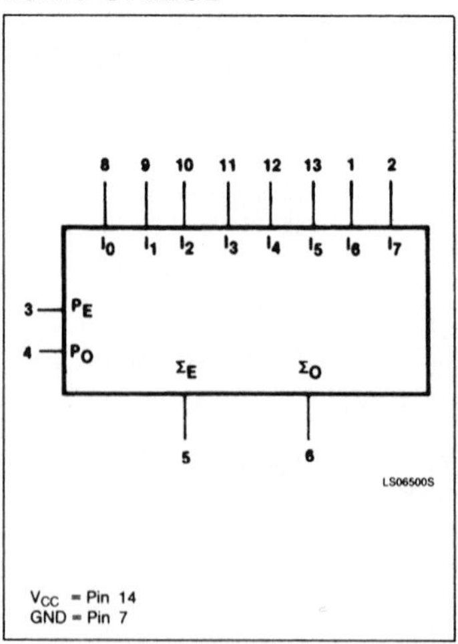

LOGIC SYMBOL (IEEE/IEC)

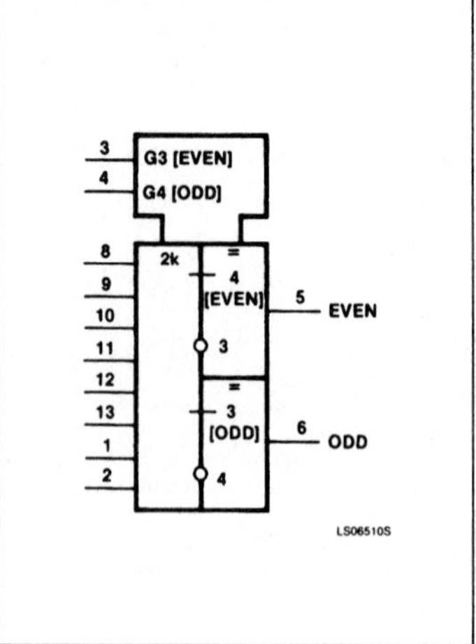

Parity Generator/Checker 74180

LOGIC DIAGRAM

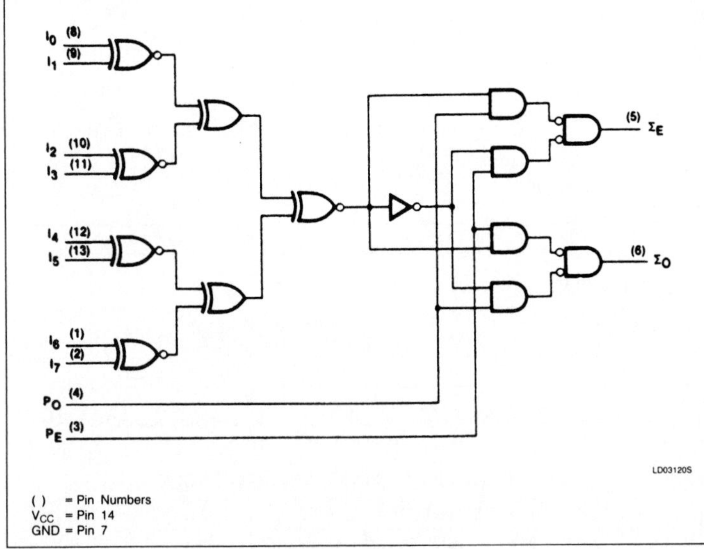

FUNCTION TABLE

INPUTS			OUTPUTS	
Number of HIGH Data Inputs ($I_0 - I_7$)	P_E	P_O	Σ_E	Σ_O
Even	H	L	H	L
Odd	H	L	L	H
Even	L	H	L	H
Odd	L	H	H	L
X	H	H	L	L
X	L	L	H	H

H = HIGH voltage level
L = LOW voltage level
X = Don't care

ABSOLUTE MAXIMUM RATINGS (Over operating free-air temperature range unless otherwise noted.)

PARAMETER		74	UNIT
V_{CC}	Supply voltage	7.0	V
V_{IN}	Input voltage	−0.5 to +5.5	V
I_{IN}	Input current	−30 to +5	mA
V_{OUT}	Voltage applied to output in HIGH output state	−0.5 to $+V_{CC}$	V
T_A	Operating free-air temperature range	0 to 70	°C

RECOMMENDED OPERATING CONDITIONS

PARAMETER		74			UNIT
		Min	Nom	Max	
V_{CC}	Supply voltage	4.75	5.0	5.25	V
V_{IH}	HIGH-level input voltage	2.0			V
V_{IL}	LOW-level input voltage			+0.8	V
I_{IK}	Input clamp current			−12	mA
I_{OH}	HIGH-level output current			−800	μA
I_{OL}	LOW-level output current			16	mA
T_A	Operating free-air temperature	0		70	°C

Signetics

74181, LS181, S181

Arithmetic Logic Units

4-Bit Arithmetic Logic Unit
Product Specification

Logic Products

FEATURES

- **Provides 16 arithmetic operations: ADD, SUBTRACT, COMPARE, DOUBLE, plus 12 other arithmetic operations**
- **Provides all 16 logic operations of two variables: Exclusive-OR, Compare, AND, NAND, NOR, OR, plus 10 other logic operations**
- **Full lookahead carry for high-speed arithmetic operation on long words**

DESCRIPTION

The '181 is a 4-bit high-speed parallel Arithmetic Logic Unit (ALU). Controlled by the four Function Select inputs ($S_0 - S_3$) and the Mode Control Input (M), it can perform all the 16 possible logic operations or 16 different arithmetic operations on active HIGH or active LOW operands. The Function Table lists these operations.

TYPE	TYPICAL PROPAGATION DELAY	TYPICAL SUPPLY CURRENT (TOTAL)
74181	22ns	91mA
74LS181	22ns	21mA
74S181	11ns	120mA

ORDERING CODE

PACKAGES	COMMERCIAL RANGE $V_{CC} = 5V \pm 5\%$; $T_A = 0°C$ to $+70°C$
Plastic DIP	N74181N, N74LS181N, N74S181N

NOTE:
For information regarding devices processed to Military Specifications, see the Signetics Military Products Data Manual.

INPUT AND OUTPUT LOADING AND FAN-OUT TABLE

PINS	DESCRIPTION	74	74S	74LS
Mode	Input	1ul	1Sul	1LSul
$\bar{A}$ or $\bar{B}$	Inputs	3ul	3Sul	3LSul
S	Inputs	4ul	4Sul	4LSul
Carry	Input	5ul	5Sul	5LSul
$F_0 - F_3$, = B, C_{n+4}	Outputs	10ul	10Sul	10LSul
$\bar{G}$	Output	10ul	10Sul	40LSul
$\bar{P}$	Output	10ul	10Sul	20LSul

NOTE:
Where a 74 unit load (ul) is understood to be 40μA I_{IH} and -1.6mA I_{IL}, a 74S unit load (Sul) is 50μA I_{IH} and -2.0mA I_{IL}, and 74LS unit load (LSul) is 20μA I_{IH} and -0.4mA I_{IL}.

PIN CONFIGURATION

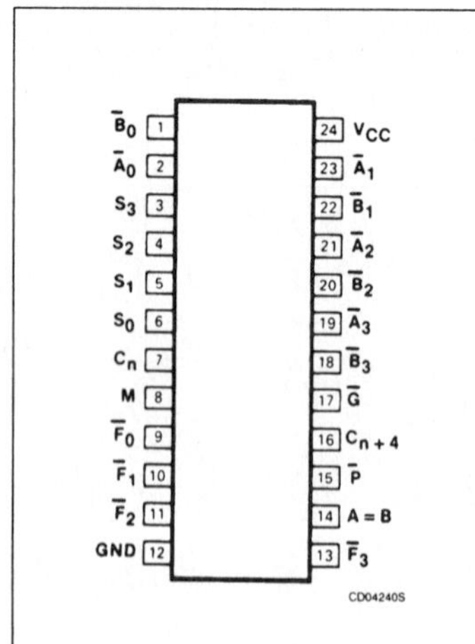

LOGIC SYMBOL

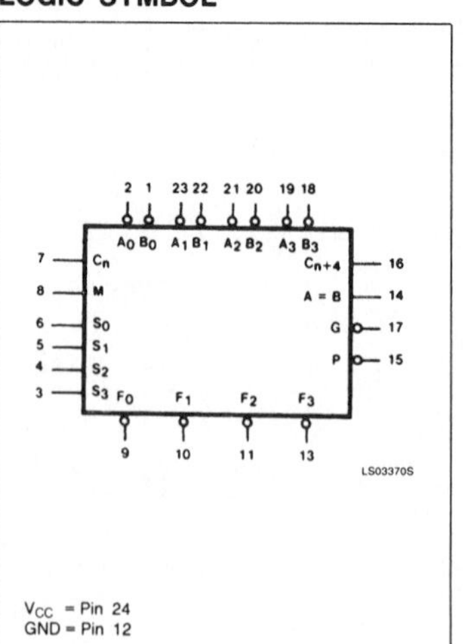

LOGIC SYMBOL (IEEE/IEC)

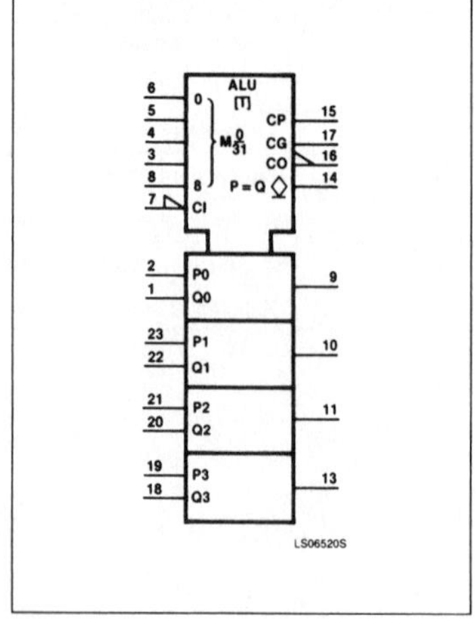

Arithmetic Logic Units

74181, LS181, S181

When the Mode Control input (M) is HIGH, all internal carries are inhibited and the device performs logic operations on the individual bits as listed. When the Mode Control Input is LOW, the carries are enabled and the device performs arithmetic operations on the two 4-bit words. The device incorporates full internal carry lookahead and provides for either ripple carry between devices using the C_{n+4} output, or for carry lookahead between packages using the signals $\overline{P}$ (Carry Propagate) and $\overline{G}$ (Carry Generate). $\overline{P}$ and $\overline{G}$ are not affected by carry in. When speed requirements are not stringent, it can be used in a simple ripple carry mode by connecting the Carry output (C_{n+4}) signal to the Carry input (C_n) of the next unit. For high-speed operation the device is used in conjunction with the '182 carry lookahead circuit. One carry lookahead package is required for each group of four '181 devices. Carry lookahead can be provided at various levels and offers high-speed capability over extremely long word lengths.

The A = B output from the device goes HIGH when all four $\overline{F}$ outputs are HIGH and can be used to indicate logic equivalence over 4 bits when the unit is in the subtract mode. The A = B output is open collector and can be wired-AND with other A = B outputs to give a comparison for more than 4 bits. The A = B signal can also be used with the C_{n+4} signal to indicate A > B and A < B.

The Function Table lists the arithmetic operations that are performed without a carry in. An incoming carry adds a one to each operation. Thus, select code LHHL generates A minus B minus 1 (2s complement notation) without a carry in and generates A minus B when a carry is applied.

Because subtraction is actually performed by complementary addition (1s complement), a carry out means borrow; thus, a carry is generated when there is no underflow and no carry is generated when there is underflow.

As indicated, this device can be used with either active LOW inputs producing active LOW outputs or with active HIGH inputs producing active HIGH outputs. For either case the table lists the operations that are performed to the operands labeled inside the logic symbol.

LOGIC DIAGRAM

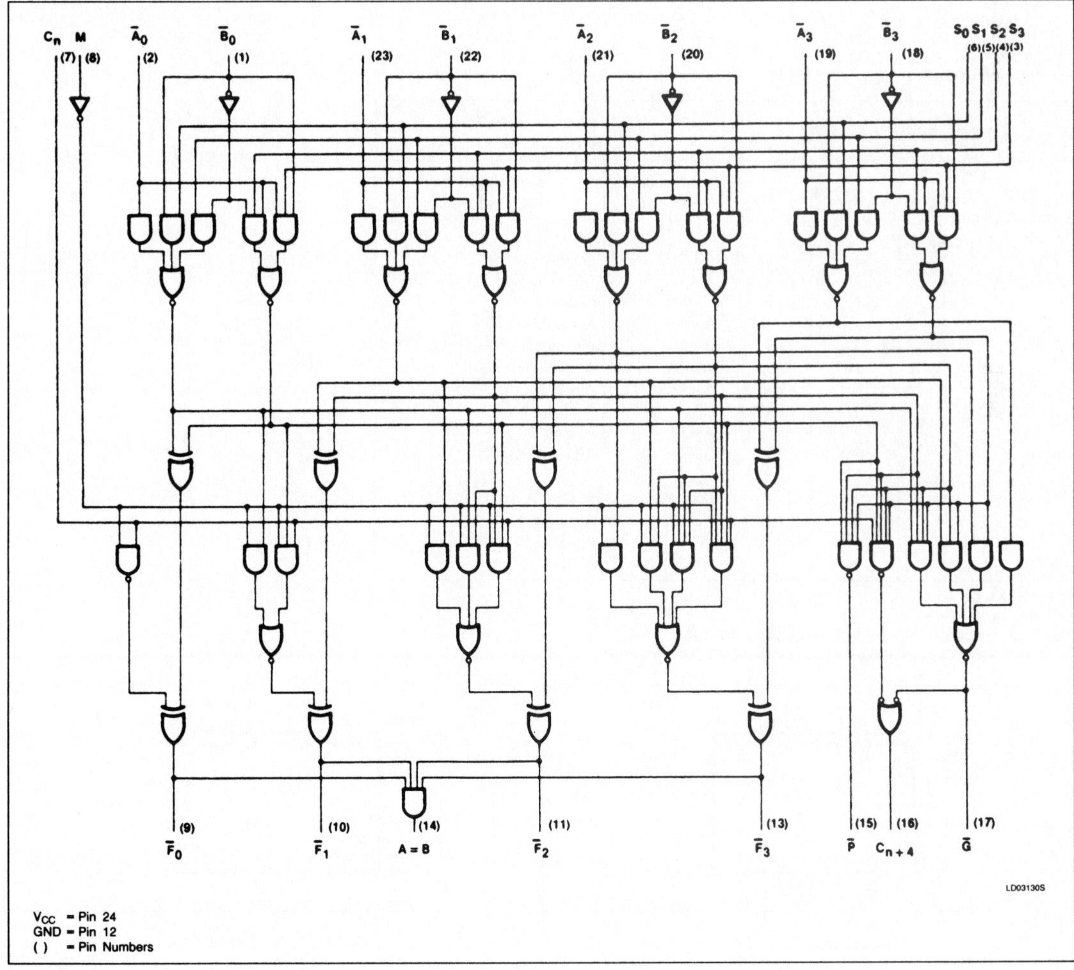

V_{CC} = Pin 24
GND = Pin 12
( ) = Pin Numbers

Arithmetic Logic Units — 74181, LS181, S181

MODE SELECT — FUNCTION TABLE

MODE SELECT INPUTS				ACTIVE HIGH INPUTS & OUTPUTS	
S_3	S_2	S_1	S_0	Logic (M = H)	Arithmetic** (M = L) (C_n = H)
L	L	L	L	$\overline{A}$	A
L	L	L	H	$\overline{A + B}$	A + B
L	L	H	L	$\overline{A}B$	$A + \overline{B}$
L	L	H	H	Logical 0	minus 1
L	H	L	L	$\overline{AB}$	A plus $A\overline{B}$
L	H	L	H	$\overline{B}$	(A + B) plus $A\overline{B}$
L	H	H	L	$A \bullet B$	A minus B minus 1
L	H	H	H	$A\overline{B}$	AB minus 1
H	L	L	L	$\overline{A} + B$	A plus AB
H	L	L	H	$\overline{A \bullet B}$	A plus B
H	L	H	L	B	$(A + \overline{B})$ plus AB
H	L	H	H	AB	AB minus 1
H	H	L	L	Logical 1	A plus A*
H	H	L	H	$A + \overline{B}$	(A + B) plus A
H	H	H	L	A + B	$(A + \overline{B})$ plus A
H	H	H	H	A	A minus 1

MODE SELECT INPUTS				ACTIVE LOW INPUTS & OUTPUTS	
S_3	S_2	S_1	S_0	Logic (M = H)	Arithmetic** (M = L) (C_n = L)
L	L	L	L	$\overline{A}$	A minus 1
L	L	L	H	$\overline{AB}$	AB minus 1
L	L	H	L	$\overline{A} + B$	$A\overline{B}$ minus 1
L	L	H	H	Logical 1	minus 1
L	H	L	L	$\overline{A + B}$	A plus $(A + \overline{B})$
L	H	L	H	$\overline{B}$	AB plus $(A + \overline{B})$
L	H	H	L	$\overline{A \bullet B}$	A minus B minus 1
L	H	H	H	$A + \overline{B}$	$A + \overline{B}$
H	L	L	L	$\overline{A}B$	A plus (A + B)
H	L	L	H	$\overline{A \bullet B}$	A plus B
H	L	H	L	B	$A\overline{B}$ (A + B)
H	L	H	H	A + B	A + B
H	H	L	L	Logical 0	A plus A*
H	H	L	H	$A\overline{B}$	AB plus A
H	H	H	L	AB	$A\overline{B}$ plus A
H	H	H	H	A	A

L = LOW voltage
H = HIGH voltage level
*Each bit is shifted to the next more significant position.
**Arithmetic operations expressed in 2s complement notation.

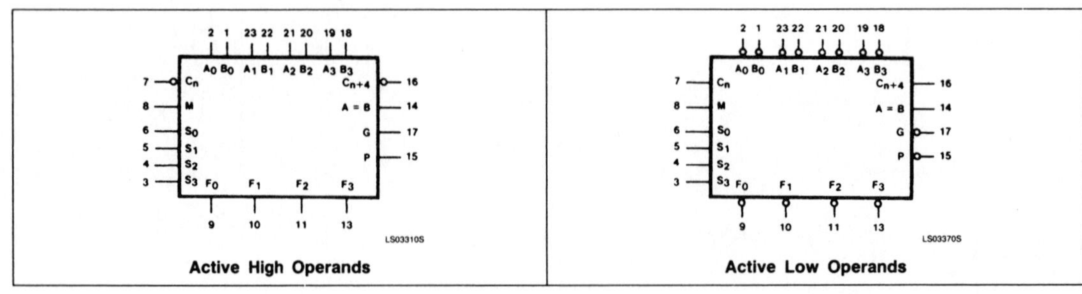

Signetics

74194, LS194A, S194
Shift Registers

4-Bit Bidirectional Universal Shift Register
Product Specification

Logic Products

- **Buffered clock and control inputs**
- **Shift left and shift right capability**
- **Synchronous parallel and serial data transfers**
- **Easily expanded for both serial and parallel operation**
- **Asynchronous Master Reset**
- **Hold (do nothing) mode**

DESCRIPTION

The functional characteristics of the '194 4-Bit Bidirectional Shift Register are indicated in the Logic Diagram and Function Table. The register is fully synchronous, with all operations taking place in less than 20ns (typical) for the 54/74 and 54LS/74LS, and 12ns (typical) for 54S/74S, making the device especially useful for implementing very high speed CPUs, or for memory buffer registers.

TYPE	TYPICAL f_{MAX}	TYPICAL SUPPLY CURRENT (TOTAL)
74194	36MHz	39mA
74LS194A	36MHz	15mA
74S194	105MHz	85mA

ORDERING CODE

PACKAGES	COMMERCIAL RANGE V_{CC} = 5V ±5%; T_A = 0°C to +70°C
Plastic DIP	N74194N, N74LS194AN, N74S194N
Plastic SO-16	N74LS194AD, N745194D

NOTE:
For information regarding devices processed to Military Specifications, see the Signetics Military Products Data Manual.

INPUT AND OUTPUT LOADING AND FAN-OUT TABLE

PINS	DESCRIPTION	74	74S	74LS
All	Inputs	1ul	1Sul	1LSul
$Q_0 - Q_3$	Outputs	10ul	10Sul	10LSul

NOTE:
Where a 74 unit load (ul) is understood to be 40μA I_{IH} and −1.6mA I_{IL}, a 74S unit load (Sul) is 50μA I_{IH} and −2.0mA I_{IL}, and 74LS unit load (LSul) is 20μA I_{IH} and −0.4mA I_{IL}.

PIN CONFIGURATION

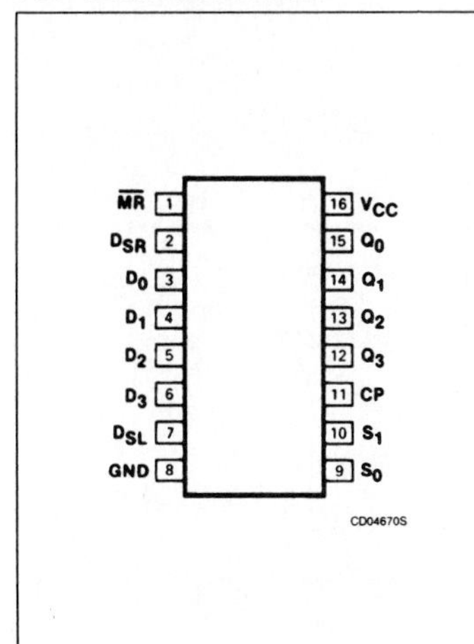

LOGIC SYMBOL

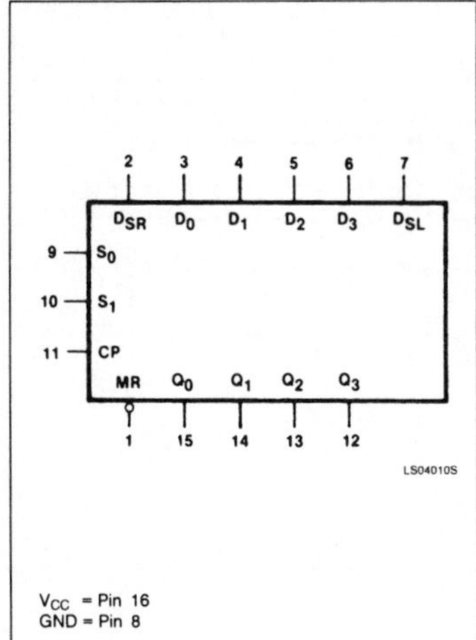

LOGIC SYMBOL (IEEE/IEC)

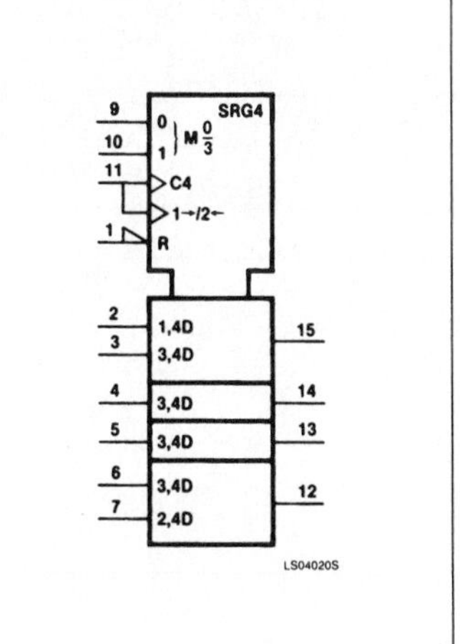

Shift Registers 74194, LS194A, S194

MODE SELECT — FUNCTION TABLE

OPERATING MODE	INPUTS							OUTPUTS			
	CP	$\overline{MR}$	S_1	S	D_{SR}	D_{SL}	D_n	Q_0	Q_1	Q_2	Q_3
Reset (clear)	X	L	X	X	X	X	X	L	L	L	L
Hold (do nothing)	X	H	l(a)	l(a)	X	X	X	q_0	q_1	q_2	q_3
Shift left	↑	H	h	l(a)	X	l	X	q_1	q_2	q_3	L
	↑	H	h	l(a)	X	h	X	q_1	q_2	q_3	H
Shift right	↑	H	l(a)	h	l	X	X	L	q_0	q_1	q_2
	↑	H	l(a)	h	h	X	X	H	q_0	q_1	q_2
Parallel load	↑	H	h	h	X	X	d_n	d_0	d_1	d_2	d_3

H = HIGH voltage level.
h = HIGH voltage level one set-up time prior to the LOW-to-HIGH clock transition.
L = LOW voltage level.
l = LOW voltage level one set-up time prior to the LOW-to-HIGH clock transition.
$d_n(q_n)$ = Lower case letters indicate the state of the referenced input (or output) one set-up time prior to the LOW-to-HIGH clock transition.
X = Don't care.
↑ = LOW-to-HIGH clock transition.

NOTE:

a. The HIGH-to-LOW transition of the S_0 and S_1 inputs on the 74194 should only take place while CP is HIGH for conventional operation.

TYPICAL CLEAR, LOAD, RIGHT-SHIFT, LEFT-SHIFT, INHIBIT AND CLEAR SEQUENCES

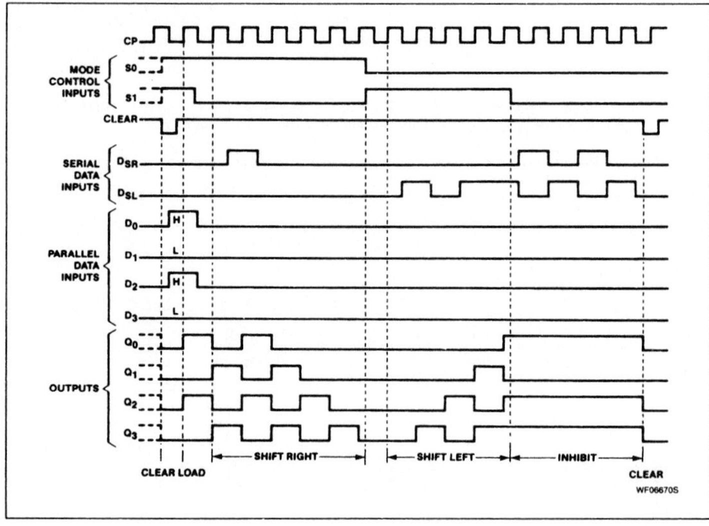

The '194 design has special logic features which increase the range of application. The synchronous operation of the device is determined by two Mode Select inputs, S_0 and S_1. As shown in the Mode Select Table, data can be entered and shifted from left to right (shift right, $Q_0 \rightarrow Q_1$, etc.) or, right to left (shift left, $Q_3 \rightarrow Q_2$, etc.) or, parallel data can be entered, loading all 4 bits of the register simultaneously. When both S_0 and S_1 are LOW, existing data is retained in a hold (do nothing) mode. The first and last stages provide D-type Serial Data inputs (D_{SR}, D_{SL}) to allow multistage shift right or shift left data transfers without interfering with parallel load operation.

Mode Select and Data inputs on the 74S194 and 74LS194A are edge-triggered, responding only to the LOW-to-HIGH transition of the Clock (CP). Therefore, the only timing restriction is that the Mode Control and selected Data inputs must be stable one set-up time prior to the positive transition of the clock pulse. The Mode Select inputs of the 74194 are gated with the clock and should be changed from HIGH-to-LOW only while the Clock input is HIGH.

The four parallel data inputs ($D_0 - D_3$) are D-type inputs. Data appearing on $D_0 - D_3$ inputs when S_0 and S_1 are HIGH is transferred to the $Q_0 - Q_3$ outputs respectively, following the next LOW-to-HIGH transition of the clock. When LOW, the asynchronous Master Reset ($\overline{MR}$) overrides all other input conditions and forces the Q outputs LOW.

LOGIC DIAGRAM

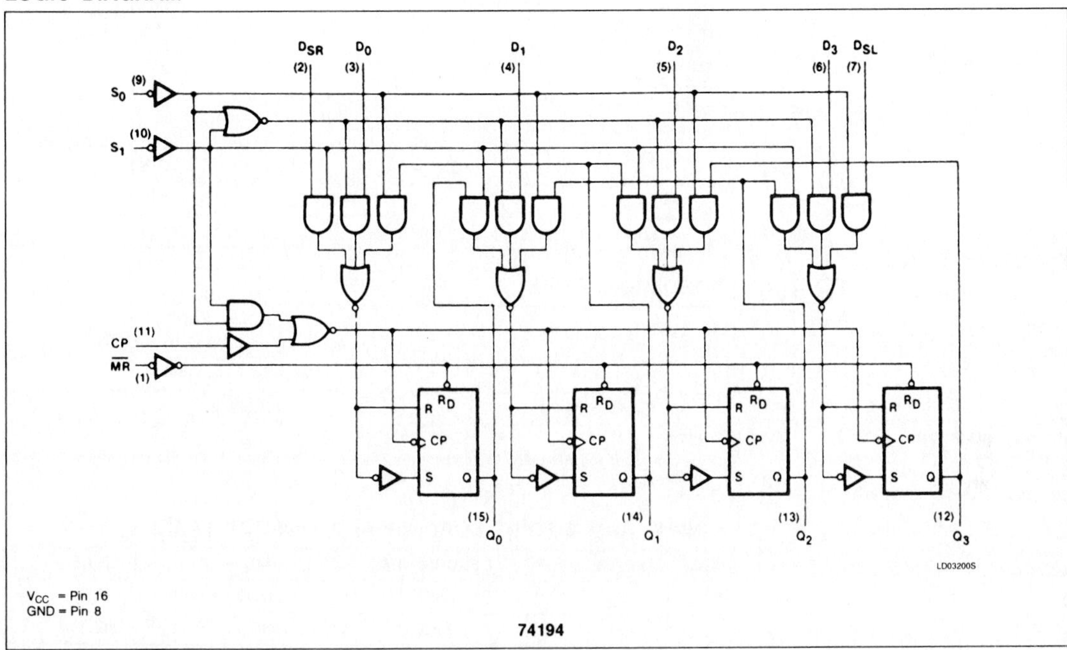

74194

LOGIC DIAGRAM

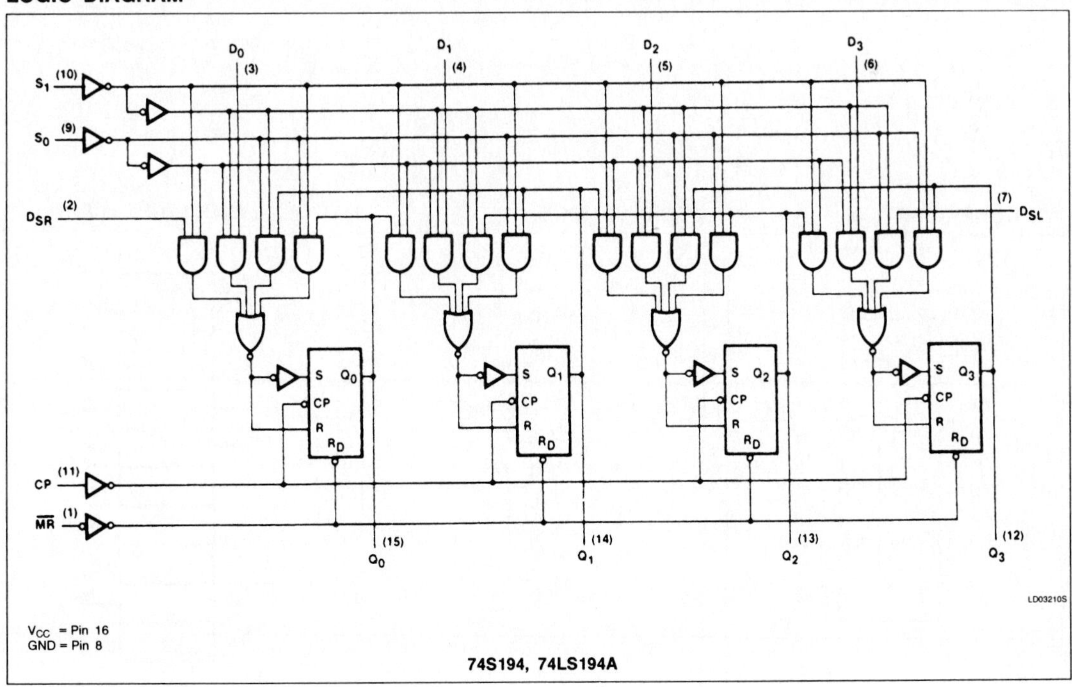

74S194, 74LS194A

Signetics

74LS244, S244
Buffers

Octal Buffers (3-State)
Product Specification

Logic Products

TYPE	TYPICAL PROPAGATION DELAY	TYPICAL SUPPLY CURRENT (TOTAL)
74LS244	12ns	25mA
74S244	6ns	112mA

FUNCTION TABLE

INPUTS				OUTPUTS	
$\overline{OE}_a$	I_a	$\overline{OE}_b$	I_b	Y_a	Y_b
L	L	L	L	L	L
L	H	L	H	H	H
H	X	H	X	(Z)	(Z)

H = HIGH voltage level
L = LOW voltage level
X = Don't care
(Z) = HIGH impedance (off) state

ORDERING CODE

PACKAGES	COMMERCIAL RANGE $V_{CC} = 5V \pm 5\%$; $T_A = 0°C$ to $+70°C$
Plastic DIP	N74LS244N, 74S244N
Plastic SOL-20	74LS244D

NOTE:
For information regarding devices processed to Military Specifications, see the Signetics Military Products Data Manual.

INPUT AND OUTPUT LOADING AND FAN-OUT TABLE

PINS	DESCRIPTION	74S	74LS
All	Inputs	1Sul	1LSul
All	Outputs	24Sul	30LSul

NOTE:
A 74S unit load (Sul) is 50μA I_{IH} and −2.0mA I_{IL}, and a 74LS unit load (LSul) is 20μA I_{IH} and −0.4mA I_{IL}.

PIN CONFIGURATION

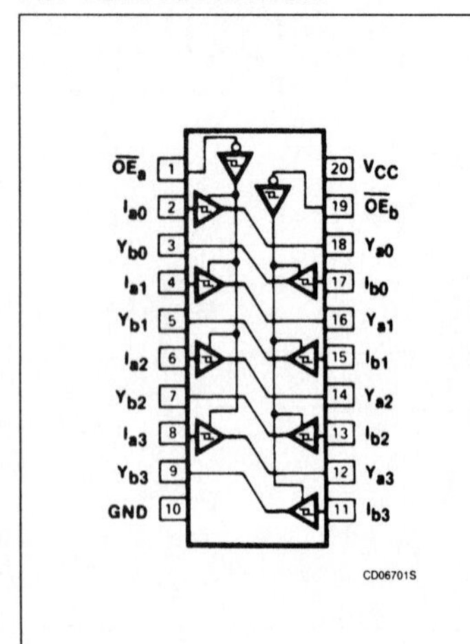

LOGIC SYMBOL

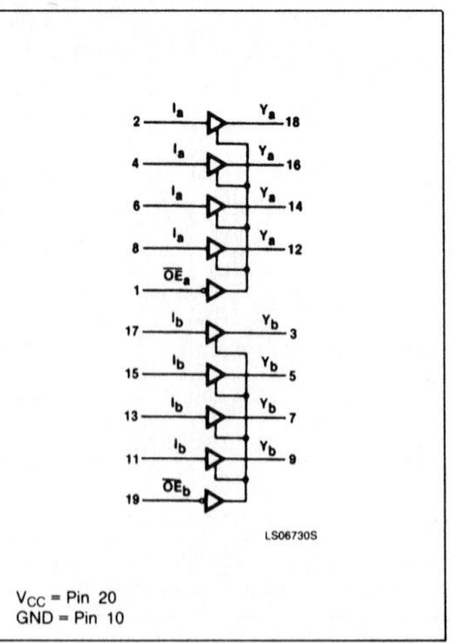

V_{CC} = Pin 20
GND = Pin 10

LOGIC SYMBOL (IEEE/IEC)

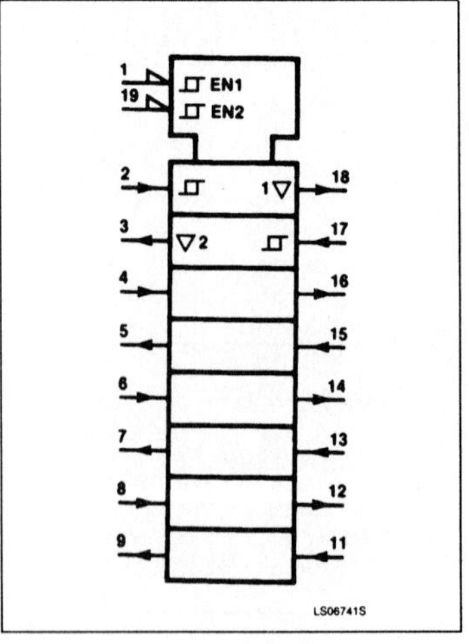

ABSOLUTE MAXIMUM RATINGS (Over operating free-air temperature range unless otherwise noted.)

PARAMETER		74LS	74S	UNIT
V_{CC}	Supply voltage	7.0	7.0	V
V_{IN}	Input voltage	−0.5 to +7.0	−0.5 to +5.5	V
I_{IN}	Input current	−30 to +1	−30 to +5	mA
V_{OUT}	Voltage applied to output in HIGH output state	−0.5 to $+V_{CC}$	−0.5 to $+V_{CC}$	V
T_A	Operating free-air temperature range	0 to 70		°C

RECOMMENDED OPERATING CONDITIONS

PARAMETER		74LS			74S			UNIT
		Min	Nom	Max	Min	Nom	Max	
V_{CC}	Supply voltage	4.75	5.0	5.25	4.75	5.0	5.25	V
V_{IH}	HIGH-level input voltage	2.0			2.0			V
V_{IL}	LOW-level input voltage			+0.8			+0.8	V
I_{IK}	Input clamp current			−18			−18	mA
I_{OH}	HIGH-level output current			−15			−15	mA
I_{OL}	LOW-level output current			24			64	mA
T_A	Operating free-air temperature	0		70	0		70	°C

TEST CIRCUITS AND WAVEFORMS

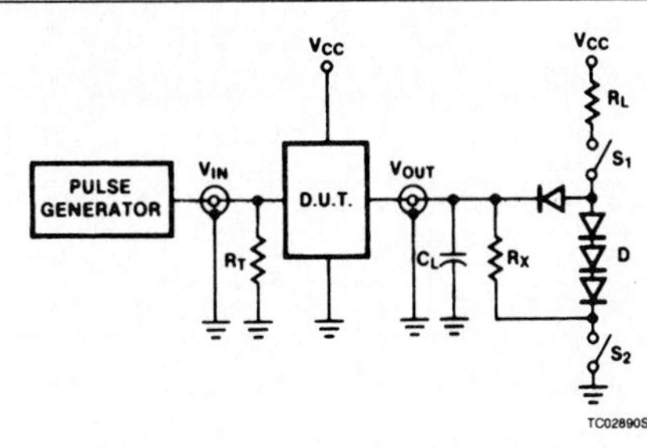

Test Circuit For 3-State Outputs

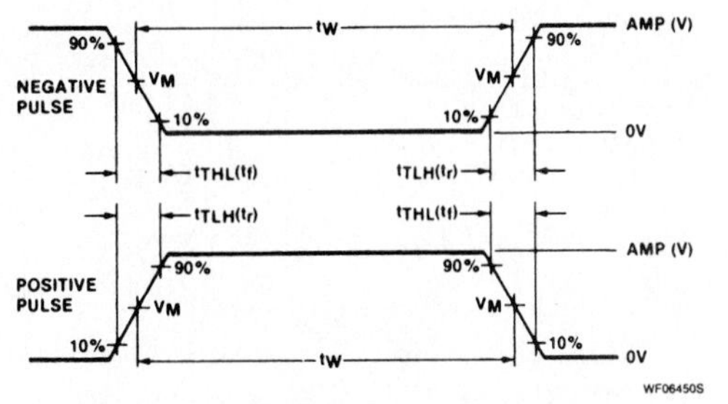

V_M = 1.3V for 74LS; V_M = 1.5V for all other TTL families.

Input Pulse Definition

SWITCH POSITION

TEST	SWITCH 1	SWITCH 2
t_{PZH}	Open	Closed
t_{PZL}	Closed	Open
t_{PHZ}	Closed	Closed
t_{PLZ}	Closed	Closed

DEFINITIONS

R_L = Load resistor to V_{CC}; see AC CHARACTERISTICS for value.

C_L = Load capacitance includes jig and probe capacitance; see AC CHARACTERISTICS for value.

R_T = Termination resistance should be equal to Z_{OUT} of Pulse Generators.

D = Diodes are 1N916, 1N3064, or equivalent.

R_X = 1kΩ for 74, 74S, R_X = 5kΩ for 74LS.

t_{TLH}, t_{THL} Values should be less than or equal to the table entries.

FAMILY	INPUT PULSE REQUIREMENTS				
	Amplitude	Rep. Rate	Pulse Width	t_{TLH}	t_{THL}
74	3.0V	1MHz	500ns	7ns	7ns
74LS	3.0V	1MHz	500ns	15ns	6ns
74S	3.0V	1MHz	500ns	2.5ns	2.5ns

Buffers — 74LS244, S244

DC ELECTRICAL CHARACTERISTICS (Over recommended operating free-air temperature range unless otherwise noted.)

PARAMETER		TEST CONDITIONS[1]			74LS244			74S244			UNIT
					Min	Typ[2]	Max	Min	Typ[2]	Max	
ΔV_T	Hysteresis ($V_{T+} - V_{T-}$)	V_{CC} = MIN			0.2	0.4		0.2	0.4		V
V_{OH}	HIGH-level output voltage	V_{CC} = MIN, V_{IH} = MIN, V_{IL} = 0.5V, I_{OH} = MAX			2.0			2.0			V
		V_{CC} = MIN, V_{IH} = MIN, V_{IL} = MAX, I_{OH} = MAX			2.4	3.4		2.4			V
V_{OL}	LOW-level output voltage	V_{CC} = MIN, V_{IH} = MIN, V_{IL} = MAX	I_{OL} = MAX				0.5			0.55	V
			I_{OL} = 12mA (74LS)				0.4				V
V_{IK}	Input clamp voltage	V_{CC} = MIN, $I_I = I_{IK}$					−1.5			−1.2	V
I_{OZH}	Off-state output current, HIGH-level voltage applied	V_{CC} = MAX, V_{IH} = MIN, V_{IL} = MAX	V_O = 2.7V				20				μA
			V_O = 2.4V							50	μA
I_{OZL}	Off-state output current, LOW-level voltage applied	V_{CC} = MAX, V_{IH} = MIN, V_{IL} = MAX	V_O = 0.4V				−20				μA
			V_O = 0.5V							−50	μA
I_I	Input current at maximum input voltage	V_{CC} = MAX	V_I = 5.5V							1.0	mA
			V_I = 7.0V				0.1				mA
I_{IH}	HIGH-level input current	V_{CC} = MAX, V_I = 2.7V					20			50	μA
I_{IL}	LOW-level input current	V_{CC} = MAX	V_I = 0.4V				−0.2				mA
			V_I = 0.5V	$\overline{OE}$ inputs						−2.0	mA
				Other inputs						−0.4	mA
I_{OS}	Short-circuit output current[3]	V_{CC} = MAX			−40		−130	−80		−180	mA
I_{CC}	Supply current[4] (total)	V_{CC} = MAX	I_{CCH}	Outputs HIGH		17	27		95	160	mA
			I_{CCL}	Outputs LOW		27	46		120	180	mA
			I_{CCZ}	Outputs OFF		32	54		120	180	mA

NOTES:

1. For conditions shown as MIN or MAX, use the appropriate value specified under recommended operating conditions for the applicable type.
2. All typical values are at V_{CC} = 5V, T_A = 25°C.
3. I_{OS} is tested with V_{OUT} = +0.5V and $V_{CC} = V_{CC}$ MAX + 0.5V. Not more than one output should be shorted at a time and duration of the short circuit should not exceed one second.
4. I_{CC} is measured with outputs open.

AC ELECTRICAL CHARACTERISTICS T_A = 25°C, V_{CC} = 5.0V

PARAMETER		TEST CONDITIONS	74LS C_L = 45pF, R_L = 667Ω		74S C_L = 50pF, R_L = 90Ω		UNIT
			Min	Max	Min	Max	
t_{PLH}	Propagation delay	Waveform 1		18		9	ns
t_{PHL}	Propagation delay	Waveform 1		18		9	ns
t_{PZH}	Enable to HIGH	Waveform 2		23		12	ns
t_{PZL}	Enable to LOW	Waveform 3		30		15	ns
t_{PHZ}	Disable from HIGH	Waveform 2, C_L = 5pF		18		9	ns
t_{PLZ}	Disable from LOW	Waveform 3, C_L = 5pF		25		15	ns

AC WAVEFORMS

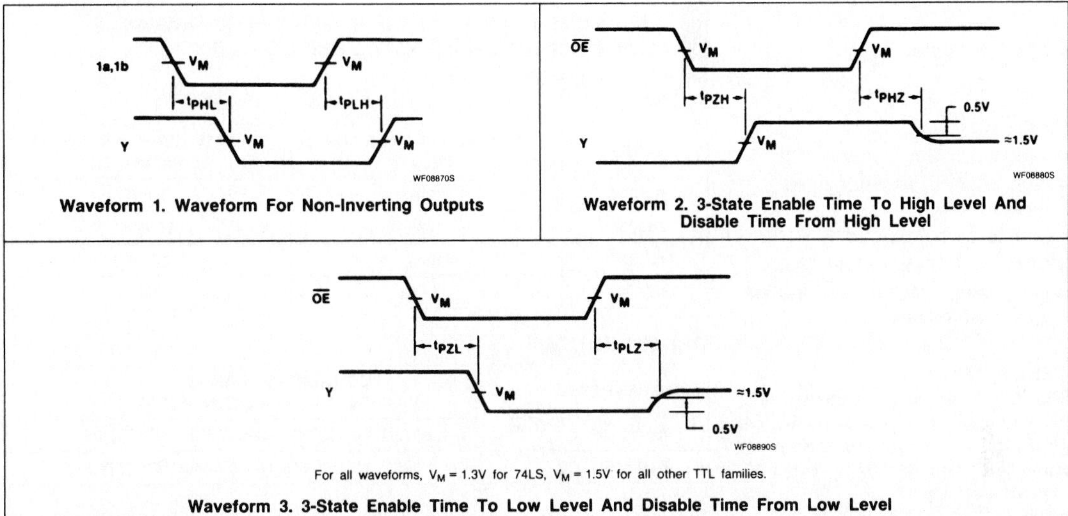

Waveform 1. Waveform For Non-Inverting Outputs

Waveform 2. 3-State Enable Time To High Level And Disable Time From High Level

For all waveforms, V_M = 1.3V for 74LS, V_M = 1.5V for all other TTL families.

Waveform 3. 3-State Enable Time To Low Level And Disable Time From Low Level

Signetics

Logic Products

74LS373, 74LS374, S373, S374
Latches/Flip-Flops

'373 Octal Transparent Latch With 3-State Outputs
'374 Octal D Flip-Flop With 3-State Outputs
Product Specification

FEATURES
- 8-bit transparent latch — '373
- 8-bit positive, edge-triggered register — '374
- 3-State output buffers
- Common 3-State Output Enable
- Independent register and 3-State buffer operation

DESCRIPTION
The '373 is an octal transparent latch coupled to eight 3-State output buffers. The two sections of the device are controlled independently by Latch Enable (E) and Output Enable ($\overline{OE}$) control gates.

TYPE	TYPICAL PROPAGATION DELAY	TYPICAL SUPPLY CURRENT (TOTAL)
74LS373	19ns	24mA
74S373	10ns	105mA
74LS374	19ns	27mA
74S374	8ns	116mA

ORDERING CODE

PACKAGES	COMMERCIAL RANGE $V_{CC} = 5V \pm 5\%$; $T_A = 0°C$ to $+70°C$
Plastic DIP	N74LS373N, N74S373N, N74LS374N, N74S374N
Plastic SOL-20	N74LS373D, N74S373D, N74LS374D, N74S374D

NOTE:
For information regarding devices processed to Military Specifications, see the Signetics Military Products Data Manual.

INPUT AND OUTPUT LOADING AND FAN-OUT TABLE

PINS	DESCRIPTION	74S	74LS
All	Inputs	1Sul	1LSul
All	Outputs	10Sul	30LSul

NOTE:
Where a 74S unit load (Sul) is 50μA I_{IH} and −2.0mA I_{IL}, and a 74LS unit load (LSul) is 20μA I_{IH} and −0.4mA I_{IL}.

PIN CONFIGURATION

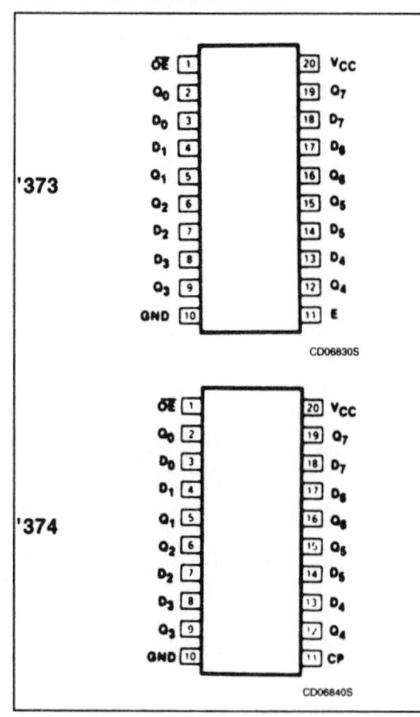

LOGIC SYMBOL

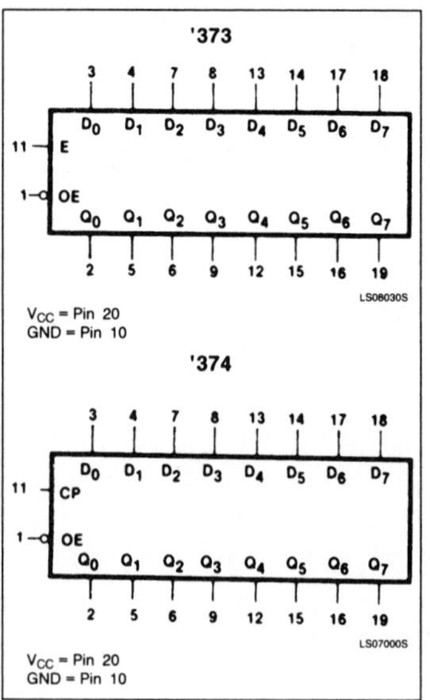

LOGIC SYMBOL (IEEE/EC)

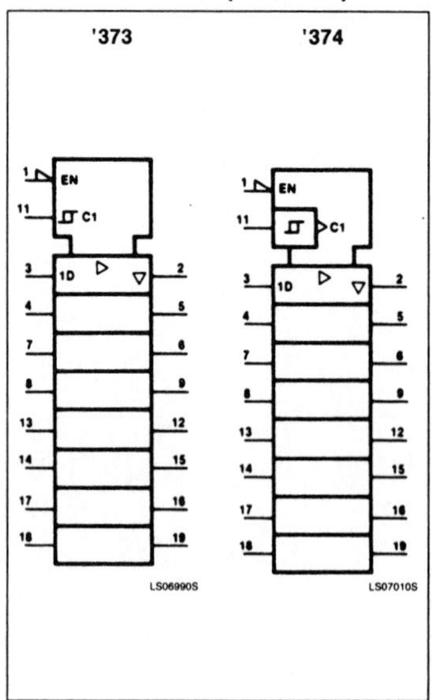

The data on the D inputs are transferred to the latch outputs when the Latch Enable (E) input is HIGH. The latch remains transparent to the data inputs while E is HIGH, and stores the data present one set-up time before the HIGH-to-LOW enable transition. The enable gate has hysteresis built in to help minimize problems that signal and ground noise can cause on the latching operation.

The 3-State output buffers are designed to drive heavily loaded 3-State buses, MOS memories, or MOS microprocessors. The active LOW Output Enable ($\overline{OE}$) controls all eight 3-State buffers independent of the latch operation. When $\overline{OE}$ is LOW, the latched or transparent data appears at the outputs. When $\overline{OE}$ is HIGH, the outputs are in the HIGH impedance "off" state, which means they will neither drive nor load the bus.

The '374 is an 8-bit, edge-triggered register coupled to eight 3-State output buffers. The two sections of the device are controlled independently by the Clock (CP) and Output Enable ($\overline{OE}$) control gates.

The register is fully edge triggered. The state of each D input, one set-up time before the LOW-to-HIGH clock transition, is transferred to the corresponding flip-flop's Q output. The clock buffer has hysteresis built in to help minimize problems that signal and ground noise can cause on the clocking operation.

The 3-State output buffers are designed to drive heavily loaded 3-State buses, MOS memories, or MOS microprocessors. The active LOW Output Enable ($\overline{OE}$) controls all eight 3-State buffers independent of the register operation. When $\overline{OE}$ is LOW, the data in the register appears at the outputs. When $\overline{OE}$ is HIGH, the outputs are in the HIGH impedance "off" state, which means they will neither drive nor load the bus.

LOGIC DIAGRAM, '373

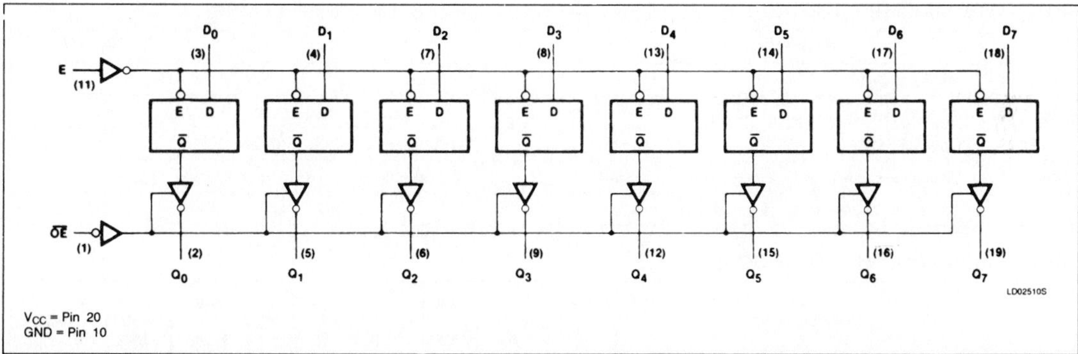

LOGIC DIAGRAM, '374

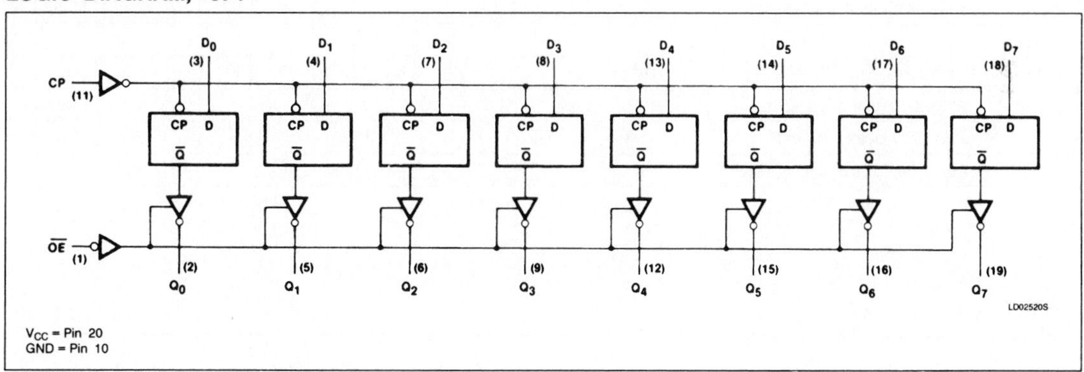

MODE SELECT — FUNCTION TABLE '373

OPERATING MODES	INPUTS			INTERNAL REGISTER	OUTPUTS
	$\overline{OE}$	E	D_n		$Q_0 - Q_7$
Enable and read register	L L	H H	L H	L H	L H
Latch and read register	L L	L L	l h	L H	L H
Latch register and disable outputs	H H	L L	l h	L H	(Z) (Z)

Latches/Flip-Flops **74LS373, 74LS374, S373, S374**

MODE SELECT — FUNCTION TABLE '374

OPERATING MODES	INPUTS			INTERNAL REGISTER	OUTPUTS
	$\overline{OE}$	CP	D_n		$Q_0 - Q_7$
Load and read register	L L	↑ ↑	l h	L H	L H
Load register and disable outputs	H H	↑ ↑	l h	L H	(Z) (Z)

H = HIGH voltage level
h = HIGH voltage level one set-up time prior to the LOW-to-HIGH clock transition or HIGH-to-LOW $\overline{OE}$ transition
L = LOW voltage level
l = LOW voltage level one set-up time prior to the LOW-to-HIGH clock transition or HIGH-to-LOW $\overline{OE}$ transition
(Z) = HIGH impedance "off" state
↑ = LOW-to-HIGH clock transition

ABSOLUTE MAXIMUM RATINGS (Over operating free-air temperature range unless otherwise noted.)

PARAMETER		74LS	74S	UNIT
V_{CC}	Supply voltage	7.0	7.0	V
V_{IN}	Input voltage	−0.5 to +7.0	−0.5 to +5.5	V
I_{IN}	Input current	−30 to +1	−30 to +5	mA
V_{OUT}	Voltage applied to output in HIGH output state	−0.5 to $+V_{CC}$	−0.5 to $+V_{CC}$	V
T_A	Operating free-air temperature range	0 to 70		°C

RECOMMENDED OPERATING CONDITIONS

PARAMETER		74LS			74S			UNIT
		Min	Nom	Max	Min	Nom	Max	
V_{CC}	Supply voltage	4.75	5.0	5.25	4.75	5.0	5.25	V
V_{IH}	HIGH-level input voltage	2.0			2.0			V
V_{IL}	LOW-level input voltage			+0.8			+0.8	V
I_{IK}	Input clamp current			−18			−18	mA
I_{OH}	HIGH-level output current			−2.6			−6.5	mA
I_{OL}	LOW-level output current			24			20	mA
T_A	Operating free-air temperature	0		70	0		70	°C

Signetics

74LS670
Register File

4 x 4 Register File (3-State)
Product Specification

Logic Products

FEATURES

- **Simultaneous and independent Read and Write operations**
- **Expandable to almost any word size and bit length**
- **3-State outputs**
- **See '170 for open collector version**

DESCRIPTION

The '670 is a 16-bit 3-State Register File organized as 4 words of 4 bits each. Separate Read and Write Address and Enable inputs are available, permitting simultaneous writing into one word location and reading from another location. The 4-bit word to be stored is presented to four Data inputs. The Write Address inputs (W_A and W_B) determine the location of the stored word. When the Write Enable ($\overline{WE}$) input is LOW, the data is entered into the addressed location. The addressed location remains transparent to the data while the $\overline{WE}$ is LOW. Data supplied at the inputs will be read out in true (non-inverting) form from the 3-State outputs. Data and Write Address inputs are inhibited when $\overline{WE}$ is HIGH.

TYPE	TYPICAL PROPAGATION DELAY	TYPICAL SUPPLY CURRENT (TOTAL)
74LS670	25ns	30mA

ORDERING CODE

PACKAGES	COMMERCIAL RANGE $V_{CC} = 5V \pm 5\%$; $T_A = 0°C$ to $+70°C$
Plastic DIP	N74LS670N
Plastic SOL-16	N74LS670D

NOTE:
For information regarding devices processed to Military Specifications, see the Signetics Military Products Data Manual.

INPUT AND OUTPUT LOADING AND FAN-OUT TABLE

PINS	DESCRIPTION	74LS
$D_0 - D_3$, W_A, W_B, R_A, R_B	Inputs	1LSul
$\overline{WE}$	Input	2LSul
$\overline{RE}$	Input	3LSul
$Q_0 - Q_3$	Outputs	10LSul

NOTE:
A 74LS unit load (LSul) is $20\mu A$ I_{IH} and $-0.4mA$ I_{IL}.

Direct acquisition of data stored in any of the four registers is made possible by individual Read Address inputs (R_A and R_B). The addressed word appears at the four outputs when the Read Enable ($\overline{RE}$) is LOW. Data outputs are in the HIGH impedance "off" state when the Read Enable input is HIGH. This permits outputs to be tied together to increase the word capacity to very large numbers.

PIN CONFIGURATION

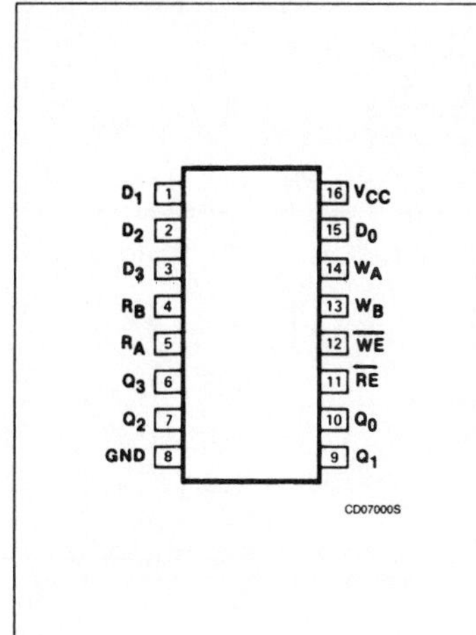

LOGIC SYMBOL

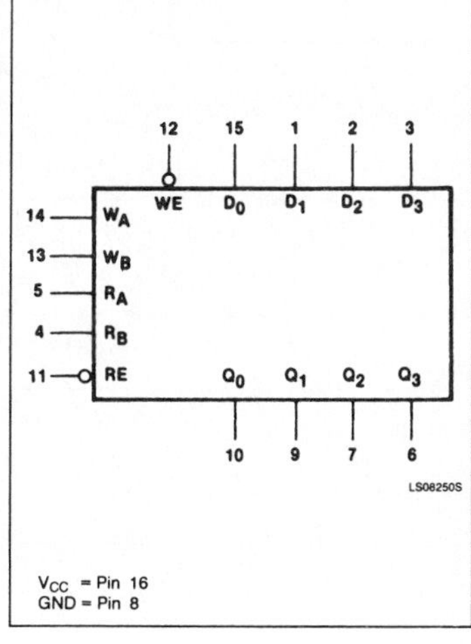

LOGIC SYMBOL (IEEE/IEC)

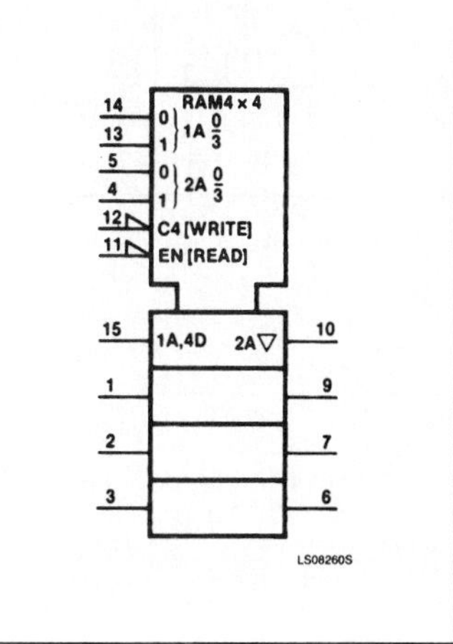

Register File 74LS670

Up to 128 devices can be stacked to increase the word size to 512 locations by tying the 3-State outputs together. Since the limiting factor for expansion is the output HIGH current, further stacking is possible by tying pull-up resistors to the outputs to increase the I_{OH} current available. Design of the Read Enable signals for the stacked devices must ensure that there is no overlap in the LOW levels which would cause more than one output to be active at the same time. Parallel expansion to generate n-bit words is accomplished by driving the Enable and Address inputs of each device in parallel.

LOGIC DIAGRAM

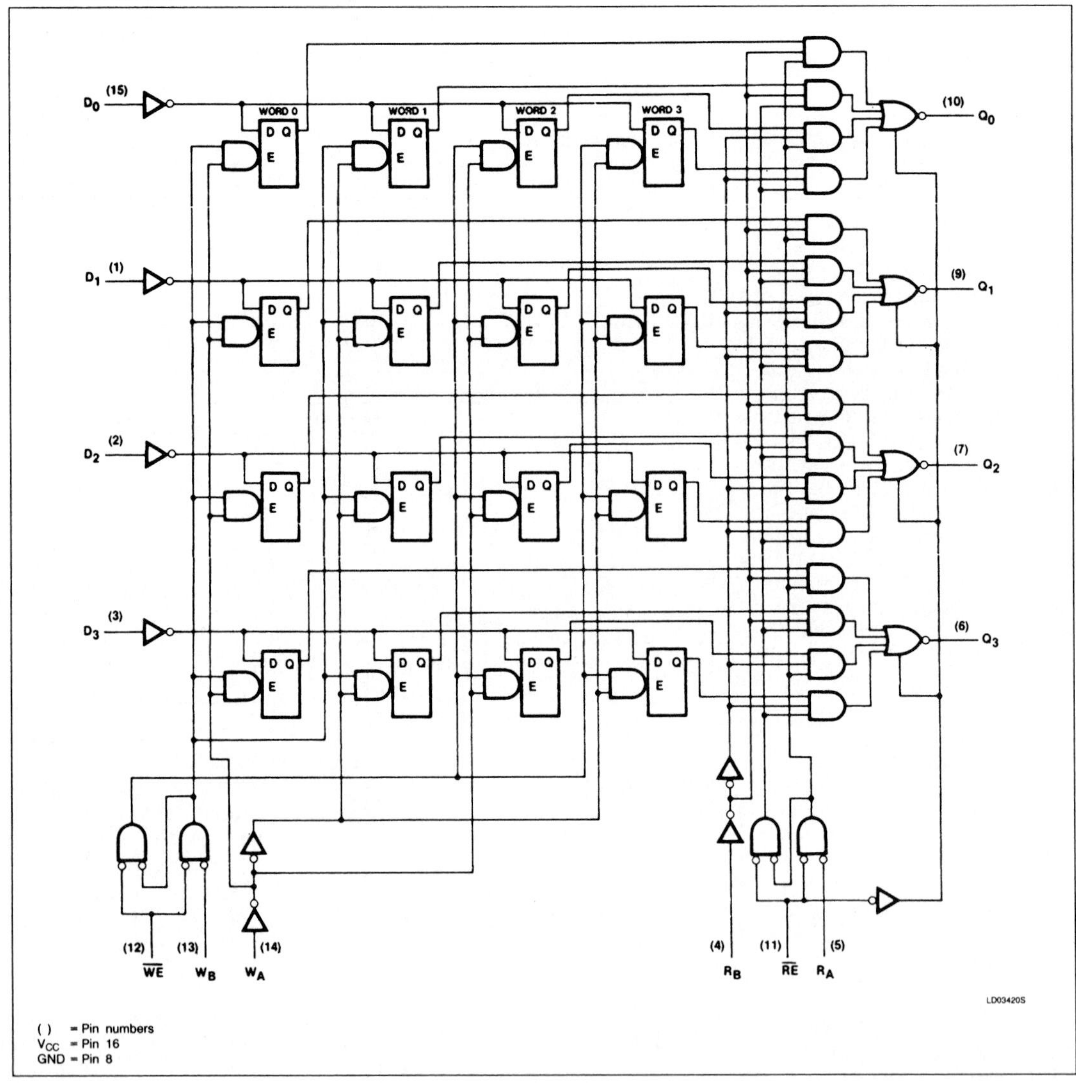

Register File 74LS670

WRITE MODE SELECT TABLE

OPERATING MODE	INPUTS		INTERNAL LATCHES[a]
	$\overline{WE}$	D_n	
Write data	L L	L H	L H
Data latched	H	X	no change

NOTE:
a. The Write Address (W_A and W_B) to the "internal latches" must be stable while $\overline{WE}$ is LOW for conventional operation.

READ MODE SELECT TABLE

OPERATING MODE	INPUTS		OUTPUT Q_n
	$\overline{RE}$	Internal Latches[b]	
Read	L L	L H	L H
Disabled	H	X	(Z)

NOTE:
b. The selection of the "internal latches" by Read Address (R_A and R_B) are not constrained by $\overline{WE}$ or $\overline{RE}$ operation.
H = HIGH voltage level
L = LOW voltage level
X = Don't care
(Z) = HIGH impedance "off" state.

ABSOLUTE MAXIMUM RATINGS (Over operating free-air temperature range unless otherwise noted.)

PARAMETER		74LS	UNIT
V_{CC}	Supply voltage	7.0	V
V_{IN}	Input voltage	−0.5 to +7.0	V
I_{IN}	Input current	−30 to +1	mA
V_{OUT}	Voltage applied to output in HIGH output state	−0.5 to $+V_{CC}$	V
T_A	Operating free-air temperature range	0 to 70	°C

RECOMMENDED OPERATING CONDITIONS

PARAMETER		74LS			UNIT
		Min	Nom	Max	
V_{CC}	Supply voltage	4.75	5.0	5.25	V
V_{IH}	HIGH-level input voltage	2.0			V
V_{IL}	LOW-level input voltage			+0.8	V
I_{IK}	Input clamp current			−18	mA
I_{OH}	HIGH-level output current			−2.6	mA
I_{OL}	LOW-level output current			8	mA
T_A	Operating free-air temperature	0		70	°C

Appendix D Glossary of terms and definitions

AC characteristics The nomenclature used by manufacturers to describe timing parameters of semiconductor ICs.

Access time The elapsed time from when the address of a memory location is sent to a memory device to when the location is accessible.

Accuracy of a DAC The difference between the calculated analog output voltage and the measured value.

Adder A logic circuit that adds binary numbers.

Address A binary number that is used to reference a particular register in a register array or a particular memory location or I/O port for read or write access.

Address foldback When multiple addresses reference the same memory location or I/O device.

Address map A rectangle that is subdivided into blocks which are identified by starting and ending addresses. These blocks are further identified by programs, data, I/O devices, or whatever occupies those addresses.

ALU Arithmetic logic unit. A logic circuit that implements various arithmetic and logical operations on binary numbers.

Analog The characteristic of being continuously variable.

Analog circuit A circuit that operates on a continuous range of values rather than a discrete (two-stated) set of values.

Analog-to-digital converter (ADC) A semiconductor IC(s) that converts an analog signal to a digital signal.

Analog signal A continuous signal without any discontinuity in the amplitude of the dependent variable.

Analysis A procedure that involves "taking apart" a circuit or trying to determine the circuit's operation.

***AND* gate** A logic gate that produces a logic 1 at its output if and only if all its inputs are at the logic 1 state.

***AND-OR* logic** Circuits that are made up of *AND* gates, *OR* gates, and inverters, or a combination of these gates.

ANSI American National Standards Institute.

Architecture The symbolic representation of a system in the form of function-blocks and connecting lines that indicate the relationship between function-blocks.

Arithmetic The branch of mathematics that deals with the four basic operations of addition, subtraction, multiplication, and division.

ASCII American Standard Code for Information Interchange.

Assembler A program that translates mnemonics into machine code.

Assembly To translate mnemonics into machine code (object code).

Astable multivibrator An oscillator. A circuit that does not have any stable states but keeps oscillating from one state to another.

Asynchronous An event that is not synchronized with the clock.

Base The number of digits in a numbering system.

BCD Binary coded decimal. A binary code used to represent decimal digits.

Binary A numbering system that is based on a set of two digits.

Binary code A set of binary numbers that are used to represent symbols such as number, letters, etc.

Bistable multivibrator A flip-flop. A circuit that has two stable states.

Bit A binary digit 0 or 1.

Bit bucket A pseudo-device that is used to collect discarded bits.

Boolean algebra The mathematics of logic.

Bootstrap loader A program that is stored in ROM which initially instructs the CPU.

Buffer A logic gate that provides no logic function but simply electronically isolates the input and output while increasing its output drive.

Buffer memory A memory (often a FIFO) that acts as a buffer between two devices.

Bus A line, or lines, that has two or more devices connected in parallel to each line.

Bus contention A condition whereby two logic circuits put binary data on the bus at the same time resulting in a conflict of logic levels.

Byte An 8-bit binary number or binary type.

Carry A digit that is carried over to the next stage in arithmetic operations.

Cascade A connection of two or more circuits where one circuit drives or triggers another.

Clear To force the state of a flip-flop or other device to the logic 0 state.

Clock A stream of pulses used to maintain the basic timing in a logic circuit.

Code A set of binary numbers that are used to represent symbols such as numbers, letters, etc.

Combinational logic Logic circuits whose outputs simply depend on the various combinations applied to the inputs.

Comparator A logic circuit that is used to compare two numbers.

Complement A term used in Boolean algebra to describe the opposite (or inverse) logic state.

Complementary MOS (CMOS) A logic family that utilizes both N- and P-channel MOSFETs in a complementary fashion.

Computer A collection of logic circuits that can be programmed to perform various arithmetic, logic and data-transfer operations.

Control bus The bus over which control signals are transmitted by the CPU to memory and I/O devices.

Control unit A special-purpose unit within the CPU that is dedicated to the execution of instructions.

Count An ordered sequence of numbers.

Counter A sequential logic circuit that is capable of producing a binary count.

CPU The central processing unit (CPU) is the section of a digital computer that processes (executes) program instructions.

Data Information that is processed by a digital system, generally in binary form.

Data bus The bus over which the CPU, memory and I/O transfer data.

DC characteristic The nomenclature used by manufacturers to describe voltage and current specifications of semiconductor ICs.

Decade counter A counter that has ten states.

Decimal A numbering system that is based on a set of ten digits.

Decimal adjust The procedure used to correct the result of a BCD addition.

Decode The process of identifying a binary code.

Decoder A logic circuit that identifies a particular input binary code by activating one of its outputs.

Decrement The process of decreasing the value of a number by 1.

Demultiplexer A logic circuit that directs binary data from a single input channel onto one of several output channels.

Difference The result of a subtraction.

Digit A symbol in a numbering system that is used to represent a value.

Digital A circuit, or a field of study, that deals with digits.

Digital-to-analog converter (DAC) A semiconductor IC(s) that will convert a digital signal to an analog signal.

Digital circuit A circuit that processes numbers.

Digital signal A signal that is composed of concatenated binary bits.

DIP Dual in-line Package. A type of package used for ICs.

Disable To deactivate or make inoperational.

Dividend The number that is being divided in a division.

Divisor The number that is divided into the dividend in a division.

Don't care A logic state that can be a 0 or a 1.

Driven logic circuit The logic circuit that serves as a load for another logic circuit.

Driver logic circuit The logic circuit that must drive another logic circuit(s).

Dynamic memory A memory that has capacitors as storage elements that tend to lose information over a period of time and therefore must be refreshed.

Dynamic one-dimensional memory cell array A one-dimensional array in which either the content of memory locations are mobile or the memory medium is mobile (dynamic).

Dynamic RAM (DRAM) A R/W memory that must have its contents refreshed. *See* Dynamic memory.

Dynamic testing Testing under rapidly changing logic level conditions.

Edge-triggered flip-flop A type of flip-flop that changes states on detecting a transition of the clock pulse that triggers it.

Electrically erasable PROM (EEPROM) A PROM that can have its contents erased electronically by the CPU. EEPROMs are also known as *read-mostly RAMs* and *nonvolatile RAMs*.

Electronic switch A transistor (bipolar or MOS) that operates in either the on (saturated) or off state.

Enable To activate or to make operational.

Encode The process of producing a binary code.

Encoder A logic circuit that produces a unique binary code when one of its inputs is activated.

End-around carry The final carry that is added to the result in unsigned subtraction.

Erasable PROM (EPROM) A PROM that can have its contents erased using ultraviolet light. The EPROM can then be programmed again. Also referred to as a *UV-PROM*.

Even parity An indication that a binary number has an even number of 1 bits.

Exclusive-*OR* gate A logic gate that produces a logic 0 at its output if and only if both inputs are equal.

Fall time The time it takes for a pulse to change from a high to a low.

Falling edge The portion of a pulse that makes transition from a 1 to a 0.

Fan-in Another term used to describe a unit load (UL).

Fan-out The number of inputs that can be safely connected to the output of a gate.

First-in first-out (FIFO) An acronym that describes the method of accessing data. That is, the first data stored is the first data that can be retrieved.

Flag A flip-flop whose logic state is used to signal the result of an operation or the request for a service.

Flat pack A type of package used for ICs.

Flip-flop A bistable sequential logic circuit that has the capability to store a bit.

Floppy disk A flexible magnetic disk that is used as a medium to store digital data.

Frequency The rate at which a periodic stream of clock pulses repeats itself.

Full adder A logic circuit that adds 3 bits.

Fully decoded addressing An addressing scheme that decodes all address lines of the address bus.

Fuse-link A fuse in a semiconductor that can be blown (programmed).

Gate An electronic circuit that implements a basic logic function such as *AND*, *OR*, *NOT*, *NAND*, and *NOR*.

Glitch An undesired pulse of very short duration.

Glue (VLSI-glue) The ICs used for interfacing various VLSI components and devices in a digital computer circuit.

Half-adder A logic circuit that adds 2 bits.

Hand assembly The act of assembling mnemonics (manually) without the use of an assembler.

Handshaking A technique used to acknowledge the transfer of data between two devices.

Hard disk A rigid magnetic disk that is used as a medium to store digital data.

Hardware The electronics of a computer system.

Hexadecimal A numbering system that is based on a set of 16 digits.

Hi-Z The high-impedance state of a device with tristate logic. *See* Tristate logic.

Hold time The time interval after the clock pulse triggers the flip-flop during which the control inputs must be stable.

IC Integrated circuit. An electronic package that integrates several circuits and components on a single silicon chip.

I/O device Devices that are used to either input or output data to or from a computer.

I/O port The electronic device used to interface an I/O device to a computer system.

I/O space Those I/O addresses that occupy the address map.

Increment The process of increasing the value of a number by 1.

Information Binary data in a digital system.

Input The signal or collection of signals that go into a circuit.

Instruction register (IR) The register within the CPU that is used to store the instruction currently being executed.

Instruction set The set of instructions that a microprocessor (MPU) is capable of executing.

Interface The electronics required to connect (interface) a device to the system.

Interrupt When the CPU stops the execution of one program at the request of an I/O device in order to service that device.

IR-decoder A decoder that decodes the content of the instruction register.

Isolated I/O A computer system that differentiates between I/O and memory addresses.

Karnaugh map A variation of the truth table organized to assist in the simplification of logic equations.

Last-in first-out (LIFO) An acronym that describes the method of accessing data. That is, the last data stored is the first data that can be retrieved.

Latch A logic circuit that can hold a logic level at its output even after the input stimulus is removed. Also used to describe the capability of a circuit to "hold" a binary number.

LCD Liquid crystal display.

LED Light emitting diode.

Linear addressing An addressing scheme that uses an address line(s) from the address bus to select and enable either an I/O port or semiconductor memory IC.

Load specifications Those voltage, current, and capacitance specifications that must be met when one semiconductor device drives another.

Logic The ability of a circuit to decide what its output(s) will be, based on its inputs and various logical rules.

Logic circuit A circuit made up of various logic gates.

Logic family The technology used to implement a group (family) of logic semiconductor ICs.

Logic function A basic logic operation such as *NOT, AND, OR, NAND,* and *NOR.*

Logic gate An electronic circuit that implements a basic logic function such as *AND, OR, NOT, NAND*, and *NOR*.

Logic level transition time The elapsed time required for the logic level output by a device to change from one logic level to the other.

Logic symbol A simple representation of a logic circuit or gate in terms of its inputs and outputs.

Logic voltage level The logic level of an input to or an output from a logic circuit.

Logical inversion The logical operation that converts a 0 (low) to a 1 (high) and a 1 (high) to a 0 (low).

Look-ahead carry A method of binary addition whereby carries from preceding stages are anticipated, thus avoiding propagation delays.

Look-up table Data stored in memory which when read is the desired translation of the input data. Also referred to as a *translation table*.

LSB Least significant bit. The rightmost bit in a binary number.

LSD Least significant digit. The rightmost digit in a number.

LSI Large-scale integration. Integrated circuits with 100 to 1000 gates per chip.

Machine code The binary bit patterns (usually bytes) that are the actual instructions executed by the CPU.

Macroinstructions Instructions that consitute a macroprogram.

Macroprogram A program that is stored in main memory and is to be executed by the CPU.

Magnetic bubble memory (MBM) A mass storage device that uses magnetic domains on a magnetic medium such as memory cells.

Magnetic tape A magnetic tape that is used as a medium to store digital data.

Magnitude The value or size of a quantity.

Main memory The working memory of the computer system.

Mask A bit, or bits, that are used to mask out other bits in a binary number.

Mask-byte A byte that is used as a mask to remove or filter out certain bits.

Mask ROM (MROM) A semiconductor memory that had its contents programmed when manufactured using a photo-masking technique.

Master-slave flip-flop A pulse-triggered flip-flop. A type of flip-flop that changes states on detecting a complete pulse.

Mass storage device A device in which programs and data can be stored on a permanent storage medium for later use.

Maxterm A logical sum that will produce a 0 for one and only one binary combination of the independent variables.

Memory cell An object that is used to store a bit of data.

Memory map A rectangle that symbolically represents all memory addresses within the system. It is further subdivided into blocks that identify specific memory chips in the system. *Also see* Address map.

Memory mapped I/O An addressing scheme that "maps" I/O devices into the memory space. That is, a system that does not differentiate between I/O and memory addresses.

Microinstructions Instructions that constitute a microprogram.

Microprocessor unit (MPU) A semiconductor IC that is used to implement the CPU, and other possible functions, of a computer.

Microprogram A program within the control unit of an MPU that instructs it as to how to execute a particular instruction from its instruction set.

Minterm A logical product that will produce a 1 for one and only one binary combination of the independent variables.

Mnemonic A memory aid for representing MPU instructions. Mnemonics symbolically indicate the MPU operation to be performed.

Modulus The number of states of a counter.

Monostable multivibrator A one-shot. A circuit that has only one stable state.

MSB Most significant bit. The leftmost bit in a binary number.

MSD Most significant digit. The leftmost digit in a number.

MSI Medium-scale integration. Integrated circuits with 12 to 99 gates per chip.

Multiplexer A logic circuit that directs several input channels containing binary information onto a single output channel.

Multiplicand The number being multiplied in a multiplication.

Multiplier The number used to multiply the multiplicand in a mulitplication.

***NAND* gate** A logic gate that produces a logic 0 at its output if and only if all its inputs are at the logic 1 state.

***NAND* logic** Circuits that are made up of *NAND* gates only.

Negative edge The portion of a pulse that makes a transition from a 1 to a 0.

Nibble A 4-bit binary number or half a byte.

9's complement An unsigned representation of a negative number in the decimal numbering system.

Node An electrical connection of two or more components.

Noise immunity The susceptibility of a circuit to the effects of noise.

Noise margin The difference between worst case output voltage and worst case input voltage.

Nonvolatile RAM An R/W memory that does not require that power be maintained to it in order for it to retain its contents. *Also see* Electrically erasable PROM (EEPROM).

***NOR* gate** A logic gate that produces a logic 0 at its output if any or all of its inputs are at the logic 1 state.

***NOR* logic** Circuits that are made up of *NOR* gates only.

Octal A numbering system that is based on a set of eight digits.

Odd parity An indication that a binary number has an odd number of 1 bits.

One-dimensional memory cell array An array of memory cells in which the cells are arranged in rows with each row constituting a memory location. Memory locations are addressed by a one-dimensional parameter, the row location.

One-shot Monostable multivibrator. A circuit that has only one stable state.

1's complement The unsigned representation of a negative binary number.

Op-code The part of an instruction that represents an MPU operation.

Open-collector TTL logic which does not have the "top" transistor of the usual totem pole output.

Operand The part of an instruction that is to be operated on by the op-code.

***OR* gate** A logic gate that produces a logic 1 at its output if any or all of its inputs are at the logic 1 state.

Output The signal or collection of signals that come out of a logic circuit.

Page of memory A pre-defined number of memory locations.

Parallel I/O device An I/O device that receives or transmits data in multiple bits (usually 8 bits).

Parallel register A register in which data is loaded and retrieved in parallel.

Parallel-to-parallel I/O port An I/O port that receives or transmits data bits (usually 8 bits) in parallel.

Parasitic capacitance An inherent capacitance that is present in all semiconductors.

Parity An indication of whether there are an even or odd number of 1-bits in a binary number.

Parity checker A circuit that checks a binary number for even or odd parity.

Parity generator A circuit that generates a parity bit for a binary number.

Partial linear addressing A memory addressing scheme

that uses linear addressing to select (enable) the semiconductor, and address decoding to select the memory location within the selected semiconductor.

Period The interval of time between periodic clock pulses.

Periodic An event that repeats itself at regular intervals.

Pipelining Using a queue to prefetch instructions so that the CPU may use the data bus and execute instructions at the same time.

Positive edge The portion of a pulse that makes the transition from a 0 to a 1.

Preset The state of a flip-flop when its output is at the logic 1 state.

Priority encoder A logic circuit that produces a prioritized binary code when one or more of its inputs are activated.

Product The result of multiplication.

Product-of-sums (POS) One of the forms that a logic equation can take in which logical sums (maxterms) are "*AND*ed" together.

Program A set of instructions to the CPU that are executed in a particular order in order to achieve a task.

Programmable Modifying something through software.

Programmable array logic (PAL) A PLD with a programmable *AND*-gate array and a preprogrammed output *OR*-gate array.

Programmable logic array (PLA) A PLD with programmable *AND*-gate and *OR*-gate array.

Programmable logic device (PLD) Semiconductor devices that can be programmed to function like various combinational logic circuits. PLDs include other programmable devices (PLEs, PLAs, and PALs).

Programmable logic element (PLE) A PLD that has a nonprogrammable *AND*-gate array with a programmable output *OR*-gate array.

Programmable ROM (PROM) A semiconductor memory that has the abiiity to be irreversibly programmed by the user using a PROM programmer.

Propagation delay The amount of delay that occurs from the time the input of a gate changes states to the time the output changes states.

Pull-up resistor A resistor used to reference (pull-up) the collector of an open-collector circuit to V_{CC} volts.

Pulse A very quick change in state from one logic level to another followed by a return back to the original state.

Pulse duration Pulse width. The time interval between the edges of a pulse.

Pulse-triggered flip-flop *See* Master-slave flip-flop.

Pulse width *See* Pulse duration.

Quantization error The inherent error of an ADC due to truncation.

Queue A FIFO that is used to sequentially store prefetched instructions. *Also see* Pipelining.

Quotient The result of division.

Random access memory (RAM) An acronym that is descriptive of the ability to access a memory location in a random manner with a constant access time. RAM is often erroneously used to mean the ability to read and write to a memory.

Read The process of obtaining or retrieving data from a register or memory location.

Read-only memory (ROM) A memory that cannot have its contents changed (written) but can only have its contents retrieved (read).

Recycle An event that causes a counter to return back to its starting (initial) state.

Refresh cycle The operation cycle performed in order to refresh the memory content of a DRAM.

Register A logic circuit that is capable of storing a single binary number.

Register array An array or collection of registers used to store several binary numbers.

Register file *See* Register array.

Reset The state of a flip-flop when its output is at the logic 0 state.

Resolution The smallest change in a magnitude that can be detected.

Rise time The time it takes for a pulse to change from a low to a high.

Rising edge The portion of a pulse that makes transition from a 0 to a 1.

Schottky technology A fast TTL-type technology.

Scratch pad memory Registers within an MPU or memory locations external to the MPU that are used to temporarily store data or intermediate results from the ALU. *See also* Register array.

Sequential logic Logic circuits whose outputs depend on the various combinations applied to the inputs and the past history of the outputs.

Serial I/O device An I/O device that receives or transmits data 1 bit at a time.

Serial I/O port An I/O port that transmits and receives serial data to an from the I/O device. *Also see* Universal synchronous/asynchronous receiver/transmitter (USART).

Serial register A register in which data is loaded and/or retrieved serially.

Set The state of a flip-flop when its output is at the logic 1 state.

Setup time The time interval before the clock pulse

triggers the flip-flop during which the control inputs must be stable.

Seven-segment decoder A logic circuit that converts a binary code to a code that activates the seven segments of an LED.

Shift register A register in which data is loaded and/or retrieved serially.

Signature analysis A dynamic diagnostic tool that counts the number of logic ones that appear at a node over an interval of time for a repetitive sequence of events. From this count a value is calculated, which becomes the signature of that node.

Sink current That current that flows into a logic circuit.

Software The instructions and programs that instruct the CPU as to the tasks to be performed.

Source current The current that flows out of a logic circuit.

SSI Small-scale integration. Integrated circuits with 1 to 11 gates per chip.

Stack memory A portion of main memory reserved and defined to constitute the stack. Its memory locations are often accessed in a LIFO manner.

Static RAM A R/W memory that retains its content without the need for refreshing, as long as power is applied to it.

Static testing Testing under DC conditions.

Subroutine A subprogram that is executed by another program.

Subtractor A logic circuit that is used to obtain the difference between two numbers.

Sum The result of addition.

Sum-of-products (SOP) One of the forms that a logic equation can take in which logical products (minterms) are "*OR*ed" together.

Switch An electronic component that stops or passes the flow of current in a circuit. Aslo refers to the capability of a logic circuit to change states.

Switching circuit A circuit that has only two states and is capable of switching between these two states.

Synchronous An event that is synchronized with the clock.

Synthesis A procedure that involves putting together a circuit or designing a circuit.

10's complement An unsigned representation of a negative number in the decimal numbering system.

Test node An electrical node that is monitored for test data.

Three-bus architecture A system that has three busses, usually the address, data, and control bus.

Three-dimensional memory cell array An array of memory cells that is constructed of duplicate planes of two-dimensional memory cell arrays. Memory locations are formed by concatenating the identically located memory cell in each plane.

Timing diagram A diagram that illustrates the operation of a sequential logic circuit by displaying the inputs and outputs of the circuit as a function of time.

Toggle The switching of a logic level from a 0 to a 1 and vice versa.

Totem pole Two bipolar transistors connected with their emitter-collectors in series with each other in the output of a TTL logic gate.

Transistor-transistor logic (TTL) A logic family that utilizes bipolar transistors.

Transition A change in logic state.

Translation table *See* Look-up-table.

Trigger A stimulus such as a pulse that is used to change the state of a logic circuit.

Tristate logic Logic circuits or gates whose outputs can exist in one of three possible states—high, low, or high impedance.

Truth table A table that shows the logic state of the dependent variable(s) for all possible combinations of the independent variable(s).

Two-dimensional memory cell array An array of memory cells in which the cells are organized into an array of rows and columns. The address of each cell within the array is an *x-y* coordinate of its row and column location.

2's complement The unsigned representation of a negative binary number.

UL Unit load. A measure of the electrical loading requirements of a logic gate.

Universal synchronous/asynchronous receiver/transmitter (USART) A programmable serial I/O port.

User memory space The area of main memory available for user programs (application programs).

UV-PROM *See* EPROM.

Vector Directed to a specific address.

VLSI Very large-scale integration. Integrated circuits with more than 1000 gates per chip.

Volatile memory A memory that loses its content when power is removed from it.

Weight The value of a digit in a number based on its position in the number.

Wired-*AND* logic The logic formed by connecting the outputs of open-collector logic gates.

Word A 16-bit binary number or a double byte. Could also be used to describe the size of binary numbers processed by a digital circuit.

Write The process of storing data into a register or memory location.

Appendix E Answers to odd-numbered problems

Chapter 1

1. Number of combinations = 100,000,000
Largest value = 99,999,999

3. 9, 1, 4, 0

5. (a) −265341 (b) −761 (c) −87 (d) −51494265

7. (a) 7 (b) 4 (c) 36 (d) 837 (e) 357

9. (a) −328761 (b) −1 (c) −78 (d) −180264979

11. (a) 6 (b) 4 (c) 90 (d) 495

13. (a) 0 1 2 3 4 5 6 7 10 11 12
(b) 144 145 146 147 150 151 152 153 154
(c) 31 32 33 34 35 36 37 40 41 42 43
(d) 67 70 71 72 73 74 75 76 77 100 101 102

15. (a) 23003 (b) 43625 (c) 100 (d) 10

17. (a) 22161 (b) 112297 (d) 64206 (e) 65535 (f) 92041

19.

(a)	(b)
0000	1100100
0001	1100101
0010	1100110
0011	1100111
0100	1101000
0101	1101001
0110	1101010
0111	1101011
1000	1101100
1001	
1010	
1011	
1100	

(c)	(d)
00100101	10100101
00100110	10100110
00100111	10100111
00101000	10101000
00101001	10101001
00101010	10101010
00101011	10101011
00101100	10101100
00101101	10101101
00101110	10101110
00101111	10101111
00110000	10110000

21. (a) 1010 (b) 1111011011
(c) 1100100001 (d) 1001100

23. (a) 110011101100011
(b) 111111111111
(c) 000110011101100
(d) 001001001000001001000
(e) No solution possible!
(f) 001101100110
(g) 010110100101

25. (a) 10101011110011011110
(b) 01101111011000101011
(c) 00100111010101001001
(d) 0111001110101100011001000101
(e) 1011101000110010
(f) 00010000000000100000001
(g) 10001000

27. (a) 254 (b) 61505 (c) 0200020
(d) 1755144 (e) 0021545 (f) 17

29. (a) 598 (b) 22548 (c) No solution possible! (d) 915 (e) 4882 (f) 02

31. (a) 1023 (b) 12130 (c) 3321
(d) 2221

33. (a) 100 (b) 1000111 (c) 01111010
(d) 1000101

35. (a) 0101 (b) 011111 (c) 01000100
(d) 000111

37. (a) 1011010 (b) 10000100
(c) 10100101

39. Quotient = 3, remainder = 2

41. (a) No parity, 101 0111
Even parity, 1101 0111
Odd parity, 0101 0111
(b) No parity, 011 1001
Even parity, 0011 1001
Odd parity, 1011 1001
(c) No parity, 100 0000
Even parity, 1100 0000
Odd parity, 0100 0000
(d) No parity, 000 0011
Even parity, 0000 0011
Odd parity, 1000 0011

Chapter 2

1. (a)

A	B	C	D	L
OFF	OFF	OFF	OFF	OFF
OFF	OFF	OFF	ON	OFF
OFF	OFF	ON	OFF	OFF
OFF	OFF	ON	ON	OFF
OFF	ON	OFF	OFF	OFF
OFF	ON	OFF	ON	ON
OFF	ON	ON	OFF	OFF
OFF	ON	ON	ON	ON
ON	OFF	OFF	OFF	OFF
ON	OFF	OFF	ON	OFF
ON	OFF	ON	OFF	ON
ON	OFF	ON	ON	ON
ON	ON	OFF	OFF	OFF
ON	ON	OFF	ON	ON
ON	ON	ON	OFF	ON
ON	ON	ON	ON	ON

(b)

A	B	C	D	L
OFF	OFF	OFF	OFF	OFF
OFF	OFF	OFF	ON	OFF
OFF	OFF	ON	OFF	OFF
OFF	OFF	ON	ON	OFF
OFF	ON	OFF	OFF	OFF
OFF	ON	OFF	ON	OFF
OFF	ON	ON	OFF	OFF
OFF	ON	ON	ON	OFF
ON	OFF	OFF	OFF	OFF
ON	OFF	OFF	ON	OFF
ON	OFF	ON	OFF	OFF
ON	OFF	ON	ON	OFF
ON	ON	OFF	OFF	OFF
ON	ON	OFF	ON	ON
ON	ON	ON	OFF	ON
ON	ON	ON	ON	ON

3. (a) $L = (A \cdot C) + (B \cdot D)$
(b) $L = A \cdot B \cdot (C + D)$

5.

A	B	C	D	E	F	G	H
0	0	1	0	1	0	1	0
1	1	0	1	0	1	0	1

7.

P	Q	R	S	W
0	0	0	0	0
0	0	0	1	0
0	0	1	0	0
0	0	1	1	0
0	1	0	0	0
0	1	0	1	0
0	1	1	0	0
0	1	1	1	0
1	0	0	0	0
1	0	0	1	0
1	0	1	0	0
1	0	1	1	0
1	1	0	0	0
1	1	0	1	0
1	1	1	0	0
1	1	1	1	1

Logic equation: $W = P \cdot Q \cdot R \cdot S$
Conclusion: The circuit functions like a four-input *AND* gate.

9.

P	Q	R	S	W
0	0	0	0	0
0	0	0	1	1
0	0	1	0	1
0	0	1	1	1
0	1	0	0	1
0	1	0	1	1
0	1	1	0	1
0	1	1	1	1
1	0	0	0	1
1	0	0	1	1
1	0	1	0	1
1	0	1	1	1
1	1	0	0	1
1	1	0	1	1
1	1	1	0	1
1	1	1	1	1

Logic equation: $W = P + Q + R + S$
Conclusion: The circuit functions like a four-input *OR* gate.

11. (a)

A	B	C
0	0	1
0	1	1
1	0	1
1	1	0

Conclusion: The circuit functions like a *NAND* gate.

(b)

A	B	C
0	0	1
0	1	0
1	0	0
1	1	0

Conclusion: The circuit functions like a *NOR* gate.

13. (a)

A	B	C	D	X
0	0	0	0	0
0	0	0	1	0
0	0	1	0	0
0	0	1	1	1
0	1	0	0	0
0	1	0	1	0
0	1	1	0	0
0	1	1	1	1
1	0	0	0	0
1	0	0	1	0
1	0	1	0	0
1	0	1	1	1
1	1	0	0	1
1	1	0	1	1
1	1	1	0	1
1	1	1	1	1

Logic equation: $X = \overline{\overline{A \cdot B} \cdot \overline{C \cdot D}}$

(b)

A	B	C	D	X
0	0	0	0	0
0	0	0	1	0
0	0	1	0	0
0	0	1	1	0
0	1	0	0	0
0	1	0	1	1
0	1	1	0	1
0	1	1	1	1
1	0	0	0	0
1	0	0	1	1
1	0	1	0	1
1	0	1	1	1
1	1	0	0	0
1	1	0	1	1
1	1	1	0	1
1	1	1	1	1

Logic equation: $X = \overline{\overline{A + B} + \overline{C + D}}$

15.

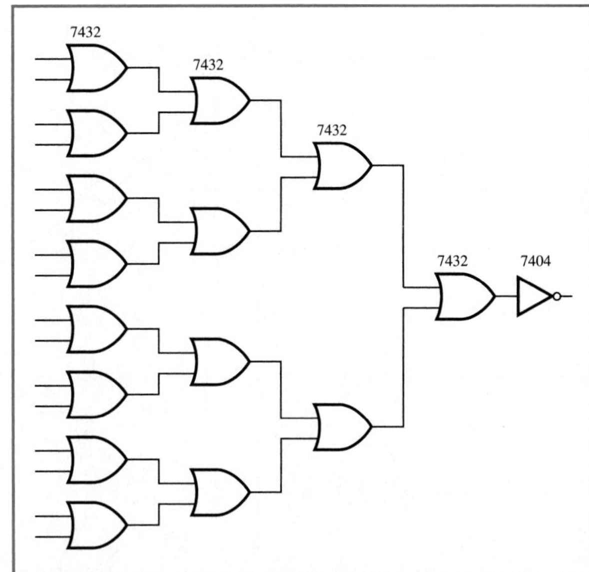

17. (a)

W	X	Y	Z
0	0	0	1
0	0	1	1
0	1	0	1
0	1	1	1
1	0	0	1
1	0	1	1
1	1	0	1
1	1	1	0

(b)

W	X	Y	Z
0	0	0	1
0	0	1	0
0	1	0	0
0	1	1	0
1	0	0	0
1	0	1	0
1	1	0	0
1	1	1	0

19. (a) $P = \overline{(W \cdot X)} + \overline{Y}$
(b) $P = \overline{(W + X)} \cdot \overline{Y}$

21. (a)

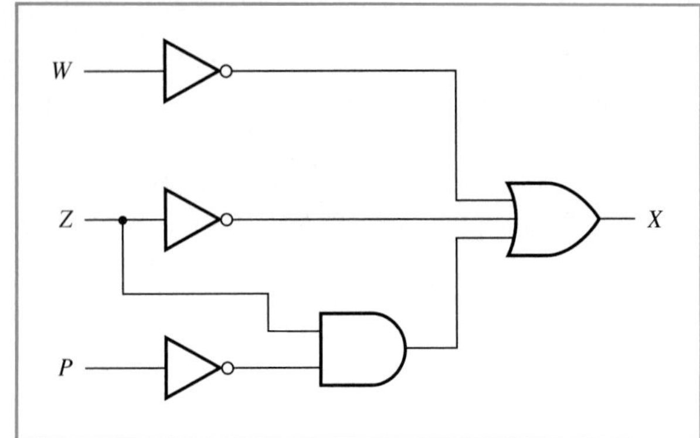
W
Z
P
X

(b)

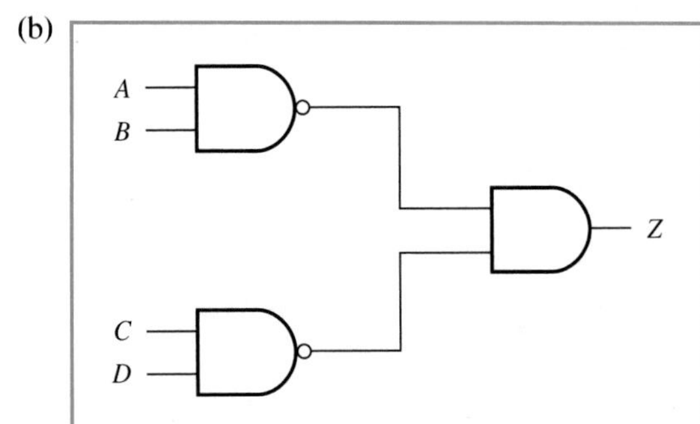
A
B
C
D
Z

(c)

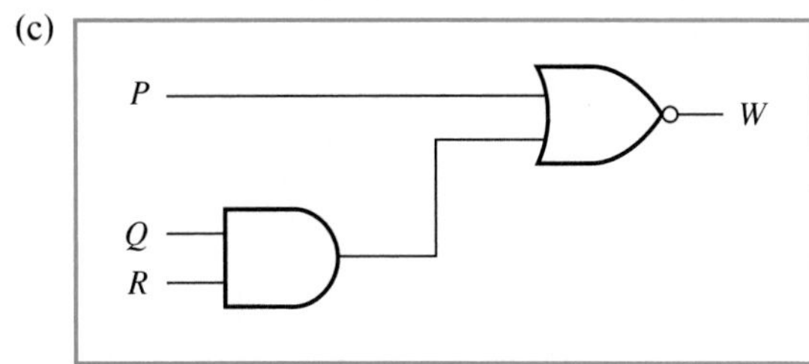
P
Q
R
W

(d)

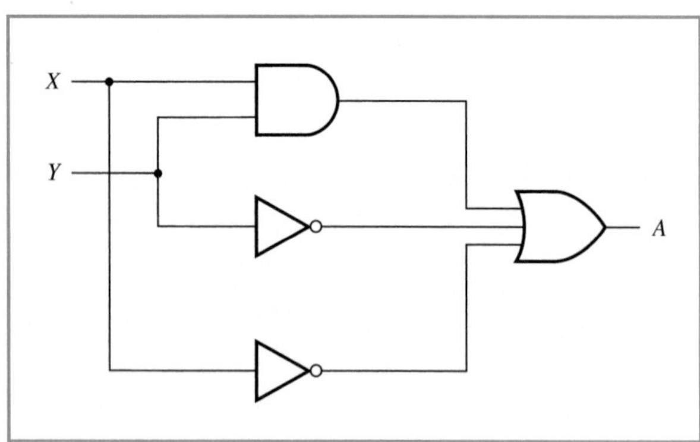
X
Y
A

23.

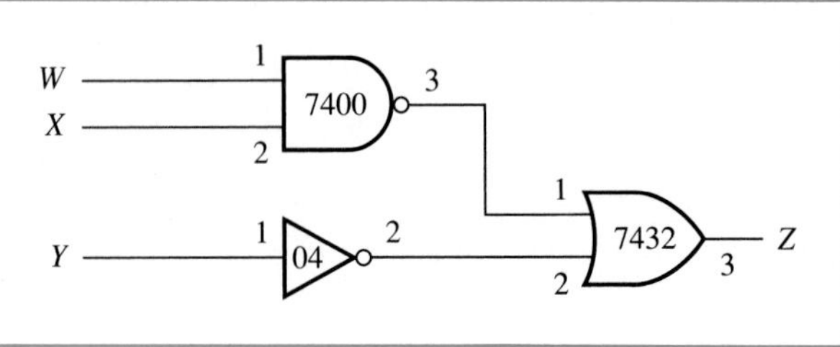
W
X
Y
7400
04
7432
Z

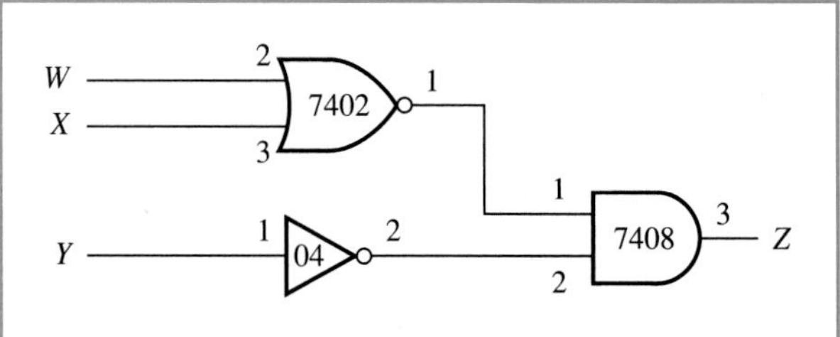
W
X
Y
7402
04
7408
Z

25.

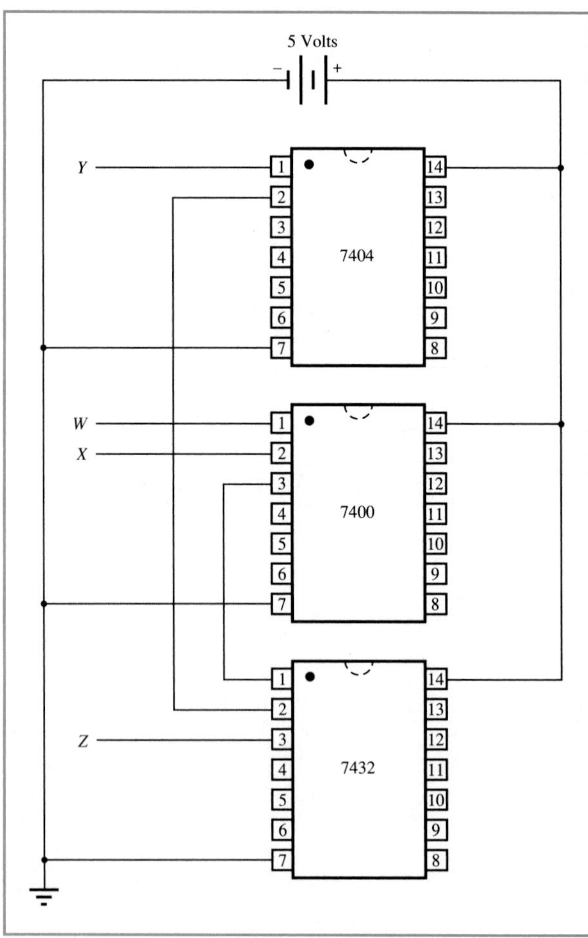
5 Volts
Y
W
X
Z
7404
7400
7432

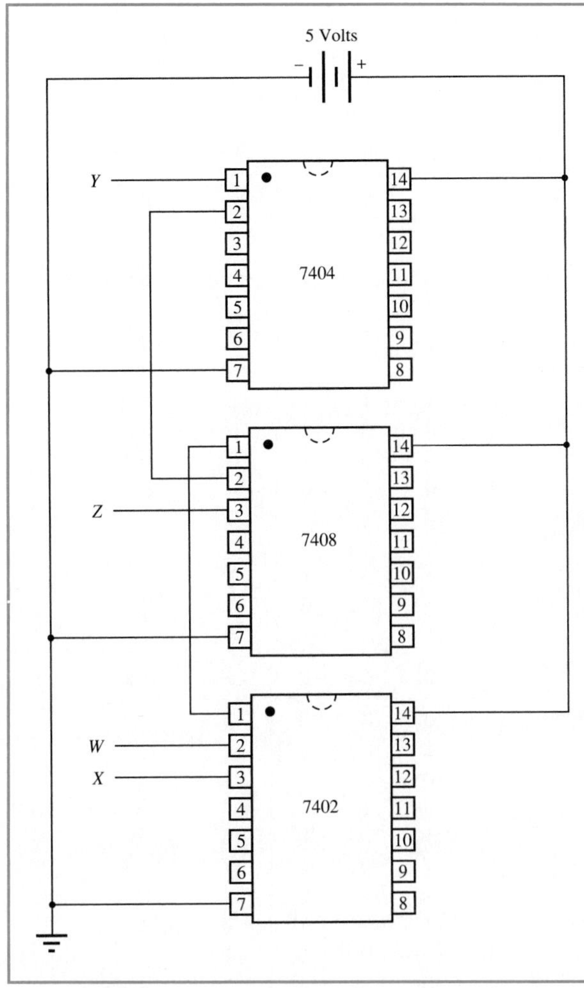

27. 97.5 mW
29. $V_{OH} = 3$ volts, $V_{OL} = 0.5$ volts, $V_{IH} = 2$ volts, $V_{IL} = 0.8$ volt
31. $I_{OH} = 2$ mA, $I_{OL} = 20$ mA, $I_{IH} = 20\ \mu$A, $I_{IL} = 500\ \mu$A
33. FO (HIGH) = 100, FO (LOW) = 40
35. 0.5 UL (HIGH), 0.3 UL (LOW)
37. $V_{NH} = 1$ volt, $V_{NL} = 0.3$ volt
39. $t_{pd} = 33$ ns

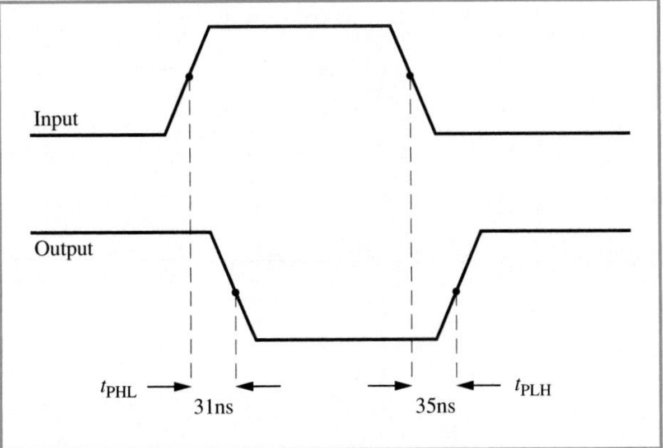

41. Output of $G3$ (Z) will be LOW
43. The output of $G3$ or the input of $G2$ could be internally shorted to ground.
45. By isolating the connection between $G3$ and $G2$, we could measure the output of $G3$ with the input A set to the LOW state. If the probe indicates a LOW then $G3$ is defective, but if the probe indicates a HIGH then $G2$ is defective.

Chapter 3

1. (a) $X = (A + B) \cdot \overline{C}$
   (b) $Z = PQ + Q + RS$

3. (a)

A	B	C	X
0	0	0	0
0	0	1	0
0	1	0	1
0	1	1	0
1	0	0	1
1	0	1	0
1	1	0	1
1	1	1	0

(b)

P	Q	R	S	Z
0	0	0	0	0
0	0	0	1	0
0	0	1	0	0
0	0	1	1	1
0	1	0	0	1
0	1	0	1	1
0	1	1	0	1
0	1	1	1	1
1	0	0	0	0
1	0	0	1	0
1	0	1	0	0
1	0	1	1	1
1	1	0	0	1
1	1	0	1	1
1	1	1	0	1
1	1	1	1	1

5. (a) Same as 3–3 (a)
   (b) Same as 3–3 (b)
7. $Z = \overline{X}\,\overline{Y} + \overline{X}\,Y$
   $S = \overline{P}\,\overline{Q}\,R + P\,\overline{Q}\,\overline{R} + P\,\overline{Q}\,R + P\,Q\,R$
9. $Z = (\overline{X} + Y) \cdot (\overline{X} + \overline{Y})$
   $S = (P + Q + R) \cdot (P + \overline{Q} + R)$
   $\cdot (P + \overline{Q} + \overline{R}) \cdot (\overline{P} + \overline{Q} + R)$

11.

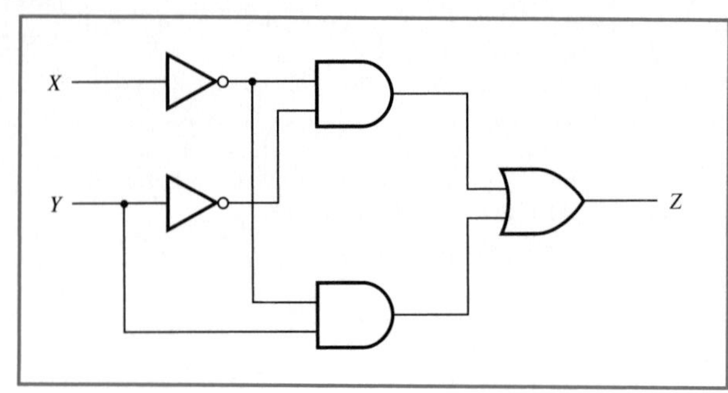

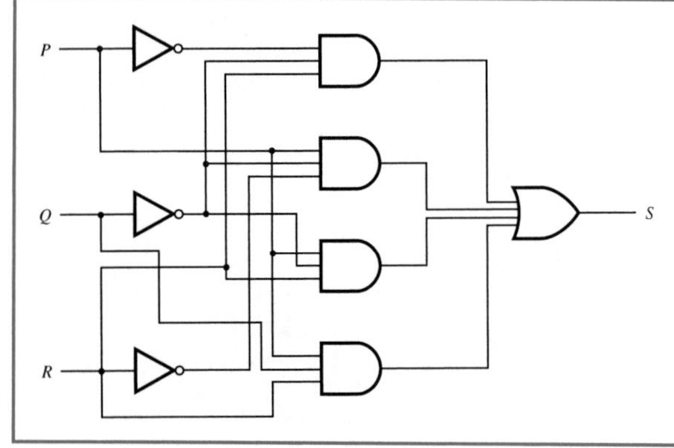

13.

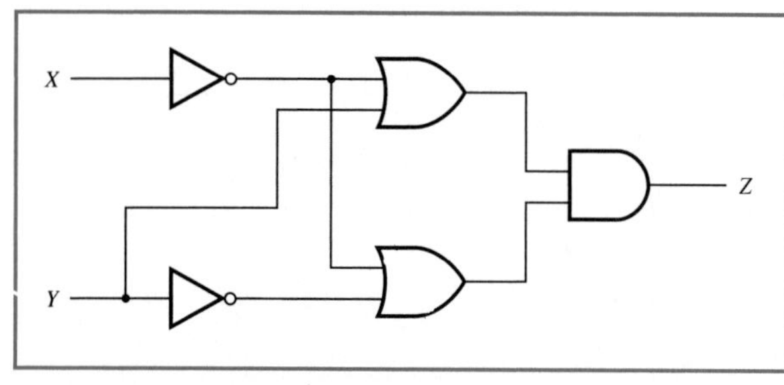

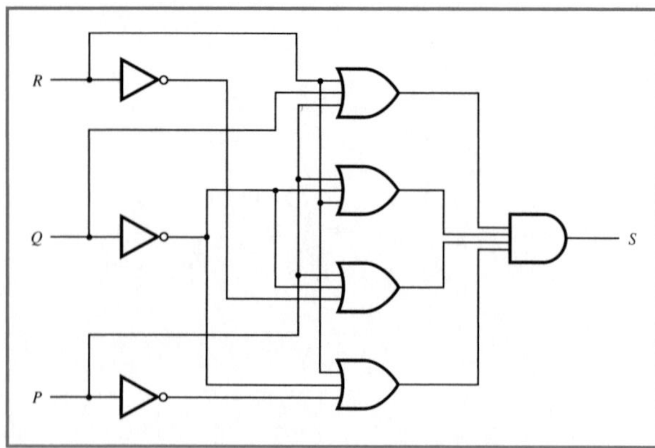

17. $Z = \Sigma_m\, 0, 1$
$S = \Sigma_m\, 1, 4, 5, 7$

19. $Z = \Pi_M\, 2, 3$
$S = \Pi_M\, 0, 2, 3, 6$

21. (a) 1 (b) 1 (c) 0 (d) 0

23. (a)

P	Q	R	S	LHS	RHS
0	0	0	0	0	0
0	0	0	1	0	0
0	0	1	0	0	0
0	0	1	1	0	0
0	1	0	0	0	0
0	1	0	1	0	0
0	1	1	0	0	0
0	1	1	1	0	0
1	0	0	0	0	0
1	0	0	1	0	0
1	0	1	0	0	0
1	0	1	1	0	0
1	1	0	0	0	0
1	1	0	1	1	1
1	1	1	0	0	0
1	1	1	1	1	1

(b)

A	B	C	D	LHS	RHS
0	0	0	0	1	1
0	0	0	1	1	1
0	0	1	0	1	1
0	0	1	1	0	0
0	1	0	0	1	1
0	1	0	1	1	1
0	1	1	0	1	1
0	1	1	1	0	0
1	0	0	0	1	1
1	0	0	1	1	1
1	0	1	0	1	1
1	0	1	1	0	0
1	1	0	0	0	0
1	1	0	1	0	0
1	1	1	0	0	0
1	1	1	1	0	0

(c)

X	Y	Z	LHS	RHS
0	0	0	0	0
0	0	1	0	0
0	1	0	0	0
0	1	1	0	0
1	0	0	0	0
1	0	1	0	0
1	1	0	1	1
1	1	1	1	1

(d)

P	Q	LHS	RHS
0	0	1	1
0	1	1	1
1	0	1	1
1	1	1	1

25. $Z = \overline{X}$ (simplified = one gate, unsimplified = five gates)
$S = \overline{Q}R + \overline{Q}P + PR$ (simplified = five gates, unsimplified = eight gates)

27. $Z = \overline{X}$ (simplified = one gate, unsimplified = five gates)
$S = (P + R) \cdot (P + \overline{Q}) \cdot (\overline{Q} + R)$ (simplified = five gates, unsimplified = eight gates)

29. (a) $A + B$ (b) 1 (c) $\overline{P} + Q\overline{R}$ (d) $\overline{R}$ (e) $A\overline{B} + AD + \overline{A}B$

31. $Z = \overline{X \cdot X}$
$S = \overline{\overline{\overline{Q}R} \cdot \overline{\overline{Q}P} \cdot \overline{PR}}$

33. *NAND* logic = one IC, one gate
AND-OR logic = one IC, one gate

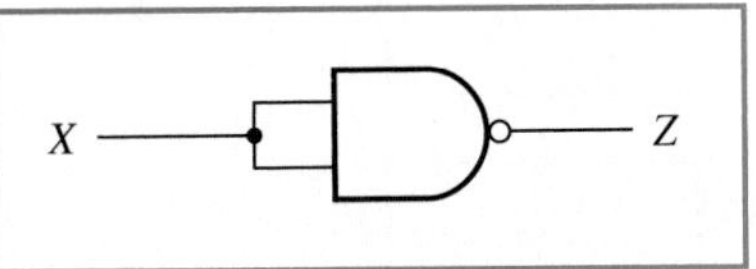

NAND logic = two ICs, five gates
AND-OR logic = three ICs, five gates

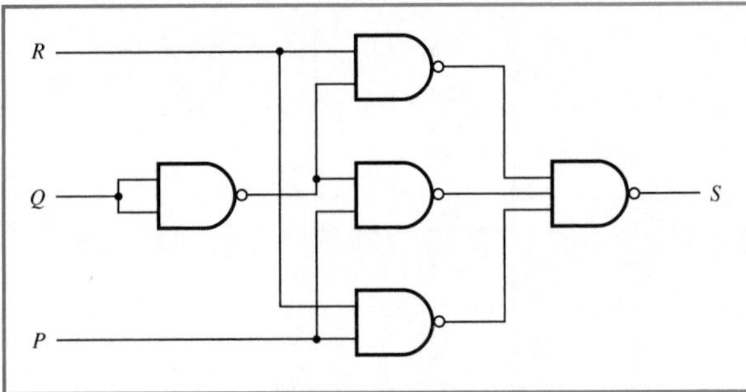

35. (a) $Z = \overline{\overline{(\overline{P} + Q)} + \overline{(P + \overline{R})}}$

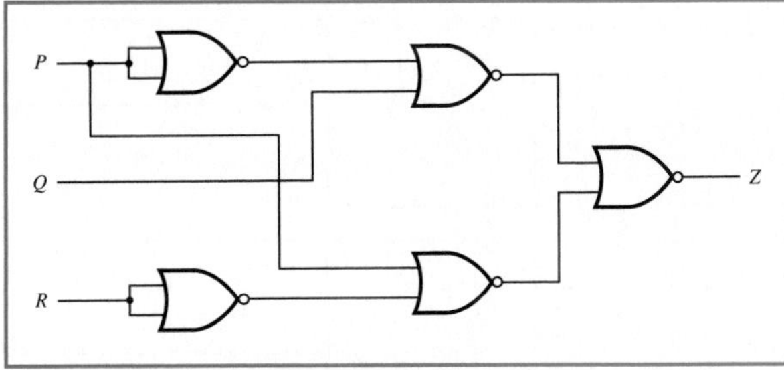

(b) $T = \overline{\overline{(\overline{X} + Y)} + \overline{(\overline{X} + Z)}}$

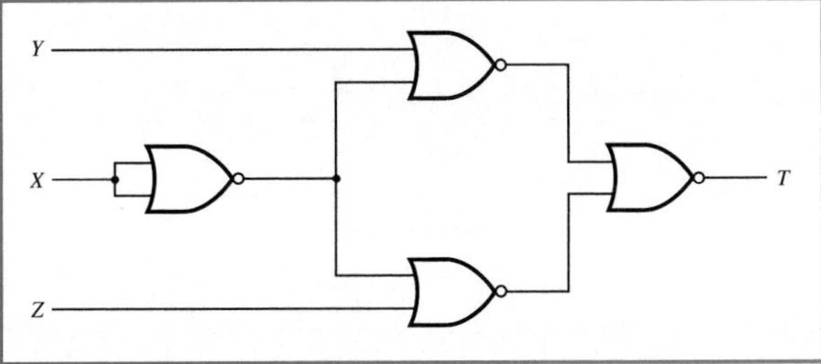

(c) $S = \overline{\overline{A} + \overline{(B + \overline{D})} + \overline{(B + C + D)}}$

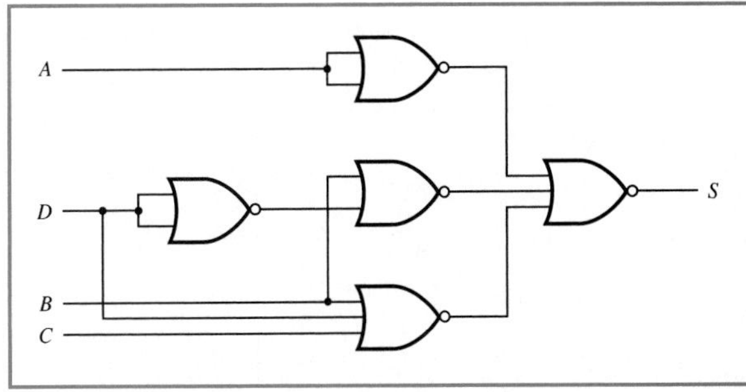

(d) $W = \overline{\overline{(P + \overline{Q} + R)} + \overline{(\overline{P} + Q + \overline{S})}}$

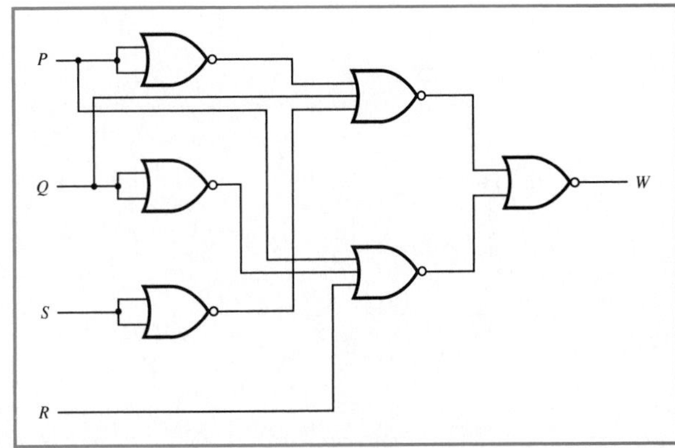

(e) $Z = \overline{\overline{(\overline{A} + \overline{D})} + \overline{(\overline{B} + \overline{C})} + \overline{(\overline{C} + D)}}$

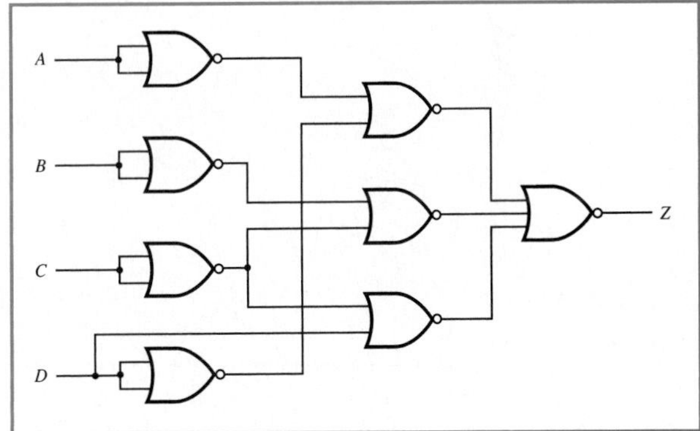

37. $Z = \overline{X}$, $S = (\overline{Q} + R) \cdot (P + \overline{Q}) \cdot (P + R)$,
$X = (\overline{A} + B) \cdot (\overline{B} + D)$
$S = (\overline{Q} + \overline{R})$

39. (a) $\overline{A} + \overline{B}$ (b) $\overline{P} + \overline{Q}$ (c) $P + R$ (d) $(\overline{A} + B) \cdot (\overline{A} + C) \cdot (B + C) \cdot (A + \overline{B} + \overline{C})$ (e) $X + Y$

41. (a) $A + B$ (b) 1 (c) $\overline{P} + Q\overline{R}$ (d) $\overline{R}$
(e) $\overline{A}B + A\overline{B} + AD$

43. (a) $Z = P\overline{Q} + \overline{P}Q$ (b) $X = B\,\overline{C} + \overline{A}\,\overline{C} + A\,\overline{B}\,C$
(c) $S = P\,\overline{Q}\,\overline{R} + \overline{P}\,Q\,R$ (d) $A = X\overline{Z} + \overline{X}Z + \overline{W}\,\overline{X}\,Y + W\,\overline{X}\,\overline{Y}$ (e) $T = PRS + QR\overline{S} + \overline{P}R\overline{S}$

45. Faults in the circuit are circled in the following table.

			Recorded			Actual		
P	**Q**	**R**	*G*1	*G*2	*G*3	*G*1	*G*2	*G*3
0	0	0	1	0	0	0	0	0
0	0	1	1	(0)	0	1	1	0
0	1	0	1	0	0	1	0	0
0	1	1	1	(0)	(0)	1	1	1
1	0	0	0	1	0	0	1	0
1	0	1	0	1	0	0	1	0
1	1	0	0	1	0	0	1	0
1	1	1	0	1	0	0	1	0

Chapter 4

1.

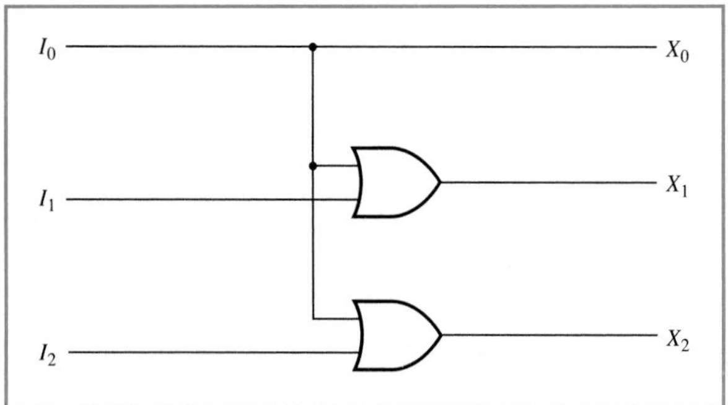

3. Convert to *AND-OR* logic, then apply DeMorgan's law and simplify.

5.

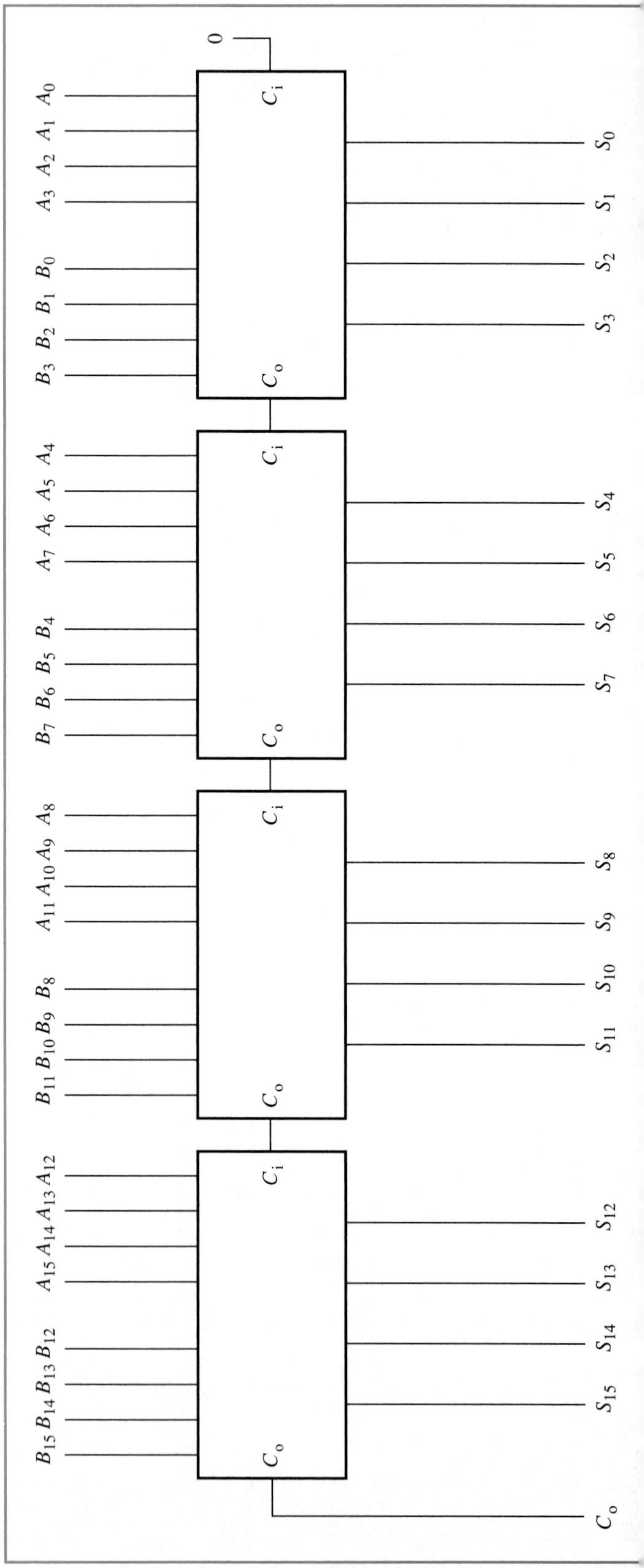

7.

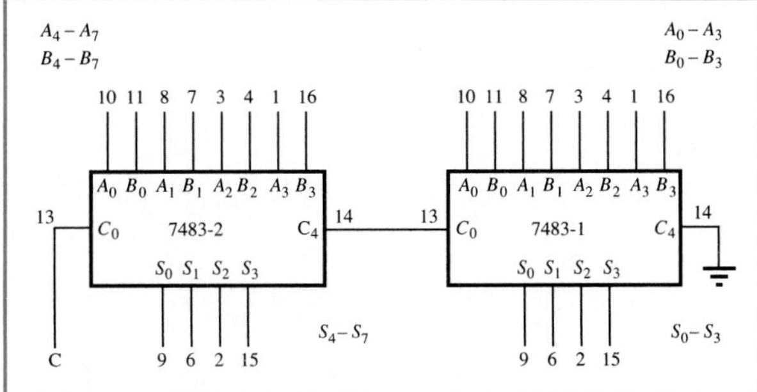

9.

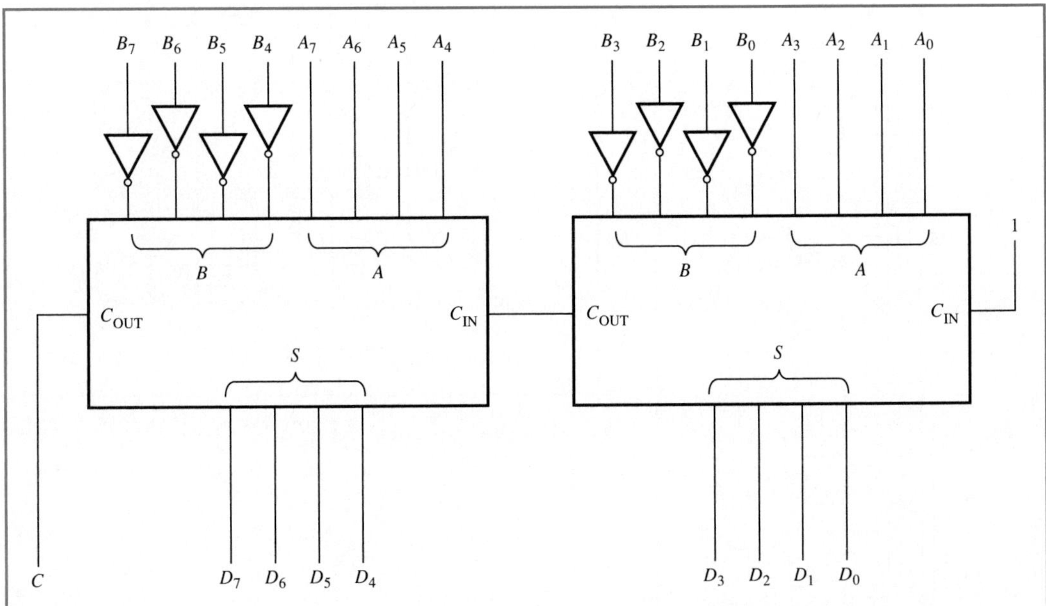

11.

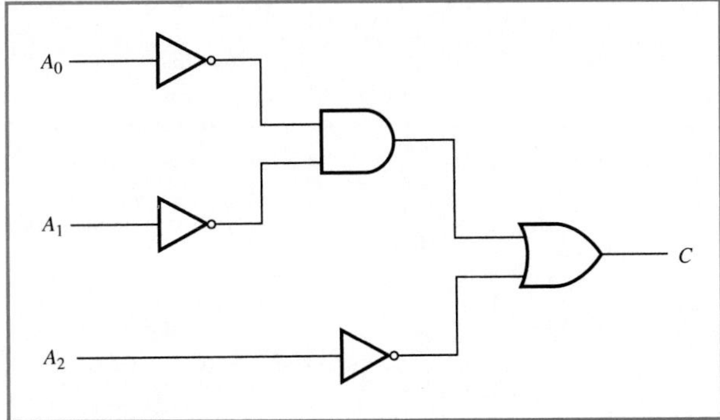

13.

I_0	I_1	I_2	I_3	X_1	X_0
0	0	0	0	1	1
0	0	0	1	1	1
0	0	1	0	1	0
0	0	1	1	1	0
0	1	0	0	0	1
0	1	0	1	0	1
0	1	1	0	0	0
0	1	1	1	0	0
1	0	0	0	0	0
1	0	0	1	0	0
1	0	1	0	0	0
1	0	1	1	0	0
1	1	0	0	0	0
1	1	0	1	0	0
1	1	1	0	0	0
1	1	1	1	0	0

15.

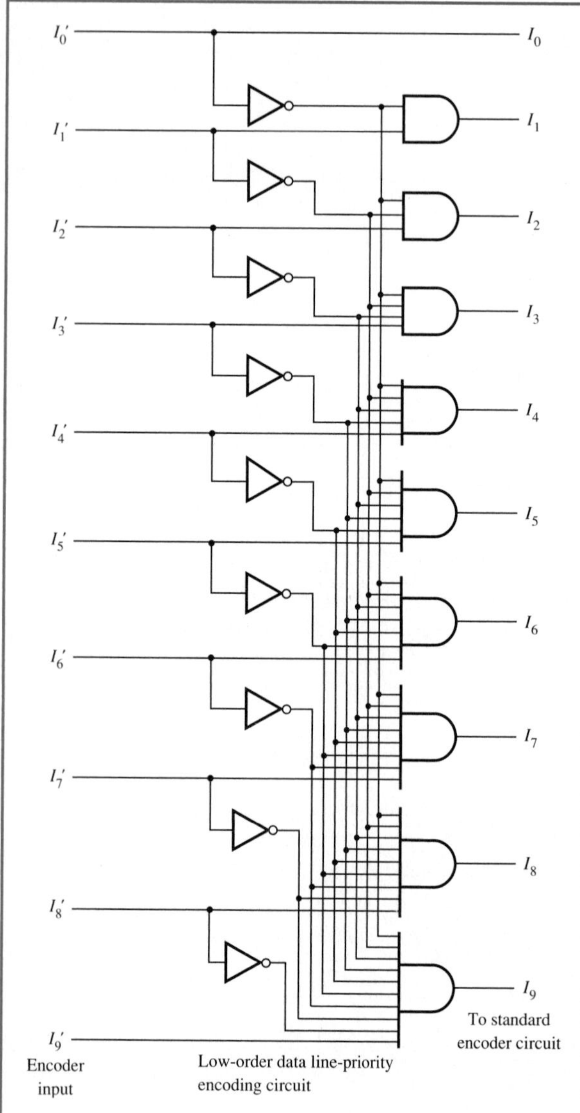

17.

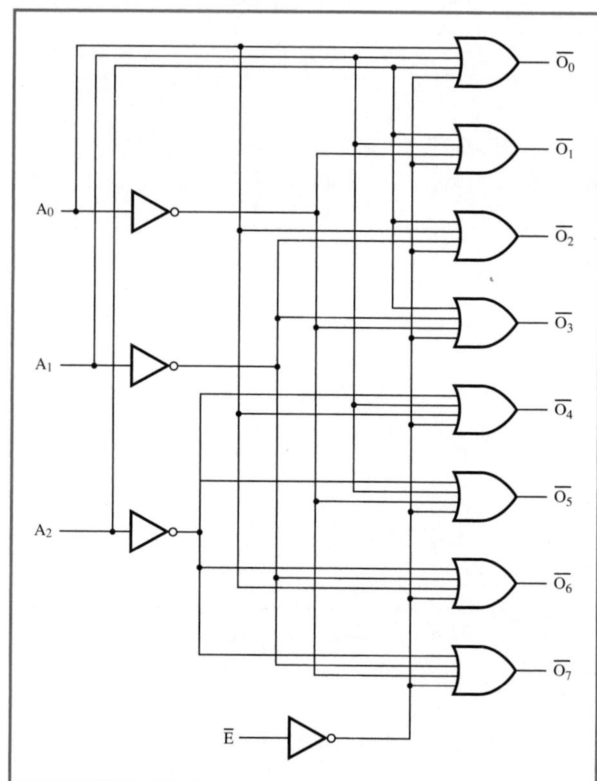

19. Priorities: Highest $\overline{7}$ $\overline{6}$ $\overline{5}$ $\overline{4}$ $\overline{3}$ $\overline{2}$ $\overline{1}$ $\overline{0}$ $\overline{F}$ $\overline{E}$ $\overline{D}$ $\overline{C}$ $\overline{B}$ $\overline{A}$ $\overline{9}$ $\overline{8}$ Lowest

21.

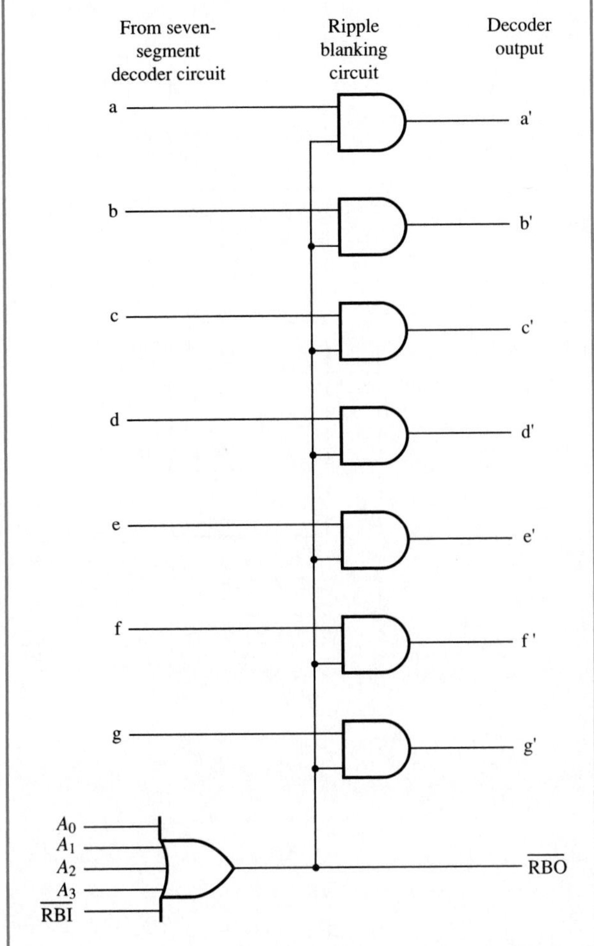
From seven-segment decoder circuit
Ripple blanking circuit
Decoder output
a
b
c
d
e
f
g
a'
b'
c'
d'
e'
f'
g'
A_0
A_1
A_2
A_3
$\overline{RBI}$
$\overline{RBO}$

23.

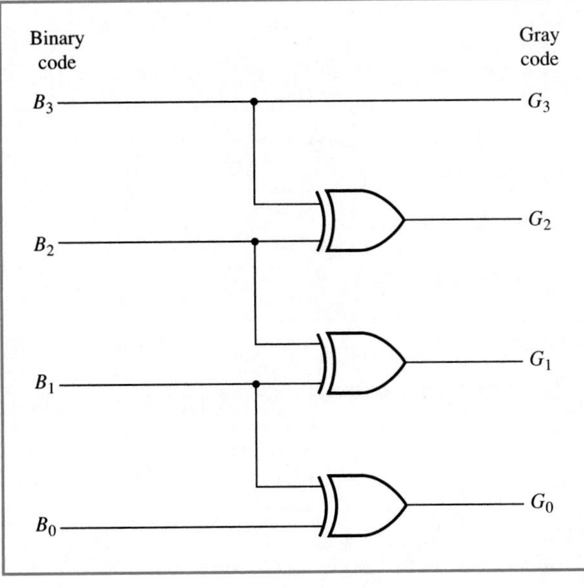
Binary code
Gray code
B_3
B_2
B_1
B_0
G_3
G_2
G_1
G_0

25.

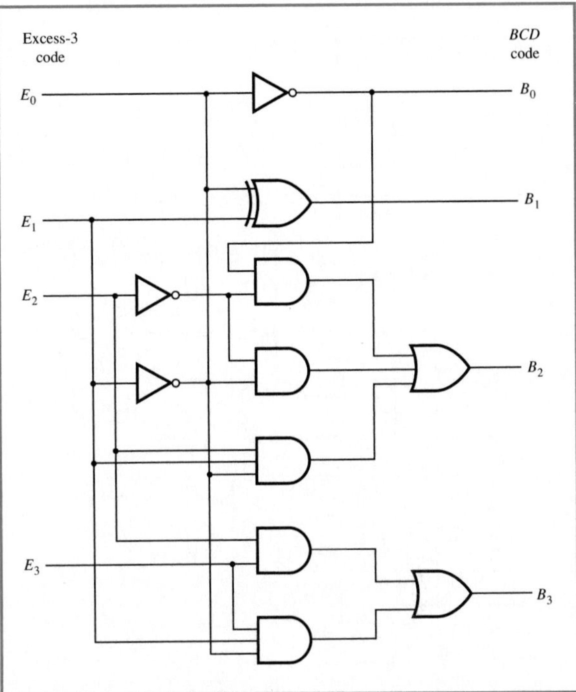
Excess-3 code
BCD code
E_0
E_1
E_2
E_3
B_0
B_1
B_2
B_3

27.

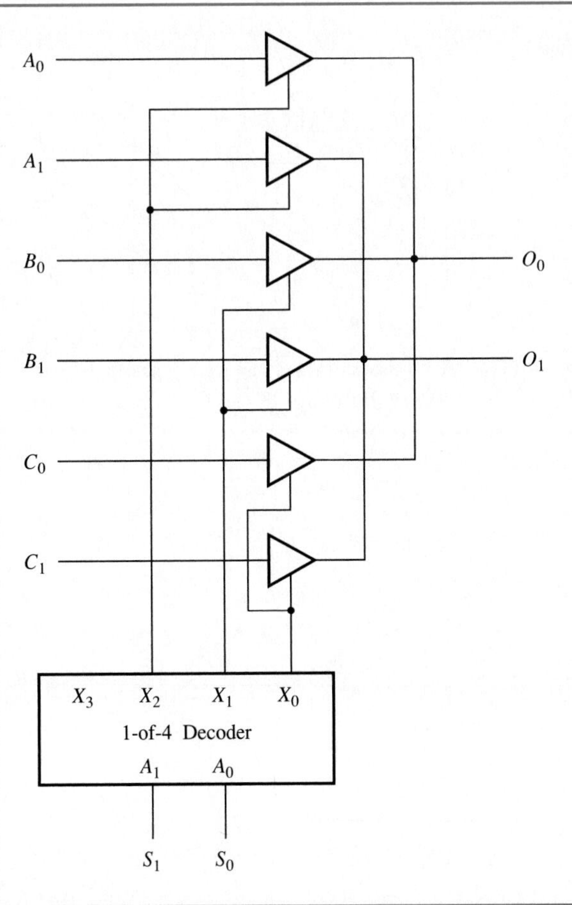
A_0
A_1
B_0
B_1
C_0
C_1
O_0
O_1
X_3 X_2 X_1 X_0
1-of-4 Decoder
A_1 A_0
S_1 S_0

29.

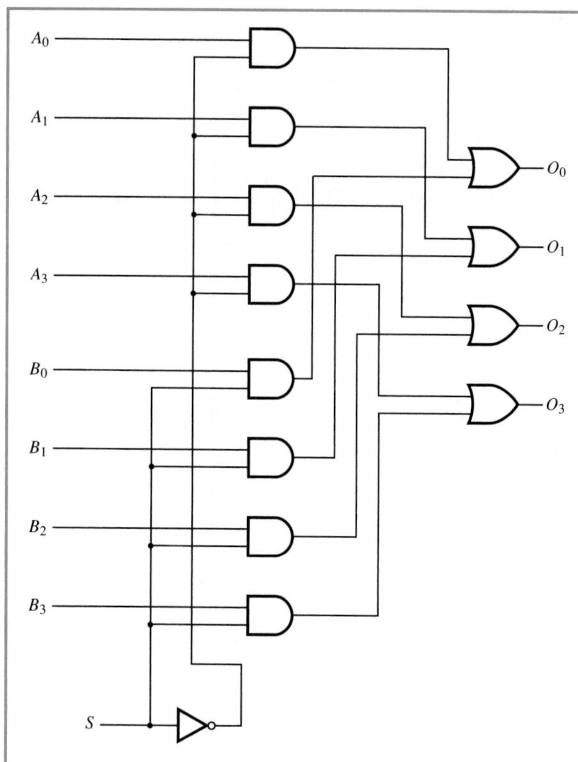

31.

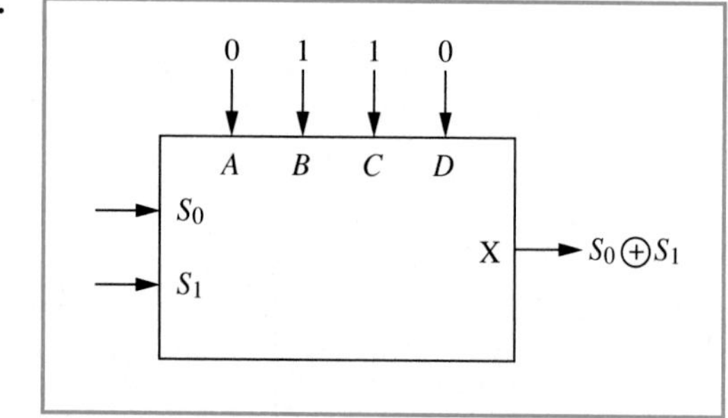

33. *AND-OR* implementation: 19 gates, seven ICs
MUX implementation: one IC

35.

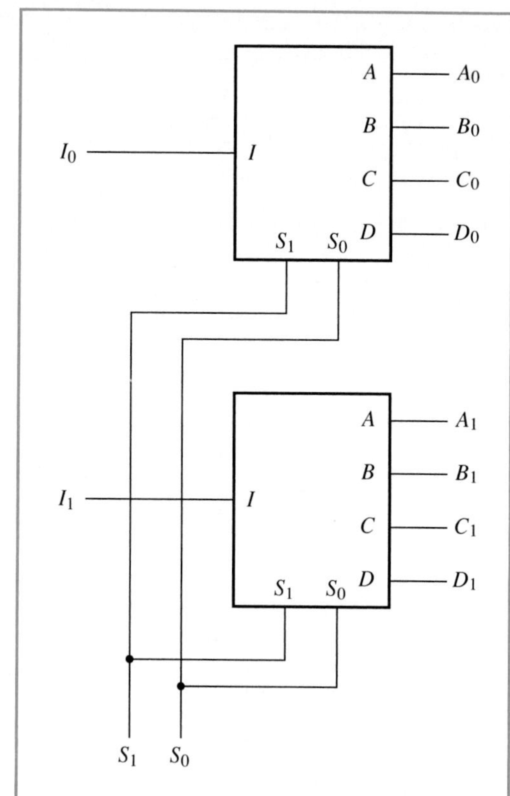

37.

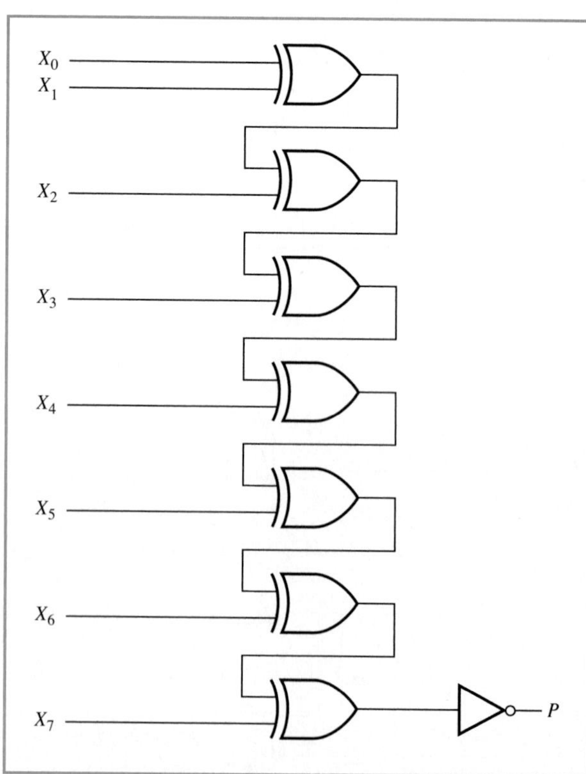

39. (a) 0001 (b) 10000 (c) 1000 (d) 1001

41. $\overline{EI}$ is permanently held high due to an internal short to V_{CC} in IC 1 or IC 2—most probable fault.

Chapter 5

1.

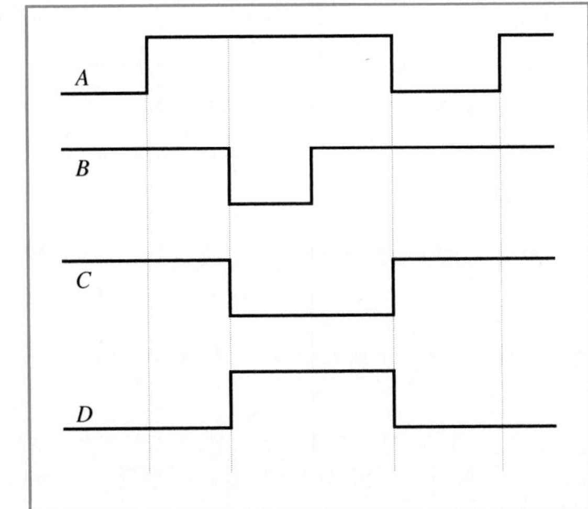

3. $\overline{S} = A$, $\overline{R} = B$, $Q = C$, $\overline{Q} = D$

5.

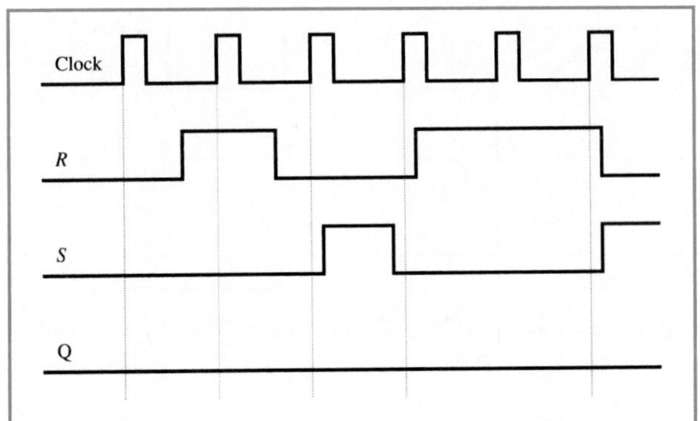

7.

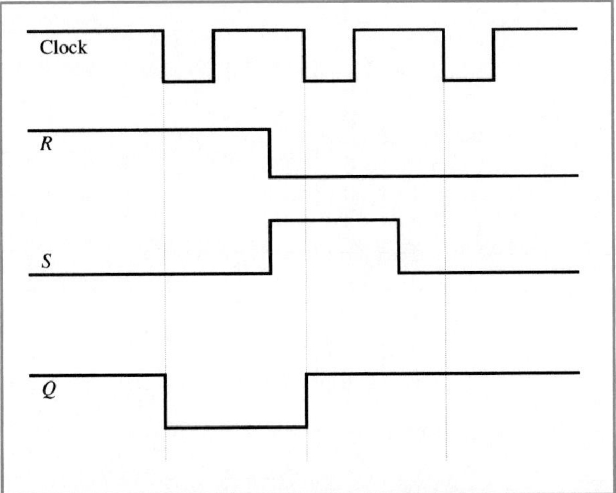

9.

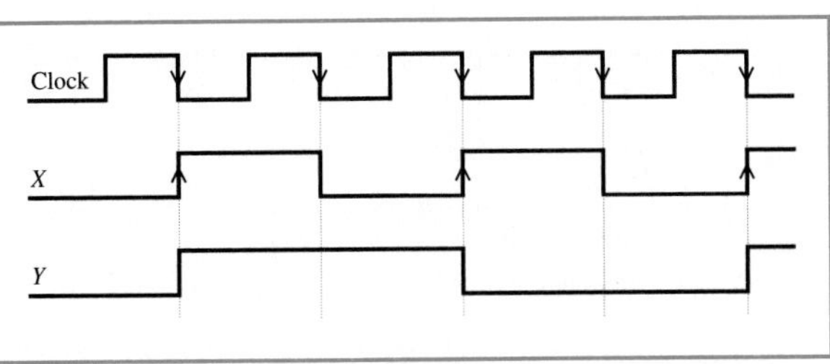

11.

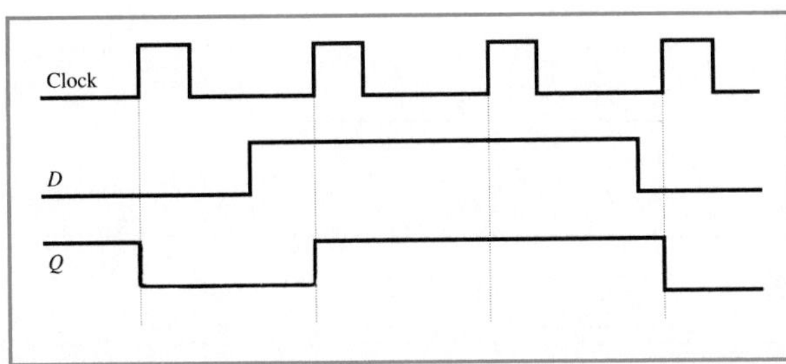

13.

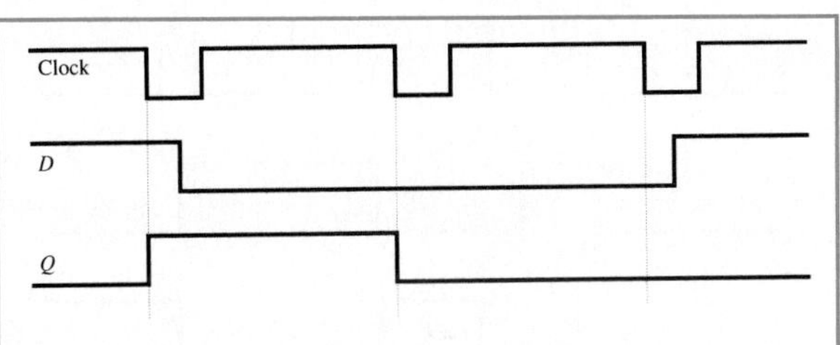

15.

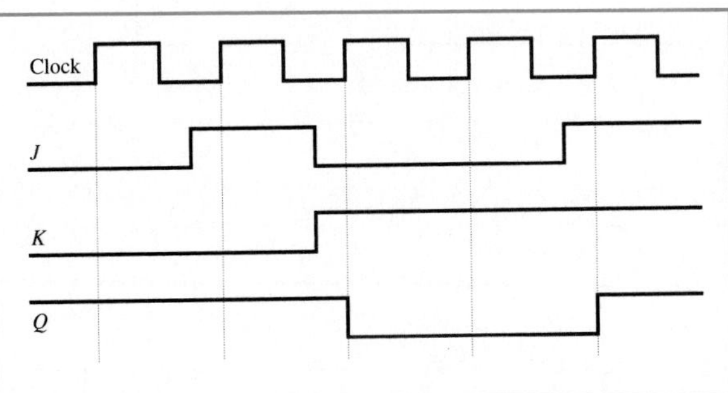

17.

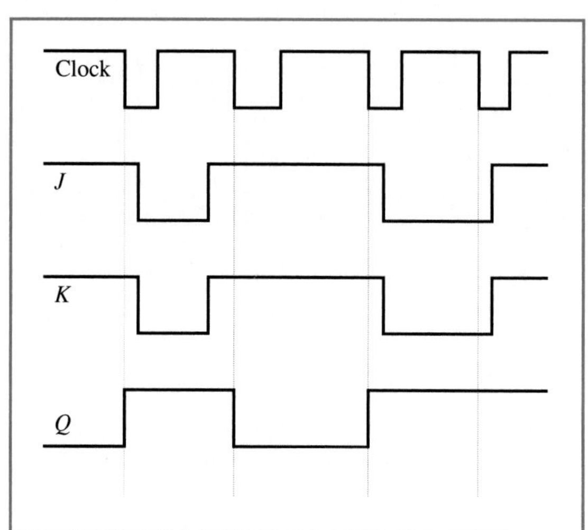

19.

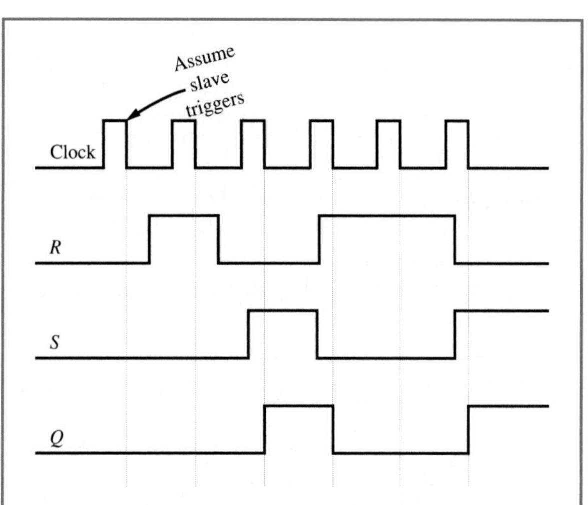

21.

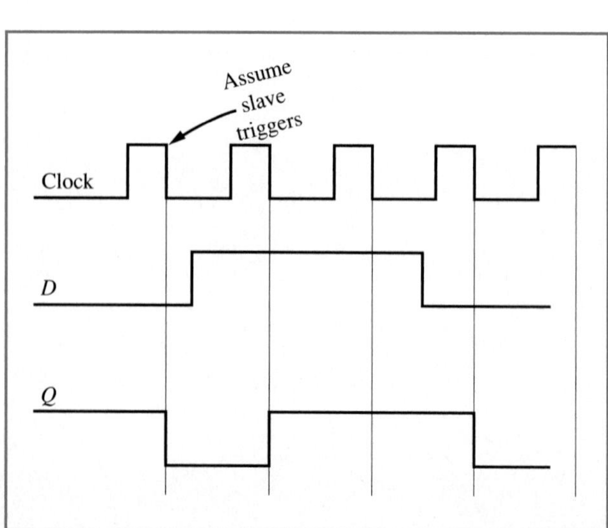

23.

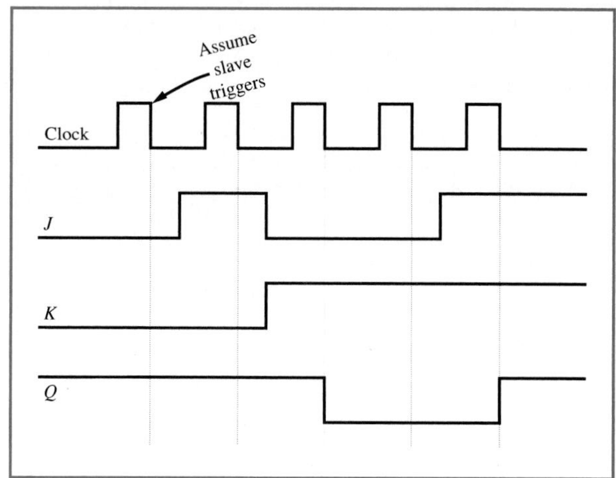

25.

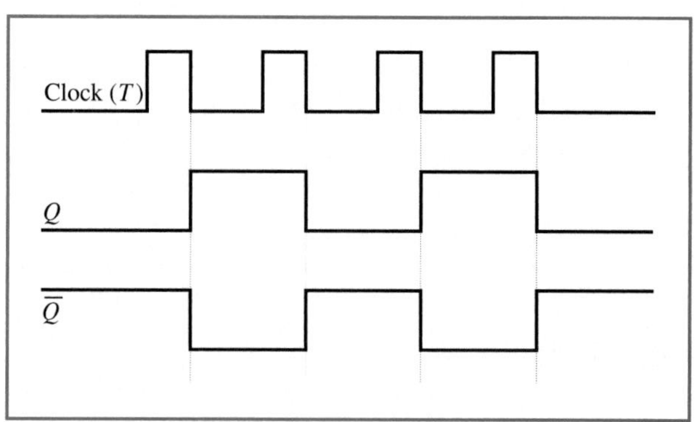

27.

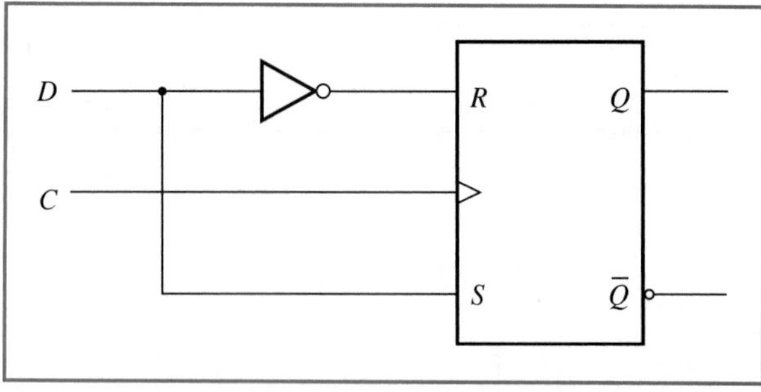

29.

Q^n	S	R	D	Q^{n+1}
0	0	0	0	0
0	0	1	0	0
0	1	0	1	1
0	1	1	1	1
1	0	0	1	1
1	0	1	0	0
1	1	0	1	1
1	1	1	1	1

31.

X	Y	Q^n	Q^{n+1}	
0	0	0	1	Set
0	0	1	1	
0	1	0	0	NC
0	1	1	1	
1	0	0	1	TOGGLE
1	0	1	0	
1	1	0	0	RESET
1	1	1	0	

$Q^{n+1} = \overline{Y}\,\overline{Q^n} + \overline{X}\,Q^n$

33.

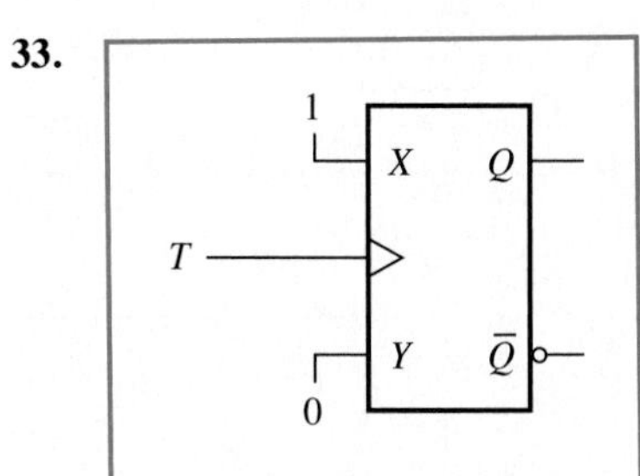

35.

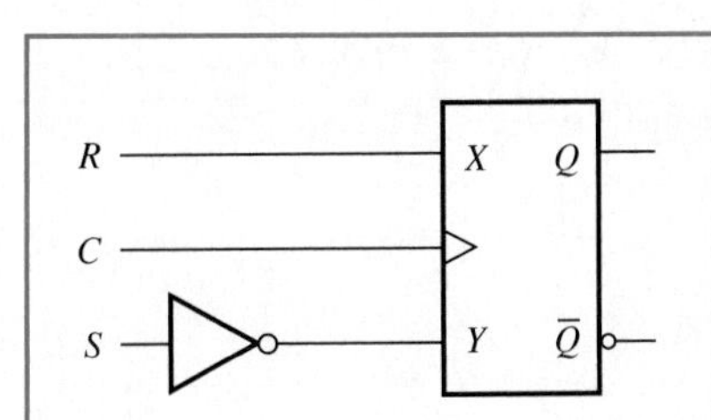

37. $R_x = 1.5\text{K},\ C_x = 1\ \mu\text{F}$

39. $R_{x1} = R_{x2} = 725$ ohms, $C_{x1} = C_{x2} = 1\ \mu\text{F}$

41.

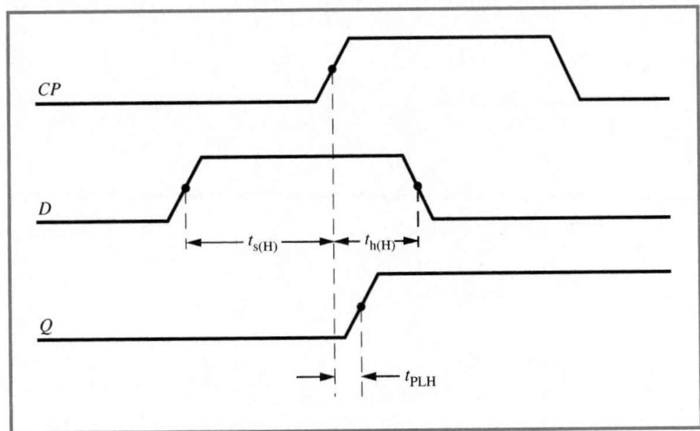

43.

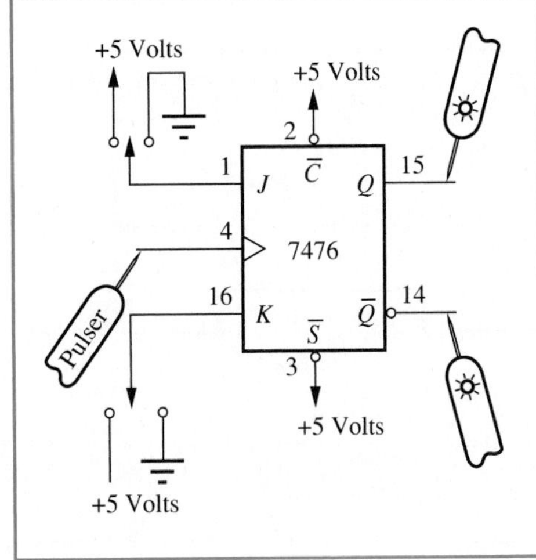

Chapter 6

1.

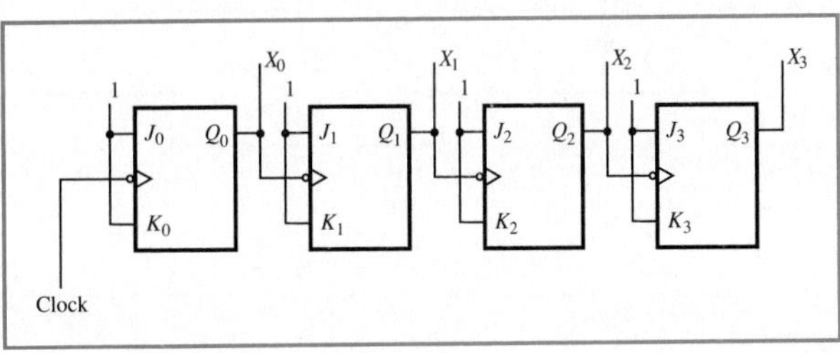

3.

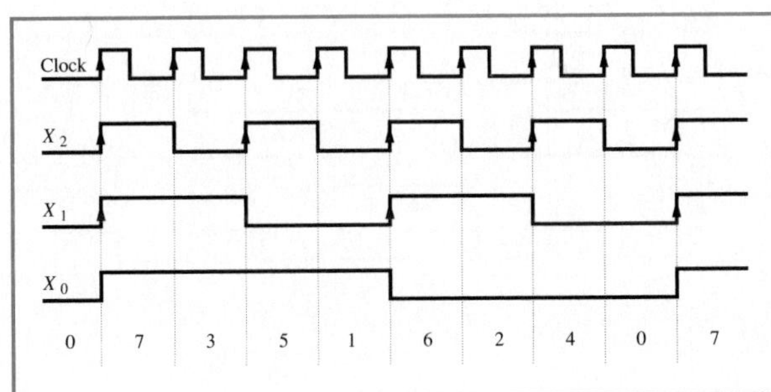

5.

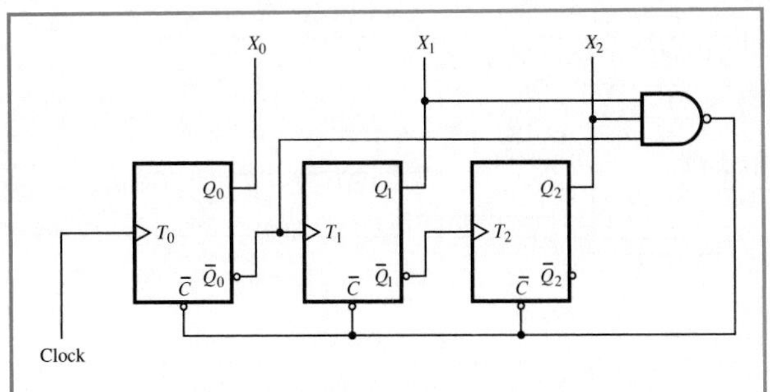

7.

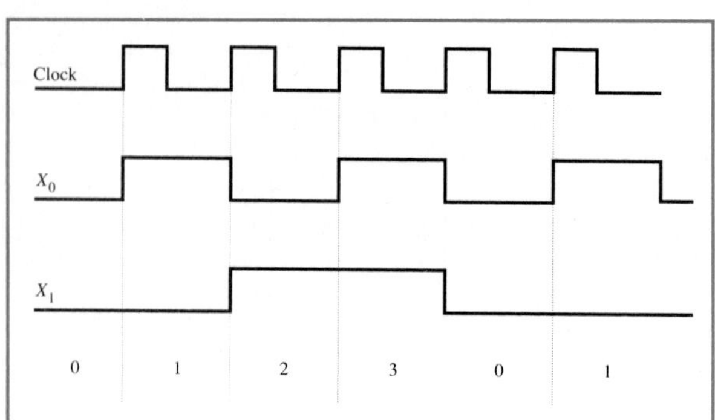

9.

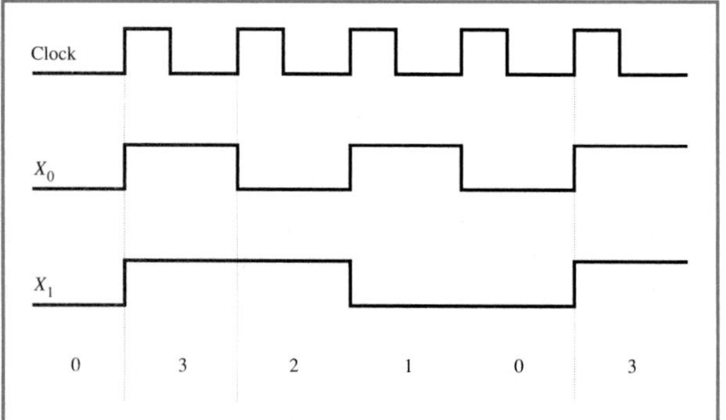

11.

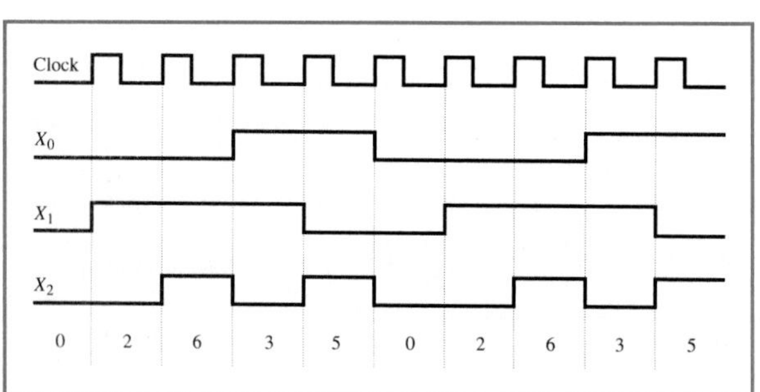

13.

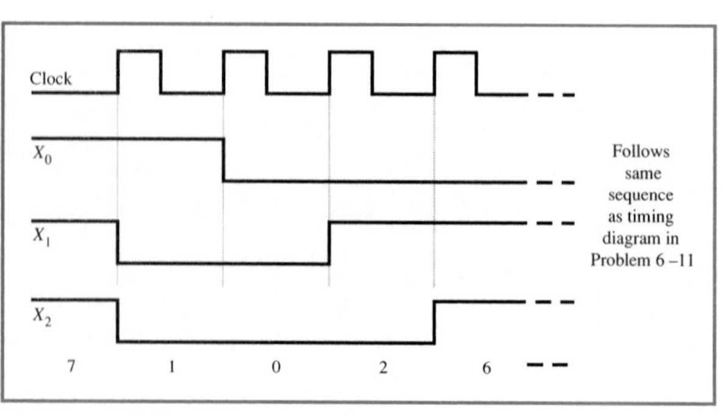

15.

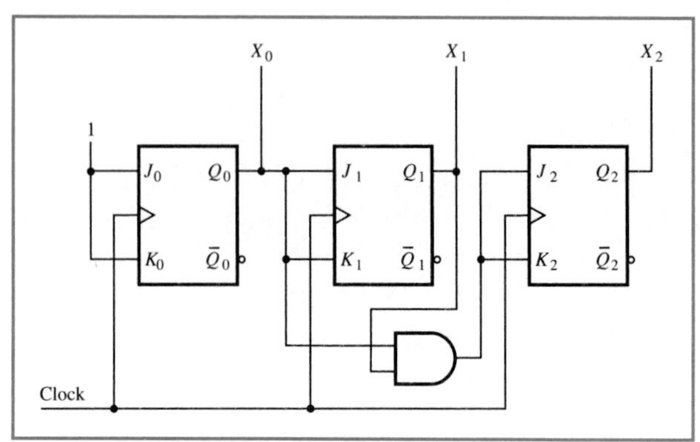

17.

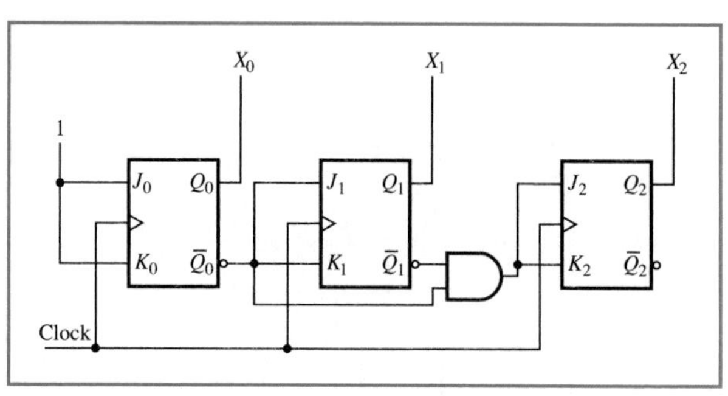

19.

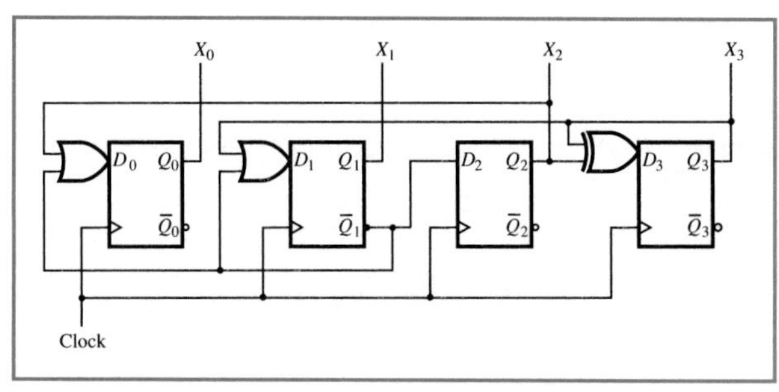

21.

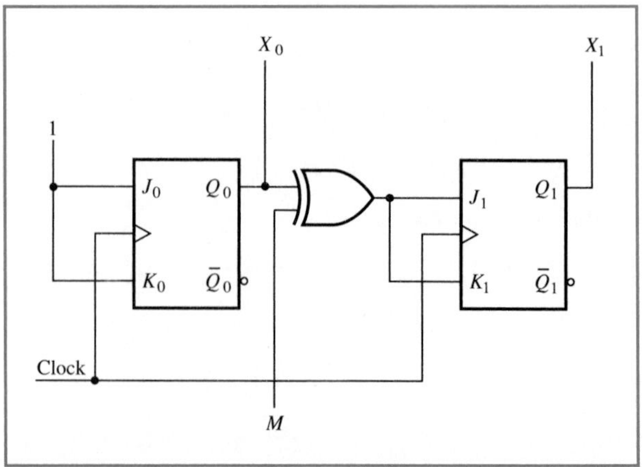

23.

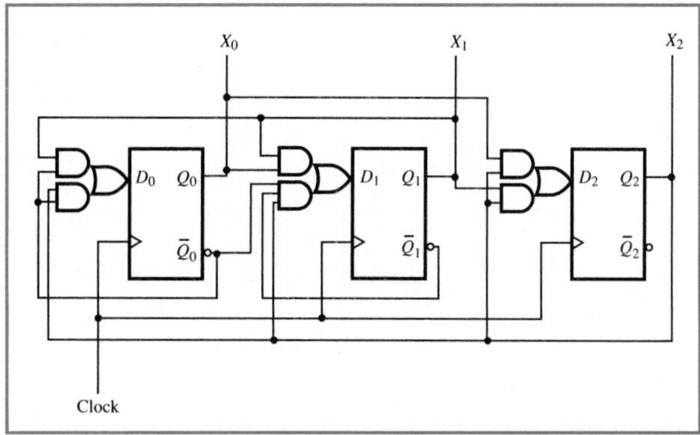

25.

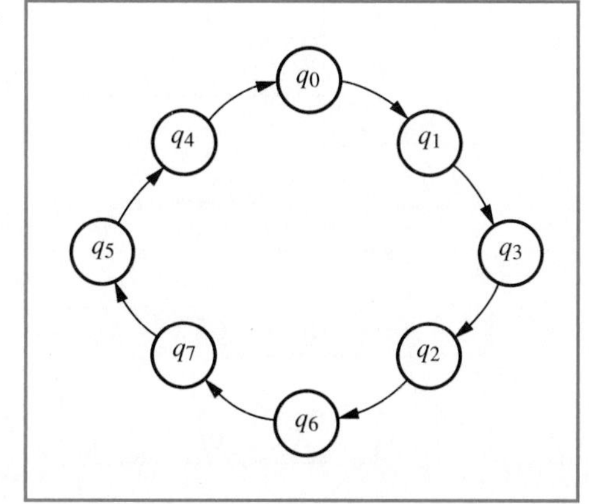

27. q_1 to q_7, q_2 to q_0, q_4 to q_F, q_5 to q_F, q_6 to q_9, q_8 to q_F, q_A to q_A, q_B to q_A, q_C to q_7, q_D to q_7, q_E to q_3

29.

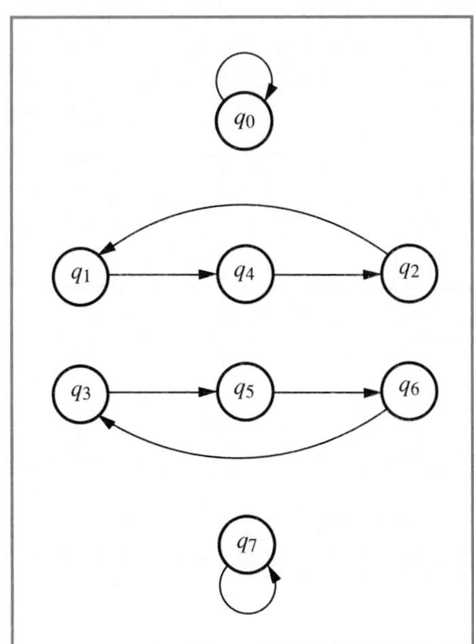

31.

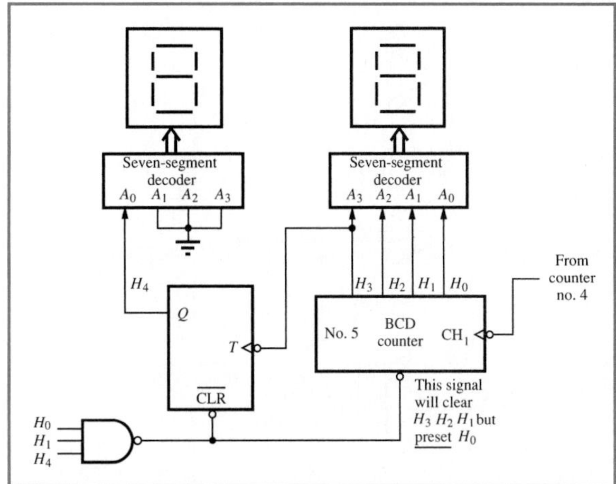

33. Output frequency = 10 Hz

35.

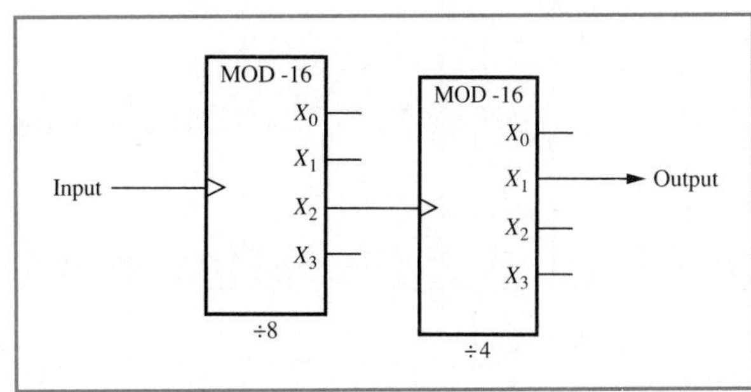

37.

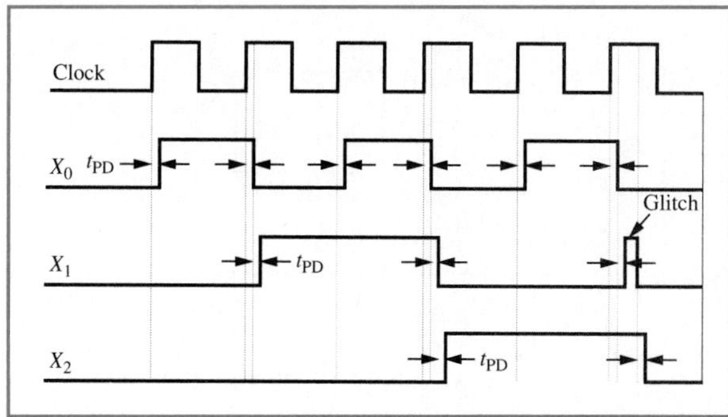

39. Possible remedies: (a) replace asynchronous counter with a synchronous MOD-8 counter, (b) enable the decoder on the negative edge of the clock pulse (see Figure 6–62)

Chapter 7

1.

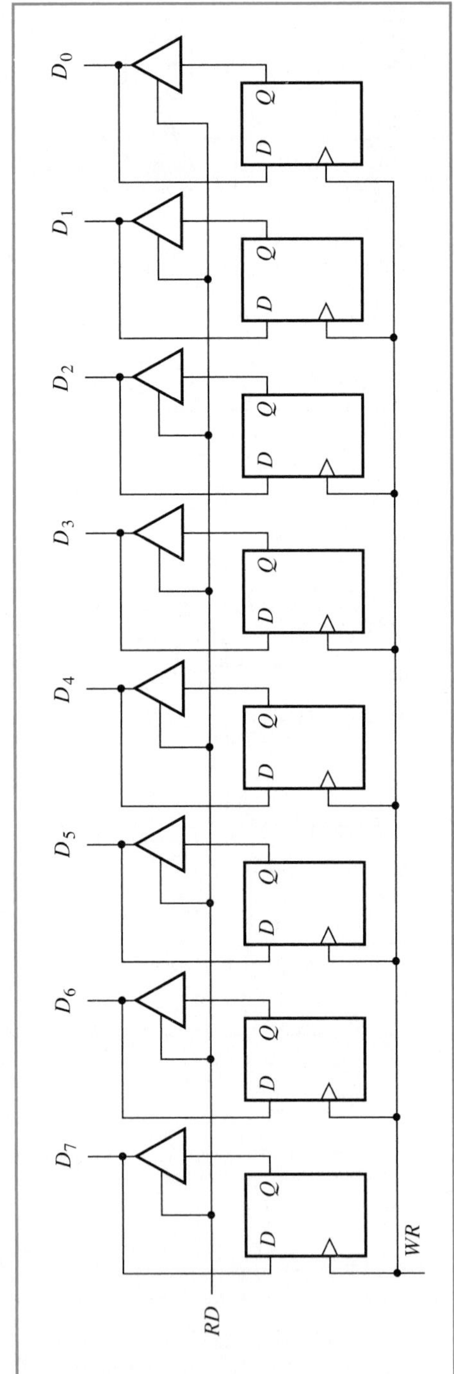

3.

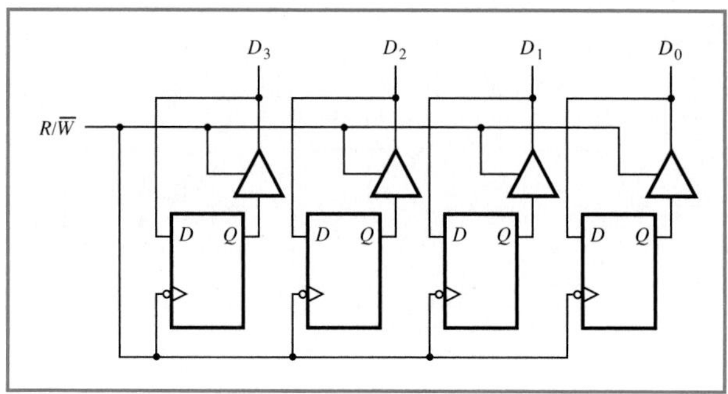

5.

Clock

I 0_A 0_B 1_A 1_B

O_3 0_A 0_B 1_A 1_B

O_2 0_A 0_B 1_A

O_1 0_A 0_B

O_0 0_A

Initial contents / Data loaded

7.

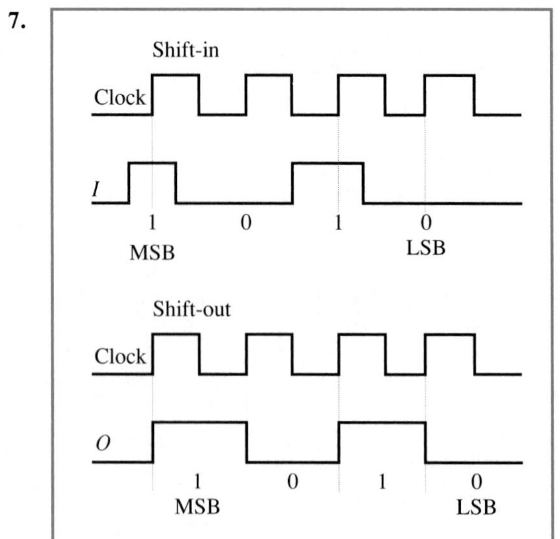

9.

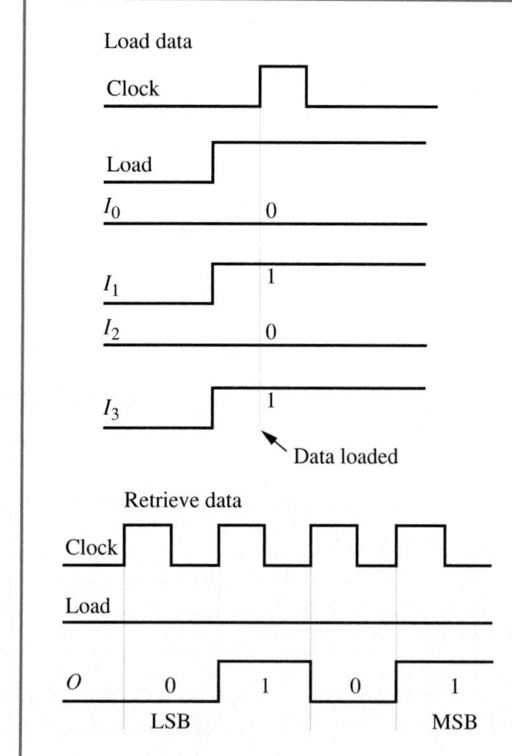

11.

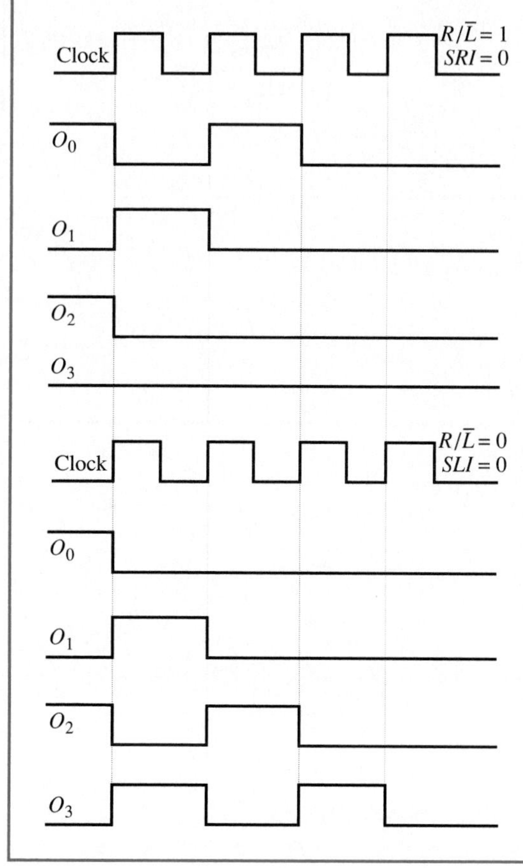

13.

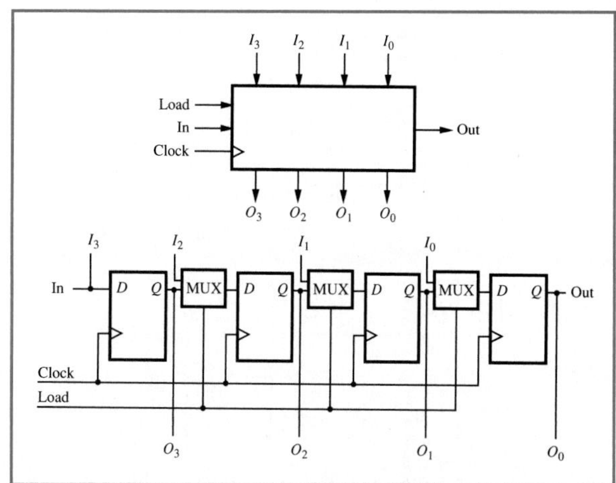

15.

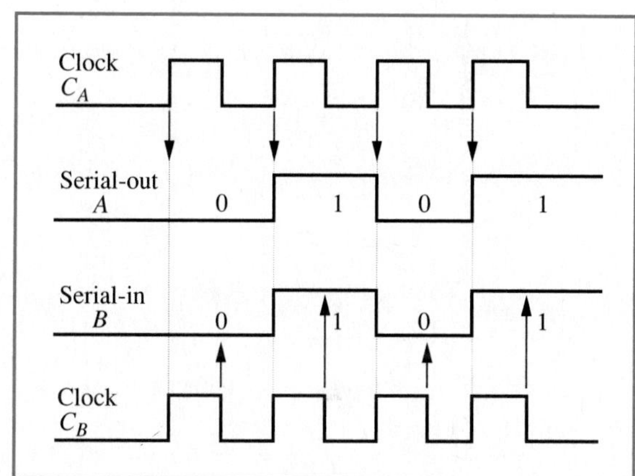

17.

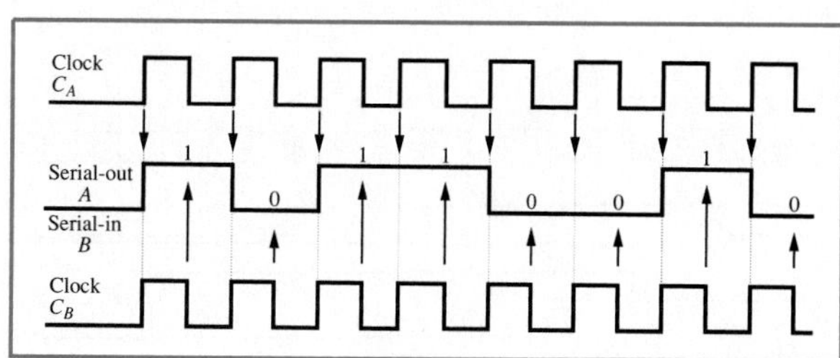

19.

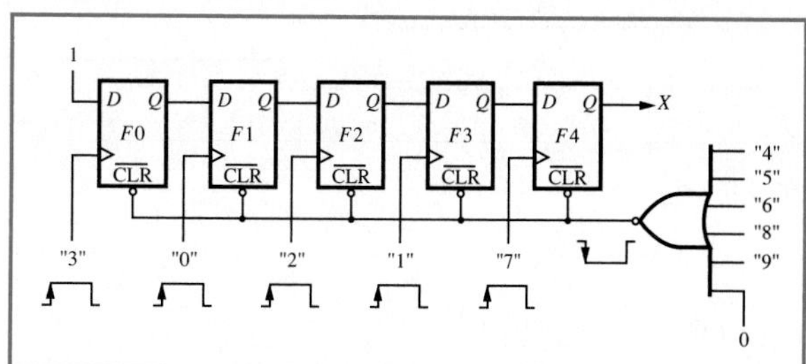

21.

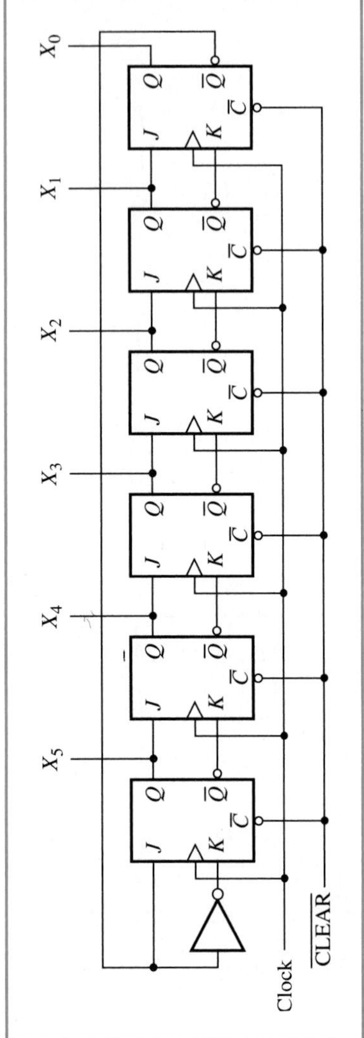

23.

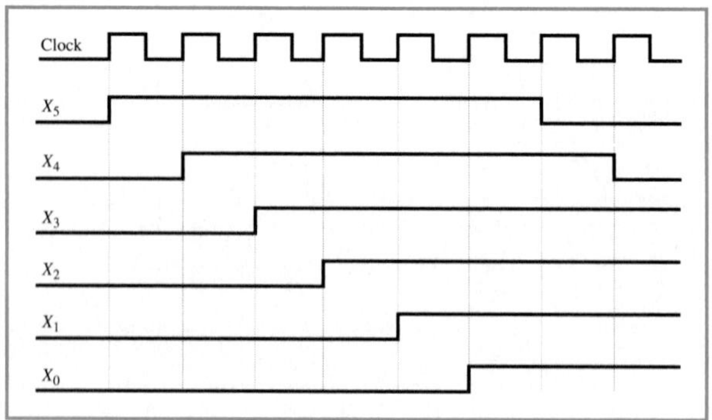

25. 0.6 μs

27.

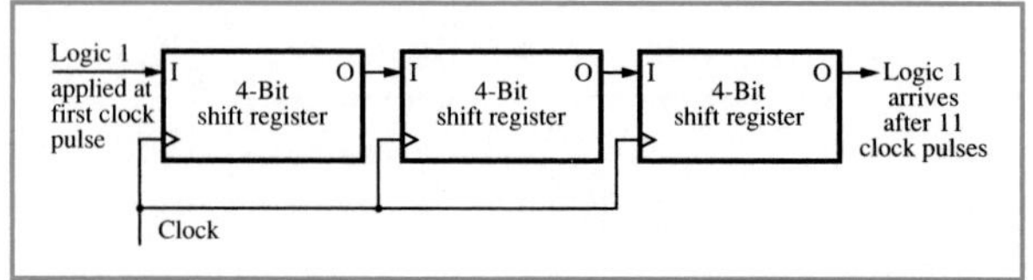

29.

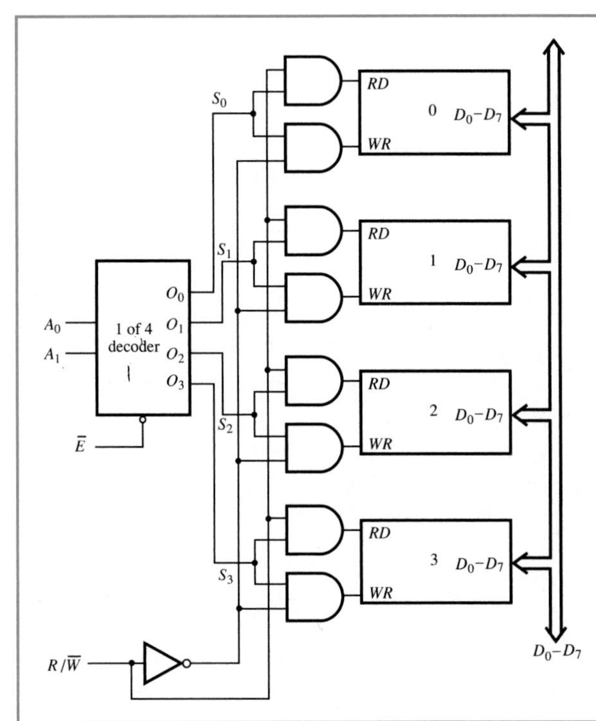

31.

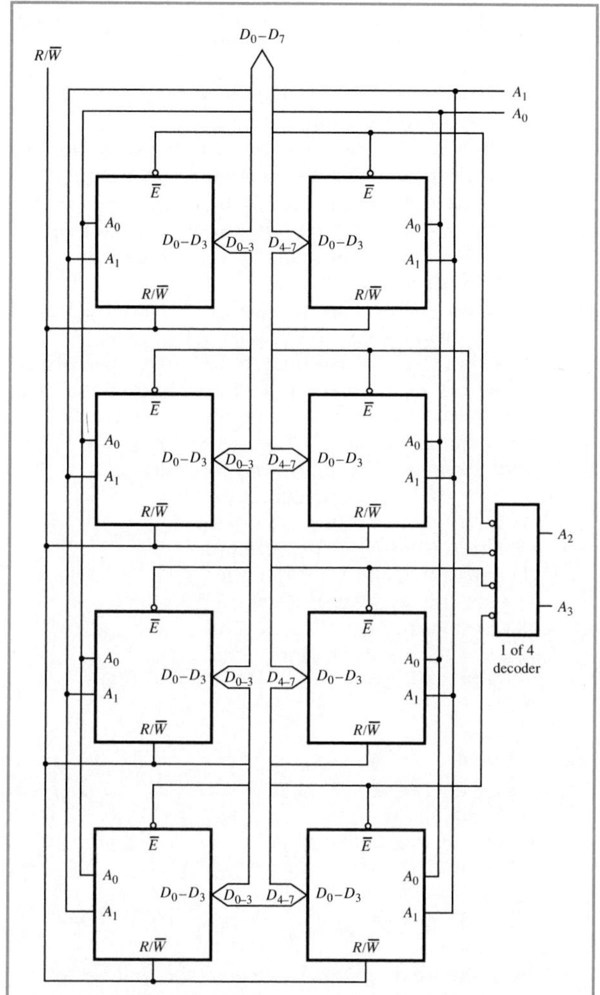

R/W̄
D0–D7
A1
A0
Ē
A0
A1
D0–D3
D0–3
D4–7
R/W̄
A2
A3
1 of 4 decoder

33.

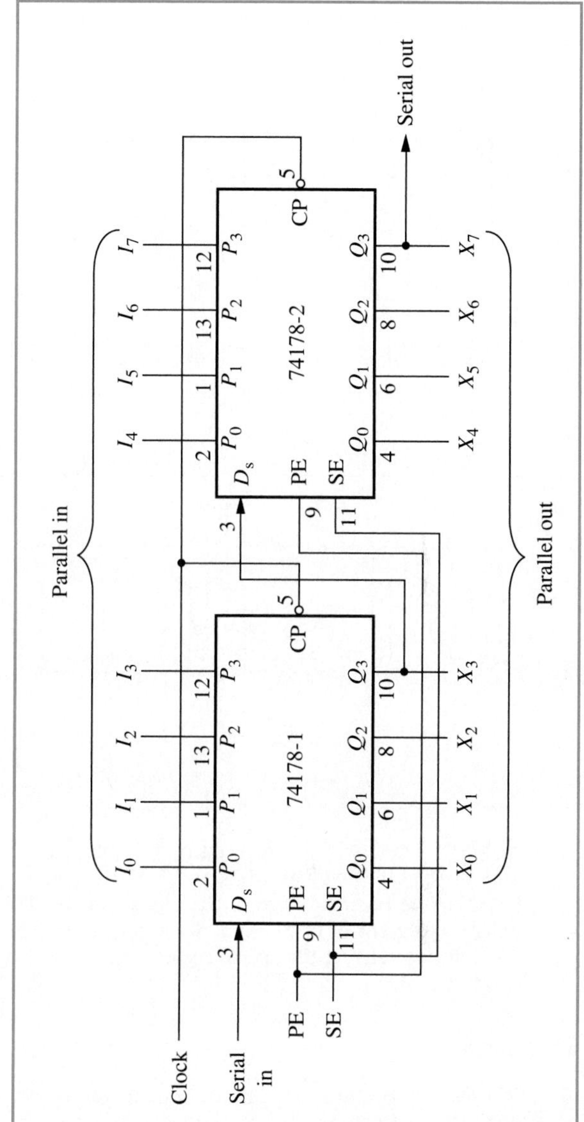

Serial out
74178-2
74178-1
CP
PE
SE
Parallel in
Parallel out
Clock
Serial in
PE
SE

35.

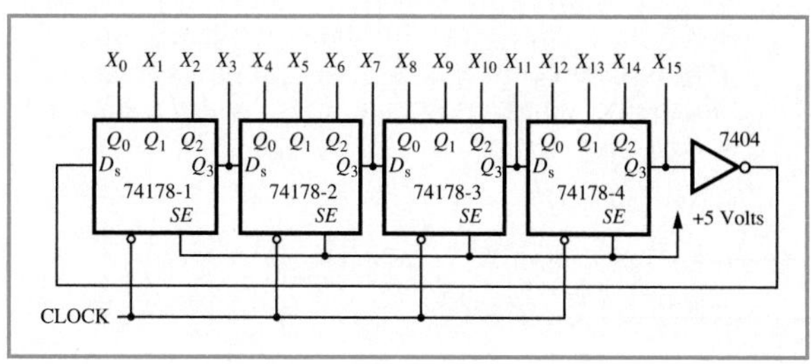

X0 X1 X2 X3 X4 X5 X6 X7 X8 X9 X10 X11 X12 X13 X14 X15
74178-1
74178-2
74178-3
74178-4
7404
+5 Volts
CLOCK

37.

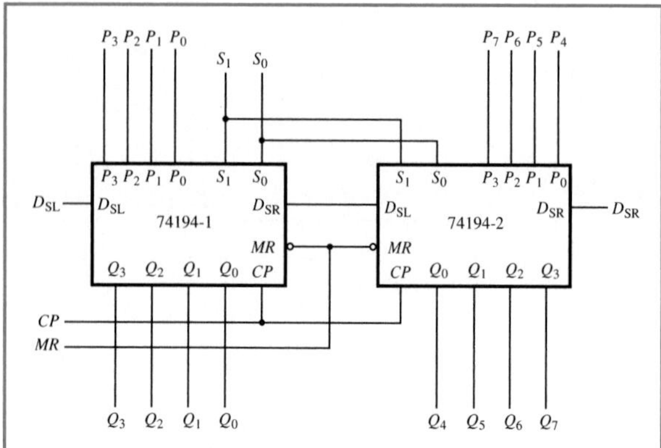

39.

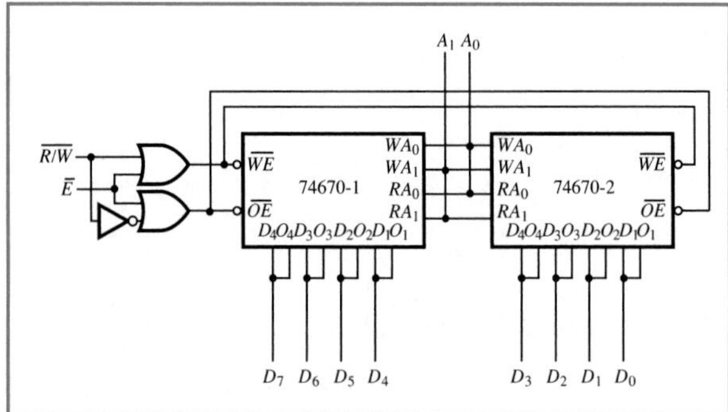

41. Problem: The 74178 is triggered by the negative edge of the clock pulses while the 74379 is triggered by the positive edge of the clock pulses. This leads to an error of one-half a clock period or 0.25 μs. Solution: Invert the clock connected to the 74379 IC.

Chapter 8

1. Program instructions are instructions to the computer that are interpreted by the computer and "tell" it which computer operation (AND, OR, ADD, etc.) is to be performed.

3. As is the case with humans, a computer's memory serves as a depository for data and instructions (a program) to achieve a task. A computer refers to this memory to retrieve and store instructions and data.

5. Main memory is the working memory of the system in that it is main memory from which the CPU works. Mass storage memory functions as a memory that can be used to store programs and data currently not being operated on.

7. Main memory locations can be accessed much faster than mass storage memory locations.

9. The five basic categories for classifying memories are: (1) organization, (2) manner of accessing, (3) alterability of memory contents, (4) if the contents of memory are retained when power is removed, (5) if the contents of memory are retained even with power applied.

11. There are 64 memory locations within the memory device, which for an *x-y* coordinate scheme requires 6 ($2^6 = 64$) parameters (columns and rows) to identify each location. These six parameters can be divided into various combinations of the columns and rows seen in Figures 8–3 and 8–4. The combinations of columns and rows are

(a) Five rows and one column
(b) Four rows and two columns
(c) Three rows and three columns
(d) Two rows and four columns
(e) One row and five columns

13. For Problem 11

Rows + Columns	= Sum
5 + 1	= 6
4 + 2	= 6
3 + 3	= 6
2 + 4	= 6
1 + 5	= 6

For Problem 12

Rows + Columns	= Sum
4 + 1	= 5
3 + 2	= 5
2 + 3	= 5
1 + 4	= 5

From the above patterns we can see that for N memory locations the possible patterns are

Rows	+	Columns	= Sum
$N-1$	+	1	$= N$
$N-2$	+	2	$= N$
$N-3$	+	$N-(N-3)$	$= N$
$N-4$	+	$N-(N-4)$	$= N$
·	+	···	= ·
·	+	···	= ·
·	+	···	= ·
$N-(N-1)$	+	$N-[N-(N-1)]$	$= N$

15.

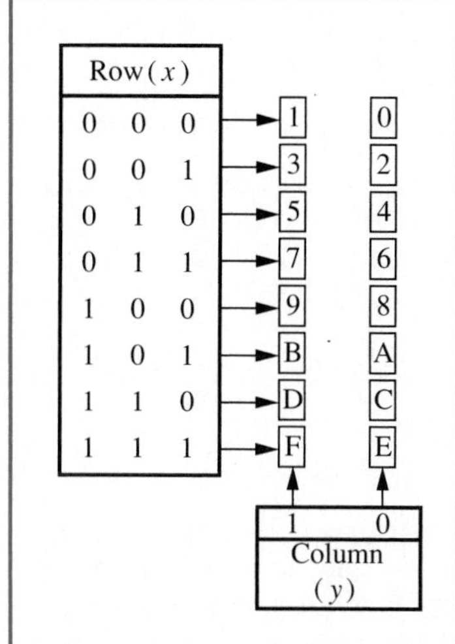

17. (a) $(256 - 10) \times 2 \times 10^{-3} = 492$ ms
(b) $(500 - 10) \times 2 \times 10^{-3} = 980$ ms
(c) $5 \times 2 \times 10^{-3} = 10$ ms

19. 2 μs

21. $2 \times 0.2\ \mu\text{s} = 0.4\ \mu\text{s}$

23. 1.2 μs

25. Yes. RAM is an acronym for random access memory and all semiconductor ROMs are accessed in this manner.

27. MROMs

29. Either PROMs or EPROMs, it would probably depend on if the company decided to purchase just an EPROM programmer or both an EPROM and PROM programmer (an engineering firm will likely have an EPROM programmer but maybe not a PROM programmer).

31. *Volatility* describes a memory that can not retain its contents without power.

33. DRAMs have a much larger memory capacity than static RAMs, hence the increased memory size justifies the additional expense for the refresh control circuity.

35. Yes. Both lose their contents when power is removed from them.

37. The CPU controls read and write operations via its read (*RD*) and write (*WR*) control signals.

39. (a) For eight address pins there will be four row address pins and four column address pins.
(b) Five row address pins and five column address pins.
(c) Six row address pins and five column address pins.
(d) Six row address pins and six column address pins.

41. For 13 address pins there will be seven row address pins and six column address pins. Therefore, there will be $2^6 = 64$ columns for a total of $128 \times 64 = 8192$ memory locations.

43. Since the CPU is the program execution function-block it must fetch program instructions from memory, hence the need for it to supply the address of the instructions in memory. In the course of executing a program the CPU may have to read and write data from or to memory and I/O, which again requires that the CPU furnish the address of the accessed memory location or I/O device.

45. The row and column select lines must be activated, for then and only then can a read or write operation be performed on those cells. To activate these select lines the memory location address must be input to the memory device.

47. Only the accessed memory cell in each plane is active and therefore only it has control of output lines.

49.

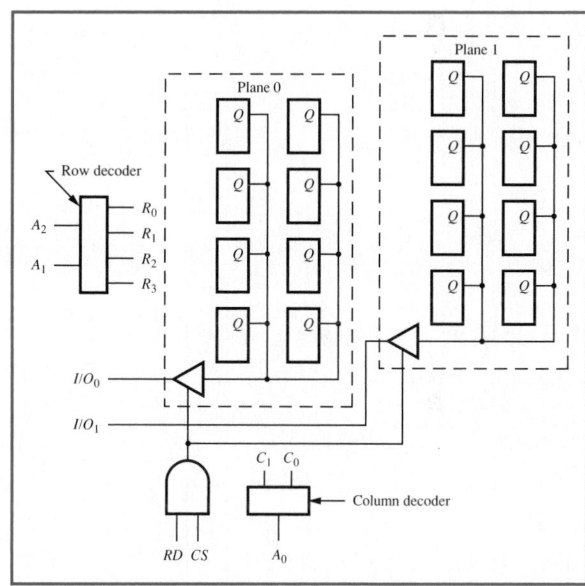

51. We will define a *page* to be the smallest number of memory locations within the memory devices that make up main memory.

53. (a) $2^{11} = 2048$
(b) One page $=$ 2048 memory locations
(c) $2^{13} = 8192$. Therefore 13 address bus lines are required.
(d) A total of 15 (11 + 4) address bus lines

55. A_{14} through A_{11} are used to select memory devices.

Decoding the linear addresses of Problem 57

Address bus lines		A_{14}	A_{13}	A_{12}	A_{11}	A_{10}	A_9	A_8	A_7	A_6	A_5	A_4	A_3	A_2	A_1	A_0
ROM-0	0800H	0	0	0	1	0	0	0	0	0	0	0	0	0	0	0
	0FFFH	0	0	0	1	1	1	1	1	1	1	1	1	1	1	1
RAM-1	1000H	0	0	1	0	0	0	0	0	0	0	0	0	0	0	0
	17FFH	0	0	1	0	1	1	1	1	1	1	1	1	1	1	1
RAM-2	2000H	0	1	0	0	0	0	0	0	0	0	0	0	0	0	0
	27FFH	0	1	0	0	1	1	1	1	1	1	1	1	1	1	1
RAM-3	4000H	1	0	0	0	0	0	0	0	0	0	0	0	0	0	0
	47FFH	1	0	0	0	1	1	1	1	1	1	1	1	1	1	1

57.

A_{14} A_{13} A_{12} A_{11}

ROM-0 *CS*

RAM-1 *CS*

RAM-2 *CS*

RAM-3 *CS*

59. $(CA > DA) = 0$, $(CA = DA) = 0$, and $(CA < DA) = 1$

61. Its access time is 12×2 ms $+ 4$ ms $= 28$ ms.

63. Just as for a magnetic tape mass storage device, the control circuitry of the access mechanism will monitor the current position of the disk (current address) and compare that address with the destination address. The address of a block of data is part of the address mark and identifies sector and track. When the current address is the same as the destination address, that data block is read.

65. The Kansas City technique is well suited for telephone transmission since logic levels are represented by a burst of sinusoidal waveforms, or tones.

67. 16.67 ms

69. Data storage on a disk is partitioned into tracks and sectors.

71.

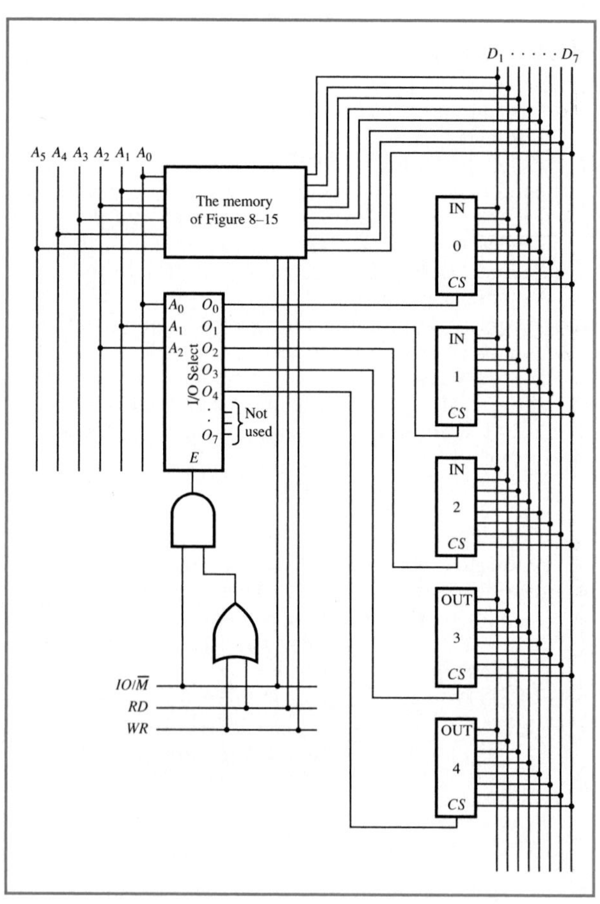

Address table for Problem 71

Address bus lines							A_5	A_4	A_3	A_2	A_1	A_0
ROM and RAM address pins									A_3	A_2	A_1	A_0
Page selector						O_n	A_1	A_0				
Memory addresses $IO/\overline{M} = 0$	ROM-0	Page 0	0 ⋮ 0	0 ⋮ F	H ⋮ H	$O_0 = 1$	0 ⋮ 0	0 ⋮ 0	0 ⋮ 1	0 ⋮ 1	0 ⋮ 1	0 ⋮ 1
	ROM-1	Page 1	1 ⋮ 1	0 ⋮ F	H ⋮ H	$O_1 = 1$	0 ⋮ 0	1 ⋮ 1	0 ⋮ 1	0 ⋮ 1	0 ⋮ 1	0 ⋮ 1
	RAM-2	Page 2	2 ⋮ 2	0 ⋮ F	H ⋮ H	$O_2 = 1$	1 ⋮ 1	0 ⋮ 0	0 ⋮ 1	0 ⋮ 1	0 ⋮ 1	0 ⋮ 1
	RAM-3	Page 3	3 ⋮ 3	0 ⋮ F	H ⋮ H	$O_3 = 1$	1 ⋮ 1	1 ⋮ 1	0 ⋮ 1	0 ⋮ 1	0 ⋮ 1	0 ⋮ 1
I/O selector						O_n				A_2	A_1	A_0
I/O addresses $IO/\overline{M} = 1$	Input device 0					$O_0 = 1$	0	0	0	0	0	0
	Input device 1					$O_1 = 1$	0	0	0	0	0	1
	Input device 2					$O_2 = 1$	0	0	0	0	1	0
	Output device 3					$O_3 = 1$	0	0	0	0	1	1
	Output device 4					$O_4 = 1$	0	0	0	1	0	0

73.

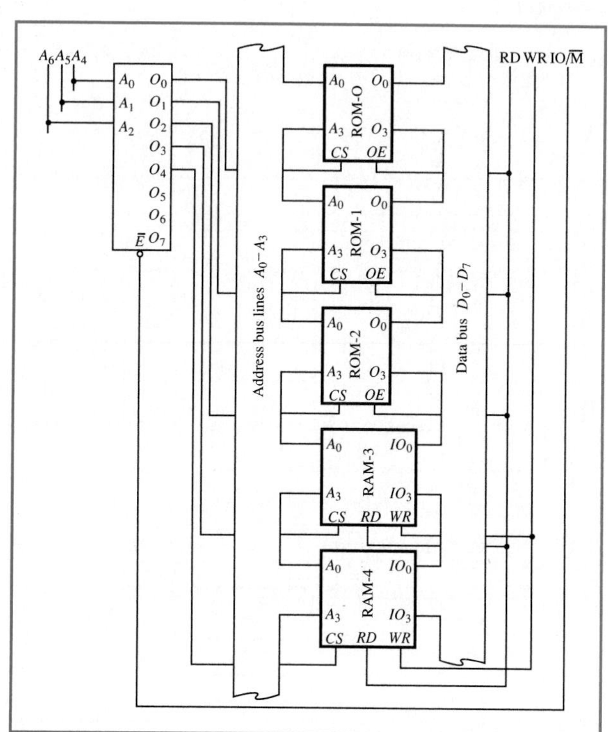

75.

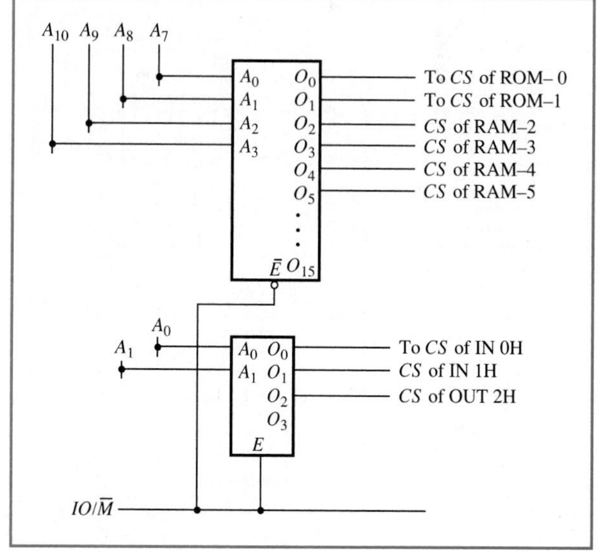

Problem 75

Address bus lines					A_{10}	A_9	A_8	A_7	A_6	A_5	A_4	A_3	A_2	A_1	A_0
ROMs and RAMs									A_6	A_5	A_4	A_3	A_2	A_1	A_0
Page select decoder				O_n	A_3	A_2	A_1	A_0							
I/O select decoder			O_n											A_1	A_0
$IO/\overline{M} = 0$	ROM-0	000H– 07FH	0	$O_0 = 1$	0 0	0 0	0 0	0 0	0 1	0 1	0 1	0 1	0 1	0 1	0 1
	ROM-1	080H– 0FFH	0	$O_1 = 1$	0 0	0 0	0 0	1 1	0 1	0 1	0 1	0 1	0 1	0 1	0 1
	RAM-2	100H– 17FH	0	$O_2 = 1$	0 0	0 0	1 1	0 0	0 1	0 1	0 1	0 1	0 1	0 1	0 1
	RAM-3	180H– 1FFH	0	$O_3 = 1$	0 0	0 0	1 1	1 1	0 1	0 1	0 1	0 1	0 1	0 1	0 1
	RAM-4	200H– 27FH	0	$O_4 = 1$	0 0	1 1	0 0	0 0	0 1	0 1	0 1	0 1	0 1	0 1	0 1
	RAM-5	280H– 2FFH	0	$O_5 = 1$	0 0	1 1	0 0	1 1	0 1	0 1	0 1	0 1	0 1	0 1	0 1
$IO/\overline{M} = 1$	IN 0	000H	$O_0 = 1$	0	0	0	0	0	0	0	0	0	0	0	0
	IN 1	001H	$O_1 = 1$	0	0	0	0	0	0	0	0	0	0	0	1
	OUT 2	002H	$O_2 = 1$	0	0	0	0	0	0	0	0	0	0	1	0

Address range for each device

77.

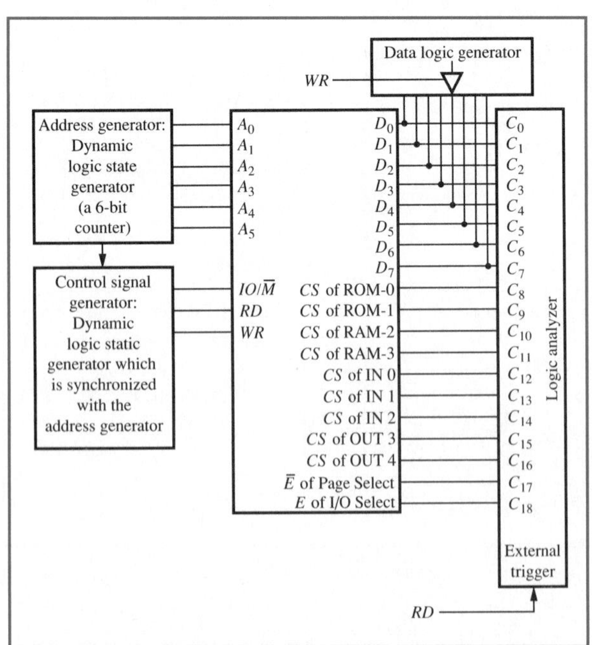

79. No. The signatures of the nodes must be known prior to using a signature analyzer. For designs being developed these signatures are not known.

Chapter 9

1. To store temporary data and intermediate data resulting from ALU operations.

3.

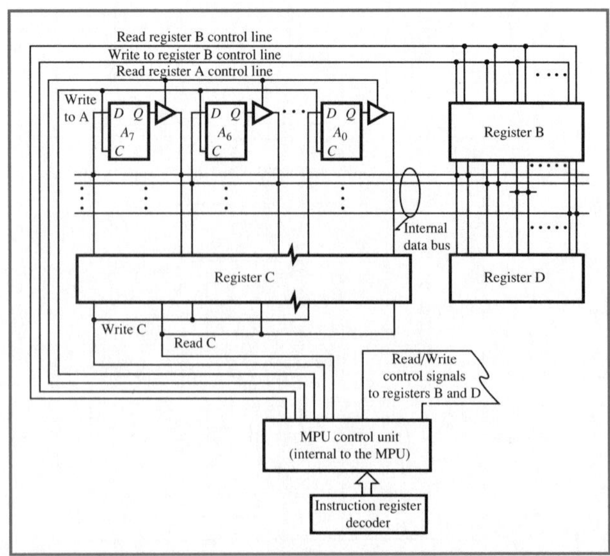

5.

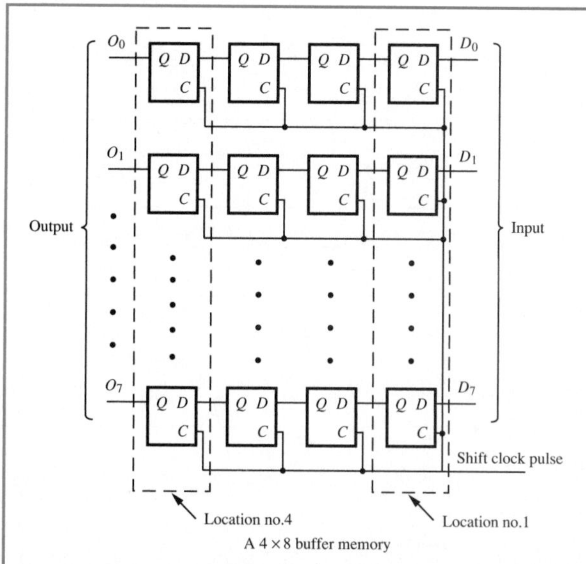

A 4 × 8 buffer memory

7. Yes
9. The MPU can prefetch instructions from memory and store them in its queue. It will fetch these instructions when the data bus is not used for instruction executions, which conserves use of the data bus. It also saves time when the MPU is ready to fetch and execute an instruction since the fetch is internal (from the queue) and the MPU will not have to wait for use of the data bus.
11. Whenever its use would save MPU time by reducing program execution time, such as in a program requiring long, complex calculations.
13. From Table 9–12 and the answer to Problem 12, we understand that to design a look-up table solution we will program the memory device so that at address 0000_2 the binary value 00111111 will be stored, which is the seven-segment code for 0. At location 0001_2 the binary code 00000110 will be stored, and so on. If say a 2716 is used to store the look-up table, then the beginning address would be 0000000000_2, the next would be 0000000001_2, and so on until address 0000001111_2, according to Table 9–12, since there are ten address pins.
15. From Table 9–2 we find that ECL is the fastest technology.
17. It has a high density with fairly fast operating characteristics and moderate power consumption.
19. Since a microcontroller performs very specific tasks, such as the control of some mechanism, its programs are also very specific and do not change. Also, these programs must be present from a cold-start. As a result, microcontroller programs are programmed in ROM and usually are relatively small. RAM is usually also very small since if any large quantities of data are to be collected, for example, if a data base was being established, then the collected data would be down loaded into mass storage. Some microcontroller systems are contained on a single chip.
21. Number of rows = 2^7 = 128 (agrees)
Number of columns = 2^4 = 16 (disagrees)
The disagreement as to the number of columns can be accounted for by considering the planes of memory cells, which are eight in number. The manufacturer considers that for each column selected there are eight memory cells (the z-coordinate), hence

$$\text{Number of columns} = 16 \times 8 = 128$$

23. Three MPU-initiated events occur for every read and write operation and these are (1) the address being output on the address bus, (2) the MPU-generated control signal being output on the control bus, (3) the data byte appearing on the data bus. Since the address being output on the address bus is the first event, all others are referenced from it.
25. The chip enable is used to enable the chip for both read and write operations while the output enable is only for read operations. Refer to the architecture of Figure 9–7.
27. Figure 9–9 states that the number of columns includes the number of memory planes (eight). Hence, the number of columns would have been stated as 128 (16 × 8) in architectures such as that shown in Figure 9–7.
29. From Figure 9–10b we see that before the write enable pin is active its logic level is a high (inactive for a write), which is the logic level for a read. Since the chip select is active (not shown) the datum is read from the addressed memory location and the datum is output during this time.
31. It is the access time parameter. The value of this parameter is compared to t_{AD} of MPU AC parameters for compatibility, where $t_{AD} > t_{ACC}$. That is, the 2764 must be fast enough to respond to MPU read commands.
33. Refer to the answer to Problem 23.
35. (a) Compared to t_{AD} of Figure 9–15a it is too large (too slow). Not compatible.
(b) $t_{DW} > t_{CC}$ of Figure 9–15b, not compatible.
(c) $t_{OE} > t_{AD} - t_{AC}$ of Figure 9–15a, therefore they are not compatible.
(d) $t_{DS} > t_{CC}$ of Figure 9–15b, not compatible.
See Table 9–6 for MPU values (Figure 9–15).
37. These enable pins ($\overline{E}_1$, $\overline{E}_2$, and E_3) serve the same function as a logic gate connected to address lines when detecting a specific logic condition, which is the process of decoding those three address bus lines. Hence, utilizing the enable pins for decoding three address bus lines results in as many as eight 74LS138s being paralleled, which allows expansion up to 64 pages of addresses.
39. The page selector will enable a memory device whenever a page of memory within its memory is addressed.
41. Not necessarily, it's up to the designer. For a small memory, of which hardware cost is of prime importance, then neither have to be prevented. It really is based on getting the most addresses from memory and convenience of programming, relative to addressing.

Address Table. Address bus table for Problem 47

Address bus lines	A_{15}	A_{14}	A_{13}	A_{12}	A_{11}	A_{10}	A_9	A_8	A_7	A_6	A_5	A_4	A_3	A_2	A_1	A_0
2716-0					$\overline{CE}$		$A_9 \leftarrow$									A_0
8417-1				$\overline{CE}$		$A_{10} \leftarrow$										A_0
8417-2			$\overline{CE}$			$\overline{A}_{10} \leftarrow$										A_0
2716-0	0	0	1	1	0	0	0 ←									0
3000H–33FFH	0	0	1	1	0	0	1 ←									1
8417-1	0	0	1	0	1	0 ←										0
2800H–2FFFH	0	0	1	0	1	1 ←										1
8417-2	0	0	0	1	1	0 ←										0
1800H–1FFFH	0	0	0	1	1	1 ←										1

43. The output of the page selector is used to detect a page of memory within a memory device by enabling that memory device. When multiple pages of memory exist within a memory device, those outputs of the page selector that address those pages are input to a single logic gate. The type of logic gate used (*AND*, *OR*, etc.) will be such that the required logic level to enable that memory device will be output by it whenever any one of the page select outputs is active. For the systems we have designed (active-lows) the logic gate used is an *AND*-gate.

45. Whenever the designer wanted simplified hardware, which also means reduced cost.

47.

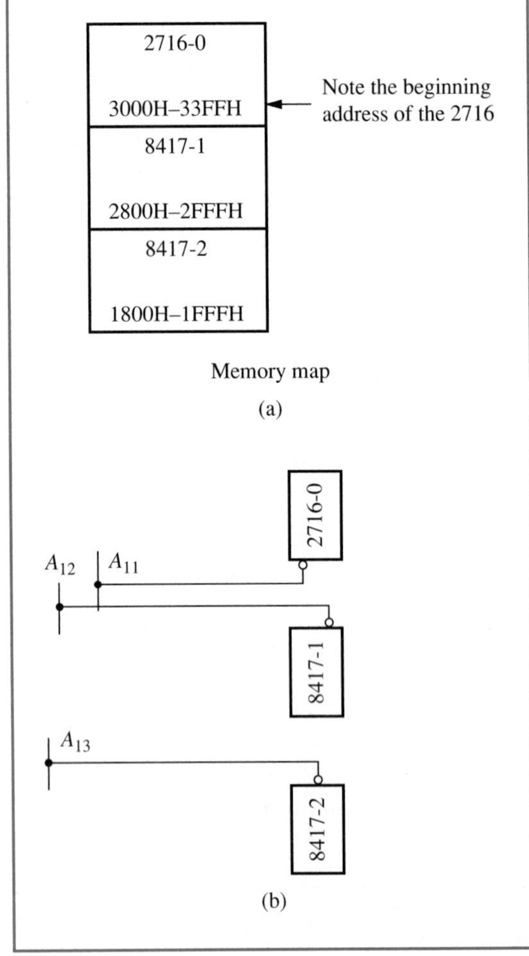

49.

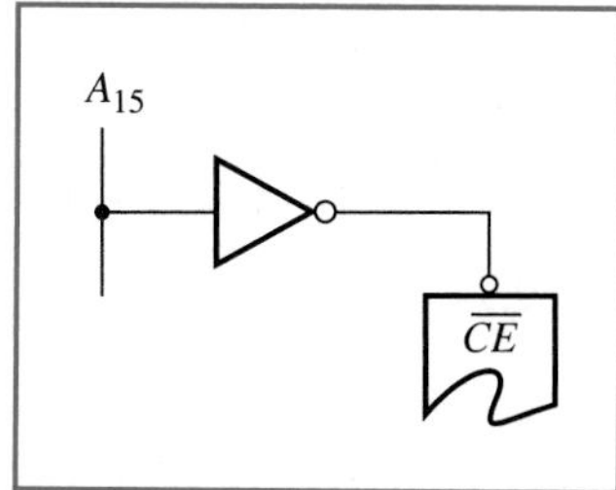

51. Since the two are pin compatible there is no physical difference. From Tables 9–4 and 9–5 we find some timing differences, but since all read and write operations will be performed according to the MPU's timing of Figure 9–15 the resultant read and write cycle time will be the same (both are AC compatible). There will be a reduction in power consumption according to Table 9–2.
53. Apply various logic level combinations to address A_0 through A_2 of the decoder and verify the output logic levels.
55. Ground the chip enable and output enable pins. Apply an address to the address pins and detect the corresponding datum output on the data bus. Since the chip enable pin was "stuck" high (internally) the data bus would be floating (Hi-*Z*) rather than having the content of the addressed memory location.
57. We would write a program that sequentially reads memory locations within 2764-0 of Figure 9–17. We would monitor the address, enable, and output pins via the data bus using the logic analyzer. When the enable pins were low we would check the output pins to verify that the proper datum appeared on the data bus. If the data bus was continuously floating we would know that the enable pins were not functioning. We need not be concerned with which enable pin is internally malfunctioning since if either fails the chip must be replaced.

Chapter 10

1. Address bus: to transmit memory and I/O addresses output by the MPU. It also transmits addresses to main memory from an I/O device having DMA capability.
   Data bus: to provide a path for the transmission and receiving of data from or to those devices connected to it.
   Control bus: the bus over which control signals are transmitted.
3. The two selectors of Figure 10–1 are the page selector and I/O selector. They allow isolation to exist between memory and I/O addresses. They are enabled with control signal $IO/\overline{M}$.
5. (a) a DMA request is made to the MPU by the I/O device with DMA capability. The MPU grants request by activating its *DMA ACK* pin and at the same time putting its address, data, and control signal pins in the Hi-*Z* state. The I/O device then takes control of the three busses and begins the DAM process.
   (b) The I/O device wishing to be served by the MPU will make the interrupt request to the MPU. When the MPU can service the I/O device it so indicates by activating its *INT REQ* pin.
7. (a) The MPU can address memory via
   PC: For fetching instructions from memory.
   SP: For accessing that portion of main memory designated as the stack.
   DC: For accessing memory data locations (registers D and C form the DC).
   (b) To address I/O, register pair *I/OPH* and *I/OPL* are used.
9. Those locations addressed by the PC contain program instructions while the content of those addressed by the SP is data (return address or register data).
11. Those MPUs that have a queue store instructions there, otherwise an instruction is loaded in the IR and then executed. Using a queue allows an MPU to prefetch instructions, thus saving time and making a more efficient use of the data bus.
13. The control unit (CU) is a special-purpose computer that is dedicated to the execution of instructions.
15. A microprogram is part of the CU and instructs the CU as to the execution of instructions. A macroprogram is stored in main memory and is executed by the CU.
17. The ALU has two inputs, *ACC*1 and *ACC*2.
19. An interrupt will cause the MPU to stop the execution of one portion of a program and jump (vector) to another and begin program execution there. A reset does the same thing. It is a reset that is singled out since a reset always jumps back to the beginning program.

21.

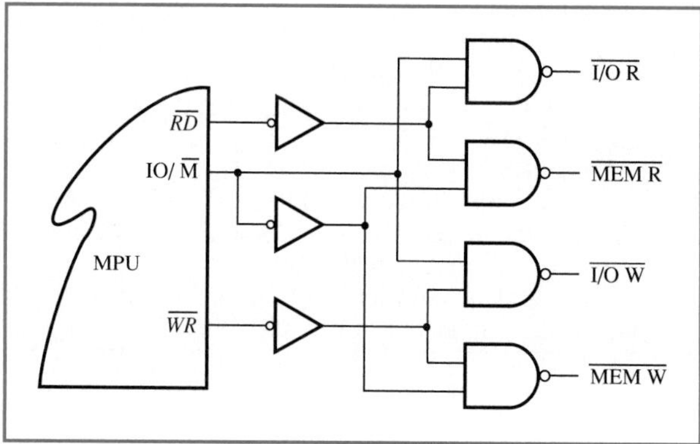

23. The MPU will output address 102CH on the address bus. At time t_{AC} the read control signal $\overline{RD}$ goes low, which enables memory. The content of memory location 102CH (*MVI A*) is loaded on the data bus before time t_{AD}. At time $t_{AC} + t_{CC}$ the MPU latches the binary instruction code for *MVI A* from the data bus and loads it in the queue, via the internal data bus. The CU increments the PC, which is 102DH (the address of the second byte of the instruction *MVI A*,50H. The second byte is fetched and loaded in the queue and again the CU increments the PC (PC = 102EH). This address is used to fetch the instruction *ADD ACC*1,*A*, which is also loaded in the queue. The queue now has 3 bytes, and they are the binary codes (machine codes) for *MVI A*,50H, and *ADD ACC*1,*A*. The CU will move those bytes which are opcodes from the queue to the IR to be executed. When byte *MVI A* is loaded in the IR and decoded by the IR-decoder, the CU will execute *MVI A*, which will cause the next byte (50H) to be loaded from the queue on to the internal data bus from which register A will latch it. When byte *ADD ACC*1,*A* is moved in the IR the CU will add the contents of register *ACC*1 and *A* and store the result in register *ACC*1, as well as setting the ALU flags accordingly.
25. From Figure 10–7, by driving the write control signal $\overline{WR}$ low.
27. *Mnemonics* are acronyms that indicate the operation of an instruction and serve as memory aids.
29. An *assembler* is a program that translates mnemonics into machine code (object code).
31. The designer will determine what must be coded in an instruction, such as operation and register codes, and the pattern with which they should appear for purposes of being decoded by the IR-decoder.
33. (a) 00000110 with $X = 0$.
(b) 00000011
(c) 00100110 with $X = 0$.
(d) 01100000
00000000 = 00H
00000101 = 05H
(e) 10011101
(f) 10101011
(g) 10111100
(h) 10100111
(i) 10110000
01111010 = 7AH (immediate data)
(j) 10100000
00111011
35. We need 3 bits to identify a second register and since only 1 bit is available (T_3) it can not be programmed with these templates.
37. An I/O port is the interface that enables an I/O device to be connected to the data bus.
39.

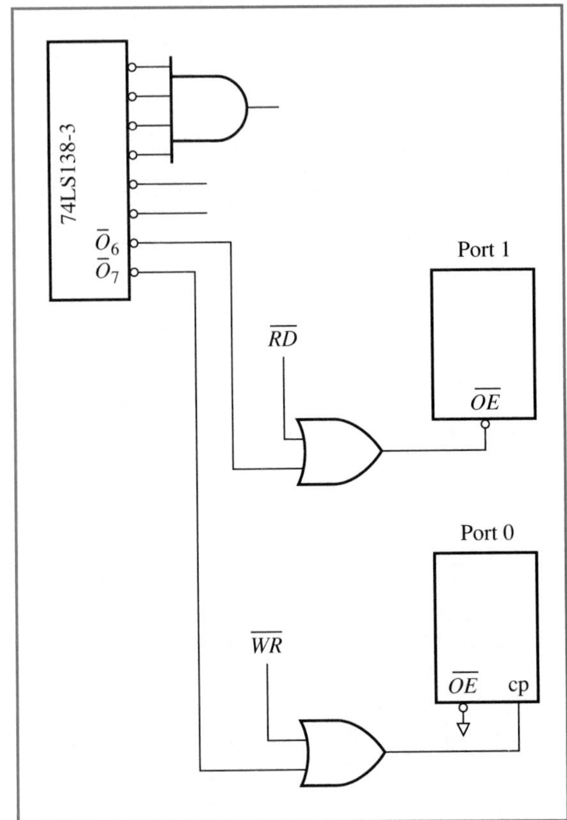

41. The *NAND*-gate provides the read/write timing as well as isolation from memory.

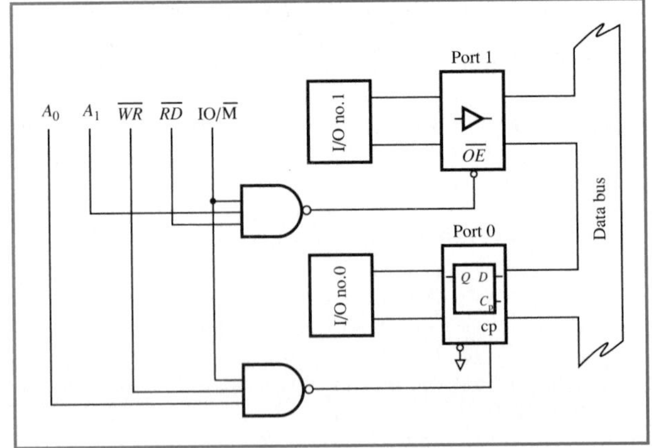

43. The 6800 and 8085 are 8-bit microprocessors while the 8086 is a 16-bit microprocessor (refer to its internal data bus size). The 8086 has an instruction queue, whereas neither the 6800 or 8085 have queues. The 8086 has two architectures in one: the execution unit (EU) and the bus interface unit (BIU). The BIU interfaces with the external data bus and therefore to memory and I/O. The execution unit executes instructions. Memory addresses are segmented by the 8086 via the segment register *CS, DS, SS*, and *ES*. The EU architecture is very similar to the 8085 architecture. The function-block identified as the EU control system is basically composed of 8085 function-blocks, IR, IR-decoder, and the control unit (CU).

45. Test the I/O select decoder (74LS138-4) by applying the I/O address for port 1 via the MPU (program the MPU to read port 1). Using a logic analyzer verify that the MPU has output the appropriate address and control signal logic levels. If any of these logic levels are not correct (but the program is correct), the MPU has failed. If these logic levels are correct, test the *AND*-gate on $\overline{E}_1$ and then test the decoder by verifying the proper output logic level of the decoders $\overline{O}_1$. If it is high then the decoder has failed, whereas if it is low, the port 1 (buffer) has failed.

47. Program the MPU to read from memory chip 2716-0, where the contents of those memory locations are known. Then, using a logic analyzer, monitor the logic levels of the address bus, control signals, and data bus. If these logic levels are correct test the page select decoder by testing for proper input logic levels to it. This requires that the *NOR*-gate be tested first. If the logic levels being input to the decoder are correct, then check the four outputs of 74LS138-3 ($\overline{O}_0$ through $\overline{O}_3$) to verify that the correct output is low. If all are high then the decoder has failed. If any one of these output is low the output of the *AND*-gate must be low, and if not, it has failed. If the output of the *AND*-gate is low and the output enable pin is also low but the data on the data bus is not correct, the 2716-0 has failed or there is a wiring error on one or more of its address pins.

Chapter 11

1. Fuse-links allow a connection to remain (not blown) or to be broken (blown). As a result it is the fuse-link which enables a user to program a PLD.

3.

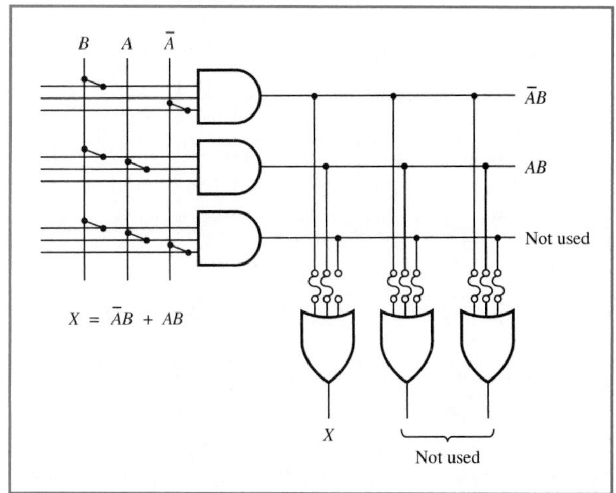

5.

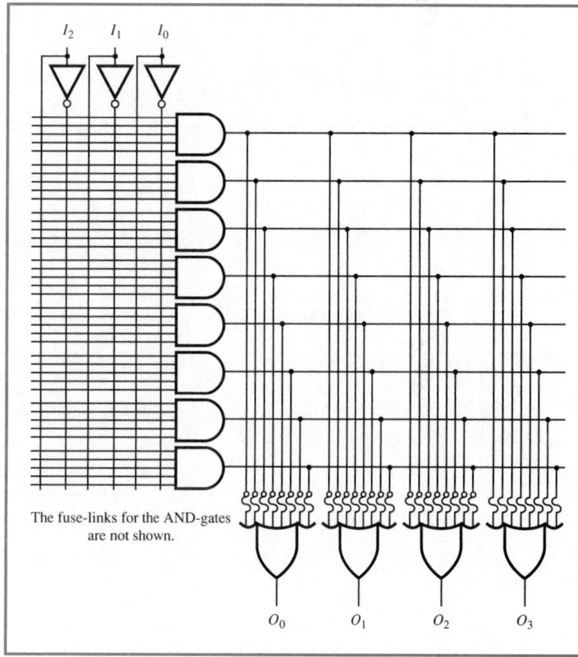

7. There must be one *AND*-gate for each minterm, which is 2^N, where *N* equals the number of independent variables (inputs).

9. $2^6 = 64$ *AND*-gates.

11. One input for each minterm. This allows each output to possibly have each minterm as part of its descriptive Boolean expression.

13.

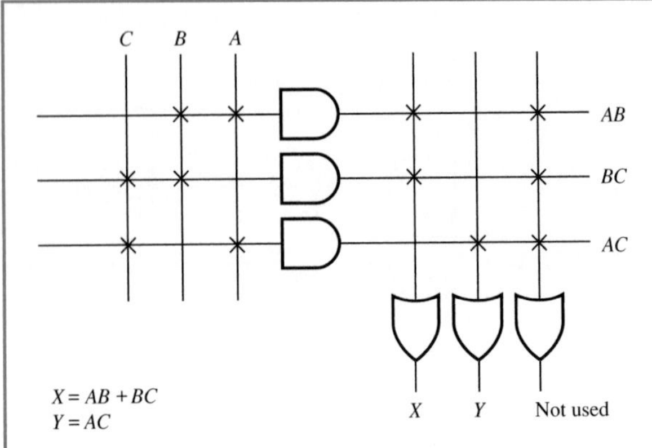

15.

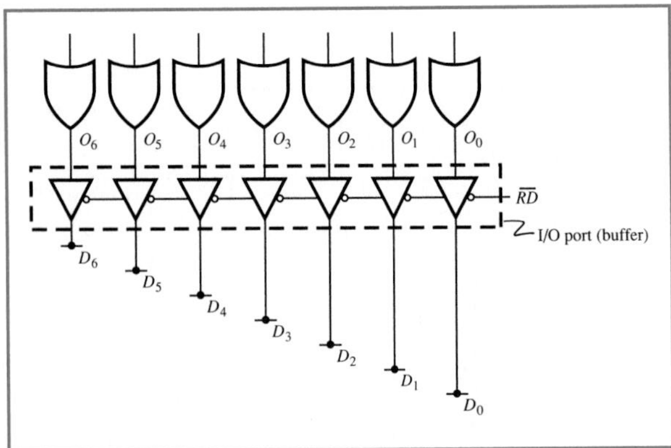

17. The *AND*-gate array is programmed by the manufacturer.

19. (a) PLE4P7 (Rather than *P* it would really be an *H*, which represents an active high.)
(b) PLE4R7

21. It can be programmed to detect SOP product terms other than just minterms.

23.

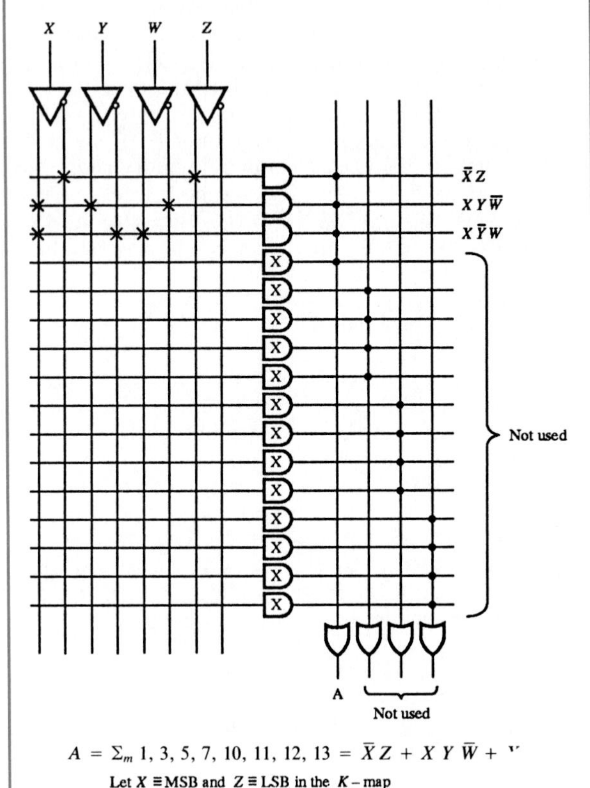

25.

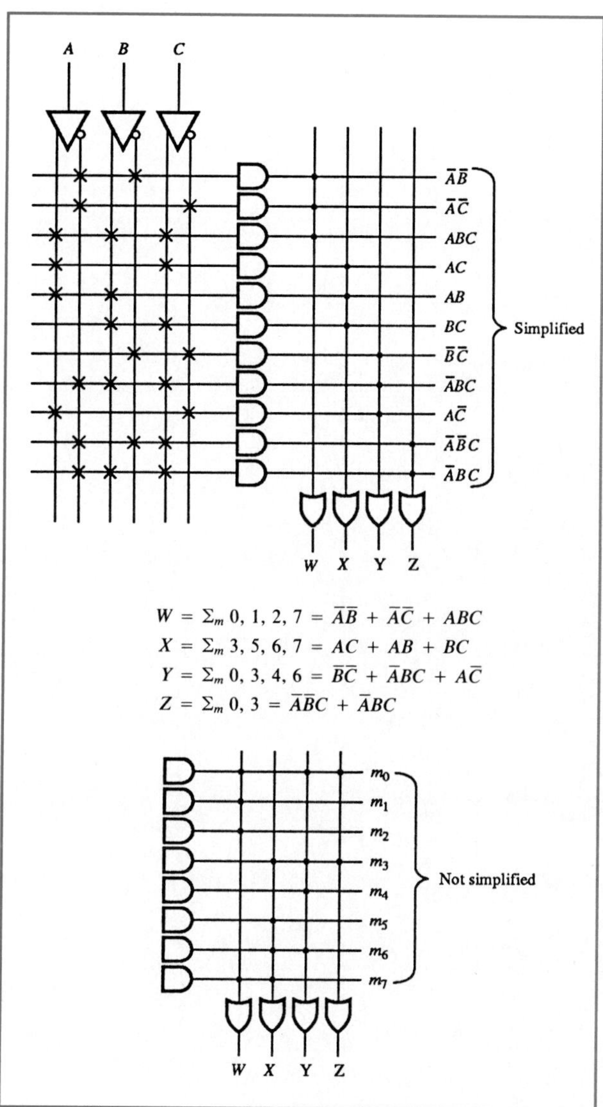

27. If the input fuse-links and *AND*-gate 1 are left intact, its output would be a logic 0, which would force the output buffer into the Hi-*Z* state. Then O_n would be an input pin.

29. Converting from POS to SOP since the outputs are *NOR*-gates.

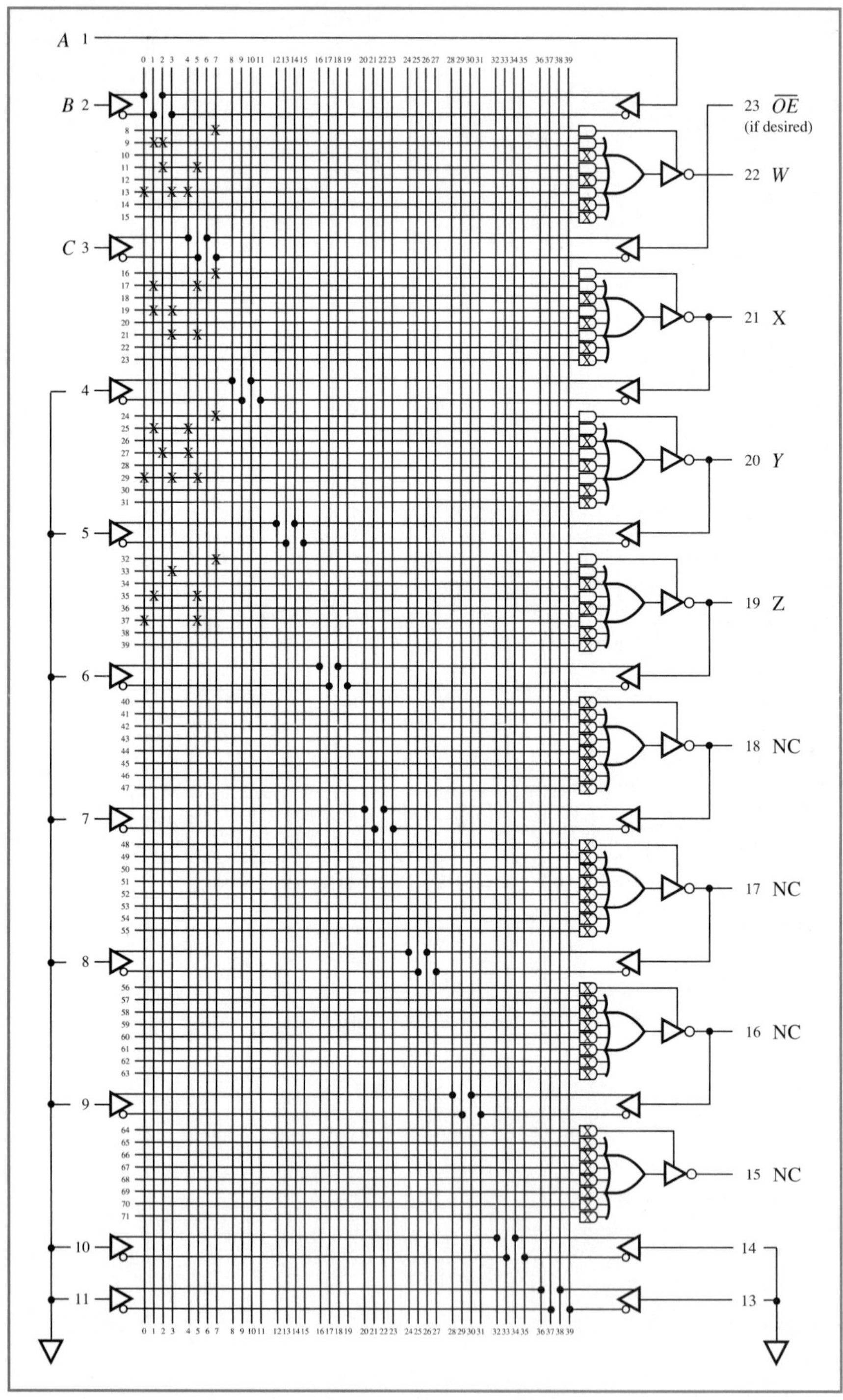

31. Design a down-counter since the outputs are inverted using SOP. Let $A = Q_3$, $B = Q_2$, and $C = Q_1$.

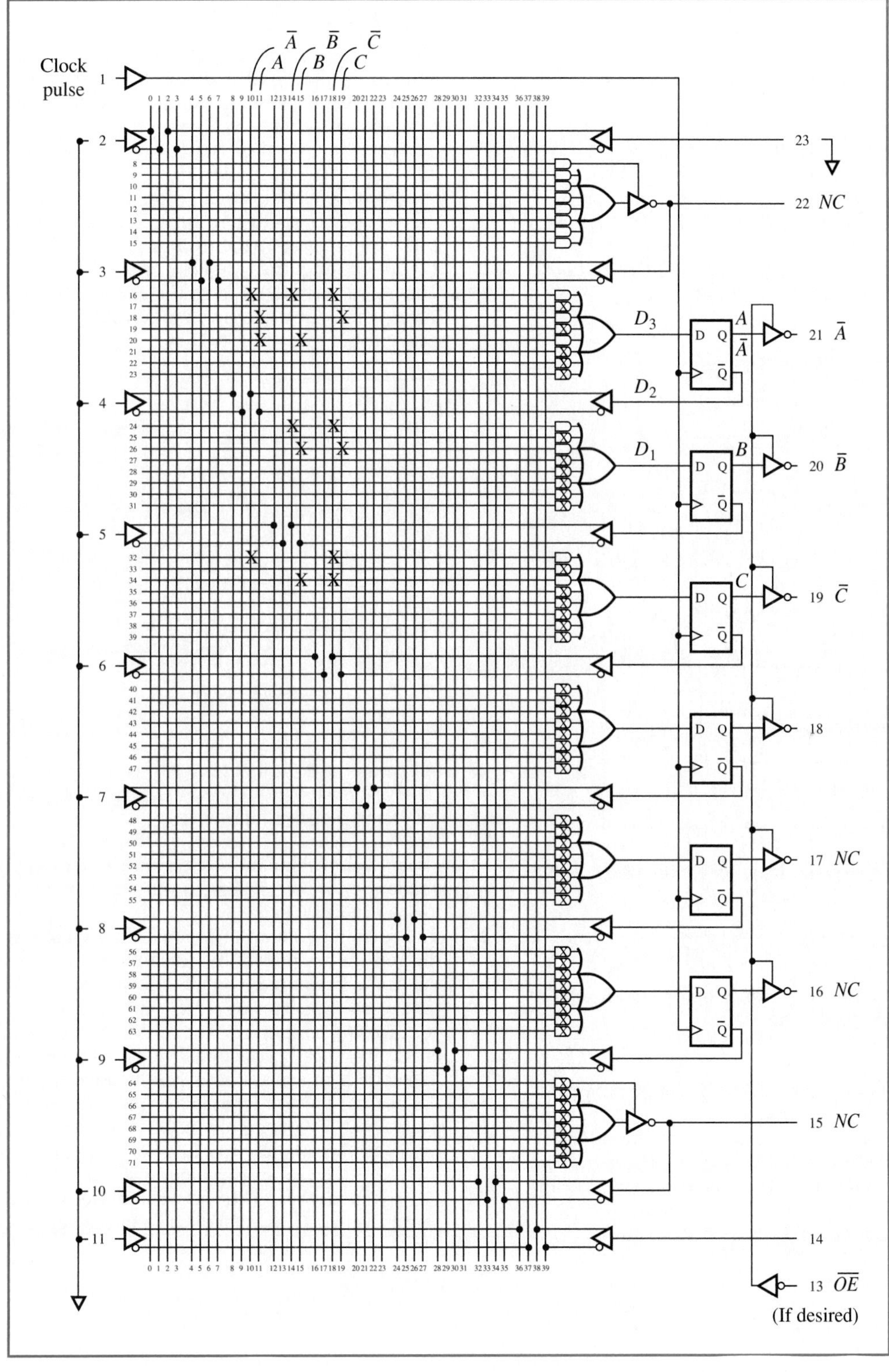

33. There need not be any relationship since a PLA designer may choose the number of product-terms and/or minterms that can be programmed. However, as a minimum 2^N (N is the number of inputs) is a guideline. Also, it is an engineering judgment as to the number of outputs (*OR*-gates) to have available on a PLA IC.

35. Since for PLDs a user does not have access to any internal connections, then all that can be monitored are external pins. As a result, apply static logic levels to input pins (A_0, B_0, etc.) and generate the logic levels indicated in the truth table of Example 11–4. For each of those input levels detect the resultant output level for outputs C, S_0, and S_1. Also compare these values to the logic levels of the truth table. If any disagree either the PLE was not properly programmed or it has failed.

37.

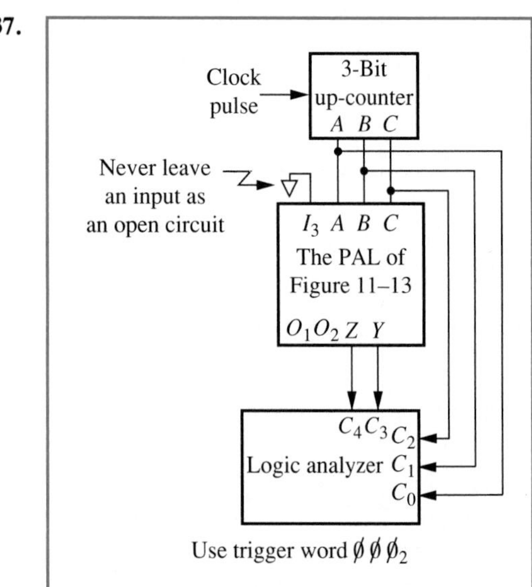

39. Since $Z = \overline{\bar{G}\bar{H} + GHJ}$, logic levels for G, H, and J are required.

Chapter 12

1. V_{IL} is a maximum voltage for a logic 0 and it just is not of sufficient magnitude.

3. An *AND*.

5. To control Q_2.

7. A totem pole is two transistors (BJT in this case) connected in series. Q_2 always turns one of the totem-pole transistors on while turning the other off. This keeps the current flow from power supply to ground at a minimum. The electronic switch has just the collector resistor to limit current.

9. If we assume an ideal BJT ($V_{be} = 0$), then $I = V_{CC}/R_1$, or $I = 5/4 \times 10^3 = 1.25$ mA. This provides a close approximation (UL = 1.6 mA).

11. Their outputs can be wired together and the result is as if these outputs had been input to an *AND*-gate (wired-*AND*).

13. These diodes would be forward-biased if, in error, a negative voltage was applied to any of the inputs. Any of these diodes being forward-biased would short that input to ground and thus protect Q_1 from drawing too much current. Hence, they are circuit protectors.

15. They can not be driven into saturation due to the base-collector Schottky diode, hence they do not build up as much charge on the collector-to-emitter parasitic capacitor which enables them to be turned off faster.

17. TTL logic has resistors integrated on the IC wafer, and because resistance is proportional to length they require a major portion of the IC wafer area. MOS resistors are MOS transistors (see Figure 12–10) and require very little area.

19. The gate is separated from the channel by an insulator, which ideally draws no current. Hence, any gate current is due to leakage.

21.

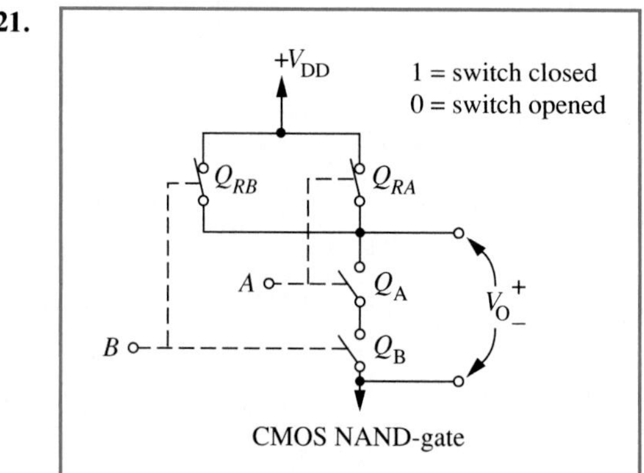

23. When the transistors are in series a negated *AND* (*NAND*) exists and when in parallel a negated *OR* (*NOR*) is the resultant logic.

25. Refer to Figures 12–12 and 12–13. By making the logic level of the control element (base or gate) dependent on the complement output element (collector or drain) logic level.

27. The row and column lines must be driven high and the sense lines pulled low for a read. For a read the sense lines are also pulled low, which provides for the on transistor to remain on by providing a path for the emitter current to flow. Since a differential amplifier is connected to both sense lines, it detects which sense line is conducting and outputs the correct voltage logic level. To write to the cell, again, both row and column lines must be driven high and then the appropriate sense line is driven high while the other is pulled low. For a logic 0 ($Q_1 = 0$ and $Q_2 = 1$) to be written into the cell, sense line for Q_1 is pulled low and the sense line for Q_2 is driven high.

29. To provide isolation (buffering) between the data pin D_n and the read and write lines of the cell.

31. The fuse-link remains if a logic 0 is to be programmed in the cell, otherwise it is blown. When a row line goes high all switches in that row with fuse-links intact are turned on and those with blown fuse-links are off. Those transistors that are turned on will short their data lines to ground, which results in a logic 0 on those data lines. Those transistors with blown fuse-links will remain off when their row line is drive high, resulting in their data lines being "pulled" high by resistor R.
33. 74L (Low-power). Increase the value of R_3 in Figure 12–7.
74H (High speed or high power). Decrease the value of R_3 in Figure 12–7.
74S (Schottky). Use Schottky transistors.
74LS (Low-power Schottky). Increase R_3 and use Schottky transistors.
35. It is a relationship between two logic circuits. The driver is the circuit whose output is connected to an input, which is the driven circuit.
37. The parameters of this table must be such that they are compatible in sign (direction or polarity) and magnitude. Voltage polarities must be the same and current directions must be the opposite. Also, $V_{OL} < V_{IL}$, $V_{OH} > V_{IH}$ and the sink current magnitude must be large enough to handle any source current requirements.
39. The magnitude of current will be preceded by a negative sign.
41. (a) 74L driven by 74C
$V_{IL} = 0.7$ volts > 0.33 volts $= V_{OL}$: OK
$V_{IH} = 2.0$ volts < 2.7 volts $= V_{OH}$: OK
$I_{IL} = -0.18$ mA and $I_{OL} = 3.6$ mA: OK
$I_{IH} = 1.0\ \mu$A and $I_{OH} = -0.36$ mA: OK
(b) 74S driven by 74HC
$V_{IL} = 0.8$ volts $> V_{OL} = 0.33$ volts: OK
$V_{IH} = 2.0$ volts $< V_{OH} = 3.84$ volts: OK
$I_{IH} = 50\ \mu$A and $I_{OH} = -4.0$ mA: OK
$I_{IL} = -2$ mA and $I_{OL} = 4$ mA: OK
(c) 74H driven by 74AC: they are compatible.
(d)74HC driven by 74S: not compatible.
43. $I_{OL} = 8.0$ mA and $I_{IH} = \pm\ 10\ \mu$A: OK
$I_{OH} = -0.4$ mA and $I_{IH} = \pm\ 10\ \mu$A: OK
They are current compatible.
45. Internal parasitic capacitance that must be charged and discharged.
47. Since they are directly connected to the address and control busses, the only propagation delay time would be that of the enable pins.
49. Some or all of the output voltage and/or current specifications (V_{OL}, V_{OH}, I_{OL}, and I_{OH}) will not be within the required limits.
51. See the answers to Problems 49 and 50. Measure logic levels and propagation delay times.
53. If too much voltage would be dropped across the resistor and thus V_{OH} of the output would not be large enough to be within specification.

Chapter 13

1. From the data sheets of Appendix C we see that the control pin of a 74125 requires a low for the buffer to be active, while a 74126 requires a high. Then the chip enable pin has an active logic level for a 74125 that is low and for a 74126 that is high.

3.

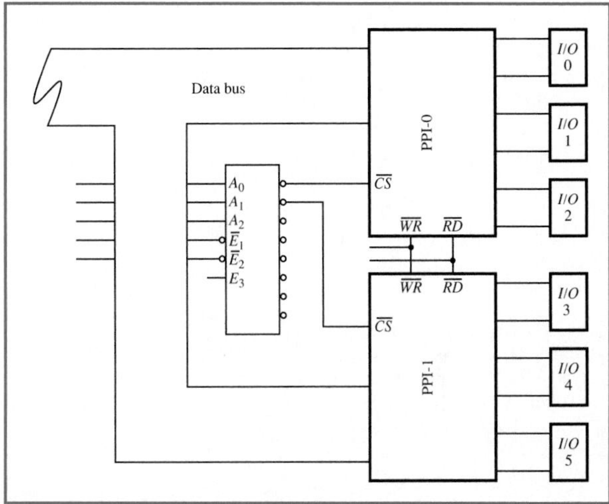

5.

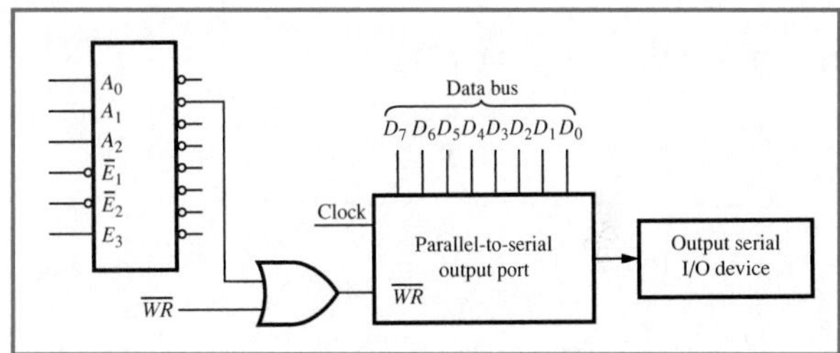

7. The serial data must be shifted, which requires a clock.
9. $e(t) < 0$ then $v_o = -V_{SAT}$
otherwise $v_o = V_{SAT}$

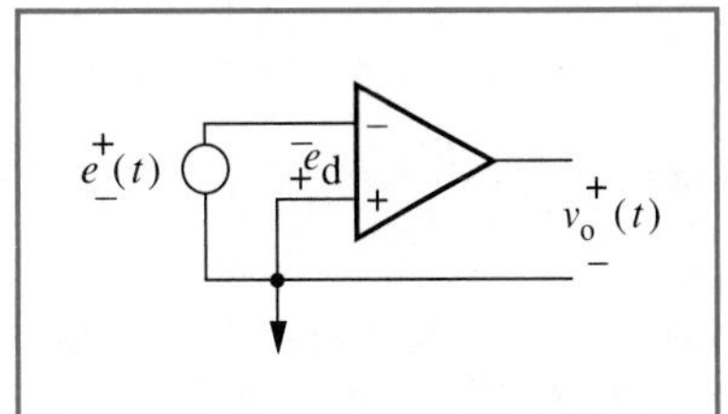

11. The maximum magnitude is V_{SAT}, which can be either positive or negative polarity.

13. $e_d = 19.5/10^5 = 195\ \mu V$ (For $V = 20$ volts), which is approximately zero volts.

15. $R_f = 183.96\text{ K} \sim 184\text{ K}$
$R_0 = R_f/2.5 = 73.58\text{ K} \sim 74\text{ K}$
$R_1 = R_f/5 = 36.79\text{ K} \sim 37\text{ K}$
$R_2 = 18.96\text{ K} \sim 19\text{ K}$

17. (a) $v_o = 1.50$ volts
(b) $v_o = 1.625$ volts
(c) $v_o = 1.406$ volts
(d) $v_o = 1.414$ volts

19. For 0000: $v_o = 0.309$ volts
For 1000: $v_o = 2.394$ volts
For 1100: $v_o = 3.958$ volts
For 1111: $v_o = 4.219$ volts

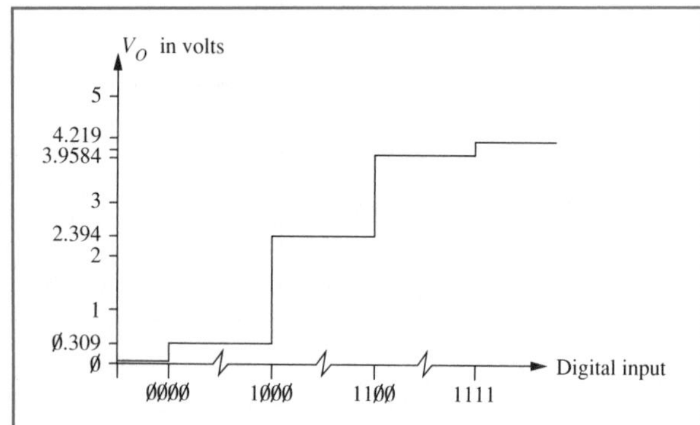

21. $R = 145\text{ K}/11 = 13.18\text{ K}$
$R_f = 10R = 131.8\text{ K}$

23. $I_f = 0.11$ mA

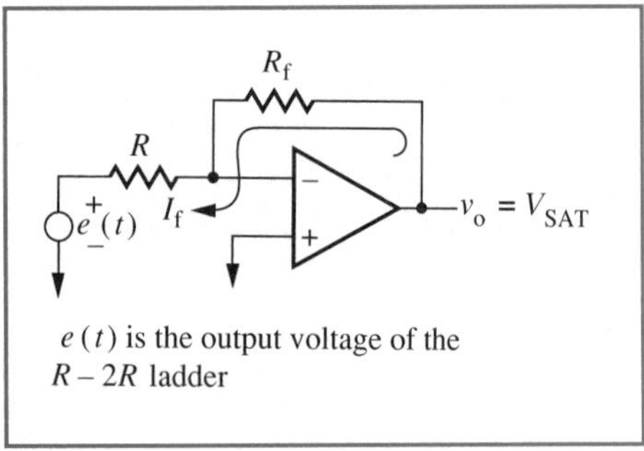

25. (a) v_o max $= -12.65$ volts
(b) v_o max $= -13.44$ volts

27. (a) QE = 0.078 volt
(b) QE = 4.882 mV

29. For a 4-bit output there would be 15 op-amps (comparators) and 16 resistors.

31. $V_{15} = 18.75$ volts

33. $n\ V_{REF}/N = V_n < e_a$
$V_{REF} < (N/n)e_a$ where e_a is its maximum value.

35. For a 4-bit ADC there are 16 counts, hence 16 clock pulses must occur between each S/H.

37.

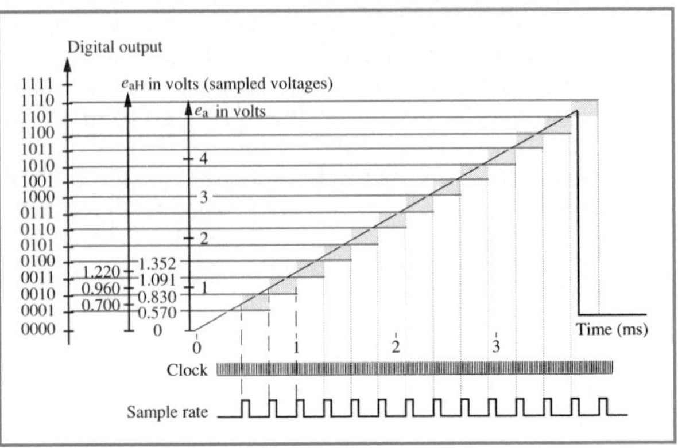

39. With a successive approximation the ADC will approximate the output in a successive fashion. Since the analog ramp voltage begins at 0 volts but the ADC begins its approximation "in the middle," the ADC will have to reduce the count by resetting the higher-weighted bits until it is approximately equal to the sample/hold value of the analog signal. When the 16th clock pulse within the sample interval occurs it triggers the EOC signal and the next approximation would begin.

41. First we must program the port. This requires that we use static logic generators to put the address of the control register on the address bus, the proper control word on the data bus, and then drive the control signals to the correct logic levels ($IO/\overline{M} = 1$, $\overline{WR} = 0$, and $\overline{RD} = 1$). Now that the port is programmed we may verify its operation by reading and writing data from and to the I/O devices. Of course this requires that the address, data (for a write), and control bus have the proper logic levels applied to them.

43. We would program the MPU to (1) Write a byte to the serial output device. (2) Read a byte from the serial input device. The MPU would generate the dynamic logic levels required ($\overline{RD}$, $\overline{WE}$, $C/\overline{D}$, and $\overline{CS}$ via the address and control busses). The clock signals *TXC* and *RXC* would be generated by some pulse-generating circuitry. Thus for a write operation the MPU would address the USART, load the data on the data bus, and then drive $\overline{WR}$ low. We would use a logic analyzer to verify that these events occurred. We also would have a channel of the logic analyzer monitoring the *TXD* to verify that the data being written is transmitted to the serial output device. For a read we would load known data into the serial input device and then read that device by again programming the MPU to do so. We would monitor the *RXD* line and busses to verify proper operation.

45. For a DAC with three inputs a 3-bit counter is used, which yields eight counts. We would have manual control over this counter by being able to advance its count when desired (we would pulse the clock input). One trace of the scope would monitor the output voltage while the other could be used to monitor the input logic levels (one input at a time).

47.

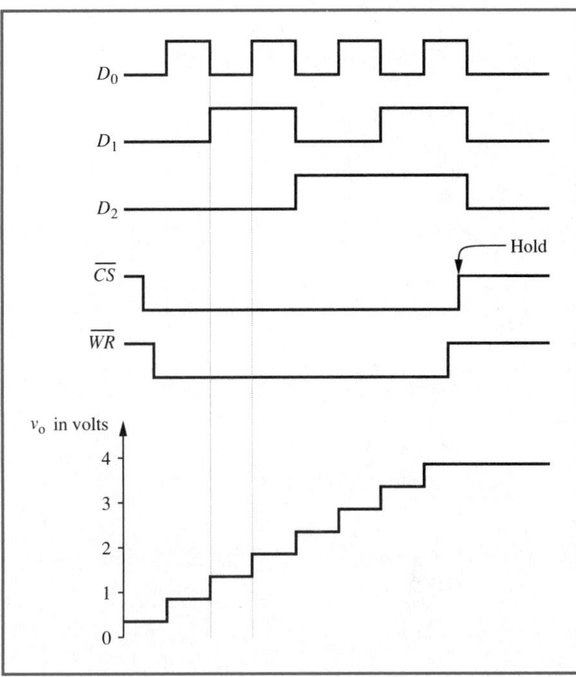

Index